McGraw-Hill Yearbook of Science & Technology

1986

McGraw-Hill Yearbook of Science & Technology

1986

COMPREHENSIVE COVERAGE OF
RECENT EVENTS AND RESEARCH AS
COMPILED BY THE STAFF OF THE
McGRAW-HILL ENCYCLOPEDIA OF
SCIENCE AND TECHNOLOGY

McGRAW-HILL BOOK COMPANY

New York   St. Louis   San Francisco
Auckland   Bogotá   Guatemala   Hamburg
Johannesburg   Lisbon   London   Madrid   Mexico
Montreal   New Delhi   Panama   Paris   San Juan
São Paulo   Singapore   Sydney   Tokyo   Toronto

McGRAW-HILL YEARBOOK
OF SCIENCE & TECHNOLOGY
Copyright © 1985 by McGraw-Hill, Inc.
All rights reserved. Printed in the
United States of America. Except as permitted
under the United States Copyright Act
of 1976, no part of this publication may be
reproduced or distributed in any form or by any
means, or stored in a data base or retrieval
system, without the prior written permission of
the publisher. Philippines Copyright, 1985,
by McGraw-Hill, Inc.

1234567890 DODO 8921098765

The Library of Congress has cataloged this serial
publication as follows:

McGraw-Hill yearbook of science and technology.
1962– . New York, McGraw-Hill Book Co.

  v. illus. 26 cm.
  Vols. for 1962– compiled by the staff of the
McGraw-Hill encyclopedia of science and
technology.

  1. Science—Yearbooks. 2. Technology—
Yearbooks. 1. McGraw-Hill encyclopedia of
science and technology.
Q1.M13   505.8   62-12028

ISBN 0-07-046181-3
ISSN 0076-2016

TABLE OF CONTENTS

FEATURE ARTICLES

Nuclear Winter	Richard P. Turco	1–13
Surface Science	John B. Hudson	14–23
The Inflationary Universe	Marek Demianski	24–33
Nuclear Electromagnetic Pulse	Mario Rabinowitz	34–47
Computer-Assisted Drug Design	A. J. Hopfinger	48–59
The Path Toward Fusion Energy	Harold P. Furth	60–71
Gene Therapy	Raju S. Kucherlapati	72–79
Accretion Tectonics	W. G. Ernst	80–88

A-Z ARTICLES 89–482

List of Contributors	485–489
Index	493–509

International Editorial Advisory Board

Dr. Neil Bartlett
Professor of Chemistry
University of California, Berkeley

Dr. Richard H. Dalitz
Department of Theoretical Physics
University of Oxford

Dr. Freeman J. Dyson
The Institute for Advanced Study
Princeton, New Jersey

Dr. George R. Harrison
Dean Emeritus, School of Science
Massachusetts Institute of Technology

Dr. Leon Knopoff
Institute of Geophysics
  and Planetary Physics
University of California, Los Angeles

Dr. H. C. Longuet-Higgins
Royal Society Research Professor,
  Experimental Psychology
University of Sussex

Dr. Alfred E. Ringwood
Director, Research School of
  Earth Sciences
Australian National University

Dr. Arthur L. Schawlow
Professor of Physics
Standford University

Dr. Koichi Shimoda
Department of Physics
Keio University

Dr. A. E. Siegman
Director, Edward L. Ginzton Laboratory
Professor of Electrical Engineering
Stanford University

Prof. N. S. Sutherland
Director, Centre for Research
  on Perception and Cognition
University of Sussex

Dr. Hugo Theorell
The Nobel Institute
Stockholm

Lord Todd of Trumpington
University Chemical Laboratory
Cambridge University

Dr. George W. Wetherill
Director, Department of Terrestrial
  Magnetism
Carnegie Institution of Washington

Dr. E. O. Wilson
Professor of Zoology
Harvard University

Dr. Arnold M. Zwicky
Professor of Linguistics
Ohio State University

Editorial Staff

Sybil P. Parker, Editor in Chief

Jonathan Weil, Editor
Betty Richman, Editor
Daniel Kaizer, Editor

Edward J. Fox, Art director

Patrick J. Aievoli, Graphic arts/production supervisor
Sherelle Ramey, Art production assistant

Joe Faulk, Editing manager

Frank Kotowski, Jr., Editing supervisor
Erin Thomas, Editing supervisor

Patricia W. Albers, Senior editing assistant
Barbara Begg, Editing assistant
Deborah Fudge, Editing assistant

Art suppliers: Eric G. Hieber,
EH Technical Services, New York, New York;
James R. Humphrey, Nanuet, New York;
E. T. Steadman, New York, New York

Typesetting and composition
by The Clarinda Company, Clarinda, Iowa

Printed and bound by R. R. Donnelley
& Sons Company, the Lakeside Press
at Willard, Ohio, and Crawfordsville, Indiana

Consulting Editors

Vincent M. Altamuro. *President, Robotics Research, Division of VMA, Inc., Toms River, New Jersey.* INDUSTRIAL ENGINEERING.

Dr. Paul M. Anderson. *President, Power Math Associates, Inc., Del Mar, California.* ELECTRICAL POWER ENGINEERING.

Prof. Eugene A. Avallone. *Formerly, Department of Mechanical Engineering, City University of New York.* MECHANICAL POWER ENGINEERING AND PRODUCTION ENGINEERING.

A. E. Bailey. *Formerly, Superintendent of Electrical Science, National Physical Laboratory, London.* ELECTRICITY AND ELECTROMAGNETISM.

Prof. T. M. Barkley. *Division of Biology, Kansas State University.* PLANT TAXONOMY.

Prof. B. Austin Barry. *Civil Engineering Department, Manhattan College.* CIVIL ENGINEERING.

Dr. Alexander Baumgarten. *Director, Clinical Immunology Laboratory, Yale–New Haven Hospital.* IMMUNOLOGY AND VIROLOGY.

Dr. Walter Bock. *Professor of Evolutionary Biology, Department of Biological Sciences, Columbia University.* ANIMAL ANATOMY; ANIMAL SYSTEMATICS; VERTEBRATE ZOOLOGY.

Robert D. Briskman. *Vice President, Systems Implementation, Comsat General Corporation, Washington, D.C.* TELECOMMUNICATIONS.

Prof. D. Allan Bromley. *Henry Ford II Professor and Director, A. W. Wright Nuclear Structure Laboratory, Yale University.* ATOMIC, MOLECULAR, AND NUCLEAR PHYSICS.

W. S. Bromley. *Consulting Forester and Association Consultant, New Rochelle, New York.* FORESTRY.

Michael H. Bruno. *Graphic Arts Consultant, Nashua, New Hampshire.* GRAPHIC ARTS.

Dr. John F. Clark. *Director, Space Application and Technology, RCA Laboratories, Princeton, New Jersey.* SPACE TECHNOLOGY.

Dr. Richard B. Couch. *Ship Hydrodynamics Laboratory, University of Michigan.* NAVAL ARCHITECTURE AND MARINE ENGINEERING.

Prof. David L. Cowan. *Department of Physics and Astronomy, University of Missouri.* CLASSICAL MECHANICS AND HEAT.

Dr. C. Chapin Cutler. *Ginzton Laboratory, Stanford University.* RADIO COMMUNICATIONS.

Dr. James Deese. *Department of Psychology, University of Virginia.* PHYSIOLOGICAL AND EXPERIMENTAL PSYCHOLOGY.

Dr. H. Fernandez-Moran. *A. N. Pritzker Divisional Professor of Biophysics, Division of Biological Sciences and the Pritzker School of Medicine, University of Chicago.* BIOPHYSICS.

Dr. John K. Galt. *Vice President, Sandia Laboratories, Albuquerque.* PHYSICAL ELECTRONICS.

Prof. M. Charles Gilbert. *Department of Geology, Texas A&M University.* GEOLOGY (MINERALOGY AND PETROLOGY).

Prof. Roland H. Good, Jr. *Department of Physics, Pennsylvania State University.* THEORETICAL PHYSICS.

Dr. Alexander von Graevenitz. *Department of Medical Microbiology, University of Zurich.* MEDICAL BACTERIOLOGY.

Prof. David L. Grunes. *U.S. Plant, Soil and Nutrition Laboratory, U.S. Department of Agriculture.* SOILS.

Dr. Carl Hammer. *Research Consulting Services, Washington, D.C.* COMPUTERS.

Dr. Dennis R. Heldman. *Vice President, Process Research and Development, Campbell Institute for Research and Technology, Camden, New Jersey.* FOOD ENGINEERING.

Dr. Ralph E. Hoffman. *Yale Psychiatric Institute, Yale University School of Medicine.* PSYCHIATRY.

Dr. R. P. Hudson. *Bureau International des Poids et Mesures, France.* LOW-TEMPERATURE PHYSICS.

Prof. Stephen F. Jacobs. *Professor of Optical Sciences, University of Arizona.* ELECTROMAGNETIC RADIATION AND OPTICS.

Consulting Editors (continued)

Dr. Gary Judd. *Vice Provost, Academic Programs and Budget, and Dean of the Graduate School, Rensselaer Polytechnic Institute.* METALLURGICAL ENGINEERING.

Dr. Donald R. Kaplan. *Department of Botany, University of California, Berkeley.* PLANT ANATOMY.

Dr. Richard F. Kay. *Department of Anatomy, Duke University Medical Center.* ANTHROPOLOGY.

Prof. Edwin Kessler. *Director, National Severe Storms Laboratory, Norman, Oklahoma.* METEOROLOGY AND CLIMATOLOGY.

Prof. Robert C. King. *Department of Ecology and Evolutionary Biology, Northwestern University.* GENETICS AND EVOLUTION.

Dr. George deVries Klein. *Professor of Sedimentology, University of Illinois at Urbana-Champaign.* GEOLOGY (PHYSICAL, HISTORICAL, AND SEDIMENTARY); PHYSICAL GEOGRAPHY.

Prof. Richard G. Klein. *Department of Anthropology, University of Chicago.* ANTHROPOLOGY AND ARCHEOLOGY.

Prof. Irving M. Klotz. *Department of Chemistry, Northwestern University.* BIOCHEMISTRY.

Prof. R. Bruce Lindsay. *(Deceased) Hazard Professor of Physics, Emeritus, Brown University.* ACOUSTICS.

Dr. Carl N. McDaniel. *Department of Biology, Rensselaer Polytechnic Institute.* DEVELOPMENTAL BIOLOGY.

Dr. Howard E. Moore. *Professor of Chemistry, Department of Physical Sciences, Florida International University.* GEOCHEMISTRY.

Dr. N. Karle Mottet. *Professor of Pathology, University of Washington School of Medicine.* MEDICINE AND PATHOLOGY.

Prof. Jay M. Pasachoff. *Director, Hopkins Observatory, Williams College, Williamstown, Massachusetts.* ASTRONOMY.

Prof. D. A. Roberts. *Plant Pathology Department, Institute of Food and Agricultural Sciences, University of Florida.* PLANT PATHOLOGY.

Prof. W. D. Russell-Hunter. *Professor of Zoology, Department of Biology, Syracuse University.* INVERTEBRATE ZOOLOGY.

Dr. Bradley T. Scheer. *Professor Emeritus, Biology, University of Oregon.* GENERAL PHYSIOLOGY.

Prof. Thomas J. M. Schopf. *(Deceased) Department of Geophysical Sciences, University of Chicago.* PALEOBOTANY AND PALEONTOLOGY.

Dr. Anthony M. Trozzolo. *Charles L. Huisking Professor of Chemistry, University of Notre Dame.* ORGANIC CHEMISTRY.

Dr. David Turnbull. *Gordon McKay Professor of Applied Physics, Harvard University.* SOLID-STATE PHYSICS.

Prof. Garret N. Vanderplaats. *Department of Mechanical and Environmental Engineering, University of California, Santa Barbara.* DESIGN ENGINEERING.

Dr. Richard G. Wiegert. *Department of Zoology, University of Georgia.* ECOLOGY AND CONSERVATION.

Dr. W. A. Williams. *Department of Agronomy and Range Science, University of California, Davis.* AGRICULTURE.

Contributors

A list of contributors, their affiliations, and the articles they wrote will be found on pages 485–489.

Preface

The 1986 *McGraw-Hill Yearbook of Science and Technology*, continuing in the tradition of its 23 predecessors, presents the outstanding recent achievements in science and technology. Thus it serves as an annual review and also as a supplement to the *McGraw-Hill Encyclopedia of Science and Technology*, updating the basic information in the fifth edition (1982) of the Encyclopedia.

The Yearbook contains articles reporting on those topics that were judged by the consulting editors and the editorial staff as being among the most significant recent developments. Each article is written by one or more authorities who are actively pursuing research or are specialists on the subject being discussed.

The Yearbook is organized in two independent sections. The first section includes eight feature articles, providing comprehensive, expanded coverage of subjects that have broad current interest and possible future significance. The second section comprises 146 alphabetically arranged articles.

The *McGraw-Hill Yearbook of Science and Technology* provides librarians, students, teachers, the scientific community, and the general public with information needed to keep pace with scientific and technological progress throughout the world. The Yearbook has successfully served this need for the past 24 years through the ideas and efforts of the consulting editors and the contributions of eminent international specialists.

SYBIL P. PARKER
EDITOR IN CHIEF

McGraw-Hill
Yearbook of
Science &
Technology

1986

Nuclear Winter

Richard P. Turco

After receiving his doctorate in electrical engineering/physics from the University of Illinois, Richard P. Turco in 1971 became research scientist and program manager, Atmospheric Chemistry and Physics, at R & D Associates, Marina del Rey, California. His studies have included the chemistry and physics of the Earth's atmosphere, the dinosaur extinction event, and Mount St. Helens eruption clouds. With his colleagues he has defined the severe environmental impacts of smoke and dust generated by nuclear war.

Nuclear weapons were invented 40 years ago. Since then, global nuclear arsenals have grown from a few bombs, each capable of destroying a modest-sized city such as Hiroshima or Nagasaki, to about 50,000 warheads carrying a total explosive power 1 million times that of the earliest bombs. The arsenals contain the destructive potential of several thousand World War II's and, with present systems and deployments, may be delivered and detonated in a matter of minutes or hours. Virtually all of civilization could be obliterated in a nuclear war.

Most people equate nuclear war with instant death by vaporization in a nuclear fireball and express little hope of surviving a nuclear conflict. Yet, some nuclear strategists and planners believe that complete physical and economic recovery, even after a full-scale nuclear exchange, could be rapid—taking only a few years or decades. This attitude is based upon the reasonable conclusion that the direct damage caused by 50,000 potential nuclear explosions—mainly due to the blast and prompt radioactive fallout—would be confined to relatively small areas around the explosion sites. Most of the Earth, and most of its people, would be spared and be able to rebuild. The longer-term worldwide effects of delayed radioactive fallout and stratospheric ozone depletion, it has been calculated, would be no more than a serious nuisance.

But all this may have changed. New scientific studies suggest that the delayed worldwide effects of nuclear war might be more dangerous to life on the planet than the nuclear bursts themselves. These studies have considered possible global climatic disturbances caused by immense quantities of dust raised by nuclear bursts and smoke generated in nuclear fires. Interestingly, with 40 years spent analyzing the effects of nuclear weapons, it was not until 1982 that anyone attempted to estimate the smoke production by a nuclear war. P. Crutzen and J. Birks made a rough calculation and found that several hundred million metric tons of smoke could be released into the atmosphere. Sunlight would be absorbed and blocked by the smoke, creating "twilight at noon" over vast areas of the Northern Hemisphere.

Nuclear winter

R. Turco, O. B. Toon, T. P. Ackerman, J. B. Pollock, and C. Sagan carried out the first quantitative estimates of smoke and dust emissions from cities and forests in a nuclear attack and calculated explicitly the related atmospheric temperature changes, radioactive fallout doses, and chemical perturbations to the upper atmosphere. In contrast to previous calculations, the findings were so startling in terms of the severity of global-scale changes in climate and environment that the aftermath of a nuclear war was metaphorically referred to as a nuclear winter. A nuclear winter involves an array of complex interconnected physical and chemical phenomena. Nuclear winter, of course, has never occurred; its conditions can only be simulated by using sophisticated modern computers. Historical and prehistorical climatic events may be used to calibrate the nuclear simulations to a limited degree. Present understanding of nuclear winter, and some of its physical consequences and implications for survival, are summarized below.

CLIMATE

The climate and weather of the Earth are driven by the energy of Sun and the thermal inertia of the oceans. Normally, most of the sunlight impinging on the Earth is absorbed in the lower atmosphere and at the surface; about 30% is reflected to space, and this portion is referred to as the planetary albedo. The absorbed sunlight heats the ground and lower atmosphere, which in turn radiate the energy back into space as thermal, or infrared, radiation. The oceans collect solar heat and store it within the top mixed layer of about 300 ft (100 m) of water. This stored heat is capable of moderating climate over a period of years, providing warmth to the atmosphere when sunlight is unavailable, and storing heat when sunlight is plentiful.

Significant changes in climate occur when the average temperature of Earth varies by only a few degrees Celsius. Following a major volcanic eruption such as Tambora (which exploded in Indonesia in 1815), there can ensue an average 1°C (2°F) global temperature decrease, frosts in summer, massive crop failures, and widespread hardship, a "year without summer." At the other extreme, the current buildup of carbon dioxide in the atmosphere may lead to a global "greenhouse" warming of about 2°C (4°F), extensive melting of polar ice fields, and flooding of coastal cities. An ice age is marked by a long-term global cooling of about 5–10°C (10–20°F). In contrast with these temperature variations, a nuclear winter could trigger a 10–30°C (20–60°F) temperature plunge lasting several months.

Meteorite winter. There is one type of natural disaster which may be comparable in magnitude to a nuclear winter. In 1980 L. Alvarez and coworkers suggested that 65 million years ago (at the Cretaceous-Tertiary boundary) a meteorite 6 mi (10 km) in diameter collided violently with the Earth. It is thought that the ancient meteor impact raised a dust cloud so dense that the Earth was cast into total frigid darkness for months—a "meteorite winter." Nearly 75% of all living species disappeared at this time, including every dinosaur. Two important lessons follow from the meteorite event. First, sudden and profound climatic disturbances can occur on the Earth when sunlight is blocked efficiently from the surface. Second, such climatic changes can prove devastating to a wide variety of life-forms, large and small. Oddly, the meteorite nightfall of 65 million years ago may have heralded the dawn of the human species by causing the ascendance of the mammals over the reptiles (dinosaurs), eventually leading to human evolution. Ironically, humankind may now face the threat of annihilation in a nuclear winter of their own making.

Clouds and surface temperatures. When sunlight is blocked from the Earth's surface by clouds of airborne dust and soot, or aerosols, land temperatures can fall rapidly. Temperature drops of 20°C (40°F) in a single night are not uncommon over deserts (of course, the heat loss is made up by the next day's warming sunlight). Heat from the soil cannot ameliorate sudden, deep temperature declines. The variation in the length of the day and maximum height of the Sun in the sky over the course of a year are responsible for the normal excursions of seasons, and the extremes between summer and winter. It is important to note, however, that the tropical and subtropical regions of the Earth, which account for about one-half of the entire surface area of the planet, do not normally experience large seasonal variations or severe winters.

Clouds and darkness do not always lead to a strong surface cooling. The effect depends on the relative visible and infrared properties of the clouds. At planetary temperatures, heat is radiated in the infrared spectrum, from about 5 to 50 micrometers. If a cloud is opaque in the infrared, heat is trapped below it for a time, just as heat can be trapped beneath a blanket. This prevents the surface from cooling rapidly.

In discussing the optical properties of clouds, terms such as opaque and opacity indicate the general capacity of the cloud to block light. The optical depth of a cloud is a quantitative measure of its opacity or opaqueness. A direct beam of radiation at a specific wavelength or electromagnetic frequency will be reduced in intensity by a factor of $e^{-\tau}$, where τ is the optical depth at that wavelength. An optical depth of 2.3 causes a factor-of-10 reduction in intensity, and of 4.6, a factor-of-100 reduction. Hence, the direct penetration of light through a cloud is very sensitive to the optical depth.

The total amount of light penetrating a cloud shows, in addition to the direct component, a diffuse, or scattered, component. In opaque clouds composed of nonabsorbing material, the scattered component can exceed the direct component. In clouds composed of absorbing material, less light can be scattered through the cloud. In the visible spectrum, pure water clouds are nonabsorbing, clouds of soil dust are weakly absorbing, and clouds of smoke (particularly sooty smoke) are strongly absorbing. In the infrared spectrum, all of these materials absorb and emit infrared radiation efficiently.

For a fixed mass of particulate matter, the optical depth also depends on the particle sizes. At visible wavelengths the greatest scattering and absorption efficiencies occur for particle sizes of about 0.1 μm. At infrared wavelengths, the efficiency is maximized at sizes of several micrometers. Natural water clouds are composed of droplets of about 10 μm in size; these clouds are therefore excellent infrared absorbers and emitters. Smoke particles generally range in size from about 0.1 to 1.0 μm, and are therefore excellent optical absorbers but weak infrared absorbers. Soil dust particles possess a wider range of sizes from less than 0.1 μm to much greater than 10 μm. Accordingly, soil dust can be effective at both optical scattering and infrared absorption. However, only the former property may be maintained by airborne dust over many days, because the larger infrared-active dust particles quickly fall out of the atmosphere under the influence of gravity.

Greenhouse effect. A greenhouse effect is created on a planet when the atmosphere is much more absorbing in the infrared spectrum than in the visible spectrum. The Earth exhibits a weak greenhouse effect mainly as a result of traces of water vapor (about 1% by mass) in the lowest mile or two of the atmosphere. Both water (H_2O) and carbon dioxide (CO_2) gases are transparent in the visible spectrum but opaque in the infrared spectrum. If the Earth were suddenly to lose its thin water-vapor shield, the surface temperature would quickly drop to $-20°C$ ($-4°F$); the planet would become a frozen, lifeless ice ball. Venus, by contrast, provides an example of an extreme greenhouse effect; the atmosphere of Venus holds about 100,000 times as much CO_2 as the atmosphere of Earth, producing a thick and greatly insulating heat blanket and a scorching surface temperature of $450°C$ ($840°F$). Energy deposited beneath a greenhouse atmosphere is carried away from the surface as radiative heat emission and as convective, sensible, and latent heat.

The surface and atmosphere energy balance for the Earth is shown in Fig. 1. Basically, sunlight is absorbed in the lower atmosphere and at the ground, and infrared (heat) radiation escapes to space. A greenhouse effect is created because sunlight passes readily through transparent gas layers and heats the surface, but thermal energy, manifested as infrared radiation, is held in by the gas blanket. Clouds play an important role in the energy balance. They reflect sunlight into space and block infrared emission to space. The infrared emission which cools the Earth originates in the middle troposphere, above which the atmosphere is quite transparent to thermal infrared wavelengths. The overall terrestrial energy balance requires an air temperature of about $-25°C$ ($-13°F$) at the altitude where infrared energy is effectively emitted to space.

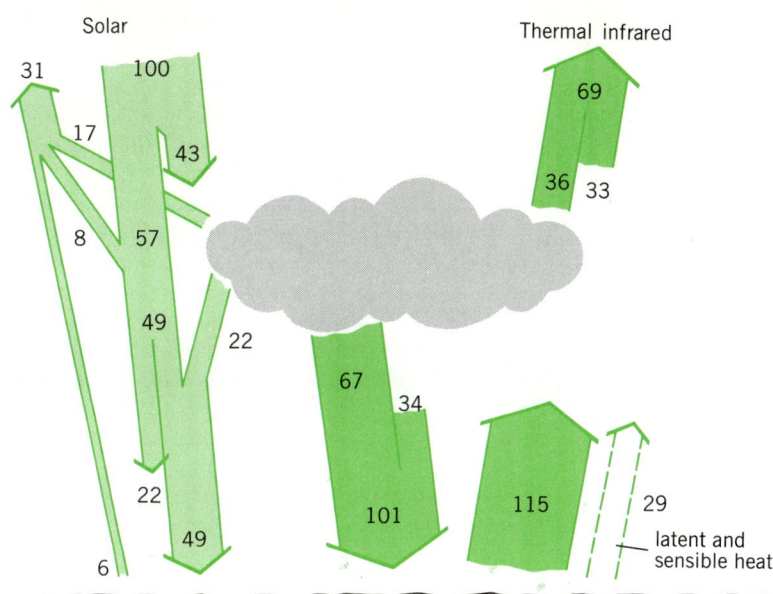

Fig. 1. The normal energy balance of the Earth's atmosphere; dimensionless units of energy are used.

When solar energy is absorbed above the greenhouse blanket, as occurs with a dense elevated layer of smoke, the normal greenhouse effect can be thwarted (also the smoky upper air layers may be abnormally heated, thereby producing unusual winds). Much of the solar energy absorbed in the smoke cloud is reradiated directly to space from the cloud top. Land surfaces receive only infrared energy that can leak through the overlying clouds and greenhouse blanket, and any residual solar radiation. As a result, the surface cools under the influence of an "antigreenhouse" effect as its initial heat, present before the darkness, leaks away to space (Fig. 2). The oceans can provide a continuing source of heat, but cannot completely ameliorate the strong cooling away from the coastlines. Nuclear winter therefore implies the reduction of sunlight at the surface, abnegation of the greenhouse effect, and subsequent cooling of the land masses.

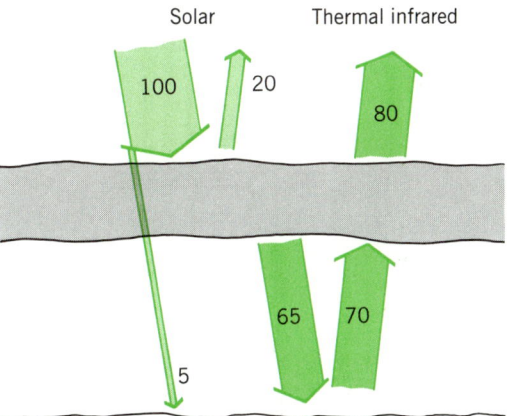

Fig. 2. Inversion of the normal energy balance when a dense cloud of smoke is placed in the middle or upper troposphere; dimensionless units of energy are used.

NUCLEAR EXPLOSIONS

Most of the explosive yield in the existing and projected nuclear arsenals of the world powers consists of warheads in the yield range 50 kilotons to 1 megaton. [The megaton (MT) is the standard measure of power of a nuclear weapon. One megaton is equivalent to 10^{15} calories, or roughly the energy released in the detonation of 1 million tons of TNT; the kiloton (KT) is equivalent in energy to 1000 tons of TNT.] The awesome destructive potential of modern nuclear weapons can be appreciated when it is recalled that the modest-sized city of Hiroshima was reduced to rubble and ashes by a 12-KT bomb. Although the stated targets of nuclear warheads are, in most instances, military facilities—for example, missile silos, airfields, submarine pens, and command centers—the destructive reach of modern warheads is so great that they would indiscriminately lay waste at least 100 mi^2 (250 km^2) around a target (Fig. 3). If a strategic military or industrial facility, near or in a city, was hit by a typical nuclear warhead, the city would sustain massive collateral damage.

The three principal physical manifestations of a nuclear explosion are the light, blast, and radioactivity. About one-third of the total energy of a nuclear explosion is emitted in a brief, intense pulse of light. The bomb light shows a spectrum similar to sunlight, but is magnified to such a degree that it can ignite fires in many common materials over a vast area around the explosion site (Fig. 3). Exposure to bomb light, or thermal flash or pulse, is measured in calories per square centimeter; at exposures of 10 to 20 cal/cm^2 from kiloton to megaton detonations, human skin sustains third-degree burns and paper, cloth, and dry vegetation ignite.

The blast, or pressure wave, is created by the sudden expansion of heated air as a nuclear device disassembles. The blast intensity can be measured by the peak increase in the static air pressure as the blast wave passes (so-called overpressure)—measured in pounds per square inch. The normal static air pressure at sea level is 14.7 psi. A peak overpressure of only 2 psi in a nuclear blast wave can destroy a wood-frame house. At shock pressures of 15 psi or greater, concrete and steel buildings are shattered. The explosion wave also generates incredibly strong winds (Fig. 3).

A nuclear explosion produces a variety of nuclear radiations. At the moment of detonation, large fluxes of high-energy neutrons and gamma rays are emitted. These radiations are dangerous, particularly in the case of low-yield explosions and enhanced-radiation weapons, or neutron bombs. The uranium and plutonium fission triggers for nuclear weapons generate up to several hundred distinct radioactive species, or radionuclides. These become attached primarily to small particles, many of which fall out of nuclear clouds within hours, contaminating the ground. A typical 1-MT weapon exploded on land will contaminate an area exceeding 100 mi^2 (250 km^2) adjacent to the burst with lethal radioactive fallout. The residual airborne radioactivity drifts around the Earth in the prevailing winds and is washed out slowly by rainfall, eventually contaminating vast regions of the planet with sublethal radioactivity. If a nuclear reactor was hit directly by a nuclear weapon, large concentrations of a potent mixture of very long-lived radionuclides—distinct from the mixture generated by nuclear fission—would be released to the environment.

The relative importance of the immediate nuclear effects varies with the altitude of detonation (often referred to as the height of burst, or hob). Explosions well above the ground (free airbursts) maximize thermal flash and blast effects and minimize the intensity of early radioactive fallout. Explosions on or very near land surfaces maximize the combined effects of blast and ground shock, create large areas of radioactive contamination, but reduce the thermal flash effect. Explosions below the surface, using penetrator warheads, maximize the ground shock but minimize the effects of blast, thermal flash, and radioactive fallout.

NUCLEAR ARSENALS AND SCENARIOS

The current worldwide nuclear arsenals contain about 50,000 warheads with a total yield of about 13,000 MT. Table 1 compares the energy of nuclear weapons with the energies of natural physical systems and events. Although the total energy bottled up in nuclear devices is quite modest in a geophysical sense, the ability to direct and focus nuclear explosives greatly amplifies their destructive capability.

Nuclear weapons are classified often as either strategic, theater, or tactical. Strategic weapons are long-distance devices such as intercontinental ballistic missiles and long-range bombers. Theater weapons are short- and medium-range missiles and aircraft. Tactical weapons are usually low-yield battlezone explosives such as artillery shells, neutron bombs, and depth charges. Most concern centers on

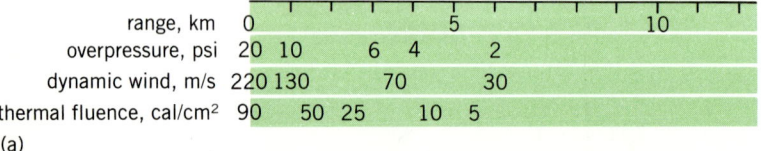

Fig. 3. The prompt effects of nuclear explosions are given for (a) 100-KT air burst detonated at a height of 1 mi (1.5 km) and (b) 1-MT air burst detonated at 2 mi (3 km). The peak overpressure is a measure of the destructive power of the blast wave. The thermal fluence is the total light energy collected on a surface of unit area facing the center of the fireball. A fluence of 10 calories/cm^2 is sufficient to cause severe skin burns and ignite fires spontaneously in a variety of materials. The thermal fluences shown assume a ground-level visibility of 12 mi (20 km).

Table 1. Comparative energies of nuclear weapons and natural physical systems and events

Source	Total energy, MT*
Nuclear arsenals	13,000
Full-scale nuclear war	5,000
Fires ignited in a nuclear war	30,000
Largest single nuclear detonation	60
Largest individual nuclear warhead	20
Krakatau explosion	100
Earthquake (8.0 Richter)	~100
Solar insolation at Earth (in 1 day)	3,000,000
Hurricane	~10,000,000
Cretaceous-Tertiary meteor impact	~100,000,000
Heat content of oceans†	~10^{11}
Solar output (in 1 second)	~10^{11}
Rotational energy of Earth	~10^{13}

*One megaton of energy is 10^{15} calories.

†The part accessible to humans as an energy source is estimated to be about 200 MT/day.

strategic and theater weapons because of their large explosive yields (50 KT to 10 MT each), although tactical weapons are troublesome because they may be the first actually used in a conflict, triggering escalation to theater and strategic weapons.

Each of the superpower alliances, NATO and the Warsaw Pact, has identified tens of thousands of potential targets for nuclear weapons should deterrence fail. The target lists include missile silos and control centers; mobile missile launchers; airbases; submarine ports; army and naval logistics stations; command, communication, and control facilities; ships at sea; military satellites; civilian airfields; key industrial areas; and economic centers. The sequence and manner in which these targets would be hit is referred to as a scenario. No one knows how a nuclear conflict would evolve; there is no meaningful theory. A growing number of strategists, recognizing the overwhelming difficulties and confusion that would arise in the initial phases of a nuclear exchange, conclude that rapid escalation to full strategic warfare is likely. War games carried out on computers suggest that thousands of megatons of weapons could promptly be detonated over a wide variety of targets. Of the 13,000 MT available, 5000 MT could readily be employed. If military systems were on alert and launch-on-warning directives were in effect, up to 10,000 MT could be detonated. Almost all of the explosions would occur as ground or air bursts.

In such nuclear wars, the implications for civilization are very grave. The immediate enormous destruction and casualties would extinguish a nation's greatness. Although the initial targets primarily would be military (silos and airbases), the fact that important military, industrial, and economic facilities are found in every major urban center makes the complete or partial destruction of up to 1000 cities possible in a global nuclear war. This destruction, in turn, implies the wrath of a nuclear winter, which may compound the ultimate costs manyfold.

SMOKE AND DUST

The urban zones of the NATO and Warsaw Pact countries sprawl over 200,000 mi² (500,000 km²), which amounts to roughly 0.1% of the total surface area of the Earth. In these urban regions are concentrated huge quantities of combustible materials such as lumber, oil, tar and asphalt, plastics, rubber, paper and cardboard, fabrics, industrial and household chemicals, and so on. At Hiroshima and Nagasaki the combination of nuclear flash and blast caused large areas of each city to burn. In a full-scale nuclear exchange, up to half of the urbanized areas in the war zones, particularly the heavily built-up city centers, could be subject to fires.

Forest, brush, and grassland encircle nearly all military bases. Estimates of forest burnoff areas in a nuclear conflict, taking into account target locations and nuclear weapons properties, range from minimum areas of 40,000 mi² (100,000 km²) to areas greater than 400,000 mi² (1,000,000 km²). The total forested area in the NATO and Warsaw Pact countries is about 7,000,000 mi² (18,000,000 km²). It is, accordingly, quite likely that 100,000 mi² (250,000 km²) or more of forest, brush, and grassland would be incinerated in a nuclear exchange.

Smoke properties. Between fires in cities and forests, perhaps 300 teragrams of smoke could be released to the atmosphere in a nuclear war. [In discussing nuclear smoke and dust emission, the adopted unit of mass is the teragram (tg); 1 tg equals 1,000,000 metric tons.] Two kinds of smoke would be distinguishable: dark sooty smoke produced by urban fires, and, light oily smoke produced by forest fires. Although both kinds of smoke contain particles with typical sizes of ~0.1 μm, sooty smoke absorbs light much more efficiently than oily smoke. Importantly, particles of 0.1-mm size are also relatively inefficient at absorbing infrared radiation and exhibit long atmospheric lifetimes. Smaller particles can diffuse rapidly to, and stick easily on, surfaces, including water droplets in clouds, whereas larger particles can settle out of the atmosphere (under the force of gravity) or collide with falling water droplets. Typical smoke particles neither diffuse nor fall rapidly, and tend to flow around objects without hitting them. Thus, smoke is an ideal light-obscuring material, and nuclear war can generate it at unprecedented rates.

Light extinction by smoke. The attenuation of light by smoke (and dust) is illustrated as a function of optical depth in Fig. 4. For smoke, the absorption and scattering components of the total optical depth are about equal. For dust, the absorption component is only about 2%. At all optical depths, sooty smoke is much more effective than soil dust in screening sunlight because soot is far more absorbing than dust. For example, at an optical depth of 3, only a few percent of the incident sunlight can penetrate a soot cloud, although more than 50% can diffuse through a dust cloud. At an optical depth of 10, less than 0.01% of the light is transmitted for soot versus

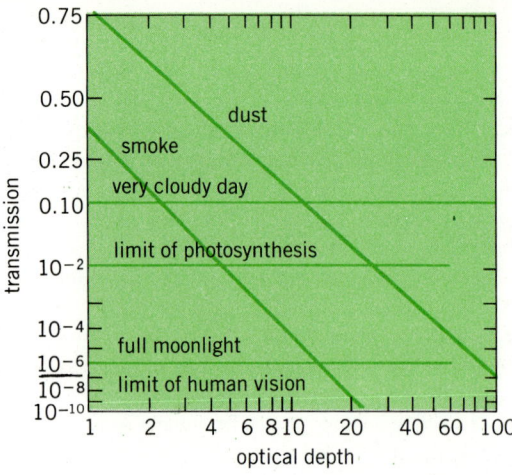

Fig. 4. The transmission of sunlight through smoke and dust clouds is shown as a function of the cloud optical depth. The transmission is defined as the fraction of incident light energy that penetrates the cloud. The transmitted light includes both direct and scattered components. In this figure, a wavelength of 550 nm (in the center of the visible spectrum) and an average sun angle of 45° from the zenith are assumed.

about 15% for dust.

If 300 tg of smoke was distributed uniformly over the Northern Hemisphere, an average optical depth of 5 to 9 would result at visible wavelengths. Even at an optical depth of 1 most of the light impinging on a smoke cloud is absorbed or scattered. In some nuclear war scenarios, light intensities below smoke clouds could be so low that photosynthesis would be impossible and human vision marginal.

Dust properties. Nuclear detonations on land generate large quantities of silicate (soil) dust. A 1-MT surface explosion can throw out several million tons of soil, producing a crater 1000 ft (300 m) across. Up to 200,000 tons of dirt and rock is liquefied and vaporized, leading to the formation of microscopic glassy spherules. Surface dust raised by the explosion shock wave and afterwinds can be sucked up by the rising fireball, adding to the dust load. Most of the dust particles are dangerously radioactive. According to measurements taken in clouds of nuclear test explosions, a 1-MT surface burst can lift up to 600,000 tons of dust to high altitude, of which up to 10% may consist of submicrometer particles. At this efficiency of injection, a full-scale nuclear conflict could inject 60 tg or more of submicrometer dust particles into the upper atmosphere. Once the dust had spread over the Northern Hemisphere, the average optical depth would be about 1. Although the reduction in sunlight beneath such a cloud is fairly modest (Fig. 4), the climatic implications are nevertheless significant, as will be explained below.

For the same optical depth, it is obvious that smoke, overall, creates a substantially larger optical impact than dust (Fig. 4). Moreover, for a given mass of material smoke is much more effective than dust in creating optical depth.

Debris injection heights. The eventual climatic impact of smoke and dust depends on their height of injection into the atmosphere. Generally, the higher an aerosol is placed above the ground, the longer it remains suspended, for two reasons. First, removal by clouds and precipitation, the natural scrubbing action of the atmosphere, is less efficient at higher altitudes. Second, above an altitude of roughly 7.5 mi (12 km; the tropopause) the atmosphere becomes very stable against vertical mixing and is therefore even less subject to cleansing.

The energy released by nuclear explosions and fires creates strong convective motions which draw air from low altitudes to high altitudes. In the case of nuclear fireballs, the test series of the 1950s and 1960s showed that, in general, megaton-yield explosions form clouds that stabilize entirely within the stratosphere, whereas 50-KT explosions form clouds that stabilize entirely within the troposphere. In actual wartime situations, with many bursts detonated over restricted areas in rapid sequence, the fireball dynamics could be significantly altered, and clouds could rise to higher altitudes than expected. There are no test data that define multiburst nuclear effects, but mathematical simulations have demonstrated enhanced cloud rise and dust-lifting capability.

Typically, the smoke produced by a large fire rises in a narrow convective column, stagnates as a cloud mass above the fire, and then spreads in the prevailing winds to form an extended stratus or cirrus cloud layer or plume. Most of the smoke is deposited well above the ground. Generally speaking, the more intense the fire, the greater the altitude of the smoke cloud. In a nuclear war, fires of unprecedented size could inject smoke well into the stratosphere, although it is expected that most of the smoke would be confined to the troposphere. During intense urban and forest fires, water vapor condensation in the smoke column occasionally produces a cumulonimbus rain cloud. The fire-induced precipitation can remove some of the smoke. Following the nuclear attacks on Hiroshima and Nagasaki, under humid maritime summer conditions, oily "black" rains poured from the smoke clouds. Nevertheless, the totality of evidence based on fire observations and plume simulations suggests that prompt washout of smoke may be relatively inefficient; in massive city fires the removal efficiency might be less than 10%.

Smoke and dust emissions. Recent estimates of the total particulate emissions in a full-scale nuclear exchange are summarized in Table 2 (the emissions are given for a number of nuclear war scenarios). It appears that significant optical perturbations could be created even with the limited use of nuclear weapons. Figure 5 illustrates the time behavior of the total smoke and dust optical depth for several of the cases listed in Table 2. The early-time optical depths (at about 1 day) do not include the effect of most of the forest fire smoke, (which is assumed to be emitted continuously over 1 week). Whenever the average hemispherical optical depth is greater than

Table 2. Smoke and dust production in a nuclear war

Case	Total yield, MT	% yield surface bursts	% yield urban industrial targets	Total number of explosions	Submicrometer smoke mass, tg	Submicrometer dust mass,[a] tg	Smoke optical depth[b]	Dust optical depth[b]
Nominal exchange	5,000	57	20	10,400	225	65	4.5	1.0
Low-yield airbursts	5,000	10	33	22,500[g]	300	15	6.0	0.2
Massive exchange	10,000	63	15	16,160	300	130	6.0	2.0
Restricted exchange	3,000	50	25	5,433	175	40	3.5	0.6
Counterforce strike[c]	3,000	70	0	2,150	0	55	0	0.8
Limited war[d]	1,000	50	25	2,250	50	10	1.0	0.1
City center attack[e]	100	0	100	1,000	150	0	3.0	0
Silos, "severe" case[c,f]	5,000	100	0	700	0	650	0	10
Future war	25,000	72	10	28,300[g]	400	325	8.0	5.0

[a]Only the stratospheric dust is counted. The tropospheric dust has a much smaller impact per unit mass injected.

[b]The initial vertical (zenith) extinction optical depth, τ_e, is given assuming that the material is distributed uniformly over the entire Northern Hemisphere. The optical extinction consists of a scattering component and an absorption component. In general, light transmission through a cloud is not simply related to τ_e, but must be calculated by using radiative transfer theory.

[c]No wildfires or urban fires are assumed to occur, which may be unrealistic.

[d]The nominal area of wildfires is reduced from 200,000 mi^2 (500,000 km^2) to 20,000 mi^2 (50,000 km^2).

[e]The central city combustible burden is taken to be 20 g/cm^2 (versus 10 g/cm^2 in the nominal case), and the net smoke emission is 0.026 g per gram of material burned (versus 0.011 g/g for city-center fires in the nominal case).

[f]A sixfold increase in the fine-dust mass lofted per megaton of yield is assumed, which is possible given the uncertainty in the nuclear data base.

[g]Complete MIRVing of existing missiles and new deployments of medium- and long-range missiles are assumed. (MIRV refers to multiple independently targeted reentry vehicles; that is, such a missile carries several nuclear warheads which separate above the atmosphere and follow trajectories to different targets.)

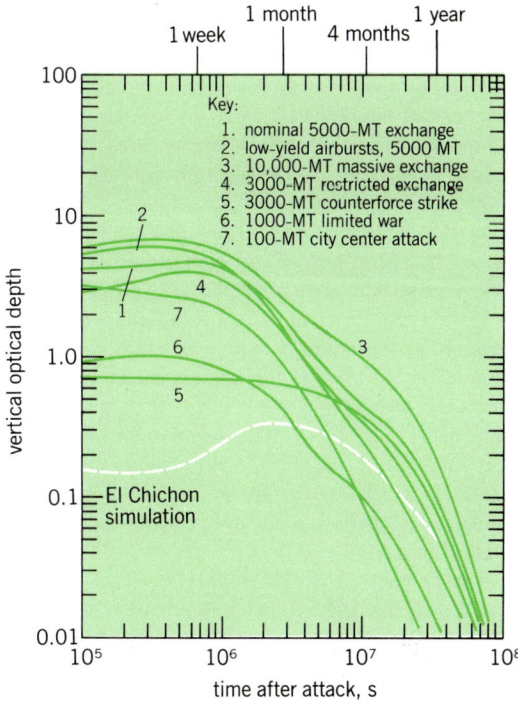

Fig. 5. Time-dependent, hemispherically averaged vertical optical depths (scattering plus absorption) of nuclear smoke and dust clouds at a wavelength of 550 nanometers. Vertical optical depths are expressed in dimensionless units. The average hemispherical optical depth of a major volcanic eruption cloud (El Chichon, April 1982) is shown for comparison.

about 1, significant optical and climatic disturbances are expected. The nuclear perturbations can persist for several months. However, longer or shorter durations are possible if fundamental changes in the debris removal rates were to occur as explained below. The optical depth perturbations illustrated in Fig. 5 lead directly to the climatic perturbations that characterize a nuclear winter.

POST–NUCLEAR WAR ENVIRONMENTS

Mathematical models have been developed which forecast post–nuclear war environments. As yet, none of the models treats adequately all of the complex physical aspects of the problem. The models are adequate, however, for obtaining a first-order description of nuclear winter. Calculations which define the essential features of a nuclear winter are reviewed below.

Light attenuation. Figure 6 shows the possible reductions in sunlight at the Earth's surface caused by clouds of debris generated in a nuclear war. Substantial reductions in light levels can persist for many weeks after a nuclear attack. Beneath dense, early smoke clouds, it may be so dark that even vision might be limited. The impact on ecosystems of reductions in sunlight varies widely with species. Because of competition, natural plant communities normally operate close to their compensation point, at which energy storage through photosynthesis just offsets energy consumption by metabolism. Primary food production by microscopic phytoplankton is the foundation of the oceanic food chain; hence, marine ecosystems are subject to collapse cascading from smaller to larger organisms. Many food crops are adapted narrowly to local climates through factors such as sunlight intensity, length of the day, duration of growing season (between frosts), and average monthly temperatures and rainfall. It follows that a variety of ecosystems and individual species could be damaged by light-level reductions as severe as those indicated in Fig. 6.

The attenuation of sunlight following a nuclear war may actually be one of the less severe environmental stresses, compared to temperature fluctuations, changes in precipitation, and radioactive fallout.

Temperature fluctuations. Changes in surface air temperatures over extended land masses for a number of nuclear war scenarios are illustrated in Fig. 7. Tropospheric smoke produces a large short-term cooling, while stratospheric dust produces a smaller but more prolonged cooling. The calculated temperature decreases are probably the greatest that would occur, and then only in continental interiors. Mean temperature decreases over oceans would probably be less than about 3°C (5°F).

These predictions do not account for patchiness in the clouds at early times or mixing of cold continental air with warm marine air at later times. Such effects lead to average land temperature decreases that are about one-half of the continental interior temperature drops shown in Fig. 7. Maximum temperature decreases of some 40°C (70°F) to an absolute temperature of around −25°C (−13°F) may be possible. In five of the cases illustrated in Fig. 7, dramatic decreases in continental air temperatures last for several months. Average temperature declines of only a few degrees Celsius in spring or early summer could destroy crops throughout the northern midlatitude belt, the world's breadbasket. The possibility of deep freezes, even in summer, emphasizes the potential peril of a nuclear winter—a time of gloomy bitter-cold days beneath black turbulent clouds, air choked with smoke and radioactivity, howling winds, and toxic snow.

In Fig. 8, calculated changes in air temperatures as a function of height and time following a 5000-MT nuclear war are shown. Several important features emerge. First, the upper atmosphere is heated strongly as the sunlight, which normally warms the ground, is absorbed in the highest smoke layers. At the same time, the ground cools in darkness. The hot clouds, like hot-air balloons, would not remain stationary (they were artificially constrained in the calculations illustrated) but would rise and expand, causing self-propelled dispersion of the nuclear debris.

The heating of dispersed nuclear clouds and the cooling of the ground below them can produce a strong temperature inversion in the lower atmosphere. Such an inversion is illustrated in Fig. 9, where perturbed vertical temperature profiles are compared to an ambient temperature profile. The troposphere usually exhibits an average decrease of temperature with increasing altitude of about 6.5°C/km, which defines the nominal stability of the atmosphere. Figure 9 suggests that, one month following a massive nuclear exchange, vast tropospheric

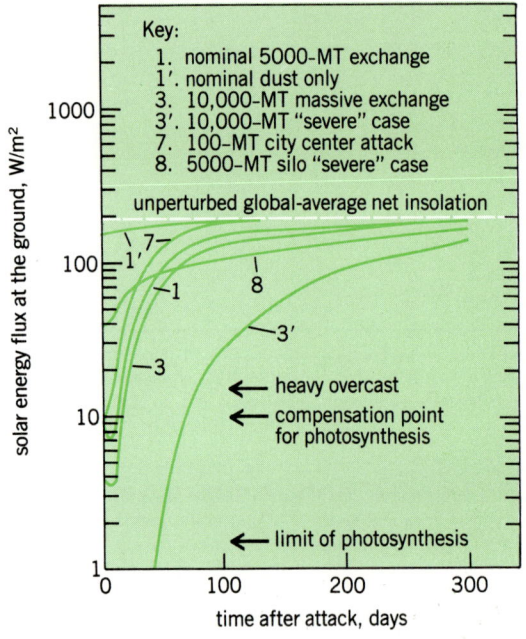

Fig. 6. The average intensity of sunlight penetrating hemispherically dispersed clouds of smoke and dust is shown for several nuclear war scenarios (see Table 2). Indicated on the figure are the approximate energy levels at which photosynthesis cannot keep pace with plant respiration (the compensation point), and at which photosynthesis ceases effectively. These limits vary widely for different species.

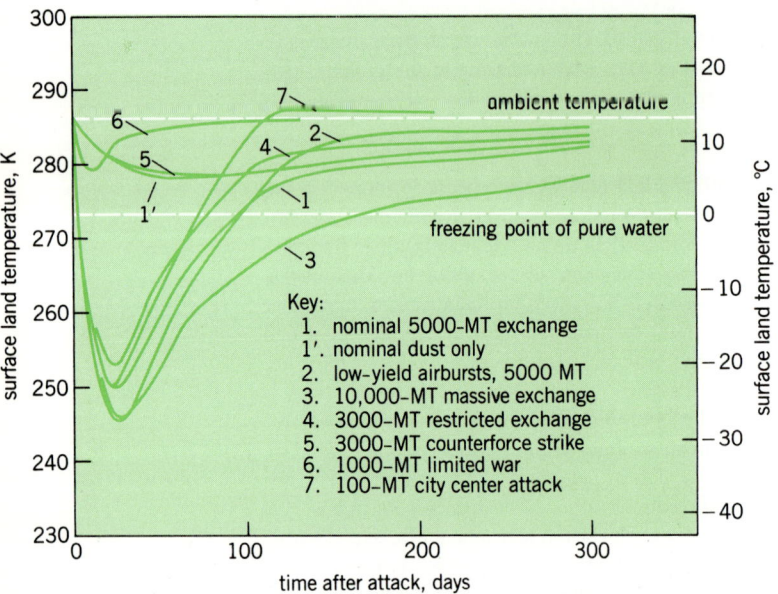

Fig. 7. Time-dependent changes in surface air temperatures over inland continental areas are shown for several nuclear war scenarios (see Table 2). The temperatures were calculated with a one-dimensional radiative-convective model. The temperature variations shown here correspond to a seasonally averaged sunlight intensity. If the war occurred in summer, larger temperature drops would be expected, and if in winter, smaller temperature drops.

air masses could be thermally stabilized over land; that is, the atmospheric temperature, rather than decreasing by 6.5°C/km, decreases more slowly or increases with increasing altitude. Even after 3 months, the ground is only receiving sufficient solar energy to drive weak convection in the lowest layers of the atmosphere. Such a smoke cloud temperature inversion is likely to increase the atmospheric residence time of nuclear smoke and dust by inhibiting vertical mixing and suppressing deep convective cloud activity.

Geographical distribution of nuclear winter. Recently, sophisticated global circulation models (GCMs) have been used to study the geographical distribution of the nuclear winter. The GCMs are not yet designed to move smoke and dust around as tracer elements or to make the required detailed radiative transfer calculations, but they are designed to reveal the initial three-dimensional perturbations in winds and temperatures caused by massive smoke injections.

Figure 10 illustrates the southward progression of freezing land temperatures for the first 10 days following a summer nuclear war in which a massive uniform smoke cloud has been generated at northern midlatitudes (30–70°N). It is found that nearly all of the land below the smoke cloud is subject to quick freezing; that is, local temperatures can fall below zero Celsius within 2 or 3 days. The surface may subsequently warm above the freezing point and possibly refreeze again in cycles of freezing and thawing. It follows that much of the midlatitude land area not shown to be solidly frozen in Fig. 10 has nevertheless suffered a quick freeze (that could devastate crops and vegetation). The area of freezing in Fig. 10 has not spread farther southward toward the Equator in part because the southern edge of the smoke cloud was artificially held at 30°N latitude. Yet, startlingly, after only 10 days all of the prominent continental land masses above 30°N latitude are predicted to be frozen.

Such theoretical studies also suggest that nuclear smoke and dust clouds could spread rapidly from northern midlatitudes into the subtropics and tropics and possibly into the Southern Hemisphere. Figure 11 illustrates a GCM prediction of the ambient and perturbed average meridional circulation of the atmosphere. Normally, the meridional motions are dominated by the powerful Hadley circulation, which rises over the warm, humid tropics, splits into two cells, and descends over the subtropics and midlatitudes in both hemispheres (Fig. 11a). Secondary circulation cells are established at higher latitudes. When a massive tropospheric soot cloud is introduced at northern midlatitudes (within the region delineated by the box in Fig. 11b) in spring or summer, heating at the southernmost edge of the cloud is intense enough to reverse the midlatitude subsidence and switch the bifurcated Hadley circulation into an unusual dynamical mode characterized by a single dominant cell with upper-level winds blowing briskly from northern to southern lat-

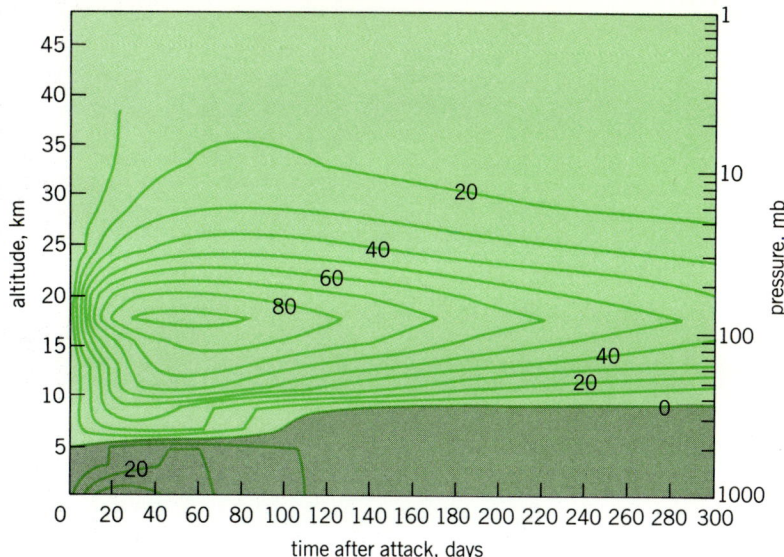

Fig. 8. Time- and altitude-dependent variations in atmospheric temperatures over Northern Hemisphere land following 5000-MT nuclear war. Contours of temperature change are given at 10°C (18°F) intervals. The shaded region shows cooling below normal temperatures; the clear region, heating above normal. The strong heating at 6–16 mi (10–25 km) is due to sunlight absorption by smoke and dust particles.

itudes. Such profound changes in the global wind system had never before been contemplated for a nuclear war.

The average meridional circulation is the residual motion of large-scale planetary wave oscillations. The GCMs predict anomalies in the planetary wave

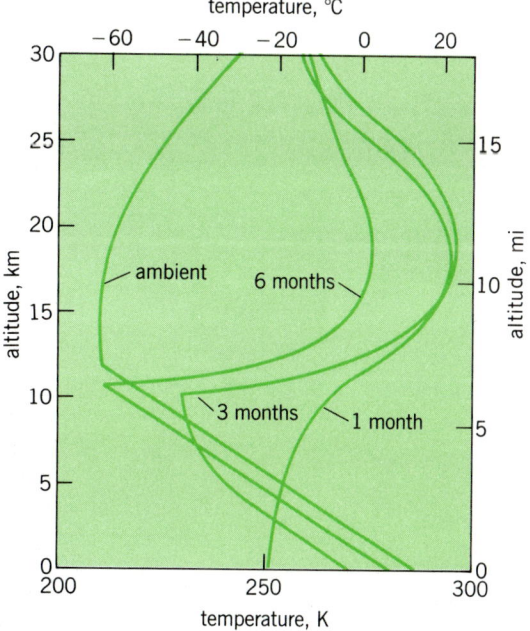

Fig. 9. Atmospheric temperature profiles before and after 5000-MT nuclear war. Temperature calculations account for radiative transfer at visible and infrared wavelengths, surface albedo, and convective adjustment, whereby atmospheric stability is maintained by mixing energy from heated air layers into cooler overlying air layers. Disregarded are large-scale dynamical responses to thermal perturbation, which could dramatically alter temperature profiles.

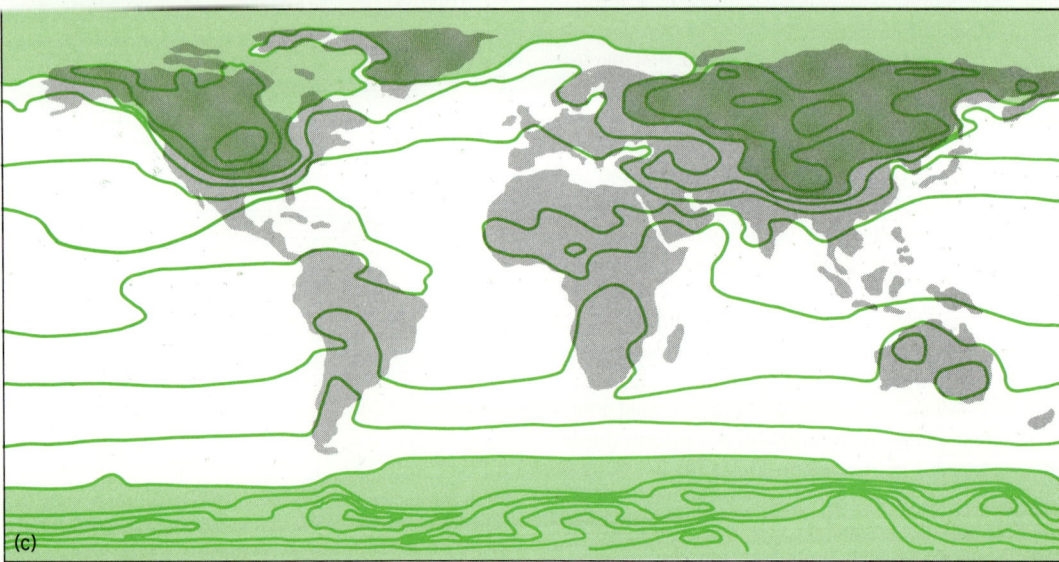

Fig. 10. Global surface temperatures are shown before and after a nuclear war in which a massive smoke cloud is injected at northern midlatitudes during the Northern Hemisphere summer. (The smoke is distributed uniformly between 30° and 70° N latitudes at altitudes from 1 to 10 km.) Freezing land temperatures are shown to progress rapidly (a, $t = 0$; b, $t = 2$ days; c, $t = 10$ days) southward over the continents during the first 10 days. Limitations of the calculations are discussed in the text. (*After C. Covey et al., Global atmospheric effects of massive smoke injections . . ., Nature, 308:21–25, 1984*)

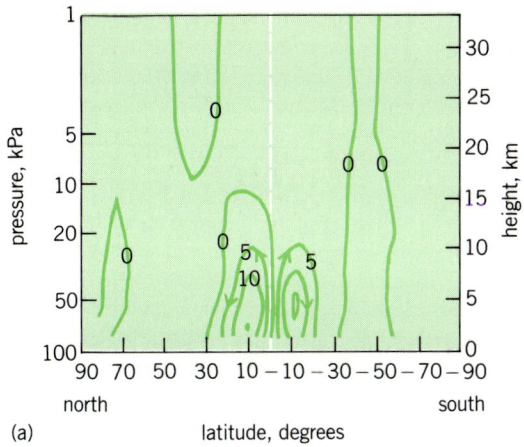 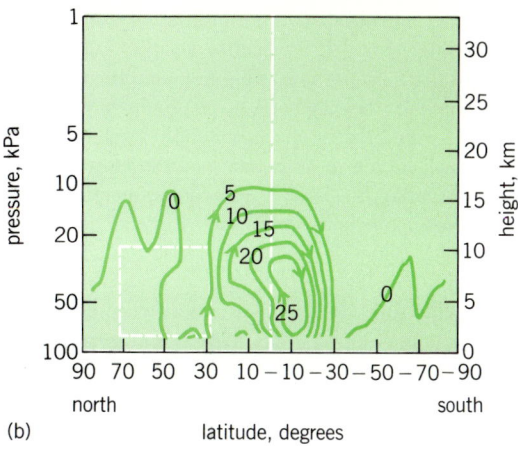

Fig. 11. The average meridional (north-south) circulation of the Earth's atmosphere is illustrated. The contours shown define the mass flow of air, in units of 10 tg/s; the winds follow these contours in the directions indicated by arrows. (a) April control. (b) April perturbed. (After C. Covey, et al., Global atmospheric effects of massive smoke injections . . ., Nature, 308:21–25, 1984)

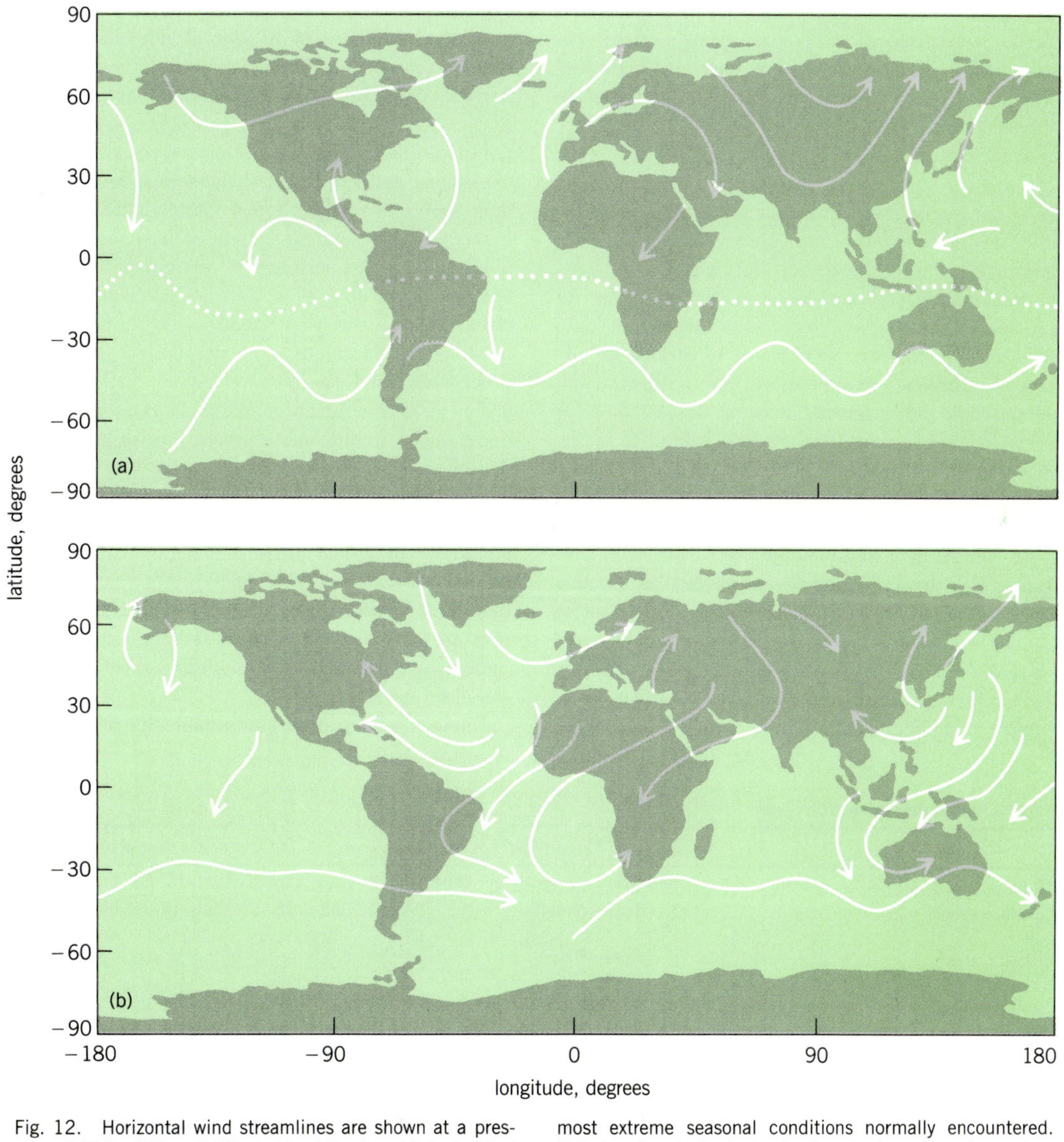

Fig. 12. Horizontal wind streamlines are shown at a pressure of 200 millibars (altitude about 7 mi or 11 km). Wind vectors correspond to (a) normal winter case (January) and (b) nuclear-perturbed spring case (April) 2–3 weeks after injection of massive smoke cloud. The dotted line in a indicates southernmost limit of southward-directed winds under most extreme seasonal conditions normally encountered. This interhemispheric barrier has broken down in nuclear winter simulation in b. (After C. Covey et al., Global atmospheric effects of massive smoke injections . . ., Nature, 308:21–25, 1984)

motions, and here, too, the preliminary results are surprising (Fig. 12). Continent-sized parcels of heated northern air seem able to penetrate deep into the Southern Hemisphere (to 40°S) in a matter of days. Essentially all of the habitable land masses of the Earth lie above 40°S and may therefore be subject to rapid blackout by nuclear-generated soot. With rapid blackouts and quick freezes, no area of the globe, north or south, may be safe from nuclear winter.

Weather effects. Weather is the day-to-day variation in temperature, winds, clouds and precipitation at a fixed locale. Climate is the average weather over extended times and areas. It is much easier to predict climate than weather. A major climatic perturbation such as a nuclear winter is expected to generate unprecedented weather fluctuations. Even a rather mild nuclear winter, characterized by an average temperature decline of only a few degrees Celsius, could produce severe weather anomalies worldwide (such as the quick freezes mentioned earlier). Nevertheless, it is currently beyond researches' capability to predict exactly where and when such anomalies might occur, or their precise intensity. Existing GCMs can only reveal exemplary weather patterns.

Common experience with weather suggests that, in a nuclear winter, storms and turbulence might increase in the lower atmosphere, driven by heat and moisture from the oceans and by strong horizontal temperature contrasts between land and seas. As prevailing winds swept masses of marine air onto cold continents, extensive bands of stratus clouds and continuous precipitation (rain, sleet, and snow) could be triggered. It is not known how far such weather activity might reach inland from the coastlines, but a 60-mi (100-km) margin probably would contain most of the activity. As land surfaces cooled rapidly, widespread fogs could develop, trapping surface heat and maintaining surface temperatures. The presence of millions of tons of nuclear debris particles in the atmosphere might also modify the microphysical properties of cloud droplets and thus the removal rate of the debris. Water injected by nuclear explosions and fires might affect atmospheric chemical and radiative processes as well.

Moreover, physical and chemical interactions between the atmosphere and the oceans might occur, interactions that could exercise important influences on short-term (1–3-year) climatic changes. All of these second-order effects require additional study.

SUMMARY

Nuclear winter is a new phenomenon associated with the massive use of nuclear weapons. A number of difficulties exist in predicting its severity. The origin and magnitude of the major uncertainties associated with nuclear winter are summarized in Table 3. Although some of the uncertainties could be reduced by further research, including experimental and theoretical work, a few elements of the nuclear winter problem, including the scenarios for nuclear warfare, will never be known precisely. Nevertheless, the overall knowledge of nuclear winter should improve significantly in the next few years as several national and international programs now focused on the problem reach completion.

In the aftermath of a general nuclear conflict, large areas of the Earth could be subject to long periods of dense overcast, hard freezes, killing frosts, toxic snowfalls, and gale-force winds; to a polluted atmosphere of poisonous gases released by explosions and fires; and to dangerous levels of radioactivity—with vast areas contaminated beyond habitation, food and water tainted by dangerous radionuclides, and millions of anticipated cases of cancer and birth defects; and to increased doses of lethal solar ultraviolet radiation. With the breakdown of transportation systems, power grids, food processing, medical care, sanitation, civil services, and central government, the world's economic interdependence would create severe hardships even in countries not directly involved in the war. Epidemics could sweep the Earth. Many tropical, subtropical, and temperate species might disappear, and countless individual organisms would perish in the nuclear devastation. Although predictions of such biological effects are still largely qualitative, scientists are now working to place understanding of the biological consequences of nuclear war on a sounder scientific basis.

Nuclear winter poses formidable barriers to the

Table 3. Uncertainty in nuclear winter

Source	Magnitude*	Nature
Nuclear scenario	Large (moderate)	Irreducible
Wildfire smoke	Large (small)	Reducible by analysis and experimentation
Urban smoke	Large (large)	Partly reducible by analysis and experimentation
Nuclear dust	Large (small)	Partly reducible by analysis
Short-term climatological responses	Moderate (moderate)	Reducible by analysis
Local weather effects	Large (moderate)	Mostly irreducible†
Long-term climatic changes	Large (large?)	Partly reducible by analysis

*First, the general magnitude of the uncertainty in the source element is given, and then, in parentheses, the contribution of the source uncertainty to the overall uncertainty in the severity of a nuclear winter.

†Precise weather prediction over a period of months is currently impossible; exemplary weather effects may be calculated with existing models, however.

establishment of effective schemes for national civil defense and postwar economic recovery. More importantly, it raises fundamental questions about nuclear strategy and policy in a world that is characterized by marginally stable political and military systems.

[RICHARD P. TURCO]

Bibliography: C. Covey, S. H. Schneider, and S. L. Thompson, Global atmospheric effects of massive smoke injections from a nuclear war: Results from general circulation model simulations, *Nature*, 308:21–25, 1984; P. J. Crutzen and J. W. Birks, The atmosphere after a nuclear war: Twilight at noon, *Ambio*, 11:114–125, 1982; P. J. Crutzen, I. E. Galbally, and C. Brühl, Atmospheric effects from post-nuclear fires, *Climat. Change*, 6:323–364, 1984; P. R. Ehrlich et al., Long-term biological consequences of nuclear war, *Science*, 222:1293–1300, 1983; S. Glasstone, and P. J. Dolan, *The Effects of Nuclear Weapons*, U.S. Department of Defense, 1977; E. Ishikawa and D. L. Swain (trans.), *Hiroshima and Nagasaki: The Physical, Medical and Social Effects of the Atomic Bombings*, 1981; National Academy of Sciences, *The Effects on the Atmosphere of a Major Nuclear Exchange*, 1984; National Academy of Sciences, *Long-Term Worldwide Effects of Multiple Nuclear-Weapon Detonations*, 1975; Office of Technology Assessment, *The Effects of Nuclear War*, OTA-NS-89, May 1979; R. P. Turco, O. B. Toon, T. P. Ackerman, J. B. Pollack, and C. Sogan, The climatic effects of nuclear war, *Sci. Amer.*, 251:33–43, August 1984; R. P. Turco et al., Nuclear winter: Global consequences of multiple nuclear explosions, *Science*, 222:1283–1292, 1983.

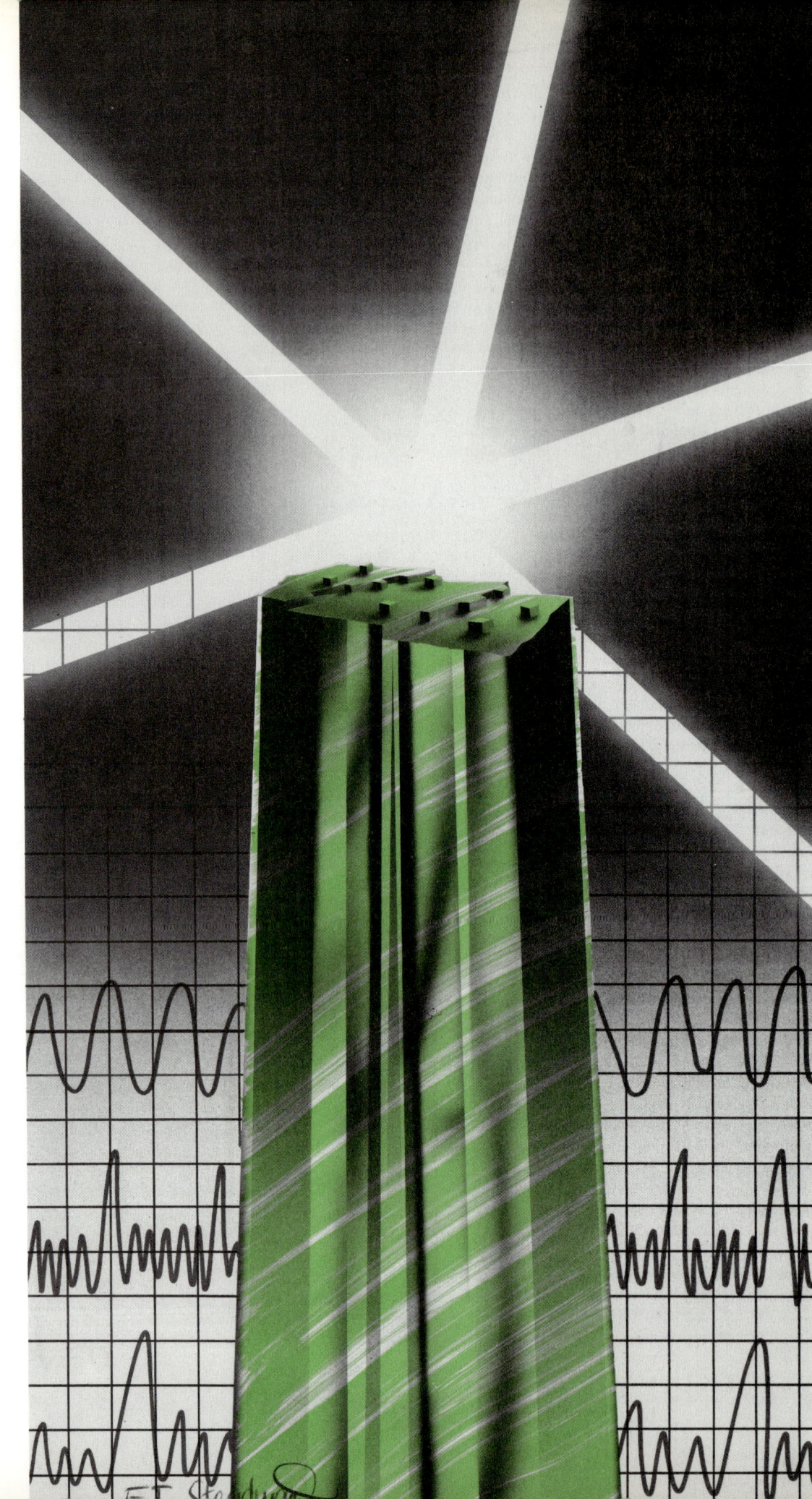

Surface Science

John B. Hudson

Recipient of a doctorate in metallurgy from Rensselaer Polytechnic Institute, John B. Hudson is professor of materials engineering there. He has been engaged in research in surface science for over 20 years, and is the author of more than 50 technical papers in the areas of gas-surface interactions and surface experimental techniques.

Modern surface science can trace its roots to the work of Irving Langmuir in the early 1900s. Then, as now, the emphasis was on the solution of practical problems, as much of his early work was concerned with the blackening of incandescent light bulbs during use. The major growth in the understanding of surfaces, however, has come primarily in the past two decades. There have been two major factors involved in this recent advance—namely, the ready availability of ultrahigh-vacuum systems, which permit the study of a truly clean surface over an extended period of time; and the development of a wide range of surface-analytical techniques based on the interaction of atoms, ions, electrons, and photons with these surfaces.

This article will begin by discussing a number of these surface spectroscopies and will describe the general picture of the surface that has emerged from their application. The discussion will then detail the application of these techniques to such technologically important areas as the catalysis of chemical reactions, the corrosion of materials, friction and wear processes, and the production and analysis of microelectronic devices.

TOOLS OF THE TRADE

As mentioned above, the recent rapid progress in surface science has been made possible in part by improvements in ultrahigh-vacuum technology and in part by the development of surface spectroscopies. Consider first the role of ultrahigh-vacuum technology. A surface exposed to the atmosphere will be struck by gas molecules at a rate sufficient to completely cover it with a layer one molecule thick in a very small fraction of a second. In order to maintain a surface at a known degree of cleanliness for the several minutes to several hours required to study its structure, composition or chemical behavior, the pressure must be reduced below 10^{-12} atm (10^{-7} Pa). Specialized systems capable of reaching this pressure range were custom-built from Pyrex glass as early as the 1930s. It was not until the development of the space program in the 1960s that new techniques for obtaining and measuring pressures in the ultrahigh-vacuum range became generally

16 Surface science

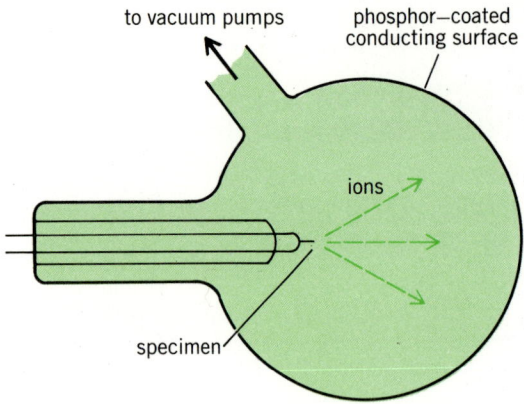

Fig. 1. Diagram of the arrangement of a field ion microscope. Ions are emitted from the tip and form a highly magnified image on the phosphor screen. (*After J. W. Blakely, Introduction to the Properties of Crystal Surfaces, Pergamon Press, 1973*)

available. It is possible now to obtain commercial systems—with stainless steel vacuum chambers, readily demountable flanges, and electronic or cryogenic pumping systems—that will operate routinely in the 10^{-13} atm (10^{-8} Pa) range over a virtually unlimited period of time.

The other major advance has been the development of techniques for determining the structure and composition of small surface areas, and of characterizing the changes that take place when these surfaces are exposed to small amounts of reactive gases. These techniques are based primarily on the processes that take place when a surface, maintained in an ultrahigh-vacuum environment, is exposed to a flux of particles, such as atoms, ions, electrons, or photons of various wavelengths. Development of these techniques began in the mid-1960s and continues today at an unslackened pace. Over this time period, about one new technique per year has been added to the list of those applicable to surface studies. The types of information attainable from these techniques have been categorized as surface atomic structure, surface composition, surface electronic structure, and surface atomic dynamics. One of the most commonly used techniques for each of these categories will be described below, and typical examples of the results obtained will be given.

Surface atomic structure. Most of the surface structure studies carried out to date have been on crystalline solids. The regular arrangement of the atoms in these materials makes the problem of structure determination tractable, as the number and kind of atomic arrangements, and the number and kind of imperfections in the atomic arrangements, are relatively small.

The general conclusion from these studies is that, for most materials, the atomic arrangement at the surface is not very different from what one would observe if the surface were formed simply by cutting the bulk crystal along the desired crystallographic direction. The most striking confirmation of this picture has come from studies using the field ion microscope (Fig. 1). The sample studied is the tip of a fine metal wire, prepared so that the very end is spherical, with a radius on the order of 10 nanometers. This tip is held at the center of a spherical glass bulb, which is coated with a phosphor material. In operation, the bulb is filled with helium at low pressure, and the tip is held at a high positive voltage. This causes the helium atoms to become ionized, and then accelerated away from the tip to the phosphor screen, where they form an image of the tip, magnified about 1 million times. This magnification is sufficient to permit resolution of the individual atomic positions on the surface.

In the field ion micrograph shown in Fig. 2a, the sample is a single-crystal tungsten wire. Each of the bright dots represents the position of an atom on the surface. In the cork ball model of this surface (Fig. 2b), the correspondence between ball position and spot positions in the micrograph is obvious. The surface of the tip is made up of atomically flat areas, called terraces, separated by steps of monatomic height called ledges. Where these ledges change di-

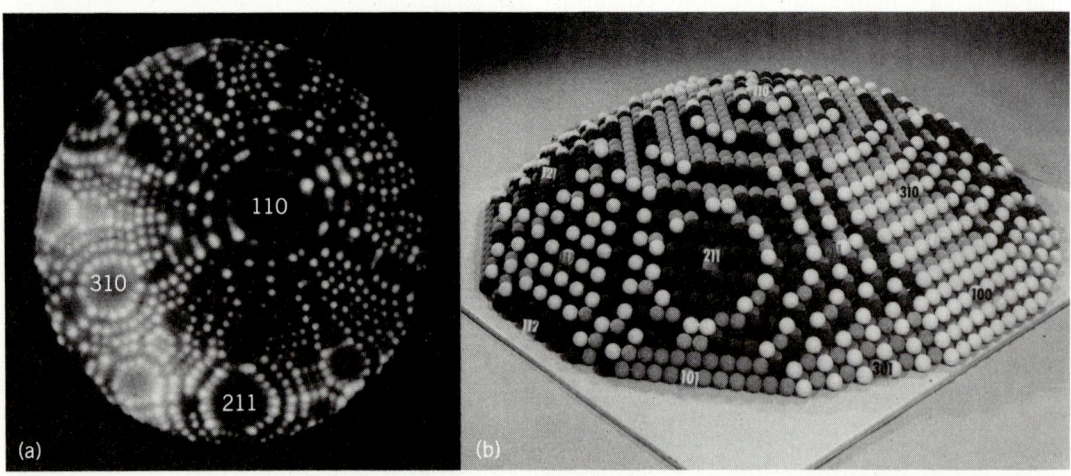

Fig. 2. Surface atomic structure: (a) field ion micrograph of a tungsten metal tip and (b) cork ball model of the same surface. (*Courtesy of Gert Ehrlich*)

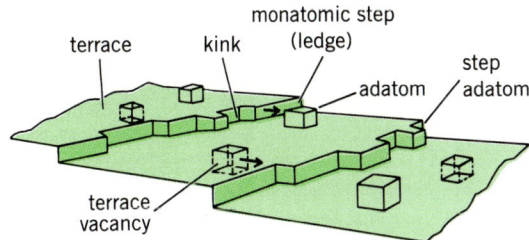

Fig. 3. Diagram of the surface of a crystalline material composed of terraces, ledges (steps), and kinks, showing a variety of surface site configurations. (*After G. A. Somorjai, Chemistry in Two Dimensions, Cornell University Press, 1981*)

rection are sites having a still different symmetry called kinks. It is possible to describe the surface of any crystalline solid in terms of these three types of sites (Fig. 3), plus defect sites arising from the termination of bulk defects in the crystal. It will be seen in the discussion of surface reactivity that the various types of defect sites, such as ledges and kinks, play a major role in determining both the overall reactivity and the selectivity for a particular type of reaction.

Another important aspect of surface atomic structure is the study of the (usually small) differences between the bulk atomic arrangement and that at the surface. This problem is studied by using low-energy electron diffraction. In this technique, the surface is bombarded by a beam of low-energy electrons (around 100–200 V). The electrons, which can be thought of in this case as behaving like radiation and having a well-defined wavelength, are diffracted from the lattice of the solid just as x-rays are diffracted from a three-dimensional crystal or as visible light is diffracted from a regular array of lines or points. Analysis of the resulting diffraction pattern provides information on the way that atoms at the surface are arranged relative to one another, and on how the near-surface arrangement differs from that in the bulk. This is true both for those atoms that make up the clean surface and for any other atoms that are present as a surface layer of some different species. (Such a layer of different species at a surface is called an absorbed layer.) The general result of these studies is that the differences between bulk and surface atomic spacing are usually small, except in compounds with primarily covalent bonding. The atomic arrangement in absorbed layers is usually some simple multiple of the interatomic distance on the clean surface, allowing the absorbed layer structure to be described in terms of the symmetry of the underlying surface.

Surface chemical composition. The technique that has provided the most information on surface chemical composition is Auger electron spectroscopy. In this technique, the surface is bombarded by a beam of electrons in the kilovolt energy range. These electrons interact with the electrons of the solid, ejecting electrons from their stable positions in the atom and leaving the atom in an excited state.

The excited atom can relax by a rearrangement of the remaining electrons. The net result of this rearrangement is that an electron having a well-defined kinetic energy is emitted from the atom. The value of this energy depends on the kind of atom involved. Measurement of the energy distribution of the electrons emitted from the surface provides information on the amount and kind of atoms in the near-surface region of the sample. A typical Auger spectrum, obtained from a clean stainless steel surface, is shown in Fig. 4. The position of the peaks along the energy axis indicates the kind of atom present in the sample. The peak-to-peak height of each of these features is proportional to the concentration of that type of atom in the near-surface region. The surface sensitivity of this technique arises from the fact that the emitted electrons, which have relatively low energies, cannot travel very far through the solid without colliding with other electrons, losing energy, and consequently not contributing to the electron current at the characteristic energy. As a practical matter, the effective depth probed by Auger electron spectroscopy is two to five atomic layers into the sample.

A variant of this technique is scanning Auger microscopy. The exciting electron beam is focused to a diameter as small as 50 nm. This focused beam can be scanned over the surface, similar to the electron beam in a television tube.

Scanning Auger microscopy can be used in several different modes. The total secondary electron current generated by the interaction of the electron beam with the surface can be collected, and used to control the contrast on a television monitor swept in sequence with the electron beam. This will generate an image of the sample on the screen, just as is done in conventional scanning electron microscopy. The image permits one to identify areas of interest for chemical analysis by Auger electron spectros-

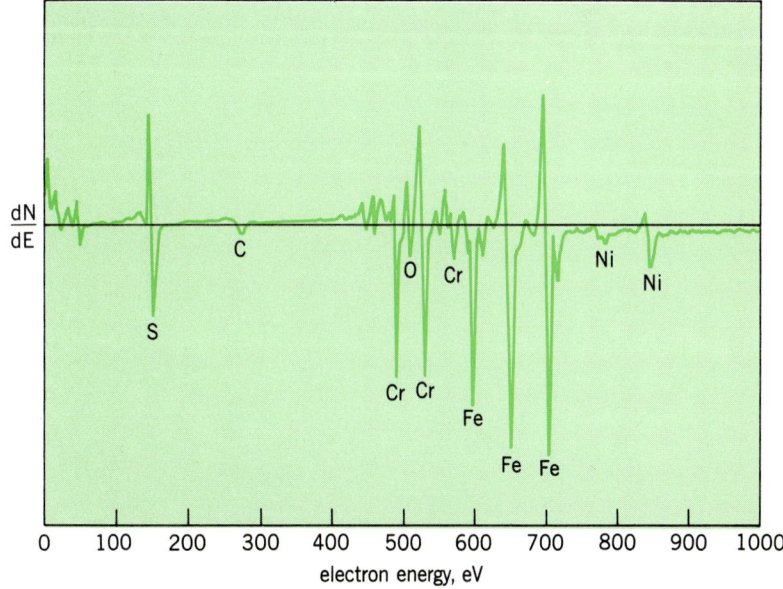

Fig. 4. Auger electron spectrum of a stainless steel surface, showing the expected peaks for the metallic species present, plus a sulfur surface impurity.

copy. Conventional Auger spectra, such as the one shown in Fig. 4, can then be obtained from areas the same size as the electron-beam diameter. The total volume of material sampled can be as small as 2×10^{-19} in.^{3} (3×10^{-18} cm^3), corresponding to roughly 100,000 atoms.

Alternatively, the associated electron energy analyzer can be tuned to an Auger energy corresponding to a particular element. If the resulting Auger electron signal is used to control the contrast on the monitor, a video image is formed in which the contrast arises from differences in the concentration of the element chosen over the surface of the sample. This technique is finding extensive use in the microelectronics industry. Examples of this application are given later.

Surface electronic structure. Surface electronic structure can be most readily studied by using ultraviolet photoelectron spectroscopy. Here the intent is to determine the binding energy of the electrons involved in chemical bonds at the surface. This is accomplished by bombarding the surface with photons, usually in the far-ultraviolet range, having photon energies from 20 to 100 eV (wavelengths of 50 to 10 nm). These photons interact with electrons in the solid, causing electrons to be ejected with a kinetic energy that is the difference between the energy of the exciting photon and the binding energy of the electron. As in Auger spectroscopy, an electron energy analyzer is used to determine the kinetic energy distribution. From this, and by knowing the initial photon energy, one can calculate the electron binding energy.

This technique is very useful in determining the energies of the electrons involved in chemical bonding. For example, Fig. 5 shows a series of spectra obtained when a clean nickel surface is first exposed to acetylene (C_2H_2) at low temperature, then heated to cause surface decomposition. Curve A shows the binding energy of electrons present in the clean surface. Curve B shows additional states associated with the bonding electrons in the acetylene molecule, and a reduction in the states associated with the clean surface. In curve C, obtained after heating to 425 K (305°F), many of the states associated with the acetylene have disappeared, leaving only a group of states associated with the bonding of carbon atoms to the nickel surface.

Surface atomic dynamics. Much of the study of the chemistry of surfaces involves measurement of the properties of the adsorbed layers mentioned in the discussion of low-energy diffraction. Some of this information is obtained from the analytical techniques already described. In the case of complex molecules on the surface, such as are formed in catalytic reactions, other techniques are required to determine the structure and orientation of the species formed on the surface. Here again electrons or photons are used as probes. In this case, however, since information is sought on the chemical structure of surface species, much lower-energy particles are used. Electron energies in the 5–10-eV range and photons in the visible or infrared range are used. These probe particles provide information on the molecular vibrational modes present at the surface and can thus differentiate between, for example, C_2H_4 molecules and the CH_2 surface fragments that might be formed in the surface decomposition of the parent molecule. These techniques, including electron energy loss spectroscopy and visible or infrared absorption spectroscopy, are similar in concept to the techniques used for chemical structure analysis in bulk gases. In the case of electron energy loss spectroscopy, a monoenergetic beam of electrons probes the surface. Some of the electrons lose energy to surface atoms or molecules, causing them to vibrate at frequencies that are characteristic of the type of chemical bonding present. These electrons are then detected, and the amount of energy loss used as a fingerprint for the type of molecule with which they interacted. Infrared and optical spectroscopies involve probing the surface with a light beam of variable wavelength (that is, varying photon energy) and noting the wavelengths at which light is absorbed due to excitation of surface species. For a given surface excitation, the energy lost by an electron in electron energy loss spectroscopy is equal to the energy equivalent of the photon absorbed in the optical experiment.

A typical result of an electron energy loss measurement is shown in Fig. 6. Here again, the adsorption of acetylene is shown, this time on an iron surface. The top spectrum is obtained from acetylene adsorbed at low temperature (120 K or −244°F). Analysis of the frequencies of the loss peaks indicates that the molecule is held without breakdown at this temperature. The second spec-

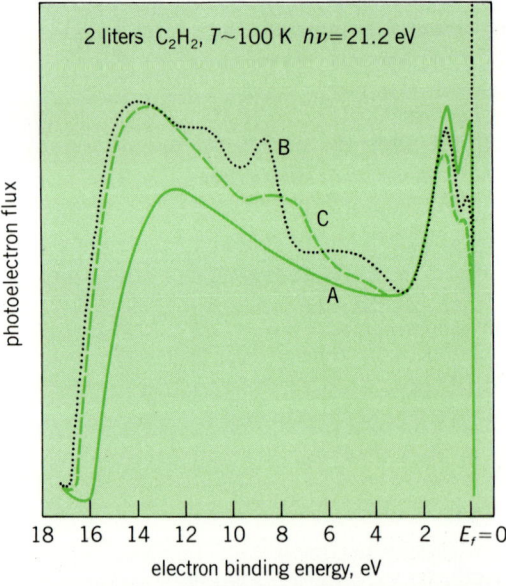

Fig. 5 Ultraviolet photoemission spectra of a clean nickel surface after exposure to acetylene gas (C_2H_2). States associated with the acetylene molecule disappear when the sample is heated. (*After J. E. Demuth, The reaction of acetylene with Ni(100) and Ni(111) surface at room temperature, Surface Science, 93:127–144, 1980*)

Surface science 19

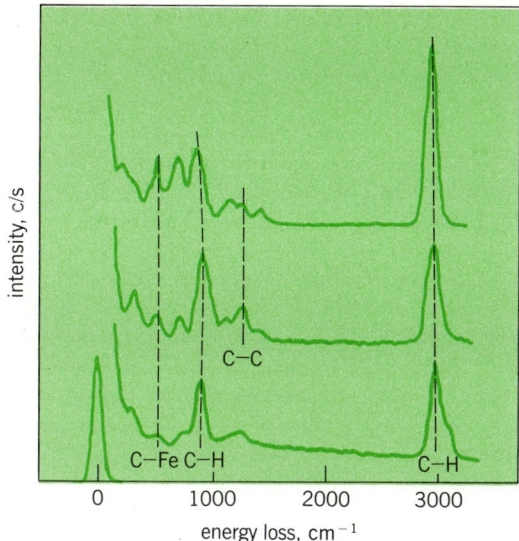

Fig. 6. Low-energy electron loss spectra of acetylene adsorbed on an iron surface, showing thermally induced decomposition at higher temperatures. (*After W. Erley et al., Vibrational spectra of acetylene and ethylene adsorbed on Fe(110), Surf. Sci., 120:276–290, 1982*)

trum shows the changes that take place when the sample is heated to 420 K (296°F). These changes are interpreted as the loss of one hydrogen atom to form a surface C_2H fragment. The bottom spectrum, taken after heating to 550 K (530°F), indicates breakage of the carbon–carbon bond to form CH fragments. The course of the surface decomposition process can thus be followed in great detail.

APPLICATIONS

Some of the ways in which the techniques described above are being applied to the solution of practical technological problems will be discussed below. It will be seen that several areas of technology are involved. Recent results obtained in such diverse areas as catalysis, corrosion, friction and wear, and microelectronic devices will be highlighted.

Heterogeneous catalysis. In any practical chemical reactor, the aim is to convert one or more reactant species into one or more desired product species, with a minimum expenditure of energy and a minimal rate of production of unwanted product species. Processes of this sort form a large fraction of the business of the chemical industry. The tonnage of chemicals passed through catalytic reactors annually vastly exceeds the tonnage of metals and ores passed through blast furnaces in the production of steel.

It has been known empirically for generations that the presence of certain solid materials in a reactor, known as catalysts, can greatly influence both the rate of the reactions taking place and the specificity for a particular product. A large fraction of the effort in surface science in recent years has been directed at understanding, on the molecular level, the role that the catalytic surface plays in the reaction. Research in catalysis has followed two general lines of attack: analysis of the structure and composition of practical catalysts, to determine the differences between a good catalyst and a poor catalyst; and studies of reaction dynamics over model catalysts, usually carefully prepared and cleaned single crystals of the catalyst material.

The analysis of practical catalysts has been carried out primarily by Auger electron spectroscopy or photoelectron spectroscopy. Typically, a catalyst is prepared by a technique that has been shown empirically to produce a good working catalyst. Its composition is then determined by one of the above techniques. The catalyst is then used in a reactor until its activity has decreased significantly. It is then removed from the reactor, and its composition again determined. Observed changes in composition provide an indication of the processes involved in the loss of activity. A typical example of such a study is shown in Fig. 7 for a cobalt molybdate catalyst used to remove sulfur from petroleum. Here photoelectron spectra are shown for an active catalyst surface, an inactive catalyst, and the bulk composition of the catalyst. The active catalyst shows

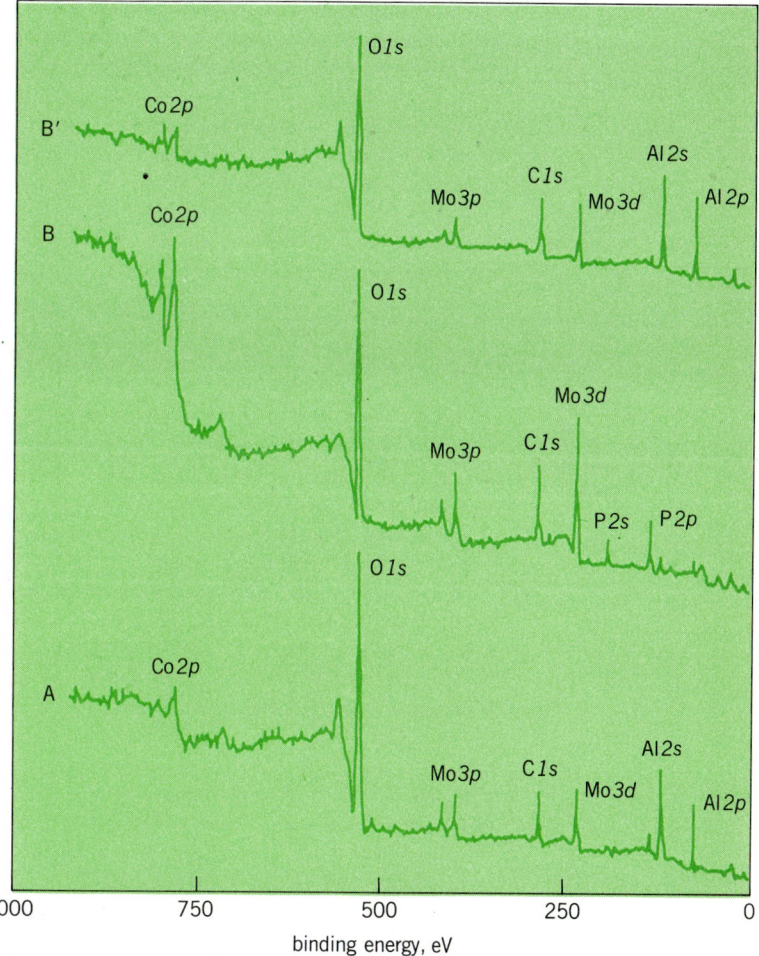

Fig. 7. Photoelectron spectra of an alumina-supported cobalt molybdate catalyst, comparing the bulk composition (B′) with the surface compositions of active (B) and inactive (A) catalysts. (*After J. S. Brinen, Proceedings of the International Conference on Electron Spectroscopy, Namur, Belgium, April 1974, Elsevier, 1974*)

high concentrations of cobalt and molybdenum, as well as a significant concentration of phosphorus. Both the inactive catalyst and the bulk composition show lower levels of cobalt and molybdenum, no phosphorus, and a relatively large amount of aluminum from the aluminum oxide support. It was deduced from these spectra that phosphorus promotes high activity, and that the loss of activity is associated with either loss of metal from the surface or agglomeration into relatively large particles.

Model catalytic studies have been made of both structure-sensitive and structure-insensitive catalytic reactions. The difference between these two types is that for the latter the reaction rate depends only on the total number of surface atoms present, and not on their distribution. In the structure-sensitive case, the details of the surface atomic arrangement are critical to the rate of reaction.

One typical structure-insensitive reaction that has been extensively studied involves the formation of methane (CH_4) from carbon monoxide–hydrogen mixtures according to the overall reaction shown below. This reaction is an important step in the man-

$$CO + 3H_2 \rightleftharpoons CH_4 + H_2O$$

ufacture of synthetic gasoline from coal. Commercially, this reaction is carried out over the surfaces of metallic catalysts, such as nickel or iron. Studies using single-crystal nickel surfaces indicated that, for this reaction, the specific rate (the amount of product formed per unit area of available surface per unit time) was essentially the same on the single crystal as for the very fine nickel particles used in common catalysts. Studies of the rate of the reaction and of the surface composition of the working catalyst enabled the detailed reaction sequence to be determined. The initial step is surface decomposition of the carbon monoxide to adsorbed carbon and oxygen. These species then react with adsorbed, dissociated hydrogen atoms to form the final products. The step that determines the overall reaction rate is the reaction between surface carbon and surface hydrogen. Efforts to improve the overall reaction rate must thus be centered on surface modifications that will speed up this step.

On the other hand, many surface reactions, notably the reactions of hydrocarbons on noble metal surfaces such as platinum, have been shown to be markedly structure-sensitive. The details of this structure sensitivity have been elucidated through reaction rate studies on stepped or kinked single-crystal surfaces. These surfaces are formed by cutting and polishing a single crystal of the desired metal a few degrees away from a close-packed crystal plane. As was shown in the discussion of the terrace-ledge-kink model of the crystal surface, such a surface, at equilibrium, will contain regularly spaced arrays of ledges and kinks. The fraction of the surface sites that are associated with ledges and kinks is controlled by the cutting angle relative to the close-packed direction. The steeper the angle, the higher the ledge and kink concentration.

Measurements of specific reaction rate as a function of the ledge and kink concentration indicate the role that these sites play in the reaction.

A typical example of such a study is shown in Fig. 8. Here the reaction rate is shown for two separate processes in a mixture of cyclohexane and hydrogen as a function of the step or kink density on the surface. One process, dehydrogenation, which involves removal of hydrogen from the cyclohexane molecule to form benzene, is independent of the defect density and thus is not structure-sensitive. The other process, hydrogenolysis, which involves opening a carbon–carbon bond of the cyclohexane ring to form n-hexane, shows a rate that increases linearly with both step density and kink density, demonstrating the structure sensitivity of this process.

Corrosion. The considerations involved in the study of corrosion processes are essentially the other side of the coin from those involved in catalysis. In both cases the initial aim is to understand the mechanism of a surface reaction sequence. In the case of corrosion, however, the next step is to determine what can be done to minimize the surface reaction rate.

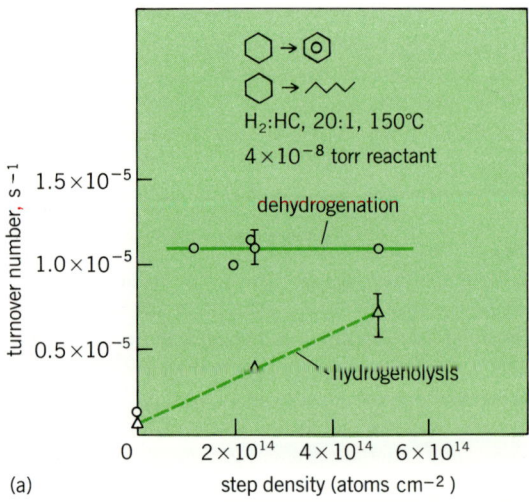

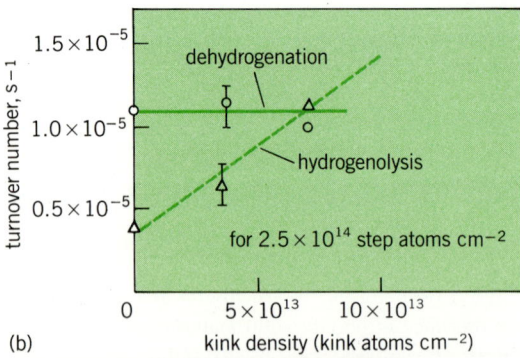

Fig. 8. Effect of (a) surface step (ledge) and (b) kink density on the reactions of cyclohexane (C_6H_{12}) on various platinum surfaces. (After D. W. Blakely and G. A. Somorjai, *The dehydration and hydrogenolysis of cyclohexane and cyclohexene on stepped (high Miller index) platinum surfaces, J. Catal.*, 42:181–186, 1976)

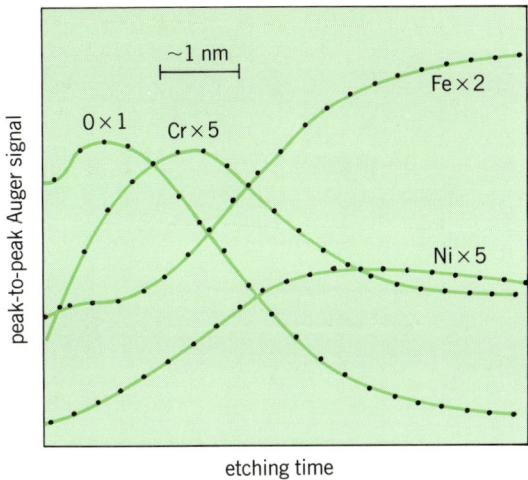

Fig. 9. Auger depth profile of a stainless steel surface, showing a very thin, corrosion-resistant chromium-enriched layer at the surface. (*After A. Joshi, Investigation of passivity, corrosion and stress corrosion phenomena by AES and ESCA, presented at the Corrosion Conference, San Francisco, 1977, and issued as an internal report by the Physical Electronics Division, Perkin Elmer*)

A great deal of attention has been paid to the question of what makes stainless steel stainless. Much of this work has involved preparing alloys of various compositions, exposing them to a corrosive atmosphere, then determining the composition as a function of depth below the surface, to determine whether a layer has been formed that protects the sample from corrosion. The most commonly used technique in these studies has been Auger electron spectroscopy. This technique, as mentioned earlier, provides chemical composition information on the topmost atom layers of the sample. If this technique is combined with a method for slowly but evenly removing material from the surface (usually bombardment of the surface with high-energy argon ions is used), then the composition can be determined as a function of depth below the original surface. Such a depth profile, for the case of a stainless steel sample, is shown in Fig. 9. In this case, a very thin layer has formed with a higher than normal chromium concentration, which protects the surface from further corrosion. This so-called passivating layer is only about 20 atom layers thick—much too thin to be seen by conventional analytical techniques.

Friction and wear. Modern surface science techniques have only recently begun to be applied to studies of friction and wear. Most of the studies to date have involved the use of techniques such as Auger electron spectroscopy to study the changes in surface composition that take place when two surfaces are in rolling or sliding contact. In cases where no wear is taking place, studies of this sort reveal the buildup of surface deposits arising from the breakdown of lubricant materials. These deposits always contain large amounts of carbon, and often have a high oxygen content as well, indicating that oxidative degradation of the lubricant takes place in the contact area, which is generally much hotter than the surroundings.

The other situation that has been investigated in some detail is that in which wear is taking place and material is being transferred from one of the contacting surfaces to the other. In one such study, a high-temperature bearing, with a steel slider running against a series of sapphire pads, was run by using molten sulfur as a lubricant (Fig. 10). The intent of the test was to determine how much material from the steel slider was transferred to the sapphire pad, and whether there were any regularities in the transfer process that could be correlated with the experimental geometry. Figure 10 was obtained by measuring the surface iron composition, using Auger electron spectroscopy, at a series of regularly spaced positions on the sapphire pad. The numbers on the figure are relative iron concentrations. The solid lines are contours of constant iron concentration. The resulting pattern shows very clearly that material transfer is greatest directly over the point at which the sapphire pad was supported from beneath by a ball bearing. The greatest rate of metal transfer is thus associated with the highest contact force between the slider and the pad.

Microelectronics. The microelectronics industry in general has been as dependent as surface science on the development of reliable high-vacuum systems. In addition, many of the operations carried out in transforming a silicon chip into an integrated circuit involve carefully controlled surface reaction processes.

As an example of the application of surface science techniques in microelectronics, a growth technique known as molecular beam epitaxy may be considered. In this process, a surface, such as a silicon chip, is prepared in ultrahigh vacuum, then

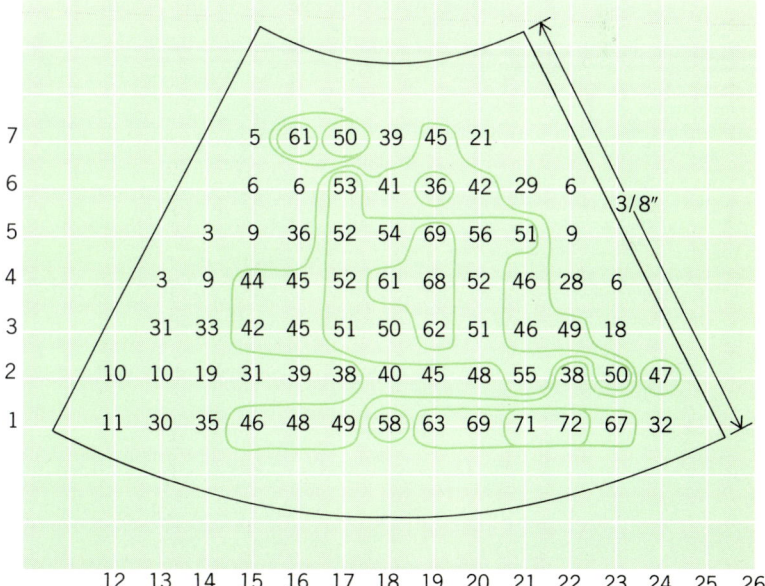

Fig. 10. Contour map of the surface concentration of iron on a sapphire bearing pad. High iron concentrations are observed at the point of highest contact pressure.

exposed to a beam of atoms of one or more other species. These beams are formed by heating a small vessel filled with the material of interest to a high enough temperature that some of the material is vaporized. The vaporized atoms pass through the vacuum and condense on the silicon chip. As a result, a layer is grown whose composition may be controlled and intentionally varied over thicknesses as little as two or three atom distances.

One very interesting application of this technique is the development of superlattice structures. These structures are formed by depositing many alternating layers of slightly different composition, each layer being only a few atoms thick. A transmission electron micrograph of such a structure, with alternating layers of silicon and silicon plus germanium, is shown in Fig. 11. Here the layer thickness is less than 10 nm (about 30 atom layers), and the change in composition takes place over a much shorter distance. These structures have unique electronic properties, and will contribute to the development of the next generation of microelectronic devices. It is also possible to change doping levels in conventional semiconductor devices over a correspondingly short distance range.

There is also a great deal of current interest in the details of the processes used to form the circuit elements in microelectronic devices. These processes typically involve coating a silicon chip with a resist material, sensitizing this resist by exposing it to light or x-rays passed through a mask to define a pattern on the surface, then using an etching process to chemically remove the resist from the exposed areas. After this process, circuit elements can be deposited on the areas from which the resist has been removed. Increasingly, these etching processes are being carried out in vacuum by using bombardment of the surface with energetic ions. The overall rate of the etching process, and the selectivity for etching exposed as opposed to unexposed areas, de-

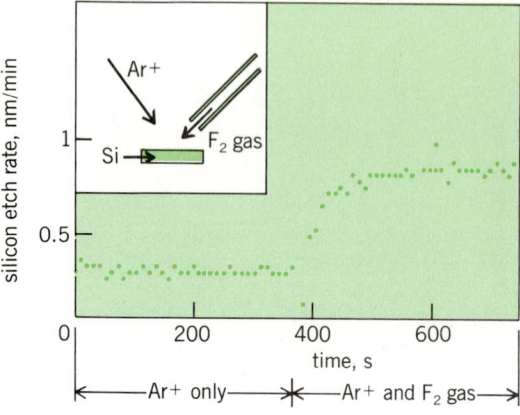

Fig. 12. Effect of the pressure of gaseous fluorine on the etch rate of silicon by argon ion bombardment. (*After J. W. Coburn and H. F. Winters, Plasma etching: A discussion of mechanisms, J. Vac. Sci. Technol., 16:391–403, 1979*)

pend critically on the details of the surface reactions involved.

These reactions are being studied by using a number of the previously described techniques. The results of one such study are shown in Fig. 12. The experimental configuration, shown in the inset, permitted the silicon surface being etched to be exposed simultaneously to controlled fluxes of fluorine gas molecules and argon ions. In a commercial apparatus, the two fluxes would not be independently controlled. The experimental setup used here demonstrates clearly the synergistic effect of the combined molecule and ion fluxes, as the etch rate increases by roughly a factor of three when fluorine gas is added concurrent with the ion bombardment.

A third area in which surface science techniques are being applied in microelectronics is in the characterization of devices, especially those that have failed in use due to defects introduced during the manufacturing process. Much of this work is done by using scanning Auger microscopy, described earlier. Figure 13 shows a series of scanning Auger micrographs of a typical microcircuit consisting of a base layer of SiO_2, overlain with a pattern of metallic silicon steps and a final metallization layer. The first picture shows the secondary electron image of the circuit pattern. The succeeding pictures show scanning Auger micrographs of the concentration distribution of silicon oxide and metallic silicon. The Auger images clearly differentiate the two oxidation states of the silicon.

THE FUTURE

There is every indication that the rapid rate of progress that has been sustained in surface science over the past 20 years will be maintained well into the future. Two areas in particular appear ready for rapid development—namely, the interaction of laser light with surfaces, and the exploitation of techniques that permit surface studies at the molecular level on surfaces outside the ultrahigh-vacuum environment.

The use of laser light is being explored both as a

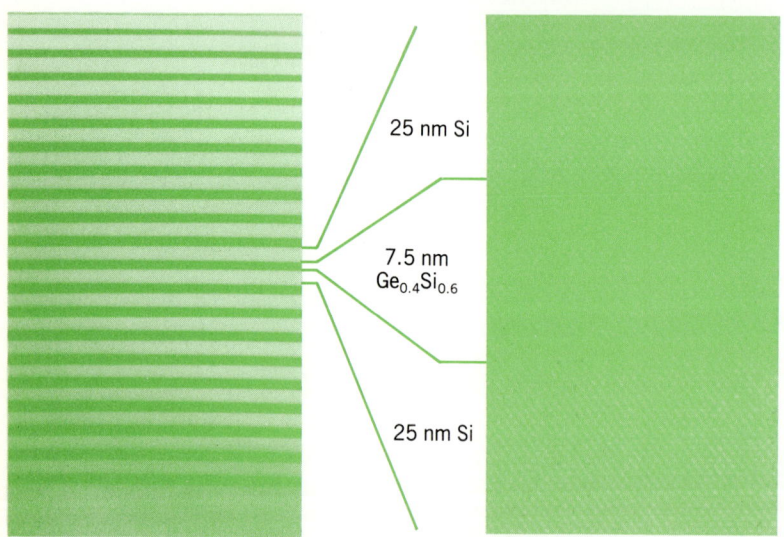

Fig. 11. Micrograph of a superlattice structure with alternating layers of silicon and silicon plus germanium. The layers are thinner than 10 nm. The concentration change occurs over a distance of roughly three atom layers. (*After J. C. Bean, Silicon MBE: From strained-layer epitaxy to device application, J. Cryst. Growth, publication pending*)

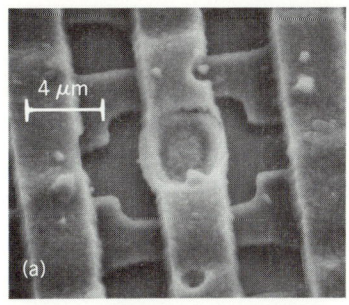 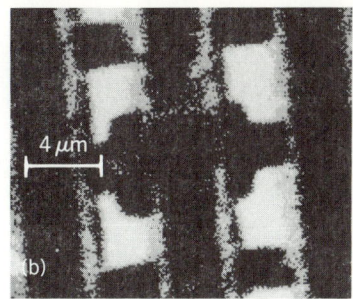 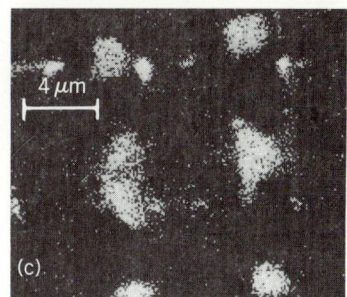

Fig. 13. Scanning Auger microscopy analysis of electronic microcircuit element: (a) secondary electron micrograph; (b) silicon (metallic) concentration map; and (c) silicon oxide concentration map. (*Physical Electronics Division, Perkin Elmer*)

means of analyzing surface structure and chemistry, and as a means of modifying structure and chemical reaction rates. The high-intensity, monochromaticity and tuneable wavelength characteristics of recently developed lasers permit measurement of surface electronic structure and of the vibrational spectra of adsorbed species over a wider range and with higher sensitivity and energy resolution than the currently used techniques, such as electron energy loss spectroscopy. The extremely high photon flux from high-intensity lasers can also induce nonlinear effects that will lead to a more detailed understanding of the surface electronic configuration of the active sites for catalytic reactions.

A great deal of effort is also being put into the use of lasers as manufacturing tools in the microelectronics industry. The two major applications to date are the use of lasers for rapid surface heating, used to control dopant profiles, and the use of tuneable-wavelength lasers to cause surface chemical reactions on well-defined areas of a surface, leading to the direct writing of circuit patterns on the semiconductor chip.

The use of lasers is also intimately linked to progress in the second area mentioned above. One of the major limitations of most of the techniques discussed so far is that they are capable of operating only at very low pressures. Many of the phenomena of interest, however, such as catalytic chemical reactions, electrochemical processes, and corrosion or wear processes, take place at atmospheric pressure or higher. Probes are thus needed that can operate in atmospheric pressure gas or through a liquid layer. The optical techniques mentioned above, using laser radiation, appear to be the best suited for this application. It is likely that a whole new set of surface spectroscopies, based on laser excitation and capable of operating over a very wide range of ambient pressures, will be developed over the next 20 years.

[JOHN B. HUDSON]

Bibliography: J. M. Blakely, *Introduction to the Properties of Crystal Surfaces*, 1973; A. W. Czanderna (ed.), *Methods of Surface Analysis*, 1975; D. A. King and D. P. Woodruff (eds.), *The Chemical Physics of Solid Surfaces and Heterogeneous Catalysis*, vols. 1–4 and following, 1982; P. A. Redhead et al., *The Physical Basis of Ultrahigh Vacuum*, 1968; G. A. Somorjai, *Chemistry in Two Dimensions*, 1981; R. Vanselow (ed.), *Chemistry and Physics of Solid Surfaces*, vols. 1–2 and following, 1978.

The Inflationary Universe

Marek Demianski

Marek Demianski teaches physics and astronomy at the University of Warsaw and the Copernicus Astronomical Center, Warsaw. He is an author of Relativistic Astrophysics. He is a member of the International Committee on Relativity and Gravitation.

How was the universe created? How has it evolved to its present state? These difficult questions have been asked since the early days of civilization. For centuries they belonged to the domain of religion and philosophy, and only very recently have viable models of the evolution of the universe been proposed.

Evolving universe. Modern cosmology began with the discovery that the Milky Way is only one of very many galaxies. In 1912 V. M. Slipher noticed that spectral lines of galaxies are slightly shifted toward the red end of the spectrum. In acoustics the effect of change of the pitch of sound as a consequence of relative motion of the source of sound and the observer, the Doppler effect, had been known for some time. Using the analogy with acoustics, Slipher interpreted the shift of spectral lines as due to relative motion of distant galaxies and the Milky Way. When it became possible to measure distances to galaxies, Edwin Hubble discovered that distant galaxies are receding with velocities proportional to their distance. He also noticed that this expansion is isotropic, and that over a sufficiently large scale galaxies are distributed uniformly. The universe was seen to be a dynamical, evolving system.

In late 1940s George Gamow, fascinated by the idea of the evolving universe, asked a profound question: how were the different chemical elements created in the course of cosmological evolution? Gamow assumed that at a very early stage of evolution the universe was very dense and very hot; the model he proposed is now known as the big bang or primeval fireball. Gamow failed to explain the observed abundance of elements, but he made an important prediction. If the early universe was indeed very hot, then even now relics of the primeval fireball should be observable in the form of thermal radiation filling all space.

Microwave background radiation. In 1964 the radio astronomers Arno Penzias and Robert Wilson were testing a new horn antenna specially designed to pick up weak signals in the microwave range. They noticed that the antenna was picking up a constant hiss that did not depend on the orientation of the antenna or the time of observation. Penzias and Wilson had discovered a microwave background radiation. When the spectrum

The inflationary universe

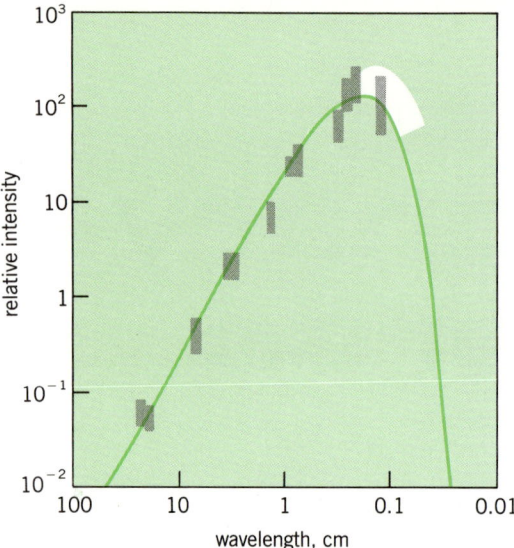

Fig. 1. Spectrum of the microwave background radiation. Bars and white region indicate measurements. Curve indicates spectrum of blackbody radiation at 2.9 K (2.9°C or 5.2°F above absolute zero). (*After J. D. Barrow and J. Silk, The structure of the early universe, Sci. Amer., 242(4):118–128, April 1980*)

of this radiation was measured (Fig. 1), it was found to have a thermal character corresponding to radiation at a temperature of 3 K (3°C above absolute zero).

An oven is a simple source of thermal radiation. It can be said that the cosmic furnace, which was initially very hot as a result of expansion, cooled to 3 K. The discovery of the microwave background radiation has provided very important cosmological information: The temperature of the microwave background radiation coming from different parts of the sky is almost the same, indicating that the universe must have been extremely uniform at the moment when the radiation was emitted.

Standard big bang model. To describe the evolution of the universe, it is necessary to construct a model and extrapolate the evolution of the universe backward in time. At present the universe is expanding, and on a large scale it is homogeneous. The universe is also uniformly filled with thermal radiation. To construct a dynamical model of the universe, it is necessary to describe how different parts of the universe interact with each other. Einstein's relativistic theory of gravity, the general theory of relativity, provides the most complete and most sophisticated framework. The general theory of relativity relates the rate of expansion of the universe with mean energy density and pressure of matter filling the universe. Thermodynamics is needed to relate the mean energy density, the mean pressure, and the temperature.

Evolution from singular state. Under general conditions this model predicts that the universe evolved from a singular state characterized by infinite density and infinite temperature. It is estimated that this singular state occurred 10 to 18×10^9 years ago.

It is quite an ambitious program to reconstruct the history of the universe starting from the singular state. Initially the universe was expanding very rapidly, and as a result of expansion it was cooling down and the density of matter was rapidly decreasing. The initial state of the matter is not known but it is reasonable to assume that at very high densities and temperatures the elementary processes were also very fast and the matter was in a state of thermal equilibrium. At thermal equilibrium the composition of matter is uniquely determined by the temperature, the masses of elementary particles, and the strength of nuclear interactions.

In the state of thermal equilibrium there should be as many particles as antiparticles. However, the observed universe is composed almost exclusively of matter rather than antimatter. In the standard big bang model, it must be assumed, therefore, that initially there was a very small asymmetry in the composition of matter and the particles slightly outnumbered antiparticles.

Properties of elementary particles. To understand the early history of the universe, it is necessary to review briefly the general properties of elementary particles. Elementary particles appearing in nature can be divided into three families: leptons, baryons, and field quanta. There are only very few kinds of leptons; the electron is the best known. Baryons are much more numerous; protons and neutrons, the elementary constituents of atomic nuclei, belong to this family. Most of the exotic particles artificially created in particle accelerators are baryons. It is now established beyond any doubt that all baryons are composed of more elementary constituents, quarks. Photons, elementary excitations of the electromagnetic field, are the best-known example of particles belonging to the third family.

First 3 minutes. The evolution of the universe can be described quite reliably starting from the moment when it was 10^{-6} s old (Fig. 2). By then the universe had expanded and cooled so dramatically that most of the heavy particle-antiparticle pairs annihilated and were transformed into photons. Matter at that epoch was composed mostly of quarks and antiquarks, electrons and positrons (antielectrons), and neutrinos.

As a result of expansion, the universe was continuously cooling. Some quark-antiquark pairs annihilated, and a very small surplus of quarks combined to form protons and neutrons. When the universe was about 10 s old, electrons and positrons annihilated, leaving, however, a small surplus of electrons to counterbalance the positive charge of protons and preserve the overall electric neutrality of the universe. About 2 min later the universe cooled sufficiently to allow protons and neutrons to combine and form light elements. The universe was not sufficiently dense and hot for long enough to allow synthesis of elements heavier than helium. The big bang model predicts that the primordial matter contained about 25% helium and 75% hydrogen. This prediction is in a good agreement with observations.

Radiation era. After an eventful first 3 min the

universe entered a relatively calm era. The mean energy density of the universe was dominated by the energy density of radiation, and the energy density of ordinary matter (baryonic matter) was practically negligible. Radiation was intimately coupled with ordinary matter, and this interaction prevented the formation of condensations. Radiation and matter were distributed uniformly. This radiation era lasted for about 3×10^5 years.

Decoupling of radiation and matter. By the end of the radiation era the temperature of the universe had dropped to about 3000 K (5000°F). At that temperature the number of high-energy photons declined to such a level that electrons could attach to protons and helium nuclei, and form atoms. Starting from that moment, radiation practically ceased to interact with ordinary matter and photons started to propagate freely. Photons of the microwave background which is observed today were emitted at that time. Almost at the same time the energy density of radiation became equal to the energy density of matter, and starting from that moment the rate of expansion was determined by the energy density of ordinary matter.

Formation of galaxies, stars, and planets. When radiation decoupled from ordinary matter, small primordial perturbations started to grow. They grew very slowly, and only after a long time did the first condensations appear. It is not clear whether galaxies formed first and then coalesced to form clusters of galaxies, or large clouds with masses typical of clusters of galaxies appeared first and later fragmented into galaxies and stars. When stars were finally formed, a complicated chain of thermonuclear reactions was initiated that transformed hydrogen and helium into heavier elements. Stars much more massive than the Sun evolved relatively quickly. At the end point of their evolution, they exploded, dispersing heavy elements. In the process of explosion, still heavier elements were synthesized. Clouds of primordial gas enriched with heavy elements condensed. The central core of one such cloud condensed to form the Sun, and the halo collapsed into a disk which fragmented into planets, among them the Earth (Table 1).

Future evolution. The future evolution of the universe depends on the mean energy density. According to general theory of relativity, there is a critical energy density. If the mean energy density of the universe is above the critical value, the gravitational attraction is so strong that it will slow down the expansion to a momentary standstill and cause the universe to recontract. If the mean energy density of the universe is equal to or is less than the critical value, the universe will expand forever. To determine the mean energy density of the universe, the universe should be weighed. This is a formidable task, however, and the mean energy density of the universe is one of the most uncertain cosmological parameters. By comparing the amount of light emitted by stars of known masses, the so-called mass-to-light ratio can be deduced. By using the mass-to-light ratio, it is possible to estimate the mean den-

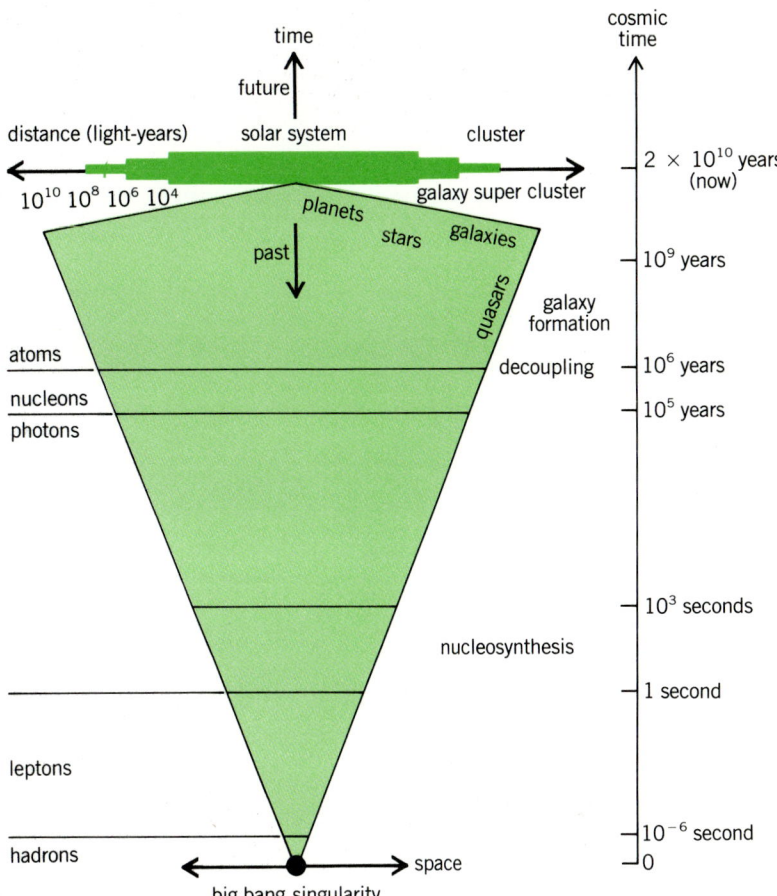

Fig. 2. Space-time diagram of the universe, tracing its history from the singularity of roughly 2×10^{10} years ago. The tinted region indicates the horizon of a hypothetical observer at the singularity. (*After J. D. Barrow and J. Silk, The structure of the early universe, Sci. Amer., 242(4):118–128, April 1980*)

sity of luminous matter. It turns out that even the Milky Way Galaxy contains much more mass than is inferred from the mass-to-light ratio. Due to this uncertainty, future evolution of the universe cannot be predicted at present. Because the universe is expanding now, it will be expanding for at least another 10^{10} years.

Unanswered questions. The standard big bang model based on a few simple assumptions describes the evolution of the universe amazingly well. There are, however, questions which are left unanswered and properties which are arbitrarily assumed. For example, the big bang model does not explain why the distribution of matter on a large scale is homogeneous; the homogeneity is postulated as an initial condition. The estimated mean energy density is very close to the critical density. Both the critical density and the mean energy density depend on time. For the mean energy density to be close to the critical density at present, it is necessary that it was almost equal to the critical density at very early times. Again this important fact is not explained in the standard big bang model.

The observed isotropy of the microwave background radiation is also a mystery. Photons of the relic radiation coming to the Earth from two patches

Table 1. Major events in the universe's history*

Cosmic time	Epoch	Redshift	Event	Years ago
0	Singularity	Infinite	Big bang	20×10^9
10^{-43} second	Planck time	10^{32}	Particle creation	20×10^9
10^{-6} second	Hadronic era	10^{13}	Annihilation of proton-antiproton pairs	20×10^9
10 seconds	Leptonic era	10^{10}	Annihilation of electron-positron pairs	20×10^9
2 minutes	Radiation era	10^9	Nucleosynthesis of helium and deuterium	20×10^9
1 week		10^7	Radiation thermalizes prior to this epoch	20×10^9
70,000 years	Matter era	10^4	Universe becomes matter-dominated	20×10^9
300,000 years	Decoupling era	10^3	Universe becomes transparent	19.9997×10^9
$1-2 \times 10^9$ years		10–30	Galaxies begin to form	$18-19 \times 10^9$
3×10^9 years		5	Galaxies begin to cluster	17×10^9
4×10^9 years			Protogalaxy collapses to the Milky Way Galaxy	16×10^9
4.1×10^9 years			First stars form	15.9×10^9
5×10^9 years		3	Quasars are born; population II stars form	15×10^9
10×10^9		1	Population I stars form	10×10^9
15.2×10^9 years			Parent interstellar cloud to the solar system forms	4.8×10^9
15.3×10^9 years			Collapse of protosolar nebula	4.7×10^9
15.4×10^9 years			Planets form; rock solidifies	4.6×10^9
15.7×10^9 years			Intense cratering of planets	4.3×10^9
16.1×10^9 years	Archeozoic era		Oldest terrestrial rocks form	3.9×10^9
17×10^9 years			Microscopic life forms	3×10^9
18×10^9 years	Proterozoic era		Oxygen-rich atmosphere develops	2×10^9
19×10^9 years			Macroscopic life forms	1×10^9
19.4×10^9 years	Paleozoic era		Earliest fossil record	600×10^6
19.55×10^9 years			First fishes	450×10^6
19.6×10^9 years			Early land plants	400×10^6
19.7×10^9 years			Ferns, conifers	300×10^6
19.8×10^9 years	Mesozoic era		First mammals	200×10^6
19.85×10^9 years			First birds	150×10^6
19.94×10^9 years	Cenozoic era		First primates	60×10^6
19.95×10^9 years			Mammals increase	50×10^6
20×10^9 years			*Homo sapiens*	1×10^5

*The precise age of the universe is not known. Times are based on an age of 2×10^{10} years.

SOURCE: After J. D. Barrow and J. Silk, The structure of the early universe, *Sci. Amer.*, 242(4):118–128, April 1980.

of sky separated by more than 30° never interacted before, but the temperature of the radiation from the two regions is almost the same. This so-called horizon problem can be illustrated by a three-dimensional space-time diagram (Fig. 3), in which scales of space and time are drawn in a nonlinear way so that the path of a light signal is represented by a line at 45° to the vertical axis. The present position of the Earth in time and space is indicated by point A. Since signals cannot travel faster than the speed of light, signals at A can be received only from inside the cone in Fig. 3 with vertex at A, which is called the past light cone of A. An event at a given point cannot be influenced in any way by events outside its past light cone. The time at which the microwave background radiation was released is indicated by a horizontal plane in Fig. 3. Radiation now reaching the Earth from opposite directions was released at points B and C, and then traveled to point A along its past light cone. Since the past light cones of points B and C do not intersect, the two points were not subject to any common influences, and it is difficult to explain how the radiation received from the two opposite directions came to be at the same temperature. In the big bang model it is simply assumed that the temperature is isotropic.

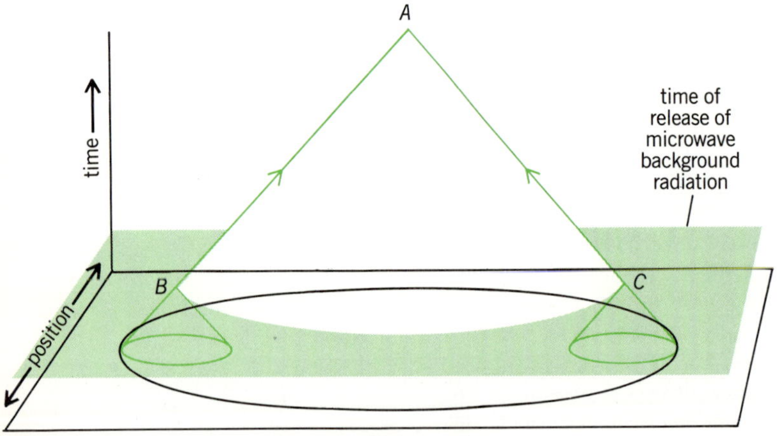

In 1981 Alan Guth proposed a new scenario of the early evolution of the universe, the inflationary model, which overcomes some of the problems of the standard big bang model. This inflationary model incorporates recent advances in the theory of elementary particles into cosmological considerations.

Fig. 3. Three-dimensional space-time diagram illustrating the horizon problem. (*After A. H. Guth and P. J. Steinhardt, The inflationary universe, Sci. Amer., 250(5):116–128, May 1984*)

Fundamental interactions and particles. There are four basic interactions (forces) in nature: gravitational, electromagnetic, weak, and nuclear

Table 2. The four fundamental interactions*

Interaction	Source	Field quantum	Relative strength now	Range, m	Manifestation	Unification energy, eV
Electromagnetic	Electric charge	Photon	10^{-12}	∞	Atomic and molecular forces, electricity	10^{11}
Weak	Leptons, mesons	W, Z bosons	10^{-14}	10^{-17}	Radioactive beta decay	10^{23}
Strong	Baryons, mesons	Pion, kaon	1	10^{-15}	Nuclear forces	10^{28}
Gravitational	Mass-energy	(Graviton)	10^{-40}	∞	Large-scale dynamics of matter	

*After J. D. Barrow and J. Silk, The structure of the early universe, *Sci. Amer.*, 242(4):118–128, April 1980.

(strong; Table 2). The gravitational interaction is universal; all kinds of matter gravitate. The gravitational interaction plays a dominant role on an astronomical scale and in cosmology, but it is negligible on an atomic scale. All charged particles interact electromagnetically, but because astronomical objects are electrically neutral the electromagnetic interaction does not play an important role in astronomy and cosmology. It is very important, however, on the atomic scale, being responsible for all chemical bonds.

It is generally believed that matter is composed of elementary constituents, quarks and leptons. Quarks interact strongly, quarks and charged leptons interact electromagnetically, and all the fundamental particles interact weakly.

Elementary particles composed of quarks and antiquarks are called hadrons. The quark model describes very well the properties of hadrons. Single quarks have not been seen, and there is good reason to believe that it is not possible to isolate a single quark from a hadron. Any attempt to pull away a single quark from a hadron involves supplying enough energy to create a quark-antiquark pair instead. All hadrons interact strongly. Strong interactions are responsible for the structure of atomic nuclei.

Weak interactions do not manifest themselves in everyday life, and the only trace of their existence is the beta decay of radioactive nuclei. When properties of weak interactions were studied more thoroughly, it was found that they violate some very fundamental symmetry principles that had been considered universal. For example, weak interactions can distinguish between left and right orientation of frames of reference.

Unification of interactions. The list of fundamental interactions leads to the question of why nature requires such diversity. About 150 years ago electricity and magnetism were considered to be separate interactions. In 1820 H. C. Oersted discovered that a current creates a magnetic field. It is known that magnetic properties are generated by microcurrents. The complete unification of electricity and magnetism by J. C. Maxwell led to new predictions, the possibility of the existence of electromagnetic waves being the most important. Because of many similarities between the gravitational interactions and the electromagentic interactions, it seemed that it should not be difficult to construct a unified theory of gravity and electromagnetism. Einstein spent the last 30 years of his life trying unsuccessfully to construct such a theory.

Weinberg-Salam model. Late in the 1960s Steven Weinberg and independently Abdus Salam proposed a theory unifying electromagnetic and weak interactions. This theory was later amended by Sheldon Glashow. According to the Weinberg-Salam model, certain new kind of particles should exist. In 1983 these particles were discovered, and their properties agree very well with predictions. This was a great success of the Weinberg-Salam model and the whole idea of unification.

Grand unified theories. Soon after Weinberg and Salam constructed the theory unifying electromagnetic and weak interactions, a more general model was proposed unifying strong, electromagnetic, and weak interactions. This class of theories is known as grand unified theories. Grand unified theories predict that at very high energies strong, electromagnetic, and weak interactions become similar. The energy required to achieve unification is estimated to be as high as 10^{14} GeV. This energy range is far above the energies obtainable in present particle accelerators, and there is no hope that it will be accessible in the foreseeable future. No astronomical objects are known that could produce particles with such enormous energies. Only at the very early stages of evolution of the universe could particles with such high energies have been created.

Spontaneous symmetry breaking. At energies higher than the grand unification energy, all the interactions except gravity should have the same strength. Below this energy, the symmetry between interactions is broken, and strong, electromagnetic, and weak interactions are distinct.

Spontaneous symmetry breaking has been known in physics for some time. There are simple examples of symmetry breaking in everyday life. Spontaneous symmetry breaking usually occurs when, as a result of a change of some external parameters, two or more symmetrical forces come into competition. One of the most common ways in which broken symmetries exhibit themselves experimentally is in phase transitions: melting, evaporation, magnetization, and so forth. Since spontaneous symmetry breaking

30 The inflationary universe

is an important concept, two examples will be discussed briefly. Imagine a donkey placed at the same distance from two haystacks. If the donkey chooses not to starve to death and approaches one of the haystacks, the symmetry becomes spontaneously broken. A more complicated example is provided by a ferromagnet. It is known that a magnet heated above 1418°F (770°C) loses its magnetic properties, and when it cools below 1418°F (770°C) its magnetic properties are spontaneously restored. To un-

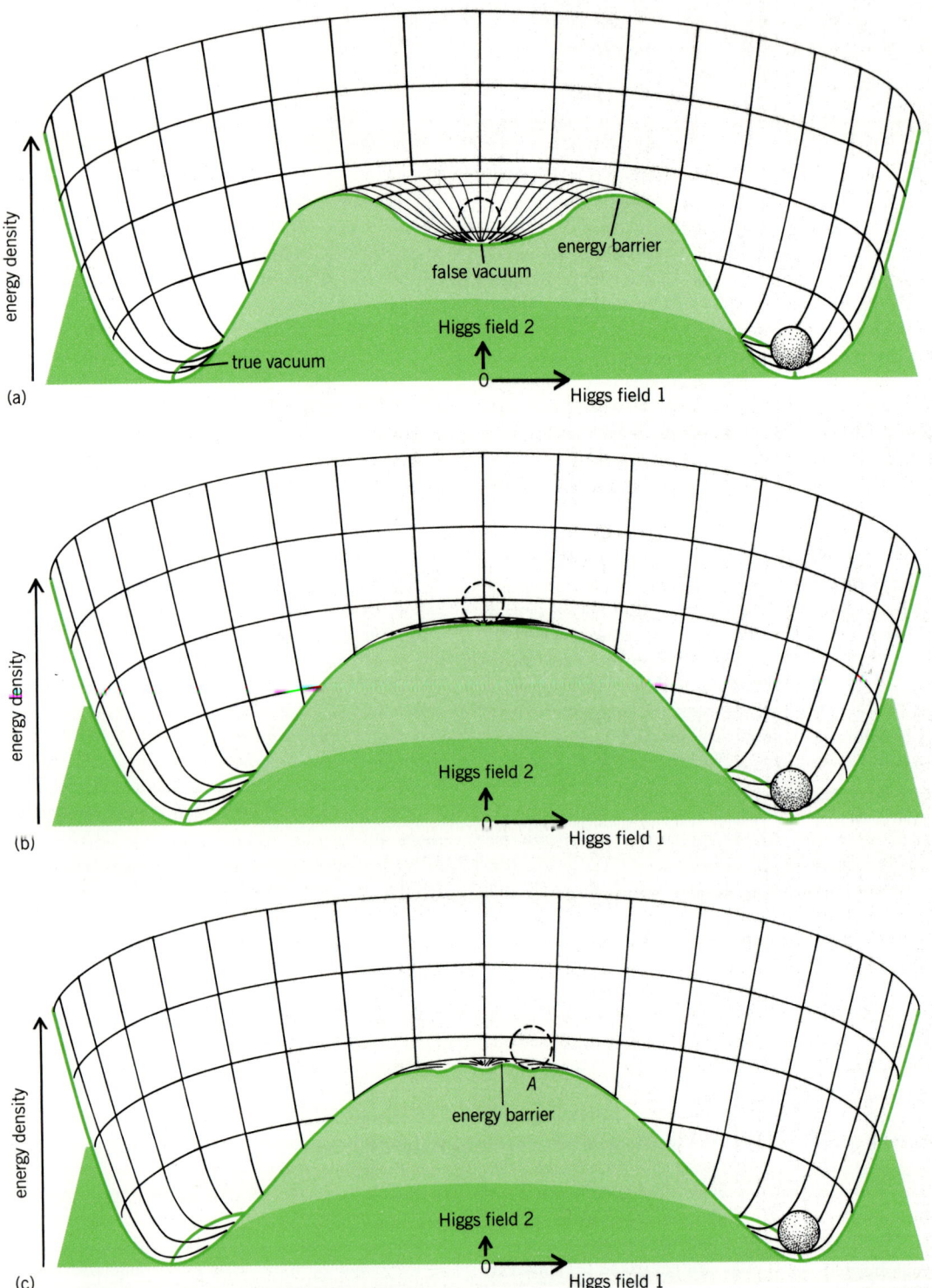

Fig. 4. Energy density of the universe as a function of two Higgs fields. (a) Original form of inflationary universe model, in which universe tunnels through energy barrier separating false and true vacuum states. (b) New inflationary model, in which there is no energy barrier and universe slowly rolls down slope from a flat plateau. (c) Variant of new inflationary model, in which false vacuum is surrounded by a small energy barrier, and universe tunnels as far as circle A, after which it slowly rolls down a very flat region. (After A. H. Guth and P. J. Steinhardt, The inflationary universe, Sci. Amer., 250(5):116–128, May 1984)

derstand the phenomenon of spontaneous magnetization, it is necessary to look at the structure of iron atoms. It turns out that because of quantum-mechanical effects it is energetically favorable for the spins of neighboring iron atoms to be parallel. The orientation of the spins is not important; it is only important that they be parallel. As a result of this tendency, spins of very many atoms of iron align themselves along a preferred direction and they form a magnetic domain. A magnetic domain is microscopic, but it is enormous on an atomic scale and it contains billions of elementary magnetic needles (magnetic moments). The division of a magnet into magnetic domains is possible because it is energetically more favorable than an arrangement with all spins pointing in the same direction. When a magnet is heated, the thermal energy of atoms increases. At a certain temperature it becomes larger than the energy of the spin-spin interactions, and the magnetic moments are free to orient themselves at random. However, when the system cools below 1418°F (770°C), the spin-spin interaction prevails, and magnetic moments spontaneously align along one of the equivalent directions in the crystal and form a magnetic domain.

Higgs fields. In the grand unified theories, spontaneous symmetry breaking is generated by including in the description a certain number of additional fields known as Higgs fields. In the symmetric phase, at energies higher than the grand unification energy, all the Higgs fields are zero. Symmetry is broken when at least one of the Higgs fields becomes nonzero, and this value of the field corresponds to a state with lowest energy density, which is called the true vacuum state (Fig. 4a).

Incorporation in big bang model. In the standard big bang model the universe cools to the grand unification energy 10^{-35} s after the bang. The 10^{-35} s is an unimaginably short time. The intervals of time at the very early stages of evolution of the universe should not be measured in seconds. There is a natural unit of time known as the Planck time which is equal to 10^{-45} s. In terms of the Planck time, the universe is quite old when it reaches the temperature corresponding to the grand unification energy.

As is well known in the special theory of relativity, it is not possible to transmit information with a velocity larger than the velocity of light. This condition restricts the size of a region in which the Higgs field could have the same value. It is not possible to communicate between adjacent regions, and therefore the Higgs field could have different values in such regions. This discontinuity of the Higgs fields across the boundary between different regions creates at least two classes of defects: pointlike defects, representing magnetic monopoles, and surfacelike defects, representing domain walls. In the Maxwell classical electromagnetic theory, there are no magnetic charges and magnetic phenomena are created by electric currents. Grand unified theories predict the existence of magnetic monopoles, which should play the role of magnetic charges. It is estimated that the mass of a magnetic monopole could be 10^{16} times larger than the mass of a proton. Magnetic monopoles and other defects could be created in copious amounts during the grand unified phase transition.

Inflationary scenario. Grand unified theories combined with the standard big bang model lead to contradictions. The predicted energy density of created magnetic monopoles is many orders of magnitude larger than the critical density. The lifetime of such a universe would be much shorter than the estimated age of the universe. To resolve the magnetic monopole problem and other conundrums of the standard big bang model, Guth proposed the inflationary scenario.

Consider a small, hot and expanding region of the very early universe. Sufficiently early, the temperature of this region should be much higher than the temperature corresponding to the grand unification phase transition. As a result of expansion, this region cools down, and at a certain moment its temperature becomes equal to the characteristic temperature of the grand unified phase transition. Some of the Higgs fields, which previously were equal to zero, acquire a nonzero value, and a new lower energy level appears. If this transition is smooth, the expanding region can be trapped at the previous energy level (the false vacuum state), which is no longer the lowest level. This situation corresponds to the well-known phenomenon of supercooling a drop of water below the freezing point.

Cosmological constant. In Guth's inflationary model, the transition to the lower energy level is prevented by a small potential barrier. The constant energy difference between the true vacuum state and the false vacuum state acts like a cosmological constant. Here a digression is appropriate. When Einstein formulated his general theory of relativity in 1916, it became apparent that this theory should be a basis of any cosmological considerations. Following Newton, Einstein thought that the universe was static. In order to describe a static space filled uniformly with matter, Einstein changed his theory slightly by introducing a cosmological constant. The purpose of this constant was to counterbalance the gravitational attraction of matter. Einstein also hoped that the modified theory would describe only a universe filled with matter and there would be no solutions corresponding to empty space. Einstein's hypothesis was immediately disproved by W. de Sitter, who found a solution of the Einstein field equations with a cosmological constant that represented empty space. The de Sitter space-time is not static, and an arbitrary region of this space expands exponentially. The small region trapped in the false vacuum state expands exponentially; the rate of expansion is proportional to the energy difference between the false and the true vacuum state. If the exponentially expanding phase lasts sufficiently long, this region could be inflated to an enormous size.

New inflationary model. According to quantum mechanics, a potential barrier is not impenetrable and the universe could tunnel through the potential barrier separating the false (higher) and the true (lowest) vacuum state (Fig. 4a). This transition is not instantaneous, and bubbles of the true vacuum state will form first. This process is random, and it creates large inhomogeneities, which are not presently observed.

A. Linde, P. Steinhardt and A. Albrecht, and S. Hawking, among others, have shown that this problem can be avoided by choosing a different type of Higgs potential. S. Coleman and E. Weinberg proposed a form of a potential which does not have a potential barrier but instead is very flat around the false vacuum state (Fig. 4b). In this model, called the new inflationary model, the universe slowly "rolls down" the potential to the true vacuum state. If this process is sufficiently slow, during the rolling-down phase the universe expands exponentially, and its size can increase enormously. When the universe reached the temperature of the grand unified phase transition, the region corresponding to the present observable universe was only 4 in. (10 cm) in diameter.

Finally the universe will roll down the steep part of the potential, oscillate around the true vacuum state, and settle down at the true vacuum state. In the process of settling down into the true vacuum state, the energy difference between the true and false vacuum state will be released mainly in the form of heat. The universe, which, as a result of exponential expansion, was cooled down practically to the temperature of absolute zero, will reheat, and copious numbers of particles and antiparticles will be produced. The expansion rate, which will be determined by the mean energy density of radiation, will be drastically reduced. The universe will continue its further evolution according to the standard big bang model.

In a variant of the new inflationary model (Fig. 4c), the false vacuum is surrounded by a small potential barrier, which is in turn surrounded by a very flat potential, as before. After the universe tunnels through the barrier, it rolls very slowly down the potential and, as in the new inflationary model, undergoes enormous expansion.

Resolution of problems. The dramatic increase in the size of the universe in the inflationary models helps to solve some of the problems of the standard big bang model.

Scarcity of magnetic monopoles. As mentioned above, magnetic monopoles are created during the grand unified phase transition. As a result of exponential expansion, the number density of monopoles decreases so drastically that at present there should be practically no chance to observe primordial monopoles. The number density of magnetic monopoles should be so small that they essentially do not contribute to the mean energy density of the universe.

Large-scale homogeneity. In the inflationary scenario the presently observed universe evolved from a region so small that light signals were able to cross it many times. In practical terms, this means that there was enough time to homogenize the system. This region was subsequently inflated to enormous size. The exponential expansion was isotropic, and it preserved homogeneity of the system. It is not surprising therefore that the microwave background radiation and matter on a large scale are distributed uniformly.

Flatness of universe. The dynamical equations describing the evolution of the universe relate the rate of expansion with the mean energy density and the curvature of space. During the inflationary phase the mean energy density remains constant but the curvature of space decreases exponentially. At the end of the inflationary phase the curvature term is reduced practically to zero and the subsequent evolution of the universe is determined solely by the mean energy density. The inflationary model predicts that the present mean energy density of the universe should be equal to the critical energy density. This also means that the universe will expand forever.

Independence from preinflationary evolution. The inflationary model has another very attractive and important feature: the present universe does not depend on the very early preinflationary phases of evolution. As a result of the enormous exponential expansion, initial inhomogeneities and anisotropies are drastically reduced, leaving no trace on the postinflationary evolution.

Small-scale structure. The distribution of matter on a small scale is highly nonuniform, and the inflationary model or any other model of the evolution of the universe should provide an explanation of how this structure originated. In the standard big bang model, it is assumed that the presently observed small-scale structure of the universe is a result of a slow growth of small primordial perturbations which are introduced as initial conditions. The standard big bang model does not provide a mechanism for generating these perturbations. The inflationary model is the only scenario providing such a mechanism. The slow transition from the false vacuum state to the true vacuum state, the process of rolling down the potential, gives only a global picture. To get a more accurate description of this transition, quantum fluctuations should be taken into account. Quantum fluctuations could slightly delay or accelerate the rolling down at random in different parts of the universe, and therefore the transition to the true vacuum state is not simultaneous. This creates slight differences in density between different regions. A. A. Starobinsky, Guth, and Hawking, among others, have shown that these density perturbations could generate the presently observed small-scale structure of the universe. There is, however, one problem: the estimated initial amplitude of density perturbations is some four orders of magnitude too big. This might indicate that the inflationary model is not compatible with observations, but it might only mean that a more sophisticated model is needed. Much effort is being devoted to clarify this problem.

Matter-antimatter asymmetry. Earth, planets, Sun, the Galaxy, and, it seems, the whole visible universe are composed predominantly of matter. Antimatter appears only sporadically. In the standard big bang model, this asymmetry between matter and antimatter is not explained but rather is related to the initial conditions. Grand unified theories provide a mechanism of creating this asymmetry. This mechanism was first suggested in the late 1960s by A. Sakharov, who pointed out that in grand unified theories the baryon number does not have to be conserved. He also pointed out that if there is a process violating the baryon conservation law, then the universe passes through an era when matter is out of thermal equilibrium and some important discrete symmetries are violated, and the asymmetry between matter and antimatter then results from a strictly symmetric initial state. It turns out that in many grand unified models these conditions are satisfied, providing a mechanism for creating the presently observed asymmetry between matter and antimatter.

Testing of grand unified theories. The grand unified theories can explain many problems facing the standard big bang model, and they provide a very general scheme of classifying and understanding the main interactions among elementary particles. The next task is to search for observational indications that the grand unified theories correctly describe the behavior of elementary particles at very high energies. Early in 1983 new families of particles predicted by the Weinberg-Salam model of unified electromagnetic and weak interactions were discovered. This was a great success. The grand unified theories predict the existence of new families of particles. The estimated mass of these particles is enormous, and there is no hope that they could be created in particle accelerators in the foreseeable future. To test grand unified theories, it is necessary to rely on indirect consequences.

In the conventional theory of elementary particles, baryon number was believed to be conserved, and since the proton is the lightest baryon it had to be absolutely stable. In grand unified theories there are processes violating the baryon number conservation law, and as a result the proton should not be absolutely stable. In the most simple model of the grand unified theories, it is estimated that the lifetime of a proton is 10^{30}–10^{32} years. The expected lifetime of a proton is some 20 orders of magnitude longer than the estimated age of the universe. The decay process of a proton is governed by quantum laws. This means that on the average protons should live some 10^{30} years. Much effort is now devoted to testing the stability of protons, and a number of experiments are in progress. Results of these experiments seem to contradict predictions of the most simple type of grand unified theory. There are, however, more complicated models, and it would be premature to abandon the idea that at very high energies all interactions in nature are similar. It may take considerable time and effort to experimentally prove or disprove this idea.

[MAREK DEMIANSKI]

Bibliography: J. D. Barrow and J. Silk, *The Left Hand of Creation: The Origin and Evolution of the Expanding Universe*, 1983; A. H. Guth and P. J. Steinhardt, The inflationary universe, *Sci. Amer.*, 250(5):116–128, May 1984; J. Silk, *The Big Bang: The Creation and Evolution of the Universe*, 1980.

Nuclear Electromagnetic Pulse

Mario Rabinowitz

Mario Rabinowitz is senior scientist in the Electrical Systems Division at the Electric Power Research Institute (EPRI) in Palo Alto, California. He is in charge of advanced projects such as superconductivity, cryogenics, amorphous metals, and the effects of EMP on the power grid. Prior to joining EPRI in 1974, he did research at the Stanford Linear Accelerator Center, Varian Associates, and the Westinghouse Research Center. Recipient of a doctor's degree in physics from Washington State University in 1963, he has taught at several universities. He holds 30 patents and has published over 25 scientific papers.

Enrico Fermi observed an electromagnetic pulse (EMP) from the first deliberate nuclear explosion in the history of civilization: the Trinity nuclear bomb test in Alamogordo, New Mexico, at 5:30 A.M. on July 16, 1945. One can only speculate whether he foresaw the different manifestations of EMP and the subtle offensive and defensive ramifications that were to evolve. Observers were stationed as close as 10,000 yd (9 km) from the tower which cradled this infant bomb. Some instrumentation failures on this test were attributed to EMP.

The electromagnetic pulse was observed with increasing interest, precision, and understanding from this first atmospheric explosion in 1945 to the last United States atmospheric nuclear tests, which were conducted over the Pacific Ocean in 1962. At 0900 h 9.029 s on July 9, 1962, Greenwich Civil Time (GCT; 11 P.M., July 8, in Hawaii), a 1.4-megaton hydrogen bomb carried by a Thor rocket was detonated 248 mi (400 km) above the Johnston Island area. The Johnston atoll (some references call it Johnson) is located at about 17°N latitude and 168°W longitude. This high-altitude nuclear test explosion, called Starfish, was previously announced to the world as an international geophysical event.

Public media stories have repeatedly stated that in Hawaii, some 800 mi (1300 km) to the east-northeast of the blast, streetlights and power lines broke down and burglar alarms started ringing as a result of the EMP from this blast. In fact, not much happened to either the power or the communications systems in Hawaii. To support a time-correlated cause-and-effect relationship, the evidential events in Hawaii would have had to occur within milliseconds after the blast. It is not documented that any such power failures occurred within even approximately 1 h of the blast. Streetlights did fail on the island of Oahu; this can readily happen since they are in an easily disrupted series circuit. The tripping of burglar alarms and downing of power lines may be apocryphal. There is no record at the Hawaiian Electric Utility of any problem on the system. Direct inquiry of Hawaiian Electric personnel who were present at the time of the blast fails to confirm the breakdown of power lines or the opening of circuit breakers. Apparently, whatever electrical disturbances occurred that day, either by coincidence or from EMP, were minor. The systems were easily restored and remained fully operational. There was no report of trouble in the telephone system in Hawaii. There is no explanation of why only Oahu should have suffered from the EMP and not the other Hawaiian islands, which are roughly at the same location from the blast. There are many other islands within a 1000-mil (1600-km) radius of the blast, including Necker (NE), Midway (NW), Howland (SSW), and Palmyra (SE), also reporting no trouble.

Evidently the military and civilian observers on both land and sea saw sufficiently uninteresting and inconsequential far-ranging EMP effects from this 248-mi-high (400-km) blast that they changed their primary test plans. Over 3 months passed before the second shot to establish a correlation between the first blast and EMP effects in Hawaii. Yet, instead of firing the next shot at the originally scheduled 500-mi (800-km) altitude, the next three blasts were fired at heights between 25 and 50 mi (40 and 80 km) over Johnston Island on October 20, October 26, and November 1, 1962 (GCT). These heights guarantee the elimination of the high-amplitude effects of EMP at distances as far as Hawaii. At these low altitudes, the steep-front EMP effect would be severely limited by line of sight, and would find no long electric power lines with which to interact.

Public media stories have also conveyed the notion that the effects of this high-altitude EMP (HEMP) came as a total surprise to the United States on July 9, 1962, but the Soviets already knew all about it. As early as 1945, Fermi recognized that EMP effects are coupled to nuclear detonations. It would be surprising if he did not envisage both the magnetic bubble and the spiraling electron mechanisms. In the early 1950s, EMP was used as a diagnostic tool for studying the time evolution of a nuclear explosion. In 1954, R. L. Garwin made theoretical calculations regarding HEMP. In 1956 magnetic-field measurements were made in conjunction with nuclear bomb tests. In 1960 the impact of EMP on military communications became an important consideration of weapon systems planners.

One may infer from these facts that large expenditures by the Department of Defense on the development of sensitive, high-speed electronic instruments such as ultrafast oscilloscopes and very fast tape recorders for measurement, recording, and storage of data were strongly motivated by the EMP phenomena. These expenditures began long before 1962. Interestingly, the EMP signal was often used to trigger oscilloscopes and other recording devices in early nuclear bomb tests.

An exhaustive 1983 Sandia Report commissioned by the U.S. Nuclear Regulatory Commission concluded that EMP poses no substantial threat to nuclear power plants. This report considered electrical equipment such as transformers, motors, switches, relays, diodes, other semiconductor devices, and the general power plant system. It predicted that EMP-induced signals at the critical equipment are generally much less than normal operating levels, and that the likelihood of individual components failing is small. Because commercial nuclear power plants are similar, conclusions that are correct for the example Watts Bar nuclear power plant are valid for other nuclear plants. There are also no essential differences between the electrical equipment at a commercial nuclear power plant and that at other commercial power plants, and although there may be differences in the electrical shielding of the buildings, it has not been determined if they are significant. No power plants and few, if any, civilian facilities are capable of sustaining the effects of a nearby nuclear explosion.

DEFINING EMP

Electromagnetic pulse is a term which encompasses a wide spectrum of electromagnetic radiation. Two basic kinds of EMP will be considered. One is a relatively slow electromagnetic pulse called magnetohydrodynamic EMP. The other is a high-altitude EMP commonly called HEMP, which gives rise to a quick pulse. However, since a high-altitude nuclear detonation also produces magnetohydrodynamic EMP, the early-time, fast-pulse EMP can be called TEMP in order to distinguish clearly between the two phenomena. The descriptions given in this section for magnetohydrodynamic EMP and TEMP will include the essentially standard unclassified specifications. In other sections TEMP will be scrutinized more closely from a first-principles point of view.

Magnetic bubble EMP (BEMP). Two similar but different magnetic perturbations have been assigned the common nomenclature magnetohydrodynamic EMP. The first is magnetic bubble EMP (BEMP). A magnetohydrodynamic bubble is produced as the nuclear bomb's ionized debris expands. A nuclear explosion achieves extremely high temperatures. For example, the steel tower used in the very small 1945 Trinity test was completely vaporized. These high temperatures and emitted x-rays produce vaporization and ionization forming a conducting plasma shell of the bomb material and any other nearby matter that initially expands at 1250 mi/s (2×10^6 m/s).

The geomagnetic flux inside this conducting shell is initially very small. If the shell were not conducting, it would simply intercept and enclose more and more of the Earth's magnetic flux as it expands. However, because it is conducting, currents are set up in this shell whose magnetic flux tends to cancel the Earth's flux (in accordance with Lenz's law). Even though the shell gets much bigger, its own currents try to limit the flux to the amount initially inside the small shell. As this magnetic bubble grows, it excludes or pushes away the Earth's magnetic field, yielding a concentration of flux outside it and almost no flux inside it (Fig. 1). As the magnetic flux density B increases outside the bubble, the magnetic pressure $B^2/2\mu$ and damping by the air act to slow down and finally stop the expanding magnetic fireball. This occurs when the magnetic pressure equals the kinetic pressure, and the magnetic field snaps back inside the bubble, producing an Alfvén wave. An initially spheroidal fireball of ionized matter probably distorts into a prolate spheroid as it expands due to the disproportionate enhancement of the magnetic field in the equatorial region of the bubble. In the far field, the effects which propagate to the Earth should decrease in strength rapidly as the distance to the source region increases. The maximum field is quite small, only 0.03 V/ft (0.1 V/m), but has a period as long as 2 to 100 s. The time delay of this pulse is of the order of 2 to 5 s after the nuclear explosion. Even small antisatellite blasts at very high altitudes may produce magnetohydrodynamic EMP effects.

Atmospheric heave EMP (AEMP). The second MHD perturbation occurs more than 10 s after the burst. It is called atmospheric heave EMP (AEMP), since it is caused by the atmospheric heave of bomb-heated ionized air across the geomagnetic field. About 70% of the energy released by a bomb appears as x-rays which photoionize the air. This forms large ionospheric current loops which have mirror images in the Earth (Fig. 2). The effects of perturbing the geomagnetic field extend out more than 1250 mi (10^6 m) from the source point and last for approximately 10^2 s. Both the field and frequency are very low, at 0.0003–0.01 V/ft (0.001–0.03 V/m) and 0.01 Hz.

The magnetohydrodynamic EMP (more so for the AEMP than the BEMP) is similar to large natural disturbances such as auroras and solar storms. The major problem, which is transformer-core saturation due to a direct-current (dc) bias on an alternating-current (ac) system resulting from solar disruptions, is well understood and easily protected by capacitor blocking of the direct current. Thus, those systems protected against solar activity are probably adequately protected against magnetohydrodynamic EMP. Since these large-scale natural disturbances occur primarily in the northern latitudes, the simple measure of incorporating similar protection in all systems of the United States should suffice to eliminate the deleterious effects of magnetohydrodynamic EMP. Thus, it appears that only the question of how to cope with TEMP remains.

Tachy-electromagnetic pulse (TEMP). To better understand why the fast EMP (TEMP) is primarily associated with a high-altitude nuclear burst, the radiation from a charged oscillating sphere will be considered. If the sphere expands and contracts (breathing mode) between a maximum radius r_2 and a minimum radius r_1, the charge on its surface is clearly accelerated. The power radiated by an accelerated charge is proportional to the square of the acceleration. Therefore, one might expect a significant amount of radiation from this oscillating (accelerating) spherical surface. In fact, because of the spherical symmetry, there is no radiation at all. Even though one can easily calculate the radiation from each element of the charged sphere, the sphere as a whole does not radiate. In this case, there is total destructive interference because of the spheri-

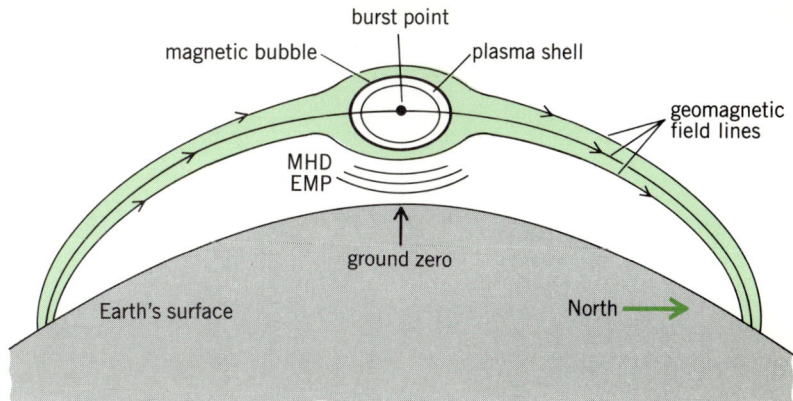

Fig. 1. Magnetic bubble EMP (BEMP); weak but long electromagnetic pulse.

cal symmetry. Figure 3 illustrates this simple spherically symmetric case for a hypothetical burst in an atmosphere with no density gradient and with no magnetic field for the Earth.

Low-altitude burst. Similarly, there is not much TEMP from a low-altitude (below 18 mi or 29 km) burst. First, the line-of-sight distance is smaller than for a high-altitude burst. Thus, the Earth gets in the way for the distant propagation of the TEMP. Second, the gamma rays produced by the bomb are emitted roughly symmetrically from the bomb. They are stopped in roughly spherically symmetric shells by colliding with electrons in the air molecules.

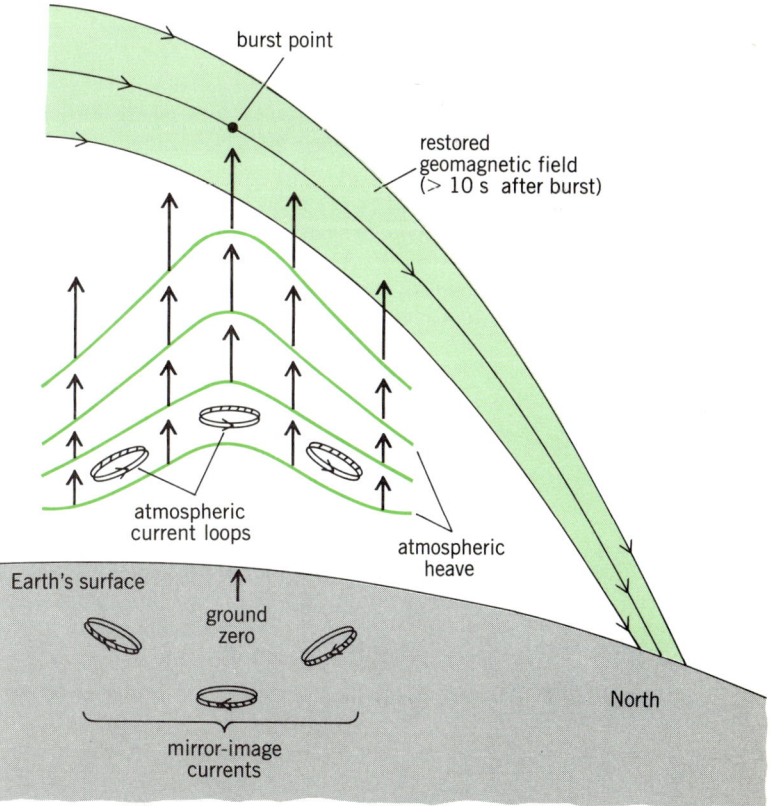

Fig. 2. Atmospheric heave EMP (AEMP).

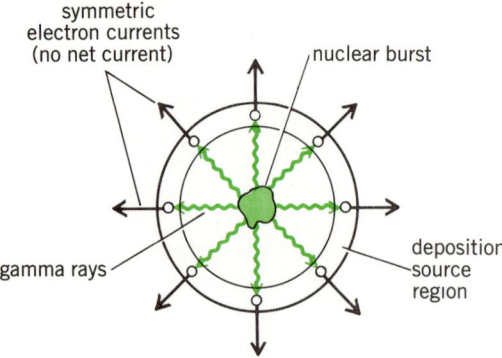

Fig. 3. No electromagnetic pulse.

By the time these electrons can spiral around (because of bending by the geomagnetic field) so that the radiation is mainly directed back toward the Earth, the coherence is lost, and many amplitude-reducing secondary electrons have been produced by collisions with air molecules. The line-of-sight distance of propagation of high frequencies is much shorter than for a low-altitude burst. Only about 10^{-7} of the prompt gamma energy is converted to EMP. So again, the region of TEMP influence is even closer to the burst point, around 10 mi (16 km).

They thus eject highly forward-directed Compton electrons at distances that are large with respect to the radius of the ionized shell of bomb debris, that is, the magnetic bubble. (The gamma rays do not interact with the Earth's magnetic field, and they travel at the speed of light.)

The Earth's magnetic field causes these electrons to spiral about the field lines. It is like the previously described oscillating sphere in the expansion phase with a twist (angular motion) added, as shown in Fig. 4, for the case of a negligible air-density gradient. The twist produces a magnetic field, whereas without it there was only a longitudinal radial electric field. Due to the higher collision frequency with the air molecules (which tends to disrupt coherence and reduces the lifetime of the Compton electrons), there is much less TEMP outside the blast region than for a high-altitude burst; and more than half of it is radiated away from the Earth. Figure 5 illustrates this case for both a magnetic field and an air-density gradient.

Surface burst. There is a hemispherical asymmetry in a surface burst that a low-altitude burst does not have (Fig. 6). Any radiation-enhancing (EMP-strength) coherence effects which occur initially are for radiation directed away from the Earth. As in the low-altitude burst, high-energy gamma rays moving out radially from the burst point produce relativistic Compton electrons in collisions with air molecules.

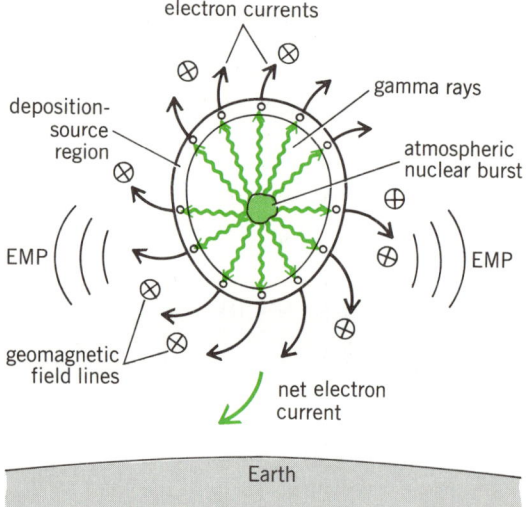

Fig. 5. Mild electromagnetic pulse.

High-altitude burst. The higher the altitude of the nuclear burst, the greater the area illuminated by the TEMP (Fig. 7). The altitude should be above 25 mi (40 km) to produce sufficient asymmetry and coverage to allow the TEMP effect to be of importance. Prompt high-energy gamma rays (ranging 0.5–2 MeV, of the order 1 MeV) are produced by a megaton nuclear explosion; they move out radially from the burst point with a rise time of about 10^{-8} s and about 10^{-3} of the bomb's energy. Those gamma rays moving toward the Earth will have increasingly more molecular collisions until they reach a height of about 19 mi (30 km). At this altitude, they will have produced an appreciable number of about 1-MeV Compton electrons by hitting and ionizing air molecules. (On the average, the energy of a Compton electron will be approximately half the energy of the gamma that produced it.) At 19 mi (30 km), the current of these high-energy electrons is roughly a maximum, since most of the gamma rays have been absorbed before reaching lower altitudes. At higher altitudes, the low atmospheric density limits the production of the Compton electrons. Few of the primary electrons are mainly scattered in the closely forward direction, and in turn their energy is absorbed as they produce secondary electrons and ions. Thus only a small fraction of the Compton electrons can participate for only a very short time in coherent effects.

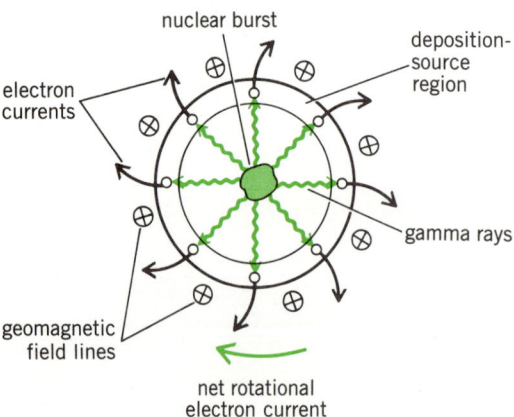

Fig. 4. Weak electromagnetic pulse.

Both the primary and secondary electrons are deflected into spiraling orbits around the Earth's magnetic field lines. It is these accelerated electrons which radiate the TEMP over large distances of the Earth's surface. The secondary electrons are an important factor in limiting the TEMP amplitude, and in decreasing it as a function of time. The affected area increases as the height of the burst increases, since the gamma rays can travel farther and cover a larger area before striking atmospheric molecules.

The maximum electric field strength E_{max} is nominally given the value of 50 kV/m, with a very short rise time of about 10^{-8} s. Although rumors from the French atmospheric tests hint that the field went as high as 30 kV/ft (100 kV/m), the United States' position in the unclassified literature has been that the peak field does not increase above the nominal value for bombs with yields in excess of 0.5 megaton. The total length of the pulse is about 10^{-6} s.

The maximum field at any point on the Earth's surface may be less than 15 kV/ft (50 kV/m) and depends upon a number of factors. Some of these factors are: (1) the height of the blast and its position in latitude with respect to the Earth's magnetic field; (2) the position in latitude and orientation of the detection point with respect to the geomagnetic field; and (3) the weapon yield and design.

Indeed, a single multimegaton weapon detonated 300 mi (480 km) over the most conducive point could blanket the entire United States with TEMP. However, more than half the United States would experience fields less than $0.6 E_{max}$. An appreciable fraction of the United States would not see fields in excess of $0.25 E_{max}$. Even in those areas where E_{max} exists, the TEMP will be able to have appreciable coupling to only a small fraction of the power lines because of orientation and interference effects. The fact that most power lines carry three phase alternating current mollifies the TEMP due to the division of the TEMP's energy between the three phases and the ground wire(s). Therefore, for these reasons as well as even stronger ones to follow, the notion that a single nuclear weapon could black out the entire United States power grid is more science fiction than science.

LIGHTNING VERSUS EMP

The electric power grid has survived and evolved in the presence of large natural electromagnetic disturbances or pulses such as solar storms and lightning. A comparison was made above between the low-frequency, low-amplitude magnetohydrodynamic EMP and the effects of auroras and solar flares. The similarity is clear. Electrical utilities have easily and successfully protected their circuits from the latter; it should be as straightforward to protect them from the former.

However, lightning and TEMP are not as similar, so it is necessary to look much deeper to see if TEMP can be handled by the power system as easily as lightning can. Lightning is the transient dis-

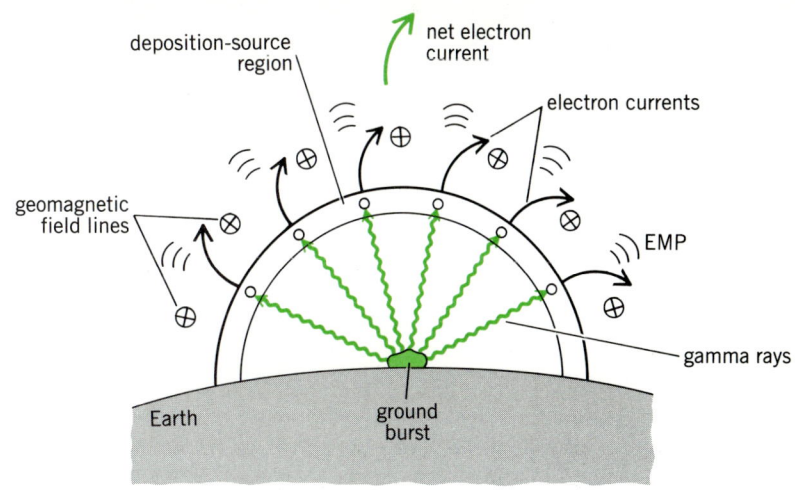

Fig. 6. Moderate electromagnetic pulse.

charge of electric charge accumulated in clouds. When a sufficiently high electrostatic field is built up, the air breaks down, forming a long lightning stroke or arc between the cloud and the earth.

A lightning stroke is made up of one or more intermittent partial discharges. Peak currents higher than 500 kA have been directly measured. Currents measured at the ground can rise to 20,000 A in about 10^{-6} s, with a maximum time rate of change in current of about 2×10^{11} A/s—not that small compared with the capabilities of EMP. The potential of charged clouds that produce lightning can be as high as 10^9 V. The current channel has a sharp temperature rise due to the rapid energy input to it. This causes the channel to expand with supersonic speed (in 10^{-5} s), producing the roughly cylindrical shock wave which becomes a part of thunder. The

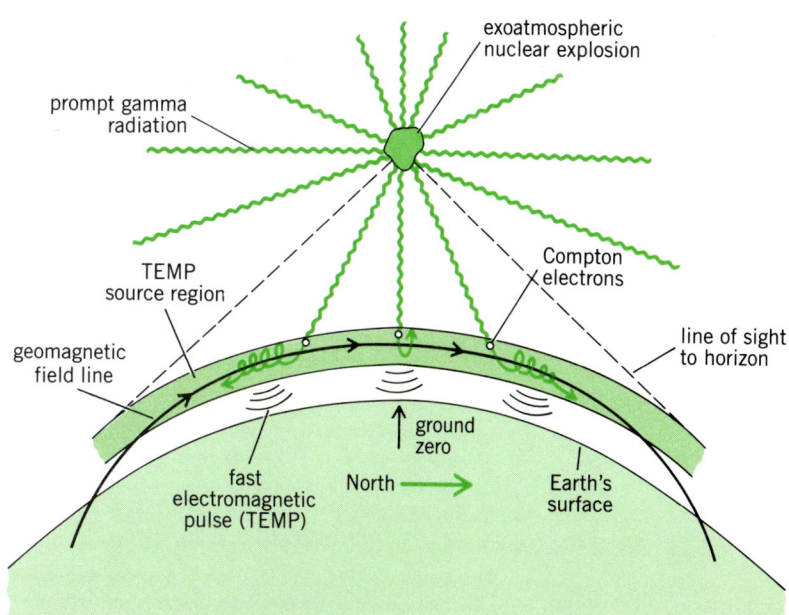

Fig. 7. Strong electromagnetic pulse.

channel temperature as measured spectroscopically (Doppler broadening) is about 54,000°F (30,000 K).

Recent measurements indicate that lightning strokes contain significant components with rise times greater than 10^{-7} s. Lightning creates electric fields in the discharge region of approximately 3×10^5 V/ft (10^6 V/m)—orders of magnitude higher than the peak TEMP field—with power levels at around 10^{12} W, and energy dissipation between 10^9 and 10^{10} joules. Instruments on orbiting satellites have detected many lightning bolts with currents as high as 10^6 A.

These lightning parameters make all the EMP numbers pale in comparison. Yet there is a significant difference between lightning and EMP which warrants looking further at EMP. Fortunately, lightning represents only a localized threat. The electromagnetic pulse has much lower power density, but is also much more global. The discussion below will examine whether this saves EMP from palling with respect to what the power system ordinarily withstands.

A direct lightning strike produces the most severe effects. The high current density (about 10^3 A/cm^2 in a stroke) delivers a high power density to the strike point, resulting in demolished structures such as exploded timber, molten metal, and charred insulation. Lightning pulses or transients can propagate along transmission lines at almost the speed of light, with circuit-limited rise times of about 10^{-6} s to peak voltages as high as 5×10^6 V, with a rate of rise of 4×10^{12} V/s. By all standards, lightning is quite a violent phenomenon with which the utility system copes routinely and with a large measure of success.

The spectacular destruction accompanying a direct lightning hit is not apt to accompany EMP. In comparison, EMP is a low-power-density disturbance, but with the subtle aspect that its effects can be gathered over a large distance, accumulated, and concentrated by electrical conductors. The steep front of the TEMP implies that small inductances and capacitances are no longer negligible. The primary effect of the TEMP is to induce overvoltages and overcurrents in the power system. It is likely that the transients produced in the system will not have as fast a rise time as that of the free-wave TEMP. If so, the newly developed zinc oxide lightning arresters may be quite effective in shunting these pulses to ground.

ENERGY IN THE TEMP

In this section the given TEMP parameters will be looked at together with nuclear bomb data provided by the Department of Defense. Implicit in the given TEMP electric field and in its given ubiquitousness within the bomb's horizon is a total energy content. By calculating this energy explicitly and comparing it with what the bomb is capable of yielding, limitations and possible inconsistencies in the given parameters may be ascertained.

The TEMP cannot have more energy than the nuclear bomb from which it came. It is not generated like fire, where the bomb acts like a small match and nature provides the materials for the energy of combustion. The Earth's magnetic field is hardly time-varying in the region of the Compton electrons and hence cannot supply energy to them. The interaction of these electrons of high velocity v and of charge e and the Earth's field of flux density B is simply the Lorentz force, Eq. (1). The field does no

$$F = e\vec{v} \times \vec{B} \qquad (1)$$

work W on the electrons, since the displacement $\vec{ds}$ is always perpendicular to the force, as shown by Eq. (2).

$$W = \int \vec{F} \cdot \vec{ds} = \int (e\vec{v} \times \vec{B}) \cdot (\vec{v}\,dt) = 0 \qquad (2)$$

In fact, the TEMP energy cannot even exceed the total energy of the prompt gamma rays, since they give birth to it by creating the Compton recoil electrons. Both the fraction of the energy of a bomb that is prompt gamma-ray energy and the parameters of TEMP are given. This energy fraction is given as about 10^{-3} (some experts use about 10^{-4}). The electric field strength E in the pulse as a function of time t is given as Eq. (3), where $E_0 = 52.5 \times 10^3$

$$E = E_0 (e^{-\alpha t} - e^{-\beta t}) \qquad (3)$$

V/m = 16.0 V/ft, $\alpha = 4.0 \times 10^6$/s, and $\beta = 4.78 \times 10^8$/s. From these parameters one can deduce that the pulse rises quickly to peak value at about 10 nanoseconds and that the pulse falls to half-value at about 160 nanoseconds.

From the time integral of the Poynting vector

$$\int_0^t \vec{E} \times \vec{H}\, dt$$

one can calculate the total energy density D delivered by the TEMP during the time t using Eq. (4).

$$\begin{aligned} D(t) &= \int_0^t \frac{E_0^2}{Z}(e^{-\alpha t} - e^{-\beta t})^2 dt \\ &= \frac{E_0^2}{z}\left[\frac{1}{2\alpha}(1 - e^{-2\alpha t}) + \frac{1}{2\beta}(1 - e^{-2\beta t}) \right. \\ &\quad \left. - \frac{2}{(\alpha + \beta)}(1 - e^{-(\alpha + \beta)t})\right] \text{J/m}^2 \end{aligned} \qquad (4)$$

$$z = 377 \text{ ohms}$$

Thus, $D(10 \text{ ns}) = 4.8 \times 10^{-2}$ J/m^2, $D(50 \text{ ns}) = 2.8 \times 10^{-1}$ J/m^2, $D(160 \text{ ns}) = 6.4 \times 10^{-1}$ J/m^2, and $D(1 \text{ }\mu\text{s}) = 8.9 \times 10^{-1}$ J/m^2.

In doing an energy balance, a 1.4-megaton bomb in precisely the Johnston Island scenario will be looked at here and throughout this article. The total energy release of this bomb is 5.9×10^{15} J. Therefore, the total gamma (Γ) energy is expressed by Eq. 5.

$$\mathscr{E}_{\Gamma \text{total}} = 10^{-3}(5.9 \times 10^{15} \text{ J}) = 5.9 \times 10^{12} \text{ J} \qquad (5)$$

The Starfish blast was at a 248-mile (400-km) height, or 248 mi − 19 mi = 229 mi (368 km) above the critical part of the atmosphere. At this height, the line-of-sight distance to the horizon is 1370 mi (2205 km). The distance on the Earth from the Johnston atoll to the bomb's horizon is 1320 mi (2125 km). Therefore, all of Hawaii is well within the area that was covered by the TEMP. A nuclear bomb similarly placed over Kansas would cover the continental United States with TEMP.

TEMP is supposed to be a nearly plane wave that is directed essentially perpendicular to the Earth and essentially undiminished by distance. If this is so, then for the area (1.4×10^{13} m^2) enclosed by the bomb's horizon, the energy delivered by the TEMP at the end of 1 microsecond is expressed by Eq. (6). About two-thirds of the gamma energy is

$$\mathscr{E}_{\text{TEMP}} = 8.9 \times 10^{-1} \text{ J/m}^2 \, (1.4 \times 10^{13} \text{ m}^2) \quad (6)$$
$$= 1.2 \times 10^{13} \text{ J}$$

directed toward outer space. Thus, the gamma energy directed toward Earth is given by Eq. (7).

$$\mathscr{E}_\Gamma = .34 \, (5.9 \times 10^{12} \text{ J}) = 2.0 \times 10^{12} \text{ J} \quad (7)$$

These two energies are compared in Eq. (8). Here

$$\frac{\mathscr{E}_{\text{TEMP}}}{\mathscr{E}_\Gamma} = \frac{1.2 \times 10^{13} \text{ J}}{2.0 \times 10^{12} \text{ J}} = 600\% \quad (8)$$

is found a clear discrepancy, since it is impossible for the energy conversion process from gamma rays to Compton electrons to TEMP to be 600% efficient, or even 100% efficient. TEMP is presented as having a magnitude of 15 kV/ft (50 kV/m) from border to border and from coast to coast, albeit with some spatial variations arising primarily due to the orientation of the geomagnetic field relative to the burst and observation points. This may reduce the discrepancy by a factor of 2 to 3, but as will be seen, what needs to be accounted for is a factor of 600.

The maximum gamma flux reaching the atmosphere occurs directly under the burst. At this point it would have an energy density of 3 J/m^2. Taking the D (1 µs) value of 0.89 J/m^2, calculated from the given TEMP electric pulse, this gives an efficiency of 30%. Even this appears high, since the kinetic energy of the Compton electrons will not be efficiently converted into radiation; a sizable amount goes into the production of secondary electrons.

The energy density of the gamma rays striking the atmosphere goes from a maximum of 3 J/m^2 to a minimum near the horizon of less than 0.1 J/m^2. Thus the density of Compton electrons goes down the farther they are produced from the blast. Even more importantly, the Compton current density is significantly reduced. With this reduction, one may expect a concomitant significant reduction in the radiated power density. This is mitigated to some extent by a concomitantly smaller production of secondary electrons, which tend to attenuate the TEMP.

From the analysis in this section, it can be concluded that at least a 28-megaton ([600%/30%] × 1.4 megaton) bomb is needed to produce the energy content in the TEMP that is implicitly claimed for a 1.4-megaton bomb, or the maximum electric field is less than 3.4 kV/ft (11 kV/m), or the TEMP is not so ubiquitous, etc. Glasstone and Dolan give a conversion efficiency of about 1% from prompt gammas to TEMP which makes the discrepancy between the Department of Defense's representative pulse and the available energy more than a factor of 600.

TEMP AND POWER-LINE FLASHOVER

Whether or not the TEMP can cause arcing between the power lines on a given tower or pole, or from the power line to ground via the tower or pole or ground wire will be examined from several points of view. It is very unlikely that the TEMP will produce arcing between power lines, that is, phase-to-phase, since it will excite all three phases similarly and at the same time. Ordinary transmission-line distances between conductors (phase-to-phase) go from 15 ft (4.6 m) for 138 kV, to 17 ft (5.2 m) for 230 kV, to 26 ft (7.9 m) for 345 kV. The corresponding nearest distances to tower are 4.9, 7.9, and 8.5 ft (1.5, 2.4, and 2.6 m). On a 13-kV distribution line, the phase-to-phase (conductor) separation is about 3.3 ft (1 m), and the closest distance from a line to the ground wire on the pole is about 5.2 ft (1.6 m). Therefore, considering a minimal gap of 2.6 ft (0.8 m), which is about half that of the smallest gap that is likely to break down, the worst case for arcing in air will be considered.

Electron time of flight. For the short distance of 2.6 ft (0.8 m) chosen, as well as for all practical purposes, electron time of flight can be taken as one criterion for deciding whether or not arcing could occur. Since 2.6 ft (0.8 m) is a small fraction of a transmission-line gap, it is considered a necessary (though not sufficient) condition for arcing that an electron have enough time to traverse this distance. (Although free electrons can be collected from midgap, the electric field is so much lower there that they cannot contribute much to an avalanche process.) In simple words, arcing cannot occur for a power line in air if the time duration of the discharge is so short that electrons do not have sufficient time to traverse an appreciable portion of the gap in an avalanche process.

When the electron mean free path between collisions is small with respect to the gap, the electron both gains and loses velocity as it traverses the gap. It gains velocity from the electric field as the field accelerates it across the gap. It loses velocity by collisions with air molecules. Its average drift velocity across the gap is given by Eq. (9), where m is the mass of the electron, e is the charge of the electron, $\overline{E}$ is the average electric field, λ is the collision mean free path of the electron, and v' is the maximum velocity of the electron.

$$\bar{v} = \int_0^{v'} \left(\frac{m}{e\overline{E}\lambda}\right) v^2 \exp\left[\frac{-mv^2}{2\lambda e \overline{E}}\right] dv \quad (9)$$

Solving this equation for the average electric field which is needed so that the electron can just traverse a gap d in time t yields Eq. (10). Now, for

$$\overline{E} = \frac{2m}{\pi \lambda e}(\bar{v})^2 = \frac{2m}{\pi \lambda e}\left(\frac{d}{t}\right)^2 \quad (10)$$

$d = (0.8$ m$)$ the following values are obtained: $\overline{E}_{10 \text{ ns}} = 1.25 \times 10^{11}$ V/m (4.1×10^{10} V/m),

$\overline{E}_{50\,\text{ns}} = 5 \times 10^9$ V/ft (1.6×10^9 V/m), $\overline{E}_{160\,\text{ns}} = 5 \times 10^8$ V/ft (1.6×10^8 V/m), and $\overline{E}_{1\,\mu\text{s}} = 1.25 \times 10^7$ V/ft (4.1×10^6 V/m).

From this it can be concluded that too high an average electric field of 5×10^8 V/ft (1.6×10^9 V/m) is needed for arcing (electrical breakdown) for discharge times of 50 ns or less. The maximum electric field in 50 ns induced by TEMP on a power line is more than an order of magnitude less than the needed field, even without charge loss.

The effects of charge loss due to field emission further work against the attainment of such high fields. To get such high average fields, the field on the line must be about 10^2 times higher, since the field is much lower in midgap. Microprotrusions on the conductor further enhance the field locally by factors of 10 to 10^4. Thus, when a limited power source such as TEMP tries to increase the field to such high levels, a substantial amount of charge will be lost by field emission which will limit the field. There is not enough local power input by the TEMP to overcome this charge drain into the air.

It is even unlikely that a high enough average field 5×10^7 V/ft (1.6×10^8 V/m) can be attained due to TEMP in 160 ns for an electron to traverse 2.6 ft (0.8 m). Since the transmission-line gaps are considerably longer than 2.6 ft (0.8 m), it can safely be concluded that they will not break down. If some distribution lines break down with discharge times of around 10^2 ns or longer, this is within the threshold of the time frame for lightning. The system with its lightning arresters has already proven its capability of shunting such surges to ground.

Electron avalanche. As energy from the TEMP is received and accumulated, the voltage of a line increases, and with this increase the electric field in the space around the line also goes up. If the electric field reaches a high enough value, the air will break down and an arc will ensue. The discharge around the line becomes an arc when it traverses the air gap and when the current in the discharge is essentially limited by only the voltage and external impedance of the circuit, that is, when the discharge becomes a small impedance element of the circuit.

There are two limiting scenarios that need to be considered in analyzing the question of whether an arc can be created by the energy delivered just by the TEMP. One is where the energy from the TEMP is accumulated over a relatively short time and a discharge initiates while the TEMP can induce high rates of rise of current and voltage in the line. This time scale is approximately 10 ns. The other is where the energy of the TEMP accumulates and feeds a discharge for a long time after there has been a significant fall in the TEMP amplitude as well as in the rates of rise of current and voltage. Of course, in the real situation, there will be energy dissipation and energy leaks such as reradiation, corona loss, and field emission.

Short time discharge. First, one can make an upper-limit calculation on the most energy that a possible discharge point on the power line could accumulate from the TEMP in a given time. The energy density D was calculated above in the plane wave hitting the Earth after different time intervals. The energy per unit area that arrived in the space surrounding the power line after 10 ns is $D = 4.8 \times 10^{-2}$ J/m^2. Most high-voltage overhead lines are three-phase with two shielding ground wires above them. So the TEMP energy divides by a factor of 5 about equally between them. This energy cannot be accumulated into a discharge point faster than the speed of light c. So, for this short time frame, an upper bound on the energy from the TEMP that can flow into the point of discharge is $\epsilon \sim D(\pi/5)(ct)^2$, which is 0.3 J. Not all of this energy can be converted into ionization of the air. The number of electron-ion pairs that can be produced by this energy input is expressed as Eq. (11), where e is the charge

$$N = \eta \frac{\epsilon}{eV_i} \qquad (11)$$

of an electron, V_i is the average ionization potential of the air [$V_i = .8\,(15.5\,V)_{N_2} + .2\,(12.5\,V)_{O_2} = 14.9\,V$], and η is the efficiency of ionization.

An electron undergoes many collisions with gas molecules, between 10^{10}/s and 10^{12}/s at atmospheric pressure. In the presence of an electric field, not all collisions produce ionization. In fact, a significant amount of the energy that the electron gains from the electric field goes into heating up the air with elastic and inelastic collisions that are not ionizing.

The ionization efficiency is given by Eq. (12),

$$\eta = \frac{V_i \alpha}{E} \qquad (12)$$

where E is the electric field producing the ionization and α is the first Townsend ionization coefficient. For air at atmospheric pressure, the low-frequency breakdown field is about 900 kV/ft (3000 kV/m) and α is 275 kV/ft (900 kV/m). This implies that for an ordinary air discharge the ionization efficiency is only 0.5%. For an impulse breakdown requiring $E = 2320$ kV/ft (7600 kV/m), α can be as high as 17,400/ft (57,000/m) and the efficiency goes up to 11%.

The maximum possible value of α, about 3×10^6/ft (10^7/m), occurs when every electron-gas collision is ionizing. If an electric field of about 3×10^8 to 3×10^9 V/ft (10^9–10^{10} V/m) is needed to produce breakdown for times of about 50–10 ns, the ionization efficiency would vary between 15 and 1.5%.

In times of about 10 ns, it is not likely that the TEMP can produce a radial electric field around the line as high as 900 kV/ft (3000 kV/m), so an upper limit for η would be 0.5%. Thus, the energy supplied by the TEMP in the first 10 ns can produce no more than $N = 6 \times 10^{14}$ electron-ion pairs.

An arc is formed by an exponential avalanche process in which one electron in ionizing a molecule results in two, and two result in four, and so forth. Thus, after going a distance δ, the total number of electrons created is $N = N_1 e^{\alpha \delta}$, where N_1 is the number of initiating electrons.

To determine the maximum excursion δ of the avalanche, recombination and regeneration effects are not relevant. Thus, Eq. (13) is obtained. The maximum

$$\delta = \frac{1}{\alpha} \ln\left(\frac{N}{N_1}\right) \quad (13)$$

excursion is obtained with $N_1 = 1$ and the smallest reasonable α, which would be 275/ft (900/m). Thus, $\delta_{max} < 0.04$ m. So, if a discharge were initiated, an energy and an avalanche consideration would limit it within a distance of 0.13 ft (0.04 m) of the line. (The electric field in midgap is about 10^{-2} lower than at the line, so it is not a region where much of an avalanche can be initiated.)

Much of the produced charge is lost by recombination, attachment, and diffusion processes. The maximum drift velocity for the electrons is about 10^5 m/s, so for a 2.6-ft (0.8-m) gap the average current would be 0.1 A if as much as 1% of the charge were collected. Even if 10^3 times as much energy went into a discharge or discharges, the average current would only be 100 A. This is an inconsequential current compared with the kiloamperes of fault current a line is equipped to handle.

Realistic worst case. A realistic worst case is where the TEMP builds up the radial electric field around the line until the average field is high enough to produce a discharge. This occurs when the average electric field strength $\overline{E}$ is about 900 kV/ft (3000 kV/m), so that the efficiency η becomes 5×10^{-3}. After a very long time, the entire energy of the pulse has been delivered with a total energy density $D = 0.9$ J/m² [see Eq. (4)] divided equally between the three phases and the two ground wires. So, roughly, the energy/length accumulated by the line is $\epsilon/\ell \sim (D/5)(ct) \sim 50$ J/m.

If it takes longer than about 10^2 ns ≡ t', for an arc to bridge the gap, this is within the threshold of the time frame for lightning, and the lightning arresters should be able to handle it. Switching surges from SF_6 breakers are also in this time frame, and the power system handles them routinely. Therefore, the maximum energy that can flow into the discharge point is expressed by Eq. (14).

$$\epsilon \sim 50 \text{ J/m } (ct') = 1600 \text{ J} \quad (14)$$

Thus the number of electron-ion pairs that can be produced by this energy input is given by Eq. (15).

$$N = \eta \frac{\epsilon}{eV_i} = 3.4 \times 10^{18} \quad (15)$$

The maximum excursion of the avalanche is given by Eq. (16). With $\alpha = 275$ kV/ft (900 kV/m) and $N =$

$$\delta_{max} = \frac{1}{\alpha} \ln N \quad (16)$$

1.2×10^{19}, δ_{max} is less than 0.16 ft (0.05 m) from the line. (It is unrealistic to expect the arc to initiate in the very low electric field region in midgap.)

Now the collection of the charge that was created by the TEMP will be considered. After the TEMP energy has discharged into the formation of the nascent arc, the field on the line will return to roughly its original value of about 520 kV/ft (1700 kV/m) and the average field value surrounding the line returns to about 15 kV/ft (50 kV/m), well below the threshold for discharge. At this average field value, the electron drift velocity is 3×10^4 ft/s (10^4 m/s). So the time interval for charge collection is about 10^{-4} s across 2.6 ft (0.8 m). Each electron makes 10^{10} collisions/s and hence suffers about 10^6 collisions in traversing the gap. This leads to substantially more attachment and recombination in traversing the gap than in the previous case. With a recombination or attachment probability of more than 10^{-5} per collision, less than 10^{-3} of the electrons cross the gap. Thus the average current will be only 5 A. Even if 10^2 times as much energy went into this long time discharge or discharges, the average current would only be 500 A, which again is an inconsequential current.

This is a negligible current that is small compared with fault currents due to lightning, and even the normal current on a transmission line. For example, a 138-kV line carrying 500 MVA has a normal root-mean-square current of 2100 A. It is equipped to handle fault currents in excess of 21,000 A.

Leaders and streamers. In the high-voltage breakdown of large gaps, arcing is preceded by discharges known as leaders and streamers. The first stage consists of the formation of a thin streamer called a pilot or leader. The leader is an ionized column advancing both by ionization at the head due to the intense electric field existing there, and by photoionization of the gas in advance of the streamer itself. The speed of propagation of this initial streamer is $\sim 10^5$–10^6 m/s.

Even if the time delay associated with the initiation of the leader is neglected, it would take more than 10^{-6} s for the leader to traverse the smallest power line gap of 1.5 m. The TEMP would be over and gone in 10^{-6} s.

NATURE OF THE TEMP

Following is a detailed discussion of the electromechanics and electrodynamics of the TEMP. To get a feeling for what is happening to the electrons in the upper atmosphere, their orbits are examined first. Equating the Lorentz force with the centripetal force on an electron, $e\vec{v} \times \vec{B} = \gamma m \vec{\omega} \times (\vec{\omega} \times \vec{r})$, where m is the electron rest mass, ω is the angular frequency, and $\gamma = (1 - v^2/c^2)^{-1/2}$. The rotational frequency ν is found to be Eq. (17).

$$\nu = \frac{\omega}{2\pi} = \frac{1}{2\pi\gamma}\left(\frac{e}{m}\right)B \quad (17)$$

Note that all the electrons with the same energy (same γ) and at the same magnetic flux density B will have the same frequency independent of where they enter the atmosphere, that is, independent of the angle that their motion makes with respect to the Earth's magnetic field. The Earth's magnetic flux density does not vary by more than a factor of 2 between its maximum value at the poles of 0.6 gauss and its minimum value at the equator of 0.3 gauss.

However, γ has a significantly larger variation. Therefore, a Larmor precession approach should not be used, since it requires that all the electrons have the same charge-to-mass ratio $e/\gamma m$, which they do not have.

It makes no difference if their orbits are large or small, or if they are circles or spirals: those electrons with the same $e/\gamma m$ will all have about the same frequency. This is why these are sometimes called synchrotron or cyclotron orbits. At a 19-mi (30-km) height, if the Earth's field is taken to be about 6×10^{-5} weber/m² (0.6 gauss), the frequency $\nu = 0.57$ MHz for 1-MeV electrons, and the period of one orbit is 1760 ns.

The orbital radius is given by Eq. (18), where Φ

$$r = \frac{\gamma mv \sin \Phi}{eB} \qquad (18)$$

is the angle between $\vec{v}$ and $\vec{B}$. Electrons of 1-MeV energy have a velocity of $0.94 c = 2.8 \times 10^8$ m/s $= 1.74 \times 10^5$ mi/s, and $\gamma = 3$. Thus, the orbital radius varies from a maximum of 260 ft (80 m) for $\Phi = 90°$, on down to 0 ft (0 m) as Φ approaches $0°$.

Power radiated per electron. The power radiated by a nonrelativistic accelerated electron is expressed by Eq. (19), where a is the acceleration and

$$P = \frac{e^2 a^2}{6\pi\epsilon_0 c^3} \qquad (19)$$

ϵ_0 is the permittivity of free space. Since the electrons are traveling near the speed of light, the radiation is enhanced by relativistic effects, and the radiated power is given by Eq. (20).

$$P = \frac{e^2}{6\pi\epsilon_0 c^3} \gamma^4 \left[a^2 + \left(\frac{\gamma}{c}\right)^2 (\vec{v} \cdot \vec{a})^2 \right] \qquad (20)$$

During the interval of maximum power radiation, most of the electrons have circular or helical orbits of roughly constant pitch and radius, where v is perpendicular to a, so $\vec{v} \cdot \vec{a} = 0$. Before losing much energy, the 1-MeV electrons from a high-altitude burst will each be radiating at most about 4×10^{-22} W. This is not much power per electron, but there are very many electrons.

To find the total power radiated by these orbiting electrons is not an easy problem. One needs to calculate the vector sum of the fields of all the electrons and then integrate over the Poynting vector. The power radiated is not the simple sum of the powers, because there are destructive and constructive interference effects. As was seen from the earlier example of the charged oscillating sphere, there was no radiation because of total destructive interference. Nevertheless, the average power radiated per electron as derived from the given TEMP characteristic can be compared with the relativistic calculation of the maximum synchrotron radiation from one electron.

As previously discussed, the power density in the TEMP is expressed by Eq. (21). The maximum

$$P/A = \vec{E} \times \vec{H} = \frac{E_0^2}{z}(e^{-\alpha t} - e^{-\beta t})^2 \qquad (21)$$

power density occurs at about 10 ns, with a value of 6.6×10^6 W/m². The total maximum power delivered to the intercepted part of the Earth is 6.6×10^6 W/m² $(1.4 \times 10^{13}$ m²$) = 9.2 \times 10^{19}$ W.

The total gamma energy hitting the atmosphere was calculated to be 2×10^{12} J. Therefore, the maximum number of Compton electrons of 1-MeV average energy that could be created is given by Eq. (22). Hence, the TEMP power per electron is Eq. (23).

$$N \sim \frac{2 \times 10^{12} \text{ J } (1 \text{ eV}/1.6 \times 10^{-19} \text{ J})}{(10^6 \text{ eV} + 15 \text{ eV})/\text{electron}} \qquad (22)$$
$$= 10^{25} \text{ electrons}$$

$$\frac{9.2 \times 10^{19} \text{ W}}{10^{25} \text{ electrons}} = 9 \times 10^{-6} \text{ W/electron} \qquad (23)$$

Therefore, the ratio of TEMP power per electron to the synchrotron radiation power from an electron in its phase of maximum radiation (early time, and $\Phi = 90°$) is Eq. (24). This is quite a large number;

$$\frac{\text{TEMP radiation}}{\text{Synchrotron radiation}}$$
$$= \frac{9 \times 10^{-6} \text{ W/electron}}{4 \times 10^{-22} \text{ W/electron}} = 2.25 \times 10^{16} \qquad (24)$$

it should be determined whether it is at all possible.

If all the electrons in the synchrotron orbits could radiate coherently, the total power from all these electrons would be about 10^{28} W, which is certainly more than the 10^{20} W that is supposed to be delivered by the TEMP. However, this is not physically possible.

The fundamental wavelength is given by Eq. (25).

$$\Lambda = \frac{c}{\nu} = \frac{3 \times 10^8 \text{ m/s}}{5.7 \times 10^5 \text{ Hz}} = 5.3 \times 10^2 \text{ m} \qquad (25)$$
$$= 1.74 \times 10^3 \text{ ft}$$

A rough estimate indicates that one way to achieve the large enhancement of radiated power to account for the given TEMP is to have about 10^9 groups of electrons, each with about 10^{16} electrons acting quasicoherently. This implies that the electrons in each group are within 330 ft (100 m) of each other, which is about $1/5\,\Lambda$. However, 330 ft (100 m) greatly exceeds the wavelengths of most of the radiated power because of relativistic forward beaming.

Ensemble energy derivation. At any instant, the energy radiated by the ensemble of Compton electrons is stored equally in the electric and magnetic fields in the oscillating electromagnetic waves. This energy cannot exceed the Earth-directed part of the gamma energy, $\mathscr{E}_\Gamma = 2.0 \times 10^{12}$ J calculated in Eq. (7). Equation (26) is thus obtained. The volume

$$\int \frac{1}{2}\epsilon_0 E^2 dV + \int \frac{1}{2}\frac{B^2}{\mu_0} dV = 2\int \frac{1}{2}\epsilon_0 E^2 dV < \mathscr{E}_\Gamma(t) \qquad (26)$$

element $dV = Ac\,dt$, and taking the time derivative of Eq. (26) gives Eq. (27), where P_Γ is the power

$$\epsilon_0 E^2 Ac < \frac{d}{dt}\mathscr{E}_\Gamma = P_\Gamma \qquad (27)$$

in the gamma flux toward the Earth, and $P_\Gamma =$

$\frac{1}{3} P_{\Gamma \text{ total}}$. ($P_\Gamma$ and area A are not independent. As the burst height decreases, A decreases and P_Γ increases.)

From Glasstone and Dolan, P_Γ can be estimated by Eq. (28), where τ_1 is about 10^{-8} s and τ_2 is

$$P_{\Gamma \text{ total}} = 1.1 \times 10^{20} \text{J/s} \, (e^{-t/\tau_2} - e^{-t/\tau_1}) \quad (28)$$

about 10^{-7} s. Therefore, combining Eq. (27) and (28) results in Eq. (29). From these parameters, the

$$\begin{aligned} E &< \left[\frac{1.1 \times 10^{20} \text{J/s}}{3\epsilon_o A c} \right]^{1/2} [e^{-t/\tau_2} - e^{-t/\tau_1}]^{1/2} \\ &< 31 \text{ kV/m} [e^{-t/\tau_2} - e^{-t/\tau_1}]^{1/2} \end{aligned} \quad (29)$$

maximum electric field $E_{\max}$ occurs at 23 ns, and is less than 6.7 kV/ft (22 kV/m). If less than 0.3 of $\mathscr{E}_\Gamma$ were converted into TEMP, then $E_{\max}$ would be less than 3.7 kV/ft (12 kV/m). Glasstone and Dolan state that only about 10^{-2} of $\mathscr{E}_\Gamma$ is converted into TEMP, which would make $E_{\max}$ on the order of 0.67 kV/ft (2.2 kV/m). Figure 8 compares the given Department of Defense TEMP electric field (curve A) with the one calculated here (curve B). Figure 9 compares the voltage induced on a transmission line by the two TEMP fields with lightning.

The briefness of the prompt gamma pulse gives it an extremely high burst of power, which results in the high value of E at short times. Insofar as the TEMP is spread out in time relative to the prompt gamma pulse, E will be further reduced in magnitude. As the Compton electrons produce secondary electron-ion pairs, the reduction of coherence and the increased conductivity and polarizability of the space will further diminish E. Both in this derivation and in the one to follow, a saturation effect due to absorption by secondary electrons is not included in the theory. As the burst point gets closer to the atmosphere or as the bomb gets very large, the intensity of the TEMP may saturate and become independent of the bomb's yield. This is another reason why these derivations overestimate E.

This derivation of an upper limit on $E_{\max}$ does not give much physical insight because it is not dependent on a model of the physical processes that give rise to it. Nevertheless, this derivation is very powerful and very general because it requires very little input. Next, the TEMP will be analyzed in more de-

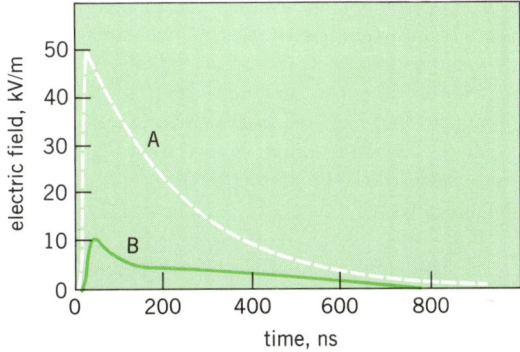

Fig. 8. TEMP electric field.

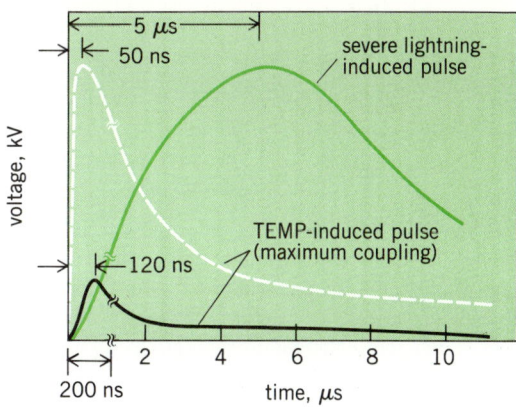

Fig. 9. TEMP-induced pulse (maximum coupling).

tail, which gives much more physical insight.

Physical-insight derivation of TEMP. As discussed above, the relativistic radiated power of an electron is given by Eq. (30), when the velocity is

$$P = \frac{\gamma^4 e^2 a^2}{6\pi\epsilon_o c^3} \quad (30)$$

perpendicular to the acceleration. The acceleration of an electron in a synchrotron orbit is described by Eq. (31). Combining Eq. (30) and (31) results in

$$\begin{aligned} \vec{a} &= \vec{\omega} \times (\omega \times \vec{r}) = \omega^2 r \\ &= \left(\frac{e}{\gamma m} B\right)^2 \left(\frac{\gamma m v \sin \Phi}{eB}\right) = \left(\frac{e}{\gamma m}\right) B v \sin \Phi \end{aligned} \quad (31)$$

Eq. 32. Substituting $v^2 = [(\gamma^2 - 1)/\gamma^2] c^2$ in Eq.

$$P = \frac{\gamma^2 e^4 B^2 v^2 \sin^2 \Phi}{6\pi\epsilon_o c^3 m^2} \quad (32)$$

(32) results in Eq. (33).

$$P = \frac{e^4 B^2 (\gamma^2 - 1) \sin^2 \Phi}{6\pi\epsilon_o c m^2} \quad (33)$$

This is the synchrotron radiated power that each electron radiates. It depends on only γ, B, and Φ. For a burst over the magnetic equator, $\Phi = 90°$ directly under the burst and everywhere to its magnetic east and west, giving the maximum value of $\sin^2 \Phi$. The angle Φ decreases the farther north or south of this burst that the Compton electrons are created.

As calculated above, the total number of Compton electrons is about 10^{25}. It was also shown that if their radiated powers add independently, the total radiated power would be a factor of 10^{16} smaller than claimed for the TEMP. The fields from these 10^{25} electrons can add coherently in two ways. If an observer is approximately the same number of wavelengths from an extended set of oscillating electrons that are somewhat locked in phase and in frequency, the fields at the observer's location will tend to add constructively. A second way in which the fields can add coherently is for groups of radiating electrons that are relatively close together to be somewhat locked in phase and frequency. Let us look at the latter mechanism.

The area-number density of Compton electrons is given by Eq. (34), that is, there are 8.8×10^{11}

$$\frac{N}{A} = \frac{1.2 \times 10^{25} \text{ electrons}}{1.4 \times 10^{13} \text{ m}^2}$$
$$= 8.8 \times 10^{11} \text{ electrons/m}^2 \quad (34)$$
$$= 8.2 \times 10^{10} \text{ electrons/ft}^2$$

electrons in cylinders of 1 m² cross section and length d. Assuming that only those electrons that are within some fraction g of the fundamental wavelength Λ can radiate together coherently, the number of coherent electrons in each group is described by Eq. (35), if $d < g\Lambda$. [If d were greater than Λ,

$$M = \left(\frac{N}{Ad}\right) d\pi (g\Lambda)^2 = \left(\frac{N}{A}\right) \pi g^2 \Lambda^2 \quad (35)$$

the relevant volume would be $(4/3)\pi (g\Lambda)^3$ instead of $d\pi (g\Lambda)^2$.]

The number of such groups is $G = (N/M)$. Thus, the total power radiated from the electrons radiating coherently in each group, and the groups radiating independently, is expressed by Eq. (36). Now, N is

$$P_T = \sum M^2 P dG = M^2 G \overline{P} \quad (36)$$
$$= NM \frac{e^4 \overline{B^2}(\gamma^2 - 1)}{6\pi\epsilon_0 cm^2} \overline{\sin^2\Phi}$$

given by Eq. (37), where f is the fraction of the

$$N = \frac{(f\mathcal{E}_\Gamma)}{[mc^2(\gamma - 1)]} \quad (37)$$

downwardly directed prompt gamma-ray energy $\mathcal{E}_\Gamma$ which is converted to Compton electrons. The value of Λ is obtained from Eq. (38).

$$\Lambda = \frac{c}{v} = 2\pi c\gamma m/e\overline{B} \quad (38)$$

Substituting Eqs. (35), (37), and (38) into Eq. (36), yields the total power radiated from all the electrons given by Eqs. (39) and (40).

$$P_T = \left(\frac{f\mathcal{E}_\Gamma}{mc^2}\right)^2 c \frac{g^2}{A} \left(\frac{\gamma}{\gamma - 1}\right)^2 \frac{2\pi^2 e^2(\gamma^2 - 1)}{3\epsilon_0} \overline{\sin^2\Phi} \quad (39)$$

$$P_T = \int (\vec{E} \times \vec{H}) \, dA = \int \frac{E^2}{Z} dA = \frac{\overline{E^2}}{Z} A \quad (40)$$

Therefore the root-mean-square TEMP electric field is expressed by Eq. (41), where $\mathcal{E}_\Gamma$ is the

$$E_{\text{rms}} = \left(\frac{f\mathcal{E}_\Gamma}{mc^2}\right) \frac{g\pi e}{A} \left(\frac{\gamma}{\gamma - 1}\right)$$
$$\cdot \left[\frac{2Zc(\gamma^2 - 1)}{3\epsilon_0}\right]^{1/2} (\sin \Phi)_{\text{rms}} \quad (41)$$

downward component of prompt gamma-ray energy, f is the fraction of $\mathcal{E}_\Gamma$ which is converted into Compton electrons, g is the fraction of fundamental wavelength within which the electrons produce coherent radiation, Z is the impedance of free space ($= 377$ ohms), and A is the area illuminated by the gamma rays.

Note that the rms value of E is with respect to space, not time. Note also that P_T and E_{rms} are independent of B. This comes from relating the coherence to the fundamental wavelength Λ. It does not mean that synchrotron radiation can occur without a magnetic field. If $\mathcal{E}_\Gamma = 2 \times 10^{12}$ J, $f < .3$, $g = .1$, $A = 1.4 \times 10^{13}$ m², and $\gamma = 3$ (1-MeV electrons), Eq. (42) results.

$$E_{\text{rms}} < (11 \text{ kV/m})(\sin \Phi)_{\text{rms}}$$
$$= (3.6 \text{ kV/ft})(\sin \Phi)_{\text{rms}} \quad (42)$$

Conclusion An illusory image of EMP has now been added to the lexicon of the established horrors of nuclear war. It is invisible, silent, and harmless to humans. Yet a pervasive and persuasive media barrage has resulted in overkill with respect to its omnipotence and ubiquitousness. It is not impotent with respect to the electric power grid, but the analyses developed above indicate that its effects are not nearly as potent as the news and magazine stories seem to indicate.

Only a small fraction of a nuclear bomb's energy release is converted to TEMP. It cannot be more than the energy content of the prompt gamma rays, which is given to be about 0.1% of the bomb's energy. But there is good reason to think that the effective energy of the TEMP might be much less, perhaps as little as 10^{-6} of the bomb's energy release. Certainly, the lack of damage to both the power and the communications systems in Hawaii from the 1.4-megaton Starfish blast does not support the view that the effects of EMP are devastating to such systems.

EMP is capable of causing damage or malfunction to unshielded electronic components. Thus, there may be damage to solid-state devices related to communications and control in the power network. Electronic logic circuits may be particularly vulnerable since they operate by transferring information as a train of pulses. It is no wonder that EMP was proper cause for alarm to military systems planners having to protect sensitive high-speed electronic circuits. Even small transients in computer circuits can lead to counting errors and flawed output. The presently known aspects of EMP have been known since at least the 1960s, and tens of billions of dollars have been spent by the military-industrial complex to harden computers, communications systems, and the electronic systems in both missiles and their silos.

Nuclear bombs are awesome. Although it is very easy to be overawed by what a nuclear bomb can do, its potential is still far from infinite. The limits of claims in either direction—both the claims of minimal effects as well as maximal effects—can be probed. Where the claims exceed what is physically reasonable, they should properly be challenged in a balanced way.

Based upon the analyses presented in this article, it appears highly improbable, if not impossible, that the EMP from a single nuclear burst could black out this nation's power grid. It would be practically impossible for the EMP to cause widespread damage to the transmission-line system. With the exception of isolated cases, it appears highly unlikely that EMP could produce extensive damage to the distribution grid nationwide. A single nuclear device exploded at high altitude will not render vital electrical services inoperable across the entire United States, as has been suggested in many media references. Thus, the government's past implicit strategy of avoiding massive retaliation in response to a single isolated nuclear explosion or other isolated catastrophe over the United States remains a sensible plan. The purpose—to avert a nuclear holocaust catalyzed by a third power or accident—remains sound and is not undermined by the unreal possibility that a single burst could cripple the United States.

[MARIO RABINOWITZ]

Bibliography: W. J. Broad, The chaos factor, *Science*, pp. 41–49, January/February 1983; C. M. Crain, *Calculation of Radiated Signals from High-Altitude Nuclear Detonations by Use of a Three-Dimensional Distribution of Compton Electrons*, Rand Rep. N-1845-ARPA, 1982; D. M. Erickson et al., *Interaction of Electromagnetic Pulse with Commercial Nuclear Power Plant Systems*, Sandia Rep. NUREG/CR-3069, SAND82-2738/2, 1983; S. Glasstone and B. J. Dolan, *The Effects of Nuclear Weapons*, U.S. Department of Defense, 1977; W. J. Karzas and R. Latter, Electromagnetic radiation from a nuclear explosion in space, *Phys. Rev.*, 126(6):1919–1926, 1962; J. R. Pierce et al, *Evaluation of Methodologies for Estimating Vulnerability to Electromagnetic Pulse Effects*, National Research Council Rep., 1984; W. E. Scharfman, E. F. Vance, and K. A. Graf, EMP Coupling to Power Lines, *IEEE Trans. Antennas Prop.*, AP-26(1):129–135, 1978; D. L. Stein, Electromagnetic pulse: The uncertain certainty, *Bull. Atom. Sci.*, pp. 52–56, March 1983; J. Steinbruner, Launch under Attack, *Sci. Amer.*, pp. 37–47, January 1984; E. Teller, Electromagnetic Pulses from Nuclear Explosions, *IEEE Spectrum*, p. 65, October 1982; A. P. Trippe, *The Threat of Electromagnetic Pulse*, National Defense, pp. 22–27, December 1984; M. A. Uman, M. J. Master, and E. P. Krider, A comparison of lightning electromagnetic fields with the nuclear electromagnetic pulse in the frequency range 10–10 HZ, *IEEE Trans. Electromag. Compat.* EMC-24 (4):410–416, 1982.

Computer-Assisted Drug Design

A. J. Hopfinger

A. J. Hopfinger is director of the department of Medicinal Chemistry at G. D. Searle & Company and holds adjunct professorships at the University of Illinois at Chicago, University of Kansas, and Case Western Reserve University. He is a former Sloan Fellow (1971–1973) and vice-chairman of the 1985 Gordon Research on Quantitative Structure-Activity Relationships. He has written extensively in the chemistry field.

The cornerstone of pharmaceutical research is the premise that the biological responses elicited from an administered compound depend upon its chemical structure; changes in chemical structure result in changes in biological responses. Researchers in the health sciences have long recognized this general Structure-Activity Relationship (SAR) and have attempted to use it to design new drug candidates. However, the word "design" means different things to different people. Also, the criterion of ease of chemical synthesis has generally been adverse to SAR-based drug design. Advocates of preferentially making those compounds most easily synthesized view drug discovery as largely a random process. Their logic is that the odds of finding a compound with acceptable biological properties are higher in making a relatively large number of easily synthesized compounds, rather than making a few target compounds more difficult to synthesize but deduced form SAR data.

Still, an enormous collection of compounds results from making relatively small numbers of substituent changes at a few sites on simple organic structures. For example, the systematic substitution of any of 20 common groups, such as CH_3, OH, or Cl, at each of the seven sites on α-naphthoic acid (I) would lead to 20^7, or

(I)

1.28×10^9 compounds. Thus, it is statistically highly unlikely that any of the relatively few compounds made on the basis of ease of synthesis in any chemical series would be optimum with respect to a particular biological end point. In other words, the vast number of compounds potentially available for synthesis and testing in a therapeutic program intrinsically demands maximum exploitation of available SAR. This is drug design.

HANSCH ANALYSIS

Drug design, as practiced today, began with the work of C. Hansch and colleagues. He first derived a mathematical model which parabolically relates biological activity to relative lipophilicity as measured by the water/1-octanol partition coefficient. Subsequently, Hansch and co-workers defined a hydrophobicity constant π to measure logarithmic changes in partition coefficient. The original mathematical model of drug action was extended to include electronic and steric physicochemical properties that could influence biological activity.

This analysis resulted in the famous Eq. (1) which relates biological response (BR) to measures of the

$$\log(BR) = k_1\pi^2 + k_2\pi + k_3\sigma + k_4E_s + k_5 \quad (1)$$

physicochemical properties of the group or groups X in a molecule. In Eq. (1), σ is the Hammett constant which is a relative measure of group acidity, E_s is Taft's steric constant which reflects group size, and π is formally defined by Eq. (2), where P is the water/1-octanol partition coefficient, and X and H

$$\pi_x = \log P_X - \log P_H \quad (2)$$

refer to the substituted and parent molecules, respectively. BR is normally expressed as $(1/C)$, where C is the concentration of compound necessary to elicit a specific biological response.

To use Eq. (1) in a design mode requires the determination of the k_i. This can be accomplished by restricting each application to a homologous series of compounds (that is, compounds having a common chemical nucleus onto which the different substituents have been attached) which are evaluated by using a common biological protocol. Intrinsic to the common biological protocol is the quantitative measurement of relative biological potency or potencies.

An initial set of compounds in a chemical series must be made and biologically tested in order to formulate a specific form of Eq. (1). The choice of the initial set of compounds, including the size of the set, dictates the particular form of Eq. (1). The k_i values are usually determined by fitting the measures of the physicochemical properties of each compound in the initial set against the corresponding measures of biological activity by using multidimensional linear regression analysis. The resulting correlation equation and affiliated measures of statistical fit are, in composite, referred to as a quantitative structure-activity relationship (QSAR). A QSAR is a recipe for quantitatively predicting a biological activity from a compound in advance of its synthesis. This capability clearly provides a means to design molecules.

The predictive reliability of any QSAR is obviously a function of the initial set of training compounds, and the composite accuracy of the property measurements. Fortunately, when a QSAR fails, the new data can be included in a reanalysis of the updated SAR table and a revised QSAR generated. This feedback loop in the evolution of a QSAR provides a means of sustaining design criteria.

At least three different strategies have been devised to maximize the SAR information realized in the generation of the initial set of compounds used to construct a QSAR. These are the methods of S. U. Free and J. W. Wilson, the Topliss tree and the Craig plot. Each approach attempts to avoid duplicate synthesis with respect to molecular properties, as well as to make possible the evaluation of the role of an individual molecular property upon biological activity. A Craig plot for aromatic substituents is shown in Fig. 1. Inspection of the plot suggests that selection of CH_3 and Cl as substituents would be a way to evaluate the role of σ on biological activity for a constant value of π. Selection of CH_2OH and $NHCHO$ could be considered a redundant choice of substituents since they both have nearly the same π and σ values, respectively.

Hansch analysis has been widely applied over the last 20 years. A representative example is the examination by Hansch and coworkers of a set of 1-aryl-2-(alkylamino) ethanol antimalarials (II). The

statistically most significant QSAR derived is Eq. (3). In Eq. (3), π_{x+y} and σ_{x+y} are the sum of the π

$$\log(1/C) = 0.31\pi_{x+y} + 0.78\sigma_{x+y} \\ + 0.13\Sigma\pi - 0.015\Sigma\pi^2 + 2.35 \quad (3)$$

$$n = 102 \quad r = 0.908 \quad s = 0.263$$

and Hammett constants, respectively, for substituents in the 1 and 8 positions of the phenanthrene ring. The sum of the constants for X, Y, R, and R' groups is $\Sigma\pi$. The number of compounds analyzed

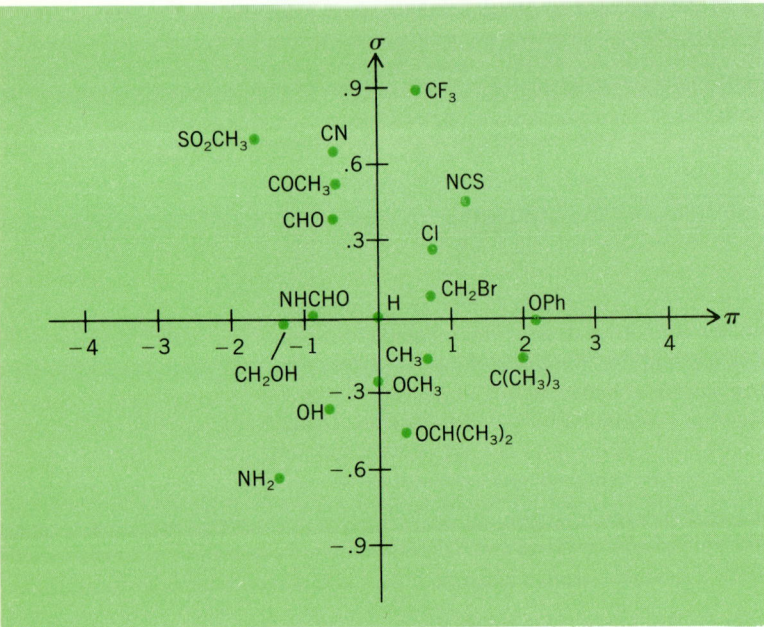

Fig. 1. An aromatic σ versus π Craig plot for certain chemical substituents.

(included in the multidimensional linear regression analysis) is n, the correlation coefficient is r, and the standard deviation is s. C is the molar concentration of compound necessary to cure malaria in 50% of a population of mice as determined in the Walter Reed screen using *Plasmodium berghei*.

Equation (3) suggests that biological potency depends upon the local relative lipophilicity of substituents X and Y, as well as the total net lipophilicity of all the substituents (X, Y, R, and R'). This is an example where competing contributions from a single molecular property (lipophilicity) distributed over a molecular framework must be fine-tuned to optimize biological potency. A further balancing act is at play through the proper selection of X and Y to take additional advantage of the role of σ as well as π upon biological potency.

There have been numerous generalizations of Eq. (1). Hansch has introduced additional physicochemical group properties as possible activity correlates. These include molar refractivity, alternate electronegativity indices to σ, and user-defined molecular features, the presence or absence of which is represented by indicator variables. Molecular properties calculated from three-dimensional computational chemistry methods and experimentally determined features, such as vibrational frequencies, have also found their way into the right-hand side of Eq. (1). It is fair to say that Eq. (1) has been conceptually generalized to include any molecular feature as a potential biological activity correlate.

MOLECULAR CONNECTIVITY INDICES

Molecular connectivity, which is based upon graph theory, provides a means of generating mathematical indices which are characteristic of the chemical composition and bonding topology of a molecule. These indices have been shown to correlate to physicochemical properties in many homologous series of compounds. Since physicochemical properties control biological activity, a correlation between molecular connectivity indices and biological activity might also be expected in some cases. This has been verified. For example, the inhibitory potency of a set of alkyl and arylalkyl derivatives of thymidine against thymidine phosphorylase in a test tube can be described by the QSAR of Eq. (4). The

$$\log (1/C) = 0.366\chi^1 - 3.364 \qquad (4)$$
$$n = 11 \quad r = 0.922 \quad s = 0.213$$

molecular connectivity index χ^1 is a measure of bond branching in a molecule.

Molecular connectivity indices can usually be computed from a chemical structure. Hence, compounds seldom have to be deleted from an analysis for lack of property measures. However, connectivity indices are entirely mathematical entities. They cannot be directly related to physical reality unless the intermediate correlations relating physicochemical properties to connectivity indices are known.

Overall, QSAR analysis based upon molecular connectivity indices has become both a complement and an alternative to classic Hansch analysis. This is particularly the case when measures of physicochemical properties required in a Hansch study are not available.

THREE-DIMENSIONAL DESIGN

The popular descriptors used in Hansch analysis, σ, π, and even E_s, do not explicitly depend upon the three-dimensional structure (shape) of a molecule. There is, however, abundant evidence that molecular shape can be of fundamental importance in controlling biological activity. For example, consider the dipeptide sweetener aspartame [(L)ASP-(L)-PHE-OCH$_3$] and its chiral isomers [(L)ASP-(D)-PHE-OCH$_3$, (D)ASP-(L)-PHE-OCH$_3$, and (D)ASP-(D)-PHE-OCH$_3$]. All four of these molecules have the same values for the Hansch type of descriptors (unless chiral indicator variables are user-defined), yet only aspartame is sweet-tasting; the other three isomers are tasteless. The way in which these compounds can ultimately be distinguished from one another is through their relative molecular shapes.

The breakthroughs made in computer technology, coupled to advances in the understanding of the theoretical basis of chemical phenomena made since 1970, have resulted in an explosive increase in three-dimensional drug design. Several new terms have arisen to describe the components of three-dimensional drug design. As in any new and rapidly expanding field, these terms are not always well defined. Consequently, it is appropriate at this point to set some definitions which can be used in a self-consistent fashion throughout the balance of the text.

Definitions. The methods and terms defined below are basic to three-dimensional drug design.

Molecular modeling. The term molecular modeling describes the generation, manipulation, or representation of three-dimensional structures of molecules and associated properties. It is not restricted to drug design studies, but can be applied to any physicochemical problem.

Computer-aided (or assisted) drug design (CADD). This includes any application of computer-based procedures for purposes of establishing structure-activity criteria. Hansch analysis and satellite approaches are under the umbrella of CADD if this definition is used.

Computer graphics. When the technology of computer graphics is used for the representation of molecular structure and properties, it is referred to as molecular graphics. This is the most powerful means of communicating information on chemical structure between people and machines. It is not surprising that sometimes molecular graphics is taken to mean drug design. However, molecular graphics is a CADD tool.

There are two types of computer graphics technologies. One is vector graphics, which consists of drawing lines (vectors) between prescribed points on the surface of a cathode-ray tube. Since the prescribed points can be set very close to one another, it is possible to generate all types of high-resolution curves, including numbers and letters. Color vector graphics can be used to differentiate molecular structures through color coding.

Raster graphics is the other computer technology used to represent molecular structure. The surface of the cathode-ray tube is divided into a matrix, and each element of the matrix, called a pixel, can be controlled from software. By requesting different pixels to be "printed" in different colors or shades, it is possible to generate space-filled graphical images like those seen on television. The resolution of this type of graphics representation increases as the number of pixels increases per unit area.

Solid objects can be portrayed by using raster graphics. Vector graphics is limited to outlining the surfaces of a solid. However, raster graphics can be used to emulate vector drawings. The quality of the raster-based vector emulation depends upon the size of the pixels. As the pixel size becomes smaller and image resolution increases, the raster representation of lines and curves becomes better.

In general, lines can be drawn faster than pixels can be turned on and off. The net result is that the time needed to draw a vector picture is normally less than that for a raster representation. The drawing time is referred to as the refresh time. If the refresh time is sufficiently small ($<1/30$ s), the human mind cannot differentiate between adjacent refresh cycles. Overall, an object will appear to move or change form in real time at these short refresh times. Real-time computer graphics, when applied to molecular representation, allows instantaneous alterations to orientation, translation, or conformation of one or more molecules. Since refresh time is faster for vector graphics, it has preferentially been used in real-time applications. However, advances in the raster technology should ultimately make real-time raster graphics a practical reality. Overall, raster technology can be expected to dominate computer graphics.

Computational chemistry. This is the quantitative modeling of chemical behavior on a computer based upon the formalisms of theoretical chemistry. Computational chemistry methods are essential to providing the most relevant estimations of molecular properties in three-dimensional drug design studies.

Molecular structure. Conformation is the three-dimensional structure of a molecule as specified by atomic coordinates of the composite atoms. Molecular shape refers to the conformation and some measure of the size of a molecule—for example, the volume encompassed by the van der Waals atomic volumes.

Conformational analysis. This is a method of computational chemistry which allows a relative energy to be associated with each conformation of a molecule. Conformational energy can be calculated by using molecular mechanics, where the molecule is thought of as a set of balls (atoms) held together by springs (bond lengths and angles). Alternatively, energies can be computed by using quantum mechanics, where molecular structure is represented by electrons flowing about nuclei fixed in space for each conformer state. Those conformations of a molecule which are low in energy (stable) are the most likely to be adopted.

Operational components. To initiate a three-dimensional drug design study, one must first build the molecule of interest on the computer. There are two major ways of building molecules currently being used:

1. Input of a description of the atom types and bonding code (connection table) for the molecule (see table). This can be done from a terminal keyboard or, as is becoming increasingly popular because of the ease of input, graphically using a "mouse," light pen, or similar input control device. The connection table representation of the molecule is translated into "rough" three-dimensional structure by using a model-builder program. The rough three-dimensional structure is then refined by carrying out conformational energy calculations.

2. A second method to build a three-dimensional molecular structure is to maintain a library of three-dimensional structure fragments. Appropriate structure fragments can be joined together to form the molecule of interest. The resulting three-dimensional model is, again, refined by using conformational energy calculations.

These two model-building methods have been combined to minimize the time and effort required for the three-dimensional assembly of virtually any molecule.

The principal part of a three-dimensional design study is the search for sterically allowable states, in

Connection table used in the three-dimensional construction of ethane

$$H_4 \diagdown \atop H_5 \diagup {}^{H_3} \! C_1 \! - \! C_2 \! {}^{H_6} \diagdown \atop \diagup H_8 \! H_7$$

Atom number	Atom symbol	Atom type*	Bonding scheme†				Charge‡
			A	B	C	D	
1	C	4	2	3	4	5	0
2	C	4	1	6	7	8	0
3	H	1	1				0
4	H	1	1				0
5	H	1	1				0
6	H	1	2				0
7	H	1	2				0
8	H	1	2				0

*The numbers in this table numerically identify the various atoms. The choice of number for an atom is arbitrary. In this case, C(sp^3) atoms are assigned the number designation 4 and the hydrogens are assigned 1.

†The numbers under A, B, C, and D refer to the atom numbers of column 1.

‡Charge refers to the net atomic charge density residing on each atom. In this example, charge density is neglected. Hence the entries are all zero.

particular, the quest for thermodynamically stable states. The intramolecular search has dominated in drug research and, as defined above, is called conformational analysis. Intermolecular energy calculations seeking stable molecular assemblies represent one of the cutting edges of current research in computational chemistry and drug design. No concise term has yet arisen to label intermolecular energy calculations.

There are two ways of searching for thermodynamically stable molecular states. One is to systematically vary each of the degrees of freedom (bond lengths and angles, torsional angles, intermolecular distances, and rotational orientations) and to correspondingly calculate the associated molecular energetics. This scanning approach allows an investigator to approximately locate all stable states for a few degrees of structural freedom. The second means of seeking stable structures is to minimize the energy as a function of the degrees of freedom. An energy minimization can be carried out for a large number of degrees of structural freedom and locates stable minima to high precision. However, minimization strategy does not necessarily locate all stable states.

Clearly the strengths and shortcomings of scanning and minimizing complement one another. Consequently, these two approaches are often used in tandem in order to optimize the efficiency of identifying thermodynamically stable molecular structures.

The conformational energy map, an example of which is shown in Fig. 2 for N-acetyl-N'-methyl-alanine amide (III), has become the standard way of

(III)

presenting information on conformational stability as a function of torsional rotations about chemical bonds. The contour lines represent specified equi-energy values above the most stable conformation, which is referred to as the global energy minimum. Stable conformations are denoted by small filled circles in this particular example.

Once the three-dimensional structures of a set of molecules have been selected for further analysis, the final task is to compare the spatial, thermodynamic, and electronic properties of these compounds. Many of the properties that can be computed from three-dimensional structure calculations are listed below:

Shape
    Interatomic distance   Surface area
    Volume   Potential field

Electronic
    Dipole moment   Wave function coefficients
    Energy levels   Bond order
    Charge density   Delocalization energy

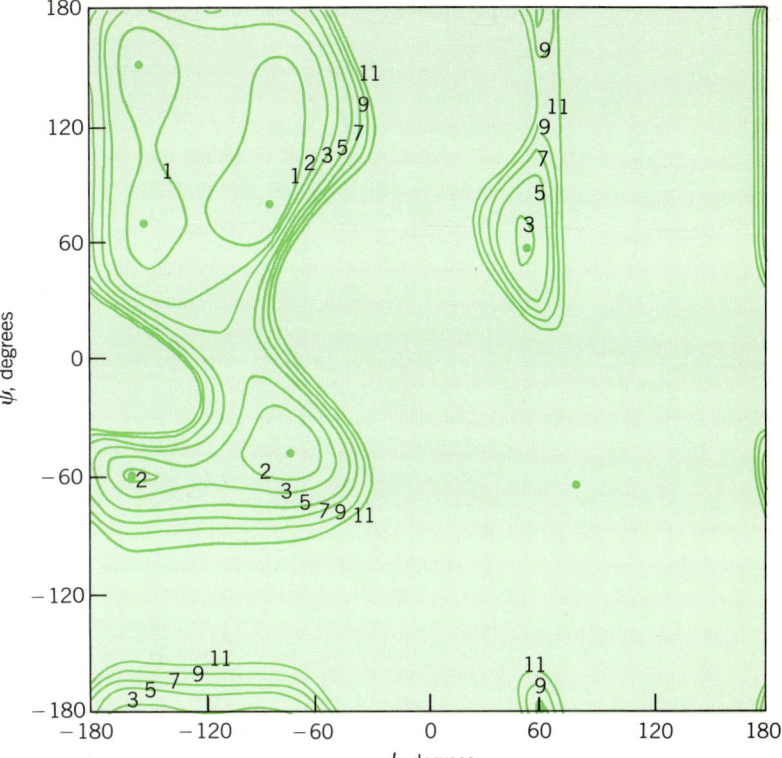

Fig. 2. Conformational energy map of N-acetyl-N-methyl-alanine amide, for a side-chain dihedral angle X^1 of 60°. Locations of minima are indicated by the dots. The contour lines are labeled with the energy ΔE in kcal/mol above the minimum-energy point at values of the backbone dihedral angles $(\phi, \psi) = (84°, 79°)$.

Thermodynamic
    Conformational energetics   Conformational entropy
                         Strain energy
    Solvation energetics   Intermolecular energetics
    Heats of formation

All of these properties, with the exception of some molecular shape descriptors, can be numerically computed. Consequently these molecular property descriptors can be used as biological activity correlates in a generalized version of Eq. (1) as mentioned earlier.

However, relative similarities or differences in molecular shape among molecules are difficult to quantify. Molecular graphics has come to the rescue, and provides a visual means of manipulating and comparing two, or more, molecular structures. Corresponding numerical fit procedures have been designed to complement the graphical capabilities. The fit procedures have been used, for example, to minimize distances between specified atom pairs or maximize common overlap molecular volumes.

Strategies. There are several philosophies, along with corresponding methods, to three-dimensional drug design. The key is how to best undertake and solve a particular design problem. There is not, and probably never will be, a definitive formula to uncover the key. The intellectual intuition and the creativity of the drug designer can be expected to remain central to solving design problems. Nevertheless, guidelines to enhance the efficiency and completeness of the drug design process are now beginning to emerge.

Two general types of problems face scientists in pharmaceutical research: identifying new chemical leads (lead discovery), and optimizing the biological profile of a lead through chemical modification (lead optimization). Each of these problems can require different drug design strategies and corresponding approaches.

Lead discovery. Lead discovery is the more difficult of the two problems. The history of lead discovery is best characterized as a set of serendipity events; that is, most new chemical leads have been chance discoveries. Attempts to design new leads prior to CADD have largely been based upon similarity comparisons of proposed chemical structures with the chemical structures of known active compounds. Two strategies for modifying a known active compound to generate a new lead have been used: (1) build in molecular rigidity through ring formation or addition of bulky substituents; (2) introduce hetero atom replacements. The reason for comparing structures in new lead design is based on the working assumption that a necessary, but not sufficient, condition for a common activity among compounds is that they share, at least in part, some common shape.

The principal shortcoming in performing molecular comparisons based upon chemical structure is in the failure to recognize that there is not an isomorphic relationship between two-dimensional chemical structure representation and the actual three-dimensional (spatial) arrangements of atoms in a molecule. The two-dimensional chemical structure notation used by an organic chemist is a convenient means of representing a molecule which, in most cases, can be built in any one of a large number of conformations. Thus, the conformational similarity-difference among molecules usually cannot be discerned from their chemical structures unless the compounds are highly homologous. Hence, the original conceptual strategies to design lead structures are valid, but the non-CADD methods of structure comparison are often inadequate for purposes of design.

Molecular graphics allows a user to rotate groups about chemical bonds, reorient molecules, superimpose molecules. Complementary structure-fitting procedures additionally allow quantitative molecule shape comparisons. Consequently, the pre-CADD strategies of introducing molecular rigidity hetero atom replacements to design new chemical leads can now be meaningfully carried out by comparing

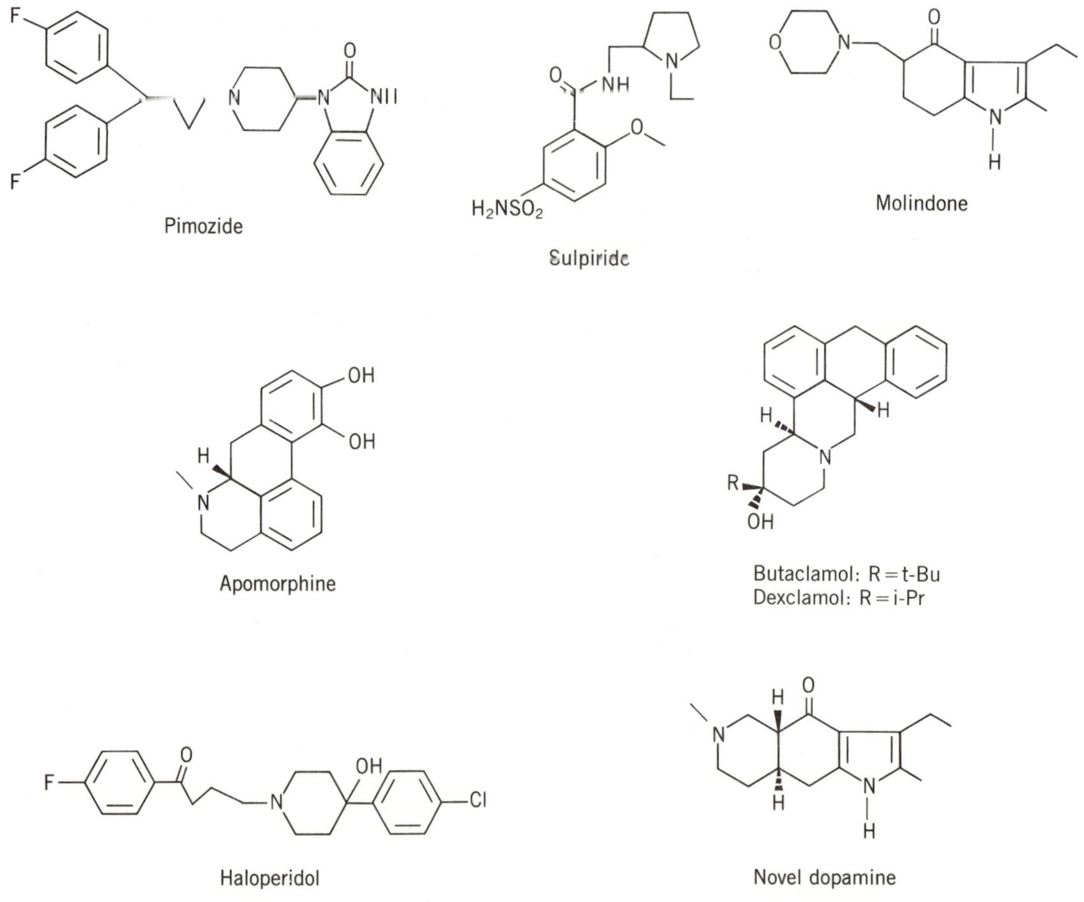

Fig. 3. Six dopamine receptor binding ligands used to predict the novel dopamine ligand structure shown.

conformations and shapes using molecular graphics and satellite computational structure-fitting methods.

There are already some noteworthy successes of new lead design based upon inclusion of the three-dimensional nature of molecular structure. For example, conformational comparison of the six dopamine receptor ligands shown in Fig. 3 led to the proposal that a set of pyrroloisoquinolines (novel dopamine structure in Fig. 3) should have the required conformation and functional groups for biological activity. It was also possible to predict that only the 4αR, 8αR configuration (the *trans* fusion of the two 6-membered rings as indicated in the novel dopamine structure) of the chiral derivatives would have the expected biological properties. This was fully confirmed. Studies with avoidance blockade tests showed that the antipsychoticlike activity of the (−) isomer of the novel dopamine is seven times more potent than that of molindone, and comparable to that of haloperidol.

A second approach to the design of new chemical leads involves explicitly considering drug receptor geometry. Most often the receptor is an enzyme and the ligand is an inhibitor or natural substrate. The geometry of the receptor is determined from x-ray crystallography. Molecular graphics again is the key ingredient that makes this design approach practical. The capability of "driving" a potential ligand into a receptor site, or piecing together a receptor-bound ligand from structural fragments based upon complementary fitting to the "receptor wall" using three-dimensional graphical display, offers the potential of generating de novo chemical leads.

Intermolecular energetics can be included as part of the ligand-receptor modeling approach. This permits approximate calculation of relative binding constants. In general, intermolecular modeling is at the cutting edge of drug design research today. The inclusion of solvent effects on ligand binding, systematic exploration of the intermolecular degrees of freedom between the ligand and receptor, allowing conformational flexibility in a receptor macromolecule, and the capacity to meaningfully include metal-ion contributions to the energetics are actively pursued research areas.

Some effective antitumor agents function by selective attack of the deoxyribonucleic acid (DNA) of tumor cells. One particular structural attack of DNA involves the insertion of a planar ring system of the attacking agent between adjacent DNA base pairs (Fig. 4). This intermolecular process is called intercalation and can lead to cytotoxicity because the intercalated agent prevents cell replication. Antitumor antracyclines, like doxorubicin (IV), are known to

(IV)

intercalate. There is strong evidence that intercalation is their molecular mechanism of antitumor action. The intercalation geometry of doxorubicin was assumed to be similar to that observed by x-ray crystallography for other known intercalators, such as ethidium (V). This intermolecular geometry (Fig.

(V)

5a) is characterized by the long axis of the ring system of the drug being nearly parallel to the two long axes of the DNA hydrogen-bonded base pairs above and below the intercalated drug. Insertion of the drug takes place along the major groove of the DNA. However, synthetic programs designed to increase the antitumor potency of anthracyclines based upon this intercalation geometry have not been successful.

Intermolecular energy calculations of the doxorubicin-DNA interaction suggests that the preferred mode of intercalation is geometrically opposite to that shown in Fig. 5a. Insertion of the anthracycline ring system takes place along the minor groove with the long axis of the anthracycline ring inserting itself nearly perpendicular to the long axes of the adjacent DNA base pairs (Fig. 5b). The crystal structure of the complex of daunomycin (VI), a doxorubicin congener, with a hexanucleotide dimer has been subsequently determined (Fig. 6). It is clear from a comparison of Fig. 5b and 6 that the predicted and observed geometries are vitually identical. Failure of the chemical synthesis programs to

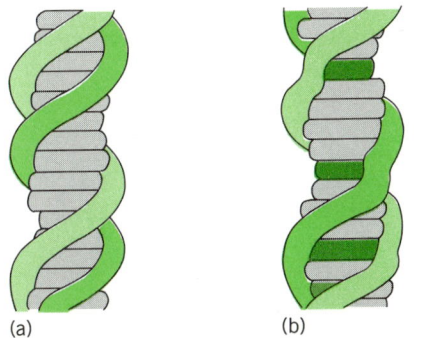

Fig. 4. Schematic representation of DNA-drug intercalation. (a) Native DNA with the base pairs shown as disks. (b) Intercalated drugs shown as shaded disks between base pairs.

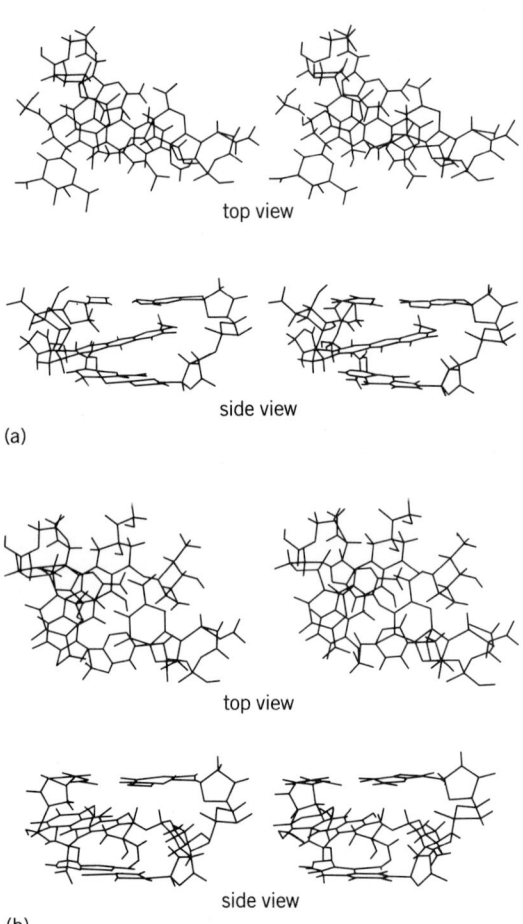

(VI)

find a more potent anthracycline analog by using the geometry shown in Fig. 5a is likely a consequence of the fact that the actual binding geometry is that shown in Fig. 5b. The molecular geometry shown in Fig. 5b can be used as a template in the design of new DNA intercalating agents.

Lead optimization. The majority of reported three-dimensional drug design studies fall into the category of lead optimization. Sometimes lead optimiza-

Fig. 5. Stereographic stick models of the intercalation geometry of doxorubicin between adjacent G-C and C-G base pairs. (a) The long axis of the anthracycline ring is almost parallel to the hydrogen-bonding base pair axes. (b) The long axis of the anthracycline ring is almost perpendicular to the hydrogen-bonding base pair axes. (*From Y. Nakata and A. J. Hopfinger, Predicted mode of intercalation of doxorubicin with dinucleotide, Biochem. Biophys. Res. Commun., 95:583, 1980*)

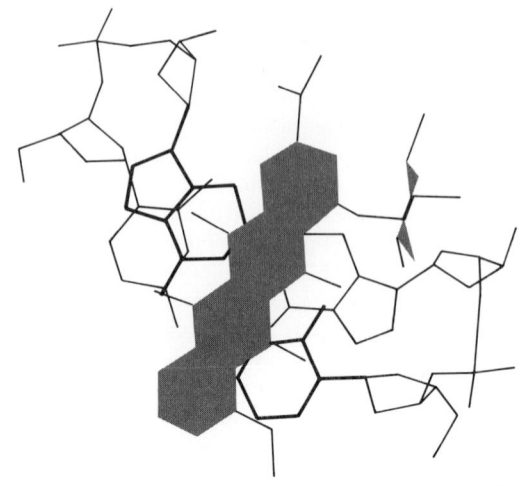

Fig. 6. Crystal structure geometry for the intercalation of daunomycin with a hexanucleotide dimer of DNA. Note that the intermolecular geometries in this figure and Fig. 5b are virtually identical. (*From Y. Nakata and A. J. Hopfinger, Predicted mode of intercalation of doxorubicin with dinucleotide, Biochem. Biophys. Res. Commun., 95:583, 1980*)

tion is not an explicit goal but rather, by hindsight, an attempt to explain the spatial origins of the specificity of biological activity in a series of compounds. However, uncovering a relationship between biological activity and molecular shape is requisite to lead optimization and, as such, can be construed as a lead optimization activity.

The discussion of lead optimization presented here is restricted to the analysis of a congeneric series of compounds for which the geometry of the receptor is unknown. If the geometry of the receptor is known, lead optimization activities parallel the corresponding new chemical lead identification strategies presented above.

The key to lead optimization based upon three-dimensional drug design is identification of the "active" shape of a molecule. The association of a unique conformation of a molecule with a corresponding biological potency is an attractive concept that has stood up to testing in several enzyme-ligand systems.

In those limited situations where the homologous compounds under study are conformationally rigid, the search for the active shape principally involves superimposing the potent compounds and the inactive compounds to identify spatial similarities and differences respectively within, and between, these two sets. The general strategy to uncover the active conformation is to seek common conformational features among potent compounds that cannot be adopted by inactive analogs. For example, the crystal structures of eight semirigid, well-known central nervous system drugs were superimposed as part of a search for the common structural components of central nervous system drugs; the common spatial location of a six-membered aromatic ring for the eight compounds was obvious (Fig. 7).

Application of three-dimensional drug design to lead optimization studies of conformationally flexible

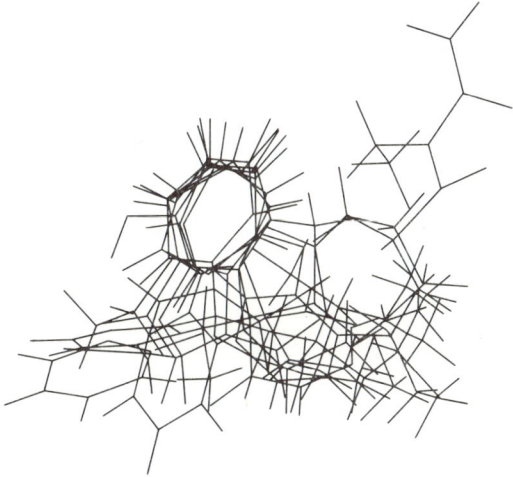

Fig. 7. Superposition of eight active central nervous system compounds in the search for common spatial features among these agents.

molecules generally requires seeking the active flexible conformational state. The predictive capacity of any such analysis is totally dependent upon the breadth and resolution of the conformational search. The search for an active conformation generally involves intramolecular conformational analyses of the molecules available in a lead set. Those conformations of a molecule which are low in energy are, thermodynamically, most likely to be adopted.

A working assumption often used in drug design studies is to restrict the search for an active conformation to low-energy conformers. The thinking behind this assumption is that nature is lazy so that the active conformation is also a low-energy state of a molecule. This idea appears to hold up in findings from enzyme-ligand crystal structures and the corresponding intramolecular energy calculations of the ligands. If this assumption is adopted, conformational analysis can be used to trim down the number of candidates for the active conformation.

The search for spatial features that are available to active analogs but not available to inactive analogs, as already described for rigid compounds, is a second strategy used to identify the active conformation of a flexible molecule.

Conformational analysis combined with differentiation of spatial features of active and inactive analogs can be an effective means of generating three-dimensional structure-activity hypotheses. Molecular graphics is a central tool used in seeking common different properties in three-dimensional structures. The ability to easily manipulate the relative translations and orientations of two or more molecules by using computer graphics provides a practical basis to generate possible three-dimensional pharmacophores (active conformations).

The implementation of the strategies described above to find an active conformation is evolving somewhat differently among drug design groups. One pathway in this evolutionary process is a set of computational and graphics procedures called mo-

lecular shape analysis which make it possible to define a relative numerical scale for characterizing molecular shape similarity. Thus, molecular shape can be treated quantitatively and used directly with other physicochemical properties, like lipophilicity, to derive QSARs.

A combined example of searching for an active conformation and also applying molecular shape analysis is the study on a set of Baker triazines (VII)

(VII)

which inhibit dihydrofolate reductases. The major conformational degree of freedom in these compounds is the torsional rotation as defined in (VII). Secondary conformational flexibility can arise through the substituents X. However, the least and most active congeners in the data base are, respectively, the $X = 2,5\text{-}Cl_2$ and $X = 3,4\text{-}Cl_2$. These two analogs contain identical sets of atoms and can only differ in conformation through θ. Figure 8 contains conformational energy plots (ΔE versus θ) of these two compounds and the unsubstituted, $X = H$, and $X = 2\text{-}Fl$ compounds. Rotation about θ is plotted on the abscissa, and the relative intramolecular conformational energy ($\Delta E = 0$ for the most stable conformation) on the ordinate. The log $(1/C)$ value is recorded in the upper right-hand corner of each plot. C is the molar concentration necessary for 50% inhibition of dihydrofolate reductase from cancer cells in laboratory culture.

There is no difference in conformational preference between $X = H$ and $X = 3,4\text{-}Cl_2$ according to the energy plots. However, $X = H$ and $X = 3,4\text{-}Cl_2$ differ in shape because of the chloro groups. A comparison of the energy maps of the $X = 3,4\text{-}Cl_2$ and $X = 2,5\text{-}Cl_2$ compounds indicates a loss in conformational stability of the minimum energy conformers located at $\theta = 250$ and $310°$. The corresponding loss in inhibition potency of the $2,5\text{-}Cl_2$ molecule is assumed to be a consequence of this conformational instability. That is, the active conformation is postulated to be near $\theta = 250$ or $310°$. Interestingly, the 2-Fl compound retains three of the four intramolecular conformational energy minima as compared to the $X = H$ and $X = 3,4\text{-}Cl_2$ compounds. The 2-Fl compound is more active than $X = 2,5\text{-}Cl_2$, but less active than $X = H$. Consequently, the minimum-energy conformer located within $\theta = 300\text{-}310°$ and destabilized for the $X = 2\text{-}Fl$ compound is deduced to be the active conformation.

One specific way to numerically compare the rel-

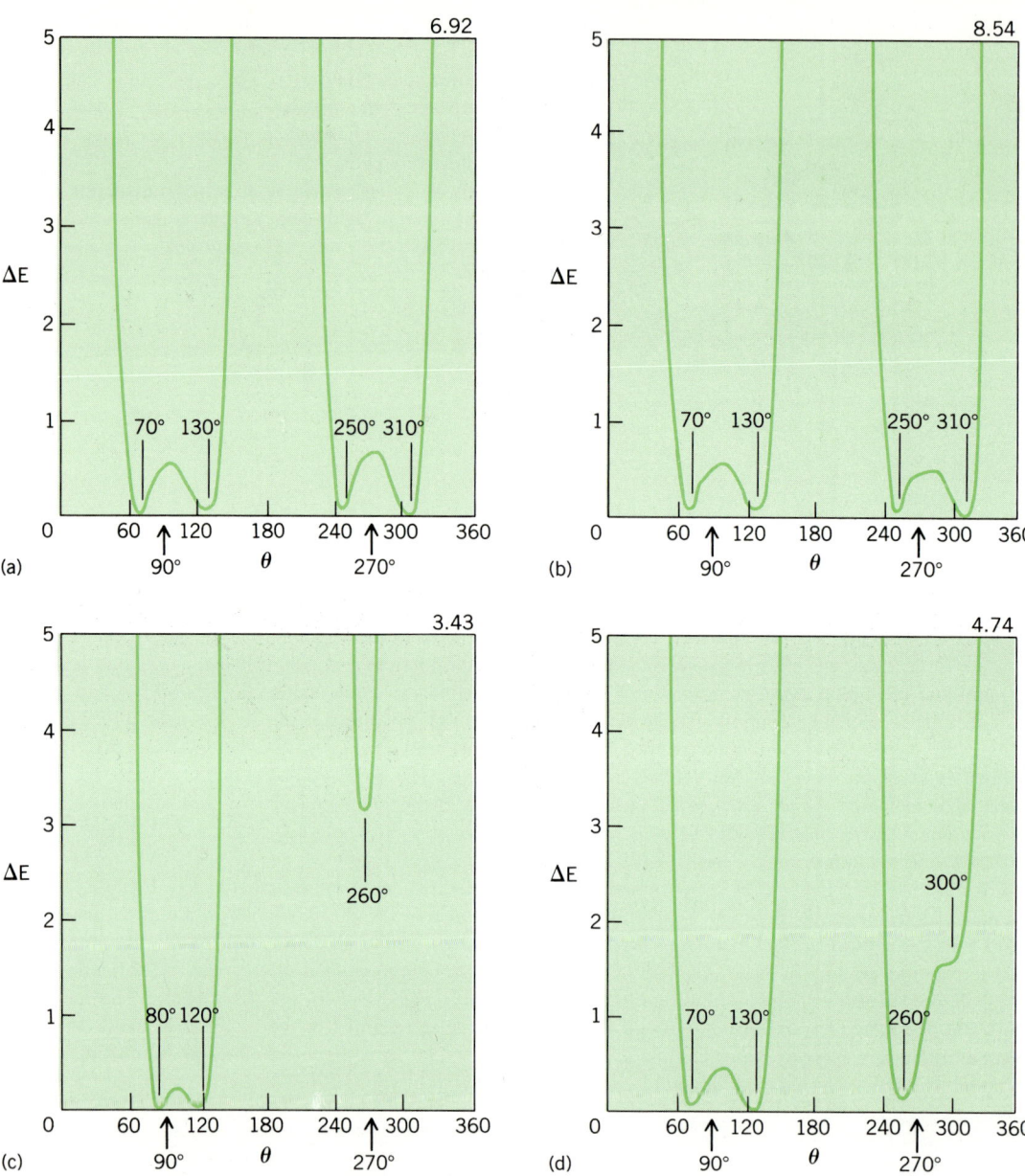

Fig. 8. Relative intramolecular energy ΔE, reported in kcal/mol, versus torsional rotation angle θ. $\Delta E = 0$ is the global energy minimum. The 90 and 270° major energy wells are defined as are the individual local energy minima. (a) X = H, (b) X = 3,4-Cl_2, (c) X = 2,5-Cl_2, and (d) X = 2-Fl. (From P. R. Andrews and E. J. Lloyd, Molecular conformation and biological activity of central nervous system active drugs, Med. Res. Rev., 2:355, 1982)

ative shape similarity of the Baker triazines is to (1) select the most active analog, X = 3,4-Cl_2, as the comparative standard; (2) freeze it into the postulated active conformation; (3) pairwise superimpose each other triazine molecule in the data base onto X = 3,4-Cl_2 by placing the pair of triazine rings in common; and (4) compute the common overlap van der Waals volume of each molecular pair.

The resulting set of relative common overlap van der Waals volumes can be used as molecular descriptors as part of an extended Hansch analysis. Application of these molecular shape analysis methods to the Baker triazines resulted in the QSAR of Eq. (5), where S_0 is the common overlap volume

$$\log(1/C) = 1.474 S_0 + 0.0224 S_0^2 + 0.378 \Sigma\pi - 17.15 \quad (5)$$

$$n = 25 \quad r = 0.960 \quad s = 0.400$$

raised to the two-thirds power, and $\Sigma\pi$ is the sum of π-substituent constants for the X substituents at the 3 and 4 positions.

Succinctly, Eq. (5) suggests that inhibition potency is controlled by the shape of the molecule and the lipophilicity of substituents at specific substitution sites. Figure 9 is a space-filling stereographic representation of the X = 3,4-Cl_2 compound in the postulated active conformation. In order to test the

Computer-assisted drug design

Fig. 9. Stereographic space-filling representation of the postulated active conformation of the Baker triazines as represented by the 3,4-Cl$_2$ compound. (*From A. J. Hopfinger, A QSAR investigation of dihydrofolate reductase inhibition by Baker triazines based upon molecular shape analysis, J. Amer. Chem. Soc., 102:7196, 1980*)

validity of Eq. (5), compound (VIII) was predicted

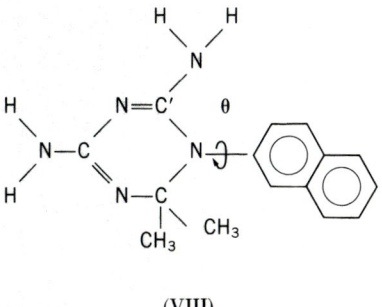

(VIII)

to have a log $(1/C)$ = 7.61 for the imposed active conformation shown in Fig. 10. The compound was synthesized and tested, and the observed log $(1/C)$ value was found to be 7.36—a successful application of three-dimensional drug design.

SUMMARY

CADD has emerged as the application of a variety of computer-based methods with the intermediate objective of identifying relationships between molecular structure and corresponding biological activity. The ultimate goal of CADD is to use these identified structure-activity relationships to predict compounds having a desired activity profile.

The methods of molecular modeling, computational chemistry, and molecular graphics which constitute CADD vary considerably in their ease of application and type of resulting chemical information. Figure 11 illustrates the major methods of CADD and suggests a hierarchical ranking of the order in which they might be applied in a drug design study.

Fig. 10. Imposed "active" conformation of compound (VIII in text) which led to the successful prediction of its dihydrofolate reductase inhibition potency. (*From A. J. Hopfinger, A QSAR investigation of dihydrofolate reductase inhibition by Baker triazines based upon molecular shape analysis, J. Amer. Chem. Soc., 102:7196, 1980*)

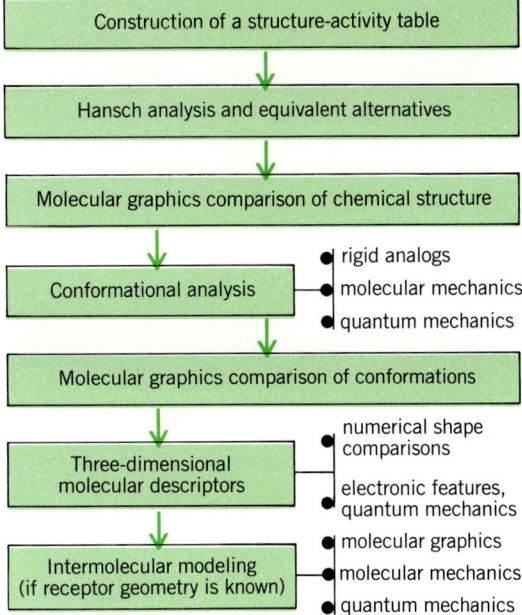

Fig. 11. Hierarchical ranking of the order in which the major methods of CADD might be applied.

Essentially, Fig. 11 summarizes the CADD methods presented and discussed in this article.

The pharmaceutical industry has been first to apply computer-based molecular design methods. Virtually every major drug company in the United States, western Europe, and Japan has CADD programs as part of its research organization. These CADD groups have not yet been in place long enough to fully ascertain their impact on the drug discovery process. However, some data, in terms of successfully designed compounds and efficiency of lead optimization, have been reported and suggest a bright future for CADD.

Other disciplines within the chemical sciences are now beginning to pay attention to molecular modeling, computational chemistry, and molecular graphics. In particular, polymer scientists appear poised to use this new technology. Quite likely CADD will become an extinct term being replaced by computer-assisted molecular design (CAMD) to reflect the general utility of this new technology.

[A. J. HOPFINGER]

Bibliography: M. Charton, *Chemtech*, no. 502, 1974, and no. 245, 1975; A. J. Hopfinger, A QSAR investigation of dihydrofolate reductase inhibition by Baker triazines based upon molecular shape analysis, *J. Amer. Chem. Soc.*, 102:7196, 1980; L. B. Kier and L. H. Hall, *Molecular Connectivity in Chemistry and Drug Research*, 1976; Y. Nakata and A. J. Hopfinger, Predicted mode of intercalation of doxorubicin with dinucleotide, *Biochem. Biophys. Res. Commun.*, 95:583, 1980; T. A. Ryan, B. L. Joiner, and B. F. Ryan, *Minitab Student Handbook*, p. 166, 1976; G. A. Segal, (ed.), *Methods of Electronic Structure Calculations*, pts. A and B, 1977.

The Path Toward Fusion Energy

Harold P. Furth

Harold P. Furth worked on controlled fusion research at the Lawrence Livermore National Laboratory from 1956 to 1967, when he became a member of the staff of the Plasma Physics Laboratory and professor of astrophysical sciences at Princeton University. He became director of the laboratory in 1981. Granted a doctorate by Harvard University in 1960, Furth received the E. O. Lawrence Memorial Award in 1974 and the James Clerk Maxwell Prize in Plasma Physics in 1983.

The world economy has three long-term energy options: fission, fusion, and solar. Some potential advantages of fusion relative to fission are that the basic fuel, deuterium, is abundant and cheap, that there is no long-term nuclear waste disposal problem, that fusion reactors are physically incapable of nuclear excursions, and that there is no need for proliferation-sensitive fuel-reprocessing plants. The use of either fission or fusion would have the advantage, relative to solar power, of minimizing demands on land area and structural materials.

While fusion is relatively free of fundamental drawbacks, in the near term it must overcome a set of extremely difficult technical challenges: (1) to make a "burning plasma" in the laboratory; (2) to develop practical power-handling technology; (3) to produce a commercially competitive power source.

International fusion research was undertaken during the 1950s with a good deal of intuitive optimism, but with no real technical basis for the belief that even the first of these challenges could be overcome. In recent years the scientific outlook has improved greatly, and reactor plasma conditions are expected to be reached by existing experimental devices during the 1980s. The prospects for achieving commercial attractiveness are still controversial, but once more there is intuitive optimism that new ideas and technologies will be found as they are needed. In particular, there is an expectation that scientific expertise in the handling of hot plasmas will continue to evolve rapidly, and will facilitate the solution of the engineering and economic problems of fusion power.

DEVELOPMENT

Nineteenth-century astronomers were able to estimate the cooling-off rate of the Sun; the surprising conclusion was that it should have turned dark long ago. In the 1920s a plausible explanation emerged. The nuclei of light atoms, such as hydrogen, were found to contain a slight excess of mass per nucleon, which might be convertible into free energy by fusing them into heavier elements, such as helium. The Sun and the stars were rightly suspected to be natural fusion reactors.

In 1934 Lord Rutherford's laboratory at Cambridge University used a 100-kV accelerator to fuse deuterium nuclei into two types of end product: helium-3 plus a free neutron, and tritium plus a free proton. The possibility of releasing practical nuclear energy on Earth then began to be mentioned, but mainly in a negative sense. Fusion reactions are less probable than ordinary nonenergy-producing collisions, and so a simple accelerator approach does not lend itself to net power production. In order to imitate the Sun's fusion furnace, it is necessary to create and confine a similarly hot body of fuel in the cold terrestrial environment. From the point of view of the technology of the 1930s, this looked like an altogether hopeless enterprise.

The reality of nuclear energy became evident during the 1940s through the explosion of fission bombs and the construction of net-power-producing fission reactors. In the case of fission energy, the fuel consists of very heavy nuclei, such as uranium. Like hydrogen, these nuclei have an excess of mass per particle, which can be turned into free energy by splitting them into lighter elements. (The lowest energy state, toward which fusion and fission reactions ultimately tend, is the nucleus of iron.)

Large-scale release of terrestrial fusion energy was demonstrated in 1952 by the explosion of a hydrogen bomb, basically a fission bomb that heats fusion fuel to stellar conditions. At about the same time, secret national programs in controlled fusion research were undertaken by the United States, the United Kingdom, and the Soviet Union. When these programs were declassified in 1958, the paths to fusion that had been discovered turned out to be remarkably similar. The basic invention was the "magnetic bottle," and bottles came in two logical shapes: tori with closed magnetic field lines, and open-ended linear "mirror machines." Other nations became involved, and the present pattern of cooperative world fusion research soon established itself.

During the next decade, major scientific progress was made, but in the light of the developing theoretical insights the prospect for early success seemed to recede. A decisive upturn took place in the late 1960s, with the experimental breakthrough of the T-3 tokamak device, a toroidal magnetic bottle developed at the Kurchatov Institute in Moscow. During the subsequent years the tokamak has become the main vehicle for high-temperature research in the world fusion program.

Since the leadership of the tokamak in approaching the reactor regime does not automatically establish its credentials as the best ultimate solution, a number of other magnetic fusion reactor concepts are being pursued in parallel: tori of different designs, as well as open-ended mirror machines. The development of high-powered lasers during the 1970s served to open up a radically different approach to fusion power: the heating and compression of small fuel pellets, from which fusion energy is to be released in the form of repetitive microexplosions. The burning fuel is held together by its own inertia, and this approach is therefore called inertial confinement fusion (ICF). Because of the relevance of inertial confinement fusion to hydrogen bomb development, much of the United States work in this field is still classified. Some potential drivers for the inertial fusion process, aside from various types of lasers, are high-powered electron and ion beams, including multi-GeV nuclei. Finally, there are yet other fundamentally different approaches to fusion, such as the use of exotic elementary particles to catalyze fusion reactions in cold fuel, but their practical potential is still uncertain.

PLASMAS AND FUSION REACTIONS

When matter is heated, it first turns into a liquid, then into a gas, and finally into a plasma. The first two transitions relate to the relaxation of interatomic bonds. The gas-to-plasma transition, which typically begins at temperatures somewhat below 18,000°F (10,000°C), consists of stripping electrons from their atoms, mainly by particle collisions. The positively charged atomic residue is called an ion. The hydrogen atom has only one electron, and therefore only a single ionization state. Heavier elements pass through a sequence of ionization states, with increasing numbers of electrons stripped from the nuclei as the temperature rises.

When two charged particles meet, they tend to deflect each other's paths, a phenomenon known as Coulomb scattering. Only when two nuclei approach to within distances of 10^{-11} to 10^{-12} cm does a strong nonelectrical force begin to develop that may cause them to fuse. The energy required to bring them that close, in the face of electrostatic repulsion, is typically in the range 10–100 keV. A plasma where the kinetic energy of random thermal motion of the nuclei is sufficient to produce frequent fusion reactions, a so-called thermonuclear plasma, should have a temperature at least in the 10-keV

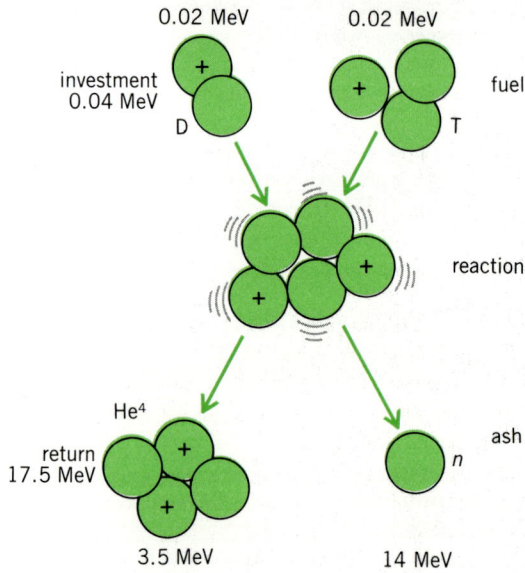

Fig. 1. D-T (deuterium-tritium) fusion reaction. Payback (ratio of energy returned to energy invested) is about 450:1.

range, corresponding to about 2×10^8 °F (10^8 °C). (One electronvolt corresponds to about 21,000°F or 12,000°C.)

Because of its huge mass and high-quality thermal insulation, the Sun is able to operate at somewhat lower temperature (about 2.5×10^7°F or 1.4×10^7°C) and can burn ordinary hydrogen as its input fuel. Terrestrial fusion reactors, for which the quality of heat insulation is a serious problem, must use faster-burning fuels, such as deuterium, as well as operating in the higher-temperature range, where fusion events occur more frequently. Fortunately, the deuterium isotope of hydrogen is sufficiently abundant on Earth so that natural water contains about 300 times the energy equivalent of gasoline.

The fastest-burning fusion fuel is a 50–50 mixture of deuterium (D) and tritium (T). The D-T reaction (Fig. 1) produces a helium-4 nucleus plus a free neutron, and releases 17.6 MeV of nuclear energy. The neutron escapes readily from the plasma region, carrying off 80% of the energy release. In a D-T reactor, the fusion neutrons will be stopped in a 24–32-in.-thick (60–80-cm) fusion reactor "blanket" (Fig. 2) made largely of natural lithium. While depositing its energy in the lithium blanket, the neutron ideally produces 1.5 new tritium atoms, so that the tritium fuel component can be self-regenerating. The problem that is being solved here is that tritium is not naturally abundant on Earth, it decays to helium-3 with a half-life of 12 years.

The helium nucleus of the D-T reaction (also called an alpha particle) receives 20% of the fusion energy release. Since it is electrically charged, the alpha can deposit its energy inside the fuel volume by colliding with other particles. If the heat insulation of the plasma is good enough, the fuel can therefore maintain its high-temperature state by means of its own fusion reactions, just as the Sun does. Entry into this desirable reactor regime is called ignition.

The required quality of thermal insulation is typically measured by the product of the plasma density n and the energy confinement time τ_E (the time that it would take for the plasma to cool off by various processes of heat leakage). The product $n\tau_E$ is called the Lawson parameter (Fig. 3). Ignition of a D-T plasma at 10-keV temperature calls for a Lawson parameter of about 2×10^{14} cm^{-3} s. Net reactor output power could be generated, of course, even when the quality of insulation is not quite that good. The Lawson break-even point (where the ratio Q of total fusion power produced to required plasma heating power is at least unity) occurs already at a Lawson parameter of 6×10^{13} cm^{-3} s. Nonmaxwellian plasma regimes with enhanced energetic-ion tails are expected to reach break-even conditions at $n\tau_E$ values as low as 10^{13} cm^{-3} s. For economically attractive electric power generation, a Q value of 10 or more is desired, and ignition ($Q = \infty$) is preferred.

Aside from D-D and D-T, some other fusion reactions, such as D-^{3}He, are of potential long-term

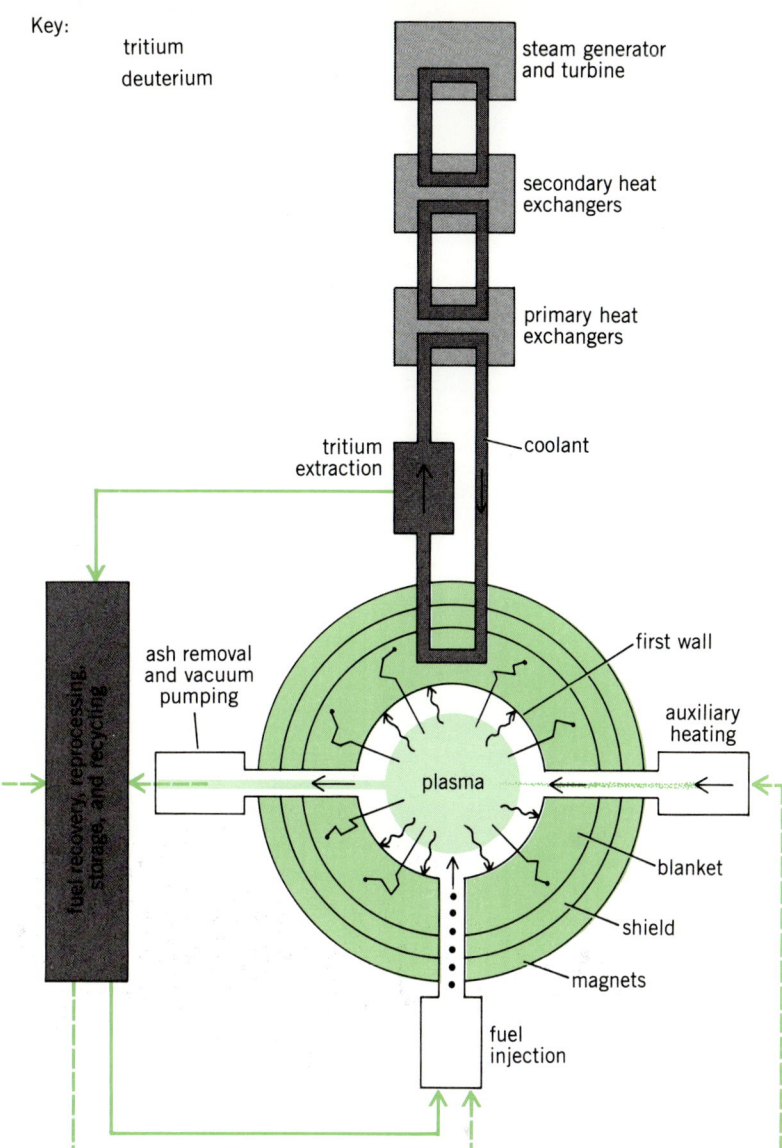

Fig. 2. Basic geometry of a magnetic fusion reactor. (*After M. A. Fischetti, Turning neutrons into electricity, IEEE Spectrum, 21(8):33–42, August 1984*)

interest, because they release most of their energy in charged particles. In that case, direct conversion of fusion energy to electric power can be considered. Another interesting possibility is to polarize the nuclei and align them so as to enhance their fusion cross section. The conventional expectation is, however, that first-generation fusion reactors will be burning ordinary D-T fuel. Most of the heat will be deposited by neutrons in the reactor blanket (Fig. 2) and will be used to generate steam and electricity in conventional ways. In this context, the high energy and penetrating power of the fusion neutrons is actually convenient, since it helps distribute the energy release over a thick volume of coolant.

FUSION SYSTEMS

Each of the approaches to fusion energy has its own set of special problems.

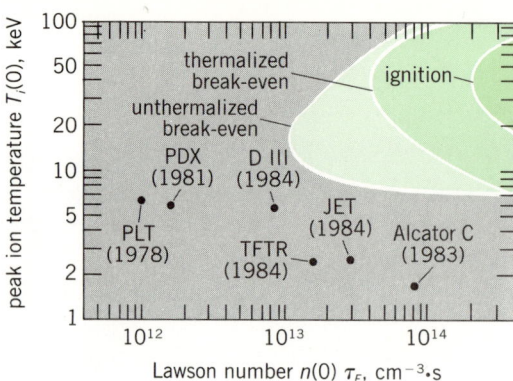

Fig. 3. Lawson diagram, comparing temperatures and Lawson parameters of current machines with conditions required for break-even and ignition. Data points for TFTR and JET represent initial values.

Toroidal magnetic confinement. Charged particles moving in a magnetic field have helical orbits (Fig. 4). If the magnetic-field lines close on themselves in toroidal geometry, and if the details of the magnetic-field structure are properly designed, a single particle cannot escape from such a bottle; however, Coulomb collisions between charged particles may allow them to hop randomly from orbit to orbit. This irreducible form of plasma leakage, known as classical diffusion, is sufficiently slow so as not to create a serious obstacle to the achievement of the $n\tau_E$ value required for ignition.

The real problem for the toroidal approach to a fusion reactor has been the transport of plasma particles and heat due to cooperative phenomena involving large numbers of particles. When the local densities of ions and electrons do not quite coincide, strong macroscopic electric fields can appear inside the plasma. Similarly, when there are net relative motions between ions and electrons, macroscopic currents and magnetic fields are generated. The interaction of these macroscopic electric and magnetic fields with the plasma can give rise to exponentially growing instability modes, turbulence, and "anomalous" transport coefficients greatly exceeding classical prediction.

Open-ended magnetic bottles. In the case of open-ended bottles (Fig. 5), the highest-priority problem has been to stop the leakage of plasma out the ends. A gyrating charged particle constitutes a magnetic dipole that opposes the ambient magnetic field; it is therefore decelerated when entering regions of stronger magnetic field, so-called magnetic mirrors, at the ends of the plasma confinement region. Mirror confinement depends on the ratio of the velocity of gyration relative to the velocity along the magnetic field. In an ordinary mirror machine (Fig. 5), a single Coulomb scattering event can cause two gyrating particles to acquire predominantly axial velocities and thus escape from mirror confinement. Even at optimally high plasma temperatures and with D-T fuel, the probability of fusion is sufficiently smaller than the probability of loss by Coulomb scattering so that the Lawson criterion for break-even can barely be met, and ignition is unattainable. An additional problem of the simple mirror machine is that the prerequisite nonmaxwellian energy distribution lends itself to the excitation of high-frequency instabilities and anomalous transport in velocity space, so that the plasma is lost on a time scale even faster than that of classical Coulomb scattering.

General magnetic confinement constraints. In magnetic confinement systems, the ratio of the plasma pressure nT (product of density and temperature T) to the Maxwell stress $B^2/8\pi$ (where B is the magnetic field strength) is called the beta value. If the plasma pressure is to be held in equilibrium, the maximum possible value of beta is unity. In general, however, the upper limit must be set at a much lower level of beta, in order to avoid a type of gross anomalous transport where the plasma escapes from confinement by bending the field lines of the magnetic bottle. For example, in a conventional tokamak, the limiting beta value is expected to be in the range 5–10%.

The combination of these plasma physics considerations with the constraints of current magnetics and materials technology determines the main characteristics of a magnetic fusion reactor. The use of superconducting magnet coils is advantageous, since the resistive power dissipation in an ordinary magnet system tends to consume an excessive fraction of the total electric power output, unless the plasma beta value is very large. Assuming the use of current superconductor technology, the magnetic-field strength is limited to about 5 teslas (50 kilogauss) at the plasma, corresponding to a Maxwell stress of about 100 atm (1500 lbf/in.^{2} or 10 megapascals) typical plasma pressure of order 10 atm (150 lbf/in.^{2} or 1 megapascal).

A representative fusion reactor plasma thus has a thermal pressure nT and energy density $\tfrac{3}{2}nT$ no greater than that of ordinary steam. The high plasma temperature T is offset by the low plasma density n, which is typically a few times 10^{14} particles per cubic centimeter. From the magnitude of n, it follows that the energy confinement time τ_E in an ignited D-T plasma must be at least of the order of 1 s. The allowed plasma pressure also limits the fusion power density to about 5 MW/m^3 (140 kW/ft^3).

These numbers indicate the general range of "conventional" present-day fusion reactor design.

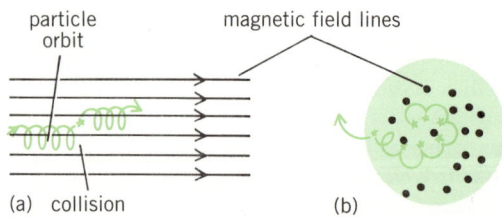

Fig. 4. Particle orbits in a magnetic field and cross-field diffusion by Coulomb collisions. (a) View across field. (b) View along field.

They reflect certain technological prejudices that could change in the future. For example, if more advanced superconducting coils were to provide a field strength of 10 T (100 kG) at the plasma, then the containable plasma pressure would be increased fourfold, and the resultant fusion power density would be increased sixteenfold. Similar increases may be achievable, even at a given magnetic field strength, in tori of advanced design, with greater rigidity against pressure-driven magnetic-field perturbations and therefore higher allowable beta values. To realize this potential in the form of higher-powered reactors, however, will call for matching advances in the technology of wall materials.

Wall materials. In current reactor design studies, the vacuum chamber inside of which low-density plasmas are produced and heated is generally assumed to be made of stainless steel. The passage of energetic fusion neutrons tends to displace the atoms of the steel wall, thus reducing its mechanical strength. In a fusion reactor with a flux of 3–5 MW of neutrons per square meter of wall surface, the expected lifetime of the wall is about 4–6 years. If the fusion power density were to be increased successfully, the lifetime of the wall material would have to be improved correspondingly.

Since the "ash" of the fusion process (that is, helium) is not radioactive, the replacement of first-wall materials gives rise to the main problem of radioactive waste disposal. By proper choice of materials,

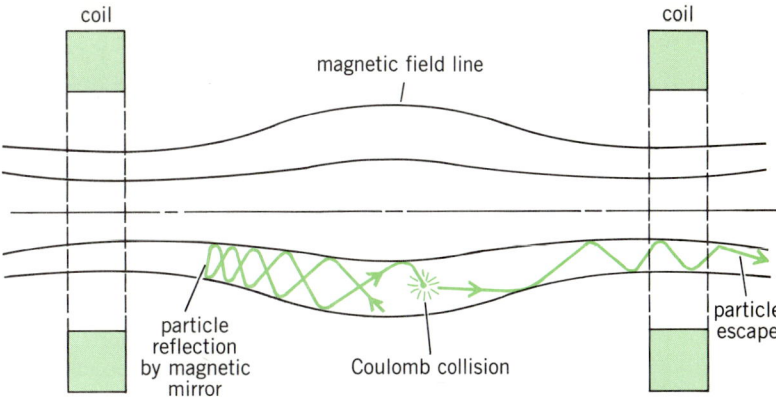

Fig. 5. Particle confinement by magnetic mirrors and escape by Coulomb scattering.

radioactivity could be minimized so that long-term burial of wastes would not be required in a fusion power economy. A second basic advantage of fusion power, that the nuclear reaction can never "run away" or fail to shut down, follows simply from the smallness of the total fuel energy that is present at any one time in a fusion reactor.

Inertial confinement. A New set of plasma physics topics and technological constraints is encountered by the inertial confinement approach to fusion. In an inertial confinement fusion reactor, the surface of a small fuel pellet (typically a frozen D-T pellet a fraction of an inch or a few millimeters in diameter)

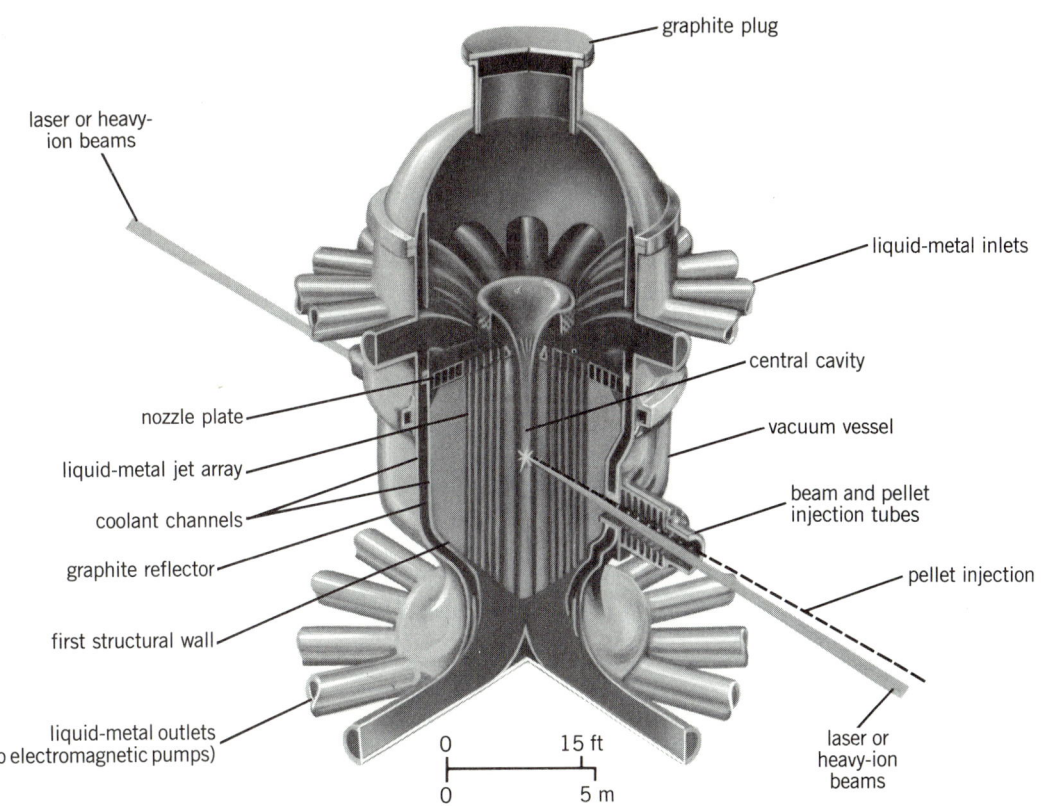

Fig. 6. Inertial confinement fusion reactor with liquid-metal blanket. (*After R. W. Conn, The engineering of magnetic fusion reactors, Sci. Amer., pp. 60–71, October 1983*)

is to be heated by means of some 10^{14} W of instantaneous power in the form of laser light or charged-particle beams. The ablation of the heated pellet surface exerts a recoil force that compresses the pellet core a thousandfold while also generating ignition-level temperatures. The ignited core then burns outward, releasing additional fusion energy before the pellet reexpands and cools.

The condition for achieving ignition is essentially the same as in the case of magnetic fusion, but the individual elements of the $n\tau_E$ product are quite different: The typical density n of the compressed pellet core is some 3×10^{25} particles per cubic centimeter, instead of 2×10^{14}, and the minimum confinement time τ_E is of the order of 10 picoseconds, instead of 1 s. Correspondingly, the plasma pressure is 10^{12} atm (10^{13} lbf/in.^{2} or 10^{17} pascals) instead of 10.

Even a well-optimized inertial-confinement-fusion compression process will be only about 10% efficient in terms of delivering energy from the driver to the hot pellet core. To achieve an overall power gain Q of about 10, the burn time must be long enough to permit a roughly hundredfold gain in the pellet itself. The required $n\tau_E$ value turns out to be about 10 times greater than the Lawson parameter for simple ignition. By specifying very large pellet compressions and final densities, this severe $n\tau_E$ constraint can be satisfied, while the pellet size is kept relatively small. In this way, the total pulsed energy required to ignite the pellet is minimized.

A basic plasma physics issue of inertial confinement fusion is whether the driver power can be absorbed efficiently at the surface of the pellet, rather than being lost by reflection or transported from the surface to the plasma interior, which would interfere with the compression process. A second critical issue is the ability to achieve high compression ratios in the face of imperfectly symmetric surface illumination. Any initial asymmetries in the compression process tend to be amplified by a Rayleigh-Taylor instability mechanism that is quite similar to ordinary gravitational instability (that is, the unstable mixing mode of a heavy fluid supported by a light fluid). One way to minimize the problem of uniform illumination has been the hohlraum approach, where the fuel pellet is placed within a cavity, the primary driver system is used to heat up the cavity walls, and the thermal x-ray emission from the walls then acts as a very uniform secondary driver.

Current estimates for the minimum energy input requirement of an inertial confinement fusion reactor are in the range of about 5 megajoules, so that the output per pulse amounts to some 500 megajoules. The pulse repetition rate is envisaged to be several per second. A promising way to capture this power would be by means of a liquid-metal curtain surrounding the pellet (Fig. 6). The curtain would typically be composed of lithium or lithium-lead eutectic, in order to meet the tritium-breeding requirement, and would serve to cushion the first reactor wall against the microexplosions, as well as protecting it against neutron irradiation damage.

PROGRESS IN MAGNETIC CONFINEMENT

During most of the 1950s and 1960s, progress toward a hot, magnetically confined plasma was impeded by various instabilities and the resultant anomalous transport. Typical plasma parameters in this period were $n\tau_E$ values below 10^{11} cm$^{-3}\cdot$s and plasma temperatures below a few hundred electronvolts. Substantial progress has been made since then, particularly by means of the tokamak confinement device.

Tokamaks. The main magnetic-field component of the tokamak (Fig. 7) is a strong toroidal magnetic field (typically 5 T or 50 kG), which is generated by an array of external magnet coils. In addition, a poloidal magnetic-field component is produced by inducing a toroidal plasma current with an air-core transformer. The two field components form helical magnetic-field lines, lying in nested toroidal magnetic surfaces. The orbit of each plasma particle is restricted to the vicinity of such a surface, except in the presence of classical collisional diffusion or anomalous transport. The outstanding experimental success of the tokamak approach has consisted in confining the plasma ion heat content about as well as the classical theory would predict. Electron heat transport in tokamaks is still anomalous, but since its magnitude is comparable to that of the classical ion heat transport, the experimentally achievable temperatures and $n\tau_E$ values are not very far below classical expectation.

An experimental tokamak plasma is produced by puffing low-density neutral gas into the toroidal vacuum chamber and then inducing a toroidal plasma current, typically 1 MA in current experiments. The electrical conductivity of the plasma improves with

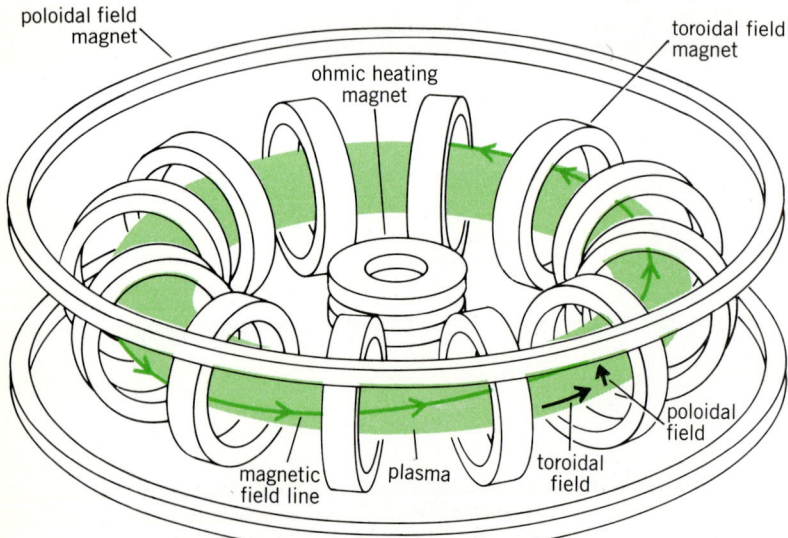

Fig. 7. Configuration of coils and fields in the tokamak magnetic bottle. (*After R. W. Conn, The engineering of magnetic fusion reactors, Sci. Amer., pp. 60–71, October 1983*)

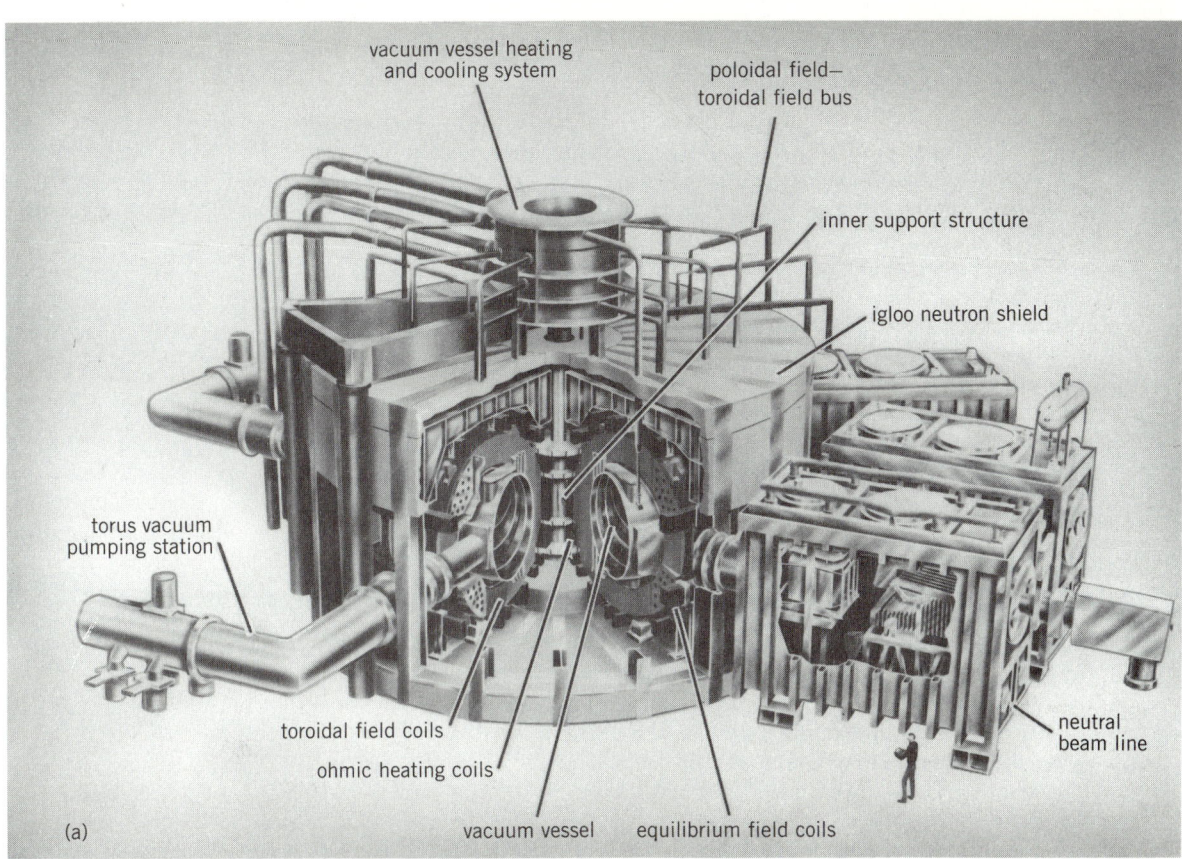

Fig. 8. Tokamak Fusion Test Reactor. (*a*) Cutaway view of components. (*b*) Fully constructed device.

rising temperature, and so the ohmic heating power associated with the plasma current tends to diminish. Typically, plasma temperatures up to 2 keV can be reached in the ohmic heating phase. For this type of heating, small-sized tokamaks with exceptionally high magnetic fields are most effective. The Alcator C device at the Massachusetts Institute of Technology, which operates at 10 T (100 kG) has achieved an $n\tau_E$ value close to 10^{14} cm^{-3} s (Fig. 3).

To advance to still higher temperatures, various forms of auxiliary heating have been developed. During the late 1970s the injection of energetic beams of neutral atoms proved effective in raising the plasma ion temperature to the 7-keV range. Experiments with deuterium plasma on the D-III tokamak at GA Technologies Inc. have achieved $n\tau_E$ values above 10^{13} cm^{-3} · s at temperatures of 6 keV. If these experiments had been carried out with D-T plasmas, the ratio Q of fusion power relative to input heating power would have been above 0.1.

The largest United States fusion research device at present is the Tokamak Fusion Test Reactor (TFTR) at Princeton (Fig. 8), which will be capable of operating with D-T fuel. The plasma current will range up to 3 MA in a 5.2-T (52-kG) magnetic field, and the plasma will be heated with 25–30 MW of neutral beams of 120-keV energy for times of order 1 s. These experiments are expected to reach 10-keV D-T plasmas with enlarged high-energy tails and $n\tau_E$ values well above 10^{13} cm^{-3} s, sufficient to produce a Q value around unity (that is, break-even). The present TFTR schedule calls for the demonstration of approximately break-even plasma conditions in a deuterium plasma during 1986, followed by a D-T demonstration in 1988.

Three other major tokamak devices will be in operation during the 1980s in Europe (JET), Japan (JT-60), and the Soviet Union (T-15). The largest of these is the JET device, which has a vacuum vessel

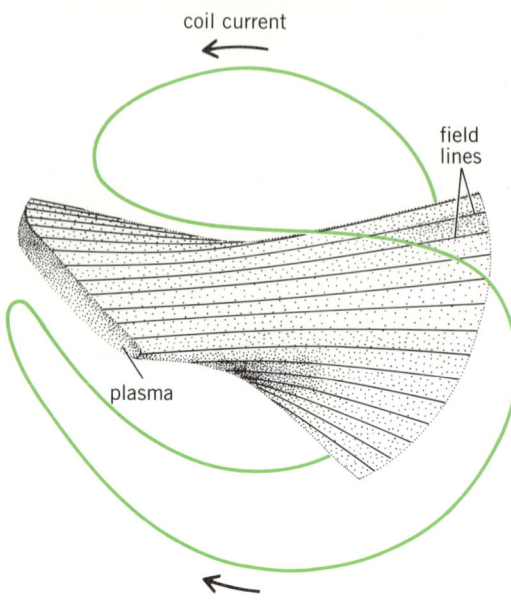

Fig. 10. Configuration of coils and fields in the minimum-B mirror machine.

with interior dimensions comparable to those expected for a practical tokamak reactor (Fig. 9). An energy confinement time of 0.6 s has been reached by JET at its full magnetic-field strength of 3.5 T (35 kG), and is expected for TFTR at its maximum field strength of 5.2 T (52 kG).

A principal objective of these large tokamak experiments is to create the scientific and technological data base for an ignited long-pulse tokamak reactor. The European, Japanese, and Soviet fusion programs have specific plans for such a facility, with construction to start around the year 1990 and operation to begin in the late 1990s. Current United States planning is oriented toward international collaboration on such an undertaking.

As tokamak experiments enter the reactor regime, increasing effort is being devoted to improving the practical engineering potential of both the tokamak and the competing approaches to a magnetic fusion reactor. In the tokamak area, the somewhat cumbersome, though effective, neutral-beam-heating method is being replaced successfully by various forms of radio-frequency (rf) heating, where plasma waves are excited by external antennas or wave guides, and damped in the plasma. The same methods have also lent themselves to driving the tokamak plasma current, thus creating the possibilty of maintaining the tokamak reactor burn in a steady state, or at least during very long pulses (many hours). The latter feature is important in order to minimize thermal and mechanical fatigue problems that might arise from shorter-pulsed operation with purely transformer-driven tokamak currents.

Major efforts are also being devoted to maximizing the beta value of the tokamak by appropriate shaping of the minor plasma cross section. The D-shaped vacuum vessel of the JET machine (Fig. 9) should lend itself to beta values above 5%. Theory predicts that a bean-shaped tokamak cross section should

Fig. 9. Interior of vacuum vessel of the JET.

permit even higher beta values, and the initial experiments along this line have been encouraging.

Mirror machines. In the case of the mirror machine approach, the problem of gross anomalous transport across magnetic field was solved in the early 1960s at the Kurchatov Institute, by means of the minimum-B configuration (Fig. 10), where the magnetic-field strength increases away from the confined plasma, not only in the direction toward the mirrors but in the cross-field direction as well. Since the plasma is diamagnetic, this geometry is energetically favorable to a stable equilibrium at the bottom of the "magnetic well."

During the following decades, the critical issue for mirror confinement was anomalous end loss caused by various high-frequency instabilities. This problem was solved in the late 1970s, along with the more fundamental problem of excessively large classical end loss, by the invention of the tandem mirror concept (Fig. 11).

The basic idea of the tandem mirror is to confine most of the reacting plasma in a long, straight central cell and to plug up the ends of this cell by means of a system of electrostatic potentials, generated by smaller, higher-energy plasmas confined in special end-plug cells. The success of this technique in stopping both classical and anomalous end losses has been documented by recent experiments. The main problems at present are the control of fine-scaled anomalous cross-field transport and the minimization of the fraction of electric power that would have to be recirculated in a reactor to maintain the end plugs. A large tandem mirror experiment, the MFTF-B, is scheduled to begin operation at Lawrence Livermore National Laboratory in the late 1980s. A principal attraction of the mirror approach is that the plasma will be maintained in true steady state.

Alternative toroidal configurations. The toroidal approach to magnetic fusion contains a number of other interesting options, aside from the development of advanced tokamak designs. The stellarator is a similiar device, but with the plasma current replaced by currents in external helical multipole windings. This configuration was invented in 1951, but at first appeared to give unfavorable confinement. Recent experiments in Japan (Heliotron E) and Germany (Wendelstein VII A) have shown that the confinement quality of stellarators can be at least comparable to that of tokamaks of the same size. The stellarator has the potential advantage of steady-state operation, but the drawback of somewhat greater geometric complexity.

Another instance where an old and formerly unsuccessful approach has recently begun to yield encouraging experimental results is the reversed-field pinch (RFP), a kind of tokamak where the toroidal field is weaker than the poloidal field and reverses its direction at the plasma edge. The reversed-field pinch may be capable of reaching very high plasma beta values and fusion power densities. Finally, there is a new generation of compact-toroid experiments (Fig. 12) which differ from conventional tori

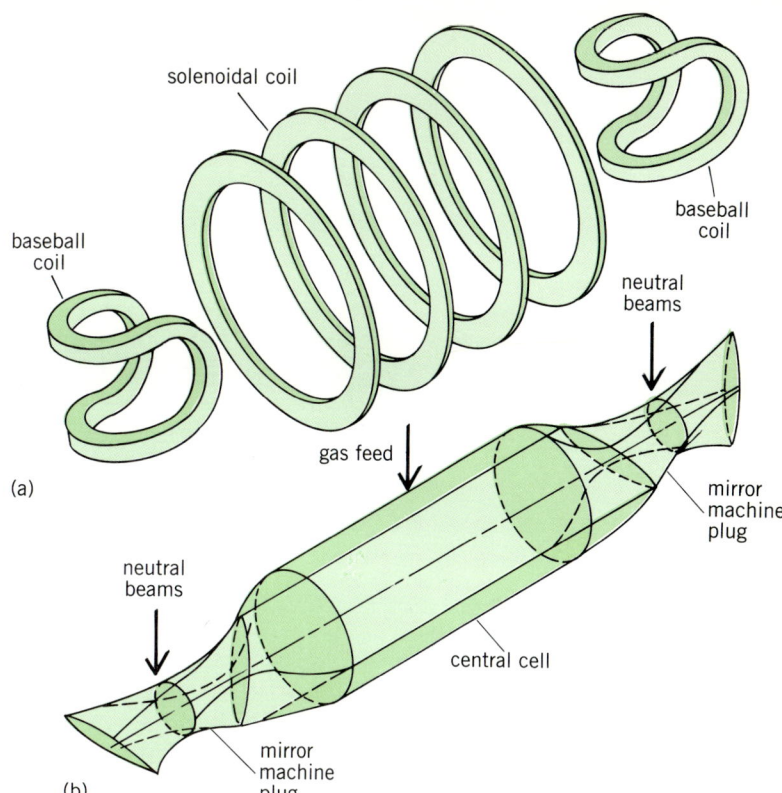

Fig. 11. Tandem mirror system. (a) Coils. (b) Plasma.

in that the plasma is not linked by toroidal-field coils. As a result, the reactor topology is simplified, and potential problems of first-wall replacement and maintenance are greatly eased.

PROGRESS IN INERTIAL CONFINEMENT

The possibility of using laser light to make an inertial fusion reactor was pointed out in the early 1960s. Ten years later, substantial experimental programs were under way in the United States, France, the Soviet Union, and Japan.

The initial expectation was that an inertial confinement fusion reactor could operate efficiently with pellet-input energies of 100 kJ or less. In the course of research, these numbers have risen by about a factor of 50. Concurrently, the most serious physical problems have been understood and solved, so that the successful ignition of D-T pellets by laser beams has become a realistic prospect (Fig. 13).

One element of recent experimental progress has been the development of shorter-wavelength laser capabilities and the associated demonstration that the main plasma physics limitations of inertial confinement fusion can be overcome by proceeding in that direction. At the natural 1-micrometer wavelength of a neodymium-glass laser, the fraction of usefully absorbed power falls to about 30% in the range of incident power fluxes of practical interest (10^{14}–10^{15} W/cm^2). When the 1-μm output is converted to a ⅓ μm wavelength by a frequency-multiplying technique, the power absorption rises to about 90%. The physical reason is that collective

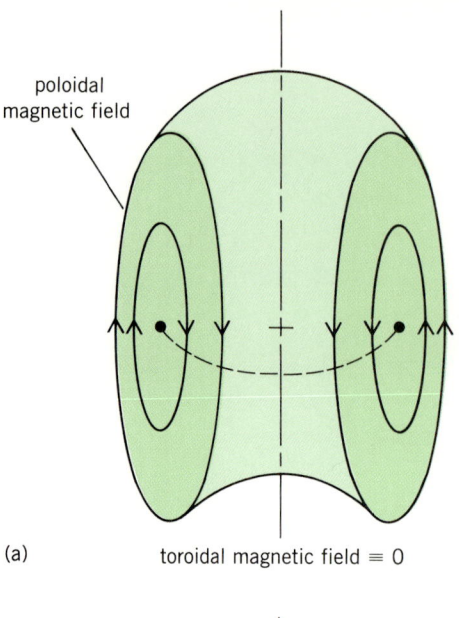

(a) toroidal magnetic field ≡ 0

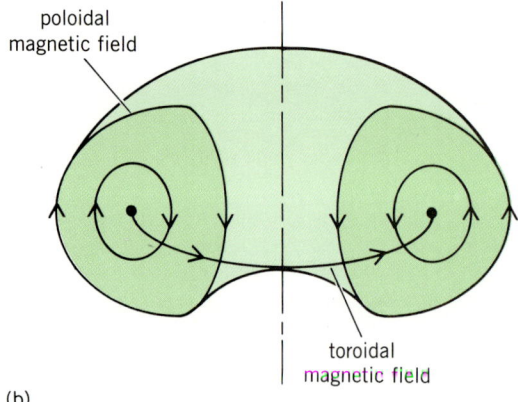

(b)

Fig. 12. Magnetic field configurations in compact tori. (a) Reversed-field θ-pinch. (b) Spheromak.

plasma effects are less easily excited when there is a larger difference between the natural frequency of plasma oscillations and the frequency of the light waves. As a result, the shorter wavelengths can penetrate to regions of high plasma density, where their energy is absorbed by ordinary collisional processes.

The same considerations explain why the acceleration of superthermal electrons decreases by orders of magnitude in the case of shorter-wavelength light. In order to maximize the compression ratio of the pellet, it is necessary to avoid preheating its core by these penetrating fast electrons. Using shorter-wavelength light, the Novette laser at Livermore (Fig. 13) has already reached density factors of 120 relative to liquid hydrogen, at temperatures up to 700 eV and pressures above 10^{10} atm (10^{11} lbf/in.^{2} or 10^{15} Pa) [Fig. 14]. Compression factors up to 1000 are expected to be achieved in the recently completed NOVA facility at Livermore, which will have a 100-kJ, 1-nanosecond output. There are similar plans for the Gekko XII laser in Osaka, Japan.

While neodymium-glass lasers with short-wavelength converters have proved to be particularly effective in achieving reactorlike pellet implosions, the efficiency and pulse repetition rate of the existing laser systems is too low for reactor purposes. A large carbon dioxide (CO_2) laser, such as the Antares at Los Alamos National Laboratory, would have much more desirable driver characteristics, but since its wavelength is 10 μm the plasma interaction problems are severe. The use of laser beams for an inertial confinement fusion reactor will depend on the development of more advanced technologies, possibly free-electron lasers.

Higher driver efficiencies appear to be readily achievable with particle-beam drivers. Electron

Fig. 13. The Novette laser installation for inertial confinement fusion at the Lawrence Livermore National Laboratory.

beams would be particularly straightforward from the point of view of source technology, but they are too penetrating to lend themselves to pellet compression. The use of light-ion beams in the energy range of 3–30 MeV is being pursued at the Sandia National Laboratory and at Osaka. Heavy-ion beams in the multi-GeV range would be advantageous for minimizing plasma effects and facilitating a sharp focus at a distance, but the magnitude of the required investment has thus far inhibited this approach.

PRACTICAL UTILIZATION OF FUSION ENERGY

While fusion fuel is an inexpensive and abundant terrestrial energy resource, the path to practical fusion power has been blocked by a series of formidable barriers. Fifty years ago, the generation of terrestrial fusion energy seemed physically impossible. Twenty years ago, it was believed to be possible in principle, but no specific reactor solution was known that had any real claim to engineering feasibility. The anomalously poor energy confinement observed in fusion experiments of that time extrapolated to power plants with unacceptably huge minimum size.

During the past decade, both magnetic and inertial fusion research have made major progress. Reactor plasma regimes will begin to be reached with existing facilities in the latter part of the 1980s, and experimental fusion reactors could then be operating around the year 2000.

Thus, the barrier of scientific feasibility seems to be fading away, and the barrier of engineering feasibility is likely to be breached by 2000. There remains the most formidable and debatable issue of all: Can the promise of cheap, inexhaustible fuel be matched by a low-cost fusion burner? In other words, can the capital investment and the associated interest payments be kept small enough, so that the cost of a kilowatt-hour of electricity will remain close to present-day prices?

A characteristic feature of fusion power is a very marked "economy of scale"; that is, a fusion reactor driving a 3000-MW electric power plant is typically estimated to cost only a few times more than a reactor driving a 300-MW plant. The desire to reduce the ideal cost per kilowatt hour strongly favors large fusion reactor units. On the other hand, there are many practical considerations relating to construction time, reactor maintenance, and system availability that strongly favor minimizing the unit size. A promising solution consists in designing multiunit fusion plants, with moderate-sized reactor units sharing conventional facilities and common auxiliary systems.

Long-range forecasts concerning the economics of high technology are inherently difficult to make. In order to produce objective estimates, as distinct from science fiction, the experts must carry out their cost analyses in terms of existing materials and designs. On the other hand, the historical record shows that such analyses tend to become outdated

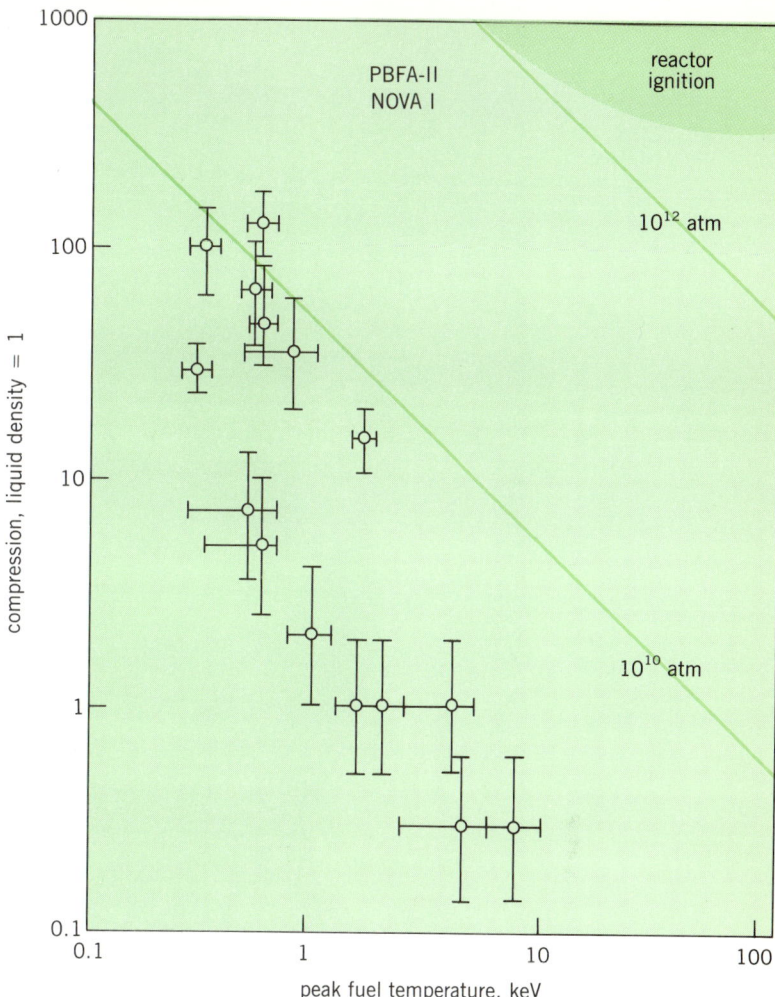

Fig. 14. Compression and temperature achieved in inertial confinement fusion devices. 1 atm = 14.7 lbf/in.^{2} = 10^5 Pa.

in a decade or two. Given past performance in fusion research, there is some basis to expect that continuing scientific advances, together with the rise of high-technology capabilities in general, will make fusion economically attractive at just about the time when it will be most needed. Fusion power could come into initial commercial use around the year 2020 and should be able to supply a major share of the world's energy requirements by around 2050.

[HAROLD P. FURTH]

Bibliography: K. A. Brueckner and S. Jorra, Laser-driven fusion, *Rev. Mod. Phys.* 46:325–387, 1974; R. W. Conn, The engineering of magnetic fusion reactors, *Sci. Amer.* 249(4):60–71, October 1983; H. P. Furth, Progress towards a tokamak fusion reactor, *Sci. Amer.* 241(2):50–61, August 1979; International Atomic Energy Association, *Plasma Physics and Controlled Nuclear Fusion Research: Proceedings of the 9th International Conference, Baltimore, 1982*, 3 vols., 1983; J. H. Nuckolls, Feasibility of inertial fusion, *Phys. Today*, 35(9):24–31, September 1982; E. Teller (ed.), *Fusion*, 2 vols., 1981.

Gene Therapy

Raju S. Kucherlapati

Raju S. Kucherlapati has been a professor at the University of Illinois College of Medicine since 1982. Earlier he was a research Fellow at Yale University, and an assistant professor at Princeton University (1975–1982). He holds a doctor's degree from the University of Illinois, Urbana.

A number of human diseases are the result of hereditary genetic disorders. In many cases, the gene involved and the exact nature of the defect that leads to the disease are well understood. For example, sickle-cell anemia is an autosomal recessive trait resulting from a single base change in the human adult β-globin gene so that there is a single altered amino acid in the globin polypeptide. Sickle-cell anemia in its homozygous state is usually lethal. Similarly, a class of disorders which are referred to as thalassemias are the result of a number of different defects, all involving the expression of human globin genes.

Because of the nature of the diseases and the nature of the defects, no satisfactory therapeutic methods are available. Understanding of the biological basis of several of these diseases has been facilitated by the ability to isolate the normal counterparts of the defective genes and to compare their structure. The recombinant DNA (deoxyribonucleic acid) methodologies which permitted the isolation of these gene sequences also permit their amplification in bacteria so that relatively large amounts of DNA can be readily obtained.

The concept of gene therapy is quite simple: if it is possible to introduce a normal copy of a gene into the appropriate cells of a person who suffers from a genetic disorder and if the introduced gene is expressed properly, the cells will be able to survive and the normal gene product will be able to alleviate the problems associated with the defective gene. Several methods of introducing foreign DNA into mammalian cells are now available, and numerous recent developments indicate that gene therapy will be a reality in a relatively short time. This article will explore the current methods for introduction of genes into mammalian cells and the study of their expression, and how these methods make it possible to initiate testing the feasibility of gene therapy.

DNA transfer into mammalian cells. Bacteria are capable of taking up DNA, and if it is homologous, the genes borne on it are expressed in the new environment. This process, bacterial transformation, permits introduction of virtually any DNA segment into bacteria. Bacterial transformation was first a tool in studying bacterial genes, and more recently has become very valuable in isolation, amplification, and characterization of genes from a number of organisms including mammals.

A method to introduce DNA into mammalian cells was described in the early 1960s. However, a reliable and reproducible gene transfer method had to await developments in the 1970s. A crucial discovery that made gene transfer possible was that DNA complexed with calcium phosphate is readily taken up and expressed by mammalian cells. The first successful gene transfer was described in 1977 by N. Maitland and J. McDougall and by M. Wigler and colleagues. Both of these groups introduced the gene coding for a pyrimidine-salvage-pathway enzyme, thymidine kinase (tk), into cultured mouse cells. The source of the gene was a relatively simple virus ubiquitous in the human population, herpes simplex virus. The success of these experiments can be attributed not only to the new method of delivering the DNA but also to the availability of a cell line which lacked the tk gene, and to a powerful selection system which, when applied to a population of cells, permits growth of cells containing and expressing the tk gene. Many of the concepts utilized in these early experiments form the basis for the current experiments leading to gene therapy.

Soon after the successful transfer of the herpes simplex virus tk gene into mammalian cells by transfection, the transfer of mammalian cellular tk genes was described. Though the mammalian cellular tk gene was not available in a purified form, it was clear from these experiments that the total cellular DNA could be used as a donor and the availability of a selectable system would permit identification of cells which have acquired and are expressing a particular gene which constitutes a small subfraction of total DNA.

Fate of foreign DNA. Several possible fates can be predicted for the transfected DNA. It may be readily lost or degraded in the cell; it may replicate autonomously and remain as an extrachromosomal element; or it may become covalently integrated into the cellular genome, thus becoming a part of the cellular DNA to be passed on to all succeeding generations of cells. All of these fates have been observed in cells transfected with DNA. All of the features important for uptake of DNA and its retention inside the cells are not well understood. Two features of DNA that would permit transfected segments autonomous replication, and thus confer extrachromosomal status, are a functional origin for the start of replication and the appropriate cell environment which recognizes this origin. If these special features are absent, the DNA is usually integrated into the cellular genome. There does not seem to be any specificity for the sites at which the integration occurs, which may have a significant impact on gene therapy. Methods and experimental approaches to target genes to specific sites in the cellular genome will be presented later.

Selectable and nonselectable It is now possible to introduce any DNA sequence into mammalian cells. For purposes of this discussion, all genes can be classified into those whose expression can be selected (selectable genes) and those whose expression cannot be selected (nonselectable genes). Selectable genes confer upon the cell a growth advantage under certain selective conditions. The herpes simplex virus tk gene falls into this category. The introduction of certain selectable genes such as tk requires cells that lack or are mutant at the locus but other selectable genes can be introduced into any cell type. A gene that codes for an enzyme called dihydrofolate reductase (dhfr) and a modified form of this gene that confers resistance to a cytotoxic drug named methotrexate fits into this second category. Since

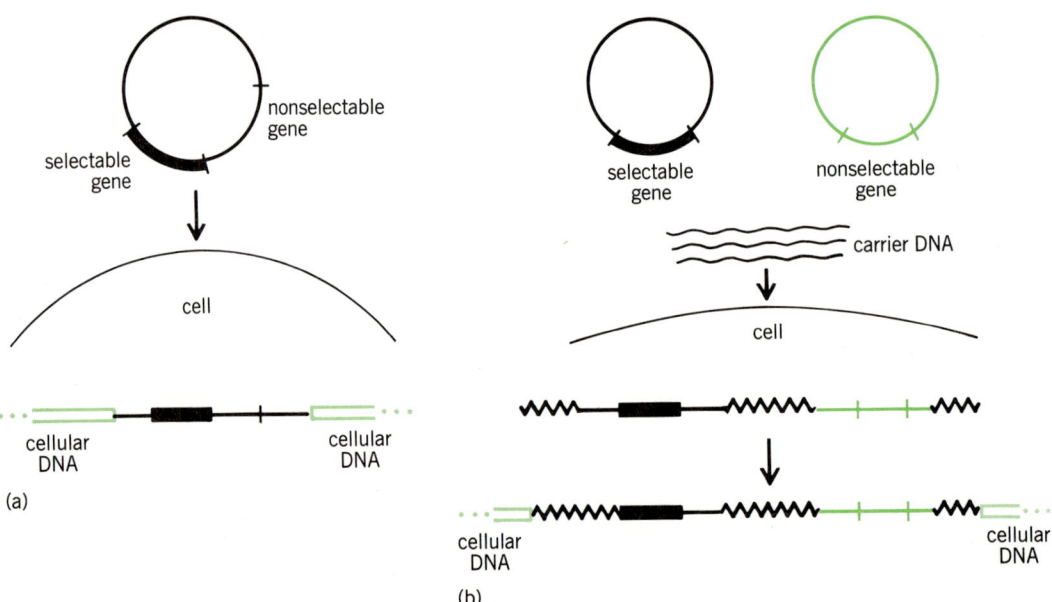

Fig. 1. Methods of introduction of nonselectable genes. (a) Selectable and nonselectable genes are covalently ligated to each other. (b) Selectable and nonselectable genes are mixed with carrier DNA before presentation to cells.

primary cells will be employed for gene therapy, and as they may have a limited life-span in culture, the so-called dominant selectable markers such as the modified dhfr gene should prove to be very useful in gene therapy.

Nonselectable genes can also be introduced into mammalian cells, by two methods. A nonselectable gene could be covalently linked to a selectable gene, and the chimeric DNA segment presented to cells (Fig. 1a). The cells can then be selected for the acquisition and expression of the selectable gene. All or most of the cells containing the selectable gene are expected to contain the nonselectable gene by virtue of the proximity of the two genes. An alternative is to mix the selectable and nonselectable genes along with nonspecific carrier DNA in excess and present this mixture to cells (Fig. 1b). It is now known that those cells that take up DNA are capable of taking up large amounts of it and the different DNA molecules are capable of joining together to form large structures, referred to as transgenomes, which contain the selectable, nonselectable, and carrier DNA molecules. This complex becomes integrated into cellular sequences carrying the nonselectable sequences along with the selectable sequences. These strategies of gene transfer thus permit introduction of virtually any DNA sequence into mammalian cells.

Other methods of gene transfer. In addition to using DNA complexed with calcium phosphate, several alternative methods to introduce foreign DNA into mammalian cells are available. These methods can be classified into direct or indirect methods of transfer. Calcium phosphate–mediated transfer is considered to be an indirect method. Other indirect methods include use of diethylaminoethyl-dextran–complexed DNA, or DNA trapped in liposomes or erythrocyte ghosts. In addition, bacteria that are harboring a plasmid can be stripped of their cell wall and the resulting protoplasts can be fused with intact cells by way of chemical agents such as polyethylene glycol. The efficiency with which the DNA is stably transferred to the recipient cells by each method varies. Since introduction of extraneous DNA should be minimized for gene therapy, the protoplast fusion method is not expected to play an important role. Since extensive preparation is required to trap DNA in liposomes or erythrocyte ghosts, those methods may not play a significant role either. Stable transfer of genes seems to be best achieved by the calcium phosphate–transfer method, where as many as one in a thousand recipient cells may be stably transfected.

DNA can also be introduced into mammalian cells by a direct method, microinjection. Purified DNA can be microinjected into cell cytoplasm or cell nucleus. Several investigators have shown that if a gene is directly introduced into the nucleus of a recipient cell, a large proportion (as much as 5% or more) of the recipient cells acquire and stably express the gene. Though this method is more time-consuming and cumbersome, it is the choice when the recipient cell number is limiting.

Microinjection has proved to be extremely useful in introducing genes into embryos. Fertilized eggs can be readily obtained from mice. Purified DNA can be introduced into the male pronucleus, and the microinjected embryos can be inserted into the uterus of a female and allowed to come to term. DNA from the tails or specific internal organs of the newborn mice can be tested for the presence of introduced DNA sequences. By using these methods, it was found that 70–90% of the injected embryos survive the trauma of injection, 10–30% of the implanted embryos come to term, and as many as 40–50% of the newborn mice contain the foreign DNA sequences. A high proportion of the mice that carry the foreign DNA have it in all cells, including germ cells. As in other gene-transfer methods, the foreign DNA becomes integrated into the cellular DNA and then can be passed on from one generation to another. Appropriate breeding of the animals permits production of mice that carry the foreign gene in a homozygous state. Since no selection schemes are necessary, virtually any DNA sequence of any origin can be readily introduced into mouse embryos which, for all practical purposes, become a part of the cellular genome.

Expression of introduced genes. In order for gene therapy to be a reality, not only is it necessary to have reliable methods of gene transfer, but the introduced genes should also be expressed in the proper fashion. When a selectable gene is introduced into cells, the fact that the cell can survive the selective pressure implies that the gene is being expressed. Direct methods of testing, which included examination of the ribonucleic acid (RNA) and protein products, confirmed this view. Examination of cells transfected with nonselectable genes revealed that in a large proportion of the cases the nonselectable gene is also expressed. The general conclusion that can be drawn from a large number of experiments is that any mammalian gene can be readily expressed in other mammalian cells.

This observation paved the way to examine the exact nature of the DNA sequences that are needed for expression in a mammalian cell. The basic strategy is to delete defined portions of the DNA segment known to carry a particular gene, introduce the modified DNA into a mammalian cell type, and assay for expression. From these and other types of experiments, a number of features of genes which are important for their expression have been identified. For genes which are transcribed by RNA polymerase II, these features are (1) a promotor located at the 5′ end of the gene, (2) transcription initiation site, and (3) poly A addition site. Also, the gene may have sequences that are removed by splicing. This information about the exact nature of the DNA sequences that are important for gene expression makes it possible to construct novel combinations of genes. For example, the promotor and transcription initiation and termination signals of a bacterial gene are not recognized by a mammalian cell, but it is possible to engineer a bacterial gene so that it carries the appropriate mammalian sequences, permitting its expression in the novel atmosphere of a mammalian cell.

Regulation of introduced genes. Genes successfully introduced into mammalian cells are not always properly regulated. For example, human globin genes can be introduced into mouse fibroblasts, where they will be expressed. Fibroblasts are inappropriate cell types for expression of globin genes. Also, globin genes in the appropriate environment (in erythroid cells) can be induced to produce higher levels of globin messenger RNA and protein, whereas similar inductions do not occur in fibroblasts.

Three features seem to be important for the properly regulated expression of genes. The first is the recipient cell type. When human globin genes are introduced into mouse erythroleukemia cells, which express their own globin genes, not only are the human genes expressed but their level of expression can be induced by a number of agents. This observation indicates that each differentiated cell may contain one or more transacting factors whose presence is necessary for the properly regulated expression of a gene. The second feature that is important for regulated expression is the presence of cis-acting sequences. Induction of some globin genes is dependent upon a segment of DNA called a cis-acting sequence located at the 5′ end of the gene. Similarly, some hormone-responsive genes are properly regulated only if this sequence at the 5′ end of the gene is retained. The knowledge of cis-acting sequences which are needed for normal and regulated levels of expression makes it possible to construct modular vectors in which parts of different genes can be mixed and matched to produce a gene that behaves in a predictable fashion when introduced into a defined cell type. A third factor that may play an important role in properly regulated gene expression is the location or site of integration of the foreign DNA. There is a fairly large body of evidence which supports the view that, in addition to intrinsic features of a gene, the nature of DNA sequences that surround the integrated sequence has profound effects on its expression. It is expected that these undesirable effects can be circumvented if the exogenous DNA sequence can integrate at the site of its cellular counterpart. Methods and experiments to achieve this goal of true gene replacement will be discussed below.

Approaches to gene therapy. Gene therapy can be effected by directly altering or replacing the defective gene, by adding a functional gene, and by indirect methods whereby the defective gene is made to produce at least a partially functional product. The indirect method, which would be useful in a limited number of cases, involves the use of suppressor transfer ribonucleic acid (tRNA) genes. In several cases, a single base change in the DNA could lead to a codon change resulting in premature termination of the protein product. This problem can be alleviated by the presence of a suppressor tRNA which recognizes this termination codon and inserts a specific amino acid at that site. Even if the amino acid is not exactly the same as the one coded by the wild type gene, it may be adequate to restore some degree of function to the protein. In such cases, all that is required of the tRNA gene is that it is expressed. Even minimal expression may be adequate to restore some function of the gene, thus affording gene therapy. This approach is of limited use because it can correct only one class of mutations.

A direct approach to gene therapy has been attempted. The strategy of this approach is to isolate the bone marrow from a mouse, treat it with calcium phosphate precipitate of DNA containing a selectable gene, and introduce the treated bone marrow into an irradiated recipient mouse. In one report, bone marrow was treated with DNA coding for the dhfr gene which confers resistance to the antitumor drug methotrexate. The researchers claimed suc-

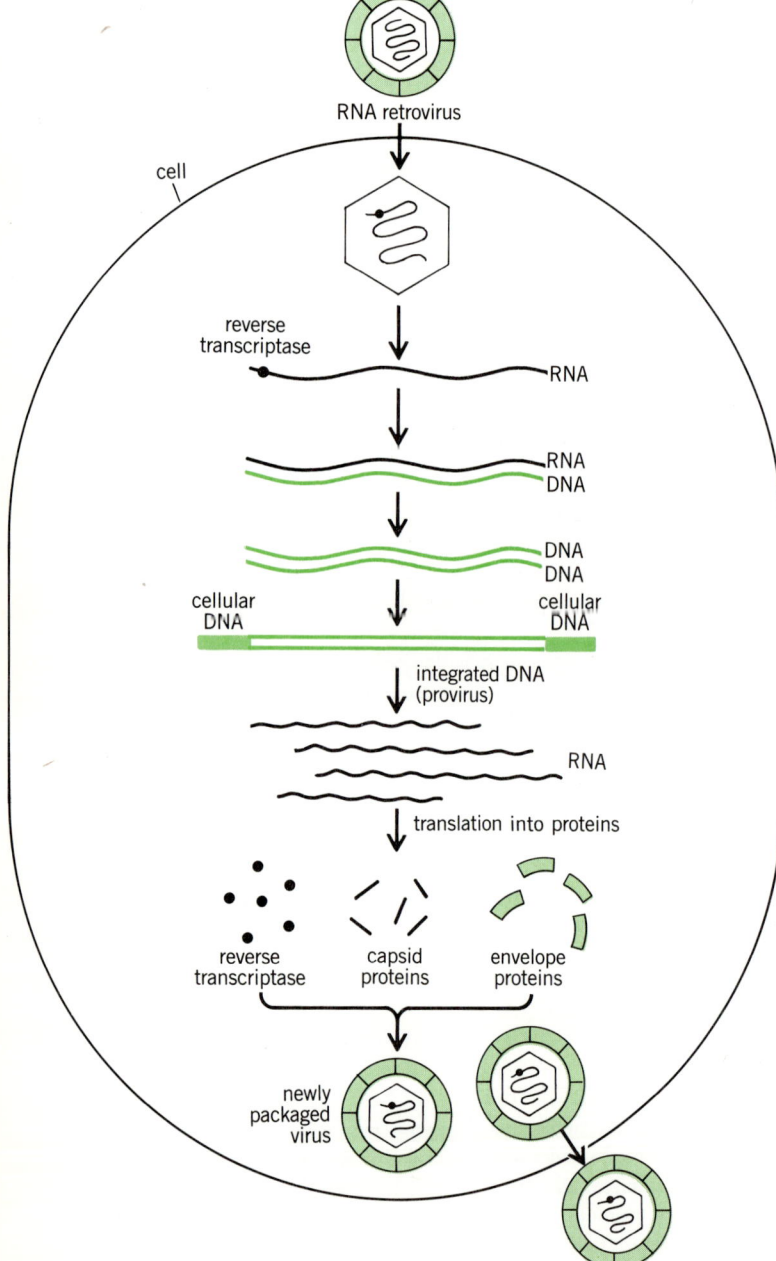

Fig. 2. Life cycle of a retrovirus; see text for details.

cessful transfer of the dhfr gene and its continued expression in the cells conferring methotrexate resistance. The same group of investigators presented evidence that the gene for herpes thymidine kinase was transferred into bone marrow cells by a similar method. These observations provide much hope for additional successes of this sort with animal models, eventually leading to similar attempts in humans. However, one possible disadvantage with this method is that stable transfer is at best about one in a thousand. Since gene therapy requires introduction of the gene into stem cells of the marrow which have continuous proliferation potential and since the stem cells constitute a minority of the population, the transfection efficiencies have to be improved for the method to be of practical significance.

An alternative direct method of gene therapy involves the use of retrovirus vectors. Retroviruses store their genetic information in RNA (see Fig. 2). The RNA is enveloped in a coat. After infection, that is, after the virus enters the cell, the virus becomes uncoated and the single-stranded RNA is used as a template for the synthesis of a complementary DNA copy (reverse transcription), which in turn is converted into its double-stranded DNA form. The double-stranded DNA becomes integrated into the cellular genome, from where it may be expressed. The integrated DNA, also referred to as the provirus, is used as a template for the synthesis of RNA, which in turn is used as a template for the synthesis of several proteins required for reverse transcription and the packaging of the RNA molecules. Newly packaged viruses are released from the cell. A considerable amount of information is available about the structure and function of the retroviral genome. Many of these viruses are identified in the mouse. The viruses have species specificity which is determined by the coat they carry. If a cell is coinfected with two viruses each having a different host range, it is possible to generate pseudotype viruses which will be able to gain entry into cells which are normally resistant to infection by the virus. The nucleotide sequence that is required for packaging of viral RNA (packaging signal) has been identified. If a virus with its packaging signal deleted is introduced into a cell along with another virus, the coat proteins produced by the first virus can be used to package the second virus. This type of strategy can be employed to introduce murine viruses into human cells.

The organization of the retroviral genome is very similar to that of many eukaryotic genes. The genes have a promoter sequence for recognition by the cellular transcriptional apparatus, gene coding sequences, splicing signals, and poly A addition signals. Though all this information is needed for the normal life cycle of the virus, it is possible to remove much of the gene coding information without affecting certain features of the life cycle such as integration and the genome's ability to initiate and terminate transcription. The deleted information can be replaced by any genetic material which can be expressed under the control of the viral sequences.

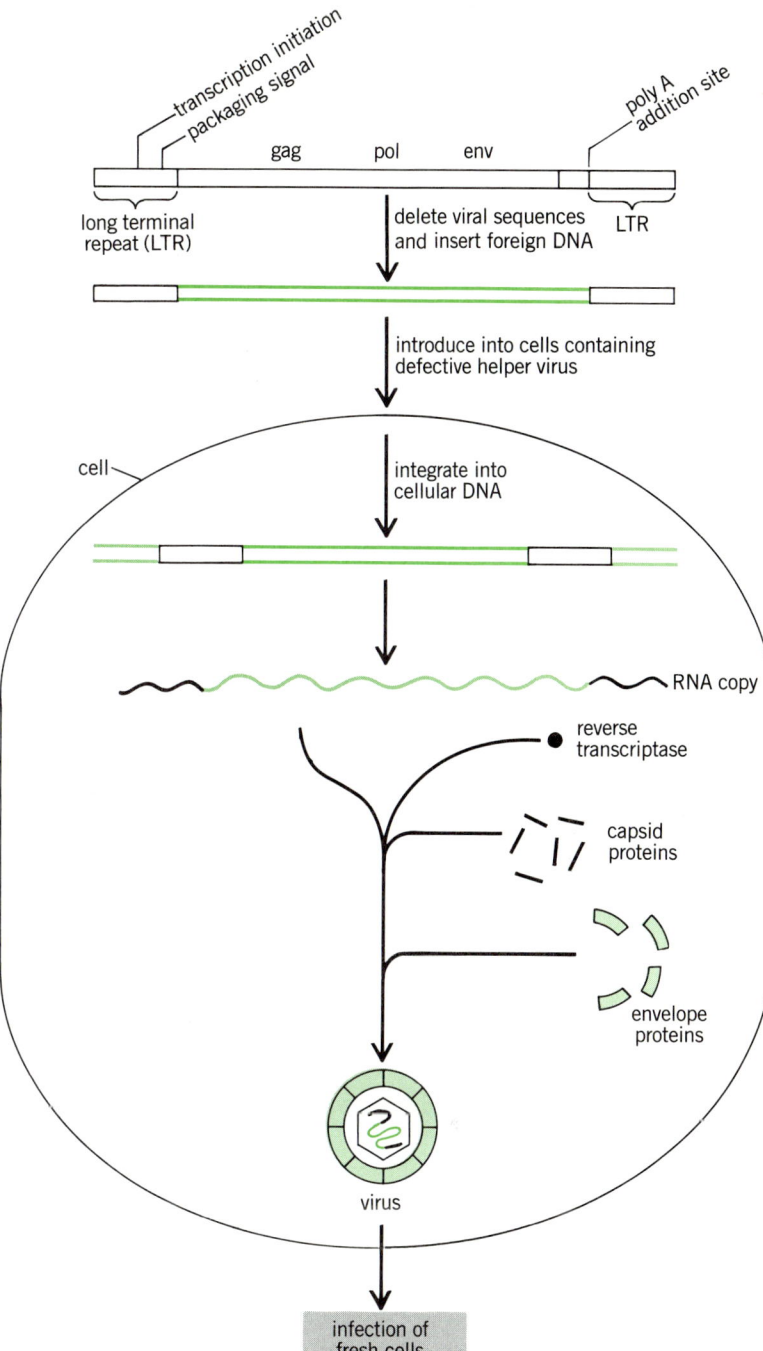

Fig. 3. RNA viruses can be used to deliver nonselectable genes. Some of the viral information can be removed and replaced with nonselectable gene sequences. Such a chimeric DNA can be introduced into cells carrying an integrated defective helper virus which provides the reverse transcriptase and the capsid and envelope proteins.

All of this information has permitted construction of retrovirus vectors which permit introduction of genes with high efficiency into mammalian cells. The general strategy is to remove the nonessential information from the DNA copy of a retroviral genome and insert one or more genes of interest in its place (Fig. 3). This chimeric molecule can be introduced into a cell line directly by DNA transfection or can be introduced into a cell line which is constitutively producing viral coat proteins. The input DNA will be utilized to produce RNA which will be packaged into intact virus particles which will be extruded into the medium. A high titer of the virus can be obtained from the medium and can be used to infect a fresh batch of cells. This method can result in 100% of the recipient cells acquiring and expressing the gene of interest.

As a prelude to gene therapy, several different types of test viruses have already been constructed. In one case, a complementary DNA (cDNA) copy of the human gene coding for the purine-salvage-pathway enzyme hypoxanthine phosphoribosyl transferase (HPRT) was inserted into a retrovirus vector. This vector was used to infect human cells derived from a patient suffering from the Lesch-Nyhan syndrome, which is characterized by a severe deficiency of the HPRT enzyme. The infected cells were shown to have acquired the ability to produce a functional HPRT protein at 5–20% of normal levels. Under the test conditions employed, these cells behave like normal cells. This demonstration of gene correction in the test tube is clearly an important step toward gene therapy in the organism. The same group of investigators have taken the system one step forward by using the virus to infect bone marrow cells from a mouse and transplanting the marrow into an appropriate recipient mouse. They were able to show that the gene was introduced into pluripotent stem cells because the recipient mice express the gene for long periods of time. Other groups of investigators have constructed similar retrovirus vectors, utilizing other genes, and showed that genes borne on such vectors can be introduced into bone marrow cells in the organism and in the test tube. Clearly this approach is ready for tests of its efficacy in correcting gene disorders in animal test systems and, if found to be consistently successful, should lead to attempts of gene therapy in humans.

Gene replacement through homologous recombination. One possible disadvantage with the retrovirus-mediated gene transfer rests with the fact that these viral sequences integrate at random sites in the genome. Such random integrations would not be able to correct dominant disorders. In addition, there are examples where viral integration has resulted in inactivation of an essential gene or inappropriate activation of genes, resulting in malignant growth. The severity of these problems, as they apply to gene therapy, is not well understood and awaits further studies. Several investigators are examining alternative approaches to gene therapy.

The most desirable method of gene therapy is gene replacement, where the defective, malfunctional or nonfunctional resident gene is replaced by its normal or active counterpart introduced by gene-transfer methods. This optimal approach could be accomplished by homologous recombination (crossing-over). Though homologous recombination is a regular feature in germ cells, it was not clear, until recently, whether chromosomes in somatic mammalian cells can also perform this reaction.

Early tests to detect homologous recombination between genetic markers located on different chromosomes in mammalian cells failed to yield positive results. The advent of high-efficiency gene-transfer systems and the use of appropriate genetic markers changed this situation (Fig. 4). Two plasmids, each of which contains a mutated or deleted mammalian selectable gene, are simultaneously introduced into cells, and reconstruction of an intact gene is monitored by the ability of the cells to survive selective pressure. In one case, deletions were introduced into the bacterial amino glycoside 3′-phosphotransferase gene which confers upon bacteria resistance to neomycin and kanamycin and upon mammalian cells resistance to an analog of neomycin called

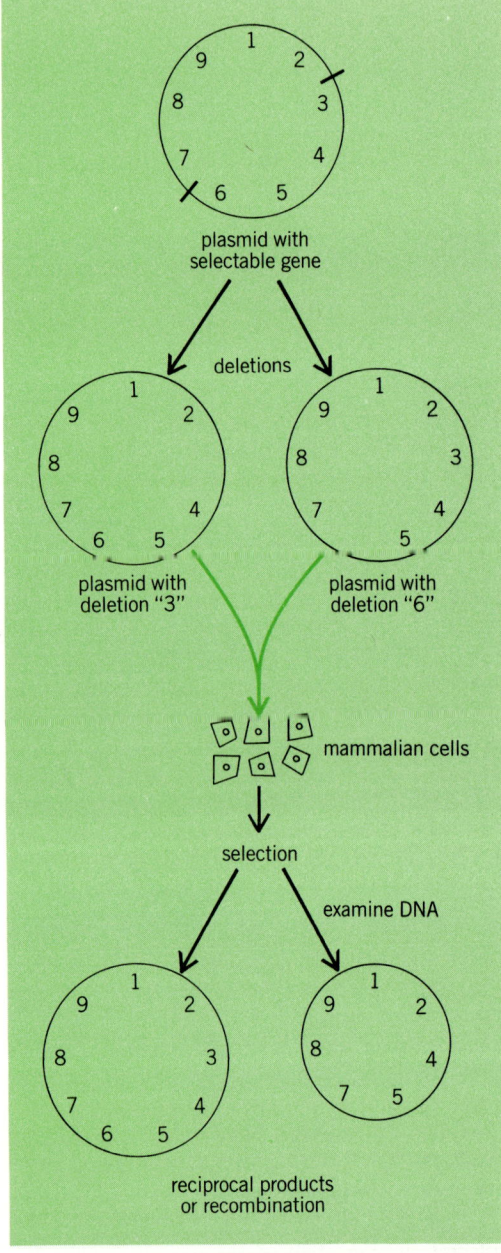

Fig. 4. A general method used to detect homologous recombination in mammalian cells. Deletions are introduced into a selectable gene and two plasmids, each of which, carrying a different deletion, is introduced simultaneously into mammalian cells. Homologous recombination can be detected by the reconstruction of an intact gene.

G418. In one plasmid, this deletion was at the 5' end of the gene, and in another at the 3' end. Each of these deletions rendered the gene inactive. When both plasmids were simultaneously introduced into mammalian cells, G418-resistant colonies arose at a high frequency. Such experiments established that somatic mammalian cells are able to mediate homologous recombination, between two introduced plasmids, at high efficiencies.

The type of recombination described above is between two introduced plasmids. For gene therapy, it is necessary to obtain recombination between a cellular sequence and its counterpart introduced by DNA transfer (Fig. 5). The following general strategy was employed to detect these types of events. A plasmid containing two selectable genes was constructed, and one of the genes was made nonfunctional by the introduction of a mutation or deletion. Such a plasmid was transfected, and cells were selected for the acquisition and expression of the intact selectable marker. Examination of the cells revealed that the plasmid had integrated into the cellular genome and for all practical purposes behaved as a cellular sequence. Homologous recombination between the "cellular" sequence and its introduced counterpart was tested by transfecting the cell line with the complementing but defective mutant. Several independent investigators using different selective systems showed that recombination between the integrated plasmid and introduced plasmid occurs. The next step is to demonstrate homologous recombination between a "true" cellular sequence and its plasmid counterpart.

Future. Although great progress has been made, no successful gene replacement experiments in organisms have been reported. Several different strategies to achieve this goal have been developed over the past few years. The technique which promises to be of immediate use is gene transfer mediated by retroviruses. This method provides a highly efficient means of introducing foreign genes into primary cells. The understanding of mechanisms of homologous recombination and identification of gene products that enhance this process are expected to make gene therapy more easily performed without concern about possible side effects from random integration of foreign DNA. One of the limitations of all current methods is that gene therapy can be effected in only those cells which are easily accessible for retrieval and replacement. This limits gene therapy to bone marrow cells. Several methods to isolate, maintain, and grow normal cells from a number of mammalian tissues are being tested. If they are successful, it may be possible to introduce the desired gene into the appropriate cell type in culture and transplant those cells to their normal environment. It is also possible to identify or develop retroviruses which will have not only species specificity but tissue specificity as well. The availability of such viruses should permit development of vectors which will allow targeting. Early identification of the genetic disorder in the fetus coupled with the available therapeutic measures and the new gene therapy methods

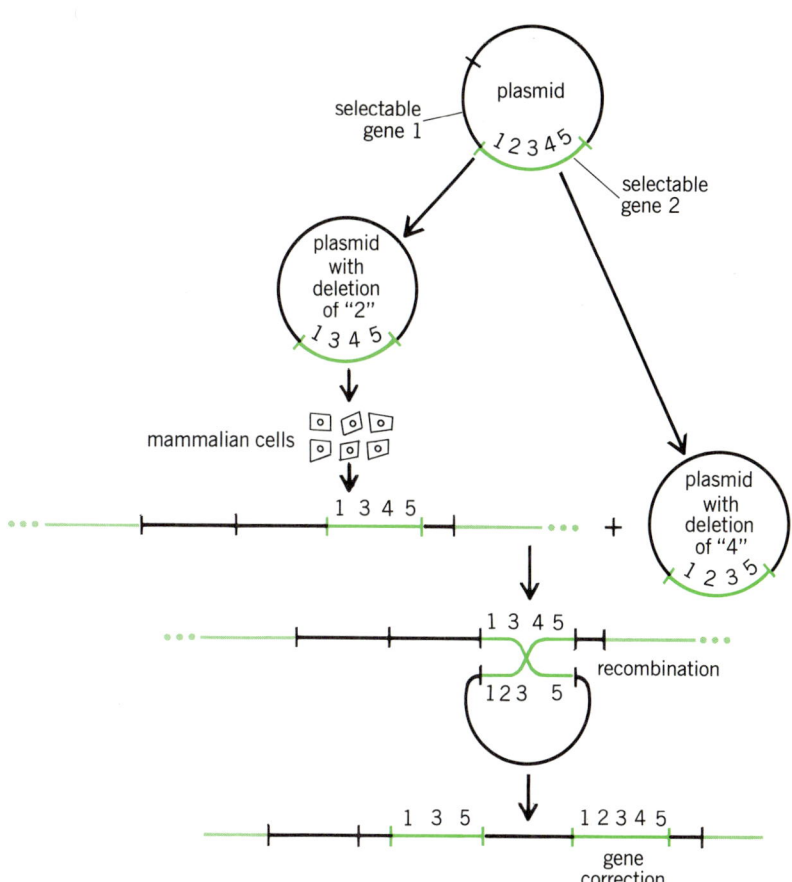

Fig. 5. Method to detect homologous recombination between a "cellular" sequence and its homolog introduced by gene transfer. A plasmid containing a deleted selectable gene is introduced into a mammalian cell and allowed to integrate into the cellular genome. A second plasmid containing a different deletion is then introduced into the cell, and homologous recombination is monitored by reconstruction of an intact gene.

might eliminate the suffering caused by these genetic disorders. [RAJU KUCHERLAPATI]

Bibliography: M. J. Cline et al., Gene transfer in intact animals, *Nature*, 284:422–425 1980; J. W. Gordon and F. H. Ruddle, Integration and stable germ line transmission of genes injected into mouse pronuclei, *Science*, 214:1244–1246, 1981; N. Hsiung et al., Co-transfer of circular and linear prokaryotic and eukaryotic DNA sequences into mouse cells, *Proc. Nat. Acad. Sci.*, 77:4852-4856, 1980; N. Hsiung et al., Introduction and expression of a fetal human globin in mouse fibroblasts, *Mol. Cell Biol.*, 2:401–411, 1982; R. K. Kucherlapati et al., Homologous recombination between plasmids in mammalian cells can be enhanced by treatment of input DNA, *Proc. Nat. Acad. Sci.*, 81:3153–3157, 1984; A. D. Miller et al., Expression of a retrovirus encoding human HPRT in mice, *Science*, 225:632–634, 1984; M. Wigler et al., Transfer of purified Herpes virus thymidine kinase gene to cultured mouse cells, *Cell*, 11:223–232, 1977; M. Wigler et al., Transformation of mammalian cells with genes from prokaryotes and eukaryotes, *Cell*, 16:777–786, 1979; R. C. Willis et al., Partial phenotypic correction of human Lesch-Nyhan (HPRT-deficient) lymphoblasts with a transmissible retroviral vector, *J. Biol. Chem.*, 259:7842–7849, 1984.

Accretion Tectonics

W. G. Ernst

After receiving a doctorate in geology from Johns Hopkins University, W. G. Ernst in 1960 joined the Earth and Space Sciences faculty at the University of California, Los Angeles, eventually becoming professor of geology and geophysics and a member of the Institute of Geophysics and Planetary Physics. He has served as president of the Geological Society of America. Author of many journal articles as well as books, he concentrates his interests in rock-forming silicates, high-pressure metamorphic terranes, and the petrogenesis of oceanic crust and mantle.

Recognition of the operation of plate tectonic processes has revolutionized geologic thought over the past two decades. Now almost universally accepted are the concepts of sea-floor spreading and continental drift. Gone are the stabilist views of the Earth as a planet with a crust fixed in geographic coordinates relative to the underlying mantle. The implications of plate tectonics have been profound; the development of the Earth is now understood much better than it was 20 years ago, including how and why rock-forming processes occur where they do. Such enhanced knowledge has influenced the ways in which earth scientists search for oil and gas, mineral deposits, and other natural resources; attempt to mitigate the potentially devastating effects of geologic hazards such as earthquakes, volcanic eruptions, and landslides, and predict the timing of their occurrences; and try to develop the environment in an orderly, conservative manner.

As illustrated in Fig. 1, plate tectonic models emphasize the formation of new lithosphere, capped by basaltic crust, at relatively simple divergent plate junctions sited over rising asthenospheric plumes. The generation of continental and island-arc sialic crust surmounting complex convergent plate junctions, where oceanic crust–capped lithosphere descends back into the deeper mantle, is depicted in Fig. 2. Magmas here are of three distinct types and sources: (1) calc-alkaline melts derived from partial fusion of the downgoing oceanic crust–capped slab; (2) incipient basaltic liquids present along grain boundaries in the asthenosphere lying directly beneath the stable, nonsubducted lithospheric plate; and (3) anatectic granitic melts produced at the base of the volatile-rich sialic crust due to advective heat transport. Although implicit in the well-recognized plate tectonic notions of conservative plate junctions (transform faults), and the consequences of oblique subduction, a new mobilistic understanding has evolved as an outgrowth of these older ideas.

The terrane concept postulates that much of the Phanerozoic continental crust bordering the Pacific Basin (and perhaps most of the Earth's crust around the globe) consists of a collage of lithotectonic units, or tectonostratigraphic terranes. As

82 Accretion tectonics

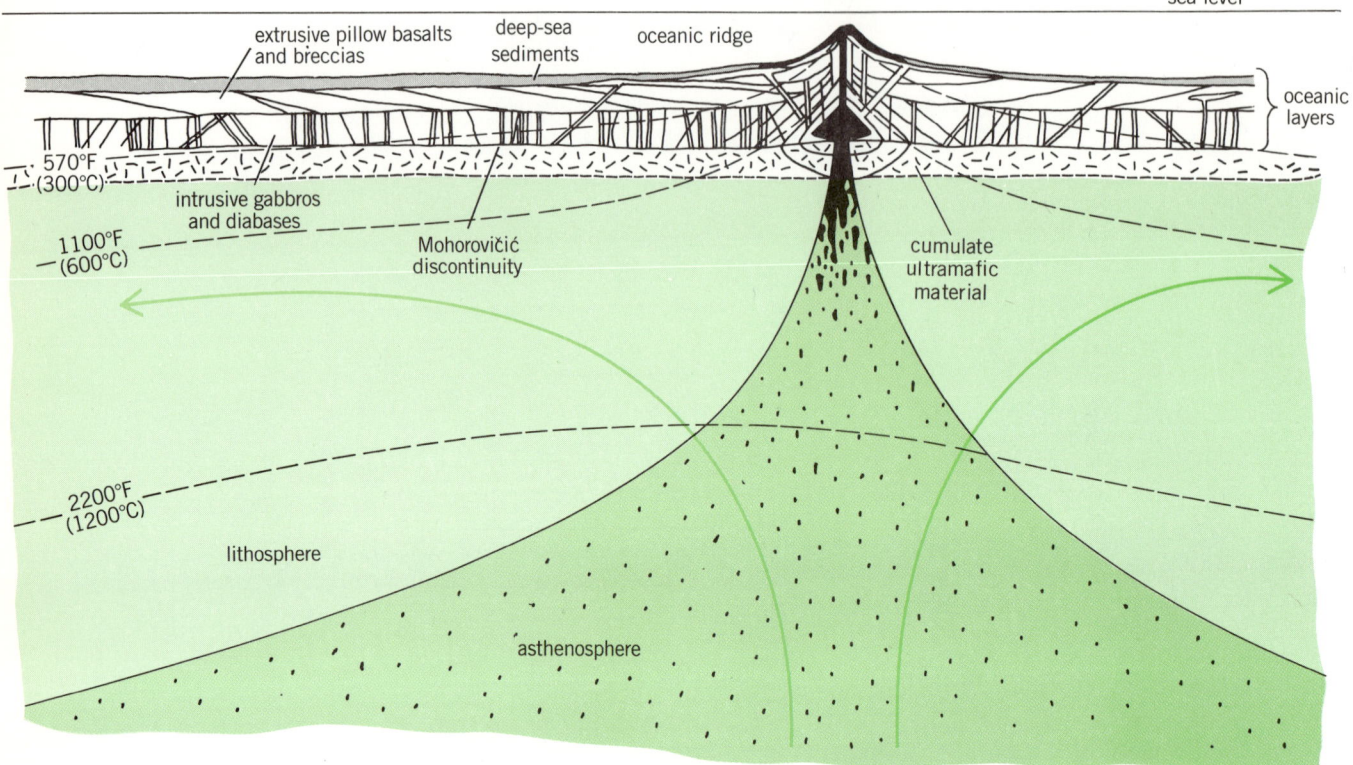

Fig. 1. Schematic diagram of an oceanic spreading center. Increasing degrees of partial melting of the rising and decompressing asthenospheric column result in the production of tholeiitic basalt magmas (shown in black). The symmetric thermal structure of a divergent plate junction is also indicated. Ultramafic material is situated below the Mohorovičić discontinuity; depleted peridotite forms the base of the section.

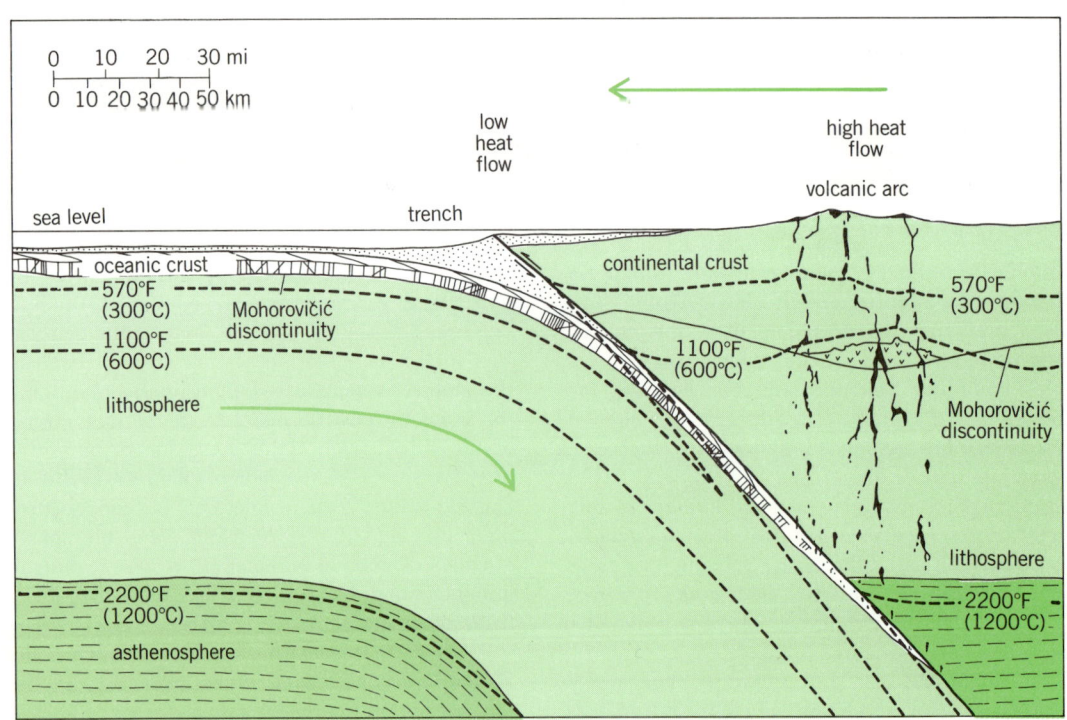

Fig. 2. Schematic diagram of a convergent plate junction sited at a continental margin. The asymmetric thermal structure of a convergent plate junction is also illustrated. Magmas, both tholeiitic and calc-alkaline, shown in black, buoyantly rise into the volcanic arc.

defined, each of these units is bounded entirely by faults, shear zones, or sutures, and possesses a unique geologic history, testified to by the particular lithic assemblages, contained fossils, and specific geochemical-geophysical characteristics. Adjacent, contrasting terranes are regarded as mutually allochthonous until, and unless, proven otherwise. Because it is difficult to imagine how individual terranes could form contemporaneously in close proximity without exhibiting geologic interrelationships, an important implication of the concept is the far-traveled nature of such allochthonous units. The bringing together, or suturing, of terranes is referred to as accretion tectonics, and is regarded by many as an important mechanism of continental growth.

Terranes are of several different varieties, depending on the dominant rock types and histories, for example, stratigraphic, plutonic igneous, disrupted (chaotic), or composite (amalgamated). The geologic evolution of any particular tectonostratigraphic entity during migration reflects its own unique history, which may involve dispersal, truncation, or amalgamation with one or more exotic terranes prior to the final collisional suturing at a growing continental margin. Following tectonic accretion, volcanic flows or sedimentary strata may cover adjacent terranes and their mutual sutures, or plutons may invade such tectonic contacts, thereby providing upper age limits regarding the time of juxtaposition.

Evidence for far-traveled terranes. As defined above, individual terranes possess geologic characteristics which contrast markedly with those of adjacent tracts; thus their geographically disparate origins seem to be required. But just how far have some of these terranes traveled, and what constraints are there on where they formed initially? As yet, these are exceedingly difficult questions to answer, but diverse kinds of data provide important constraints.

Packets of rock which must have been generated in very different geographic environments are now juxtaposed. Three well-studied examples are: (1) the contact along the San Andreas Fault in the central California Coast Ranges between the Franciscan relatively high-pressure, low-temperature graywackes and argillites of a late Mesozoic subduction zone complex on the east, against coeval Salinian granites intruding older high-grade metamorphics on the west; (2) rock sections exposed in Nelson and

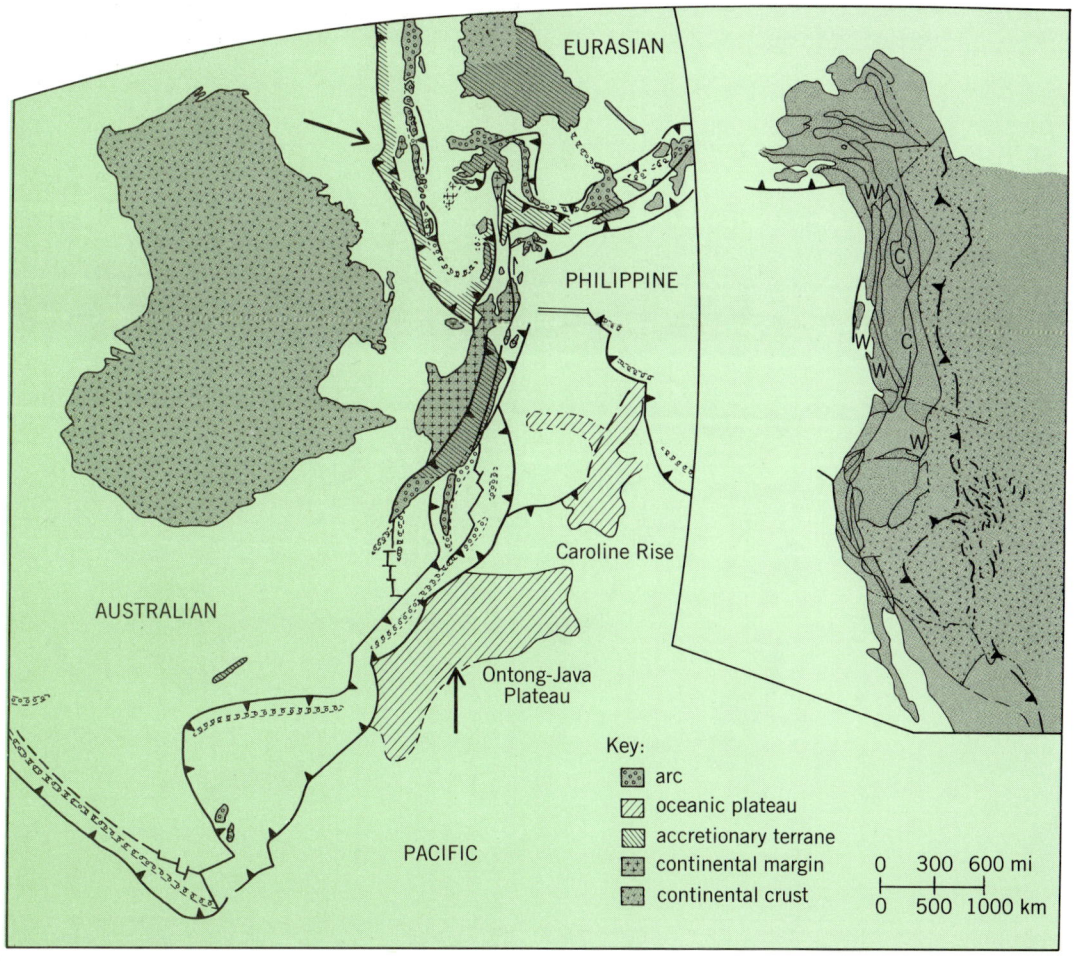

Fig. 3. Modern transpressive margin of the Indonesian archipelago with Australian, Pacific, Philippine, and Eurasian plates designated, compared at the same scale with suspect terranes of western North America. W stands for Wrangellia, and C for the Cache Creek terrane. *(After E. A. Silver and R. B. Smith, Comparison of terrane accretion in modern Southeast Asia and the Mesozoic North American Cordillera, Geology, 11:195–202, 1983)*

Otago, South Island, New Zealand, along the Alpine Fault; on the west, highly metamorphosed lower Paleozoic miogeoclinal strata intruded by mid Paleozoic to late Mesozoic mafic to felsic plutons, against Permian and lower Mesozoic volcanogenic arc and forearc strata of the Brook Street—Murihiku composite terrane, and amalgamated, roughly contemporaneous quartzofeldspathic strata of the feebly recrystallized Torlesse terrane on the east; and (3) Siluro-Devonian and younger granites, schists, and included carbonate blocks of the serpentinite-rich Kurosegawa tectonic zone, Shikoku, southwestern Japan, emplaced adjacent to the more northerly high-pressure, low-temperature Sanbagawa schists of Permian and Mesozoic age. In these and many other cases, the juxtaposition of deep-seated intrusive igneous complexes and weakly recrystallized country rocks must have occurred subsequent to thermal relaxation of the former, for the low-grade metamorphics which formed in a very different geologic environment show no evidence of high-temperature overprinting.

Long-distance migration is also suggested by the bringing together of tropical and temperate water fauna represented by fossil assemblages in now-adjacent terranes. Although offsets based on paleontological assemblages are equivocal in many instances, a persuasive case has been made for the northward translation of Permian strata containing verbeekinid fusilinids relative to schwagerinids in the westernmost coterminous United States, British Columbia, and southeastern Alaska. The former represent a cosmopolitan tropical fauna of Eurasian, or Tethyan, affinities, whereas the latter apparently were confined to temperate North American waters. Such faunal contrasts now mark much of this ancient continental margin, with contemporaneous inboard strata generally carrying schwagerinids, and the outboard, allochthonous Cache Creek–type terranes, verbeekinids. The interpretation of large-scale northward drift of the verbeekinid-bearing tectonostratigraphic terranes seems reasonable.

Perhaps the most compelling evidence concerning the disturbed and far-traveled nature of exotic terranes is based on paleomagnetic data. For western segments of the North American continent, where remnant magnetism has been investigated, large clockwise rotations of the host rocks are indicated since their times of formation. Even more significant is the fact that for many terranes, paleomagnetic latitudes measured for such rocks are consistently low. Wrangellia (Fig. 3), for instance, is thought to have drifted northward a minimum of 1800 mi (3000 km) since the Triassic. A few microterrane specimens have even been shown to possess remnant magnetic fields appropriate to near-equatorial or Southern

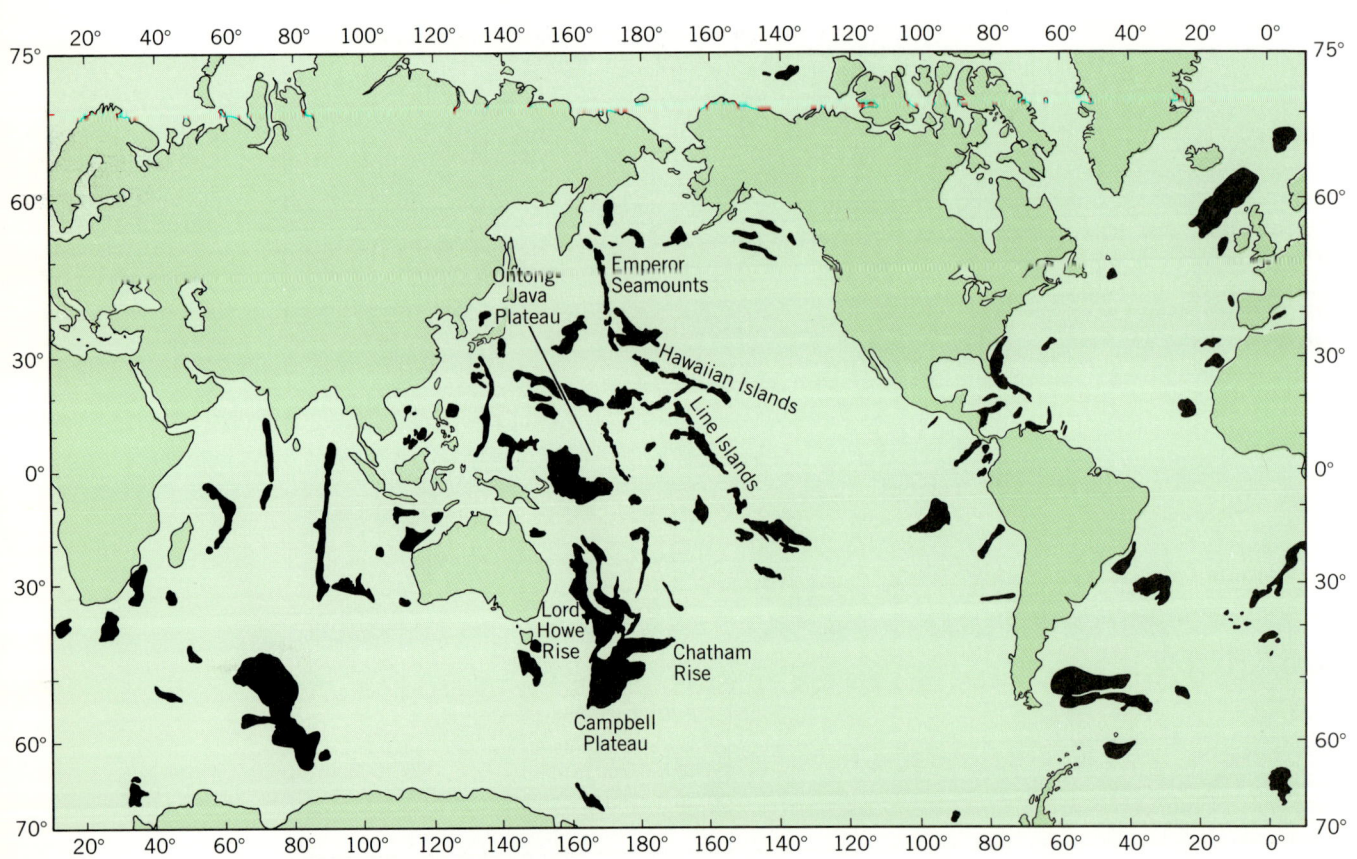

Fig. 4. Present-day distribution of oceanic plateaus (in black), both of continental (sialic) and oceanic (simatic) affinities. (After A. Nur and Z. Ben-Avraham, Oceanic plateaus, the fragmentation of continents and mountain building, J. Geophys. Res., 87:3644–3662, 1982)

Hemispheric sites of origin.

The aggregate weight of geologic, paleontologic, and geophysical evidence thus suggests that certain terranes have migrated considerable distances prior to their incorporation in a Circumpacific continental margin. Clearly, accretion tectonics is an important aspect of crustal evolution.

Modern examples of allochthonous terranes. Because of the differential motions of major lithospheric plates, segments of continental and oceanic crust on any particular plate are moving with respect to crust on each of the other plates. Allochthonicity is a fundamental aspect of all classical plate tectonic models; hence the terrane concept should not be considered as a fundamentally new idea. What is novel is the fuller realization that many unrelated portions of continental crust have been brought together by sea-floor spreading, subduction, and transform fault mechanisms, either acting singly or in combination. Three modern examples of obviously far-traveled terranes are presented below: the first from a midplate regime, the second from a transform shear system, and the third from a combined convergent and strike slip (= transpressive) plate junction.

Oceanic plateaus of Pacific Basin. Except in the vicinity of spreading systems, oceanic crust is overlain by about 3 mi (5 km) of sea water; the crust itself is another 3 mi (5 km) thick. Within each ocean, and especially the Pacific, the topography is relieved by a number of submarine plateaus of various dimensions scattered about the basin. These blocks are characterized by depths to the Mohorovičić discontinuity of approximately 9–15 mi (15–25 km) or more. Some, such as the Lord Howe and Chatham rises, the Campbell Plateau, and the Ontong-Java Plateau, appear to constitute microcontinents, and doubtless consist at least in part of sialic material. Others, for example, the Line Islands, and the Hawaiian Island–Emperor Seamount chain, represent tholeiitic basalt complexes which record the passage of the Pacific plate over fixed mantle hot spots. Still others are defunct spreading centers and isolated, intraplate volcanic islands. Many of the oceanic plateaus illustrated in Fig. 4 are of unknown crustal type.

What is clear is that, providing the present relative plate motions continue, most of these topographic anomalies will arrive at a transform- or convergent-type plate junction in the near geologic future. Depending on the local geometry, some of the oceanic prominences will be sheared off and incorporated within the subduction or transpression zone. Certainly the larger masses of sialic material will be tectonically decoupled from the descending slab because of their buoyancy and sutured against the stable, nonsubducted plate.

Salinian block of coastal California. Cenozoic dextral strike slip of great magnitude along the San Andreas Fault has been known since 1953. This break represents but one strand of a major, anastomosing, right-lateral shear system. The deformational zone has resulted from encroachment of the North American continental crust–capped plate toward and over the East Pacific Rise through consumption of the Cocos-Farallon plate, and the consequent change in motion in the overriding continental crust relative to the Pacific plate.

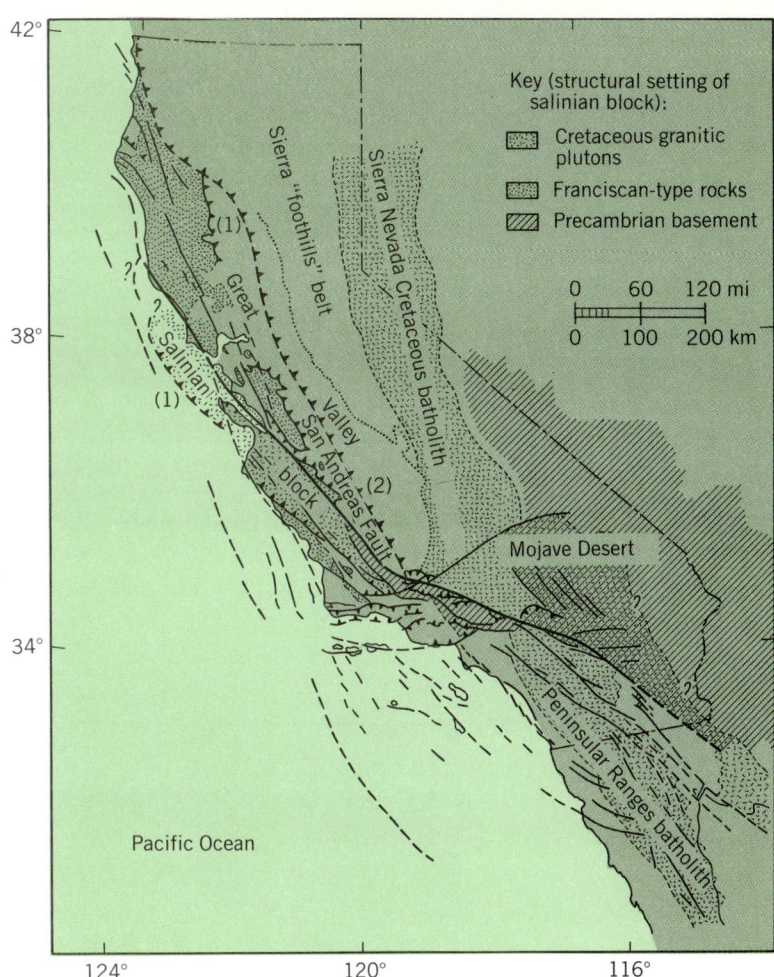

Fig. 5. Offset basement terranes of western California. Thrust faults are shown with barbs on the upper plate. The two most important are the fault contact between the Franciscan trench complex and the Great Valley forearc basin deposits; and the tectonic contact between Sierran and Franciscan basement. Other features shown are the San Andreas Fault (SAF) and Peninsular Ranges batholith (Pen. Ra. bath) (*After B. M. Page, Migration of Salinian composite block, California, and disappearance of fragments, Amer. J. Sci., 282:1694-1734, 1982*)

Because the San Andreas transects the western margin of North America at a low angle, northward drift has duplicated the lithotectonic belts which had been previously constructed during late Mesozoic subduction. As shown in Fig. 5, sialic crust west of the San Andreas has moved northward at least 300 mi (500 km) relative to the eastern side because, for this distance, Late Cretaceous Salinian granitic plutons from the originally inboard volcanic-plutonic arc are juxtaposed on the west against the initially outboard high-pressure, low-temperature Franciscan trench complex. Although of latest Jurassic and principally Cretaceous age, the Franciscan exhibits no indication of thermal upgrading which would be expected if these contrasting terranes had formed side by side. The total amount of northward drift of the Salinian block, at a minimum, 300 mi (500 km), is yet debated; geologic and paleomagnetic evidence indicates that the total northward displacement may be as great as 1500–3500 mi (2500–5600 km).

Indonesian Archipelago. Indonesia represents an intricate collage of convergent and conservative plate boundaries, the crustal expressions of which consist of backarc basins, island arcs, forearc basins, and trenches, involving continental and oceanic crust–capped plates. This transpressive junction complex defines the collisional boundaries of the Australian, Eurasian, Philippine, and Pacific plates.

As shown in Fig. 3, the mosaic of crustal fragments present here include truly far-traveled exotics such as the Caroline Rise, the Ontong-Java Plateau, and numerous scraps of oceanic crust (ophiolite) sequestered within highly tectonized suture zones.

Moreover, the northward-migrating Australian continental crust is in the process of colliding with Asia, due to the subduction of intervening oceanic lithosphere. Much of Indonesia, however, consists of compressed and laterally translated island-arc materials formed during oblique subduction. That most such sialic terranes are not truly exotic, and thus unrelated, is indicated by their systematic spatial relationships, as manifested by telescoped lithotectonic assemblages in the sequence trench, forearc basin, volcanic arc, and backarc basin.

Inferred ancient allochthonous terranes and accretion tectonics. The recognition of far-traveled terranes in the geologic record is more difficult compared to the examination of modern mobile belts. The problem is especially severe in tracts of pre-Jurassic continental crust, because the oceanic crust and spreading systems which could provide constraints regarding lithospheric plate dynamics are totally lacking. Post-Paleozoic mountain belts are

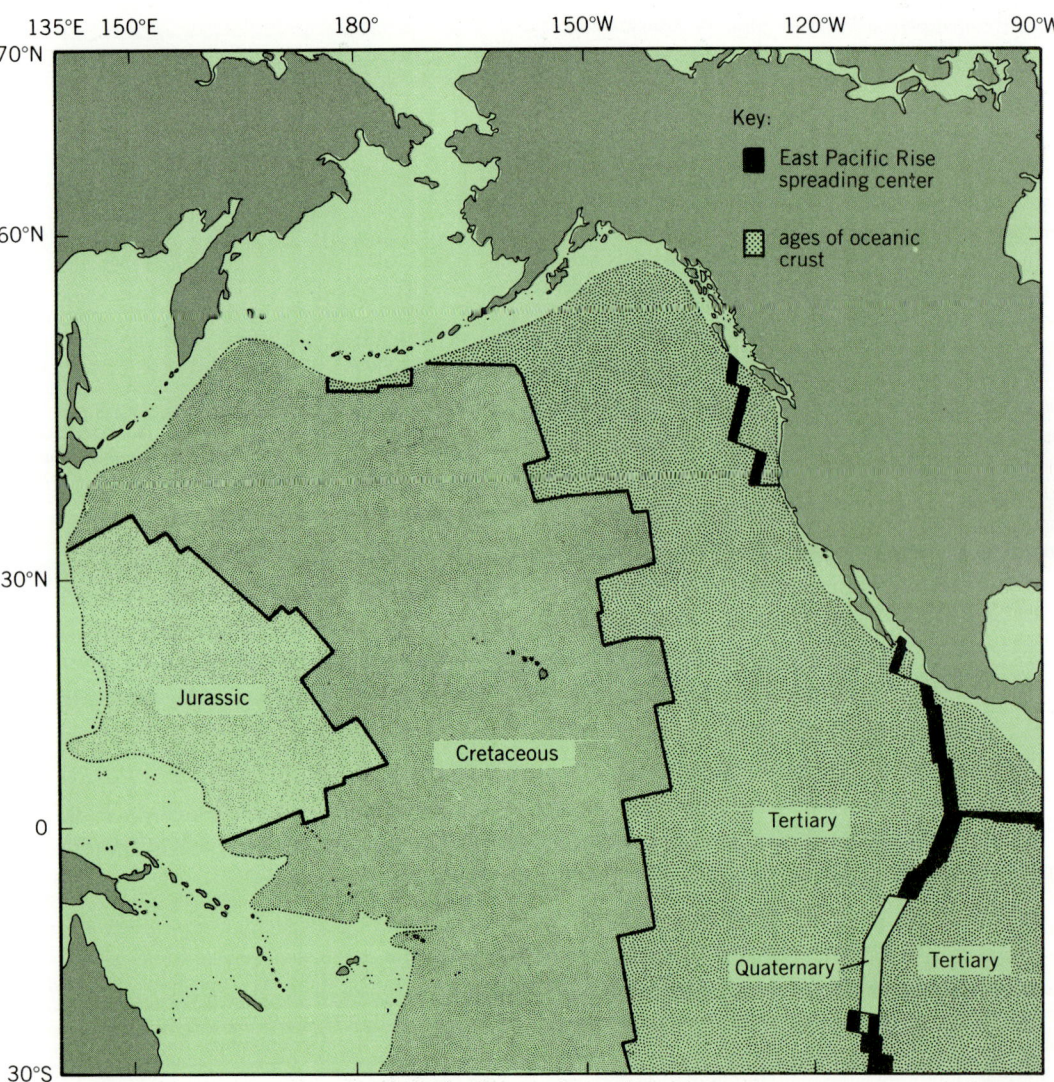

Fig. 6. Consumption of eastern limbs of paleo-Pacific oceanic crust due to the overriding by the North American plate. (*After W. G. Ernst, California blueschists, subduction, and the significance of tectonostratigraphic terranes, Geology, 12:436–440, 1984*)

the most readily understood, and three relatively unambiguous examples are presented below.

Western North America. A comparison of the accretionary margin of western North America and the Indonesian archipelago is presented in Fig. 3. These regions exhibit important similarities. Certain western cordilleran lithotectonic belts, such as Cache Creek and Wrangellia, clearly are far-traveled, whereas many others are suspect. Yet other tracts consist of calc-alkaline volcanic-plutonic belts such as the Idaho, Sierra Nevada, and Peninsular Ranges batholiths which evidently were produced in place predominantly by lithospheric underflow. The entire section has been assembled during later Phanerozoic time, seaward of the stable North American craton. Accretion tectonics is the term applied to the growth of this sector of the Circumpacific but, similarly to Indonesia, this process obviously involved the generation of autochthonous or parautochthonous lithotectonic belts at the Pacific margin of North America as well as the addition of exotic sialic microcontinental blocks, and tectonic slices of oceanic terranes. As illustrated in Fig. 6, approximately 6000 mi (10,000 km) of oceanic crust–capped lithosphere has descended beneath the westward-encroaching North American continent since Jurassic time, whereas a few thousands of kilometers of northward drift of the various paleo-Pacific plates occurred during this interval. Obviously subduction was an important factor in the growth of this continental margin.

South Island, New Zealand. On a much smaller scale, South Island, New Zealand, displays the effects of accretionary tectonics prior to its Late Cretaceous rifting and northeastward drift away from the Antarctic-Australian portion of the Gondwana supercontinent. Nine contrasting lithotectonic entities are shown in Fig. 7, some of which are almost surely exotic. For instance, the Permo-Triassic Brook Street–Murihiku terrane consists of a landward calc-alkaline volcanic and plutonic arc and a genetically related coeval, seaward forearc basin, the Southland syncline. On its Pacific side, this belt is bordered by the roughly contemporaneous Dun Mountain–Maitai terrane, which consists of an ophiolitic tectonic melange overlain by deep-water flysch sediments. Yet further outboard lies the Permo-Triassic Caples volcanogenic graywacke terrane.

Although these three complexes, the mutual contacts of which are well-described fault zones, conceivably could be related to one another, and juxtaposed intimately by differential slip along a transpressive plate junction, the Pacificward Torlesse belt of quartzofeldspathic sediments appears to represent a voluminous Permian and Mesozoic deposit of unknown but possibly distant provenance. Curiously, its contact with the Caples terrane is an indistinct and enigmatic suture hidden within a regionally metamorphosed tract known as the Haast Schist belt.

Taiwan. This sector of the Asiatic margin marks the impingement of Eurasian and Philippine Sea lithospheric plates, and records interactions involving

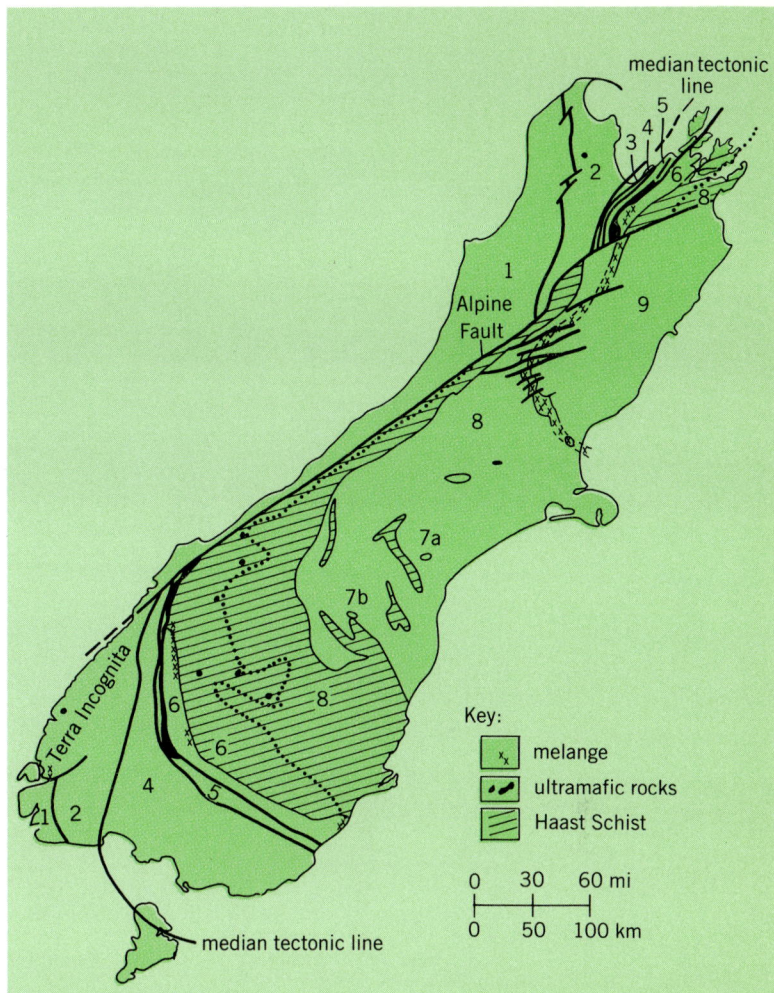

Fig. 7. Provisional terrane map of South Island, New Zealand. Terranes are indicated by numbers as follows: (1) Buller; (2) Golden Bay; (3) Drumduan; (4) Brook Street–Murihiku; (5) Dun Mountain; (6) Caples; (7a) Kakahu; (7b) Akatarawa; (8) older Torlesse; and (9) younger Torlesse. Dotted line locates the cryptic suture between Caples and older Torlesse terranes. (*After D. G. Bishop et al., Provisional terrane map of South Island, New Zealand, Proceedings of the Circum-Pacific Terrane Conference, Stanford University, pp. 24–31, 1983*)

rifting, drifting, and tectonic accretion since approximately mid Mesozoic time. The major lithotectonic belts are illustrated in Fig. 8. Mesozoic basement consists of the landward, relatively high-temperature low-pressure, granite-invaded Tailuko calc-alkaline belt, and a seaward, apparently coeval high-pressure low-temperature argillite and serpentinite Yuli blueschist belt. These two lithotectonic entities, juxtaposed along a major fault, contain tectonic lenses of ophiolitic materials, now largely recrystallized to greenstones and amphibolites; the ophiolites are probably far-traveled paleo-Pacific oceanic crust which has been tectonically inserted into the surrounding lithologies by convergent plate motion. The entire Yuli belt is suspect and could be allochthonous inasmuch as its source area has not yet been recognized.

The Mesozoic basement is overlain by a cover series of Cenozoic slates. Descent of the Asiatic margin along an east-dipping subduction zone and col-

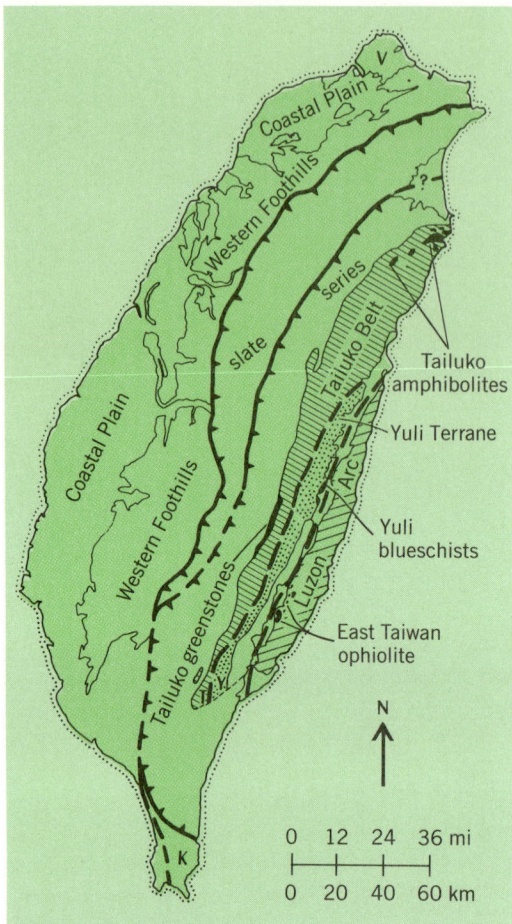

Fig. 8. Major lithotectonic belts of Taiwan slightly modified. Tectonically accreted portions of the island include the exotic Tailuko ophiolites (greenstones, serpentinites, and amphibolites), the calc-alkaline Luzon Arc (= Coastal Range), and olistostromal debris representing the East Taiwan ophiolite. Of possible allochthonous nature is the Yuli suspect terrane. (*After W. G. Ernst, C. S. Ho, and J. G. Liou, Rifting, drifting, and crustal accretion in the Taiwan sector of the Asiatic continental margin, American Association of Petroleum Geologists, publication pending*)

lision with the approaching andesitic Coastal Range terrane (the Luzon Arc) in Plio-Pleistocene time led to the continentward thrusting of the cover series; hence it is now parautochthonous relative to the underlying basement. Volcanogenic units of the outboard, exotic Coastal Range grade laterally and upward into an olistostromal unit which contains pebbles and slide blocks of the East Taiwan ophiolite; these latter are remnants of the intervening Eurasian oceanic crust which was consumed prior to impaction of the continental margin with the far-traveled Coastal Range. Thus Taiwan, although chiefly formed in place or nearly so, contains abundant lithologic units which reflect the tectonic accretion of exotic terranes.

True continental growth. Accretion tectonics is the process whereby allochthonous terranes are brought to and sequestered at an evolving continental margin. Accretion, of course, means growth; hence continental growth obviously can occur through accretion tectonics. Allochthonous terranes must possess a source, however, and inasmuch as the migration of far-traveled crustal blocks does not actually generate new continental material, accretion along one continental margin through this mechanism requires the spallation or calving of terranes from another. Continents are dominantly sialic, in contrast to the simatic ocean basins; hence the volumetrically insignificant addition of ophiolitic slices to continental margins through the process of accretion tectonics results in only minor crustal growth of the continents.

Real increase in the mass of continental material, as opposed to reshuffling of preexisting sialic fragments, involves the production of new sialic crust, as is being generated today in island arcs and continental margins sited above subduction zones. This growth of sialic crust and of continents themselves is chiefly a result of petrologic processes accompanying convergent plate motions. The recycling of old crustal materials as well as the rifting, drifting, and suturing of exotic terranes complicates this picture and results in local regimes dominated by accretion tectonics, and others by the disjunction of preexisting sialic crust. True growth through geologic time, however, appears to be caused chiefly by subduction.

[W. G. ERNST]

Bibliography: P. J. Coney, D. L. Jones, and J. W. H. Monger, Cordilleran suspect terranes, *Nature*, 288:329–333, 1980; M. Hashimoto and S. Uyeda (eds.) *Accretion Tectonics in the Circumpacific Region*, 1983; J. B. Saleeby, Accretionary tectonics of the North America Cordilleran, *Annu. Rev. Earth Planet. Sci.*, 15:45–73, 1983; E. R. Schermer, D. G. Howell, and D. L. Jones, The origin of allochthonous terranes: Perspectives on the growth and shaping of continents, *Annu. Rev. Earth Planet. Sci.*, 12:107–131, 1984; E. A. Silver and R. B. Smith, Comparison of terrane accretion in modern Southeast Asia and the Mesozoic North American Cordillera, *Geology*, 11:198–202, 1983.

A-Z

Acarina

Of the arthropod decomposers in natural soil systems, mites in the suborder Oribatida are perhaps the most evolutionarily successful and ecologically significant. In recent years their role as rate regulators in energy flow and nutrient cycling, processes that are directly accomplished primarily by soil microorganisms, has been accepted by many workers, and their evolution is beginning to be examined more critically. Of particular interest to evolutionary studies are recent fossil and cytological discoveries, and new hypotheses on phylogenetic relationships. Some of these could start new investigations on the evolution of sex and parthenogenesis.

Slow early evolution. Over the course of the last half century, scattered studies on Old World remains of oribatid mites have pushed back the fossil record from the Tertiary, when many extant genera and even species already existed, to the Mesozoic, when some extant families and genera existed. In 1979 M. Hammer and J. Wallwork added biogeographic evidence of the great age of certain oribatid mites, which have poor dispersal ability, when they related present distributions of genera to the major timed events of continental history suggested by theories of plate tectonics. They suggested that one extant species, *Mucronothus nasalis*, has existed since before the breakup of the supercontinent Pangea some 200 million years ago (m.y.a.). Recently, W. Shear and coworkers presented preliminary results on investigations of the oldest known terrestrial animal fossils from North America. These fossils, which were from lenses of plant remains in Middle Devonian (about 380 m.y.a.) mudstone in southern New York, included an extraordinarily well-preserved specimen of the extant family Ctenacaridae (illustration *a*), one of several families in the earliest known derivative cohort of oribatid mites, the Palaeosomata. This fossil predates the previously oldest known oribatid mite remains by almost 200 m.y.

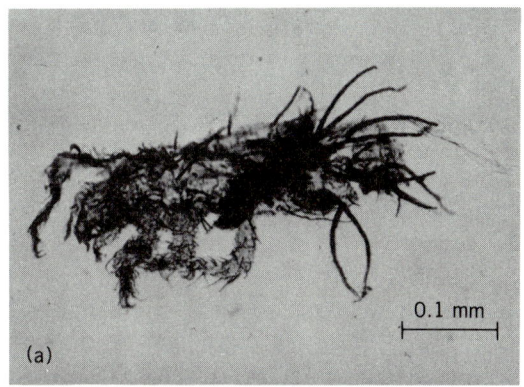

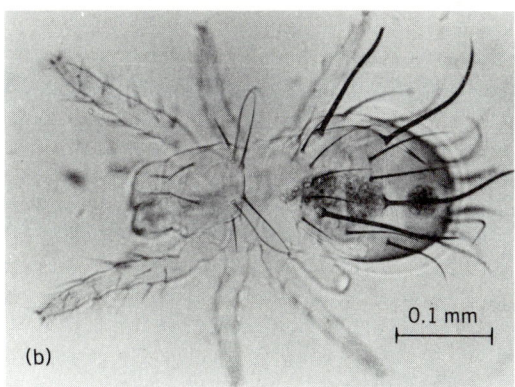

Comparison of a fossil and modern mite. (*a*) Oribatid mite (family Ctenacaridae) from Devonian mudstone at Gilboa, New York (*from W. Shear et al., Early land animals in North America: Evidence from Devonian age arthropods from Gilboa, New York, Science, 224:492–494, 1984*). (*b*) Ctenacarid mite from North Carolina (genus *Beklemishevia*) (*Courtesy of R. A. Norton*).

and indicates that these animals were already present when terrestrial floras, and the intricate association of plant roots, dead plant tissue, microorganisms, invertebrate animals, and mineral particles composing soil, began to develop. By extrapolating from the known feeding habits of extant ctenacarid mites (illustration b), which differ little from the Devonian specimen, the role of this fossil species in such early associations was that of a particle-feeding fungivore. As such, it was the only identified animal preserved in the material which was not a fluid-feeding predator.

Many later derivative oribatid mite groups have retained a similar feeding strategy, although some are now fungivores on the surfaces of trees or herbaceous plants, where microorganisms abound. Other groups developed the ability to tear off and ingest particles of the substrate on which most soil fungi grow, that is, the decomposing plant debris (leaves, wood). Growing competition for fungal hyphae and spores on the surfaces of plant remains may have influenced this development; other arthropods, such as the springtails (Collembola) and representatives of the mite suborder Prostigmata, are also known from terrestrial Devonian deposits and were probably also fungivorous. It is clear, however, that the ability to eat and utilize fungal tissue was not lost during the evolution of these groups; most extant oribatid mites are catholic feeders that consume both types of material.

Chromosomes and parthenogenesis. Important evolutionary questions can be raised by considering recent developments in the poorly studied field of oribatid mite cytogenetics. A team of researchers led by W. Helle has studied chromosome complements of several species in the most highly derived cohort, Circumdehiscentiae, by staining squashed eggs and examining mitotic divisions. All of the five species studied were haplo-diploid (haploid males, diploid females), with $n = 8$ or 9. This finding contrasts sharply with prior assumptions of chromosome complements and sex determination mechanisms in oribatid mites. Earlier Russian and French workers, by studying stained sections of gonadal tissue, had determined that a wide variety of oribatid mites were diplo-diploid, (mostly $2n = 18$). The sex determination mechanism of such a system is generally heterogametic, comparable to that of humans, but no differences between male and female chromosomes were noted in these studies. These early workers found no evidence of haplo-diploidy. The genera of Circumdehiscentiae with haploid males studied by Helle's group (*Oppia, Humerobates, Orthogalumna*) and those found by earlier workers to have diploid males (*Achipteria, Euzetes, Galumna, Xenillus, Poroliodes, Damaeus*) do not represent mutually exclusive evolutionary lineages. As yet, no single species has been studied with both egg and gonadal techniques, but assuming the data are correct, sex determination and male ploidy level in the Circumdehiscentiae seem to have evolved in some mosaic pattern unknown in other animal groups. There is still no evidence of haploid males in the older lineages of oribatid mites, but few have been studied, and none by the egg-squash method.

One problem encountered in working with many early derivative cohorts is that males are often rare or nonexistent in populations. Parthenogenesis, or the development of eggs without fertilization by males, is a widespread phenomenon in mites, and thelytokous parthenogenesis, in which unfertilized eggs develop into diploid females, is pervasive in most non-Circumdehiscentiae oribatid cohorts. On the basis of sex-ratio studies and laboratory cloning, some generalization can be made. The oldest cohort, the Palaeosomata, are mostly bisexual (only the family Palaeacaridae appears to be thelytokous); the cohort Enarthronota have a mixture of thelytokous and bisexual families; the Parhypochthonata are totally thelytokous; the Mixonomata are mixed, but predominantly thelytokous; and the Euptyctima are mixed according to family. The remaining cohort, the Nothronata, or Desmonomata, which comprise the cohort most closely related to the predominantly bisexual Circumdehiscentiae, are thelytokous with exception of the bisexual family Hermanniidae.

The predominance of thelytokous reproduction in so many major groups of oribatid mites poses an interesting evolutionary question. It has long been the consensus that thelytokous parthenogenesis is a reproductive strategy which constitutes an evolutionary dead end. However, the cohort Nothronata, exclusive of the bisexual Hermanniidae, have almost 30 genera and close to 400 known species, with probably several times the latter number yet to be described. It seems irrational to think that each extant species of such a large group is a terminal representative of a separate lineage which was previously bisexual, as the dead end theory would suggest. Thelytokous lineages supposedly cannot radiate in evolutionary time since genetic and phenotypic variability, the raw materials of evolution, are limited by the cloning process.

Another type of parthenogenesis which is common in many other mite groups but unknown in oribatid mites is arrhenotoky, in which unmated females lay unfertilized eggs producing haploid males. This phenomenon is especially characteristic of mite groups which often colonize new habitats, since it allows an unfertilized female to produce her own prospective mate. It has apparently evolved several times independently in mites with haplo-diploid sex determination, and its absence in oribatid mites is unexplained, except by their general lack of a colonizing habit. Those species which Helle's group found to be haplo-diploid are not known to lay eggs without first mating. Although no cytological evidence yet exists, it may be that fertilization is required to stimulate the development of male eggs, after which the paternal chromosome complement is extruded; this phenomenon, called parahaploidy, is known in other mite groups.

Relationships with Astigmata. Another recent development in the study of oribatid mite evolution has challenged the traditional limits on the group. In studying origins of the mite suborder Astigmata, B. OConnor has suggested these limits are too narrow. For the past quarter century, most workers have ranked the Astigmata as one of three suborders in the mite order Acariformes (along with the suborders Prostigmata and Oribatida). Perhaps the foremost biological pecularity of this large suborder (69 families, 785 genera) is that practically all members have facultative or obligatory relationships with other animals. They include significant agricultural pests, which may utilize insects or other invertebrate animals for phoretic transport; important parasites of humans and domestic and wild vertebrate animals; and still other large groups for which there is very little biological information, such as the feather mites of birds. In contrast to the opinions of most previous workers that the Astigmata are an early derivative lineage from the common ancestor of all Acariformes, OConnor contends that the Astigmata are in fact a highly derived group. The relatively simple body form and unsclerotized cuticle exhibited by the adults of most members are not ancestral conditions but neotenic ones, derived by maintaining in their own evolution not the characteristics of the adults of their ancestors but the characteristics of their immatures. Such neotenic trends have apparently been common in the mites as a whole and are probably the source of much of the so-called regressive evolution in the group. According to OConnor's hypothesis, the closest relatives of the Astigmata must be sought among the oribatid mites, and indeed among one of the more highly derived oribatid mite cohorts, such as the Nothronota or Mixonomata.

It has often been said that the oribatid mites are the only major mite group in which great diversity in form and species richness have evolved in the absence of parasitism or other associations with plants, insects, or vertebrates. If OConnor is correct, this is untrue; all major mite lineages have at one or more points in their evolution taken advantage of these rich resources. Current studies are being undertaken to identify more closely the Astigmata's closest relatives among the higher oribatid mites, but it seems likely that they will prove to be one of the groups which have predominantly thelytokous reproductive strategies. Like the Circumdehiscentiae, which have apparently also evolved from thelytokous ancestors, the Astigmata are mostly bisexual organisms and thelytoky, when it occurs, seems to be restricted to terminal branches of evolutionary lineages, as standard theory would suggest. Also, as in the Circumdehiscentiae, both diplo-diploidy and haplo-diploidy are known in the Astigmata, with distributions which at present seem rather mosaic. Unlike the Circumdehiscentiae, arrhenotokous parthenogenesis has evolved among the haplo-diploid groups, and possibly more than once.

For background information see ACARINA in the McGraw-Hill Encyclopedia of Science and Technology.

[ROY A. NORTON]

Bibliography: M. Hammer and J. Wallwork, A review of the world distribution of oribatid mites in relation to continental drift, Biol. Skrifter, Copenhagen, 22:1–31, 1979; W. Helle et al., Chromosome data on the Actinedida, Tarsonemida and Oribatida, in D. A. Griffiths and C. E. Bowman (eds.), Acarology VI, vol. 1, 1984; B. M. OConnor, Phylogenetic relationships among higher taxa in the Acariformes, with particular reference to the Astigmata, in D. A. Griffiths and C. E. Bowman (eds.), Acarology VI, vol. 1, 1984; W. Shear et al., Early land animals in North America: Evidence from Devonian age arthropods from Gilboa, New York, Science, 224:492–494, 1984.

Acid rain

Scientists, legislators, and a growing segment of the public have expressed concern that the deposition of acid precipitation falling as rain, fog, mist, snow, or sleet may have adverse consequences for human health, agricultural crops, water resources such as lakes, ponds, and streams, and forest ecosystems. Declines in forest growth, dieback of tree crowns, and other symptoms of forest tree stress and tree death have been recorded in certain sections of West Germany, Scandinavia, other European locations, Canada, and the United States. Whether acid precipitation is the direct or indirect cause of forest stress is a contemporary issue of intense scientific and public interest and debate. The central issue is whether acid precipitation influences the growth or health of forests or individual tree species, and whether it changes the species composition of forest communities or destroys certain tree species, associated plants, or animals over significant forest areas.

In the absence of air pollutants, the pH of precipitation is presumed to be dominated by carbonic acid formed from atmospheric carbon dioxide. In pure rain, this would produce a pH of approximately 5.6. However, large areas of the Northern Hemisphere are currently receiving precipitation which is more acidic (lower pH value) than the carbon dioxide equilibrium would predict. The pH of precipitation presently falling on forests in north and central Europe and in the northeastern United States and adjacent portions of Canada is commonly in the range 3.0–5.5; individual storm events have been recorded with pH values as low as 2.0. European areas receiving precipitation between pH 4.0 and 4.5 have increased considerably in the past few decades and now comprise substantial portions of northern and central regions. Precipitation falling in central New Hampshire in the early 1960s was found to have pH values between 4.0 and 4.5 and led to numerous determinations substantiating the widespread occurrence of pH 5.0–5.5 precipitation throughout the United States east of the Mississippi River, and pH 4.0–5.0 in certain northeastern locations, particularly the Adirondack region of New York. In western sections of the United States, scattered pockets of pH 4.0–5.0 precipitation have been recorded in Tucson, the Los Angeles basin, the San Francisco Bay area, Spokane, and the Willamette Valley–Portland area.

The scope of these areas affected by acid rain and the concern they have aroused make it important to explore the difficult and complicated questions they raise. A review of the most significant hypotheses advanced to describe the possible interaction between acid precipitation and forest systems comprises the rest of this article.

Forest variability. Forest ecosystems are extremely variable. Forests differ in soil type, climate, topography, species composition, age, and health; occur at different elevations above sea level; and are located on slopes facing in all directions. Trees in a given forest may be the same age or many different ages. Forests may be reproduced by seed, sprouting, or planting. Forests may be subject to various levels of management intensity; forest trees exist in a continuum of management activity ranging from unmanaged and managed natural forests, to plantation forests, to specialized forests (seed orchards, Christmas tree plantations, nurseries), to urban and suburban forests. The response of these different forests to influence by acid precipitation varies with forest type and location.

The development and health of forest ecosystems involves integration, over time (as trees are very long-lived), of all the stresses to which the ecosystems are exposed. These stresses are extremely variable and may be living (insects, fungi, bacteria, viruses), nonliving (lightning, drought, low temperature), natural (volcanic eruptions, hurricanes), and anthropogenic (fire, harvesting, air pollution). Documentation of acid precipitation damage and the separation of this damage from damage caused by these other factors is extremely difficult.

Air pollutants. The compounds that pollute the troposphere 6 to 12 mi (3.6 to 7.2 km) above the earth are numerous and chemically varied. Materials may be solid, liquid, or gas, and may be organic or inorganic. Some pollutants are released directly into the atmosphere, while others result from chemical reactions in the atmosphere (acid rain). Pollutants are transferred to water or land ecosystems in wet deposition (rain, fog, snow, sleet), in dry deposition (particles), and by direct gas exchange. Regional air pollutants, that is, those transported tens, hundreds, or even thousands of miles from their point of release, are judged to be of particular significance to the expansive natural forests of the temperate latitudes. Regional air pollutants of particular importance include photochemical oxidants, particularly ozone, and a variety of trace metals, particularly heavy metals, for example, lead. Ozone and trace metals are judged to be the most important regional-scale pollutants deposited in forests in the temperate zone.

Chemicals (primarily sulfur dioxide and nitrogen oxides) that lead to the formation of acid precipitation are released into the atmosphere in major industrial and population areas. Conversion of these oxides into sulfuric and nitric acids occurs in the atmosphere, followed by deposit to earth hundreds of miles from the point of release. Acid precipitation, therefore, qualifies as a regional pollutant. Whatever impact acid rain may have on forest ecosystems will be associated with the influence of other pollutants at a given location, since forests are exposed to multiple pollutants concurrently and sequentially. Interaction of influences of multiple pollutants may be extremely important.

Though adverse effects of acid precipitation on forest ecosystems have not been conclusively proven by existing evidence, it cannot be concluded that adverse effects are not occurring. Numerous legitimate hypotheses for adverse effects, worthy of scientific evaluation, have been proposed and are discussed below.

Forest soils. Natural processes that occur in the maturation of both forests and the soils in which they grow cause the soils to become extremely acidic (pH $<$ 5.0). This is particularly true in coniferous forests, or those dominated by coniferous species, growing in cool, moist climates. Since most forest soils in the eastern United States are acidic, they should not be sensitive to further reduction of the pH of their upper horizons by acid deposition. Forest soils potentially subject to acidification are judged rare and would be characterized by a pH above 5.5 and have low buffering capacity.

Acid deposition may cause net increases, no change, or net decreases in the nutrient (elements required for plant growth) content of soils. Deposition of nitric acid may add nitrogen to soils, while deposition of sulfuric acid may contribute sulfur. Accumulation or uptake of atmospherically deposited nitrogen would probably be accompanied by a tree growth increase in most cases. Excess deposition of nitric acid, however, could "saturate" forest soils with nitrate and cause an increase in the loss (leaching) of other positively charged nutrient elements (cations). Since the forest sulfur requirement is only about 7% of that for nitrogen, and since atmospheric sulfur inputs usually exceed those of nitrogen, forests in most polluted regions are past the point of receiving nutritional benefits for sulfur input. As with nitrogen, however, sulfur deposition in excess of the forest ecosystem capacity to retain it causes accelerated nutrient leaching from the soil. Many forest soils, however, have a high capacity for sulfate retention and are therefore protected from this leaching loss.

As the pH in the surface horizon of forest soil drops below pH 5.0 during natural soil formation, aluminum and other metals begin to be solubilized and leached into soil solutions and transported to deeper horizons of higher pH and clay content where they precipitate. Extremely acid soils, which are low in positively charged nutrient elements but high in hydrogen ion (H^+) and aluminum ion (Al^{3+}), may experience increased Al^{3+} leaching in response to increased sulfate and nitrate leaching. Possibilities for mobilization of toxic trace metals such as lead or cadmium also exist, if soils are subject to sufficiently high acid inputs. In addition to the possibility that increased soil aluminum or increased trace metals could be toxic to tree roots or soil organisms, these metals may be transported to contaminate surface water (streams, ponds, lakes) or groundwater (wells, aquifers).

Forest trees. Numerous hypotheses have been advanced to explain potential mechanisms of adverse acid precipitation effect on forest trees. Forest canopies (leaves and twigs) present a large collecting surface to the atmosphere and can "accumulate" large amounts of rain and fog. In high-elevation forests in the northeastern United States, eastern Canada, and many locations in Europe, forest leaves are subject to frequent fog events of particularly low (acid) pH. Wax layers that cover leaves may be eroded by acid precipitation. Nutrient elements and organic chemicals may be leached from leaf tissue during precipitation events. Tree roots, even though belowground, may be injured by soluble aluminum, trace metals such as lead, cadmium, or nickel, or other chemical or physical changes in the soil around roots. Beneficial root associations with specialized soil fungi and bacteria may be disturbed or eliminated. Acid precipitation may also influence reproductive processes of trees including flower production, pollination, seed production, seed germination, and seedling survival.

In addition to these possible direct effects of acid precipitation on trees, there is also a very important indirect potential interaction. Acid rain may alter the ability of insects or microorganisms (fungi, bacteria) to influence the health of trees. If this alteration increases insect or microbial activity, greater tree injury or disease may result.

Conclusion. Presently, tree mortality (death) and morbidity (disease) and growth rate reductions in European and North American forests occur where regional air pollution, including acid deposition, is generally high. Numerous separate hypotheses for adverse forest effects from acid deposition, worthy of testing, have been proposed. However, it may be inappropriate to consider the effects of any regional pollutant on forests in isolation since under natural conditions forest ecosystems are exposed to multiple air pollutants simultaneously or sequentially, and interactive and accumulative influences are important. The growth reductions and decline symptoms of the forests of West Germany are dramatic and should warn all nations that the resiliency of forest ecosystems has limitations. Until the cause of this decline is more clearly understood, natural resource science can neither reject nor indict any single cause, but extreme caution must be exercised to protect invaluable forest resources.

For background information see AIR POLLUTION; FOREST ECOLOGY; PH in the McGraw-Hill Encyclopedia of Science and Technology.

[WILLIAM H. SMITH]

Bibliography: W. H. Smith, *Air Pollution and Forests*, 1981; W. H. Smith, Ecosystem pathology: A new perspective for phytopathology, *For. Ecol. Manag.*, 9(3):193–214, 1984; Society of American Foresters, *Acid Deposition and Forest Ecosystems*, Report of Task Force on the Effects of Acid Deposition on Forest Ecosystems, 1984; U.S. Environmental Protection Agency, The *Acidic Deposition Phenomenon and Its Effects*, Critical Assessment Review Papers, vol. 1: *Atmospheric Sciences*, Publ. EPA-600-8-83-016AF, vol. 2: *Effects Sciences*, Publ. EPA-600-8-83-016BF, 1984.

Acoustic microscope

The acoustic microscope uses very short-wavelength sound waves, focused in a liquid, to image samples. Short wavelengths are desired because the resolution of the microscope improves with shorter wavelengths. However, because sound attenuation increases with shorter wavelengths (higher frequencies), it becomes impossible to operate a very high-resolution microscope with conventional liquids. Fortunately, there is one liquid which possesses almost no attenuation of sound: superfluid helium at temperatures less than 0.2 K (0.4°F above absolute zero). By using this fluid, an acoustic microscope has been constructed which uses wavelengths as short as 30 nanometers, 20 times shorter than red light. The corresponding resolution of the microscope is 20 nm, which is approximately 10 times better than the capabilities of very fine optical microscopes and is beginning to compare with the resolution of electron microscopes.

Principles of operation. The heart of the acoustic microscope is the acoustic lens, which is typically made from sapphire (aluminum oxide, Al_2O_3). A spherical depression is polished at one end of the sapphire, and the other end is polished flat. A thin-film zinc oxide (ZnO) transducer is deposited on the flat end. When the transducer is excited with microwaves, acoustic plane waves are generated in the sapphire, and propagate to the spherical depression and into a liquid. Since liquid sound speeds are typically much less than sound speeds in solids, the acoustic plane waves refract across the spherical interface and come to a tight focus in the liquid. If a sample is placed near the focus, part of the acoustic radiation will be reflected from the surface and return to the spherical depression. The acoustic signal will again become plane-wave radiation, and the "echo" can be detected by the thin-film transducer. To form an image, the microwaves are repeatedly pulsed while the lens is mechanically scanned across the sample. The echo intensity is recorded as a function of position, and the result is displayed on a cathode-ray tube.

Sound attenuation problem. The resolution of the acoustic microscope is determined by the focal spot size. The spot size is diffraction-limited, and the resolution is approximately equal to $F\lambda$, where F is the F-number of the lens (aperture diameter divided by focal length) and λ is the acoustic wavelength in the liquid. Since $\lambda = c/f$, where c is the sound speed and f is the frequency, the resolution is improved by using liquids with low sound speeds and by increasing the frequency. A fundamental limitation occurs, however, when the sound attenuation in the liquid path becomes so large that the signal-to-noise ratio is unacceptable. In most fluids the acoustic loss is given by Eq. (1), where ρ is the

$$\alpha = \frac{2\pi^2}{\rho c}\left[\frac{4}{3}\eta + \eta' + K\left(\frac{1}{c_v} - \frac{1}{c_p}\right)\right]f^2 \quad (1)$$

mass density, c is the sound speed, η is the shear viscosity coefficient, η' is the dilatational viscosity coefficient, K is the thermal conductivity coefficient, and c_v and c_p are the specific heats at constant volume and pressure, respectively. Thus, increasing the frequency to obtain shorter wavelengths and higher resolution quickly increases the attenuation. Because of this attenuation, operation of the acoustic microscope in normal liquids (typically water) is extremely difficult when the wavelength is much below optical wavelengths (approximately 500 nm).

Superfluid helium at temperatures less than 0.2 K (0.4°F above absolute zero), however, has negligible attenuation of microwave-frequency sound. The dominant form of attenuation for low temperatures is given by Eq. (2), where f is the frequency and T is

$$\alpha = (4.88 \times 10^{-3})fT^4 \quad \text{dB/cm} \quad (2)$$

the temperature in kelvins. It is obvious that the sound attenuation of superfluid helium given in Eq. (2) is vastly different from that of normal fluids given in Eq. (1). The attenuation in the superfluid is due to the scattering of the sound waves (phonons) by the thermal vibrations (thermal phonons) in the fluid. As the temperature is lowered, the energy and number of thermal phonons decreases, and the attenuation disappears as given by Eq. (2).

Cooling of liquid helium. To use the short acoustic wavelengths desired for high-resolution microscopy, Eq. (2) demands that the acoustic microscope be cooled down to extremely low temperatures, typically 0.1 K (0.2°F above absolute zero). Helium-4 liquefies at 4.2 K (−452°F) at its saturated vapor pressure. The simplest method of cooling below this temperature is to pump on the liquid and cool by evaporation. Below temperatures of about 1.0 K (1.8°F above absolute zero), however, the vapor pressure of the helium is nearly zero, and the cooling power becomes negligible. An improvement can be made by using the isotope helium-3, but the lowest achievable temperature by this method is approximately 0.3 K (0.5°F above absolute zero). To achieve reasonable cooling powers at even lower temperatures, a dilution refrigerator is required.

The first operational dilution refrigerator was made in 1965. The refrigerator relies on the finite solubility (6.4%) of ^{3}He in ^{4}He at temperatures approaching zero. This finite solubility gives rise to a nondiminishing, effective vapor pressure of ^{3}He in a ^{3}He-^{4}He solution. The effect allows cooling by pumping on the ^{3}He through the ^{4}He (instead of pumping through a vacuum as in a normal evaporation refrigerator).

Imaging. An example of an acoustic image taken in superfluid helium at approximately 0.07 K (0.13°F above absolute zero) is shown in Fig. 1. The sound frequency is 4.2 GHz and the wavelength in helium is 57 nm. The lateral resolution is better than 50 nm. The object is a bipolar transistor on a silicon integrated circuit manufactured by TRW, Inc. Visible in the images are 2-micrometer-wide, 0.5-micrometer-thick aluminum lines making connections to the emitter, base, and collector of the transistor. The crystal grain boundaries of the aluminum can easily be seen.

Depth of focus. The superfluid helium microscope exhibits a very narrow depth of focus. At 4.2 GHz, the depth of focus is approximately 130 nm, as a result of the wide-angle acoustic lens and short wavelengths used. Because of the short focal depth and the considerable vertical relief on the integrated circuit, it is not possible to focus simultaneously on the silicon substrate and the top of the aluminum lines in Fig. 1. To present the entire integrated circuit in a single picture, three separate images are taken at three different focal positions. The images are digitally stored and superimposed electronically to form the composite micrograph shown in Fig. 1.

Contrast. By looking carefully around the base region of the transistor, a dark ring can be seen encircling the base and emitter connections of the device. This ring is the border of the ion-implanted region for the base of the transistor. This feature had not been previously seen by the manufacturer using other forms of imaging. There is a slight surface displacement in the implanted region, and the acoustic microscope's sensitivity of topography provides the contrast needed to image the border.

The contrast in the superfluid helium acoustic microscope arises primarily from topographical features on the surface of the sample. Because of the large acoustic impedance mismatch between helium and virtually any solid, sound incident from the liquid has almost a unity reflection at the surface. The transmitted fraction of sound normally incident from helium into a sample is given by Eq. (3), where Z_{He} and Z_{solid} are the acoustic impedances of the helium

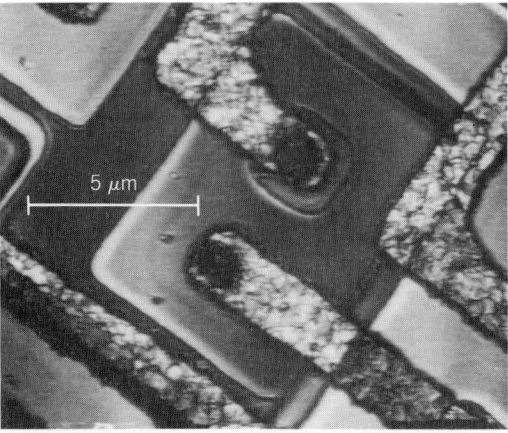

Fig. 1. Acoustic micrograph of a bipolar integrated circuit.

$$T \approx \frac{4Z_{\text{He}}}{Z_{\text{solid}}} \qquad (3)$$

and solid, respectively. The acoustic impedance Z is equal to ρc, where ρ is the mass density and c is the sound speed. Therein lies the problem with transmitting sound in and out of helium; it is very light ($\rho = 0.145$ g/cm^3 or 0.145 times the density of water) and the sound speed is very slow (238 m/s or 780 ft/s). For example, at a helium-sapphire interface, the transmitted acoustic signal is 25 dB down from the incident signal. (For transmitting and receiving sound through the acoustic lens interface, a quarter-wave matching layer is used to enhance transmission.)

Although the previous arguments demonstrate that the contrast comes from the surface features of the sample, there are several mechanisms which can provide contrast. First, the short depth of focus of the microscope makes it possible to differentiate small height changes on the sample. This feature alone makes the acoustic microscope a complementary tool to electron microscopes, which have a long depth of focus. Second, the acoustic microscope is sensitive to tilted surfaces which specularly reflect all or part of the focused beam so that it is not received. Finally, because the imaging uses coherent radiation, interference effects can be seen with the acoustic microscope, allowing perception of changes in sample height as small as 2 nm.

Capability. Figure 2 demonstrates the current capability of the acoustic microscope. The sound frequency is 8.0 GHz and the corresponding wavelength in helium is 30 nm. The lateral resolution is 20 nm. The sample is a *Myxococcus* bacterium. This image is a composite of three images taken at three different focal planes, just as in Fig. 1. While this bacterium would normally be imaged in a transmission electron microscope and would have to be stained with heavy atoms to be visualized, such sample preparation is not necessary for acoustic microscopy. The contrast seen in acoustic images of these bacteria appears to be far better than for the transmission electron microscope images.

Ultimate limit of resolution. At frequencies above 20–30 GHz, Eq. (2) for the attenuation of sound in helium no longer represents the dominant form of attenuation. At the higher frequencies, the scattering of sound from the zero-point motion of the helium (quantum noise) is dominant. However, this form of attenuation can be eliminated by pressurizing the helium to more than 20 atm (2 megabars) and maintaining low temperatures. Under these conditions, acoustic radiation with wavelengths as short as 1 nm can propagate long distances without loss.

The generation and detection of acoustic waves with these very short wavelengths will provide a technological challenge. The frequencies correspond to 100–200 GHz (millimeter waves). Fortunately, much of the equipment necessary for those high frequencies is being developed presently for communications and radar use. Advances in these areas will directly aid any program to develop a 100–200-GHz acoustic microscope. The absence of apparent physical limits for sound wavelengths as short as 1 nm makes the "ultimate" acoustic microscope an enticing research area.

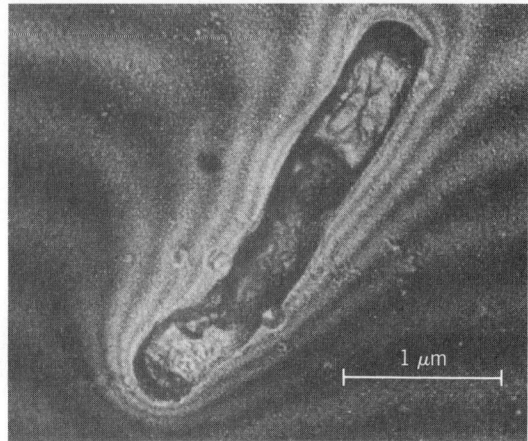

Fig. 2. Acoustic micrograph of a *Myxococcus* bacterium.

For background information *see* ACOUSTIC MICROSCOPE; LIQUID HELIUM; LOW-TEMPERATURE PHYSICS in the McGraw-Hill Encyclopedia of Science and Technology. [JOHN S. FOSTER]

Bibliography: J. S. Foster and D. Rugar, High resolution acoustic microscopy in superfluid helium, *Appl. Phys. Lett.*, 42(10):869–871, 1983; B. Hadimioglu and J. S. Foster, Advances in superfluid helium acoustic microscope, *J. Appl. Phys.*, in press.

Acoustic transients

Acoustics traditionally deals primarily with steady-state situations, that is, with sound waves characterized by discrete frequencies or by a random superposition of frequencies. Since the early 1970s, acoustic transients—particularly impulsive transients characterized by an abrupt rise time and short duration—have attracted the attention of researchers. Individual transients embody a continuous spectrum of frequencies, the pressure p being represented by an inverse Fourier transform over all frequencies as given by Eq. (1), where ω is the cir-

$$p(t) = \frac{1}{2\pi}\int_{-\infty}^{\infty} p(\omega)e^{-i\omega t}\,d\omega \qquad (1)$$

cular frequency and t is the time.

Recent studies of transients fall into two widely separated areas of acoustics: psychoacoustics, specifically the hearing loss experienced by workers in noisy industrial environments; and the use of sound waves as a diagnostic technique for determining the mechanical properties of elastic or fluid continua, and the location and nature of inhomogeneities. Research in these areas was stimulated, respectively, by two developments: the discovery that impulse noise is more damaging to the auditory system than steady noise of comparable level, a danger aggravated by the fact that a brief sound pulse is perceived subjectively as less loud than the equivalent steady sound; and the availability of computers and the development of the fast Fourier transform for rapidly extracting the frequency response of a linear system from its transient response to an impulsive excitation, thereby shifting the burden of measuring frequency response from multiple discrete-frequency measurements performed over the desired frequency range to signal processing of a single transient response. The practical importance of this shift derives from the fact that in the field, over the time interval required by multiple measurements, conditions gradually changing with time may contaminate the observed frequency dependence; furthermore, the time and hence the expense of measurements are drastically reduced by substituting a single transient measurement.

These two developments have in turn stimulated applied research. Awareness of the hearing loss produced by transients has generated interest in the physical processes underlying the radiation of sound pulses by industrial machinery with the purpose of formulating silencing techniques. The increasing use of sound pulses as a diagnostic tool has resulted in the development of various novel sound pulse generators.

Sound pulses as diagnostic tools. Provided the underlying theory is adequately developed, the frequency dependence of a system's response provides the basis for computing various characteristic parameters. The required frequency dependence is determined by taking the Fourier transform of the transient response. Recording, for example, the transient sound pressure $p(t)$, the frequency dependence is given by Eq. (2), where T is the duration of the

$$\tilde{p}(\omega) = \int_0^T p(t) e^{i\omega t} \, dt \qquad (2)$$

signal. Representative examples of the information obtained in this manner are: (1) the modal response of a bounded system, for example, an impact-excited bell or an explosively insonified reverberant room; (2) the dispersion curves of group velocities of an effectively infinite waveguide, for example, a layered geological formation insonified by borehole soundings; and (3) the echo time history obtained by echo-ranging a structure submerged in the ocean, or by performing tomography of a fetus in the womb, or by nondestructive testing of a metal casting for the purpose of detecting cracks.

The extraction of the frequency dependence of the system's response is based on the convolution theorem whereby the convolution integral given by Eq. (3) transforms into a product of Fourier transforms

$$f_1(t) = \int_{-\infty}^{\infty} f_2(t - t') f_3(t') \, dt' \qquad (3)$$

given by Eq. (4). Here f_3 is the forcing function,

$$\tilde{f}_1(\omega) = \tilde{f}_2(\omega) \tilde{f}_3(\omega) \qquad (4)$$

while f_2 is in the nature of a Green's function which embodies the system's properties. This desired information is therefore contained in the ratio $\tilde{f}_1/\tilde{f}_3$. In the case of explosive echo ranging, $\tilde{f}_1$ if the backscattered pressure measured near the source, $\tilde{f}_2$ is proportional to the square root of the scattering cross section, and $\tilde{f}_3$ is the incident pressure measured near the target. The scattering cross section A is computed as a function of frequency in the form of Eq. (5), where R is range. Another familiar example

$$A = 4\pi R^2 \, |\tilde{f}_1|^2 |\tilde{f}_3|^{-2} \qquad (5)$$

is a multimodal structure. In this case, $\tilde{f}_1$ is the vibratory velocity response, $\tilde{f}_2$ the mechanical admittance, and $\tilde{f}_3$ the exciting force. The mechanical impedance therefore is given by Eq. (6).

$$Z(\omega) = \tilde{f}_3/\tilde{f}_1 \qquad (6)$$

The forcing function $f_3(t)$ must display a spectrum encompassing the frequency range of interest. If this includes high frequencies, the excitation must display, by virtue of the initial value theorem, a sharp rise time. For example, excitation of the higher modes of a bell requires a blow with a metal hammer rather than a rubber mallet. If low frequencies are of interest, the signal processed must encompass the trailing end of the system response. For example, in echoranging, the entire echo must be captured before contamination by surface or bottom reflections sets in.

While explosive sources are still used, mechanically, electrically, and hydraulically actuated transducers have recently been developed because they produce a highly repeatable pulse. For underwater echo ranging, a representative hydraulic source consists of a piston, accelerated by hydrostatic pressure, whose motion is abruptly stopped by an anvil, thereby producing a condensation wave. A widely adaptable impulse generator consists of computer-controlled piezoelectric transducers energized by an electric signal adjusted to generate a specified pressure-time history which typically requires that the signal suppress ringing of the normal modes of the transiently energized crystals.

Hearing loss caused by sound pulses. The practical importance of this problem results from the large number of industrial workers—1,000,000 in the United States—regularly exposed mainly to impact noise. A valid criterion for estimating the cumulative hearing loss from relatively steady noise is the time integral of the acoustic energy reaching the ear. This procedure, which models hearing damage as a fatigue phenomenon, is not conservative when the incident noise embodies short sound pulses whose level lies substantially above the average noise level. The reason for the relatively greater damage inflicted by noise displaying a brief rise time is that the muscle contraction reflex in the middle ear, which shields auditory nerves from potentially harmful noise, requires 300–500 milliseconds, that is, a time interval orders of magnitude longer than the duration of pulses produced by representative metal-forming operations. This inadequacy of the ear's protective mechanism parallels the familiar visual experience associated with the iris's inability to contract rapidly enough to shield the eye from a bright flash, as opposed to, for example, steady bright sunshine. The shortcoming of the ear's defense mechanism is compounded by the fact that the integration time of the auditory nervous system is 20–100 ms, thereby leading to a subjective underestimation of the intensity of transients shorter than 20 ms. A quantitative measure of the ear's vulnerability to transients is provided by comparing two noisy environments: In an aircraft, where the difference between peak noise levels and the dB(A) noise level is a modest 10 dB, the time-integrated sound energy has been shown to constitute a reliable criterion for predicting hearing loss as a function of years of exposure. [dB(A) is a sound pressure scale weighted to reflect the lesser sensitivity of the human ear at low frequencies.] In metal-working shops with punch presses, where a representative difference between peak and dB(A) noise levels is 20 dB, a comparable hearing loss occurs for dB(A) noise levels 10–20 dB lower than in aircraft. Permissible maximum dB(A) levels should be accordingly lower to provide metal workers with comparable protection.

Sound pulses are radiated by metal-forming operations whereby kinetic energy of a moving part, such as a flywheel, is abruptly transformed into work on the material being formed. Among the various transient-noise mechanisms embodied in these operations, the most important is generally the sudden acceleration or deceleration of a moving surface, a phenomenon analyzed for a translating sphere over a century ago by G. R. Kirchhoff. This pulse-generating mechanism is also present when a metal surface is rapidly deformed or even fractured. The abrupt accelertion of a boundary produces a condensation wave embodying precisely half the work performed on the acoustic medium by the moving boundary. The other half, stored as kinetic energy in the near field of the moving boundary, is transformed into sound energy associated with the rarefaction wave radiated when the moving boundary is abruptly decelerated. To reduce the amplitude of sound pulses, accelerations must be kept at the lowest level compatible with the function of the moving part.

A similar conclusion is reached with regard to transients radiated by air ejected from the gap between two planar surfaces moving toward each other. Another type of transient, namely, ringing of an impact-excited structure, once again displays a pressure amplitude proportional to peak acceleration. Consequently, metal-forming operations should be performed as slowly as practical so as to achieve an inherently less harmful acoustic environment. Other recent silencing efforts consist in adapting familiar protective measures, such as damping treatment, absorptively lined machinery enclosures, and mechanical isolators, to sources of transient noise.

For background information see ACOUSTIC NOISE; FOURIER SERIES AND INTEGRALS; HEARING (HUMAN); NONDESTRUCTIVE TESTING; PHYSIOLOGICAL ACOUSTICS; PSYCHOACOUSTICS; SONAR in the McGraw-Hill Encyclopedia of Science and Technology.

[MIGUEL C. JUNGER]

Bibliography: A. Akay, A review of impact noise, *J. Acoust. Soc. Amer.*, 64(4):977–985, 1978; P. Bruel, Do we measure damaging noise correctly?, *Noise Contr. Eng.*, 8(2):52–60, 1977; C. J. Mazzola, J. D. Birdwell, and M. Athans, On the application of modern control theory to improving the fidelity of an underwater projector, *J. Acoust. Soc. Amer.*, 66(3):739–750, 1979; E. J. Richards et al., On the prediction of impact noise, pt. 1: Acceleration noise, *J. Sound Vib.*, 62(4):547–575, 1979, pt. 2: Ringing noise, 65(3):419–451, 1979, pt. 3: Energy accountancy in industrial machines, 76(2):187–232, 1981.

Acoustical materials

A porous solid saturated with a fluid or a gas often has unusual acoustic properties compared to those of a nonporous solid. A nonporous isotropic solid has two distinct sound modes, each of which has its own speed of propagation: (1) a longitudinal mode characterized by oscillatory motion of the substance along the direction of propagation; and (2) a transverse mode characterized by motion along any direction perpendicular to the direction of propagation. If a given porous solid has the property that the fluid part and the solid part each form a connected cluster throughout the sample, there are in general two distinct longitudinal modes as well as a transverse mode. Examples of such systems are naturally occurring sedimentary rocks, gels (which are mostly fluid but are supported by a tenuous solid network), packed snow, and a variety of artificially manufactured samples that are formed, typically, from a fusing of some granular material. Roughly speaking, sound can propagate through the fluid at one speed or through the solid at another speed, although each of the true normal modes of propagation is a combination of the motion of the two constituents. From a theoretical point of view, such systems are best described by a formalism which treats the displacement of the solid part and of the fluid part separately and on an equal footing, although the two motions are coupled. A consequence of this theory is that one of the longitudinal modes, called the slow wave, corresponds to a great deal of relative motion between the fluid and solid constituents; the two components are 180° out of phase with each other.

Low-frequency behavior. At low frequencies, where the viscosity of the fluid is a significant effect, the slow wave is overdamped because of the large drag force that the solid exerts on the fluid. In this case the evolution of the slow mode is governed not by a wave equation with a well-defined speed of propagation but rather by the heat equation. The spatial and temporal evolution is diffusive in character. As a simple example, if the fluid is suddenly pressurized by the injection of additional fluid into a porous solid, then the fluid slowly expands through the porous network in a manner mathematically equivalent to the way that heat is conducted in an ordinary thermally conducting medium. Other examples of this diffusive slow wave are fairly commonplace.

High-frequency behavior. The situation can be quite different for acoustic disturbances at high frequencies. If the frequency is high enough, the viscous skin depth δ, given by Eq. (1), where η is the

$$\delta = \sqrt{2\eta/\rho_f \omega} \qquad (1)$$

fluid viscosity, ρ_f is the fluid density, and ω is the frequency of the sound, becomes small compared to the sizes of the pores in the solid. In this limit the fluid behaves as if it were nonviscous except in a thin boundary region of size δ along the walls of the pores. The consequences for acoustic propagation are dramatic: The fast and the slow compressional waves, as well as the transverse wave, are propagatory in the sense that a pulse of finite duration of any one of these can propagate more or less undistorted with a well-defined speed. Of course, the different modes have different sound speeds.

Theoretical prediction of sound speeds. The theory has been developed to the point where it is possible to characterize a given sample by independent measurements and thereafter predict the speeds of the three modes when the sample is saturated with an arbitrary fluid. The relevant input parameters are ρ_f and ρ_s, the densities of the fluid and of the solid; K_f and K_s, the bulk moduli of the fluid and of the solid; ϕ, the porosity or volume fraction of the fluid. The bulk and shear moduli K_b and N of the dry porous solid must also be known. These can be determined by measuring the transverse speed in the dry material V_T (dry), given by Eq. (2), and the

$$V_T(\text{dry}) = \sqrt{\frac{N}{(1-\phi)\rho_s}} \qquad (2)$$

longitudinal speed in the dry material, given by Eq. (3). Finally, information about the pore geometry is

$$V_L(\text{dry}) = \sqrt{\frac{K_b + \tfrac{4}{3}N}{(1-\phi)\rho_s}} \qquad (3)$$

specified through the tortuosity parameter α. This parameter can be determined by saturating the porous sample with a nonviscous, highly compressible fluid; in practice this means using superfluid helium-4. The speed of the slow wave is then given by Eq. (4), where V_f, given by Eq. (5), is the speed of

$$V_{\text{slow, lim}} = \frac{V_f}{\sqrt{\alpha}} \qquad (4)$$

$$V_f = \sqrt{K_f/\rho_f} \qquad (5)$$

sound in the fluid alone. In addition, α can be measured by purely electrical means. If the solid is insulating and the fluid has a conductivity σ_f, the porous fluid-saturated sample has a conductivity σ which is proportional to σ_f; that is, $\sigma = x\sigma_f$. It can be shown that $\alpha = \phi/x$.

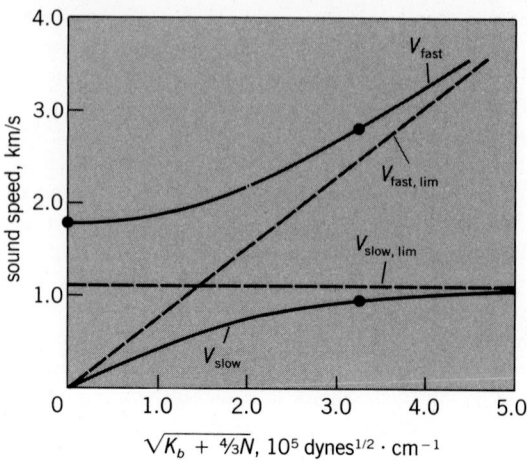

Speeds of fast and slow waves in a porous medium consisting of glass and water as functions of frame moduli K_b and N. The ratio K_b/N is held fixed, with $K_b = 1.80\ N$, while $K_b + \tfrac{4}{3}N$ is varied. Values of input parameters are: $\phi = 0.38$, $\alpha = 1.79$, $\rho_s = 2.48\ \text{g/cm}^3 = 2.48 \times 10^3\ \text{kg/m}^3$, $K_s = 4.99 \times 10^{11}\ \text{dynes/cm}^2 = 4.99 \times 10^{10}\ \text{Pa}$, $\rho_f = 1.00\ \text{g/cm}^3 = 1.00 \times 10^3\ \text{kg/m}^3$, $K_f = 2.25 \times 10^{10}\ \text{dynes/cm}^2 = 2.25 \times 10^9\ \text{Pa}$. 1 km/s = 3281 ft/s. Broken lines represent the limiting values of the speeds in the case that the porous frame is much stiffer than the fluid.

Dependence on frame moduli. It is informative to plot the speeds of the fast and slow compressional waves as functions of the "stiffness" of the porous skeleton, $K_b + \tfrac{4}{3}N$, always assuming the validity of the high-frequency approximation, of course. This is done in the illustration for a porous medium consisting of glass and water. The ratio K_b/N is held fixed while $K_b + \tfrac{4}{3}N$ is varied. There is also a transverse mode whose speed is not shown; its speed is given simply by Eq. (6). This kind of plot

$$V = \sqrt{N/[(1-\phi)\rho_s + (1-1/\alpha)\phi\rho_f]} \qquad (6)$$

has approximate validity for a porous solid formed by sintering a collection of loose glass beads. As the sintering progresses, the contacts between the individual beads stiffen and the frame moduli K_b and N grow from an initial value of zero appropriate to a suspension. The data points are for two different samples of glass beads and water at a fixed porosity. In one case the beads are loose and unfused ($K_b = N = 0$), whereas in the other the beads were first fused in an oven. The porosity did not measurably change after sintering. It is clear that there is a radical change in the acoustic properties of the system due to a relatively minor change in the microstructure (the formation of small welds between the beads).

Limiting values. The broken lines in the illustration represent the limiting values of the speeds in the case that the porous frame is much stiffer than the fluid (K_b and N are much greater than K_f). The expressions for the speeds of the modes simplify. One of the modes corresponds to sound propagating in the fluid; its speed is reduced from V_f to $V_f/\sqrt{\alpha}$ as in Eq. (4) due to the fact that the sound must wind its way around the tortuous pore space, and is

Speeds of sound for water-saturated glass bead samples*

Porosity (φ), %	Tortuosity (α)	Slow wave speed		Fast wave speed		Shear wave speed	
		Theory	Experiment	Theory	Experiment	Theory	Experiment
10.5	3.84	0.71 [0.77]	0.58	5.17 (5.16)	5.15	3.04 (3.09)	2.97
16.2	3.02	0.81 [0.86]	0.70	4.82 (4.84)	4.83	2.74 (2.81)	2.68
21.9	2.40	0.89 [0.97]	0.88	4.35 (4.32)	4.60	2.57 (2.65)	2.57
26.6	2.00	0.97 [1.06]	0.94	3.89 (3.83)	3.98	2.20 (2.28)	2.21
33.5	1.75	1.01 [1.13]	0.99	3.23 (3.10)	3.19	1.75 (1.82)	1.68

*All speeds are in km/s; 1 km/s = 3281 ft/s.

thus given by Eq. (7). The fast wave corresponds to

$$V_{\text{slow,lim}} = \sqrt{\frac{K_f}{\alpha\,\rho_f}} \qquad (7)$$

sound propagating through the solid frame, dragging along some of the fluid as it goes. Its speed is given by Eq. (8).

$$V_{\text{fast,lim}} = \sqrt{\frac{K_b + {}^{4}\!/_{3}N}{(1-\phi)\rho_s + (1 - 1/\alpha)\phi\rho_f}} \qquad (8)$$

Comparison with experiment. With the measured values of α and of the dry speeds (in a given sample), it is now possible to calculate, with no adjustable parameters, the speeds of the fast longitudinal, slow longitudinal, and transverse waves in the water-saturated sample. A comparison of theoretically calculated speeds with the experimentally measured values for a series of fused glass bead samples is presented in the table. The input data are the dry speeds, listed in parentheses, and the tortuosity α, deduced from the superfluid helium-4–saturated data. For comparison, the theoretical speed of the slow wave for an infinitely rigid frame, equal to $V_f/\sqrt{\alpha}$, is given in square brackets. The agreement between theory and experiment is quite reasonable. Thus, effort now has shifted to the problem of understanding the input parameters to the theory.

For background information *see* CONDUCTION (HEAT); ELASTICITY; LIQUID HELIUM; SOUND in the McGraw-Hill Encyclopedia of Science and Technology.   [DAVID LINTON JOHNSON]

Bibliography: M. A. Biot, Theory of elastic waves in a fluid-saturated porous solid, I. Low-frequency range, *J. Acoust. Soc. Amer.*, 28:168–178, 1956; M. A. Biot, Theory of propagation of acoustic waves in a fluid-saturated porous solid, II. Higher frequency range, *J. Acoust. Soc. Amer.*, 28:179–191, 1956; D. L. Johnson, Recent developments in the acoustic properties of porous media, in D. Sette (ed.), *Proceedings of the Enrico Fermi Summer School "Frontiers of Physical Acoustics,"* 1985; D. L. Johnson et al., Tortuosity and acoustic slow waves, *Phys. Rev. Lett.*, 49:1840–1844, 1982; D. L. Johnson and T. J. Plona, Acoustic slow waves and the consolidation transition, *J. Acoust. Soc. Amer.*, 72:556–565, 1982.

Adenohypophysis hormone

Since the initial report by M. Raben in 1958, human growth hormone has been widely employed in the treatment of short stature secondary to growth hormone deficiency. Unlike many other polypeptide hormones, growth hormone appears to be highly species-specific. Thus, while every mammalian growth hormone so far evaluated appears capable of stimulating growth in the rat, only primate growth hormone is biologically active in humans. The practical consequences of this species specificity has been that clinical supplies of growth hormone have been limited by the availability of human cadaver pituitary glands which, until the advent of recombinant deoxyribonucleic acid (DNA) technology, constituted the sole source of human growth hormone.

Human growth hormone is a single polypeptide chain (Fig. 1) with a molecular weight of approximately 22,000. It comprises 191 amino acids, with one disulfide bond between Cys-53 and Cys-165, and a second disulfide bond between Cys-182 and Cys-189. The growth hormone present in pituitary extracts actually constitutes a family of growth hormone molecules, probably resulting from posttranslational modifications of the parent 22,000-dalton protein. The biological actions of growth hormone include (1) stimulation of somatic growth, typically reflected by epiphyseal growth, increased body weight, nitrogen retention, and increases in total body DNA, ribonucleic acid (RNA), and protein; (2) increased mobilization of fat; (3) acute insulinlike activity; (4) chronic insulin-resistant glucose intolerance; and (5) mild lactogenic activity. The growth-promoting actions of the hormone are believed to be mediated by the somatomedins, a family of growth hormone–dependent, insulinlike peptides, while the other effects presumably reflect direct actions of growth hormone.

Because of it size (191 amino acids), human growth hormone has not lent itself to synthesis by standard methodologies of protein chemistry. On the other hand, since pituitary tissue is highly enriched for messenger RNA (mRNA) for human growth hor-

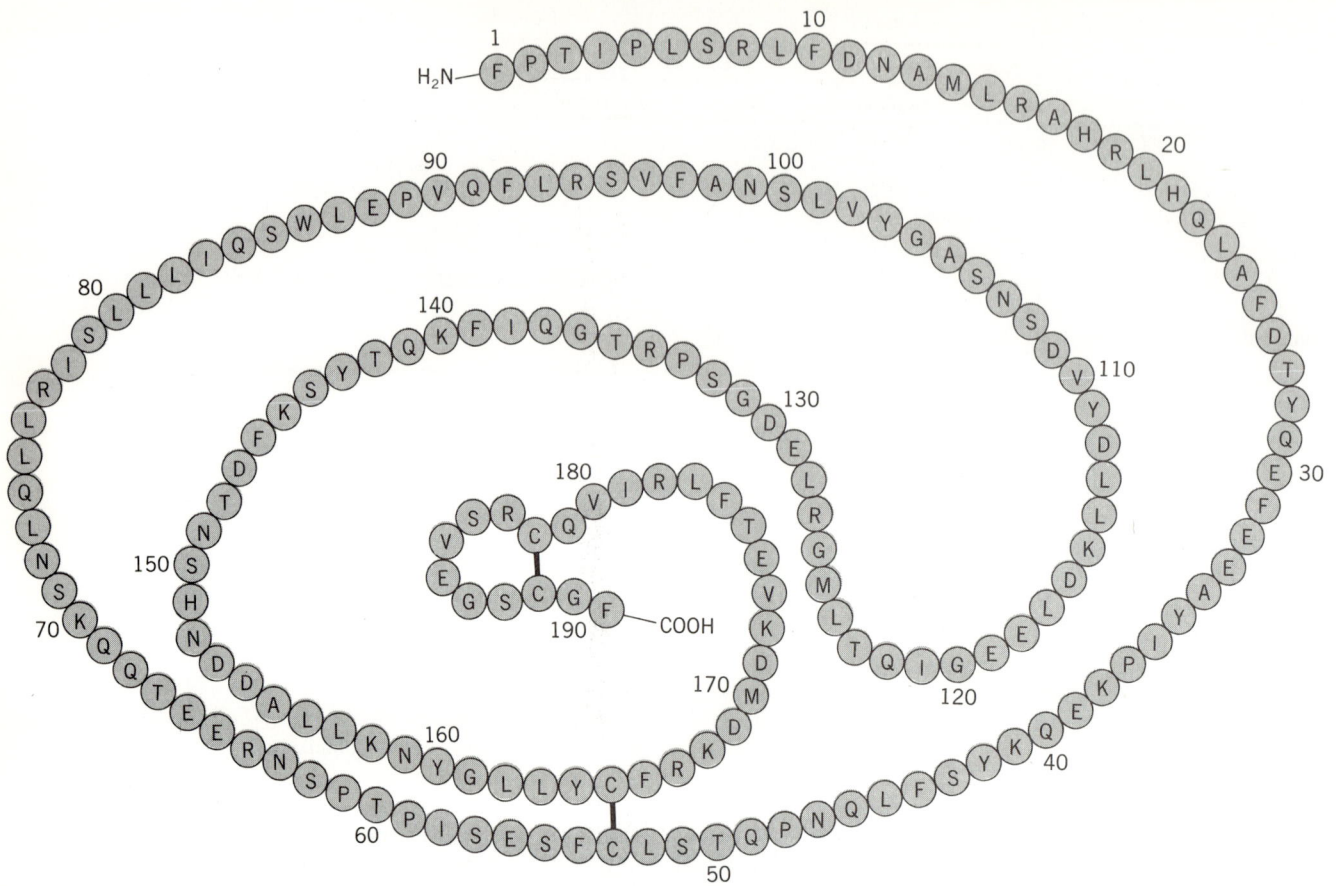

Fig. 1. Amino acid sequence of human growth hormone. A = alanine; C = cysteine; D = aspartic acid; E = glutamic acid; F = phenylalanine; H = histidine; I = isoleucine; K = lysine; L = leucine; M = methionine, N = asparagine; P = proline; Q = glutamine; R = arginine; S = serine; T = threonine; V = valine; Y = tyrosine. (After J. L. Geuriguian, E. D. Bransome, Jr., and A. S. Outschoorn, eds., Hormone drugs, Proceedings of the FDA-USP Workshop on Drug and Reference Standards for Insulins, Somatotropins, and Thyroid-Axis Hormones, U.S. Pharmacopeial Convention, Inc., Rockville, Maryland, 1982)

mone, and since the protein is not glycosylated, human growth hormone has proven to be an excellent candidate for production via recombinant DNA technology (Fig. 2).

Synthesis. To assemble the complete DNA sequence for human growth hormone, a synthetic-natural "hybrid" gene was constructed. This strategy was based upon the existence of an *Hae*III restriction endonuclease site in the sequence coding for amino acids 23 and 24 of human growth hormone. Poly(A)-RNA was first prepared from human pituitaries, and a double-stranded complementary DNA (cDNA) was then produced by reverse transcription. The cDNA was treated with *Hae*III and electrophoresed on an 8% polyacrylamide gel, yielding a DNA fragment of 551 base pairs, including the complete coding sequence for amino acids 24–191. Terminal deoxynucleotidyl transferase was then employed to add approximately 20 deoxycytosine residues per 3′ terminus, and the DNA fraction was cloned in an appropriate DNA vector, pBR322, which had been cleaved with the enzyme *Pst*I, and extended with deoxyguanine residues. This vector was then employed to transform *Escherichia coli* χ 1776.

Additionally, a DNA fragment coding for amino acids 1–24 was chemically synthesized. An ATG codon, specifying the amino acid methionine, was added immediately prior to the TTC codon for phenylalanine, normally the first amino acid in human growth hormone. The methionine acts as an initiation codon, and promotes peptide synthesis in *E. coli*. Since *E. coli* contains enzymes which normally cleave initiator methionines from nascent polypeptides, as evidenced by the fact that most bacterial proteins do not contain N-terminal methionine residues, it was initially expected that *E. coli* would be capable of direct expression of human growth hormone 191. Thus the DNA fragment coding for amino acids 1–24 plus methionine was assembled, cloned in pBR322, and used to transform *E. coli* 294.

After the DNA fragments for amino acids 24–191 and methionine-1–24 were separately cloned, the fragments were repurified and linked together to yield the DNA sequence for Met-1–191. This annealed DNA sequence was cloned in a plasmid, and used to transform *E. coli* χ 1776. Lac promoters were employed to initiate transcription in the direction of a tetracycline resistance gene (Tc)R. Colonies were selected for growth on tetracycline, since antibiotic resistance was dependent upon transcription from the lac promoter reading through the human growth hormone gene sequence into the tetracycline-resistance gene. Transformed colonies were then screened by hybridization with ^{32}P-labeled cDNA probes, and the existence of the correct human growth hormone DNA sequence was verified.

Purity. Once the *E. coli* have been successfully transformed, they can be grown by a batchwise procedure, harvested, lysed, and the human growth hormone extracted from the bacterial cytosols. The biosynthetic met–human growth hormone has been purified by conventional biochemical separation techniques, and the complete amino acid sequence, including the correct disulfide bond formation, has been confirmed. Contrary to initial hopes, the N-terminal methionine has not been cleaved by the bacteria, presumably because the structure of the human growth hormone molecule prevents bacterial enzymes from working properly. Nevertheless, the secondary and tertiary structures of met–human growth hormone and pituitary human growth hormone have been found to be identical, and greater than 99% purity has been achieved for the met–human growth hormone. This level of purity compares favorably with that of pituitary preparations, and the met–human growth hormone has the additional advantage of being entirely free from other pituitary polypeptides or peptide fragments. Bacterial endotoxin contamination is less than 1 part in 10 million, and the polypeptide retains both immunological and biological activity after greater than 1-year storage at 2–8°C.

Biological activity. Initial biological studies demonstrated that met–human growth hormone and pituitary human growth hormone were equivalent in their ability to stimulate weight gain and tibial growth in hypophysectomized rats, indicating that the presence of an N-terminal methionine does not affect the biological activity of human growth hormone. In 1982 the first trials of met–human growth hormone in humans were begun at Stanford University. Twenty-two healthy adult male volunteers were randomly assigned, in a double-blinded, crossover study, to four daily injections of 0.125 mg/kg body weight of either pituitary human growth hormone or met–human growth hormone. In all parameters tested, including stimulation of somatomedin production, lowering of blood urea nitrogen, raising of serum triglycerides, lowering of serum cholesterol, and induction of insulin resistance and glucose intolerance, the two human growth hormone preparations proved identical (Fig. 3).

Therapeutic use. On the basis of these studies, the Food and Drug Administration has approved several multicenter investigations of recombinant DNA–derived met–human growth hormone in children with growth failure. Studies are already under way in the United States and Europe on the use of met–human growth hormone in children with growth hormone deficiency and short stature associated with Turner syndrome (a chromosomal disorder of phenotypic females, in which the second X chromosome is either missing or structurally abnormal).

The therapeutic implications of recombinant DNA technology are staggering, and this is certainly true for human growth hormone. Critical limitations upon supplies of pituitary-derived human growth hormone have restricted its clinical use to the treatment of growth hormone deficiency, a rare disorder, in which endogenous production of pituitary growth hormone is compromised. Even here, the dosage of human growth hormone used clinically has been sharply limited by scanty supplies. However, with the advent of recombinant DNA–derived human growth hormone, the potential therapeutic uses of human growth hormone have greatly expanded, and now include (1) treatment of growth hormone deficiency with higher doses of human growth hormone than previously possible; (2) treatment of non-growth hormone–deficient forms of short stature, such as Turner syndrome, intrauterine growth retardation, and normal short children; (3) acceleration of wound healing; and (4) treatment of osteoporosis,

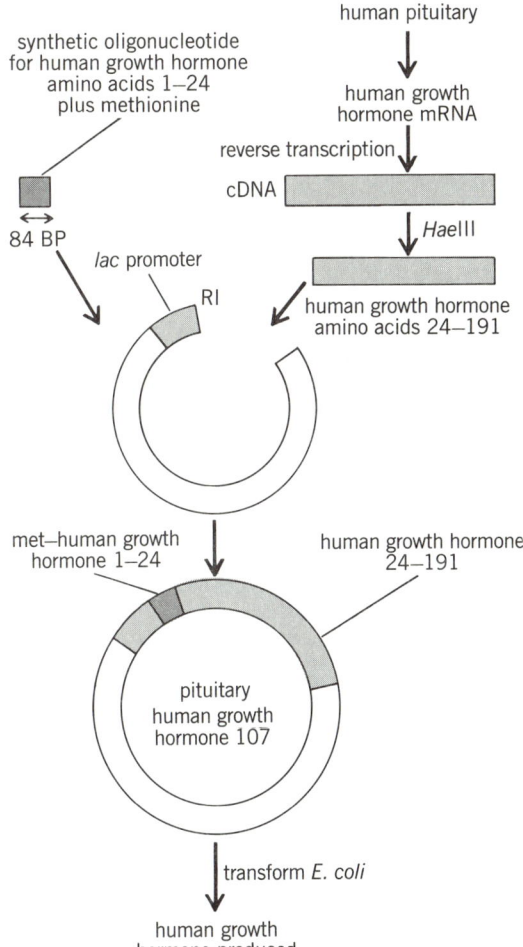

Fig. 2. Production of met–human growth hormone in *Escherichia coli* by recombinant DNA technology. A synthetic-natural "hybrid" gene was constructed, cloned, and expressed in *E. coli*. (*After J. T. Watson, J. Tooze, and D. T. Kurtz, Recombinant DNA: A Short Course, Scientific American Books, W. H. Freeman, 1983*)

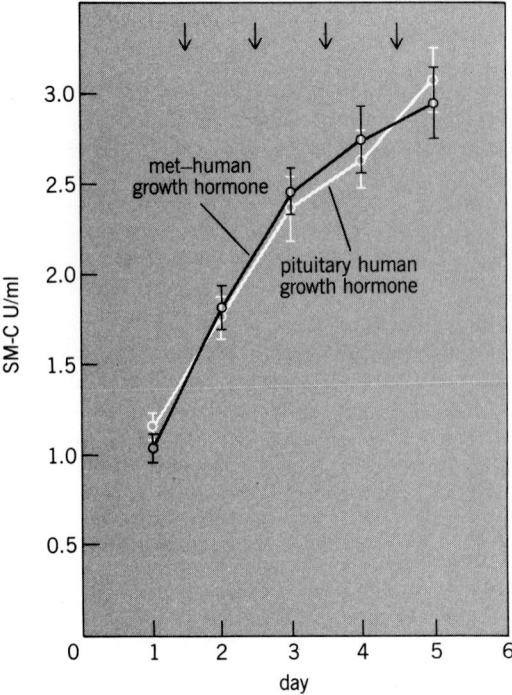

Fig. 3. Increase in plasma somatomedin levels following daily injections of pituitary human growth hormone or met–human growth hormone at a dose of 0.125 mg/(kg)(day). (After J. L. Geuriguian, E. D. Bransome, Jr., and A. S. Outschoorn, eds., Hormone drugs, Proceedings of the FDA-USP Workshop on Drug and Reference Standards for Insulins, Somatotropins, and Thyroid-Axis Hormones, U.S. Pharmacopeial Convention, Inc., Rockville, Maryland, 1982)

poorly healing fractures, and cachexia. On the other hand, these potential therapeutic benefits must be weighed against possible complications of human growth hormone therapy, which include diabetogenic effects, antigenicity of met–human growth hormone and abuse of human growth hormone in the treatment of normal children.

For background information see ADENOHYPOPHYSIS HORMONE; GENETIC ENGINEERING; PITUITARY GLAND in the McGraw-Hill Encyclopedia of Science and Technology.

[RON G. ROSENFELD]

Bibliography: D. V. Goeddel et al., Direct expression in *Escherichia coli* of a DNA sequence coding for human growth hormone, Nature, 281:544–548, 1979; R. L. Hintz et al., Biosynthetic methionyl human growth hormone is biologically active in adult man, Lancet, 1:1276–1279, 1982; K. C. Olson et al., Purified human growth hormone from *E. coli* is biologically active, Nature, 293:408–411, 1981; R. G. Rosenfeld et al., Recombinant DNA–derived methionyl human growth hormone is similar in membrane binding properties to human pituitary growth hormone, Biochem. Biophys. Res. Comm., 106:202–209, 1982.

Amebas

Free-living amebas are single-celled animals belonging to the subkingdom Protozoa, the unicellular eukaryotic organisms. They are a large group, abundant worldwide and found in soil, fresh-water, and marine environments. The major group or phylum of Protozoa to which the free-living amebas belong is Sarcomastigophora, the amebas and flagellates.

Free-living amebas play an important role in soil microbiology and in aerobic decomposition of organic matter. They are also important in the percolating-filter and activated-sludge processes of sewage and wastewater treatment. In the laboratory, cell biologists use several species of free-living amebas (such as *Acanthamoeba*, *Naegleria*, and *Tetramitus*) to study eukaryotic cellular differentiation (cell development). Free-living amebas have also been used as bioassay organisms for environmental pollution by pesticides and other toxicants.

Most free-living amebas are harmless to humans. However, in recent years some amebas belonging to the genera *Naegleria* and *Acanthamoeba* have been shown to cause fatal disease in humans and animals. Infection involves mainly the central nervous system, but may also affect the eye and other organs.

Nonpathogenic amebas. Free-living amebas are largely nonpathogenic organisms. They live in aqueous or soil habitats, where they feed on bacteria and other unicellular microorganisms. Free-living amebas vary greatly in size and, generally, they are considered to be asymmetrical in shape. They travel by means of cytoplasmic extensions known as pseudopodia (Fig. 1). Amebas feed by surrounding and engulfing their prey with these cuplike pseudopodia, a process referred to as phagocytosis. Amebas may also engulf small amounts of fluids by pinocytosis.

An elastic cell membrane envelops the ameba and its cytoplasm. Within the cytoplasm are a nucleus, contractile vacuole, food vacuoles, and other organ-

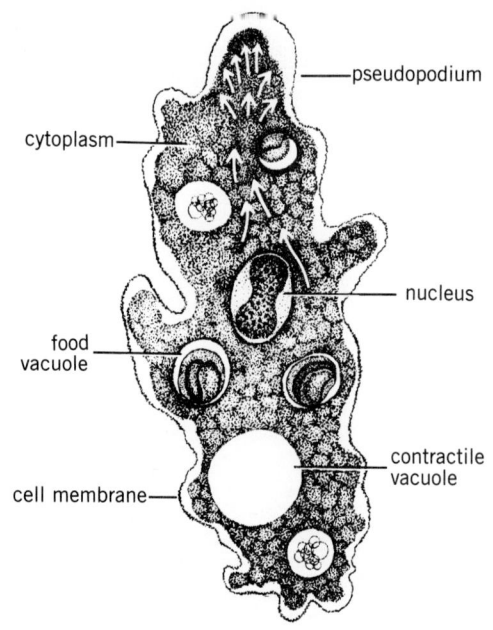

Fig. 1. A generalized free-living ameba; arrows show the direction of cytoplasmic flow.

elles typical of a eukaryotic cell (Fig. 1). Contractile vacuoles are used to maintain water balance and occur in fresh-water and soil species, but are absent in marine amebas.

Free-living amebas reproduce by asexual binary fission. In this process, the ameba divides into two equal-sized daughter cells. Many free-living amebas also produce a resting stage, called a cyst, in which a resistant wall forms around the cell. The cyst enables the organism to survive unfavorable conditions, such as drying. An ameba emerges from the cyst once environmental conditions are again favorable.

Some free-living amebas have a flagellate stage in their life cycle. These are known as ameboflagellates. The flagellate form is usually a transient phase and, depending upon the species of ameba, may or may not divide. The ability of free-living amebas to encyst and to enflagellate has been exploited in the laboratory by scientists to study various biochemical events in eukaryotic cellular differentiation.

Pathogenic amebas. Certain members of the free-living amebas *Naegleria* and *Acanthamoeba* are lethal pathogens, capable of producing fatal disease in humans and animals. *Acanthamoeba* species generally cause a long-term infection in chronically ill or debilitated persons, some of whom may be undergoing immunosuppressive therapy. Invasion of the central nervous system appears to be by way of the bloodstream with amebas originating from a lesion of the skin, lungs, or kidneys. Serious eye infection also has been caused by certain *Acanthamoeba* species. The species of *Acanthamoeba* indicted in human infection are strains of *A. castellanii*, *A. culbertsoni*, *A. polyphaga*, and *A. astronyxis*.

In contrast to *Acanthamoeba* infection, the ameboflagellate *N. fowleri* causes a rapidly fatal human disease known as primary amebic meningoencephalitis. Since the majority of described human infections by free-living amebas have been caused by *N. fowleri*, the remainder of this article will be restricted to *N. fowleri* and the disease it produces.

Primary amebic meningoencephalitis was first detected in humans in Australia in 1965. Since then, human infections have been described from around the world. Primary amebic meningoencephalitis typically occurs in previously healthy children or young adults with a recent history of swimming and diving in fresh-water lakes or pools; infection usually follows breathing in of water containing amebas or flagellates of *N. fowleri* (Fig. 2).

Amebas gain entrance to the brain by way of the nasal mucosa and cribriform plate, a bony structure separating the brain from the nasal passages. Within the brain amebas provoke inflammation, hence the term meningoencephalitis (inflammation of brain membranes and tissue), and cause extensive tissue damage. The disease is rapidly fatal, usually producing death within 3 days after the onset of symptoms.

Symptoms begin with severe frontal headache, fever, and loss of appetite, followed by nausea and

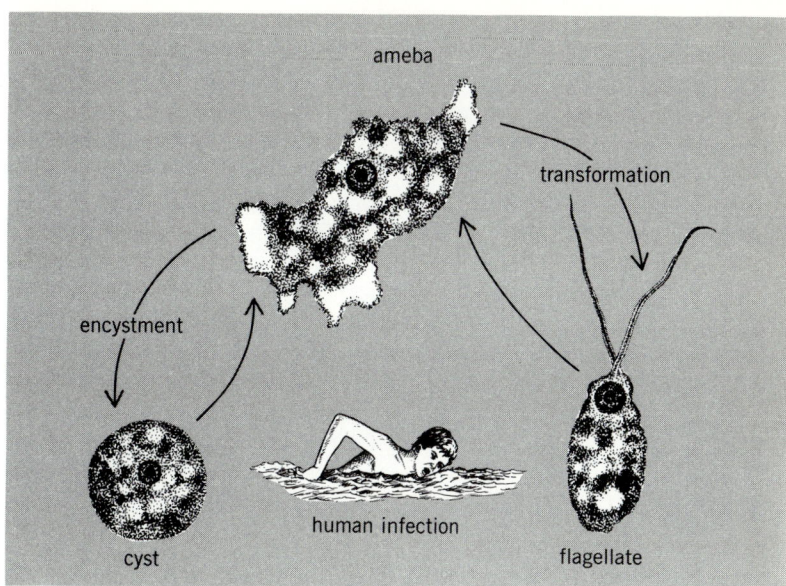

Fig. 2. Life cycle of *Naegleria fowleri*.

vomiting. Involvement of the olfactory part of the brain may affect the sense of smell or taste and may be noted early in the course of the disease. Vision may be affected. The victim may become confused, irritable, and restless, and may become irrational before lapsing into coma. Generalized seizures also may be present.

Primary amebic meningoencephalitis is diagnosed by identifying living or stained amebas in cerebrospinal fluid. A history of swimming within the preceding week is strongly suggestive of possible naeglerial infection. *Naegleria fowleri* amebas may be grown by placing some of the spinal fluid on agar which has been spread with a lawn of bacteria. The bacteria serve as a source of food for the growing amebas.

At present there is not a completely satisfactory treatment for primary amebic meningoencephalitis. The antibiotics used to treat bacterial meningitis are not effective in naeglerial disease. The antifungal agent amphotericin B, a considerably toxic drug, was used to treat the only two known survivors of primary amebic meningoencephalitis. However, others have not survived naeglerial infection even though they were given amphotericin B. A more effective drug or combination of drugs is needed for treating primary amebic meningoencephalitis.

Recently, unique suckerlike structures, referred to as amebostomes, have been described for the amebas of *N. fowleri* (Fig. 3). Amebostomes are functional structures used for engulfment, but they may also be involved in pathogenesis and virulence.

Another pathogenic species, *N. australiensis*, has been isolated from environmental sources. Although this ameba produces deaths in experimentally infected mice, it has not yet been recovered from a human infection. Nonetheless, this species of ameba, and perhaps others, given the right conditions, may be able to cause serious disease in humans.

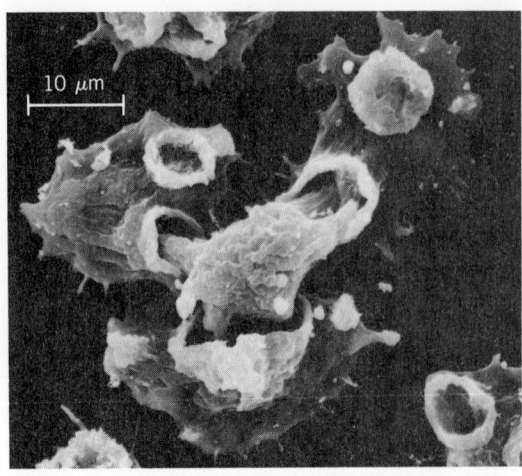

Fig. 3. Three *Naegleria fowleri* amebas using suckerlike structures (amebostomes) to attack and engulf portions of a fourth, presumably dead, ameba.

For background information *see* AMEBA; MEDICAL PARASITOLOGY; PROTOZOA; SARCOMASTIGOPHORA in the McGraw-Hill Encyclopedia of Science and Technology.

[DAVID T. JOHN]

Bibliography: D. T. John, Primary amebic meningoencephalitis and the biology of *Naegleria fowleri*, *Annu. Rev. Microbiol.*, 36:101–123, 1982; D. T. John, T. B. Cole, Jr., and F. M. Marciano-Cabral, Sucker-like structures on the pathogenic amoeba *Naegleria fowleri*, *Appl. Environ. Microbiol.*, 46:12–14, 1984; F. C. Page, *An Illustrated Key to Freshwater and Soil Amoebae*, 1976; F. L. Schuster, Small amebas and ameboflagellates, *Biochem. Physiol. Protozoa*, 1:215–285, 1979.

Antenna

The remarkable growth in the channel capacity of geostationary orbit communications satellites since their inception two decades ago has come about largely from more efficient use of two finite resources, such as the available radio-frequency (rf) spectrum and the geostationary orbital arc. The most dramatic efficiency improvement is due to satellite antennas which have multiple-shaped beams so that the available frequency bands can be reused many times at the satellite.

The most prominent example of multiple-shaped beams is the INTELSAT VI series of satellites scheduled for first launch in 1986. This satellite (Fig. 1) is nearly 39 ft (12 m) tall and 12 ft (3.6 m) in diameter, and will weigh about 4000 lb (1800 kg). Its solar cells will produce 2 kW of power. It will relay voice, television, and data signals by using analog and digital modulations. The satellite capacity exceeds 33,000 equivalent voice channels, and the effective bandwidth is 3200 MHz in spite of the fact that the total allocated bandwidth is only slightly more than 1000 MHz. As explained below, this is achieved by incorporating multiple, shaped antenna beams to reuse portions of the allocated frequency bands as many as six times. For comparison, INTELSAT I *(Early Bird)*, launched in 1965, had a capacity of only 240 voice channels.

Satellite operation. To better appreciate how multiple beams improve bandwidth and orbit efficiency, it is useful to review the basic operation of a geostationary communications satellite. In spite of its apparent complexity, such a satellite is simply a frequency-channelized repeater in orbit in the equatorial plane at an altitude of 22,000 mi (36,000 km). The orbit period is 24 h so the satellite appears stationary from the Earth. The Earth subtends an angle of about 18° as seen from the satellite; more than one-third of the Earth's surface lies within this field of view. For signal relay, the satellite's uplink antenna receives signals from a number of earth station antennas. For frequency-division multiple access (FDMA), each Earth-to-satellite transmission is assigned a unique frequency slot within the overall uplink band which, for the typical C-band allocation, is 500 MHz wide, that is, 5.925–6.425 GHz. Onboard the satellite, the signals are amplified, frequency-translated to the 3.7–4.2-GHz downlink band, and grouped by frequency into a number of transponder channels whose typical bandwidth is 36 MHz. For conventional frequency modulation (FM), this transponder bandwidth can typically accommodate either 900–1200 voice channels or 1–2 television channels. Amplifiers in each channel boost the signal level, typically to valves in the 4 to 10-W range. An output multiplexer combines the signals onto a common waveguide for transmission to Earth by the downlink antenna. For time-division multiple access (TDMA) operation, users share a common band but are assigned unique time slots for burst transmissions.

Traffic growth beyond the capacity of a single satellite requires the use of more satellites, the adoption of new frequencies, more efficient spectrum use, or combinations of these. Additional satellites are obviously expensive, requiring duplication of satellite and earth facilities. New frequency allocations require international coordination and are difficult to obtain because the spectrum is a finite resource for which many applications compete. The prevalent bands in use are 500 MHz wide at the 6-GHz (uplink) and 4-GHz (downlink) C band and 500 MHz at the 14-GHz and 12-GHz Ku band. Although these bands have been recently expanded to each be about 1000 MHz wide, heavy capital investment in existing earth facilities motivates system owners to use existing bands as much as possible. The adoption of completely new bands, such as those allocated at 20 and 30 GHz, is inevitable, but technology risk, propagation uncertainties, and high costs associated with the development of new equipment encourage maximum exploitation of existing bands.

Frequency reuse. The same frequencies can be used more than once at the satellite by incorporating multiple independent antenna beams. Figure 2 depicts a satellite transmit antenna connected to two independent sets of transponders, A and B, each using the same allocated frequency band. Beam A is shaped and pointed to radiate efficiently over a northwestern coverage area on the Earth. Beam B covers a southeastern region. The beam patterns for the satellite receive antenna are identical to those for the transmit antenna so the illustration (Fig. 2) could apply for either the uplink or the downlink since an antenna's patterns are the same for receive and transmit modes.

Consider the beams at the right of Fig. 2 for the transmit or downlink case. Each beam must radiate relatively strong signals, that is, have high gain, over its intended main beam coverage area, but its gain in the direction of the other coverage area, that is, its sidelobe levels, must be low enough that earth antennas there do not receive an appreciable amount of its signal relative to the intended beam for that area. The cofrequency undesired signal received by an Earth antenna will appear as interference noise which adds on a power basis to its receiver's thermal noise and degrades the link. Since both beams occupy the same frequency spectrum, the satellite antenna sidelobe discrimination is the only mechanism for controlling interference. The beam isolation is the ratio of desired signal power to undesired power as measured at the receive location, and it should be at least 27–33 dB. This means that the desired signal power at the receive location should be at least 500 to 2000 times that of the undesired power. Similar values would apply to the uplink beam isolations where, for example, satellite receive beam A must be highly responsive, that is, have high gain, to uplink signals from the Earth antennas in the northwestern area but should have low gain to signals coming from the southeastern area. Uplink interfering signals are carried through the satellite, and their powers add to those of the downlink interfering signals to further degrade the link. Frequency reuse can easily be extended to more than two beams, provided all beams can be isolated from each other.

Antenna polarization is another degree of freedom that permits almost a doubling of spectrum use over a common coverage area by using two spatially coincident but orthogonally polarized beams. Polarization refers to the orientation of the electric field vector of the radio transmission. Orthogonal linear polarizations are typically used for United States domestic satellite systems and, for historical reasons, orthogonal circular polarizations are used at C band for the INTELSAT system. The polarization isolation must also be at least 27–33 dB. For N isolated coverage areas, it is theoretically possible to achieve a $2N$-fold increase in effective bandwidth by using both spatial and polarization isolations.

The INTELSAT system uses both techniques to achieve a sixfold reuse of parts of the 4- and 6-GHz bands. The inset in Fig. 1 illustrates the shaped beam coverages for the Atlantic Ocean region. Here, the Western Hemisphere (hemi) shaped beam covers North and South America with a circularly polarized beam of, say, right-hand sense. The sense of polarization is right hand if the electric field vector rotates clockwise in a plane as viewed from behind. The eastern hemi beam covers Europe and Af-

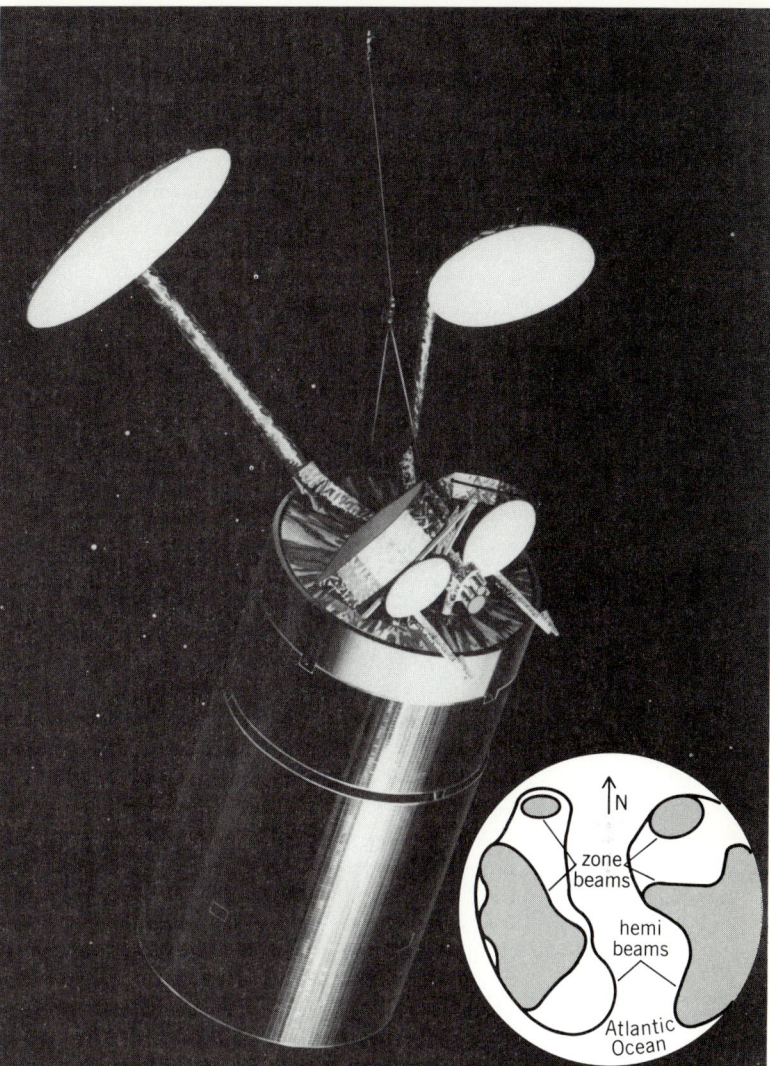

Fig. 1. INTELSAT VI satellite. The large reflectors and feed arrays form shaped hemispheric and overlaid cross-polarized zone beams for a sixfold reuse of C-band frequencies. (F. Taormina, Hughes Aircraft Co.)

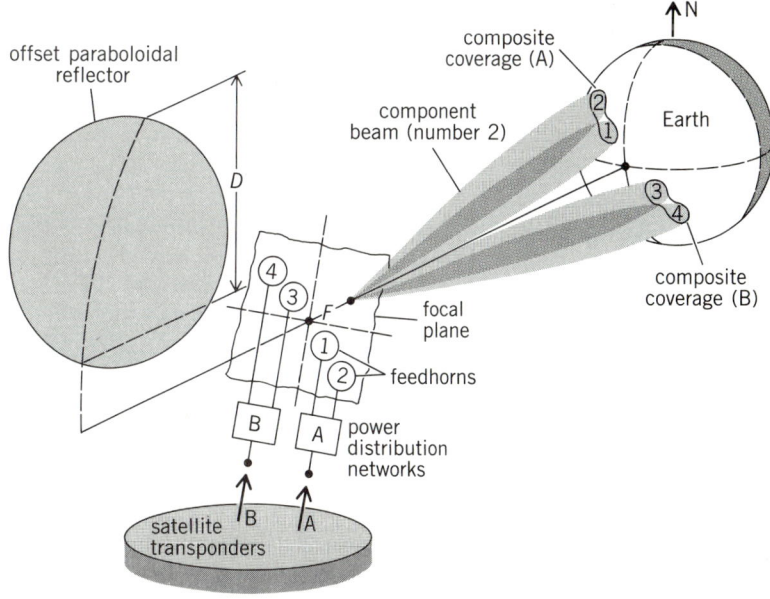

Fig. 2. Beam shaping with component beams from an offset paraboloid.

rica, and it has the same sense of polarization as the western hemi beam. These two beams are isolated only by the antenna amplitude patterns which are designed to suppress sidelobe radiation from each beam into the other coverage area. Overlaid on the hemi beams are four zone beams, located and shaped to cover specific INTELSAT earth stations. These zone beams are orthogonally polarized relative to the hemi beams but copolarized to each other. For this example, they would all be left-hand circularly polarized. They are isolated from the hemi beams by polarization orthogonality, but they are isolated from each other only by their sidelobe suppression. The patterns described apply to both the uplink and downlink antennas, but all corresponding polarization senses for each would be opposite.

Antenna designs. The most prevalent satellite antenna implementation to achieve multiple shaped beams is the offset-fed paraboloid with multiple focal region feeds. This antenna is illustrated in Fig. 2 for the simple example where four feed horns are used to form two shaped beams. The offset section of a paraboloid has projected aperture diameter D. While the principles of beam shaping and beam pointing would apply to a symmetrical paraboloid, the aperture blockage caused by the feeds would make it impossible to achieve low sidelobes.

A transmitting feed at point F would produce a beam whose maximum intensity is along the focal axis. Feeds 1 and 2 located below and to the right of the axis in the focal plane would, if independently excited with rf energy, produce component beams 1 and 2 pointed toward the northwest as shown; feeds 3 and 4 would produce component beams 3 and 4 squinted or "scanned" toward the southeast. The angle between the beam maximum and the focal axis is proportional to the lateral feed displacement, in the focal plane, away from the focus (F). The quality of the beam degrades with increasing scan angle due to phase aberrations. The half-power beamwidth of each component beam θ_c is the angular extent between the points where the radiated power intensity is one-half that of the peak. It is approximately related to the aperture diameter by Eq. (1), where λ is the wavelength at the oper-

$$\theta_c = 65\lambda/D \quad (1)$$

ating frequency ($\lambda = 0.3/f$ meters, where f is the frequency in GHz). The beam associated with a particular feed horn is called a component beam because it is not generally excited alone; rather it is fed from a power distribution network that coherently (such as on a voltage basis) excites a number of feeds. Each component beam is a building block for the composite shaped beam. For example, in Fig. 2 the voltage sum of component beams 1 and 2 forms shaped beam A, and component beams 3 and 4 sum to form shaped beam B. By locating a number of feeds in the focal region and by properly designing the feed distribution network to excite each feed with the proper amplitude and phase, it is possible to build up coverage patterns of considerable complexity. The design challenge is to choose feed positions and the amplitudes and phases of their excitations to concentrate efficiently the radiated power from the satellite over the desired coverage area while suppressing the radiation in the directions of other beams reusing the same frequencies. Many systems do not reuse frequencies via spatially isolated beams, but beam shaping is important in order to maximize the efficiency with which the antenna concentrates its radiated power over a particular coverage region.

The design process begins with the specification of the coverage areas which, in the INTELSAT case, might be determined by the locations of a number of earth stations. For a domestic or regional satellite, it might be defined by the geographical boundaries of a country or group of countries. For United States systems, it might be the contiguous United States (CONUS) with the possible addition of spot beams pointed toward Alaska, Puerto Rico, and Hawaii. The angular maps of the coverage areas are enlarged to take into account the apparent motion of earth locations due to satellite antenna-pointing errors and attitude variations. Typical pointing accuracies for satellite designs are now around 0.1°.

The fundamental antenna design parameter is the aperture diameter which established the optical resolution of the antenna. Optical terms are used for antennas because many of the principles of telescope design apply even though the microwave wavelengths are much longer than those of visible light. For frequency reuse, this resolution refers to the minimum allowable spacing σ between the edges of adjacent coverage regions which are to be isolated by a specified amount, say 27 dB. A reliable rule of thumb is that, for 27 dB isolation between two re-

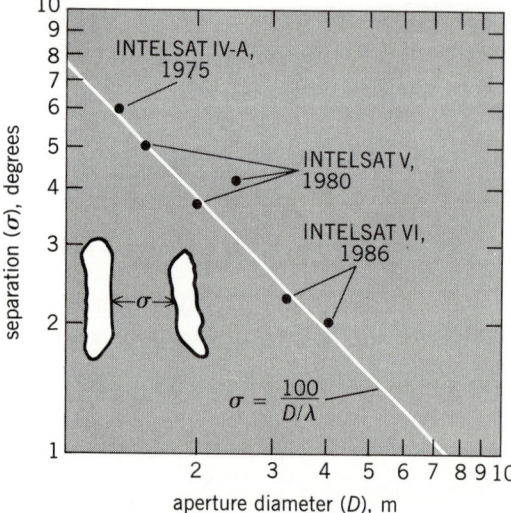

Fig. 3. Relation between aperture diameter D and minimum angular spacing σ between the edges of adjacent coverage regions for 27 dB isolation at frequency of 4 GHz. White line gives rule of thumb. Data points are results of extensive computer computations.

Fig. 4. INTELSAT VI 4-GHz feed array. (F. Taormina, Hughes Aircraft Co.)

gions separated by σ degrees, the ratio of the antenna diameter to its operating wavelength must be at least that given by Eq. (2).

$$D/\lambda \geq 100/\sigma \qquad (2)$$

Once the diameter is determined, feed locations and other geometrical parameters are found, and the major task is to find the optimum set of feed excitations to form the beams. This is done with the aid of large computer programs. The resolution rule calls for a guard spacing between the edges of adjacent coverage regions of at least 1.5 component beamwidths, and it has proven to be very consistent despite more than a decade of extensive development of computer programs by many competing organizations. Figure 3 illustrates the rule applied to more than a decade of INTELSAT shaped-beam designs. It also makes the point that for more reuses, implying closer beams, the antenna diameter must be increased. This makes the antenna one of the most significant subsystems on a typical satellite. For example, the INTELSAT VI satellite in Fig. 1 uses a 10.5-ft diameter (3.2-m) reflector for its 4-GHz hemi-zone antenna. This is the large reflector shown in the upper left of the figure. The feed array contains 146 horns and is 5.75 by 6.2 ft (1.6 by 1.9 m; Fig. 4). The 6-GHz receive antenna in the upper right has a similar design but is about two-thirds as large as the 4-GHz antenna. The total weight of all antennas on INTELSAT VI is nearly 660 lb (300 kg), and this represents nearly 18% of the total satellite mass.

Prospects. Large reflectors will have to be able to be unfurlable or erectable in space because the available launch vehicles cannot accommodate a rigid reflector larger than about 15 ft (4.5 m) in diameter. For a fixed field-of-view, the scan aberrations may dictate more complex multiple reflector designs such as Cassegrain or Gregorian designs. The requirement to achieve high power at the 20- and 30-GHz bands to overcome propagation losses may lead to the use of phased array antennas having thousands of elements. Weight limitations may dictate the use of very lightweight radiating elements, such as printed-circuit antennas. These may radiate directly over the field of view, or they may be used with multiple reflector systems. Phased arrays offer the possibility that the rf power can be distributed over all the elements, allowing a few high-power amplifiers, such as are now typical, to be replaced with many small monolithic microwave integrated circuit (MMIC) amplifiers. Such an antenna may be appropriate for a system that uses scanning or hopping beams to distribute a form of TDMA traffic. This type of antenna and TDMA system will be demonstrated in the NASA Advanced Communications Technology Satellite (ACTS) program scheduled for launch in 1987.

For background information *see* ANTENNA (ELECTROMAGNETISM); COMMUNICATIONS SATELLITE in the McGraw-Hill Encyclopedia of Science and Technology.   [DANIEL F. DIFONZO]

Bibliography: D. DiFonzo, Antennas: Key to communications satellite growth, *Microwave Systems News*, 8(6):83-91, June 1978; F. Taormina, et al., INTELSAT IV-A communications antenna: Frequency reuse through spatial isolation, *IEEE Conference on Communications*, June 14–16, 1976; P. Neyret Antenna technology at INTELSAT (English trans.), *(Journées Internationales de Nice sur les Antennes* (JINA '84), November 13–15, 1984.

Anthropology

Study of the early ancestry of humans includes many active fields, three of which are reported on in this article. These are the origin of bipedal locomotion, the early manufacture of stone artifacts, and the appearance of modern humans. The oldest evidence for bipedal locomotion lies in a series of footprint trails found in Africa, dated about 3.5 million years ago; the meaning of these trials and other anatomical findings that relate to bipedality are discussed in the first section. The second section reports on another major advance, the making of stone artifacts by humans or humanlike creatures. These tools reveal a great deal about the behavior of human ancestors about 1.8 million years ago. The final section describes the origin of modern human types, which took place less than 50,000 years ago; the transition from archaic to modern human forms was not only a biological process, but also a cultural and a technological one.

BIPEDAL LOCOMOTION

Humans are easily distinguished from other primates because they have big brains, hands adapted for refined manipulation instead of locomotion, and peculiar bipedal postures and locomotor habits. These functional-morphological complexes are readily discernible from bones alone. Therefore, if a complete series of properly dated fossils existed, anthropologists would be able to trace each complex to its roots and to infer the selective forces behind it. During the past 25 years, and especially since 1974, a wealth of fossils have been collected that pertain to the evolution of human bipedalism. The most dramatic evidence comes from Pliocene sites at Laetoli, northern Tanzania, through efforts of Mary Leakey, and at Hadar, northern Ethiopia, where Donald Johanson and Maurice Taieb directed research.

Footprint trails. The oldest indisputable evidence for the existence of humanoid bipedalism is the two parallel footprint trails made by three individuals in a 3.5-million-year-old moist volcanic ash at Laetoli site G. The footprints show that one species of the Hominidae had a virtually human foot morphology and gait pattern during the Pliocene. They also show that bipedalism preceded the development of stone tool technology by at least 1 million years.

Because the hominid trials at site G extend northward over 89 ft (27 m), it is clear that the creatures were not merely momentarily bipedal and then resumed quadrupedal postures. One trail (G-1) was made by a single individual (also designated G-1). The trail close to the right (east) of it (G-2/3) was made by individuals G-2 and G-3. G-3 had overprinted and partly obliterated the tracks of G-2.

The Laetoli G hominids were short by comparison with average adult modern humans. The statures of G-1 and G-3 can be computed from foot lengths on the trials if it is assumed that they had the same relationship between foot length and stature that modern humans do. Thus G-1 was about 50 in. tall (1.3 m), and G-3 was about 56 in. (1.4 m) tall. Assuming that they were adults and different sexes, they exhibit the same degree of sexual dimorphism in size (female 89% of male value) that is shown by modern humans and common chimpanzees. However, it is likely that the stature estimates for the Laetoli G hominids are too high because small-brained Pliocene hominids would lack comparable cranial height and their lower limbs, spines, or both could have been relatively shorter than in modern humans.

The morphological features that are revealed by the Laetoli prints indicate that the feet were indistinguishable from those of modern humans and were used similarly for support during bipedal walking. The great toe was aligned with the four lateral toes, and it often left a deep impression that is like the final toeing-off prior to swing phase by people walking in moist sand on the beach. The lateral toes did not project beyond the tip of the great toe, and they generally left only faint impressions in the ash. The best prints of G-1 reveal a prominent medial longitudinal pedal arch. This arch is less apparent in prints of G-3, perhaps because G-2 had compacted the ash where G-3 subsequently stepped. It appears that each step began with a substantial heel strike. Initially, weight was borne on the heel and lateral side of the sole. Then, as in normal human walking, weight shifted medially onto the ball of the foot and finally onto the great toe prior to toe-off.

The feet of G-1 and G-3 were relatively broad. But similar length-breadth relationships can be found in many modern humans who have never worn restraining footgear.

Gait analyses. Many characteristics of gait for the Laetoli G hominids can be inferred from the trials. The major difference between an analysis on the fossils and those analyses performed in modern human gait laboratories is that there is no direct evidence for the speed of their walking. However, even this can be approximated by careful comparisons with results from modern humans.

The average step lengths were similar in G-1 and G-3. They indicate that both individuals walked slowly while making the trails. The interstep stride widths, that is, the degree to which the feet are placed apart, of the two individuals were quite variable. Generally, however, G-1 employed a stride width that is narrower than the average stride width of modern laboratory subjects, while the average stride width of G-3 is greater than that of G-1, and than the average stride width of most modern experimental subjects. But it is comparable to those of modern elderly subjects, who are taller and have longer lower limbs than G-3 did.

The degrees to which G-1 and G-3 angled their feet with reference to the direction of travel are dramatically different from one another. Whereas G-1 always toed-out, G-3 toed-in or placed its feet straight forward. Indeed the right foot of G-1 was so markedly out-toed that the possibility must be considered that the limb had suffered a trauma at some point in G-1's life. This alone could explain the slow pace of G-1 and the other hominids, if they were walking together. Further, the slow pace, deep heel, and great toe impressions, and variable stride widths of G-1 and G-3; the outwardly angled feet of G-1; and the wide stride width of G-3 could indicate that the individuals were walking on irregular terrain and perhaps were carrying loads (for example, dependent youngsters). Studies on modern subjects have shown that loads, including obesity, fatigue, and old age, can cause individuals to slow down, toe-out, and spread their feet more widely when they walk.

Hadar hominids. The earliest hominid foot bones come from Hadar, Ethiopia. Several dates have been proffered for the Hadar Formation, but the bulk of the fossils are probably somewhat younger than the Laetoli prints. One of the most striking features of the collection is the consistent prominent downward curvature of the shafts of the second to fifth finger and toe bones. This is characteristic of arboreal apes and monkeys; it seems unlikely that such curved toes could be parts of feet that made the Laetoli footprints. Prominent impressions of the toe tips would be expected if the creatures walked with their toes extended, or deep knuckle prints if their toes were flexed, during bipedal walking in a pliable moist substrate like the Laetoli ashfall.

Although the toes of Hadar feet were curved downward, the joints between them and the bones of the sole would have permitted bipedal locomotion. Their big toes were robust. But it cannot be determined whether at least some of the Hadar feet had divergent (apelike) versus aligned (human) big toes.

The Hadar knee joints are more human than their feet. Many features indicate that they were fully extended during bipedal standing, unlike apes, which commonly exhibit notable knee flexion when they stand upright.

The pelvis of the most famous Hadar specimen, "Lucy," is a mixture of human and nonhuman features. The sacrum is broad and short like human sacra (ape sacra are long and narrow). The blades (ilia) of the hipbones are also broad and short, but instead of bending forward to form a human basin-like structure they remain flared sideways as in the apes. This could mean that the rump muscles were still used to extend the hips powerfully during climbing and that the human mechanism for balancing the pelvis over the supporting lower limb during bipedalism had not developed fully. Nonetheless, other clearly human features of the hipbones place the Hadar pelvis closer to human than to ape pelves.

Numerous other anatomical features of the Hadar hominids suggest that although they were bipedal on the ground, they continued to utilize trees, probably for security at night, and for feeding and escape from predators during the day. The mechanics of vertical climbing are in some ways similar to those of propelling the body forward during bipedal walking.

In brief, numerous features indicate that the Hadar hominids were derived rather recently from small-bodied forest dwellers that frequently climbed tree trunks and vertical vines in order to eat and that probably also moved bipedally on horizontal boughs. These creatures have been termed the hylobatians because living gibbons (Hylobatidae) engage in similar behaviors, albeit with the added specialization of rapid athletic arm-swinging locomotion which probably was not developed markedly in human ancestors. The alternative suggestion that human ancestors were quadrupedal chimpanzeelike knuckle walkers is denied by numerous hand bones from Hadar that show none of the special features that enable African apes to engage in this peculiar behavior.

[RUSSELL H. TUTTLE]

STONE ARTIFACTS

The earliest known stone artifacts come from East Africa, and date to just over 2 million years age. The most informative archeological sites contain the earliest stone artifacts together with fragmentary animal bones. These kinds of sites provide insights into the lifeways of early hominids. Recent major research breakthroughs, from the discovery of microwear polish on stone artifacts to cut marks on fossilized bone, provide a greater understanding of functions of stone tools and their role in the lives of early hominids. In the same strata as the stone artifacts are fossils from various populations of hominids, including the earliest known species attributed to the genus *Homo*.

Age and geographical distribution. The first discoveries of well-documented early stone artifacts were made by Louis Leakey and Mary Leakey at Olduvai Gorge in northern Tanzania (Fig. 1) beginning in 1959. Potassium-argon and fission-track dating showed that the lowermost strata with artifacts at

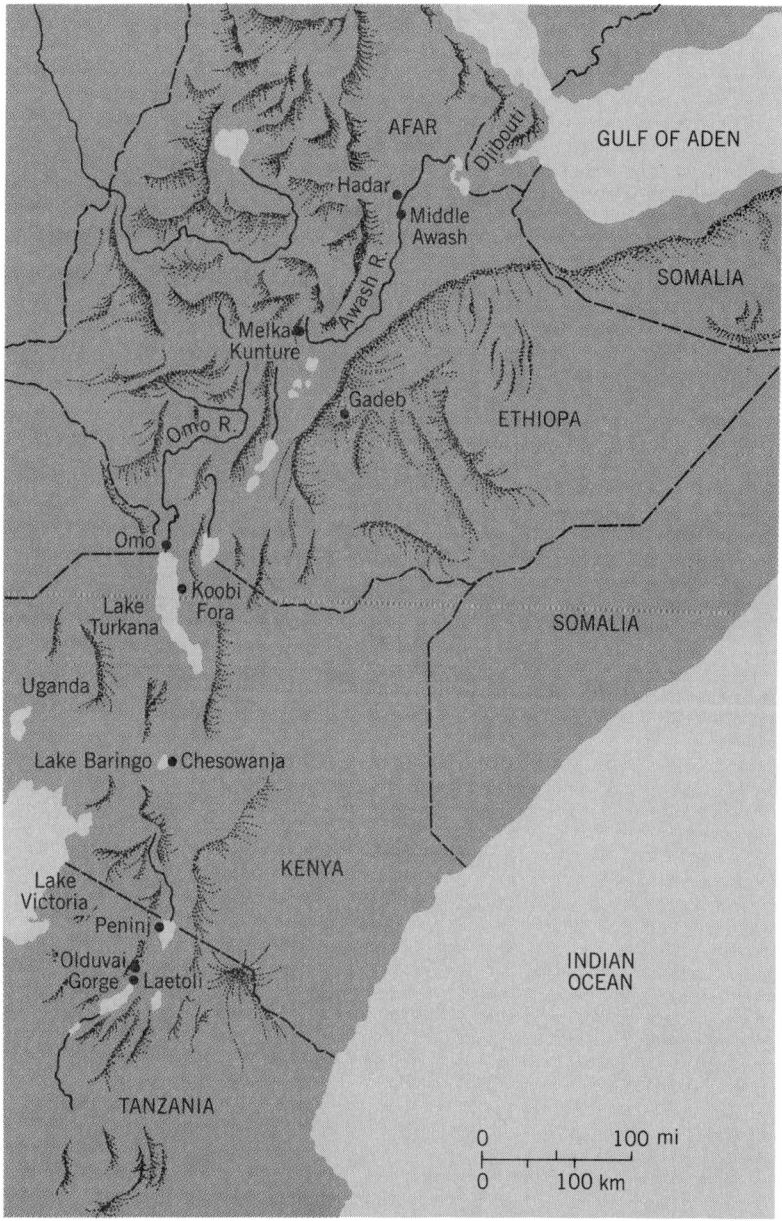

Fig. 1. Map of east Africa showing the location of the Oldowan artifact bearing localities found in the Rift Valley and surrounding terrain. In the same strata are also preserved the remains of early hominids.

Olduvai were 1.8 million years old. Subsequent paleomagnetic determinations have confirmed the age. More recently, large-scale field studies conducted in new or poorly known areas have found well-dated stone artifact occurrences of a similar character in sedimentary basins located along the floor and on the hilly flanks of the Eastern Rift Valley, which bisects the Kenyan and Ethiopain highlands. Absolute age determinations indicate that these new discoveries fall within the time interval between approximately 2.1 and 1.5 million years ago, but still earlier stone artifacts dated to 2.5–2.7 million years have recently been described from Hadar in Ethiopia. Moreover, this enlarged sample of artifact-bearing localities illustrates the geographical range of environments that early hominids occupied during this time range.

In addition to Olduvai Gorge (Tanzania), archeological sites containing the earliest stone artifacts at Chesowanja, Lake Baringo (Kenya), at Koobi Fora and Omo, Lake Turkana (Kenya and Ethiopia), and in the adjacent areas of Central Afar Depression, Hadar, and the Middle Awash (Ethiopia; Fig. 1) document early hominid occupation and activities on lake shores or close to the banks of rivers draining into ancient lakes on the Rift Valley floor. In contrast, archeological sites found at Melke Kunturé and Gadeb (Ethiopia) indicate hominid activities and hominid movements into habitats at higher elevation (6560 ft or 2000 m) associated with ancient river valleys situated on the Rift volcanic uplands. Occurrences of the very early stone artifacts outside of the Eastern Rift Valley are rare. Potentially informative early stone artifact occurrences have been reported along the Atlantic littoral at Sidi Abderrahman near Casablanca in Morocco, as well as at Ain Hanech in Algeria. Comparisons are difficult, however, because of the paucity of excavated sites. Similarly, at the cave site of Swartkrans situated on the high Plateau of South Africa near Johannesburg, early stone artifacts from the basal levels may be on the order of 2 million years old, but these have not yet been described in detail.

Form and function. Although the best-known very early excavated sites are separated by hundreds and even thousands of miles within the East African Rift Valley, the stone artifacts so far recovered exhibit broadly similar characteristics. Two basic classes of manufactured "tools" are present: fist-sized cobbles or blocks of stone from which chips or flakes have been removed, and the chips or flakes of stone themselves (Fig. 2). These tools were fashioned by striking repeatedly one cobble or pebble, which functioned as a hammerstone, against another. No arbitrarily imposed tool design is apparent, but early hominids must have had a rudimentary knowledge of the fracturing properties of stone. By simple stone percussion flaking practices, opportunistically shaped cobbles ("cores") were being manufactured in what appear to be most efficient solutions to the production of large numbers of smaller sharp-edged chips or flakes. In addition, there are many stone artifacts ("hammerstones" and "anvils") which show battering or pounding but not intentional flaking. There is also a residue of unmodified cobbles and slabs of stone (manuports) that were clearly transported to sites by early hominids but show no obvious traces of flaking and use.

The manufacture of sharp-edged tools made from durable materials like stone arose out of new sets of activities requiring simple cutting, scraping, and piercing devices. These activities marked the beginning of an awareness in the relationship between form and function by early hominids. Recent research has established direct evidence for tool function. Microscopic traces of use were discovered on some of the Oldowan stone tools excavated from several archeological sites at Koobi Fora dated to 1.5 million years. Distinct microwear polishes identified along the edges of stone tools made from fine-grained siliceous rocks (chert and chalcedony) document that they were used variously for cutting up meat, shaping wood, and cutting up plant materials. Independent corroborative evidence for the butchery of animal carcasses and thus the consumption of meat by early hominids comes from microscopically detectable cut marks found on fossilized animal bones recovered from archeological sites at Koobi Fora and Olduvai Gorge. The characteristics of the

Fig. 2. Stone artifacts recovered from the earliest known excavated site near the Gona River, Hadar, Ethiopia, dated to approximately 2.5 million years ago. (a) "Core" (chopper). (b) Small sharp-edged chips or flake tools. (From J. W. K. Harris, Cultural beginnings: Plio-Pleistocene archaeological occurrences from the Afar, Ethiopia, Afr. Archaeol. Rev., 1:3–31, 1983)

cut marks showed that they were made by early hominids using sharp-edged stone chips or flakes. Cut marks are present on bones of animals that ranged in size from elephant, giraffe, and hippopotamus to a small antelope. The position of the cut marks indicates that early hominids were not only interested in bones with substantial quantities of meat. Slicing marks on less meaty bones suggest the deliberate removal of animal skins, and perhaps the detachment of sinews and tendons to be used as cord. Also, evidence of pounding and bashing marks on significant numbers of bones indicates that weightier stone artifacts such as cores, hammerstones, or manuports were probably used for breaking open bone to obtain marrow.

Behavioral implications. The fossil record shows that by 5.5 million years ago divergence and speciation among hominoid groups had already taken place and that early protohuman creatures (hominids) had emerged in East Africa. Paleontological and paleoenvironmental evidence from Hadar (Ethiopia) and Laetoli (Tanzania) dating between 3 and 4 million years show that early hominids lived in savannas. They were small in stature, possessed ape-sized brains, but were fully bipedal. Apart from this uniquely human mode of locomotion, little is known about their behavior. Admittedly, with the freeing of the hands there may well have been a rudimentary stage of simple tool use (the use of sticks and unmodified stones, for example) that has not been preserved or is no longer visible in the archeological record. The appearance of stone tools just over 2 million years ago, however, provides the earliest tangible evidence for the beginnings of culturally elaborated behavior. Moreover, in numerous instances, Oldowan stone tools together with broken-up and cut-marked animal bones occur as concentrated patches across ancient landscapes. These concentrations reveal features of a truly human pattern, as they appear to be ancient campsites where food was repeatedly carried in for processing and perhaps consumption. There is nonhuman primate that produces such campsites.

In assessing the adaptive significance of stone tools, it is important to realize that the earliest known evidence for their manufacture and use coincides with climatic changes from moister to drier and more arid conditions. Furthermore, during this period of environmental transition there appears in the East Africa fossil record shows that there were several kinds of early hominids. These are the robust australopithecine, *Australopithecus boisei*; a more gracile form, *A. africanus*; and *Homo habilis*, which exhibits significant brain expansion and a reduction in the size of the molars or back teeth compared to the others.

In view of the marked environmental change, competition for already existing food resources or the search for alternative food resources was probably heightened between 2.5 and 2 million years. Direct evidence linking artifact use with butchery and processing of wild animals at ancient campsites shows that some hominid populations diversified their food resource base to include meat in the diet. Sharp-edged tools were crucial for cutting through the thick hide or skin of an animal to gain access to the meaty portions. Furthermore, the tools facilitated the rapid removal of parts of the carcass, flesh, or other useful materials, so that early hominids could quickly leave the scene if necessary and thus avoid competition with other carnivores. Therefore, in various ways stone tools can be viewed as having a significant effect on the survival of early hominids who were partly dependent on obtaining meat.

It is generally assumed, based upon modern ethnographic analogies of present-day hunters and gatherers, that early hominids consumed principally plant foods. By adopting a diet that also included a high protein source such as meat, however, some populations of early hominids may have greatly enhanced their reproductive success compared to others with less nutritious diets obtained perhaps from less predictable sources.

Once meat eating and stone tool use became established patterns of behavior, they had a profound effect on the foraging habits of early hominids. In this regard, sites containing stone artifacts and the bones of many different kinds of animals are particularly informative. There is an overall tendency for sites of this nature to be situated on floodplains close to stream banks. In addition to the advantages of shade, shelter, and the proximity to plant foods, there was a local source of cobblestones available for making stone tools in nearby stream conglomerates. However, stone tools were made from various kinds of stone raw materials. Therefore, out of necessity early hominids walked over widely dispersed areas of the landscape, transporting stone raw materials or discarding stone artifacts at sites up to 12.5 mi (20 km) from their source.

It seems likely, based upon the evidence for stone tool cut marks overlying carnivore gnaw marks, that early hominids were scavengers rather than hunters of wild game. Furthermore, there is good evidence to show that early hominids sometimes postponed consumption of food by transporting scavenged portions of meat away from the place where an animal had died or been killed by carnivores. They were then transporting and concentrating meat together with stone tools for processing and consumption elsewhere on the landscape. The behavioral reconstruction implies not only cooperative foraging activities in transporting meat and stone from widely separated localities, but also perhaps provisioning behavior and sharing food among social groups of early hominids. In contrast, nonhuman primates are solitary foragers who feed as they move and who rarely share food with other members of the troop.

It is generally assumed, but uncertain, that this novel shift in foraging behaviors, including meat eating and stone tool use and manufacture, is linked to the appearance of a larger-brained hominid, *H. habilis*, in the fossil record approximately 2 million years ago. It seems likely that natural selection began to favor various mental abilities reflected in these innovative adaptive patterns of behavior that began the trend toward brain enlargement and reorganization.

Stone tools should be thought of as part of a complex set of adaptations to life on the African savanna, which also favored a shift toward a higher protein meat diet, transportation of food and raw materials, and other behaviors by early hominids that formed the basis upon which human patterns of behavior and elaborations of culture were built.

[J. W. K. HARRIS]

ORIGIN OF MODERN HUMANS

Recent years have seen a renewal of interest in the evolutionary processes involved in the origins of modern humans and their adaptive patterns across the Old World. There is continued controversy as to the roles of specific human groups (such as the Neandertals of Europe and western Asia) in the ancestry of regional early modern human populations, but there is increasing consensus that the origin of modern people was associated with a major advance in human adaptive effectiveness. This shift is reflected in both the paleontological and archeological records, and is being actively investigated by an international community of anthropologists. The discussion here represents in many ways an interim report on ongoing research, research that is based on analyses during the last few years of both new discoveries of upper Pleistocene (128,000 to 10,000 years ago) remains and specimens known for most of the twentieth century.

Timing of transition. The evolutionary transition, from archaic humans (archaic *H. sapiens*, such as Neandertals,) to modern humans (*H. sapiens sapiens*), and from the adaptive pattern of the former (the Middle Paleolithic or Middle Stone Age) to that of the latter (the Upper Paleolithic or Later Stone Age), took place at different times across the Old World. This shift is here referred to as the upper Pleistocene transition. The transition started first in sub-Saharan Africa sometime around 50,000 B.P. (B.P. = years before present) and occurred last in western Europe between 31,000 and 35,000 B.P. The earliest modern humans appeared in the Near East and north Africa around 40,000 B.P., in eastern Europe about 36,000 B.P., in Australia between 40,000 and 32,000 B.P., and probably about the same time in east Asia. The arctic regions of Eurasia were first occupied about this time, but it was at least 10,000, and probably closer to 20,000, years later before people colonized the Americas via the Bering land bridge between Siberia and Alaska. The geographical spread of modern humans was a complex process, in which anatomically modern people both displaced (probably through competition for resources) and interbred with regional archaic *H. sapiens* populations. This produced a remarkably homogeneous (for their geographical dispersion) set of human populations that differed primarily in minor aspects of facial shape, bodily proportions, and presumably dermal features (such as hair color and texture, and skin pigmentation) inherited from their regional archaic ancestors.

The anatomical and associated cultural changes indicate a host of behavioral shifts, all of which were related to a significant increase in human cultural activities. The human cultural and biological spheres appear to have contributed equally toward the emergence of this new and successful adaptive pattern.

New subsistence pattern. Prior to the upper Pleistocene transition, human subsistence appears to have been largely opportunistic, with extensive gathering of plant foods (when and where available) and some hunting of small and medium-sized mammals and of the young of larger herbivores, combined with extensive scavenging of carcasses left by large mammalian carnivores. The associated technology in stone, bone, and wood was elaborate and patterned, but contained few special-purpose tools, few or no hafted points or implements, and no effective projectiles. This pattern was replaced by one with increased emphasis on hunting of larger game with less scavenging, and was facilitated by the use of hafted projectile points, using polished bone for the first time in addition to flaked stone, and by a significantly greater number of task-specific tools. This new pattern appears as a logistically organized system, in which information processing (indicated by the explosion of "art" and related notational systems) became an integral part of the system, and locations within the landscape (sites) started to be used for special-purpose activities.

Anatomical changes. Anatomical changes of the upper Pleistocene transition included a sharp reduction in massiveness of the trunk and limbs, an increase in stature, and a reduction in facial size and anterior tooth dimensions. The decrease in lower-limb robustness (to about half of its former level) indicates both lower levels of peak stress placed on the legs and a reduced ability to resist fatigue failure from repetitive activities. It therefore implies fewer strength-demanding tasks and a reduced level of habitual activity; reasonably continuous movement during waking hours probably shifted to more intermittent activity. The reduction of arm, hand, and upper trunk massiveness (also to about half of its previous level) indicates a marked decrease in the strength needed for manipulative activities. These are changes which more effective technology and planning would have allowed and which may have been energetically advantageous for the early modern human populations. The increased stature contributed to this by allowing longer strides and thus further reducing energetic expenditures. The decrease in the rate of attrition on the front teeth relative to the cheek teeth reflects a marked reduction in use of the front teeth as a vise, as may the decreases in facial and anterior dental size. The vise function of the teeth may have been taken over by the more elaborate technology.

This transition also saw the advent of the modern

human gestation length of about 9 months, decreased from the 11–12 months expected for *H. sapiens* given the developmental patterns (normal nonhuman primate perinatal developmental changes take place among modern humans about 3 months postnatally). Archaic *H. sapiens* had pelvic apertures large enough for the expected 11–12-month gestation periods, whereas early modern humans, similar to living humans, had smaller pelves that required the birth of less mature infants. Even though the earlier births of modern humans made their newborns more vulnerable to cold, infection, and injury, the earlier births may have been demographically, energetically, and neurologically advantageous once the cultural system was sufficiently sophisticated to keep a reasonable number of the altricial newborns alive.

Cultural changes. Indications of cultural sophistication are evident in the technological, notational, subsistence, and robusticity changes listed above. Early modern humans were also more effective at keeping warm, as indicated by the appearance of constructed hearths and elaborate shelters and by their spread into previously unoccupied arctic regions. Social systems became more elaborate, since their living sites were larger and more internally differentiated. Their burials included recognizable grave goods for the first time and indicate some social role differentiation. Geographical zones with identifiable artistic and technological styles emerged. Many of these cultural features appeared in relatively simple form immediately after the upper Pleistocene transition, but they evolved in highly elaborate forms, similar to those known for modern hunter-gatherers, in 10,000 to 15,000 years; by contrast, little or no perceptible cultural evolution occurred during the last 50,000 to 75,000 years prior to this transition. Clearly a more elaborate and differently organized human adaptive system emerged across the Old World between 50,000 and 31,000 B.P.

Neurological aspects. Interestingly, there was no change in human brain size or, to the extent that can be determined from impressions on the insides of skulls, in neurological organization between archaic and modern *H. sapiens*. Both archaic and early modern *H. sapiens* had brains that were, on the average, slightly larger than those of living humans, probably due to their larger average body masses. The only change in the brain was a shift toward higher and more rounded brains. This shape alteration was a product of slightly earlier brain growth among modern humans, due probably to the earlier exposure of their infants to the environment, and perhaps to the richer cultural environment to which they were exposed as infants.

The origins of modern humans across the Old World was therefore a complex biological and cultural process that saw the emergence of a technologically sophisticated and organizationally elaborate adaptive system similar to that of modern human hunter-gatherers. Within a relatively short period of time, early modern humans and their associated cultural systems replaced and absorbed the preceding archaic humans to become the dominant terrestrial mammal. The upper Pleistocene transition, although modest in some respects, was the beginning of the cultural evolutionary process that led, by the end of the upper Pleistocene (about 10,000 B.P.), to the spread of people into most inhabitable regions of the world and to the planned exploitation of a wide spectrum of natural resources. This developed, perhaps inevitably, into the direct control of resources and the emergence of agriculture, sedentation, and complex social systems.

For background information *see* ANTHROPOLOGY; ARCHEOLOGY; AUSTRALOPITHECUS; DATING METHODS; FOSSIL MAN; PHYSICAL ANTHROPOLOGY; PRIMATES in the McGraw-Hill Encyclopedia of Science and Technology.

[ERIK TRINKAUS]

Bibliography: J. W. K. Harris, Cultural Beginnings: Plio/Pleistocene archaeological occurrences from the Afar, Ethiopia, *Afr. Archaeol. Rev.*, 1:3–31, 1983; R. L. Hay and M. D. Leakey, The fossil footprints of Laetoli, *Sci. Amer.*, 246:50–57, 1982; G. Ll. Isaac, The foodsharing behavior of protohuman hominids, *Sci. Amer.*, 238:90–108, 1976; F. E. Johnston (ed.), *Pliocene Hominid Fossils from Hadar, Ethiopia*, *Amer. J. Phys, Anthrop.*, 57:373–719, 1982; L. Keeley and N. Toth, Microwear polishes on early stone tools from Koobi Fora, Kenya, *Nature*, 293:464–65, 1981; R. Potts, Home bases and early hominids, *Amer. Sci.*, 4:338–47, 1984; F. H. Smith and F. Spencer (eds.), *The Origins of Modern Humans*, 1984; E. Trinkaus, *The Shanidar Neanderthals*, 1983; E. Trinkaus (ed.), *The Mousterian Legacy*, British Archaeological Reports, vol. S164, 1983; E. Trinkaus and W. W. Howells, The Neanderthals, *Sci. Amer.*, 241(6):118–133, 1979; R. H. Tuttle, Evolution of hominid bipedalism and prehensile capabilities, *Phil. Trans. Roy. Soc. Lond. B*, 292:89–94, 1981; R. H. Tuttle, Kinesiological inferences and evolutionary implications from Laetoli bipedal trails G-1, G-2/3, and A, in M. D. Leakey and J. M. Harris (eds.), *The Pliocene Site of Laetoli, Northern Tanzania*, in press; M. H. Wolpoff, *Paleoanthropology*, 1980; J. Wymer, *The Palaeolithic Age*, 1982.

Antibacterial agents

Carboxylic acids have recently been found to be effective antibacterial agents. When administered orally to humans, they are active against a wide variety of bacteria.

Chemistry. This group of antibacterial agents consists of synthetically produced, weak, organic acids. It has sometimes been called the quinolines or the quinoline carboxylic acids; those names cover all members of the group except for ofloxacin, which is a pyridino carboxylic acid (see illustration). In this article, the carboxylic acids have been classified into two subgroups—those which are nonfluorinated and those which contain a fluoride. The former are

Chemical structures of carboxylic acid antibiotics.

older compounds and are considerably less active than the latter. Within the latter group there are sometimes only very small chemical differences; norfloxacin is, for example, the demethyl variant of pefloxacin, which in living subjects is partly metabolized to norfloxacin. Ciprofloxacin resembles norfloxacin, but in the 3-position the ethyl group in norfloxacin is replaced by cyclopropyl, a modification which seems to markedly increase the antibacterial activity.

Mode of action. The carboxylic acids affect nucleic acid synthesis in metabolically active bacterial cells and are bactericidal at concentrations close to the inhibitory ones. The exact mechanism by which their activity is exerted is still unclear, but studies on nalidixic acid indicate that it interferes with the enzymes responsible for building up the complete double-stranded deoxyribonucleic acid (DNA) molecule and also, at high cytoplasmic concentrations, affects messenger ribonucleic acid (mRNA). Toxicological studies have shown that the carboxylic acids do not affect nucleic acid synthesis of mammalian cells.

Antibacterial activity. The nonfluorinated carboxylic acids are narrow-spectrum antibiotics, mainly active against gram-negative aerobic bacilli. *Escherichia coli*, *Klebsiella* spp., *Enterobacter* spp., *Proteus mirabilis*, *Salmonella* spp., and *Shigella* spp. are normally susceptible to 10 mg/liter or less of these agents, while the susceptibility of indol-positive *Proteus* spp. and *Serratia* spp. varies. *Pseudomonas aeruginosa* and other *Pseudomonas* species, as well as aerobic gram-positive bacteria and anaerobes are resistant.

Relatively small and clinically insignificant differences have been observed among the carboxylic acids. The fluorinated compounds are considerably more active than the nonfluorinated ones. Against the gram-negative aerobes mentioned above, enoxacin, norfloxacin, ofloxacin, and pefloxacin are at least 10 times and ciprofloxacin about 100 times more active than the nonfluorinated compounds. They are also generally active against *Pseudomonas* spp., indol-positive *Proteus* spp., *Citrobacter* spp., and *Acinetobacter* spp., all of which are pathogens characterized by a high degree of resistance to many other antibacterial agents. The activity of the fluorinated carboxylic acids against gram-positive bacteria varies. Ciprofloxacin is the most active and inhibits streptococci (including enterococci) and staphylococci at concentrations below 1 mg/liter, while the other compounds are less active and not likely to affect these strains at concentrations achievable outside the urinary tract. Due to its higher intrinsic activity, ciprofloxacin is also the only carboxylic acid active against anaerobic bacteria, rendering it one of the most active antibacterial agents available.

Resistance. Resistance to carboxylic acids is chromosomal and appears to be nontransferable between bacterial cells. In fact, these antibiotics seem to be able to inhibit transfer of plasmids between bacteria. In cultures, resistance can be induced at a rate of 10^{-7} or 10^{-8} with nalidixic acid, while it seems to occur 10 to 100 times less often with the other compounds studied—cinoxacin, norfloxacin, and ciprofloxacin. The fluorinated compounds are likely to be active against many bacterial strains which have developed resistance to the nonfluorinated ones, although a reduced susceptibility to one of the carboxylic acids will affect the activity of all members of the group. However, due to the high intrinsic activity of the fluorinated compounds, they will in many cases be sufficiently active against strains resistant to the nonfluorinated ones. Importantly, there is no cross-resistance between carboxylic acids and other groups of antibiotics.

Pharmacokinetics. With the exception of ciprofloxacin, the carboxylic acids have been evaluated in humans only as oral agents. The degree of gastrointestinal absorption varies and is generally delayed and reduced by food. In subjects who have fasted, quinolines are well absorbed; the absorption of norfloxacin and ciprofloxacin is, however, incomplete and does normally not exceed 50% of the dose. The binding to serum protein (probably albumin) varies but is consistently below 50%. Metabolism of the carboxylic acid is common: nalidixic acid is metabolized to hydroxylnalidixic acid, which is antibacterially active; pefloxacin is metabolized to norfloxacin, pefloxacin N-oxide, oxonorfloxacin (which is also a metabolite of norfloxacin), and oxopefloxacin, of which only norfloxacin is active. Similar metabolites have been identified with all carboxylic acids studied, but the degree of metabolism seems to vary.

In addition to metabolism, glucoronide conjugation takes place for many of the compounds, reducing the urinary excretion of antibacterially active drug. With nalidixic acid, only about 15% of the dose is excreted as nalidixic acid or hydroxylnalidixic acid; the proportion can be increased to 25–30% by administration of the drug with sodium citrate, which increases the urinary pH and reduces glucoronidation. With the fluorinated compounds, 70% or more of the dose has been recovered as unmetabolized drug or active metabolites in the urine when ofloxacin and pefloxacin were tested. The corresponding figure for norfloxacin and ciprofloxacin is about 35%, while no human data have been found for enoxacin.

The route of excretion seems to be mainly renal in humans. The serum half-life is 2–4 hours in subjects with normal renal function, and increases considerably in patients with renal failure. Penetration to peripheral compartments seems to be excellent for all carboxylic acids studied. In deep peripheral compartments, concentrations of about 50% of those obtained in serum are achieved. With pefloxacin, cerebrospinal fluid concentrations of about 50% of the peak serum concentrations have been reported in patients with meningeal inflammation. Animal experiments indicate that in kidney tissue, concentrations which are about 10 times higher than those in serum are achieved.

Clinical use. The nonfluorinated carboxylic acids are well documented in treatment of urinary tract infections; in addition, rosoxacin has been used in the treatment of gonorrhea. The fluorinated compounds, of which only norfloxacin is well documented, also seem to have their main indication in treatment of urinary tract infections. Since they are bactericidal and since high kidney tissue concentrations are achieved, they should be suitable for treatment of both lower and upper urinary tract infections. Their antibacterial spectrum makes them the first antibiotics which can be used orally in treatment of urinary tract infections caused by *Pseudomonas aeruginosa* and other pathogens which would otherwise require the use of injectible antibiotics. Although clinical studies comparing the various fluorinated carboxylic acids are lacking, it can be assumed that they will be equally effective in the treatment of urinary infections.

Another possible indication for the fluorinated carboxylic acids is treatment of intestinal infections, for example, salmonellosis, shigellosis, campylobacter infections, and yersinosis. Respiratory tract infections caused by *Hemophilus influenzae* should be treatable with all of the fluorinated compounds, but those caused by pneumococci are probably susceptible only to ciprofloxacin due to the higher antibacterial activity of that compound. Ciprofloxacin also has a potential to become the first orally available antibiotic for treatment of systemic infections caused by *Ps. aeruginosa* and multiply resistant organisms which today require injectible antibiotic treatment. Documentation of such indications, however, is still lacking.

Safety. All carboxylic acids cause nonspecific neurological symptoms such as dizziness, headache, blurred vision, and even increased intracranial pressure. The frequency of these reactions seems to be related mainly to the plasma concentration achieved, and they are reversible. Gastrointestinal reactions, mainly nausea, occur rather frequently. In beagle puppies whose mother had been given high doses of carboxylic acids, arthritis was noted. No reports exist on similar adverse reactions in humans. Crystalluria may occur in patients given very high doses. In urine from healthy subjects receiving high norfloxacin doses, crystals were observed when the pH of the urine was between 7 and 8, the interval at which norfloxacin has a poor solubility.

For background information *see* ANTIBACTERIAL AGENTS; ANTIBIOTIC; ANTIMICROBIAL AGENTS in the McGraw-Hill Encyclopedia of Science and Technology.

[S. RAGNAR NORRBY]

Bibliography: N. X. Chin, and H. C. Neu, Ciprofloxacin, a quinoline carboxylic acid compound active against aerobic and anaerobic bacteria, *An-*

timicrob. Agents Chemother., 25:319–326, 1984; N. X. Chin, and H. C. Neu, In vitro activity of enoxacin, a quinoline carboxylic acid, compared with those of norfloxacin, new β-lactams, aminoglycosides, and trimethoprim, Antimicrob. Agents Chemother., 24:754–763, 1983; G. C. Crumplin, and J. T. Smith, Nalidixic acid: An antibacterial paradox, Antimicrob. Agents Chemother., 8:251–258, 1975; G. Montay, Y. Goeffon, and F. Roquet, Absorption, distribution, metabolic fate, and elimination of pefloxacin in mice, rats, dogs, monkeys, and humans, Antimicrob. Agents Chemother., 25:463–472, 1984; M. Rylander, and S. R. Norrby, Norfloxacin penetration into subcutaneous tissue cage fluid in rabbits and efficacy in vivo, Antimicrob. Agents Chemother., 23:352–355, 1983; R. Ziegler et al., Studies on absorption and excretion of ciprofloxacin (Bay no. 9376) in healthy male volunteers, 23d Intersciences Conference on Antimicrobial Agents and Chemotherapy, Amer. Soc. Microbiol. Abst. 851, 1983.

Archeology

Research advances in nuclear physics and chemistry have repeatedly been found applicable to problems in other sciences. Techniques used in archeology include radiocarbon dating, neutron activation analysis, mass spectroscopy, and the Mössbauer effect.

Radiocarbon dating. After the radioisotope carbon-14 was first artificially produced in the cyclotron, it was found that it could also be detected in biogenic carbon in nature. Carbon-14 is continually generated by cosmic rays bombarding nitrogen in the Earth's atmosphere, and is then transmitted as labeled carbon dioxide to plants and animals. There is always a rather steady natural level of ^{14}C in biogenic carbon, but the radioactive decay of the nuclide, once the biogenic carbon is separated from the natural pool on the death of the plant or animal, permits a radiocarbon date, or age, to be calculated.

To make this age calculation, the proportion of ^{14}C in the sample must be estimated. This ratio, $^{14}C/^{12}C$, is normally about 10^{-12}, and has been determinable until recently, only through a measurement of the radioactivity of the ^{14}C. Since 1976, however, a radically new method, employing electrostatic acceleration, has been developed. Here, charged carbon ions are brought to energies of several MeV, then separated according to their masses and measured. Thus, the ratio $^{14}C/^{12}C$ is determined directly, the Van de Graaff accelerator serving as a high-energy mass spectrometer. This method, routinely available today for ^{14}C dating, has two major advantages over conventional counter techniques: it is much faster, requiring only hours instead of days; and it can handle samples which are minute (a few milligrams) compared to those needed for the conventional procedure used previously. On the other hand, accelerators are very expensive and have substantial maintenance and operating expenses.

Parallel to the development of the accelerator, the conventional proportional counter for ^{14}C was refined and miniaturized to the point where it too could date 10 mg of carbon. Although the miniature counter is slower than the accelerator (3 months versus a few hours), it is also far simpler to operate and much cheaper. Miniature counter equipment is now installed in several ^{14}C laboratories and, like the accelerator, is available commercially to archeologists.

It is to be anticipated that further development of the accelerator method will lead ultimately to the possibility of routine ^{14}C dating of milligram or submilligram samples (single plant seeds, hairs, pollen, plant fibers used as temper in pottery, smoke on the ceilings of cave dwellings, food remains in cooking pots, and so forth), thereby greatly broadening the range of materials of interest to archeology that can be dated.

Neutron activation analysis. A totally different nuclear method in archeology is neutron activation analysis: this is analytical chemistry based upon the observation that many chemical elements become radioactive, emitting characteristic gamma radiation, when they are exposed to a current of neutrons. For example, minute traces of cobalt present in an excavated potsherd absorb neutrons, converting some of the normal ^{59}Co nuclei to ^{60}Co. When the specimen is later counted with a special gamma-ray detection device, called a germanium diode, the presence of the two characteristic gamma rays of radioactive ^{60}Co at 1173 and 1332 keV signals the presence of cobalt in the specimen: the number of these gamma rays measures the quantity of the cobalt.

There are several advantages to neutron activation as an analytical technique: it is nondestructive, can be applied to small specimens, can measure as many as 25 or 30 chemical elements simultaneously, and has a very wide dynamic range—that is, it can usually handle elements at concentrations ranging from very large (percent levels) down to ultratrace (parts per billion). But the method does require a powerful neutron source such as a nuclear reactor, highly specialized equipment, and computer facilities for data processing.

Neutron activation analysis is useful to the archeologist because the pattern of chemical element distribution in the sample (its compositional profile or fingerprint) often characterizes that sample, reflecting its geological source. Thus, archeological ceramics (potsherds); stone such as flint, chert, or obsidian used for blades or points; semiprecious gems like jade and turquoise; construction materials like sandstone and limestone; and metals like native copper can all be analytically described by compositional profiles which relate to their origins, rather than to their shaping, decoration, travels, or burial spots. This information permits the archeologist to gain insight into ancient trade routes, cultural contact areas, and even economic colonialism.

Approximately 40,000 analyses of archeological ceramics and other materials, for the most part carried out to answer specific research questions, are being combined and reduced to a data bank for archeological research at the SARCAR (Smithsonian Archaeometric Research Collection And Records) unit of the Smithsonian Institution. In many cases, the ceramic material itself will be archived as well. This data bank, already used in a number of comparative studies, is expected to grow in value as archeological sites continue to be looted and destroyed and access to them by archeologists is further restricted.

Mass spectroscopy. Because it depends upon the phenomenon of isotopy—the presence in nature of atoms of the same chemical element having virtually identical chemical properties but differing nuclear masses—the use of mass spectroscopy as an archeological tool can also be termed nuclear archeology. Mass spectroscopy has long been used in correcting the results of conventional ^{14}C dating and, as noted above, Van de Graaff high-energy mass spectrometers are now themselves performing carbon dating.

Lead has four stable isotopes, of mass 204, 206, 207, and 208. Because the relative proportions of these four differ quite widely in different natural lead deposits, the isotopic signature, measurable by mass spectroscopy, can be most helpful to the archeologist in determining the origin, or tracing the movements, of lead itself, lead alloys such as bronze, glass containing lead, and lead compounds like galena (a pigment used and traded among the American Indians). Even in silver, such as that used in coinage in archaic Greece, there is sufficient lead alloyed to permit identification of the silver mines, located among the Aegean Islands, that provided the metal for coinage.

In a similar way, the isotopic ratios of oxygen, carbon, and strontium isotopes contained in marble used in ancient sculpture and architecture have been used to locate the particular quarries in which those marble blocks were cut. A more recent development is the use of isotope ratios in carbon, nitrogen, and strontium to give indications as to primitive diets. For example, aboriginals who consumed much maize, or much fish, tend to show the carbon isotope ratios characteristic of those foods in the collagen of their bones, and this even after millennia of burial.

Mössbauer effect. The Mössbauer effect involves the emission and absorption without recoil of certain low-energy gamma rays. One of the elements whose radioisotopes emit such gamma radiation is iron, and research employing the Mössbauer technique has produced a wealth of information on the structure and bonding processes operating in crystalline compounds of iron. The connection of this nuclear phenomenon to archeology arises from the fact that almost all ancient ceramics contain iron, in the form of various oxides and silicates, which often produce the gray or bright red colors of the pottery fabric. If an unfired raw clay is examined by the Mössbauer technique, the iron present is found in one or more molecular forms which can be identified and at least roughly estimated. But as the firing proceeds to higher and higher temperatures, these compounds of iron alter themselves or interact chemically with other components of the clay to produce new forms.

When the firing has attained its maximum temperature and the pottery is subsequently cooled for removal from the kiln, the final high-temperature distribution of iron compounds is permanently "frozen in." It is then relatively easy to measure its Mössbauer pattern and to relate this pattern to the highest firing temperature that was attained, and to the important technological question of whether the firing took place under oxidizing (excess of air) or reducing (deficiency of air) conditions. All these data are obtained by comparing the Mössbauer patterns to those found in controlled laboratory firing calibration experiments. Thus the archeologist, thousands of years after the fact, can estimate to about $\pm 100°F$ ($\pm 50°C$) the firing temperatures attained in ancient pottery kilns. This kind of information is a clue to the sophistication of the technology in use at that time. Open brushwood fires generally yield lower-fired pottery than closed-in kilns, and even in the latter the design, fuel used, and airflow conditions are important clues to the technology.

Nuclear measurements in archeology are now routinely included in the design of current and future archeological field projects, and the substantial data banks already on hand ensure that the interdisciplinary interaction generally called archeometry should continue to grow.

For background information *see* ACTIVATION ANALYSIS; MASS SPECTROMETRY; MASS SPECTROSCOPE; MÖSSBAUER EFFECT; RADIOCARBON DATING in the McGraw-Hill Encyclopedia of Science and Technology.

[GARMAN HARBOTTLE]

Bibliography: J. E. Ericson and T. K. Earle (eds.), *Contexts for Prehistoric Exchange*, 1982; R. Gillespie et al., Radiocarbon measurement by accelerator mass spectrometry: An early selection of dates, *Archaeometry*, 26:15–20, 1984.

Architectural acoustics

Organists and organ builders typically emphasize the necessity for especially reverberant spaces for their instruments and resulting music. However, an examination of the acoustics problem and experience with organs indicates that the way in which the reflective sound field decays is more important than the length of time that it persists. In other words, the qualitative nature of reverberation is of greater import to a satisfactory audition of the music than the quantitative extent.

Reverberation time. The persistence of the reverberant sound field is estimated for proposed auditoriums through the equation below, developed by

$$T = \frac{0.049V}{A}$$

W. Sabine, which gives the length of time T in seconds for a reflected sound field to decrease in sound

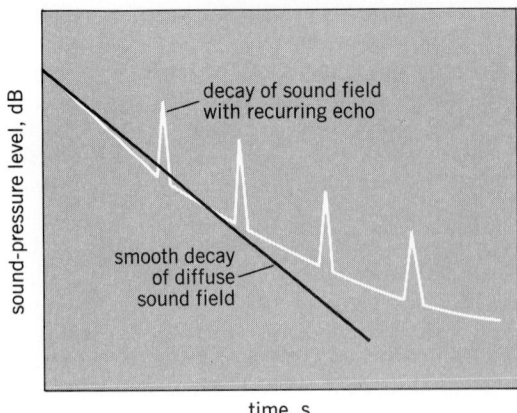

Comparison of reverberation decay curves for diffuse and nondiffuse sound fields.

pressure level by 60 dB, or from a rather loud sound level down to virtual inaudibility. Here V is the volume of the room in cubic feet, and A is the absorption in the room in sabins. If the volume is expressed in cubic meters, the formula constant becomes 0.161. The formula assumes that the reverberant field is perfectly diffuse; that is, it contains equal sound energy density at all points. Furthermore, the reverberant field decays smoothly, and long-delayed images (echoes) do not occur. The illustration shows the difference between the change in sound pressure level with time for diffuse and nondiffuse sound fields.

In most rooms the major absorption is offered by the audience. Thus, if particularly long reverberation times are desired for organ music, the Sabine formula indicates that a larger volume per person is required than for other music or speech where less lively acoustic environments are judged to be acceptable. However, auditoriums with small seating capacities will typically contain a small volume per person, no more than 200 ft^3 (6m^3) per person. As room size is increased, the volume per person rises so that an auditorium with 2500 persons may have at least 250–300 ft^3 (7–8.5 m^3) per person. If all room surfaces are hard and reflective, the reverberation time will range from about 1 s for the smaller space to almost 2 s for the larger space. However, organists and organ builders usually ask for 3–4 s reverberation time. Obviously, so lively an acoustic environment is not possible unless there is a larger volume per seat than is required for any other auditorium use.

The table lists four examples of European churches from the Gothic and Renaissance periods, including the Thomaskirche in Leipzig where Bach spent his last years. Even he did not have a space where the occupied reverberation time was 2 s. The data illustrate the necessity of having huge volumes if long reverberation times are to be realized. The table also lists four contemporary churches to illustrate that the reverberation time increases with the increase in volume per seat.

Of course, a typical suburban church does not have an extraordinary volume and is not going to sound like a cathedral.

V. O. Knudsen proposed an optimum midfrequency reverberation time of 1.7 s for a volume of 100,000 ft^3 (2832 m^3) and up to 2 s for a volume of 1,000,000 ft^3 (28,320 m^3). These suggestions are supported by the tabular data. Mendelssohn played a recital of Bach organ music on Bach's original instrument at the Thomaskirche in 1840 and commented about the end of the Toccata in F, "The Toccata sounded as if it were going to make the church tumble down . . . what a giant Bach was!" Schumann also praised the concert, tacitly approving the room acoustics.

Reflections. Even though it is generally agreed that a significantly long reverberation time is necessary to impart fullness to the musical sound, there are also concomitant requirements for good audition of organ music. Contrapuntal music, which is dominant in organ literature, demands musical clarity which, like speech intelligibility, requires the early

Occupied reverberation times for selected churches at octave bands from 125 to 2000 Hz

Church	Volume ft^3	Volume m^3	Seating capacity	Volume per person ft^3	Volume per person m^3	125 Hz	250 Hz	500 Hz	1000 Hz	2000 Hz
Historical churches										
Thomaskirche, Leipzig	640,000	18,125	1,800	355	10.0	1.8	1.9	1.9	1.8	1.7
Dom, St. Blasii, Braunschweig	565,000	16,000	1,100	514	14.6		3.2		2.4	
Münster, Weingarten	1,553,640	44,000	1,200	1,295	36.7		6.6		5.5	
St. Margaret's, Westminster, London	257,000	7,280	1,000	257	7.3			1.9		
Contemporary churches										
Dreifaltigkeits-kirche, Hamburg	247,170	7,000	470	526	14.9		2.6		2.4	
St. Mary's Cathedral, San Francisco	2,175,000	61,600	2,400	906	25.7	4.2	4.1	2.4	2.2	2.5
First Lutheran Church, Gainesville, Florida	126,120	3,570	410	308	8.7	1.6	1.1	1.2	1.1	1.2
First Presbyterian Church, Stamford, Connecticut	220,000	6,230	550	400	11.3	1.9	2.5	2.1	1.9	1.5

arrival of strong first-order reflections at the seats. Therefore, the development of a fugue whose voices can be heard entering and leaving needs high-level first-order reflections arriving at every seat no more than 50 milliseconds after the sound is heard directly from the instrument. These early images coalesce with the direct sound to form that which is perceived as a single, reinforced entity, thereby promoting clarity. To achieve the 50-ms limit, the path of travel of sound along the reflected route must not be more than 44 ft (13 m) longer than the direct path to a seat.

Continuing reflections are required to give a sense of space and fullness to the music. However, as the reverberant field continues to die away, it must not contain reflected images strong enough to overlay and mask the direct sound and early reflections that are imparting clarity. The illustration shows that smooth decay of a diffuse reverberant field avoids the masking and accompanying loss of clarity, while a field with long-delayed first-order images, that is, echoes, will confuse subsequent direct sound and early reflections. The echoes cause a "muddiness"; the music is no longer definitive and articulate. It is more important, therefore, to establish a proper pattern of decay than to have a precise reverberation time for a space for organ music.

Architectural design. Since the decay pattern and diffuseness of the reverberant field are controlled by the shape of a room, the reflective surfaces must be appropriately related to an instrument, one another, and the seating. An organ, like any other musical source, should stand immediately in front of a hard, stiff reflecting surface and be raised high enough to project sound to every seat. Side walls need to be close enough to all seats, including those at mid-width, to generate the strong early reflections. In smaller rooms, horizontal ceiling surfaces may be available for the same purpose. Well-diffused continuing reflections to complete the decay come primarily from cross-reflections between side walls and to a lesser degree from ceiling-to-wall multiple reflection paths. Diffusion is encouraged by convex, pyramidal, zigzag, and other surface modeling of moderate scale which scatters the sound radiation to make it an unintelligible blur of lower-level sound which will not mask the earlier definitive musical sounds. Rear wall surfaces must be treated with such surface modeling to prevent intelligible echoes from being returned to front seating. The steeply pitched roofs with minimal wall surfaces found in many contemporary churches are not conducive to producing and sustaining the kind of protracted sound field desired for the organ.

The essentials for good acoustics for the organ are thus (1) adequate room volume, preferably at least 400 ft^3 (11 m^3) per person, in order to obtain an adequate reverberation time; (2) hard, stiff reflective room surfaces to promote maximum reverberation; and (3) room shape to generate loud, very early reflections to every seat and continuing diffuse reflections of lower level.

Acoustics for the organ is subject to more variation than for any other instrument because of the wide range of architectural settings for the organ. Chamber groups produce an intimate sound best heard in smaller rooms, whereas the large orchestras require larger spaces if the music is not to become overwhelming during loud passages. Similarly, a small organ in a small room has an intimate charm, while a big cathedral instrument in a generous volume of space can achieve grandiosity.

For background information see ARCHITECTURAL ACOUSTICS; MUSICAL ACOUSTICS in the McGraw-Hill Encyclopedia of Science and Technology.

[BERTRAM Y. KINZEY, JR.]

Bibliography: L. L. Beranek, *Music, Acoustics, and Architecture*, 1962; Committee on Acoustics and Architecture, American Guild of Organists, *Acoustics of Worship Spaces*; L. Cremer and H. A. Muller, *Principles and Applications of Room Acoustics*, vol. 1, 1978; L. Keibs and W. Kuhl, Zur Akustik der Thomaskirche in Leipzig, *Acustica*, 1(2):49–58, 1951; V. O. Knudsen and C. M. Harris, *Acoustical Designing in Architecture*, 1950, reprint 1978.

Automobile

Up to the mid-1970s the use of electronics in automobiles was confined to the radio and occasional use on tachometers and alternator regulators. Apart from these minor applications, the control of the many vehicle systems had not significantly changed for 40 years and was carried out through traditional mechanical and electromechanical technology.

The introduction of increasingly severe exhaust emission regulations in the United States during the 1960s and 1970s made it progressively more difficult to achieve the required emission levels by conventional carburetor fueling and contact-breaker ignition control. Fortunately, at the same time the development of semiconductor technology was making it possible to develop electronic controls for fueling and ignition at an acceptable cost and with an acceptable accuracy and reliability. These controls were developed in the 1960s for use in racing cars, but were not then considered practical or cost-effective for mass-produced vehicles.

Electronic control was applied first to the ignition system and shortly thereafter to the fuel system, so that by 1975 a significant proportion of cars sold in the United States were using electronic engine control to meet the emission standards. Since then, electronics has spread to other systems of the car and is expected to be used for virtually all system control and communication functions in automotive vehicles by the mid-1990s.

Engine controls. Since the relatively simple engine control systems of 1975, the increased severity of emission regulations in the United States and Japan has forced the use there of exhaust-gas catalysts to reduce the major pollutants of carbon monoxide (CO), hydrocarbons (HC), and the oxides of nitrogen (NO_x) to acceptable levels. The use of such catalysts makes it mandatory that the engine be run on nonleaded gasoline and that the air-fuel mixture used in the engine be held at or very close to the stoichiometric air-fuel ratio of 14.7:1; otherwise the catalyst fails to operate correctly. To achieve this consistently has required the development of a sensor mounted in the exhaust system which detects the exhaust-gas oxygen level and produces a sudden voltage change as the air-fuel ratio fed to the engine passes through stoichiometry. This voltage is then fed back to the fuel control system, which usually employs a separate electromagnetic fuel injector for each engine cylinder, to modify the amount of fuel being injected and thereby control the air-fuel ratio. Rapid changes in engine speed and power output which cannot be effectively controlled by feedback from the exhaust-gas sensor are accommodated by measuring inlet manifold air pressure or mass flow together with engine speed. Then these two parameters are used to access semiconductor-stored lookup maps which specify optimum fueling and ignition conditions for the whole range of engine speeds, loads, and transient conditions.

In Europe, however, because of the much less severe emission constraints and the much higher price of gasoline, work has concentrated on operating the engine under a "lean-burn" condition, in which air-fuel ratios down to 18:1 or more are used to reduce fuel consumption dramatically while maintaining adequate drivability. This has been achieved by the use of open-loop control from look-up maps similar to those referred to above but without the exhaust-gas oxygen-sensor feedback. However, development efforts currently under way to produce a lean-burn exhaust-gas oxygen sensor could result in a convergence of these two control approaches. Other developments concerned with sensing the efficiency of the actual combustion process in the cylinders of the engine by means of pressure or ionization sensor measurements offer the possibility of much improved feedback control of ignition and fueling in the future.

Transmission control. In recent years much work has been done to develop systems which control the transmission gear ratio between the engine and the rear wheels continuously during operation to give the optimum power matching between engine and drive wheels. This generally requires that the operation of the accelerator pedal controls the transmission gear ratio and engine speed so that the engine is always run at the lowest speed possible consistent with producing the power required to propel the vehicle at the desired speed and acceleration. The continuously variable transmission is the ideal theoretical device for this. However, in practice the relatively low efficiency of the available tapered pulley/flexible belt–driven systems sacrifice a major part of the efficiency gained by the optimized wide-open throttle operation of the engine which they make possible. Future continuously variable transmission designs are likely to overcome many of these problems, but electronically controlled, continuously variable transmissions at the current level of development are now being introduced into European cars. Integration of electronic controls for engine and transmission to provide a fully optimized single power-train control system is engaging considerable numbers of engineers in the automotive and electronics industries. It should result in future power trains of improved efficiency which meet all the legislative constraints and fuel economy requirements.

Electronic instrumentation. Many electronic display technologies have been investigated for automotive use, but only two have been shown to be technically and economically viable: vacuum fluorescence and liquid crystals. Vacuum fluorescent instruments use a special form of flat vacuum tube to produce complex multielement displays in a range of colors. This technology has the same disadvantage as all light-emissive displays in an automotive vehicle in that it can be "washed out" when in direct sunlight. The liquid-crystal technology, however, separates the functions of display segment control from illumination and, with back illumination and color filtering in a semireflective operating mode, can provide complex and lively displays of excellent quality under all lighting conditions. Vacuum fluorescent technology has been used in most production instruments since its introduction in the United States in 1979, but it is being overtaken by liquid-crystal technology. The latter has the potential of providing a much more compact low-cost display if it becomes possible to mount the associated electronic circuits directly on the glass of the liquid-crystal display device.

The use of digital displays for such functions as vehicle speed indication is very simple and has been shown to be advantageous in various ergonomic studies. Similar benefits have been demonstrated for diagnostic displays of various important vehicle system conditions such as bulb failure, brake pad wear, liquid levels, and external temperature. Further development of these features is certain over the next 10 years as is the wider use of trip computers which can display information on distance traveled, average speed, fuel consumption, and expected driving range for a tank of fuel. Voice synthesis and voice recognition also seem destined to play a part in driver-vehicle interaction.

Traffic and road information. The development of increasingly sophisticated communications systems has also received much attention recently. Systems currently available range from the occasional short broadcast of information on regular broadcast transmissions, to frequently repeated information on a special channel which can be automatically received by a suitably equipped standard receiver. A system with these features is available in some European countries; similar systems exist in other parts of the world.

Vehicle routing systems which rely on communication with the vehicle by means of inductive loops in the road, are also under intensive development.

These permit the vehicle driver to enter the coordinates of a destination into a central computer which then transmits back to the driver information on the best route, usually in the form of directions on which way to turn at each road junction. Clearly, much better use of available roads is possible by such means. Navigational systems using information from satellites have also been developed for use in vehicles.

Multiplex wiring. Among the most unreliable elements in any car are the electrical connections and switches. One way of overcoming some of these problems, as well as space and installation problems associated with the increasingly complex wiring looms, is to use multiplex wiring. The multiplex ring-main operates by using a single large-gauge circuit to distribute electrical power to all parts of the vehicle, with a second signal circuit carrying switching signals to control an electronic transistor switch at each electrical load within the vehicle. Usually a group of four electronic switches is contained in a module, together with their electronic signal decoding circuits, and the module is placed close to a group of electrical loads such as a lamp cluster. The signals controlling these electronic switches are generated by appropriate dashboard switches.

The signal transmission circuit may employ a single wire, a twisted pair of wires, a coaxial cable or, in the long-term future, an optical fiber. Each method has advantages and disadvantages with respect to the system sensitivity to external interference, the information capacity of the system, the ease with which the system may be connected to the multiplex cable, and its cost. The system can easily incorporate a diagnostic system which permits the driver to receive warning information on whether the electrical component being switched is actually taken the correct current or is open- or short-circuited.

Such systems have been technically possible for some years, but their application has been delayed by the nonavailability of a low-cost, low-saturation-voltage-drop transistor switch. However, this problem has been overcome with the latest generation of these devices, and extensive use of multiplex wiring in vehicles can be expected in the future.

Other applications. Apart from the systems described above, electronics currently has, and will have increasingly in the future, an impact on a number of functions related to driver convenience, comfort, and safety. These include antilock brakes; automatic temperature and air-conditioning control; monitoring, warning, and control of seat belts, and other safety systems; speed control; driver diagnostics for fluid levels, bulb failure, brake pad wear, and tire pressure; electronic headlamp dim-dip; active or modulated rate and deflection control of the suspension; and vehicle collision warning radar.

Most of these are relatively straightforward applications of existing electronics and are currently available in some form. Others, such as active suspension control and vehicle warning radar, involve sophisticated control systems and transducers and are not likely to appear in mass-produced cars until the end of the 1980s. All, however, will contribute toward the further improvement of the convenience, performance, and safety of the vehicle over the next 10 years, in which as many dramatic developments in automotive electronics are likely to occur as over the last 10 years.

For background information see AUTOMOBILE; AUTOMOTIVE TRANSMISSION; AUTOMOTIVE VEHICLE; CONTROL SYSTEMS; ELECTRONIC DISPLAY; INTEGRATED CIRCUITS; INTERNAL COMBUSTION ENGINE; LIQUID CRYSTALS; MICROPROCESSOR in the McGraw-Hill Encyclopedia of Science and Technology.

[M. H. WESTBROOK]

Bibliography: Proceedings of the 4th International Conference on Automotive Elecronics, London, IEE Conf. Publ. 229, November 1983; *Proceedings of the International Congress on Transportation Electronics (Convergence 82)*, Dearborn, Michigan, SAE Publ. P111, October 1982; Society of Automotive Engineers, *Electronic Engine/Drivetrain Control*, SAE Publ. SP540, 1983; Society of Automotive Engineers, *An Update on Automotive Electronic Displays and Information Systems*, SAE Publ. P123, 1983.

Barnacle

The most familiar barnacles can be loosely divided into two categories—sessile or acorn barnacles which may dominate rocky shores, and stalked or goose barnacles which are usually oceanic, hanging down attached to floating objects such as logs or ship bottoms.

Most barnacles hold their fertilized eggs within their shells until they hatch as planktonic first-stage nauplius larvae. These do not feed but swim in the sea, surviving on the remnants of egg yolk until molting after about a day to second-stage nauplii. The nauplius larvae then feed on phytoplankton (minute drifting plant life) and increase in size, passing through four further molts over a period of about a month to reach the sixth nauplius or metanauplius stage.

The sixth-stage nauplius undergoes a further molt or indeed a metamorphosis to become a very different larva—the cypris larva, so named because of its similarity to a genus of ostracod crustaceans. The cypris is a nonfeeding specialized settlement stage which swims for up to three weeks or so in the case of littoral sessile barnacles but probably longer in the case of oceanic goose barnacles, relying on the stored fat reserves built up during nauplius development. The cypris selects a settlement site after alighting on a hard substratum, cements itself permanently, and metamorphoses into a recognizable barnacle.

Barnacle nauplii have three pairs of jointed appendages (listed in order from front to back): the uniramous (one-branched) antennules or first antennae, the biramous (two-branched) antennae or sec-

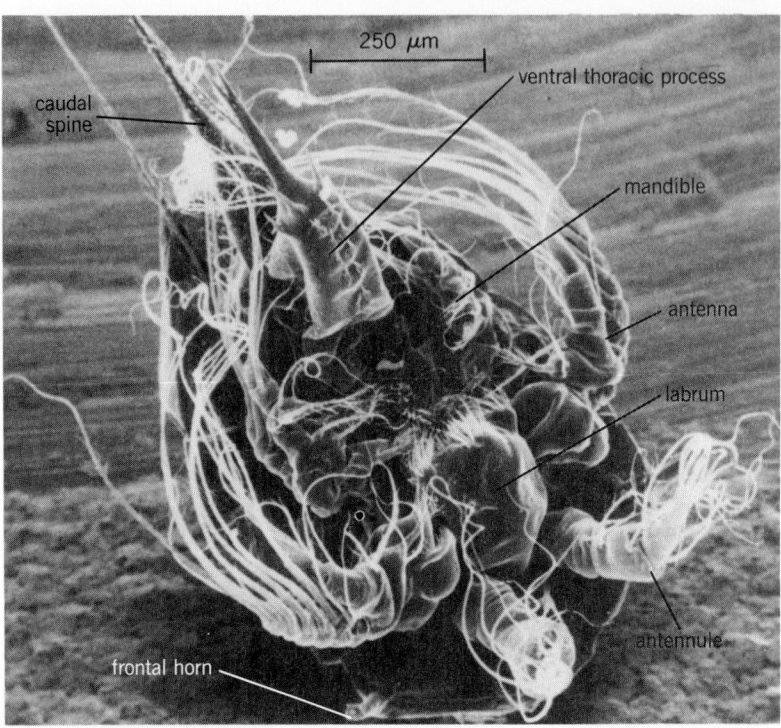

A scanning electron micrograph of the sixth-stage nauplius of the sessile barnacle *Semibalanus balanoides*. (From P.S. Rainbow and G. Walker, The feeding apparatus of the barnacle nauplius larva: A scanning electron microscope study, J. Mar. Biol. Ass. U.K., 56:321-326, Cambridge University Press, 1976)

ond antennae, and the biramous mandibles. Extending posteriorly are the ventral thoracic process (also called the abdominal process) and the caudal (dorsal) spine (see illustration).

Nauplius larvae. Swimming and feeding occur simultaneously. The antennules are usually held forward to counterbalance the posterior process and spine but may be moved in rhythm with the antennae and mandibles. The antennae and mandibles project ventrolaterally and have very long setae on each branch. They are the major locomotory and feeding organs beating in metachronal rhythm—the mandibles first, followed by the antennae and then (if at all) the antennules. The antennae, and to a lesser extent the mandibles, rotate slightly during each backward propulsive stroke so that in midstroke each antenna with setae forms a rigidly extended flattened plane vertical to the axis of the body. On the return forward recovery stroke, the antennae and mandibles are flexed partly passively and partly by muscular action so that the plane of setae is reduced in area, closing like a fan. In this way, resistance to water movement is reduced and the larva continues moving forward.

Feeding. The beating of the antennae and mandibles also brings about feeding. In the case of the larvae of sessile balanoid barnacles, which have been best studied, the propulsive backstroke of the double sweep net formed by the two pairs of limbs creates swirls which cause phytoplankton particles to be carried in toward the ventral surface of the body. The base and inner branches (endopodites) of each mandible have setae projecting in, and food particles are trapped in a "basket" between these setae and the body. As the mandibles move forward again (recovery stroke), the broad blades of the limbs close against the body and produce a forward current of water along the ventral surface of the body. Food particles are therefore moved forward along the body and pass under the labrum, the flap overlying the mouth. A midline groove along the ventral surface and setae on this surface and on the labrum, together with further setae on the mandibles, help prevent escape of the particles. The forward stroke of the antennae is still exerting a suction of water forward ventrally as the mandible beat reverses to begin the propulsive backstroke, thus causing the basal setae of the mandibles to direct the particles further in toward the mouth. The stout jaw-like processes (gnathobases) on the bases of the antennae finally push the particles into the mouth under the labrum, and they are sucked in by muscular expansion of the esophagus. Secretion from glands in the labrum may assist in tangling and holding the food particles during transport to the mouth.

Periodically, balanoid nauplii will stop and curl up so that the base of ventral thoracic process touches the labrum. Simultaneously, the long setae of the mandibles and occasionally the antennae are run across the dorsal caudal spine and the spiny dorsolateral surfaces of the ventral thoracic process. These long setae may also interlock and run through each other. This curling and wiping action has been interpreted as a step in the transfer of food, but at least in balanoid nauplii it appears to be only a cleaning process, the particles wiped off being inaccessible to the mandible endopodite setae even during curling.

Variation in feeding. This description of the feeding process of nauplius larvae of balanoid barnacles almost certainly does not apply to all other barnacle nauplii. The nauplius larvae of the north Atlantic balanoid species *Semibalanus balanoides* (previously named *Balanus balanoides*) feed only on diatoms, which are relatively large, as opposed to smaller phytoplankton types such as flagellates. On the other hand, larvae of oceanic goose barnacles *Lepas* and of other sessile barnacle genera including *Verruca* and *Chthamalus* feed and survive well on flagellates in the laboratory, as do nauplii of the warm-water species *B. perforatus*. Dietary differences appear to be correlated with differences in limb setation—those larvae feeding on flagellates having finely feathered setae capable of filtering the very small flagellates, but such setae being absent in diatom-filtering larvae.

Balanoid barnacles are thought to be relatively advanced evolutionarily, and the absence of the fine-mesh filter is therefore probably a secondary specialization to allow efficient feeding on the rich seasonal diatom bloom of temperate seas. The barnacle *S. balanoides*, for example, restricts its nauplius development to the spring months when diatoms are present in abundance. Smaller flagellates are the dominant phytoplankton of warmer and oceanic waters, and the genera *Lepas* and *Chthamalus* and the species *B. perforatus* are better served by being adapted to feed thereon. The flagellate filtering nauplius larvae of *B. perforatus* have possibly reevolved from a diatom-feeding ancestral *Balanus* species, for a fine-mesh filter (a fringe of setae) is now located on the antennal expodite (outer branch). In all other (more primitive) nauplii capable of feeding on flagellates (such as *Lepas*, *Verruca*, and *Chthamalus*), the flagellate filter is located on the antennal endopodite (inner branch).

The oceanic nauplii of the lepadid goose barnacles are very large and slow-moving compared to the nauplii of the barnacles of coastal waters, including *Balanus*. *Lepas* sixth-stage nauplii may reach 10 mm (0.4 in.) in total length and swim at two limb beats per second (about 4 mm or 0.2 in. per second) at 68°F (20°C). Nauplii of balanoid barnacles, on the other hand, reach about 1 mm (0.04 in.) in length and beat six times per second at the same temperature. It is quite probable that the large nauplii of goose barnacles with correspondingly large setal areas on antennae and mandibles are close to the maximum size limit for the nauplius pattern of locomotion. The posterior body processes are greatly extended and may be necessarily so to give stability in the presence of the large limb areas. The antennules beat regularly and may be needed to maintain forward movement during the recovery phase of the large antennae. Very little net forward movement is actually produced, and the major role of the beat of the antennae and mandibles is feeding. The limited forward locomotion will prevent repeated filtering at one location, but is not sufficient to move the nauplii to any significant degree in the water column as is typical for most zooplankters, including balanoid nauplii which undergo diurnal vertical migrations.

The feeding process of lepadid nauplii is also different from that of balanoid nauplii. Small food particles are filtered on the fine filters on the antennal endopodites, and further food particles adhere to the long antennal setae before they are wiped against the spines of the ventral thoracic process. In contrast to the situation in balanoid nauplii, these wiped particles are positioned so that they come into contact with the bristles of the labrum on curling and can be transferred forward to the mouth by the action of mandibular and antennal setae. In addition the forward food transfer currents observed in balanoid nauplii appear to be absent in large goose barnacle nauplii.

Some stalked barnacles and parasitic barnacles have nonfeeding nauplii drifting in the plankton for a period using yolk reserves provided by the parent, as do the first-stage nauplii of balanoids.

Cypris larvae. The cypris larvae of all barnacles are nonfeeding and swim by means of six pairs of thoracic limbs. These are homologs with the elongated ventral thoracic process in the nauplius stages and with the six pairs of thoracic cirri in adult barnacles. After alighting on a substratum, the cypris larvae walk by means of the antennules which are much altered from the swimming appendages of the nauplii. Antennae and mandibles are much reduced and play no role in cypris locomotion.

For background information *see* BARNACLE; CRUSTACEA; FEEDING MECHANISMS (INVERTEBRATE) in the McGraw-Hill Encyclopedia of Science and Technology.

[P. S. RAINBOW]

Bibliography: J. H. Lochhead, On the feeding mechanism of the nauplius of *Balanus perforatus* Bruguiere, *J. Linn. Soc. (Zool.)*,, 39:429–442, 1936; J. Moyse, Some observations on the swimming and feeding of the nauplius larvae of *Lepas pectinata* (Cirripedia:Crustacea), *Zool. J. Linn. Soc.*, 80:323–336, 1984; P. S. Rainbow and G. Walker, The feeding apparatus of the barnacle nauplius larva: A scanning electron microscope study, *J. Mar. Biol. Ass. U.K.*, 56:321–326, 1976.

Bioluminescence

Recent studies of the patterns and functions of bioluminescence in coastal marine organisms have shown that while representatives from at least nine groups produce light (Table 1), it is estimated that only 1–2% of all coastal species worldwide are bioluminescent. This contrasts with oceanic and midwater species, where in certain groups more than 75% of the species and 85–95% of the individuals may be bioluminescent. The reason for this difference is not known, but it may be related to the more variable environment of the coastal marine waters with regard to light, temperature, currents, physical surroundings, and other factors.

From these studies it was concluded that most luminescence seen in coastal waters occurs in re-

Table 1. Principal marine coastal organisms that produce light

Group	Organism
Bacteria	Gram-negative rods, free living and symbiotic
Dinoflagellates	Unicellular, flagellated algae
Cnidarians	Jellyfish, colonial hydroid, sea pansy
Ctenophores	Comb jelly, sea walnut
Annelids	Fireworms, scaleworms, tubeworms
Arthropods	Crustaceans, ostracods, shrimp
Mollusks	Squid, clam
Echinoderms	Brittle stars
Chordates	Fishes

sponse to contact stimulation and that it is generally displayed in one of two different patterns—as fast flashes (<2 s) or as slow glows (>5 s). Interestingly, both are viewed as being primarily defensive, aimed at deterring potential predators. But bioluminescence is recognized as having at least two other types of functions, which may be classed as offensive (predation) and communication (Table 2). It is hypothesized that even where other functions do occur, the light signals are displayed so as to thwart predators from using these signals to locate or capture the emitting organism.

Strategies for defense. The three main defensive strategies, described below, are to frighten, to act as decoy, and to camouflage.

Flash and frighten effect. Dinoflagellates provide a good example of this strategy; they respond to physical stimulation by emitting bright rapid (0.1-s) flashes readily visible to the dark-adapted naked eye. Light emission comes from subcellular organelles where dinoflagellate luciferase and luciferin are localized; triggering is believed to involve a membrane action potential and an intracellular pH change. Dinoflagellate flashing is generally believed to serve a defensive role, either directly by frightening, startling, diverting, or temporarily blinding the predator, or indirectly by attracting a predator of the predator (alarm effect). Experimental results are consistent with a direct effect, and selection for the indirect effect involves complex constraints.

Many ctenophores and cnidarians are similar to dinoflagellates in their flashing behavior, but all differ in their biochemical basis; triggering of the flash is believed to involve a change in intracellular calcium instead of pH, and the luciferin and luciferase are completely different structurally.

Decoy effect. A glowing object in the ocean generally appears to attract feeders or predators. This might be turned to the advantage of the prey in escaping predation if an organism could create a decoy light to attract the predator and then slip off under the cover of darkness. In several cases this is exactly what is done. In the waters off Japan there is a small ostracod crustacean that synthesizes its luciferin and luciferase in two separate glands. Upon excitation, these glands squirt their contents into the sea, where the bioluminescent reaction occurs as the organism swims away; the predator has only light to eat.

Some organisms, such as tubeworms and boring pholad clams, produce luminescent "clouds," a luminous exudate that is projected away from the "house" and the vital body parts by the flow stream of these filtering organisms. The light may be intense enough to confuse and drive out smaller invading organisms, perhaps offering these up as morsels to would-be predators.

Other organisms sacrifice more than light: in scaleworms and brittle stars, a part of the body may be automized (broken off) and left behind as a luminescent decoy to divert the predator. These animals illustrate Morin's distinction between flash and glow. The scales or arms exhibit frightening flashes while still attached to the animal, but after being jettisoned the light changes form to an attracting glow.

Camouflage effect. A unique method for evading predation from below is to camouflage the silhouette by emitting light which exactly matches the color and intensity of the background light. By analogy with countershading in reflected light, this has been called counterillumination. Imagine an airplane in the sky. If it could emit light from its bottom surface matching the sky behind, it would be invisible from below. It is not necessary to match the background exactly in order to conceal an object; it is only necessary to disrupt its appearance so it is no longer recognizable. With bioluminescence, this would then be called disruptive illumination. Many marine organisms utilize disruptive illumination, including many shallow-water fishes with either gut-associated glandular light organs (carrying luminous bacteria as a light source) or ventral photophores.

Strategies for offense. These same luminous signals—glow and flash—may also be used for predation. A luminous decoy may be used as a lure to attract potential prey. The coastal flashlight fish, *Photoblepharon*, utilizes luminescence to attract positively phototactic plankters on which the fish then feeds. In this case, the light organ of the fish

Table 2. Functions of bioluminescence

Function	Strategy	Method
Defense (predator avoidance)	Frighten, startle	Flashing
	Decoy, divert	Glow, luminous cloud, sacrificial lure
	Camouflage	Disrupt or obscure silhouette
	Vision	Emit light to see predator
Offense (predation aid)	Frighten, startle	Flashing
	Lure	Glow to attract phototactic prey
	Camouflage	Disrupt or obscure silhouette
	Vision	Emit light to see prey
Communication	Courtship	Flash, glow, luminous cloud
	Spacing	
	Interspecific advertising (for mutualism)	

may be used also as a light source (flashlight) to aid in vision for locating and capturing the prey. The ventral glow used to camouflage the silhouette and escape predators may be used offensively by the same animal; it can allow closer approach to prey. A similar argument can be made with regard to flashes that frighten, stun, or temporarily blind: organisms so affected are more susceptible to predation by the flashing animal itself.

Strategies in communication. Information exchange between individuals of a species may involve bioluminescent signals. Organisms known to use such signals include crustaceans, squid, fishes, and some annelids. The best-known cases involve using the light for courtship and mating. In the Bermuda fireworm a truly extraordinary display occurs as the males and females engage in mating, shedding eggs and sperm in the ocean in a circle of brilliant luminescence. The actual role of luminescence is not certain, but it may function to help mates to locate one another. The midshipman fish, which occurs in shallow water along the west coast of the United States, may also use luminescence in courtship. Luminescence probably plays a role in the spacing of fishes in coastal waters. Flashlight fish, for example, may use their lights to form and maintain the typical large aggregations (50–200 fishes).

Light emitted by one species which is detected and acted on by a second for the mutual advantage of both falls into a separate category. Morin has referred to this as advertisement. The light of luminous bacteria growing on fecal pellets or on invertebrates as parasites would attract a fish or crustacean to consume the material to the benefit of both.

Bacterial luminescence. Except for those involved in specialized light-organ symbioses, the luminous bacteria, which emit light continuously via an enzymatic pathway related to cell respiration, represent an evident exception to the generalization that light signals are displayed for defense. These signals function primarily for interspecies communication by advertisement. As microorganisms, the bacteria emit too little light to be seen individually. They grow as saprophytes on dead animals, and as commensals on the surfaces of live ones. They occur as major marine enteric bacteria; the intestines of many fishes and crustaceans are brightly luminous, and fecal pellets may be very bright—meaning that they can be readily recycled by night-feeding coprophagous organisms. Luminous bacteria also are parasitic and may infect and ultimately kill crustaceans and other organisms. In that case also, the luminescence of the infected organism would make it more visible by night to predators; this may be viewed as advertisement which allows the bacteria to be cycled into a high-nutrient enteric environment.

Luminous bacteria are also cultured as true symbionts in special light organs by higher organisms such as fishes and squid. At least 10 different families of fishes possess such organs. The host utilizes the light for its purposes (which may include any or all of the three listed above), while the bacteria are provided with a niche and nutrients.

Diversity and evolution. Bioluminescence is believed to have arisen independently many times during evolution, for there appear to be some 30 different luminescent systems among extant species. Inasmuch as luminescence is a nonessential function, it might have been evolved and then been lost many times; its loss in any given group would not necessarily have led to extinction of that group.

Although many examples of patterns and the functions of bioluminescence in coastal marine organisms have been described, there are many more which are still not understood. Of the some 30 biochemically different luminous systems, only 7 have been characterized chemically in any detail. Knowledge of behavior is also still rudimentary; much research remains to be carried out in order to understand the origin, evolution, biochemistry, and function of bioluminescence in living organisms.

For background information see BACTERIAL LUMINESCENCE; BIOLUMINESCENCE; ECOLOGICAL INTERACTIONS in the McGraw-Hill Encyclopedia of Science and Technology.

[J. W. HASTINGS]

Bibliography: J. W. Hastings, Biological diversity, chemical mechanisms and evolutionary origins of bioluminescent systems, *J. Mol. Evol.*, 19:309–321, 1983; P. J. Herring (ed.), *Bioluminescence in Action*, 1978; J. G. Morin, Coastal bioluminescence, *Bull. Mar. Sci.* (London), 33:787–817, 1983; K. H. Nealson, *Bioluminescence: Current Perspectives*, 1981.

Biomechanics

Biologists have long been concerned with the relationship between form and function in organisms; one such function has been loosely referred to as "support." Recent exchanges between biologists and mechanical engineers have led to the emergence of a more sophisticated and quantitative view of materials, structure, and support as the field of comparative biomechanics. Because the participating biologists have been interested in marine organisms, and also because the greatest diversity among organisms occurs in the marine environment, work has focused on these plants and animals. This article describes the biomechanics of soft, sessile marine organisms of the rocky intertidal zone.

If the measure of mechanical stress is considered to be the maximal force to which a given object might be subjected, the most stressful habitats of any organisms must be at the edges of oceans. In a storm hitting a rocky headland, water may be thrown more than 30 ft (10 m) into the air; it is a simple matter to calculate that the water current against the shore must have had an initial speed of about 35 mi/h (15 m/s), neglecting air resistance. This figure, in fact, has been confirmed by direct measurements using recording and transmitting equipment attached to the rocks. While such a speed is slow compared to that of, say, a hurricane, water is nearly a thousand times denser than air and so the forces generated are comparably greater. A fist extended from an automobile window into air at hurricane speed will be exposed to a force of a pound or two; in the water current of 35 mi/h (15 m/s) the comparable force would be about 200 lb.

Stratagems to reduce mechanical stress. The soft, flexible organisms with which coastal rocks are commonly covered must manage to endure such pounding. Several stratagems are recognized which, it must be emphasized, are neither mutually exclusive nor exhaustive. Most importantly, all of these devices are not merely possible for flexible organisms—they positively depend on flexibility.

Weathervaning. An organism may be arranged so that the area exposed perpendicular to the direction of flow is minimized; the exposed surface will incur much less drag if oriented parallel to, rather than normal to, local currents. While flows may vary from moment to moment in both magnitude and direction, a design using sufficiently flexible materials can easily be arranged to reorient as necessary in the manner of a flag or weathervane. Tall sea anemones, gorgonian corals (sea whips), sea pens, and many kinds of large algae use this device.

Proximity to substratum. Reorientation in severe currents which brings the surface of the organism closer to that of the substratum will reduce the forces on it. At the substratum itself, as at any solid-fluid interface, the speed of flow with respect to the solid is zero; the speed increases in the direction away from the substratum and asymptotically approaches the free-stream velocity. The thickness of this gradient region, the boundary layer, varies with local circumstances; but it is always present. Planar, fenestrated sea fans as well as the organisms already mentioned use the scheme; and again it is flexibility which creates the opportunity to reduce drag.

Reduction of exposed area. Another useful reorientation is the convergence of the otherwise diverging portions of an organism with the result that the total area exposed to rapid flow is reduced. A familiar case is an isolated tree in a strong wind; and, indeed, the magnitude of the effect has been evaluated in pines, and its structural basis investigated in palms. Among marine organisms, large, laminate algae have been given the most attention in this regard.

Streamlining. Streamlining, in the usual sense of ellipsoidal form with rounded upstream end and pointed downstream end, is relatively uncommon in sessile coastal organisms; it has been noted mainly as a feature of large, actively locomoting forms such as fishes, squid, shrimp, sea turtles, cetaceans (dolphins and whales), and diving birds. (For small organisms, streamlining is much less useful as they face a different fluid-mechanical regime.) But at least one algal thallus has now been shown to be well streamlined. It is elliptical and about four times as long (ranging up to 12 in. or 30 cm) as thick. The thallus is tethered to a rock by a short and, as is mandatory for streamlining, very flexible stalk attached to the upstream end.

Reduction of volume. A less obvious device is the reduction of surface area by reducing the overall volume of the organism. In the short term, at least, volume is constant for most organisms. Many coelenterates, though, are soft, filled with sea water, and capable of body wall contraction and vomiting of a large fraction of their functional volume back into the littoral. The most impressive are some tall (12 in. or 30 cm) anemones that can reduce both volume and height by around 90% in a few seconds.

Movement with the current. Still another mechanism is movement with the local current until the speed and acceleration of surging water has ebbed from the maxima. One cannot pull with much force on a nontaut string, and a long enough organism will take many seconds to reverse direction in the bidirectional flows induced by surface waves hitting a shore. The scheme is known in several very large algae.

Nutrition and stress. Two of the common ways that inhabitants of the rocky intertidal zone obtain energy are through photosynthesis and passive suspension feeding (that is, where flow is ambient, not pumped). In either case, the yield must be closely related to the area of surface exposed. For suspension feeding, the consideration is particularly difficult—surface is effective in filtration to the extent that it is normal to flow. The usual solution to the divergent requirements of nutrition and stress minimization is to take advantage of the intermittent character of very rapid flows and arrange to minimize drag with increasing "singlemindedness" as flows become more rapid. In all cases so far measured, drag increases as velocity increases. But it is noteworthy how slight is the rate of increase as organisms make increasing use of the devices mentioned at higher speeds.

Drag. For an ordinary object in the size and speed range of present interest, drag increases in proportion to the square of speed: the exponent of proportionality is 2.0. For a streamlined object, the exponent is lower, 1.5. For the streamlined algal thallus mentioned above, the exponent is similar but a little lower yet, 1.2. For a cabbagelike alga, a sea pen, and a small, branched colonial hydroid, the exponents are still lower, 0.7 to 0.9. Curiously, the exponents for small pine and holly trees in a wind tunnel are in this last range as well. Flexibility is crucial: when the hydroid is artificially stiffened, its exponent rises to 1.8, almost the same as the ordinary object. The lowest value now known, 0.5, was obtained for another alga, the sargassum weed, which is capable of particularly drastic convergence of its branches.

Absence of erect forms. There are no marine analogs of trees in the rocky intertidal. The nearest, perhaps, is the so-called sea palm; but it is far more flexible than any proper tree and appears treelike only under fairly calm conditions. As noted earlier, however, the forces in marine habitats can greatly exceed any occurring in air. A quantitative threshold seems to separate responses to the flows in the two media. "Trees," from tree ferns to bamboos, re-

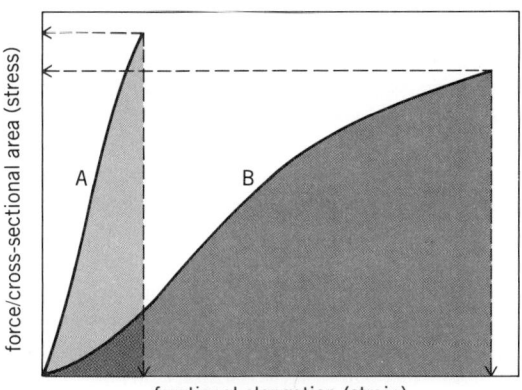

The response of two materials to stretching. The maximum height of each curve gives the strength of the material; the horizontal distance gives the material's extensibility; the slope of each curve indicates the tensile stiffness (elastic modulus); the area under a curve is the material's toughness. Material A may be stronger, but it is much less tough (is more brittle) than is B.

main erect until the point of destruction; only stony corals (quite a different story—breakage seems in part a reproductive device) do the same in water. It seems that among flexible forms, only nonerect designs are suitable for habitats with rapid flows of water.

The absence of continuously erect forms from rocky coasts reflects a basic mechanical as well as biomechanical problem. It is far cheaper in material to withstand a given load if that load is tensile (pulling) than if it is compressive, bending, or torsional. While marine forms may stand erect in low flows, and many must do so to feed or photosynthesize, bending through appropriate flexibility circumvents the more mechanically awkward forces.

Strength vs. toughness. The question arises as to whether flexible organisms are really strong, or whether they are merely specialists at avoiding drag. The question masks a distinction between the force (properly, the force per unit cross section, or stress) needed to break something, or its strength; and the work (the work per unit volume) needed to break something, or its toughness (see illustration). Work is force times distance; those items which must be stretched to be broken are less brittle—in short, tougher. Concrete and hard coral are strong but not very tough; a sharp blow cracks either. Wood and bone, the axial material of a sea whip, and the stalk of one alga are only a little less strong but over 10 times tougher. But some flexible, biological materials which take stresses in tension are 10 to 1000 times tougher yet; among these are several other algal stipes, cellulose, tendons, and silk. Toughness amounts to a resistance to the propagation of cracks and tears, and in this property the flexible organisms of exposed rocky coasts particularly excel.

For background information see BIOMECHANICS; FEEDING MECHANISMS (INVERTEBRATE) in the McGraw-Hill Encyclopedia of Science and Technology.

[S. VOGEL]

Bibliography: M. A. R. Koehl, How do benthic organisms withstand moving water?, *Amer. Zool.*, 24:57–70, 1984; S. Vogel, Drag and flexibility in sessile organisms, *Amer. Zool.*, 24:37–44, 1984; S. Vogel, *Life in Moving Fluids*, 1981; S. A. Wainwright et al., *Mechanical Design in Organisms*, 1976.

Biomedical engineering

New technology, particularly in the area of microprocessors and microcomputers, has opened up possibilities for new types of aids for the blind and partially sighted. It is not only the dramatic reduction in price of such components but also the much lower development costs which are of great significance for aids required in small quantities.

People without vision have obvious problems with mobility and reading. It is less obvious that blindness results in a serious lack of privacy; for instance, personal financial information may have to be read to a blind person. The general public assumes that blind people are those without sight; however, over 85% of the legally blind in most developed countries have some useful vision.

Television reading aids. The principal technological development for those with residual vision has been closed-circuit television reading aids (Fig. 1). These devices usually provide variable magnification, enhanced contrast, and image reversal (white letters on a black background), but the disadvantages include the high cost and the weight of the machines (typically 55 lb or 25 kg). Nevertheless, they have proved to be very useful for those whose needs have not been met by conventional optical aids.

Tape recorders. For users with no useful residual vision, aids for access to written information have either audio or tactual output. Probably the most useful technical aid is the cassette tape recorder. The main disadvantages are the user's inability to vary the speed and the lack of a good indexing system. The variable-speed problem can be overcome by using compressed-speech modules which correct the pitch changes resulting from increasing the speed. This is done by sampling the signal and discarding some of the samples; the electronic processing links the remaining samples without unpleasant transients. The indexing problem is not so easily solved. What is required is an inexpensive system so that the blind user can input a number or keyword and the machine will then automatically find the information; such systems exist but are prohibitively expensive. A less satisfactory solution is to record index terms on another track from that carrying the sound, but such a system is laborious to use.

Synthetic speech. A variety of devices with speech output are now commercially available. The simplest use spelled speech where the output is character by character; the advantage is the very low cost, but the quality is unacceptably low for any application involving prolonged listening. The qual-

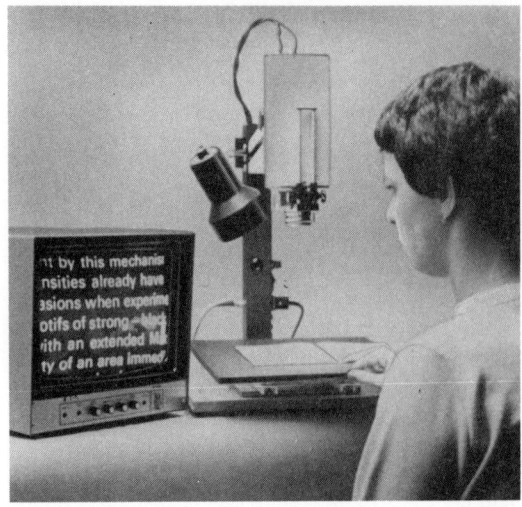

Fig. 1. Closed-circuit television reading aid. (*J Heathcoat and Co.*)

ity of full-vocabulary synthetic speech, where the machine approximates a human speaker, is dependent on the size of the computer program, but high-quality synthetic speech systems are still expensive.

The major breakthrough has been in the medium-quality systems. There are now a large number of full-vocabulary speech-output devices with computer interfaces with relatively low prices. These devices are very useful for blind people who want a file readout and are prepared to tolerate the robot-like speech. However, many blind people want to use a microcomputer in a word-processing mode; the available speech terminals, with cursor information, are very expensive—presenting a large increment in cost for a small increment in performance.

A reading machine with optical character recognition and synthetic-speech output permits a blind person to have direct access to printed books, but costs are prohibitive, making it cheaper to employ a sighted reader.

Braille. The best-known communication medium for the blind is braille, where dots are embossed on paper or plastic. The system was developed over 150 years ago by a blind Frenchman, Louis Braille. Braille utilizes a 6-dot cell giving 64 possible combinations. One of the disadvantages of braille is the considerable bulk, which is typically 20 times that of the print version. To reduce this bulk, 190 contractions and abbreviations are used which result in a 25% saving in space. There is an acute shortage of people skilled in transcribing braille, so a number of computer-based systems have been developed to translate text to contracted braille. Such systems permit a typist with no knowledge of braille to produce documents in both ink-print and contracted braille from a single typing operation (Fig. 2).

Computer-based systems are significantly cheaper than manual transcription for information which already exists in digital form. For instance, major banks in the United Kingdom use an automated system to produce statements of account in braille. Since the system is totally automated, it also minimizes the risk of an error in the braille version.

Braille has not been superseded by other forms of nonvisual media despite numerous predictions to the contrary, and it is still supreme in its use for reference and technical material. Another important aspect is that a blind person can write braille without having to invest in expensive equipment.

A number of systems have been developed for storing braille digitally on cassette or floppy disk. The braille is output on a transitory display such as an array of pins which can be raised to represent the braille characters. These devices typically are beyond the financial reach of most blind people for use at home.

Reading machines. Devices have been developed for converting printed characters to some form of tactual output. Most of these devices do not recognize the characters, but present a tactual display which has to be recognized by the human reader. The most widely used is the Optacon, which gives an enlarged tactual image of the letter being scanned (Fig. 3). The advantage of this device is that it can be used on any printed or typewritten material. The disadvantages are the considerable training and practice required to reach speeds of 50 words per minute, and the high cost.

Large print. Although most of the legally blind have residual vision, it is only very recently that modern technology has been used to produce reading material for this group. The most notable development has been the application of laser printers for the fast production of large or "clear" print. A laser printer is capable, with special computer programs, of producing good-quality print of any size at speeds up to four pages per second. Another method of producing clear print is to use a matrix printer with up to 12 passes for each line; the matrix is offset each pass and the end result is a very clear bold image, but the printing speed is slow (typically 13 characters per second).

Electronic mobility aids. The traditional mobility aids for the blind are the long cane and the guide dog. The long cane has the advantages of simplicity and cheapness; it can be a very effective mobility aid with the proper training, but it does not provide the user with any warning of obstacles at head height. Electronic aids can be used as a supplement to a long cane by providing an audio or tactile warning of such obstacles.

The electronic aids are usually obstacle detectors or environmental sensors. The former indicate only whether there is an obstacle within range; for instance, the device may give out five different musical notes depending on the distance of the nearest object at which the device is being pointed. The environmental sensors give an image of the immediate environment as a complex stereophonic audio signal. These devices give a more complete "picture," but require significantly more training for effective

Fig. 2. Braille and ink-print text-processing system. (*RB Aids for the Blind*)

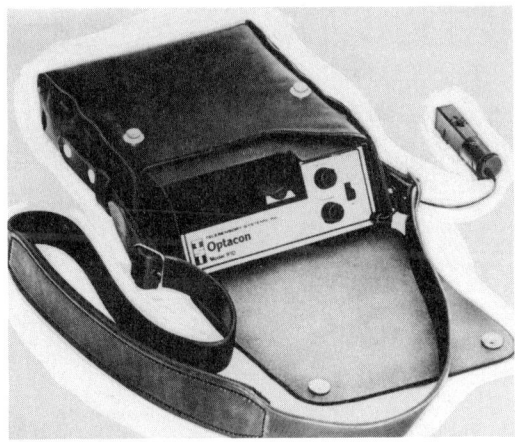

Fig. 3. Optacon reading aid. (*Telesensory Systems*)

use. There are few commercially available mobility aids, however. The chief problems with existing aids concern identifying the information required by the blind pedestrian and determining the optimal nonvisual display of this information.

For background information *see* BIOMEDICAL ENGINEERING; BLINDNESS, CHARACTER RECOGNITION; MAGNETIC RECORDING; MICROCOMPUTER; MICROPROCESSOR; VOICE RESPONSE in the McGraw-Hill Encyclopedia of Science and Technology.

[J. M. GILL]

Bibliography: J. M. Gill, *International Survey of Aids for the Visually Disabled*, 1984.

Blastocystis

Blastocystis hominis, an intestinal protozoan parasite of humans, is a newly discovered disease agent in the sense that it languished for 70 years labeled as a harmless intestinal yeast and the disease that it caused was attributed to other etiological agents. It was almost totally neglected as a subject for study. A few researchers were puzzled by its multiplicity of forms and questioned the yeast classification, but failure to find intracellular *B. hominis*, a life cycle, a thick-walled cyst form, or intermediate hosts hampered efforts. Following studies conducted in the late 1960s, *B. hominis* has become generally recognized as a protozoan and as a cause of diarrhea of humans.

Morphology. *Blastocystis hominis* cells as seen in intestinal contents are spherical, about 10–15 micrometers in diameter, with only a glistening membrane as envelope, and appear empty under phase microscopy or by stained smear except for a few peripheral granules. The apparently empty membrane-bound central body comprises most of the volume of the cell, as seen by light microscopy. Electron microscopy, however, reveals membrane-bound granules, some of which develop into progeny.

The *B. hominis* organelles include one or more nuclei, mitochondria, Golgi apparatus, microbodies

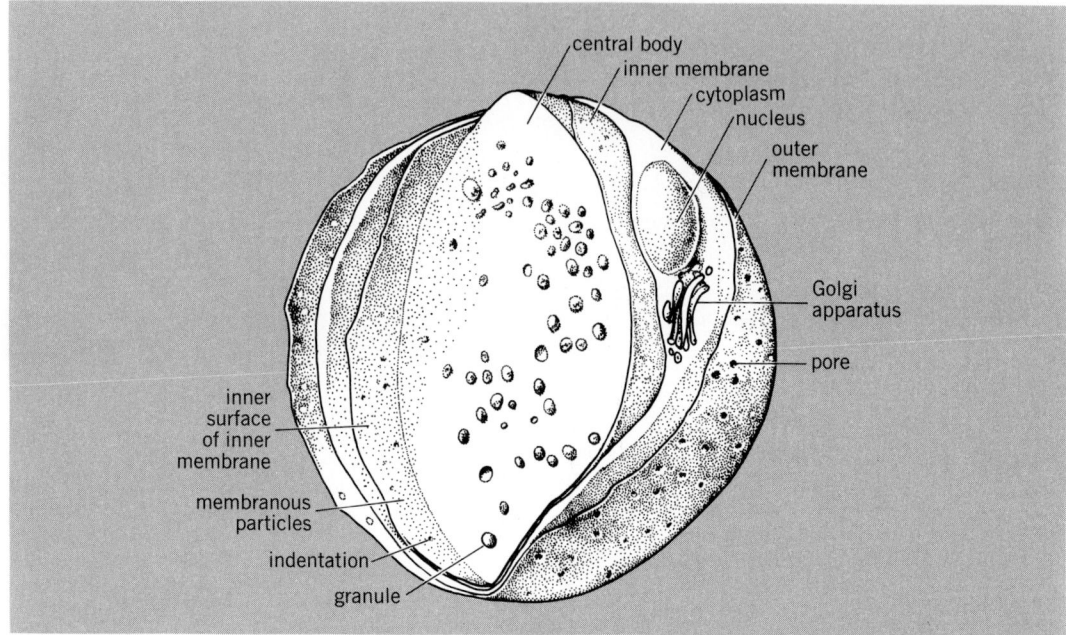

Fig. 1. Schematic of *Blastocystis hominis* showing inner and outer membranes with pores and particles, also the central body and organelles in peripheral cytoplasm.

(resembling glycosomes), rough and smooth endoplasmic reticulum, polyribosomes, and vacuoles (see table). They are squeezed between the central body membrane and outer cell membrane in a narrow peripheral band of cytoplasm (Fig. 1). In most cultures, *B. hominis* cells are the same as they are in intestinal contents, but under special conditions they can be made to grow extremely large, as giant cells of 300–400 μm. If the cells were less transparent, at this size they could be seen in suspension with no magnification.

Classification. Although *B. hominis* is a protozoan, classification within the Protozoa presents problems since *B. hominis* has no constant intracellular form. It has been seen intracellularly in intestinal mucosal cells of infected germfree guinea pigs. There is no intricate life cycle involving different hosts or host tissues, and no secondary or alternate host (Fig. 2). These lacks were largely responsible for the long mystery of *B. hominis* classification, and permit the organism to fit rather nicely into a yeast classification, given its presence in the gastrointestinal tract as uniform, glistening, spherical, and (to some researchers) yeastlike cells.

Endodyogeny and endosporogeny have been demonstrated in *Blastocystis*, in addition to the more usual simple binary fission. *Blastocystis hominis* is now classified in the phylum Protozoa subphylum Sporozoa, class *Blastocystea*, and order *Blastocystida*. Fusion of cells, probably indicative of sexual reproduction, has been observed microscopically in slide cultures. *Blastocystis hominis* is placed with the coccidia and plasmodia, along with *Toxoplasma gondii* and *Pneumocystis carinii*, in the subphylum Sporozoa.

Protozoan properties of *Blastocystis hominis*

Structural	Physiological
No cell wall, but a thin membrane with vesicles and pores	Strict anaerobe
Mitochondria with typical protozoan morphology functioning anaerobically; aerobic pathways absent	Similar to *Entamoeba histolytica* in requiring live bacteria for growth, unless subjected to axenization under carefully controlled conditions
Golgi apparatus, with typical protozoan morphology	Does not grow on any medium used in cultivation of bacteria or fungi or on agar media
Supports a stable bacteroid endosymbiont	Optimum temperature 98.6°F (37°C); no growth at 75°F (24°C)
Exhibits pseudopodial activity for feeding, not locomotion	Ingests bacteria and other particulate matter
Reproduction by binary fission, endosporulation, or endodyogeny, not budding	Preference for neutral pH; killed in strongly acid environment
Specialized organelle, the central body, active in reproduction	Resistant to highest concentrations of amphotericin B
Definite eukaryotic nucleus with distinct crescentic peripheral chromatin; well-defined double nuclear membranes	
Well-demarcated smooth and rough endoplasmic reticulum	

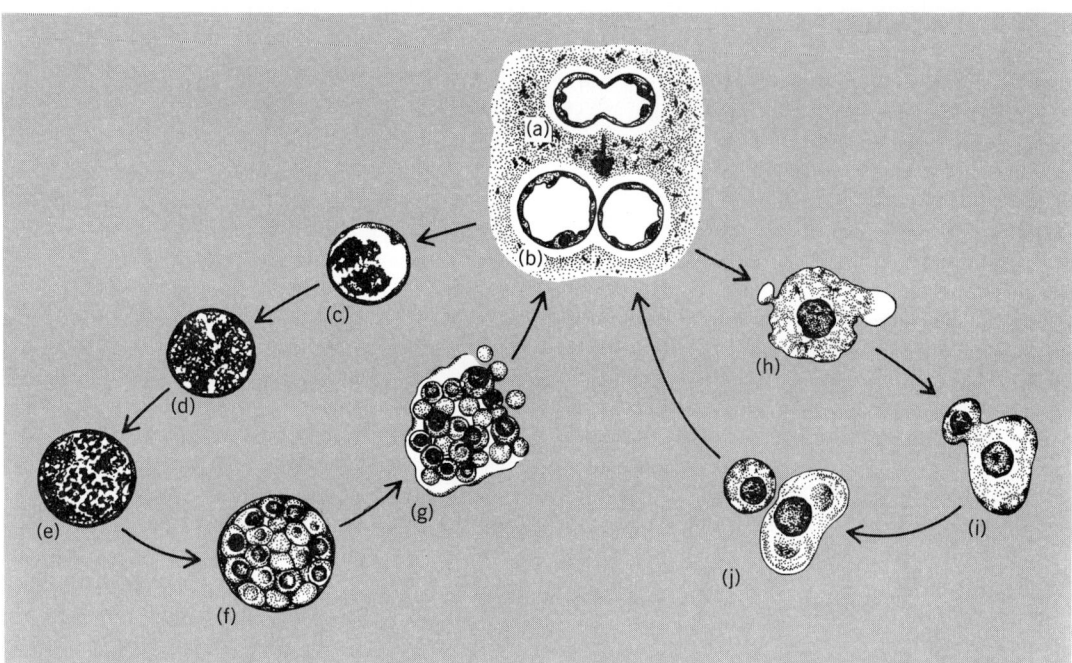

Fig. 2. Schematic of different stages in *Blastocystis hominis* life cycle. (*a, b*) Reproduction by binary fission. (*c–g*) Developing sporocyst as "granule" cell. (*h–j*) Ameba cell in process of cutting off daughter cell.

Culture. *Blastocystis hominis* is cultured anaerobically in glass tubes with a base slant of coagulated Locke's solution–egg mixture, overlaid with Locke's solution with 25% added serum. Most strains require the presence of bacteria for continual culture transfers. With antibiotics included in the medium, a few strains have been coaxed to grow without bacteria (axenic growth).

An amebalike, convoluted form of *B. hominis* is seen in culture as well as in severe cases of human diarrhea. It is distinguished from leukocytes by the concomitant presence of the central body form, and by culture of samples of diarrheal fluid, which develop the classical spherical central body form. This central body, a seemingly empty space occupying the greater volume of the center of the cell, is in truth not a vacuole but an organelle where endosporogeny occurs. Thin-section electron microscopy reveals a characteristic crescentic chromatin body in the nucleus which also distinguishes *B. hominis* from leukocytes.

Mitochondria. An intriguing aspect of *B. hominis* that has not attracted much attention is the presence of hundreds of mitochondria in the typical large culture cell (Fig. 3). *Blastocystis hominis* is strictly anaerobic; in the presence of even a little air the organism self-destructs within a few minutes, developing holes in the cell membrane through which the cell contents escape.

Rapid change in mitochondrial numbers and shape leads researchers to believe that *B. hominis* mitochondria are fully functional. Although 10 or so of the enzymes restricted to the usual oxidative mitochondrion have been sought, none has been found. The *B. hominis* mitochondrion probably has unique enzymatic pathways.

Pathogenicity. Currently there is active investigation of *B. hominis* by physicians and clinical parasitologists as a cause of diarrhea in humans. A few physicians did not wait for the reclassification of this strange organism from yeast to protozoan to seriously regard it as an agent of human disease. For decades they have diagnosed infections in humans caused by

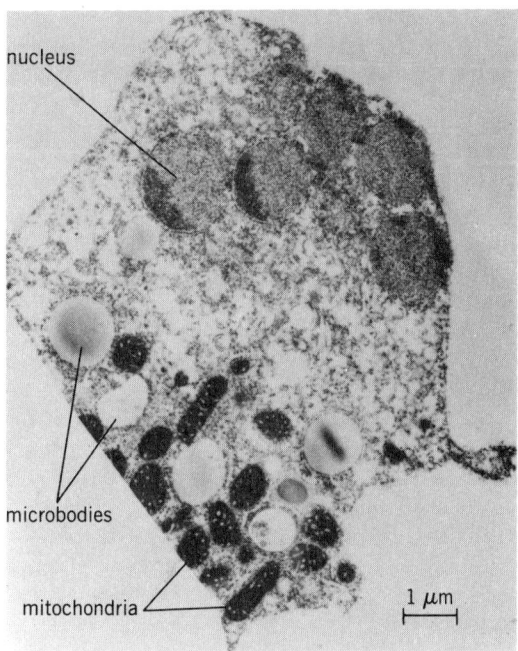

Fig. 3. Half of giant ameba cell from culture, showing five and perhaps six nuclei, numerous mitochondria, microbodies, and other inclusions.

B. hominis, and have successfully treated these infections with different quinoline compounds, particularly iodochlorhydroxyquin. Since use of the latter compound is now restricted, metronidazole (Flagyl) is most often prescribed.

For background information *see* DIARRHEA MEDICAL PARASITOLOGY; PROTOZOA; SPOROZOA; YEAST in the McGraw-Hill Encyclopedia of Science and Technology. [CHARLES H. ZIERDT]

Bibliography: H. K. Tan and C. H. Zierdt, Freeze-etch studies of granular and vacuolated forms of *Blastocystis hominis*, *Z. Parasitenk.*, 44:267–278, 1974; C. H. Zierdt, *Blastocystis hominis*, a protozoan parasite and intestinal pathogen of human beings, *Clin. Microbiol. Newslett.*, 5:57–59, 1983; C. H. Zierdt, *Blastocystis hominis*, an intestinal protozoan parasite of man, *Pub. Health Lab.*, 36:147–160, 1978; C. H. Zierdt and H. K. Tan, Ultrastructure and light microscope appearance of *Blastocystis hominis* in a patient with enteric disease, *Z. Parasitenk.*, 50:277–283, 1976.

Breeding (plant)

New genetic engineering technology has great potential for the improvement of crop species, both as a means for introducing genes into plants and as a way of identifying valuable traits in plants generated by conventional plant breeding. Gene transfer among plants is the breeder's major activity. However, the gene pool from which one may select those genes or characteristics is confined to those species or varieties with which the plant to be improved can be naturally mated. Gene transfer via genetic engineering could widen this gene pool to include incompatible organisms also. Herein genetic engineering is defined as gene transfer between species that do not normally exchange deoxyribonucleic acid (DNA). Through modern molecular biology, genes can be isolated from any organism, and various gene transfer methods can be used to introduce these DNA sequences into plant cells. Recent advances in plant tissue culture are making the regeneration of whole plants from these transformed cells possible in increasing numbers of species.

In addition, genetic engineering techniques may shorten the time required for crop improvement since preexisting techniques are so laborious. Often breeders may find a desirable gene in a species that is cross-compatible but carries with it many undesirable characteristics in the cross. Many generations of backcrossing the hybrid to the desirable parent, always selecting for the desirable gene in each generation of the progeny, must follow in order to cull the unwanted characteristics.

An example is the necessity for oat breeders in the midwestern United States to regularly incorporate stem rust– and crown rust–resistance genes into their varieties, since new races of these pathogens are frequently emerging. Occasionally resistance-conferring genes come from alien species that are cross-compatible to the crop only with some difficulty. Cytogenetic testing of progeny and many generations of backcrossing are required before a commercially viable variety can be released. If the transfer of a specific gene could be made in a single generation, much of the time and effort required for such an endeavor could be eliminated.

Principles of plant genetic engineering. Three major breakthroughs in the manipulation of DNA have made the genetic engineering of plants possible. The first is the process known as molecular cloning, in which DNA from any organism is cut into small fragments and introduced into a bacterial cell to amplify individual sequences. Each cell receives a different fragment, and progeny of individual bacteria are clones, each containing the same inserted sequence. Clones encoding the desired genes and regulatory regions are identified and assembled. The second breakthrough is the ability to transfer these DNA fragments to plant cells where they are stably integrated to form a permanent part of the plant genome, a process called transformation. Third is the ability to regenerate whole plants from the individual transformed plant cells (see illustration).

The theoretical advantages of the molecular methods of genetic engineering include not only the precision with which gene transfers can be made (single genes can be transferred without the necessity for time-consuming backcrossing to eliminate deleterious genes), but also the ability to use protein coding regions derived from almost any source—plant, animal or microorganism. In addition, the same gene could potentially be used in many plant species. Some disadvantages of the genetic engineering approach are the need to isolate the agronomically important genes and obtain proper gene expression, the difficulty of dealing with traits that are the result of the interaction of many genes, and the need to regenerate whole plants from transformed cells.

The kinds of genes that are receiving the most attention for transfer are those involving the determination of seed quality characteristics, herbicide resistance, disease resistance, and secondary plant products. Seed quality could be improved by modifying the amino acid composition of seed storage proteins in cereals and legumes; increased lysine and tryptophan levels in cereals, and increased methionine levels in legumes, would result in more valuable feed products since these foods are naturally low in these essential amino acids. Commercialization of an herbicide-resistance gene would also be attractive in the present agricultural system which stresses crop uniformity and chemical weed control. Genes for herbicide resistance are being found and isolated from resistant plants (crop or weedy species) as well as from mutagenized bacteria, but cloning a disease-resistance gene remains an elusive goal because resistance mechanisms are not well understood.

Using genetic engineering techniques, the possibility exists to construct entirely new genes not presently available in any organism. This could be done by random or directed mutagenesis of existing genes or by actual chemical synthesis of novel DNA sequences. Although genes have already been successfully modified by mutagenesis, de novo synthesis of novel genes is still in its infancy.

Although methods have recently been hypothesized to delete or suppress specific preexisting gene sequences, none has yet been put into practice in plants. An example of the usefulness of gene deletion is the elimination of a gene necessary for the production of trypsin inhibiter in soybean seed. Feed and food processing costs of soy products would be lower without this antinutritional protein.

Transformation methods. Several methods of delivery of the gene construct to cells have been developed. They can be divided into those using biological vectors (bacteria or viruses which insert DNA into a plant cell) and those using direct methods of DNA transfer.

The most widely used method of plant cell transformation uses the *Agrobacterium tumefaciens* Ti (for tumor-inducing) plasmid. Infection of plant cells with *Agrobacterium* leads to the insertion of a portion of the Ti plasmid (called the T region) into the plant genome. The T region normally encodes genes that induce tumors on plants; however, it has been shown that these oncogenes can be replaced with other DNA sequences without impairing the transfer functions of the system. The Ti system has been used extensively to transform dicotyledonous plants, but its value for monocots is not yet clear.

Another biological vector system that shows promise is the virus-based vector. Many viruses infect plant cells and generate large quantities of transcribed and translated gene products. These viruses can move from cell to cell, thus causing a systemic infection from a single wound site. In some cases, such as Cauliflower Mosaic Virus, purified DNA copies of the viral genome have been shown to cause symptoms when rubbed onto a plant. Thus one could potentially insert a foreign gene into a cloned viral genome, inoculate plants, and thus obtain plants that express large quantities of the foreign gene product. Such a system would eliminate the need for regeneration but could be thwarted for a number of reasons. First, the host range of the virus may be limiting. Also the virus would have to be infectious, but not pathogenic. In addition, stability of the autonomously replicating viral DNA would have to be assured. Finally, these viral vectors may pose an environmental problem, if they spread to other species.

A number of methods of transformation that do not require the use of biological vectors are currently under development. These include direct uptake of DNA into protoplasts (plant cells lacking cell walls) mediated by treatments such as polyethylene glycol, polycations, polyvinyl alcohol, calcium ions, high pH, or osmotic shock. In addition, some researchers are experimenting with the use of lipid vesicles (liposomes) to enclose the DNA for uptake into protoplasts. Finally, a great deal of interest has been focused on techniques for microinjection of

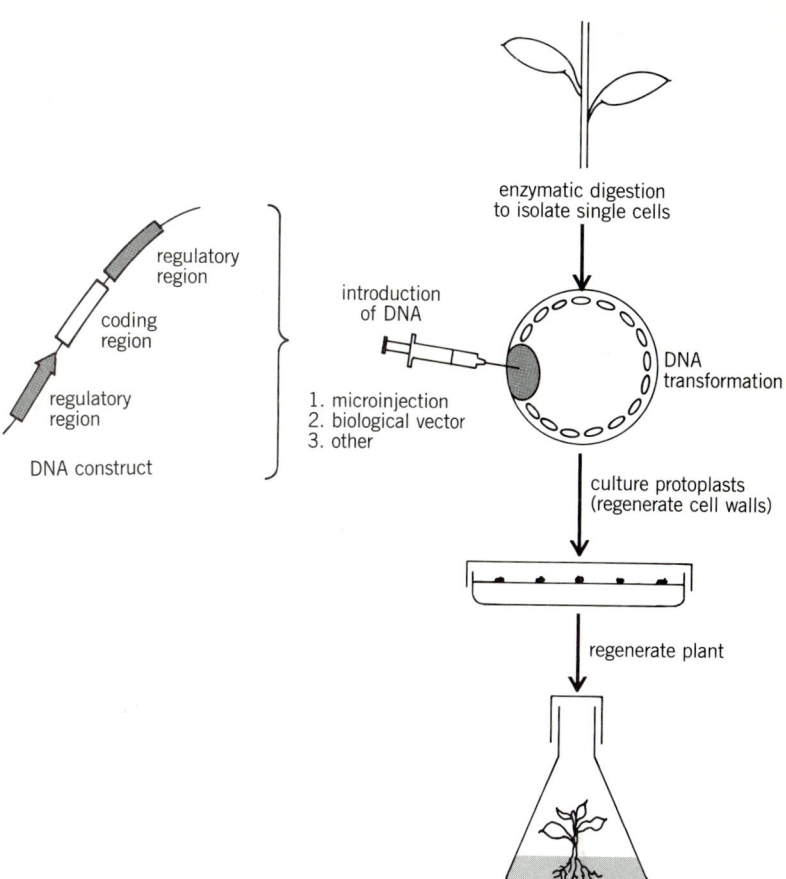

Steps in plant transformation. Protoplasts can be prepared from plant leaves or from cultured cells by enzymatic digestion of the cell wall. The DNA construct is then introduced into the cell by using biological vectors or mechanical means. The transformed protoplasts can regenerate cell walls in culture. Fertile plants can be regenerated from this transformed callus tissue.

DNA into cells which could make it possible to deliver DNA directly into the desired cell compartment (nucleus, cytoplasm, or organelle). In each of these cases, stable transformation of the cell would require integration of the foreign DNA into the plant genome by cell mechanisms that are not yet well understood.

Most of the transformation methods described above require the ability to culture the plant cells and to regenerate whole plants from cultured tissue. But ability to regenerate varies considerably from species to species. Some plants such as tobacco and petunia can be easily transformed, cultured, and regenerated, whereas others such as corn or soybean are extremely difficult to regenerate.

Foreign gene expression in plants. In one of the first examples of plant genetic engineering, T-DNA regulatory sequences have been used to express genes in plants. Here, a bacterial gene encoding an enzyme that confers resistance to the antibiotic kanamycin was attached to regulatory sequences from the nopaline synthase gene, one of the genes located in the T-DNA of the Ti plasmid. The hybrid promoter-gene construction was transferred from *Escherichia coli* to *Agrobacterium tumefaciens*, which was then used to transform protoplasts of *Nicotiana plumbaginifolia*, a relative of domesticated tobacco. Because growth of plant cells is inhibited by kanamycin, callus tissue could be selected by its growth in the presence of the antibiotic. Progeny of plants regenerated from callus transformed with the kanamycin-resistance gene have been tested, and antibiotic resistance was inherited in a 3:1 ratio among the offspring—this is the kind of inheritance pattern expected if the originally transformed plants were heterozygotes.

While expression of antibiotic resistance is not an agronomically important trait, demonstration of the ability to express a bacterial gene in a plant is a significant step in the development of techniques for the genetic engineering of plants. Furthermore, since the kanamycin-resistance gene confers a dominant selectable phenotype on transformed cells, a second, nonselectable gene can be linked to the antibiotic-resistance gene which then acts as a marker for the identification of transformed cells or tissue.

Other applications for plant improvement. Recently, breeders have begun correlating allozymes (enzyme variants encoded by different alleles at the same locus) and morphological markers with yield components to mark chromosomal segments that carry desirable genes. A major shortcoming of this strategy is the paucity of known markers in most plants. Restriction enzyme digestion of total plant DNA followed by DNA:DNA hybridization using a low-copy-number sequence probe (a radiolabeled DNA sequence) can be used to create a new class of chromosome markers. These markers could be constructed from any DNA sequence present in low copy number in a plant genome, and could enable ready detection of heterozygous individuals; unlike morphological markers, they do not require expression of a protein. If such a DNA marker is consistently correlated with a useful characteristic that cannot be assayed easily or directly in the whole plant, finding genotypes carrying the trait could be made easier by first screening unknown genotypes for the presence of the marker DNA.

Other uses of marker DNA (small random DNA fragments from the digestion of plant nuclear DNA with restriction enzymes) may be important. For example, they may be useful in characterizing new varieties or hybrids of seed crops unambiguously for legal proof of germplasm ownership or invention even if they are not linked to any known gene. DNA hybridization techniques will also provide sensitive diagnostic screens for the presence of viral infection in perennial crops and for seed-borne pathogens.

Finally, microinjection will make nuclear, chromosomal, and organellar transplants possible in plants. The most immediate application this may have in crop improvement is the ability to create cytoplasmic male sterile (cms) genotypes in a single generation. Presently, cms is transferred to a new line by repeatedly backcrossing the plant with the desired nuclear genome (and subsequent hybrid progeny) to that with male sterile cytoplasm, effectively replacing the nuclear genome of the original cms plant with that of another. This takes many generations to complete, but is essential to the production of hybrids in all crops that cannot be readily emasculated in the field. Transplantation of the desired genome to an enucleated cell bearing cms cytoplasm followed by regeneration of that cell to a plant would accomplish the nuclear substitution in a single generation.

Conclusion. The first crops bearing recombinant DNA are now being tested for stability, inheritance, and proper expression of the transformed DNA. Although not all of the techniques required for successful gene transfer via genetic engineering have been refined for every crop species, knowledge is increasing so rapidly that the commercial impact of genetically engineered plants may be felt by the end of the decade.

For background information *see* BREEDING (PLANT); GENETIC ENGINEERING; SOMATIC CELL GENETICS in the McGraw-Hill Encyclopedia of Science and Technology.

[HOLLY HAUPTLI; CATHERINE M. HOUCK]

Bibliography: A. Caplan et al., Introduction of genetic material into plant cells, *Science*, 222:815–821, 1983; R. C. Gardner and C. M. Houck, Development of plant vectors, in T. Kosuge and E. W. Nester (eds.), *Plant-Microbe Interactions*, 1984; R. B. Horsch et al., Inheritance of functional foreign genes in plants, *Science*, 223:496–498, 1984; T. Kosuge, C. P. Meredith, and A. Hollaender (eds.), *Genetic Engineering of Plants: An Agricultural Perspective*, Basic Life Sciences, vol. 26, 1983.

Cell differentiation

Multicellular organisms develop into diverse forms by the generation of different cell types from a single cell, a process that can be thought of as a series of developmental decisions. Understanding the genetic and molecular events of this process of differentiation is fundamental to understanding development.

Lineage of Caenorhabditis elegans. The free-living soil nematode *Caenorhabditis elegans* is an ideal organism with which to address cell differentiation because every cell division, cell migration, and programmed cell death, from the first cleavage division of the zygote to the adult form, has been traced. Understanding the entire cell lineage of an organism that generates many of the cell types found in higher organisms is a landmark in developmental biology (Fig. 1). Further, the lineage provides a detailed map which precisely identifies the relationship between diverse cell types. Patterns of division, programmed cell deaths, and terminal differentiations within the lineage are constant from one individual to another. This means that at any given time during development each cell not only has a predictable past but also a predictable future, a reproducible position, and a defined and invariant group of neighbors. Further, the lineage has provided a spatial and temporal pattern indicating where developmental

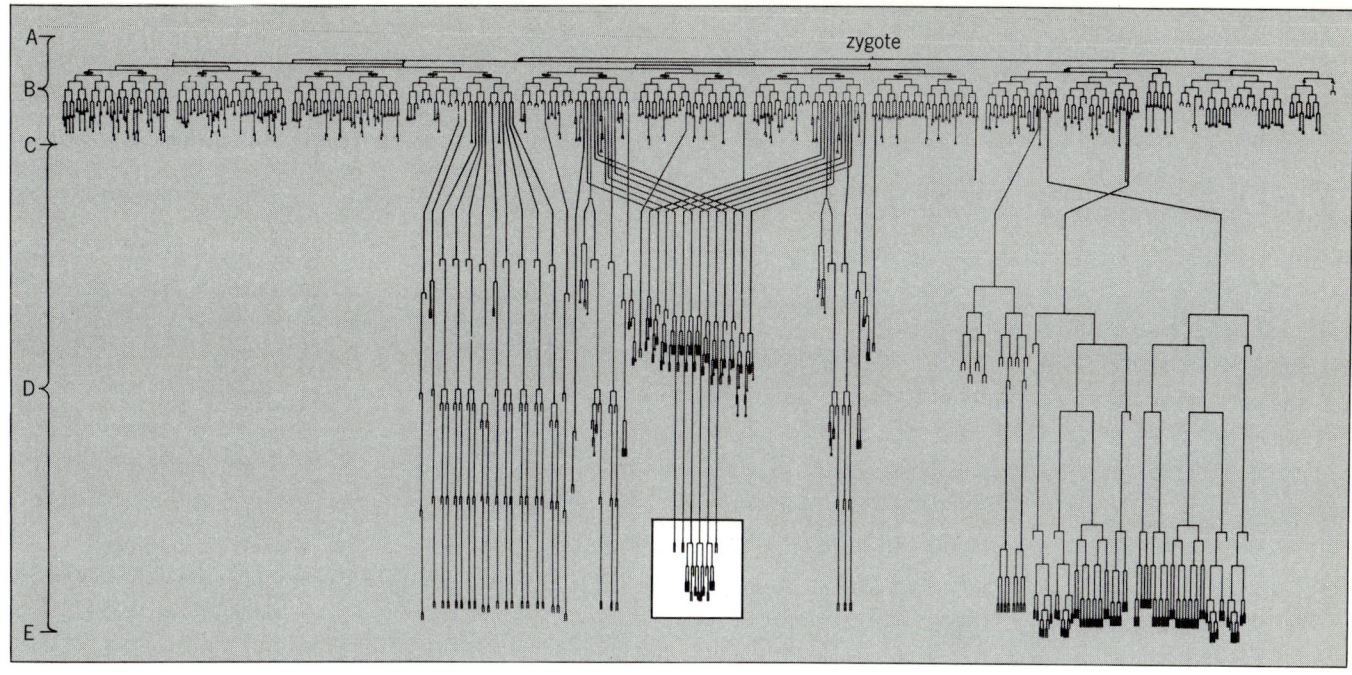

Fig. 1. Complete cell lineage of Caenorhabditis elegans. Each vertical line represents a cell, and each horizontal line represents a cell division. The capital letters to the left of the lineage indicate various times during the life cycle of the nematode. A = fertilization, B = embryogenesis, C = hatching, D = larval forms, and E = adult form. The portion of the lineage within the box is enlarged in Fig. 3. (After C. J. Kenyon, Pattern, symmetry and surprises in the development of Caenorhabditis elegans, Topics Biochem. Sci., 8:349–351, 1983)

decisions probably occur during development.

In C. elegans the first developmental decisions occur during the first unequal cleavage divisions which create the six embryonic blast cells, AB, MS, E, C, D, and P4 (Fig. 2). The divisions following embryonic blast cell formation are essentially equal. Embryonic blast cells E, D, and P4 generate only one cell type; however, embryonic blast cells AB, MS, and C produce both ectodermal (neurons and epidermis) and mesodermal (muscle and gonad) descendants. Thus, the developmental decisions separating the cell types generated by precursors AB, MS, and C occur after embryonic blast cell formation.

The adult nematode is only 1 mm (0.04 in.) long and can be easily grown in the laboratory on a diet of Escherichia coli grown on agar plates. Since the animal is transparent, individual or groups of cells can be watched as they divide, migrate, and subsequently differentiate. The majority of cell divisions take place during the first half of embryogenesis (Fig. 1). At hatching, the hermaphrodite is composed of 558 somatic cells; postembryonic divisions add 401 somatic cells, so an adult contains 959 cells. 131, or one in eight, of all the somatic cells produced die almost as soon as they are formed; the identity and approximate times of these deaths are predictable. Most cells move only a short distance after their births, but a few migrate long distances. The lineage, then, in addition to creating cells of the correct cell types, also places them in their approximate positions for normal development to proceed.

The lineage provides a basis from which to approach the genetic and molecular events which lead to cell determination and subsequent differentiation. To address this question, two approaches have been widely utilized: cell ablation, by use of a laser microbeam, and the creation of genetic mutations which alter the lineage division patterns and, many times, cell fates.

Selective cell ablation. In selective cell ablation experiments, an animal is oriented under a compound microscope, with Normarskioptics so that individual cells can be identified. After a cell is exposed to several low-energy pulses with a diameter of about 0.5 micrometer (the diameter of an average nucleus in C. elegans is 2.0 micrometers), death is not instantaneous, but much like a programmed cell death. No apparent damage to other cells in the animal is usually observed.

The results of such ablation experiments show overwhelmingly that the only cells missing are those that are normally generated by the ablated precursor (notable exceptions are discussed below). The determination and subsequent differentiation of a cell are not affected by the ablation of any of its neighbors; that is, a cell whose neighbor is ablated does not compensate for the loss of its neighbor by changing its own developmental fate. Such experiments have revealed that development in C. elegans is largely mosaic, with cell determination proceeding in a cell-autonomous fashion. Determination, then, in the vast majority of cases is controlled by a cell's lineage rather than by intercellular interactions with its neighbors.

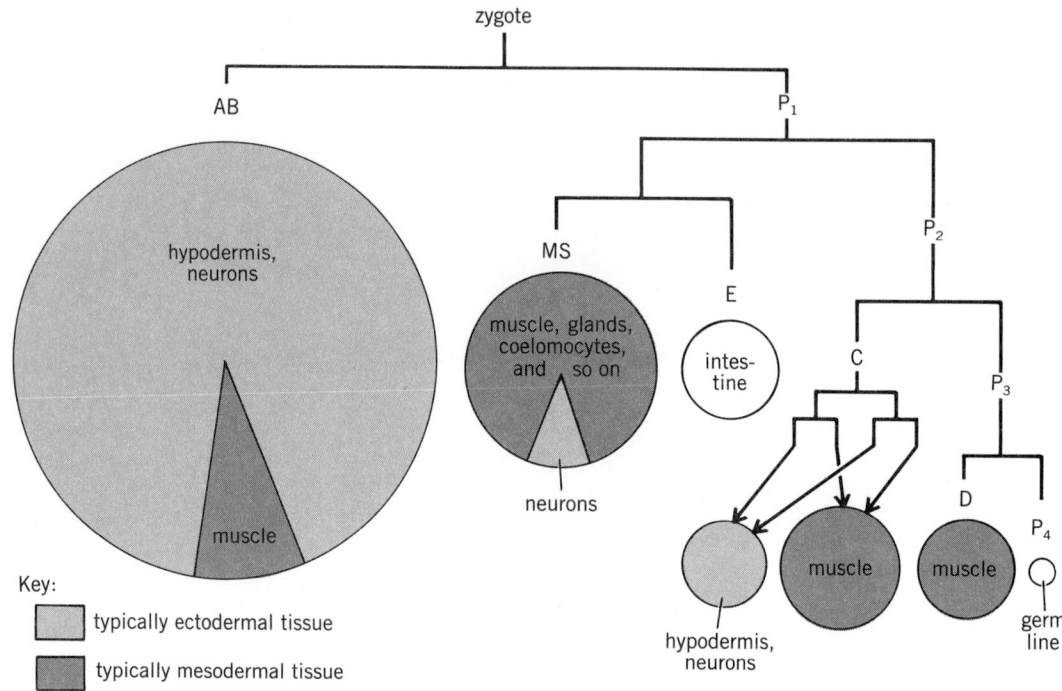

Fig. 2. Generation of embryonic blast cells and a summary of cell types derived from them. (After J. E. Sulston et al., *The embryonic cell lineage of the nematode Caenorhabditis elegans*, Dev. Biol., 100:64–119, 1983)

In a small number of cases, a cell has been observed to reproducibly adopt the fate of its neighbor after the removal of that neighbor by laser ablation. These limited cases of developmental regulation or cell replacement are seen only within discrete groups of neighboring cells called equivalence groups. These groups of cells are known to make specialized structures within the nematode.

One equivalence group consists of six midventral, linearly arranged precursor cells in the region of the future vulva or egg-laying apparatus which are designated from anterior to posterior as P3.p, P4.p, P5.p, P6.p, P7.p, and P8.p. Normally, P5.p, P6.p, and P7.p divide several times to form the 22 cells of the vulva; P3.p, P4.p, and P8.p divide only once and become part of the epidermis. The fate normally adopted by P6.p is designated 1°, that of P5.p and P7.p, 2°, and that of P3.p, P4.p, and P8.p, 3°. If cells P5.p, P6.p, and P7.p are killed by laser ablation, precursor cells P3.p, P4.p, and P8.p abandon their normal fates and instead divide several additional times to form vulval cells (Fig. 3).

P5.p, P6.p, and P7.p will only divide to form a vulva in the presence of a nondividing cell in the gonad, the anchor cell. When the anchor cell is ablated, P5.p, P6.p, and P7.p undergo only one division and, like their equivalent precursors, become part of the epidermis (Fig. 3).

The cellular interactions described above demonstrate that the six precursor cells are equivalent in that each can adopt either a vulval or epidermal fate. Such examples of cellular regulation during development are rare in *C. elegans*.

Mutation and cell fate. To date, over 500 genes have been identified and mapped onto one of the six pairs of chromosomes carried by *C. elegans*. Nematodes with mutations that identify specific genes are easily maintained and manipulated by standard genetic means.

The cell lineages and developmental fates of cells in *C. elegans* can be affected by mutation. One such cell lineage mutant identifies the gene *lin-12*. Mutations in the *lin-12* locus result in a variety of phenotypic abnormalities, all of which indicate that the normal function of the *lin-12* gene product is to

		P3.p	P4.p	P5.p	P6.p	P7.p	P8.p
(a)	cell fates	epidermis		vulva			epidermis
	wild-type (normal) fates	3°	3°	2°	1°	2°	3°
(b)	fates after selective cell ablation	2 or 1°	2 or 1°	ablated	ablated	ablated	2 or 1°
	anchor cell ablated	3°	3°	3°	3°	3°	3°
(c)	lin-12(d)	2°	2°	2°	2°	2°	2°
	lin-12(o)	3°	hybrid	1°	1°	1°	3°

Fig. 3. Lineages and fates of midventral precursor cells (see Fig. 1) P3.p, P4.p, P5.p, P6.p, P7.p, and P8.p in (a) wild type, (b) wild type after selective cell ablation, and (c) *lin-12* mutants.

effect a developmental decision between alternative cell fates.

There are two classes of alleles of the *lin-12* locus: semidominant mutations designated *lin-12(d)*, and recessive mutations designated *lin-12(0)*. The normal or wild-type form of the gene is designated *lin-12(+)*. When a semidominant mutation is present, only one copy of the mutant allele is needed to express the mutant phenotype. The semidominance of the *lin-12(d)* mutations suggests that they may result either in novel gene activity or in an increased level of gene activity. When a recessive mutation is present, two copies of the mutant allele are needed to express the mutant phenotype. Recessive *lin-12(0)* alleles are thought to result in the loss or decreased activity of the *lin-12* function.

In wild-type hermaphrodites, two gonadal cells, Z1.ppp and Z4.aaa, have naturally variable fates. Either Z1.ppp or Z4.aaa becomes the anchor cell; the other cell becomes the ventral uterine precursor cell, which divides several times to form the ventral wall of the uterus. Mutations in the *lin-12* locus affect the fates of Z1.ppp and Z4.aaa. Hermaphrodites carrying a semidominant mutation, *lin-12(d)*, lack an anchor cell, and both Z1.ppp and Z4.aaa become ventral uterine precursor cells. In contrast, hermaphrodites homozygous for a recessive mutation, *lin-12(0)*, have two anchor cells and are lacking ventral uterine precursor cells. One likely interpretation of these results is that a high level of *lin-12* activity—*lin-12(d)*—causes both Z1.ppp and Z4.aaa to become ventral uterine precursor cells; the absence of *lin-12* activity—*lin-12(0)*—causes both cells to become anchor cells.

Both semidominant and recessive *lin-12* mutations affect the equivalence group responsible for formation of the vulva. In hermaphrodites, the elevated gene activities produced by *lin-12(d)* alleles affect the midventral precursor cells so that all cells within the equivalence group express the 2° fate (Fig. 3). Elimination of gene activity by *lin-12(0)* alleles results in the expression of a non-2° fate.

In the case of *lin-12(d)* mutations, the expression of the 2° fate by all cells of the midventral equivalence group and the lack of an anchor cell leads to a phenotype in which multiple small, nonfunctional protrusions, or pseudovulvae, are observed along the ventral midline. Eggs produced have no means of exit; they accumulate within the body of the hermaphrodite, where they hatch and devour the mother from the inside out. In the case of *lin-12(0)* mutations, the midventral precursor cells which adopt the 1° fate form a single large protrusion in the normal position of the vulva. The hermaphrodites often explode as young adults; survivors are usually sterile.

The state of activity of the *lin-12* gene specifies cell fates. A high level of *lin-12* activity results in the expression of one fate, while a low level results in the expression of an alternative fate. Cell ablation experiments in wild type have shown that cellular interactions influence the fates of many of the cells affected in *lin-12* mutants, which indicates that they must have more than one developmental potential before the interactions have occurred. The interactions, then, restrict their developmental potential. It is believed that cellular interactions specify the level of *lin-12* activity in each of these cells and that the level of *lin-12* activity controls the expression of other genes responsible for effecting specific cell fates. The molecular mechanism by which *lin-12* effects the developmental decision between alternative cell fates is unknown.

Future studies will involve the genetic identification and molecular characterization of more genes with functions similar to those of *lin-12*. Understanding the molecular events which lead to the expression of *lin-12* and other similar genes is paramount to the elucidation of the mechanisms underlying cellular differentiation.

For background information see CELL DIFFERENTIATION, SENESCENCE, AND DEATH; CELL LINEAGE; FATE MAPS (EMBRYOLOGY) in the McGraw-Hill Encyclopedia of Science and Technology.

[LIZABETH A. PERKINS]

Bibliography: I. S. Greenwald, P. W. Sternberg, and H. R. Horvitz, The *lin-12* locus specifies cell fates in *Caenorhabditis elegans*, Cell, 34:435–444, 1983; J. Kimble, Alterations in cell lineage following laser ablation of cells in the somatic gonad of *Caenorhabditis elegans*, Dev. Biol., 87:286–300, 1981; J. E. Sulston et al., The embryonic cell lineage of the nematode *Caenorhabditis elegans*, Dev. Biol., 100:64–119, 1983; J. E. Sulston and J. G. White, Regulation and cell autonomy during postembryonic development of *Caenorhabditis elegans*, Dev. Biol., 78:577–597, 1980.

Chaotic behavior

This article discusses the notions of order, chaos, and noise as they occur in deterministic dynamical systems.

Throughout history, sequentially using magic, religion, and science, people have sought to perceive order and meaning in a seemingly chaotic and meaningless world. This quest for order reached its ultimate goal in the seventeenth century when newtonian dynamics provided an ordered, deterministic view of the entire universe epitomized in P. S. de Laplace's statement: "We ought then to regard the present state of the universe as the effect of its preceding state and as the cause of its succeeding state." In everday life, the predictable swing of a long, massive pendulum or the regular motion of the Sun, Moon, and planets provides reassurance of a mechanistic newtonian order in the world.

But if the determinism of Laplace and Newton is totally accepted, it is difficult to explain the unpredictability of a gambling game or, more generally, the unpredictably random behavior observed in many newtonian systems. Commonplace examples of such behavior include smoke that first rises in a smooth, streamlined column from a cigarette, only to abruptly burst into wildly erratic turbulent flow

138 Chaotic behavior

Transition from order to chaos (turbulence) in a rising column of cigarette smoke. The initial smooth streamline flow represents order, while the erratic flow represents chaos.

(see illustration); and the unpredictability of the weather.

At a more technical level, flaws in the newtonian view had become apparent by about 1900, leading J. H. Poincaré to remark, "Determinism is a fantasy due to Laplace," and J. C. Maxwell to assert, "The true logic of this world is the calculus of probabilities." The problem is that many newtonian systems exhibit behavior which is so exquisitely sensitive to the precise initial state or to even the slightest outside perturbation that, humanly speaking, determinism becomes a physically meaningless though mathematically valid concept. But even more is true. Many deterministic newtonian-system orbits are so erratic that they cannot be distinguished from a random process even though they are strictly determinate, mathematically speaking. Indeed, in the totality of newtonian-system orbits, erratic unpredictable randomness is overwhelmingly the most common behavior.

Examples. These notions will be illustrated through three simple mapping systems which retain many features of general dynamical systems but which are not encumbered by extraneous detail.

Highly orderd motion. First, consider the highly ordered motion represented by a discrete mapping of points around a circle of unit circumference in which each point is carried one-fifth way around the circle upon each iteration. After n iterations, a point initially at angular position θ_n given by Eq. (1), where the angle θ has been normalized such that

$$\theta_n = (0.2)n + \theta_0 \tag{1}$$

$\theta + 1 = \theta$. This mapping is obviously determinate since each θ_0 uniquely determines an "orbit" of θ_n iterates. The "motion" here is also highly ordered because, from Eq. (1), all orbits are periodic with $\theta_{n+5} = \theta_n$, because the entire circle of points is mapped as a rigid body upon each iteration, and because the angle of rotation is the precisely computable, terminating rational number, $1/5 = 0.2$. Moreover, except for a displacement around the circle, all orbits are identical. This mapping system is thus fully ordered, strictly deterministic, and totally dull. In more general terms, the monotony of the streamline flow in the illustration may be contrasted with the greater variety of behavior of the turbulent flow. Ordered motion while comfortably simple is also rigidly constrained motion in which all or almost all capricious options have been discarded.

Ordered motion. Next, consider the less constrained but still ordered motion generated by the mapping which rigidly rotates the whole circle of points through the larger angle $\alpha = \sqrt{2}$ upon each iteration. Here the result of n iterations is given by Eq. (2), where again $\theta + 1 = \theta$. Like the mapping

$$\theta_n = (\sqrt{2})n + \theta_0 \tag{2}$$

of Eq. (1), except for a displacement around the circle, all orbits are identical. However, here there is one exciting difference. Because the angle of rotation α is now irrational, the θ_n iterates of each θ_0 in Eq. (2) spread out uniformly around the circle, almost as if their positions were selected from a uniform random distribution. In fact, it is possible to rigorously prove for Eq. (2) that Eq. (3) is satisfied,

$$\lim_{N \to \infty} \left(\frac{1}{N}\right) \sum_{n=0}^{N} f(\theta_n) = \int_0^1 f(\theta) d\theta \tag{3}$$

where f is any reasonable well-behaved function. Equation (3), a version of the ergodic theorem, is merely a sophisticated way to assert that all θ values occur with equal frequency and that the mapping of Eq. (2) therefore passes one test for randomness. However, this slight randomness in the mathematically deterministic Eq. (2) has appeared only because a sacrifice has been tacitly made for it. Indeed, the introduction of an irrational angle of rotation into Eq. (2) which yields a uniform θ distribution simultaneously also renders Eq. (2) not fully determinate from the physical viewpoint. Specifically, humans can never compute the precise decimal representation for $\sqrt{2}$. Thus, let ϵ be the error in the determination of $\sqrt{2}$; then the error $\Delta(n)$ in θ_n will be at least $\Delta(n) = n\epsilon$. Since this error clearly grows without bound with iteration number n, physical determinism in Eq. (2) is completely lost when the error $\Delta(n)$ is about unity. Nonetheless, physical determinism over human time scales can be maintained, provided the error ϵ in the determination of $\sqrt{2}$ is sufficiently small.

It is now possible to proceed through a sequence of maps which become more "random" and less "physically determinate." For them, the error $\Delta(n)$ in each θ_n would grow as some polynomial in n, but the notions of order and physical determinism could still be maintained, at least over human time scales, by using humanly available accuracy. The great divide occurs when the error growth becomes exponential, yielding mathematically determinate systems which pass every test for randomness. Here the mapping equations require such massive input of data strings in order to maintain accuracy that, informationally speaking, they are not computing or determining the solution but rather merely copying out the solution being given to them. This case will be illustrated with a final example.

Chaotic motion. Suppose now that the points on the circle of unit circumference move according to the "multiplicative" mapping equation (4), where

$$\theta_n = 2^n \theta_0 \qquad (4)$$

$\theta + 1 = \theta$ as before. For Eq. (4), orbits starting with slightly differing θ_0 values clearly separate exponentially with iteraton number n. In consequence, Eq. (4) exhibits that extreme sensitivity of final state to precise initial state characteristic of all chaotic systems; indeed, Eq. (4) is perhaps the simplest system which can serve as a paradigm for chaos.

However, Eq. (4) is no less mathematically deterministic than are Eqs. (1) and (2), since a θ_n set exists and is unique for each given θ_0. To see how the θ_n iterates of Eq. (4) can nevertheless be unpredictably random and nondeterministic, consider the consequences for determinism of the exponentially separating orbits in Eq. (4). Assume, for example, that the spatial angle θ_0 is known to an accuracy of 10^{-31}, far beyond the accuracy available to contemporary science; even then, after only 100 iterations, the error in θ_{100} is of order unity, and the last vestige of determinism in Eq. (4) has been lost.

This issue may be further clarified by computing a full set of θ_n iterates of Eq. (4), each having an accuracy of at least 2^{-1}, that is, at least one-binary-digit accuracy. Then let 2^{-m} be the accuracy of θ_0 required to compute a given θ_n. In consequence, Eq. (5) must be satisfied. Thus $m = n + 1$ or, for

$$2^{-1} = 2^n 2^{-m} \qquad (5)$$

large n, $m \cong n$. Now the situation becomes clear. In order to assure n informational bits of solution output, one per θ_n iterate, it is necessary to include a total of $m = n$ informational bits into the binary representation of θ_0. Equation (4) is thus seen to be merely a "copy" algorithm rather than a "deterministic" one. Alternatively stated, Eq. (4) is merely a fixed "language translation" algorithm which translates a "German" θ_0 into an "English" θ_n sequence.

Even greater insight into these matters can be obtained by writing the θ_0 of Eq. (4) in binary notation as a digit string of zeros and ones; typically it might be given by Eq. (6). Then, each iterate θ_n specified

$$\theta_0 = 0.11101100000100\ldots \qquad (6)$$

by Eq. (4) may be obtained by merely moving the "decimal" in Eq. (6) n places to the right and dropping the integer part. Now, the string of zeros and ones in Eq. (6) may be regarded as specifying a coin-toss sequence. Moreover, the set of all binary θ_0 sequences is in a one-to-one correspondence (essentially) with the set of all possible random coin-toss sequences. Consequently, almost all binary θ_0 sequences pass every humanly computable test for randomness. In short, they are random. A straightforward derivation then proves that almost all θ_0 yield random θ_n sequences despite the fact that Eq. (4) is mathematically deterministic.

Equation (4) is mathematically deterministic provided that θ_0 is given or somehow determined mathematically; however, in Eq. (4), randomness arises precisely because the infinite digit string for θ_0 is, in general, as random as a coin toss and therefore not computable by any finite algorithm. Moreover and equally important, the exponentially sensitive dependence of final state upon initial state in Eq. (4) transforms randomness of the θ_0 digit string into randomness of the θ_n sequence. Indeed, the θ_n sequences of Eqs. (1) and (2) lack full randomness only because they lack this sensitive dependence upon initial state.

Equation (4), as mentioned above, is the epitomy of chaos just as Eq. (1) is the epitome of order. In addition, Eq. (4) exhibits that richness of behavior inherent to all chaotic systems: everything that can happen, does happen. In Eq. (4), for example, the full orbit is known once the θ_0 digit string of Eq. (6) is known. But here all digit strings are possible and, in fact, each digit string actually occurs for some θ_0. Thus, chaos results in richness of opportunity rather than meaninglessness.

Systems in the real world. The use of deterministic, ordered mathematical models to describe behavior of systems in the real world is so familiar that only chaos need be discussed. Consider first the evolution of life on Earth. Were this evolution deterministic, the governing laws of evolution would have had built into them anticipation of every natural crisis which has occurred over the centuries plus anticipation of every possible ecological niche throughout all time. Nature, however, economizes and uses the richness of opportunity available through chaos. Random mutations provide choices sufficient to meet almost any crisis, and natural selection chooses the proper one.

One can consider the problem that the human body faces in defending against all possible invaders. Again, nature appears to choose chaos as the most economical solution. Loosely speaking, when a hostile bacterium or virus enters the body, defense strategies are generated at random until a feedback loop indicates that the correct stategy has been found. A great challenge for chaos researchers is to mimic nature and to find new and useful ways to harness chaos.

Another matter for consideration is the problem of predicting the weather or the world economy. Both these systems have much in common with Eq. (4), although their governing equations are certainly more complicated. Nonetheless, like a θ_n sequence of Eq. (4), the weather and the economy are chaotic and can be predicted more or less precisely only on a very short time scale. Nonetheless, by recognizing the chaotic nature of the weather and the economy, it may eventually be possible to accurately determine the probability distribution of allowed events in the future given the present. At that point it may be asserted with mathematical precision that, for example, there is a 90% chance of rain 2 months from today. Much current work in chaos theory seeks to determine the relevant probability distributions for chaotic systems.

Finally, many physical systems exhibit a transition from order to chaos, as exhibited in the illustration, and much current work studies the various routes to chaos. Examples include fibrillation of the heart and attacks of epilepsy, manic-depression, and schizophrenia. Physiologists are striving to understand chaos in these systems sufficiently well that these human maladies can be eliminated.

Noise. Reduced to basics, chaos and noise are essentially the same thing. Chaos is randomness in an isolated system; noise is randomness entering this previously isolated system from the outside. If the noise source is included to form a composite isolated system, there is again only chaos.

For background information see ELECTRICAL NOISE; STOCHASTIC PROCESS; TURBULENT FLOW in the McGraw-Hill Encyclopedia of Science and Technology.

[JOSEPH FORD]

Bibliography: G. J. Chaitin, Randomness and mathematical proof, *Sci. Amer.*, 232(5):47–52, May 1975; J. Ford, How random is a coin toss?, *Phys. Today*, 36(4):40–47, 1983; L. P. Kadanoff, Roads to chaos, *Phys. Today*, 36(12):46–53, 1983; I. Peterson, Pathways to chaos, *Sci. News*, 124:76–77, 1983.

Cnidaria

Animals in the phylum Cnidaria (Coelenterata) are characterized by microscopic stinging capsules called nematocysts. Nematocysts serve most importantly in feeding and in defense. There are several different types of nematocysts; differences in their structure are important taxonomic characters used to distinguish species of coelenterates. Nematocysts in the tentacles of coelenterates are directly responsible for prey capture, but the functions of the various types of nematocysts and their effects on prey have been mostly speculative. Recent research has shown how different nematocysts function in prey capture and that the kinds of prey captured may depend on the nematocyst types of the predator.

Nematocyst structure. Nematocysts are classified according to the structure of the thread, which is coiled within the nematocyst capsule and everts upon discharge. Some types have short threads with closed tips and adhere to surfaces. Most nematocysts have very elongate threads, many times the length of the capsule, which are open at the tip. These have been presumed to inject toxin through the hollow thread into animals that the coelenterates contact, whether for feeding or for defense. Most of these elongate threads have spines along their length that differ in size among different species.

Nematocyst effects on prey. Most people are familiar with large stinging jellyfish such as *Physalia*, the Portuguese man-of-war, which plagues swimmers along the southern United States coast, and *Chrysaora*, the sea nettle, which is a pest in Chesapeake Bay. These jellyfish inject toxin into swimmers. Such examples probably have led most people to expect all nematocysts to kill prey by functioning as tiny syringes, but the natural prey of most coelenterates are only a few millimeters (only about 1/20 to 1/4 in.) in length, and the nematocysts are less than 1 mm (only about 1/5000 to 1/50 in.) in length. Most prey are crustaceans, which have a hard external armor (exoskeleton) that would have to be pierced by a hollow thread 1/1000 to 3/1000 mm (1/25,000 to 1/8000 in.) in diameter. One type of nematocyst, the stenotele, can pierce the exoskeleton of the crustacean prey of *Hydra*. Stenoteles penetrate by the impact of three stylets, which form a pointed arrowhead, and then the thread everts through the hole.

The natural diets of siphonophores and the effects of their nematocysts upon prey have been well studied. Siphonophora is an order of colonial jellyfish consisting of nonvisual predators that feed with a tentacle "net" suspended in the water. *Physalia* is the only species to float at the water surface. Species in two of the three Siphonophora suborders, Calycophorae and Physonectae, have nematocysts concentrated in nematocyst batteries, one battery on each of several side branches of the tentacles. Calycophoran species have the same four types of nematocysts in these batteries, but the nematocysts differ in size among different species. Calycophoran diets consist mainly of small copepod crustaceans. Physonect species uniformly have three or four types of nematocysts in the batteries; again nematocyst sizes differ among species. Physonect nematocysts generally are larger than those of calycophorans, and physonect diets consist of large copepods and other zooplankton. Species in these two suborders that have larger nematocyst batteries (with more and larger nematocysts) capture larger prey.

The effect on prey of these two groups is dramatic—scanning electron microscopy shows that the conspicuous spines on the threads adhere to the prey (Fig. 1a). The nematocyst threads wrap around and entangle the prey (Fig. 1b). Only physonects have penetrating stenotele nematocysts in the batteries, and the stenoteles are outnumbered 20:1 to 200:1 by other nematocyst types; no evidence indicates penetration of the crustacean prey by other nematocyst types. Thus, prey capture by calycophoran and physonect siphonophores appears to be primarily by entanglement, rather than by penetration and injection of a toxin. Tentaculate species of another group of gelatinous zooplankton, the phylum Ctenophora, capture the same crustacean zooplank-

Cnidaria

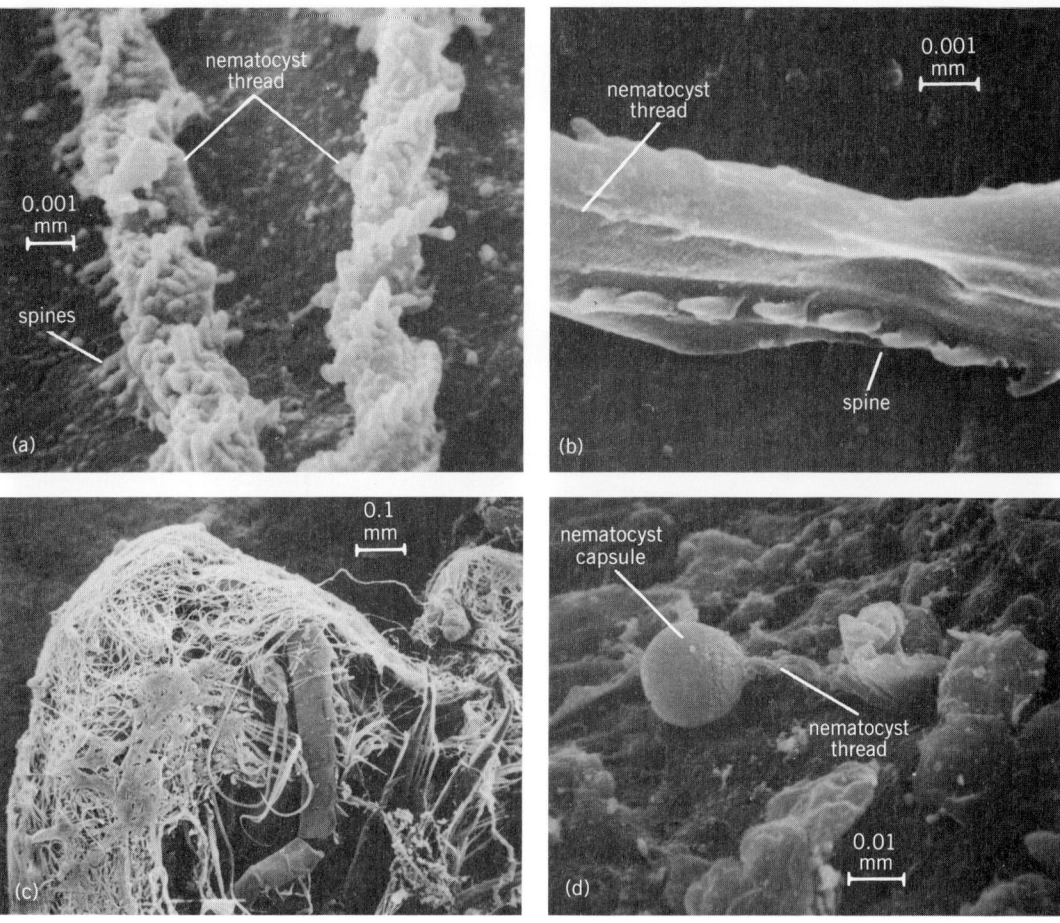

Fig. 1. Scanning electron micrographs of nematocysts. (a) Nematocyst threads of calycophoran siphonophore; spines are adhering to the prey surface. (b) Thread of cystonect siphonophore penetrating a fish larva. (c) Copepod entangled in adhering nematocyst threads after capture by a physonect siphonophore. (d) Nematocyst of cystonect siphonophore penetrating a fish larva.

ton prey by using adhesive colloblasts.

Siphonophores in the suborder Cystonectae, including *Physalia*, in contrast to Physonectae and Calycophorae, eat only soft-bodied prey; larval fishes constitute 90–100% of their diets. Cystonects all have the same type of nematocyst in their tentacles (not in batteries). The threads of their nematocysts lack spines or have minute spines (Fig. 1c), in contrast to the conspicuous spines along calycophoran and physonect nematocyst threads. Cystonect nematocysts clearly penetrate their soft-bodied prey (Fig. 1d), but have not been seen to penetrate or adhere to crustaceans. Thus cystonect siphonophores seem to be unable to entangle prey and may be limited to prey which their nematocysts can penetrate.

Dietary differences among siphonophores are related to the differences in nematocyst structure. The above results cannot eliminate the possibility that the extreme dietary specificity of cystonect siphonophores is due to a chemical stimulus necessary for nematocyst discharge, which might be present in the mucus coating soft-bodied prey but which is lacking in crustaceans.

Other nematocyst functions. The nematocyst batteries serve calycophoran and physonect siphonophores in two other capacities related to prey capture. (1) The swimming activity of most calycophoran and physonect siphonophores spreads their tentacles in a three-dimensional array. The diameters of the nematocyst batteries are greater than those of the tentacles; hence the nematocyst batteries act to increase drag at the ends of the tentacle branches and cause the tentacles to be drawn out. The batteries aid in tentacle extension and enable the spread of tentacles to be greater and the tentacles finer (and therefore less conspicuous to prey) than is possible with tentacles of uniform thickness. (2) The nematocyst batteries may serve as lures to attract large predatory zooplankton into the tentacles of the siphonophores. The nematocyst batteries of some physonect siphonophores resemble small zooplankton: those of *Agalma okeni* resemble copepods (Fig. 2a), those of *Athorybia rosacea* resemble fish larvae (Fig. 2b), and those of *Athorybia lucida* resemble larvaceans. Each type of lure is moved in the water; large zooplankton (potential prey for the siphonophores) could be attracted to the lures visu-

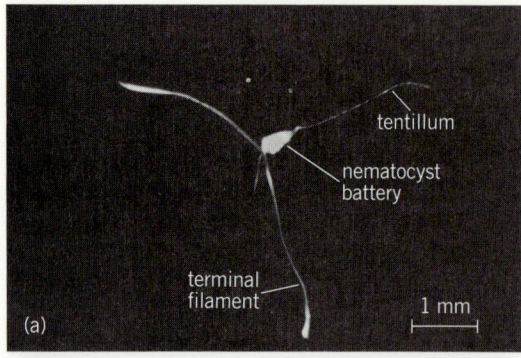

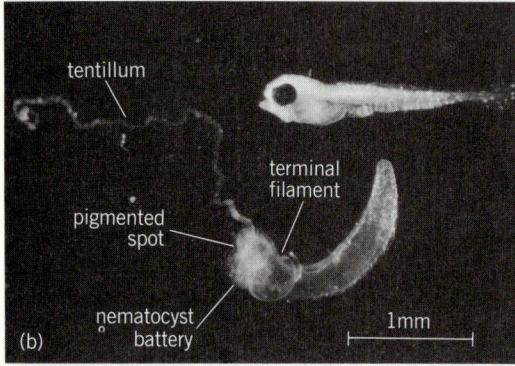

Fig. 2. Light micrographs of nematocyst batteries. (a) Battery that resembles a copepod, from the physonect siphonophore *Agalma okeni*. (b) Battery that resembles a larval fish (top), from the physonect siphonophore *Athorybia rosacea*. Two pigmented spots at the enlarged "head" resemble eyes.

ally or by the vibrations created in the water.

Siphonophores seem that they would be passive feeders, because they are nonvisual predators which drift in the water with their tentacles spread and depend on prey to swim into contact with them. Distinct dietary differences exist among the many species; these differences are related to the size, number, and thread structure of the nematocysts. These data, plus observations on nematocyst batteries that may act as lures to attract prey, suggest that morphological and behavioral mechanisms related to feeding by coelenterates are unexpectedly sophisticated.

For background information *see* COELENTERATA; FEEDING MECHANISMS (INVERTEBRATE) in the McGraw-Hill Encyclopedia of Science and Technology.

[JENNIFER E. PURCELL]

Bibliography: L. Muscatine and H. M. Lenhoff (eds.), *Coelenterate Biology: Reviews and New Perspectives*, 1974; J. E. Purcell, The functions of nematocysts in prey capture by epipelagic siphonophores (Coelenterata, Hydrozoa), *Biol. Bull.*, 166:310–327, 1984; J. E. Purcell, Influence of siphonophore behavior upon their natural diets: Evidence for aggressive mimicry, *Science*, 209:1045–1047, 1980; P. Tardent and T. Holstein, Morphology and morphodynamics of the stenotele nematocyst of *Hydra attenuata* Pall. (Hydrozoa, Cnidaria), *Cell Tissue Res.*, 224:269–290, 1982.

Coastal engineering

The Thames Barrier is a unique flood defense system that protects London. The project involved the installation in a major waterway of large concrete and mechanical structures within extremely close tolerances. The barrier became operational in 1982.

Historical background. Records show that land alongside the tidal section of the Thames has been subjected to flooding for many centuries due to increasing high water levels and gradual land settlement in the southeast corner of Great Britain.

Coordinated flood defense construction was authorized about 100 years ago. In 1928 lives were lost in London when floodwaters rose above the tops of the defense levels. In the next surge flood in 1953, banks protecting low-lying land in the outer estuary were breached and 300 lives were lost.

Feasibility studies. A government commission reported in 1953 on the severe flood risk to London, and detailed studies were undertaken.

The cause of the surge floods was established to be a random combination of meteorological conditions and a high spring tide. Land levels relative to sea levels recorded over 200 years showed surge levels increasing by about 2½ ft (¾ m) per 100 years (Fig. 1).

Numerous schemes were investigated during 1956 to 1969, and a barrier was judged the most desirable form of defense. Meeting the requirements of navigational interests was onerous, and the alternatives investigated varied from massive structures with clear openings of up to 1400 ft (427 m) to much smaller structures with clear openings of 200 ft (61 m). Reduced navigation requirements following progressive closure of upriver dock systems led to the choice of the smaller type defense. In 1972, a barrier, located 8 mi (13 km) from London Bridge, and based on the novel concept of the rising sector gates developed by engineers in 1968, was authorized by parliamentary act.

Hydraulic-model studies were used to investigate the effect of the barrier on tidal propagation and silt movement, river-bed protection requirements for failure of one gate to close, possible vibration problems on the gates, and techniques for partial gate closure to avoid reflected waves from tidal flow.

Characteristics of the Thames Barrier. The width of the river at the chosen site is 1700 ft (520 m). The barrier has 10 waterway openings contained between the south and north abutments and nine piers. Six openings are for navigation and have rising sector gates. The center four have 200-ft (61-m) clear openings and the others 103-ft (31.5-m) openings. One 103-ft (31.5-m) opening at the south end and three at the north end have conventional falling radial gates. The control building, standby generator building, workshop, and substation are located on the south bank, and there is a further substation adjacent to the north abutment. Living quarters for the duty shift and a tourist facility are also provided on the south bank. Figure 2 is a photograph of a practice low-tide barrier closure held on June 19, 1984. The leading edges of the center pair of 200-ft (61-m) rising sector gates are just about to rise above the river's surface. A typical arrangement of a rising sector gate and its relationship with the piers and sill is shown in Fig. 3.

Under normal conditions the box-girder rising sector gate is housed in the concave recess in the top of the prestressed-concrete hollow-box sill spanning the concrete piers. Each sill contains two 10-ft- (3-m-) diameter subways which carry duplicated power and control cables and service pipework. Thus the services in one subway are duplicated by the services in the other subway.

Each gate is moved by the operation of hydraulic cylinders acting through the triangular-shaped rocking beams connected to the gate ends by link arms. The gate ends are supported on solid-lubricant spherical bearings which allow the gate assemblies to rotate from the river-bed position to the defense and maintenance positions.

The main operating cylinders are installed at two levels, and the actuating hydraulic power packs are housed at the upstream end of each pier under the large shell-roof structure. Three actuating power packs are provided on each of the piers adjacent to the main spans, one pack for each gate mechanism and one pack as standby. The downstream building houses lift machinery.

The installed machinery is a powerful and effective method of gate operation which is simple and economic to maintain. Its power is such that the gates can be moved satisfactorily with only one in operation, although normally the hydraulic cylinders at each end of a gate will be operated in unison. An additional mechanism installed at each end of a rising sector gate, known as the shift-and-latch mechanism, is used to lock the gates in various positions. This mechanism also has other functions.

Operation of the gates is controlled from control panels in local control rooms situated on each pier, or from control panels located in the central control room situated near the top of the control building on the south bank.

The power supply for operation of the machinery and the control systems is from three separate sources: the National Grid on the south side of the river, the National Grid on the north side of the river, and from any one of three 1.5-MW generators installed in the generator building on the south bank.

River traffic through the barrier is controlled by navigation lights installed at the upstream and downstream ends of each pier. The lights are linked to and operated from the Port of London Authority control tower situated near the south end of the barrier.

Experience in flood control. Since the barrier became operational two years ago, there have been no occasions where closure has been necessary to protect London against flooding. In February 1984, the storm warning system indicated the possibility of a

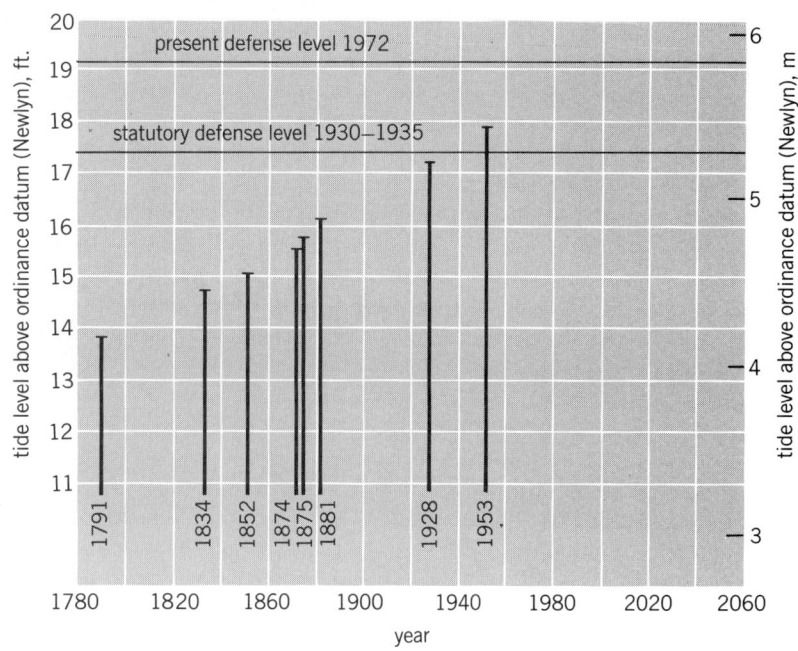

Fig. 1. Tide levels at London Bridge.

Fig. 2. Low-tide barrier closure.

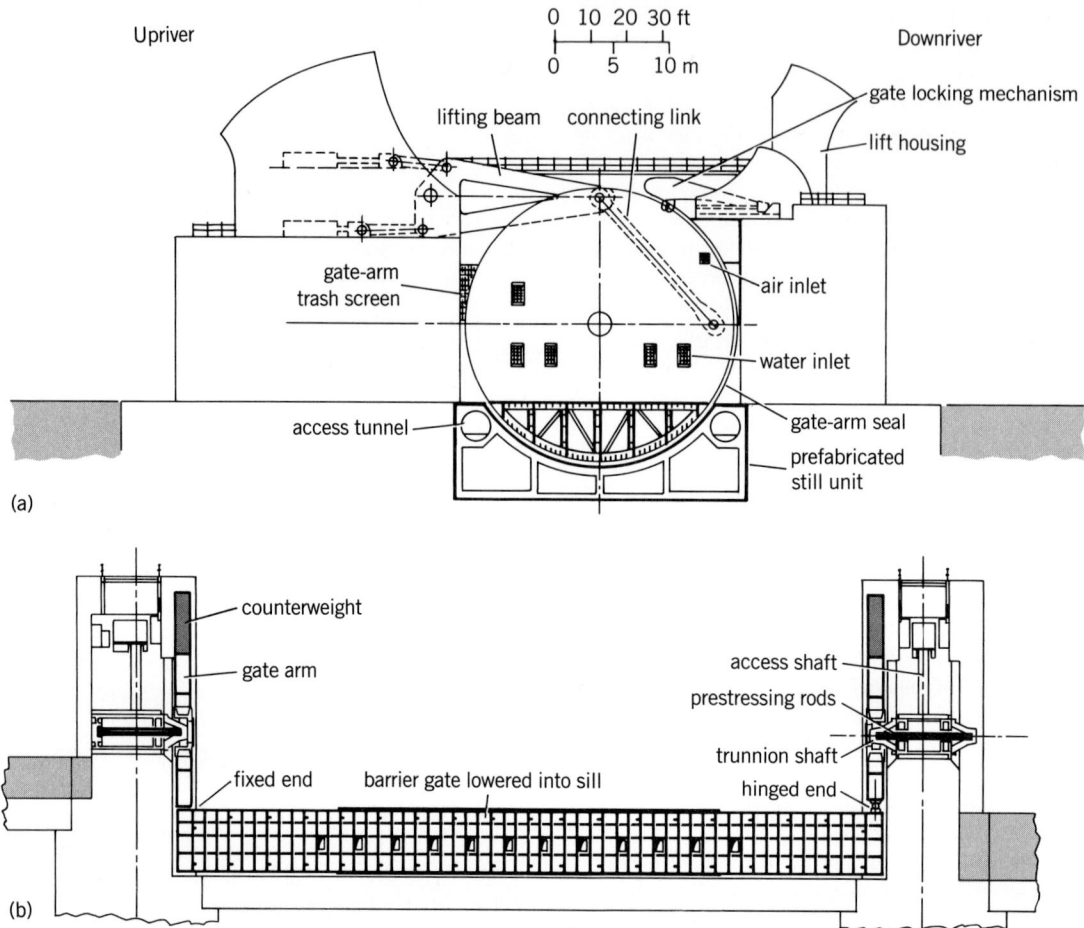

Fig. 3. Typical arrangement of rising sector gate. (a) Cross section through rising sector gate. (b) Cross section on center line of gate looking downriver.

significant surge tide but, in the event, the surge dissipated before the high tide.

Once a month, since becoming operational, the barrier has been closed at varying states of the tide in order to provide experience for the operational team and to test various modes of closure. Also, as part of a regular maintenance program, each individual gate is moved into the defense position once every month in order to ensure that all the operating machinery and associated systems are in a constant state of readiness for emergency closure.

For background information see COASTAL ENGINEERING in the McGraw-Hill Encyclopedia of Science and Technology.

[PETER A. COX]

Bibliography: J. R. Grice and E. A. Hepplewhite, Design and construction of the Thames Barrier cofferdams, paper presented at the Institution of Civil Engineers, London, May 1983; R. G. R. Tappin, P. J. Dowling, and P. J. Clark, Design and model testing of the Thames Barrier gates, paper presented at the Institution of Structural Engineers, London, April 1984; *Thames Barrier Design*, Proceedings of Conference held at the Institution of Civil Engineers, London, October 1977; *The Thames Barrier*, six papers presented at the Institution of Mechanical Engineers, June 1983.

Communications satellite

At the beginning of the commercial satellite era, early in the 1960s, the satellites were small and feeble, and the earth stations communicating through them required huge dish antennas more than 100 ft (30 m) in diameter. As the available power has increased on later generations of satellites, and as the sophistication of certain kinds of digital communications has increased, it has become possible to consider satellite communications directly to and from portable, and even hand-held, units.

Strengths and weaknesses of satellites. Communications to mobile and portable devices pushes the state of the art in communications. An indication of how it can be done is given by considering the inherent strengths and weaknesses of satellite communications. There are three inherent weaknesses in this mode of communications: (1) Bandwidth (the range of frequencies available for sending signals) is severely limited, particularly in comparison with fiber-optic systems. (2) The power available at the satellite for relaying signals is also severely limited. When considered with the great distance from geostationary orbit to the Earth (over 22,300 mi or 36,000 km), the power limit means that only a modest number of information bits can be sent per second to any receiver that does not have a large dish antenna. (3) There is a delay of about 0.3 s between the sender and the addressee. For voice communications, this tends to throw off the normal reaction time of give-and-take conversation, and sometimes makes conversations slow and awkward.

Against those negatives may be set two strong advantages: (1) Satellite signals can be received at, or sent from, any unobstructed spot in an entire nation. Therefore, on the very first day of operation of a new satellite, it can serve the most remote areas of a country as easily as it can major cities. (2) A satellite signal travels a direct line of sight between the satellite and the user. Thus there is a direct, exact relationship between the distance of the user from the satellite and the time delay of the user's signal. That allows, in principle, locating a user with great accuracy by measuring the time delay of a signal sent to the user from several satellites, and applying trigonometry.

These two advantages of satellites have been used in the design of military positioning systems, the U.S. Navy's Transit system and the more recent Global Positioning System (GPS) of the Department of Defense. Both systems use medium altitude satellite orbits. In the GPS system as many as 24 satellites, orbiting in three different planes inclined at angles of 60° to each other, all send signals simultaneously, in the same frequency bands. The signals are distinguished by coding, using pseudorandom noise (PN) codes. A PN code is a specific sequence of binary ones and zeros, known both to the sender and to the receiver. In the GPS, each satellite uses its own particular PN code, and that code is compared with a stored set of codes in each user receiver.

Direct communication with mobile users. At present a number of companies are proposing to supply direct communications to and from mobile users via satellite. Some of these companies plan to do so for the same services that are available over the telephone system or by cellular radio: voice and low-speed digital data. A difficulty in such systems is that they use only one of the two strong advantages of satellites (direct access to any point in a country). They also suffer from the necessity to support a relatively high power output for the long duration of a telephone call. A voice channel requires the equivalent of about 5000 information bits per second to provide minimum acceptable voice quality. The power output needed to send information bits, either from the user equipment or from the satellite, is proportional to the number of bits per second sent, other things being equal. In practice, this means that such user equipment has to be connected to a vehicle power source, and that a typical satellite can support only a small number (roughly 24,000) customers. The inherent 0.3-s time delay of satellite transmission is a further difficulty, though one that customers have become used to in many overseas telephone calls through the INTELSAT system. An operational example of such mobile systems is Inmarsat, which links ships at sea to shore telephone networks by way of satellites. The Inmarsat ship terminals are relatively bulky and expensive, because they must use gyro-stabilized platforms carrying dish antennas about 3 ft (1 m) in diameter in order to transmit enough bits per second for voice and data.

RDSS. Another approach to mobile satellite communications makes use of the inherent advantages of satellites, and avoids the disadvantages, to provide a special set of services quite different from voice telephone systems. The Federal Communications Commission voted in July 1984 to issue a Notice of Proposed Rulemaking that would allocate spectrum in the microwave bands for such services. They will be called Radio Determination Satellite Services (RDSS). In the RDSS a continuous spread-spectrum signal is sent from a central ground station through a satellite relay in geostationary orbit. That signal is relayed to cover a large area. Within the signal are embedded interrogation timing marks repeated many times per second. The user receiver-transmitters (transceivers), using spread-spectrum decoding techniques, receive the interrogations. A user who wants to know position allows his or her transceiver to respond to the interrogation. The response is picked up by at least two of the RDSS satellites and relayed to the central ground station. There powerful computers combine the known time of the interrogation with the measured times of response and with the terrain height (called up from a digital data base) in the vicinity of the user. The four pieces of data are combined to provide a highly accurate measurement of position. The position is then sent to the user, addressed by the user's unique digital transceiver identification code, as part of the continuous outbound data stream. The transmission is completed by the return of a positive acknowledgment. Because the RDSS uses all of the available bandwidth for just one channel, it can measure times, and therefore positions, very precisely (typically to several feet or a few meters.)

The value of the RDSS basic service is enhanced by a number of other features, all relatively simple to add because of the power and size of the central ground computer and its data base. The user may be given navigational guidance, for example, to a precise street address, using directions no more complicated than those on a highway sign. For example, a digital message "Turn right in 1.45 miles at Murphy Street" may be flashed to the transceiver of a particular user. Information on the positions of all the vehicles, or any particular vehicle, in a fleet of trucks, police cars, taxis, or delivery vans may be returned to the fleet dispatch headquarters.

The RDSS can also be used to relay telegraphic messages and data between any two transceivers. It can provide emergency location information and verify through its digital message capability that an alarm is genuine. It can warn users in boats and aircraft against collision with shoals or terrain. And it can interconnect at the central ground station with the entire telephone system for all digital services. And because the user transceiver responds with only a short burst signal, the transceiver can be handheld and powered by penlight cells.

It seems likely that all the possible types of satellite services to mobile users will eventually be implemented, and ultimately that the marketplace will determine which services are economical to provide through satellites, and which are better handled by competitive systems such as cellular radio.

For background information *see* COMMUNICATIONS SATELLITE; SATELLITE NAVIGATION SYSTEMS in the McGraw-Hill Encyclopedia of Science and Technology.

[GERARD K. O'NEILL]

Computer-aided design and manufacturing

Computerized manufacturing systems often involve a model of the process response. The model is used to predict the changes in the manufactured part that can be expected from variations in operating parameters, or vice versa. Thus, if a correction or control is needed in the manufactured part, the adjustments necessary in the operating parameters can be calculated. For example, a rolling mill may be producing a strip that is momentarily thicker than allowed by product specifications. This could be caused by, among other things, a fluctuation in the incoming thickness or a fluctuation in strip temperature. To correct this problem of increased thickness, the rolls may be moved closer together, or a tensile stress may be increased on the strip before or after the rolls. A process model can be used to tell how much the rolls must be moved or how much stress must be added.

In some approaches the relationship between process parameters and process result involves elaborate empirical data bases, developed simply from process experience. In other cases a discrete adjustment is made in the proper direction, without regard to a quantitative estimate provided by a model. That is, if the rolled strip is too thick, the rolls are moved closer together by a given amount. If that is not enough, the rolls are moved a further increment, and so on. However, the most sophisticated control systems involve a process model.

The process model involves properties of the material being processed, properties of lubricants and coolants, and, in some cases, the properties of the tooling. In fact, the development of such properties

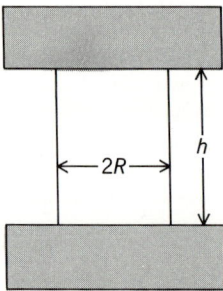

Fig. 1. Schematic representation of the forging of a cylindrical billet.

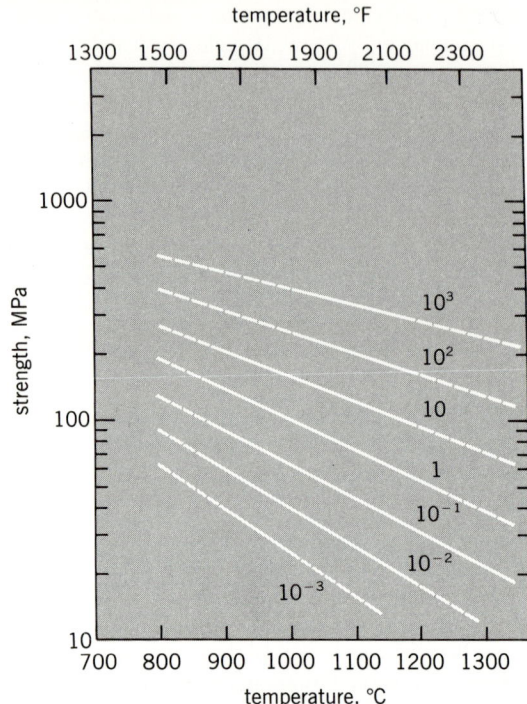

Fig. 2. Strength (flow stress) of mild steel as a function of elevated temperature and strain rate. The solid portion of each curve indicates the extent of the data range. The numbers on the curves represent the true strain rate $\dot{\epsilon}(s^{-1})$. 1 MPa = 145 psi.

can be the most difficult step in establishing a reliable computer-aided manufacturing process. An elementary example is the open die forging, or upsetting, of a round metal bar, as shown in Fig. 1. A reasonable estimate of the average forging load or force F is given by Eq. (1), where R is the bar ra-

$$F = \pi R^2 \sigma_0 \left[1 + \frac{2\mu R}{3h} + \frac{1}{3}\left(\frac{\mu R}{h}\right)^2 \right] \quad (1)$$

dius, h is its height, σ_0 is the strength of the metal, and μ is the coefficient of friction between the bar and the tooling. The forging force depends on the shape of the billet (R/h) and will therefore vary as the forging deformation ensues. However, at any point in the process, Eq. (1) gives the force needed for continuing deformation. The value of the force allows assessment of whether a given billet can be forged on a given press, or how much deformation a given press energy can produce (force × distance = work or energy).

Equation (1) shows the great dependency of force on material strength. The strength, in turn, is a function of strain, strain rate, $\dot{\epsilon}$, temperature, and the basic composition and microstructure of the metal being forged. Figure 2 shows some data for mild steel in the hot-working range. The strain rates vary from those of routine laboratory mechanical testing (10^{-3}/s) through those of hydraulic pressing (perhaps 10^{-1}/s) up to those of rapid mechanical pressing and high-energy rate forming (10/s and higher). Moreover, the strength is very sensitive to such variations, and it is necessary to obtain data at the relevant strain rate and temperature. In some cases it is possible to summarize such wide-ranging data in terms of a constitutive equation. The constitutive equation could be directly incorporated into Eq. (1). Where no simple equation is practical, data such as shown in Fig. 2 may be directly stored in the computer.

At high deformation rates and large deformations, significant temperature increases can occur, thus complicating the concept of deformation temperature. In a process like that shown in Fig. 1, the adiabatic temperature increase from the deformation, ΔT_d, can be estimated from Eq. (2), where $\bar{\sigma}_0$

$$\Delta T_d = \frac{\bar{\sigma}_0}{C\rho} \ln\left(\frac{h_1}{h_2}\right) \qquad (2)$$

is average strength, C is heat capacity, ρ is density, and h_1 and h_2 are the initial and final billet heights, respectively. The adiabatic ΔT_d only applies to very rapid deformation rate. At slower deformation rates some deformation heat would be lost to the environment and the temperature increase would be less than ΔT_d in Eq. (2).

Data such as that given in Fig. 2 can be generated by mechanical testing machines which allow wide ranges of deformation rate and accurate temperature control. A considerable amount of time can be consumed generating such data. For rough estimates on common alloys, data in the open literature may suffice. However, for less common alloys or for subtle heat-to-heat differences on a given alloy, there is little substitute for new testing to provide a sophisticated data base for the computerized control system. At present such testing can be costly as well as lengthy, and there is a great need for rapid, inexpensive test techniques suitable for deformation rates and temperature ranges pertinent to manufacturing.

Equation (1) also involves the friction coefficient. Friction coefficients are difficult to measure reliably and, in some cases, rough estimates are obtained by back-calculating from equations similar to Eq. (1). However, more sophisticated approaches also are available. Thick lubricant films can involve friction coefficients as low as 0.01–0.03. At the other extreme, the friction coefficient approaches 0.5 as the metal sticks to the tooling. Lubrication with light liquids can involve friction coefficients in the 0.10 range. Such friction produces heating. The overall amount of friction heating is usually small compared to that involved in Eq. (2). However, it is concentrated at the tooling interface, and the lubricant and adjacent surfaces can get very hot. Such heating can grossly alter the value of the friction coefficient. Thus, in practice a value of friction coefficient that is accurate for normal processing is used, while it must be understood that vastly different values can be encountered during lubricant breakdown.

There are other materials data needed, beyond the scope of Eq. (1). Fracture or workability data, and related criteria, are necessary. Elastic constants are needed in many cases. It is important that ultrasonically measured elastic constants be used, because low-strain-rate elastic measurements are not accurate at high temperatures. The resistance of the material to various instabilities (necking, buckling, shear band development, vibrations, and so on) may have to be incorporated into the model. Surface descriptions such as heat-transfer coefficients and wear and oxidation rates may be involved as well.

As more experience is gained with computer-aided manufacturing systems, the data needed for common alloys will become generally available. Even so the concerns voiced above will continue to apply to the processing of new materials.

For background information *see* COMPUTER-AIDED DESIGN AND MANUFACTURING; STRENGTH OF MATERIALS; STRESS AND STRAIN in the McGraw-Hill Encyclopedia of Science and Technology.

[ROGER N. WRIGHT]

Bibliography: G. Dieter (ed.), *Workability Testing Techniques*, 1984; S. Kalpakjian, *Mechanical Processing of Metals*, 1967; A. T. Male, Variations in friction coefficients of metals during compressive deformation, *J. Inst. Met.*, 94:121–125, 1966.

Computer networking

A person with a computer can now send a letter across the United States by using a computer network for less cost than through the mail system. With a new computer network technology called packet switching, a letter sent coast to coast costs as little as one-half cent per 200 words. More significantly, a computer letter will arrive within seconds.

Even these costs will continue to decrease. The cost of electronic circuitry, the critical ingredient of computer and computer networking equipment, is falling at a rate of over 20% per year, so that circuits that cost $1000 today may cost only $328 in 5 years. Computers and computer networking facilities are also becoming easier for the layperson to use, and standards for exchanging information among diverse computers are being defined.

The future is likely to bring worldwide interconnected computer networks resembling the telephone system of today. Like the telephone system, these networks will have local components called local area networks (LANs), akin to a building or campus telephone system called the Private Branch Exchange (PBX), and components in the public domain called wide area networks (WANs). Access to computer networks will become as common as access to the telephone system, as prices reflect the circuit cost decreases and networking facilities become easier to use. Using computer networks, people will not only send letters but have access to a variety of information available from a new breed of electronics entrepreneur, the information provider.

Development. In a sense, the functions of the telegraph system developed during the nineteenth century more closely resembled a computer network than today's telephone network. The telegraph system, like the computer network, does not encourage direct interaction between the sender and receiver. A telegraph message, like a computer letter sent over a packet network, is stored at a number of intermediate stops along its path to the recipient.

In the 1950s it became useful to transfer computer information over the telephone system. At that time, computer information was usually first punched into computer cards or placed on magnetic tape before being sent long distances between two communicating card reader-punch machines or between two communicating magnetic tape machines. In the early 1960s, computers were redesigned to communicate directly without placing information on cards or magnetic tape. However, costs for circuits were high then, and the use of the telephone system for computer data was restricted by law. Not until the Supreme Court's "Carterfone decision" in 1969 was it deemed legal to attach equipment not owned by the telephone company to the telephone system.

The first wide area networks were government networks and a few private ones like airline reservation systems. One of these was to become the research vehicle for many other future wide area networks: the U.S. Department of Defense Advanced Research Projects Agency's (ARPA) network called the ARPANET. It was envisioned in a 1964 report as a packet network for computer sharing between government workers, and its initial version was completed in 1969. It became very useful as a test bed for computer networking principles. Instead of being used primarily for computer sharing, ARPANET rapidly became much more widely used as a vehicle for electronic mail between researchers in various computer science fields, including computer networking.

During the 1970s, wide area computer networks of two types, private and public, became available. Private networks, designed by computer manufacturers and owned by corporations and government organizations, provided information workers with data needed for daily operations. Digital Network Architecture (DNA) from Digital Equipment Corporation, and Systems Network Architecture (SNA) from International Business Machines Corporation were two prominent examples of these designs.

Public wide area networks, designed and owned by telephone companies, gave people access to information and services from companies such as Dow Jones, which provided electronic financial information, and Mead Data, which provided electronic news. Datapac in Canada, Transpac in France, and Telenet and Tymnet in the United States were public networks through which these kinds of services were available.

During the 1970s local area networks became the subject of significant research, primarily as a result of the growing interest in moving the processing power of the computer to the user rather than asking the user to come to the computer room or share a large computer with other users. A number of different concepts, based on two physical network configurations, were proposed. Work was undertaken on a

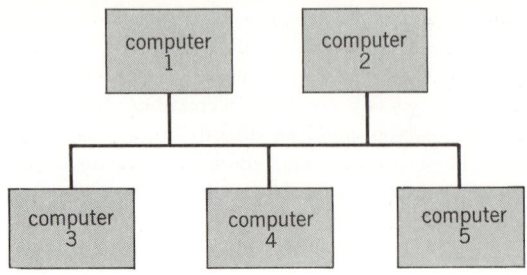

Fig. 1. Bus configuration.

bus configuration (Fig. 1), using a mathematically based scheme called Carrier Sense Multiple Access (CSMA) to determine which computer has access to the bus at any given time.

A second configuration called the ring configuration (Fig. 2) was found to have the advantage of natural ordering. That is, each computer can determine which computer is before it and which follows it. For example, if the electrical signal flows clockwise in Fig. 2, computer 1 follows computer 5 and precedes computer 2; computer 2 follows computer 1 and precedes computer 3; and so on. This property allows a "token" or a permit to circulate to control access to ring at any given time. Until the ring is needed for message transmission, the token continues to circulate. Any station may remove it, send a message, and return it to the ring at any time. Thus, only one station will send a message at a given time. Both the ring and the bus were discovered to have the practical disadvantage that they could not be wired into a building in the traditional way. Traditionally, building wire extends outwardly radially from wiring centers in a "star" configuration. Researchers then discovered that the ring could be made more reliable by wiring it as a star (Fig. 3). This configuration has replaced the pure ring in practical LANs. Other combinations have been developed from these two.

In the early 1980s the number of both public and private networks grew substantially. Almost every major noncommunist country had one or more public networks, and over 10,000 private networks were operational. The number and variety of information services available through computer networks also grew dramatically. By 1984 over 600 different data sources in the United States provided information almost instantly through computer networks. The technological differences between public and private

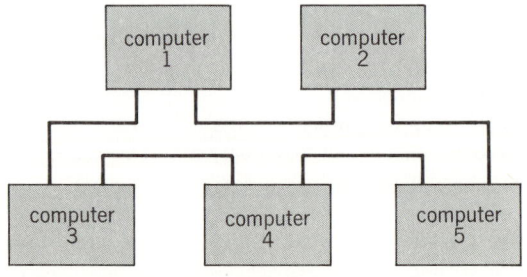

Fig. 2. Ring configuration.

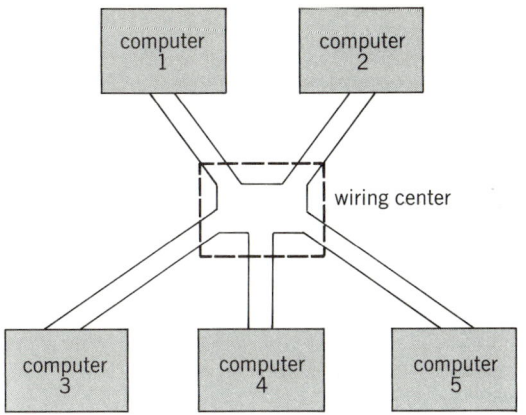

Fig. 3. Star-ring configuration.

networks were fading, and interconnection of both kinds of networks permitted information exchange between organizations. Despite these advances, relatively little information in the form of images, graphics, or voice was transferred through computer networks.

Principles. The principles of computer networking are, in many ways, similar to the principles of vehicular traffic flow through the network of highways, bridges, and tunnels.

Three of the important problems of computer networking are addressing, routing, and flow control. Addressing is analogous to assigning names and numbers to the populated areas (city and town names) and routes between them (U.S. 70, I80, and so forth). The advantages of assigning designations that give information about the characteristics of the route (Interstate 80 is a better route than U.S. 70 which is better than Vt. 4) or the populated area (a city has more services than a town which has more services than a village) are obvious. Frequently, computer networks use a similar concept. A given node (analogous to a populated area) in a computer network may be designated, implicitly or explicitly, as a host, switching node, or terminal node. Similarly, a link (analogous to a piece of route connecting two adjacent populated areas) may be designated as a high- or low-capacity link.

Routing of messages in a computer network is similar to routing of cars and trucks from one location to another. Frequently, travelers unfamiliar with an area obtain the best route from one city to another through a travel service. A centralized routing service of this sort is used in many computer networks. The public network, Tymnet, has used this approach in the past, and private networks using SNA employ it in some small networks.

Another approach to traveling is simply to determine the best route from a map or from knowledge of geography. This normally works well, but can make traveling difficult if information about the terrain or road conditions is outdated. This technique is called source routing because the source (sender) determines the route the message will follow. Finally, a traveler may have no specific route in mind but make route decisions at each intersection. The network analogy to this semirandom approach is called fully distributed routing. This technique has been used in the ARPANET.

Flow control, used in computer networks, minimizes the ill effects of congestion. It has no universal analogy in the highway system. Data that enter a computer network at a high speed, say 8000 characters (a basic element of computer information, capable of representing a single letter or number) per second, may need to share the same telephone link with data entering from an unrelated terminal at a rate of 2000 characters per second. If the link which must accommodate the two data flows is able to carry only 6000 characters per second, there must be some technique that limits the two flows in a way which is fair to both. This is much like the problem of two highways coming together at a tunnel that is not able to accommodate the traffic from both highways. When this happens in a data network, flow control reduces the two data flows and may even limit the amount of data entering the network from the terminals. In the highway example, reducing the number of cars that enter the system would be akin to asking certain people to stay home since the tunnels are crowded.

Prospects. The future holds promise for tremendous growth and improvement in computer networks, both local area networks and wide area networks. It is likely that these networks will become as important economically and militarily as the highways and airways of today.

Computer networks will become more flexible, easier to use, and will perform better. Most importantly, computer networks will be easier to access from personal computers and terminals. As computer networks become as easy to use as the dictionary, the telephone book, and the automated bank teller, they will become important to a more diverse population of users.

The computer network will continue to improve its performance. This will be partly due to the improving underlying technologies, computers, and transmission systems, but will also be influenced by a demand for more rapid network response consistent with the human brain's reaction time.

The computer network will become more flexible so that it is easily able to accommodate a wide variety of information, either separately or mixed. For example, voice messages and pictures will be sent over the network in the same way that text messages or text documents can be sent today. These messages or documents may also have voice, text, graphics, and photographs all integrated into one information package. Computer networks will gradually change to accommodate, at lower costs and more rapidly, all of the information that can be sent through the mail today.

For background information *see* DATA COMMUNICATIONS in the McGraw-Hill Encyclopedia of Science and Technology.

[ROBERT F. STEEN]

Bibliography: P. Baran, F. Boehm, and P. Smith,

On Distributed Communications, vols. 1–11, Rand Corp., August 1964; A. S. Tanenbaum, *Computer Networks*, 1981.

Cryosphere

Snow and ice phenomena are attracting the attention of a widening range of scientific disciplines. The scale of glaciological research has broadened primarily from a local activity to a global framework, and from a concern with current processes to the study of past and potential future changes in snow and ice conditions and related climate. These developments have been supported by advances in the application of remote-sensing techniques for mapping snow and ice, in the numerical modeling of glacier and ice-sheet motion, and in the paleoclimatic interpretations of ice cores extracted from ice sheets in the polar regions as well as from high-elevation ice masses in the tropics.

Remote sensing. Until the advent of satellites, snow cover was surveyed at snow courses and sea ice was observed by ships and aircraft. The early satellites provided visible and infrared images which are still the mainstay of the operational mapping of global snow and ice extent. However, clouds obscure large areas of the Earth, and their presence complicates the detection of snow and ice due to their similar spectral signatures in the visible and, for low clouds, infrared wavelengths. Sensors detecting thermal microwave radiation emitted by the surface have provided the first year-round, all-weather information on polar sea ice. The utility of this technique is a result of the contrast of approximately 100 K (180°F) in brightness temperatures between sea ice and open ocean. Data collected by the 1.55-cm-wavelength electrically scanning microwave radiometer (ESMR) on *Nimbus 5* showed a large year-to-year variability in Antarctic sea ice, and identified a previously unknown area of low ice concentration and open water (polyn'ya) during several winters in the Weddell Sea. As sea ice ages, and especially when it survives a summer melt season in the central Arctic Ocean, brine drainage causes its salinity to decrease markedly, and this in turn affects the dielectric coefficient and the emissivity at microwave frequencies. *Nimbus 7* carries a scanning multichannel microwave radiometer (SMMR), which provides data from vertical and horizontal polarizations for a range of frequencies. These data enable not only the total ice concentration to be calculated but also the fraction of multiyear ice and the ice surface temperature.

Parallel studies using the SMMR data over the continents in winter show that snow extent, snowpack water content, and snowmelt progression can be mapped routinely. This information, together with airborne surveys of natural terrestrial gamma radiation which are used over the Midwest to give more detailed mapping of snow water content, will be increasingly useful in snowmelt runoff and flood forecasting.

Remote sensing from aircraft and satellites is also being used to map the elevation and thickness of glaciers and ice sheets. Radar (and, in the future, laser) altimeters enable precise mapping of the ice surface, currently to within ±3 ft (±1 m) elevation with a spatial resolution of about 1–3 mi (2–5 km). Up to now, only the northern parts of Antarctica and southern Greenland equatorward of 72° latitude have been so mapped due to the orbital configuration of satellite tracks. Corresponding data on bedrock elevation have been obtained by airborne radio-echo sounding at frequencies of 60 and 300 MHz over much of Greenland and about half of Antarctica. The measurements reveal clearly the bedrock terrain, including subglacial mountains and areas below sea level, as well as internal-layer echoes which are continuous over wide areas and appear to indicate changes in the acidity of the ice resulting from the deposition of volcanic sulfate aerosols (Fig. 1).

Borehole data. Direct measurements in ice sheets are now available from boreholes. Two deep ice cores to bedrock have been extracted in Greenland (at Camp Century, 77°N, and Dye-3, 65°N) and one at Byrd (80°S) in Antarctica, each spanning a record of more than 100,000 years, together with approximately 100 cores of 330 ft (100 m) depth or more; the latter include a 1-mi (2-km) core from the Soviet station Vostok and a 3000-ft (900-m) core from Dome C, both in Antarctica. These records show rapid changes in crystal structure, particle concentration, and chemical constituents (particularly carbon dioxide trapped in air bubbles) at the Holocene/

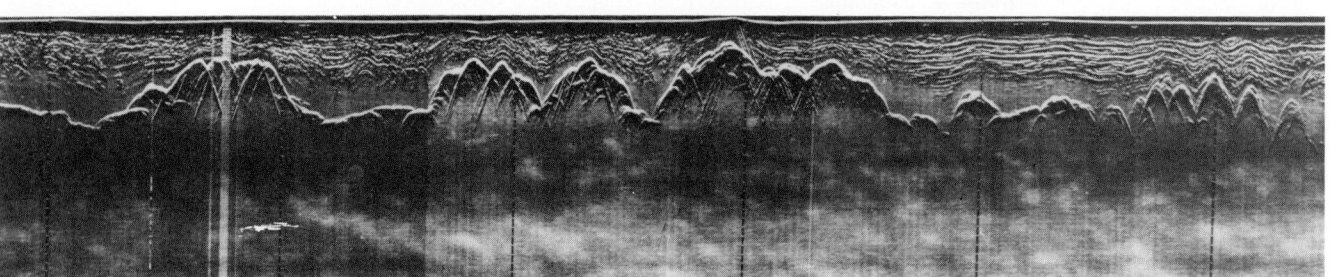

Fig. 1. Trace of airborne radio-echo sounding in Victoria Land, Antarctica, near Dome C (74°30′S, 123°E). Subglacial mountains rising to within 3900–4900 ft (1200–1500 m) of the ice-sheet surface, bedrock plains at 8200–9800 ft (2500–3000 m) depth, and internal-layer echoes are visible. The vertical markers are at 18-mi (30-km) intervals. (*World Data Center-A for Glaciology* [*Snow and Ice*])

Wurm-Wisconsin boundary (approximately 10,500 years ago). For example, atmospheric aerosol levels, which were large during glacial times due to the aridity of the continents and stronger wind circulation, subsequently decreased by an order of magnitude. Atmospheric carbon dioxide levels increased dramatically by some 50–100 ppm at the end of the last glacial interval when they approached or exceeded the preindustrial level of about 275 ppm (the current value is 340 ppm). Global climate models suggest that such a shift corresponds to a mean global air temperature difference of 0.5–1°F (0.3–0.6°C; 7–15% of the total glacial/postglacial change), although the change in atmospheric CO_2 may have been in part a response to the postglacial changes in climate and ocean conditions. It is especially intriguing that abrupt, short-term CO_2 changes of similar magnitude seem to have occurred during the glacial period, pointing to chemical changes in the world ocean as a possible cause.

Numerical models. Remote-sensing information and borehole data have contributed substantially to understanding the motion and past histories of ice sheets and glaciers. Numerical models have been used, for example, to simulate ice sheet history over North America. It has been shown that a realistic time scale of ice growth and decay during the last glacial cycle can be modeled, provided that the effects of ice-albedo feedback, isostatic bedrock response to the ice load, and decreasing snow accumulation over the central part of the ice sheet are incorporated into the model. Figure 2 shows the difference in phase between the climate, itself forced by astronomical variations in solar radiation and amplified by ice-sheet–albedo feedback, ice volume, and bedrock depression. One model of the Antarctic ice sheet suggests that its marine segment in West Antarctica should undergo a long-term cyclical pattern of buildup and shrinkage before renewed growth. The geological evidence for such a pattern is unclear, although the West Antarctic ice sheet appears to have retreated appreciably between 17,000 and 7000 years ago as world sea level rose during the melting of the northern continental ice sheets.

On a smaller scale, modeling has been used to study the possible catastrophic retreat of the Columbia Glacier in southwestern Alaska. This glacier, now grounded on a moraine, calves into the shipping lane for the oil terminal port of Valdez, and it is the only tidewater glacier in Alaska not to have undergone major retreat during this century. Modeling projections indicate that accelerated retreat and much increased iceberg discharge may be imminent.

Glacier retreat. On a global scale, glacier retreat during the last 100 years appears to be contributing about 0.02 in. (0.5 mm) per year to the observed global sea-level rise of 0.05 in. (1.2 mm) per year. Ocean thermal expansion may account for the remainder. However, information on glacier volume changes is extremely limited in its spatial and temporal coverage. A related question for the future is the stability of the West Antarctic ice sheet to any global warming. Much of the West Antarctic ice is grounded well below sea level (Fig. 3) and to some extent retained by the Ross Ice Shelf, so that a sea-level rise or increased basal ice melt could cause a rapid retreat of the present grounding line leading to a substantial shrinking of the ice sheet, accompanied by a 20–23 ft (6–7 m) rise in world sea level, within a few centuries. At present, the grounding line of the Ross Ice Shelf appears to be in dynamic equilibrium, or may even be advancing due to isostatic uplift below the ice shelf. Geophysical observations and modeling studies of this problem are continuing.

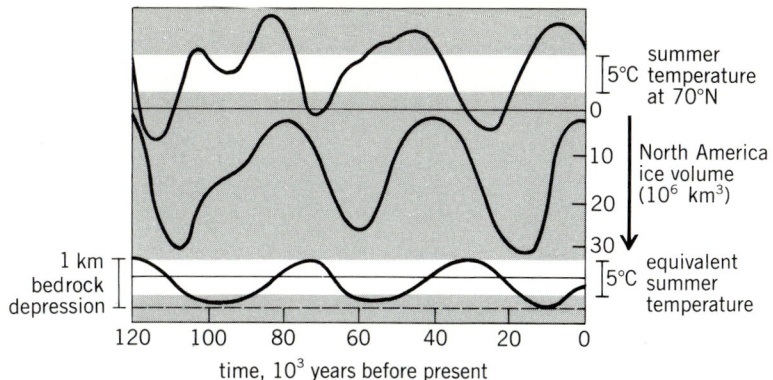

Fig. 2. Phase difference between climate (including the effect of ice-sheet–albedo feedback), ice volume, and bedrock depression, in a numerical simulation of the Northern Hemisphere ice sheets. An ice volume of 3×10^5 mi^3 (10^6 km^3) is approximately equivalent to 12.5 ft (3.8 m) of world sea level, a difference of 5°C-9°F; 1 km-0.6 mi. (*After W. F. Budd and I. N. Smith, The growth and retreat of ice sheets, Sea Level, Ice, and Climatic Change, Int. Ass. Hydrolog. Sci. Publ. 131, pp. 369–409, 1981*)

Arctic sea ice. A further concern for the future is the stability of the Arctic sea ice cover. This persists year-round at present, although its extent increases by a factor of 2 from the late summer minimum (2.7×10^6 mi^2 or 7×10^6 km^2) to the spring maximum (5.8×10^6 mi^2 or 15×10^6 km^2). There is sedimentological evidence that a quasipermanent annual ice cover has persisted for at least the last 5×10^6 years, even during warm intervals. Nevertheless, it has been estimated that a 9°F (5°C) increase in summer temperature might be sufficient to remove the pack ice each summer, although it would reform in the winter. The question is complicated by the nature of the thermal and salinity structure of the Arctic Ocean surface waters. Fresh-water runoff from Eurasia and North America and summer ice melt maintain a shallow layer of low salinity above the halocline, and this water readily refreezes in the autumn. At present, convective overturning does not penetrate below 660 ft (200 m), and therefore warmer waters of Atlantic origin below 660–980 ft (200–300 m) have no effect on the surface regime. The oceanographic information and the geological record suggest a picture of stability, unless drastic climatic or other changes in runoff are invoked. Proposals in the Soviet Union for diverting Siberian riv-

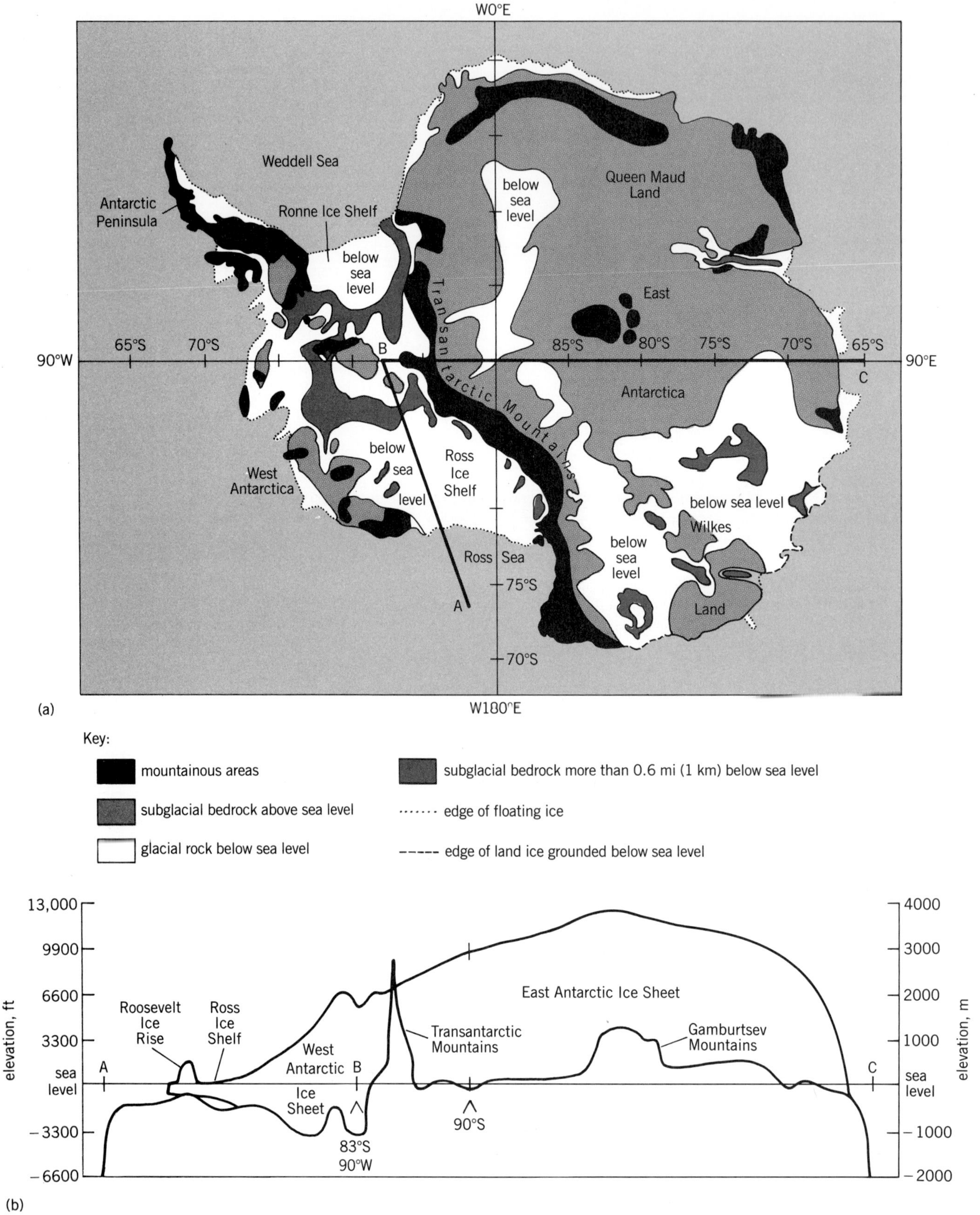

Fig. 3. Subglacial topography of Antarctica. (a) Surface map. (*After C. R. Bentley, The Antarctic ice sheet: Diagnosis and prognosis, in Proceedings of the Carbon Dioxide Research Conference: Carbon Dioxide, Science, and Consensus, Sec. IV, pp. IV3–IV73, CONF-820970, U.S. Department of Energy, 1983*) (b) Cross section. (*After J. T. Hollin and R. G. Barry, Empirical and theoretical evidence concerning the response of the Earth's ice and snow cover to a global temperature increase, Environ. Int., 2:437–444, 1979*)

ers toward central Asia seem unlikely to affect salinities sufficiently to disrupt the present Arctic sea ice cover. However, a doubling of atmospheric carbon dioxide, by late in the twenty-first century, if accompanied by a warming of 9–18°F (5–10°C) in high latitudes as calculated by present climate models, would be a cause for concern. Changes in the Arctic sea ice cover would cause significant changes in the climatic regime of middle and high latitudes, although the nature of such changes cannot yet be predicted with any confidence.

For background information *see* GLACIOLOGY; GREENHOUSE EFFECT, TERRESTRIAL; REMOTE SENSING; SEA ICE; SNOW SURVEYING in the McGraw-Hill Encyclopedia of Science and Technology.

[R. G. BARRY]

Bibliography: D. L. Clark and A. Hanson, Central Arctic Ocean sediment texture: A key to ice transport mechanisms, *Glacial-Marine Sedimentation*, pp. 301–330, 1983; D. J. Drewry (ed.), *Antarctica: Glaciological and Geophysical Folio*, 1983; G. de Q. Robin (ed.), *The Climatic Record in Polar Ice Sheets*, 1983; H. J. Zwally et al., *Antarctic Sea Ice 1973–1976: Satellite Passive-Microwave Observations*, NASA SP-459, 1983.

Cryptosporidium

Cryptosporidium is a protozoon that completes its life cycle on intestinal and respiratory epithelial cell surfaces of mammals, birds, and reptiles. It is a genus in the family Cryptosporidiidae, suborder Eimeriina, order Eucoccidiida, subclass Coccidia, class Sporozoa, phylum Apicomplexa. The suborder Eimeriina contains 13 families with over 1500 named species. The great majority of these species belong to the genera *Eimeria* and, to a lesser extent, *Isospora*. Both are intracellular parasites which primarily infect the intestinal tract of vertebrates. *Toxoplasma*, a tissue cyst–forming coccidia, is another important member of the suborder Eimeriina.

Cryptosporidiosis is a common infection which, to date, has been described in 20 different species of animals, including humans; in at least 12 of them the infection was associated with some illness. The infection was first observed in mice in 1907, but the significance of the disease in humans and ruminants has only recently been recognized.

Biology. The life cycle broadly follows that of other enteric coccidia, with asexual followed by sexual endogenous stages (Fig. 1). Distinct features by which *Cryptosporidium* achieves greater reproductive potential include cyclic development of the schizogeny and autoinfection; the latter feature allows it to maintain persistent infection in immunologically compromised hosts.

The organism has been observed in the gastrointestinal tract of clinically healthy mice, rabbits, guinea pigs, raccoons, cats, and birds; and in clinically ill calves, lambs, goats, piglets, deer, humans, monkeys, snakes, and immunologically deficient foals. The infection has also been observed in association with upper respiratory tract infections of

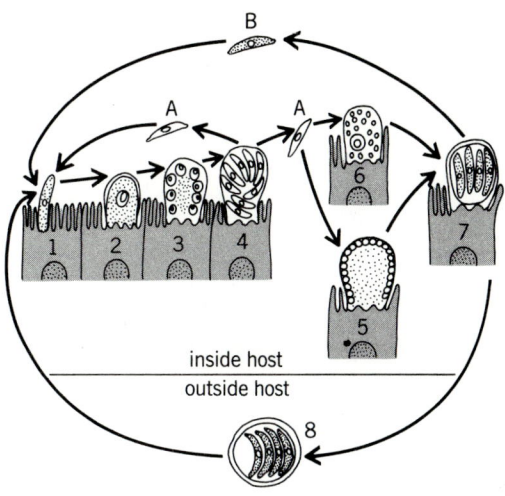

Fig. 1. Diagram of the life cycle of *Cryptosporidium*. Asexual cycle of the endogenous stage: (1) a sporozoite or merozoite invading a microvillus of a small-intestinal epithelial cell; (2) a fully grown trophozoite; (3) a developing schizont with eight nuclei; (4) a mature schizont with eight merozoites. Sexual cycle: (5) microgametocyte with many nuclei; (6) macrogametocyte. (7) A mature oocyst containing four sporozoites without sporocyst. (8) Oocyst discharged in the feces. Merozoite (A) released from mature schizont; sporozoite (B) released from mature oocyst, the segment of the life cycle responsible for autoinfection which in immunodeficient host can maintain infection indefinitely.

birds. Based on recent transmission experiments, *Cryptosporidium* was shown to lack host specificity; it is, therefore, a potential zoonosis. Organisms from some species can infect a wide variety of other animals, with or without causing illness. Thus, isolates from calves, humans, deer, lambs, and goats readily infect lambs, calves, goats, and piglets, causing enterocolitis; and mice, rats, guinea pigs, chickens, dogs, cats, and foals without causing illness. Infection is diagnosed either histologically (by intestinal biopsy) or by demonstration of oocysts (3–4 micrometers) in Giemsa-stained or modified acid-fast stained fecal smears. The oocysts are extremely resistant to the action of disinfectants commonly used in laboratories and hospitals.

Cryptosporidiosis has most commonly been studied in calves, and more recently in humans. Studies in animals show that the lower small intestine is the organ most severely affected; in very young animals, however, the entire bowel may be infected.

Cryptosporidium, unlike other coccidia, does not invade the epithelial cytoplasm, but remains physically outside the cell boundaries and firmly embedded in the brush-border membrane (Figs. 2 and 3). The most common mucosal changes observed are stunted, fused, and swollen villi coated with immature absorptive cells; and the lamina propria is moderately infiltrated with macrophages, neutrophils, and eosinophils. The infection has a marked effect on the level of membrane-bound digestive enzymes, and diarrhea results from brush-border maldigestion and malabsorption. Relapses of the disease have been reported several weeks after apparent recov-

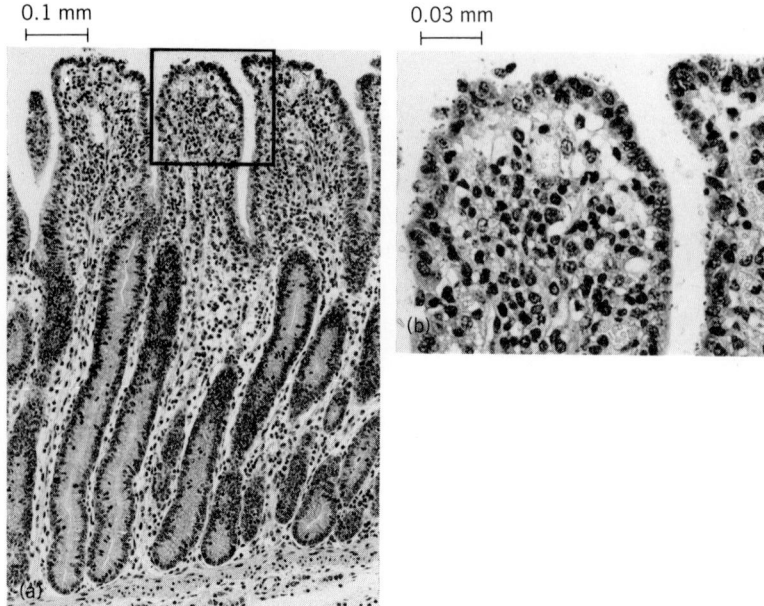

Fig. 2. Photomicrographs of a histological section from the midileum of a calf infected with *Cryptosporidium* of human origin. (*a*) Note the partial villous atrophy, and increased cellular infiltration of the lamina propria. (*b*) Higher magnification of the area marked in rectangle in *a*, showing numerous organisms in different stages of the life cycle embedded in the brush border of stunted enterocytes.

ery. The disease has also been described in artificially reared lambs and deer calves and in suckled goat kids and lambs. *Cryptosporidium* was reported to cause moderate to severe upper respiratory tract infection in turkeys and other birds.

Some species, although readily infected, appear to possess an innate resistance; in rats, mice, and guinea pigs, for instance, the infection is asymptomatic with a long incubation period. Ruminants, on the other hand, are susceptible and can become ill if infected at an early age. Though the evidence against host specificity is strong, some differences between animal isolates do exist, the extent of which is not fully understood.

Medical and veterinary significance. Cryptosporidiosis is economically most significant in calves. The incidence of diarrhea attributed to this infection in calves is 20–40% of all diarrheal cases, and is equal to or slightly less than the incidence of rotavirus infection.

The first clinical case of cryptosporidiosis in humans was reported in 1976. The disease manifests itself as two clinically distinct entities: a self-limited diarrhea in immunologically normal humans; and a persistent, life-threatening diarrhea in immunologically compromised individuals, particularly those with acquired immunodeficiency syndrome (AIDS). The infection causes diarrhea more commonly among young children, and in adults it is often associated with "traveler's diarrhea." The incidence of diarrhea in the human population is thought to be between 2 and 10%. The symptoms vary from severe diarrhea, vomiting, abdominal pain, fever, and loss of body weight, often requiring hospitalization and intensive care up to 2 weeks, to mild diarrhea. The recovery in severe or mild cases is spontaneous. Sources of infection are other humans, contaminated food and water, and domestic animals (particularly calves). In immunologically compromised patients, cryptosporidiosis causes protracted diarrhea, which is usually fatal; it tends to be further complicated by other conditions associated with AIDS.

Cryptosporidium readily propagates in laboratory animals, 8-day-old chicken embryos, and continuous human and mouse lung-cell lines, which should facilitate further research on the biology of the organism, the host-immune response to the infection, and control of the disease. At present there is no treatment against the infection despite exhaustive search for an effective antimicrobial agent, and prevention by vaccination is not yet available.

For background information see COCCIDIA; DIARRHEA; EUCOCCIDA; MEDICAL PARASITOLOGY; PROTOZOA; SPOROZOA in the McGraw-Hill Encyclopedia of Science and Technology.

[SAUL TZIPORI]

Bibliography: W. L. Current et al., Human cryptosporidiosis in immunocompetent and immunodeficient persons, *New Eng. J. Med.*, 308:1252–1257, 1983; S. D. Pitlik et al., Human cryptosporidiosis: Spectrum of disease, *Arch. Intern. Med.*, 143:2269–2275, 1983; S. Tzipori, Cryptosporidiosis in animals and humans, *Microbiol. Rev.*, 47:84–96, 1983; S. Tzipori, The relative importance of enteric pathogens affecting domestic animals, in Cornelius and Simpson (eds.), *Advances in Veterinary Science and Comparative Medicine*, vol. 29, in press.

Data-base management systems

Data-base management technology has been one of the most active areas of research and development in computing since the early 1970s. A data-base management system (DBMS) provides an environ-

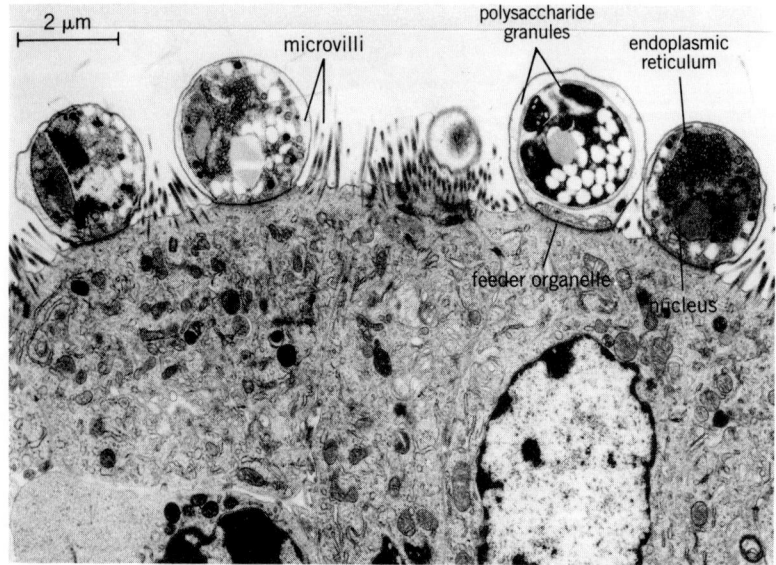

Fig. 3. Electron micrograph of ultrathin section taken from the midileum of the calf shown in Fig. 2. Four macrogametes are shown at various stages of their development, the second from right being the most immature stage. Note the reduced number of microvilli of infected cells ($\times$ 10,400).

ment in which distinct data files can be stored and accessed by means of a single user interface. Such systems can thus provide for integration of data, shared access to data by different users, and nonredundant storage of data. Current developments in the technology address several major issues: improvement in the user interface to data-base management systems; capability to provide rapid response from extremely large data bases; and application of the technology to new domains, for example, to support knowledge bases for applications in artificial intelligence.

Current-generation data-base systems are designed to handle files containing occurrences of records of identical internal structure. Systems are characterized by the manner in which records can be logically related, that is, how the user can view relationships among different types of records. Three main classes of systems are commercially available, distinguished by the data model supported: hierarchical systems, in which data records are organized into tree structures; network systems, in which data records can be related in more general structures permitting a given record to have more than one parent; and relational systems, in which data can be viewed as being stored in two-dimensional tables (relations.)

Relational data models and languages. A row in a relational table corresponds to a record occurrence in standard data-processing terminology. A column in a relational table corresponds to values of a particular data item. In relational terminology, a particular row in a relation is referred to as a tuple. The columns, which are named, are referred to as attributes, and can be thought of as the names of data items contained in the relation. A key feature of the relational model is that relationships (associations) among tuples in different relations are represented by data values of attributes common to two or more relations. More precisely, since attributes may have different names in different relations, common attributes are those that draw their values from a common domain of values.

In 1969–1973, the foundation was laid for much of the subsequent theoretical and practical work in relational systems. Two principal issues have been investigated: the data-base design problem, and the design of data manipulation languages to operate on a data base composed of a set of relations.

Relational data-base theory provides a useful framework for deciding which attributes should be grouped together in relations. Basically this involves issues relating to the functional dependency of the values of one attribute on the values of one or more other attributes. Rules have been proposed governing the desirability of particular groupings of attributes into relations, in order to construct relations in a variety of normal forms. The importance of this issue is that if a data base is organized into relations in a desirable normal form, undesirable side effects that might occur when tuples are added, deleted, or modified can be eliminated.

There are two classes of data manipulation languages for operating on a data base of relations. Relational algebraic languages are low-level procedural languages providing for tuple selection, projection of a relation (selection of specified columns), join of two relations (concatenation of two tuples based on, usually, the existence of identical values in two common attributes, in which case it is referred to as an equijoin), and the union and difference of compatible relations. Algebraic languages may include, for convenience, other operations that combine some of these five primitive operations. Relational calculus languages are higher-level nonprocedural languages; statements in these languages can be mapped to the fundamental algebraic operations. A language is said to be relationally complete if it provides for the construction of all relations derivable from some finite set of base relations by the application of the primitive algebraic operations. The output of any algebraic or calculus expression is always a derived relation, and the arguments for an operation may be entire relations.

Relational software systems. In recent years several relational software systems have emerged for use on large mainframes and superminicomputers. All provide relational calculus language interfaces, and most also provide access to their data bases from programs written in languages such as COBOL, PL/1, and FORTRAN, in which data-base language statements can be embedded. The following list summarizes these systems as of 1983.

System name	Host hardware
SQL/DS	IBM 370
RELATE/3000	HP 3000
INGRES	DEC VAX
ORACLE	DEC PDP 11, VAX; IBM 370
INFO	Prime 50; DEC VAX, PDP; IBM 370
REXCOM	Prime 50
MRDS/LINUS	DPS8; DPS8M
NOMAD	IBM 370
ENCOMPASS	TNS 11

More recently there has been a surge of releases of data-base software for microcomputer environments. Many of these systems are termed relational by their vendors, but are not relationally complete in the sense defined above. The following systems are claimed as relational; in some cases the database management system is incorporated in a software product containing other functions.

CONDOR Series 20, Level 3	METAFILE ORACLE
dBASE II, III	R:base 4000
FMS-80	SYMPHONY
Knowledge Manager	10-BASE
MDBS III	

Relational hardware systems. The commercial systems listed above are designed to run on general-purpose computer hardware. Two special-purpose data-base machines, designed to run as back-end processors for general-purpose main frames or work stations, are commercially available—the Britton-Lee IDM 500 and the Intel iDBP. Both may be regarded as relational engines which implement storage and retrieval operations on physical data bases through a user language interface resident in an attached (general-purpose) host machine.

Development trends. One major trend is related to improving the friendliness of the user interface to data-base systems. Several natural language interfaces are currently commercially available, permitting the user to access a data base with (restricted) English language commands. In the same category is the recent emergence of systems incorporating integrated data-base, spreadsheet, and text processing functions, which provide powerful application development systems. *See* NATURAL LANGUAGE PROCESSING.

An area of intense current research activity, driven in part by rapid advances in semiconductor technology, is the design and development of experimental data-base machines. The opportunity to develop new computer architectures specialized to increasing performance in accessing very large data bases, to providing improved protection mechanisms, and to extending the functionality of systems is being explored in the United States, France, Italy, West Germany, and Japan. The development of such machines is an important factor in the thrust to produce the fifth generation of computing systems designed to exploit artificial intelligence applications supported by large knowledge bases.

For background information *see* ARTIFICIAL INTELLIGENCE; DATA-BASE MANAGEMENT SYSTEMS in the McGraw-Hill Encyclopedia of Science and Technology. [ALFRED G. DALE]

Bibliography: C. J. Date, *An Introduction to Database Systems*, 3d ed., 1981; D. K. Hsiao (ed.), *Advanced Database Machine Architecture*, 1983; W. Kim, *Relational Database Systems*, ACM Computing Surveys, 2(3):185–211, 1979; J. D. Ullman, *Principles of Database Systems*, 1980.

Deoxyribonucleic acid (DNA)

Deoxyribonucleic acid (DNA) is the universal carrier of the hereditary information in living organisms. The genetic information encoded in the DNA molecule has to be expressed in a carefully controlled manner, and also has to be duplicated and transmitted accurately into daughter cells during cell division. The molecular architecture of the DNA molecule ensures that both these functions can be performed.

It has been known since 1953, when J. Watson and F. Crick published their classical work, that DNA exists as a double-helical molecule in which two long DNA single strands pair with each other and coil around each other to generate the right-handed double helix (B-DNA). During the last 5 years, however, new insights into the molecular structure of the DNA double helix have been obtained, primarily as a result of two methodological developments. First, new techniques for rapid chemical synthesis of DNA oligonucleotides with defined nucleotide sequences make available pure DNA material in high quantity. Second, the technique of recombinant DNA technology (that is, the recombination and cloning of DNA sequences in laboratory cultures) allows biochemical characterization of individual DNA segments of natural origin. Using chemically synthesized oligonucleotides of defined sequence, x-ray crystallography studies in the laboratory of A. Rich led to the discovery by A. H.-J. Wang and his colleagues of a left-handed DNA double-helical molecule called Z-DNA. The discovery of Z-DNA led to an enhanced awareness of the conformational flexibility of the DNA double helix and, indeed, new variations of the DNA conformation have since been described. The spectrum of structural DNA polymorphism ranges from subtle but distinct deviations from the classical right-handed B-DNA conformation to the radically different left-handed Z-DNA. Current research emphasis centers on the question of the biological function and physiological role of alternative DNA structures like Z-DNA within living cells.

Molecular structure of Z-DNA. The Z-DNA structure is formed by Watson-Crick–type base pairing (hydrogen bonding) between complementary nucleotide sequences of the individual single strands of DNA. These two DNA strands are coiled around each other in a left-handed helix, and this represents a major difference between the classical right-handed B-DNA and Z-DNA (Fig. 1). In addition, Z-DNA differs from B-DNA in other helical parameters, such as helix diameter, position of the helical axis, distance of phosphates from the helical axis, and internal conformation of base and sugar moieties. The phosphate molecules that are part of the sugar-phosphate backbone of the DNA strand are arranged in a zigzag pattern (hence the abbreviation Z-DNA). As a result, shorter distances can be found between phosphates, and this is assumed to be the cause of the reduced stability of Z-DNA. It is therefore expected that, within a living cell, Z-DNA can exist only if it is stabilized by specific cellular factors. An understanding of such factors, which determine the equilibrium between B and Z forms of DNA, is necessary for both physicochemical and biological investigations of the biological role of Z-DNA within living cells.

Stabilization of Z-DNA. The nucleotide composition of a given DNA segment determines the ease with which such a segment can assume the left-handed Z-DNA conformation. An alternating sequence of purine and pyrimidine bases has the greatest potential for forming Z-DNA, while other sequences will form Z-DNA less readily. Four factors have been found to help stabilize the Z-DNA conformation within a DNA segment: solvent conditions, chemical modification of DNA, binding of proteins, and supercoiling of DNA.

Solvent conditions. Unusual solvent conditions of high ionic strength led to the first recognition of a left-handed conformation of DNA by F. M. Pohl and T. M Jovin. While most ions can stabilize Z-DNA only at nonphysiologically high concentrations, there are some ionic compounds (Mg^{2+}, polyamines) that are effective at physiological concentrations.

Chemical DNA modification. Chemical modification, such as methylation of cytosine residues, can facilitate the salt-induced transition from B-DNA to Z-DNA. Such DNA modifications either can be introduced artificially into the DNA in laboratory experiments or can be of natural origin.

Protein binding. Protein binding may be a major component in the intracellular recognition and stabilization of Z-DNA. Proteins that can interact specifically with Z-DNA have been described in extracts from eukaryotic systems.

Negatively supercoiled DNA. The topological stress of a negatively supercoiled region of DNA can cause Z-DNA formation within that segment (Fig. 2). Circular DNA molecules, such as plasmids, are often "underwound," which means that the two DNA strands that form the double helix are coiled around each other fewer times than in the topologically unrestrained condition of a linear double-helical molecule; the two DNA strands are therefore linked with each other fewer times than they would be in a relaxed state. A normal double-helical B-DNA conformation can establish itself in an underwound circle only if a compensatory superstructure (negative supercoiling) is introduced. Deviations from the B-DNA double helix geometry can reduce the degree of supercoiling, and therefore Z-DNA formation is energetically favored in a negatively supercoiled DNA circle. Since, in natural conditions, DNA is found most frequently in the underwound, negatively supercoiled state, topological strain may well represent a major physiological factor for Z-DNA formation within the living cell. Laboratory experiments have already shown that DNA supercoiling is indeed extremely effective in stabilizing Z-DNA segments.

Function of Z-DNA. Although Z-DNA formation within living cells represents an exciting possibility for novel molecular genetic mechanisms, no direct evidence for occurrence of Z-DNA in intact cells has yet been obtained. However, a substantial body of data from laboratory experiments is available for the characterization of Z-DNA, and also indirect evidence has been obtained for Z-DNA formation in DNA of natural origin. Such experiments were performed owing to the existence of tools that allow Z-DNA recognition; these tools are used in physicochemical techniques (sedimentation analyses, two-dimensional gel electrophoretic studies), biochemical approaches (nuclease digestion), and immunochemical studies (use of antibodies that specifically recognize and bind to Z-DNA). By using these techniques, it was shown that factors stabilizing Z-DNA, such as ionic strength, chemical DNA modification, DNA binding by proteins, and DNA supercoiling, can have additive effects, thereby making the formation of Z-DNA a likely event in living organisms.

Perhaps a Z-DNA segment within the genome of

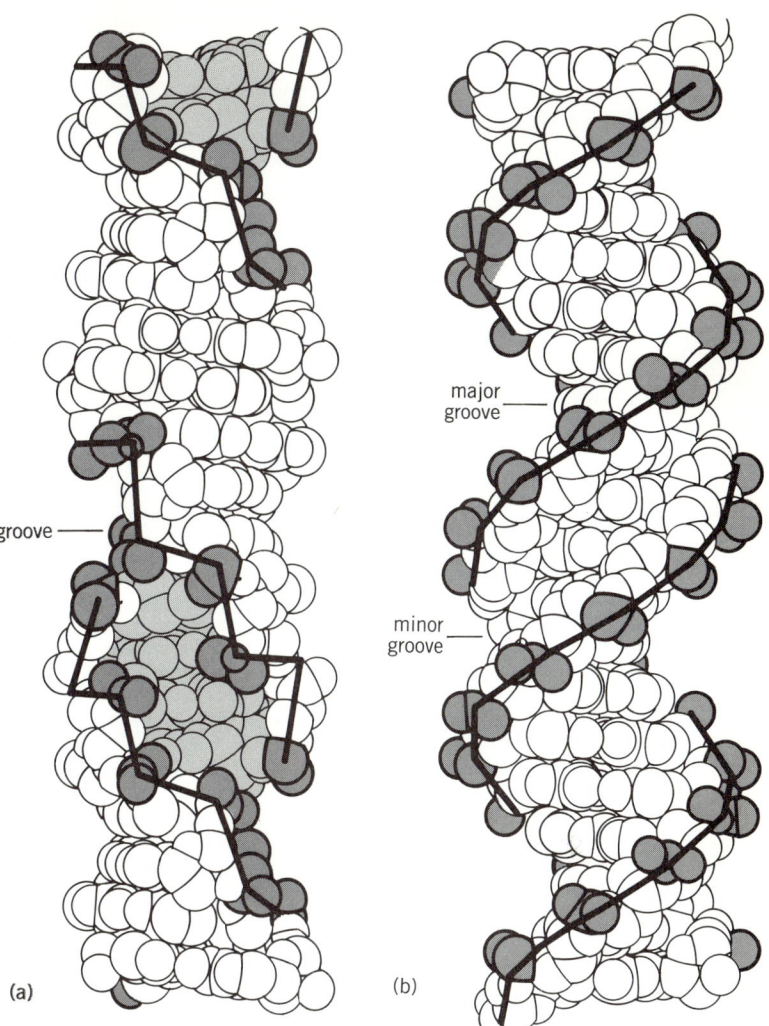

Fig. 1. Van der Waals's drawings of the (a) left-handed Z-DNA and (b) right-handed B-DNA double helices. The concave major groove of B-DNA forms the outer convex surface of Z-DNA. The thick black lines are drawn to show the positions of the phosphates along the sugar-phosphate backbones. This line follows a zigzag course in Z-DNA, indicating alternating positions of the phosphate residues. (*From A. H.-J. Wang et al., Molecular structure of a left-handed double helical DNA fragment at atomic resolution, Nature, 282:680–686, 1979*)

a cell represents an important signal for the binding of specific proteins. Such proteins may be involved in regulatory processes of genome function. The effectiveness of DNA supercoiling in stabilization of Z-DNA allows one to postulate that localized generation of topological stress within a genomic segment may be associated with Z-DNA formation. Topological stress will occur in the genome most easily where local denaturation of the DNA double helix occurs, as is the case during initiation of transcription, replication, and recombination. This notion is supported by recent experimental findings that identified the potential for Z-DNA formation within regions that regulate the expression of defined genes. This potential for Z-DNA formation was detected in laboratory conditions, however, and it remains to be seen whether it is realized within the living cell.

The identification in cellular extracts of proteins that can selectively bind to Z-DNA segments may suggest that these proteins also bind Z-DNA in liv-

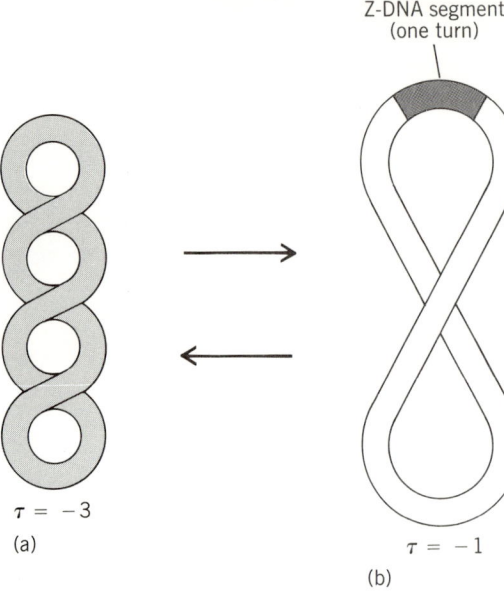

Fig. 2. Z-DNA formation within a negatively supercoiled circle of DNA. (a) A DNA circle in its underwound state containing three negative superhelical turns ($\tau = -3$) is in equilibrium with (b) the same circle in which two negative superhelical turns have been removed due to the introduction of a Z-DNA segment of approximately one left-handed helical turn in length. Such a transition of B-DNA into a Z-DNA segment is energetically favored in a negative superhelical DNA circle. (*From A. Nordheim et al., Supercoiling and left-handed Z-DNA, Cold Spring Harbor Symp. Quant. Biol., 47:93–100, 1983*)

ing organisms. Such proteins may be involved in regulating transcription of genes and also in events of genome recombination and rearrangement. Elucidation of the natural occurrence of Z-DNA, and an understanding of the potential biological role of this unusual DNA conformation, awaits further experimentation and is an area of active scientific investigation.

For background information see DEOXYRIBONUCLEIC ACID (DNA). GENE ACTION in the McGraw-Hill Encyclopedia of Science and Technology.

[ALFRED NORDHEIM]

Bibliography: A. Nordheim et al., Negatively supercoiled plasmids contain left-handed Z-DNA segments as detected by specific antibody binding, *Cell*, 31:309–318, 1982; F. M. Pohl and T. M. Jovin, Salt-induced co-operative conformational change of a synthetic DNA: Equilibrium and kinetic studies with poly(dG-dC), *J. Mol. Biol.*, 67:375–396, 1972; A. Rich, A. Nordheim, and A. H.-J. Wang, The chemistry and biology of left-handed Z-DNA, *Annu. Rev. Biochem.*, 53:751, 1984; A. H.-J. Wang et al., Molecular structure of a left-handed double helical DNA fragment at atomic resolution, *Nature*, 282:680–686, 1979.

Echolocation

Recent behavioral studies have shed new light on the performance of the sonar systems of echolocating bats, and the accuracy with which these animals can use information contained in the echoes of their sonar pulses to determine the location and motion of objects in space.

Target detection. Several important studies have been conducted on the big brown bat (*Eptesicus fuscus*), an insectivorous species that is widely distributed through the temperate regions of the Northern Hemisphere. Like most other members of its family, the Vespertilionidae, the fundamental component of its echolocative cries or sonar pulses sweeps downward from a high starting frequency of about 60 kHz to a much lower terminal frequency around 20–25 kHz. A second harmonic is also present. When the bat is searching for prey in open areas, the low-frequency terminal portion of the pulse may be maintained for several milliseconds at an almost constant frequency. When trying to intercept a flying insect or avoid an object in its flight path, *Eptesicus* increases the repetition rate of its sonar pulses from perhaps 10/s to as high as 200/s at the instant of prey capture. As it does this, it shortens the duration of each sonar pulse, from several milliseconds down to 1 ms or less, by eliminating the almost constant frequency terminal portion and producing a steeply frequency-modulated pulse.

The point at which a bat responds to a target by increasing its pulse repetition rate has often been used as an estimate of the distance at which it first detects the target. By using this technique, it was estimated that bats like *Eptesicus* can detect a sphere 1.2 in. (32 mm) in diameter at a distance up to about 3 ft (1 m). It now appears that those estimates were too conservative and that the bat often detects a target well before it begins to increase its pulse repetition rate. It was discovered through an operant conditioning experiment using a two-choice psychophysical procedure that *Eptesicus* can detect echoes from a 0.2 in.-diameter (4.8-mm) stationary nylon sphere at 9.5 ft (2.9 m), and from a 0.7-in. (19.1-mm) sphere as far away as 16.7 ft (5.1 m). Calculations indicate that the intensity at the bat's ear of the echo from these spheres is about equal to the most sensitive threshold of human hearing (0 dB sound pressure level with respect to 2×10^{-5} newton/m^2).

Target direction. Similar conditioning experiments have yielded important new insights into the ability of *Eptesicus* to measure the direction, or azimuth, of a sonar target. In order to do this, the animal must determine the direction from which the echo arrives by comparing certain aspects of the acoustic stimulus arriving at its two ears. The binaural cues which can theoretically be used to localize a sound source in the horizontal plane include differences in sound intensity at the two ears due to a shadowing effect of the head on the ear farthest from the sound source, differences in time of arrival or phase of the acoustic waveform due to different path lengths that sound must travel to each ear, and frequency-dependent (spectral) differences arising from the directional sensitivity of each ear.

Two-choice psychophysical experiments in which *Eptesicus* was conditioned to distinguish between closely spaced or more distantly spaced pairs or arrays of vertical rods reveal that this bat is able to discriminate horizontal angles as small as 1.5°. The physical cues upon which this discrimination is based are not known with certainty, but some evidence suggests that, using their broadband sonar pulses, these bats may be able to detect extremely small interaural time-of-arrival differences, perhaps as small as a microsecond. If so, they could use this information to localize a sound source in the horizontal plane to within 1 or 2°. It is not clear how the nervous system could detect and code a 1-microsecond disparity in arrival time at the two ears, and the use of some other cue cannot be completely ruled out. Well-trained human subjects are unable to detect interaural time disparities in broadband clicks of less than about 10 μs.

Binaural cues in the acoustic stimulus that are potentially useful for the lateral localization of a sound source or sonar target are not helpful for vertical localization. Yet bats which pursue flying insects must do so in three-dimensional space where vertical localization is as important as horizontal localization. Recent psychophysical measures of the big brown bat's ability to resolve the direction of a sound source in the vertical plane show that it can do this to within about 3°. In humans, cues for vertical localization depend on the fact that the acoustic properties of the pinna vary according to the vertical angle of the sound source. The ear of echolocating bats contains a leaf-shaped cartilaginous structure which projects upward at the base of the pinna in front of the auditory canal. This structure, the tragus, has a characteristic shape in different species (Fig. 1). Its function, until recently, was unknown, but in recent experiments in which it was bent down and glued to the fur the acuity for resolving vertical angles dropped from 3° to about 12 or 14°, showing that in its normal position the tragus considerably improves the bat's ability to localize the source of an echo in the vertical plane. It presumably does this by altering, according to its vertical angle of incidence, either the path length or spectral properties of the sound reaching the middle ear. The tragus may thus have evolved as a means of improving the vertical resolution of bat sonar systems.

Target velocity. Many bats use echolocation to capture moving prey. In addition to knowing the position of a moving target at a given instant, it is advantageous to have information about the relative velocity of the prey. For some time it has been known that the Old World insectivorous horseshoe bat (*Rhinolophus*), which emits sonar pulses with a long-duration constant-frequency component, can detect the upward Doppler shift imposed on the echo from an approaching target, and compensates for this shift by lowering the frequency of its emitted pulses. It is not clear, however, whether or not horseshoe bats actually obtain information regarding the relative velocity of the target from measurement of the echo's Doppler shift. Target velocity could

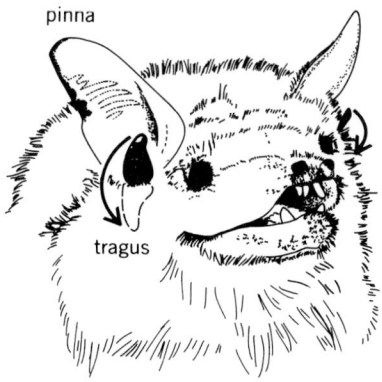

Fig. 1. Head of the big brown bat (*Eptesicus fuscus*) showing the tragus and the pinna. (*After B. Lawrence and J. Simmons, Echolocation in bats: The external ear and perception of the vertical positions of targets, Science, 218:481–483, 1982*)

also be determined by measuring the target's range (distance) on successive sonar pulses and then computing the rate at which the range is increasing or decreasing. It can be shown that the narrow-bandwidth pulses such as the long-duration constant-frequency pulses emitted by *Rhinolophus* are better suited for frequency measurement, whereas the short, broadband frequency-modulated pulses often emitted by *Eptesicus* are capable of providing more accurate range information. Target range can be determined by measuring the time that elapses until the echo returns.

The first direct experimental evidence that at least some echolocating bats are capable of measuring target velocity was recently obtained in experiments on the fish-catching bat (*Noctilio leporinus*). This is a neotropical species that has evolved an ability to use its sonar to detect the protruding dorsal fins of small fish swimming just below the surface of the water or the ripples which they produce. Although it also feeds on flying insects, it frequently flies low over inland lakes and streams or coastal waters searching for fish, which it catches by dipping its disproportionately large feet into the water and spearing the fish on its sharp claws. As they search for prey, fishing bats emit moderately long-duration (8–16-ms) sonar pulses which either are mostly constant-frequency at about 60 kHz or contain an initial constant-frequency portion followed by a downward frequency sweep terminating around 30 kHz.

Captive fishing bats were trained to catch pieces of fish moving toward them or away from them at a controlled velocity over the surface of a pool. They were then conditioned to associate a food reward with a target moving at a certain velocity which they had to distinguish, on the basis of its velocity, from either a faster or a slower moving target. By noting the performance of the bats as the velocity difference between the two types of targets was reduced, it was found that they could discriminate target velocity differences as low as 16 in./s (40 cm/s), independent of the absolute speed or direction of target movement (Fig. 2). When the same bats were

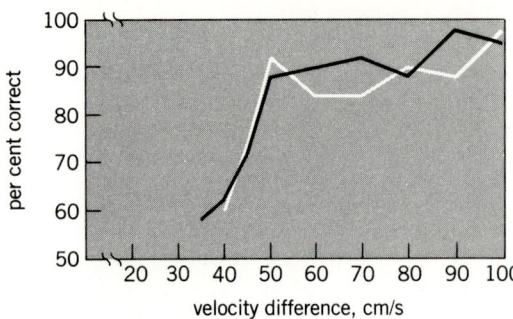

Fig. 2. Performance of two fish-catching bats, *Noctilio leporinus*, in discriminating, during flight, a target moving toward them at 10 in./s (25 cm/s) from a target moving toward them at a higher velocity. 1 cm/s = 0.4 in./s. (*After J. Wenstrup and R. Suthers, Echolocation of moving targets by the fish-catching bat, Noctilio leporinus, J. Comp. Physiol., A155:75, 1984*)

then trained to discriminate between the presence or absence of an artificial upward Doppler shift in the simulated echoes of their sonar pulses, they could not discriminate frequency shifts smaller than 570 Hz. Since a moving target would have to have a relative velocity of 67 in./s (170 cm/s) in order to produce a detectable Doppler shift, fishing bats must rely on range-rate information, presumably obtained from the frequency-modulated portion of their echolocative pulses, in order to discriminate velocity differences as small as 16 in./s (40 cm/s). It remains to be seen whether other echolocating bats also measure relative velocity in this way or if some constant-frequency–emitting species utilize Doppler shift information for this purpose.

For background information *see* BAT; DOPPLER EFFECT; EAR; HEARING (HUMAN); PHONORECEPTION; SONAR in the McGraw-Hill Encyclopedia of Science and Technology.

[RODERICK A. SUTHERS]

Bibliography: J. A. Simmons et al., Acuity of horizontal angle discrimination by the echolocating bat, *Eptesicus fuscus*, *J. Comp. Physiol.*, 153:321–330, 1983; J. A. Simmons and S. A. Kick, Physiological mechanisms for spatial filtering and image enhancement in the sonar of bats, *Annu. Rev. Physiol.*, 46:599–614, 1984; J. J. Wenstrup and R. A. Suthers, Echolocation of moving targets by the fish-catching bat, *Noctilio leporinus*, *J. Comp. Physiol.*, A155:75, 1984.

Eclipse

Eclipses of the Sun are an awesome sight. From antiquity to the present, the blotting out of the Sun in the middle of the day has been considered a powerful omen. Current studies of eclipses take advantage of the dark sky in the daytime to study parts of the Sun that cannot otherwise be well seen. These studies are tied in with studies of the Sun from space satellites to provide a fuller picture of the Sun's atmosphere. A burgeoning field is the linking of knowledge about the Sun's atmosphere with studies about the atmospheres of other stars.

Solar chromosphere and corona. A solar eclipse occurs when the Moon crosses in front of the Sun during its monthly revolution around the Earth. Though the Moon is about 400 times smaller than the Sun, it is also about 400 times closer, and so blocks the bright solar disk, the photosphere (the sphere from which the light comes), to ±10%. As the photosphere is completely hidden, a reddish rim becomes visible, the chromosphere (color sphere). This chromosphere is made of tiny jets of gas, each perhaps 600 mi (1000 km) across by 6000 mi (10,000 km) high, which rise and fall with a period of about 15 min, resembling blades of grass waving in the wind. The chromosphere is reddish because most of its radiation is concentrated in spectral lines of emission from several elements, chiefly the reddish radiation typical of hydrogen. The spectral lines flash into view just before totality and remain visible for only 5 or 10 s, giving the flash spectrum. The same recurs at the end of totality.

When the chromosphere is hidden by the advancing edge (limb) of the Moon, the total phase of the eclipse begins. This phase may last from no time at all up to somewhat over 7 min, depending on the positions of Sun, Earth, and Moon. During totality, the corona, a ghostly white irregular halo of light around the Sun, is seen. Most of the light seen at this time is photospheric sunlight scattered toward the Earth by electrons in the corona (inner part) and by dust in interplanetary space (outer part). A dozen or so spectral lines are observable in the innermost corona. They come from iron and other elements so heated that approximately half of their electrons are stripped off. This is one of several signs that the coronal temperature is extremely high, about 2×10^6 kelvins (2×10^6 °C or 3.6×10^6 °F).

A major question now being addressed is how the corona is heated to this high temperature. A theory of heating by acoustic (sound) waves was dominant until the mid-1970s, though the theoretical details of how these waves would dissipate their energy in the corona was never satisfactorily formulated. But the flux of acoustic waves passing through the chromosphere, when measured with *OSO-8* (*Orbiting Solar Observatory 8*) proved to be over 100 times too low. New theories had to be formulated.

The new theories were largely based on the observation, made from the 1973–1974 *Skylab* missions, that the corona as seen on the solar disk by its x-radiation was strongest over sunspot regions. This indicated that coronal heating was somehow linked to the solar magnetic field, which is strongest in sunspots. Several new theories involving magnetic and electric fields have been advanced and are under theoretical and observational study.

Shortly after *OSO-8*, NASA's *Einstein Observatory* x-ray satellite studying distant stars discovered x-radiation, presumably from coronas, about stars of all temperatures. Since only relatively cool stars like the Sun had been expected to have acoustic waves generated, the surprise observation indicated a need to provide a nonacoustic-wave mechanism for coronal heating in any case.

Recent eclipse expeditions. The apparent superposition of the Moon and Sun is so precise that it is visible from a region on Earth less than 200 mi across. As this region sweeps across the Earth, it traces out a path of totality that is hundreds of miles wide and thousands of miles long. Partial phases are visible for over a thousand miles to the sides of this path.

The three longest and therefore most favorable total eclipses in recent years occurred in 1973 in Africa, in 1980 in Africa and India, and in 1983 in Indonesia. The U.S. National Science Foundation (NSF) mounted major expeditions to each, with a dozen American observatories, colleges, and universitites participating, and many other countries also sent expeditions. Since the shape of the solar corona changes with the phase of the sunspot cycle, a link between the solar magnetic field and the corona has long been known. These three eclipses took place at different phases in the sunspot cycle, allowing study of the corona at different parts of the cycle.

The most recent maximum of the sunspot cycle took place in 1979–1980. At solar maximum, so much solar activity is present on the surface of the Sun that coronal streamers go off in all directions. When they are seen in projection against the plane of the sky, the corona looks relatively round, as it did in 1980 (Fig. 1). The NSF group's site on that occasion was in south India, in the countryside south of Hyderabad, using an observing platform at the Japal-Rangapur Observatory. Totality lasted approximately 2 min 10 s.

At the NSF site, astronomers from the Sacramento Peak Observatory studied the motions of gas in the corona by making Doppler-shift studies of the emission spectral line in the green at 530.3 nanometers using a Fabry-Perot interferometer. Kitt Peak National Observatory observers also studied coronal

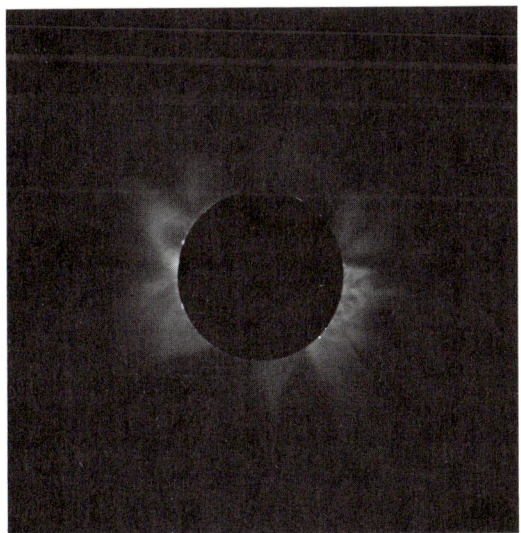

Fig. 1. The Sun at the total solar eclipse of February 16, 1980, photographed from India through a filter radially graded in density so that the bright interior parts of the corona can be shown on the same film as the faint exterior parts. (*High Altitude Observatory/NCAR and Southwestern at Memphis*; *J. L. Streete and L. B. Lacey*)

Fig. 2. The Sun at the total eclipse of June 11, 1983, photographed from Java through a radially graded filter. (*High Altitude Observatory/NCAR*; *R. R. Fisher, T. Baur, L. B. Lacey, and M. McGrath*)

motions, using spectroscopy. A group from the University of Hawaii took the spectrum of solar prominences over the entire visible spectrum, data that have since been reduced in conjunction with a Williams College team. A joint Williams-Hawaii experiment studied coronal loops with high spatial and time resolution to look for predicted oscillations resulting from surface magnetohydrodynamic waves on the solar surface, one of the leading contenders among solar-heating models. Scientists from the High Altitude Observatory and Southwestern College at Memphis took photographs of the corona through a filter graded radially in density to study the coronal intensity (Fig. 1), and also observed coronal polarization. University of Minnesota scientists made infrared studies to search for an interplanetary dust ring around the Sun. For the minutes just before and just after totality, astronomers from Johns Hopkins University and Gettysburg College observed shadow bands, ripples of shadow on the Earth's surface caused by turbulence in the Earth's upper atmosphere. Indian scientists observed from an adjacent site, and Japanese scientists were also located in the region.

Other scientists observed the eclipse from Africa, from which rockets were sent up into totality, and from an airplane flying from Africa toward India. A coronal transient eruption was seen to change in shape and disappear over the observing period.

After study of the data, many of the same groups joined the NSF expedition to the 1983 total solar eclipse, which crossed the Indonesian island of Java. The sunspot cycle was much advanced toward minimum, so the corona was irregular in shape as only a few streamers were visible and those mainly at solar equatorial latitudes (Fig. 2). The Sacramento Peak Observatory and the solar group of the Kitt Peak National Observatory (now combined as the National Solar Observatory) again studied tem-

Fig. 3. The total eclipse of 22/23 November 1984. (*Courtesy of Jay M. Pasachoff*)

peratures and velocities in the corona. The Hawaii group again took prominence spectra; prominences with different characteristics are visible on the Sun at different times. The Williams-Hawaii group had indications of coronal loop oscillations from the 1980 data, and carried out an improved version of the experiment. Iowa State University researchers searched for dust in interplanetary space with infrared observations. High Altitude Observatory scientists again used their coronal camera to get radial filter photographs (Fig. 2) and polarization maps. Groups from East Carolina University and North Carolina State University studied the effect of the eclipse on the Earth's atmosphere. A team led by a Naval Research Laboratory scientist spread out across Java to study the size of the Sun, discussed below. Indian, Japanese, British, and other scientists were also in Java for the eclipse. Chinese observers went to Papua New Guinea.

Other eclipses also took place in the period, but were of lesser interest because of shorter durations or inferior weather forecasts. The Pacific Ocean eclipse of 1977, the United States–Canadian eclipse of 1979, the Siberian eclipse of 1981, and the Papua New Guinean eclipse of 1984 (Fig. 3) fell into that category.

Annular eclipses. Sometimes the Moon's angular diameter is too small to cover the Sun. An annulus—a ring—of photospheric light remains visible, and an annular eclipse occurs. Since the photosphere is so bright, the corona cannot ordinarily be seen at an annular eclipse.

The 1984 annular eclipse that crossed the southeastern United States was so close to total that many of the phenomena ordinarily limited to total eclipses could be seen. The Baily's beads, bright bits of photospheric sunlight shining through valleys on the edge of the Moon, were visible all around the Sun. In fact, an unbroken annulus, without lunar mountains projecting upward through it, was never visible. The weather turned out to be clear from New Orleans through the Georgia–North Carolina border. The skies were sometimes so clear that even the corona could be briefly seen or photographed for a minute or so surrounding the 10 s of annularity. Shadow bands were also seen. Observers northeastward, through Virgina, were clouded out. The partial phases were visible throughout the country, wherever it was not cloudy.

The major scientific experiment that can be carried out at an annular eclipse involves observations of Baily's beads to monitor the Sun's size. Over several centuries, measurements of the Sun's size have led some to believe that it might be shrinking slowly, or perhaps oscillating. The rate of change possible under the observations is such that significant changes in the solar size and therefore brightness could have occurred in times short on a geological scale. It is known, however, from fossil and other records that such major changes did not occur. It is possible or probable that the remaining uncertainties in the observations are sufficiently large that any shrinkage is unmeasurably small. Many professional and amateur astronomers were deployed in the path of annularity, especially near its edges, to monitor the Baily's beads. The data will be compared with past observations to look for long-term effects. Photographic and video observations are now replacing visual observations.

Safety in observing. Many people remain misinformed about the possible hazards of observing an eclipse. It has been advanced that solar eclipses are hazardous to the eye, which is true only in extreme cases. Most professional observers glance briefly at the Sun with the unaided eye with no harm. Only prolonged staring at the Sun is hazardous. The rule should be never to stare at the Sun at any time, eclipse or not. And the Sun should not be observed through a telescope or binoculars that concentrate the image, unless special filters are used.

To observe an eclipse safely, simply do not stare at the Sun. The partial phases can be seen by glancing through a special solar filter or by making a pinhole camera. Solar filters can be made by taking black-and-white (not color) film, holding it up to light so that it is fogged, and then developing it. If the Sun cannot be seen through two thicknesses of this fogged and developed film, then one thickness will do. Commercial solar filters made of aluminized Mylar are inexpensively available. A pinhole camera can be constructed by punching a hole 0.1–0.2 in. (2–5 mm) in diameter in a piece of cardboard. If the cardboard is held up, an image of the solar crescent will be projected onto the ground or onto a second piece of cardboard. In this case, the observer watches the image by facing away from the Sun (the Sun should never be observed through the

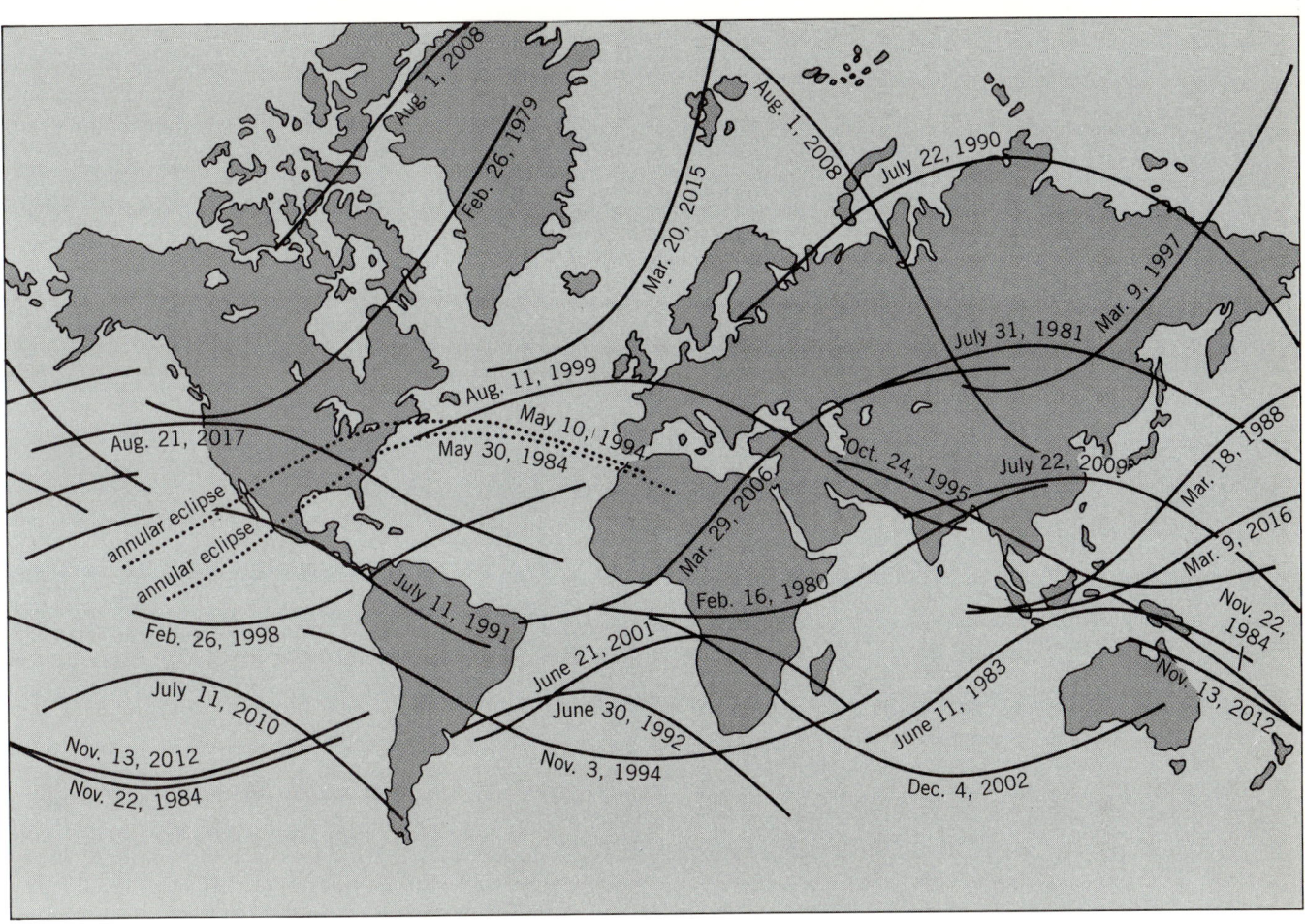

Fig. 4. Paths of total eclipses, 1979–2017. The next total eclipse to cross the United States mainland will be in 2017. (After B. Brewer, Earth View, Inc.)

hole in the cardboard) and can watch indefinitely.

During totality at a total solar eclipse, the solar photosphere is entirely covered so the solar corona can be observed safely without any eye protection or filters for the duration of totality. Television gives no idea of what it is like to observe a total eclipse, and students should be encouraged to see an eclipse directly rather than on television.

Future eclipses. Solar eclipses occur regularly around the world, though it is often necessary to travel to observe them (Fig. 4).

The next accessible total eclipse will not occur until 1988. On November 12, 1985, a total eclipse will be visible only in the ocean off Antarctica. The brief annular eclipse on October 3, 1986, will be visible only over the ocean near Greenland, and on March 29, 1987, an annular eclipse will cross Africa; both will have very brief total sections. (The partial phases are much more widely visible; the 1986 partial phases will be visible in the United States.) The September 23, 1987, annular eclipse will be visible in China and Okinawa.

The March 17–18, 1988, total eclipse will be favorable for observers. Totality crosses Indonesia, (prime sites may be in Sumatra) and the southern Philippines. Maximum totality is 3 min 46 s.

Annular eclipses will cross Somalia and the ocean on September 11, 1988, and Antarctica and the ocean on January 26, 1990. Partial eclipses will be visible in Hawaii and northwestern North America on March 7, 1989, and in southeastern Africa and Madagascar on August 31, 1989.

The total eclipse of July 22, 1990, will largely cross the northern coasts of Europe and Asia, and will cross United States territory in Alaska's Aleutian Islands, where totality will last 2 min 20 s.

The January 15–16, 1991, annular eclipse will cross Australia and New Zealand.

The favorable total eclipse of July 11, 1991, will cross the island of Hawaii with over 4 minutes of totality. Totality will also be visible in Mexico and in Central and South America.

The annular eclipse of January 4, 1992, will be almost completely over the Pacific Ocean, though the setting eclipsed sun will be visible at sunset in Los Angeles and San Diego. The June 30, 1992, total eclipse will be very low in the sky in Uruguay.

The May 10, 1994, annular eclipse will cross the United States over El Paso; Oklahoma City; Tulsa; Wichita; St. Louis; Cleveland; Detroit; Buffalo and Albany; and Portland (Maine). Annularity will be over 5 min in duration.

For background information *see* ECLIPSE; SUN in the McGraw-Hill Encyclopedia of Science and Technology.

[JAY M. PASACHOFF]

Bibliography: R. G. Giovanelli, *Secrets of the Sun*, 1984; D. H. Menzel and J. M. Pasachoff, *A Field Guide to the Stars and Planets*, 2d ed., 1983; R. W. Noyes, *The Sun, Our Star*, 1982; J. M. Pasachoff, *Contemporary Astronomy*, 3d ed., 1985.

Ecological interactions

The chemical diversity of plants was discovered centuries ago by herbalists and physicians who explored the "virtues" or medicinal properties of different types of plants. The underlying chemical bases for medicinal activity were of little concern to early investigators, and the assumed purpose for the pharmacological activity was to render service to humankind in the form of drugs, euphorics, and other curatives. This anthropocentric justification pervaded botanical chemistry well into the nineteenth century. From these early chemical studies, patterns began to emerge—primarily, the constancy of production of particular chemicals within a species regardless of locality, and the irregularity in distribution of chemical types among plant families.

Secondary plant products. In 1922 F. Czapek noted the sporadic occurrence and inconstant appearance of naturally occurring plant alkaloids in closely related species and proposed that the formation of such substances is not a general process of all cell plasma but is of a secondary character. Since that time, naturally occurring plant products that play no apparent role in the internal physiological process of plant life have been grouped together as secondary products. General features of secondary products are: (1) taxonomic limitation in the occurrence of the substances; (2) formation by specific enzymes or from a precursor formed by a specific enzyme; (3) metabolic regulation of the concentration by biosynthetic and catabolic pathways; (4) formation in specialized cells or organs; and (5) lack of an apparent primary function in the plant producing the substance.

Glucosinolates or mustard oil glycosides, such as sinigrin, are typical secondary plant products. They are restricted almost exclusively to the closely related families Cruciferae (mustards), Capparidaceae (capers), and Resedaceae (mignonettes). When plant tissues are damaged, volatile isothiocyanates (mustard oils) are released as a result of hydrolysis of the glucosinolates, a reaction catalyzed by a specific group of enzymes called myrosinases. In cabbage, sinigrin is hydrolyzed to allylisothiocyanate, as shown in the reaction below; this product is the major component of the odor of cooked cabbage. In intact leaves, myrosinases and their glucosinolate substrates are compartmentalized and stored separately within the plant. The glucosinolates are not known to perform any physiological functions in the plants.

$$CH_2=CH-CH_2-C\begin{matrix}N-O-SO_3^-\\ \\ S\text{-glucose}\end{matrix} \xrightarrow{\text{myrosinase}}$$

Sinigrin (allylglucosinolate)

$$CH_2=CH-CH_2NCS + \text{glucose} + HSO_4^-$$

Allylisothiocyanate

The apparent lack of any physiological function for such a vast array of plant chemicals resulted in speculation as to what common function such structurally different substances may share. One early suggestion was that these substances represented metabolic waste products, stored in the plant only by virtue of the fact that plants lack means of actively excreting them. However, the structural complexity of many plant secondary products, as well as the accumulation of such compounds despite the existence of catabolic pathways, is inconsistent with a waste-product hypothesis. By the end of the nineteenth century, a more attractive hypothesis had been advanced. After studying the relations between snails and the plants they eat, E. Stahl formulated a general hypothesis that ascribed the chemical variation among plants to the "struggle" between plants and animals.

Coevolution. The importance of insects in particular as selective agents in determining plant chemistry in the parallel evolution of insects and plants was not emphasized until G. Fraenkel suggested that "reciprocal adaptive evolution" determined patterns of host plant utilization by insects; that is, plant chemicals can act as repellents to most insects but at the same time can act as attractants for those few species of insects that can feed on a particular plant species. P. Ehrlich and P. Raven coined the term coevolution to describe the stepwise procedure by which plants elaborate chemical defenses and insects evolve mechanisms of resistance or tolerance to those defenses. According to Ehrlich and Raven, who based their assumptions on circumstantial evidence in the form of feeding patterns of various groups of butterflies, insect-plant coevolution is a five-step process: (1) Angiosperms produce new secondary substances as the result of mutation or recombination. (2) By chance, some of the new substances affect the suitability of the plant as food for insects. (3) Plants with such newly required resistance "escape" to a large extent from insect herbivory and are free to undergo evolutionary diversification. (4) By mutation or recombination, insects evolve resistance to the new compounds. (5) Insects enter a new adaptive zone and undergo their own evolutionary diversification, in some cases using the erstwhile toxic or repellent compounds as attractants.

R. Whittaker and P. Feeny likened the process to an "evolutionary arms race," in which plant chemical defenses are countered by insect resistance, which in turn prompts escalation and development of more advanced chemical "weaponry" by the plant, and so on. Umbelliferone (7-hydroxycoumarin; I) is a common secondary compound found in plants, but lacks appreciable toxicity to lepidopteran herbivores. But some members of the Rutaceae (citrus family) and Umbelliferae (carrot family) have developed a specific enzyme which prenylates umbelliferone to form furanocoumarins such as psoralen (II). The double bond in the newly formed furan ring can, in the presence of ultraviolet light, form cycloadducts (III) with thymine in deoxyribonucleic acid (DNA) and cause mortality in many insects and other organisms, including lepidopterans.

Whittaker and Feeny introduced the term allelochemical to describe chemicals whose principal functions involved individuals other than the one producing the chemical. Although much of the theory with respect to plant-animal coevolution was derived from studies with herbivorous or plant-feeding insects, the principles apply for the most part equally well to vertebrate as well as invertebrate herbivores.

Quantitative vs. qualitative toxins. The mode of action by which plant secondary substances affect the suitability of plants as food for herbivores depends largely upon the chemical structure of the substance. Basically, there is a continuum of activity. At one extreme there are large-molecular-weight compounds, often of amorphous structure, that act by complexing with proteins and reducing the digestibility of nutritive quality of the plant material. Tannins, resins, and proteases fall into this category. Since these compounds increase in effectiveness with increasing concentration and characteristically occur in plants at high concentration (ranging from 5 to 60% of the dry weight), they are referred to as quantitative toxins. At the opposite extreme are chemicals, typically of low molecular weight, that act upon specific biochemical target sites in the herbivores. Typical toxins of this type include glucosinolates; cyanogenic glycosides, which upon hydrolysis liberate cyanide, a substance that inhibits respiratory electron transport; and furanocoumarins, which in the presence of ultraviolet light can cross-link strands of DNA and interfere with template formation (see above). These and others are referred to as qualitative toxins.

In 1976, Feeny introduced the idea that the two types of defenses are associated in nature with certain ecological features of plants. In long-lived or "apparent" plants such as trees, quantitative defenses provide protection against generalized herbivores that over time are eventually able to locate their hosts. In contrast, short-lived or "unapparent" plants are associated with highly host-specific herbivores that are resistant to the qualitative toxins produced for defense in their particular host plants; thus, unapparent plants depend upon escape in time and space to avoid herbivory.

Cost of production. In that metabolic energy is required to synthesize secondary products as well as, in some cases, to acquire certain elements essential for growth (for example, nitrogen, a plant nutrient, is a characteristic component of plant alkaloids), it has been assumed that there is an inherent "cost of production" for defensive compounds. While estimating these costs has proved difficult, evidence has amassed to suggest that plants employ many strategies to reduce metabolic expenditures on defense. Many plants produce low or negligible concentrations of defensive compounds until herbivory begins, at which point production of large quantities begins de novo or translocation from other plant parts takes place. Many species of forest trees frequently subjected to outbreaks of foliage-feeding caterpillars, such as oaks (*Quercus* spp.), produce leaves in years following defoliation that are significantly higher in quantities of phenolics and other digestibility-reducing substances. Insects feeding on such foliage experience reduced growth rates and decreased fecundity.

Resistance. The means by which herbivores develop resistance to allelochemicals is dependent upon the structure and mode of action of the chemicals. Resistance can be behavioral, physiological, or biochemical. Behavioral resistance is the failure of a herbivore to ingest sufficient quantities of an allelochemical to suffer lethal effects. Aversive conditioning is a prime example and is a major form of behavioral resistance in large mammalian herbivores, which consume a large variety of foods each in small quantities, thereby reducing the probability of ingesting a toxic dose. Any food that causes physiological disturbance is subsequently dropped from the diet. Another form of behavioral resistance, commonly displayed by insects, is selective feeding—avoiding plant parts that are high in a particular toxin or feeding in such a manner as to avoid consuming large quantities of toxin. Squash beetles (*Epilachna borealis*), for example, feed in a manner called trenching. They cut a circular trench in the upper epidermis of the surface of leaves of squash plants (Cucurbitaceae) and feed only on the leaf area inside the trench. The trenching prevents the leaves from translocating cucurbitacins, extremely toxic and bitter triterpene derivatives, into the area consumed by the beetles.

Physiological resistance involves the failure of an allelochemical to reach its target site within the body of a herbivore. Many wood-feeding insects, for example, contain surfactants that interfere with the protein-complexing ability of tannins. Nicotine, a neurotoxic alkaloid produced by the tobacco plant *Nicotiana tabacum*, is toxic only in its neutral or uncharged form (IV); in the hemolymph or blood of

Manduca sexta, the tobacco hornworm, the pH is such that over 90% of the molecule exists only in ionized form (V). Since an insect's nervous system is protected by an ion-impermeable sheath, nicotine is unable to penetrate the nervous system and the insect suffers no ill effects.

The most widespread mechanism of resistance is biochemical—enzymatic inactivation of a toxin within the body of a herbivore. Both vertebrate and invertebrate herbivores possess a suite of enzymes associated with cell membranes in particular organs, called mixed-function oxidases or polysubstrate monooxygenases, which function in converting lipophilic (fat-soluble) substances into hydrophilic (water-soluble) substances by oxidative reactions; this facilitates solubilization and subsequent excretion. The transformation from lipophilic to hydrophilic is generally associated with detoxification since many sensitive sites, such as cell membranes, are lipid-rich and susceptible to attack by lipophilic foreign substances.

Use of chemicals by insects. Not only are herbivores able in many instances to overcome the defenses of their food plants, but they also can exploit the erstwhile toxins to their own ends. Many insects use allelochemicals in their host plants, toxic to unadapted species, as host-recognition cues. *Pieris brassicae*, a European cabbageworm that feeds exclusively on plants in the Cruciferae, possesses contact chemoreceptor cells that are sensitive to mustard oil glycosides, the compounds characteristic of its cruciferous host plants. These receptors are capable of detecting sinigrin (a mustard oil glycoside) at concentrations as low as $10^{-5}M$. Adult *P. brassicae*, called large white butterflies, possess similar receptors on fine hairs on their forelegs; female butterflies "tap" potential host plants to ensure that they are suitable before ovipositing.

In addition to using plant secondary products for purposes of recognition, many herbivorous insects use allelochemicals for defense against their own enemies. For example, the pipevine swallowtail *Battus philenor* feeds as a caterpillar only on plants in the family Aristolochiaceae. These plants as a rule contain aristolochic acid, a plant poison capable of inducing nausea and vomiting in vertebrates. The pipevine swallowtail butterfly is unaffected by the aristolochic acid and sequesters or stores the compound from its food plant in fatty tissues in its body (see illustration). These compounds are also present in the wings of the adult butterflies after metamorphosis. The aristolochic acid in the body of the butterfly still possesses the same toxicological properties that it does in the plant and protects the pipevine swallowtail from attacks by birds and other potential predators.

Evolutionary history. This stepwise interaction between plants and herbivores, the "coevolutionary arms race," is believed to be at least partially responsible for the tremendous diversity of plants and animals, particularly insects, on Earth. Almost half of all known species—approximately 500,000—are herbivorous insects. The interaction between insects and plants is presumed to have arisen early in Cretaceous times, coincident with the rise to dominance of angiosperm plants, and is still very much in evidence today as herbivorous insects foil attempts by plant breeders to develop permanently resistant varieties of crop plants. The metabolic adaptations possessed by herbivorous insects for detoxifying plant allelochemicals may well have preadapted them for detoxifying synthetic organic insecticides, such as DDT and other chlorinated hydrocarbons, introduced during this century for insect control. Resistance to synthetic organic insecticides developed rapidly after their introduction and continues to interfere with human attempts to effect chemical control of insects in crop plants.

For background information *see* ECOLOGICAL INTERACTIONS; GLYCOSIDE; PLANT PHYSIOLOGY in the McGraw-Hill Encyclopedia of Science and Technology.

[MAY BERENBAUM]

Bibliography: W. J. Bell and R. T. Carde (eds.), *Chemical Ecology of Insects*, Sinauer Associates, Sunderland, Massachusetts, 1984; M. Berenbaum,

Battus philenor, the pipevine swallowtail: (a) caterpillar and (b) adult. As a larva, this insect can consume leaves of *Aristolochia* without succumbing to the toxic effects of the aristolochic acid contained in the foliage. Instead, the insect stores the toxin, which protects it from enemies. (*Photos by J. Sternburg*)

Coumarins and caterpillars: A case for coevolution, *Evolution*, 37:163–179, 1983; P. Ehrlich, and P. Raven, Butterflies and plants: A study in coevolution, *Evolution*, 18:586–608, 1964; P. Feeny, Plant apparency and chemical defense, *Rec. Adv. Phytochem.*, 10:1–40, 1976.

Electrical power engineering

Electrical power projects in third world countries face a completely different set of parameters from those in the industrial world. Success in the less developed areas owes as much to imagination, flexibility, patience, and sympathy with different cultures and business procedures as to the need for high standards of engineering skills and equipment.

Availability of finance. Third world countries are very often dependent on aid from the developed countries and their agencies. Electrical power projects need to be identified and examined for practical and economical viability prior to search for finance. As discussed below, an aid project is most beneficial when it includes the encouragement of local manufacture and training. A world economic recession reduces the availability of soft funds and tends to widen the gulf between the developed and third world countries.

Urgency. Third world countries have to export to live. Be it jute, tea, sugar, or a small cottage industry, some kind of processing is usually involved and electricity is a great and convenient aid for increasing productivity. The social needs should not be neglected; for example, electric light allows children to study in the evenings. Supply has never outstripped demand. The fact that labor is cheap leads some economists to suggest that development of electricity is not essential, but this does not stand up to analysis. Many processes are impossible without electric power. The demand for labor in, for example, agriculture can be highly variable. Pumped water is often required at the same time as intensive labor is needed in the fields. If water is pumped manually, this decreases the labor for planting which then lowers productivity compared to that achieved when using an electrically driven pump. For all these and many other reasons, the urgency of establishing supplies to towns and rural areas is very apparent.

Existing system. The existing electrical power system is often very run down (Fig. 1) due to lack of investment. Though there is a temptation to demolish the whole system and start again, this would leave many people without supplies for a considerable period. Rehabilitation can be by small sections or by paralleling and then removing the old equipment. Power generation plant can usually be added without interruption of supplies, provided the transmission equipment has sufficient capacity for the additional plant.

It is not uncommon to find very few records of the existing system. This may be due to bad storage, wars, or lack of management. An important preliminary function is to know "what is there" before the system can be improved. This may involve many months of collecting field data. Often students of the local university or polytechnic can be used to do this, acquiring at the same time useful practical experience.

Fig. 1. Distribution point in Old Dhaka, Bangladesh.

Environmental conditions. In general, third world countries usually have much hotter and wetter climates than developed countries, and these severe conditions must be allowed for in the design of the equipment. Other unusual hazards are, for example, birds building nests across transformer insulators or plants growing rapidly toward a conductor. Local experience is invaluable and should be considered very carefully before the automatic use of well-tried methods used in developed countries. A good example is the method used in Bangladesh to drive ground rods into the earth.

The problem of providing ground continuity to a depth of 20 ft (6 m) or so has been solved in this labor-orientated country in the following manner (Fig. 2). A sturdy bamboo frame is erected adjacent to the chosen grounding point, at which site an initial excavation is made to a depth of approximately 1.5 ft (0.5 m). This hole is filled with water. A 20-ft-long (6-m), 1.5 in.-diameter (380-mm) galvanized iron pipe is placed vertically in the excavation and held by a laborer perched at the top of the bamboo frame with his or her palm over the top of the pipe. A second laborer uses part of the scaffold as a pivot for a bamboo lever to raise the water pipe. The pipe is permitted to drop under its own weight, during which fall the upper laborer releases his or her palm seal. In this manner, water is lifted up through the pipe, carrying with it the silt directly from the opening at the pipe bottom, and the slurry is ejected from the top each time the pipe falls. By reciprocat-

Fig. 2. The local method of driving ground rods into the earth in Bangladesh.

ing the level, raising and lowering the pipe, and ejecting silt from the ground through the top of the pipe, the two laborers succeed in driving the pipe to a depth of 20 ft (6 m) in 2 h.

Infrastructure. Getting customs clearance and the transport of equipment to the final point of erection can be fraught with difficulties. In Bangladesh barge transportation (Fig. 3) is often the only way to get heavy loads inland, and then it is only possible in the flood season for as little as 2 months in the year. Little can be taken for granted. For example, such points as the availability of local contractors capable of doing the work and of adequate supplies of cement for the civil element of the work must be checked out prior to the commencement of the project.

Local manufacture. For an aid project to give the maximum benefit, it must not only supply the immediate needs but also stimulate self-reliance in the developing country. If it is a case of importing steel poles or stimulating local manufacture of concrete poles, the latter should have preference. Time, quality control, and patience may be needed, but the end result can be very rewarding.

Maintenance and spare parts. Equipment should be maintained locally and, preferably, spares manufactured locally. An electrical system may technically be best protected by fuses, but replacement fuses may not be available. When this happens, they are generally shorted out, and then a local fault in the system is not isolated and can shut down a vast number of consumers. In one incident, an 11-kV overhead conductor, glowing red hot, sagged, touched the low-voltage line, and caused a major shutdown; a week was required to reestablish the supply. Power stations have been shut down for a year or more due to lack of spares. The original fault could have been avoided, in some cases, by properly trained staff.

Training. For nationalistic reasons, most power utilities no longer have significant expatriate staff in key positions. Some countries have been very successful in filling the gaps, while others have been struggling. Training is an essential element of any progress. A power station is useless if it cannot be operated, a point not always fully appreciated until the expatriate engineer has finished his or her part of the operation.

Prospects. Well-engineered electrical power projects can be of immense assistance to developing countries. There are also projects which, in relation to the capital finance, have not realized their full potential through insufficient planning and consideration of local conditions. The trends would be encouraging if it was not for lack of funds to undertake worthwhile projects.

For background information see ELECTRIC DISTRIBUTION SYSTEMS; ELECTRIC POWER GENERATION; ELECTRIC POWER SYSTEMS; RURAL ELECTRIFICATION in the McGraw-Hill Encyclopedia of Science and Technology.

[DAVID A. JONES]

Bibliography: D. A. Jones, Power engineering for the developing countries, *IEE Electr. Power*, 28:35–38, 1982; D. A. Jones, The role of the consultant in developing countries, *IEE Proc.*, A, 128:18–25, 1981.

Electrical units

Two electrical units, usually the ampere and farad, have to be derived by measurement in terms of the SI base units of mass, length, and time, and from these two units all other electrical units are derived in turn. The present situation with regard to the farad is satisfactory, whereas that of the ampere is

Fig. 3. A transformer being transported by barge in Bangladesh.

not, but new measurement techniques are in hand to improve this situation.

Farad and derived units. The farad is arrived at by calculating the capacitance between electrodes forming the boundaries of a region of space, and the Thompson-Lampard electrostatic theorem shows how this capacitance can be derived from a single length measurement on a properly configured capacitor. The electrical permittivity of free space, ϵ_0, must be known, but it can be calculated as $\epsilon_0 = 1/\mu_0 c^2$ from the velocity of light, c, and the defined value of the magnetic permeability of free space, $\mu_0 = 4\pi \times 10^{-7}$ henry/m.

The SI value of a capacitor C thus derived can be used to measure the SI value of a standard resistor R at a frequency f from the relationship $R = 1/2\pi f C$. This process is fairly satisfactory for the present needs of science and technology in that the SI value of the mean of a set of resistors found in this way will be known for a year or so afterward to an accuracy of about 0.1 part per million. The mean value can be expected to change at a rate which is predictable from past SI realizations and which is of the order of 0.05 ppm/year.

The unit of inductance, the henry, can be derived from the farad by equating the impedance $2\pi f L$ of an inductance L to the impedance $1/2\pi f C$ of a capacitance C with an accuracy of a few parts per million, which is adequate for present needs of inductance measurement.

Once derived, the units have to be preserved for use in terms of maintained standards, and here also new techniques are now available.

Resistance unit and quantum Hall effect. The recent discovery by K. von Klitzing that the quantum Hall effect provides an accurate and unchanging way of preserving a resistance unit offers the possibility of setting up or recovering the SI ohm in any laboratory without the need to rederive it in terms of the base SI units. As the quantum Hall effect appears to depend only on the unchanging value of the combination $(h/e^2)/n$ of the fundamental physical constants h (Planck's constant), e (the charge on the electron) and an integer n, the resulting resistance unit would be unchanging and accurately the same in different laboratories. The quantum Hall effect occurs in certain semiconductor devices at temperatures of a few kelvins when they are in a magnetic flux density B of a few teslas. For particular values of B, in the vicinity of B_0/i, where B_0 is a flux density proportional to the surface carrier concentration of an individual device, the voltage drop along the direction of the current through the device is vanishingly small, while the ratio of the voltage across the device to this current (that is, the Hall resistance) becomes equal to $(h/e^2)/n$. Typically $n = 4$, for which value the quantum Hall resistance is 6453.2 ohms. Present indications are that the value of the quantized Hall resistance is independent of whether the device is based on silicon or gallium arsenide technology or of the particular device of either kind used, to at least 0.1 ppm. Although the physical phenomena involved are quite different, the quantum Hall effect could be used to maintain a resistance unit in the same way as a voltage unit is preserved by relating it to the ratio $2e/h$ of the ac Josephson effect. Any national standards laboratory can reproduce this voltage unit with an accuracy of 0.05 ppm or better, but the relationship of this unit to the SI volt is uncertain at present to a few parts per million. The realizations of the volt or ampere now in progress are aimed at improving this situation. *See* HALL EFFECT.

Ampere. In the past, the definition of the ampere in terms of the force of $2\pi \times 10^{-7}$ newton/m between parallel conductors a meter apart, each of which carries a current of 1 A, has been implemented fairly directly by using a current balance to measure, by weighing, the force between two current-carrying coils, one of which is suspended from the balance. The accuracy of this method is limited both by the need to make hundreds of length measurements to determine the geometry of the coil system and by the small force obtained in relation to the convective forces on the coils which are heated by the current. Recent evidence discussed below indicates that the accuracy of a current balance determination of the ampere is probably not better than 10 ppm, and this is at least a factor-of-10 worse than present-day needs. Once the ampere has been derived by this or a better technique, it can be preserved in perpetuity in terms of the ratio of a Josephson effect voltage standard to a quantum Hall effect resistance standard.

Indirect determinations. Over the past few years there have been several opportunities to derive the ampere with greater accuracy indirectly from relationships between the values of physical constants measured in various ways. Some of these ways involve the base SI units directly through force measurement, while others are in terms of electrical units so that agreement on the values of these constants will be obtained only if the correct SI ampere is used, and if not, the process can be inverted to yield a realization of the SI ampere. While there is not complete agreement between the values derived from various possible relationships, a unit some 7 ppm larger than that derived from current balance measurements is indicated. The uncertainty in the unit derived in this way is about 1 ppm, which is a considerable improvement on that afforded by the use of a current balance. It is still most desirable, however, to have a direct method of realizing the ampere, preferably with still lower uncertainty, both to improve accuracy still further and to allow measurements of the physical constants to revert to their proper role of testing the accuracy of understanding of physical theory. To this end, alternatives to the current balance are being examined.

Direct determinations. One approach which is yielding a reproducibility of 0.3 ppm is to suspend a coil from a balance so that part of it is in a strong magnetic field, make two separate measurements, and combine the results. First (illustration *a*), as in the current balance, the force F produced by a current I is measured by opposing it with a mass M in

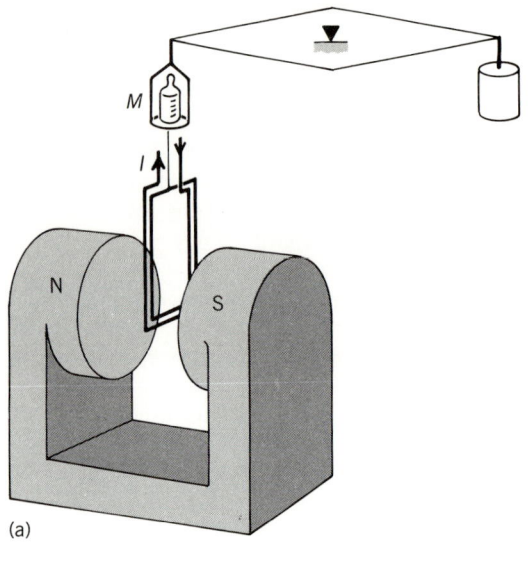

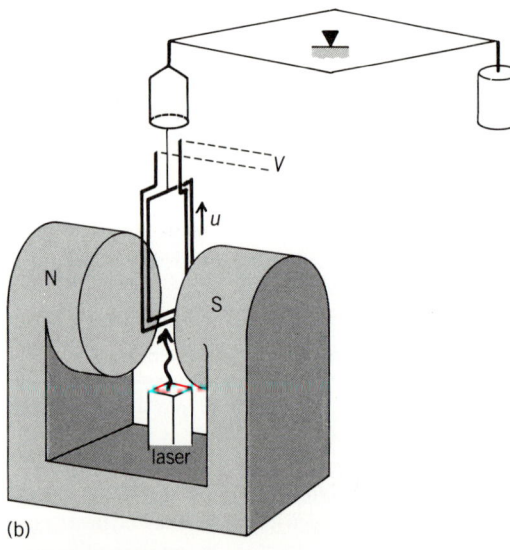

Apparatus to determine the ampere. (a) Measurement of force produced by current I in a coil suspended partially in a strong magnetic field. (b) Measurement of voltage V induced in the same coil when it moves vertically with velocity u measured by laser interferometer.

the Earth's gravitational acceleration g. Second (illustration b), the voltage V induced when the coil moves vertically with velocity u through the position that it occupied for the weighing is measured. Equality of the electrical and mechanical power units then gives the simple equation $IV = Mgu$.

The current I can be measured in terms of the voltage drop that it produces across a resistor whose value is known in SI units from a calculable capacitance and frequency, and both this voltage drop and the voltage V can be related to a non-SI voltage unit, thus calibrating this unit in SI volts. As neither the dimensions of the coil nor the strength or distribution of the magnetic field need be measured, the single dimensional measurement needed is that to establish the velocity of the moving coil. The velocity measurement can be carried out with great ac-

curacy by using a laser interferometer, and the weighing observations have a much lower uncertainty than in the case of the current balance because a much larger force is generated with a much lower power dissipated in the coil.

Another method which is yielding results of comparable accuracy is to measure the force exerted on the electrode surface of a capacitor when the voltage to be calibrated in SI units is applied. Since, as discussed above, an accurate value for the SI ohm can be derived, this is an equivalent route to the ampere. Usually one electrode is suspended from a balance, and the change in force and capacitance is determined when the other electrode is displaced, but in the one determination which is yielding results of part-per-million accuracy the necessary force per unit area is ingeniously determined directly as a hydrostatic pressure on a pool of mercury, and no capacitance measurements need be made.

Kilogram. Finally, it is possible to consider a future development of SI. If the quantum Hall effect is proved capable of maintaining a resistance unit in terms of h/e^2 in a similar way to that in which a voltage standard is maintained by the Josephson effects, it is conceivable that at some future date the quantum Hall effect and Josephson effects together with the present meter and second could provide an alternative basis for the SI units. The kilogram would then be a unit derived from an ampere determination in reverse instead of a preserved artifact. This situation would be philosophically more satisfying, but an improvement of more than a factor of 10 in the present determinations of the ampere is needed to make the suggestion practicable.

For background information see ATOMIC CONSTANTS; CAPACITANCE MEASUREMENT; CURRENT BALANCE; ELECTRICAL UNITS AND STANDARDS; INTERFEROMETRY; JOSEPHSON EFFECT; PHYSICAL MEASUREMENT in the McGraw-Hill Encyclopedia of Science and Technology.

[BRYAN P. KIBBLE]

Bibliography: B. P. Kibble, R. C. Smith, and I. A. Robinson, The NPL moving coil ampere determination, *IEEE Trans.*, IM32:141–143, 1983; K. von Klitzing, G. Dorda, and M. Pepper, New method for a high accuracy determination of the fine structure constant based on quantized resistance, *Phys. Rev. Lett.*, 45:494–497, 1980; G. J. Sloggett et al., in *Proceedings of the 2d International Conference on Precision Measurement and Fundamental Constants, June 1981*, NBS Spec. Publ. 617, pp. 469–473, 1984; B. N. Taylor, Is the present realisation of the absolute ampere in error?, *Metrologia*, 12:81–83, 1976.

Electrical utility industry

The year 1984 marked the first since the economic depression of the 1930s in which capital expenditures of the utility industry in the United States actually declined. That decline is due primarily to the cancellation or completion of fossil and nuclear generating units on which construction had previously begun, and the absence of new construction starts. Capital expenditures for generating capacity, which dominate utility budgets, will probably decline, due to the same reasons, until the late 1980s.

The second trend that emerged strongly during 1984 was the "phasing-in" of the cost of new generating units coming into service. In most jurisdictions, utilities are not permitted to put capital construction costs incurred during any given year into the rate base, and are thereby prevented from earning a return on that investment. The entire cost of the project is taken into the rate base when the facility begins commerical service and becomes "used and useful." Because of the huge cost of a modern generating units—ranging up to several billion dollars for a nuclear station—this can cause rate increases of, in extreme cases, up to 50%. To avoid the shock of this to the consumer, some regulators have now begun to phase the cost of such units into the rate base over a number of years.

Ownership. Ownership of electrical utilities in the United States is pluralistic in nature, being shared by investor-owned corporations; customer-owned cooperatives; and publicly owned agencies on the municipal, district, state, region, and federal levels. The investor-owned utilities, however, dominate in all essential measures. Investor-owned utilities serve 76.7% of 98.5 million electricity customers in the United States. Publicly owned utilities serve only 13.3% of electricity customers, slightly more than the 10.1% served by the cooperatives. Although federal utilities do serve some few customers on the retail level, they are primarily wholesalers, selling to other utilities that distribute the energy to the ultimate customer.

The cooperatives are the fastest-growing sector of the industry, their growth being driven by the continuous electrification of farms, and the shift of population away from urban and suburban centers into the rural areas where cooperatives operate.

The noticeable discrepancy between the percentage of customers served by the cooperatives and the much smaller percentage of capacity owned by them comes about because the cooperative sector comprises primarily distribution companies that buy their power from others at wholesale rates and distribute it to their member customers. Some cooperatives do generate electricity to serve their members, and some, such as the Basin Electric Power Cooperative in Bismarck, North Dakota, serve no retail customers at all, but generate and transmit the power to member distribution cooperatives over a wide area.

The combination of tremendous financial pressure on investor-owned utilities arising out of the burgeoning need for new capacity and consumer resistance to rate increases, plus antitrust action by the courts against the private utilities, has led them to join in partnership with each other and with cooperatives and publicly owned utilities in the construction and operation of large new units. Municipal bodies and cooperatives which themselves seldom have the resources necessary to build a large-scale, efficient fossil or nuclear plant have therefore eagerly bought into large efficient new units to gain access to the lower-cost energy that they produce. The importance of this cooperation was underscored in 1984 when the costs of the Seabrook nuclear plant threatened to bankrupt Public Service of New Hampshire, the major shareholder. Only the financial guarantees of the National Rural Electric Cooperative Agency, acting on behalf of a consortium of New England cooperatives that are partners in the plant, enabled construction to continue.

United States electric power industry statistics for 1984*

Parameter	Amount	Increase compared with 1983, %
Generating capacity, MW		
Conventional hydro	64,619	0.5
Pumped-storage hydro	15,149	5.6
Fossil-fueled steam	461,190	0.7
Nuclear steam	72,750	8.3
Combustion turbine and internal combustion	56,700	−0.9
TOTAL	670,410	1.01
Energy production, $\times 10^3$ GW	2457.7	5.7
Energy sales, GW		
Residential	779,700	3.8
Commercial	556,100	4.3
Industrial	838,100	4.6
Miscellaneous	80,900	4.0
TOTAL	2,254,800	4.2
Revenues, total, $\times 10^6$ dollars	138,500	7.1
Capital expenditures, total, $\times 10^6$ dollars	41,150	−10.5
Customers, $\times 10^3$		
Residential only	86,600	1.5
TOTAL of all classes	98,450	2.6
Residential usage, kWh/customer	9,000	4.3
Residential bill, cents/kWh (average)	7.4	5.7

*From 35th annual electric utility industry forecast, *Elec. World*, 198(9):49–56, September 1984. Extrapolations from monthly data of the Edison Electric Institute; and 1984 annual statistical report, *Elec. World*, 198(4):49–72, April 1984.

Capacity additions. Utilities had, at the end of 1984, a total generating capacity of 670,400 MW, having added 12,820 MW during the year. This produced a capability—that is, actual ability to generate power at the time of peak summer demand—of 624,600 MW. The difference comprises that capacity that is unavailable because of the need for maintenance, failure, or derating from nameplate capacity (see table).

The composition of the capacity additions during 1984 was 1170 MW of hydroelectric, 5685 MW of fossil-fueled steam, 5680 MW of nuclear, 78 MW of diesel and combustion turbines, and 210 MW of alternative sources such as geothermal, wind, refuse-fueled, or solar.

The composition of total plant capacity by type of generation at the end of 1984 was 461,190 MW or 68.8% of fossil-fired steam; 79,770 MW or 11.9% of hydroelectric; 72,750 MW or 10.9% of nuclear; 51,700 MW or 7.7% of combustion turbines; and 5000 MW or 0.7% internal combustion engines, essentially all diesel (see illustration).

Fossil-fired capacity. All units entering commercial service during 1984 were coal-fired, as dictated by the Fuels Use Act of 1974. Utilities continue to convert older oil-fired capacity to coal. These remaining units are concentrated primarily in the Northeast, with 69%; the Southeast, with 23%, and the rest in the Southwest, with 8%. Utilities now plan to convert 5640 MW from oil to coal, representing 30 units, from 1985 to 1990.

Environmental constraints on the pollutants in flue gas are driving utilities to investigate alternative combustion processes. One is the fluidized-bed boiler, in which pulverized coal is burned together with limestone in a bed that is kept turbulent, or fluidized, by passing air through it. This process removes most of the sulfur and particulates from the hot effluent gases. A 40-MW unit, the largest in actual operation in the United States, is successfully running in tests on the Tennessee Valley Authority system.

A number of developmental projects to gasify coal and use the cleaned gas as a fuel are also in progress. At a site in Coolwater, California, a gasifer supplying clean gas to a combustion turbine has operated at roughly a 70% capacity factor since startup in the spring of 1985. The exhaust of the gas turbine generates steam to drive a steam turbine in a combined-cycle arrangement that gives high relative efficiency (60%). If no insurmountable problems occur during the test period, utilities have shown great interest in pursuing this development although, at current oil and natural-gas prices, the process is not fully competitive.

Utilities spent $8.7 billion for construction of fossil-fired units in 1984. Investor-owned utilities spent $6.3 billion; publicly owned units, $1.7 billion; cooperatives, $552 million; and federal agencies, only $204 million.

Nuclear power. The cancellation of units already announced or under construction continued in 1984, with eight units totaling 9040 megavolt-amperes (MVA) joining the list of those canceled. Because the average completion time from date of Nuclear Regulatory Commission (NRC) approval to entrance into commercial service has now stretched to as much as 14 years, units begun in previous years continue to come on line. During 1984 four units with an aggregate capacity of 5680 MVA began commercial service. The total capacity of all nuclear units now in service in the United States is 72,750 MW, provided by 85 units in 56 individual generating plants.

All four of the units entering service in 1984 were boiling water reactors (BWR). This brings the totals of the United States to 32 BWRs and 53 pressurized water reactors (PWRs).

Utilities currently have under construction, but less than 50% complete, a total of nine nuclear units aggregating 10,535 MW. There are an additional 41 under construction and more than 50% complete, aggregating 44,524 MW.

Utilities spent $15.9 billion on nuclear construction in 1984. Investor-owned spending accounted for $13.2 billion, state and municipal utilities for $1.1 billion, cooperatives for $445 million, and federal agencies $1.1 billion.

Combustion turbines. Combustion turbines have historically been installed by utilities for use at times of peak demand to supply up to 10–15% of

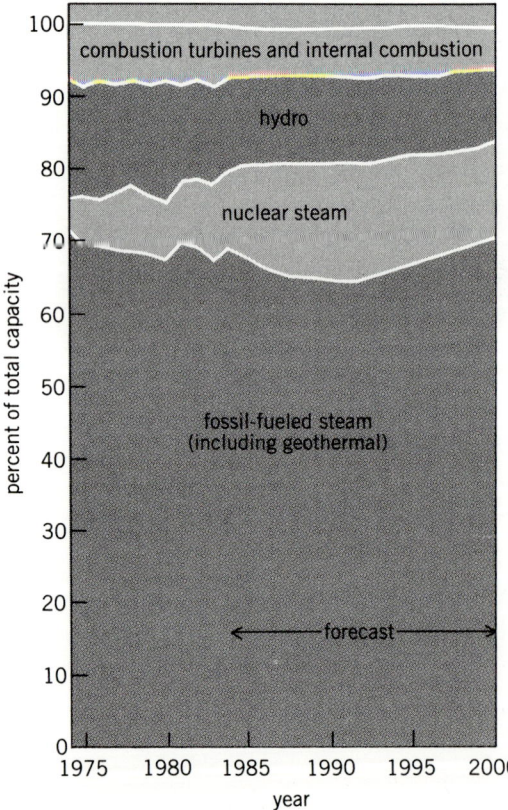

Probable mix of net generating capacity. (*After 35th annual electric utility industry forecast, Elec. World, 198(9):49–56, September 1984*)

that peak. The low capital cost of such machines, currently about $300/kW compared to $700/kW for coal-fired units, more than offsets their high fuel consumption when used only for about 200 h per year in this way. Combustion turbines, which use either natural gas or oil as fuel, have heat rates of about 13,000 Btu/kWh (3.8 joules of heat per joule of electric energy) compared with less than 9000 Btu/kWh (2.6 joules of heat per joule of electric energy) for a modern fossil unit. Further, they have quick-start capability which permits them to be brought up to full load in 3–8 min and provides flexible capability for emergency conditions. They are also used to provide startup power for generating stations that have experienced a complete shutdown, as during a blackout of an entire area or system.

Because of government restrictions on fuel, and because there are already excess reserves of steam, hydroelectric, and nuclear units at peak periods, the percentage of peak capacity represented by gas turbines is decreasing.

Although individual units as large as 125 MW have been installed, units most favored are in the 25–50-MW range. In 1984, utilities added only 75 MW of combustion turbine capacity. Due to retirements, total industry capacity of combustion turbines actually decreased from 51,992 to 51,700 MW, and such turbines are now 7.8% of total capacity installed.

Combustion turbines are also used in combination with steam turbines in highly efficient combined-cycle operation. In this mode the 900–1000°F (482–538°C) gas turbine exhaust produces steam in a heat recovery boiler which supplies a steam turbine. Efficiencies of this cycle may run as high as 60%, compared with 34–35% for the most efficient current steam cycles. In this mode the combustion turbine will contribute about 70% of the total output.

Utilities spent $68.1 million on combustion turbines in 1984.

Hyroelectric installations. Utilites brought 370 MW of conventional hydroelectric capacity into service in 1984, raising the total now installed to 64,619 MW. Hydro units of this type, that is, hydroelectric turbines driven by impounded water or by the natural flow of a river, constitute 9.6% of total installed capacity of all types of utility system in the United States.

Future plans call for an additional 3600 MW of capacity residing in 200 units to be built over the next 10 years, though locations for the required dams are becoming increasingly difficult to find and to license. The few major sites suitable for high dams are, in general, in areas where the environmental effects of the resulting lake are unacceptable. Utilities spent $603 million on this type of installation during 1984. Investor-owned companies spent $337 million; state and municipal utilities, $186 million; cooperatives, $6.7 million; and federal agencies, $73.4 million.

The future of hydroelectricity may rest with small-scale projects that either can be incorporated into existing dams or can be powered by the normal flow of the river. These units carry ratings of 1.5–50 MW. Such units have become especially popular in the West, where the extensvie irrigation canal systems often have locks that create heads of 15–25 ft (4.5–7.5 m), adequate to power small tube- or bulb-type hydroelectric units. New England and New York State also have numerous small falls that formerly were used to power mills and factories and which now could be developed as hydroelectric sites. New York State, for instance, plans to develop about 20 such sites in the next 10 years, with capacities of 1–35 MW. New England plans 25 sites with capacities of 1–25 MW each.

Pumped storage. Pumped storage represents one of the few possible methods for storing large amounts of energy from electrical generators. Water is pumped from a body of water on which the generating unit is located into a reservoir some distance above. This is normally done during off-peak periods when large, efficient base-load units that would ordinarily be shut down for lack of demand are available. During the subsequent peak demand period, this water is released through the pumps, which can be reversed to act as turbines, recovering only about 65% of the energy originally expended but reducing the need for the equivalent capacity to be provided at peak. This type of generation is especially attractive when coupled to nuclear plants with their high capital cost and low operating cost.

Utilities installed an additional 800 MW of pumped storage capacity during 1984, raising total installed capacity in the United States to 15,149 MW, or about 2.3% of the nation's total capacity.

Utilities spent $241 million for pumped storage faciltes in 1984. The investor-owned companies spent $236 million, the major portion of that total. Federal agencies spent only $2.6 million, and other publicly owned utilities, $2.0 million.

Renewable energy sources. Utilities have actively sought energy sources that provide alternatives to the combustion of coal, oil, gas, and uranium. At the end of 1984, utilities had a total of 1690 MW of capacity other than the conventional types listed above. These include geothermal, wind, solar, waste, and refuse-fired units. Utilities have planned an additional 3383 MW of such units for the 1983–1992 period. These consist of 2040 MW of geothermal, 357 MW of solar, 273 MW of wind, and the rest waste- or refuse-fired capacity. More than 87% of all such new capacity planned is located in the state of California.

Rate of growth in demand. Some credible forecasters have begun to project extremely low or even negative growth in demand over the next 10–20 years. These projections are predicated to a large degree on a rise in electricity prices as rates adjust to absorb the large additions to the rate base that new, expensive units represent as they come into service. For example, the just-completed Shoreham unit of the Long Island Lighting Company would raise rates roughly 50% upon entering the rate base.

Coupled with this negative price elasticity that will encourage conservation, continuing stable prices for competitive energy sources such as coal and oil would also exert downward pressure on use.

A long-term decreasing trend also naturally arises from the mix of demographic factors that characterize a maturing society such as the United States. Population is growing at a decreasing rate, and passing beyond the years of peak consumption. This effect will be compounded by the price-induced conservation. Because of the high technological content of the utility industry's plant, the preponderance of skilled labor employed, the high cost of capital engendered by unresponsive regulation, and escalating fuel costs, electricity prices should rise at or slightly more than the rate of inflation over the coming decade, somewhat moderating growth, also.

Some utility forecasts, however, point to the fact that the normalized rate of growth, even throughout the past 10 years, has closely tracked the growth in the gross national product; and further, that the increase in productivity required to sustain that growth can be achieved only through further electrification of commerce and industry. It is the balance of all these factors that has produced a consensus forecast of 2.5–3% growth per year.

The overall national declining pattern of annual growth (from 7.2% before 1974 to an average of 3.4% in the period 1974–1984) has had a major effect on reserve margins—that is, the excess of installed capacity over demand—on a national basis. A rule of thumb is that average national reserve margin should be about 25%, and it is now about 36%. Though utilities are delaying the construction of, or canceling, many major generating units, some plants started years ago will be finished and will come into service. Margin will not decline to the 25% level until 1990.

Usage. Sales of electricity, following a rise of 3.1% in 1983, rose another 4.2% in 1984. Total national usage was 2.255×10^{12} kWh. The rise was due almost entirely to a surging economy, since a mild winter and summer contributed little.

Commercial sales held up well in 1984, rising 4.3% to a total of 556.1×10^9 kWh.

Industrial sales rose vigorously, about 4.6%, reflecting the strong economic recovery, and reached 838.1×10^9 kWh consumed.

Residential sales rose only slightly less strongly than industrial and commercial, gaining 3.8%, with 779.7×10^9 kWh consumed. This usage is dependent on housing starts, which continued at a moderate annual level of 1.7 million units.

Electric heating for residences continues to make gains. More than half of all new homes constructed in the United States in each of the last 10 years have been electrically heated, and in 1984 heating energy sales topped 172.4×10^9 kWh. The emergence of the heat pump, now taking almost a third of new electric installations, has made electric heating competitive in most areas with other fuels. Electric heating is now used in almost 20% of all housing stock in the United States, second only to gas heating.

Residential use per customer rose slightly to 9000 kWh from 1983's use of 8826 kWh. Continued rate increases, however, boosted the average residential rate to 7.4 cents/kWh and the annual average bill per residential customer to $666.

Total revenue for all classes of service for the entire industry was $138.5 billion.

Fuels. Consumption of coal by utilities in 1983, the last year for which good figures are available, rose 5.3% to 625.2×10^6 tons (568.4×10^6 metric tons). Oil dropped again 2.0% to 215.5×10^6 bbl (38,934 m^3). Essentially all the shift from oil was absorbed by nuclear and hydroelectric units since consumption of natural gas also dropped 9.8% to 2.910×10^{12} ft^3 (83.4×10^9 m^3).

The same type of shift was seen in actual energy generated by fuel. Total generation output increased from 2.3582×10^{12} kWh, to 2.4577×10^{12} kWh in 1984, while that portion generated by other than hydroelectric increased from 1.9320×10^{12} kWh to 1.9782×10^{12} kWh. Coal generated 1.2594×10^{12} kWh of the total; oil accounted for 144.5×10^9 kWh; gas was used to generate 274.1×10^9 kWh; and nuclear power put out 293.7×10^9 kWh. The rest was primarily hydroelectric, with a minor contribution from other sources such as geothermal.

Distribution. Distribution capital expenditures for 1984 amounted to $6.042 billion, and an additional $2.16 billion was spent maintaining existing plants. During the year, 10,900 mi (17,558 km) of three-phase equivalent overhead lines and 5870 three-phase equivalent miles (9458 km) of underground lines came into service at voltages ranging from 4.16 to 35 kV. The majority of this mileage was at 15 kV, which accounted for 7900 three-phase equivalent miles (12,725 km) of overhead and 3933 three-phase equivalent miles (6335 km) of underground circuitry. The percentages for overhead construction held by other voltage classes were 9.5%, 15.8%, and 2.3% for 35-, 25-, and 4-kV, respectively. For underground construction, the equivalent percentages are 11.2%, 20.0%, and 1.8%. During 1984, utilities energized 13,420 MVA of distribution substation capacity and expended $734 million for substation construction.

Transmission. Utilities spent $4.0 billion in capital accounts for transmission lines in 1984. During the year, they spent $903 million for overhead lines at 345 kV and above, and $932 million for overhead circuits of 220 kV and below. For underground transmission construction, which can cost, on average, eight times more than equivalent overhead construction, capital expenditures amounted to $51.5 million at voltages of 69–161 kV, but only $800 million for circuits at 220 kV and above. Utilities installed 3000 mi (4832 km) of overhead lines at 345 kV and above, but 5380 mi (8666 km) at 220 kV and below.

The picture is completely different in the underground sector, because current cable technology costs favor the lower voltages. In 1974, only 4 mi (6.4 km) of cables operating at or above 230 kV came into commercial service, and 30 mi (48 km) at or below 161 kV.

Utilities brought 40.4 gigavolt-amperes of transmission substation capacity into service in 1984 and spent $1.20 billion for substation construction. Maintaining existing transmission plant cost $600 million.

Capital expenditures. Total capital expenditures in 1984 dropped from 1983's $45.99 billion to $41.15 billion as a result of combined utility efforts to curtail construction. Of this total, $27.6 billion went for generating facilities, $4.2 billion for transmission, $7.58 billion for distribution, and $1.77 billion for miscellaneous facilities such as headquarters buildings and vehicles. Total assets held by the investor-owned segment of the industry were $345.3 billion at the end of 1984. Municipals held about $34 billion in assets and cooperatives about $35 billion.

For background information *see* ELECTRIC POWER GENERATION; ELECTRIC POWER SYSTEMS; ENERGY SOURCES; TRANSMISSION LINES in the McGraw-Hill Encyclopedia of Science and Technology.

[WILLIAM C. HAYES]

Bibliography: Edison Electric Institute, *Statistical Yearbook of the Electric Utility Industry*, 1983; 1984 annual statistical report, *Elec. World*, 198(4):49–72, 1984; 35th annual electric utility industry forecast, *Elec. World*, 198(9):49–56, 1984; 23rd steam station cost survey, *Elec. World*, 197(11):49–50, 1983.

Electronic displays

The application of large flat-panel displays made from liquid crystals, electroluminescent, gas-discharge and plasma materials is expanding rapidly. Flat-panel displays are not replacing the cathode-ray tube, but are finding new applications in the commercial marketplace (such as truly portable televisions and personal computers) which cannot utilize the cathode-ray tube. The cathode-ray tube is improving in performance with higher resolution and color in product applications such as computer-aided design, and 1000-line cathode-ray tubes in full color are now readily available from several domestic and foreign sources.

The market enthusiasm for flat-panel displays in personal computers is very strong. Several technologies are competing with liquid crystal displays, electroluminescent displays, and gas-discharge or plasma-display panels being the leading contenders. The emphasis is not on color but on ergonomics, compatibility with existing cathode-ray tube formats, and economy in size and power.

Flat-panel portable color TV displays. Flat-panel color television displays have evolved more rapidly than had been predicted. High-quality color and black-and-white demonstrator displays have been made by several Japanese companies (Fig. 1). The primary breakthrough has been due to the application of thin-film transistors which have been made

Fig. 1. Pocket-sized color television using 2-in.-diagonal (5-cm) liquid crystal display with 52,800 polysilicon thin-film transistors deposited on a glass substrate. (*Seiko*)

by using amorphous or polycrystalline silicon as the semiconductor.

Several problems have been encountered in attempting to enlarge liquid crystal displays beyond the watch or calculator size—all of which have been overcome in the latest color television designs. First, liquid crystal displays in general are not matrix-addressable for television-size arrays. In other words, the individual picture elements (pixels) cannot be addressed by a simple orthogonal array of row and column leads. Liquid crystal material is not sufficiently nonlinear. To accomplish the addressing, individual switching elements such as thin-film transistors are added at each pixel. Second, the color is not sufficiently saturated from simple inclusion of dye mixes; therefore, filters are added at each pixel triad for the red, green, and blue colors. Third, the contrast is low. To correct this, the twisted nematic mode is used with polarizers in order to enhance the contrast and provide some immunity to variations in ambient lighting. Fourth, liquid crystal displays do not in themselves emit light but depend upon reflected light. Consequently, they are dim and become dimmer because of the polarizers and filters. The dimness is overcome by adding a back white lighting source that uses incandescent or electroluminescent lamps to augment the reflected light. The performance in terms of color content, viewing angle, and resolution is good—approaching that of a cathode-ray tube at orders of magnitude less power and a fraction of the volume for displays in the 2–5-in. diagonal (5–13-cm) size category.

In the final configuration of the color liquid crystal display (Fig. 2) each row lead is dedicated to a color. Consequently, a color liquid crystal display requires three times more row leads than a comparable black-and-white liquid crystal display. A simple thin-film transistor is constructed in the back

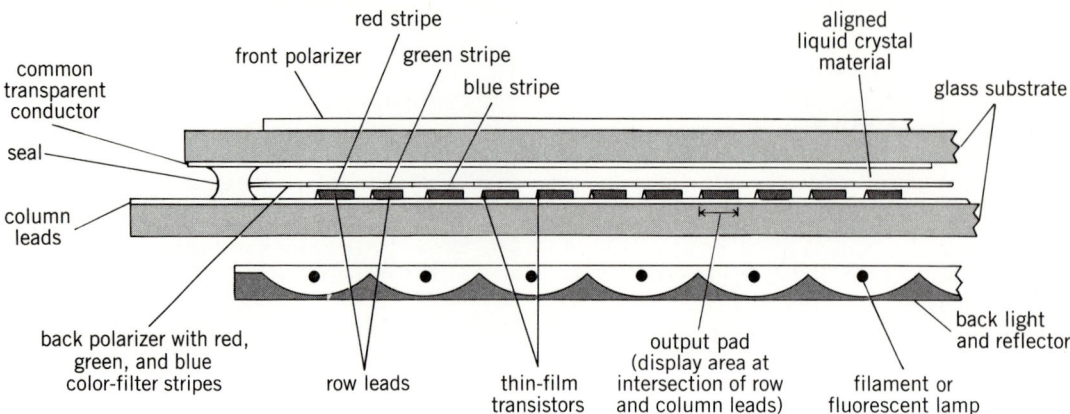

Fig. 2. Basic construction of a color liquid crystal display.

plane at the intersection of each row and column lead with a transparent output pad connected to the drain of the thin-film transistor. The output pad faces the common transparent front electrode and rotates the intermediate liquid crystal material when charged by the thin-film transistor and connecting row and column leads. The row leads typically control the thin-film transistor gate, and the column leads supply the charge through the thin-film transistor source. Rotating the surface-aligned liquid crystal material stops the transmission of light through the crossed polarizers and color filters. When at rest and aligned with the upper and lower glass surfaces, the optically active liquid crystal material rotates polarized light from the first polarizer, allowing it to pass through the second crossed polarizer.

A display of this nature has a potential cost problem in high-volume production. The primary problem is the production yield of the thin-film transistor array. Additionally, the complexity of the display caused by the required filters, polarizers, and line drivers further increases the cost. The thin-film transistor array requires one thin-film transistor at every pixel. Present designs utilize a 240 by 360 pixel array. A complete television picture utilizes 480 by 360 pixels, 30 times per second (U.S. Standard), with two-to-one interlace. A significant economy in construction and operation is realized in the smaller television display by reducing the row leads by a factor of two and combining the interlace fields onto the same pixels. The thin-film transistor is typically made from five deposition, masking, and etching cycles—a manageable number since all the dimensions are large compared to those encountered in metal oxide semiconductor (MOS) technology. Here, however, all or nearly all of the pixels must work or the whole display is lost. The polycrystalline and amorphous-silicon semiconductor technology has been advanced by the solar energy industry.

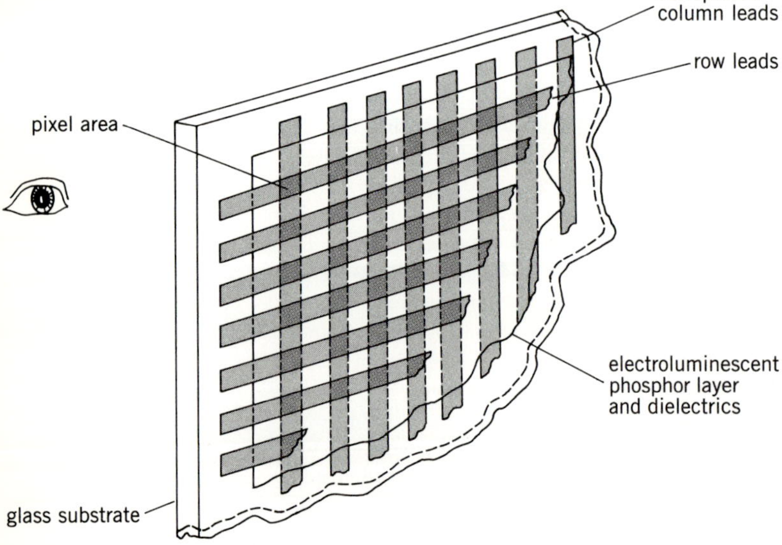

Each of the array leads in the display requires a line driver for addressing the individual pixels. For liquid crystal displays, the line drivers are made from complementary MOS (CMOS) chips with typically 32 line drivers per chip and approximately 20 chips per display. The line driver chips are mounted on a separate board and attached by wire bonding or other means to the liquid crystal display leads around the periphery of the liquid crystal glass substrate. The chips, mounting, testing, and attachment to the display cost as much as the glass portion of the display. This cost may be reduced in the future by making the line drivers from thin-film transistors with appropriate shift registers, at the same time and on the same glass substrate as the pixel thin-film transistors. The enthusiasm for this approach is tempered by the reality that adding this much complexity to the display glass may reduce its yield to the point where no net cost gain is realized. Each line driver would require 10 or more thin-film transistors.

The liquid crystal technology discussed here is ideally suited for the portable television market. There are several factors which indicate that this form of liquid crystal display could do for the portable television what the transistor did for the porta-

Fig. 3. Construction of a thin-film alternating-current electroluminescent display.

ble radio. First, the liquid crystal display is readable in direct sunlight without any increase in power. The lack of this feature can be very disappointing to owners of portable televisions made with cathode-ray tubes. No other display technology has the same immunity to high ambient illumination as the liquid crystal display. Second, the liquid crystal technology is the lowest power consumer of any known technology. The back light takes a significant amount of power, but still the net power is lower than any other approach. Additionally, the overall display dimensions such as viewing area and volume are ideally suited for portable applications. It remains to be seen if the price can be made acceptable to the marketplace.

Flat-panel computer displays. The display for a personal computer has a great effect on its portability and its desktop area coverage (footprint). There is a strong interest in the application of flat-panel displays to improve these factors. The first applications of liquid crystal displays have been with limited character formats.

Two-thousand-character liquid crystal displays have been demonstrated and are expected to be in computer products by 1985. The liquid crystal display's sunlight readability and low power make it ideal for portable computers; however, without some augmentation such as thin-film transistors or thermal addressing, the viewing angle and general ergonomics are marginally acceptable in the larger sizes. Augmented liquid crystal displays for personal computers are expected.

Electroluminescent display technology is also finding applications in personal computers. The alternating-current (ac) electroluminescent display has the simplest construction of all flat-panel displays, requiring only one glass substrate (Fig. 3). The thin film, less than 1 micrometer thick, is made of ZnS:Mn transparent polycrystalline phosphor and is sandwiched between two dielectrics. Film and dielectrics are in turn sandwiched between two sets of orthogonal leads. The display is truly solid-state as there is no moving fluid or gas cavity.

The first flat-panel personal computer used an electroluminescent display. The thin-film ac electroluminescent display has the advantage of being ergonomically superior to any other monochrome display, with good power efficiency and sunlight readability. The sunlight readability of an electroluminescent display is not as good as that of a liquid crystal display. An electroluminescent display requires more power because it emits light. The readability of the electroluminescent display is definitely superior to that of any other monochrome display. The electroluminescent display can operate over a wide range of ambient lighting, temperature, and viewing angle. It has sufficient nonlinearity for direct addressing without augmentation for all conventional computer terminal sizes.

Plasma-display panels are also used in computer terminals where the emphasis is on the footprint area and not on portability. Plasma-display panels have the lowest luminous efficiency and are heaviest in comparison with liquid crystal or electroluminescent displays. They are used in military computer terminals where ruggedness in construction is important and ample power is available. Photonics has made large plasma-display terminals (up to 3 ft or 1 m in diameter) which are sunlight-readable.

It is expected that liquid crystal displays will be used in lap computers and portable computers where less than 2000 characters are required. Electroluminescent displays will dominate where 2000 characters and standard computer graphics are required over a wide range of temperature and lighting with optimal ergonomic features. The plasma-display panels will be used where larger-than-typical computer terminal formats are required and where flatness is important.

For background information see CATHODE-RAY TUBE; ELECTROLUMINESCENCE; ELECTRONIC DISPLAY; INTEGRATED CIRCUITS; LIQUID CRYSTALS in the McGraw-Hill Encyclopedia of Science and Technology.

[LAWRENCE E. TANNAS, JR.]

Bibliography: L. E. Tannas, Jr. (ed.), *Flat-Panel Displays and CRTs*, 1984.

Electronics

Recent improvements in the production of electronic circuit boards include surface-mounted technology and elastomeric connectors for connecting components.

Surface-mounted devices. The technique of mounting electronic circuit components and their electrical connections on the surface of a printed board (rather than through holes) is known as surface-mounted technology. Surface mounting is gaining favor in the electronics industry because smaller components and smaller circuit assemblies not only reduce size and weight and improve performance but also can be manufactured at lower cost.

Surface mounting is becoming well accepted by the proponents of both high-density hybrid construction on ceramic printed circuit boards and standard printed wiring board construction for a variety of reasons. Producers of hybrid circuits find that the surface-mounted chip carriers improve component handling, component testing, and production yield, thereby lowering cost. Builders of printed wiring boards and assemblies find that surface-mounted components, which are smaller than their counterparts, provide smaller and lighter assemblies with fewer copper wiring layers. Size and weight reduction of ceramic printed circuit or organic wiring boards are important factors in improving military, avionics, and space electronics, as well as medical devices—for example, implanted pacemakers and insulin pumps.

The surface-mount technology also reduces interconnecting lengths between circuit components, a requirement for increased circuit speed. Surface-mounted technology leads to reduced assembly costs by offering a way to attach components to printed

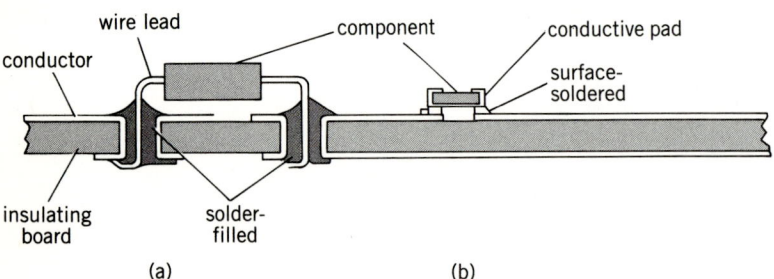

Fig. 1. Comparison of attachment of electronic components by (a) through-hole mounting and interconnecting of components and (b) surface mounting.

circuits without having to straighten component leads and without having to insert through holes in the printed circuit board and to cut, cinch, and solder leads. Instead, the components can be placed on a film of printed sticky solder paste by a robot or a pick-and-place machine, then soldered by heating the assembly. This simpler, more automated assembly technique and smaller-sized assembly lead to lower costs, better electrical performance, and easier repair as well as reduction in size and weight.

Early packages. The electronic components on most printed circuit boards are attached to conductor paths by wires. Each wire is fed through a hole in the printed circuit board, cut, clinched over, and then soldered to hold the component in place and to make the electrical connection to the conductive path on the printed circuit board (Fig. 1). The so-called through-the-board method of mechanical attachment is a carry-over from technology prevalent before the advent of printed circuits, when it was necessary to wrap the conductor lead around a metal terminal and connect a wire from one terminal to the next.

Most components were designed and produced at a time when terminals were used to mechanically and electrically hold the component and its interconnecting wires. Components such as resistors and capacitors were encased in plastic, and tinned copper wires extended from the cases to make the connections. Diodes were encased in glass tubes that were hermetically sealed to low-expansion-alloy wires. Generally, transistors were individually molded in plastic or hermetically sealed in metal cases, and the wires were either wrapped around the terminals or, for through-hole connections, attached by solder to the conductive paths on a printed circuit board.

These components with wire connections, originally made for connection to terminals, were used with the introduction of the integrated circuit. When more input-output terminals became necessary, the dual–in-line package became popular. It was standardized to have a row of metal terminals along each of two sides of the package; the terminals were spaced 0.1 in. (2.5 mm) apart in a row, with a minimum of 0.3 in. (7.5 mm) space between rows. The integrated circuit industry has been doubling the number of interconnected components on a monolithic chip each year since 1959, and was producing chips with up to a million components in 1984. The vast increase in components interconnected on a chip required an increase in the number of input-output terminals, albeit at a lesser rate; even so, the dual–in-line package outgrew an efficient size for packaging integrated circuits.

Hybrids. In the meantime, the design and production of very small electronic circuitry was developed by the hybrid industry. In hybrid technology, uncased active components such as diodes, transistors, and integrated circuit chips were mounted directly to the surface of a thin (sputtered) film or a thick (fired) film of conductive metal on a ceramic printed circuit board. Passive components, such as resistors or capacitors, were either formed directly on the substrate or fabricated on miniature ceramic chips, then surface-mounted. The resultant hybrids were about one-tenth the size of either the similar electronic assemblies having cased passive components, or the dual–in-line active components mounted and interconnected on a printed wiring board or on a terminal board with wire-wrapped or welded-wire connections.

With the advent of more complex integrated circuits, however, it became more difficult to completely test the unpackaged chips, and the manufacturing yield of hybrids dropped to unacceptable levels simply due to the statistics of yield. For instance, if 10 components, each with an individual yield of 90%, are assembled, the resultant assembly will have a yield of 0.9^{10}. This means that 35% or only one in three assemblies will be satisfactory without repair. On the other hand, if the components have a 99.9% yield, then the resultant assembly will have a yield of 0.999^{10}, which equals 99% yield or 99 satisfactory assemblies out of 100.

Chip carrier. Since packaged integrated circuits are much easier to handle and test without damage, the hybrid industry looked for a small package. The chip carrier package became available in the early 1970s. This package was standardized by industry in several forms. The basic package is rectangular with input-output connections spaced 0.05 in. (1.25 mm) apart in one family and 0.04 in. (1 mm) apart in another family, on all four sides of the package. The chip carrier package is about one-third the area of the dual–in-line package with the same number of terminals.

The 0.05-in. (1.25-mm) centered package has six variations (Fig. 2), all of which have the same mounting and electrical connection pattern onto the mating surface of the printed circuit board site for surface mounting. The leadless types A, B, C, and D are typically ceramic packages with hermetically sealed metal or ceramic lids. When clip leads are added to the leadless types, the chip carriers become the leaded type B package, which is generally a plastic package.

The chip carrier is one of many surface-mounted packages adapted for active or passive devices. Other popular packages include the small-outline

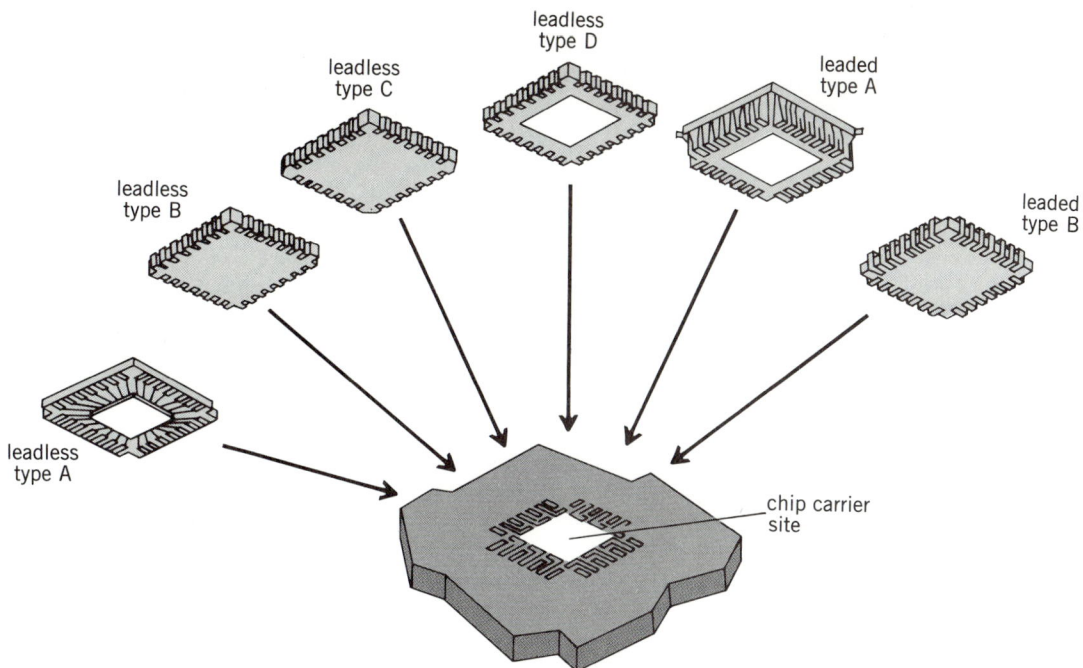

Fig. 2. A family of chip carriers showing surface-mounting compatibility. (*After Institute for Interconnecting and Packaging Electronic Circuits, Guidelines for Surface Mounting and Interconnecting Chip Carriers, IPC-CM-78, November 1983*)

(SO) transistor, or integrated-circuit leaded package, and the four-sided leaded flat pack (QUAD). Devices available in surface-mounted form include resistors, capacitors, inductors, potentiometers, and crystals, with many others becoming available.

The new generation of surface-mounted components are much smaller than their older leaded counterparts. Even so, the continuous increase in the number of components on a chip with resultant increase in input-output connections has already generated the need for even smaller and more efficient chip carriers. The current trend is toward carriers with spacings of 0.02 and 0.025 in. (0.5 and 0.6 mm) and increasing numbers of input-output connections (Fig. 3). There are, however, problems to be solved. The leadless chip carrier package can easily be soldered onto a printed circuit board with high reliability in adverse environmental conditions if the printed circuit board matches the temperature coefficient of expansion of the surface-mounted chip carrier. In temperature-extreme environmental cycling conditions, the coefficient of expansion of a ceramic circuit board is matched to the ceramic chip carrier. The standard glass/epoxy/insulated-copper-conductor printed wiring board has a higher temperature coefficient of expansion than ceramic-cased components, and eventually the solder joint fails from metal fatigue caused by the different expansion characteristics.

Two generic solutions are currently being tested to improve organic-insulated printed wiring boards. One solution involves the creation of a printed wiring board having the same temperature coefficient of expansion as the ceramic chip carrier, that is, 3.3×10^6 ppm/°F (6×10^6 ppm/°C). Among the many materials being tested are composites of insulating organic materials, such as epoxy or polyimide, that have reinforcement with materials of low temperature coefficients of expansion. Many reinforcing materials are being tested such as quartz, carbon, aramid fibers, or composite metal planes such as copper-Invar-copper insulated with appropriate resins. With the proper quantities of each material, a coefficient of expansion can be produced that matches that of the ceramic chip carriers, ceramic resistors,

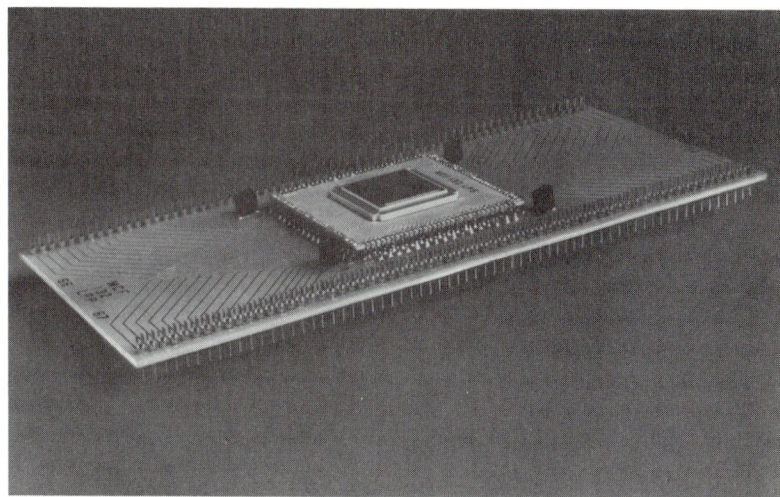

Fig. 3. Size reduction by the 132-connection, 0.025-in.-spaced (0.6-mm) chip carrier mounted on a test board with 132 leads in dual-in-line format.

or capacitor chips. An alternative solution, especially for large chips that heat up by internally generated power, is to mount the chip by compliant leads to the printed wiring board. In this case, when the component expands and contracts because of heat generated from turning the component on and off, or as the different materials stretch or bend, the lead can flex without breaking or affecting the electrical connection.

Deriving the solutions to these problems is one of the greatest challenges facing electronic packaging engineers because the fast-moving solid-state industry is moving simultaneously toward more components on a chip, higher numbers of inputs and outputs, and higher-speed circuits. The systems engineers are generating more complex systems because consumers want smaller size and weight, higher reliability, and lower-cost products that will survive under the severe conditions imposed by avionics, outer space, automotive engine compartments, and human implantation.

[JOHN A. BAUER]

Elastomeric connectors. Elastomeric connectors are electronic connectors composed of conductive silicone rubber or of metal lines supported by insulating silicone rubber. The rubber is made conductive by incorporating carbon, silver, nickel, or other conductive particles in the silicone matrix. The amounts, sizes, and shapes of these particles are chosen to give a balance of electroconductivity, physical properties, and ease of fabrication.

Types. In today's elastomeric connectors, conductive layers are alternated with nonconductive layers to produce connectors in strips of various sizes. Typically, these layered elastomeric connectors have 250 conductive layers per inch (100 per centimeter).

In the other major type of elastomeric connector, metal traces are positioned on a nonconductive silicone rubber core to produce the metal-elastomeric connector. Typically, this is copper-nickel-gold laid down on a polyester or polymide film that is wrapped around and sometimes adhered to the silicone core.

Another version of this is the metal-on-elastomer connector, which utilizes all-silicone construction under the metal lines, usually having a very thin scrim cloth in the outer layer to maintain thermal and mechanical stability. These metal patterns can be as fine as 0.001-in. (0.025-mm) lines and 0.001-in. (0.025-mm) spaces, or 500 connections per inch (200 per centimeter), with an accuracy of 0.1% if necessary.

A third type of elastomeric connector exists. It technically is not a connector, but a pressure-sensitive pad that acts as a switch. A composition of conductive particles and elastomeric can be produced where the mixture is nonconductive and insulating until local pressure is applied, at which time the particles are squeezed closer together and it becomes conductive with a resistance of approximately 1 ohm.

Composition. Silicone rubber is predominantly used because of its flexibility at temperatures of $-67°F$ ($-55°C$) or lower and long-term stability at $347°F$ ($175°C$) with the ability to withstand $572°F$ ($300°C$) for short periods without degradation. It does not degrade in most chemicals, ultraviolet radiation, or ozone, oxygen, or other gases normally present in the atmosphere, including moisture. The silicone rubber used maintains its dielectric constant of 2.7–3.0 and a very low dielectric loss over a wide range of temperatures and frequencies. It has a volume resistivity greater than 10^{12} ohm-cm even when saturated with water.

Uses. The carbon-filled layered elastomeric connectors have become the connector of choice for liquid crystal displays. This application began in 1973 with digital watches. These connectors are also used with liquid crystal matrix displays in portable computer terminals, in automobiles, calculators, clocks, airplane cockpits, gasoline pumps, torque wrenches, counters, digital voltmeters, and in an increasing variety of electronic equipment.

The carbon-filled layered elastomeric connectors have a resistance between 500 and 5000 ohms, so their use is limited to high-impedance circuits like those in liquid crystal displays.

For applications that have low-impedance requirements, silver-filled layered elastomeric connectors were developed. These are produced in the same way as carbon-filled layered elastomeric connectors and have the same pitch. But their resistance typically is 0.1 to 5 ohms, depending on the height and width of the connector itself and the electronic pads to which it contacts.

Fig. 4. Silver-filled layered elastomeric connector being used to connect two printed wiring boards.

These silver-filled layered elastomeric connectors are being widely used to connect displays such as electroluminescent, electrophoretic, gaseous-discharge, and vacuum-fluorescent. They are also being used in large quantities to interconnect printed wiring boards (Fig. 4) and chip carriers to printed wire boards. Because of their versatility and ease of use, these silver-filled layered elastomeric connectors are used in test and burn-in fixtures for electronic components.

The metal-elastomer and metal-on-elastomer connectors have even lower resistances than the silver-filled layered elastomeric connectors in the range of 0.020–0.200 ohm. These connectors are used where minimum resistance is necessary: connection of printed wiring boards, chip carriers, displays, flat-screen television, medical implants, outer lead bonding, and bumped chips. They also have found acceptance as switch elements where as many as 24 simultaneous contacts are made.

In most elastomeric applications, connections are made between the elements to be connected by clamping the mating surfaces to the compliant conductive elastomeric elements. One of the major design constraints in using elastomeric connectors is the necessity for maintaining the elements in a deflected configuration. As shown in Fig. 5, when held in constant strain, silicone rubber loses only 50% of the original stress in 50 years at 77°F (25°C). Long-term compression set or stress relaxation that is found in all materials is not a significant problem.

In some electronic equipment, housings have been designed so that the pressure is applied when the unit is assembled.

Advantages. The popularity of elastomeric connectors stems from their versatility, shock absorbance, inertness to the environment, ease of custom fabrication, gas-tight seals at contact points, redundancy, and use as a zero insertion force connector.

The metal-elastomer and metal-on-elastomer types can be constructed with such accuracy that their electrical properties, such as capacitance and inductance, can be tailored to precise predetermined characteristics. Very short metal-path lengths allow their use in the gigahertz range without producing signal distortion.

Extensive testing in both field and laboratory applications shows that carbon and silver layered elastomeric connectors, and meta-elastomer and metal-on-elastomer types all exclude atmospheric contaminants and are totally unaffected by long exposure to high concentrations of SO_2, NO_2, H_2S, and similar corrosive gases at high humidities and temperatures.

Other advantages in using elastomeric conductors are their ease of mounting and dismounting—nothing to solder or unsolder—shock and vibration resistance, and their ability to take up or make up for differences in tolerances.

Another useful attribute is redundancy. Because the connector layers can be made so thin, 0.001 to 0.002 in. (0.025 to 0.050 mm), contacts can be made to pads 0.010 in (0.25 mm) and below. As a result of this range of 250–500 connections per inch (100–200 per centimeter), there are always two or more conductive paths connecting opposing pads, and two or more insulating layers between adjacent pads. Where the contact pads are on centers less than 0.010 in. (0.25 mm) or minimum resistance is required, the metal traces on the elastomer can be placed in one-to-one correspondence with the substrate pads. This alignment of connector and circuit pad is also required when a unit is made to a matched impedance.

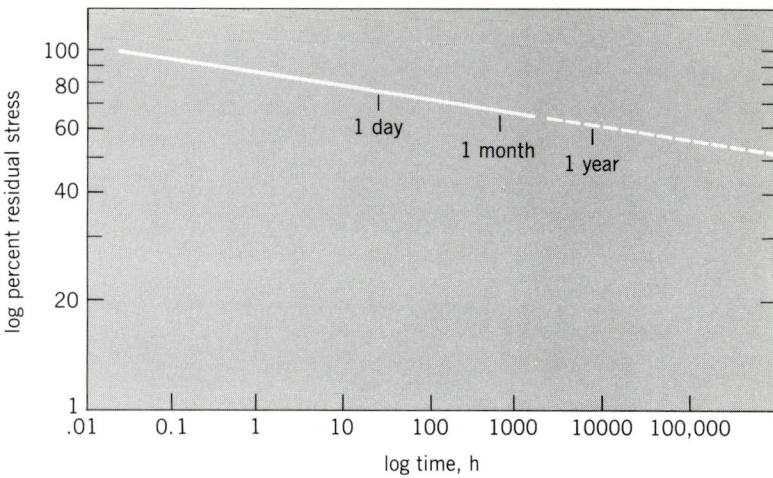

Fig. 5. Retained stress as a function of log time of a layered elastomeric connector held at constant strain.

The connective density, redundancy, and environmental seal attributes promise a bright future for elastomerics as electronic connections become denser and the devices are down-sized. Elastomerics come into their own as connections densities rise to 20 per inch (8 per centimeter) and higher.

Future applications. As a result, future uses of elastomeric connectors will include applications where the high density of connections, unsolderable substrates, very hostile environments, or the need to mount and dismount the connector periodically will make them the unit of choice.

There have been several experimental and prototype applications that should find wide use in the near future. These include use as battery contacts, contacts to thermal printheads, sensors in automobiles and other equipment, large flat-screen television, medical implants (there has been an extensive study showing that if properly designed, elastomeric connectors can form a matable contact that is impervious to long-term immersion in salt water and body fluids even while conducting current).

Future applications also include contact to the outer leads of tape-automatic-bonded integrated circuit chips in a test fixture or as permanent connections to a circuit, and removable connection between integrated circuit chips and substrates. Also, connections to leadless-pad grid arrays have been made in prototype quantities.

Because the nature of elastomeric connectors is

inherently zero insertion force, they are finding increasing use in surface-mounting applications.

Several investigations involve elastomeric connectors soldered to ceramic substrates such as a chip carriers and to printed wire boards in surface-mounted configurations. This eliminates the problems normally found when materials of different thermal expansion are rigidly connected together.

For background information see CIRCUIT (ELECTRONICS); ELECTRONIC DISPLAY; INTEGRATED CIRCUITS; LIQUID CRYSTALS; POLYMER; SILICONE RESINS in the McGraw-Hill Encyclopedia of Science and Technology.

[LEONARD S. BUCHOFF]

Electrostatics

When liquids are sprayed using mechanical forces such as hydraulic, pneumatic, or centrifugal forces, the spray droplets are usually slightly charged. Electrostatic forces may be used either alone or in conjunction with mechanical forces to generate sprays so highly charged that the electrostatic forces act upon droplets to significantly affect trajectories and influence depositional patterns. Several very important industrial processes have been developed which rely upon the electrical properties of highly charged sprays.

There have been many developments involving modification of various types of spray nozzles so that electrostatic charging of droplets occurs. These nozzles have found uses in many industries. In crop spraying, water-based and oil-based pesticides can be sprayed and charged, and efforts have been made to control droplet size distributions to improve droplet depositional characteristics. Crop spraying has become so efficient that ultralow volumes of chemical per unit area can now be used, ultimately reducing environmental pollution. In addition, techniques have been developed that enable liquids which are extremely good electrical insulators to be sprayed by using charge-injection and electrostatic forces alone. At the other end of the spectrum, ion beams and submicrometer-sized droplets of liquid-metal sprays may be produced and used for special surface treatment of certain materials and in the fabrication of microelectronic devices.

Paint spraying. In the paint-spraying industry, electrified spinning disks, cups, or bells are used extensively. Paint, fed to the center of a disk or the bottom of a cup, is driven by centrifugal force to the spinning edge where it is dispersed. Rotational speeds of 20,000 revolutions per minute and above are used; the droplets are sometimes impelled by air blowing toward the workpieces. By controlling rotational speed and the flow rate and rheological characteristics of the paint, an array of ligaments can form along the edge of the spinning nozzle and each ligament will break up into droplets of identical size (Fig. 1). The spray has an extremely narrow size distribution; centrifugal forces usually dominate the dispersion process. Aqueous or oil-based liquids may be dispersed and charged by rotary nozzle systems. Viscous paints which have high solids content and utilize very little solvent may be successfully dispersed to achieve uniform film thickness of about 20 micrometers. Spinning-nozzle paint sprayers are often programmed to move on preset paths with respect to workpieces so that optimum results are obtained.

Agricultural spraying. The agricultural sprayer developed by S. E. Law is a hybrid system in which an air blast atomizes water-based pesticide and a charging electrode charges the droplets. An annular airstream passes a ring electrode which is positioned a fraction of an inch (a few millimeters) away from the nozzle orifices. The airflow keeps the electrode dry so that liquid short-circuit paths do not develop between the electrode and the grounded parts of the nozzle. Sprays are polydisperse (droplets have a spread in size) with a volume mean diameter in the range 30–50 μm and with charge-to-mass ratios of about 10^{-3} coulomb/kg at liquid flow rates of approximately 0.03 fl oz/s (1 ml/s). The Law nozzle requires a high-pressure air supply and is thus most suitable for tractor-mounted operations. The nozzle operates with no significant electric field between itself and ground and relies on airflows to direct charged droplets toward the crop and thence on space-charge and image-charge depositional forces. Field and laboratory tests with the Law system have shown that spray deposition onto some crops, such as cabbage, may be increased sevenfold in comparison with uncharged spray.

An example of a hand-held sprayer which uses electrostatic forces to both atomize and charge an oil-based pesticide is the Electrodyn. Torch batteries within the handle of the unit drive a power supply which maintains an annular slit nozzle at a potential of about 20 kV. The pesticide spray has a very narrow size distribution. The Electrodyn is claimed to be so effective that only ultralow volumes of chemical are required, 7 fl oz/acre (500 ml/hectare) or even less being satisfactory. An important advantage of an ultralow volume system is that refills are very infrequent. A disadvantage of the Elec-

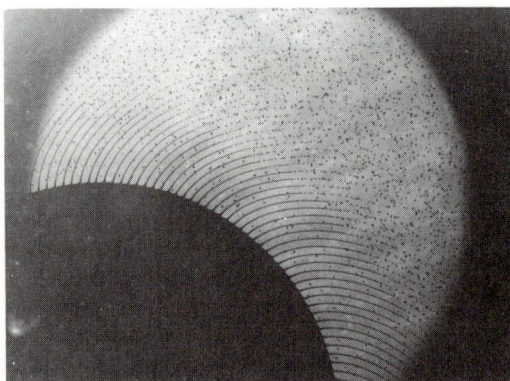

Fig. 1. High-speed flash photograph of ligament-mode liquid dispersion from the edge of a spinning disk. (*From A. G. Bailey, Electrostatic spraying of liquids, Phys. Bull., 35:146–148, 1984*)

trodyn is that only oil-based liquids can be satisfactorily sprayed. It is questionable whether droplets with a very narrow size distribution will distribute over all surfaces of a complex foliar target.

A novel method of directing pesticide to a target in a selective manner has been developed. Two side-by-side airfoil-shaped linear nozzles are used, and liquid dispersion is by air blast (Fig. 2). Induction charging electrodes close to the nozzles are energized by a 6-kV direct-current supply. One nozzle is supplied with pesticide and the other with water so that dual spray clouds are produced. Space-charge forces due to the charged water droplets tend to direct charged insecticide droplets toward the target and reduce losses due to drift away from the target. In field tests an insect repellent aerosol was generated as a low-level cloud with the water aerosol cloud above it. Cattle beneath the dual clouds were subjected to a substantial deposition increase relative to more conventional uncharged spraying, and a reduction in environmental contamination was inferred. The dual nozzle was also tested in enclosed spaces for insect control. It was shown that with a lower cloud of water aerosol the upper insecticide aerosol penetrated upper regions of a barn with satisfactory deposition in out-of-reach areas.

Liquid-metal spraying. Liquid metals may be sprayed by using slit nozzles or capillary tubes. Microscopic and submicroscopic droplets may be produced, and under some conditions metallic ions may be directly pulled from liquid-metal surfaces by electrostatic disruptive forces. Metallic ion and droplet sources have considerable potential for industrial applications such as ion implantation in the semiconductor field and for thin-film production during the manufacture of hybrid microcircuits. The treatment of many surfaces by metallic ion or droplet beams can enhance corrosion-resistance properties.

A field-emission deposition process is under development in the United Kingdom, and spray tests with copper, silver, silicon, aluminum, gold and various alloys have been carried out. A gold sprayer uses a capillary tube with a centrally positioned pointed wire which becomes wetted by the molten gold (Fig. 3). Extractor electrodes operated at a few thousand volts set up an electric field at the gold-wetted point and disrupt the surface to generate a spray of gold droplets which solidify upon impact onto a target. Field-emission deposition coatings adhere exceptionally well to surfaces, and the spraying technique shows great promise for the microelectronics industry. Electrostatic focusing and deflection electrodes should allow maskless pattern generation to be developed.

Liquid-insulator spraying. For a liquid to be satisfactorily sprayed by electrostatic forces, free surface charge must build up on the liquid surface. As insulating liquids are relatively free of charge, charge-injection techniques have been developed. The simplest charge-injection sprayer incorporates an electrode with sharp points or edges within the

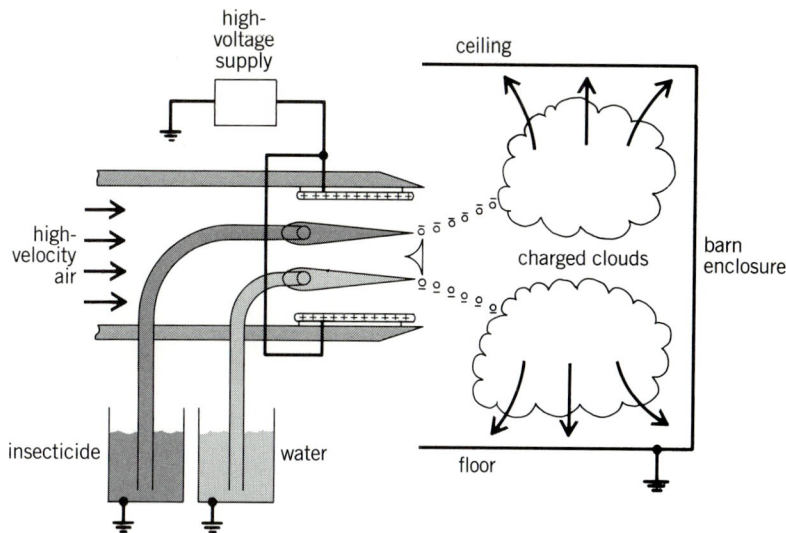

Fig. 2. Airfoil nozzle for producing dual sprays. (*After I. I. Inculet, K. J. Hodgson, and J. G. Milward, Cross current dual air-foil electrostatic spray nozzle, IEEE–Industry Applications Society, Conference Record, Mexico City, pp. 1050–1054, 1983*)

liquid-feed system. By raising the potential of the electrode by several thousand volts, the high field set up at the sharp surfaces leads to the generation of free charge. Thoriated tungsten emitters have been used to enable silicone oil, Freon, and liquid hydrogen to be sprayed. Hydrogen isotope spraying opens up interesting possibilities of producing microparticles for controlled thermonuclear fusion.

Insulating hydrocarbon fuels may be dispersed by using charge-injection methods. Hydrocarbons rang-

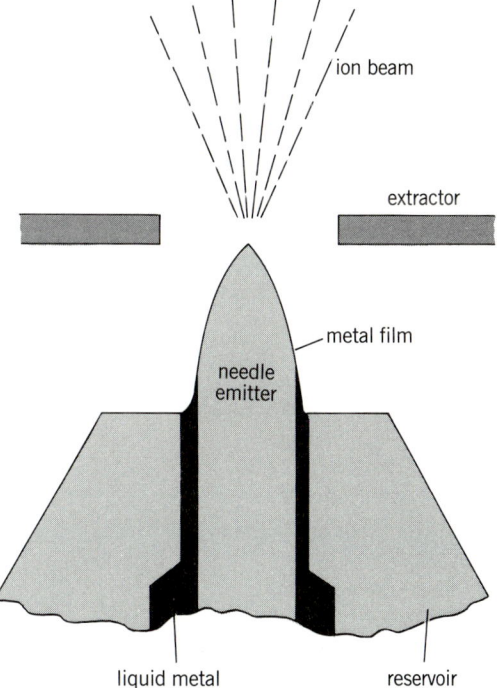

Fig. 3. A liquid-metal field-emission ion source. (*After P. D. Prewett and D. K. Jeffries, Liquid metal field-emission ion source and their applications, Inst. Phys. Conf. Ser. 54, pp. 316–321, 1980*)

ing from the nonviscous Jet A fuel to a rubber extender oil of viscosity 80 centipoise (0.08 Pa · s) have been successfully sprayed. Higher-viscosity insulating oils are more difficult to disperse with electrostatic forces alone. An electrically charged spray of fuel oil may be manipulated by electrostatic fields so that electrostatic flame control is possible. Electrostatic spray nozzles also have the important advantage of being free of moving parts.

Other applications. An important and well-developed application of electrostatics is the ink-jet printer, which utilizes a charged jet of ink broken up into charged drops which are deflected to a paper target. Another device that utilizes a single jet of liquid is a microsprayer, developed for lubrication purposes. Lubricants are sprayed onto workpieces during machining and other workshop operations. Lubricant is applied only where it is required, and there is virtually no environmental pollution problem from sprayers of this type. Work has been carried out recently in the United States in which copious quantities of charged water spray have been injected into dust-laden atmospheres to scavenge out the dust. Highly promising results have been reported.

For background information *see* ATOMIZATION; ELECTROCHEMICAL PROCESS; ELECTROSTATICS; ION IMPLANTATION; METAL CASTING; PAINT in the McGraw-Hill Encyclopedia of Science and Technology.

[A. G. BAILEY]

Bibliography: A. G. Bailey, Electrostatic spraying of liquids, *Phys. Bull.*, 35:146–148, 1984; IEEE–Industry Applications Society, *Conference Record*, Mexico City, 1903, and Chicago, 1984; S. Singh (ed.), *Electrostatics 1983*, Inst. Phys. (Lond.) Conf. Ser. 66, 1983.

Elementary particle

During the past two decades understanding of the fundamental structure of matter has increased profoundly. The picture that has emerged from an extraordinary series of experimental discoveries and theoretical advances has great beauty and simplicity. This picture is described, in large part, by the so-called standard model, which is discussed in the first part of this article. The second part of the article discusses technicolor forces, which were hypothesized to explain a poorly understood aspect of the standard model, the breakdown of symmetry in the unified electroweak interaction.

STANDARD MODEL

The constituents of all matter are quarks and leptons, pointlike particles of spin $\hbar/2$ (where $\hbar$ is Planck's constant divided by 2π). If the very weak gravitational interaction is ignored, the forces among these particles seem to be of three types—weak, electromagnetic, and strong. These forces were once thought to be unrelated, but according to present understanding all three arise from the exchange of vector (spin-$\hbar$) particles, called gauge bosons. Furthermore, the electromagnetic and weak interactions are not separate phenomena. They are described by the unified electroweak theory of S. Glashow, S. Weinberg, and A. Salam. This theory predicted the existence and properties of the weak neutral currents discovered at the European Center for Nuclear Research (CERN) and Fermilab in the early 1970s and, more dramatically, of the very massive charged $W^{\pm}$ and neutral Z^0 gauge bosons that are partners of the familiar, massless photon of electromagnetism. The recent discovery at the CERN proton-antiproton collider of these heavy bosons, with masses on the order of 80 to 100 GeV, is simultaneously a tremendous experimental achievement and a satisfying confirmation of the basic correctness of the now-standard unified electroweak model.

This unified electroweak theory and the theory of strong interactions, called quantum chromodynamics (QCD) because the names of colors are used to label the charges of the theory, form the now widely accepted standard model of particles and their interactions.

Quarks. In the 1950s and 1960s the profusion of strongly interacting particles (called hadrons) led to the suggestion by M. Gell-Mann and G. Zweig that the bewildering number of these particles could be understood simply if the particles were built from three basic constituents, generically called quarks, and distinguished by the labels u, d, s. The discovery at Brookhaven National Laboratory of the particle called the omega-minus, consisting of three s quarks, was an impressive confirmation of the basic idea. Subsequently, pioneering experiments at the Stanford Linear Accelerator Center (SLAC) in the late 1960s on deep inelastic scattering of electrons colliding with protons indicated that protons (and neutrons) contained pointlike charged constituents. There were thus two distinct reasons for believing in the composite nature of hadrons: one as an explanation of the patterns of hadron types and masses, and one related to their internal structure as probed by high-energy collisions.

Quantum chromodynamics. While theorists struggled with the complexity of the strong interactions, numerous experiments, with neutrinos as well as electrons, showed that when probed over short distances, that is, by high-energy collisions, the quarks inside hadrons behaved as if they were nearly free particles. In the early 1970s quantum chromodynamics was proposed as the theory of strong interactions. It attributes the strong forces among quarks to the exchange of massless vector gauge bosons called gluons, analogous to photon exchange in quantum electrodynamics.

According to quantum chromodynamics, quarks possess the remarkable property of asymptotic freedom, which means that the overall strength of the coupling $a(r)$ between quarks becomes weaker as the available energy increases, or equivalently as the distance r between colliding particles becomes smaller. The potential between quarks is given by $a(r)/r$, and the weakening of the coupling strength as particles come closer together is a quantum-mechanical effect. The corresponding quantum-mechanical effect in electromagnetism is known as vacuum polarization and varies with distance in the opposite way.

In quantum chromodynamics it is further believed that the force between quarks colliding at low energies or widely separated in space is so strong that two quarks can never be completely separated. Quantum chromodynamics thus permits a qualitative understanding of the almost-free quarks confined inside hadrons, on the one hand, and possibly of the absence of free, isolated quarks on the other. It is a rich and complicated theory from which it is difficult to extract precise predictions, but there is increasing evidence for its basic validity.

Successes of the model. Over the past 10 years, particularly following the dramatic discovery at SLAC and Brookhaven of the J/φ particle and its related states, and more recently of the upsilon at Fermilab and its related states at the Cornell Electron Storage Ring (CESR), the reality of quarks as the fundamental constituents of hadronic matter has been established beyond doubt. The numerous massive psi and upsilon particles imply the existence of additional quarks, called c and b. The discovery of the tau lepton at SLAC in 1975 added to the lepton class of particles. And finally, the pattern of electroweak interactions has led to the grouping of quarks and leptons into families or generations: the u and d quarks, plus the electron and its neutrino, form the first family; the c and s quarks, plus the muon and its neutrino, form the next; the b quark and the t quark, plus the tau lepton and its suspected neutrino, form a third.

The electroweak theory has been thoroughly tested at low energies, and with the discovery of the $W^\pm$ and Z^0 vector bosons it becomes ever more firmly established. Meanwhile, quantum chromodynamics continues to provide the best theoretical description of the strong interactions. Even though quantitative tests have proved to be elusive, there are many observed phenomena that can be easily understood only in the framework of quantum chromodynamics. An example is the clustering of hadrons produced at high energies into highly collimated jets, which provided the first direct evidence for gluons at the electron-positron collider PETRA in Hamburg, Germany.

Possible extensions of the model. The standard model (the unified electroweak theory, plus quantum chromodynamics) provides a wide-ranging synthesis, but it cannot be the final answer. It does not predict the number of families of quarks and leptons, or their masses or mass ratios. Also not determined is the exact relation between the strengths of the weak and electromagnetic interactions, specified by an empirical angle θ_w. Further, many features of the weak currents are left unconstrained by the theory. Finally and perhaps most importantly, an essential but as yet poorly understood aspect of the theory is the mechanism that causes the W and Z gauge bosons to acquire large masses while leaving the photon massless.

Higgs symmetry-breaking mechanism. This last puzzle—the breakdown of the symmetry called gauge invariance, known as the Higgs phenomenon—presents an especially important and exciting challenge. There are general arguments based on theoretical consistency that the mass, or energy, scale where the Higgs symmetry-breaking mechanism must begin to reveal itself can be no more than a few TeV. Interest is heightened by the fact that almost nothing is known about what will be found there. The Higgs sector has traditionally been described by a set of scalar (spinless) fields with interactions arranged to break gauge invariance and to generate the W and Z gauge-boson masses as required by experiment. However, there are reasons to regard this device with some skepticism and to anticipate that the actual underlying physics will be rather different.

Of course, the Higgs phenomenon could occur at a mass scale considerably less than 1 TeV, and such a light Higgs particle should certainly be searched for. However, the theory as conventionally formulated suggests that such a low mass scale is unstable against quantum corrections. Specific models that address this problem do indeed conclude that a few TeV is the natural mass scale of the Higgs sector.

Grand unification. Whereas the strong interactions become weaker at higher energies, the electroweak interactions become stronger. It is thus natural to ask at what energy, if any, these very different forces become equal and, therefore, perhaps appear as parts of a single unified theory, with quarks and leptons on an equal footing. It is remarkable that such an equality turns out to be possible, but only at extraordinary energies of order 10^{14} GeV, beyond the reach of imaginable collision energies in a laboratory.

The grand unification of quarks and leptons implies the existence of very massive gauge bosons, causing transitions between those two classes of particles, in addition to the known gauge bosons of the standard model. One beautiful consequence of this enlarged theory is the understanding of the exact equality of the magnitudes of the electronic and protonic electrical charges. Another is the prediction of proton decay at an exceedingly small but possibly measurable rate. Other consequences of grand unification are the successful prediction of the weak angle θ_w, and the possible existence of nonzero masses for neutrinos and of very massive (10^{16}-GeV) magnetic monopoles.

Supersymmetry. Another area of theoretical speculation, beyond grand unification, is supersymmetry. In the world as it is understood at present, particles of integral spin appear disjoint from those with half-integral spin. But the procedure of putting quarks and leptons, very different in their interactions, though all spin ½, on an equal footing through grand unification, suggests the possibility of going one step further and unifying bosons (integral-spin particles) and fermions (half-integral-spin particles). This is the aim of supersymmetric theories, in which every particle has a supersymmetric partner, a sparticle, of identical properties except for the opposite kind of spin. Since no identifiable supersymmetric pairs have been seen so far, the symmetry must be broken. Sparticle partners, if they exist, must have much greater masses than the known particles. Bizarre as supersymmetry may seem, there are enough attractive theoretical reasons, most compelling the possibility of incorporating gravity into a unified description of all elementary forces, to keep the idea alive and to continue the experimental search for its manifestations.

In the past two decades, then, there have been encouraging signs that physicists are on the right track. Nonetheless, there are many unanswered questions, including: whether there are more quarks and leptons; whether there are other, still heavier, gauge bosons; what is the nature of the Higgs sector; whether neutrinos are massless, and if so, why; how particles acquire mass; and whether there are supersymmetric partners of the known particles.

[THOMAS W. APPELQUIST]

TECHNICOLOR FORCES

Technicolor is the popular name given to a hypothetical fifth force proposed to explain the origin of the breakdown of symmetry in the unified electroweak interaction. The technicolor interaction is based on a gauge theory, and it is much like the ordinary color force that is responsible for binding quarks, antiquarks, and gluons into the strongly interacting particles called hadrons. The principal difference is that the characteristic energy scale of technicolor is about 500 GeV, or 2500 times larger than the energy scale of ordinary color. If the technicolor proposal is correct, many new particles, called technipions, will be found in experiments to be carried out over the next 10 years or so.

Masses of gauge bosons. As discussed above, according to the standard model, the strong, electromagnetic, and weak forces are described by gauge theories in which elementary, spin-$\frac{1}{2}$ fermions (the quarks and leptons) possess a generalized charge and interact through the exchange of spin-1 gauge bosons which couple to this charge. The most familiar example of this construct is electromagnetism whose gauge boson, the massless photon, couples to electrically charged quarks and leptons. It is the photon's masslessness that is responsible for the infinite range of electromagnetic interactions, that is, the inverse-square-law force found by C. A. Coulomb. The gluons, the gauge bosons of the strong interactions between quarks, are also massless. However, their peculiar gauge properties result in the permanent confinement of quarks (and gluons themselves) inside hadrons. Thus, a single quark, antiquark, or gluon can never be isolated as can an individual electron or a colorless hadron such as the proton. The typical distance over which quarks are confined, and hence the range of strong interactions, is approximately 10^{-15} m. According to Heisenberg's uncertainty principle, this distance corresponds to a characteristic strong-interaction energy scale of $hc/(2\pi \times 10^{-15}$ m$) = 200$ MeV. Here, h is Planck's constant and c is the speed of light. Unlike all the other gauge bosons, the $W^\pm$ and Z^0, the gauge bosons of the weak interaction, are very heavy, with masses of 82 GeV and 93 GeV, respectively. These large masses are responsible for the most obvious features of the weak interaction, namely that it is weak and has a range of about 10^{-18} m, 1000 times shorter than the strong interaction. This range corresponds to a characteristic weak-interaction energy scale of about 250 GeV. This is a very high energy indeed, and one that will be important in the discussion of technicolor.

The key ingredient in the unification of the electromagnetic and weak forces into a single electroweak interaction is symmetry. The theory is based on the gauge symmetry group SU(2) $\times$ U(1). This group requires that there be four gauge bosons—the photon, W^+, W^-, and Z^0. Furthermore, this group, together with the threefold color charge of quarks, implies that quarks and leptons belong to electroweak families. These are the fermion doublets, now repeated in three generations. If this electroweak symmetry were exactly respected by nature, the photon, $W^\pm$, and Z^0 would have the same mass, precisely zero, and the weak and electromagnetic interactions would have approximately the same strength and be of infinite range (or, possibly, partially confining). It is apparent, therefore, that electroweak symmetry is broken. The central unresolved problem raised by the standard model is to find the basic dynamical mechanism responsible for this symmetry breaking and the consequent gross difference between the mass of the photon and that of the weak gauge bosons.

Symmetry breaking mechanisms. In their model of electroweak symmetry breaking, Weinberg and Salam sidestepped this issue by the stratagem of introducing a set of four hypothetical fundamental spin-0 particles, usually called Higgs bosons (after P. W. Higgs, who introduced them in the context of gauge theories in 1964). The Higgs bosons can couple through the electroweak interaction to the W and Z and, through this coupling, they may give mass to the weak bosons. The Higgs bosons do not couple to the photon, and so it is left massless. What is bothersome about this scenario for generating W and Z masses is that it is based on arbitrary assumptions about the nature of Higgs bosons and their self-interactions. In fact, an equally plausible assumption about Higgs boson dynamics would leave the electroweak symmetry intact, with the W, Z, and photon all massless. This freedom to choose arbitrarily the interaction of Higgs bosons may be traced directly to the assumption that they are fundamental particles, composed of nothing smaller and more elementary.

The clue to what might be the underlying dynamics can be found in the observation that electroweak symmetry still gets broken, even if elementary Higgs bosons are not included in the theory. The agency for this breaking is the ordinary color interaction of quarks. This interaction binds quarks into composite spinless mesons, the familiar pions. Then, because of the electroweak doublet structure of quarks, the pions act as Higgs bosons with just the right couplings to give mass to the W and Z while leaving the photon massless. The problem with this natural and economical picture is that the strong-interaction energy scale is only 200 MeV and this would lead to W and Z masses 2500 times smaller than observed.

The way out is clear, and was proposed independently by Weinberg and L. Susskind in 1979. They assumed the existence of a new strong gauge interaction, now called technicolor, with a characteristic energy scale of approximately 500 GeV. Its fundamental constituents are spin-½ technifermions, interacting through the exchange of gauge technigluons. As are the ordinary quarks, technifermions and their antiparticles are permanently confined by their strong interaction into bound states, known as technihadrons. The large technicolor energy scale implies that most (but not all) technihadrons weigh 1000–3000 times as much as their ordinary hadronic counterparts. Most important, if technifermions exist in the same electroweak doublet family structure as do quarks and leptons, then certain spin-0 technihadrons play precisely the role of elementary Higgs bosons and automatically break electroweak symmetry. In other words, technicolor dynamics guarantees the correct pattern and magnitude of photon, $W^\pm$, and Z^0 masses.

Technipions. In many respects, the technicolor scenario for electroweak symmetry breaking is identical to the elementary-Higgs one. The most obvious difference is that, with technicolor, there exists a rich spectrum of technihadrons with masses in the range 1000–3000 GeV. Unfortunately, these particles' high mass and complicated decay modes may make them inaccessible to experiments at existing and even planned high-energy accelerators. However, it was realized in 1979 (by E. Eichten, K. Lane, and S. Dimopoulous) that technicolor models generally require the existence of technihadrons which are anomalously light compared to the technicolor scale. These are called technipions, by analogy with the ordinary pions which are light compared to the usual strong-interaction scale. They are tightly bound, spin-0 states of a technifermion and an antitechnifermion. And they may be of two general types: those that do not carry the ordinary color quantum number and those that do. It is the hallmark of technicolor that, unlike the situation in elementary-Higgs models with arbitrary self-interactions, technipion masses are calculable with some confidence. Technipions thus provide the means for testing the technicolor idea.

Colorless technipions have electric charge 0 (denoted by P^0) or ± 1 ($P^\pm$), and they are very light indeed. Their masses are estimated to lie in the range 5–50 GeV. If they are light enough, the $P^\pm$ may be detectable in experiments at the positron-electron (e^+e^-) storage rings PEP and PETRA. Otherwise, they should be found in the decays of Z^0 (to $P^+ + P^-$) and $W^\pm$ (to $P^\pm + P^0$). These searches will be carried out at the proton-antiproton colliders at CERN and Fermilab and at the e^+e^- colliders SLC and LEP now under construction at SLAC and CERN. Another way to search for charged technipions is in the decay of mesons containing the top quark t. If the $P^\pm$ mass is at least 5 GeV less than the t mass, then $t \rightarrow P^+ + b$ (the bottom quark) is expected to be the major decay mode of t. There are recent indications of the discovery at CERN of the t quark, having a mass of about 45 GeV and with no sign of decay to $P^+ + b$. If this proves correct, then the existence of $P^\pm$ with mass less than about 40 GeV probably is excluded.

Colored technipion masses are expected to range from 100 to 300 GeV. They may have a variety of electric charges, depending on the details of the model. Some of them, called leptoquarks, may decay into a charged lepton plus a quark (that is, hadrons), thus providing a very striking signature for their presence. Others are expected to decay into mesons containing heavy (top or bottom) quarks or antiquarks. These complicated decay modes may be difficult to identify. In any case, the relatively large masses of colored technipions preclude their being produced in even the large accelerators now under construction. A definitive search will require the very large proton-(anti)proton collider, known as the Superconducting Super Collider, that has been proposed for construction in the United States.

Superconducting Super Collider (SSC). The primary mission of the SSC is to uncover and study the physics of electroweak symmetry breaking. Whether or not that physics is described by the technicolor scenario, it is known that the appropriate energy scale is within a factor of 2–4 of 1 TeV. In a hadron machine such as the SSC, the really high-energy collisions occur between individual quarks or gluons inside the colliding protons. Since these constituents share the proton's energy and momentum, proton energies 5–10 times greater than a few TeV are required to achieve the desired collision energies. Thus, the total machine energy proposed for the SSC is 40 TeV. If its construction is approved, the SSC should be ready for experiments in 1993 or so. Although a long time off, the experimental program at the SSC promises to yield exciting and, probably, surprising new insights to the fundamental interactions.

For background information see ELEMENTARY PARTICLE; FUNDAMENTAL INTERACTIONS; GLUONS; QUANTUM CHROMODYNAMICS; QUARKS; WEAK NUCLEAR INTERACTIONS in the McGraw-Hill Encyclopedia of Science and Technology.

[KENNETH LANE]

Bibliography: E. Farhi and R. Jackiw (eds.), *Dynamical Gauge Symmetry Breaking*, 1982; M. K. Gaillard, Toward a unified picture of elementary particle interactions, *Amer. Sci.*, 70:506–514, 1982; C. Rebbi, The lattice theory of quark confinement, *Sci. Amer.*, 248(2):54–65, February 1983; G.

't Hooft, Gauge theories of the forces between elementary particles, *Sci. Amer.*, 242(6):104–138, June 1980.

Ethylene

Ethylene, one of the simplest organic molecules, has been known for some time to be a plant hormone. Its physiological effects include ripening, senescence, abscission, and germination—all processes which are primarily catabolic. Ethylene is a "stop sign." While the mode of action of ethylene is still under intense scrutiny by plant physiologists and biochemists, the details of how ethylene is synthesized by plants to transmit the stop signal are now becoming known.

ACC synthesis. Through experiments in which the ability of hypothetical precursors, in radioactive form, to be converted into radioactive ethylene was examined, the amino acid methionine was found to be a likely biosynthetic precursor to ethylene. Furthermore, it was shown that carbons 3 and 4 of methionine specifically become the two carbons of ethylene. The carboxyl group was found to become CO_2. The production of ethylene had long been known to require oxygen, but the exact enzymatic mechanism by which methionine might be converted to ethylene was obscure.

The issue was clarified greatly by the discovery that methionine is converted by plants to a new compound which accumulates in the absence of oxygen. When tissue containing this compound is exposed to oxygen, ethylene is produced. Through chemical studies, this substance was shown to be 1-amino cyclopropanecarboxylic acid (ACC). The enzymatic synthesis of this compound from methionine is explained by a two-step process (Fig. 1), the first of which involves conversion of methionine to *S*-adenosylmethionine, a compound well known to participate in a number of enzymatic reactions. The next step is unique, however; *S*-adenosylmethionine is converted by an enzyme called ACC synthase to ACC. The mechanism which has been proposed for this transformation includes Schiff base formation from *S*-adenosylmethionine and pyridoxal phosphate, which is known to be a required cofactor. Deprotonation, ring closure to give an ACC-pyridoxal Schiff base, and hydrolysis of this species yield ACC and regenerate the cofactor. With the exception of the ring closure reaction, which is chemically very reasonable, ample precedent exists for these conversions in other pyridoxal enzymes. The levels of ACC synthase and therefore ACC are controlled by other hormones such as auxin, and they are involved in an intricate feedback loop with ethylene which regulates the biosynthesis. ACC synthase has now been isolated, and further study should firmly establish its mechanism.

Ethylene synthesis. The conversion of ACC to ethylene by plants is also a unique transformation, and the enzyme which executes it has thus far defied

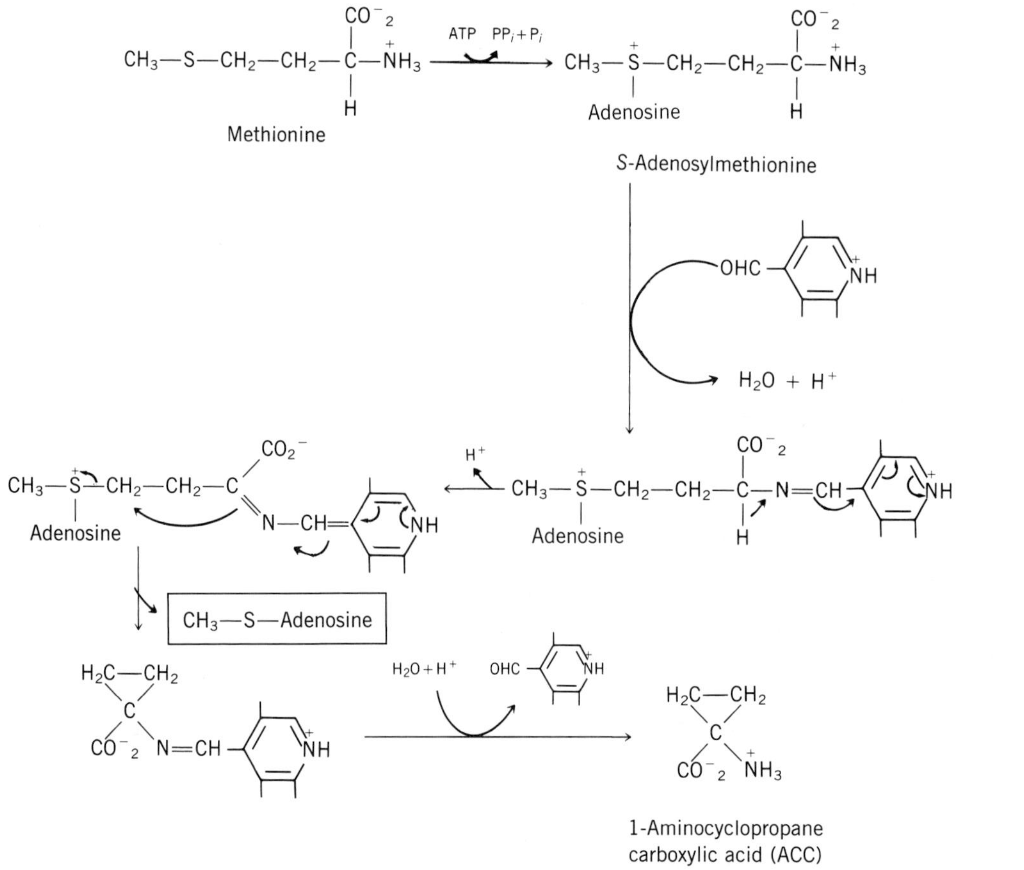

Fig. 1. Pathway for the conversion of methionine to ACC.

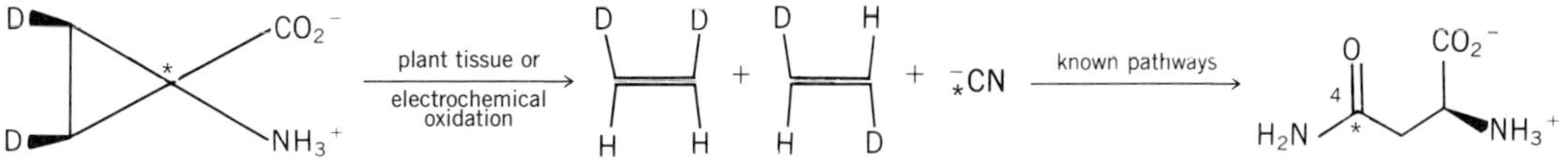

Fig. 2. Processing of isotopically labeled ACC derivatives.

isolation. Again the techniques of isotopic labeling have been useful in studying this process in intact plant tissue. Both radioactive and stable isotopes have been used. ACC labeled in its methylene carbons was prepared by anaerobically providing the appropriate radioactive methionine to plant tissue. On resubmission to the tissue in the presence of oxygen, radioactive ethylene is produced from labeled ACC. While establishing which carbons of ACC become ethylene, this experiment does not reveal how it occurs. Stereospecific hydrogen isotope labeling was used to answer the question. ACC derivatives were synthesized in which two hydrogens, on either the same or opposite side of the cyclopropane ring, were replaced by deuterium. These compounds are processed by plant tissue to the deuterium-substituted ethylenes. From either cyclopropane isomer, equimolar mixtures of the deuterated ethylenes are produced (Fig. 2). This loss of stereochemistry of the ACC molecule in the ethylene produced from it indicates that the two carbon-carbon bonds of ACC that must be cleaved to convert it to ethylene are not broken at the same time. This sort of behavior is characteristic of a free-radical reaction. Free radicals derived from ACC have been examined by electrochemical oxidation, and this treatment converts ACC to ethylene. That there are some similarities in the biosynthetic reaction and the electrochemical model is shown by oxidation of the deuterium-substituted ACC derivative. The same loss of stereochemistry is observed.

This model also yields cyanide from carbon 1 of ACC. Though the fate of carbon 2 of methionine during ethylene biosynthesis had never been determined, the mechanism of ACC synthase requires that it become carbon 1 of ACC. It was important then to determine the fate of this atom during ethylene biosynthesis. ACC derivatives isotopically labeled at carbon 1 were prepared biosynthetically from radioactive methionine and chemically from carbon-13 glycine. When supplied to plant tissue, these are converted to ethylene. The location of the isotopic label proved interesting. It was found in carbon 4 of the amino acid asparagine. A rationalization of this result was possible with knowledge of cyanide metabolism in plants, in which a major product is asparagine. ACC is thus converted to cyanide, which is then incorporated into asparagine. The amount of ethylene formed from ACC and the amount label-incorporated into asparagine are the same. This was determined by enzymatic or chemical degradation to give labeled CO_2 from carbon 4 of asparagine.

Other substrates. Other compounds which are substrates for the ethylene-forming enzyme have also recently been prepared (Fig. 3). Both methyl- and ethyl-substituted ACC derivatives are processed by plant tissue into propylene and butene, respectively. As in many enzymes, only specific stereoisomers are substrates for this conversion, which is somewhat surprising since the natural substrate has a plane of symmetry and therefore no stereochemistry. These compounds also inhibit the biosynthesis of ethylene from ACC, but not very powerfully. Because of ethylene's importance in regulating catabolic processes, it is likely that compounds which inhibit ethylene biosynthesis will continue to be sought as preservatives of agricultural products or plant growth regulators. See PLANT METABOLISM.

For background information see AUXIN; ENZYME; METHIONINE; PLANT HORMONES; PLANT METABOLISM in the McGraw-Hill Encyclopedia of Science and Technology.   [MICHAEL C. PIRRUNG]

Bibliography: D. O. Adams and S. F. Yang, Ethylene biosynthesis: Identification of 1-aminocyclopropane carboxylic acid as an intermediate in the conversion of methionine to ethylene, Proc. Nat. Acad. Sci. USA, 76:170–174, 1979; M. Lieberman, Biosynthesis and action of ethylene, Annu. Rev. Plant Physiol., 30:533–591, 1979; G. D. Peiser et al., Formation of cyanide from carbon 1 of 1-aminocyclopropane-1-carboxylic acid during its conversion to ethylene, Proc. Nat. Acad. Sci. USA, 81:3059–3063, 1984; M. C. Pirrung, Ethylene biosynthesis, 2. Stereochemistry of ripening, stress, and model reactions, J. Amer. Chem. Soc., 105:7207–7209, 1983.

Fertilizer

The production of sufficient food for the world's population requires the use of fertilizers. Environmental problems can be caused by the loss from agricultural fields of nutrients not completely utilized by crops, particularly nitrogen (N) and phosphorus (P). Thus, it is important that these potential problems be recognized and best management practices be utilized to minimize them. The four primary environmental concerns about fertilizer use are: accelerated eutrophication of surface waters, the nitrate (NO_3) content of drinking water, increased emission of nitrous oxide (N_2O) causing depletion of stratospheric ozone, and increased cadmium (Cd) content of soils as a result of phosphorus fertilization.

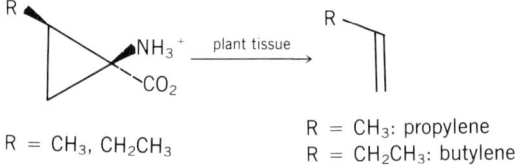

Fig. 3. Processing of alkyl-substituted ACC derivatives in plants.

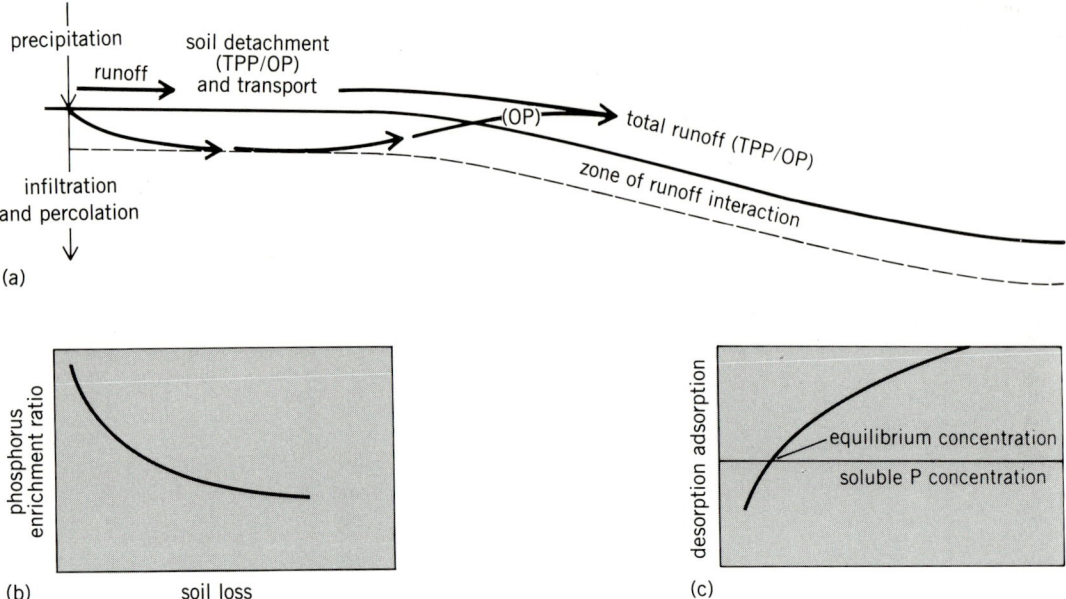

Fig. 1. Conceptual view of phosphorus transport in runoff from agricultural land. (a) Overall process. (b) Total particulate phosphate (TPP) transport. (c) Soluble phosphate (orthophosphate, OP) transport. (After T. J. Logan and J. W. Adams, The Effects of Reduced Tillage on Phosphate Transport from Agricultural Land, Lake Erie Wastewater Management Study, Corps of Engineers, Buffalo District, Technical Report Series, 1981)

Eutrophication is a natural process whose rate is usually limited by phosphorus and sometimes nitrogen. There is no question that accelerated eutrophication is a serious water-quality problem in some areas and that the use of fertilizers has contributed to this problem. The maximum safe NO_3-N concentration in drinking water that has been accepted throughout the world is 10 micrograms per milliliter. There have been very few confirmed human health problems caused by NO_3 in drinking water, but because subsurface drainage water from fertilized agricultural land often exceeds 10 μg NO_3-N ml^{-1}, this potential problem has received considerable attention.

The concerns for increased N_2O emission and the cadmium content of soil resulting from fertilizer use have received much less attention than the other problems because they are perceived to be smaller.

Phosphorus. Reduction in eutrophication rate of surface waters is most dependent upon reduction in phosphorus input. The two major sources of phosphorus to waters are sewage discharges and nonpoint (that is, from diffuse sources) contributions from agricultural lands. The control of phosphorus from nonpoint sources is assuming more importance in areas where current eutrophication problems exist. Although nonpoint sources include urban runoff, forest and agricultural land runoff, outlets from subsurface drainage, and atmospheric deposition, most nonpoint phosphorus comes from fertilized agricultural land.

Most of the phosphorus lost from land is by surface runoff because the sorption of phosphorus by soil particles tends to concentrate added phosphorus near the soil surface where it is susceptible to runoff and erosion. Runoff from precipitation detaches and transports sediment containing phosphorus and also removes some dissolved phosphorus (Fig. 1). During soil detachment and transport, there is a selective removal of clay-size particles which contain higher concentrations of phosphorus than the larger sand and silt particles. Thus the soil removed in the runoff contains a higher concentration of phosphorus than the original surface soil (Fig. 1b). As soil loss increases, the phosphorus enrichment ratio decreases and approaches unity under highly erosive conditions. Only a fraction of the particulate phosphorus in runoff is available to algae, so all of the sediment phosphorus lost to water does not contribute to eutrophication. Research studies have indicated that about 20% of the particulate phosphorus in sediment is bioavailable, with a range of 5–50%.

The processes which control the transport of dissolved inorganic phosphorus from soils by runoff water are shown in Fig. 1c. In soil-water systems, an equilibrium is reached between phosphorus in the solid phase and phosphorus in solution. The solution concentration where phosphorus is neither adsorbed nor desorbed from the solid phase is known as equilibrium phosphorus concentration (EPC_0). The EPC_0 is a characteristic of the soil which is a function of natural soil chemical properties and previous fertilizer history. There is a direct relation between the EPC_0 of a soil and the amount of dissolved inorganic phosphorus which is lost in surface runoff water.

Phosphorus fertilization increases the phosphorus content of surface soils because only a very small percentage of added phosphorus fertilizer is harvested with the crop in most agricultural systems.

Thus phosphorus fertilization will increase the phosphorus lost in runoff unless erosion is reduced. Phosphorus fertilization also increases the fraction of the absorbed sediment phosphorus which is bioavailable.

Considerable effort has been expended in the past few years to quantify the loss of phosphorus to drainage water from forest and agricultural lands. Generally, the annual loss from forest or native grassland is only 0.09–0.18 lb per acre (0.1–0.2 kg·ha^{-1}). The losses of phosphorus from cultivated land is much larger, and the range of values measured is also large. Most of the values for annual particulate phosphorus loss are between 0.9 and 16 lb per acre (1 and 18 kg·ha^{-1}), and dissolved inorganic phosphorus from 0.46 to 1.8 lb per acre (0.5 to 2 kg·ha^{-1}). In general, the larger the watershed, the smaller the amount of phosphorus lost in drainage water per unit area.

The best way to minimize phosphorus losses from agricultural land is through soil conservation practices which reduce sediment losses. For example, T. J. Logan and J. W. Adams have reported that in most studies reviewed by them, particulate phosphorus losses were reduced by as much as 90% by no-till or conservation tillage as compared to conventional plowing on the same soil. However, the losses of dissolved inorganic phosphorus were increased by these same cultural practices. Another way of minimizing phosphorus losses to water is by providing good subsurface drainage systems. Because of the phosphorus absorption capacity of most subsoils, only a very small amount of phosphorus is lost from agricultural subsurface drainage. However, subsurface drainage improvement tends to increase the nitrogen lost to surface waters. Finally, it should be pointed out that only a small percentage of the phosphorus added to soils is lost to drainage waters as most remains in place attached to the soil particles.

Nitrogen. When fertilizer nitrogen is added to soils, the nitrogen may be harvested with the crop, incorporated into soil organic matter, lost to drainage waters, or denitrified (converted to a gas). The factors controlling the distribution among these possible pathways are complex and vary greatly between soils and locations.

The nitrogen fertilizer which is harvested with the crop seldom exceeds 50%. Some perennial forage crops are much more efficient than 50%, but most fertilizer nitrogen is added to annual cultivated crops. Some of the nitrogen not harvested with the crop is incorporated into the soil organic matter. Using ^{15}N experiments to determine what happens to the applied nitrogen, researchers have found that 30% or more of the nitrogen applied to a crop is present in the organic fraction of the soil at the end of the experiment. However, very few agricultural soils are increasing in total nitrogen content, so the fertilizer nitrogen found in the organic fraction is replacing, for the most part, a portion of the nitrogen which was lost from soil organic matter through mineralization (conversion of organic to inorganic forms). Thus the amount of fertilizer nitrogen remaining in soils as organic nitrogen is of minimal importance in long-term nitrogen balances. The range of nonutilized fertilizer nitrogen lost through denitrification in the field is from near zero to 100%. However, approximately 15% of the fertilizer nitrogen applied is believed to be denitrified in agricultural fields.

The nitrogen not harvested with the crop or not lost through denitrification is subject to loss in drainage water; it is this loss that is the greatest potential problem to both ground and surface waters.

In contrast to phosphorus, fertilizer nitrogen lost through surface runoff is not considered to be a significant problem. The addition of nitrogen fertilizer does frequently result in a greater loss of nitrogen via surface runoff, but the increase is usually insignificant with regard to fraction of the total nitrogen lost. The effect of nitrogen fertilizer on increased surface loss is usually limited to the first rainfall following fertilizer application. Much of the nitrogen entering surface waters via surface runoff is organic nitrogen, which is less of a eutrophication problem than inorganic nitrogen readily available to algae.

There is a very large range in the quantities of nitrogen which are lost from agricultural fields to surface waters through subsurface drains. Values as high as 218 lb per acre (245 kg·ha^{-1}) per year have been measured, but the most common values are 18–46 lb per acre (20–50 kg·ha^{-1}). There are many factors controlling the quantitites of nitrogen lost through subsurface drains, but one of the most important is fertilizer application rate. This is illustrated in Fig. 2, where it can be seen that the recommended rate of 100 lb per acre (112 kg·ha^{-1}) per

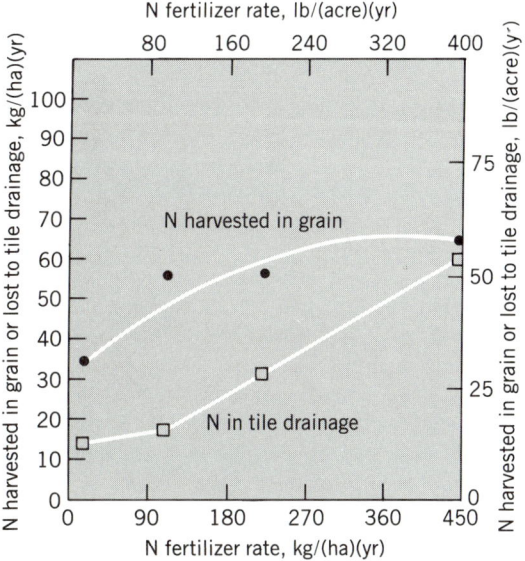

Fig. 2. Effect of nitrogen application rate on corn with reference to nitrogen utilized in the grain and nitrogen lost in the drainage water. (*After R. G. Gast, W. W. Nelson, and G. W. Randall, Nitrate accumulation in soils and loss in tile drainage following nitrogen applications to continuous corn, J. Environ. Qual., 7:258–261, 1978*)

year resulted in only a small increase in loss of nitrogen through tile lines as compared to a very low rate of fertilizer application, but that rates higher than recommended caused very significant increases in nitrogen losses. It is not necessary for agricultural fields to have subsurface drains for nitrogen applications to increase the entry of nitrogen into surface waters, since nitrogen content of seepage water and springs can be related to upslope activity.

It cannot be assumed that high nitrogen in tile effluents from the subsurface drainage system is a result of overfertilization since some soils mineralize large quantities of nitrogen when they are drained; nor can it be assumed that fertilization at recommended rates will not have an influence upon nitrogen content of drainage waters. It is unusual for nitrogen fertilizer to be 100% effective for crop utilization, and all unutilized nitrogen can potentially be lost to drainage water.

The effect that increased nitrogen fertilizer usage is having upon contents of surface waters varies with location. In many areas of the United States there apparently has been little change in the past decades, while other areas have observed significant increases in nitrogen contents of streams. Areas which have not shown significant increases in stream nitrogen contents are generally those with relatively high rainfall and a low percentage of the land under cultivation.

The contamination of groundwater with high concentration of nitrate resulting from fertilizer use has been a major concern. There is evidence that nitrate is accumulating in groundwater in some irrigated agricultural areas. Many studies have shown a relationship between fertilizer application levels and nitrate in the soil profile above the groundwater and below the rooting zone. Irrigation water management practices also have a large influence on nitrate leaching, with the largest movement of nitrogen being in areas with the largest amount of irrigation water moving below the root zone. It is clear that the threat to groundwater from fertilizer application can be minimized by using nonexcessive rates of fertilizer nitrogen and irrigation water. Most scientists currently believe that these management factors can be used to control the potential threat to groundwater in agricultural areas.

Nitrogen fertilizer and the atmosphere. The fixation of N_2 from the atmosphere, utilization of fixed nitrogen by plants and animals, and return of nitrogen to the atmosphere as N_2 and N_2O through denitrification constitute an important life cycle. The denitrification step has long been recognized as a problem in fertilizer efficiency because of loss of potentially crop-utilizable nitrogen. Only recently has the N_2O been recognized as a potential problem through reactions which result in loss of ozone from the stratosphere. Nitrous oxide evolution has been measured from a number of cropland soils throughout the world, with most of the measurements falling in the range of 1.8–8.9 lb per acre (2–10 kg·ha^{-1}) per year. This amount is greater than that evolved from nonagricultural land, but is not necessarily altogether a result of nitrogen fertilizer use. This is because an increase in nitrogen fixation through the cultivation of a legume crop not requiring nitrogen fertilizer also results in an increase in N_2O evolution. The effect that any increased evolution of N_2O may have upon the protective layer of ozone is not clear at this time. It is clear that any fertilizer practice which maximizes fertilizer nitrogen efficiency will minimize N_2O evolution problems.

Cadmium accumulation. Cadmium has been recognized as posing some risk to humans through entry into the food chains. Because phosphate rock contains as much as 980 µg Cd g^{-1}, some cadmium is added to soils through application of phosphorus fertilizers. This caused some to question whether cadmium might be accumulating in heavily fertilized soils to the extent that crops might absorb dangerous quantities. Investigators of this potential problem have concluded that the cadmium levels which are added to soils in phosphorus fertilizer pose little or no potential health threat.

For background information *see* EUTROPHICATION; FERTILIZER; NITROGEN CYCLE in the McGraw-Hill Encyclopedia of Science and Technology.

[J. W. GILLIAM]

Bibliography: R. G. Gast, W. W. Nelson, and G. W. Randall, Nitrate accumulation in soils and loss in tile drainage following nitrogen applications to continuous corn, *J. Environ. Qual.*, 7:258–261, 1978; J. W. Gilliam, T. J. Logan and F. E. Broadbent, Fertilizer use in relation to the environment, in F. C. Boswell et al. (eds.), *Fertilizer Technology and Usage*, 1984; G. F. Lee, R. A. Jones, and W. Rast, Availability of phosphorus to phytoplankton and its implications for phosphorus management strategies, in R. C. Loehr, C. S. Martin, and W. Rast (eds.), *Phosphorus Management Strategies for Lakes*, 1980; T. J. Logan and J. W. Adams, *The Effects of Reduced Tillage on Phosphate Transport from Agricultural Land*, Lake Erie Wastewater Management Study, Corps of Engineers, Buffalo District, Technical Report Series, 1981.

Fertilizing

With extensive changes in farming practices and increased fertilizier use, many changes in techniques of fertilizer application have occurred in recent years. The techniques vary depending on the characteristics of the fertilizer, the crop being grown, soil properties, and environmental conditions. Therefore, much of the development of new techniques concerns improvement of fertilizer effectiveness and efficiency for the crop, soil, and climate where fertilizers are applied.

The changes in techniques include: (1) application of fertilizers in bands in the soil rather than by mixing throughout the soil; (2) application of fertilizer-pesticide mixtures instead of separate applications of these materials; (3) injection of fertilizers below the soil surface in reduced tillage systems; (4) application of fertilizers carried in irrigation water;

and (5) foliar application of fertilizer solutions sprayed on the crop plants.

Fertilizer materials. The fertilizer nutrients come in different forms and react differently. The optimum method of application varies with the characteristics of the different nutrients and the materials being applied.

Nitrogen. Nitrogen (N) may be applied as ammonia (NH_3) which is a gas, or as urea ($CO[NH_2]^2$), ammonium (NH_4^+) or nitrate (NO_3^-), which are solids, or in solution. Ammonia is obtained as a pressurized liquid which immediately vaporizes as it emerges from the applicator, so it must be injected into the soil and cannot be surface-applied. Recently a "cold-flow" method has been developed in which the NH_3 is supercooled to liquid form and metered into the soil ahead of an incorporating implement. In the soil, the NH_3 reacts readily with water to form NH_4^+. Urea is applied as a dry, granular material or in solution; in moist soil the naturally occurring enzyme, urease, reacts readily with the added urea to form NH_4^+. Ammonium from either of the above sources or ammonium added as a sulfate, chloride, or nitrate salt reacts with the cation-exchange complex of the soil to become adsorbed on the surfaces of clay particles. However, ammonium is also biologically oxidized to nitrate (NO_3^-) in moist, well-drained soil. Nitrate, whether produced in the soil or added as a fertilizer, is a negatively charged anion that is not held on the cation-exchange complex in the soil. Nitrate is water-soluble and mobile in the soil solution, moving readily as the soil water moves. With excess soil water, leaching of nitrate from the soil may occur. Also, excess water in the soil may result in anaerobic conditions, with loss of nitrates due to denitrification.

Phosphorus. Phosphorus (P) is most commonly applied to soils as an orthophosphate (PO_4^{3-}), but it may be applied as a polyphosphate or in the mineral form, apatite. Orthophosphates react rapidly in soils to form much less soluble compounds. The degree of reaction and the products formed depend largely on the soil pH and the kinds and amounts of precipitating cations present. Because of the phosphorus reactions in soil, phosphorus is immobile in most soils and the added phosphorus must be placed where plant roots will come in contact with it. Therefore, phosphorus fertilizers are often applied in bands within the rooting zone in the soil rather than as broadcast applications. Rarely is more than 20% of the added fertilizer phosphorus taken up by the crop plants during the first growing season. Instead, the phosphorus remains in the soil and is slowly available in succeeding seasons.

Potassium. Potassium (K) is usually applied as a chloride (KCl), sulfate (K_2SO_4), or nitrate (KNO_3), salts that are soluble in water. The cation (K^+) reacts with the soil cation-exchange complex and is held as an exchangeable cation that is relatively immobile in the soil but is readily available to plants.

Secondary elements. Calcium (Ca), magnesium (Mg), and sulfur (S) are of two types. The cations, Ca^{2+} and Mg^{2+}, are the primary cations on the cation-exchange complex of the soil and are only slightly mobile in the soil. Calcitic and dolomitic limestones are commonly broadcast-applied to reduce soil acidity. Sulfur exists in different forms, many of which are oxidized to the anion sulfate form (SO_4^{2+}) which is relatively mobile in soils.

Micronutrients that are present in soils and are essential for crop growth include zinc (Zn), copper (Cu), manganese (Mn), boron (B), and iron (Fe). However, the amounts required are relatively small, and these minor elements are often mixed into other fertilizers. Because of the different characteristics of the materials applied and the needs of the crops, the methods of application vary. They may be broadcast on the soil, banded for row crops, or sprayed as a foliar application.

Soil applications. Different methods of application of fertilizers to soils have been developed to increase their effectiveness.

Band versus broadcast applications. Soil applications of fertilizers may be made in bands at or below the soil surface, or may be broadcast on the soil surface and subsequently plowed under or disked in. Band applications result in localized placement of fertilizer nutrients and generally less reaction of the added nutrients with the soil. Thus, the availability of the added nutrients to the crop plants is often higher than for fertilizers that are mixed with the soil. This is especially important for the relatively immobile nutrients, phosphorus and potassium. However, it is essential that the band of high nutrient availability be located where the plant roots will be active in nutrient uptake when the growing plants require the nutrients.

Fertilizers combined with pesticides. Combining insecticides or herbicides with fertilizers for application can improve the farmer's efficiency and sometimes increase the effectiveness of the pesticide, especially herbicides. Pesticides can be added to liquids (solutions and suspensions) and to dry fertilizers. However, not all pesticide-fertilizer mixtures are compatible. To be effective, the pesticide must mix freely with the fertilizer carrier, stay in suspension in liquid fertilizers without undue agitation, and be effective in controlling the pest for which it is intended. Since not all fertilizer-pesticide combinations meet these requirements, there are limitations on this practice. But where the practice meets these requirements, it can be used very effectively.

Placement for reduced tillage systems. Reduced tillage systems such as chisel plowing and no-till plowing have been developed to reduce costs and to provide better control of soil erosion. However, different methods of nitrogen fertilizer application must be developed and used for such systems. Present systems generally involve plowing and disking to prepare the seedbed.

With reduced tillage, much of the phosphorus and potassium from fertilizer applications and from crop residues returned to the soil remain near the soil surface. The nutrients in this position may be relatively unavailable to crop plants during periods when the surface soil is dry or cold. Furthermore, studies indicate that a more acidic layer develops near the soil surface with continuous use of no-till production of crops. This is due to the production of organic acids during the decomposition of the crop residues and the leaching of calcium from this layer. Because of this stratification within the soil, it may be desirable to inject phosphorus and potassium fertilizers below the soil surface and to apply lime periodically on the soil or to plow the soil every 5–10 years with a moldboard plow.

Application with irrigation water. In many areas, irrigation is used to supply water to the growing crop; adding fertilizers to the irrigation water can be a convenient and effective method of fertilizer application. There are several methods of irrigation: (1) Flood irrigation involves applying the water from a ditch or pump at the high end of the field so the water flows by gravity in a sheet or in furrows to the lower end of the field. This results in nonuniform distribution of water, and of fertilizer if it is added. (2) Sprinkler irrigation involves pumping water to a series of aboveground sprinklers that are either solid-set or movable. Distribution of water from these sprinklers can be affected by plugging or other malfunctions of the nozzles or by wind which causes drifting of the sprayed water. (3) Trickle or drip irrigation involves delivery of water under low pressure through pipes or tubing to drippers or emitters at a slow rate. The emitter lines and tubing may be placed below the soil surface. Clogging of orifices by particles or organic slime is a major problem with this system.

Adding nitrogen fertilizers in the irrigation water has become a widely used practice. Urea and nitrates are very soluble in water and move down in the soil with the applied water. Although most ammonium sources are soluble, the NH_4^+ is held on the cation-exchange sites near the soil surface and is subject to volatilization losses.

Application of phosphorus and potassium fertilizers in irrigation water is not generally recommended. Precipitates of phosphorus in the water cause plugging in the system. The restricted movement of both phosphorus and potassium in most soils results in retention of these nutrients very near the soil surface, where their availability to plants is restricted when the soil surface is dry.

Foliar applications. Plants are capable of absorbing fertilizer nutrients through aerial plant parts, so foliar application of fertilizer solutions can be effective. However, because severe leaf burn results if excessive amounts of fertilizer are applied at one time, the amount per application is limited. Foliar fertilization has been used primarily for applications of minor elements that are needed by the crop plants in only small amounts.

Foliar applications of the minor elements—iron, zinc, manganese, boron, copper, and molybdenum—are effective in correcting deficiencies. Such applications are used extensively for deciduous fruit trees, grape vineyards, and citrus and also for other crops such as beans, potatoes, peas, sugarbeets, corn, soybeans, grain sorghum, and pineapples. Foliar fertilization of these crops is often especially desirable where soil applications are difficult or, even in massive doses, not effective.

Foliar application of the major elements—nitrogen, phosphorus, potassium, and sulfur—has not been practiced extensively. Applications of nitrogen and phosphorus together with minor elements are made regularly for pineapple and sugarcane in Hawaii, and foliar applications of these major elements can be effective in colder regions or seasons. Recent research involving field grain crops such as soybeans, corn, and wheat has shown that foliar applications of these major elements made during the seed-filling period will sometimes, but not consistently, result in very significant yield increases.

In field grain crops, during the seed-filling period most of the photosynthate (sugar) produced by photosynthesis in the leaves is channeled to the developing seeds. Growth of vegetative plant parts, including the roots, is very limited during this period, and nitrogen fixation in the nodules on the roots of leguminous plants slows and stops. Although nutrient uptake from the soil continues, retranslocation of several nutrients, especially N, P, K, and S, from the leaves and other plant parts to the developing seed is a major source of these nutrient elements to the developing seeds. This results in loss of the elements in the leaves during the seed-filling period. As this nutrient depletion in the leaves progresses, photosynthesis in the leaves gradually slows and stops.

Some nutrient elements, especially calcium but also magnesium and manganese, are not effectively retranslocated from the leaves and other plant parts to the developing seeds. The concentration of calcium in the seeds is very low, and the concentration in the leaves increases during the seed-filling period.

Although research does not justify general use of foliar fertilization on field grain crops during seed-filling, it has provided some useful guidelines: (1) Foliar fertilization applications containing urea should not be made when the sun is shining brightly. Instead, spraying should take place in the evening or early morning to avoid serious leaf burn. (2) Urea is the most effective form of nitrogen for foliar applications, but the amount of urea-nitrogen that can be applied in one spraying without causing serious leaf burn does not exceed 18–23 lb N/acre (20–25 kg/ha). (3) Excessive amounts of some urea decomposition products such as cyanate, biuret, and cyanamid, which may be found in some urea fertilizers, cause serious leaf burn and decrease yields. (4) Best yield increases of soybeans have resulted from foliar applications which included N, P, K, and S in a ratio of 10–1–3–0.5 (similar to the ratio in soybean seeds). (5) Foliar applications of N, P,

K, and S during seed filling have not delayed leaf senescence appreciably.

For background information *see* AGRICULTURAL SOIL AND CROP PRACTICES; FERTILIZER; FERTILIZING; IRRIGATION (AGRICULTURE) in the McGraw-Hill Encyclopedia of Science and Technology.

[JOHN J. HANWAY]

Bibliography: J. J. Hanway, Foliar fertilization of soybeans during seed-filling, in F. T. Corbin (ed.), *World Soybean Research Conference II Procedings*, 1980; R. D. Hauck (ed.), *Nitrogen in Crop Production*, American Society of Agronomy, 1983; G. E. Richards (ed.), *Band Application of Phosphatic Fertilizers*, Agriculture Production Department, Olin Corp., 1977.

Food manufacturing

Continuous thermal processing and aseptic packaging for preserving foods is a growing industry trend. Compared to in-can processing, continuous thermal processing offers reduced energy and packaging costs and favors product quality retention.

Aseptic processing. Food preservation by thermal sterilization is accomplished in one of two ways. The food is either heated to achieve sterility, placed in a presterilized container, and sealed; or packaged in a nonsterile container and then heated. Sterility is described as a heat treatment yielding an inability of microorganisms and their spores to grow under normal storage conditions. The first method is referred to as aseptic processing or ultrahigh-temperature (UHT) processing, and the second as conventional canning.

Aseptic processing equipment. For continuous thermal sterilization of foods, it is necessary to have equipment to perform the following processes: feeding raw product, preheating product, providing constant product flow, heating for sterilizing, providing time for sterilizing (holding), cooling product, storing aseptic product, and packaging aseptic product (see illustration).

The raw product feed and preheating are usually accomplished with conventional food-handling equipment—a supply pump and standard heat exchangers. The product has not been sterilized and temperatures are relatively low through this stage. Constant product flow, maintained by the use of positive displacement pumps which accurately meter product through the system, is required to assure adequate exposure of product to sterilizing heat through the sterilizer and holding section.

Commercial sterilizers are classified as direct-heating or indirect-heating units. Direct-heating units are either steam injectors or infusers. Indirect-heating units are tubular, plate, or scraped-surface heaters. The source of heat for all sterilizers is steam or hot water.

For direct heating, steam and product are mixed, causing the steam to condense and yield its latent heat to the product. Since no barrier exists between product and heating media, rapid heat transfer occurs resulting in extremely fast product temperature

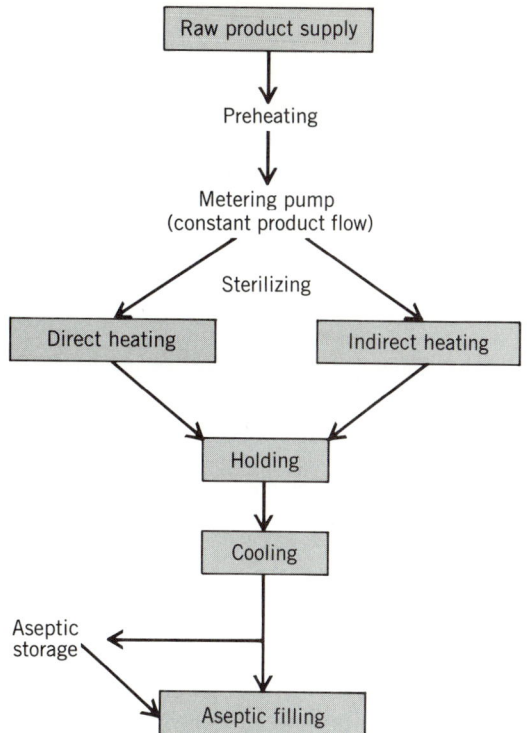

Flow chart of continuous-flow aseptic processing.

rise. Vacuum equipment provides for partial cooling and removal of the excess moisture derived from dilution of the product by condensing steam. When the direct-heating unit is a steam injector, steam is injected directly into the product flow, and all of the steam condenses by maintaining required pressure with a flow restriction device (usually an air-operated valve). When a steam infuser is used, product is introduced into a vessel with a steam atmosphere. As with steam injection, adequate pressure must be maintained.

For indirect heating, product and heating media are separated by a barrier through which heat is transferred to the product. The barrier is usually a wall of a tube, a plate, or a scraped surface. Indirect heating requires a large surface–to–product volume ratio in order to yield rapid heating; however, rapid heating for indirect sterilizers is slow relative to direct units. To achieve a large ratio, tubular heaters have small tube diameters; more plates may be added to increase the ratio for plate heat exchangers.

Indirect systems provide energy-saving benefits by regenerative heating and cooling. However, this is offset by the fact that indirect heaters tend to form product deposits on hot surfaces which may severely limit the operating time and make cleaning more difficult. After heating to the scheduled sterilizing temperature, product is maintained at this temperature in a holding section or holding tube, which is a pipeline section located immediately downstream from the sterilizing unit. Its design ensures that all product is held at or above the sterilization temperature for at least the scheduled process time.

Direct sterilizing occurs very rapidly compared to indirect. In the direct systems, virtually all of the thermal effects on product occur in the holding tube. In indirect systems, substantial thermal effects may occur during heating, and some additional effects may occur during cooling. Aseptic processing (direct and indirect) occurs at higher temperatures and shorter times than processing by in-can retorting. This increased process temperature combined with decreased time can give the required lethal effect on microorganisms and their spores. It simultaneously reduces the thermal effect on chemical reactions associated with loss of product quality, for example, nutrient degradation.

Storage of product between processing and packaging is sometimes desirable. This is accomplished in an aseptic surge tank.

Equipment sterilization. Processing systems are usually sterilized by pumping heated water (250°F or 121°C) under pressure through the system. This includes piping that interconnects the processing system with aseptic tanks or fillers. All potential product-contact surfaces must be brought to sterile conditions according to the scheduled process. Once the equipment sterilization cycle is completed, the processing system downstream from the holding tube is cooled and readied for product.

Aseptic tanks and fillers must also be sterilized, and they usually have their own specific sterilization cycles. An aseptic tank may be steam-sterilized. The condensate of steam contacting the tank, as with any product contact surface, must be of food-grade quality. When the specified temperature is achieved inside the tank, timing of the sterilization period begins. Once sterilization is completed, the cooldown is obtained by displacing the steam with sterile air, always maintaining a positive pressure in the tank.

Aseptic fillers also have sterilization cycles, but the lengths of the cycles and the sterilizing chemicals vary for each manufacturer.

Equipment operation. Upon completion of equipment sterilization, processing conditions are established. All equipment beyond the holding section is cooled to the processing level. When product is introduced to the system, it forces out the sterilizing water. When the system is operating on pure product, filling may then begin.

It is possible for failures to occur at any time during processing. Examples of such failures are loss of processing temperature or pressure, or an indication of nonsterile product from any controlled parameter. Should a failure occur, the system must be resterilized. At the completion of the process run, all product-contact surfaces are completely cleaned of product, and then the next equipment sterilization cycle is possible.

Aseptic packaging. There are five basic categories of aseptic packaging: fill-and-seal, form/fill/seal, erect/fill/seal, thermoform/fill/seal, and blow-mold/fill/seal.

Although fill-and-seal systems may use cans and glass bottles, few glass aseptic systems exist due to weight and cost factors. The chief benefit for the environment is that glass is reusable, but return systems tend to be costly and it is difficult to create consumer interest. All returned glass containers must be washed, sterilized with steam, filled with sterile product, and then sealed with a presterilized closure. Cans are used in a manner similar to glass except that they are not reused. The new cans are sterilized with superheated steam or hot air, filled with sterile product, and sealed with lids. In appearance, can and glass packages produced by continuous aseptic processing are not different from those produced by traditional canning methods.

In the form/fill/seal aseptic packaging systems, a roll of incoming material is sterilized and then formed, filled, and sealed in an aseptic environment. A common example involves the use of coextruded paperboard composed of polyethylene, paper, and aluminum foil. To sterilize the laminate, hydrogen peroxide, ethanol, or some other chemical sterilizing agent is used. After sterilization the roll stock is shaped into a tube, filled with product, and formed into individual packages.

In the erect/fill/seal system, a knockdown blank of incoming material is erected, sterilized, and then filled and sealed. Coextruded paperboard is also used to make knockdown blanks. The units are formed in a location separate from that of the filler. The flat preformed packages are unfolded, sealed at the bottom, treated with chemical sterilant, filled with product, and sealed at the top. Unlike packages which are made from tubes without head space, these units have head space to reduce splashing during filling, a condition which applies to all types of open filling.

Pouches made from plastic laminates are another example of erect/fill/seal packaging. They vary in size and lack rigidity, requiring some support container, but can be used in both canning and continuous aseptic systems.

Thermoform/fill/seal systems use roll stock, which is sterilized, thermoformed, filled, and sealed in an aseptic environment. Thermoforming is the process of shaping containers from heated roll-stock material by using plug-type formers.

The fifth type of aseptic packaging, blow-mold/fill/seal, utilizes an extrudable material which is blow-molded. The blowing is done with sterile air or nitrogen. Within a sterile environment, the gas is vented, the sterile product is metered in, and the package is sealed.

For background information *see* FOOD MANUFACTURING; HEAT EXCHANGER in the McGraw-Hill Encyclopedia of Science and Technology.

[K. R. SWARTZEL]

Bibliography: D. M. Adams et al. (eds.), *International Conference on UHT Processing and Aseptic Packaging*, Department of Food Science, North Carolina State University, 1979; D. S. Hsu, *Ultra-High-Temperature (U.H.T.) Processing and Aseptic Packaging (A.P.) of Dairy Products*, Damana Tech. Inc., New York, 1970; P. Jelen, Review of basic technical principles and current research in UHT Processing of foods, *Can. Inst. Food Sci. Technol. J.*, 16(3):159–166, 1983; J. R. Russo and R. Bannar, Aseptic packaging: How far will it go?, *Food Eng.*, 3:49–59, 1981.

Gene

The term transposable element designates a type of deoxyribonucleic acid (DNA) sequence that is able to alter its chromosomal location by a unique mechanism. The movement of transposable elements requires element-encoded enzymes that recognize short, highly specific sequences repeated in inverted order at each end of the element. Transposable elements vary markedly in size and structure, and have been found in virtually every organism examined for their presence. Here the history of their discovery will be summarized briefly, and their structure and function in different organisms will be discussed.

History. Transposable elements were first identified in maize by Barbara McClintock in the early 1940s. The two related elements that McClintock first understood to transpose were initially detected because of their ability to cause chromosome breakage at a specific site on the short arm of chromosome 9. The chromosome breakage or dissociation site was designated the Dissociation (Ds) locus and proved to be transposable. However, chromosome breakage and transposition of Ds required the presence of an additional genetic element designated the Activator (Ac) locus. This locus was also found to be transposble, but differed from the Ds locus (or element) by its ability to promote its own transposition. In subsequent studies, McClintock and other maize geneticists identified and analyzed additional, genetically distinct families of transposable elements and elucidated the in behavior. Maize transposable elements, termed controlling elements by McClintock, were found to be the causative agents of unstable mutations in maize and have also been shown to promote a variety of chromosomal rearrangements, including deletions, duplications, inversions, and translocations. Several of the maize transposable elements have recently been isolated; their structure is discussed below.

Two decades after their identification in maize, transposable elements were detected in bacteria as insertion sequences (IS) that cause mutations and as the agents which mobilize the drug-resistance markers frequently found on plasmids. Bacterial IS elements and transposable elements (transposons) have been studied extensively, yielding detailed information about the mechanism of transposition. During the past several years, transposable elements have been identified in other organisms, including yeast, the nematode *Ceanorhabditis elegans*, and the fruit fly *Drosophila melanogaster*. Insertion sequences which bear a structural resemblance to transposable elements have been identified in an even larger variety of organisms, although direct evidence of their ability to transpose is either minimal or lacking.

Structure. Transposable elements vary so widely in structure that it is perhaps best to begin by de-

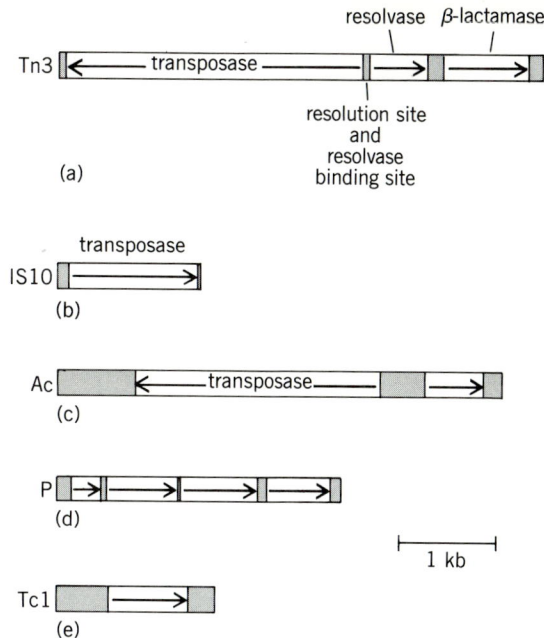

Fig. 1. The protein coding sequences of each transposable element are represented by open boxes; arrows designate the polarity of each sequence. The name of the enzyme encoded is given where known. Each element has a terminal inverted repetition sequence of the following length: (a) bacterial transposon Tn3, 38 bp; (b) bacterial insertion element IS10, 23 bp; (c) maize-controlling element Ac, 11 bp; (d) *Drosophila* transposable P element, 31 bp; (e) *Caenorhabditis elegans* transposable element Tc1, 54 bp.

scribing a bacterial transposon whose structure and function are understood in some detail. The bacterial transposon Tn3, whose structure is represented diagrammatically in Fig. 1a, will be used for this purpose. Tn3 is a relatively small bacterial transposon. It is 5.0 kilobase pairs (kb) in length and contains three genes. Two of the genes encode proteins which participate in the transposition process, and the third is the passenger β-lactamase gene that confers drug resistance on cells harboring the transposon. The element is bounded by 38-base-pair (bp) terminal inverted repetitions (IR) that contain the specific recognition sites required to mobilize the sequence lying between them.

Bacterial IS elements are some of the simplest elements capable of autonomous transposition. They resemble Tn3 in encoding transposition functions, but lack the passenger genes characteristic of more complex transposons. IS elements vary in length from less than 1 kb to more than 5 kb, and have a coding capacity of one to three proteins. All have terminal IRs which vary in length from less than 10 base pairs (bp) to more than 40 bp. The structure of the 1.4-kb IS10 is shown in Fig. 1b as an example of a simple bacterial IS element. Several of the transposable elements isolated from higher organisms strongly resemble bacterial IS elements (Fig. 1c–e). The 4.6-kb Ac element of maize bears a striking resemblance to the part of Tn3 that encodes its transposition functions. The 2.9-kb P element of *D. melanogaster* and the 1.6-kb Tc1 element of *C.*

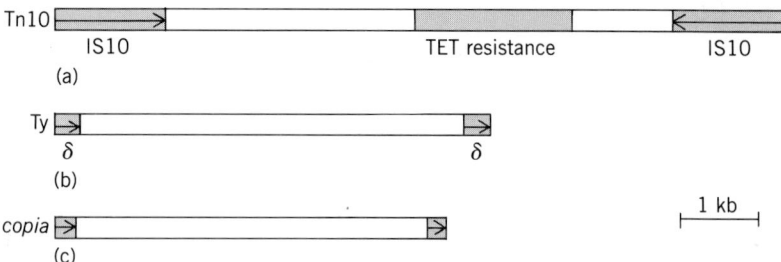

Fig. 2. Structure of transposable elements from (a) bacteria, (b) yeast, and (c) Drosophila. The short, direct terminal repetitions of the Ty element are designated δ sequences.

marker flanked on each side by a copy of the IS10 element.

The structure of two rather similar families of eukaryotic transposable elements is represented by a yeast Ty element and the *copia* element of *Drosophila* (Fig. 2b–c). This type of element ends with rather long, direct repetitions, each of which terminates in short IRs. The yeast Ty element is about 6.3 kb, of which the terminal 330 bp comprise directly repeated sequences. The 5-kb *copia* element of *D. melanogaster* has a 276-bp terminal direct repeat. The *copia* element strongly resembles integrated retroviral sequences in other organisms, and there is growing evidence that it occurs in viruslike particles, as well as in the genome.

Even very large sequences can be mobilized if they are bracketed by transposable elements. Examples of such elements are the 100-kb transposable element of *D. melanogaster*, which comprises a mobile *white* locus bounded by a structurally unique class of transposable elements termed the foldback elements. A similar 30-kb sequence bounded by Ds elements has been described in maize. It is also known in both *Drosophila* and maize that chromosome segments bounded by P or Ds elements, respectively, can undergo duplications, inversions, and more complex rearrangements.

It has become increasingly evident that there are many elements that are unable to promote their own transposition, but can be mobilized by a related element's transposition enzymes. Many such elements appear to be defective transposable elements derived from functional ones by various mutations, especially deletions. As an example, one of the maize Ds elements that has been studied appears to have arisen directly from an Ac element by a short deletion in its transposase gene. Similarly, many defective P elements in *Drosophila* appear to have arisen by various internal deletions. Other mobilizable sequences have been studied, however, which are virtually unrelated to the mobilizing elements, except for the terminal IRs. For example, a very small 0.4-kb Ds element has been isolated whose resemblance to the mobilizing Ac element is confined almost entirely to its 11-bp IRs. These observations, taken together with direct mutational analysis in bacteria, indicate that the sequence information required by the transposition enzymes resides in the very ends of the mobile element.

elegans also resemble bacterial IS elements in structure.

Bacterial transposons are somewhat more complex and commonly larger than IS elements. Transposons were generally identified as carriers of mobile marker genes, most commonly encoding enzymes that confer drug resistance on the bacterium within which they reside. Tn3 is a small transposon that resembles an IS element containing a single extra gene. The structure of Tn10 is given as an example of a more complex transposon (Fig. 2a). It comprises a sequence containing a tetracyline-resistance

Function. The transposable element for which the most functional information has been accumulated is Tn3. Tn3 encodes two proteins which participate in the transposition process, the transposase and the resolvase enzymes. Tn3 undergoes replicative transposition by a mechanism that involves the replication of the element and the union of the donor and recipient DNAs (Fig. 3). The element's transposase is involved in the formation of this structure, termed a cointegrate. Tn3, like all known transposable elements, generates a short, direct repetition at the site of insertion. This is presumed to reflect the unequal or staggered cutting of the two DNA strands at the

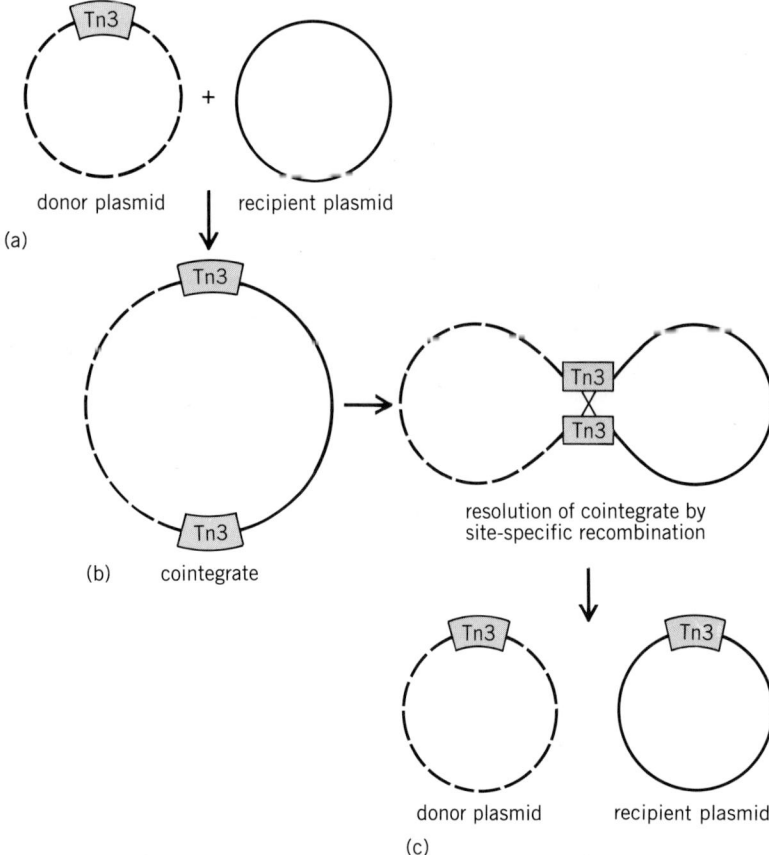

Fig. 3. Transposition of Tn3. (a) The first step in transposition is the fusion of the donor and recipient DNA [represented as plasmids] (b) with concomitant replication of the element to form a cointegrate structure; resolution of the cointegrate involves a site-specific recombination event between the two transposons, catalyzed by the element-encoded enzyme resolvase, to generate (c) two plasmids, each carrying a copy of the transposon.

insertion site, followed by repair synthesis to fill in the gap created upon ligation of the transposable element to the insertion site. The element's resolvase separates the cointegrate by recombination of the specific sequences located between the resolvase and transposase genes and termed the resolution sites. The end products are two separate DNA molecules, each containing a copy of the transposable element. In addition to its role in transposition, Tn3's resolvase acts as a transcriptional repressor to control expression of both the transposase and resolvase genes. It does this by binding to the sequence containing both the resolution site and the promoter regions of both genes.

Less is known about the transposition and regulation of other transposable elements. All known transposable elements generate a short, direct repetition at the site of insertion, albeit of different lengths for different elements. This suggests a common mechanism for cleaving DNA at the insertion site. There is evidence that some transposable elements in both maize and bacteria undergo a nonreplicative type of transposition in which the element is removed from its former insertion site and inserted into a new one. There is also evidence that element-encoded functions are regulated at both transcriptional and translational levels in bacteria, and that the developmental timing as well as the tissue specificity of transposition are regulated in higher organisms.

Summary. Transposable elements are ubiquitous in living organisms. They share enough features of structure, organization, and the transposition mechanism to suggest the existence of underlying universal principles unique to the movement of transposable elements. The elements of bacteria and eukaryotes are not simply interchangeable, however. Each element carries genes that are expressed and regulated correctly in the organism in which it resides, but which will not necessarily allow it to function in an unrelated organism. The growing awareness that transposable elements not only can alter the expression of genes within which or near which they insert, but also can rapidly and dramatically restructure genomes, invites speculation about their role in the evolutionary history of organisms.

For background information see BACTERIAL GENETICS; DEOXYRIBONUCLEIC ACID (DNA); GENE; MUTATION in the McGraw-Hill Encyclopedia of Science and Technology.

[NINA V. FEDOROFF]

Bibliography: K. O'Hare and G. M. Rubin, Structures of P transposable elements and their sites of insertion and excision in the *Drosophila melanogaster* genome, *Cell*, 34:25–35, 1981; R. F. Pohlman, N. V. Fedoroff, and J. Messing, The nucleotide sequence of the maize controlling element Activator, *Cell*, 37:635–643, 1984; K. Ruan and S. W. Emmons, Extrachromosomal copies of transposon Tc1 in the nematode *Caenorhabditis elegans*, *Proc. Nat. Acad. Sci. USA*, 81:4018–4022, 1984; J. A. Shapiro (ed.), *Mobile Genetic Elements*, 1983.

Genetic mapping

The human gene map may be the fastest-growing body of knowledge in human biology. It has been estimated that the human nuclear genome contains 50,000 structural genes (those that code for protein and are present in single copies). These genes are distributed on the 24 different human chromosomes (22 autosomes and the X and Y chromosomes).

The assignment of genes to specific chromosomes and regions of chromosomes is carried out using a combination of techniques, including classical mendelian genetics, somatic cell genetics, and recombinant deoxyribonucleic acid (DNA) technology. Classical family studies alone produce limited information because of the long human generation time, the small size of human sibships, and the small number of genetic markers. New information is limited to linkage relationships; only a few genes have been assigned to specific chromosomes by family studies alone, primarily by exploitation of abnormal chromosomes, X linkage, or gene dosage. A large number of genes, however, have been localized by using somatic cell genetics, a methodology that employs human-rodent cell hybrids that retain only specific numbers and combinations of human chromosomes. The boom in recombinant DNA technology has also produced many new genetic markers which have, in turn, led to a multitude of new gene assignments as well as the prospects for many more.

Somatic cell hybridization. Development of somatic cell hybridization resulted in a significant increase in the number of human genes assigned to specific chromosomes and regions of chromosomes. This mapping methodology is based on the construction of proliferating hybrid somatic cell lines that carry reduced numbers of human chromosomes. To produce the hybrids, human cells are fused with mouse or Chinese hamster cells that carry selectable markers (Fig. 1). The only cells that grow are those hybrid cells that express the human gene which complements the rodent cell growth deficiency. For unknown reasons, human chromosomes are preferentially lost from human-rodent hybrids, resulting in cells that contain various numbers and combinations of human chromosomes. The cells are cloned and their chromosomal content is determined. Panels of hybrid cells containing various combinations of human chromosomes on constant rodent backgrounds have been assembled in a number of laboratories. Any human gene product that is expressed in cell culture (for example, enzymes or cell surface markers) can be assigned to a specific human chromosome by correlating the presence or absence of the genetic marker with the retention or loss of a specific human chromosome, respectively. Hybrid cells containing deleted or rearranged human chromosomes are useful for regional chromosome assignments.

Recombinant DNA technology. Cloned human genes can be mapped to human chromosomes by us-

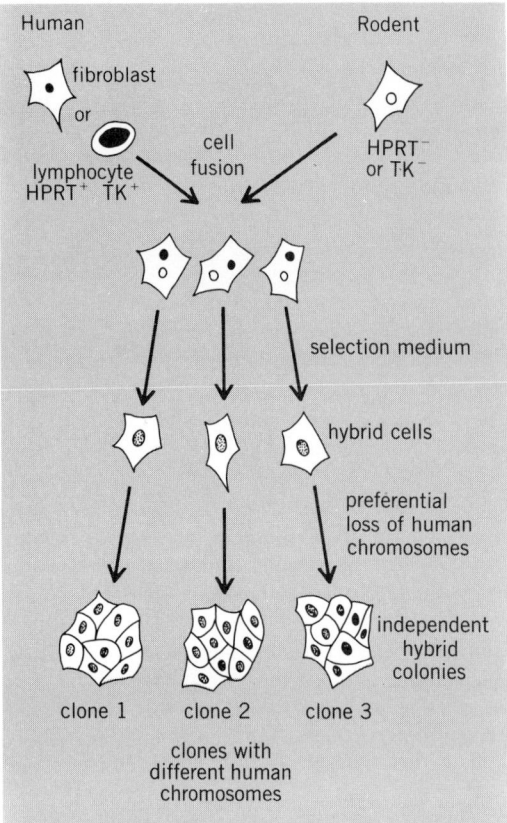

Fig. 1. Scheme for production of somatic cell hybrids for use in human gene mapping. Human cells carrying the normal genes for the enzymes hypoxanthine phosphoribosyl transferase (HPRT) and thymidine kinase (TK) are fused, by using polyethylene glycol or sendai virus, to rodent cells deficient in one of those enzymes. The cells are then grown in a selection medium that kills unfused cells. Different human chromosomes are subsequently lost from each hybrid; clones of these hybrid cells allow mapping of human genes. (*After T. B. Shows, A. Y. Sakaguchi, and S. L. Naylor, Mapping the human genome, cloned genes, DNA polymorphisms, and inherited disease, Adv. Human Genet., 12:341–452, 1982*)

ing somatic cell hybrids. First, the cloned gene is radiolabeled with ^{32}P; this radiolabeled gene is called a DNA probe. Next, DNA is isolated from human-rodent hybrid cells, cut into fragments by using a restriction endonuclease, and separated on the basis of size by agarose gel electrophoresis. Restriction enzymes recognize specific short sequences of DNA and generate double-stranded cuts at those sites. The DNA is transferred by using capillary action from the gel to nitrocellulose filter paper by the method of Southern blotting. The DNA is bound to the filter and then hybridized with the radiolabeled DNA probe. The hybridized blots are exposed to x-ray film, which is developed to reveal specific bands corresponding to DNA fragments that are bound to the probe (Fig. 2). As with the enzyme and cell surface markers, the concordance of a specific hybridization band with a specific human chromosome assigns the cloned gene to that chromosome.

Anonymous DNA fragments. Most of the DNA in the human genome, and thus in recombinant DNA libraries, is anonymous, that is, undefined and of unknown function. The technology to isolate and map these anonymous probes, however, is available. They can be identified during mapping studies using somatic cell hybrids and localized to specific chromosome bands using in-situ hybridization. Thus, not knowing the coding capacity of a single-copy DNA probe does not diminish its usefulness as a genetic marker.

DNA polymorphisms. Restriction-fragment-length polymorphisms are exciting new molecular tools for gene mapping. These polymorphisms result from variations in DNA sequence that can be detected because they either generate or erase restriction enzyme cutting sites. Restriction polymorphisms are inherited as traits and, like genes, have different forms or alleles. A polymorphism located close (or linked) to a deleterious gene will tend to be inherited along with that gene. These polymorphisms are not necessarily linked to lesions in specific genes, but are preexisting DNA sequence variations present in the population. Therefore, the polymorphism allele linked to a particular disease locus in a family is related only to whichever allele was present on the ancestral chromosome where the mutation arose. The polymorphisms produced by one enzyme at two or more alleles can be identified, since different-size fragments can be separated by agarose gel electrophoresis and can be detected with specific DNA probes using Southern blotting, and the form of the polymorphic sequence carried by an individual can be determined.

Huntington's disease. A recent example of human gene mapping illustrates the value of combining the techniques of mendelian genetics (family studies), somatic cell genetics, and recombinant DNA (restriction-fragment-length polymorphisms). Using family studies, James Gusella and coworkers identified an anonymous, polymorphic DNA marker that is tightly linked to the gene for Huntington's disease. Huntington's disease is a dominantly inherited, progressive neurodegenerative disorder which has its onset during the third to fifth decade of life and results in death within about 15 years. Gusella and his colleagues used DNA samples taken from members of a large American family with a history of Huntington's disease to find a DNA marker closely linked to the Huntington gene. After testing only 11 DNA probes, they found one that was present in all affected family members. The marker has four alleles and the affected individuals all carried the same form, whereas those free of the disease carried other alleles. To confirm their results, Gusella's group studied the DNA of the members of a very large Venezuelan family in which there are 100 living victims of the disease and 1100 individuals who are at risk of developing it. As in the American family, it was found that all of the affected Venezuelan family members carried the same form of the polymorphic marker, although it was a different allele than in the American family. The odds are 100 million to 2 that the marker is linked to the Huntington's gene. The marker was then mapped by Susan Naylor to the short arm of chromosome 4 using somatic cell hybrids. Until this significant discovery by Gusella and his colleagues, there was no clue to the location of the Huntington gene and no reliable method for presymptomatic or prenatal diagnosis of this devastating genetic disorder. There is still no effective method of treatment.

In-situ hybridization. The method of in-situ hybridization has been improved so that it can be used to map single-copy genes in addition to repeated sequences. In this technique, a radioactively labeled DNA or RNA molecule is hybridized to fixed metaphase chromosome spreads. The preparations are coated in the dark with photographic emulsion, and after a suitable exposure the latent images that result from the decay of the radiolabeled probe are developed. The grain distribution over the chromosomes is analyzed to determine the chromosomal localization of the specific gene being probed. The chromosomes are stained (banded) to facilitate their specific identification. The combination of in-situ hybridization, somatic cell hybridization, and Southern blotting is a powerful method to accurately determine the chromosomal distribution of single-copy genes and related genes within multigene families.

Flow sorting. The technology of flow cytometry and chromosome sorting has improved tremendously in the past few years. This technology utilizes an instrument in which cells or particles (in this case, chromosomes) in suspension and stained with a fluorescent dye are passed in single file through a narrow laser beam. The fluorescent signals emitted when the laser excites the dye are amplified by complex electronic devices and transmitted to a computer which can be programmed to instruct the flow cytometer to sort cells or particles having specific characteristics into collecting vessels. Chromosomes can be sorted to about 90% purity, and single-chromosome DNA libraries can be prepared. Unfortunately, with present technology, human chromosomes 9 through 12 cannot be clearly resolved, but the use of somatic cell hybrids containing individual chromosomes in this group on a background of larger or smaller rodent chromosomes will enable these chromosomes to be isolated and their DNA cloned.

In addition, DNA can be isolated directly from sorted chromosomes and analyzed for the presence of a specific gene by restriction endonuclease analysis and Southern blotting. For example, the localization of the hemoglobin β and insulin genes to the short arm of chromosome 11 were confirmed by using an X/11 translocation that could be distinguished from the normal chromosome 11 by flow cytometry and sorting. A rapid gene mapping system has been developed in which chromosomes are sorted directly onto nitrocellulose filter paper, the DNA is denatured, radioactive DNA probes are tested as in Southern blotting, and chromosome assignments are made.

New genetic markers, cloned genes, and infor-

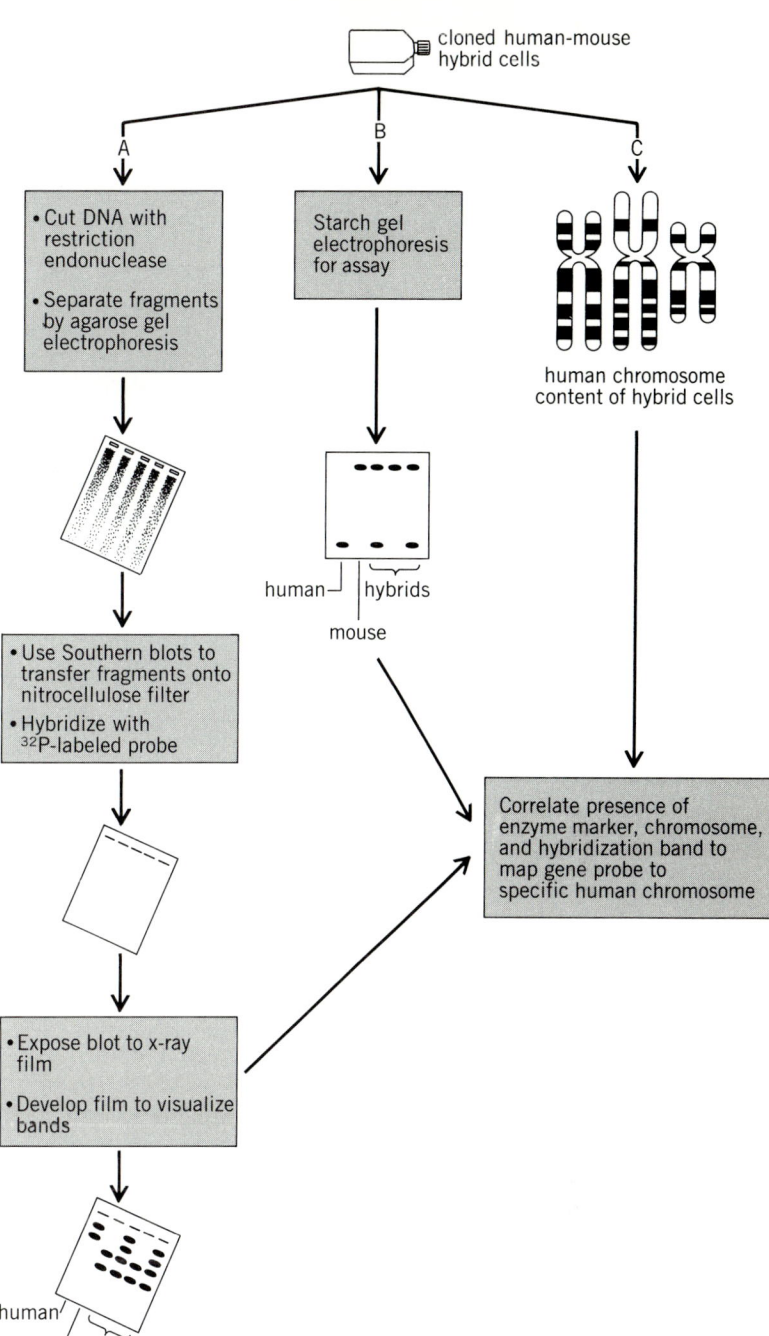

Fig. 2. Mapping cloned genes in human-rodent somatic cell hybrids by Southern blotting. Hybrid cells are used for (A) preparation of DNA, (B) extraction for assessment of enzyme markers, and (C) karyotyping. Correlation of the presence of human-specific bands on the blots with the results of the enzyme marker analysis and karyotyping is used to assign a gene to a particular chromosome. (*After T. B. Shows, A. Y. Sakaguchi, and S. L. Naylor, Mapping the human genome, cloned genes, DNA polymorphisms, and inherited disease, Adv. Human Genet., 12:341–452, 1982*)

mation about their location on the human genetic map have led to practical applications, such as identification of carriers of particular genetic disorders, prenatal and postnatal diagnosis of individuals with genetic diseases, and identification of disease loci. Further understanding of the human genetic map will lead to greater insight into gene regulation and function and to strategies for therapy of genetic diseases.

For background information *see* GENETIC ENGINEERING; GENETIC MAPPING; SOMATIC CELL GENETICS in the McGraw-Hill Encyclopedia of Science and Technology.   [SUSANNE M. GOLLIN]

Bibliography: J. F. Gusella et al., A polymorphic DNA marker genetically linked to Huntington's disease, *Nature*, 306:234–238, 1983; R. V. Lebo et al., High-resolution chromosome sorting and DNA spot-blot analysis assign McArdle's syndrome to chromosome 11, *Science*, 225:57–59, 1984; T. B. Shows, A. Y. Sakaguchi, and S. L. Naylor, Mapping the human genome, cloned genes, DNA polymorphisms, and inherited disease, *Adv. Human Genet.*, 12:341–452, 1982; B. U. Zabel et al., High-resolution chromosomal localization of human genes for amylase, proopiomelanocortin, somatostatin, and a DNA fragment (D3S1) by *in situ* hybridization, *Proc. Nat. Acad. Sci. USA*, 80:6932–6936, 1983.

Genetics

Histones are basic proteins found associated with deoxyribonucleic acid (DNA) in chromosomes; they play a major role in the structural and functional organization of the genetic material. There are five classes of histones, each defined by its electrophoretic mobility and by its content of the amino acids lysine and arginine. The H1 class is richest in lysine; H3 and H4 are richest in arginine; and the H2A and H2B classes are slightly rich in lysine. The H1 class contains up to six distinct subtypes that differ in primary structure. The other histone classes are highly conserved and contain fewer subtypes. These are attributed primarily to modifications made after synthesis of these molecules and not to their primary structure.

Gene expression may be controlled at the level of transcription into ribonucleic acid (RNA), of processing and transporting the transcript, of translation of the transcript into protein, or of posttranslational modification of the protein product. In their search for the means by which the expression of genes is controlled, molecular biologists theorize that histones, especially H1, are an integral part of the total process involved in promoting or repressing gene transcription.

Histones and chromatin structure. Genes are located on chromosomes in linear arrays, as sequences of the nucleotides that constitute DNA. Chemically chromosomes are known to contain proteins and RNA in addition to DNA; the term chromatin is applied to the aggregate of proteins and nucleic acids of the chromosome. The proteins of chromatin are divided into two groups, the histones and the nonhistones. While histones are a group of well-defined basic proteins, the nonhistones include a large number of uncharacterized proteins. The RNA of chromatin is a minor component (about 3%) representing nascent chains at various stages of transcription.

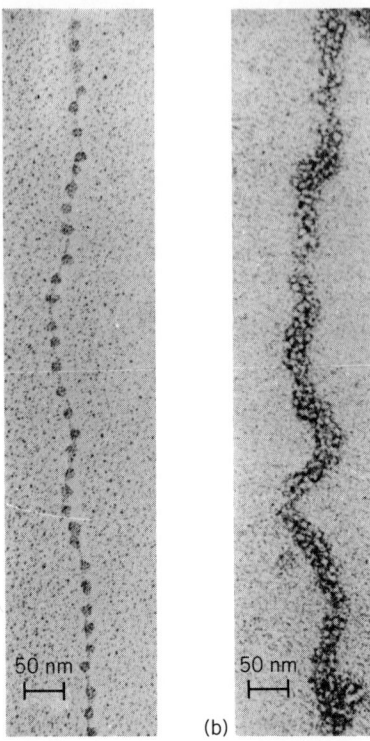

Fig. 1. Electron micrograph of decondensed chromatin (a) in the beads-on-a-string form showing beads (11 nm in diameter) connected by linker DNA (*courtesy of Victoria Foe*), and (b) showing the native structure of chromatin as a fiber (30 nm in width) of condensed nucleosomes (*courtesy of Barbara Hamkalo*).

Roger Kornberg proposed a structure for chromatin in 1974 based on dissection of chromatin by nuclease digestion and on the results of x-ray diffraction analysis. The basic unit of chromatin structure consists of a repeating subunit called a nucleosome, composed of DNA and histones. Each nucleosome core has been found to be composed of an octamer of histone molecules, two each of H2A, H2B, H3, and H4, surrounded by 146 base pairs of superhelical DNA. The nucleosome core particles, or beads, are connected by a stretch of linker DNA approximately 60 base pairs in length (Fig. 1a). The linker DNA plus the nucleosome bead constitutes the entire nucleosome, which therefore contains about 200 base pairs of DNA. Current evidence suggests that the 146 base pairs surrounding the histone core make 1¾ turns around the octamer.

Individual nucleosome beads can be detached from the linker DNA by the action of a bacterial enzyme, micrococcal nuclease. This enzyme degrades only the DNA between beads; the remainder is protected by the bound octamer of histones. One molecule of H1 histone is located just outside the nucleosome bead and associates simultaneously with both ends of the linker DNA as it enters and exits the nucleosome, spanning 165 base pairs (two complete turns) of DNA (Fig. 2). Nonpolar amino acids are clustered into a globular form in the middle of the H1 molecule; the positively charged polar ends, consisting of lysine and proline residues, bind with the negatively charged phosphate groups of the linker DNA. In this position, the various subtypes of H1 molecules are thought to control the superstructure of chromatin, influencing the space between nucleosome beads.

An average-sized protein molecule contains about 400 amino acids and therefore requires about 1200 DNA base pairs in the gene that codes for it. Since there is one nucleosome per 200 base pairs, an average gene will contain six nucleosomes.

Chromatin in the living cell. In the nondividing, metabolically active cell, chromatin is in a contracted state and rarely assumes the beads-on-a-string configuration described above. When chromatin is prepared by gentle means, most of the chromatin is seen as a fiber with a diameter of about 30 nanometers (Fig. 1b). The binding of chromatin into clusters of nucleosomes in the 30-nm fiber is apparently a function of histone H1: it binds the DNA that enters and leaves the nucleosome into a regular repeating array (Fig. 3). H1 histones exhibit reversible cooperative binding to DNA, in that H1 molecules tend to bind in clusters; any one H1 molecule has a strong tendency to bind next to another bound H1 molecule rather than on some DNA region relatively devoid of H1 molecules.

Since it has been demonstrated that the various subtypes of H1 histone differ in their ability to condense DNA, it is reasonable to assume that the structural subtypes of H1 histone may behave differently from one another in their pattern of packing the nucleosomes in chromatin. H1 histones also vary in the presence or absence of a phosphorylation site, a factor which could drastically influence DNA configuration. This site is subject to phosphorylation by hormonal induction and is not related to condensation occurring during the cell division cycle.

Active versus inactive chromatin. Only a portion of the genome operates in a given tissue of a multicellular organism. Maximally, only 2 to 8% of the DNA of the genome is represented by messenger RNA (mRNA) in the cytoplasm. Further, each tissue has a unique set of mRNAs. A similar situation exists in oocytes and embryos, where sets of genes are switched on or off during the various stages of oogenesis and development. In the genome as a whole, DNA is transcribed to provide cytoplasmic messenger, ribosomal, and transfer RNAs, as well as a heterogeneous class of nuclear RNAs. Thus, some regions of chromatin are actively being transcribed, while others are quiescent. It is believed that active chromatin retains its complement of core histones in many if not all instances, even though these structures seem to disappear from active chromatin in the electron microscope. It is probable that core histones are modified in active chromatin, and it is known that H1 histone is lacking from active chromatin, leading to the decondensed state associated with gene activation.

Core histones may be acetylated and deacetylated in minutes in the living cell. Acetylation serves to

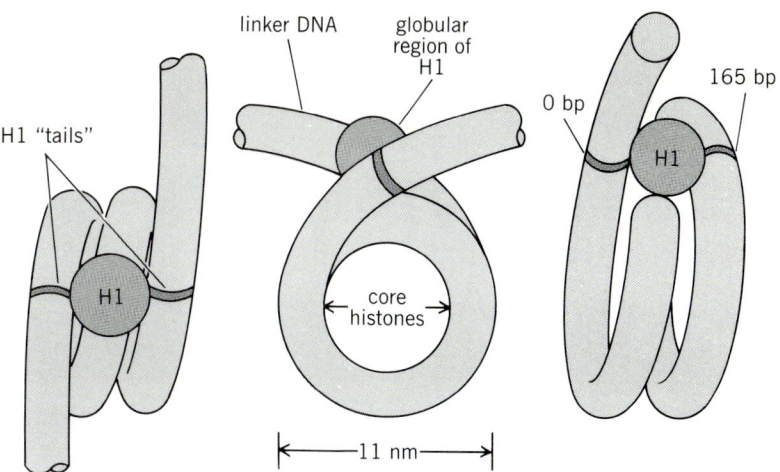

Fig. 2. Diagram of a model of the nucleosome showing the globular portion of the H1 histone molecule and the two lysine-rich tails binding across two complete turns of superhelical DNA. The H1 histone molecule spans 165 base pairs, including 146 in the nucleosome bead and 19 of the linker DNA. The core histone octamer occupies the central space. (*After J. Allan et al., The structure of histone H1 and its location in chromatin, Nature, 288:675–679, 1980*)

neutralize the charge on the lysine residue, thus loosening the association between histone and DNA and making the DNA strand more accessible to RNA polymerases and to DNases. Histone acetylation is enhanced in regions of active chromatin, and active chromatin is more susceptible than inactive chromatin to DNase I, an enzyme that causes "nicks" in DNA. Changes in the core histone through acetylation seem to parallel gene activation, though the mechanics of the movement of enzymes concerned with transcription through the modified nucleosomal core are not understood.

A number of reports have documented the depletion or absence of H1 histone from active chromatin, leading to the change from the 30-nm inactive chromatin fiber to the actively transcribed 11-nm beads-on-a-string configuration. Vadim Karpov and associates recently reported a depletion not only of H1 histone but of core histones as well, leading to linearized DNA in the actively transcribed region of the *Drosophila* genome coding for heat shock proteins. In this study, the depletion of H1 occurred first in the sequence of changes. The absence of histones correlated well with high nuclease sensitivity and the disappearance of the nucleosome pattern in micrococcal nuclease digests of active chromatin. The loss of H1 and core histones is assumed to be cooperative and reversible as the genes are activated and repressed. It is possible in this case that RNA polymerase molecules reversibly displace H1 and core histones from DNA, leading to chromatin decondensation and gene activation.

Summary. It is apparent that histone molecules do not contain sufficient information within themselves for the initiation of the schedule of chromatin dynamics during the program of gene activation. It is equally apparent that drastic alterations in chromatin structure accompany gene activation. During

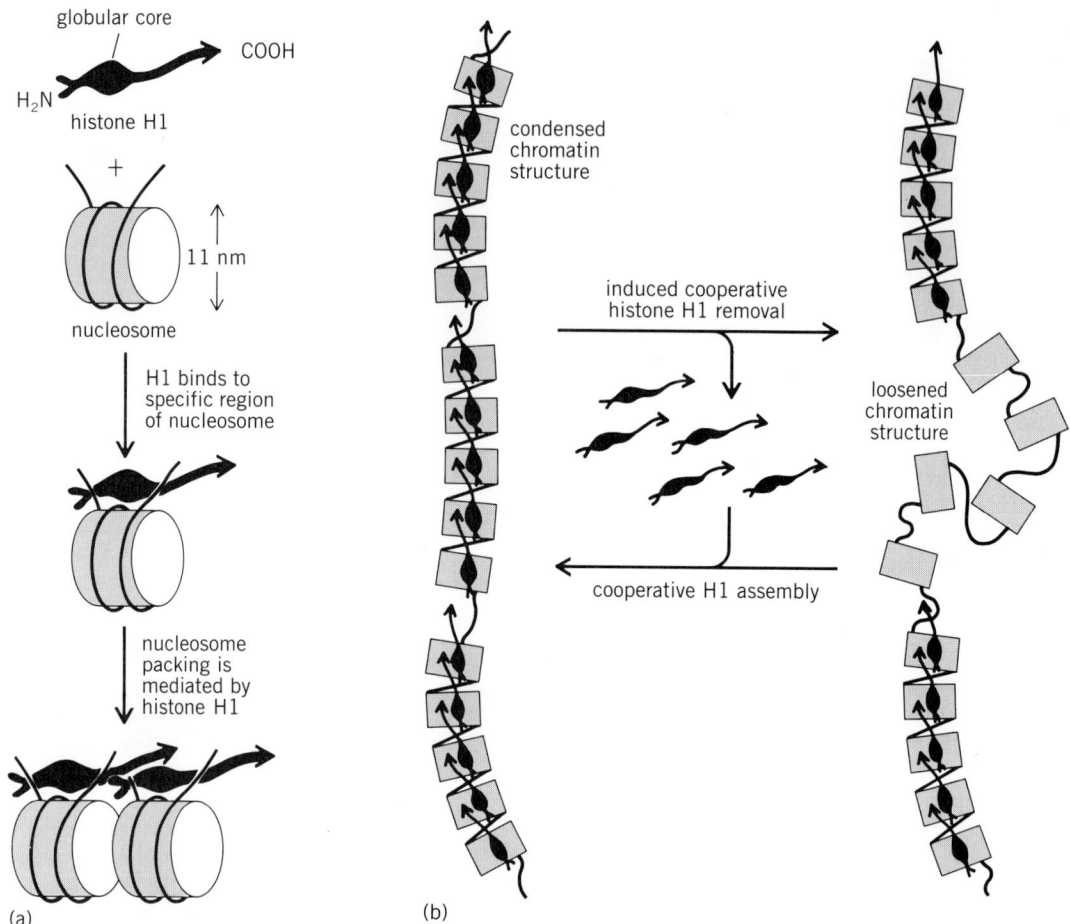

Fig. 3. H1 histone (a) binding and nucleosome packing and (b) reversible cooperative binding to DNA forming condensed and loosened chromatin. A chromatin unit is depicted suddenly being decondensed upon receiving some external regulatory signal. Chromatin relaxation of this type is thought to underlie the process of gene activation. (After B. Alberts et al., Molecular Biology of the Cell, Garland Publishing, 1983)

development of the sea urchin (and other species) the subtype of H1 produced is drastically changed from the morula (late cleavage) stage to the gastrula. Such shifts have been correlated with different nucleosome repeat lengths in the total chromatin, presumably reflecting overall changes in transcription patterns at different stages of development. The existence of heterogeneity among H1 molecules provides a mechanism for control of the physical state of chromatin at different times in the life history of a cell. R. D. Cole has recently likened the role of H1 histone in gene activity to that of the disulfide bridge in stabilizing the folding of a protein. Assuming that the various subtypes of H1 differ in their ability to condense DNA, and perhaps chromatin as well, they could act to stabilize patterns of cell activity and cell differentiation dependent on specific gene readout by opening and closing segments of the genome to transcription. The system which sets in motion the changes in H1 and core histones requisite for transcription of DNA remains unknown.

For background information see CHROMOSOME; GENE; GENE ACTION; NUCLEOPROTEIN in the McGraw-Hill Encyclopedia of Science and Technology.

[JOHN W. BROOKBANK]

Bibliography: R. J. Arceci and P. R. Gross, Histone variants and chromatin structure during sea urchin development, Develop. Biol., 80:186–209, 1980; R. D. Cole, A minireview of microheterogeneity in H1 histone and its possible significance, Anal. Biochem. 136:24–30, 1984; V. L. Karpov, O. V. Preobrazhenskaya, and A. D. Mirzabekov, Chromatin structure in hsp 70 genes, activated by heat shock: Selective removal of histones from the coding region and their absence from the 5′ region, Cell, 36:423–431, 1984; R. D. Kornberg, Structure of chromatin, Annu. Rev. Biochem., 46:931–954, 1977.

Geochemical prospecting

Recent advances in geochemical prospecting include the application of fluid inclusion analysis and its use in mineral exploration, and biological methods in mineral exploration.

Fluid inclusion analysis. The use of fluid inclusions in the examination and evaluation of mineral deposits has been a common form of study for several decades. The point of such study is, however, largely oriented toward determinations of the physicochemical parameters, such as temperature, pressure, and fluid composition, present during mineralization. While informative, the data produced from this approach are of only indirect use in terms of the discovery of new deposits, or of new pay zones within known deposits. Since 1970, however, fluid inclusion analysis techniques, both classical and newly developed, have emerged as geochemical prospecting tools of considerable potential.

Unlike many other geochemical methods, fluid inclusion analysis is often ambiguous and complex. Individual differences between deposits will be reflected in the character of fluid inclusions formed, and a clear awareness of the particular nature of the deposit sought is necessary to avoid misinterpretation and the loss of time, money, and effort.

Nature of inclusions. Fluid inclusions are discrete bodies of fluid (with or without solid phases) that are trapped within crystals of most minerals [both ore and non-ore (or gangue)] normally associated with mineral deposits. This trapping may occur during crystal growth, or during later fracturing and subsequent "healing over" of the crystal. Inclusion size may range from less than 5 micrometers to over 10 mm, but the bulk of inclusions lie between 1 mm and 10 micrometers.

The fluids most commonly found as inclusions are liquid H_2O and CO_2 and vapor-phase H_2O, CO_2, sulfur dioxide (SO_2), and hydrogen sulfide (H_2S). Solid phases may also precipitate from included fluids; these are referred to as daughter minerals. Since most ore-forming fluids are brines, the most commonly seen daughter minerals are chloride salts, especially halite (NaCl) and sylvite (KCl).

The utility of these inclusions lies in their identity as direct samples of the fluids present at some time during the deposition of an ore body. The validity of this identity hinges upon the degree to which the inclusion has behaved as a closed and isochemical system after its formation. Any leakage, exchange of components with its surroundings (that is, the host crystal), volume changes, or other open-system behavior alters the conditions within the inclusion and changes the parameters recorded therein.

Methods of analysis. There are a number of methods used in the analysis of fluid inclusions. Many of the classic methods involve optical techniques, using doubly polished sections of rocks or minerals thin enough to transmit light. Most common among these methods are the determination of homogenization and freezing temperatures.

Since the fluids involved in the formation of many types of ore deposits are brines at elevated temperatures (up to several hundred degrees Celsius), cooling to ambient temperatures results in significant thermal contraction of the fluid. If the inclusion was completely liquid at the time of trapping, this contraction will result in the formation of a vapor bubble within the inclusion. The temperature at which the fluid expands to completely fill the inclusion, causing this bubble to disappear, is called the homogenization temperature and is presumably close to the original fluid temperature. A pressure correction must be applied to homogenization temperatures for deposits formed at depth, usually determined by estimation of the thickness of rock overlying the deposit at the time of formation. The temperature of freezing is of interest because it reflects the salinity of the fluid (expressed as equivalent weight percent NaCl), which directly affects the ability of the ore fluid to transport (and therefore deposit) metals.

Other optical methods in frequent use include measurement of overall inclusion abundance, which reflects the inclusion density within the host mineral, inclusion size, and the type of inclusions present. The last usually involves the number and character of phases present within inclusions, which may be broken down into a variety of classes (Fig. 1). Daughter minerals may also be identified optically, providing insights into fluid chemistry.

Other analytical methods involve actually break-

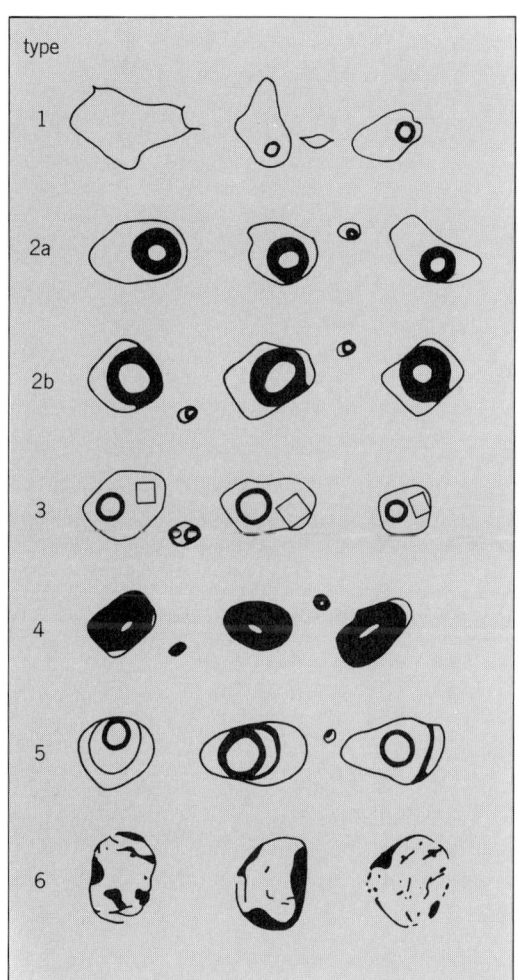

Fig. 1. Classification scheme for fluid inclusions in granites from southwestern England. Types 1 and 2 are aqueous inclusions with varying proportions of liquid and vapor. Type 3 fluid inclusions contain cubes of halite. Type 4 are gas-rich, and type 5 contain gas plus liquid CO_2 and aqueous solution. Type 6 are solid inclusions with minor vapor—perhaps solidified melt inclusions. (*After D. H. M. Alderton and A. H. Rankin, The character and evolution of hydrothermal fluids associated with the kaolinized St. Austell Granite, southwest England, J. Geol. Soc. Lond. 140:297–309, 1983*)

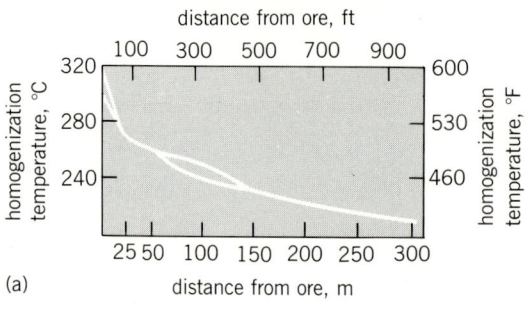

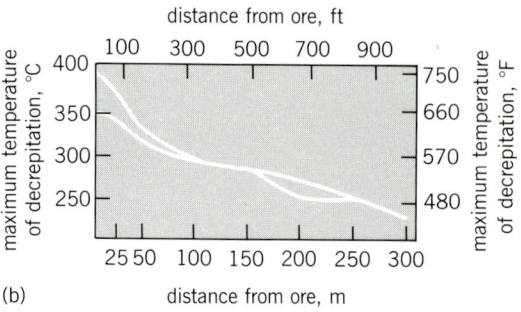

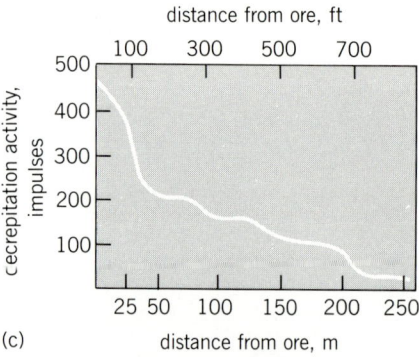

Fig. 2. Variance of (a) homogenization and (b) decrepitation temperatures, and (c) decrepitation activity with distance from the ore body at the Sandonoky poly metallic vein deposit in the Soviet Union. Branches in the curves represent two sets of data. (*After N. P. Ermakov, and A. G. Kuznetsov, The use of thermobarogeochemistry methods in the search for hidden ore deposits, Fluid Inclus. Res., 4:122–125, 1971*)

ing open the inclusion and releasing the fluid within. This is generally done by rapidly heating the host crystal until the fluid pressure within the inclusion exceeds the mechanical strength of the crystal, a process called decrepitation. Decrepitation temperatures may deviate considerably from homogenization temperatures, the magnitude of deviation depending on a host of factors such as rate of heating, the physical characteristics of the host crystal, and the geometry of the inclusion. Once decrepitated, analyses performed may be as straightforward as measurement of sample weight loss during decrepitation or of the number of decrepitations occurring relative to temperature, a characteristic termed decrepitation activity. The fluids released may also be directly analyzed for composition, whether chemical or isotopic, by a host of methods ranging from wet chemical to plasma emission spectroscopy.

Exploration methods. One of the most common uses of fluid inclusions as prospecting tools in recent years has revolved around the search for aureoles surrounding ore deposits. These include any kind of anomalous characteristic which goes beyond the immediate extent of, and changes progressively toward, the ore body, thereby providing a method of remote detection and a directional vector indicating the location of the ore body.

Characteristics typically used in this way are homogenization temperature, weight loss on heating, and decrepitation temperature and activity. Soviet researchers, in particular, have reported a number of instances in which distinct halos of anomalous temperatures or decrepitation activities are detectable to several meters or tens of meters around an ore body. In almost all cases, the temperatures and activity increased dramatically with closer proximity to the ore body. In one case, analysis revealed an outer halo with a decrepitation activity of 5–7 times background levels, and an inner halo of 15–20 times background, resulting in a bull's-eye pattern with the ore body at the center (Fig. 2). In addition, exploration companies frequently use homogenization temperature analyses from field or drill core samples to determine if conditions are conducive to formation of a specific type of deposit, and so are able to eliminate unpromising prospects early.

Specific types of inclusions may also offer clues as to the location of a deposit. For example, high-vapor inclusions have been noted to increase in abundance immediately adjacent to some types of deposits. This is often referred to as steaming, or streaming through aureoles. Another instance is found in a British study of tin-tungsten-copper-bearing granites in Cornwall, where high-CO_2 inclusions were found only in association with tungsten-bearing vein swarms. As a result, these inclusions could conceivably be used as an effective exploration tool in detecting these vein systems.

Physical characteristics of inclusions also have considerable potential in mineral exploration. For example, overall inclusion abundance and size have been shown to correlate well with copper soil anomalies at a large copper deposit in Indonesia. The causal connection behind this correlation may be the dependence of both the degree of inclusion formation and of copper mineralization on the density of small-scale fractures in the host rock. These fractures provide passage for the solutions which are responsible for both inclusions and mineralization. This link between inclusion abundance, in particular, and the degree of mineralization has been demonstrated for several deposits of different types.

A more direct approach is the analysis of the fluids themselves. Detection of a metal anomaly in included fluids may be the most elegantly straightforward method yet devised of revealing the passage of metalliferous fluids through a specific area, permitting discovery of the mineralizing fluid conduits, which may lead to ore deposits. Work of this type on uranium-bearing inclusions has already been performed. This type of analysis grows more attractive as new methods are evolved that provide ever-decreasing detection levels and require smaller sample sizes. Isotopic compositions of inclusion fluids may also prove useful as a method of detecting changes in fluid character in the vicinity of a mineral prospect. [DAVID I. NORMAN]

Biological methods. Recently there have been interesting developments in the application of biological methods to mineral exploration. The two main branches of this field are biogeochemistry and geobotany. In geobotany, certain indicator plants tend to occur in mineralized soils over ore deposits. In biogeochemistry, the trace-element content of vegetation is used to identify areas that contain economic mineralization.

Bacteria. Perhaps the most novel application of indicator methods has been work on bacterial species that thrive in mineralized areas. This application is based on the research of Alexander Fleming in the 1930s on bacterial resistance to penicillin and to toxic metals. However, studies on bacterial distributions in marine sediments from the New York Bight first alerted researchers to this possible use of bacteria in mineral exploration. It was found that bacteria naturally resistant to penicillin tended to occur in those sediments which had the highest concentration of mercury and cadmium from industrial pollution.

Following this lead, researchers cultured soils collected over three base-metal and gold deposits in the western United States. The soils with a higher content of trace metals contain species of *Bacillus* with increased penicillin resistance. These soils apparently favor the development of *Penicillium* spp., which are metal-tolerant molds that produce penicillin and a variety of other antibiotics. The dominant species of *Bacillus* over the deposits, *B. cereus*, is resistant to these antibiotics and can also use the mycelium of the mold as a source of nutrients. Whether this will emerge as a practical exploration method will depend on much more extensive field trials. For example, there will be little point in measuring the population of *B. cereus* in soil samples if it is easier to determine the concentration of one or more indicator elements in the soils. The method will be most effective if it can outline mineralization in areas where there are no readily detectable geochemical anomalies. Tentative evidence of this promise has been found in 1000 × background counts of *B. cereus* over a large porphyry molybdenum deposit at Pine Grove Creek, Utah, that is covered by almost 3300 ft (1000 m) of transported overburden.

Trees. In the spruce forests of northern Saskatchewan, there are several potential advantages in using biogeochemical methods in the search for gold. The country is mostly covered by a blanket of glacial till. Plant roots penetrate this till to draw nutrients from material closer to the bedrock source of the mineralization. These roots create a highly corrosive micro-environment within which gold and other metals can be dissolved and absorbed by the plant. Access to this swamp and insect-ridden region during the summer to sample soils for conventional geochemical surveys can be tedious. Trees can be sampled in winter, when travel by snowmobile is easier.

Individual plants show great variability in the trace-metal content of their different organs. In one 25-year-old jack pine the gold content of the ash of 17 different parts was measured. Gold in this one tree varied by an order of magnitude, from 14 ppb in the inner trunk to 140 ppb in the scales of the outer bark. This might indicate that the outer bark is the best material to sample. However, collecting bark samples can be time-consuming; twigs are easier to collect and have an intermediate content of gold. In this region of Canada, alders are the most sensitive to gold in the soils, followed by balsam fir, jack pine, birch, spruce, and willow.

A very significant factor in the recent advances in biogeochemistry is that far more sensitive and rapid analytical techniques are now available. Neutron activation analysis was used for gold in the ash of the alder twigs. This permits gold contents as low as 5 ppb to be detected in ash samples weighing only 0.35 oz (1 g). Analytical sensitivity is obviously of paramount importance for an element as rare as gold. However, it is an advantage for all biogeochemical surveys, since the sampler does not wish to carry away a mass of vegetation from each sample site. It has been shown that ashed alder twigs can very effectively outline zones of gold mineralization in northern Saskatchewan. Because gold occurs as rather rare particulates, it is irregularly distributed in rocks and soils. The alders, however, effectively sample several cubic meters of soil and till, and thus provide a more representative sample.

Soil. Exploration geologists are generally more interested in the ground than in the vegetation that grows on it. In the recent hunt for gold, forest litter or humus is the predominant sample medium for routine surveys within forested regions. In the Canadian Shield, much of this activity was stimulated by the discovery of the Hemlo gold deposit in Ontario in 1981. This appears to be one of the largest single deposits of gold ever found in North America. The deposit is transected by Canada's main east-west highway, and it is exposed below a moderate cover of till. Along with the gold at Hemlo, there are concentrations of antimony, mercury, thallium, molybdenum, and barium two to three orders of magnitude greater than background levels in rock and soil. In such cases the analysis of vegetation need not be restricted to the metal commodity being sought; often it is better to use another, indicator element. For example, antimony is usually more mobile in soil and overburden than gold. Anomalies for antimony in vegetation may, therefore, be more widespread than those for gold, allowing wider sampling intervals and cheaper costs.

Termites. Animals as well as plants have a role in exploration. Termite mounds, for example, have proven useful in the Leopard mine in Zimbabwe. Termites must have water, and during the dry season the only source of water for them in the mine is underground fissures. In this mine the water-carrying passages extend down 200 ft (60 m) to the water table. To clear these passages, the termites must remove material from the fissures, which is then deposited at the surface in mounds. Many mineral deposits are formed along faults and fissures, so that, in effect, the termites are very effective miners. The termite mounds over the Leopard mine had more than 3 ppm of gold. The overburden in this area is Kalahari sand, which effectively blankets exploration of the underlying rock, since it is not of local origin but has drifted in from elsewhere. This preliminary conjecture was eventually followed up and was successful, as is attested by the discovery and operation of the Termite Gold Mine, Zimbabwe.

Remote sensing. Much of the future development of biological methods of exploration will be from the air or outer space. Remote sensing by aircraft or satellite depends to a large extent on the appearance of vegetation. This may occasionaly reveal the presence of economic mineralization. However, it is more likely that remote sensing will reveal the faults and other structures, or hydrothermal alteration that are associated with mineral deposits. At the outset it was hoped that *Landsat* imagery alone would lead to the discovery of mineral deposits. However, in modern practice this imagery is incorporated into a package of remote-sensing methods, which usually include a variety of airborne geophysical measurements. One of the most interesting applications of airborne geobotanical sensing was the mapping of the Sokli carbonatite complex in Finland. Carbonatite rock contains a relatively high content of phosphorus compared to the surrounding rocks, and so stimulates the growth of plants. Prior to airborne mapping, the size of the complex was thought to be 8 mi^2 (20 km^2). After mapping, this was revised to 60 mi^2 (150 km^2).

Apart from sensing the physical properties of vegotation from above, they may also be sampled from aircraft. The Airtrace and Surtrace systems of airborne geochemistry are based on the principle that metal-bearing organic compounds accumulate on leaf and stem surfaces. In the Surtrace method a vacuum probe carried by a helicopter is drawn through the vegetation, which collects the surface microlayer of the plants. Within the helicopter this material is cleaned and sorted in a cyclone and impacted on a moving roll of high-purity tape. Back at the laboratory the tape is drawn through a laser, volatilizing the plant material, which then enters the plasma of an argon plasma emission spectrometer. The result is a multielement profile of the composition of the vegetation along the track of the helicopter. It has been shown that these profiles can identify changing bedrock compositions and the presence of mineralization.

For background information *see* GEOBOTANICAL INDICATORS; GEOCHEMICAL PROSPECTING; REMOTE SENSING in the McGraw-Hill Encyclopedia of Science and Technology.

[EION M. CAMERON]

Bibliography: R. R. Brooks, *Biological Methods of Prospecting for Minerals*, 1982; D. Carlisle et al. (eds.), *Biological Systems and Mineral Exploration*, Rubey Colloquium Series, vol. 5, 1984; A. R. Chivas and R. W. T. Wilkins, Fluid inclusion studies in relation to hydrothermal alteration and mineralization at the Koloula Copper Prospect, Guadalcanal, *Econ. Geol.* 72:153–169, 1977; A. H. Rankin et al., Determination of U:C ratios in fluid inclusion decrepitates by inductively coupled plasma emission spectroscopy, *Mineralog. Mag.* 46:179–186, 1982; E. Roedder, Fluid inclusions as tools in mineral exploration, *Econ. Geol.*, 72:503-525, 1977; E. Roedder and A. Kozlowski (eds.), *Fluid Inclusion Research: The Proceedings of COFFI*, 1971.

Gold, geochemistry of

Gold is a very widespread but geochemically scarce element found in all types of terrestrial igneous, sedimentary, and metamorphic rocks and in meteoritic and lunar samples (Table 1). In spite of its scarcity, the element generally occurs in nature as a free (native) metal phase, either gold or electrum, which is an alloy of gold and silver. This is because of its noble nature, that is, its failure to form oxides or sulfides; the only anion with which it bonds to form important minerals is tellurium (Table 2).

Geochemical abundance and compositions. Naturally occurring gold and electrum alloys contain up to about 50 wt % silver. Copper and mercury are commonly present in the range 0.01–0.1% and only rarely reach levels of 2–5% or higher; the copper content generally rises with the gold content. Traces of iron, nickel, platinum, and palladium may also be present, and rare alloys rich in platinum and antimony have been found. Although gold and silver form a complete solid solution above 300°C (570°F), nearly all naturally occurring gold-silver alloys have

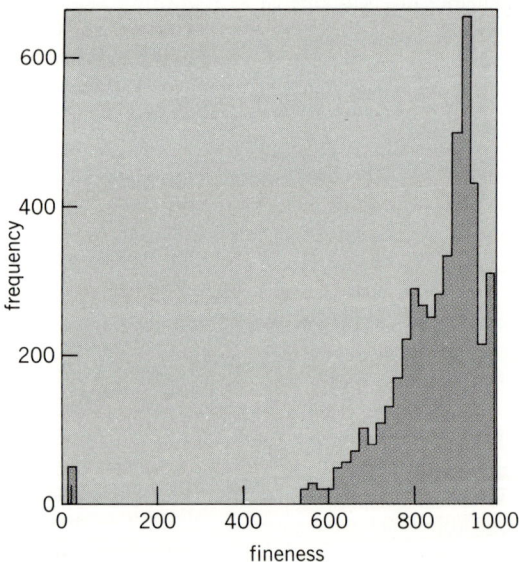

Fineness of naturally occurring gold and gold-silver alloys on the basis of more than 4600 analyses from worldwide localities.

Table 1. Geochemical abundance of gold in rocks and waters

Rock type	Range, ppm	Mean, ppm
Cosmic	—	0.0004
Lunar rocks: basalts	0.00001–0.0007	0.0001
breccias	0.0004–0.0057	0.0021
Meteorites: chondrites	0.0170–1.6000	0.2225
achondrites	0.0005–0.2800	0.0310
siderites	0.0550–8.7440	1.2050
tektites	0.0003–0.0242	0.0081
Earth's crust	—	0.0035
Igneous rocks		
Ultrabasic (dunite, peridotite, pyroxenite, and so on)	0.0002–0.7800	0.0114
Basic intrusive (norite, gabbro, and so on)	0.0003–0.6800	0.0230
Basic extrusive (basalt and so on)	0.0001–0.2300	0.0174
Intermediate—intrusive (granodiorite, syenite, monzonite, and so on)	0.0001–0.3500	0.0075
Intermediate—extrusive (andesite, dacite, trachyte, and so on)	0.0001–0.0650	0.0129
Acid—intrusive (granite, aplite, and so on)	0.0002–2.9000	0.0114
Acid—extrusive (rhyolite, obsidian)	0.0001–0.1130	0.0037
Sedimentary rocks		
Sandstone, arkose, conglomerate	0.0003–2.5000	0.0523
Shale, mudstone, siltstone, argillite	0.0001–0.8000	0.0080
Pyritic black shale, and so on	0.0025–2.1000	0.1320
Limestone, dolomite	0.0002–0.0889	0.0070
Evaporites (halite, gypsum, and so on)	0.0005–0.0850	0.0220
Metamorphic rocks		
Quartzites	0.0002–1.1500	0.0317
Gneisses, granulites	0.0002–0.3000	0.0031
Schists	0.0001–3.7000	0.0186
Marble	0.0002–0.1000	0.0132
Amphibolites, greenstones	0.0001–0.1000	0.0071
Natural waters		
Ocean	0.00001–0.044 ppb	0.011 ppb
Streams and rivers	0.0002–4.7 ppb	0.01 ppb
Hot springs	0.0004–2.2 ppb	0.5 ppb

compositions of approximately 50–100 wt % gold. The distribution of compositions in terms of fineness ($[Au/(Au+Ag)]$ wt % $\times$ 1000) shown in the illustration suggests that the gold/silver ratios are controlled by natural buffers and that there may well be a miscibility gap in the silver-rich portion of the system at low temperatures.

Economic gold-bearing deposits. Gold is recovered commercially from many ore deposits, but it is a principal metal in only the types summarized in Table 3. In order for gold to be recovered as the principal metal from lode deposits—those formed within the host rocks as veins, replacement, or disseminations—the concentration, or grade, of the ore must be at least on the order of 3 ppm (about 0.1 troy oz per ton or 3 g per metric ton), a value more than 1000 times the average crustal abundance.

Table 2. Important gold-bearing minerals

Mineral name	Formula
Gold	(Au,Ag); Au > 80%
Electrum	(Au,Ag); Au = 50–80%
Calaverite	$AuTe_2$
Krennerite	$AuTe_2$
Petzite	Ag_3AuTe_2
Sylvanite	$AuAgTe_4$

Placer deposits, containing gold grains weathered out of lode deposits, are widespread and can frequently be profitably worked at much lower concentrations because the unconsolidated sediments may be processed at much less cost. Most modern placer deposits are localized and small, but the Witwatersrand District of South Africa, the world's principal producing area and reserves, contains a series of gold-bearing deltaic placer deposits that formed approximately 2 billion years ago.

The most important gold-bearing lode deposits consist either of veins deposited by hot, aqueous fluids (hydrothermal solutions) or of plate- to lenslike bodies formed as a result of hot-spring activity. The veins are commonly associated with acidic to intermediate intrusive igneous rocks that provided the heat and some of the fluids and metals; frequently, metal has been leached from the surrounding rocks by convecting groundwater which accounts for most of the fluid passing through the veins. The hydrothermal fluids sometimes cause broad alteration zones in which microscopic to submicroscopic gold occurs. Veins also occur in moderate- to high-grade metamorphic terrains, and are often attributed to hydrothermal fluids expelled during metamorphism or tectonic deformation.

Gold-bearing submarine massive to disseminated

Table 3. Major types of gold-bearing deposits

Type	Mode of occurrence	Examples
Hydrothermal-epithermal	Quartz-rich veins and disseminations related to intrusive igneous bodies. Probably formed as a result of subsurface boiling within subaerial hot-spring systems. Gold is commonly only a minor constitutent relative to Cu, Zn, Pb, Ag; it is commonly associated with FeS_2, FeAsS, or tellurides.	Kirkland Lake, Canada; Hollinger, Canada; Sunnyside, Colo.; Creede, Colo.; Kalgoorlie, Australia; Carlin, Nev.; Getchell, Nev.; Cripple Creek, Colo.
Metamorphic	Quartz-rich veins in moderate- to high-grade metamorphic terrains. Also present are pyrite (FeS_2) and carbonates and trace amounts of Pb, Zn, Cu sulfides.	Dahlonega, Ga.; Mother Lode, Calif.; Yellowknife, Canada
Volcanogenic exhalative deposits	Plate- to lens-shaped bodies that are produced by submarine volcanism or hot-spring activity. Gold may occur as an accessory to massive Cu-Zn-Pb-bearing iron sulfide ore or be disseminated in fine-grained siliceous rocks	Hemlo Dist., Canada; Kuroko Dist., Japan; Haile-Brewer, N.C.
Porphyry-type	Disseminated to stock-work veins around intermediate-type igneous intrusions. Gold is a by-product of Cu-Mo (+Zn, Pb) mining.	Bingham Canyon, Utah; Chuquicamata, Chile; Bougainville Papua, New Guinea; Ok Tedi, Papua, New Guinea
Placer and paleoplacer deposits	River and beach deposits containing gold grains and nuggets that have weathered out of deposits of the types listed above. Uranium oxides are also present in the deposits greater than 1 billion years old	Witwatersrand Dist., S. Africa; Elliot Lake (Blind River), Canada; Jacobina, Brazil

sulfide ores and exhalative hot-spring deposits appear to represent different parts of a spectrum of deposits associated with volcanic activity. The massive to disseminated ores are iron sulfide–dominant and commonly contain significant and economic quantities of $Cu \pm Zn \pm Pb$ (copper, zinc, and lead) sulfides. They form when metal- and sulfur-charged hydrothermal fluids are vented onto the sea floor along spreading and subduction plate tectonic boundaries. Some stages of hydrothermal emanation on the sea floor contain little sulfide and precipitate, dominantly silica with trace amounts of dispersed gold and iron sulfides. Both types of submarine deposits tend to precipitate in sea floor depressions adjacent to the vents and hence produce tabular to platelike bodies. Porphyry-type deposits, which have now become major by-product gold producers, have formed as a result of shallow intrusion of intermediate magmas. Mineralization occurs as disseminations and within many generations of fractures generated by successive phases of steam explosion.

Gold transport and deposition. Gold accumulates in lode deposits when hydrothermal solutions leach trace amounts (Table 1) from large volumes of surrounding country rocks and transport it to a site of deposition, where abrupt chemical or physical changes within a restricted zone (ore deposit) cause it to precipitate. Gold is most soluble under oxidizing conditions and in the presence of bisulfide and chloride ligands, which complex the gold to form $AuHS$, $Au(HS)_2^-$, and $AuCl_2^-$. The common association of arsenopyrite with gold deposits suggests that a sulfarsenide-gold complex may also be important. At the site of deposition, these solutions give up much of their dissolved gold as a result of one or more of the following processes: boiling, cooling, pH changes, and reduction of the solution by carbon compounds or sulfide minerals. Depending upon the Ag and Te content of the solutions, the gold may precipitate as native gold, electrum, or gold tellurides.

When erosion uncovers a lode (hydrothermal) gold deposit, the noble character of the gold and electrum grains allows them to resist destruction by chemical weathering for extended periods while the surrounding minerals are dissolved or eroded away. Thus, soil horizons overlying marginal lode deposits may become enriched to ore grade. Nonetheless, chloride from rainfall, cyanide from plants, and thiosulfate from the partial oxidation of sulfide minerals are all thought to contribute to minor gold solubility in the weathering environment. This minor redistribution may account for the presence of rims of nearly pure gold on electrum grains. Silver, copper, and tellurium are much more soluble than gold under these conditions and are carried away in solution. Ultimately, erosion transports the gold grains to sites of placer deposition, where they are concentrated by fluvial processes as a result of their high density (19.3 g cm^{-3} or 11.2 oz/in.^{3} for pure gold to 15.5 g cm^{-3} or 8.96 oz/in.^{3} for silver-rich electrum; the density of common rock ranges from 2.8 to 3.0 g cm^{-3} or 1.6 to 1.7 oz/in.^{3}).

For background information see GOLD; GOLD ALLOYS; ORE AND MINERAL DEPOSITS; SOLID SOLUTION in the McGraw-Hill Encyclopedia of Science and Technology. [J. R. CRAIG; J. D. RIMSTIDT]

Bibliography: R. W. Boyle, *The Geochemistry of Gold and Its Deposits*, 1979; A. Lewis, Gold geochemistry, *Eng. Min. J.*, 183:56–60, 1982; K. H. Wedepohl (ed.), *Handbook of Geochemistry*, vol. 2, 1972.

Granite

Recent studies of granite have yielded information concerning rock crystallization as well as characterization of Proterozoic anorogenic granite, a type fundamentally different from most granites, which is generally associated with orogenic cycles.

CRYSTALLIZATION

The variety of minerals and the spectrum of mineral textural relationships in granitic rocks are attributable to variability in the physical and chemical conditions that prevailed during the crystallization of parental silica-rich magmas (naturally occurring silicate liquids) in the crust of the Earth. The crystallization process is influenced by the composition of the magma, the rate of cooling, and the depth at which crystallization occurs. Laboratory studies have played an important role in gaining an understanding of the factors that control crystallization of granitic magmas. Some new insights into the origins and evolutionary development of granitic rocks have recently come from high-temperature, high-pressure experimental studies of phase equilibria and crystal growth kinetics in chemical systems which serve as models for natural granitic magmas.

Phase equilibrium studies. These describe the changes in stable mineral assemblages that occur with changes in pressure, temperature, and composition. Additionally, equilibrium investigations provide a reference point for interpreting the results of dynamic crystallization experiments which are conducted to investigate the factors that influence crystal nucleation and growth in silicate melts. These kinetic studies have focused on either: (1) the quantitative determination of crystal nucleation rates (the number of crystal nuclei formed in a unit volume per unit time) and crystal growth rates (the size increase of a crystal per unit time) of specific minerals as a function of temperature, or (2) the textural variations produced during crystallization of a simulated magma in response to changes in cooling rate.

One recent phase-equilibrium study of model granite and granodiorite systems containing H_2O reemphasized the important influence of water on the sequence of mineral crystallization in granitic magmas. The results obtained in this study provide a detailed example of the physicochemical effects of H_2O, the most common and abundant volatile component of granitic magmas (others include CO_2, Cl_2, and F_2), on mineral stabilities in multicomponent chemical systems which approach the chemical complexity of natural silica-rich magmas. Experiments which were conducted at 800 megapascals (8 kilobars) for this study simulate conditions occurring

Crystallization sequences inferred from results of experiments with the synthetic granite and granodiorite compositions at 800 MPa*

	Water content of the system in wt % →			
		Granite		
1.0	4.0	5.8	8.0	H_2O-saturated
Orthopyroxene	*Orthopyroxene*	*Orthopyroxene*	Biotite	Biotite
Plagioclase	*Clinopyroxene*	*Biotite*	*Clinopyroxene*	*Clinopyroxene*
Alkali feldspar	*Biotite*	*Clinopyroxene*	Alkali feldspar	Alkali feldspar
Clinopyroxene	*Alkali feldspar*	*Orthopyroxene (resorbed)*	Quartz	Quartz
Quartz	Plagioclase	Alkali feldspar	Plagioclase	Epidote
Biotite	*Orthopyroxene (resorbed)*	Quartz	*Clinopyroxene (resorbed)*	*Clinopyroxene (resorbed)*
Orthopyroxene (resorbed)	Quartz	Plagioclase	Epidote	Plagioclase
Clinopyroxene (resorbed)	*Clinopyroxene (resorbed)*	*Clinopyroxene (resorbed)*		
	Epidote ?	Epidote		
		Granodiorite		
1.0	2.0	4.0	10.0	H_2O-saturated
Plagioclase	Plagioclase	Plagioclase	Hornblende	Hornblende
Orthopyroxene	*Orthopyroxene*	*Orthopyroxene*	Biotite	Biotite
Quartz	Quartz	Hornblende	Plagioclase	Plagioclase
Alkali feldspar	Biotite	*Orthopyroxene (resorbed)*	Epidote	Epidote
Biotite	*Orthopyroxene (resorbed)*	Biotite	*Hornblende (resorbed)*	*Hornblende (resorbed)*
Orthopyroxene (resorbed)	Alkali feldspar	Quartz	Quartz	Quartz
Epidote	Epidote	Alkali feldspar	Vapor	Alkali feldspar
		Hornblende (resorbed)	Alkali feldspar	
		Epidote		

*Italics indicate phases that are not present in the solidus phase assemblage under equilibrium conditions. Temperature scale is relative to each crystallization sequence and is not correlative between crystallization sequences for different H_2O contents. Biotite is underlined to emphasize the changes of mineral crystallization sequence with H_2O content.

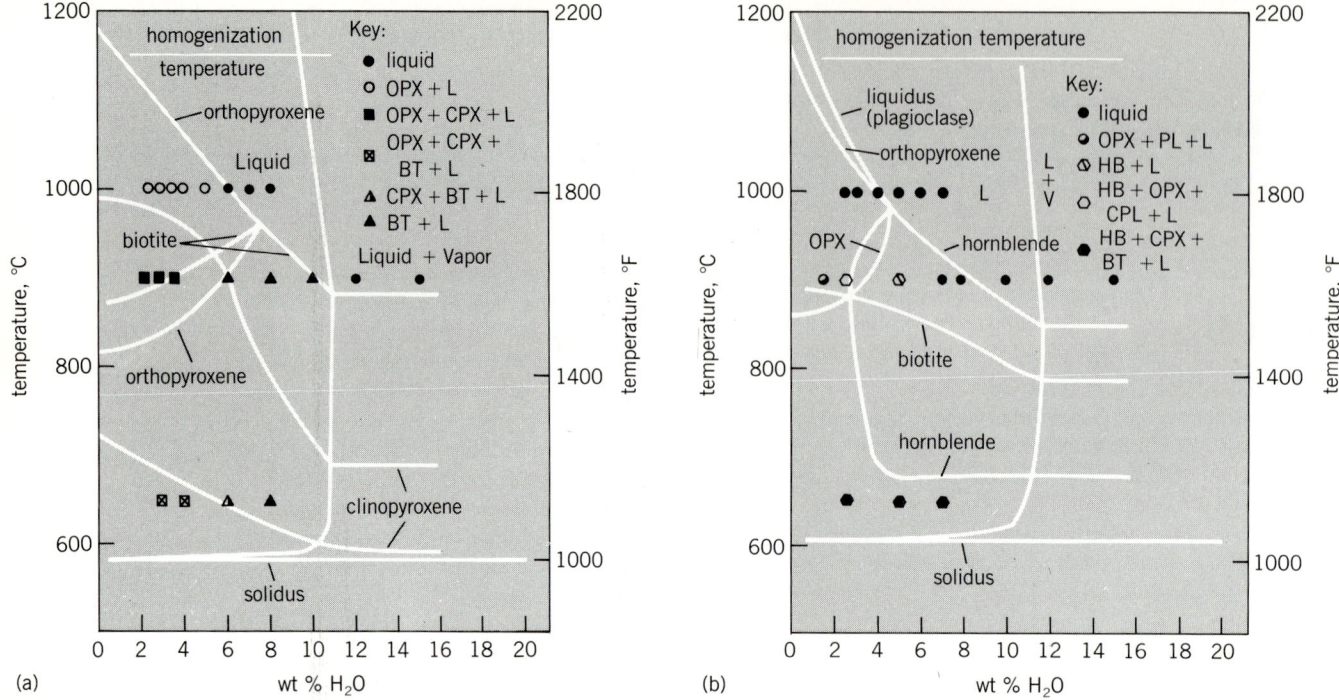

Fig. 1. Results of dynamic crystallization experiments at 800 MPa (8 kb) with (a) synthetic granite and (b) granodiorite compositions. Symbols indicate the phase assemblage observed in the experiments. Phase assemblage boundaries shown are those determined in phase equilibrium experiments and are included for comparative purposes. L = liquid, OPX = orthopyroxene, CPX = clinopyroxene, BT = biotite, HB = hornblende. (After M. T. Naney and S. E. Swanson, The effect of Fe and Mg on crystallization in granitic systems, Amer. Mineralog., 62:966–978, 1977)

at a depth of 16 mi (26 km).

Results obtained from 800-MPa (8-kb) experiments with hydrous synthetic granite and granodiorite compositions provide information on the combinations of temperature and H_2O content for which assemblages of minerals are stable at constant pressure. Phase diagrams based on the experimental data define the limits of stability of silicate liquid, aqueous vapor, or a specific mineral with respect to temperature and H_2O content. The synthetic rocks are completely molten at approximately 1200°C (2200°F). If the silicate liquid is progressively cooled and equilibrium is maintained, a specific sequence of mineral phases will crystallize.

Depending on H_2O content, the first minerals to crystallize are orthopyroxene or biotite in the granite, and plagioclase feldspar or hornblende in the granodiorite. The crystallization of different minerals at liquidus temperatures and subsequent differences in mineral assemblages on further cooling reflect the compositional differences between the granite and granodiorite.

The solubility limit of water in a magma system is marked by the appearance of an H_2O-rich vapor phase. This solubility limit is readily seen on phase diagrams for both compositions as a line which separates regions having phase assemblages containing silicate liquids not saturated with H_2O from those containing H_2O-saturated silicate liquids. In the H_2O-saturated region the silicate liquid has dissolved all of the H_2O it is capable of holding at the pressure being imposed on the system; therefore additional water exists as a separate vapor phase.

Progressive cooling and crystallization of assemblages containing H_2O-vapor-saturated melts or melts not saturated with H_2O (vapor-absent melts) produce contrasting sequences of mineral precipitation. The mineral crystallization sequences shown in the table are those predicted by the phase diagrams for constant-composition (constant-H_2O-content) cooling paths. A variety of mineral crystallization sequences are possible in the vapor-absent regions of these diagrams due to the complex intersection of stability curves. In contrast, the crystallization sequence in the vapor-saturated portions of the diagrams is unchanged with variations in H_2O content. Geologic evidence suggests that early crystallization in many natural granitic magmas takes place under H_2O-vapor-absent conditions. Selected phase assemblage boundaries determined in these experiments are shown in Fig. 1.

The conditions of crystallization of several minerals are particularly noteworthy. Orthopyroxene and clinopyroxene are stable high-temperature phases in the granitic compositions investigated. However, at lower temperature they react with the silicate liquid to produce other ferromagnesian minerals. The limited high-termperature stability of these minerals is consistent with the rare occurrence of pyroxene in natural granitic rock. Biotite, a hydrous mineral, was found to crystallize from compositions containing small amounts (<1 wt %) of H_2O. In contrast, hornblende, another hydrous mineral, only precipitates from synthetic granodioritic compositions containing more than 3 wt % H_2O. A third hydrous phase, epidote, commonly thought to be a metamorphic mineral, was observed to coexist with silicate liquid over a small temperature interval in both the granite and granodiorite systems. This suggests that epidote can crystallize directly from granitic magma.

The solidus mineral assemblage is produced by the complete crystallization of the silicate melt. This final product of crystallization contains the same assemblage of minerals in both the hydrous synthetic granite and granodiorite compostions (epidote, biotite plagioclase, alkali feldspar, quartz, and H_2O-rich vapor), although the proportions of minerals and their compositions may differ. However, the textural relationships among the minerals can be expected to reflect differences in temperature-H_2O cooling paths. In principle, the mineral crystallization sequence of a granitic rock—and by inference the H_2O content of the parent magma system—could be interpreted from the textural relationships among the constituent minerals.

Equilibrium is probably rarely maintained throughout the cooling history of a natural granitic magma as evidenced by the frequent occurrence of chemically zoned minerals and textures indicative of incomplete reaction. After the complete solidification of a granitic magma (approximately 600°C or 1100°F in the model granitic systems of the experiments), the cooling granitic rock can be expected to undergo recrystallization, which would modify original textural features. Therefore, at least some of the primary textural information which could be used to interpret the cooling history of a granitic magma will be lost. Despite difficulties in applying results of phase equilibrium studies to natural granitic rocks, data obtained from these experiments permit detailed analysis of, and quantitative constraints on, models for granitic magma chemical fractionation and crystallization.

Cooling experiments. To better understand the development of textures in granitic rocks and the influence of nonequilibrium processes on granitic magma crystallization, a number of workers have conducted programmed cooling experiments. In these experiments a granitic composition is heated to a temperature above its liquidus to form a homogeneous silicate liquid which is subsequently cooled continuously or in steps to promote crystallization. The crystallization of synthetic granitic compositions which have been used in the equilibrium experiments described above have been studied. Hydrous granitic liquids were subjected to single undercooling steps of 50–200°C (90–360°F). The term undercooling represents ΔT, the temperature difference between the liquidus temperature and the temperature at which crystal growth takes place. Additional multistep cooling experiments were performed with undercoolings up to 450°C (810°F). Although large ΔT experiments are more appropriate for modeling a magma during volcanic eruption, these experiments provide an extreme example of the effects of nonequilibrium crystallization. The results of these experiments contrast markedly with the idealized equilibrium crystallization sequences derived from the phase relationships.

The results obtained in the cooling studies are summarized in Fig. 1. Particularly noteworthy is the delayed appearance or absence of framework silicate minerals—plagioclase feldspar, alkali feldspar, and quartz—from the crystallizing mineral assemblages. The model granitic liquids were allowed to crystallize at conditions hundreds of degrees below the temperatures at which these minerals first crystallize under equilibrium conditions, yet no quartz or alkali feldspar was observed in the experimental products. The precipitation of plagioclase feldspar was suppressed 250°C (450°F) below the anticipated crystallization temperature, based on the equilibrium phase relations. This tendency for the suppression of framework mineral crystallization in rapidly cooled granitic liquid may in part explain the enrichment of ferromagnesian chain and sheet silicate minerals (for example, hornblende and biotite) at the margins of many granitic rock bodies, where magma was cooled by older, preexisting rocks.

Nucleation and crystal growth. Quantitative measurements of crystal nucleation and growth rates in model magma systems have permitted application of theory developed in materials science to problems of geologic interest.

Crystallization experiments have been conducted with hydrous synthetic granite and granodiorite compositions which do not contain iron or magnesium but are otherwise chemically similar to the synthetic compositions investigated in both of the previously described studies. In these experiments quantitative measurements were made of crystal nucleation density (a practical measure of nucleation rate) and growth rates for plagioclase feldspar, alkali feldspar, and quartz. Some of the results are summarized in Fig. 2. This diagram illustrates the variation in nucleation density (number of crystal nuclei per unit volume) and growth rate (size increase per second) for each mineral with changes in undercooling.

For small undercooling (50°C or 90°F), the crystallization of melt containing 3.5 wt % H_2O is dominated by the growth of alkali feldspar. The relatively high growth rate and low nucleation density (small number of crystals) of alkali feldspar produce a few large alkali feldspar crystals. When undercooling is increased to 100°C (180°F), the growth rate of alkali feldspar continues to increase together with the number of crystals. At this temperature the nucleation densities of the feldspars are approximately equal, but the rate of alkali feldspar growth is more than 200 times faster than the rate of plagioclase feldspar growth. Quartz has not begun to crystallize. At undercooling in the range 200–250°C (360–450°F), the growth rates of all three minerals have peaked and nucleation densities differ by a factor of 10. Crystallization at this large undercooling will produce approximately equal quantities of

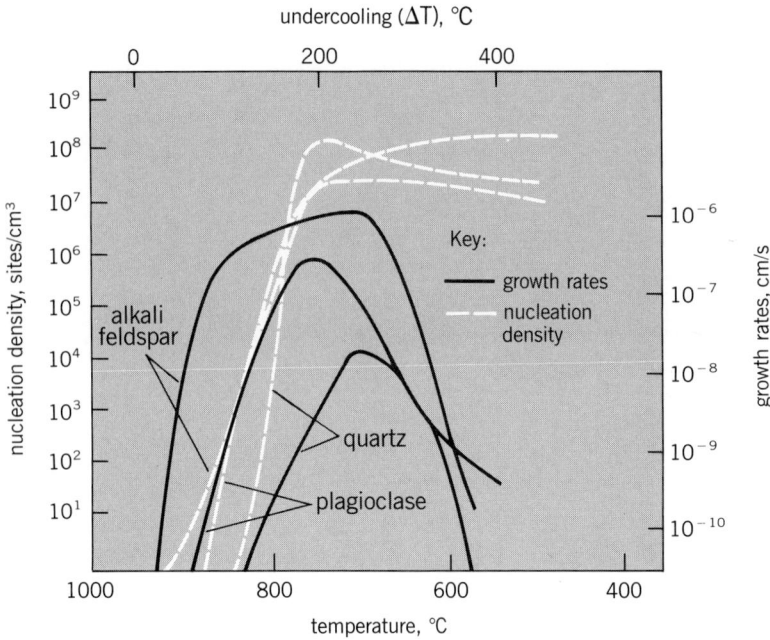

Fig. 2. Nucleation and crystal growth rate curves for an iron- and magnesium-free synthetic granite containing 3.5 wt % H_2O at 800 MPa (8 kb). °F = (°C × 1.8) + 32°; 1°C = 1.8°F. (After S. E. Swanson, Relation of nucleation and crystal-growth rate to the development of granitic textures, Amer. Mineralog., 62:966–978, 1977)

alkali feldspar, plagioclase feldspar, and quartz crystals growing at comparable rates. These conditions produce a fine-grained aggregate of crystals of approximately equal size.

The relationships between nucleation density and growth rate curves determined in these experiments provide a new explanation for the origin of granitic rock textures composed of large alkali feldspar crystals (phenocrysts) set in a matrix of equigranular, smaller crystals. These experiments indicate that the crystallization kinetics of alkali feldspar, plagioclase feldspar, and quartz in some granitic magma compositions can produce a bimodal size distribution of crystals from a continuously cooled magma. Previous explanations have required multistage processes in which large crystals are produced in a slowly cooling magma. The magma environment subsequently changes (for example, rises into a cooler region of the Earth's crust) to permit faster cooling and crystallization of the fine-grained matrix (groundmass).

The knowledge gained from integrated studies of phase equilibria and crystal growth provides new insights into the crystallization of granitic magmas and new interpretations for the development of granitic rock textures. Some of these new interpretations differ markedly from less quantitative but long-standing earlier explanations which have become the conventional wisdom.

[MICHAEL T. NANEY]

PROTEROZOIC ANOROGENIC GRANITE

Granitic magmas are characteristic of the Earth as a planet, representing an extreme product of planetary differentiation toward sialic crust formation. However, many types exist due to a wide range of possible origins. Most granitic magmas have some link to plate convergence and thus are typically associated with orogenic cycles. Yet a fundamentally different type of granite, one that is much hotter, drier, and enriched in a variety of incompatible elements, has an origin in intraplate continental regions not experiencing orogenesis. Termed anorogenic, this form of magmatism reached unprecedented proportions during the middle Proterozoic, attesting to a unique stage in the Earth's evolution.

Tectonic setting. In stark contrast to the Phanerozoic, the Proterozoic tectonic history of the Earth is characterized by long (about 600–800 million years) periods of stability interrupted by distinct and shorter (about 200 m.y.) episodes of orogeny. The latter represents a time of widespread deformation, metamorphism, and coeval igneous activity resulting in a major growth of the continents through transfer of material from the mantle. On a worldwide basis, these events occurred at 1800 ± 150 m.y., 1150 ± 100 m.y., and 550 ± 100 m.y. Currently, there is much debate as to what degree these are ensimatic (Wilson cycle) in origin.

During the intervening anorogenic periods, the newly formed Proterozoic crust continued to undergo considerable modification involving deep-seated crustal melting leading to widespread intrusion of high-level, potassium-rich rapakivi granites into the upper crust. Coeval ignimbritic volcanism was not uncommon. In many areas, this magmatic activity coincided with intrusion of large volumes (massifs) of rocks ranging from gabbro to anorthosite and swarms of diabase, indicating that this thermal disturbance was rooted in the mantle. Chemically intermediate, high-alkali intrusions (jotunite, mangerite, syenite) form a third component in some complexes and have been inferred to originate from a mixed mantle–lower crustal source.

In comparison to orogenic cycles, the addition of new mantle material to the crust was not large. However, in regions affected, these granites constitute up to 40% of the exposed crust, representing a magnitude of crustal reorganization without parallel in the history of the Earth. Moreover, the periodicity of this form of magmatism was not uniform even in the anorogenic periods of the Proterozoic. The major pulses occurred between the orogenies of around 1800 m.y. and 1150 m.y. Included is the 1400–1500-m.y.-old anorogenic province of North America (Fig. 3) which involves hundreds of rapakivi granite–mangerite–anorthosite complexes distributed in a belt extending across the entire continent from Labrador to California. Examples older than 2000 m.y. are uncommon, and there is a distinct lessening of such activity after 900 m.y. Phanerozoic anorogenic granites certainly exist but are not as densely distributed across such large regions of a continent, nor do they occur with anorthosite.

Essential features. There are several characteristic attributes of anorogenic granites that clearly make them distinct from their orogenic counterparts, including intrusive style, mineralogy, whole-rock

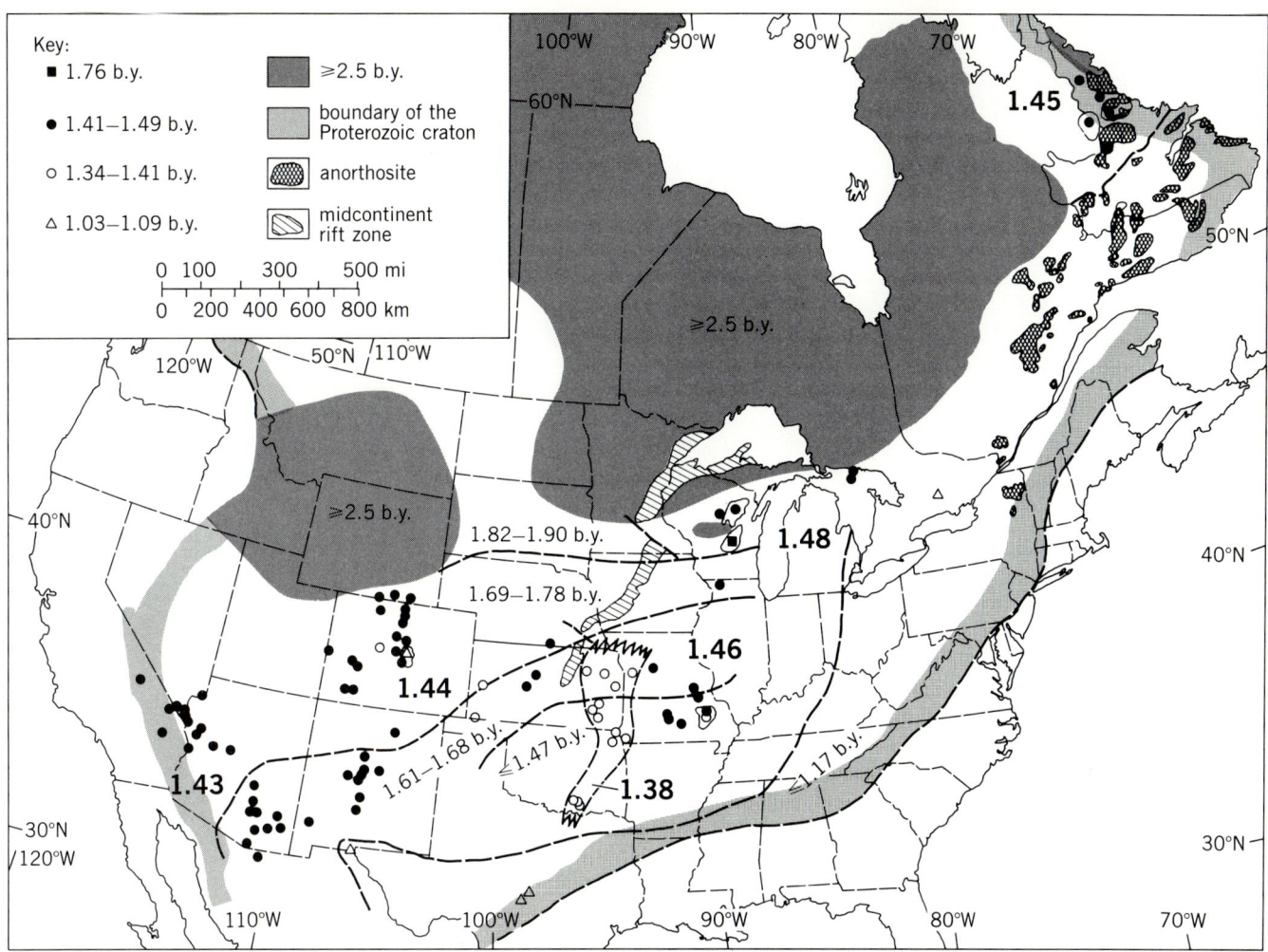

Fig. 3. Distribution of Proterozoic anorogenic complexes of North America. Large numbers represent the average age (b.y.) of the anorogenic event (U-Pb). Small numbers identify the orogenic crustal province (b.y.). (After J. L. Anderson, Proterozoic anorogenic granite plutonism of North America, Geol. Soc. Amer. Mem., 161:133–154, 1983)

composition, and isotopic composition.

Intrusive style. These plutons are posttectonic, usually circular intrusions emplaced high in the crust. Ring dikes and coeval volcanism are associated with the more shallow examples. Passive stoping and permissive emplacement are the dominant mechanisms of intrusion. Associated anorthosite and mangerites are invariably older; late, usually tholeiitic, diabase swarms are not uncommon.

Mineralogy. Modal compositions range from monzogranite to syenogranite, attesting to the predominance of alkali feldspar over plagioclase. The remaining mineralogy varies with bulk composition—peralkaline suites contain sodic amphibole and pyroxene ± biotite, fluorite; metaluminous suites contain biotite ± hornblende (commonly hastingsitic), sphene, fluorite; and peraluminous suites contain muscovite ± biotite, monazite. The Fe-Ti oxide mineralogy ranges from ilmenite dominant to magnetite dominant with related changes in composition of the mafic mineral phases (see below).

Whole-rock composition. Enrichment in a range of incompatible large-ion lithophile elements (LILE), including K, Ba, Rb, U, Th, rare-earth elements (REE), Ta, and Nb, is uniformally widespread for a given level of silica (Fig. 4). Some enrichment also occurs for Fe, Ti, Zr, Sn, and F. Likewise, Ca, Mg, and Sr are other compatible elements that tend to run lower than that of orogenic suites. Economic concentrations of U, Th, Sn, Ta, and Nb occur locally in pegmatite and vein deposits.

Isotopic composition. There is considerable variation in the Sr and Nd isotopic composition in these suites. For example, initial $^{87}Sr/^{86}Sr$ ratios for 1400-m.y.-old North American granites average 0.7051, but with range from 0.701 to 0.718. Likewise, initial ϵ_{Nd} averages -0.45 (ϵ_{Nd} for a chondritic earth = 0) but ranges from -5.3 to $+4.8$.

Conditions of crystallization. Thermobarometric estimates for emplacement conditions for anorogenic granite indicate intrusion depths to be shallow; generally these are less than 5 mi or 9 km (2.5 kb or 250 MPa) with many less than 2 mi or 4 km (1.0 kb or 100 MPa). Hence, the fact that several complexes

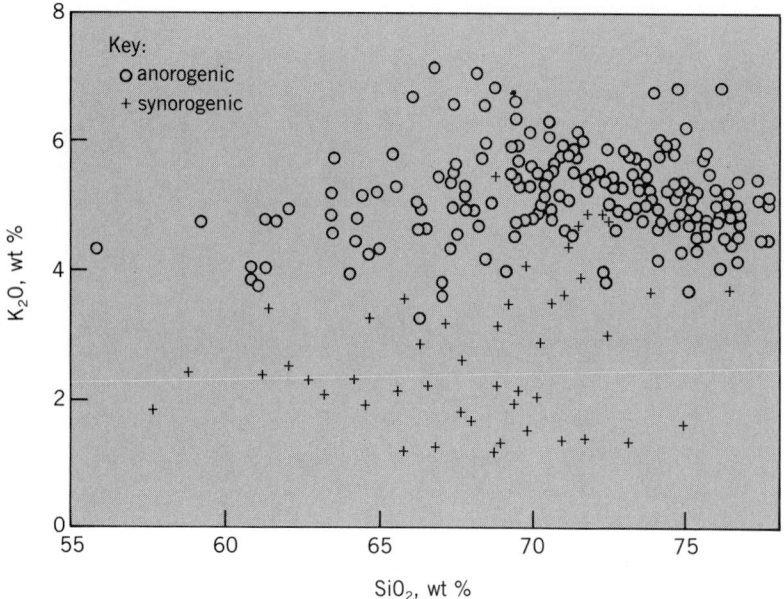

Fig. 4. Contrasting potassium enrichment of Proterozoic anorogenic and synorogenic granites.

drier, fayalite-bearing granitic suites. Such high temperatures are consistent with low partial pressure of water (PH_2O), which, where rigorously constrained, is on the order of 20–40% of the total pressure. This is consistent with the textural inference that the hydrous phases, mica and amphibole, usually crystallized after most of the other minerals, as implied by their interstitial, anhedral nature.

Given this range of pressure, temperature, and PH_2O, the intensive parameter that exhibits the largest variation is the oxygen fugacity (fO_2). Calculated values of fO_2 (Fig. 5) span three orders of magnitude from $10^{-15.6}$ bar ($10^{-10.6}$ Pa) [at 740°C, or 1360°F] near that of the experimental quartz-fayalite-magnetite (QFM) buffer upward to $10^{-12.8}$ bar ($10^{-7.8}$ Pa) [at 740°C or 1360°F]. Internally consistent mineralogic changes demonstrate the validity of this remarkable variation. The more reduced granites (ilmenite series) contain ilmenite as the principal Fe-Ti oxide, and associated mafic minerals (biotite, hornblende, pyroxene, olivine) are iron-rich with Fe/(Fe+Mg) ratios typically in excess of 0.80 up to 0.99. Examples include the Wiborg rapakivi massif of Finland, the Wolf River batholith, and the Pikes Peak batholith of Colorado. The more oxidized granites (magnetite series) contain magnetite as the principal Fe-Ti oxide and mafic minerals with lower Fe/(Fe+Mg) ratios, generally in the range 0.4–0.6. This includes many of the 1.4–1.5-b.y.-old granites of North America (for example, St. Francois complex, the Silver Plume and St. Vrain batholiths of Colorado, and the Gold Butte granite of Nevada). An extreme case is represented by the Hualapai granite where, despite the high Fe/Mg of the rock, the biotite is quite phlogopitic with the ratio Fe/(Fe+Mg) ranging down to 0.27. Many of these high-fO_2 granites contain sufficient amounts of magnetite to generate striking anomalies on regional aeromagnetic compilations.

contain an eruptive phase (for example, Kiruna and Dala porphyries of Sweden, Montello batholith of Wisconsin, and the St. Francois complex of Missouri) is not surprising. Calculated crystallization temperatures are usually in excess of 720°C (1330°F) [for example, 720–790°C (1330–1450°F) for the Wolf River batholith of Wisconsin, 740–760°C (1360–1400°F) for the Silver Plume batholith of Colorado, and 730–790°C (1350–1450°F) for the Hualapai granite of Arizona] and range up to 940°C (1720°F) [820°C–940°C or 1500–1720°F for the Albany granite of western Australia] for some of the

Magma evolution. Trace element and isotopic studies have confirmed that most of these granitic magmas were generated from a crustal source. An exception may be some of the peralkaline, riebeckite granites that exhibit a sodic enrichment trend on line with mangerite-syenite suite. This latter magma lineage shown in Fig. 5 appears to originate from a mixed mantle-crustal source with increased crustal involvement during evolution toward felsic members. These magmas types are volumetrically less significant relative to the potassic granites that make up the bulk of most anorogenic granite complexes.

The cogenetic association of anorthosite-mangerite-rapakivi granite appears to represent not a comagmatic lineage but a sequence of melting events initiating in the upper mantle and progressing upward into the lower crust. The Sr and Nd isotopic data preclude significant mantle component in most of the granite magma systems. Moreover, the age of the source is constrained also to be Proterozoic in age, rather than Archean, with a crustal residence age of only 200–800 m.y. This crust, formed during an earlier period of Proterozoic orogenesis, includes

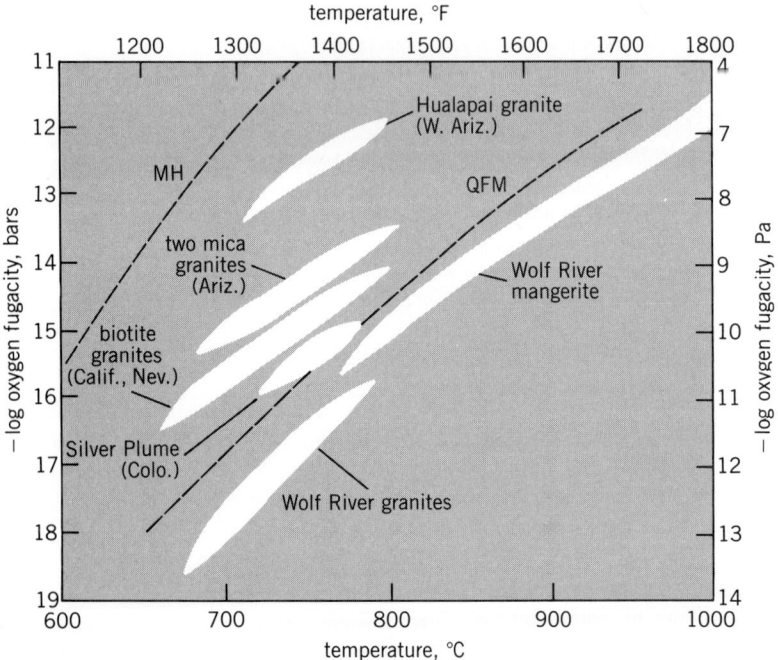

Fig. 5. Estimated conditions of crystallization of 1400–1500-m.y.-old Proterozoic anorogenic granites of North America. MH = magnetite-hematite buffer, QFM = quartz-fayalite-magnetite buffer.

a mixture of metaigneous to metasedimentary material with a range of oxidation state but with low water content due to dehydration during the preceding period of metamorphism and deformation. A relatively dry crustal source is incapable of large degrees of melting. The distribution of rare-earth elements has been demonstrated to be consistent with 20 ± 10% melting of a quartzofeldspathic source. Thus, the granitic magma generated is forced to remain near a minima composition and correspondingly is enriched in incompatible large-ion lithophile elements, Fe/Mg, and F. Comparison of normative composition to the pressure-sensitive granite minima indicates melt generation greater than 27–36 km (>7–10 kb or 700–1000 kPa). This is not unlike that inferred for crystallization of some anorthosite-mangerite suites, implying that they may have provided the heat prerequisite for the thermal event. The state of oxidation of the crustal source is apparently inherited by the derived granitic magma and carried through, albeit with some probable modification, to the final stages of crystallization and solidus mineral stability. The metaluminous suites were likely generated from a metaigneous source, one approximating the andesitic (or tonalitic) character of the orogenic terrain. By comparison, the anorogenic two-mica suites exhibit a general lower enrichment of large-ion lithophile elements and greater concentration of the compatible elements, including Ca, Mg, and Sr. An increased contribution of metasedimentary component not only would increase the alumina saturation of the derived melt but would also have a greater water budget facilitating a greater degree of melting and dilution of the overall high abundance of incompatible large-ion lithophile elements.

Tectonic considerations. The Proterozoic, southward-younging orogenic belts of North America do not appear to be intracratonic in origin and point to an ensimatic (Wilson cycle) evolution in construction of the final Precambrian craton. Yet elsewhere several of the orogenic belts of south Africa, Australia, and South America do indeed appear to be intracratonic, or ensialic, prompting the widespread disagreement about the transitory nature of the Archean versus the Phanerozoic. The significant orogenic hiatuses of the Proterozoic and coincident and vast periods of anorogenic igneous activity, never to be repeated in the Phanerozoic, point to a major, time-dependent difference.

The Proterozoic, with its few ensimatic orogenic "pulses," generated a major growth and thickening of the continents. Subcontinental, lithospheric mantle remained trapped under the newly stabilized cratons never to experience the continuing depletion typical of the suboceanic mantle that exists today. The model presented here is that the root of the Proterozoic anorogenic plutonism was critically double-edged. First, the subcontinental mantle with a near original allotment of heat-producing elements (including K, U, and Th) was prone to thermal and gravitational instability and resultant diapiric upwelling. Depending on the pressure-temperature path of the thermal plume, formation of mantle-derived mafic melt would be possible, and this would certainly further aid in transfer of heat to the lower crust. Second, the newly formed Proterozoic crust was largely undifferentiated and contained a significant low-melting fraction, making the lower crust highly susceptible to thermal perturbations from the mantle below. The upwelling would also generate a form of extensional strain on the crust that, if integration into an evolving worldwide plate system was possible, could lead to a continental rift system. In most situations, however, rifting did not occur. After a period of voluminous intrusion and volcanic outpouring tied to these subcontinental hot spots, the event waned, having accomplished sufficient transfer of heat and sialic component out of these thermally unstable regions of the mantle and lower crust. After 800 m.y. of such activity, considerable reorganization of the crust and mantle led to a more stable configuration, one with a mantle being less enriched and a lower crust more mafic and more refractory. Thus, anorthosite formation ceased with the beginning of the Phanerozoic. Likewise, the anorogenic granite plutonism of the Proterozoic was never to be repeated on such continent-wide dimensions.

For background information see GRANITE; GRANODIORITE; PETROGRAPHY; PETROLOGY in the McGraw Hill Encyclopedia of Science and Technology.

[J. LAWFORD ANDERSON]

Bibliography: J. L. Anderson, Proterozoic anorogenic granite plutonism of North America, *Geol. Soc. Amer. Mem.*, 161:133–154, 1983; A. J. Baer, Proterozoic orogenies and crustal evolution, *Geol. Soc. Amer. Mem.*, 161:47–58, 1983; A. Kroner, Proterozoic mobile belts compatible with the plate tectonic concept, *Geol. Soc. Amer. Mem.*, 161:59 74, 1983; M. T. Naney, Phase equilibria of rock-forming ferromagnesian silicates in granitic systems, *Amer. J. Sci.*, 283:993–1033, 1983; M. T. Naney and S. E. Swanson, The effect of Fe and Mg on crystallization in granitic systems, *Amer. Mineralog.*, 65:639–653, 1980; B. K. Nelson and D. J. DePaolo, Rapid production of continental crust 1.7–1.9 b.y. ago: Nd and Sr isotopic evidence from the basement of the North American midcontinent, *Geol. Soc. Amer. Bull.*, in press; S. E. Swanson, Relation of nucleation and crystal-growth rate to the development of granitic textures, *Amer. Mineralog.*, 62:966–978, 1977.

Grass

One of the primary goals of grass taxonomists is to prepare a treatment of the grass family, Gramineae (or, equally correctly, Poaceae), that reflects the probable evolutionary relationships among its members—related genera should be in the same tribe, and closely related tribes in the same subfamily. Relationships among genera or tribes are inferred from their similarities: those that are alike in many features are considered more closely related than those with only a few similarities. Not all features are of equal importance in assessing similarity. Sharing a feature that appears to have evolved only a few times within the family is considered more significant than the common possession of a feature that seems to have evolved independently in several different lines. A major concern of taxonomists is evaluation of this taxonomic importance of various features. It is accomplished by comparing the distribution of each feature with the distribution of other features of the family. Features that have patterns of occurrence similar to those of other features are considered more informative than others. Features that have contributed most to the current understanding of grass taxonomy are discussed below. This is followed by a brief synopsis of the distinguishing features of the subfamilies recognized today. Many of the structures mentioned, and some of their major variants, are illustrated in Figs. 1–4. Representative tribes and genera for each subfamily are listed in the table.

Historical developments. The current era in grass taxonomy began in 1931, when N. Avdulov published an article showing that data on chromosome number and size did not support division of the family into two subfamilies, as was accepted by most grass taxonomists at that time. Some tribes were chromosomally uniform, but others were very variable. Moreover, examination of previously published data from other features (such as epiblasts, starch grains, and seedling leaves) showed that the data displayed a pattern of variation very similar to the chromosomal data. Avdulov proposed a new taxonomic treatment for the family, one that incorporated the newly available data. In it he recognized three, rather than two, subfamilies and revised the limits of several tribes. In a paper published independently a year later, H. Prat presented a similar conclusion from his study of the epidermes of leaf blades from a wide range of grasses.

After publication of these two papers, many previously ignored characters were reevaluated and many new characters studied. This process continues today, since many grasses, especially tropical grasses, are still poorly known. There are also several genera and tribes for which the best taxonomic disposition is still controversial. New techniques reveal new features which may be taxonomically useful, but this can be established only by examining a large number of grasses representing a wide range of tribes.

Fossils have contributed relatively little so far to the understanding of the taxonomy of the family as a whole, partly because the taxonomic value of the anatomical features that can be seen in fossil grasses has only recently been appreciated. Equally important, before scanning electron microscropy was developed, the techniques for examining fossil grasses did not reveal sufficient detail to be informative. Although insufficient data are available to contribute significantly to the taxonomy of the family as a whole, J. Thomasson's work on fossil Stipeae has already clarified the history of this tribe in the Americas. Further studies should provide similar insight into the history of other tribes.

The following overview of the characters used in grass taxonomy is restricted to those features that are particularly important at a tribal or subfamily level, although some are also useful at lower levels. Because the constancy of a character may vary from group to group, characters important in some tribes may not be important in others. For example, all members of the Paniceae have two florets per spike-

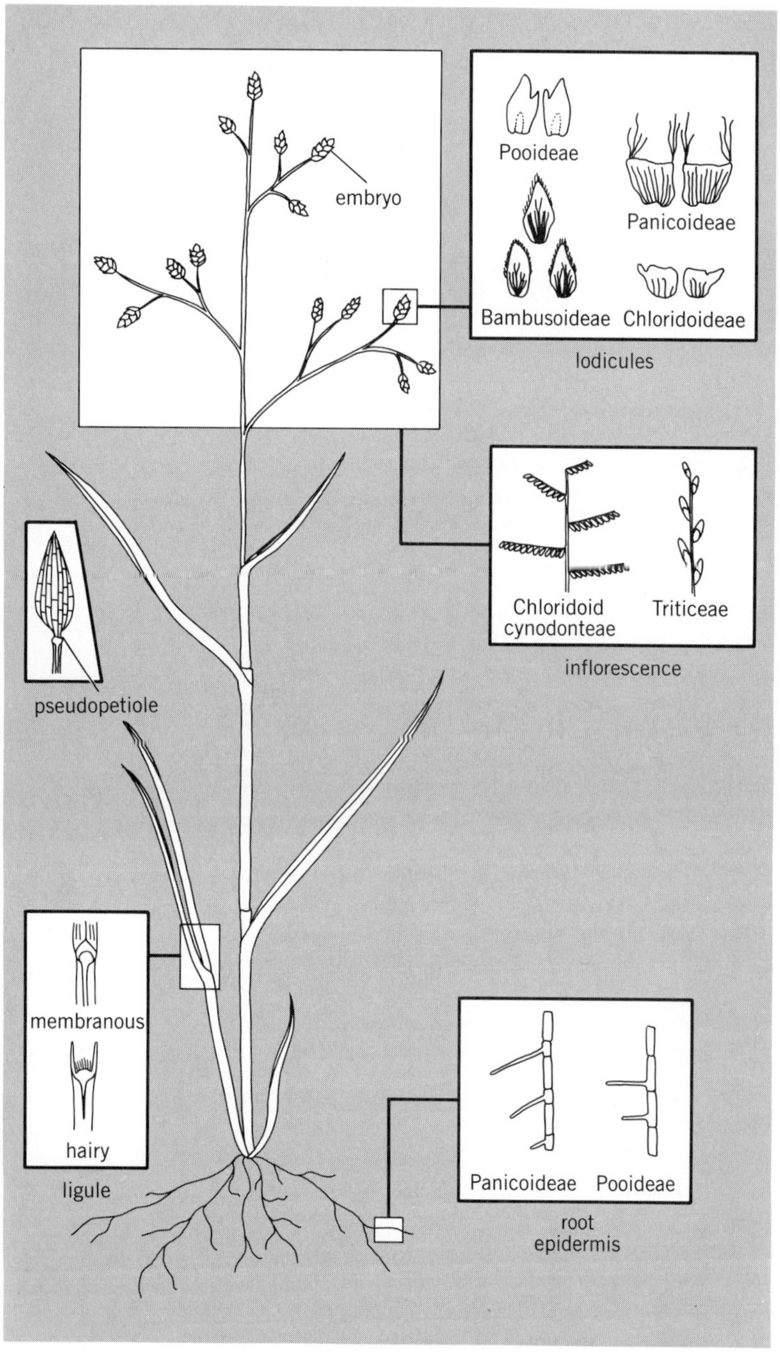

Fig. 1. Diagram of a grass plant, showing the location of some of the variants of the structures mentioned in the text.

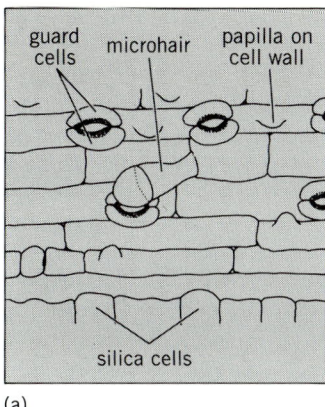 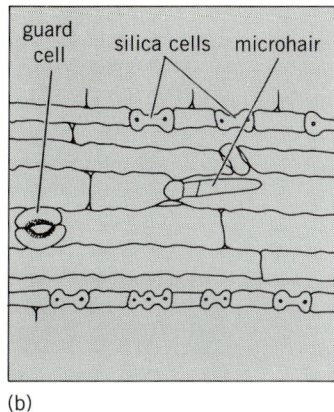 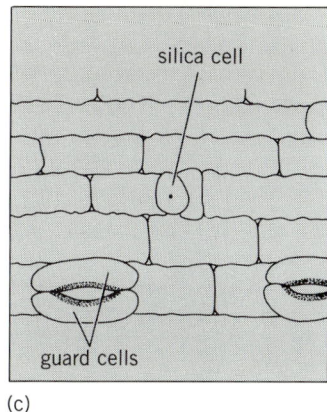

Fig. 2. Taxonomic variation in lower leaf epidermes. (a) Chloridoideae (*Austrochloris*) with bicellular microhair, papillose cell walls, and triangular guard cells. (b) Panicoideae (*Panicum*) with bicellular microhair, dumbbell-shaped silica cells, and triangular guard cells. (c) Pooideae (*Deschampsia*) with round silica cells paired with another short cell, and parallel-sided guard cells. (*After H. Clifford and L. Watson, Identifying Grasses, University of Queensland Press, 1977*)

let, all members of the Stipeae and Aristideae have one floret per spikelet, but in the tribe Poeae the number of florets in a spikelet varies from one to eight. Thus the number of florets per spikelet helps describe the Paniceae, Stipeae, or Aristideae, but is not so helpful in characterizing the Poeae.

Morphological characters. Taxonomically important characters that can be seen with a hand lens are: angle and size of the first seedling leaf blade; presence or absence of woody tissue; nature of the ligule, whether hairy, membranous, or a combination; presence or absence of a pseudopetiole and cross venation in the blade; structure of the inflorescence, whether spikelike, paniculate, or a combination; the number of florets in a spikelet; location of sterile or staminate florets; the number of veins on the lemma; number of stamens and style branches; and size of the embryo relative to the caryopsis (grain). Equally important, but needing a microscope for examination are: number, shape, and vasculature of the lodicules; presence or absence of an epiblast on the embryo; and morphology of the starch grains in the endosperm.

The functional significance, if any, of most of these characters is unknown. One exception is the photosynthetic pathway. The C_4 pathway is more efficient than the C_3 pathway at high temperatures and light intensities. Not surprisingly, therefore, grasses with the C_4 pathway are most abundant in hot deserts and grasslands; none occurs in arctic or alpine habitats. The more usual situation exists with respect to lodicules. These structures help to expose the anthers and styles to the wind by pushing the lemma and palea apart, but there are no data indicating the membranous-winged lodicules of the Pooideae are more or less efficient in this regard than the truncate lodicules of the Panicoideae. Obviously, both kinds are sufficiently effective for survival.

Anatomical characters. Root epidermes may have alternating long and short cells, with only the short cells developing into root hairs (Panicoideae), or all cells may be similar (most other subfamilies). Leaf epidermes are a source of several taxonomic characters (Fig. 2): distribution and appearance of cells filled with silica; shape and position of the guard cells; presence and shape of bicellular microhairs; presence or absence of papillae on the undifferentiated long cells; and sinuosity of the side walls of these long cells. Three aspects of the embryo are particularly important (Fig. 3): whether the scutellum is fused to the coleorhiza, whether the vascular strand to the scutellum and coleoptile diverge at the same point, and whether the embryo has overlapping leaf margins. As with the morphological character, the functional significance of these features is unknown.

Leaf-blade cross sections (Fig. 2) are the source of many taxonomic characters, most of which are related to the presence or absence of C_4 photosynthesis. Grasses lacking the C_4 pathway have randomly arranged chlorenchyma; more than four cells between adjacent vascular bundles; and two unicellular layers of cells around each bundle, the vascular bundle sheaths, neither one of which has starch plastids. Grasses with the C_4 pathway have more or less radially arranged chlorenchyma; fewer than four chlorenchyma cells between bundles; and one (two in Aristideae) bundle sheath of large cells containing specialized starch plastids. C_4 grasses in which the initial four-carbon compound is converted to malate have starch-containing sheath cells that are elongated parallel to the vascular bundle; in those forming aspartate, the cells are cuboid or elongated perpendicular to the bundle.

Other taxonomically interesting features of the leaf cross section include the size of the bulliform cells, and the presence or absence of arm cells and fusoid cells.

Cytological characters. As Avdulov showed, and subsequent evidence supports, chromosome number and size vary between the tribes. Tribes that are largely tropical such as Paniceae and Andropogoneae tend to have smaller chromosomes than tribes common to temperate or boreal regions. Chromosome base number also varies among the tribes and, to some extent, between the subfamilies. Another feature of interest is the persistence of the nucleolus into mitotic metaphase. This appears to occur only in non-pooid grasses.

Physiological characters. The most frequently used physiological character is the presence or absence of C_4 photosynthesis. This is usually constant at the tribal level, as is the subtype present, that is, whether malate- or aspartate-forming. Other physiological characters of interest are the amino acid composition of the caryopses and nature of the storage products in the endosperm.

Taxonomic overview. There is no taxonomic treatment with which all taxonomists concur. Considerable variation exists in each subfamily, and particular genera or tribes may lack a feature common in their subfamily or have a feature that is more common in another subfamily. Moreover, there is no objective means of determining which treatment is best in critical cases. The answer depends, in part, on one's taxonomic philosophy.

The treatment presented here (see table) would be acceptable to most western agrostologists. Some major areas of controversy are indicated, but some tribes and genera whose placement is dubious have been ignored. The most substantial disagreement is with the Soviet taxonomist N. N. Tsvelev, who recognizes only two subfamilies, Bambusoideae and Pooideae, but many more tribes than most western taxonomists.

Bambusoideae. The Bambusoideae includes several herbaceous species in addition to the woody bamboos. Distinguishing features of the subfamily include florets with three large vasculated lodicules, six stamens, and three stigmas; seedlings in which the coleoptile does not elongate and the first leaves lack blades; broad leaf blades with a pseudopetiole and cross venation; mesophyll with arm and fusoid cells; leaf midribs with two rows of vascular bundles; bicellular microhairs with tapering distal cells; embryos with vascular strands diverging from the same location, an epiblast, unfused scutellum, and overlapping leaf margins. The chromosomes are small. Photosynthesis is C_3.

Bambusoid grasses grow primarily in humid tropical and subtropical regions, but some of the Oryzeae (for example, wild rice) grow in temperate regions. The woody bamboos have retained many primitive reproductive characters, but are evolutionarily advanced in their vegetative features. The Oryzeae is sometimes treated as a separate subfamily, Oryzoideae.

Centostecoideae. Centostecoid grasses resemble herbaceous bambusoids in having broad leaf blades. Differences include the two truncate lodicules, three stamens, and two stigmas; midribs with only one vascular bundle or one major bundle with a few, much smaller, lateral bundles; the absence of arm or fusoid cells; and unspecialized leaf-epidermal cells almost square in outline rather than rectangular. The embryos differ from those of the Bambusoideae in having well-separated vascular strands.

Centostecoid grasses are found primarily in tropi-

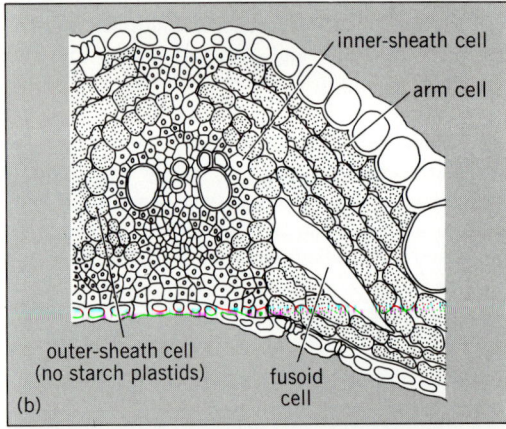
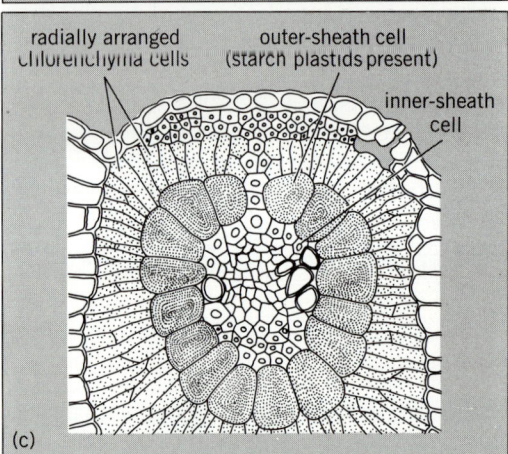

Fig. 3. Taxonomic variation in leaf blade cross sections. (a) Pooideae (*Poa*), showing vascular bundle with double sheath. (b) Bambusoideae (*Maclurolyra*), showing arm cells, fusoid cells, and double bundle sheath. (c) Cynodonteae (*Bouteloua*), showing two bundle sheaths (the outer consisting of large cells with starch plastids), and radially arranged chlorenchyma cells. (*After F. W. Gould and R. B. Shaw, Grass Systematics, 2d ed., Texas A&M University Press, 1983*)

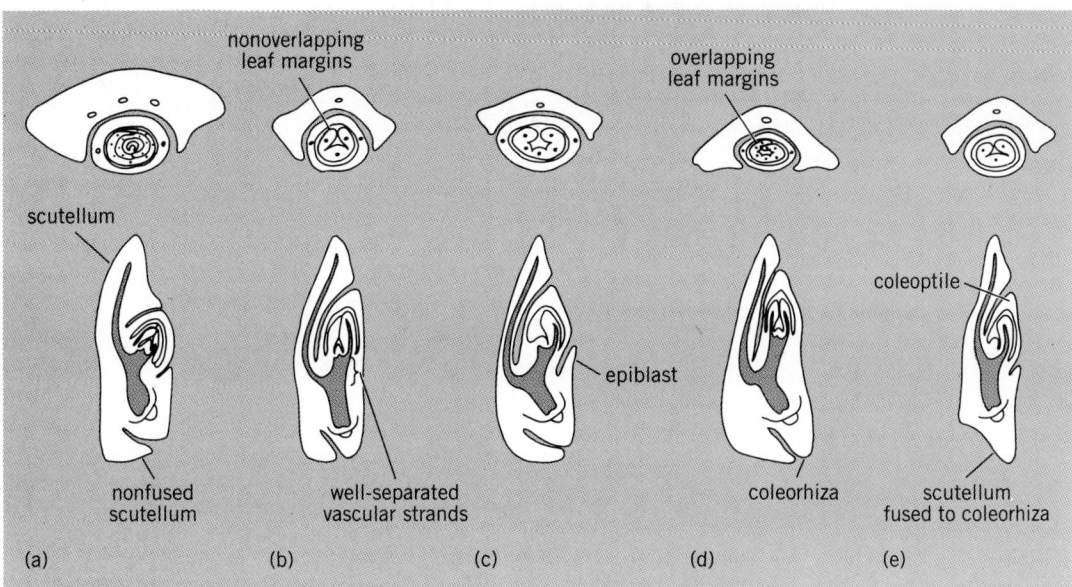

Fig. 4. Taxonomically important features of grass embryos. The upper part of each diagram is a cross section through the embryo, and the lower part is a longitudinal section. (a) *Poa pratensis* (Pooideae); (b) *Panicum clandestinum*, (Panicoideae); (c) *Eragrostis pectinacea* (Chloridoideae); (d) *Dendrocalamus strictus* (Bambusoideae); and (e) *Phragmites australis* (Arundinoideae). (After J. R. Reeder, The embryo in grass systematics, Amer. J. Bot., 44:756–768, 1957)

cal forests and warm woodlands. They are sometimes treated as a tribe, Centosteceae, of the Arundinoideae.

Arundinoideae. The Arundinoideae is hard to characterize. As recognized here, all of the Arundinoideae have an embryo with well-separated vascular strands, no epiblast, unfused scutellum, and nonoverlapping leaf margins, but in their other characteristics they are more variable. In most members the ligule consists of hairs or a hair-fringed membrane, and two vascular bundle sheaths are present. Arm cells are present in a few genera, but not fusoid cells. Most genera have only the C_3 photosynthetic pathway, but some of the Danthonieae and all of the Aristideae have the C_4 pathway. Bicellular microhairs are present in the leaf epidermes of some, but not all, members.

Further work may lead to reinterpretation of the Arundinoideae, but the available data do not support any particular alternative. The most controversial feature is the inclusion of Aristideae here, rather than in the Chloridoideae.

Chloridoideae. All members of this subfamily have the C_4 photosynthetic pathway and associated anatomical features. Other characteristic features of chloridoid grasses are seedlings with broad leaf blades; lemmas with only three veins; ligules of hairs or a hair-fringed membrane; two fleshy, truncate lodicules; bicellular microhairs with an inflated distal cell; and embryos with well-separated vascular strands, an epiblast, unfused scutellum, and nonoverlapping leaf margins. The chromosomes are small.

The Chloridoideae has its greatest representation in xeric tropical and subtropical regions. The Cynodonteae is sometimes treated as two tribes, the Chlorideae and Eragrosteae, or even as three tribes when including the Sporoboleae. C. Campbell has summarized the reasons for treating them as a single tribe.

Panicoideae. All panicoid grasses have spikelets with two florets, one of which is staminate, sterile, or reduced. In addition, the spikelets are usually dorsally compressed or terete and disarticulate below, rather than above, the glumes. This unusual combination of characters led to the Panicoideae being recognized earlier than the other subfamilies, and more recent data confirm this interpretation. Panicoid grasses have broad seedling leaf blades; two truncate lodicules; bicellular microhairs with a narrow distal cell; and embryos with well-separated vascular strands, no epiblast, an unfused scutellum, and overlapping leaf margins. The chromosomes are small. Both C_3 and C_4 grasses occur in the Paniceae, but other tribes of the subfamily are entirely C_3 or C_4. The subfamily occurs primarily in hot, strongly seasonal regions.

Pooideae. The Pooideae as conceived here is a much more restricted group than in the past. Most of its members have narrow seedling leaf blades; membranous ligules; laterally compressed spikelets with several florets and disarticulation above the glumes; lemmas with five veins; and two nonvasculated, membranous lodicules. The embryos have poorly separated vascular strands, an epiblast, a fused scutellum, and nonoverlapping leaf margins. The chromosomes are large. Photosynthesis is C_3 in all members of the subfamily, which is the most common grass subfamily in cool temperate regions.

For background information *see* CYPERALES;

Representative tribes and genera in each subfamily of grasses*

Subfamiles and representative tribes	Representative genera
Subfamily: Bambusoideae	
Tribe: Bambuseae	*Bambusa, Arundinaria, Chusquea, Dendrocalamus* [woody bamboos]
Phareae	*Pharus, Leptasis* [herbaceous bamboos]
Oryzeae	*Oryza* (rice), *Zizania* (wild rice), *Leersia* (cutgrass)
Subfamily: Centostecoideae	
Tribe: Centosteceae	*Centosteca, Chasmanthium*
Subfamily: Arundinoideae	
Tribe: Danthonieae	*Danthonia* (oatgrass), *Molinia, Schismus*
Arundineae	*Arundo* (giant reed), *Phragmites* (reedgrass), *Cordateria* (pampasgrass)
Aristideae	*Aristida* (three-awn)
Subfamily: Chloridoideae	
Tribe: Cynodonteae	*Eragrostis* (lovegrass, stinkgrass), *Cynodon* (bermudagrass), *Chloris* (windmillgrass), *Muhlenbergia* (muhly), *Bouteloua* (grama), *Sporobolus* (dropseed)
Aeluropodeae	*Distichlis* (saltgrass)
Subfamily: Panicoideae	
Tribe: Paniceae	*Panicum* (panicgrass), *Setaria* (bristlegrass), *Digiteria* (crabgrass), *Paspalum* (paspalum), *Echinochloa* (barnyardgrass)
Andropogoneae	*Andropogon* (bluestem), *Saccharum* (sugarcane, ravennagrass), *Zea* (maize), *Themeda*
Subfamily: Pooideae	
Tribe: Poeae	*Poa* (bluegrass, meadowgrass), *Festuca* (fescue), *Puccinellia, Bromus* (bromegrass)
Aveneae	*Avena* (oats), *Danthonia* (oatgrass), *Phleum* (timothy)
Triticeae	*Triticum* (wheat), *Hordeum* (barley), *Secale* (rye), *Elymus* (wheatgrass, wildrye), *Agropyron* (crested wheatgrass)
Meliceae	*Melica* (melic), *Glyceria* (mannagrass)
Stipeae	*Stipa* (needlegrass, feathergrass), *Piptochaetiun, Achnatherum*

*Common names are given in parentheses.

PLANT TAXONOMY in the McGraw-Hill Encyclopedia of Science and Technology.

[MARY BARKWORTH]

Bibliography: C. S. Campbell, The subfamilies and tribes of grasses in the southeastern United States, *J. Arnold Arbor.*, April 1985; H. T. Clifford and L. Watson, *Identifying Grasses*, 1977; F. W. Gould and R. B. Shaw, *Grass Systematics*, 1983; N. N. Tsvelev, *Grasses of the Soviet Union*, Smithsonian Institution of Libraries, 1983.

Gravitational radiation

The effort to detect gravitational radiation in the laboratory continues as evidence for its existence mounts from the pulse arrival timing measurements of the binary pulsar PSR 1913+16. The laboratory detection of gravitational waves will open a new observational window on previously inaccessible astrophysical events such as stellar core collapse leading to the formation of white dwarfs, neutron stars or black holes, and the coalescence of compact binary stars.

Nature of gravitational radiation. Gravitational radiation is a distortion of the space-time geometry that appears as a gravitational tidal force (Fig. 1) which propagates at the velocity of light, carrying with it energy. The weakness of the interaction of gravitational radiation with matter gives it great penetrating power, which will allow astronomers to probe the inner reaches of astrophysical sources impenetrable by electromagnetic radiation. This feature also makes it extremely difficult to detect gravitational radiation in the laboratory. For example, the strongest predicted impulsive gravitational wave from a source located in the Galaxy would change

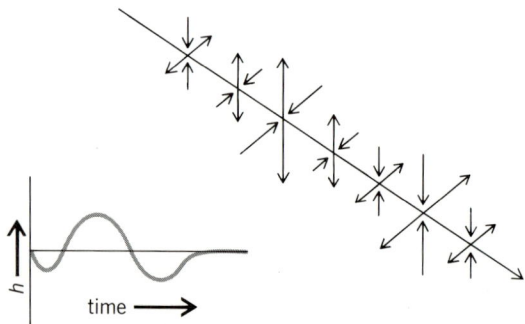

Fig. 1. Representation of a propagating burst of gravitational radiation and the associated lines of force. The force is a gradient which is transverse to the direction of propagation. The graph shows the strength of gravitational wave *h* as a function of time. (*After P. Meystre and M. O. Scully, eds., Quantum Optics, Experimental Gravitation, and Measurement Theory, Plenum Press, 1983*)

the separation of two bodies initially separated by 1 ft by approximately 10^{-18} ft.

Gravitational radiation is emitted by a system of accelerating masses which has a time-varying quadrupole moment, there being no analog of the familiar electromagnetic dipole radiation due to the fact that all mass has the same sign. For example, the linear acceleration of an isolated mass would not generate radiation, but a rotating rod or nonspherically symmetric collapse of an extended mass would generate gravitational radiation.

The laboratory generation and detection of gravitational radiation is a logical experiment to consider in analogy with the early experiments of H. Hertz which proved the existence of electromagnetic radiation. Communication with gravitational waves is conceivable; for example, using a very large rotating rod as the source, communication directly through the Earth or the center of the Galaxy would be possible due to the penetrating power of the radiation. Unfortunately, calculations of the flux of gravitational radiation that could be generated with present technology show that it would be many orders of magnitude weaker than could ever be detected. This leads the experimentalist to abandon the hope of generating and detecting gravitational waves in the laboratory and to seek natural sources of radiation. Fortunately, cataclysmic astrophysical events may generate gravitational waves which are strong enough to be detectable.

The essential feature of any astrophysical source of gravitational radiation is that there be a very large changing quadrupole moment of the distribution of mass such as in a rapidly rotating compact binary star or in an unsymmetrical stellar core collapse. It is very difficult to estimate the amount of asymmetry in such an event which makes the predictions of the possible strength of the emitted waves very uncertain. Figure 2 is an estimate of the strength and frequency of the strongest waves that may be bathing the Earth. The strength of a gravitational wave is given by h, the dimensionless amplitude, which is roughly the amount of strain or the fractional change of the dimension, induced in a detector. The frequency of a burst source of radiation is taken to be the inverse of the burst duration. Bursts of amplitude 10^{-20} to 10^{-23} with a frequency near 1 kHz may occur once per month. Much stronger burst signals emitted by a supernova in the Milky Way Galaxy probably occur only once every 10 to 30 years.

Experiments to detect gravitational radiation are designed to be sensitive to bursts of radiation in the kilohertz range or to continuous sources in the 100–10,000-Hz band. There are also experiments which use the radio ranging of satellites to detect the very low-frequency waves, in a band around 10^{-4} Hz, which could be generated by some main-sequence binary star systems or may be the background radiation left over from the big bang. Before these experiments are discussed, evidence will be examined for the existence of gravitational radiation from observations of the binary pulsar PSR1913+16.

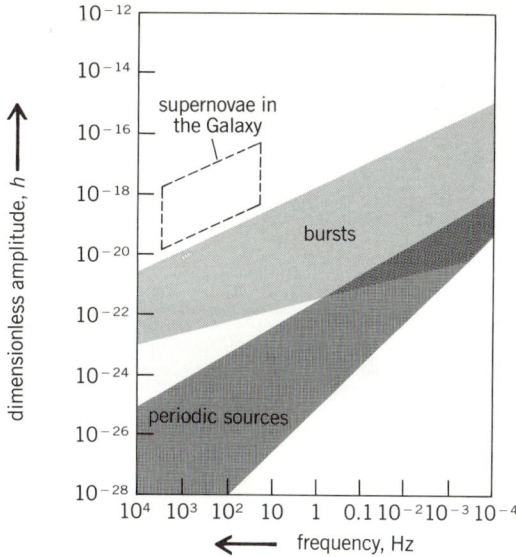

Fig. 2. Estimate of the strength and frequency of the gravitational radiation bathing the Earth from astrophysical sources. (*After P. Meystre and M. O. Scully, eds., Quantum Optics, Experimental Gravitation, and Measurement Theory, Plenum Press, 1983*)

Binary pulsar source. The binary pulsar PSR1913+16 furnishes firm evidence for the existence of gravitational radiation. The system contains a pulsar, which is rotating about its axis with a period of 0.059 s, in a binary orbit with a compact companion star. By observing the arrival times of a large number of pulses from 1974 up to the present, a full set of orbital parameters for the binary system has been determined. The orbital period has been measured to be 27906.98163 ± 0.00002 s or approximately 7 h 45 min. Of most interest is the rate of change of the orbital period $(-2.40 \pm 0.09) \times 10^{-12}$ s s^{-1}, which indicates that the binary system is losing energy and speeding up as the two stars spiral inward.

Einstein's theory of gravity predicts that two objects in mutual orbit such as the binary pulsar will emit gravitational radiation at a rate which depends upon the masses of the two stars, the orbital period, and the orbital eccentricity. The gravitational radiation carries away energy, causing the binary orbit to slowly decay. From the extensive observations of the system, it has been possible to independently determine these parameters, allowing a calculation of the rate of orbital decay due to the loss of energy by the emission of gravitational radiation. The predicted rate of decay of the orbit is 1.00 ± 0.04 times the measured value, which leads to the conclusion that gravitational radiation does indeed exist as predicted by Einstein's theory of gravity. The evidence for the existence of gravitational radiation from these observations provides the impetus to continue the effort to make the laboratory reception of gravitational radiation routine.

Experimental detection efforts. The original experiments to detect gravitational radiation were pi-

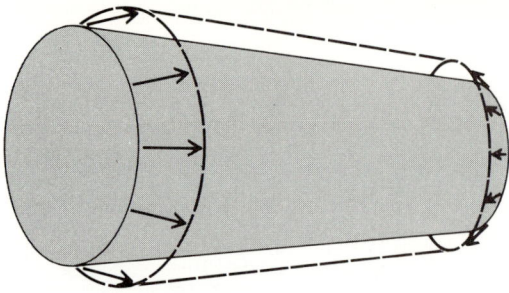

Fig. 3. Cylindrical-bar gravitational radiation antenna, strained in its lowest longitudinal mode by a passing burst of gravitational radiation. (*After P. Meystre and M. O. Scully, eds., Quantum Optics, Experimental Gravitation, and Measurement Theory, Plenum Press, 1983*)

oneered by J. Weber in the late 1950s and 1960s. His detectors were large aluminum cylinders (3100 lb or 1400 kg) isolated from environmental noise and instrumented with sensitive strain-monitoring devices to monitor their lowest-frequency longitudinal modes (Fig. 3). Although the other resonant modes of the detector interact with a gravity wave, this mode, with the two end faces of the cylinder moving in opposition, has the largest cross section for the absorption of gravitational radiation. After Weber's report of coincidences of events between his pair of detectors in Maryland and Illinois, several others joined the search. All of these experiments, the most sensitive of which was capable of measuring a strain of 10^{-16}, failed to detect any events.

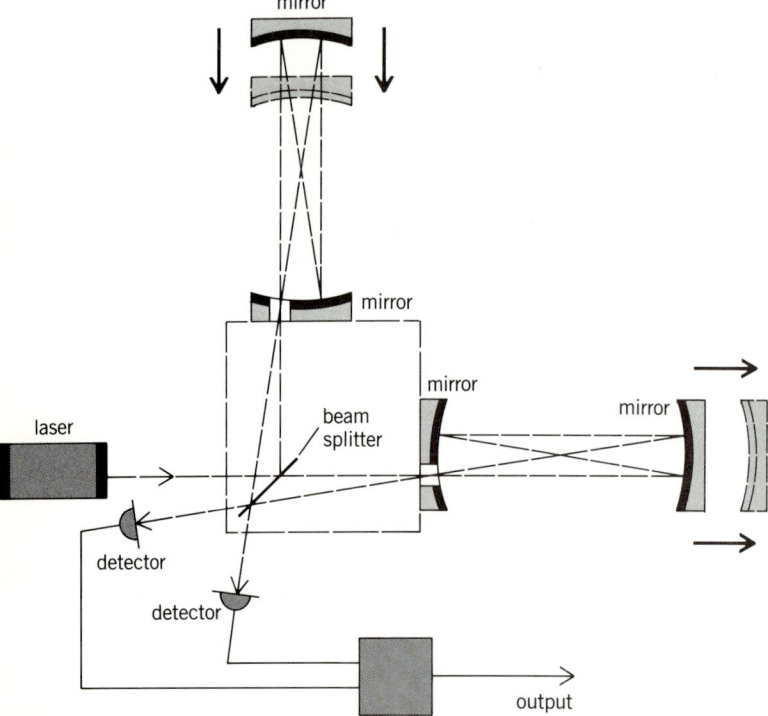

Fig. 4. Laser interferometer detector. A gravitational wave will cause an alternate increase in the separation between the mirrors in the two arms of the interferometer, as indicated by the broken-line arrows. (*After P. Meystre and M. O. Scully, eds., Quantum Optics, Experimental Gravitation, and Measurement Theory, Plenum Press, 1983*)

It became apparent that vast improvements in detector sensitivity were needed. This could be achieved by using antennas cooled to near absolute zero with improved electromechanical motion transducers, so work on room temperature detectors was abandoned in the late 1970s and efforts on a new generation of detectors were started.

Improvements in resonant-bar detectors. The biggest improvement in the second generation of detectors would come from operating them in a cryogenic environment. This serves to reduce the brownian motion of the antenna, which was the limitation of the room temperature detectors. The magnitude of the brownian motion is proportional to the temperature and inversely proportional to the mechanical quality factor, or Q. The Q determines the number of cycles of oscillation of a resonant system for the amplitude of oscillation to be damped. It was found that in addition to the obvious advantage gained by lowering the antenna temperature, a factor of 10^4 or better improvement in Q could be achieved at low temperatures in specific alloys of aluminum, niobium and in single-crystal sapphire and silicon.

The other area in which tremendous strides have been made is in the design of the electromechanical transducers which convert the motion of the antenna to an electrical signal. The first generation of room temperature detectors had a number of piezoelectric crystals glued directly to the bar. The voltage produced in these crystals by their deformation from the vibration of the cylinder was amplified with a transistor amplifier and recorded to be later analyzed. This method has been superseded by the use of transducers in which an additional resonant diaphragm is mounted to the end of the cylinder. If the frequencies of the cylinder and the diaphragm are closely matched, most of the energy is transferred from the cylinder to the diaphragm, and due to the smaller mass of the diaphragm its amplitude of vibration is enhanced. In most designs the motion of the diaphragm is made to modulate a parameter of an electrical circuit, either an inductance or a capacitance. This parametric coupling of the bar to the electronics affords greater sensitivity than the original method of electromechanical transduction. In addition, the cryogenic cooling of transducers reduces the electrical analog of brownian noise, the Johnson-Nyquist noise. Recent research has led to a technique which may isolate the detector from the noise of the amplifier which affects the bar by acting back through the transducer. This is called a back-action-evasion measurement strategy.

There has also been considerable progress in the development of ultralow-noise amplifiers for use on these detectors. SQUIDs (super-conducting quantum interference devices) are amplifiers which depend upon the Josephson effect, the tunneling of superconducting electron pairs through a weak link joining two superconductors. These devices have nearly achieved the ultimate theoretical limit to the sensitivity of any amplifier. The SQUIDs which have been implemented on the second generation of detectors have not yet reached this limit because of conflicting requirements to make an extremely low-noise device and a device which will serve as a useful amplifier. Work is continuing to develop and implement extremely low-noise SQUIDs.

In 1982 the most sensitive cryogenically cooled detector yet developed achieved a strain sensitivity of about 3×10^{-18} and collected data for a total of 74 days. Unfortunately, this was the only detector of its type to operate during that period, so it was not possible to veto extraneous impulses in a coincidence experiment with another detector.

Laser interferometers. Concurrent with the effort on resonant-bar gravitational radiation detectors, work was progressing on a different type of detector which measures the separation between a pair of free masses with a laser interferometer technique. There are several variations on the original design, but all systems currently under development have three masses arranged in the shape of an L with mirrors mounted on them (Fig. 4). A gravity wave with some component of its propagation direction perpendicular to the plane of the interferometer will cause the separation between the masses in the two arms to alternately increase and decrease, and this will produce a shift in the interference fringe pattern at the output of the interferometer. The use of high-power, frequency-stabilized lasers, high-reflectivity mirrors, and sophisticated isolation from the ground vibration has led to unprecedented sensitivities for such systems. Although the most sensitive of these systems has yet to achieve a sensitivity better than the best room temperature bar detectors, the interferometers have the advantage of being broader-band detectors than the bar type of antenna. In practice, the bar detectors will probably be limited to a range of frequencies of a few hundred hertz centered on the resonant frequency of the cylinder. The interferometers could search in a large range of frequencies, between about 100 and 10,000 Hz for continuous sources as well as for the more dispersed burst sources.

Work is continuing on this type of detector by constructing longer baseline interferometers. The longest is 130 ft (40 m). Efforts are under way to make more stable lasers and to isolate the free masses from the vibrations of the ground in the laboratory.

Radio and laser ranging of spacecraft. Radio ranging of NASA *Voyager* spacecraft has been used in a search for gravitational radiation of very low frequency, around 10^{-4} Hz, such as could be emitted by a binary system of two main-sequence stars. A strain sensitivity of 10^{-14} to this sort of signal was achieved in these experiments, but no signals were detected. However, it was possible to place an upper limit to the stochastic background of gravitational radiation left over from the density perturbations in the big bang. Work is continuing to improve upon these results by using laser ranging of spacecraft and possibly sending a series of dedicated satellites to stable points in the solar system to form a very long-baseline (10^5 mi or more) laser interferometer detector.

Status and prospects. Since the end of the era of first-generation Weber bar detectors, most of the research in the field has focused on the development of new technology to be used in the following generation of detectors. The unprecedented sensitivity which is being sought has required orders-of-magnitude improvement in such areas as transducer design, low-loss materials, low-noise amplification devices, and vibration isolation. This long task is just now beginning to come to completion with the assembly of a number of second-generation detectors.

Hopefully the work proceeding at present will prove the feasibility of the new-generation detectors and lead to large-scale implementation of the new technology. Plans are now being formulated to build two 3-mi (5-km) baseline laser interferometer detectors in the United States and to construct large arrays of bar detectors. With these instruments, a new window on some of the most exotic and energetic events in the universe will open to astronomers.

For background information *see* BROWNIAN MOVEMENT; GRAVITATION; INTERFEROMETRY; PULSAR; RELATIVITY in the McGraw-Hill Encyclopedia of Science and Technology.

[MARK F. BOCKO; DAVID H. DOUBLASS]

Bibliography: S. P. Boughn et al., Observations with a low-temperature, resonant mass, gravitational radiation detector, *Astrophys. J.*, 261:L19–L22, 1982; K. S. Thorne, Experimental gravity, gravitational waves, and quantum nondemolition: An introduction, in P. Meystre and M. O. Scully (eds.), *Quantum Optics, Experimental Gravitation, and Measurement Theory*, 1983; J. Weber, The search for gravitational radiation, in A. Held (ed.), *General Relativity and Gravitation*, vol. 2, 1980; J. M. Weisberg and J. H. Taylor, Observations of postnewtonian timing effects in the binary pulsar PSR 1913+16, *Phys. Rev. Lett.*, 52:1348–1350, 1984.

Hall effect

The quantum Hall effect is a new physical phenomenon. At low temperatures and in high magnetic fields the ratio of the Hall voltage to the current through a two-dimensional electron system is quantized to $h/(ve)^2$, where h is Planck's constant, e is the electronic charge, and v is either integral or a rational fraction. Concomitant with the quantization of the Hall resistance, the resistivity of the two-dimensional system vanishes as the absolute temperature approaches zero. Depending on the value of v, the related phenomena are termed either the integral quantum Hall effect or the fractional quantum Hall effect. Though the phenomena differ only in the value of v, the physical mechanisms responsible for their existence are quite different. While the integral quantum Hall effect results from the unique quantization conditions for independent electrons of a two-dimensional system in a high magnetic field, the fractional quantum Hall effect, according to its present understanding, seems to indicate the condensation of the electrons into a novel, incompressible electron liquid with factionally charged quasi-particles.

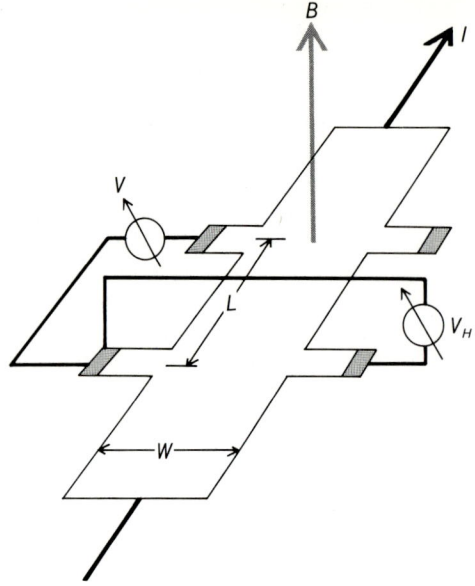

Fig. 1. Hall experiment. Two-dimensional specimen carrying current *I* is placed in a magnetic field *B*, perpendicular to the plane of the specimen. The Hall voltage V_H and longitudinal voltage V are determined. W and L are geometry factors. (*After H. L. Stormer and D. C. Tsui, The quantized Hall effect, Science, 220:1241–1246, 1983*)

Two-dimensional electron systems. In two-dimensional electron systems, the motion of the carriers is restricted to a plane. Such systems are generally established at the interface between two different materials. The electrons are confined to this interface by a strong electric field creating a narrow potential well approximately 10 nanometers wide. In such a narrow well, the carrier motion perpendicular to the plane is quantized, and only a discrete set of states are quantum-mechanically allowed. At low temperature (less than about 10 K or 18°F above absolute zero) a situation is attained where only the lowest of these states (the ground state) contains electrons, leaving no degree of freedom for perpendicular motion, while the motion in the plane remains unrestrained.

The carriers in silicon MOSFETs (metal-oxide-semiconductor field-effect transistors), electrons on the surface of liquid helium and, more recently, carriers in modulation-doped semiconductor heterostructures are examples of such two-dimensional carrier systems. *See* SEMICONDUCTOR DEVICES.

Hall experiment. In the Hall experiment (Fig. 1), a two-dimensional specimen carrying an electric current *I* is placed in a uniform magnetic field *B* which is perpendicular to the plane of the specimen and hence to the current. The Hall voltage V_H is measured across the flow of electric current. The Hall effect is normally employed to determine the electron concentration *n* of the specimen by using Eq. (1) for the Hall resistance ρ_{xy}. Generally, the resistivity of the material, given by Eq. (2), is determined simultaneously by measuring the voltage

$$\rho_{xy} = V_H/I = B/(ne) \qquad (1)$$

$$\rho_{xx} = (V/I) \times (W/L) \qquad (2)$$

drop *V* along the current path. Here *W* and *L* are geometry factors shown in Fig. 1. The combination of the measurements of V_H and V yields direct information on the transport properties of a given specimen.

Integral quantum Hall effect. For a two-dimensional electron system of fixed carrier concentration *n*, a strictly linear relationship between Hall resistance and magnetic field is expected, according to Eq. (1). However, the results of low-temperature (about 4 K or −452°F), high-field (about 10 teslas) experiments on two-dimensional systems deviate considerably from such a simple relationship. Instead of a linear field dependence of ρ_{xy} on *B*, a staircaselike dependence is observed, where flat plateaus alternate with steep vertical sections (Fig 2). The steps are found to be quantized to the values in Eq. (3) to much better than 1 part in 10^6. Con-

$$\rho_{xy} = h/(ie^2) \qquad i = 1, 2, 3, \ldots \qquad (3)$$

comitant with the appearance of plateaus, the resistivity ρ_{xx} seems to vanish, being finite only in the regime of the vertical section in ρ_{xy}. From current decay times in circular two-dimensional geometries under the condition of the quantum Hall effect, resistivities ρ_{xx} of less than 10^{-10} ohm/□, equivalent to three-dimensional resistivities as low as 10^{-18} ohm-meter, have been inferred. Such values are more than six orders of magnitude lower than the resistivity of any nonsuperconducting material. The sum of these observations is termed the integral

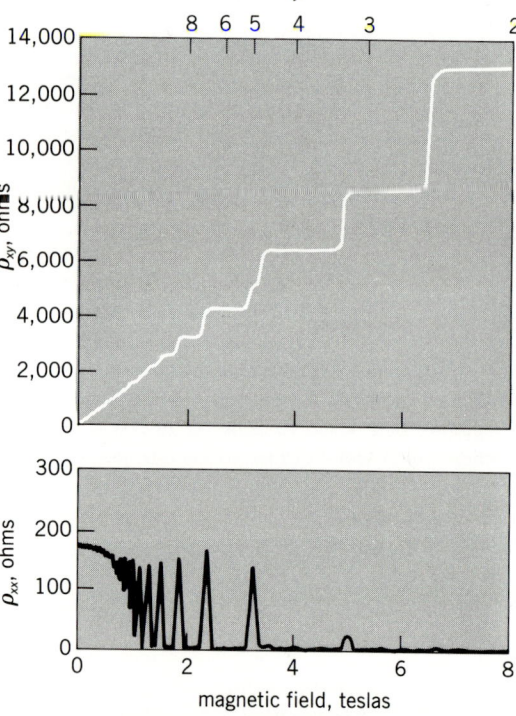

Fig. 2. Hall resistance ρ_{xy} and resistivity ρ_{xx} of a two-dimensignal electron system displaying the integral quantum Hall effect as a function of magnetic field. The top scale indicates the filling factor *ν*. (*After M. A. Paalenen, D. C. Tsui, and A. C. Gossard, Quantized Hall effect at low temperatures, Phys. Rev., B25:5566–5569, 1982*)

quantum Hall effect (sometimes also called the normal quantum Hall effect).

Applications. Applications of the integral quantum Hall effect as a novel resistance standard are presently being considered. Furthermore, high-precision measurements of ρ_{xy} allow a determination of the fine-structure constant α, given by Eq. (4) where c

$$\alpha = e^2 c/(2h) \quad (4)$$

is the speed of light, which is independent of quantum electrodynamic theories and might ultimately help to decide to what degree the electron is truly a point charge.

Origin. A high magnetic field perpendicular to the plane of a two-dimensional electron system completely quantizes the energy spectrum of its carriers. This spectrum consists of a set of discrete levels (Landau levels) at energies E_j given by Eq. (5) (neglecting spin) where $\hbar = h/2\pi$, and ω_c, given by

$$E_j = (j + \tfrac{1}{2})\hbar\omega_c \quad j = 0, 1, 2, \ldots \quad (5)$$

Eq. (6), is the cyclotron frequency with the carrier

$$\omega_c = eB/m^* \quad (6)$$

mass m^* (Fig. 3). Each Landau level can hold up to s electrons per unit area, given by Eq. (7), while the

$$s = eB/h \quad (7)$$

gaps between such levels contain no state at all. At low temperatures and at a given field B_0 a two-dimensional system with fixed carrier concentration n occupies ν Landau levels, where ν, given by Eq. (8),

$$\nu = n/s = nh/eB_0 \quad (8)$$

is called the filling factor. Generally ν is not integral; hence, the highest level is only partially filled. However, at a set of field values given by Eq. (9), ν becomes integral ($\nu = i$) and the classical equation Eq. (1) yields exactly Eq. (10).

$$B_i = nh/(ie) \quad i = 1, 2, 3, \ldots \quad (9)$$

$$\rho_{xy} = B_i/(ne) = h/(ie^2) \quad (10)$$

The concomitant loss of resistivity is a result of the lack of possibility for electron scattering to occur: Since all occupied Landau levels are completely filled, carrier scattering cannot occur within a given Landau level but only via states in higher, empty Landau levels. At low temperatures, the energies to scatter into higher Landau levels are not available. Hence, a suppression of carrier scattering and a loss of resistivity result.

The above derivation of ρ_{xy} and ρ_{xx} does not account for the finite width of the Hall plateaus, nor for the wide stretches of vanishing resistance which are the truly outstanding features of the quantum Hall effect. Explanation of a finite width of the effect requires the existence of localized states as they are present in any real two-dimensional system. In contrast to the delocalized states in the Landau levels, which contribute to the electronic transport, localized states *trap* electrons. As the magnetic field is varied, electrons are periodically pumped between delocalized and localized states. The localized states act as an electron reservoir for the delocalized states which determine the transport properties of the two-dimensional system. This buffer action of the localized states extends the field range over which the transport coefficients ρ_{xy} and ρ_{xx} assume their quantized value. In this sense, the imperfection of the two-dimensional system and the resulting development of localized states is essential for the observation of the quantum Hall effect.

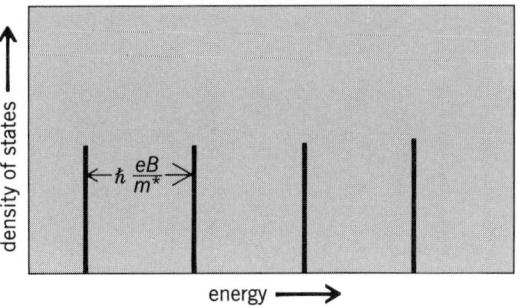

Fig. 3. Energy scheme of a two-dimensional electron system in a high magnetic field B. Only discrete energies separated by gaps of width $\hbar(eB/m^*)$ are allowed (spin is neglected).

Fractional quantum Hall effect. According to the above explanation of the origin of the integral quantum Hall effect, ρ_{xy} is given by Eq. (3) and ρ_{xx} approaches zero exclusively for integral-filling of Landau levels. However, more recently the same phenomena have been observed also for fractional occupation of Landau levels, that is, for values of the filling factor ν given by Eq. (11), with Hall resistivities given by Eq. (12), where p and q are integers

$$\nu = p/q \quad (11)$$

$$\rho_{xy} = h/(p/q)e^2 \quad (12)$$

(Fig. 4). These experimental findings are termed the fractional quantum Hall effect (sometimes also called the anomalous quantum Hall effect). These observations require temperatures below about 1 K (2°F above absolute zero) and a two-dimensional electron systems of exceptionally high quality. Fractions with odd denominators q have been found exclusively, including several representatives of the $q = 3, 5, 7$, and 9 series. In some cases, their quantization has been demonstrated to better than 3 parts in 10^5.

Origin. The fractional quantum Hall effect cannot be accounted for on the basis of the discreteness of the energy spectrum of two-dimensional electrons in a high magnetic field, in analogy to the integral quantum Hall effect. At fractional values of the filling factor ν, the top Landau level is partially occupied and hence cannot give rise to plateaus in ρ_{xy} and vanishing ρ_{xx}. However, since the integral quantum Hall effect is directly connected with the existence of a discrete energy spectrum with wide gaps in between containing no states, it is concluded from the fractional quantum Hall effect that at low temperatures gaps of a new kind appear at fractional ν. Such gaps must be of many-particle or-

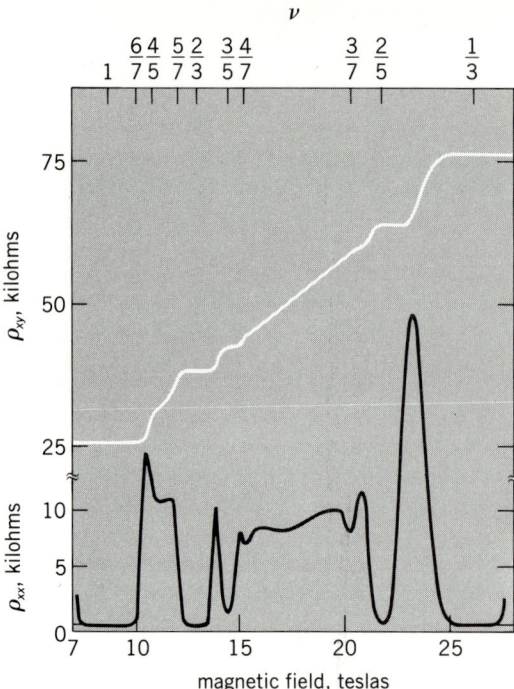

Fig. 4. Hall resistance ρ_{xy} and diagonal resistivity ρ_{xx} of a high-mobility, low-density two-dimensional electron system displaying the fractional quantum Hall effect as a function of magnetic field. Temperature of the sample is 90 milli-Kelvins. The top scale indicates the fractional filling factor ν. (After A. M. Chang, et al., Higher-order states in the multiple-series, fractional, quantum Hall effect, Phys. Rev. Lett., 53:997–1000, 1984)

igin resulting from a strong correlation between the motion of the individual carriers.

Initially the formation of an electron crystal (Wigner lattice) was assumed to cause the fractional quantum Hall effect. This interpretation is presently believed to be unlikely since electron transport through such a crystal is expected to show nonlinearities at low temperature which have not been observed. Furthermore, theoretical calculations show that electron crystallization does not preferentially occur at any particular (rational) filling factor in contrast to the experimental data.

The presently most widely accepted interpretation of the fractional quantum Hall effect is based on a recent theory by R. B. Laughlin, who proposed the formation of a novel electron quantum liquid. He succeeded in designing a many-particle wave function, reminiscent of wave functions used to describe superfluid helium, which describes a state that exists at $\nu = 1/q$, $q = 1, 3, 5, \ldots$, in agreement with the experimental observation of exclusively odd denominators. For small q (q less than about 9), the wave function describes an incompressible quantum liquid whose energy is lower than that of an electron crystal and hence is expected to form the ground state of a two-dimensional electron system in a high magnetic field. For q greater than about 9, a transition to a not yet observed electron crystal is predicted. The liquid ground state is separated from its excitations by a small but finite energy gap, similar to the case of superconductivity. This gap precludes low-lying excitations and causes resistanceless electron transport at temperatures approaching absolute zero. The quasiparticles, describing the excitations, have fractional charge given by Eq. (13) and are

$$e^* = e/q \qquad (13)$$

responsible for the observation of rational quantum numbers ν in the Hall resistance ρ_{xy}. Ground states at filling factors $\nu = p/q$ with p not equal to 1 are more difficult to design, and are believed to develop out of the primary states at $\nu = 1/q$ following a hierarchical scheme.

Research areas. The fractional quantum Hall effect is presently a very active field of research. Experimentally, the precise determination of the size of the many-particle energy gaps, the search for new quantum numbers, the transition from a liquid to a crystal, and studies of the influence of localized states on the fractional quantum Hall effect are pursued in many laboratories. The formation of two-dimensional electron systems in novel material combinations allowing for yet lower carrier scattering rates will be of utmost importance in carrying out many of these studies.

Theory focuses on finding appropriate descriptions for higher-order states at $\nu = p/q$ ($p \neq 1$) and determining their gap energies, on the interaction between quasiparticles and the effect of disorder, and on finding a description of the new state at $\nu = 1/q$ at finite temperatures.

For background information see DE HAAS–VAN ALPHEN EFFECT; HALL EFFECT; LIQUID HELIUM; NONRELATIVISTIC QUANTUM THEORY in the McGraw-Hill Encyclopedia of Science and Technology.

[H. L. STORMER]

Bibliography: K. von Klitzing, The fine-structure constant α: A contribution of semiconductor physics to the determination of α, in J. Treusch (ed.), *Festkorperprobleme* (*Advances in Solid State Physics*), vol. 21, pp. 1–23, 1982; R. B. Laughlin, Anomalous quantum Hall effect: An incompressible quantum fluid with fractionally charged excitations, *Phys. Rev. Lett*; 50:1395–1398, 1983; H. L. Stormer, The fractional quantum Hall effect, in P. Grosse (ed.), *Festkorperprobleme* (*Advances in Solid State Physics*), vol. 24, pp. 25–44, 1984; D. C. Tsui, Quantum Hall effect: Factional quantization, in J. Chadi (ed.), *Proceedings of the 17th International Conference on Physics: Semiconductors*, San Francisco, 1984.

Heart

The calcium channel blockers are a structurally diverse group of compounds which are known to specifically antagonize or inhibit certain calcium-dependent cellular functions. From the clinical standpoint, the significance of these drugs lies in their ability to prevent or lessen the severity of angina pectoris (chest pain), which is often associated with coronary artery disease and heart attacks. However,

Table 1. Use of calcium channel blockers in cardiovascular disease syndromes

Disease syndrome	Calcium channel blocker*
Coronary artery spasm	DZ, NF, VP
Chronic stable angina pectoris	DZ, NF, VP
Supraventricular tachycardia	VP
Hypertension	DZ, NF
Hypertrophic cardiomyopathy	VP
Afterload reduction in congestive heart failure	NF
Protection of ischemic myocardium	DZ, NF, VP
Other vascular disorders	DZ, NF

*DZ = diltiazem; NF = nifedipine; VP = verapamil.

the subject of this article is a second property of calcium channel blockers: their ability to prevent irreversible myocardial cell damage secondary to episodes of oxygen deprivation (ischemia) in heart muscle and to myocardial infarction.

Clinical agents. The calcium channel blockers as a therapeutic class were first described by German investigators in the early 1960s. Some of these agents were first studied in the United States approximately 15 years later, and it was not until 1980 that the first calcium blocker was tested in the clinical setting there. Today, there are three calcium channel blockers available for clinical use in the United States. The first to be approved was verapamil, which has both antiarrhythmic and antianginal efficacy. Nifedipine and diltiazem have also been approved for clinical use as antianginal agents. Although these three agents have different characteristics both in the experimental and in the clinical setting, all are effective antianginal agents and are potentially useful for the treatment of other cardiovascular disease syndromes (Table 1). In addition, there is a group of second-generation calcium channel blockers which is presently under active experimental and clinical investigation. Perhaps the most noteworthy of these is bepridil hydrochloride.

Angina pectoris. In order to understand the mechanisms of action of the calcium channel blockers, it is important to briefly review the pathophysiology of angina pectoris. Angina pectoris is generally the result of an imbalance between oxygen delivery to the cardiac muscle and the demand by the working muscle for oxygen. Most individuals who have fixed blockages of the coronary arteries secondary to atherosclerosis develop angina when coronary blood flow cannot be further increased to meet an increasing demand for oxygen caused by an increase in heart rate, ventricular wall tension, or cardiac contractility. In these individuals, angina is typically precipitated by exercise. Prinzmetal's angina, or variant angina, develops during an imbalance caused by a primary reduction in coronary flow. Usually, this situation is the result of an inappropriate increase in coronary vascular tone localized within the large vessels of the coronary circulation, and often superimposed on a background of atherosclerotic coronary artery disease. However, coronary spasm can occur in individuals with little or no coronary artery disease. In either case, the occurrence of angina pectoris can be interpreted as an indication of myocardial ischemia and impending irreversible muscle damage.

Mechanisms of action. The calcium channel blockers can prevent irreversible myocardial damage by several possible cardiovascular hemodynamic mechanisms. Because contraction of cardiac muscle is a calcium-dependent process, calcium channel blockers which reduce the availability of calcium ion to the myocardial contractile proteins can prevent myocardial ischemia by reducing myocardial demand via a direct effect to decrease myocardial contractile activity. Myocardial oxygen demand can also be reduced by reduction of heart rate. Since conduction of the atrial action potential through the fibers of the atrioventricular node of the heart is a calcium-dependent process, calcium channel blockers can reduce heart rate by delaying atrial-ventricular conduction. Myocardial oxygen demand can also be decreased by a reduction in peripheral vascular resistance or afterload. This mechanism contributes substantially to the antianginal efficacy of nifedipine; both diltiazem and verapamil effectively reduce afterload but to a lesser extent (Table 2). The most important mechanism by which all three agents achieve their potent antianginal efficacy is through direct coronary vasodilation, which increases oxygen supply to the ischemic myocardium.

In addition to these classical hemodynamic explanations of the beneficial effects of the calcium channel blockers in the prevention of angina pectoris, there are now a number of other mechanisms which appear to contribute importantly to these beneficial effects and which may also contribute to the apparent efficacy of these drugs for the prevention of myocardial ischemia. The illustration shows the potential consequences of reduced coronary flow. The direct vasodilator effect of the calcium channel block-

Table 2. Primary hemodynamic mechanisms contributing to the antianginal efficacy of the calcium channel blockers*

Calcium channel blocker	Decreased heart rate	Negative inotropism	Systemic vasodilation	Coronary vasodilation
Diltiazem	+	+	+	+++
Verapamil	+	++	++	+++
Nifedipine	−	−	+++	+++

*Symbols +, ++, +++ indicate minimal, moderate, and substantial degrees of contribution, respectively; − indicates no contribution.

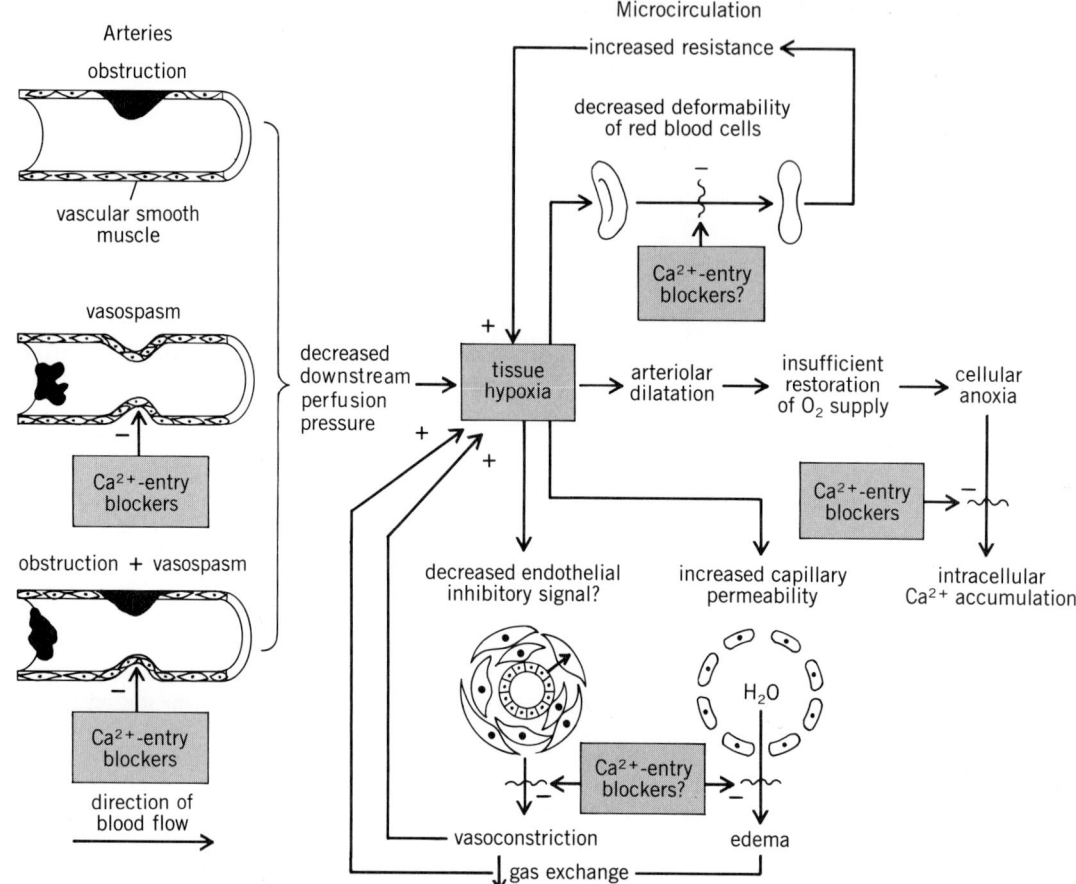

Events secondary to a decrease in perfusion pressure at the level of the larger arteries which may lead to and contribute to the maintenance of tissue ischemia at the level of the microcirculation, and which potentially may result in tissue infarction. The possible sites of action of calcium channel blockers (Ca^{2+}-entry blockers) to inhibit (−) specific events are indicated. (After P. M. Vanhoutte and R. A. Cohen, Calcium-entry blockers and cardiovascular disease, Amer. J. Cardiol., 52:99A–103A, 1983)

ers can act to reverse the primary event when vasospasm is involved. Development of tissue hypoxia and anoxia results in changes in the microcirculation which may include cellular damage due to a reduction in the high-energy intermediates used to support membrane integrity and ionic gradients. Under these conditions, cells tend to take up calcium, which leads to irreversible cell damage. This process has been shown to be inhibited by calcium channel blockers.

There are also a number of positive feedback loops which are set off during tissue hypoxia which further contribute to the development of tissue damage. Blood viscosity is partially determined by red cell deformability, which becomes important in the microcirculation where the red cells must pass through capillaries having internal diameters smaller than the diameters of the red cells. Decreased metabolic support secondary to hypoxia results in reduced red cell deformability, which may increase apparent blood viscosity and resistance to flow through the capillaries, thereby further reducing the delivery of oxygen to the tissue. Some calcium blockers can reduce the viscosity of hypoxic blood by preserving red cell deformability.

The ability of calcium blockers to inhibit the aggregation of platelets can also contribute to their ability to preserve ischemic tissue. Tissue hypoxia also increases capillary permeability and subsequent fluid loss to the interstitial space. This state of tissue edema further reduces the delivery of oxygenated blood to the tissue and potentiates the hypoxic condition. Some calcium blockers are thought to interfere with this process.

Finally, it is now known that the intact endothelial lining of blood vessels releases a factor that causes the surrounding vascular smooth muscle cells to relax. It has been suggested that hypoxia may cause damage to the endothelium, resulting in a reduced release of endothelium-derived relaxing factor. The net effect of this situation would be an increase in vascular smooth muscle tone which is countered by the calcium channel blockers.

Experimental studies. The evidence directly demonstrating the efficacy of the calcium channel blockers for the prevention of myocardial ischemia comes from experimental studies on the reduction of infarct size. All three antianginal agents have been shown to reduce the size of an experimental myocardial infarction secondary to artificial ligation of a coronary artery. Possible mechanisms for these effects are: (1) improvement of the supply-demand ratio as described above; (2) prevention of unopposed uptake of calcium in the ischemic myocardial cell, thereby deterring cell death; and (3) coronary vasodilation, supporting the development of collateral circulation to the ischemic area. It is interesting to note that the protective effect of the calcium channel blockers in some cases (for example, nifedipine) appears to be dose-dependent. Low doses of nifedipine that have minor systemic effects appear to provide the desired protection, but higher doses may actually increase infarct size secondary to hypotension and reduced perfusion pressure as well as increased heart rate due to reflex stimulation.

The protective effect of verapamil in the myocardium is also dose-dependent. Low doses have no effect on coronary flow or myocardial ischemia, whereas moderate doses reduce infarct size without affecting heart rate or left ventricular function. High doses have a greater beneficial effect on infarct size but also depress left ventricular function and cause atrioventricular block. Verapamil appears to selectively reduce contractility in the ischemic myocardium without significantly affecting normal regions. It is unclear whether this effect is of significant benefit.

Diltiazem has also been shown to reduce myocardial infarct size in experimental models and to increase subepicardial blood flow in ischemic areas after experimental myocardial infarction. In addition, pretreatment with diltiazem protects myocardial cell mitochondrial function as well as ventricular function during reperfusion after experimental coronary ligation.

Summary. The calcium channel blockers are extremely beneficial in the prevention of angina pectoris. Experimental data suggest that these agents will also be of benefit in the preservation of the myocardium in certain potential acute ischemic syndromes. However, further clinical testing under highly controlled conditions will be necessary to establish the efficacy of calcium channel blockers as therapy for the protection of the ischemic myocardium.

For background information see ANGINA PECTORIS; CARDIOVASCULAR SYSTEM; CARDIOVASCULAR SYSTEM DISORDERS; HEART (VERTEBRATE); HEART DISORDERS in the McGraw-Hill Encyclopedia of Science and Technology.

[STEPHEN F. FLAIM]

Bibliography: L. R. Bush et al., Effects of diltiazem on extent of ultimate myocardial injury resulting from temporary coronary artery occlusion in dogs, *J. Cardiovasc. Pharmacol.*, 4:285–296, 1982; A. G. Ellrodt and B. N. Singh, Clinical applications of slow channel working compounds, *Pharmacology and Therapy*, 23:1–43, 1983; S. F. Flaim and R. Zelis (ed.), *Calcium Blockers: Mechanisms of Action and Clinical Applications*, 1982; A. Fleckenstein, *Calcium Antagonism in Heart and Smooth Muscle*, 1983.

Herbicide

Herbicide resistance in plants is a relatively new phenomenon, even though herbicides have been used extensively for the past 30 years. The development of resistance in a variety of organisms to other pesticides and pharmaceuticals has been an established principle for some time. As early as 1956, it was predicted, based on data from other biological systems, that resistance to herbicides in plants would develop. However, it was not until 1970 that specific data even suggested that this was a potential problem. Subsequent research has confirmed the occurrence of herbicide resistance in plants.

A number of factors led to the delay in acceptance of herbicide resistance in plants as a valid concept. The primary reason is probably because higher plants have a much longer life cycle and are less prolific than insects and microorganisms. Therefore, it requires several years for herbicide resistance in plants to be expressed in the field, whereas those organisms with shorter life cycles and those that are more prolific can express resistance in a relatively short time. Furthermore, certain agronomic practices, such as crop rotation with or without the use of different herbicides, prevent or delay the development of herbicide resistance.

Herbicide resistance. Herbicide resistance is a decreased response of the population of a plant species to a herbicide; that is, a formerly susceptible species becomes resistant. Where this occurs, there are two distinct biotypes of the species present in the population, resistant and susceptible. They are essentially identical in appearance, but differ in a specific biochemical or biophysical characteristic. The resistant biotype occurs in very low numbers relative to the susceptible biotype, because it is less competitive. However, the resistant biotype survives and grows normally at herbicide doses effective for control of the susceptible biotype and this results in greatly increased numbers of the resistant biotype after application of a particular herbicide for several years on the same area of land. There is no evidence to suggest that a mutation occurring after the herbicide application is involved.

The terms resistance and tolerance are often used interchangeably, and there are no uniform and consistent definitions. However, weed scientists are increasingly using resistance to denote the conditions described above, while tolerance refers to the variability of response within or among plant species to a single herbicide application.

Distribution. Triazine has been investigated more than any other class of herbicides, since the initial evidence of herbicide resistance was found with this class. By 1982 triazine resistance had been confirmed in at least 18 genera and 30 species of weeds located in at least 23 states in the United States, four provinces of Canada, and seven countries of Europe. Herbicide resistance has also been indicated with several other herbicides, including bentazon, chlorfenprop, 2,4-D, diquat, diuron, paraquat, phenmedipham, and picloram.

Prevention. The occurrence of herbicide-resistant weeds as a serious practical problem is limited because of the nature of resistance and the availability of alternate control methods. Few natural weed populations have developed herbicide resistance, which has usually occurred in the field only after several years of successive applications of the same herbicide or class of herbicides. Also, since resistance is generally herbicide-specific—cross resistance between chemically unrelated groups of herbicides is rare—the resistant biotype can usually be controlled by another herbicide. In addition, there are several alternate weed control methods. Cultivation or other tillage methods will control the resistant biotype. Crop rotation increases the number of herbicides available and may also change crop management practices, (time of planting, maturity and harvest date, cultivation methods, and irrigation) enough to make the environment unfavorable for the resistant biotype. In perennial crops, control of resistant biotypes is somewhat more difficult and is limited to the use of a variety of herbicides and intensive cultivations.

Mechanism of resistance. Triazine resistance has provided a model for determining the mechanism of herbicide resistance in plants. Early research demonstrated that the difference between resistant and susceptible biotypes was not due to differences in absorption, translocation, or degradation of the triazine molecule. Rather, in susceptible biotypes, triazine herbicides act in blocking photosynthesis by binding to a specific component in the photosynthetic electron transport system; the triazine molecule fails to bind in the resistant biotype. It also appears that triazine resistance is maternally inherited and controlled by a single gene.

Potential benefits. The phenomenon of herbicide resistance in plants can be applied beneficially. Perhaps the greatest potential benefit is the possible transfer of resistance from a resistant plant to a susceptible plant with the objective of developing herbicide-resistant crops. The transfer of resistance from a resistant plant to a closely related crop plant can now be accomplished by conventional plant-breeding methods. Considerable progress has been made by using this method to transfer triazine resistance from the resistant weed biotype of wild *Brassica campestris* to rapeseed cultivars of Polish rape (*B. campestris*) and Argentine rape (*B. rapus*). Protoplasts from resistant and susceptible lines have been fused and whole plants with decreased susceptibility developed. In the future the transfer of herbicide resistance may also be possible with genetic engineering techniques.

Cell and tissue culture methods also show promise for developing plants with decreased susceptibility to herbicides. Intact plants, callus cultures, or cell cultures are treated with relatively high herbicide levels, and whole plants are regenerated from the cells that survive. This approach is limited by the fact that whole plants cannot be regenerated from cells of many species. However, herbicide-tolerant tobacco plants have been developed for several herbicides by using this approach.

For background information *see* BREEDING (PLANT); HERBICIDE in the McGraw-Hill Encyclopedia of Science and Technology.

[FLOYD M. ASHTON]

Bibliography: J. Gressel, Triazine herbicide interaction with a 32000 Mr Thylakoid protein—Alternative possibilities, *Plant Sci. Lett.*, 25:99–106, 1982; J. S. Holt and S.R. Radosevich, Differential growth of two common groundsel (*Senecio vulgaris*) biotypes, *Weed Sci.*, 31:112–120, 1983; H. M. LeBaron and J. Gressel (eds.), *Herbicide Resistance in Plants*, 1982; K. E. Steinback et al., Identification of the triazine receptor protein as a chloroplast gene product, *Proc. Nat. Acad. Sci.*, 78:7463–7467, 1981.

Hydrogenic ions

At very high temperatures, normal matter becomes ionized as its atomic electrons become free of their atomic nuclei. Ionized media have long been of interest to astronomers concerned with the nature of the interiors of stars. The recent development of directed-energy devices such as high-energy laser beams and particle beams has made it possible to produce highly ionized matter in the laboratory. Such matter can now be studied under controlled conditions so that its possible use in applications such as fusion energy and x-ray lasers can be investigated.

There are many different types of particles found simultaneously in highly ionized matter. Aside from the nuclei and electrons of which all atoms and ions are made, the most simple particles in ionized matter are the hydrogenic ions. These atomic particles consist of a single electron (with electric charge $-e$) electrically bound to a nucleus of charge $+Ze$ to form a two-particle system. The hydrogen atom corresponds to $Z = 1$, which is electrically neutral. If the atomic number Z is larger than 1, the net charge of the system is positive and it is called a positive ion. Hydrogenic ions have been produced with Z between 2 for hydrogenic helium and 92 for hydrogenic uranium.

Ion structure. If the size and structure of the nucleus itself are ignored, nonrelativisitic quantum mechanics describes hydrogenic ions as having an infinite discrete set of allowed electron binding energies labeled by a positive integer n called the principal quantum number. The binding energy increases with the square of Z and decreases with the square of n. The value $n = 1$ is the lowest possible one and is associated with the ion being in its ground state. As no energy level is lower, the ground-state ion cannot decay and is stable. The excited states with n larger than 1 can decay, however, and are thus unstable.

Each possible energy level of the ion corresponds to a set of elliptical electron orbits described by classical mechanics. The different members of such a set have values of electron angular momentum 0, $\hbar$, $2\hbar$, ..., $\ell\hbar$, ..., $(n-1)\hbar$, where $\hbar$ is Planck's quantum unit of action divided by 2π and the integer ℓ is called the angular momentum quantum number. The average radius of an electron orbit with quantum numbers n and ℓ about a nucleus of atomic number Z increases according to $3n^2 - \ell \cdot (\ell + 1)$ and decreases linearly with Z. Hence the ground state of the ion corresponds to the smallest possible orbit for the given value of Z. As the nuclear charge increases, the orbit size decreases. The eccentricity of the orbit depends on the quantum number ratio ℓ/n.

Excited ion lifetimes. The lifetime of an excited hydrogenic ion can be theoretically determined once the quantum nature both of particle motion and of the electromagnetic field is considered. The results show a trend for the lifetime to increase with n and with ℓ, while decreasing strongly with the atomic number Z. Many of the excited states of highly charged ions decay in times considerably shorter than 1 nanosecond, making such ions particularly scarce in stars and in the laboratory.

Relativistic effects. The time-averaged electron orbital velocity within the ion is proportional to Z/n. The ground state of an ion with Z equal to 60 has its electron orbiting at about half the speed of light. When particles travel at such speeds, the effects of the theory of relativity are important, and the above conclusions of nonrelativistic quantum theory are in error. Relativistic quantum mechanics then describes the electron's motion, and relativistic quantum field theory describes its interaction with the ionic electromagnetic field. Among the consequences are a restructuring of the ion's energy levels and changes in the ion's radius and lifetime. As relativistic quantum field theory is one of the fundamental underpinnings of modern physics, considerable research on highly charged hydrogenic ions is being carried out to check the theoretical predictions as precisely as possible.

The ground state of hydrogenic uranium has an average electron orbit radius of 5×10^{-13} m while the radius of the uranium nucleus itself is about 10^{-14} m. Thus the structure of the nucleus is fairly unimportant for the properties of even the most highly charged hydrogenic ions.

Partially ionized media. Because matter on Earth is both relatively cold and essentially uncharged, nuclei of charge Ze are normally found with Z electrons orbiting around them. If the temperature were to be considerably increased, the atoms in matter would move much faster relative to one another and electrons would be removed from atomic nuclei during atom-atom collisions. At extremely high temperatures such as in the center of stars, all electrons are free of their nuclei and a completely ionized plasma is formed, the fourth pure state of bulk matter. At a somewhat lower temperature than required to form such a plasma, only one or a few electrons are usually found around each nucleus and a partially ionized plasma is formed containing fairly large amounts of hydrogenic ions. The temperature required for this situation increases with Z and is about 300,000 K (540,000°F) for ionic helium, the hydrogenic ion of lowest charge. Thus hydrogenic ions are not found on Earth in matter existing in a state of thermal equilibrium.

Hydrogenic helium. Significant amounts of ionic helium do exist in certain helium-containing gas discharge devices designed to operate at gas pressures considerably lower than that of the atmosphere. These devices function far from thermodynamic equilibrium in the sense that the temperature of the ions is very different from that of the walls of the vessel containing them. Practical devices such as gas lasers are based upon the ease of producing hydrogenic helium in gas discharges. The unusually large electron binding energy of this ion plays an essential role in the chemistry of such discharges, resulting in the excited-state population inversions needed for laser action.

Production of highly charged ions. Hydrogenic ions with Z larger than 2 can now be produced in nonequilibrium situations, and applications are being pursued. Various approaches for ion production are employed, with transient plasmas being produced by either high-energy particle beam impact or high-power laser beam incidence on a surface of a solid or within a gas. Hydrogenic carbon ions ($Z = 6$) have been produced by using each of these alternatives. The particle beam approach has produced hydrogenic calcium. Still another way of producing hydrogenic ions has been to pass a high-energy beam of ions having more than one electron through a thin layer of solid material, with the remaining electrons being stripped off the fast ions during collisions with the atoms in the solid. With this approach, beams of hydrogenic uranium ($Z = 92$) were produced in 1984.

The recent development of ways to produce highly charged ions makes feasible the pursuit of new technology based on the extension of techniques developed for devices using hydrogenic helium ions. Also in the offing are x-ray lasers and fusion-energy devices using high-energy beam impact on small solid pellets containing fusible materials. Observation of the x-ray spectra emitted by hydrogenic ions is an important diagnostic tool for assessing the properties of the plasmas that are produced.

For background information see ATOMIC STRUCTURE AND SPECTRA; BEAM-FOIL SPECTROSCOPY; NONRELATIVISTIC QUANTUM THEORY; RELATIVISTIC QUANTUM THEORY in the McGraw-Hill Encyclopedia of Science and Technology.

[JAMES E. BAYFIELD]

Hypervalent species

Most stable compounds of main-group elements follow the G. N. Lewis octet rule, with no more than eight valence shell electrons involved in bonding four or fewer groups to the central atom. The electrons are present either as unshared pairs or as pairs forming a chemical bond (traditionally represented by a single line joining two atoms in the written structure). If more than four groups are joined to the central atom, it may be termed hypercoordinated. This may occur with electron-deficient bonding using only the standard octet of electrons in molecules such as CH_5^+ or the carboranes; or it may occur with electron-rich bonding using 9 or 10 electrons in bonding five groups to the central atom (or 11 or 12 electrons in a 6-coordinate species). The latter type of bonding, in hypercoordinate molecules such as PF_5, is called hypervalent. The term hypervalent also includes species which are not hypercoordinated, such as SF_4, IF_3, and XeF_2, in which unshared pairs of electrons supplement those involved in bonds to ligands to bring the total to more than eight. Recent reports of the first examples of hypervalent first-row elements (B, C, and F) have changed the perspective from which hypervalent bonding is viewed.

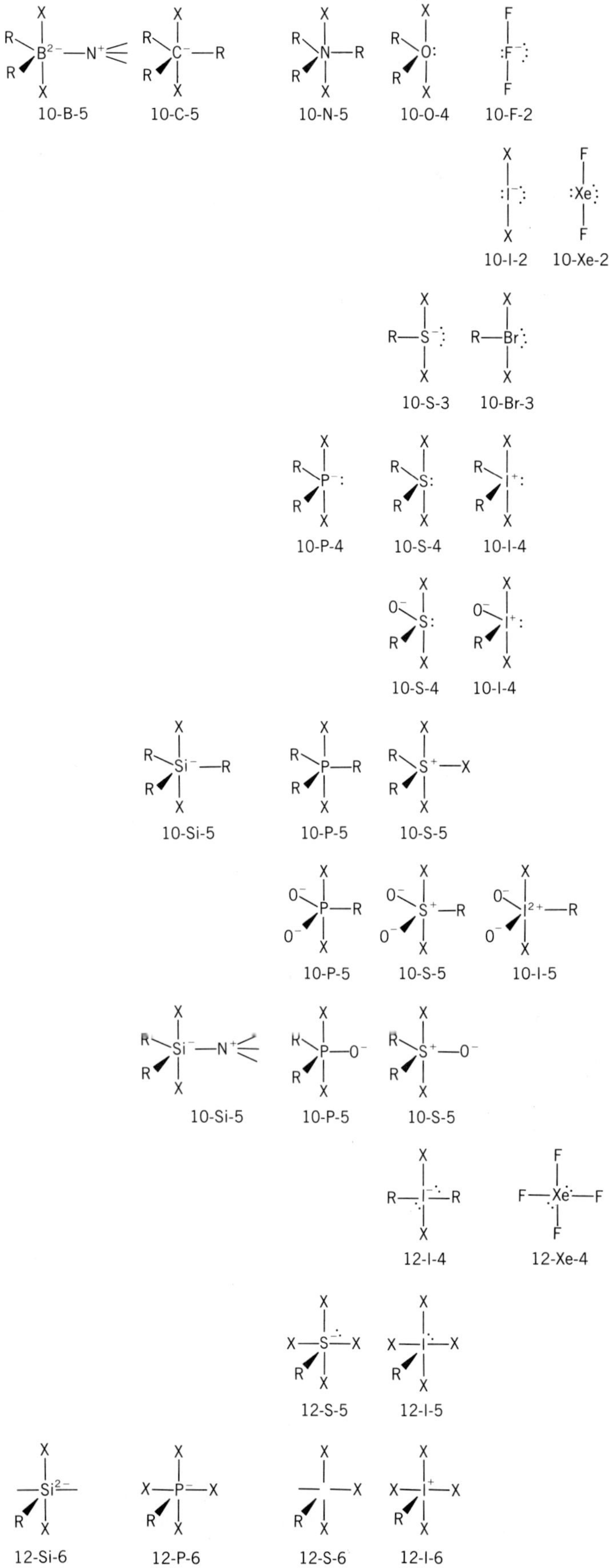

Hypervalent transition states. Hypervalent species may be formed from an appropriately substituted normal-valent substrate by the approach of a nucleophile, which has an unshared pair of electons available to form a bond to the substrate. In many cases, such as the approach of nucleophile Y^- to CH_3X (I) in reaction (1), the formation of the C—Y bond is synchronous with the departure of the leaving group (X^-) to form another normal-valent compound, reaction product CH_3Y (III). The transition state (II) for the reactions uses 10 electrons in bonding five ligands to the central carbon. This hypervalent transition state formally expands carbon's valence octet to 10—formally in the sense that while quantum-mechanical calculations may assign the extra electrons to the ligand atoms rather than to the central atom, the 10 electrons are involved in the five bonds of the transition state (II). The geometry of the energy maximum for the reaction [transition state(II)] is trigonal bipyramidal (*TBP*), while (I) and (III) have tetrahedral (*Td*) geometry. Transition states for associative nucleophilic displacements on other nonmetallic central atoms may also generally be classified as hypervalent.

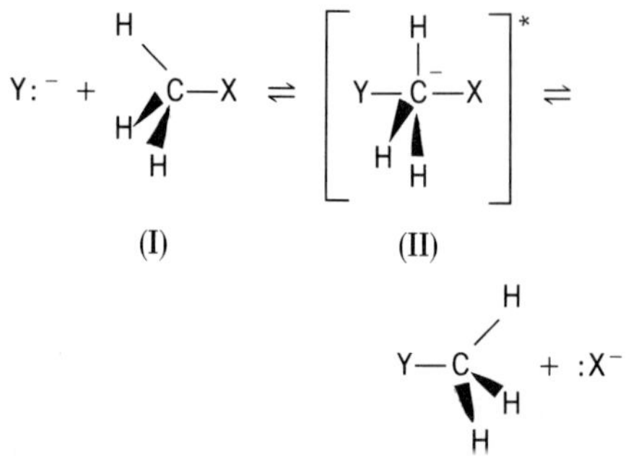

The magnitude of the increase in energy on going from (I) to (II), the activation energy of the nucleophilic displacement, depends strongly on the nature

Fig. 1. Partial "periodic table" of hypervalent nonmetallic species, with *N-X-L* designations.

of the five ligands to carbon in (II). In this example, with three hydrogens attached to a central carbon, (II) is an energy maximum. With other groups attached to the central atom, or with a different central atom, the *TBP* geometry can be stabilized relative to the *Td* geometries sufficiently to make the hypervalent *TBP* species an energy minimum. The term hypervalent is used to describe such *TBP* species with electron-rich bonding.

Isolable, stable *TBP* hypervalent species such as PCl_5 have been known since the nineteenth century, as have related species, called pseudo trigonal bipyramidal (ψ-*TBP*), which have electron pairs instead of one or more of the equatorial ligands. These include compounds such as SF_4, IF_3, and XeF_2 with, respectively, 1, 2, or 3 equatorial lone pairs. These species may be classified by using the *N-X-L* system—*N* is the number of electrons involved in bonding *L* ligands to the central atom *X*. Transition state (II) of reaction (1) is a 10-C-5 species, SF_4 a 10-S-4 species, and XeF_2 a 10-Xe-2 species. Figure 1 shows a systematic listing of a variety of hypervalent species, including cationic, anionic, and electrically neutral isovalent species in each of several of the horizontal rows. Both 10-X-L and 12-X-L types of compounds are shown. The octahedral 12-X-L compounds are in several respects similar to the *TBP* compounds, which will be discussed in more detail.

The two-electron reduction of an 8-P-4 phosphonium cation (IV) to a 10-P-4 phosphoranide (V), reaction (2), results in a change in geometry from *Td*

$$\begin{array}{c}L_3\diagdown \\ L_4\diagup \end{array}\!\!P^+\!\!\begin{array}{c}\diagup L_1 \\ \diagdown L_2\end{array} \xrightarrow{2e} \begin{array}{c}L_3\diagdown \\ L_4\diagup\end{array}\!\!\overset{\displaystyle L_1}{\underset{\displaystyle L_2}{P:^-}} \qquad (2)$$

$$\text{(IV)} \hspace{3cm} \text{(V)}$$

to ψ-*TBP*. In this sense, the extra electron pair is clearly involved in bonding.

Molecular orbital description of hypervalent bonding. For many years, chemists have considered the stability of 10-electron compounds of main-group elements from the second row (or higher rows) of the periodic table to result from the use of *d* orbitals to accommodate the expansion of the valence octet. In fact, the electron occupancy of *d* orbitals is low because of their high energy in such species, and the bonding scheme (Fig. 2), termed hypervalent by Musher, is gaining wider acceptance as a more useful approximate description of the bonding than the traditional dsp^3 hybridization scheme. Only *p* orbitals are used in this picture to form the three-center four-electron hypervalent bond to the two apical fluorines of the 10-Xe-2 compound XeF_2. The electron density placed on the fluorines by two-electrons in ψ_2 reflects an ionic contribution to the XeF bonds and suggests that the octet expansion is not to 10 electrons at the central atom, but is a formal expansion, not requiring *d* orbitals to provide a useful approximate description of the bonding.

Some researchers use the term hypervalent to describe compounds such as sulfur dioxide [reaction

$$\underset{\text{(VI}a\text{)}}{\overset{\ddot{\text{S}}}{\underset{\ddot{\text{O}}\cdot\quad\cdot\ddot{\text{O}}\cdot}{\diagup\!\!\diagdown}}} \longleftrightarrow \underset{\text{(VI}b\text{)}}{\overset{\overset{++}{\text{S}}}{\underset{-\text{O}\quad\text{O}-}{\diagup\!\diagdown}}} \qquad (3)$$

(3)]. The first resonance structure (VI*a*), with p_π-d_π double bonds, would be a 12-electron sulfur species, but the second one (VI*b*) is an 8-S-2 species. The term hypervalent is more commonly used to describe sigma delocalized electron-rich bonding of the type illustrated in Fig. 2.

Hypervalent first-row elements. The general acceptance of the dsp^3 hybridization picture has inhibited attempts to prepare compounds of hypervalent first-row elements, since *d* orbitals (or vacant *s* orbitals) for such elements are so high in energy. Recent successes in preparing a 10-C-5 species, a 10-F-2 species, several 10-B-5 species, and a 12-B-6 species have shown that to accommodate the distribution of electrons in the scheme of Fig. 2, the proper choice of ligands can make single minimum hypervalent compounds of first-row elements accessible. Indeed, the compounds of hypervalent hydrogen, such as the strongly bonded bifluoride anion with its symmetrical hydrogen bond, have been known for many years. These are 4-H-2 species with a formal expansion of the filled shell, which has two electrons in it for H and He, to four electrons. In 1957 the HF_2^- molecule was described with a molecular orbital scheme very similar to Fig. 2 using the hydrogen 1s orbital, rather than a *p* orbital on the central atom.

For background information see CHEMICAL BONDING; COORDINATION CHEMISTRY; MOLECULAR ORBITAL THEORY; PERIODIC TABLE; RESONANCE (MOLECULAR STRUCTURE) in the McGraw-Hill Encyclopedia of Science and Technology.   [J. C. MARTIN]

Bibliography: D. Y. Lee and J. C. Martin, Compounds of pentacoordinate (10-B-5) and hexacoordinate (12-B-6) hypervalent boron, *J. Amer. Chem. Soc.*, 106:5745–5746, 1984; J. C. Martin, "Frozen" transition states: Pentavalent carbon et al., *Science*, 221:509–514, August 5, 1983; J. C. Martin and E. F. Perozzi, Isolable oxysulfuranes in organic chemistry, *Science*, 191:154–159, 1976; J. I. Musher, Hypervalent molecules, *Angew. Chem. Int. Ed. Engl.*, 8:54–69, 1969.

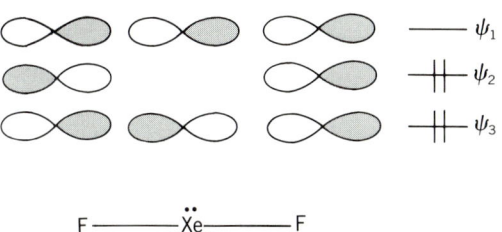

Fig. 2. Musher hypervalent bonding scheme illustrated for XeF_2.

Immobilized cell technology

Immobilized biocatalysts were defined at the 1st Enzyme Engineering Conference in 1971 as those agents "physically confined or localized in a certain defined region of space with retention of their catalytic activity and, if possible or even necessary—their viability, and which can be used repeatedly and continuously." The techniques which were first employed to achieve the immobilization of enzymes have been more recently extended to whole, primarily microbial, cells, and have led to the efficient industrial production of a number of biological compounds. The process of whole-cell immobilization should prove effective in achieving high-cell-density, continuous reactor systems. However, care must be taken when choosing an immobilization technique, since the relative productivity of any specific immobilized cell system will vary with the technique employed.

Rationale for whole-cell immobilization. The immobilization of whole cells offers a number of advantages over enzyme immobilization. These include: circumventing the need for expensive and laborious enzyme isolations; achieving an increase in biocatalyst rate and stability; exploiting multienzyme synthetic pathways; and in-place cofactor regeneration. However, consideration of the option of whole-cell immobilization must also take into account certain disadvantages: the addition of cell membranes and walls, as well as support membranes, can accentuate diffusion limitations; byproducts may be formed from extraneous biosynthetic pathways; and less enzyme per unit area or volume of catalyst particle can be achieved when compared to immobilized enzyme systems. The first two disadvantages can sometimes be remedied by postimmobilization treatment of the immobilized cells. The severity of the third problem can be lessened by increasing intracellular enzyme concentrations through standard selection procedures or through genetic engineering techniques.

Methods of immobilization. These methods can be classified into four groups: entrapment within a support matrix; attachment to the surface of an insoluble support; formation of macroscopic flocs or pellets without the use of a carrier; and retention of the biocatalyst through the use of semipermeable membranes. Regardless of the method, the support must be nontoxic to the organism, nonbiodegradable, and capable of attaching high levels of microorganisms.

Entrapment. Of these methods, entrapment procedures are the most widely reported. Usually the cells are mixed with a solution of prepolymer, which is then allowed to polymerize. The first of these polymers to be tested was polyacrylamide. However, it was found that in some cases free radicals formed, and the heat evolved during the polymerization process led to either inactivation of the desired biosynthetic pathway or the loss of cell viability. In spite of these limitations, polyacrylamide beads have been used to immobilize an impressive diversity of microorganisms.

Polymerization of some natural polymers was found to occur under suitably mild conditions. Calcium alginate and κ-carrageenan (both are isolates from seaweed) are noteworthy examples. Further treatment of these natural polymer matrices through chemical crosslinking can result in enhanced stability of the catalyst particles.

Insoluble support. Immobilization of cells onto the surface of the support can be accomplished through either covalent coupling or adsorption processes. In covalent coupling, the support surface is derivatized with a bifunctional reagent which, when brought in contact with the cell, reacts with one of the many reactive sites on the cell surface. Presumably, this method creates a stable bond between the support and the cell, unaffected by variations in the microenvironment surrounding the catalyst particle (for example, pH, ionic composition, and strength). Adsorption processes exploit ionic and hydrogen bonding forces between, predominantly, the hydroxy groups on the cell surface and the active sites on the support particle. Ion-exchange resins have been used extensively to achieve this type of immobilization. This mechanism is also responsible for the immobilization of cells to such support materials as glass, coal, wood chips, and stones. An interesting extension of these general adsorptive processes is biospecific adsorption. The use of immobilized lectins or antibodies can result in specific, stable bonds between the cell and the support.

Flocculation and pelletization. This is often a natural part of the microbial growth cycle of mycelial organisms. The degree of agglomeration and the size of the resulting particles are dependent on variables such as agitation rate, dissolved oxygen concentration, media composition, pH, and the concentration additives. Flocculated cells quite often achieve dimensions equivalent to carrier-immobilized cells. The same types of reactor systems can therefore be used for both flocculated and carrier-immobilized cells.

Semipermeable membrane. The final method used for the immobilization of cells is their containment in semipermeable membranes. This can be accomplished by two very different procedures. One method is to emulsify cells in a mixture of hydrocarbon solvent, a surfactant, and a membrane stabilizer, resulting in the microencapsulation of cells within a liquid membrane. Typically, 500–600 cells can be contained in emulsion droplets 20–40 micrometers in diameter. The other method employs semipermeable material—such as dialysis tubing, ultrafiltration membranes, or hollow fibers—to retain the biocatalyst within the reaction chamber while maintaining a viable cell population within the reactor.

Industrial applications. The first application of immobilized cell technology was the production of vinegar using a mixed culture of living microbial cells immobilized on wood chips. This process used a naturally occurring film to achieve immobilization. In 1964, the first industrial process using an artificially produced immobilized cell system employed heat-treated whole cells for the production of prednisolone by steroid dehydrogenation of hydrocortisone. It was not until the middle to late 1970s that another industrial process was developed using polyacrylamide-entrapped *Escherichia coli* cells to convert ammonium fumarate to L-aspartic acid on an industrial scale. Later this same procedure was used to immobilize *Brevibacterium ammoniagenes* to produce L-malic acid. Further testing of the support materials led to the finding that the immobilization of *E. coli* in κ-carrageenan beads would result in a fivefold increase in stability and a fifteenfold increase in productivity over a polyacrylamide-immobilized *E. coli* system. Similar results were found when *B. ammoniagenes* was immobilized in κ-carrageenan.

Another industrial application of immobilized whole-cell technology is the production of high-fructose syrups. This application received a tremendous boost when the soft drink industry replaced approximately half of the sugar content of these drinks with high-fructose syrups.

These four processes, however, are the only current industrial applications of immobilized cell systems. This lack of large-scale processes may be the result of any of several factors: poor operating stability, unfavorable economics, or simply the efficiency of well-established fermentations. It should be further noted that all of these industrial processes utilize only a single enzyme to effect the desired transformation. Thus, while whole cells are immobilized, these biocatalysts behave as simple immobilized enzyme systems.

The production of products involving multienzyme systems, or cofactors, such as the production of enzymes and antibiotics, will likely be included in the next generation of immobilized whole-cell industrial reactors. Future applications of immobilized whole-cell technology will likely be in the food, fine chemical, and pharmaceutical industries where relatively high-value products are of interest. These applications will require that further work be conducted in the areas of biocatalyst stability, mass transport characteristics, and reaction kinetics, where a detailed understanding of the effects of immobilization on these phenomena will lead to reliable process control strategies and system optimizations.

For background information *see* ENZYME; ION EXCHANGE; POLYACRYLATE RESIN in the McGraw-Hill Encyclopedia of Science and Technology.

[JOHN LLOYD GAINER; BLAIR OKITA]

Bibliography: P. S. J. Cheetham, Developments in the immobilization of microbial cells and their applications, in E. Wiseman (ed.), *Topics in Enzyme and Fermentation Biotechnology*, 1980; I. Chibata and L. B. Wingard, Jr., *Immobilized Microbial Cells*, 1983; S. Fukui and A. Tanaka, Immobilized microbial cells, *Annu. Rev. Microbiol.*, 36:145–172, 1982.

Immunogenetics

A basic puzzle of immunogenetics has been the means by which individuals generate enough different antibody (or immunoglobulin) molecules to specifically combine with the vast array of different antigens (estimated at 10^6–10^7). The specificity with which each variety of antibody binds to a particular antigen is analogous to the way in which a key fits a particular lock, and these different binding specificities are related to the deoxyribonucleic acid (DNA) sequences of the genes encoding the immunoglobulin molecules. There must exist, then, genetic mechanisms which can preserve and transmit, from one generation to the next, a standard set of immunoglobulin genes but which also can produce, in each individual, immunoglobulin genes diverse enough to produce 10^6–10^7 different structural variants.

Historically, two theories have attempted to explain the diversity of antibodies: (a) the germ line theory, contending that the individual carries a separate gene for every different type of antibody produced, and (b) the somatic mutation theory, proposing that there are very few such genes but that they mutate rapidly during development, so that different B lymphocytes express mutant variants of the original immunoglobulin genes. Recent scientific consensus, described below, incorporates parts of both theories as well as a system of gene rearrangement, whereby the linear sequences of the DNA encoding the immunoglobulin genes are altered.

Antibody structure. An antibody, or immunoglobulin (Ig), molecule (Fig. 1) is a protein containing

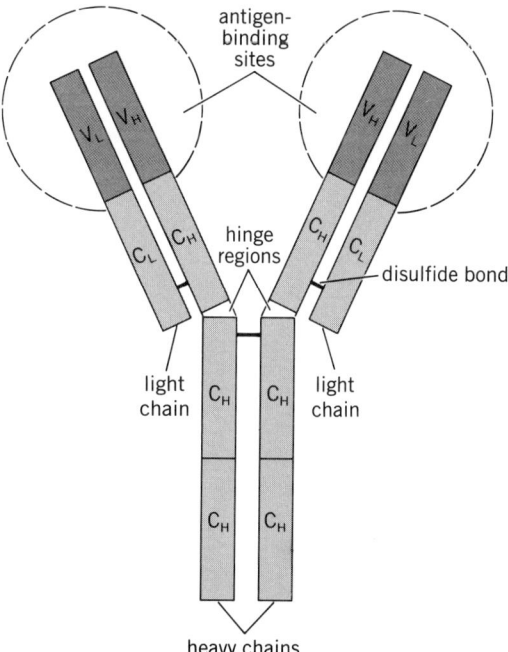

Fig. 1. Typical immunoglobulin molecule (IgG). Variable regions are darker-tinted; other regions show constant amino acid sequence. Abbreviations are explained in the text.

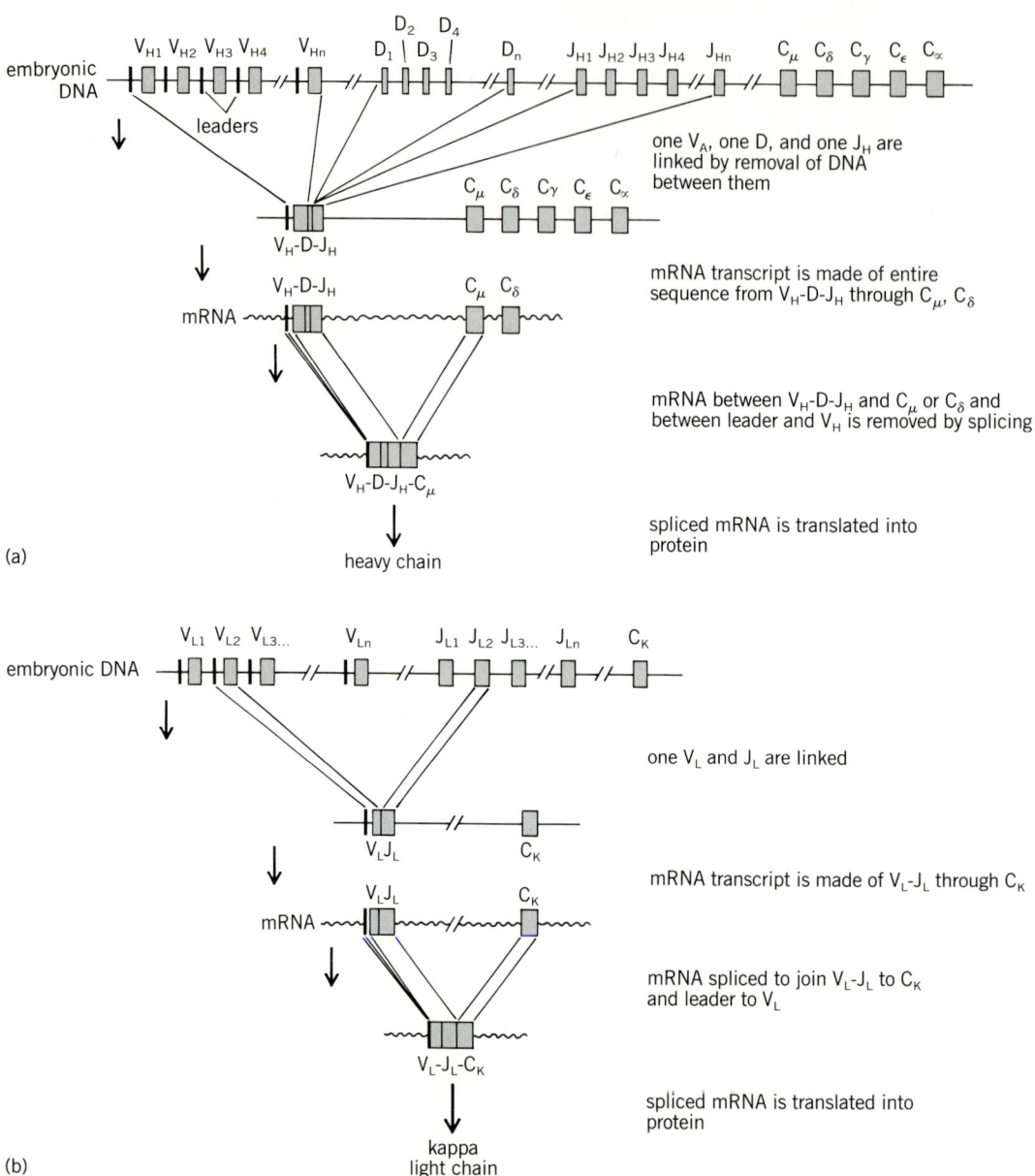

Fig. 2. DNA rearrangement and mRNA splicing in synthesis of (a) heavy chains and (b) kappa light chains. Abbreviations are explained in the text.

two identical light chains and two identical heavy chains linked by covalent disulfide bonds. Molecules of three antibody classes (IgG, IgD, IgE) consist of only one such basic unit. The remaining two classes differ in that IgM molecules contain five basic units and IgA molecules have either one or two basic units.

Each chain contains one or more constant regions in heavy chains (C_H) or light chains (C_L) and a variable region (V_H or V_L) which contributes to the antigen-binding sites of the immunoglobulin molecule. The constant regions of the light chains exist in two forms, kappa and lambda (C_k and C_λ), while the constant regions of the heavy chains exist in five forms, each associated with a particular immunoglobulin class (C_α = IgA, C_δ = IgD, C_ϵ = IgE, C_γ = IgG, and C_μ = IgM). In a class of immunoglobulins, variable regions are extremely diverse within each individual, while constant regions are consistent. For example, the constant regions of the IgM molecules within an individual vary little, but the variable regions are very different.

There are three types of antibody diversity: (1) idiotypes—the various forms, within the individual, of the antigen-binding sites of the molecule; (2) isotypes or classes—differences within the individual among heavy-chain constant regions determining the physiological functions of the immunoglobulin molecule; and (3) allotypes—minor structural differences which vary between individuals in the manner of classical mendelian genes. In addition, a particular variable region may be the same in antibodies of different classes—for example, there may be IgM and IgG molecules with identical variable regions.

Genetic specification of antibody molecules. Each immunoglobulin chain, heavy and light, is encoded not by a single gene but by a series of genes known as a multigene family. For heavy chains, there are several clusters of genes: variable genes, known as V_{H1-n}, diversity segments (D_{1-n}), joining segments (J_{H1-n}), and constant-region genes (C_μ, C_δ, C_γ, C_ϵ, and C_α) [Fig. 2a]. Each V_H gene is preceded by a leader sequence that encodes a portion of the chain important for synthesis but that is removed when the molecule becomes functional. While all of these genes are present in the early embryo, those cells destined to become B lymphocytes lose some of them during development so that only one V_H gene, one D segment, and one J_H segment remain. These are fused, by removing the DNA between them, to form a V_H-D-J_H combination encoding the heavy-chain variable region. The number of V_H genes in humans has been estimated at 100–1000, of D genes at 20, and J_H genes at 4. If one from each group is chosen at random, there are 8000–80,000 different combinations (100 or 1000 × 20 × 4) available to encode a heavy-chain variable region (estimates vary among species). Since there is also some variability in the exact sites at which these genes may be joined, the number of possibilities is actually higher.

Light chains are similarly encoded in the DNA as multigene families (Fig. 2b), although they differ from the heavy chains in lacking diversity (D) sequences. Kappa and lambda genes differ in their organization of V, J, and C genes. As in heavy-chain synthesis, during development of the B-cell lineage, some DNA is lost and a single V_L and J_L gene are joined in each cell to form a V_L-J_L unit encoding the light-chain variable region. Kappa light-chain genes are located on a different chromosome than lambda light-chain genes (heavy-chain genes are on still another chromosome). The numbers of genes encoding light chains have been estimated at 200 V_L genes and 5 J_L genes for kappa chains, but only 2 V_L and 2 J_L genes for the less complex lambda chains. Thus there are about 1000 different light-chain variable regions possible.

With about 1000 different types of light-chain variable regions and 8000–80,000 different heavy-chain variable regions, the total number of different immunoglobulin binding regions (combining heavy- and light-chain variable regions) ranges from 8×10^6 to 8×10^7.

After V_H-D-J_H or V_L-J_L joining, the DNA encoding the variable region is still separated from the constant gene or genes by an intervening stretch of DNA. This entire sequence, including the intervening DNA, is transcribed into messenger ribonucleic acid (mRNA), but the intervening RNA sequence is subsequently removed by splicing to produce an mRNA which carries the V_H-D-J_H-C_H or V_L-J_L-C_L genes directly linked to each other in the correct sequence for translation into protein. In similar fashion, the mRNA sequences between the leader segments and the V_H or V_L genes are spliced out.

Both kappa and lambda light chains are produced by this means. In heavy chains, however, this method for uniting the variable and constant genes may exist only for the IgM and IgD classes. IgM and IgD are the earliest immunoglobulin classes produced by B lymphocytes. In fact, naive unstimulated B lymphocytes may express both IgM and IgD on their surfaces. The initial union of V_H-D-J_H units with C_μ and C_δ genes is probably similar to the process in light chains. An mRNA transcript including

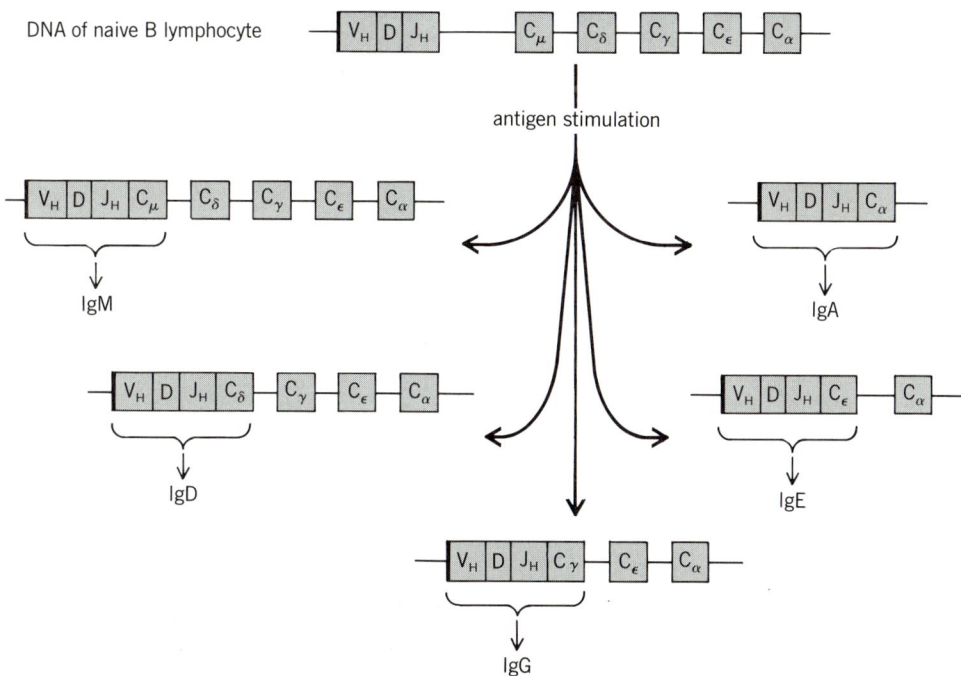

Fig. 3. Class switching in heavy chains by DNA rearrangement. Abbreviations are explained in the text.

the leader, the V_H-D-J_H unit, the intervening segment, and the C_H gene is produced (possibly including both C_μ and C_δ), then spliced to remove the RNA between V_H-D-J_H and either C_μ or C_δ. The spliced mRNA can then be translated to produce the entire IgM or IgD heavy chain. Production of IgG, IgE, or IgA heavy chains, however, requires further DNA rearrangement.

Class switching. The same heavy-chain variable region may be united with different heavy-chain constant regions. The DNA rearrangements of variable regions occur during embryonic development; each B lymphocyte expresses a rearranged variable region (V_H-D-J_H or V_L-J_L) sequence before making contact with the appropriate antigen. In heavy chains the V_H-D-J_H unit is nearest to the C_μ and C_δ genes (Fig. 3) and becomes linked to these genes, probably via mRNA splicing as discussed above. Upon stimulation by antigen, however, additional DNA rearrangements may occur (Fig. 3). The DNA encoding C_μ may be excised, leaving the V_H-D-J_H unit to combine with the next C_H gene in the sequence (C_δ). Thus, the cell involved in this class switch would stop secretion of IgM and produce only IgD, but since the V_H-D-J_H combination is the same, the variable region would be unchanged. If longer sequences were deleted, such as $C_\mu + C_\delta$, the V_H-D-J_H segment would then combine with the C_γ genes to produce IgG (again with the same variable region). Longer deletions, in turn, would permit the synthesis of IgE (C_ϵ) and IgA (C_α). Whether any particular B lymphocyte lineage can undergo a series of class switches (IgM to IgD, then to IgG, and so on) or only a single switch (IgM to IgD, or IgM to IgG, or IgM to IgE, and so on) is unknown. However, since the constant regions of the heavy chains determine the physiological reactions in which the immunoglobulin molecule can function, class switching permits the production of a variety of functionally different antibodies directed against the same antigen.

For background information see ANTIBODY; IMMUNOGLOBULIN; IMMUNOLOGY in the McGraw-Hill Encyclopedia of Science and Technology.

[ROGER MELVOLD]

Bibliography: T. Honjo, Immunoglobulin genes, *Annu. Rev. Immunol.*, 1:499–528, 1983; J. Klein, *Immunology: The Science of Self-Nonself Discrimination*, 1982; A. Nisonoff, *Introduction to Molecular Immunology*, 1982; M. Zaleski et al., *Immunogenetics*, 1983.

Information processing (psychology)

One of the most widely used data collection techniques in psychological, sociological, consumer, and medical research is introspective reports about people's thoughts and feelings. It is thus important to establish the accuracy of introspective reports. This is no easy task, because there are usually no independent sources against which verbal reports can be checked. For example, if a person expresses a feeling of remorse, it is very difficult to come up with evidence that challenges this claim.

Nonetheless, a considerable amount of work has been done on the validity of introspective reports. Much of this work is controversial, and due to the complexity of the issues psychologists are far from answering all the remaining questions. Several major issues and recent discoveries can, however, be outlined.

Inaccuracy due to lying. Introspective accounts about mental processes may be inaccurate because people purposefully distort their accounts; that is, they may report that their thoughts are different from what they know them to be. An ardent suitor might convey better intentions than he knows he actually has, or a defendant pursuing an insanity plea might report more bizarre thoughts than were actually experienced. In socioscientific research, investigators attempt to minimize dishonest reporting by reducing the motive to misrepresent one's thoughts (for example, by making the responses private and anonymous), or by including scales that attempt to measure the extent of misrepresentation, such as the lie scale of the Minnesota Multiphasic Personality Inventory (MMPI).

Inaccuracy due to lack of access. More fundamentally, introspective reports might be inaccurate because the thoughts and feelings under scrutiny are inaccessible to conscious awareness. Even an honest, forthright attempt to report one's thoughts will be inaccurate to the extent that the thoughts are inaccessible. This possibility raises one of the most debated and fundamental questions in psychology, namely the nature and extent of the unconscious.

Virtually all psychologists agree that there are limits to what is accessible via introspection. Indeed, if the mind were totally accessible, there would be no need for the burgeoning field of cognitive science; each person would need only introspect to answer complex questions about mental functioning. At the same time, few psychologists would argue that mental processes are totally inaccessible. (Some radical behaviorists would argue that the mind is epiphenomenal and unworthy of scientific investigation, but few would deny the existence of the mind or deny that it is at least partially accessible via introspection.) It thus becomes necessary to specify precisely what sorts of thoughts and feelings are and are not accessible.

Threatening versus nonthreatening thoughts. One of the earliest psychologists to address the question of what is accessible to awareness was Sigmund Freud. Freud argued that most of a person's mental life is unconscious, and thus people's reports about their thoughts and feelings should be considered suspect. He characterized unconscious thought as childlike and irrational, and due to basic sexual and aggressive drives. People are motivated to repress these thoughts, Freud argued, because the thoughts are highly threatening (that is, anxiety-provoking) to the conscious, more rational self. According to this view, psychologists interested in assessing personality cannot rely on an individual's introspective reports. Instead, projective tests such as the Rorschach must be used to bypass a person's repressive forces.

Freud's ideas remain influential, but they have never been fully accepted by experimental psychologists, who are convinced only by empirical data, not clinical case studies. In recent years, however, the role of the unconscious has been gaining increasing acceptance in this discipline as well. The type of unconscious processing that is studied tends to be different than that described by Freud; indeed, recent research suggests that Freud may have been too conservative about the limits of awareness.

Recent experimental work. Several experiments have demonstrated that people can respond to meaning and can form emotional responses outside of awareness; two such studies will be described here for illustrative purposes. One technique that has been widely used to study unconscious processing is dichotic listening. Subjects wearing headphones are instructed to pay close attention to a verbal message that is played to one ear, called the attended channel. Just as all voices are tuned out at a cocktail party except the one to which a person is listening, messages played to the other ear (the unattended channel) are not consciously heard. Several experiments have demonstrated, however, that although people cannot report the message played to the unattended channel, this message can influence their behavior. For example, D. Mackay presented people with ambiguous sentences in the attended channel. One such sentence was, "They threw stones at the bank," with the ambiguity centering on whether a river bank or a financial bank was meant. For some subjects the word "river" was simultaneously played to the unattended channel, while for others the word "money" was played. Although no subject could report the word played to the unattended channel, when "river" was played subjects tended to interpret the sentence as referring to a river bank, and when "money" was played they tended to interpret the sentence as referring to a financial bank. The words played to the unattended channel registered at a nonconscious level, affecting the meaning attached to the sentence consciously heard.

W. Kunst-Wilson and R. Zajonc demonstrated that affective preferences can develop unconsciously. Subjects were shown differently shaped polygons at extremely fast exposure times, so fast that subjects could not report what they had seen. The polygons were each shown five times in this way. Then, subjects were shown polygons at long enough durations that they could be consciously seen. On these trials some of the polygons were ones that subjects had "seen" earlier at the fast exposure times, and other were ones they had never seen before. Subjects could not report at better than chance levels which ones they had been exposed to before. When the subjects were asked how much they liked the polygons, however, the ones seen earlier were preferred significantly more than the new ones. Affective preferences developed for stimuli that were apparently never consciously seen.

While controversial, these and other experiments indicate that people can respond to meaning and can form preferences unconsciously. This work casts doubt on Freud's argument that only anxiety-provoking material is unconscious; the words used by Mackay and the polygons used by Kunst-Wilson and Zajonc can hardly be said to be threatening to subjects. A picture of the mind is emerging in cognitive psychology that portrays consciousness as the focus of a person's attention at any given time, with a good deal of other information being processed nonconsciously. Thus, unconscious processing may be an asset of an efficient and capable mental system, and not limited to material that is anxiety-provoking. The threatening-nonthreatening distinction, then, appears not to fully explain what is accessible and what is not.

Lower-order versus higher-order processes. Another distinction concerns lower-order versus higher-order mental processes. Few people argue that lower-order processes such as sensation and perception are accessible. A request to explain how one perceives depth, for example, is likely to be beyond answering. It seems that many higher-order processes are accessible, such as the processes underlying judgments, choices, evaluations, and problem solving. R. Nisbett and T. Wilson have demonstrated, however, the access even to many higher-order processes is limited. In these studies subjects often made inaccurate reports about the determinants of their preferences for other people, consumer goods, and literary passages. Similarly, the two experiments reviewed above demonstrate people's inability to specify the processes underlying their interpretations of an ambiguous sentence and preferences for polygons, processes that are typically considered to be higher-order. Nisbett and Wilson argued that people make inferences about the causes of their responses which are often based more on cultural norms about causality than on introspectively gained information.

Mental processes versus mental states. A further distinction that may elucidate what is accessible and what is not concerns mental processes versus mental states. Nisbett and Wilson argued that people have introspective access to the states resulting from mental processing, such as sensations, emotions, and plans, but not to the cognitive processes that produced these states. For example, subjects in the Kunst-Wilson and Zajonc experiment were presumably making accurate reports about their preferences for the polygons, but would be considerably less accurate in specifying why they preferred the ones seen earlier. Similarly, A. Ericsson and E. Simon argued that reports about mental processes will be inaccurate if it is necessary for people to interpret their responses, rather than simply to report them. Thus, consistent with Nisbett and Wilson's argument, they argued that subjects will be considerably less accurate at answering why they responded the way they did than in giving the response itself. This distinction is in need of some refinement, however. There is some evidence that people can be unaware of mental states as well as mental processes. Furthermore, the difference between a mental state and a mental process is not always clear.

Conclusion. Considerable work needs to be done to specify when introspective reports will be valid. Investigators should be particularly wary, however, of reports in situations where people are motivated to lie, reports about anxiety-provoking material, reports about lower-order mental processes, and reports requiring substantial interpretation and inference on the part of the respondents, such as reports about the causes of their responses. When possible, independent means of assessing people's mental processes should be used, such as observing their behavior.

For background information see COGNITION; CONSCIOUSNESS; INFORMATION PROCESSING (PSYCHOLOGY) in the McGraw-Hill Encyclopedia of Science and Technology.

[TIMOTHY D. WILSON]

Bibliography: A. Ericsson and H. Simon, Verbal reports as data, *Psychol. Rev.*, 87:215–251, 1980; D. Fiske, When are verbal reports veridical?, *New Dir. Meth. Soc. Behav. Sci.*, 4:59–66, 1980; D. Lieberman, Behaviorism and the mind: A (limited) call for a return to introspection, *Amer. Psychol.*, 34:319–333, 1979; R. Nisbett and T. Wilson, Telling more than we can know: Verbal reports on mental processes, *Psychol. Rev.*, 84:231–259, 1977.

Insect control, biological

A World Health Organization (WHO) scientific group has defined genetic control as "the use of any condition or treatment that can reduce the reproductive potential of noxious forms by altering or replacing the hereditary material."

Genetic control of agricultural pests was introduced in 1954 against the screwworm (*Cochliomyia hominivorax*), a pest of livestock, but has not been widely employed against any other species except fruit flies because chemical methods of pest control have been more cost-effective. However, in recent years attention has been directed to alternative methods of pest control, including genetic control, because of the problem of widespread resistance of pests to chemicals and public concern over environmental pollution by pesticides. Compared with chemical control, genetic methods are unparalleled in their specificity and safety because only one species in the ecosystem is affected. Genetic control is achieved by rendering the pests infertile.

Available genetic techniques. There are a wide variety of techniques currently available for rendering pests infertile. These are as follows:

Sterile-insect technique. This aims at the release of reared insects, which have been made sterile by exposure to ionizing radiations or chemosterilants, into the environment occupied by the wild population. This is also referred to as sterile-insect release method.

Genetically altered insects. Four methods have been used in which the insects have been modified genetically or carry deleterious genes.

1. Inherited sterility. Also called inherited partial sterility, delayed sterility, or F_1 sterility, inherited sterility is produced by lowering the radiation dose considerably so that the released insects are only partially sterile rather than completely sterile. The radiation dose can be adjusted so that when the released males and females interbreed, practically no progeny are produced, but when they outcross with insects in nature the egg hatch is decreased and the resulting progeny are almost completely sterile.

2. Hybrid sterility. Currently available only for the suppression of the tobacco bud worm (*Heliothis virescens*), this approach involves the release of backcross insects derived from crossing *H. subflexa* females with *H. virescens* males. These fertile females attract and mate with *H. virescens* males either in the laboratory or in nature, and they will continue to produce fertile females and sterile males indefinitely. Hybrid sterility has also been found in some mosquito species complexes, such as *Anopheles gambiae*. The sterile hybrid males produced in the laboratory could then be released.

3. Chromosomal translocations. These are produced when chromosomes are broken by radiation and allowed to rejoin in abnormal combinations. The translocations can be useful because they cause inherited partial sterility and can link an unusual combination of genes together. One application of these abnormal linkages is in the production of genetic sexing systems in which insecticide-resistance genes are linked to the Y chromosome so that the females in a batch can be selectively killed, leaving only males for release. Another special kind of translocation is by using compound chromosomes which have the remarkable property of causing complete sterility when the carriers mate with the wild insects, and yet the insects from the compound chromosome strain are partially fertile. All these kinds of translocations have been intensively studied in the Australian sheep blowfly (*Lucilia cuprina*).

4. Insects harmless to humans. A different concept of genetic control is not to suppress the population by sterility, but to introduce genes which render the insect population harmless to humans. There is considerable interest with regard to rendering mosquitos in the wild genetically unable to transmit human disease. Among agricultural insects, a similar concept has been tested in the Hessian fly in the United States where genes causing inability for the pest to survive on local wheat strains have been introduced by male releases.

Other techniques. The mating behavior of insects may be used to transfer a lethal or sterilizing agent to members of the natural population, and this method might be particularly useful for suppressing insects that are difficult or costly to rear in large numbers. This technique has been successfully tested for the control of tsetse flies. The flies are trapped in cages which are impregnated with chemosterilants, and after exposure the flies are released. Techniques such as cytoplasmic incompatibility, conditional lethals (such as temperature sensitivity), inability to diapause, and many others are available in some species of insects and are being further investigated.

Screwworm. The eradication of the screwworm by the first major use of the sterile-insect technique was carried out on the island of Curaçao in 1954, in Florida in 1959, and on the island of Puerto Rico in 1975, and was an outstanding achievement in this field. The program of eradication of the screwworm from the southwestern United States and of suppression in northern Mexico now appears to be successful after some earlier difficulties. The present goal of this Mexican–United States program is to establish and maintain a barrier of sterile flies across the isthmus of Tehuantepec in Mexico by 1985.

Mediterranean fruit fly. The control of the Mediterranean fruit fly (medfly; *Ceratitis capitata*) is also showing encouraging results. The sterile-insect technique for medfly suppression was originally tested in 1969 on the Italian island of Capri, and has also been tested in several other Mediterranean and Latin American countries. After the medfly was introduced into Costa Rica in the mid-1960s, a program based in that country was planned to halt the northward movement of the medfly across Central America and to eventually eliminate the pest north of the Panama Canal. Unfortunately the program was abandoned in 1971, and subsequently the Mediterranean fruit fly spread through Central America. It was first detected in Mexico in January 1977, and a major sterile-insect technique program was started there. This program has been highly successful in that it has checked the movement of the pest in the north of the country and eradicated the pest from an area covering more than 7.5 million acres (3 million hectares) in Mexico.

When the medfly was accidentally introduced into California in the 1980s, the sterile-insect technique was an important component of an integrated pest management program which succeeded in eradicating this pest from California.

Currently, a program to eradicate the medfly from the Nile delta and valley in Egypt (about 6 million acres or 2.5 million hectares) is being planned by the Egyptian government, the Food and Agriculture Organization, and the International Atomic Energy Agency. If successful, this would eliminate the significant losses due to medfly that occur each year in the production of fruit and vegetables in Egypt.

Other fruit flies. The medfly is not an important pest in the Far East, but two other species of fruit flies, the melon fly (*Dacus cucurbitae*) and the oriental fly (*D. dorsalis*), are of considerable economic importance. As there are no fruit flies in mainland Japan, the transport of untreated host fruit from infested areas is prohibited. Programs to eradicate these two species from certain islands which lie south of Japan were undertaken and have been successful. In 1980 Japan began a large-scale project to eradicate the melon fly from the entire Okinawa prefecture by utilizing the sterile-insect technique.

Lepidopteran species. It is widely recognized that lepidopteran species, that is, butterflies and moths, are the most important pests of many major crops, forests, and stored products. A sterile-insect technique trial to suppress the corn earworm (*Heliothis zea*) on the island of St. Croix was conducted in 1968–1969, but poor results were obtained. On the other hand, the tobacco hornworm (*Manduca sexta*) there was suppressed during field trials in 1971–1972. Suppression of the codling moth (*Cydia pomenella*) to a low level has been achieved in several field trials in western Canada and the United States.

The largest program to date utilizing the sterile-insect technique against lepidopteran species is not aimed at suppression or eradication, but uses the technique as a quarantine method to prevent the establishment of the pink bollworm (*Pectinophora gossypiella*) in the San Joaquin valley of California.

The inherited sterility method has been tested for the suppression of the cabbage looper (*Trichoplusia ni*) and the gypsy moth (*Porthetria dispar*).

The tobacco budworm is a major pest in the United States. In 1977 a pilot project was initiated on St. Croix to test the hybrid sterility method against this pest. The results indicated that the feasibility of *Heliothis* management by releasing hybrids will depend upon the number of native moths in the natural population and the cost of rearing and distributing the insects. The eventual application of this method will depend largely on economic factors.

Other species. The sterile-insect technique has been developed to the point of large-scale implementation for dealing with onion flies, olive fruit flies, stable flies, horn flies, tsetse flies, ticks, and mosquitos.

In the field of public health, two notable attempts have been made to eradicate *Anopheles albimanus* in El Salvador by releasing chemosterilized males. The preliminary attempt, a small-scale effort, was successful, but a larger trial faced several difficulties even when the released sterile males were preceded and accompanied by the use of larvicides.

Population replacement and the introduction of favorable genes such as those controlling insecticide susceptibility and those preventing the development of human-disease pathogens offer better prospects; methods for mass-producing harmless male mosquitos with these characteristics are well on the way to being perfected. The method for producing batches consisting of males involves only the translocation of resistance genes to the Y chromosome and exposure of the resulting mosquitos to solutions of the insecticide sufficient to kill the susceptible females.

Refractoriness to the rodent malaria parasite *Plasmodium yoelii nigeriensis* was introduced into a susceptible population of *A. gambiae* by releasing males of a refractory strain (one in which the parasite is unable to develop). The ultimate hope is that it will be possible to do this in the field situation with mosquitos refractory to *P. falciparum* and *P. vivax*, and thus help to eliminate malaria.

Factors affecting application and success. A basic requirement for all genetic control techniques, if they are to be cost-effective compared with other methods of control, is the ability to rear the required number of vigorous insects in sufficient numbers and at a reasonable cost. The successful use of any of these techniques will require that the insects released should be able to compete with members of the natural population for mating. Many ecological factors, the dynamics of the pest population, and the distribution and dispersal of the released insects are some of the important factors that influence the successful application of genetic control methods in the field.

For background information *see* ENTOMOLOGY, ECONOMIC; INSECT CONTROL, BIOLOGICAL in the McGraw-Hill Encyclopedia of Science and Technology.

[R. PAL]

Bibliography: E. F. Knipling, *The Basic Principles of Insect Population Suppression and Management*, Agr. Handb. 512, U.S. Department of Agriculture, 1979; L. E. LaChance, *Genetic Methods for the Control of Lepidopteran Species Status and Potential*, International Organization for Biological Control, 1985; R. Pal and L. E. LaChance, Methods for control of insects of medical and veterinary importance, *Annu. Rev. Entomol.*, 19:269–291, 1974; *Sterile Technique and Radiation in Insect Control*, proceedings of a symposium, Neuharberg, June 29 to July 3, 1981, IAEA, Vienna, 1982.

Integrated services digital network

The costs associated with solid-state circuitry have been rapidly decreasing, leading to unprecedented changes in the telecommunications industry. In the past, communications network design with analog technology was optimized for voice transport. Since the early 1960s, the worldwide communications system has been evolving toward use of the most advanced digital technology for both voice and nonvoice applications. The resulting system has been named the integrated services digital network (ISDN).

Characteristics. Integrated services digital network is a generic term referring to the integration of communications services transported over digital facilities such as wire pairs, coaxial cables, optical fibers, microwave radio, and satellites.

Digital technology. ISDN provides end-to-end digital connectivity between any two (or more) communications devices. Information enters, passes through, and exits the network in a completely digital fashion. Digital technology has played an important role in the growth of the telecommunications industry. In 1962 the arrival of pulse-code modulation (PCM) marked the beginning of the evolution of the digital network. Pulse-code modulation is a sampling technique which transforms a voice signal with a bandwidth of 4 kHz voice signal into a 64-kilobits-per-second (kbps) digital bit stream, the fundamental building block for data transmission rates in the existing telecommunications industry.

Many aspects of telecommunications are improved with digital technology. For example, digital technology lends itself to very large-scale integration (VLSI) technology and its associated benefits of miniaturization and cost reduction. In addition, computers operate digitally. The digital transport of ISDN provides for human-to-human, computer-to-computer, and human-to-computer interactions. The network is capable of transporting voice, data, graphics, text, and even video information over the same equipment.

Service integration and standard interface. The customer has access to a wide spectrum of communications services by way of a single access link. This is in contrast to existing methods of service access, which segregate services onto specialized lines. ISDN, on the other hand, integrates services on a single access line to the network (Fig. 1).

Associated with integrated access and ISDN is the concept of a standard interface. The objective of a standard interface is to allow any ISDN terminal to be plugged into any ISDN interface, resulting in terminal portability, flexibility, and ease in operation. The interface separates the network from the customer's equipment. The customer's perspective of the service is through the interface. In fact, ISDN is recognized solely by the service characteristics (performance, bandwidth, and protocol) offered at the user-network boundary, rather than by any internal network components. This promotes competition among telecommunications providers since internal network attributes, such as transmission technology (fiber, twisted pair, satellite, and so forth), are not specified but are left up to the discretion of the telecommunications service provider.

Through a standard interface, customers can select their particular service needs. The benefits of standard interfaces are readily visible in the United States in the electrical industry, where almost any electrical outlet can be used by any electrical device. On the other hand, the international electrical industry is a notable example of the difficulties and inconveniences associated with nonstandard interfaces.

Signaling. In addition to service integration and a standard interface, ISDN provides an advanced capability in the form of a signaling control channel. This control channel is a common channel, transported out-of-band, that provides advanced digital signaling capabilities associated with the integrated service access.

Digital signaling is important for several reasons. First, with the increase in communications involving computers, machine-readable signaling information as opposed to human-audible signals allows computers, which do not understand voice messages, to react to the varied states which occur in a typical call. Second, with digital signaling, the time it takes to set up a call is reduced to the order of a few seconds. Third, additional messages are easily introduced when required for new service applications.

Out-of-band signaling offers advantages as well. Out-of-band means that the messages do not use the same channel as the customer information but are carried independently. This allows messages to be sent to the users while the call is active without interfering with the call, such as the identity of a caller in a call-waiting message.

Thus, out-of-band digital signaling provides features unheard of in the existing telephone network. An example is security where the identity of the calling party, sent to the called party in machine-readable form, can prevent invalid users from accessing private computers.

In summary, then, the goal of ISDN is to provide a versatile, multiservice end-to-end digital network with internationally standardized customer-network interfaces and a digital out-of-band control channel.

Standards. In order to establish ISDN, it is necessary to define a set of interfaces between the customer premises equipment and the telecommunications transport network. An important goal is to make these interfaces internationally compatible. Internationally compatible interfaces benefit the users of the network and the network providers, as well as the telecommunications equipment vendors. This is particularly true for multinational corporations and international equipment vendors.

In order to define internationally compatible interfaces for ISDN, an international team of experts in customer premises equipment and telecommunications planners has been formed. These experts operate under the auspices of the International Telegraph and Telephone Consultative Committee (CCITT), an operational entity of the International Telecommunication Union (ITU) which is a specialized treaty agency of the United Nations. This team meets many times annually, and is scheduled to issue recommendations every 4 years. The first set of ISDN recommendations was issued in 1984. These recommendations are widely accepted as the standard for implementing ISDN. Through this cooperative effort, ISDN will reflect the natural evolution of telecommunications and computers and will take maximum advantage of their increasing synergies.

Interfaces. Currently, three ISDN interfaces are being defined in the CCITT. Each is an outgrowth of the existing digital network capabilities. Since the existing network encodes voice into a 64-kbps bit stream, the building block of ISDN is a data transmission rate of 64 kbps.

The first interface is called the basic interface, operating at a 192-kbps rate, providing two 64-kbps channels plus a 16-kbps packet-oriented digital channel. (The remaining 48 kbps are overhead providing structure to the channels.) The 64-kbps

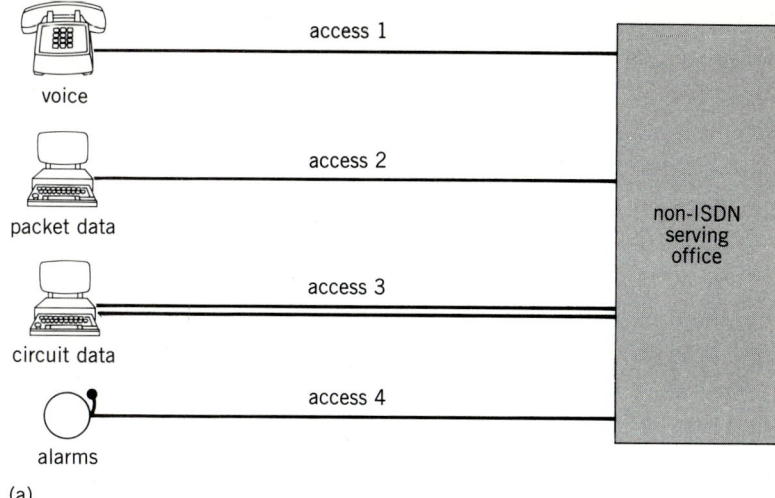

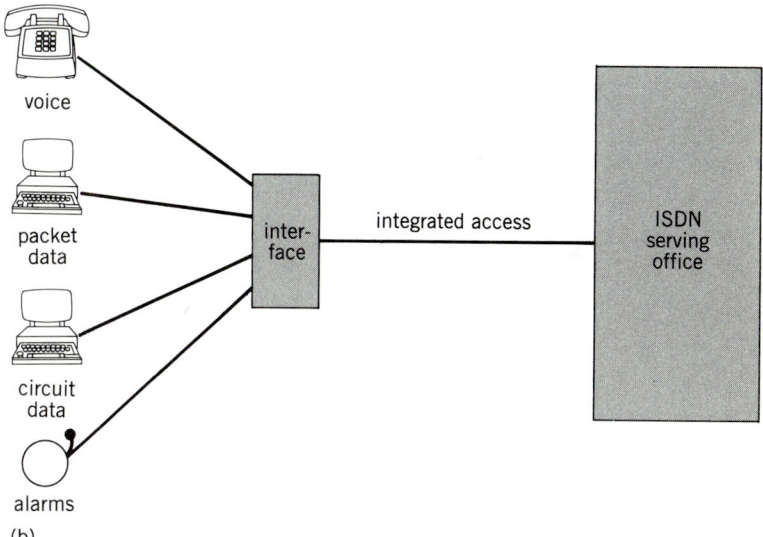

Fig. 1. Comparison of segregated and integrated access. (a) For segregated access customer-network interfaces differ whereas (b) for integrated access there is one customer interface.

channels, referred to as B channels (for bearer channels), transport 64 kbps of customer information, either voice or data. The 16-kbps channel, referred to as a D channel (for delta channel), carries signaling and packet data.

The second interface at 1.544 megabits per second (Mbps) can provide access to digital private branch exchanges (PBXs) or other large communications systems. This interface consists of twenty-four 64-kbps channels, and is an outgrowth of existing multiplexed digital carrier communication. One of these 64-kbps channels, called a D channel, operates as a signaling and packet data channel. The other twenty-three channels are B channels, as in the basic access. This interface is flexible in that several of the 64-kbps B channels can be combined to provide higher-speed services. Two recognized speeds are 384-kbps and 1.536-Mbps data rates.

The third interface is a wide-band interface for

fiber optic access to the user. Likely candidates for this interface are speeds of 45 or 139.264 Mbps, with a packet or message-oriented control channel.

Additional work is under way in the CCITT to standardize bearer service subrates, such as 8, 16, and 32 kbps. These standards, together with those described above, will provide a wide spectrum of customer data rates for numerous applications. As computers become more ubiquitous and as communications costs decrease, the need for varied bandwidths will grow. ISDN is designed to provide bandwidth on demand, which means that the user will be able to specify the data rate desired on a call-by-call basis. Also, since different applications require different performance levels, performance on demand will be another key component of ISDN.

Communications network. In addition to integrated-access transport from the customer to the switching node, ISDN must provide digital switching, interoffice transport, and signaling functions. Switching functions include circuit switching, packet switching, channel switching, and wide-band switching.

Circuit switching is the technology used in the existing voice network where end-to-end paths are dedicated to a single call during the call duration. Packet switching is intended for data calls which, unlike voice, tend to be "bursty" in information content. Packet-switched networks bundle customer data into self-contained messages that use the transport facility only during the transmission of the packet and share the medium with packets from other users.

Leased lines, while strictly not a switching function, are fundamental to ISDN. They are information pipelines dedicated to two (or more) users, and are generally used for applications which require large amounts of data to be transported between fixed points. In ISDN, many of the leased-line applications will migrate to a new switching class called channel switching. Channel switching is the capability of switching leased lines in real time under customer control. This allows the users to reconfigure their private network in real time. Wide-band switching is applied to high data rates, such as those required for video applications.

B channels are carried over the integrated-access line to the ISDN serving office. Here, they are routed over the circuit-switched network, are circuit-switched to a packet-switched network, or travel over the leased line or channel-switched network. The first case is likely for most voice calls, the second for interactive data which are statistically multiplexed on customer premises, and the third for applications with very long holding times such as host-to-host communications.

The D channel is the integrated-access control link between the customer and the ISDN serving office, ending at the office. It carries signaling information and packet data. The packet data are routed to a packet-switched network, while the signaling information is either interpreted by the local switch or sent into the network over a signaling network (Fig. 2a). It is envisioned for the future that the transport networks will be integrated (Fig. 2b).

Applications and prospects. ISDN is designed to accommodate multiple applications over a single integrated-access line. It is capable of serving many existing and emerging applications. These include: (1) telemetry, such as remote meter reading, energy management, and alarm sensors such as fire alarms and burglary alarms; (2) interactive packet data, including word processors, data-base access, telex, and videotext; (3) bulk data, such as computer-

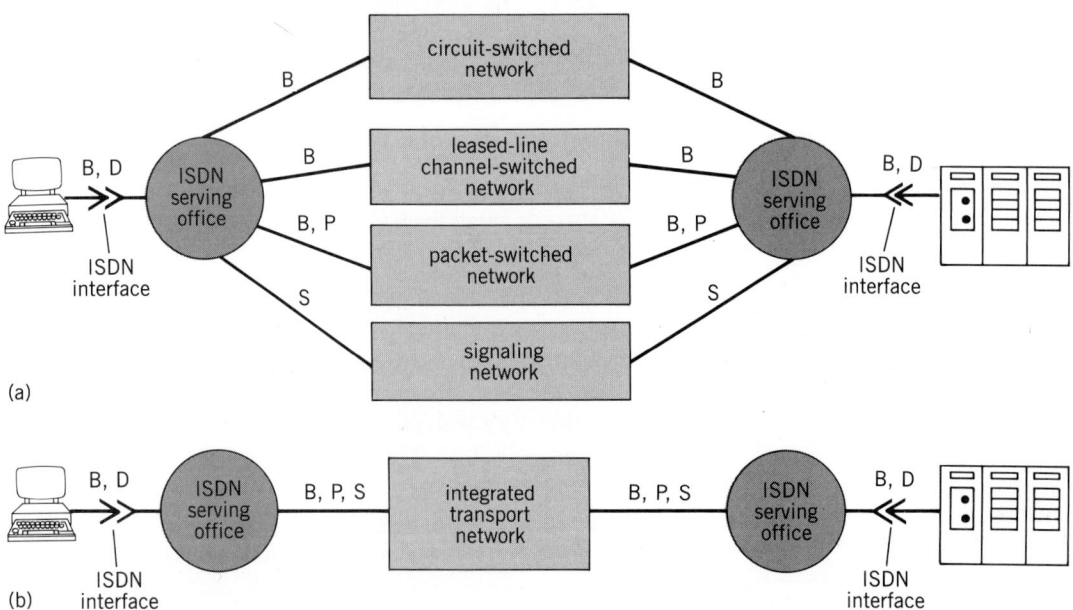

Fig. 2. Examples of integrated services digital network capabilities provided through (a) separate transport networks and (b) an integrated transport netowrk. D = D channel, B = B channel, P = packet data, and S = signaling information.

aided design and manufacturing, graphics, slow-scan and freeze-frame video, and facsimile; (4) voice applications, such as high fidelity and secure voice; and (5) teleconferencing applications. *See* TELEPHONE SERVICE.

By providing an integrated network transport capability, ISDN caters to the varied communications needs of the user. The key to ISDN is its ability to provide transport for all of these applications on a service-independent basis. As new applications arise, ISDN, by its design, will be ready to provide connectivity, so that the user will have a flexible, intelligent, low-cost, advanced telecommunications system.

ISDN provides a strong telecommunications infrastructure rather than one leading to a proliferation of incompatible communications networks. Since it is a public network concept, it will provide all members of the population with new telecommunications capabilities. Because of these characteristics, the integrated services digital network will be the gateway into the "Information Age."

For background information *see* DATA COMMUNICATIONS; ELECTRICAL COMMUNICATIONS; PULSE MODULATION; SWITCHING SYSTEMS (COMMUNICATIONS) in the McGraw-Hill Encyclopedia of Science and Technology.   [ROBERT M. WIENSKI]

Bibliography: I. Dorros, Telephone nets go digital, *IEEE Spectrum*, 20(4):48–53, April 1983; Special issue on integrated services digital network, *IEEE Commun. Mag.*, vol. 22, no. 1, January 1984.

Laser welding

Laser welding has become established as an important production joining process. Although probably no more than 10% of metal fabrication plants are using the process, laser welding has had significant impact in high-technology areas such as electronics, aerospace, and defense technology. Also laser welding is beginning to show economical advantages where high production rates are needed, such as in the automotive and the appliance industries. Ease of operator training and facility for automation have enhanced its acceptance. For many applications, lasers are used for welding simply because they solve fabrication problems by enabling welds to be made with low heat inputs or with dissimilar alloys. Continued improvements in laser technology and ancillary equipment are broadening the application of laser welding. The development of fast-axial-flow carbon dioxide (CO_2) lasers has provided compact systems that are more suitable where production space is limited. Face-pumped Nd:YAG (neodymium:yttrium-aluminum-garnet) lasers are anticipated to provide low-divergence beams of 1000 W or more. Use of quartz fiber optics for light transmission and robotics is expected to further automation.

Laser types. Both solid-state and gas lasers are used for welding. Most solid-state lasers employ neodymium as the lasing medium in either YAG or glass matrices, and are usually optically pumped with krypton or xenon lamps. Pulsed Nd:YAG la-

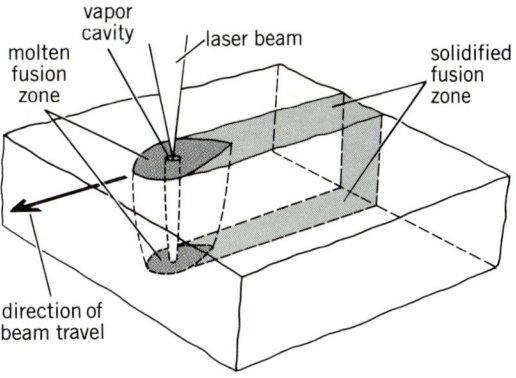

Fig. 1. Deep-penetration laser welding. (*After American Society for Metals, Metals Handbook, 9th ed., vol. 6, 1983*)

sers are the most frequently used solid-state lasers for welding. Pulsed ruby lasers and pulsed Nd:glass lasers are used for some spot-welding applications. Also seam welds have been made with continuous-wave Nd:YAG lasers. Carbon dioxide (CO_2) gas lasers are used more extensively for seam welding. The CO_2 gas molecules, which are contained in a $He-N_2-CO_2$ gas mixture, are excited by an electrical discharge. Carbon dioxide lasers can be pulsed for both seam welding and spot welding.

Weld types. Depending on the power densities and weld speeds used, either conduction-limited or deep-penetration welds can be made with lasers. Typically, welds made with solid-state or CO_2 lasers at less than 1000 W are conduction-limited. High-power CO_2 lasers can be used to produce high-speed conduction-limited welds. In conduction-limited welding, the beam energy is delivered to the surface of the part being welded, and spreads inward by conduction and, in some cases, by convection in the molten pool. Deep-penetration welding takes advantage of a laser's ability to rapidly vaporize a metal to initially form a vaporized "keyhole" that permits the laser energy to be deposited below the surface (Fig. 1). As a deep-penetration weld advances, the vaporized column or keyhole is carried along with a surrounding liquid pool. Deep-penetration welding, also termed keyhole welding for through thickness welds, is highly efficient and can produce the same weld penetration with an order of magnitude less energy than would be required by conventional arc welding processes. The depths of keyhole welds are usually at least four times their width. Few solid-state lasers have been built that produce greater than 400 W average power, whereas CO_2 lasers are available with outputs as large as 20,000 W. Consequently, deep-penetration welding is the realm of CO_2 lasers. High power CO_2 lasers are also used to produce high speed conduction-limited welds.

Laser selection. Selection of a laser for welding is based on an evaluation of the application. Principal factors to be considered are the alloys to be welded, whether spot welds or seam welds are required, weld penetration needed, heat sensitivity of the assembly, and production rates desired. The

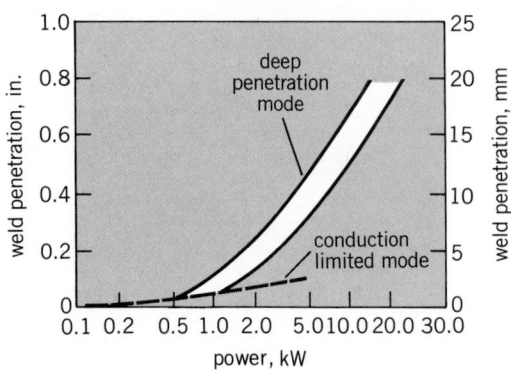

Fig. 2. Effect of power on maximum weld penetration. (*After American Society for Metals, Metals Handbook, 9th ed., vol. 6, 1983*)

shorter-wavelength light produced by solid-state lasers (for example, 1.06 micrometers for Nd) is absorbed more efficiently by metals than is the longer-wavelength light (10.6 micrometers) of CO_2 lasers. This is an important consideration in conduction-limited welding of highly reflective metals such as aluminum, copper, gold, and silver. The high average powers presently available only from CO_2 lasers are required for either deep weld penetration (greater than about 0.08 in. or 2 mm) or high welding speeds (greater than about 4 in./s or 100 mm/s). Figure 2 indicates the maximum weld penetration that can be expected for both conduction-limited and deep-penetration welding as a function of laser power. The high peak powers afforded by pulsed lasers, particularly pulsed solid state lasers, are advantageous for spot welding. Also, pulsed Nd:YAG lasers are highly efficient for welding heat-sensitive parts. For alloys for which weld cracking is a problem, continuous-wave lasers are generally preferred, because pulsed heat sources often increase the propensity for cracking.

Medium-power (300–2000-W) CO_2 lasers normally operate as fundamental-mode lasers, where the laser beam energy is concentrated in a single spot whose flux density distribution is approximately gaussian. Most Nd:YAG lasers used for welding, however, operate as multimode lasers where the beam is broken into parallel bundles of light. As a consequence, a multimode beam diverges more rapidly than a fundamental-mode beam and cannot be focused to as small a spot. Because Nd:YAG laser beams are typically not as well collimated as medium-power CO_2 laser beams, they are more difficult to transmit large distances. Two recent developments are expected to overcome this problem and enhance the capabilities of Nd:YAG lasers in automated, multiple-work-station environments. The development of face-pumped slab lasers is anticipated to provide Nd:YAG lasers that produce most of their energy in the fundamental mode. Also, the ability to transmit light from Nd:YAG lasers several meters by quartz fiber optics has been demonstrated. Robotic work stations that utilize fiber optics to transfer laser power have become available.

Advantages. The most intrinsic characteristic of a laser as a welding heat source is high intensity. This characteristic affords high-efficiency welding, where most of the heat is actually used to produce a fusion weld. In more conventional welding processes, such as the various forms of arc welding, only a fraction of the deposited heat results in melting. Other advantages stem from the ability to optically focus and manipulate the beam, the inertness of light, and the reproducibility of welding parameters. Advantages of laser welding include: high processing speed; very accurate positioning of welds; minimal damage to heat-sensitive parts; deep, narrow welds; low thermal distortion; narrow heat-affected zones; ability to weld in open atmospheres; wide latitude to permissible alloys and alloy combinations; ease of automation and time sharing; ease of changing position, shape, and intensity of heat source; no problems with magnetic materials; minimal charge buildup compared with electron-beam welding; no x-ray generation; noncontaminating heat source; accessibility to normally difficult-to-reach locations; and an excellent working environment.

Comparison with electron-beam welding. Laser welders and electron-beam welders are often compared, because they both produce focused, high-intensity heat sources. Since light beams can be manipulated by optics, for example, to achieve multiple-work-station automation, lasers have some distinct advantages compared with electron-beam welders. Also, since electron-beam welding normally requires a vacuum chamber, maintenance and setup costs are typically higher than for laser welding. These potential advantages of laser welding are offset by the complex manner in which photon energy is absorbed by metals. The fraction of the energy of the laser beam that is absorbed depends on the composition of the alloy, surface roughness, oxide films, surface temperature, and the wavelength of light. Depending on these factors, absorptivity varies from a few percent to 70–80%. By contrast, most of the electron-beam energy incident on a surface is absorbed, and that amount lost to backscatter doesn't vary much with the material. This not only is important from the perspective of process efficiency, but is of concern because variations in surface roughness, oxidation, temperature, and so forth can result in problems regarding weld reproducibility. During keyhole welding, effective absorptivity can increase significantly because the beam is being delivered in a hole rather than at a surface. But, the evaporation of metal during keyhole welding can introduce a further complexity in that the beam may become absorbed or scattered by the resulting vapor plume, which for some laser-metal vapor interactions may become a plasma. Attenuation of the beam by the plume can significantly influence the efficiency (absorption can exceed 50%) and reproducibility of the process. Approaches to minimizing the plume or plasma effects include directing a jet of inert gas into the keyhole and welding in vacuum.

Applications. Laser welding in the automotive and appliance industries takes advantage of the ease of automation of lasers or the high welding speeds afforded by high-power CO_2 lasers. Laser welds are used on such diverse parts as transmission gears, automobile underbodies, and pollution control devices. In the electronics industry, laser welding is used to minimize heat damage and produce hermetic welds. Typical applications include closure welds on relay cans and microcircuits, attachment of contacts, and joining of electrical feed-throughs. These welds are often made with pulsed Nd:YAG welders. Similar applications include closure welds on batteries and heart pacemakers. Aerospace applications include laser welding of turbine blades and high-pressure gas vessels.

For background information *see* GEOMETRICAL OPTICS; LASER; LASER WELDING in the McGraw-Hill Encyclopedia of Science and Technology.

[J. L. JELLISON]

Bibliography: ASM Committe on Laser Beam Welding, Laser beam welding, *Metals Handbook*, 9th ed., vol. 6: *Welding, Brazing, and Soldering*, pp. 647–671, 1983; J. Mazumder, Laser welding: State of the art review, *J. Metals*, 34:16–24, 1982.

Lightning

Although lightning has received serious scientific study for over 200 years, many important questions still remain about the physics of the discharge and the mechanisms of lightning damage. In recent years, new experimental techniques have been used to determine physical properties of lightning on time scales ranging from tens of nanoseconds to several seconds. Methods have also been developed to artificially trigger lightning with small rockets, and thus it is now possible to study both the physics of the discharge and the interactions of lightning with structures in a partially controlled environment.

The hazard. Lightning is a transient, high-current electric discharge that occurs in the atmosphere and that has a length ranging from hundreds of meters to several tens of kilometers. The vast majority of lightning discharges are produced by thunderclouds, and most flashes remain within the cloud or are from cloud to cloud or from cloud to air. About one-third of the discharges are from cloud to ground, and it is this type of lightning that is the primary hazard to people or structures. Altogether, about 40 million cloud-to-ground flashes strike the continental United States each year, and the resulting loss of life and property make lightning one of the nation's worst weather hazards.

Besides its many deleterious effects, lightning has some unique benefits. The chemical effects of lightning may have played an important role in the prebiotic synthesis of amino acids, and today lightning is still an important source of fixed nitrogen, a natural fertilizer, and other nonequilibrium trace gases in the atmosphere. Lightning-caused fires have long dominated the dynamics of forest ecosystems throughout the world, and the electric charge that is transferred to the ground by lightning is a major component in the global circuit of atmospheric electricity.

Cloud-to-ground lightning. Almost all cloud-to-ground flashes begin within the cloud with a process called the preliminary breakdown. After several tens of milliseconds, the preliminary breakdown initiates an intermittent, highly branched discharge that propagates horizontally and downward, and is called the stepped leader. When the tip of any branch of the stepped leader gets close to the ground, the electric field just above the ground becomes very large and one or more upward propagating discharges form at the surface and then rise to meet the leader. When contact occurs between an upward junction discharge and the stepped leader, the first return stroke begins. The return stroke is an intense, positive wave of ionization that starts at or just above the ground and propagates up the partially ionized leader channel into the cloud. The upward speed of a return stroke is typically about one-third the speed of light. The peak currents in return strokes have been measured during direct strikes to instrumented towers and range from several to hundreds of kiloamperes with a typical value being 10 to 40 kA.

After a pause of 40 to 80 ms, a new leader begins in the cloud, the dart leader, and this leader then propagates without stepping down the previous return-stroke channel. When the dart leader contacts the ground, a subsequent return stroke propagates from the ground back to the cloud. Most cloud-to-ground flashes contain two to four return strokes, and each of these effectively neutralizes a different portion of the cloud charge. Visually, lightning often appears to flicker because the human eye can just resolve the time intervals between different return strokes. In 20–40% of all flashes to ground, a dart leader propagates down just a portion of the previous return-stroke channel and then forges a different path to ground. In these cases, the lightning actually strikes the ground in two places, and the channel has a characteristic forked appearance that can be seen in many photographs.

Recent research has provided new information about the physics of lightning. This includes remote measurements of the electric and magnetic fields produced by lightning, which can be used to infer various properties of the discharge; the sources of radio-frequency noise, which can be used to trace the geometrical development of lightning channels as a function of time; and the use of small rockets to trigger lightning artificially under thunderstorms.

Electromagnetic fields. It has become clear that the electric and magnetic fields that are radiated by different lightning processes have different but characteristic signatures that are reproduced from flash to flash. For example, Fig. 1 shows three of the many impulses that were radiated by a typical cloud-to-ground discharge at a distance of about 30 mi (50 km). The trace in Fig. 1*a* shows a cloud impulse that was radiated during the preliminary

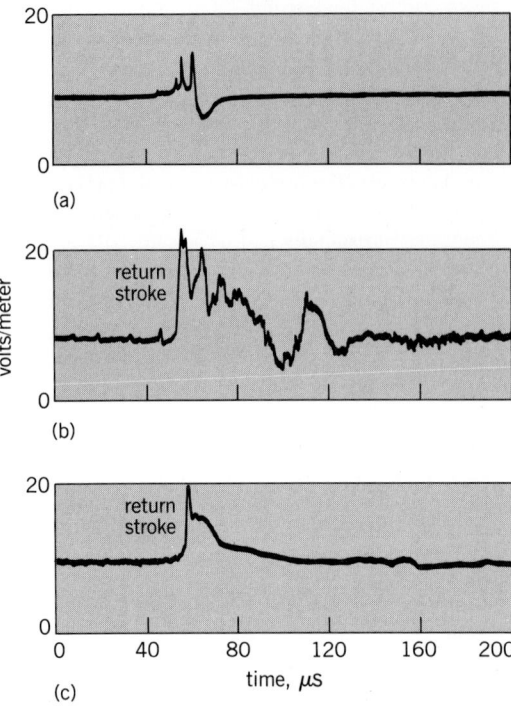

Fig. 1. Examples of some electric field impulses that were produced by a cloud-to-ground flash at a distance of about 30 mi (50 km). (a) Radiated during the preliminary breakdown. (b) Due to the first return stroke. (c) Due to a subsequent return stroke.

breakdown; the trace in Fig. 1b shows the waveform that was radiated by the first return stroke; and the trace in Fig. 1c shows a subsequent return stroke. The small pulses that precede the first return stroke in the trace in Fig. 1b were radiated by individual steps of the stepped leader just before attachment occurred at the ground. The characteristics of these newly measured signatures have been put to use in the detection and location of cloud-to-ground flashes. For example, operating throughout North America and other parts of the world, there are now extensive networks of lightning direction finders that can discriminate between the shapes of the return-stroke fields and other processes and that can provide accurate locations of the ground-strike points.

A theoretical model has been developed that describes the shapes of the electric and magnetic fields that are produced by return strokes as a function of distance. One particularly important result of this work is the prediction that during the first few microseconds of the stroke, that is, just after the attachment process has been completed, the waveform of the distant or radiation field is proportional to the channel current, as shown by the equation below, where E is the vertical electric field that is

$$E_{\text{RAD}}(t) = -\frac{\mu_0 v}{2\pi D} I(t - D/c)$$

measured at the ground at time t, μ_0 the permeability of free space, v the return-stroke velocity, c the speed of light, and D the horizontal distance to the flash. A typical first return stroke will produce a peak field of about 8 V/m at a distance of 60 mi (100 km), and the stroke velocity near the ground is typically on the order of 3×10^8 ft/s (10^8 m/s). With these values, the equation predicts a peak current of about 40 kA, a value in good agreement with currents that have been measured in direct strikes to towers.

The microsecond and submicrosecond structure of return-stroke fields has been measured, and it has been found that a typical first stroke produces an electric field front that rises in 2–8 μs to about half of the peak-field amplitude. This front is followed by a fast transition to peak whose mean 10–90% rise time is about 90 nanoseconds. Subsequent stroke fields apparently have fast transitions very similar to first strokes, but fronts that last only 0.5 to 1 μs and that rise to only about 20% of the peak field.

The submicrosecond field components must be produced by submicrosecond components in the current, but very few currents that have been measured during direct strikes to instrumented towers show components as fast as 90 ns. Perhaps an enhanced upward connecting discharge from a tall tower or the electrical characteristics of the tower itself lengthen the tower rise time to a value that is substantially larger than a stroke to normal terrain. More measurements of lightning currents and fields with fast time resolution will be necessary before the behavior of lightning currents is understood under all conditions.

The current rise time is a very important property of lightning because, if a rapidly varying current interacts with an inductive load, the voltage on that load will be substantially larger than that for a slow current. Most of the standard surge waveforms that are used to verify the performance of protectors on power and telecommunications circuits specify that open-circuit voltage should have a rise time of 0.5, 1.2, or 10 μs and that the short-circuit current should have a rise time of 8 or 10 μs. These values are substantially slower than those implied by the electromagnetic fields; therefore, it is possible that the degree of protection that is provided by devices that have been tested to these standards will not be adequate for direct lightning surges.

Lightning radio sources. The radio-frequency (rf) noise that is generated by lightning in the high-frequency (HF) and very high-frequency (VHF) bands appears in the form of discrete bursts, and within each of these bursts there are hundreds to thousands of separate pulses. If the difference in the time of arrival of each pulse is carefully measured at four widely separated stations, the location of the source of each pulse can be computed, and the geometrical development of the burst can be mapped as a function of time. Unfortunately, the physical processes that produce HF and VHF radiation in lightning are not well understood. It has been reported that the pulses in most bursts are produced by a regular progression of source points; therefore the bursts may

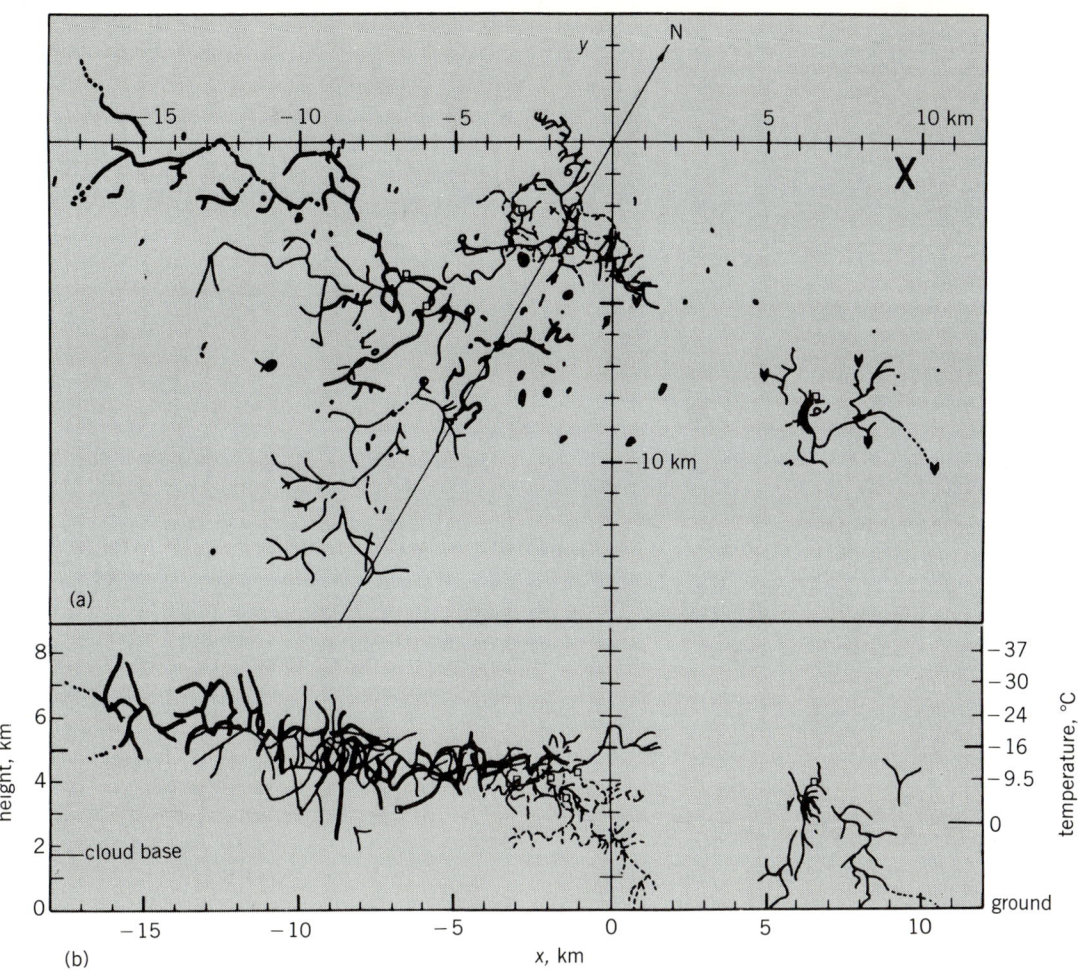

Fig. 2. Geometrical reconstructions of six successive lightning discharges in South Africa. (a) Plan view of the channels shown from above. (b) Elevation view that is a projection of the same channels on a vertical plane parallel to the x axis. 1 km = 0.6 mi. °F = 9/5 (°C) + 32. (Courtesy of D. E. Proctor)

be produced by new ionization processes and extensions of old channels.

If the source location of each rf pulse within a burst is plotted, the width of the associated radio image of the channel ranges from about 300 ft (100 m) to more than 0.6 mi (1 km). Figure 2 shows the paths of the central cores of six successive lightning discharges that were reconstructed geometrically. By combining reconstructions such as these with measurements of the associated electric field changes at the ground, it has been inferred that incloud channels usually have a net negative charge and that the average line charge density is about 0.9 coulomb/km. Also, by dividing the length of a channel segment by the time required for that portion to develop, it has been determined that the average velocity of streamer formation ranges from 9×10^4 mi/h to 18×10^5 mi/h (4×10^4 m/s to 8×10^5 m/s) with a mean of $3.1 \pm 2.7 \times 10^5$ mi/h ($1.4 \pm 1.2 \times 10^5$ m/s).

The geometrical forms of intracloud discharges range from concentrated knots or stars a few kilometers in diameter to extensive branched patterns up to 50 mi (90 km) in length. Successive discharges in a storm have been reported to form an interconnected system, and some flashes seem to extend the paths of earlier discharges. In one case, two flashes that were separated by just 1.6 s produced tortuous channels that ran parallel to each other for almost 1.2 mi (2 km), but the channels remained about 1000 ft (300 m) apart.

Recent observations of lightning with radio interferometers have provided results similar to those obtained with time-of-arrival methods, and perhaps interferometric methods will be able to provide unambiguous three-dimensional reconstructions of the discharge processes in the future.

Artificial triggering of lightning. The last development to be described here is the artificial triggering of lightning by small rockets. This technique is particularly important because it provides, for the first time, the capability of studying both the physics of the discharge process and the interactions of lightning with structures and other objects in a partially controlled environment. Although rockets were first used to study atmospheric electricity in the eighteenth century, the first artificial initiation of lightning was attempted in the early 1960s. The

Fig. 3. Example of a rocket-triggered lightning flash in New Mexico. (*Courtesy of P. Hubert*)

technique has subsequently been improved and is now being used to investigate a variety of lightning problems in France, Japan, and the United States.

When a thunderstorm is overhead and the electrical conditions are favorable, the technique consists of firing a small rocket that carries a grounded wire aloft. If the rocket is fired when the surface electric field is 3 to 5 kV/m, then about two-thirds of all launches will trigger a lightning discharge. Most triggers occur when the rocket is at an altitude of only 300 to 900 ft (100 to 300 m), and the first stroke in the flash usually propagates upward into the cloud. The majority of the subsequent strokes follow the first stroke and the wire to ground; but in about one-third of the cases, the subsequent strokes actually forge a different path to ground. These latter events are now called anomalous triggers. The first stroke in a triggered discharge is not like natural lightning, but subsequent strokes appear to be almost identical to their natural counterparts.

An example of a lightning that was triggered by this method is shown in Fig. 3. The upward branching in this photograph was produced by a leader that propagated upward from the wire, and the bright, straight section of channel near the ground shows the path of the wire just before it exploded as a result of the lightning current.

Triggered lightning is now being used to investigate the luminous development of lightning channels, characteristics of lightning currents, velocities of return strokes, the relation between currents and fields, mechanisms of lightning damage, the performance of lightning protection systems, and many other problems.

Among the more important results to date have been a direct experimental verification of the existence of submicrosecond fields and currents during return strokes and the general validity of the equation (given above) developed from the theoretical model. The main benefit of the rocket triggering technique is that it is now possible to cause lightning to strike a known place at a known time, thus enabling controlled experiments to be performed. The triggering wire guides the lightning to a point where a variety of sensors can measure the physical properties of the discharge and its deleterious effects directly. All cameras and data-recording equipment can be turned on and be fully operational just before the rocket is fired. In most geographic regions, the frequency of overhead storms limits the total number of triggers to a few tens of events per year, but the quality of these events can be quite high.

Future research. New experimental techniques now provide an opportunity to investigate many of the important questions that remain about the physics of lightning on Earth. Spacecraft observations have shown that lightning may be present in the atmospheres of Jupiter, Venus, and Saturn, and the upcoming *Galileo* probe will carry a lightning detector to Jupiter. In the future, perhaps a study of lightning in atmospheres that are radically different from Earth will lead to a better understanding of lightning on Earth and offer even more challenging questions for the future.

For background information *see* CLOUD PHYSICS; LIGHTNING; LIGHTNING AND SURGE PROTECTION in the McGraw-Hill Encyclopedia of Science and Technology. [E. PHILIP KRIDER]

Bibliography: R. Fieux et al., Research on artificially triggered lightning in France, *IEEE Trans. Power Apparat. Sys.*, PAS-97:725–733, 1978; E. P. Krider et al., Lightning direction-finding systems for forest-fire detection, *Bull. Amer. Meteorol. Soc.*, 61:980–986, 1980; D. E. Proctor, Lightning and precipitation in a small multicellular thunderstorm, *J. Geophys. Res.*, 88:5421–5440, 1983; M. A. Uman, *Lightning*, 1969, reprint 1984; M. A. Uman and E. P. Krider, A review of natural lightning: Experimental data and modelling, *IEEE Trans. EMC*, EMC-24:79–112, 1982; C. D. Weidman and E. P. Krider, Submicrosecond risetimes in lightning return-stroke fields, *Geophys. Res. Lett.*, 7:955–958, 1980 (see also C. D. Weidman and E. P. Krider, Correction, *J. Geophys. Res.*, 87:7351, 1982).

Liquid helium

In the last few years, studies of superfluid helium-3 (^{3}He) have revealed many unexpected properties which are unique to this fluid and differ radically from those of the other known quantum liquid, superfluid helium-4 (^{4}He). Two areas of particular interest are the properties of quantized vortices in superfluid ^{3}He and the collective modes of motion in this system.

QUANTIZED VORTICES IN HELIUM-3

The first experiments on the behavior of superfluid phases of liquid ^{3}He during rotation have been carried out during the past few years. Several unexpected vortex properties in the anisotropic superfluid A and B phases of liquid ³He have been discovered, and vortices in superfluid ³He are now known to differ radically from those in superfluid ⁴He in several important respects. Vortices in the A phase of superfluid ³He, or ³He-A, can occur in several different quantum states; the ones identified as the vortices seen in the experiments (1) are multiply quantized, (2) possess continuously distributed vorticity, and (3) are nonaxial. Vortices in the B phase of superfluid ³He, or ³He-B, (1) display a phase transition of the vortex core, (2) are spontaneously magnetized, (3) possess an electric dipole moment or a spontaneous supercurrent along the vortex axis, and (4) exhibit a unique superfluid core structure. Some of these features are shared by the A- and B-phase vortices alike. Because of all these fundamentally new, rich physical phenomena, the quantized vortex lines in superfluid ³He rank among the most interesting objects in condensed matter physics.

Quantum liquids do not rotate as a solid body, as do classical fluids such as water. Rather, rotation perforates the superfluids in units of singular lines, called vortex lines. Circulation is quantized around any path in space embracing the vortex axis. Vortices are well known from the earlier-discovered superfluids, superfluid ⁴He and superconductors. In ⁴He the superfluid order parameter ψ is a complex-valued scalar function. The distribution of the order parameter in the linear quantized vortex state along z with topological charge m is given by Eq. (1).

$$\psi(\vec{r}) = C(r)e^{im\phi} \quad (1)$$

Here m denotes the number of circulation quanta of the superfluid velocity field around the vortex axis, and r, ϕ, and z denote the cylindrical coordinates, with z along the vortex axis. The amplitude of the order parameter $C(r)$ tends to zero on the vortex axis [that is, $C(r = 0) = 0$], in order to prevent the superflow from having infinite energy. Hence, superfluidity is broken in the ⁴He vortex on the vortex axis and the core of the quantized vortex line in ⁴He is normal (Fig. 1).

The superfluid velocity field $\vec{v}_s$ for this pure phase vortex is potential flow, except at $r = 0$, and is given by Eq. (2), where $\hbar$ is Planck's constant h

$$v_s = (\hbar/m_4)r^{-1} \quad (2)$$

divided by 2π and m_4 is the mass of the ⁴He atom.

Because of the symmetries of the scalar vortex state in Eq. (1), there exists only a single type of an axially symmetric vortex in the classical superfluid ⁴He with m quanta of circulation. Therefore, phase transitions for the ⁴He vortex cannot occur, unless axial symmetry is broken. As discussed below, the axially symmetric vortex structures are much richer in superfluid ³He-B. The multicomponent order-parameter tensor of this state allows several possible phase transitions in the vortex core structure, both of first and second order, due to breaking of discrete internal symmetries without breaking of axial symmetry.

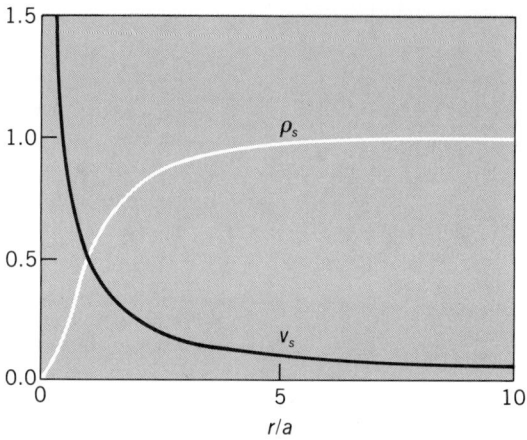

Fig. 1. Normalized superfluid density ρ_s and superflow velocity v_s around a vortex line in superfluid ⁴He. The superfluid density tends to zero on the vortex axis; hence the vortex core is normal. The characteristic distance a is of the order of the interatomic spacing.

Cooper pairing in superfluid ³He. The coherent superfluid state of liquid ³He results from pairing of the constituent ³He fermion atoms. The frictionless motion of the pairs is responsible for the superfluidity in ³He. In liquid ³He the pairing occurs only below the very low temperature T_c of 0.0026 K.

In superconductors the pairing occurs between electrons, and the resulting entities are known as Cooper pairs. The Cooper pairs in ³He are essentially different from their counterpart in superconductors: The electrons in a superconductor are paired with opposite spins, producing magnetically neutral "molecules" with zero spin ($S = 0$). However, the ³He atoms form magnetic molecules with (nuclear) spin $S = 1$. Due to the coherence of the condensate, the superfluid phases of liquid ³He are liquid antiferromagnets. Moreover, the shape of the wave function for a Cooper pair in ³He is not spherically symmetric. In the coherent condensate this property is magnified, which serves to produce orbital anisotropy axes for the whole superfluid, as in a liquid crystal. Hence, superfluid ³He is a superfluid antiferromagnet and a liquid crystal. *See* SUPERCONDUCTIVITY.

The anisotropy of the Cooper pairs is a consequence of the nonzero value of the orbital quantum number L, which serves to characterize the relative orbital motion of the ³He atoms in the Cooper pairs. The orbital momentum in superfluid ³He is $L = 1$, while $L = 0$ for the Cooper pairs in superconductors, where the motion of the electrons is isotropic.

However, for ³He the two quantum numbers S and L are not sufficient to describe the internal motion of the Cooper pairs. The projections μ and ν of the spin and orbital momenta in certain directions must be specified as well. Thus, various superfluid phases can occur, which differ in the quantum numbers μ and ν. The three superfluid states of liquid ³He which are known to exist in stationary experiments are the A phase ($\mu = 0$ and $\nu = 1$), the A_1

phase ($\mu = \nu = 1$; this phase occurs only in the presence of an applied magnetic field), and the B phase, which corresponds to the total angular momentum $J = 0$, where $\vec{J} = \vec{L} + \vec{S}$. The B phase may alternatively be regarded as a mixture of three substates, each with equal weights: ($\mu = 1$, $\nu = -1$), ($\mu = -1$, $\nu = 1$), and ($\mu = 0$, $\nu = 0$). New superfluid phases of liquid ^{3}He with other sets of quantum numbers exist in the cores of quantized vortex lines.

Continuous vortices in the A phase. The superfluid A phase shares many properties with ordinary liquid crystals. The A phase is a liquid crystal with uniaxial anisotropy, like the nematics. The direction of this anisotropy axis is denoted by the vector $\vec{l}$; it is also the axis of quantization of the orbital motion of the ^{3}He atoms in the Cooper pairs. This liquid-crystal behavior of the A phase is apparent in the observed "textures" (that is, the spatial distribution of the orbital anisotropy axis $\vec{l}$). Various textures result due to the competition of different orienting effects on the $\vec{l}$ vector, such as boundaries and superflow. Superflow orients $\vec{l}$ along the superfluid velocity as a result of the anisotropy of the superfluid density. Such an anisotropy of superflow exemplifies simple superposition of the superfluid and liquid-crystal-like properties of ^{3}He-A.

The most striking property of the A phase is, however, a consequence of the important conjugation of superfluid and liquid-crystal-like characteristics: a nonuniform $\vec{l}$ texture in the A phase serves to give rise to continuous vortex superflow. In the presence of a nonuniform $\vec{l}$ texture, the superflow cannot be irrotational; that is, Eq. (3) holds. In contrast to superfluid ^{4}He, the circulation of superflow in this case is not quantized, but is arbitrary. This follows from the specific symmetry breaking that occurs in the A phase: although the gauge and rotational symmetries are separately broken, the combined gauge-orbital symmetry is nevertheless retained.

$$\vec{\nabla} \times \vec{v}_s \neq \vec{0} \quad (3)$$

This produces several intriguing effects, in particular during rotation. Since the A phase can support arbitrary vorticity, it could rotate with the vessel like a solid body. This would, however, not be favorable for the $\vec{l}$ texture. In fact, in a bulk sample of ^{3}He-A, the A phase accommodates rotary motion in a manner which is intermediate between the ordinary rotation of a classical viscous liquid and that of superfluid ^{4}He (Fig. 2). The superfluid velocity field $\vec{v}_s$ and the $\vec{l}$ vector texture form periodic structures, which are continuous everywhere. This arrangement serves to imitate the rotation of a classical liquid on the average. The circulation of the superfluid velocity $\vec{v}_s$ along the boundary of a primitive hexagonal vortex lattice cell is $2\,h/M$, where $M = 2m_3$ is the mass of a Cooper pair of ^{3}He atoms. Hence, each lattice cell constitutes a vortex with two quanta of circulation. In contrast with the quantized vortex in ^{4}He, there occurs no singularity on the vortex axis; the superfluidity of the A phase is not broken anywhere. (Singular vortices can also exist in the rotating A phase under specific conditions.)

Magnetic vortices in B phase. Prior to the experiments on the rotating B phase, it was considered to be a dull superfluid, because it lacks the intrinsic coupling of the superfluid and liquid-crystal-like properties of the A phase. The quantized vortex lines in the B phase were thought to resemble the classical vortex lines in superfluid ^{4}He. The vortex lines in ^{3}He-B can indeed only be singular, with quantized superflow around the vortex core, whose size is of the order of the superfluid coherence length ξ. However, the large size of the vortex core, hundreds of times larger than that in ^{4}He, and the intrinsic coupling of the liquid-crystal-like and magnetic properties serve to make the B-phase vortices quite exciting.

The Cooper pairs in the B phase have the quantum number $J = 0$. In superconductors, too, $J = 0$, and here the Cooper pairs are completely isotropic; the orbital and spin motions in superconductors are, moreover, separately isotropic since $L = S = 0$. However, the Cooper pairs in ^{3}He-B are isotropic only under specific combined rotations of the spin and orbital spaces, but they are anisotropic under separate rotations. The B phase is said to have broken relative spin-orbital symmetry, which gives rise to the coupling of the magnetic and liquid-crystal-like behaviors.

This broken symmetry is reflected in a special degeneracy of the B-phase states. This is commonly described with use of the orthogonal matrix $R_{\alpha i}$. The state labeled with $R_{\alpha i}$ is obtained from one with $J = 0$ through a rotation of the spins of the Cooper pairs by the matrix $R_{\alpha i}$; such a state can be characterized by the quantum number $\tilde{J} = 0$, where $\tilde{J}_i = L_i + S_\alpha R_{\alpha i}$. The weak spin-orbital interaction partially lifts this degeneracy: states with the minimal spin-orbital energy correspond to rotation of spins around an arbitrary axis $\vec{n}$ through the fixed "magic" angle $\theta_0 = \arccos(-1/4) \simeq 104°$. This angle is universal for the B phase and has been seen in many nuclear magnetic resonance (NMR) experiments. The vector

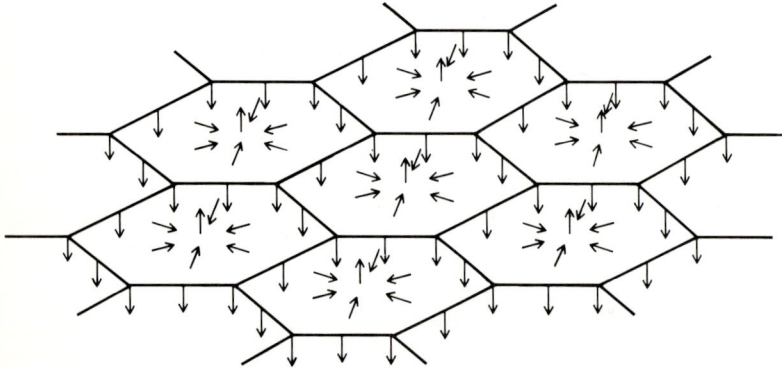

Fig. 2. Periodic hexagonal $\vec{l}$ texture (arrows) in the rotating A phase. Each cell is a continuous vortex with two quanta of circulation of the superfluid velocity along the boundary (solid lines).

$\vec{n}$ which specifies the remaining degeneracy is the common axis for the spin and orbital anisotropy. The NMR absorption in superfluid ³He-B is very sensitive to the orientation ("texture") of the $\vec{n}$ vector.

Quantized vortex lines in ³He-B can be detected through their orienting signature on the $\vec{n}$-vector texture. This is because the superflow field around the vortices bends the $\vec{n}$ texture. This effect was seen in the NMR experiments on rotating ³He-B. In an axial magnetic field a series of peaks are observed in the transverse NMR signal, corresponding to spin waves localized on the $\vec{n}$ texture. Such waves are formed in the cylindrical sample because perturbations caused by the boundaries in the B phase penetrate deep into the bulk liquid. During rotation the splitting $\Delta \nu$ of these peaks increased appreciably, indicating the presence of a new orientational effect on the $\vec{n}$ vector, due to the quantized vortex lines.

The most unexpected result from this experiment was in the temperature dependence of $\Delta \nu$, which exhibited a jump discontinuity at a temperature $T = 0.6 \, T_c$ (at a pressure of 29.3 bars or 2.93 megapascals). This discontinuity in the NMR spectrum manifested features of a first-order phase transition. The transition was seen only under rotation, but the transition temperature did not depend on the angular velocity of rotation Ω. Such a behavior is consistent only with a phase transition inside the vortex core. The vortex core possesses a nontrivial structure, which is different above and below the transition temperature.

Further NMR experiments showed that there exists a gyromagnetic effect: The NMR splitting $\Delta \nu$ was different when the ³He cryostat was rotated clockwise or counterclockwise. The same change could be brought about by reversing the magnetic field direction. Hence, the effect possesses odd parity with respect to both the rotation speed and magnetic field and can therefore be measured, although it is rather small. Quite unexpectedly it was found that the gyromagnetism also undergoes a jump discontinuity at the same temperature $T = 0.6 \, T_c$, where the vortex structure changes.

Only a magnetization which is concentrated inside the vortex core and which changes in magnitude at the core transition can produce such an effect. The magnetic moment of the ³He-B vortices arises because the superflow around the vortex induces an internal rotation of the Cooper pairs inside the core with the orbital momentum $\vec{L}$ directed along the vortex axis $\hat{\Omega}$. Due to the rigidity of the quantum state of the Cooper pairs, this produces a spin momentum given by Eq. (4) and hence also a magnetization given by Eq. (5), where γ is the gyromagnetic ratio. The interaction of the magnetic moments of the vortices with an external magnetic field $\vec{H}$ gives rise to an additional gyromagnetic energy

$$S_\alpha = -R_{\alpha i} L_i \quad (4)$$

$$M_\alpha = \gamma S_\alpha \sim -R_{\alpha i} \Omega_i \quad (5)$$

$-\vec{M} \cdot \vec{H}$, proportional to $H_\alpha R_{\alpha i} \Omega_i$, which essentially depends on the vortex core structure. (Magnetic vortices have also been conjectured in the neutron superfluid in rotating pulsars).

Both the phase transition inside the vortex core and the magnetic moment of the vortices indicate a complicated vortex core structure. The core structures were recently classified by using symmetry arguments: It was found that an axially symmetric vortex in ³He-B with unit quantum of circulation may be in five different states, distinguished by the discrete symmetry of the vortex core. These states are labeled as o, u, v, w, and uvw; in all of them the vortex possesses a magnetic moment. The most symmetric, the o vortex, has a normal core with superfluidity destroyed on the vortex axis, as in the ⁴He vortex. In the u, v, and w vortices the discrete symmetry is partially destroyed, while in the uvw vortex it is completely broken. The vortices v and w have striking properties: they have a superfluid core, which contains a mixture of the A phase and a new superfluid phase with ferromagnetically ordered spins of the Cooper pairs. Moreover, the v vortex possesses a spontaneous electric polarization, caused by the parity violation in the core, while the w vortex has a spontaneous superflow along the vortex core axis. The uvw vortex has the properties of both the v and w vortices. Detailed numerical calculations show that the v vortex is energetically more advantageous than the other vortices, at least at low pressures. It is probable that just this vortex has been observed at temperatures T below $0.6 \, T_c$.

The most general expression for the order parameter of an axially symmetric vortex state in ³He-B with unit quantum of circulation is given by Eq. (6).

$$A_{\alpha i} = \Delta(T) \sum_{\mu \nu} a_{\mu \nu} \lambda_\alpha^\mu \lambda_i^\nu \quad (6)$$

Here $\Delta(T)$ is the scalar magnitude of the superfluid ³He-B energy gap and the indices α and i refer to Cooper pair spin and orbital spaces, respectively, and λ_α^μ and λ_i^ν denote cylindrical coordinates in the spin and orbital spaces, with the former given by Eq. (7). The quantities $a_{\mu \nu}$, given by Eq. (8), contain in general the nine complex-valued amplitudes

$$\lambda_\alpha^\mu = R_{\alpha i} \lambda_i^\mu \quad (7)$$

$$a_{\mu \nu} = C_{\mu \nu}(r) \, e^{i(1-\mu-\nu)\phi} \quad (8)$$

$C_{\mu \nu}(r)$ as functions of the radial distance r from the vortex axis.

Each function $C_{\mu \nu}$ describes the amplitude of Cooper pairing of the ³He atoms into a respective general p-wave superfluid state, characterized by the quantum numbers μ and ν, referring to the projection of the Cooper-pair spin and orbital momenta onto the corresponding coordinate spaces, spanned by the vectors λ_α^μ and λ_i^ν. For example, $C_{\mu \nu}(r)$, with the projections $\mu = 0$ and $\nu = +1$ of pair spin and orbital momenta, represents the A-phase Cooper pairing amplitude.

For the v vortex, Eq. (8) may be expanded as Eq. (9), where all the nine functions $C_{\mu\nu}$ ($\mu,\nu = +, 0,$

$$a_{\mu\nu} = \begin{pmatrix} C_{++}e^{-i\phi} & C_{+o} & C_{+-}e^{i\phi} \\ C_{o+} & C_{oo}e^{i\phi} & C_{o-}e^{2i\phi} \\ C_{-+}e^{i\phi} & C_{-o}e^{2i\phi} & C_{--}e^{3i\phi} \end{pmatrix} \quad (9)$$

$-$) are real. The resulting vortex structure in the space of the nine real functions $C_{\mu\nu}(r)$ is illustrated in Fig. 3. The upper curves represent the five pairing amplitudes, $C_{\mu\nu}(r)$ with $\mu + \nu$ even, that also occur in the maximally symmetric o vortex; the lower curves display the four additional pairing amplitudes, $C_{\mu\nu}(r)$ with $\mu + \nu$ odd, that exist in this vortex structure. This vortex solution has a superfluid core with A phase (C_{o+}, and its dual phase C_{o-}) and a ferromagnetic so-called β phase (C_{+o}, and its dual phase C_{-o}) at the vortex axis. In Fig. 3 the real pairing amplitudes $C_{\mu\nu}(r)$ are scaled so that the bulk ³He-B order parameter at $r = \infty$ is described by $C_{+-}(\infty) = C_{oo}(\infty) = C_{-+}(\infty) = 1$. Radial distances r from the vortex axis are measured in units of the Ginzburg-Landau coherence length given by Eq. (10). The scale is linear in r for r less

$$\xi_{GL}(T) = \sqrt{3/5}\,\xi_o/(1 - T/T_c)^{1/2} \quad (10)$$

than $5\xi_{GL}$, and varies as $1/r$ for r greater than $5\xi_{GL}$, so that the structure of an isolated vortex line can be represented for $0 \le r \le \infty$.

Point vortices in momentum space. The structure of the vortex core in the B phase is intimately related to the properties of the spin-½ ³He quasiparticle spectrum inside the core, and in particular to the topology of the nodes in the energy gap. For superconductors with spin singlet pairing, the energy gap is a scalar, but for the spin triplet paired superfluid states of liquid ³He, the energy gap becomes a 2 × 2 matrix in the ³He quasiparticle spin space. The nodes of the energy gaps for an arbitrary order parameter $A_{\alpha i}$ may be found from a consideration of the determinant of the 2 × 2 gap matrix $\hat{\Delta}(\hat{p})$ given by Eq. (11), where $\hat{p}$ denotes the ³He

$$\det(\hat{\Delta}(\hat{p})) = -A_{\alpha i}\hat{p}_i A_{\alpha j}\hat{p}_j = \Delta_\uparrow(\hat{p})\Delta_\downarrow(\hat{p}) \quad (11)$$

quasiparticle momentum direction on the Fermi sphere. If, for a given $\hat{p}$, the determinant vanishes, then for this $\hat{p}$ at least one of the gaps, that for ³He quasiparticles with spin projection up ($\Delta_\uparrow$) or with spin projection down ($\Delta_\downarrow$), is zero.

The nodes in the energy gap define all the superfluid properties of ³He, including the peculiar continuous vorticity of superflow. These nodes of the gap are just those points on the Fermi surface around which the phase of det ($\hat{\Delta}(\hat{p})$) changes by $2\pi N$, where N denotes an integer ($N = 2$ in the A phase). Such point vortices on the Fermi surface are called boojums, because they resemble point singularities of the same name in real space that may occur on the surface of a container. These vortices in momentum ($\vec{p}$) space and the more common vortices in position ($\vec{r}$) space have basically the same origin; they are vortices in the order-parameter field $\hat{\Delta}(\hat{p}, \vec{r})$, and thus they may transform to each other by the deformation of the order parameter in both $\vec{p}$ and $\vec{r}$ space. Due to these transformations, singular vorticity of the quantized vortex in $\vec{r}$ space may be smoothed by "flaring out" of the vortex in $\vec{p}$ space. This is what occurs in the B-phase v vortex.

The singly quantized o vortex has no nodes in the gap anywhere across the core. Hence, there are no vortices on the Fermi surface and vorticity in the singly quantized o vortex is strictly singular. That is, the circulation of superflow along any path embracing the vortex axis is 2π, irrespective of the path chosen, and the superfluid velocity has a singularity on the vortex axis, where the gap vanishes. In other words, $\hat{\Delta}(\hat{p}, \vec{r} = 0) = 0$ all over the Fermi surface, giving rise to the normal state at the vortex axis.

The situation is fundamentally different for the v vortex, where this singularity is resolved through the formation of vortices in $\vec{p}$ space, with the result that superfluidity is not broken on the vortex axis. Topological analysis predicts that in order to escape

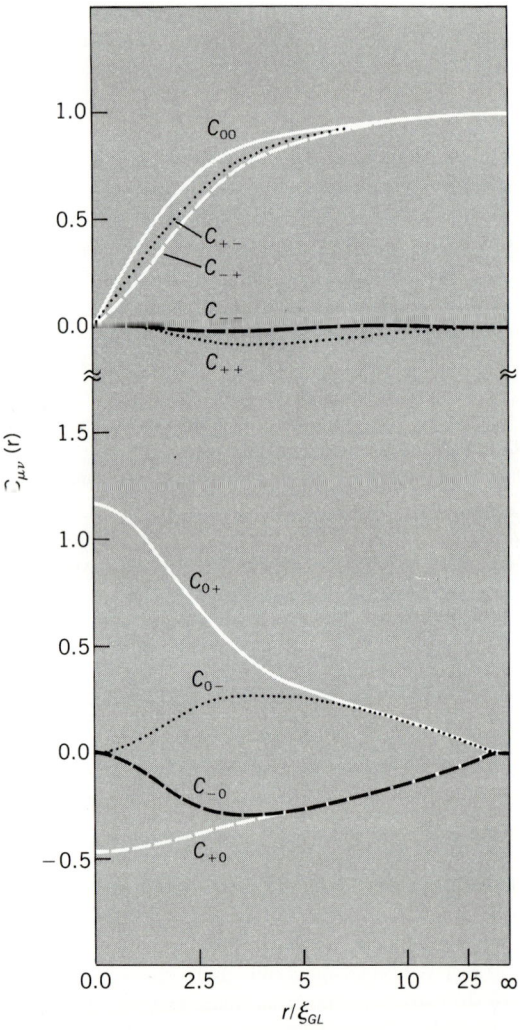

Fig. 3. Calculated structure of the v vortex in superfluid ³He-B. This spontaneously ferromagnetic superfluid core vortex structure is observed in nuclear magnetic resonance (NMR) experiments on rotating bulk superfluid ³He-B.

the singularity on the vortex axis, two pairs of boojums associated with the opposite topological charges $N = +1$ and -1 appear at the Fermi surface for some finite distance from the vortex axis. The distance at which the nodes in the ^{3}He quasiparticle excitation spectrum first appear is called the vortex core radius (r_{core}), since everywhere in the core region inside this radius the superflow is nonpotential.

As the distance r from the vortex axis is decreased to values less than r_{core}, these point vortices move continuously to the poles of the Fermi surface, while covering the whole Fermi surface on their way. On reaching the vortex axis at $r = 0$, the vortices fuse together on the poles. There they produce antipodal nodes of the gap with net topological charges $N = +2$ and -2, which correspond to the superfluid A phase on the vortex axis.

Figure 4 illustrates the computed magnitude of the energy gap $|\det(\hat{\Delta}(\hat{p}))|$ on cross sections of the Fermi surface for the v vortex in the weak-coupling limit shown in Fig. 3. Here the energy gap is drawn in the plane which contains all the four nodes. These cross sections are given for several different distances r from the vortex axis. This figure displays the process of splitting of the vortices on the antipodal points with the topological charges $N = +2$ at $r = 0$ into two pairs of vortices, each with $N = +1$, at $r > 0$. As r approaches r_{core}, the vortices move toward the equator, where the opposite topological charges annihilate each other at $r = r_{core}$. For r greater than r_{core}, the energy gap is strongly anisotropic, but nodes no longer exist. As r approaches infinity, the energy gap approaches that of the isotropic B phase.

[MARTTI M. SALOMAA]

COLLECTIVE MODES IN HELIUM-3

The dynamic behavior of a many-body system on a macroscopic scale is governed by its collective modes. These modes consist of weakly damped and therefore long-lived coherent motions of a large fraction of the particles in the system. In a normal liquid, for example, there exists only one well-defined collective mode: ordinary longitudinal sound. By contrast, superfluid ^{3}He can support a rich variety of collective modes, some of a type never before observed. The existence of these modes is intimately

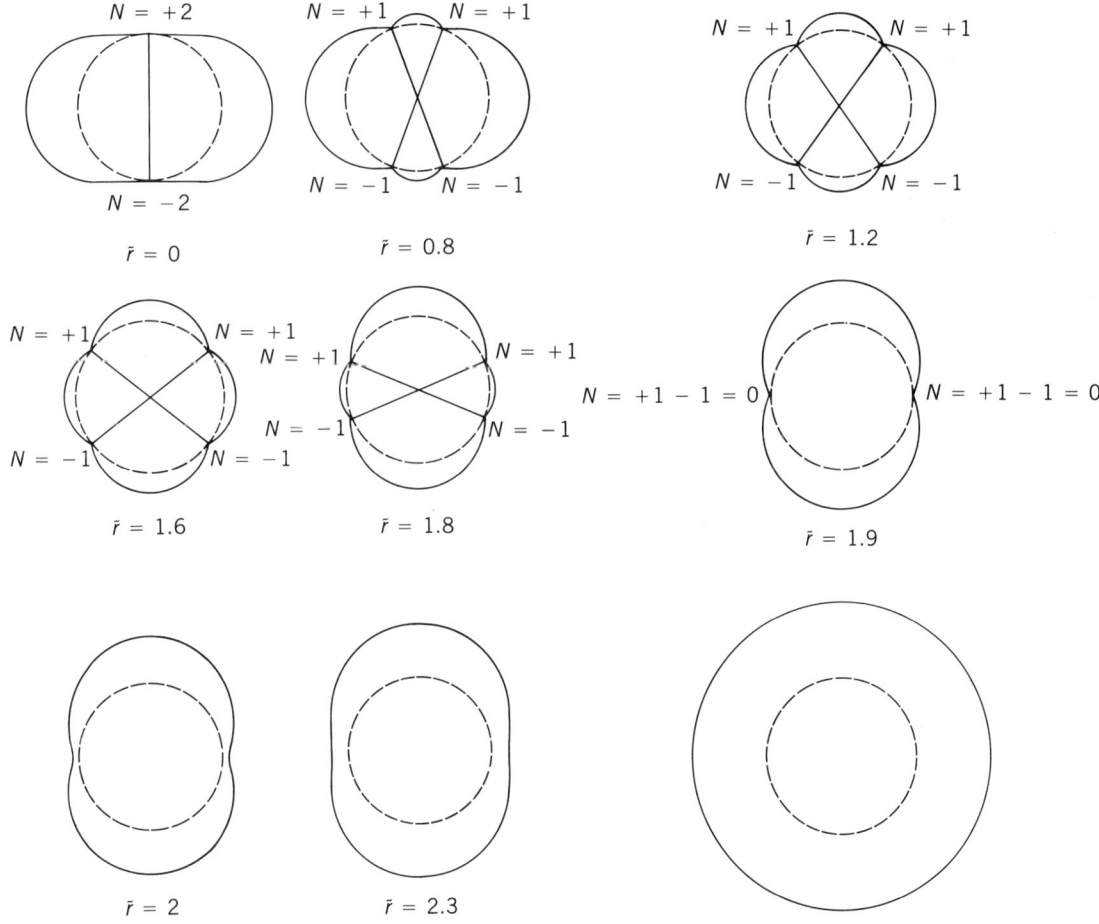

Fig. 4. Computed magnitude of the ^{3}He quasiparticle energy gap $|\det(\hat{\Delta}(\hat{p}))|$ for the superfluid core vortex in Fig. 3 as a function of the ^{3}He quasiparticle momentum and the normalized radial distance $\tilde{r} = r/\xi_{GL}$ from the vortex axis, illustrating the transfer of vorticity from position space to momentum space, that is, a point vortex on the Fermi surface.

related to the peculiar type of macroscopic order believed to be present in the superfluid phases of ^{3}He. The study of these phenomena has added new insights to the understanding of many-particle systems in general.

Macroscopic order. As shown by J. Bardeen, L. N. Cooper, and J. R. Schrieffer in 1957, any system of Fermi particles with attractive interaction forces is unstable below a critical temperature T_c against formation of correlated pairs. These Cooper pairs, being essentially Bose particles, all occupy the same lowest-energy quantum state, the so-called condensate.

As discussed above, the Bardeen-Cooper-Schrieffer (BCS) concept successfully explains the properties of the Fermi liquid ^{3}He, as well as the Fermi system of conduction electrons in superconductors for which it was originally invented. In contrast to conventional superconductors, where the Cooper pairs are formed in an isotropic spin singlet state, the Cooper pairs in liquid ^{3}He are in a relative angular momentum state $L = 1$ and consequently in the triplet spin state $S = 1$, as required by the antisymmetry of the pair wave function.

As the magnetic quantum numbers S_z and L_z may take on the values 0, +1, and −1, the Cooper-pair wave function is a three-by-three matrix of complex components, $A_{\mu\nu}$. Any such state is necessarily anisotropic; that is, the matrix $A_{\mu\nu}$ defines preferred directions in spin space and orbit space, just as an ordinary molecule is characterized by a spatial structure. The remarkable feature distinguishing Cooper pairs from molecules is that they are necessarily oriented in precisely the same way throughout the sample. Any deviation from uniformity costs additional energy. Thus, $A_{\mu\nu}$ may be considered to be an order parameter.

The 18 real variables necessary to specify the matrix $A_{\mu\nu}$ may be considered as collective variables of the system. They are capable of performing various oscillations about the equilibrium state, the collective modes. It turns out that the modes are quite distinct for the different phases of liquid ^{3}He.

B phase. The order parameter in the B phase is given by Eq. (12), where $\delta_{\mu\nu}$ equals 1 for $\mu = \nu$

$$A_{\mu\nu} = \delta_{\mu\nu} e^{i\phi} \Delta_\nu \qquad \mu,\nu = 1, 2, 3 \qquad (12)$$

and is zero otherwise. In equilibrium, $\Delta_1 = \Delta_2 = \Delta_3 \equiv \Delta$, where Δ is the (isotropic) energy gap in the single-particle spectrum; that is, Δ is the minimum energy that has to be spent in order to excite a particle out of the condensate of bound Cooper pairs. To break a Cooper pair into two single particles requires at least the energy 2Δ.

Quite generally the collective modes can be grouped into two categories, the so-called gapless modes, that is, those with eigenfrequency tending to zero as the wavelength becomes infinitely large, and modes with nonzero eigenfrequency in this limit. According to a theorem by J. Goldstone, there appears a gapless collective mode for every continuous symmetry that is broken in the ordered state. In the B phase there are two such symmetries: the order parameter is not invariant under gauge transformations ($\phi \to \phi'$) and under relative rotations of spin and orbit space ($\delta_{\mu\nu} \to R_{\mu\nu}$, where $R_{\mu\nu}$ is a rotation matrix). The corresponding modes are a density wave giving rise to the so-called zero-, second-, and fourth-sound modes under various conditions and three spin-orbit waves. With the exception of second sound, all of these have been observed. Similar modes exist in superconductors, antiferromagnets, and liquid crystals.

The unique property of ^{3}He-B is the existence of well-defined collective modes with finite frequency below the pair-breaking threshold 2Δ. These modes consist of a quadrupolar distortion of the equilibrium gap structure as described, for example, by $\Delta_1 = \Delta_2 = \Delta(1+\delta)$, $\Delta_3 = \Delta(1-2\delta)$. They may be viewed as molecular vibrations of a "giant molecule," the condensate. If the equilibrium state of the Cooper pairs is a state of total angular momentum $J = 0$, these excited states correspond to $J = 2$. As first shown in 1963, the frequency ω_{sq} for imaginary δ is given by Eq. (13), while the frequency ω_{rsq} for real δ is given

$$\hbar\omega_{sq} = \sqrt{12/5}\,\Delta \qquad (13)$$

by Eq. (14). These modes have come to be known

$$\hbar\omega_{rsq} = \sqrt{8/5}\,\Delta \qquad (14)$$

as the squashing and the real squashing modes, respectively. At finite temperatures T, the modes are damped by interaction with the thermal excitations, but the relaxation rate tends rapidly to zero as $\exp(-\Delta/T)$ as T approaches 0.

The squashing mode was first observed experimentally in 1976 through its strong coupling to sound waves. It causes a huge peak in the ultrasound absorption at approximately the temperature T at which the gap function $\Delta(T)$ is equal to $\sqrt{5/12}\,\hbar$ times the sound frequency (Fig. 5). At these low temperatures, ultrasound waves propagate

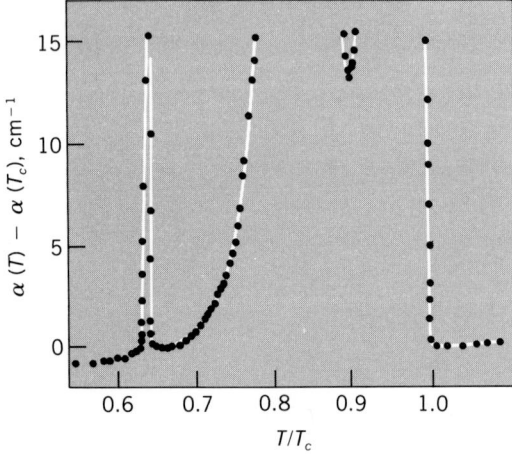

Fig. 5. Sound attenuation α in ^{3}He-B as a function of temperature T, showing pair-breaking, squashing, and real squashing states in order of decreasing temperature. (*After R. W. Giannetta et al., Observation of a new sound-attenuation peak in superfluid ^{3}He-B, Phys. Rev. Lett., 45:262–265, 1980*)

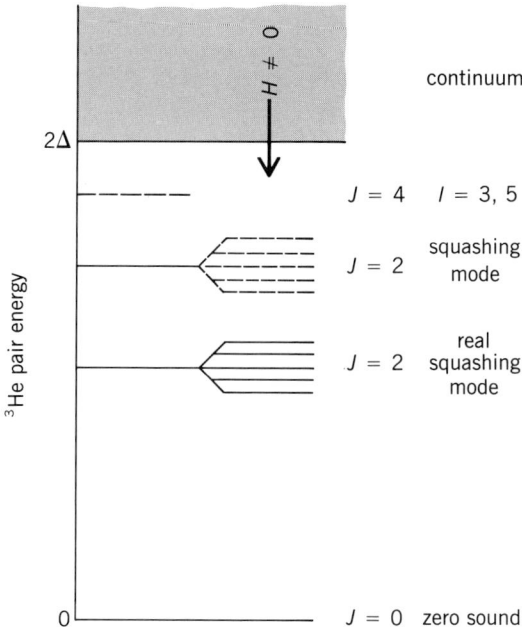

Fig. 6. Schematic energy level diagram for the Cooper-pair excited states of ^{3}He-B. The broken lines indicate levels for which there is not yet experimental evidence. (*After W. P. Halperin, Acoustic order parameter mode spectroscopy in superfluid helium-3-B, Physica, 109B and 110B:1596–1621, 1982*)

as collisionless "zero" sound, a gapless mode characteristic of Fermi liquids. The real squashing mode, detected in 1980, has a coupling strength to sound waves four orders of magnitude smaller than that of the squashing mode.

The analogy with a molecule carries even farther: a fivefold splitting of the $J = 2$ modes in a magnetic field, reminiscent of the Zeeman splitting in molecular spectroscopy, was predicted in 1979 and verified experimentally, for the real squashing mode, shortly thereafter (Fig. 6).

The extremely long lifetime of the squashing modes sufficiently far below the transition temperature allows large-amplitude excitation to be achieved. This makes nonlinear mode coupling an interesting possibility. Although preliminary experimental and theoretical studies indicate the existence of saturation effects and self-induced acoustical transparency similar to nonlinear resonant optical systems, this field remains to be fully explored.

Additional collective modes have been predicted, but experimental confirmation is ambiguous.

A phase. The order parameter in the A phase has a separable form in spin and orbit, given by Eq. (15). The unit vectors $\vec{d}$ and $\vec{m}, \vec{n}$ specify pre-

$$A_{\mu\nu} = \Delta_o d_\mu (m_\nu + n_\nu) \quad (15)$$

ferred directions in spin space and orbit space, respectively. The $\vec{m}$ and $\vec{n}$ are orthogonal such that $\vec{m}, \vec{n}$ and $\vec{l} = \vec{m} \times \vec{n}$ form an orthonormal triad.

Here the gap $\Delta \vec{p}$ in the energy spectrum of particles with momentum directed along $\vec{p}$ is anisotropic and even vanishes along the direction $\vec{l}$.

Hence, even for excitation energies well below $2\Delta_o$, there is a small probability for excited Cooper pairs to decay. Nevertheless, there exist two well-defined excited resonance states of the Cooper pairs. Actually these were observed shortly after the discovery of superfluid ^{3}He in ultrasound experiments in 1973. As pointed out shortly thereafter, these modes can be visualized as oscillations of the preferred axes $\vec{m}$ and $\vec{n}$ against each other (the clapping mode) and out of the common plane (the flapping mode), with calculated frequencies given by Eqs. (16) and (17). The coupling of these modes to zero

$$\hbar\omega_{cl} = 1.23\Delta_o(T) \quad (16)$$

$$\hbar\omega_{fl} \cong (4/5)^{1/2}(T/T_c)\Delta_o(T) \quad (17)$$

sound is very sensitive to the orientation of the gap axis $\vec{l}$ with respect to the sound propagation direction, which makes ultrasound an ideal probe of the $\vec{l}$-vector field.

Finally, the five broken symmetries of the A phase (gauge symmetry and symmetry with respect to rotations of $\vec{l}$ and $\vec{d}$) give rise to a density wave, two orbits waves, and two spin waves.

For background information *see* FERMI SURFACE; LIQUID CRYSTALS; LIQUID HELIUM; LOW-TEMPERATURE ACOUSTICS; NONRELATIVISTIC QUANTUM THEORY; NUCLEAR MAGNETIC RESONANCE (NMR); PHASE TRANSITIONS; QUANTIZED VORTICES; SUPERCONDUCTIVITY; SYMMETRY LAWS (PHYSICS) in the McGraw-Hill Encyclopedia of Science and Technology.

[PETER WÖLFLE]

Bibliography: P. J. Hakonen et al., NMR experiments on rotating superfluid ^{3}He-A and ^{3}He-B and their theoretical interpretations, *J. Low Temp. Phys.*, 53:425–476, 1983; P. J. Hakonen et al., Magnetic vortices in rotating ^{3}He-B, *Phys. Rev. Lett.*, 51:1362–1365, 1983; J. B. Ketterson et al., Probing collective modes of superfluid ^{3}He-B with zero sound, in *Quantum Fluids and Solids, 1983*, AIP Conf. Proc. 103, pp. 288–305, 1983; D. M. Lee and R. C. Richardson, Superfluid ^{3}He, in K. H. Bennemann and J. B. Ketterson (eds.), *The Physics of Liquid and Solid Helium*, vol. 2, pp. 287–380, 1977; M. M. Salomaa and G. E. Volovik, Symmetry and structure of quantized vortices in superfluid ^{3}He-B, *Phys. Rev. B.*, 31:203–227, 1985; M. M. Salomaa and G. E. Volovik, Vortices with ferromagnetic superfluid core in ^{3}He-B, *Phys. Rev. Lett.*, 51:2040–2043, 1983 and 52:2008, 1984; P. Wölfle, Collisionless collective modes in superfluid ^{3}He, *Physica*, 90B:96–106, 1977; P. Wölfle and D. Vollhardt, *The Superfluid Phases of Helium 3*, 1985.

Local-area networks

Recent progress in local-area networks (LANs) has provided opportunities for interconnecting personal computers for resource sharing and remote computation. This article describes a strategy for reducing the complexity and cost of personal computer interconnection. The key to this strategy is to strive for maximum compatibility with existing user applications, software, and hardware. Furthermore, compatibility will necessarily be achieved at some sacrifice in efficiency, and compatibility can be achieved only by careful and detailed planning of interfaces at all levels of the network architecture. The approach to interconnection should be motivated by application and software requirements and not by the traditional emphasis on hardware connections. There are different considerations involved in networking computers depending upon whether the objective is to provide intra-local-area network, inter-local-area network, or local-area network–to–mainframe (where individual personal computer-to-mainframe is a special case) communication, or some combination of these. *See* COMPUTER NETWORKING.

Both software and hardware considerations are important in networking. Software encompasses operating system and application software concerns. Hardware elements include interface boards, communication media (including building wiring), and communication facilities. Network standards, involving the standards activities of both the International Standards Organization (ISO) and the Institute of Electrical and Electronic Engineers (IEEE), play an important role in both software and hardware considerations. Although this article focuses on personal computer interconnection, the approaches and concepts are applicable to networking all types of computers.

Strategy for interconnection. The only way to minimize the investment and time involved in interconnection is to maximize compatibility at all pertinent network layers. This means that the user may have to pay for some overhead and services that are neither desired nor needed, but this is the price of compatibility.

The recommended strategy for interconnecting computers—an approach that is frequently overlooked—is to start with the users' application and software requirements and employ these as the baseline for identifying the appropriate hardware, topology, software packages, and network control. The reason for preferring this approach to one that starts with hardware specifications is that the characteristics of the application lead naturally to considering the possible needs of the system in the following areas:

1. The choice between multiple services (for example, voice and data) and a single service (for example, data only) has implications for the choice between broadband and base-band communications, respectively.

2. The choice between multiple simultaneous access to files and records by two or more users and a system in which each user employs files which are unique to that user's application has implications for the level of file and record locking which may have to be provided.

3. The importance of interoffice communication (for example, memo distribution) has implication for the use of electronic mail.

4. The need for access to a multiplicity of programs and data bases has implications for the degree of resource sharing which may be required and the number and diversity of software packages which may have to be provided.

Operating system considerations. The choice of operating system will, in large measure, determine the functionality of the network. For example, EtherMail requires DOS 2.0 or 2.1 in order to provide electronic mail service on a network of IBM personal computers.

Thus, if a user wants to network existing computers, the user's operating system will determine the types and brands of software and hardware which can be used in the network. If, on the other hand, the user does not have microcomputers installed and is considering acquiring them, the user's options for networking and choice of microcomputers will be significantly constrained by the marketing decision which the network vendor previously made concerning compatibility between the vendor's products and existing microcomputers.

In addition to specifying operating system compatibility, the network vendor will also specify random-access memory (RAM) size and other resources which are required in order to run the network software and the microcomputer operating system. Thus it is apparent that by merely identifying major applications the user who stresses compatibility above all else can quickly gauge the range of network functionality and performance available.

Software interconnection considerations. Although, as indicated above, the user must exercise great care in planning for the use of software in a single local-area network, that problem seems straightforward when contrasted with the challenge of providing for inter-local-area network and local-area network–to–mainframe communication. In the case of inter-local-area network communication, the strategy of maximizing compatibility can be achieved by using the following rule: implement all seven layers of the ISO architecture (application, presentation, session, transport, network, link, physical) in the interconnected local-area networks. The reason for this approach is that, in general, wide-area network (WAN) vendors have implemented the ISO Open Systems Interconnection (OSI) architecture, so it makes sense to avoid compatibility problems by adapting to the technology used by the dominant forces in the industry. Unfortunately, the standardization effort for local-area networks is in a more primitive state of development than the ISO model.

The IEEE standards effort, although making good progress, has essentially addressed only layers 1 and 2 of the ISO model. Even assuming the existence of seven-layer ISO software in a microcomputer local-area network, which would be difficult to implement given the amount of code required in relation to the speed and capacity of microcomputers, a gateway would still be required to resolve differences in message format, speed, and addressing between the local-area network and the wide-area network. In short, the market has not matured to the point where it is easy to interconnect multiple microcomputer networks operating through a wide-area network. A totally different situation exists relative to interconnecting minicomputers and mainframes to a wide-area network. Here, the technology and impetus exist within the Department of Defense community, for example, to develop standard protocols (Internet Protocol, Transmission Control Protocol) for interfacing subscriber communities, through hosts and front-end processors, to the Defense Data Network.

Local-area network–to–mainframe interconnection, particularly personal computer to mainframe communication, requires special treatment because, in this case, the user must deal with an organization and system—the data-processing department and its computers—which is outside the user's own organization and control. From a software standpoint, there may be a requirement to install programs in the mainframe in order to effect such operations as file transfer. Obviously, it will be necessary to make arrangements with computer center management for the installation of this software and its subsequent checkout. An example of this situation is the file transfer program which operates under IBM's VM/SP (virtual machine/system product) operating system, which must be installed with the IBM 3270PC (personal computer) in order for the user to interact with an IBM mainframe. The important point is that networking decisions should not be made in isolation and that while users can dictate interface requirements within their own organization, they must conform to the computer center's interface if they want communication with the mainframe. This illustration again points up the fact that cost avoidance in network interconnection requires conformance to someone else's standard.

Hardware interconnection considerations. A good procedure for interconnection planning is to identify and analyze all the interfaces which are involved in the interconnection. This means the user must be concerned with seemingly mundane matters but, in actuality, it is essential that these aspects be considered in order to achieve a low-cost interconnection. These physical or hardware considerations include the following:

1. Sufficient adapter board slots must be available in a microcomputer so that communication boards can be installed. For example, the five expansion slots in the IBM PC can fill rapidly to accommodate a floppy disk, color monitor, printer, memory expansion, and extender card for an expansion unit, thus requiring an expansion unit, in order to accommodate communication boards.

2. A wiring plan with local-area network taps should be provided for new buildings, so that massive cable pulling can be avoided in the future.

3. The trade-off must be recognized between the use of low-speed, low-quality asynchronous dial-up communication, with no wiring installation required, versus high-speed, high-quality synchronous communication, with wiring installation of dedicated or leased lines required (for example, the trade-off between using an asynchronous board in the IBM PC for dial-up communication versus using a binary synchronous board for dedicated communication).

4. It is necessary to ascertain the adequacy of input/output ports in mainframe controllers or front-end processors in order to accommodate hard-wired personal computer–to–mainframe connections [for example, the adequacy of ports on an IBM 3274 controller to accommodate a coaxial cable connection from an IBM 3270 PC, where four ports are required on the controller in order to provide four concurrent VM/CMS (virtual machine/conversational monitor system) sessions and windows on the 3270 PC; or the adequacy of line sets on an IBM 3705 to accommodate a dedicated telephone cable connection to support bisynchronous communication on the IBM PC].

One important conclusion which can be drawn from the above considerations and from an examination of available data communications facilities and services is that dial-up asynchronous communication is still dominant, because it is highly accessible and compatible and inexpensive. Also, the advent of smart modems has permitted a speed increase from 300 to 1200 bps for this type of communication.

From a hardware standpoint, intra-local-area network communication planning must address the following items:

1. Acquiring and installing an adapter board for each personal computer which will be part of the local-area network [for example, the adapter board which can be installed in an expansion slot of an IBM PC in order to implement PCnet for the purpose of providing the network's CSMA/CD (carrier sense multiple access/collision detection) protocol].

2. Determining whether a file/print server is required in the network (as in EtherLink) or is not required (as in PCnet).

3. Identifying communication media requirements, whether coaxial cable, as in PCnet, or twisted pair, as in Omninet.

4. Ascertaining maximum network cable length, maximum segment cable length (that is, between repeaters), maximum distance between microcomputers, and maximum number of microcomputers on the network.

5. Identifying which devices in the network are sharable—printer, hard disk, floppy disk, modem—and what restrictions apply to their use.

As mentioned above, software is not commercially available for interconnecting personal computer local-area networks with wide-area networks, except for the special case of connecting individual personal computers to mainframes; the situation is entirely different with respect to interconnecting minicomputer and mainframe local-area networks to wide-area networks, where rapid progress is being made, as in the Defense Data Network. Therefore, without the required software, the matter of achieving hardware interconnection and compatibility of personal computer local-area networks with wide-area networks is a moot point. An illustration of the inherent incompatibility which exists between the two types of networks is that Ethernet, the most prevalent type of local-area network, does not use the international standard, HDLC (high-order data link control), for its data link protocol, because HDLC is for point-to-point, multipoint, and loop communication whereas Ethernet uses broadcast transmission.

Status of interconnection technology. As discussed above, the major strategy for achieving a low-cost interconnection of networked computers is to maximize compatibility of hardware and software between connected devices and systems. This will result in a performance penalty; however, this is to be expected, since there is no way to simultaneously maximize performance and minimize cost. As also discussed, whereas intracomputer networking (for example, intra-local-area network communication) is fairly straightforward, inter-local-area network communication, particularly the netting of personal computer local-area networks to wide-area networks, is complex. The major reason is that at this point in the evolution of networks, personal computer local-area networks are marketed primarily for office applications, whereas wide-area networks have been developed primarily for large-scale computation, data-base management, and message communication. Furthermore, the two standards efforts—the ISO model for wide-area networks and IEEE 802 for local-area networks—have not produced standards which are entirely compatible. However, future efforts will undoubtedly be directed at reconciling the two approaches.

For background information *see* DATA COMMUNICATIONS; DIGITAL COMPUTER; MICROCOMPUTER; MULTIACCESS COMPUTER in the McGraw-Hill Encyclopedia of Science and Technology.

[NORMAN F. SCHNEIDEWIND]

Bibliography: V. E. Cheong and R. A. Hirschheim, *Local Area Networks: Issues, Products and Developments*, 1983; *Defense Data Network Program Plan*, Defense Communications Agency, January 1982, revised May 1982; L. E. Jordan and B. Churchill, *Communications and Networking for the IBM PC*, 1983; J. Nelson, 802: A progress report, *Datamation*, 29(9):136–148, September 1983; N. Schneidewind, Interconnecting local networks to long-distance networks, *Computer*, 16(9):15–24, September 1983.

Loricifera

The first specimen, an adult, of the newest phylum in the animal kingdom, the Loricifera, was found in 1974 in a sample of coarse sand collected from the continental shelf off North Carolina. During the following years a few larval specimens were found by a second investigator. However, it was not until the latter, R. M. Kristensen, found both larvae and adults in a single sample of coarse sand off the northern coast of France in 1982 that the significance was realized. At that time 38 phyla of living animals were known, and only two, the Pogonophora and the Gnathostomulida, had been discovered during this century. The existence of the thirty-ninth phylum was announced in 1983.

Nannaloricus mysticus, the only species of Loricifera currently described, is part of an ecological assemblage of microscopic invertebrates called the meiobenthos. These animals are well adapted for life between the sediment particles that make up most of the ocean floor. Meiobenthic organisms commonly have ways of attaching to grains of sand which make it difficult for them to be removed from the sediment. Although relatively few studies have been conducted on the meiobenthos of coarse sediments, the discoverers of the Loricifera believe that a new extraction technique—rinsing the sand sample briefly in fresh water to induce an osmotic shock causing the organisms to release their grasp on the grains—contributed to the discovery.

External morphology. The adults (Fig. 1) of *N. mysticus* reach a maximum length of 235 micrometers (0.009 in.). Immature stages of the animal, called Higgins larvae (Fig. 2), are as small as 120 μm (0.005 in.). The life history of the Loricifera includes several larval instars which grow by molting and finally undergo metamorphosis to form a preadult stage which soon molts into the adult. The body of the loriciferan consists of three regions. The anterior region is the head, or introvert. It consists of a thinly cuticularized spherical area with a mouth cone protruding from the center of a series of rings of appendages, or scalids. The expanded bases of eight oral styles surrounding the buccal canal cause an increase in the diameter of the mouth cone near its base, and the mouth cone then constricts as it joins with the spherical portion of the head. This gives the mouth cone a distinctive biconical appearance. Following the head is a thorax. Both head and thoracic regions are capable of being withdrawn, or inverted, into the thicker-walled cuticle, or lorica, of the third region, the abdomen.

The head of the adult has nine rings of appendages, or scalids. The first row consists of eight blunt clavoscalids which tend to project anteriorly. In the female all clavoscalids are similar, but in the male the two ventral clavoscalids remain as in the female while the remaining six divide into three branches, each thereby giving the appearance of 20 clavoscalids in the first ring. The scalids of the next eight rings project posteriorly. Elongate, spinelike scalids in the second ring become progressively reduced in the posterior rings until those of the eighth and ninth rings are small and toothlike in appearance. Following the head is a relatively short thoracic region which may consist of two regions. The posterior region bears a ring of basal plates from which specialized scalids project posteriorly over the edge of the third region of the body, the abdomen. The cuticle of the abdomen is much thicker and forms a lorica consisting of six longitudinal plates. The anterior margin of these plates forms hollow spinous projections resulting in a crown-shaped appearance. The midventral plate appears hinged to the adjacent plates. No appendages are associated with the adult lorica. A few small sensory structures called flosculae, clusters of four to seven micropapillae surrounding a pore, are present on the dorsal and ventral portions of the lateral plates. Many other glandular pores are present on the lorical plates. The anal opening is slightly dorsal to the caudal end of the lorica. The presence of reproductive pores cannot be clearly demonstrated, but they are thought to be subterminal as well.

Higgins larva. Although the several larval stages have the same body regions as in the adult, these regions have some significant differences in their morphology. The mouth cone is less complex and lacks oral styles. Eight anteriorly directed clavoscalids, all similar in structure, make up the first ring of head appendages. As in the adults, the remaining rings are posteriorly directed and become smaller, less spine-shaped, and reduced to toothlike structures in the last of only seven rings. A single middorsal scalid appears to be modified into a special sensory structure reminiscent of the dorsal antennae of some rotifers. The thoracic region of the Higgins larva is more prominent than it is in the adult. Its appendageless cuticle is divided into a series of plates by several longitudinal and transverse folds. When the head is withdrawn into the abdomen, the thoracic plates, including two distinctive midventral plates, collapse over the withdrawn head and close off the anterior region of the lorica.

The larval lorica is not as strongly cuticularized as it is in the adult. The anterior three-fourths is made up of four large plates, each with a series of longitudinal folds. The caudal region of the lorica has a series of five dorsally displaced plates surrounding the anus. Unlike the lorica of the adult, there are a series of prominent appendages associated with the larval lorica. At the anterior margin of the lorica are two appendages, one on either side of the ventral midline. Each consists of three elongate rami; the outer two are spinose, but the inner ramus has a clawlike tip. These appendages are a combination of locomotory and sensory organs used in crawling over the sand grains. A second pair of uniramous setae, presumed to be sensory, are situated on either side of the dorsal midline at the beginning of the caudal region of the lorica; a third pair of similar setae occur on the lateral caudal plates. The most prominent feature of the Higgins larva is the pair of large caudal locomotory organs, or toes. The toes are connected to the caudal region by ball-and-socket joints. Near their origin the toes enlarge to form a broad, flat structure with a series of cuticular folds that appear to be able to reverse their orientation as in a venetian blind. This portion of the toes gradually tapers into an elongate flexible tube terminating with a small pore. The animal can creep along the substrate by using its toes in a manner

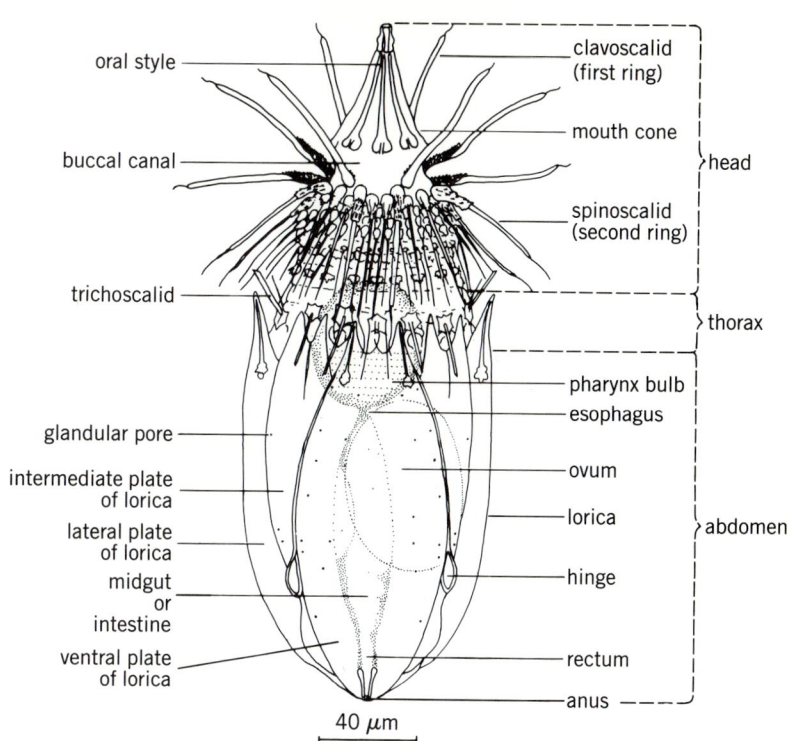

Fig. 1. *Nannaloricus mysticus*, ventral view of adult female. (*After R. M. Kristensen, Loricifera, a new phylum with Aschelminthes characters from the meiobenthos, Z. Zool. Syst. Evolutionsforsch., 21(3):163–180, 1983*)

similar to the rotifers, thereby suggesting that the glandular material seen at the base of each toe constitutes an adhesive organ. Otherwise, the caudal appendages can move rapidly to propel the loriciferan in a swimming mode.

Internal morphology. The alimentary canal begins with a cuticularized foregut composed of a flexible buccal canal with a large muscular pharynx bulb at its base. The pharynx bulb can be moved from a recessed position in the abdomen to the head by a series of six retractor muscles. Near the anterior end of the pharynx bulb are three hollow accessory stylets associated with a hexaradially arranged salivary gland. The lumen of the pharynx is triradiate; the cuticular lining of each of the radii secretes a row of five placoids to which muscles of the pharynx attach. The contractile elements of the pharynx are myoepithelial cells with true cross-striated myofibrils. A short esophagus follows the pharynx and continues as a simple midgut, a short rectum, and finally a cuticle-lined anal cone or hindgut.

The sexes are separate; the principal external difference is the structure of the first ring of head appendages in the adult. Internally, paired saccate ovaries, usually with only a single egg, and testes are visible in the adults. A small seminal receptacle may be present in the female.

The nervous system consists of a very large dorsal brain. Each ring of scalids is innervated separately. The clavoscalids, or first ring of head appendages, are associated with eight circumoral ganglia which

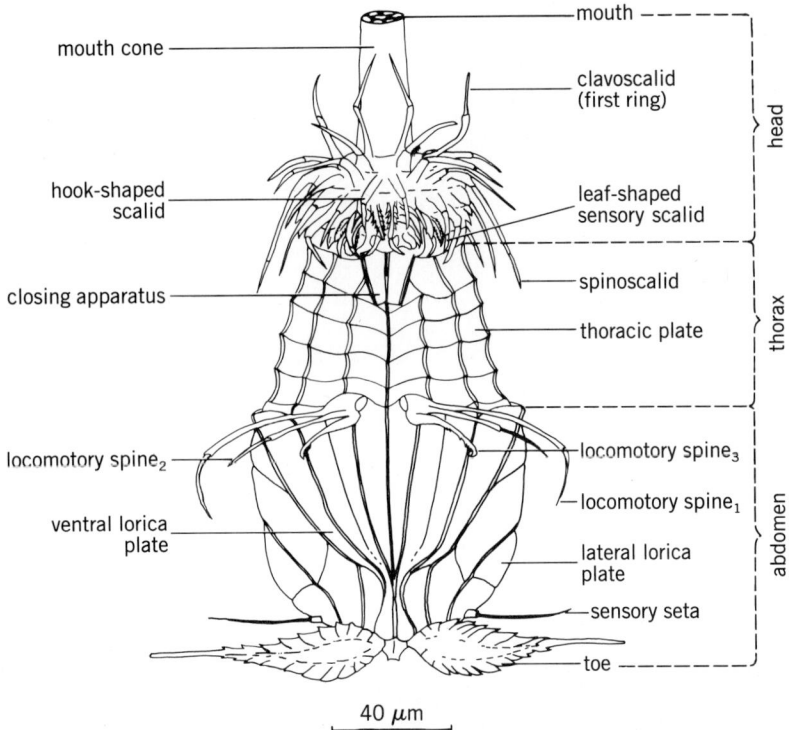

Fig. 2. *Nannaloricus mysticus*, ventral view of Higgins larva. (*After R. M. Kristensen, Loricifera, a new phylum with Aschelminthes characters from the meiobenthos, Z. Zool. Syst. Evolutionsforsch., 21(3):163–180, 1983*)

are nearly fused with the brain. A large ventral ganglion in the thorax innervates the special scalids in that region. At least one and possibly more ventral ganglia are present in the abdomen, and a single ventral caudal ganglion innervates the floscular system, the only sensory structures on the adult lorica. The presence of a pair of protonephridia is suspected. Other features of the internal anatomy of the adult and larval stages must await further study.

Taxonomy and phylogenetic implications. Although only one species of loriciferan has been described, at least six others are in the process of being described. Differences in structure and appearance of the lorica of both adults and immature stages, the structure and arrangement of scalids and elements of the mouth cone, and the structure and arrangement of the appendages of the larval lorica have been noted in the new species.

Phylogenetically, the Loricifera appear to be most closely related to the Kinorhyncha and the Priapulida. However, they share several characteristics with the Nematomorpha, Rotifera, and even the Tardigrada.

For background information *see* KINORHYNCHA; NEMATOMORPHA; PRIAPULIDA; ROTIFERA; TARDIGRADA in the McGraw-Hill Encyclopedia of Science and Technology.

[ROBERT P. HIGGINS]

Bibliography: R. M. Kristensen, Loricifera, a new phylum with Aschelminthes characters from the meiobenthos, *Z. Zool. Syst. Evolutionsforsch.*, 21(3):163–180, 1983.

Lyme disease

Lyme disease is a complex multisystem human illness caused by the tick-borne spirochete *Borrelia burgdorferi*. The disease usually begins with the characteristic skin lesion, erythema chronicum migrans (ECM), that may be accompanied by headache, stiff neck, fever, myalgias, arthralgias, malaise, fatigue or swelling of the lymph nodes. These early clinical manifestations may last for several weeks and may be supervened by meningoradiculitis, meningoencephalitis, myocarditis, or migrating musculoskeletal pains. Later, intermittent attacks of arthritis may occur. The arthritis may become chronic and result in the destruction of bone and cartilage in the large joints, especially the knees.

Development of the skin lesions is a unique clinical marker. The lesion is usually characterized by a small red papule or macule, often considered to be the site of tick bite, that 3–23 days later spreads centrifugally, with a hardened center up to 2 cm (0.8 in.) wide and usually flat borders. The centers of early lesions may also be vesicular or necrotic. The centers clear as the lesions expand. Common sites for these itching, burning, painful lesions are thigh, groin, and axilla. Within several days after onset of the initial lesion, annular secondary lesions appear in almost half the affected individuals. These lesions, generally smaller, less migratory, and lacking hardened centers, may be located anywhere on the body except on palms and soles. During or soon after resolution of the initial and secondary skin lesions, new lesions may develop in the form of small red circles and blotches (2–3 cm, or about 1 in., in diameter) that do not migrate but may persist for several weeks.

Lyme disease was first described in 1977 as a new clinical entity following investigations of unusual clusters of arthritis among children and adults in East Haddam, Lyme, and Old Lyme, Connecticut. Although 13 of 51 individuals had skin lesions preceding arthritic complications, the disease was considered to be distinct from the previously reported cases of the skin lesion in Wisconsin and Connecticut, or from those in Europe where the lesion had been recognized since 1908 and associated with the ixodid tick, *Ixodes ricinus* (Fig. 1). More recent clinical investigations, however, suggest that the European ECM also represents a complex syndrome with the one or more diagnostic skin lesions followed in many cases with neurological, cardiac, and arthritic complications. Lymphocytic meningoradiculitis (Bannwarth's syndrome), lymphocytoma (lymphadenitis benigna cutis), and acrodermatitis chronica atrophicans are the other disorders that appear to be associated with the European ECM.

Etiologic agent. The etiologic agent in Europe as well as in the United States remained elusive until 1981 when, during a survey of spotted fever group rickettsiae in the ixodid tick *Ixodes dammini* from Shelter Island, New York, a spirochete was discovered that reacted strongly in indirect immunofluorescence with the convalescent sera of Lyme disease patients. Of 126 adult *I. dammini*, 77 (61%) contained spirochetes in their midgut that could readily be isolated in a medium suitable for growth of relapsing fever spirochetes. Antigenically similar if not identical spirochetes were subsequently discovered in the European tick vector, *I. ricinus*. Since then, the significance of these microorganisms as the cause of Lyme disease in the United States and in Europe has been confirmed by the recovery of spirochetes from blood, cerebrospinal fluid, and skin lesions of affected individuals. Investigations into the genetic and phenotypic characteristics of such isolates have shown that these microorganisms represent a new species of *Borrelia*, for which the name *Borrelia burgdorferi* has officially been proposed.

Borrelia burgdorferi is a helically shaped bacterium 0.18–0.25 micrometer by 4–30 micrometers (Fig. 2). An outer membrane surrounds the protoplasmic cylinder, which consists of the peptidoglycan layer, cytoplasmic membrane, and the enclosed cytoplasmic contents. Seven periplasmic flagella (axial filaments or axial fibrils) are attached subterminally in a row parallel to the organism's long axis and overlap at the central region of the cell. These flagella are probably responsible for the organism's rotational and translational movements. *Borrelia burgdorferi* is gram-negative and stains moderately well with Giemsa. It is chemoorganotrophic and uses carbohydrates such as glucose as energy and carbon sources. Like certain tick-borne relapsing fever spirochetes, it grows well in modified Kelly's medium between 34 and 37°C (93 and 98°F) and has a generation time of 11–12 h at 35°C (95°F).

Geographic distribution. In the United States, Lyme disease has so far been reported from three geographic areas: the Northeast and Midwest, where *I. dammini* is the principal vector, and the West, where *I. pacificus* carries the spirochete. Isolated cases from Texas, Arkansas, and Georgia suggest that other species of ticks such as *I. scapularis*, *Amblyomma americanum*, and even *Dermacentor variabilis* may be involved as vectors in the Southeast. Although spirochetes have been demonstrated in these latter species, only *A. americanum* has been shown to be an additional vector of *B. burgdorferi*.

In Europe, Lyme disease and related disorders have been reported from practically all countries within the distributional area of *I. ricinus*, which includes Scandinavia and middle and eastern Europe and most of southern Europe. Cases have also been recorded in England, where *I. ricinus* occurs abundantly, and in Australia, where the arthropod vector has not as yet been identified.

Epidemiology. Prevalence of infected ticks varies. On Shelter Island, New York, where the first isolate of *B. burgdorferi* originated, up to 100% of *I. dammini* were found to carry spirochetes. In California, on the other hand, not more than 2% of *I. pacificus* were infected, and in Switzerland up to 36% of *I. ricinus* were positive for spirochetes.

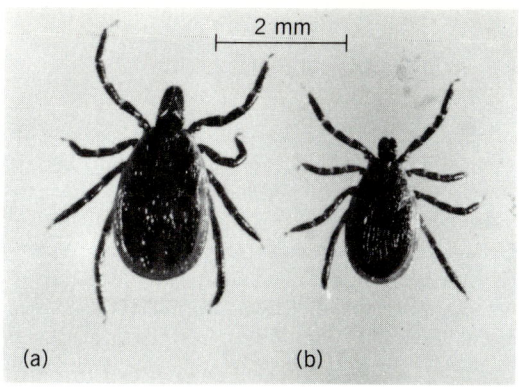

Fig. 1. *Ixodes dammini*, (a) female and (b) male, the principal vector of the Lyme disease spirochete in the northeastern and midwestern United States.

Development of *B. burgdorferi* in its tick vectors and the modes of transmission to humans are still under investigation. In the majority of infected ticks, spirochetal distribution is limited to the midgut. Such ticks, it is speculated, may be capable of transmitting spirochetes by regurgitation. A generalized infection including the tissues of the salivary glands and of the ovary occurs in a few ticks only. These ticks undoubtedly transmit via saliva. They have also been shown to be capable of passing spirochetes via eggs to their offspring.

The fact that many persons with Lyme disease in the United States as well as in Europe contract the disease without apparent exposure to ticks suggests that other hematophagous (blood-feeding) arthropods, such as biting flies and mosquitoes, serve as vectors of *B. burgdorferi*.

As yet, little is known concerning the natural history of *B. burgdorferi*, particularly the sources for infecting ticks. Spirochetes have been recovered from the blood of white-tailed deer (*Odocoileus virginianus*), a raccoon (*Procyon lotor*), and deer mice (*Peromyscus leucopus*). The latter is a preferred host of immature *I. dammini* and is considered a poten-

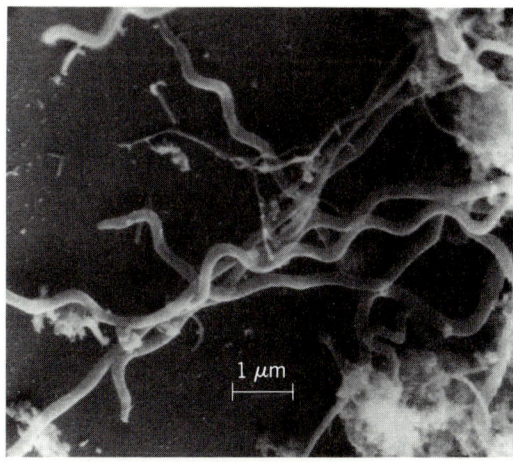

Fig. 2. Scanning electron micrograph of *Borrelia burgdorferi* in modified Kelly's medium.

tial reservoir. The New Zealand white rabbit (*Oryctolagus cuniculus*) and the hamster (*Cricetulus* sp.) appear to be useful animal models for laboratory research.

Diagnosis. Diagnosis of Lyme disease can readily be made based on clinical observation, especially if it includes the development of the skin lesion following tick bite and exposure in an endemic area. Histological demonstration of silver-stained spirochetes in skin biopsy specimens, and cultivation of the organism from blood, cerebrospinal fluid, and skin biopsies also are useful tools. In the absence of the skin lesion, the diagnosis may be established serologically by the indirect immunofluorescence or enzyme-linked immunosorbent assay with cultured spirochetes; the latter appears to be more sensitive in detecting immunoglobulin M (IgM) antibodies which are usually present during the third through sixth week after onset. Both tests, however, detect immunoglobulin G (IgG) antibodies that reach maximum titers months later during arthritic, neurologic, or cardiac sequelae. Because of the close relationship of *B. burgdorferi* to other spirochetes, particularly those causing relapsing fever and syphilis, cross reactions do occur but may be eliminated through absorption of sera with *B. hermsii* or *Treponema phagodenis*, respectively. When the illness is treated early with appropriate antibiotics, a detectable serological response to *B. burgdorferi* does not develop.

Treatment. Antibiotics are effective in the treatment of Lyme disease. Treatment with oral tetracycline is recommended. In children, phenoxymethyl penicillin is effective. Erythromycin should be given to individuals who are allergic to penicillin.

For background information see ACARINA; IXODIDES; SPIROCHETE in the McGraw-Hill Encyclopedia of Science and Technology.

[WILLY BURGDORFER]

Magnetic field

Observations have established the existence of magnetic fields in essentially all astronomical objects. Planets, stars, the gaseous disks of galaxies, and even the space throughout clusters of galaxies are filled with magnetic fields. The fields are so strong as to influence the dynamical behavior of the gases. The fields are the agent by which most of the fast particles and the superheated x-ray coronas of stars and galaxies are produced.

On astronomical scales, the magnetic fields embedded in the electrically conducting gases of stars and galaxies take on an independent existence that is wholly unlike their subservient relation to electric currents in the terrestrial laboratory. The astronomical magnetic field rides free with the churning gas, and is maintained and amplified by the vorticity, and dissipated in the current sheets formed by the internal strains. Energy from the dissipation of magnetic fields produces the explosive flare phenomena of the Sun and other stars. Stresses within magnetic fields heat the outer atmosphere of stars to temperatures of millions of degrees, so that each star emits x-rays. Opposition of magnetic stresses to the convective transport of heat leads, in some way not yet clear, to the formation of sunspots and, by inference, to the gigantic star spots that appear on some faint red dwarfs. Magnetic fields of the planets accelerate and store fast particles, forming the Van Allen radiation belts around Earth, the radiation zones around Mercury, Jupiter, and Saturn and, it is expected, around Uranus and Neptune. Even Venus and Mars, which have little or no magnetic field of their own, live in the environment of the outstretched magnetic field of the Sun.

Solar wind. The solar magnetic field, with an average value of a few gauss (1 gauss equals 10^{-4} tesla) at the Sun, is extended outward through the solar system by the solar wind to form the heliosphere, and meets the interstellar medium somewhere beyond the planets. The magnetic field carried outward in the solar wind sweeps the charged particles (electrons and ions) before it, dominating space throughout the solar system. Even the galactic cosmic rays are swept back by the fields in the wind, so that the cosmic-ray intensity at Earth is much reduced and varies oppositely to the mean level of activity of the Sun. The field has a spiral form (Fig. 1) as a consequence of the Sun's rotation. At the orbit of the Earth, the field is inclined approximately 45° to the radius, and it is primarily in the azimuthal direction beyond the Earth's orbit.

Cosmic rays. The cosmic rays fill the galaxy, trapped and stored by the galactic magnetic field of $2-3 \times 10^{-6}$ G ($2-3 \times 10^{-10}$ T) embedded in the interstellar gas. It appears that the cosmic rays, produced by supernovae and supernova remnants, build up inside the galactic magnetic field until they are able to push their way out in a body and escape into intergalactic space. Hence the energy density

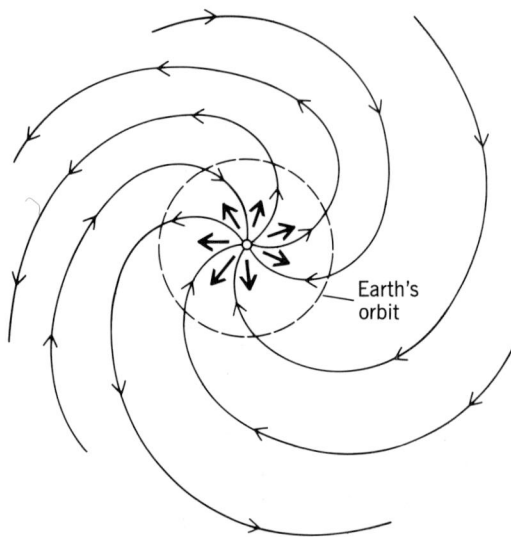

Fig 1. Spiral magnetic lines of force in the equatorial plane of the Sun pulled outward by the radial solar wind (short arrows).

(pressure) of the galactic cosmic rays is about equal to the energy density (pressure) of the galactic magnetic field.

Field of the Earth. It was first pointed out by W. M. Elsasser in 1945 that the magnetic field of Earth must originate in the liquid iron-nickel core as a consequence of the slow convection and the associated nonuniform rotation of the core (Fig. 2). The core is presumed to rotate more rapidly near the axis of Earth, by 0.004–0.04 in./s (0.1–1 mm/s). This nonuniform rotation shears the dipole field (whose lines of force lie in the merdional planes and whose strength is 0.6 G (6×10^{-5} T) at the surface of Earth at the north and south poles, and 5 (5×10^{-4}) in the core T) in the manner sketched in Fig. 3 to produce a strong east-west (azimuthal) field in the core, with a strength variously etimated to be 20 to 200 G (2×10^{-3} to 2×10^{-2} T). The cyclonic convection within the rotating core interacts on a small scale with the azimuthal field, producing many small loops throughout the core (Fig. 4) whose coalescence through resistive diffusion leaves a net circulation of field in each meridional plane, and reinforces the dipole component with which the process began (as may be seen by comparison of the meridional fields of Figs. 2 and 4). The convection that accomplishes this is evidently powered by slow dendritic condensation of the liquid iron and nickel into the solid phase forming the central solid core.

The past history of the magnetic field of Earth is recorded in lavas and in the new ocean floor that spreads outward from mid-ocean ridges. The rocks take up a residual magnetism with a direction and strength determined by the field at the time of their solidification. From studies of these rocks, as well as of kiln-fired bricks, it can be shown that the field is maintained in a quasisteady state (with fluctuations in strength of approximately 30% on time scales of 2000 years) for periods of 10^5–10^7 years. Such quasisteady epochs terminate with an abrupt (approximately 10^3 years) reversal of the dipole field, followed by a new quasisteady state with the opposite polarity. It is presumed that the reversals are caused by some disturbance in the convective pattern in the core, although the precise mechanism is not known.

There is a widespread belief that the increase of the cosmic-ray intensity during the reversal (when the weakened geomagnetic field turns back fewer cosmic rays) has been responsible for the abrupt demise of various species, such as the dinosaurs. The idea evidently has great appeal, but quantitatively the net effect of removing the geomagnetic field would be to increase the cosmic-ray intensity at sea level to approximately the present level in Denver, Colorado. Thus, something more lethal must have been responsible for the demise of ancient species.

Solar field. In the Sun, where the outer one-quarter of the radius is subject to convective overturning, the same principles that apply to the generation of the quasisteady field in the liquid core of Earth lead to an oscillating dipole field, with a pe-

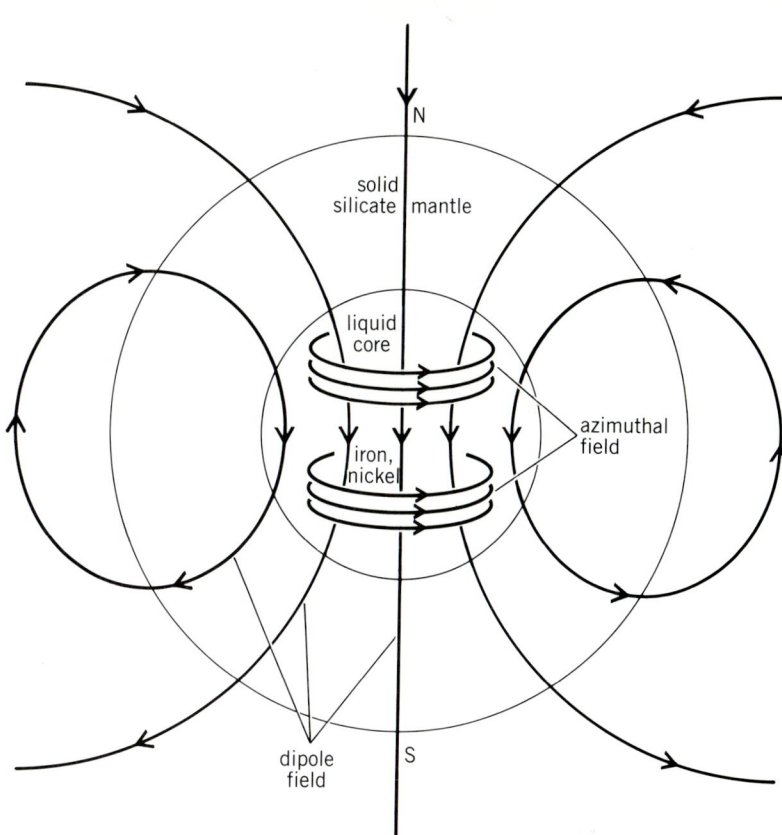

Fig. 2. Interior of the Earth. The small central solid iron-nickel core is not shown. Lines of force of the dipole and azimuthal magnetic fields are indicated.

riod comparable to the 22-year observed oscillation of the magnetic field of the Sun. The azimuthal field, generated from the dipole field by nonuniform rotation, is again the stronger field, estimated to be 2–10 kilogauss (0.2–1 T) in the lower half of the convective zone. It is the azimuthal field, bursting locally through the surface of the Sun, that is responsible for the large active regions, with their sunspots and flares and the hot, dense x-ray–emitting corona above.

The activity of magnetic regions on the Sun is attributed to the continual deformation of the field by the churning gas beneath the surface. A magnetic field subject to any but the simplest strains has no equilibrium, instead developing internal shear planes (current sheets) where the very high electric current density is dissipated. If the background gas is tenuous, the shear planes may accelerate the local electrons and ions and generally heat the gas to temperatures of 10^6 K or more. Sometimes the dissipation between distinct regions of field occurs explosively, producing the flare phenomenon, with temperatures ranging up to 10^8 K. At such times hordes of fast particles and large volumes of gas and field are ejected into space. The total energy output of a large flare may be as high as 10^{32} ergs (10^{25} joules).

Stellar fields. Most other stars have as much magnetic activity as the Sun, evident in their continual variable x-ray emission. Some stars such as the small, faint (dM) red dwarfs may be very much more active, with cool spots—star spots—covering as much as half of one side of the star, and flares a thousand times more energetic than the largest flares on the Sun.

There is a class of hot stars called magnetic stars which have general surface fields so strong (1–35 kG or 0.1–3.5 T) that they were the first astronomical fields to be directly measured. For the most part these stars have dipole fields set at an angle to the axis of rotation. It is not clear whether the fields are primordial (that is, carried in with the interstellar gas that formed the star) or are the product of a present-day dynamo effect, along the lines described above for Earth and the Sun.

The strongest fields are found in some of the collapsed stars, that is, white dwarfs and neutron stars. Magnetic fields in white dwarf stars are difficult to detect because of the absence of sharp spectral lines. However, some white dwarfs have fields so strong (10^4–10^5 kG or 10^3–10^4 T) that they produce measurable circular polarization of the emitted light. Neutron stars have fields of the order of 10^9 kG (10^8 T), first inferred from the fact of their emission of intense pulses of radiation with each rotation, and later verified through direct measurement of the cyclotron frequency of trapped particles. The extreme fields of collapsed stars are produced, it is assumed, by the compression of original stellar fields (of perhaps 1 kG or 0.1 T) with the collapse of the star.

Galactic fields. The magnetic fields of spiral galaxies are observed to lie more or less along the spiral arms. It is conjectured that the magnetic fields are primordial, compressed from some (so far undetected) fields of 10^{-10} G (10^{-14} T) in intergalactic

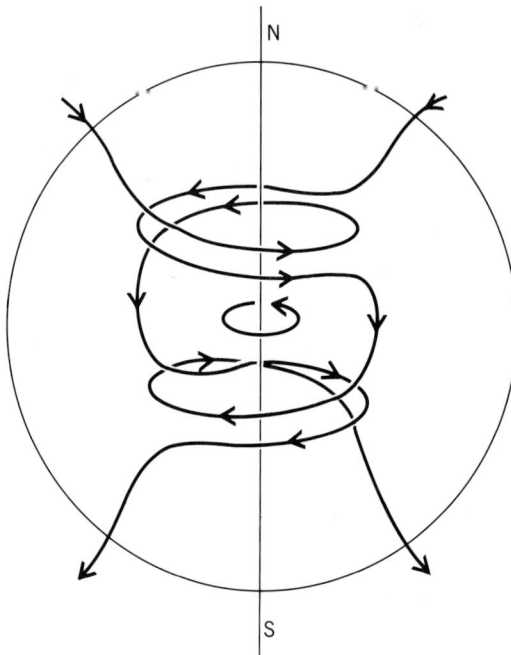

Fig. 3. Shearing of the dipole field by the more rapid rotation of the inner part of the liquid core of Earth, producing a strong east-west field in the core.

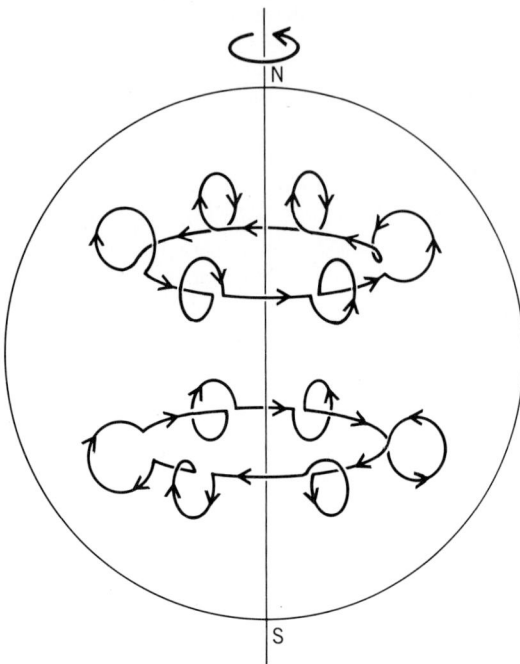

Fig. 4. Small loops twisted in the east-west magnetic field in the Earth's core by rising cyclonic (rotating) convective cells in the liquid iron.

space into their present form and intensity (several microgauss) with the formation of the galaxy. A closer look suggests serious difficulty with this idea in accounting for the simple outward extension of the fields along the open spiral arms of some of the neighboring galaxies so late in their history (at ages of 10^{10} years) when the inner part of the galaxy has rotated ten or a hundred more revolutions than the outer part. Serious theoretical difficulties arise also in understanding how a field is maintained for 10^{10} years in the face of so many dynamical instabilities. Quantitative application of the same dynamo principles that account for the magnetic field of Earth and the Sun suggests that the magnetic fields of galaxies may be maintained in the present form by the motions within the gaseous disks of the galaxies.

Quantitative study. These theoretical conclusions are based on extensive quantitative study and modeling of the mean-field dynamo equations, for which the fluid motions are prescribed by other, generally elementary, considerations. In particular, the turbulent mixing and diffusion of magnetic fields, which plays an essential role in their generation and decay, is treated in only the crudest way, by an effective "eddy diffusion" coefficient. The formal solution of the dynamical equations of motion to deduce quantitatively the unseen motion of the fluid within the liquid core of Earth and within the convective zone of the Sun and other stars is extremely difficult. It can be shown that no conceivable electronic computer can handle the complete calculation. Only the natural object—planet or star—acting as a precise analog computer can determine the fluid motion by direct "computation." The "readout" is restricted to what can be observed at the visible surface. On this basis, it appears that the general ideas (outlined above) are correct, but the specific details are not clear in many cases. For instance, it is not known precisely why the field of Earth reverses occasionally. Close examination of the surface of the Sun reveals that the magnetic field there is broken into intense (1 kG or 0.1 T) separate fibrils (with diameters of 120 mi or 200 km) unlike the continuum fields found elsewhere. The mean field, typically 5 G (5×10^{-4} T) in quiet regions and perhaps 100 G (10^{-2} T) in active regions, depends on the spacing of the fibrils. This effect does not alter the mean-field dynamo effect of the fluid motions, but it demands explanation of the intense fibril state. Until that is forthcoming for the fields at the surface of the Sun, the state of the field deep in the Sun and in other stars can only be conjectured.

For background information *see* COSMIC RAYS; GALAXY; GEOMAGNETISM; SOLAR WIND; STELLAR MAGNETIC FIELD; SUN in the McGraw-Hill Encyclopedia of Science and Technology.

[EUGENE N. PARKER]

Bibliography: E. F. Borra, J. D. Landstreet, and L. Mestel, Magnetic stars, *Annu. Rev. Astron. Astrophys.*, 20:191–220, 1982; F. Krause and K. H. Rädler, *Mean-Field Magnetohydrodynamics and Dynamo Theory*, 1980; A. M. Soward (ed.), *Stellar and Planetary Magnetism*, 1983; H. K. Moffatt, *Magnetic Field Generation in Electrically Conducting Fluids*, 1978; E. N. Parker, *Cosmical Magnetic Fields*, 1979; E. N. Parker, Magnetic fields in the cosmos, *Sci. Amer.*, 249(2):44–55, August 1983.

Marine sediments

Recent advances in the study of marine sediments have resulted in new insights into transport of sediments in active margins and valuable information on the nature of eolian sediments and deposition.

Sediment drifts in active margins. Transport of sediment within active-margin settings has been attributed traditionally to three dominant mechanisms: density-driven turbidity and debris flows, background fallout of relatively fine-grained material from the overlying water column, and mass wasting. Models of sediment deposition in active margins suggest that these mechanisms are specific to particular tectonomorphic environments: turbidite deposition is dominant in the forearc and backarc basins, trench-slope basins, and trench floor; hemipelagic sediments drape the trench-slope and the backarc-basin region distant from the volcanic arc; and mass wasting is prevalent at the base of steep tectonic scarps along both trench walls. This classification of mechanisms seems fairly restrictive considering the widespread tectonic variations observed in intraoceanic and continental arc-trench systems.

Contemporaneous with the development of active-margin sediment-depositional models, investigations of deep-ocean abyssal-plain and passive-margin geologic environments have revealed a widespread influence of abyssal circulation on sedimentation. Wind-driven and thermohaline currents are capable of sculpting vast areas of the continental slope and abyssal floor, resulting in sediment drifts and deep-sea unconformities covering as much as thousands of square kilometers. Sediment drifts are accumulations of sediment deposited on the sea floor by ocean currents. Such drift deposits have only rarely been described in active-margin settings, though there exists no a priori explanation for this discrepancy.

Indeed, it seems likely that sediment drifts are common features in active-margin settings, especially in intraoceanic island arcs where ocean currents pass through narrow passages between arc islands. One example of this occurs in the eastern Sunda arc of Indonesia, where a sediment drift is recognized at the trench-slope break of the collision-modified intraoceanic forearc (Fig. 1). Geologic and geophysical characteristics of this drift are defined by closely spaced seismic reflection and coring data.

Sunda arc sediment drift. The Sunda arc sediment drift is located in an elongate, roughly triangular-shaped slope basin that occupies a structural depression between Sumba Ridge and the Sawu-Timor Ridge (Fig. 1). A narrow, incised gap between Sumba and Sawu Islands breaches Sumba Ridge at 1150 m (3770 ft) water depth. This sill gap is elevated above the adjacent floor of the forearc basin to the northeast (Sawu Basin), and provides a mid- to deep-water connection between Sawu Sea and the Indian Ocean.

The sill gap is connected to the northeast apex of the slope basin via an erosional spillway (Fig. 1). Structural divergence of oblique-trending Sumba Ridge and Sawu-Timor Ridge produces a westward widening of this west-southwest–trending slope basin. This widening alleviates the bathymetric constriction and forms a catch basin downstream from the sill gap. A convex-upward slope of the basin sea floor forms a smooth, continuous mound with moat-like boundary channels around the margin. The boundary channels, outlined by the 1500-m (4900-ft) bathymetric contour, trend parallel to the axis of the basin at the base of the slope along Sumba Ridge and Sawu-Timor Ridge. The convex-upward sea floor is most pronounced in the northeast portion of the basin near the sill gap and diminishes to the west. Along the northern margin of the slope basin, the slope of Sumba Ridge is smooth and relatively free of major submarine canyons; consequently, apart from the erosional spillway associated with the sill gap, there are no bathymetric corridors for significant downslope transport of material from the ridge.

Seismic data show that the slope basin is filled with more than a kilometer (0.6 mi) of sediment (Fig. 2). The thickness of the basin sediment fill decreases to the west, where it merges with thin sequences of trench-slope sediment. Truncation of the moderately continuous internal seismic reflections of the sediment-drift sequence indicates erosion of specific areas prior to subsequent deposition. These complicated variations in reflection truncations and external body geometry are in marked contrast to the

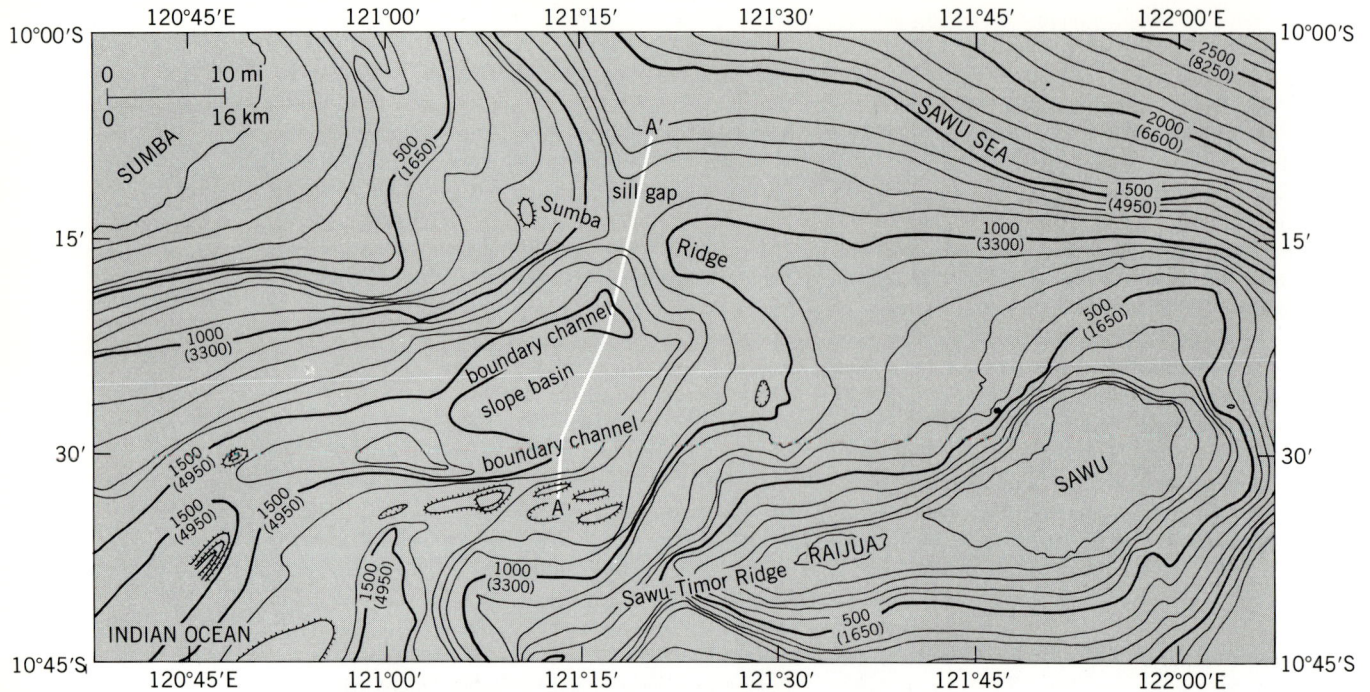

Fig. 1. Map of part of the eastern Sunda arc of Indonesia showing bathymetry and tectonomorphic features described in text. Sediment drift is located in the slope basin. A-A' shows the location of the seismic reflection profile collected with a watergun sound source. Depth contour intervals in 100 m (330 ft).

planar, continuous reflections commonly associated with basinal turbidites.

A piston core collected from the sediment drift (piston core 1, Fig. 2) suggests that the drift is built of late Quaternary reworked hemipelagic sediments interbedded with thin silt layers that appear to be classic contourites. These sediments represent accumulations of material eroded off Sumba Ridge by ocean currents. A piston core taken in the sill gap (piston core 2, Fig. 2) contains well-consolidated early Pliocene calcareous ooze, suggesting the erosion of as much as a kilometer (0.6 mi) of overlying material by ocean currents. A third piston core taken on Sumba Ridge near the erosional sill gap

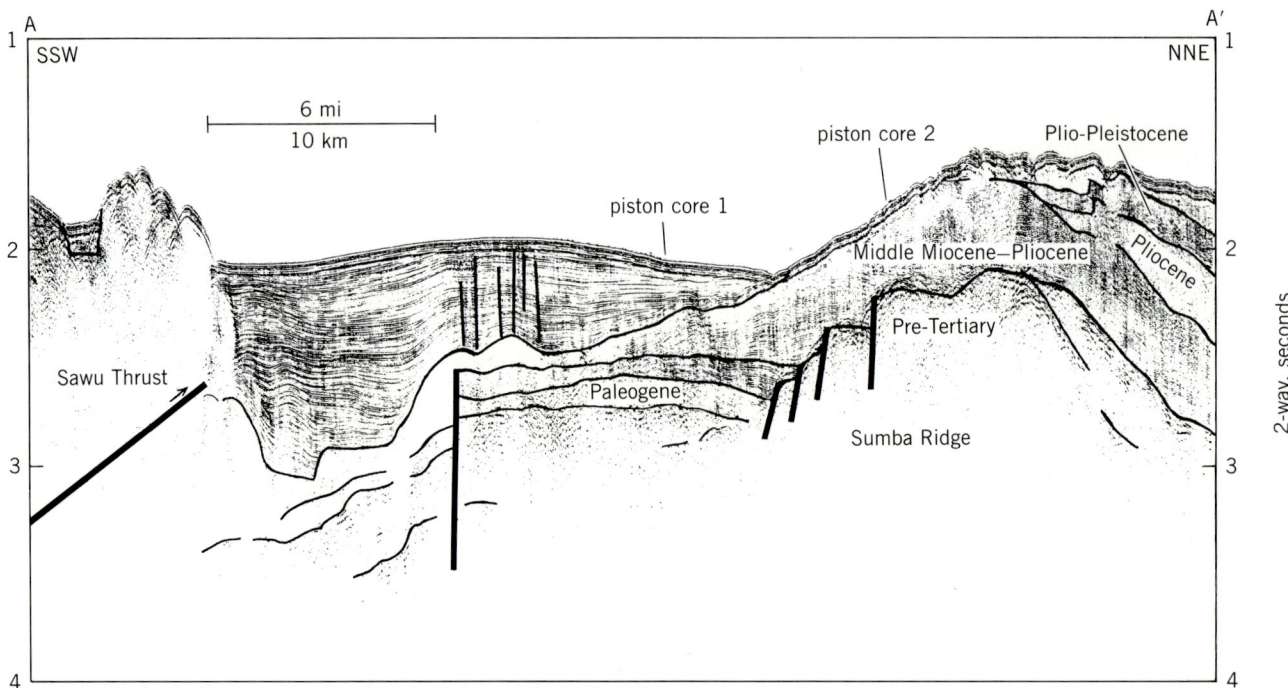

Fig. 2. Seismic reflection profile collected with watergun sound source on the R/V *Kana Keoki* across the slope-basin sediment drift, eastern Sunda arc, Indonesia. Location of seismic line A-A' is shown in Fig. 1. Vertical exaggeration, ×12.

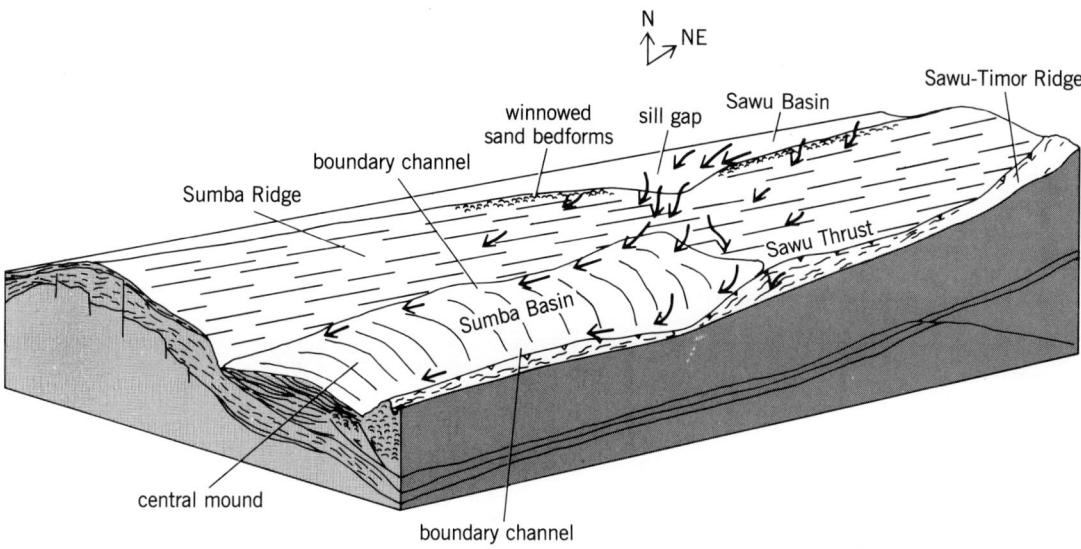

Fig. 3. Physiographic diagram modeling current circulation and sediment drift deposition in eastern Sunda arc, Indonesia.

contains current-winnowed, reworked foraminiferal sands. Seismic data indicate that these current-winnowed sands are configured in large (>30 m or 100 ft high) asymmetric sediment waves.

Physical oceanographic data suggest that Southern Intermediate Water and Pacific Ocean Deep Water are transported from the western Pacific, over the sill gap between Sumba and Sawu islands, and into the Indian Ocean. The restricted nature of the sill gap accelerates current activity through the gap, causing erosion of the ridge and transportation of the eroded material by outflow currents to the slope-basin sediment drift. A model of this depositional mechanism is shown in Fig. 3.

Likely depositional settings. Many factors influence the relative importance of abyssal circulation as a depositional mechanism in active margins. The most likely type of active-margin setting in which to find sediment-drift deposits are intraoceanic active margins located between adjacent ocean basins. Volcanic island arcs characteristic of this type of active margin provide a barrier to interbasinal water mass exchange. The small volume of the island arc allows mid- to deep-water masses to breach the arc through gaps in the volcanic edifices. Examples of this type of active-margin setting include the Marianas arc, the eastern Sunda arc, and the Aleutian arc.

Uplifted segments of complex active-margin forearcs provide further obstructions to current quiescence, and can steer and locally accelerate sluggish current flow patterns. The scale of this type of deep circulation accentuation depends on the extent of forearc disruption.

Sediment-drift deposits are more likely to occur along the western sides of the ocean basins or adjacent to major corridors for global thermohaline circulation. These intensified thermohaline currents will more effectively erode arc material (either the volcanic edifices or the submarine passages between them), and redeposit the fine-grained erosional detritus in sheltered regions away from the current-swept submarine slopes.

Although sediment drifts are not directly responsible for the formation of current-related accumulations, it is probably easier to recognize sediment drifts in immature arcs that lack the large volume of turbiditic detritus that might mask or bury drift deposits.

Sediment transport and deposition by ocean currents is certainly a more common depositional process in active margins than has been recognized in the past. With further oceanographic studies, the relative role that ocean current activity plays in active-margin sedimentation processes will become better defined. In addition, results of these oceanographic studies will be used by field geologists to identify similar sediment-drift deposits in the ancient rock record.

[AUDREY W. MEYER; DONALD REED]

Eolian sediments and processes. During the past 15 or 20 years, oceanographers have used three categories of data to address broad-scale questions about the history of climate. The categories include sedimentary microfaunal and microfloral assemblages, commonly followed by multivariate statistical or empirical mathematical analyses; the accumulation rate of the biogenic components such as calcium carbonate and opal in pelagic sediments; and the oxygen and carbon isotopic composition of planktonic and benthic foraminifera. From these data can be drawn reasonable inferences for past sea-surface temperature and salinity, the location of oceanographic frontal zones, sea-surface biological productivity, deep-ocean carbonate dissolution, the volume of continental ice cover, and the fractionation of carbon between the deep and shallow waters. More recently it has been shown that a fourth category of data can be compiled which allows definition of the nature of past atmospheric circulation, a primary climatic variable. This information comes from analyses of eolian dust deposited on the deep-sea floor.

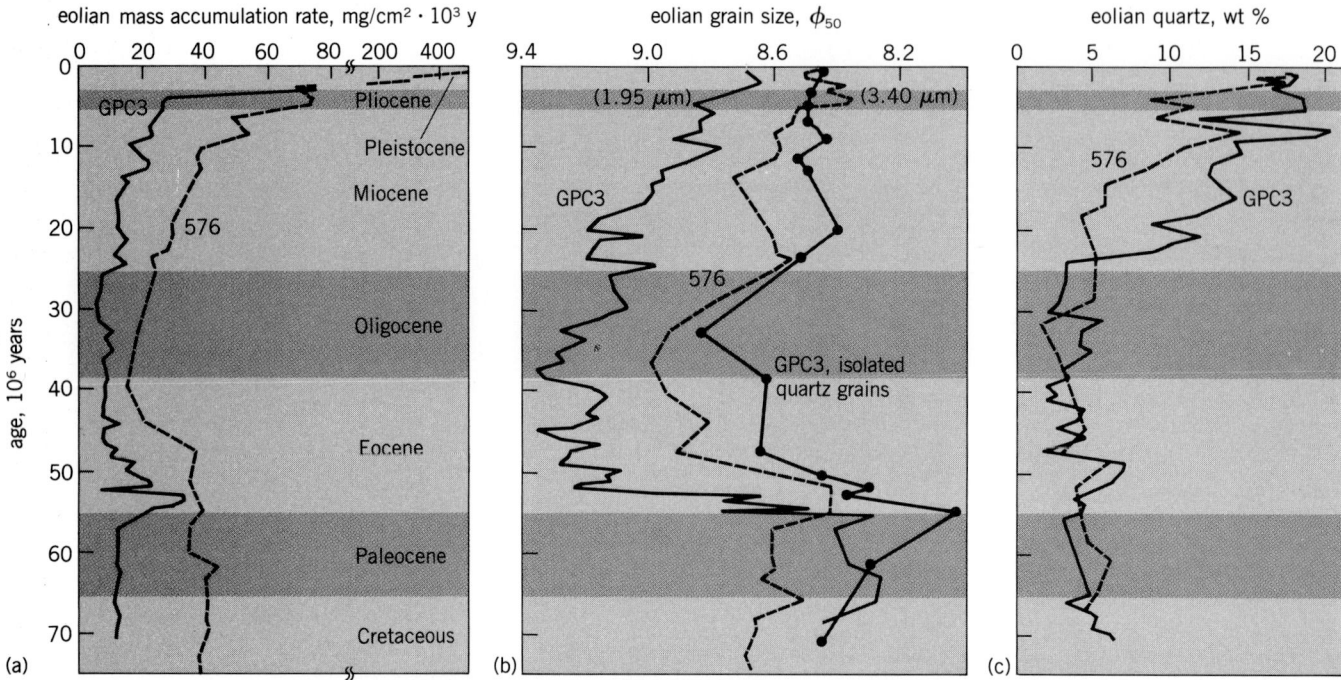

Fig. 4. A 70-million-year record of North Pacific eolian sediments, based on data from piston core LL44-GPC3 and DSDP Hole 576 (a) mass accumulation rate, (b) grain size, (c) mineralogy. θ_{50}-median grain size, the diameter of those grains at the 50th percentile of the total (100%) distribution.

Windblown dust is an important component of pelagic sediments. Dust grains are raised from arid and semiarid continental surfaces (commonly by spring dust storms), lifted to the upper troposphere, and transported by the zonal winds. Large grains quickly settle out of the air, but fine-grained minerals, generally less than 10 micrometers in diameter, remain suspended indefinitely and can travel global distances. These mineral grains are removed from the atmosphere and delivered to the sea surface by rainout.

Continentally derived dust is composed of illite and quartz, with lesser amounts of feldspar, kaolinite, chlorite, and smectite. Volcanic glass is also transported long distances by wind. Far from any continent—that is, seaward of the realm of hemipelagic deposition which extends hundreds of kilometers offshore and equatorward of the region of ice rafting—the mineral component of pelagic sediments is essentially all eolian. Eolian materials make up over 50% of the siliceous and pelagic (red) clays underlying the North Pacific westerlies, about 5% of the calcareous sediments beneath the northern tradewinds, and less than 0.5% of the calcareous sediments beneath the Southern Hemisphere tradewinds.

Observations. The present pattern of eolian deposition as determined by the mineralogy of surface sediments in the Pacific reflects the east-west zonal winds, even where the boundary currents of oceanic gyres flow north or south. This observation illustrates how rapidly small particles are removed from the sea surface by the combined effects of vertical transport in fecal pellets and the settling of large amorphous aggregates, thus preserving sea surface input patterns in the sediments of the ocean floor.

Three types of data can be derived from the eolian component extracted from pelagic sediments: mineralogy, mass accumulation rate, and grain size. Dust mineralogy is determined by the regional geology of the eolian source region, and can be used to characterize or identify those sources. The flux of dust to the sea floor, or the mass accumulation rate, commonly expressed in units of $mg/cm^2 \cdot 10^3 y$, is a quantitative measure of dust supply, and reflects the relative aridity of the source; more-arid regions supply more dust. The grain size of the dust that is in equilibrium with the transporting wind—that is, dust usually more than 2000 km (1200 mi) downwind from the source—provides a quantitative estimate of the intensity of the zonal winds. This is achieved by calculating Stokes settling in air and the lifting motions of atmospheric turbulence which depend on wind speed, and comparing the results. Eolian mass-accumulation-rate values for the Holocene are in excess of 1000 $mg/cm^2 \cdot 10^3 y$ in the Northwest Pacific westerlies, 10–40 $mg/cm^2 \cdot 10^3 y$ in the north equatorial region, and only 1–3 $mg/cm^2 \cdot 10^3 y$ beneath the Southern Hemisphere tradewinds, a range of three orders of magnitude. The diameter of the eolian grains (median grain size) now accumulating in the surficial sediments is 2.4–3.0 μm in the Northern Hemisphere and 5–7 μm in the Southern Hemisphere.

Interpretation. Two studies illustrate the types of information that can be derived from the eolian component of pelagic sediments. The first is a long-term (70 million years) record of eolian sediment yield to the North Pacific. Detailed information exists from two apparently continuous cores of pelagic clay that span the entire Cenozoic: LL44-GPC3, a pelagic clay core raised from 30.3°N, 157.8°W, and Deep Sea Drilling Project Site 576 from 32.4°N, 164.3°E. These cores are separated by 3600 km (2160 mi). The temporal pattern of mineralogy, mass flux, and grain-size variations from Site 576 and GPC3 match closely, illustrating the broad-scale uniformity of eolian processes in the North Pacific over the past 70 million years (Fig. 4). Site 576, closer to Asia, exhibits a greater eolian mass accumulation rate and slightly coarser grain diameter.

These data show that large-scale changes in the atmospheric circulation factors controlling dust transport have occurred during the past 70 million years. An important result of this work was documentation of a sudden decrease in the intensity of atmospheric circulation that occurred during the early Eocene. Both cores show that a significant reduction in eolian grain size occurred then (Fig. 4b). Atmospheric circulation remained less intense until 35–40 million years ago, when wind intensity began a long-term overall increase, culminating in the maximum associated with the Plio-Pleistocene glaciations. The post-Eocene data agree with the present understanding of Cenozoic polar cooling and evolution of the cryosphere beginning about 38 million years ago in the Antarctic. Similarity of the two grain-size curves for the Paleocene and latest Cretaceous implies that atmospheric circulation at that time was as vigorous as that of the past few million years. This conclusion is contrary to the general impression of the Cretaceous as a time of warm and equitable climates, characterized by sluggish ocean circulation.

Eolian mass-accumulation-rate values were smallest during the middle Tertiary, a time of relatively humid continental climates. Dust transport began to increase during the early Miocene, and during the Plio-Pleistocene it increased by up to an order of magnitude (Fig. 4a) with the onset of Northern Hemisphere glaciation, emphasizing the general correlation between glacial ages and global aridity.

The second data set reveals the history of atmospheric circulation and dust flux during the past several glacial and interglacial cycles. Piston core KK75-02, raised from 5465 m (17,930 ft) at 38.6°N, 179.3°E, contains a 750,000-year record of this information. Eolian mass-accumulation-rate values average about 400 mg/cm² · 10³y and show relative maxima approximately every 110 kiloyears (Fig. 5). In general, times of accumulation maxima correspond to interglacial periods. The similarity of the eolian mass-accumulation-rate pattern to the 105-ky ice-volume signal is not surprising as the advance and retreat of continental ice would dominate the source area climate in mid to high latitudes. Eolian flux in Deep Sea Drilling Project Hole 503B from the eastern equatorial Pacific, 4.0°N, 95.6°W,

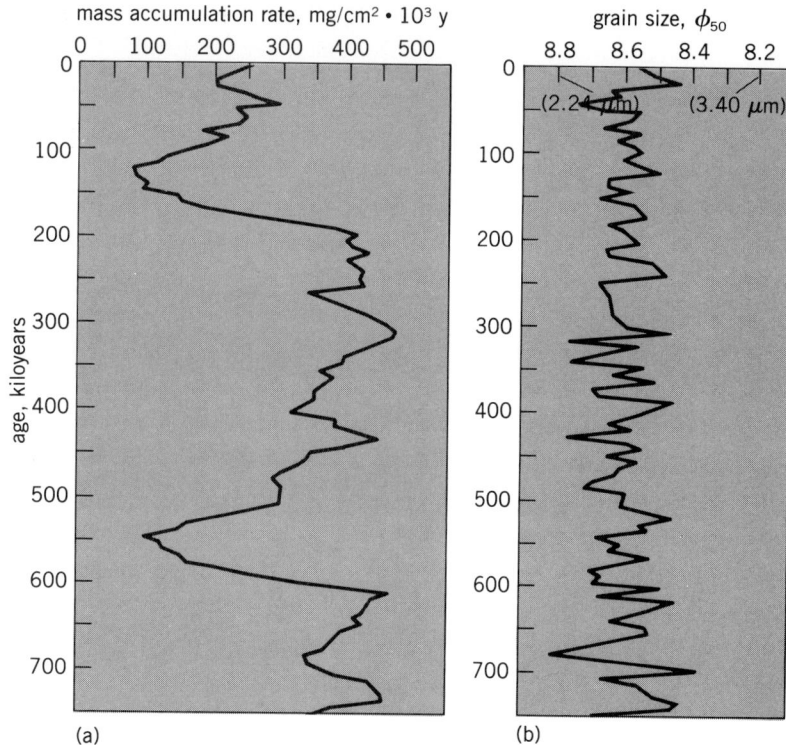

Fig. 5. A record from the North Pacific westerlies, core KK75-02, of quaternary fluctuations in (a) eolian mass accumulation and (b) grain size. θ_{50}-median grain size, the diameter of those grains at the 50th percentile of the total (100%) distribution.

also indicates arid interglacials and more humid glacial periods in Central American and northern South American source regions.

The grain size of eolian dust in KK75-02 (Fig. 5a) shows high-frequency fluctuations, which indicate an average change in wind intensity of 22%. The amplitude of the inferred wind fluctuations changed markedly about 250 ky ago from 27% before that time to 17% more recently. Spectral analysis of the KK75-02 eolian grain-size data reveals three significant peaks, corresponding to 104, 41, and 23 ky. These periods correspond to the orbital cycles calculated by M. Milankovitch of eccentricity, obliquity, and precession, and document that atmospheric circulation responds to orbitally forced stimuli as do other paleoclimatic proxy indicators.

These and other studies demonstrated the long-suspected importance of eolian transport to the deep sea, and have provided the beginning of an understanding of how atmospheric circulation differs with geologic time and geographic location.

[DAVID K. REA]

Bibliography: D. A. Gillette, Production of dust that may be carried great distances, in T. L. Péwé (ed.), *Desert Dust: Origin, Characteristics, and Effect on Man*, Geol. Soc. Amer. Spec. Pap. 186, pp. 11–26, 1981; W. Hamilton, *Tectonics of the Indonesian region*, U.S. Geol. Surv. Prof. Pap. 1078, 1979; C. D. Hollister and B. C. Heezen, Geologic effects of ocean bottom currents: Western North Atlantic, in A. L. Gordon (ed.), *Studies in Physical*

Oceanography, vol. 2, pp. 37–66, 1972; T. R. Janecek and D. K. Rea, Eolian deposition in the northeast Pacific Ocean: Cenozoic history of atmospheric circulation, *Geol. Soc. Amer. Bull.*, 94:730–738, 1983; M. Leinen and G. R. Heath, Sedimentary indicators of atmospheric activity in the Northern Hemisphere during the Cenozoic, *Palaeogeogr. Palaeoclimatol. Palaeoecol.*, 36:1–21, 1981; D. K. Rea, M. Leinen, and T. R. Janecek, Geological approach to the long-term history of atmospheric circulation, *Science*, vol. 227, 1985. J. B. Sangree and J. M. Widmier, Interpretation of depositional facies from seismic data, *Geophysics*, 44:131–160, 1979; K. Wyrtki, Physical oceanography of the southeast Asian waters, *Scientific Results of Marine Investigations of the South China Sea and the Gulf of Thailand 1959–1961*, Scripps Institution of Oceanography, La Jolla, California, NAGA Rep., vol. 2, 1961.

Medical imaging

Nuclear imaging in medicine encompasses one new technique and two older ones.

Nuclear magnetic resonance (NMR), now also called magnetic resonance imaging (MRI), is a relatively new diagnostic technique based on the precessional properties of the tiny magnetic moments of atomic nuclei. Hydrogen nuclei in tissue are magnetic and can absorb stimulating electromagnetic radiation. Subsequent emissions by such nuclei are detected and can be made to form images.

Positron emission tomography (PET) is an older technique that provides images by identifying the sites of positron-emitting radionuclides introduced in the body. When positron decay occurs, each positron quickly annihilates close to where it is stopped, and emits two simultaneous back-to-back (180°) gamma rays. A straight line in space is defined by this 180° pattern, which includes the small volume, almost a point, in which the positron annihilates, and passes through the positron-bearing organ. The intersections of these lines are "points" which delineate the image of that organ.

Nuclear medicine imaging techniques belong to a field of considerable maturity. Gamma-emitting radionuclides are introduced in tissue or organs to be viewed, and they have gamma energies specifically recognized by the detector devices in the form of total energy peaks. In this way, background noise is lowered very substantially, and good images are produced by scanning or by cameralike detectors.

Magnetic resonance imaging. MRI has been introduced in very recent times, but its success has been remarkable. It would not be unreasonable to expect that this new technique could become the preferred method of medical imaging because of its excellent image detail and contrast, nonuse of ionizing radiation, absence of any injurious effects to patients, and its noninvasive nature. There is no doubt that MRI is already competitive with the best imaging techniques provided by x-ray computerized tomography (CT). MRI probably represents the biggest breakthrough in medical imaging since the discovery of x-rays by W. Röntgen in 1895.

The MRI method is most commonly used in imaging the presence of hydrogen through detecting the nuclear magnetic dipole moment of the proton, the nucleus of the hydrogen atom. Other nuclei may also be viewed, such as ^{7}Li, ^{13}C, ^{19}F, ^{23}Na, and ^{31}P, although the intensities of the observed signals are much weaker for these nuclei. Probably phosphorus and sodium will assume greater importance as technique improves.

[ROBERT HOFSTADTER]

The principles underlying MRI were first enunciated in 1973 by Paul Lauterbur and Peter Mansfield. The potential application of this noninvasive, hazard-free technique to diagnostic medicine was clear from these early demonstrative studies. Exploitation, however, required the experiment to be scaled up over two orders of magnitude, from the test tube–sized samples of conventional NMR instruments to human whole body dimensions. This goal was first achieved by Raymond Damadian with the publication of the first crude human thoracic image in 1977. Industrial interest was triggered, investment was large, and subsequent progress rapid. Limited Food and Drug Administration approval was granted for head and whole-body studies in 1983; commercial MRI instrumentation is currently available.

Principles. NMR is based on spin properties of atomic nuclei. Since nuclei carry a positive charge, this spin has associated with it a current loop and hence a magnetic moment. When placed in a static magnetic field, these nuclear magnetic moments have a tendency to align themselves with the field. However, just as in the electronic case, the magnetic, or Zeeman, energy is quantized and only certain states are permitted. For a spin-½ nucleus such as the proton, there are two possibilities, corresponding to alignment parallel or antiparallel to the field direction. Transitions between these levels can be induced by application of electromagnetic radiation whose frequency exactly matches the energy splitting; hence, this is a resonance phenomenon. This frequency is given by the Larmor equation, $\omega = \gamma B$, where ω is the angular (Larmor) frequency, B is the magnetic field strength, and γ is a constant known as the magnetogyric ratio. The γ is a small quantity or, put another way, the interaction is a weak one. This has the advantage of safety since, for practically attainable field strengths, frequencies correspond to the nonionizing radio-frequency region of the electromagnetic spectrum. However, the weakness of the interaction also means that the difference in populations of the energy levels is low, and the method therefore suffers from a fundamental lack of sensitivity. This can be alleviated in part by operating at high magnetic field strengths. In conventional NMR instruments, for example, field strengths up to 14 tesla (140,000 gauss) are used. For human imaging applications, fields in the range

0.05–2.5 T (500–25,000 G) are the norm, the upper limit being dictated by a combination of practical and safety considerations.

In order to detect the NMR signal, a short intense pulse of radio-frequency irradiation is applied. This has the effect of deflecting the bulk magnetization, arising from the excess of spins parallel to the magnetic field direction, into an orthogonal plane where it precesses at the Larmor frequency and can be detected by the electromotive force induced in a receiver coil. The method of spatial encoding is based on the Larmor expression. If the applied magnetic field B is made to vary in space, the frequency distribution or NMR spectrum will reflect this same spatial variation. The simplest dependence that can be imposed is a linear one, and this is what is used in practice.

If an object is placed in such a magnetic field, planes of spins which are at right angles to the gradient direction will resonate at the same frequency, and the spectrum will be a one-dimensional projection of the object normal to this direction. By viewing the object in different gradient orientations, a series of projections are built up which can be combined by using a suitable algorithm, for example, the filtered back-projection method, to yield an image. This technique of projection reconstruction was initially attractive because of its obvious similarity to reconstruction methods used by x-ray computerized tomography instruments. It has now been largely superseded by Fourier-based techniques which employ a static field gradient in one dimension and gradients which are applied for varying times or, more effectively, whose amplitudes are varied in a stepwise linear fashion to achieve spatial resolution in the other dimensions. The serial recording of spectra obtained in different gradients leads to a relatively long imaging time, typically in the range 30 s–5 min. More efficient spatial encoding schemes are available, though they are technically more demanding. One such is the echo-planar imaging technique of Mansfield. By using this method, a complete image can be obtained in a time as short as 10 ms. This has enabled remarkable real-time studies of cardiac motion to be performed.

Image contrast. Many factors determine the contrast in an NMR image. The signal is directly proportional to the number of contributing nuclei, so that a proton image would be expected to reflect differences in proton density. Protons come in many different guises, but only those which are in mobile environments give rise to NMR signals which are sufficiently prolonged to be observed. For practical purposes, this means the protons of water and, to a lesser extent, those of mobile lipids. On average, the human body contains about 70% water, with a 15% variation between the different soft tissues.

Other factors, however, are also important in determining contrast. When the bulk magnetization is perturbed by the excitation pulse, it recovers toward its equilibrium value in an exponential fashion with a time constant T_1 known as the spin-lattice relaxa-

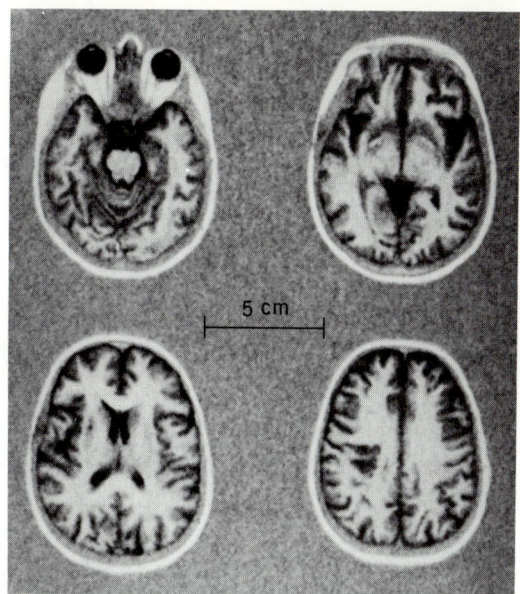

Fig. 1. Multislice magnetic resonance imaging (MRI) axial head scans (slice thickness 0.25 in. or 6 mm) obtained at a frequency of 21 MHz using an inversion recovery method with time interval of 400 ms. (*Siemens Ltd., 1984*)

tion time. T_1 values show rather large tissue variation, perhaps 200–300% depending on the frequency, and are thus a valuable source of tissue contrast. The variation reflects the state of the tissue water, in particular the relative proportions of free water to that bound to macromolecular structures. Pulse sequences have been designed to exploit T_1 contrast. In the inversion recovery method, for example, the magnetization is first inverted and then inspected after a certain time interval. Different tissues recover at different rates and so give rise to signals of different strength.

The NMR signal itself decays with a time constant T_2 known as the spin-spin relaxation time. This reflects the strength of the internuclear interactions which give rise to variations in the local magnetic field. Pulse sequences such as the spin-echo method have been designed to elicit T_2-dependent contrast. Generally, however, images will simultaneously display elements of spin-density, T_1, and T_2 contrast in differing proportions, depending on the nature of the imaging sequence. Attempts at standardization have been made based on absolute determinations of these parameters, but success has been mitigated by difficulty of measurement, frequency dependence of T_1, and the multicomponent nature of the T_2 process.

Considerable effort has been expended in the development of contrast-enhancing agents. They all contain unpaired electrons and rely on the relative strength of the electronic magnetic moment (some three orders of magnitude greater than its nuclear counterpart) to increase the NMR relaxation rates. The gadolinium ion, with seven unpaired electrons, is particularly attractive and has proved an effective intravenous contrast agent at dose levels of 0.1 millimole kg^{-1}. In the head, it gives rise to a characteristic ring enhancement of tumors at sites of blood-brain barrier breakdown.

Clinical applications. The full diagnostic potential of MRI has yet to be realized. Clinical experience is, however, being rapidly accumulated; many hospitals can claim in excess of 1000 patient examinations. In the majority of cases, these have been head studies, where the ability of MRI to discriminate between white and gray matter is particularly remarkable (Fig. 1). As a general rule, most lesions are associated with a prolongation of the relaxation times, both in the lesion itself and in the surrounding edema. The white-gray discrimination is especially useful in cases of demyelinating diseases, and MRI has become the method of choice for the diagnosis of multiple sclerosis. It has also been successfully applied to studies of stroke where changes can be seen as early as a few hours following the onset of symptoms: vascular lesions of various types, trauma, and, in some, organic brain diseases.

One of the key applications that provided the stimulus for much of the early development is the detection of tumors. MRI is particularly useful in the region of the posterior fossa since it does not suffer from bone artifacts as does x-ray computerized tomography. The unique ability to switch between axial, sagittal, coronal, or intermediate plane orientations under electronic control is an invaluable aid in establishing the full extent of lesions. Whole-body studies are also gathering momentum with cardiovascular applications foremost. Exquisite visualization of cardiac anatomy is available from electrocardiogram-gated images. The principal vessels are generally well delineated due to the lack of signal from flowing blood, enabling screening for the presence of atherosclerotic plaques. Most of the other principal organs have also been studied.

It is important to exclude patients with ferromagnetic surgical implants, such as certain types of aneurysm clip, from MRI examination. In addition, those fitted with cardiac pacemakers not only should be excluded from examination but also must be kept outside the 5-gauss (0.5-millitesla) field contour, since fields substantially greater than this can switch these devices into a synchronous mode of operation. With the exception of these patients, MRI promises to be a safe and effective alternative to the use of x-ray computerized tomography for medical diagnosis.

[PETER G. MORRIS]

Positron emission tomography (PET). The PET technique was invented by G. L. Brownell in 1953 and lately has undergone a strong resurgence because its capability for resolving small volumes in organs has improved, and also because functional behavior, such as metabolic activity within the brain, for example, can be observed in addition to structural detail. Although the figure of merit for resolution may be something less than 0.4 in. (1 cm) in a given direction and therefore is not as good as MRI, where 0.04–0.08 in. (1–2 mm) may be observed in immobile tissue, the ability to study physiologic or metabolic function is probably better than in any other noninvasive method. This advantage arises from the capability of PET of choosing radioelements from almost anywhere in the periodic system, and from the extraordinary sensitivity of this method to the imaging of small volumes identified by specific pharmaceuticals. Dynamic studies involving short time intervals are becoming increasingly possible, and this may be another advantage of the PET modality.

In most PET facilities one or more rings of gamma-ray detectors, made of small scintillator crystal modules, surround the patient undergoing examination. Since the two annihilation gamma rays, each of energy 512 keV, are emitted simultaneously, timelike coincidences in the detectors are requirements for acceptable signal events. By using an algorithm associated with back-to-back (180°) emission, the regions of intersection of acceptable

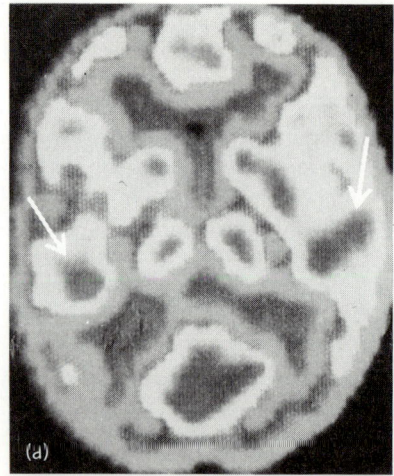

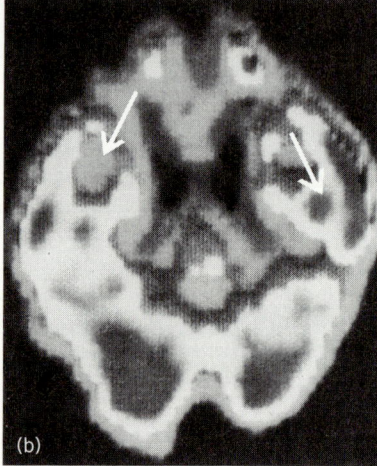

Fig. 2. Positron emission tomography (PET) images of the brain of a young man who was asked to concentrate on a Sherlock Holmes story: (*a*) while listening to story, and (*b*) during memory processing. (*From J. C. Mazziotta et al., Tomographic mapping of human cerebral metabolism: auditory stimulation, Neurology, 32:921–937, 1982*)

events may be correlated by computer techniques to construct a clear image of the positron-bearing organ. In this way it is possible to look for normal or abnormal behavior. In an attempt to improve signal-to-noise ratio, PET methodology has borrowed a time-of-flight measuring procedure used heretofore only in sophisticated nuclear physics studies.

Studies have shown how to exhibit small regions in the brain which are activated, or "light up," metabolically speaking, by body activity or even mental effort. Figure 2 shows PET images of a normal subject's brain while he is listening to a Sherlock Holmes story and asked to remember specific details of the story. The images are obtained by injecting deoxyglucose labeled with the positron-emitting isotope ^{18}F, which allows for the quantitative measurement of glucose utilization in the living human brain. The dark areas indicate increased glucose utilization. In Fig. 2a, high levels of glucose utilization are seen on both the left and right side of the brain in the areas of the auditory cortex (indicated by arrows) and reflect the fact that the subject is listening to verbal information. Since the subject's eyes are open, the visual cortex at the bottom of the image also demonstrates a relatively increased level of glucose uptake. Figure 2b demonstrates increased glucose utilization in the middle portion of the temporal lobe in areas including the hippocampus and parahippocampus (indicated by arrows). These structures are thought to be important in processing of memory tasks and may reflect this component of the subject's study.

Nuclear medicine imaging. Nuclear medicine imaging was developed by B. Cassen in 1950, and has evolved into a valuable technique of medical diagnosis. It is concerned with the noninvasive study of dysfunction in internal organs by using gamma rays emitted from radioelements introduced in the body and captured by scintillation detectors. For this reason the subject is sometimes called scintigraphy. Early studies concentrated on thyroid images using ^{131}I. The original devices were scanning instruments that used single or multiple detectors, but current modalities use large crystal assemblies that behave as "cameras." H. Anger did the pioneering work in this field in 1957, using planar slabs of thallium-activated sodium iodide [NaI(Tl)], viewed by photomultipliers placed in a lattice type of array. This formulation is known as the Anger camera, and is used in almost every major hospital or medical care center.

The resolution attainable by this method is better than 0.4 in. (1 cm). Tomographic studies can be made by translation and rotation of the camera around the patient. Figure 3a–d shows four images of the liver taken at 15-min intervals after administration of technetium aminodiacetic acid to a normal individual. The radioelement in this case is ^{99m}Tc. Bile is being sent into the gall bladder and then into the gut in normal fashion, only a little showing in the gut at the time of exposure. Figure 3e–h shows an acute cholecystitis, and a gallstone blocks the

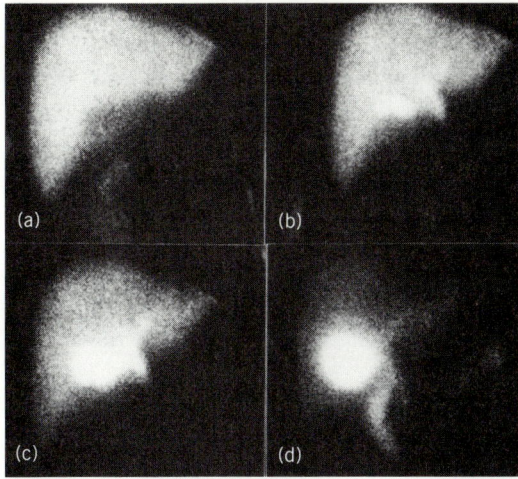
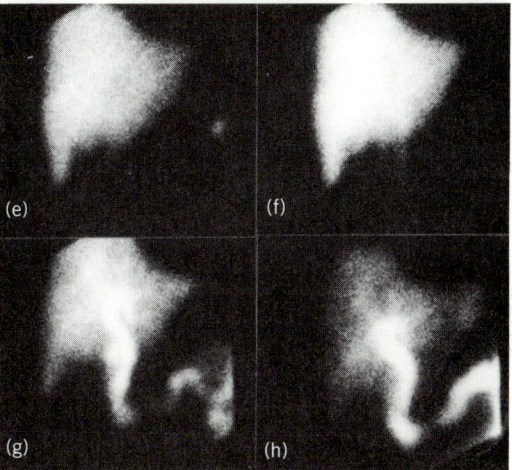

Fig. 3. Liver and gall bladder behavior observed with nuclear medicine imaging techniques. (a–d) Images taken at 15-min intervals showing normal behavior. (e–h) Abnormal case representing a gallstone blockage. (*Ross McDougall, Stanford University Hospital*)

duct between the liver and the gall bladder so that the gall bladder does not fill. The differences between the two sets of images are readily apparent. With larger scanning cameras, whole-body images may be observed and are used to study bone diseases.

For background information see COMPUTERIZED TOMOGRAPHY; GAMMA-RAY DETECTORS; NUCLEAR MAGNETIC RESONANCE (NMR); NUCLEAR MEDICINE; RADIOACTIVE TRACER; RADIOISOTOPE (BIOLOGY) in the McGraw-Hill Encyclopedia of Science and Technology.

[ROBERT HOFSTADTER]

Bibliography: G. L. Brownell et al., Positron tomography and nuclear magnetic resonance imaging, *Science*, 215:619–626, 1982; P. C. Lauterbur, Image formation by induced local interactions: Examples employing NMR, *Nature*, 242:190–191, 1973; *Magnetic Resonance in Medicine* (journal), Society of Magnetic Resonance in Medicine; P. G. Morris, *NMR Imaging in Medicine and Biology*, 1985; I. L. Pykett et al., Principles of nuclear magnetic resonance imaging, *Radiology*, 143:157–168, 1982; M. M. Ter-Pergossian, M. E. Raichle, and B. E. Sobel, Positron emission tomography, *Sci. Amer.*, 243(4):170–181, 1980; J. S. Waugh (ed.), *Advances in Magnetic Resonance*, suppl. 2, 1982; P. N. T. Wells (ed.), *Scientific Basis of Medical Imaging*, 1982.

Meter (unit)

Since the early 1970s, frequency-stabilized lasers have proved to be outstanding tools for length measurement. One of their noteworthy contributions to science was to enable a determination of the value of the speed of light c 100 times more accurate than any previous measurement. This was performed by measuring both the frequency f and the wavelength λ of the radiation emitted in the near-infrared by a laser whose frequency was stabilized on an absorption line of methane. By using the elementary relation $c = \lambda f$, a value of c was deduced equal to 299 792 458 ± 1.2 m/s. While the most difficult part of the experiment was to measure f, it could be performed with a high accuracy. The less accurate part of the experiment was the measurements of λ. As the accuracy of the latter was limited essentially by the realization of the meter, this initiated the process that led to the change of definition of the meter.

Need for redefinition. When the first laser determination of the speed of light was carried out in 1972, the best reference available to measure λ was the wavelength of a spectral line of radiation emitted by a standard discharge lamp filled with krypton-86. The definition of the meter had been based on this standard since 1960.

The frequency reproducibility of the stabilized laser used in these measurements had been checked by the optical beats method and was better than a few parts in 10^{11}. This guaranteed a similar reproducibility of the wavelength. However, there was no reason to expect that the measurement of the wavelength could be performed with an accuracy definitely better than 1 part in 10^8, the reproducibility expected from the krypton lamp. Repeated measurement of λ showed a reproducibility of a few parts in 10^9, a little better than expected but still limited by the krypton lamp and not by the laser.

There was no hope of significantly improving the performance of the krypton-86 radiation. Emitted by a conventional discharge lamp, this radiation was subject to Doppler broadening, first-order Doppler and pressure shifts, and so forth. In contrast, the expected wavelength reproducibility of a few stabilized lasers was better than 1 part in 10^{10}. To measure such a wavelength efficiently, a standard reproducible within better than 1 part in 10^{10} was required. Obviously, a better reference than the krypton radiation standard was needed.

Alternatives. It was tempting to choose the wavelength of one of these stabilized lasers as a new standard and to base upon it a new definition of the meter that would have been very similar to the former one. But this raised the question of which stabilized laser should be chosen. None of them is likely to maintain prominence for more than a few years, and it is not desirable to change such a standard each time a new device supersedes the performances of the former one.

Another approach was to use the speed of light itself as a standard defining the meter. The speed of light is a fundamental constant of physics. As such, it intervenes in many relations between physical quantities. Conversely, the reliability and coherence of these relations give confidence in the everlasting and universal character of the speed of light. This is exactly what is expected from a standard.

The measurement of c had demonstrated two points. First, a better reference, in other words a new definition of the meter, was needed. Second, instead of measuring λ in terms of the standard wavelength of krypton-86, it was more efficient to assign a conventional value to c and deduce the value of λ from the measured frequency f and the elementary relation $\lambda = c/f$.

Fixing a conventional value for the speed of light would have been an indirect, rather esoteric way of redefining the meter. It seemed to be more sensible to redefine the meter in such a way that a fixed, conventional value of c be implied in the definition itself.

Final steps. Various wordings of a definition along these lines were proposed. They were discussed in many national and international bodies whose advice was collected by the Comité International des Poids et Mesures (CIPM) and its Comité Consultatif pour la Définition du Mètre (CCDM). All these wordings were recognized to be equivalent from the scientific point of view. Finally, the CCDM and CIPM expressed their preference for the wording that appeared the best from the viewpoint of clarity and simplicity. Following their advice, the 17th Conférence Générale des Poids et Mesures (1983), which is the highest authority in these matters and brings together representatives of the governments of all countries adhering to the Convention du Mètre, decided to abrogate the former definition of the meter and to replace it by the following (translated): "The meter is the length of the path traveled by light in vacuum during a time interval of 1/299 792 458 of a second."

Among the consequences of this new definition, the most obvious is that the speed of light becomes exactly equal to 299 792 458 m/s. This stems from the conventional meaning attached to the symbol m, representing the meter.

Continuity of measurements. One may wonder why such an ugly number has been retained. As the definition of the meter is a matter of international agreement on some arbitrary choice, why not take this opportunity to round the value of c to 300 000 000 m/s? The answer is simple. Doing so would reduce the size of the meter by some 0.7 mm, almost 1 part in 1000. Sizes specified in "new meters" would differ from sizes specified in "old meters" by an amount that would become significant as soon as some precision was required. All precise data involving the meter would become ambiguous, all precision mechanical measuring instruments would have to be changed, and all mechanical industries would be endangered. Such a change would, to say the least, be unwieldy.

This illustrates a general rule for a change in the definition of any unit: the size of the unit should not change; the new definition should define the same quantity, within the uncertainty of the best realization of the former definition.

This rule ensures the continuity of measurements. It has been applied to the meter since its origin. The meter has been successively defined in terms of the terrestrial meridian (1791), the "Mètre des Archives" (1799), the "Prototype international du mètre" (1889), the standard wavelength of krypton-86 (1960), and now the speed of light (1983). Each of these definitions is compatible with the former one, within the accuracy available at the time of the change.

The price to pay for continuity is either careful adjustment of the new material standard (1799, 1889) or choice of a proper, usually unattractive, number to be included in the new definition (1960, 1983).

Realization. The new definition of the meter seems to suggest that length should be measured by using the time-of-flight method that is in current use for satellite ranging. The time interval t needed for a light pulse to travel the unknown distance l is measured and l is deduced from the relation $l = ct$ (or $2l = ct$ if the distance is traveled in both directions). This method works quite well for long distances in free space. It is not yet usable for laboratory or workshop distances.

In laboratory practice, the typical instrument for accurate measurements of length is the Michelson interferometer. It allows measurement of the unknown length l by means of its ratio p to the wavelength λ of a monochromatic radiation, for instance, the radiation emitted by a stabilized laser. Assuming that the frequency f of the laser is known, its wavelength is $\lambda = c/f$ and the unknown length l may be expressed as $l = pc/f$.

Actually, the wavelength of any wave is defined as the path traveled by the wave during one period $T = 1/f$, hence the relations $\lambda = cT$ and $l = pcT$ are valid. Thus the interferometric method is just a time-of-flight method in which the wave itself is used as a clock to measure the duration pT of the travel: the "number of wavelengths" p is also the "number of periods" needed to travel the distance l.

Measuring the frequency f of a laser is not an easy task. Between the standard cesium frequency defining the hertz, around 10 GHz, and the visible frequencies, around 500 THz, the ratio is of the order of 50 000. This gap has been bridged by few laboratories. Actually, up to now, only one laboratory has pushed this exercise of "frequency synthesis" up to visible frequencies, the National Bureau of Standards at Boulder, Colorado. This achievement played a key role in the decision to promulgate the new definition. It had been backed by wavelength comparisons between visible lasers and infrared lasers whose frequency was well established. This last method is easier, but less powerful in terms of accuracy.

Whatever the method used, measuring the frequency of a laser in the visible or near-infrared cannot be considered as routine work, nor is it likely to become so for some years to come.

Fortunately, a few stabilized lasers are reliable enough to reproduce the same frequency, hence the same wavelength, in the visible, within a few parts in 10^{10} or 10^{11}. Their frequency or wavelength has been measured, and need not be measured again each time they serve as a reference. They may be used as "secondary wavelength standards," although there no longer exists a primary one. Values of their frequency and wavelength have been recommended by the CIPM, together with an estimate of the uncertainty attached to their realization, assuming good laboratory practice. The same recommendation also allows the use of the conventional discharge lamps (krypton-86, mercury-198, cadmium-114) provided that a larger uncertainty, of the order of a few parts in 10^9 or 10^8, may be tolerated.

Benefits of redefinition. Thus the new definition entails some major benefits. First, more accurate measurements are possible. Second, they are not linked to the use of a unique, specified standard; any laser whose frequency can be accurately measured may be used as a wavelength standard. Third, this way of operating relies upon the frequency standard, the cesium-beam standard, whose present accuracy of realization is of the order of a few parts in 10^{14}; some room is left for progress. Future progress in stabilized lasers and in the measurement of their frequency will benefit the accuracy of length measurements without requiring a further change in the definition of the meter.

For background information *see* INTERFEROMETRY; LASER; LASER SPECTROSCOPY; PHYSICAL MEASUREMENT in the McGraw-Hill Encyclopedia of Science and Technology.

[P. GIACOMO]

Bibliography: Bureau International des Poids et Mesures, *Comité Consultatif pour la Définition du Mètre*, 7th session, 1982; Documents concerning the new definition of the metre, *Metologia*, 19:163–177, 1984.

Microbial interference

Microorganisms simultaneously compete with and depend upon other microorganisms for their survival. Pure populations of microorganisms are found only in cultures; in nature, mixed populations predominate. Because of frequent overlapping of growth requirements in such populations, competition between the constituents is inevitable. The mechanisms by which microorganisms compete for carbon and energy sources and essential chemicals are varied and complex, and involve direct antagonism of one microorganism by another, as well as mutually beneficial synergistic arrangements between microbial species. Interest in such relationships derives, in large part, from the fact that each has the potential for enhancing or impairing health under appropriate conditions.

There are both direct and indirect mechanisms by which one microorganism inhibits growth of another. The most familiar example is the elaboration of antibiotics by various fungi and bacteria. The many other mechanisms are in evidence throughout the environment, but particularly so among the complex microbial populations that constitute the normal flora of the human skin and mucosal surfaces. Largely as a result of such mechanisms, certain species of microorganisms predominate on these body surfaces, while others are noteworthy for their absence.

Direct inhibition. One of the best-characterized direct mechanisms of microbial interference is the release of chemical substances that have toxic or inhibitory effects on other microbial populations. It was precisely this phenomenon that A. Fleming explored in his work leading to the isolation of penicillin. This form of competition is widespread in nature as well as among the populations that compose the normal flora of human body surfaces. In the latter situation, bacteriocins are the toxic chemicals. These are bacterially derived, proteinacious antibiotics that have a high molecular weight and—in contrast to classical antibiotics—generally have limited spectra of activity. Their production by resident viridans streptococci (α-hemolytic streptococci) is thought to represent an important mechanism by which colonization of the oropharynx by pathogenic bacteria such as *Streptococcus pneumoniae*, *S. pyogenes*, and gram-negative bacilli is prevented. It is also likely that bacteriocin production is the principal means by which indigenous bacteria of the bowel and lower genitourinary tract suppress potential pathogens that attempt to colonize these two sites. That this mechanism is responsible for suppressing pathogenic microorganisms on skin surfaces is suggested further by studies demonstrating a reduced incidence of secondary infections in individuals with eczematous or varicose ulcerations that are colonized by antibiotic-producing bacteria.

In the intestine, production of toxic short-chain fatty acids by the indigenous flora might be an important means by which pathogenic microorganisms are discouraged from intruding into established intestinal ecosystems. There are probably many other metabolic end products of the indigenous flora that are either directly toxic to potential pathogens or inhibit growth of pathogenic microorganisms indirectly by lowering local oxidation-reduction potentials.

In addition to excreting metabolic by-products and other inhibitory substances, microbial populations change their environment by extracting nutrients. In nature, there is intense competition for limited supplies of chemicals from which biological molecules are synthesized, and for available energy sources. In situations where different microbial populations compete for the same nutrients, the potential exists for more efficient scavengers of such nutrients to suppress growth of less efficient microorganisms. Although this type of interference appears to be widespread in nature, it has not been shown to be an important mechanism by which indigenous microorganisms suppress potential pathogens on human body surfaces.

Some of the most innovative studies of microbial inhibitory mechanisms have considered the effect of resident microorganisms on adherence of nonresident microorganisms to epithelial surfaces. *Candida albicans*, for example, has been shown to adhere twice as well to oral epithelial cells of germ-free mice as to those of conventional animals (that is, animals with fully developed, normal microbial floras). Similarly, adherence of *Actinomyces viscosus* (a cause of root surface caries and periodontal bone loss) to hydroxyapatite has been shown to be suppressed by various bacterial species found among the normal flora of the human oropharynx. In one instance, adherence of *A. viscosus* appears to be impaired because its attachment sites partially overlap those of indigenous microorganisms, whereas in another case a bacteriocin (produced by *S. mutans*) appears to be responsible for inhibition of adherence.

Electron micrographs of mucosal surfaces in the colon reveal a carpet of indigenous microorganisms so complete that it is difficult to conceive of a means by which potential pathogens might find access to the underlying mucosal surface, short of burrowing through this carpet. Such steric hindrance, if important, would obviously be less evident in more sparsely populated surfaces such as the skin or small-bowel mucosa.

Indirect inhibition. Aside from their ability to inhibit nonresident microorganisms directly through the mechanisms described above, indigenous microorganisms might obtain the same result indirectly by stimulating host immune or clearance systems. Comparison of germ-free and conventional animals reveals a variety of immunological deficiencies in animals lacking indigenous microbial associates, suggesting that indigenous microbes are, at least in part, responsible for the full maturation of the immunological defenses of higher animals.

Since many bacteria are destroyed in laboratory cultures by deconjugated bile acids, deconjugation of bile acids has been proposed to be an indirect means by which the normal gastrointestinal flora contributes to resistance to intestinal pathogens. Contrary to this hypothesis, some intestinal pathogens resist the toxic effects of bile acids, and other (sensitive) pathogens avoid exposure to the acids by colonizing areas of the upper small intestine where concentrations of deconjugated bile acids are low.

Finally, the normal flora appears to heighten host clearance mechanisms, at least to the extent that it stimulates intestinal peristalsis. In this regard, it is likely that indigenous microorganisms limit the capacity of intestinal pathogens to establish themselves by facilitating their expulsion from the gastrointestinal tract.

Microbial assassins. Microorganisms are consummate parasites. As a result, their plight as occasional victims of parasitism has generally been overlooked. The best-characterized examples of parasitism between microorganisms involve bacteriophages. Bacteriophages are ubiquitous viruses that are frequently highly virulent for their host bacterium, and yet completely nonpathogenic for animals. Within minutes of their first encounter with susceptible bacteria, virulent, lytic phages will have multiplied a hundredfold and, in the process, caused the production of enzymes that attack the cell wall of bacteria, weakening it until it bursts to liberate new and equally virulent bacteriophages. In theory, any parasite behaving in this fashion should have little difficulty in effecting the total destruction of its host bacteria within a relatively short period. For these reasons, there was great hope early in this century that bacteriophages might be used to treat human infections. Unfortunately, in spite of hundreds of experiments originating as early as 1918 and continuing to as late as 1970, no viable clinical application of this phenomenon has ever been achieved.

Parasitization of one microorganism by another is by no means limited to the association between the bacteriophage and its host. Viruses that parasitize nematodes and protozoa have been identified. Bacteria (bdellovibrios) that prey upon other bacteria, and bacteria that parasitize human trematodes have also been identified. Although the destruction of schistosomes by bacteria inoculated into schistosome-infected laboratory animals has led to speculation that selected bacteria might someday be used to eradicate human schistosomiasis, the successful clinical application of live microbial "assassins" has yet to be achieved.

Potential importance to humans. Competitive inhibition of exogenous microbes by the normal flora is one of the many important defenses against infection of higher animals. The indigenous microorganisms of the oropharynx (particularly viridans streptococci) suppress growth of such diversified respiratory pathogens as *Streptococcus pneumoniae*, *Neisseria meningitidis*, *Staphylococcus aureus*, *Mycobacterium tuberculosis*, *Legionella pneumophila*, and gram-negative bacilli. More importantly, numerous investigators have shown that antibiotic-induced elimination of such inhibitory bacteria is rapidly followed by the emergence of many of these same pathogens in the oropharynx.

A similar situation seems to prevail in the gastrointestinal tract. In numerous experiments with laboratory cultures, resident microorganisms of the gastrointestinal tract have been shown to inhibit growth of important pathogens such as *Vibrio cholerae*, *Salmonella*, and *Shigella*. Furthermore, suppression of these same resident bacteria (by antibiotics) reduces substantially the resistance of higher animals to invasion by the intestinal pathogens. In addition to suppressing growth of intestinal pathogens, the resident microflora of the bowel appear to protect the host by degrading bacterial toxins, a function that almost certainly accounts for the increased resistance of conventional animals (as compared to germ-free controls) to *Clostridium botulinum*. There have also been reports of antagonism of *Neisseria gonorrheae* in laboratory cultures by various microbial species within the resident flora of the human endocervix.

Capitalizing on interference. Recognizing the importance of endogenous inhibitory microorganisms as a defense against infection, numerous investigators have sought to increase resistance to infection by supplementing the normal flora with nonpathogenic bacteria that are more effective inhibitors of particular pathogenic microorganisms. The most extensive efforts in this regard have been devoted to attempts to control epidemic staphylococcal infections. These studies have demonstrated that interference between strains of *Staphylococcus aureus* can be exploited to curtail epidemics of staphylococcal disease in nurseries and to interrupt cycles of recurrent furunculosis in older persons. In these studies the nonpathogenic 502A strain of *S. aureus* has been the primary replacement agent used to prevent colonization by more virulent strains of staphylococci. Unfortunately, nonpathogenicity has been a relative, rather than an absolute, characteristic of this bacterium in that instances of disease caused by *S. aureus* 502A have been reported.

Investigators in South Africa have also attempted to eliminate carriage of multiresistant pneumococci by inducing nasopharyngeal colonization by *Streptococcus faecalis*. In spite of the fact that this bacterium inhibited growth of the multiresistant organism in laboratory cultures, it did not eradicate carriage of the pneumococcus. *Streptococcus faecalis* appears unable to actually displace the pneumococcus in humans.

In addition to these attempts to increase the effectiveness of the normal flora as a barrier to infection by introducing new inhibitory microorganisms into its ranks, some investigators have sought to manipulate the normal flora through a process of selective decontamination. According to this process, antimicrobial agents that suppress aerobic microorganisms while leaving the obligate anaerobic flora undisturbed are administered to infection-prone leukemia patients. This is done with the knowledge that obligate anaerobes rarely cause infections in leukemic patients, and in fact might constitute a limited barrier to colonization by aerobic bacteria that are responsible for the majority of infection seen in these patients. Such attempts to capitalize on the colonization resistance of the anaerobic constituents of the normal flora, while suppressing potential aerobic pathogens, have already met with some success in clinical trials. However, the ultimate question of whether selective decontamination enhances survival in such patients has not been resolved.

Microbial synergism. In the intense competition for vital elements, microorganisms occasionally find it advantageous to ally themselves with other microorganisms. These synergistic relationships have long been recognized in nature, but have only recently been appreciated as playing a part in human infections. Traditionally, such infections have been regarded as disorders caused by single microorganism, requiring a single, specific therapy. It is now clear, however, that in certain mixed infections a high degree of synergism exists between pathogens, dictating that these processes be viewed as complexes of interacting microorganisms rather than the simple sum of independent infections.

The means by which one microorganism assists another in consummating infection are both subtle and sinister. An otherwise weak pathogen, for instance, may succeed in establishing itself in a given host, because another, more virulent microorganism has previously suppressed the host's immunologic defenses. This type of synergism is probably at least partially responsible for the bacterial superinfections that complicate influenza—that is, influenza viruses induce a number of immunologic defects which facilitate secondary bacterial invasion of respiratory tissues.

One organism may also predispose to infection by another by promoting its dissemination to new susceptible hosts. This phenomenon is best illustrated by simultaneous outbreaks of viral and bacterial upper respiratory infections in newborn nurseries. In this situation, airborne dissemination of bacterial pathogens appears to be a direct consequence of the sneezing and coughing that accompany the viral respiratory infections.

In rare instances, synergistic relationships involve provision of essential elements from one microorganism to another. This form of synergism is well illustrated by the delta agent, a defective virus that replicates only in the presence of the hepatitis B virus. Interestingly, in the course of obtaining elements vital to replication from the hepatitis B virus, the delta agent simultaneously inhibits synthesis of hepatitis B virus products.

Finally, an otherwise nonpathogenic or weakly pathogenic microorganism may increase in virulence because of factors obtained from another microorganism. An example of this form of synergism is the association between *Corynebacterium diphtheriae* and a bacteriophage responsible for induction of diphtheria toxin production. As a result of the activity of the bacteriophage, avirulent strains of *C. diphtheriae* are converted to virulent strains.

For background information see ANTIBIOTIC; BACTERIOPHAGE; ECOLOGICAL INTERACTIONS; MEDICAL BACTERIOLOGY; MEDICAL PARASITOLOGY; TOXIN; VIRULENCE in the McGraw-Hill Encyclopedia of Science and Technology.   [PHILIP MACKOWIAK]

Bibliography: A. G. Fredrickson and G. Stephanopoulos, Microbial competition, *Science*, 213:972–979, 1981; P. A. Mackowiak, Microbial synergism, pts. 1 and 2, *N. Engl. J. Med.*, 298:21–26, 83–87, 1978; P. A. Mackowiak, The normal flora, *N. Engl. J. Med.*, 307:83–93, 1982; K. Sprunt and W. Redman, Evidence suggesting importance of role of interbacterial inhibition in maintaining balance of normal flora, *Ann. Intern. Med.*, 68:579–590, 1968.

Microorganism

This article reports on two fields of active research in medical microbiology. The first section discusses the essential role of iron in microbial infections and how deeper understanding of iron's effect on the metabolism of both host and invader may lead to clinical techniques to curb some infections. The second section discusses IgA1 proteases, a class of microbial enzymes associated with virulence in several bacterial species; work in this area provides insight into the human mucosal immune system that may also yield clinically useful techniques to stem some bacterial pathogenicity.

ROLE OF IRON IN INFECTION

During the past half-century, clinical and laboratory observations have revealed a vigorous competition for iron between bacterial, fungal, protozoan, and neoplastic (tumor-inducing) invaders and their vertebrate hosts. The hosts possess an array of mechanisms to withhold the growth-essential metal; the ability of invaders to overcome these mechanisms is an important component of virulence. High amounts of iron not only support growth of the invaders but also impair the host's chemotactic responsiveness, phagocytic capacity, and bactericidal activity of granulocytes and monocytes. Iron-withholding host defense is compromised in diverse conditions that involve aspects of iron overload in specific tissues. Such conditions include excessive exogenous iron via ingestion, injection, or inhalation; destruction of iron-storage cells (as in hepatitis); excessive destruction of erythrocytes (as in clinical episodes of various hemoglobinopathies, malaria, bartonellosis, leukemias, and lymphomas); and decreased synthesis of transferrin (as in kwashiorkor and jejunoileal bypass).

Host iron-withholding proteins. Powerful iron-binding proteins of the transferrin class are stationed by vertebrates at potential sites of invasion. Known examples include conalbumin in egg white, transferrin in plasma and perspiration, and lactoferrin in such exocrine secretions as tears, nasal exudate, saliva, bronchial mucus, gastrointestinal fluid, hepatic bile, cervical mucus, seminal fluid, and milk. Lactoferrin also is a major protein component of the specific granules of circulating neutrophils. It is released on degranulation of these cells in a septic area. After combining with iron in the invaded site, the metal-saturated protein is ingested by macrophages.

Each of the transferrin class of proteins consists of a single polypeptide chain of about 680 amino acids plus an oligosaccharide moiety; the molecular weight is around 77,000 daltons. The carbohydrate may serve a protective structural role, guarding against proteolytic destruction in various body fluids. Two metal-binding sites are present in the polypeptide; each is composed of three tyrosyl and two histidinyl residues. The sixth ligand is satisfied by a carbonate ion. Iron-binding activity of conalbumin and transferrin requires alkaline or neutral pH, whereas lactoferrin is highly active at the acidic pH values that develop in sites of invasion.

Emergency mechanisms of iron withholding. Vertebrate hosts invaded by microorganisms or neoplastic cells promptly lower their plasma iron. In a typical series the mean value in normal persons was 17.8 micromolar (range 12–27 μM), whereas individuals with active infections or neoplasia had a mean of 5.5 μM (range 2–12 μM). Iron saturation of transferrin likewise was lowered from a normal mean of 30% (range 25–50%) to a mean of 15% (range 10–25%). This shift in iron metabolism is enhanced as the clinical condition worsens, and returns to normal as the patient improves.

The metal is diverted to liver and spleen by a block in its normal release to plasma transferrin from macrophages that have acquired the iron from decaying erythrocytes. The excess metal in the phagocytic cells combines with intracellular apoferritin to derepress additional ferritin synthesis. Thus, the metal can be stored in ferritin (rather than excreted) until the danger of invasion has ceased. The shift in metal metabolism is induced by a monokine hormone termed leukocytic endogenous mediator (LEM). This 13,000–16,000-dalton peptide has also been called endogenous pyrogen, lymphocyte-activating factor, and interleukin-1. It is formed mainly by monocytes; production begins quite early in the course of the invasion.

Additionally, LEM stimulates liver cells to synthesize such "acute-phase" reactive proteins as C-reactive protein, amyloid A protein, fibrinogen, haptoglobin, and ceruloplasmin. The hormone also causes zinc to be shifted from plasma to liver, the hypothalamic temperature set point to be raised from 37°C (98.6°F) to at least 38.5°C (101.3°F), neutrophiles to be released from the bone marrow storage pool, and granules to be discharged from the neutrophiles at septic sites. Moreover, LEM induces production of a lymphokine termed interleukin-2 which, in turn, stimulates the formation of transferrin receptor proteins on the membranes of T lymphocytes. Such receptors are essential for acquisition of transferrin-bound iron by the T cells. An important function of the metal is catalysis of ribonucleotide reductase, whose activity is needed for the synthesis of deoxyribonucleic acid (DNA) required for the clonal expansion of these defense cells.

Capture of host iron by invaders. Pathogenic bacteria, fungi, and protozoa must be able to capture host iron in order to survive and multiply in body tissues or fluids. To acquire the metal, bacteria and fungi generally release high-affinity, low-molecular-weight chelators, then bind and transport into their cytoplasm the resulting iron chelates by means of specific membrane-bound permeases. Most widely studied in the past 33 years have been hydroxamate and catechol siderophores. The former contain one or more oxidized peptide bonds, the latter one or more 2,3-dihydroxybenzoyl groups. An important component of virulence is the ability of the pathogen to utilize a siderophore that can compete with host iron-binding proteins for the metal. In some cases, synthesis of "virulence" siderophores is coded by specific "virulence" plasmids. Although transformed and neoplastic animal cells have been shown to be able to utilize specific low-molecular-weight peptides as siderophores, most observations have been concerned with the capacity of these cells to markedly increase their rate of iron acquisition by increased synthesis of transferrin iron receptors. The latter, located in the cytoplasmic membrane, are glycoproteins of about 90,000 daltons.

Alteration of iron-withholding host defense. Hundreds of studies on humans and other vertebrates have shown that procedures or conditions that suppress the ability of hosts to withhold iron markedly increase morbidity and mortality of diseases caused by bacteria, fungi, protozoa, and neoplastic cells. As a corollary, procedures or conditions that enhance the ability of hosts to withhold iron significantly decrease morbidity and mortality of these diseases. Although the most obvious and most frequently demonstrated mechanism whereby excess iron enhances infection and neoplasia is by serving as a nutrient for the invading microbial or neoplastic cells, additional modes of action may contribute to a weakening of host defense. For example, the metal may inhibit monocytes and macrophages by suppressing the phagocytosis-associated metabolic burst as well as by inactivating peroxide.

On the other hand, if hosts become truly iron-deficient from severe or prolonged starvation or bleeding, susceptibility to infection and neoplasia can become intensified rather than diminished, presumably because the metal is needed to catalyze various aspects of the humoral and cell-mediated immune systems. Nevertheless, the present state of knowledge about iron-withholding defense suggests that methods and agents for strengthening such defense now can be safely and effectively developed for use in human and veterinary medicine.

A number of available or potential methods and agents for strengthening iron-withholding defense are now apparent. These include avoidance of excess dietary iron, especially in the presence of such enhancers of iron absorption as ethanol or ascorbic acid; elimination of administration of parenteral iron, except in those who are truly iron-deficient (hemoglobin value ≤ 10 g/dl); induction of brief hypoferremia and fever by means of purified LEM; appropriate use of either low-molecular-weight or protein iron chelators; and prevention or correction of underlying hyperferremic or hypotransferrinemic defects by suitable vaccines, chemotherapeutic agents, or restoration of adequate protein nutrition.

[EUGENE D. WEINBERG]

IgA1 PROTEASES

IgA1 proteases (originally named IgA proteases) are a family of bacterial enzymes whose only known substrate is human IgA1 immunoglobulin. Antibodies of the IgA isotype are secreted onto mucosal surfaces and therefore are central elements in the immune defense of tissues lining the oral cavity, respiratory tree, intestine, and genital tract; the bacteria that secrete IgA1 proteases typically colonize or infect such tissues. For this reason the IgA1 proteases are tentatively regarded among the virulence factors—attributes of microorganisms that may initiate, sustain, or complicate infectious disease.

IgA1 protease–excreting bacteria include several major pathogens, such as *Neisseria gonorrhoeae*, *N. meningitidis*, *Hemophilus influenzae*, *Streptococcus pneumoniae*, *S. sanguis*, and certain *Bacteroides* and *Capnocytophaga* species in the oral cavity. These bacteria are responsible for a variety of human infections, including sexually transmitted diseases, pneumonia, bronchitis, otitis media, and infections of the oral cavity, and are generally unable to colonize or infect other animals. In contrast to many opportunistic bacteria that cause disease in persons whose immune capacity is reduced, these pathogens commonly infect previously healthy individuals.

Organization of mucosal immune system. Secretory IgA antibodies in colostrum, milk, and most secretions of epithelial membranes are dimeric immunoglobulins secreted by differentiated plasma cells lying in the lamina propria beneath the epithelium. These large proteins pass through the epithelial cells by a unique receptor-mediated transport system that involves a polypeptide chain synthesized by the epithelial cell itself, and called secretory component. Secretory component becomes covalently bound to the IgA dimer, and both appear together in secretions as secretory IgA, which then has a molecular weight of approximately 380,000 daltons. Secretory IgA has antibody specificity for viral, bacterial, and environmental (including food) antigens.

There are two isotypes of IgA in normal individuals, designated IgA1 and IgA2, and both are present in secretions and serum. The primary structure of the IgA heavy chain is similar in both isotypes, with the exception that IgA1 has a unique sequence of 13 amino acids in the hinge region. This localized structural difference between IgA1 and IgA2 heavy chain is of major importance in understanding IgA1 proteases, because these enzymes all cleave IgA1 in the 13-residue segment corresponding to that deleted in IgA2. For this reason human IgA2 proteins are completely resistant to attack by IgA1 proteases. The relative role of the two isotypes in mucosal antibody defense is not presently known, although the percentage of IgA2 among IgA proteins in secretions is higher than it is in serum.

Characteristics. As work has progressed in studying the enlarging number of these enzymes, several consistent features have been noted. The enzymes are extracellular, being secreted free into the culture medium with little if any accumulation within the bacterial cell or in the periplasmic space. the proteases are large, exceeding 100,000 Daltons. IgA1 proteases apparently cleave a single peptide bond in the IgA1 heavy-chain hinge peptide. and invariably proline contributes the carboxyl group to the sensitive bond. As will be discussed below, bacterial isolates differ in which bond they cleave, but within a given isolate the enzyme specificity remains invariant despite repeated passage of the strain in culture. Most importantly, each bacterial species releasing these enzymes is a recognized human pathogen, and nonpathogenic but taxonomically related species are enzyme-negative. For example, *Neisseria gonorrhoeae* and *N. meningitidis* are the only two pathogens in their genus and are the only species within the genus that excrete IgA1 protease; the same is true for the genus *Hemophilus*, where *H. influenzae* is both the only pathogen and the only protease-positive species. Several lines of evidence (for example, susceptibility to metal-chelator and other inhibitors, serologic analysis, and genetic analysis) indicate that the known IgA1 proteases differ in structure and biochemical characteristics despite their singular requirement for human IgA1 as substrate. Detailed insight into the molecular evolution of this enzyme family is not yet available.

Types. By examining many strains among genera of pathogenic bacteria, it has become clear that IgA1 proteases differ in exactly which peptide bond they cleave in IgA1 (see table). The IgA1 hinge-region domain has the primary structure (amino code: P = proline, T = threonine, S = serine):

```
←        —TPPTPSPSTPPTPSPS—        →
Amino    |     |     |     |    Carboxy
terminus 225   230   235   240  terminus
```

Residue numbers are based on the published primary sequence of IgA1 protein. The location of the peptide bond cleaved by several IgA1 proteases has been identified, and among *Neisseria* and *Hemophilus* species enzymes of differing specificities are found. Each microorganism, with very few exceptions, elaborates the single enzymatic specificity and this remains constant in subculture. With each species, enzymes cleaving different bonds are assigned different type numbers.

Several laboratories have observed that a protease type is not randomly distributed within a genus, but

Bonds cleaved by various IgA1 proteases

Microorganism	IgA1 peptide bond cleaved*
Streptococcus sanguis	227–228
Streptococcus pneumoniae	227–228
Neisseria gonorrhoeae	235–236
	237–238†
Neisseria meningitidis	235–236
	237–238
Hemophilus influenzae	231–232
	235–236
	237–238

*Numbers refer to primary sequence.
†Cleavage site not yet directly verified by sequence analysis.

rather correlates with the capsular polysaccharide type (serotype) of the strain. For example, *Neisseria meningitidis* serogroup A strains yield only type 1 enzyme, and serogroups X and Y only type 2. In *Hemophilus*, capsular serotypes B and D yield only type 1 enzyme, while serotypes C and E yield type 2. The explanation for the correlations of protease type and capsule polysaccharide structure is uncertain, but genetic studies have provided some evidence that the genes specifying protease, and those for capsule structure, are near one another in the bacterial genome.

Synthetic substrate analogs. Attempts to circumvent the requirement of human IgA1 protein as substrate have not been successful. A series of synthetic peptides ranging in length from 2 to 34 amino acid residues, all primary sequence analogs of the IgA1 hinge region, were not hydrolyzed, implying that structural elements of IgA1, such as carbohydrate residues or regions of the protein distant from the cleaved hinge itself, are needed for enzyme activity. Although they are not substrates, certain of these hinge-region analogs competitively inhibit the enzymes, and experiments are presently being done to improve this inhibition enough that these analogs may be clinically useful in treating or preventing bacterial disease. During colonization or infection by some of these bacteria, particularly the *Neisseria* species, antibodies develop that are themselves inhibitors of the IgA1 proteases. In secretions these antibodies are of the secretory IgA type; in serum they are IgG.

Genetics and infectivity. Several laboratories have recently undertaken studies of the molecular biology and genetics of IgA1 protease in *Neisseria* and *Hemophilus*. Synthesis of active enzymes is not plasmid-mediated in either genus, and the gene (designated *iga*) encoding the protease of both these pathogens has been cloned into *Escherichia coli* by using chromosomal DNA incorporated into plasmid vectors. Screening of bacterial colonies of lytic plaques has been aided by the development of an overlay technique using agar that contains insolubilized human IgA1 proteins. The enzyme is expressed in *E. coli* at considerably lower activity levels than that of the parent strain. At present it is apparent that the *N. gonorrhoeae* enzyme is secreted extracellularly in *E. coli*, but that of *H. influenzae* accumulates in the periplasmic space, in contrast to its extracellular secretion in its natural state. In both species the cloned gene product is larger than the native enzyme, indicating the presence of a leader sequence, but the role of this extra peptide length in the secretory process itself has not yet been defined. The ability of cells to excrete this large macromolecule is quite unusual.

Cloned DNA which has been modified to eliminate production of active enzyme has been introduced into native *Hemophilus* and *Neisseria* species by the technique of homologous transformation. IgA1 protease–negative transformants are readily identified by using the overlay screen mentioned above. These strains are otherwise isogenic and are presently being used to examine the role of the enzymes in the infectious process. Because of the extreme substrate specificity of the enzymes, human tissues and cells are required for appropriate models of infectivity by these pathogens.

In Southern blotting experiments using segments of the *iga* gene as homologous probes, the chromosomal DNA of *Hemophilus* and *Neisseria* species that are not human pathogens have no segments homologous with the probe, consistent with their lack of enzymatic activity as discussed earlier. The mutual homology of the IgA1 protease genes in *Hemophilus* and *Neisseria* is low but demonstrable. Among *Hemophilus* capsular serotypes there is a high degree of homology when using the *iga* gene probe despite the aforementioned minor differences in the precise peptide bond cleaved by the various *Hemophilus* enzymes.

Summary. IgA1 proteases are extracellular enzymes secreted by bacteria pathogenic for humans, and are not found in nonpathogenic species within the same genus as the pathogens. The enzymes specifically attack immunoglobulins of the IgA1 isotype which provide antibody function at mucosal surfaces. Secreted IgA2 antibodies that differ in structure from IgA1 at the enzyme-susceptible hinge peptide are resistant. The precise specificity of these proteases is linked genetically to capsular serotype of infective bacteria, and the enzymes are encoded by chromosomal DNA. There is presently no direct evidence that the enzymes are involved in the ability of these bacteria to colonize or infect.

For background information *see* IMMUNOGLOBULIN; MEDICAL BACTERIOLOGY; VIRULENCE in the McGraw-Hill Encyclopedia of Science and Technology.

[ANDREW G. PLAUT]

Bibliography: J. Bricker et al., IgA1 proteases of *Haemophilus influenzae*: Cloning and characterization in *Escherichia coli* K-12, *Proc. Nat. Acad. Sci. U.S.A.*, 80:2681–2685, 1983; C. A. Dinarello, Interleukin-1, *Rev. Infect. Dis.*, 6:51–95; M. Kilian et al., Molecular biology of *Haemophilus influenzae* IgA1 proteases, *Mol. Immunol.*, 20:1051–1058, 1983; G. K. Lee, The anemia of chronic disease, *Semin. Hematol.*, 20:61–80, 1983; T. K. Low et al., Primary structure of a human IgA1 immunoglobulin, *J. Biol. Chem.*, 254:2850–2858, 1979; P. C. McNabb and T. B. Tomasi, Jr., Host defense mechanisms at mucosal surfaces, *Annu. Rev. Microbiol.*, 35:477–496, 1981; J. B. Neilands, Iron absorption and transport in microorganisms, *Annu. Rev. Nutr.*, 1:27–46, 1981; A. G. Plaut, The IgA1 proteases of pathogenic bacteria, *Annu. Rev. Microbiol.*, 37:603–622, 1983; E. D. Weinberg, Iron withholding: A defense against infection and neoplasia, *Physiol. Rev.*, 64:65–102, 1984.

Mountain systems

Uplift of individual mountains and mountain systems has attracted much attention in the last few years. Most motions of the Earth's crust work so slowly that their study is largely academic, but in certain regions crustal movements are so rapid that they result in earthquakes, landslides, floods, and tsunamis (giant sea waves).

The world's highest mountain range, the Himalayas, is actively rising. The uplift is causing episodic landslides, catastrophic floods, and extremely destructive earthquakes, which can damage or destroy the vital dams and irrigation systems on which India and Pakistan rely for much of their agriculture and thus their vital foodstuffs.

Geophysical monitoring. The mountain uplift is being monitored in part by an array of seismic stations that permit the geophysicists to pinpoint sites of maximum activity (often hidden far below the surface). The motion is being monitored also by geodetic releveling; this technique calls for precise determination of the elevation of benchmarks, and then the survey is repeated every few years to determine height differences (if any).

Over one 4-year period (1976–1980) in the Shanan area of the Himachel Pradesh province of India, vertical shifts have been determined ranging from 0.23 to 1.91 in. (5.8 mm to 48.6 mm). Although this uplift of up to 0.5 in. (13 mm) per year may not sound serious, over centuries and millennia such uplifts can lead to catastrophic changes in the landscape. For instance, a critical threshold effect is observed often. Slow tilting of a mountain slope may cause no immediate effect, but at a certain point the slope reaches a threshold of stability and the entire mountain side collapses, burying adjacent valleys, towns, highways, and people. In the northeastern Himalayas and Shillong area, an extremely earthquake-prone region, a joint Indian–United States project has been started to monitor the crustal shocks. An array of 50 accelerographs will provide data on this active seismic belt, and will furnish warnings of dangerous crescendos of activity.

Geological studies. The progressive uplift of the Himalayas is also being studied by geologists. The rate of uplift cannot be measured directly by geological methods, but as the mountains are uplifted the rocks are rapidly attacked by rain and rivers and eventually, as the mountains get higher, by snow or ice. Streams and rivers carry the eroded debris down to adjacent plains or into the sea. The rate of accumulation of sediment here tells how much of the mountains is being eroded, and approximates the rate of mountain uplift.

Bangladesh straddles the combined delta of the Ganges and Brahmaputra rivers. In this area geologists have been studying deltaic deposits in connection with a search for hydrocarbons. In correlating the samples from well borings for natural gas, they found that fossil pollen was particularly useful. But something unexpected also emerged. Much of the fossil pollen in the delta was carried down by the annual flood from the Himalayas. The pollen from different types of vegetation is very distinctive, particularly so from the flora which grows at different altitudes. By establishing the date of each assemblage of altitude-indicating pollen, the rate of Himalayan uplift can be deduced.

The Ganges-Brahmaputra delta started to build out into the Bay of Bengal following the final collision between the Indian and Asiatic plates nearly 30 million years ago (m.y.a.). At that time a river system arising in southern Tibet began flowing to the south, and eventually into the Indian Ocean. Gradually the mountain belt folded and crumpled; heating below was added as the Indian plate pushed in, under the edge of the Tibetan Plateau. Melting occurred, and molten granite injected into the heart of the folded mountains. Uplift now followed, and by 5 m.y.a. the young Himalayas were approaching 6600 ft (2000 m) in elevation. By 2 m.y.a. the pollen shows evidence of high mountain species, the broad-leaved evergreens and trees like pines and rhododendron that are typical of 9800 ft (3000 m) today in Nepal and Bhutan. Since 2 m.y.a. the Himalayan uplift has been dramatic. The uplift has accelerated to rates up to 0.4 in. (10 mm) a year, so that many peaks are now at over 26,000 ft (8000 m), including Mount Everest, the world's highest mountain.

For background information *see* EARTH DEFORMATIONS AND VIBRATIONS; GEODESY; MOUNTAIN SYSTEMS; PALYNOLOGY; PLATE TECTONICS in the McGraw-Hill Encyclopedia of Science and Technology.

[RHODES W. FAIRBRIDGE]

Mud volcano

Mud volcanoes are extrusions of deep-seated sediments that have risen through overlying layers to erupt on the surface, and these sedimentary volcanoes have been described in many land areas, including Pakistan, Afghanistan, Timor, New Zealand, Trinidad, and Texas. The eruptions can be violent, but are more often relatively passive seeps. The most spectacular examples can have the classic conical shape (Fig. 1) of igneous volcanoes with basal diameters as great as 0.6 mi (1 km) and heights of over 330 ft (100 m). More typically, however, mud volcanoes are smaller mounds and seeps, or groups of these features, where the extruding mud

Fig. 1. Photograph of a cone-shaped mud volcano which rises over 165 ft (50 m) above the surrounding plain, located south of Chandragup, West Pakistan. (*From R. E. Snead, Active mud volcanoes of Baluchistan, West Pakistan, Geogr. Rev., 14:546–560, 1964*)

is combined with large amounts of water and gas, producing a fluid mixture that readily flows and accumulates in sheets and ponds.

The recent development of long-range side-scan sonar equipment for deep-sea exploration has enabled researchers to map large areas of the sea floor with a level of resolution not previously available, and has resulted in the location of numerous mud volcanoes in oceanic plate convergence regions. This sea-floor mapping suggests a tectonic origin for many of the sea-floor mud volcanoes, an origin somewhat different from that of most of the mud volcanoes that have been studied on land.

Overpressured sediments. Periods of rapid deposition of sediments, particularly in a marine environment where pelitic, water-saturated sediments are common, can produce thick beds that are quickly buried by often relatively impermeable overburden. The deeper beds are compacted as the weight of the overburden forces expulsion of interstitial water from these beds. Normally the fluids are expelled at a sufficiently rapid rate so that the interstitial water remains at hydrostatic pressure regardless of the overburden weight. With compaction, however, clay minerals in the sediments become more closely packed and can substantially reduce the permeability of the formation. In many cases this and related effects will cause the pressure of the interstitial water in these deep beds to become high compared to surrounding beds, sometimes elevating to close to the overburden weight. In this situation the interstitial water pressure will prevent complete compaction of the sediments, causing the overlying beds to partially float on the overpressured fluids. At the same time, these rapidly deposited sediments will often have a high content of organic matter, which will decay through time and produce natural gas. The addition of this gas to the beds can further elevate the pressure to the point where it surpasses the overburden pressure and will then find a path to the surface and erupt, producing a mud volcano.

Mud volcanoes typically have associated seeps of natural gas, and have often been cited as an indicator of petroleum generation potential. Because the overpressured sediments will tend to rise to the surface through some zone of weakness in overlying layers, the distribution of mud volcanoes can also be an indicator of large faults or anticlinal structures.

Mud volcanoes caused by overpressured sediments have been known to erupt violently or as passive seeps, either spasmodically or periodically, and in submarine locations (as evidenced from surface observations of ejecta and discolored water) as well as on land. Violent eruptions from the Mangaehu Stream mud volcano in New Zealand, for instance, had fountaining to heights of 400 ft (120 m) in 1908.

In recent years, increasing attention has been given to the influence of tectonic stresses on the formation of overpressured sediment beds and mud volcanoes. Any horizontal compression of sediment layers will tend to increase the lithostatic pressure derived from the weight of the overburden. When overlying beds partially "float" on overpressured beds, the undercompacted sediment grains are held apart by the high fluid pressure, producing a decoupling of the sediment layers that can provide a preferred zone for thrust faulting. This compressional tectonic activity can then augment the already overly high fluid pressures, further promoting the chances of overpressuring to the point where diapirism and mud volcanism may occur.

Side-scan sonar sea-floor swathmapping. The recent development of high-resolution side-scan sonar sea-floor mapping systems has provided the means to produce images of large portions of the sea floor that provide reconnaissance capabilities comparable to the usage of aerial photographs over land areas. The SeaMARC II system in use at the Hawaii Institute of Geophysics is the newest of these systems. It is a hydrophone array towed from a ship which transmits a focused beam of approximately 12-kHz sound in the water that reflects from features on the sea floor. The returning echoes are received by the same array and computer-processed to produce a sonar image of the sea floor in a 6-mi-wide (10-km) swath along the ship's track. The SeaMARC II system also has the capability to measure the angle of incidence of reflected echoes returning to the hydrophone array. When the direction and range of a reflector relative to the sonar array are known, the position and depth of the reflector are easily calculated. The combination of many of these reflector positions is then used to produce a bathymetric map of the ensonified area of the sea floor. This bathymetry gives the dimensions of the sea-floor features in a SeaMARC II image.

The SeaMARC II system can map up to 1200 mi^2 (3000 km^2) of deep-ocean sea floor each day. Thus a sea-floor area the size of Connecticut could be completely surveyed by SeaMARC II in 4 days. All linear features, such as faults or erosion channels, can be identified in SeaMARC II images, and two-dimensional features on the scale of a football field will have recognizable shape.

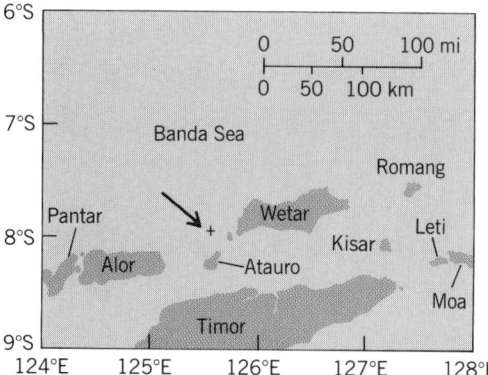

Fig. 2. Location of Wetar Island in the eastern Indonesian archipelago. The oceanic crust of the Banda Sea is converging south relative to the Indonesian islands and is being subducted beneath the overriding islands. Arrow indicates location of mud volcanoes in Fig. 3.

Mud volcano

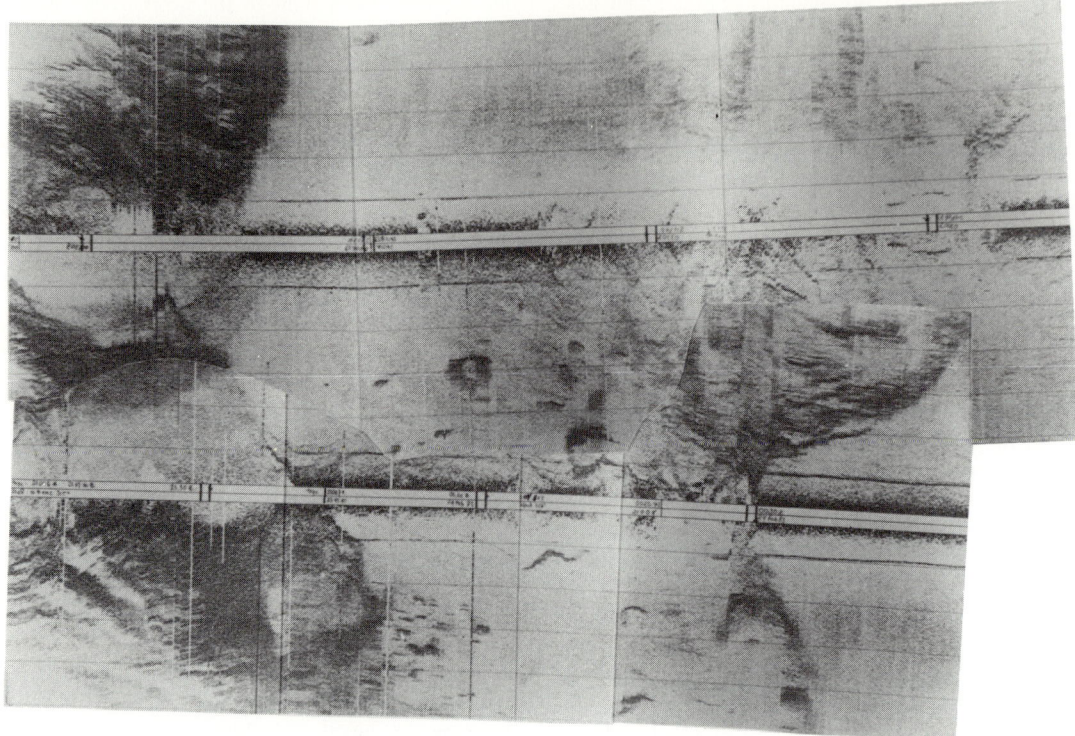

Fig. 3. SeaMARC II side-scan sonar image of mud volcanoes in 9900-ft-deep (3000-m) water north of Wetar Island.

This sea-floor mapping tool has enabled marine geologists to observe vast areas of sea floor with a resolution that was previously unknown. One of the many significant discoveries of early SeaMARC II and other sonar mapping studies has been the identification of large numbers of mud volcanoes in island arc and trench regions of the sea floor. The first published observation of a single mud volcano on the sea floor was the result of an earlier low-resolution study using the British GLORIA side-scan system near Barbados. During the first SeaMARC II survey near the island of Wetar in the eastern Indonesian archipelago (Fig. 2), large numbers of sea-floor mud volcanoes were observed. Figure 3 is a representative image of the Indonesian marine mud volcanoes. In this image the darkness of features is proportional to the strength of bottom reflectors, so that dark portions of the image are indicative of rough-textured bottom and light areas are smooth sediments. The shapes in the sonar image are geometrically correct, so that it can be interpreted much the way an aerial photograph is interpreted on land. The smooth sediments of the Banda Sea are converging against the island slopes from the right

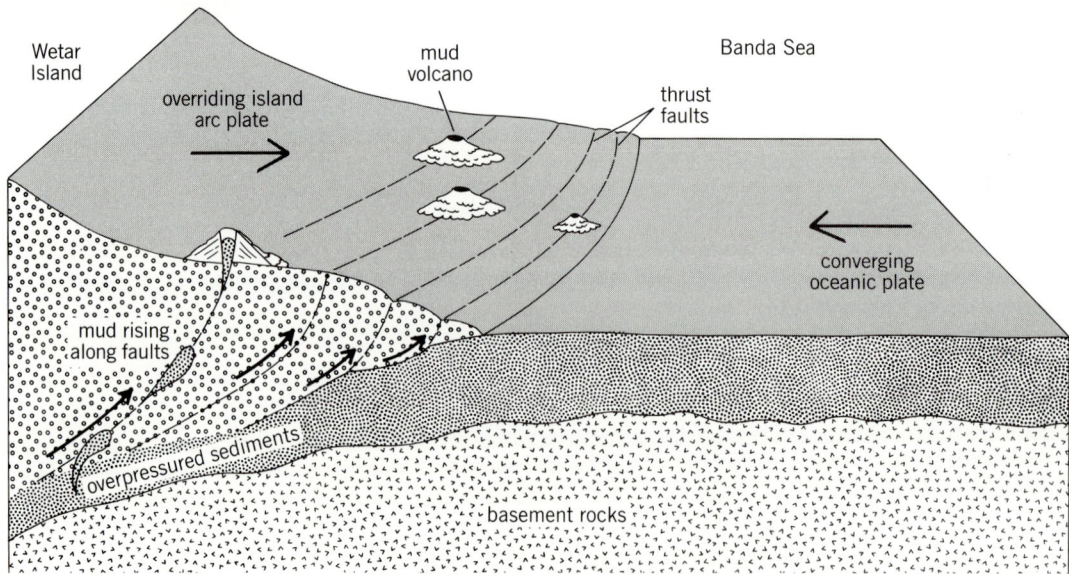

Fig. 4. Sketch of the processes producing convergence zone mud volcanoes such as seen in Fig. 3.

side of the sonar image, causing deformation and compaction of the sediments on the lower slope of the island. As the oceanic sediments are thrust beneath the lower island slope sediments, a combination of the increased overburden weight from the overriding sediments and the compaction from the converging plates causes overpressuring of deep sediment layers and subsequent mud volcanism on the island side of the deformation front.

The mud volcanoes in this image vary in size from basal diameters of over 0.6 mi (1 km) for the two largest vents, to many smaller eruptive centers with diameters of less than 330 ft (100 m).

The most interesting thing about the location of these features is their concentration in what is known as the deformation front of the overriding forearc in a plate convergence zone (Fig. 4). These mud volcanoes are probably the result of the overpressurization of sediments being carried into and down the subduction front. The compression of these sediments is thus a combination of nearly horizontal tectonic compression as the plates converge, and the sudden increase in overburden pressure as the subducting plate is overridden by the forearc.

Sea-floor mud volcanoes in island arc deformation fronts have now been observed by sonar mapping near Barbados, in eastern Indonesia, and in the Mariana and Bonin island arcs. Their distribution suggests that many subaerial mud volcanoes may also be caused by horizontal tectonic compression. It is hoped that further study of sea-floor mud volcanoes may help determine strain and fault distribution in active convergent margins.

For background information see DIAPIRIC STRUCTURES; MARINE GEOLOGY; MUD VOLCANO; OCEANIC ISLANDS; PLATE TECTONICS in the McGraw-Hill Encyclopedia of Science and Technology.

[DONALD M. HUSSONG]

Bibliography: M. K. Hubbert and W. W. Rubey, Role of fluid pressure in mechanics of overthrust faulting: 1. Mechanics of fluid filled porous solids and its application to overthrust faulting, *Geol. Soc. Amer. Bull.*, 70:115–166, 1959; M. F. Ridd, Mud volcanoes in New Zealand, *Amer. Ass. Petrol. Geol.*, 54:601–616, 1970; R. E. Snead, Active mud volcanoes of Baluchistan, West Pakistan, *Geogr. Rev.*, 14:546–560, 1964; A. H. Stride, R. H. Belderson, and N. H. Kenyon, Structural grain, mud volcanoes and other features on the Barbados Ridge Complex revealed by GLORIA long range side-scan sonar, *Mar. Geol.*, 49:187–196, 1982; G. K. Westbrook and M. J. Smith, Long decollements and mud volcanoes: Evidence from the Barbados Ridge Complex for the role of high pore-fluid pressure in the development of an accretionary complex, *Geology*, 11:279–283, 1983.

Multiport network analyzers

In order to design microwave systems such as radar and communications or navigation aids, it is necessary to know and to be able to measure various properties which characterize microwave circuits. These include input and output impedances, attenuation of passive circuits, gain of amplifiers, and similar quantities which can be given the generic title of scattering parameters. Methods for measuring these quantities at a fixed frequency have long been known, and use, for instance, bridge or slotted-line techniques. In recent years very wide-band systems (covering perhaps a 10:1 range of frequency) have been developed. To make measurements on such systems one frequency at a time by manual methods is prohibitively time-consuming. Automatic network analyzers, using wide-band tunable sources and receivers under computer control, have been developed, enabling measurements to be made more rapidly. They are, however, complex and expensive, and more recently the use of multiport couplers has resulted in simpler systems.

A multiport network analyzer is a linear passive microwave network having five or more ports which is used for measuring power and the complex reflection coefficient. These parameters are measured at one port when a signal is applied to a second port, and the three or more remaining sidearm ports are terminated with power detectors. The power and the reflection coefficient at the measurement port are calculated from the sidearm power detector readings. Usually four sidearm detectors are used so that the network has six ports in all, and is called a six-port reflectometer (Fig. 1).

A typical multiport network analyzer consists of components such as directional couplers or probes that couple in different ways to the incident and reflected waves in a transmission line. A simple arrangement consisting of a directional coupler and three voltage probes is shown in Fig 1a. The probes are inserted in a transmission line with a spacing equal to 60° at the frequency of operation. The spacing is not critical, and such an arrangement will cover about an octave in frequency. Broadband designs that cover a decade in frequency are shown in Fig. 1d and 1h. They consist of 3-dB couplers or quadrature hybrids Q and power dividers D.

The idea of using multiple sidearms with power detectors dates back at least to 1947, when A. L. Samuel described an impedance meter consisting of four probes inserted in a transmission line, where the probes were spaced one-eighth wavelength apart and terminated with diode detectors. The concept was given new importance in 1972–1973, when G. F. Engen and C. A. Hoer developed a general theory describing any circuit with six ports. Since then, over 120 papers have been published on the theory and application of six-port reflectometers, vector voltmeters, and network analyzers.

Theory. It has been shown that the observed power at sidearm i of an n-port reflectometer (with n equal to or greater than 3) can be written as in Eq. (1), where Γ_2 is the reflection coefficient of the ter-

$$P_i = |A_i b_2|^2 |\Gamma_2 - q_i|^2 \quad i = 3, \ldots, n, \, n \geq 3 \quad (1)$$

mination being measured at port 2, and b_2 is the emerging wave into the termination. The parameters A_i and q_i are constants which are functions only of the network parameters of the n-port reflectometer

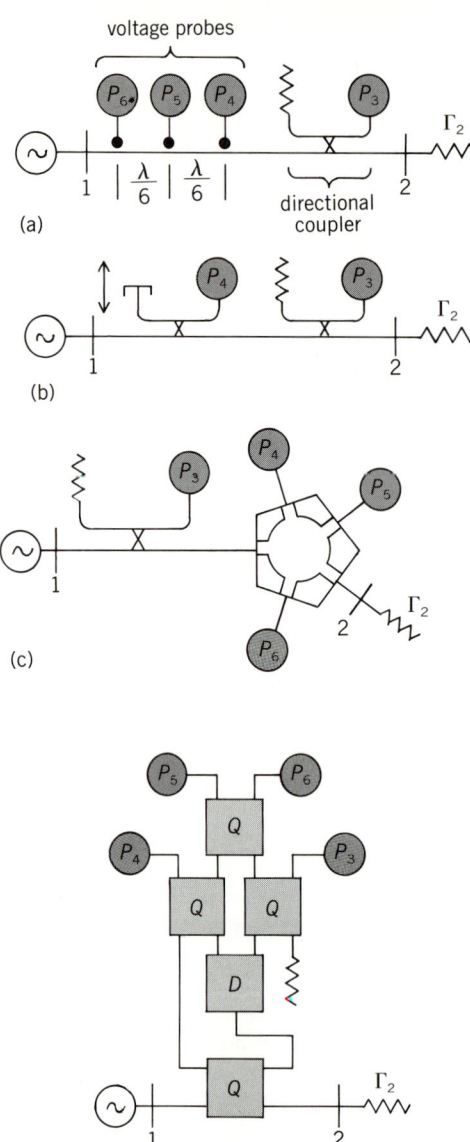
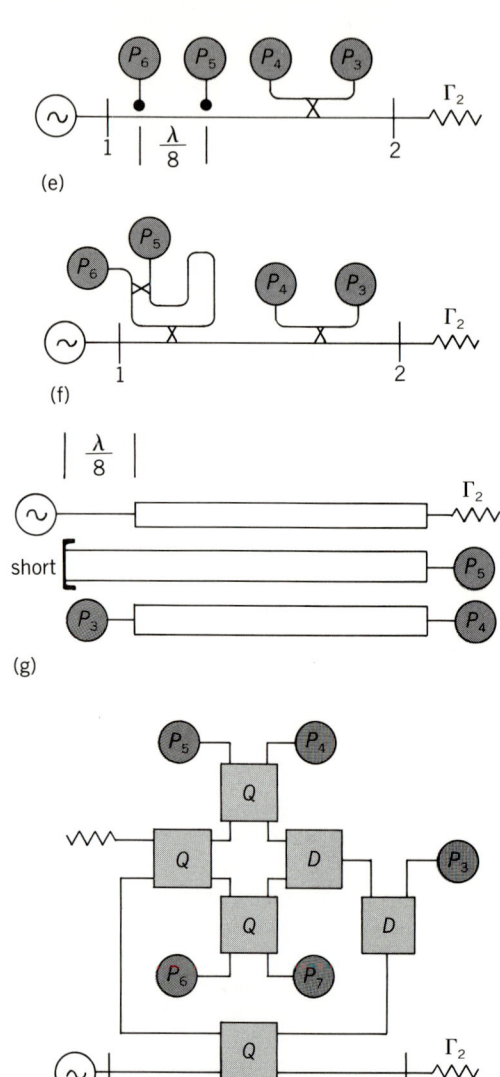

Fig. 1. Examples of multiport reflectometers. P = power detector, Q = quadrature hybrid, D = power divider. Numbers indicate ports. (a) Narrow-band reflectometer with three voltage probes spaced 60° apart to give an optimum q_i spacing of 120° with $|q_i| = 1$. (b) Reflectometer similar to that in a, in which Griffin movable short simulates several probes. (c) Riblet symmetrical five-port junction with a directional coupler, with q points same as in a and octave bandwidth. (d) Engen broadband approximation of reflectometer in a. (e) Engen narrow-band reflectometer with two voltage probes plus a four-port coupler. (f) Griffin broadband version of e. (g) El-Deeb microstrip five-port reflectometer. (h) Hoer broadband seven-port reflectometer with q_i spacing of 90° and $|q_i| = 1$.

and of the reflection coefficient and efficiency of the sidearm detectors. The A_i and q_i are independent of the reflection coefficient of the generator and the termination. These constants must be determined by some calibration procedure such as those described below before a set of equations like Eq. (1) can be solved for Γ_2 and $|b_2|$.

An approximate but useful illustration of how a multiport reflectometer obtains phase information from power measurements can be obtained by assuming that one of the sidearm outputs, say P_3, is proportional mainly to the emerging wave $|b_2|^2$. Most multiports are designed to approach this approximation. Then for that detector, P_3 can be expressed as in Eq. (2), where B_3 is a new constant similar to

$$P_3 = |B_3 b_2|^2 \qquad (2)$$

A_i. Taking the ratio of each remaining detector output power to P_3 gives Eq. (3), where $K_i = |A_i/B_3|^2$.

$$\frac{P_i}{P_3} = K_i |\Gamma_2 - q_i|^2 \qquad i = 4, \ldots, n \qquad (3)$$

Since q_i is fixed for each sidearm of a given multiport reflectometer, this equation is that of a circle of radius $\sqrt{P_i/P_3 K_i}$ centered at q_i on which Γ_2 must lie. It shows that the power readings from the sidearm of a multiport reflectometer can be thought of as giving the radii of circles whose intersection gives the desired value of Γ_2.

For a five-port reflectometer with three detectors,

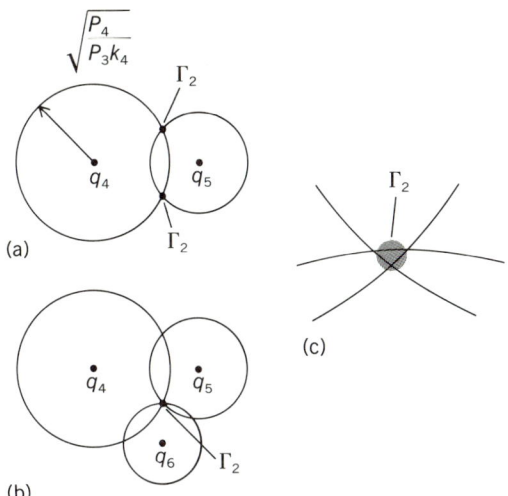

Fig. 2. Process for obtaining reflection coefficient Γ_2 (a) at the intersection of two circles for a five-port reflectometer, and (b) at the intersection of three circles for a six-port reflectometer. (c) Enlargement of b, showing range of values of Γ_2.

only two circles are obtained which intersect in two possible values for Γ_2 (Fig. 2a). Therefore the value of Γ_2 obtained from a five-port reflectometer is not unique, and a prior estimate of Γ_2 is necessary to be able to choose the correct value. It is possible to design a five-port reflectometer so that one choice for Γ_2 has a magnitude greater than 1, and this choice can be discarded for passive terminations.

For a six-port reflectometer with four detectors, three circles are obtained which intersect in only one point (Fig. 2b), thereby determining Γ_2 uniquely.

Least-squares solution. If the intersection of the three circles in Fig. 2b is examined more closely, it will be found that because of measurement error the circles do not intersect in a point but give a range of values for Γ_2 (Fig. 2c). Four or more equations like Eq. (1) have a nonlinear least-squares solution which yields not only Γ_2 but an estimate of the standard deviation σ_2 of Γ_2. This σ_2 is a measure of the spread in the values of Γ_2 shown in Fig. 2c and is a valuable indicator of the quality of the measurement. It can be used to tell the operator when the multiport needs to be recalibrated or that problems exist such as nonlinearity or instability in the detectors, or harmonics or frequency drift in the signal source.

Linear solution. Although Eq. (1) is nonlinear in Γ_2, four equations like Eq. (1) yield the rather simple equations (4) and (5) for the reflection coefficient

$$\Gamma_2 = \frac{\Sigma f_i P_i}{\Sigma g_i P_i} \quad i = 3, \ldots, 6 \quad (4)$$

$$P_2 = \Sigma h_i P_i \quad i = 3, \ldots, 6 \quad (5)$$

and the net power P_2 at the measurement port. Here the g_i and h_i are real constants, and the f_i are complex constants which can be calculated from the $|A_i|$ and q_i or determined directly from some calibration procedure. The linear solution is easier to implement on a small computer, but gives no estimate of σ_2.

Two-port measurements. Two six-port reflectometers have been used quite extensively in the circuit shown in Fig. 3a to measure all the scattering parameters of a reciprocal two-port device. The scattering parameters are complex numbers that describe the relationship between the incident voltage wave amplitudes a_1 and a_2, and the emergent voltage wave amplitudes b_1 and b_2. They are determined by the characteristics of the two-port device and are independent of the a and b waves. The radio-frequency signal from the source is divided into two channels by the power divider and is applied to both reflectometers at the same time. The ratio b_1/a_1 is seen by reflectometer 1 as a reflection coefficient Γ_1. Likewise reflectometer 2 measures $\Gamma_2 = b_2/a_2$. These two reflection coefficients are related to the scattering parameters, S_{11}, S_{22}, S_{12}, and S_{21}, of the two-port device under test by Eq. (6), where Δ is

$$\Gamma_2 S_{11} + \Gamma_1 S_{22} - \Delta = \Gamma_1 \Gamma_2 \quad (6)$$

defined by Eq. (7). Three equations like Eq. (6) are

$$\Delta = S_{11} S_{22} - S_{12} S_{21} \quad (7)$$

needed to solve for S_{11}, S_{22}, and Δ. Then S_{12} is obtained from Eq. (7). These equations are generated by measuring Γ_1 and Γ_2 for three different settings of the phase shifter ϕ. The values of phase do not need to be known since they do not appear in the equations; they simply generate different sets of Γ_1 and Γ_2.

Usually more than three different phase settings are used so that a least-squares solution for the S_{ij}

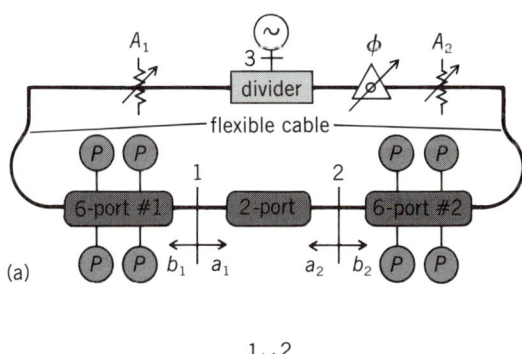

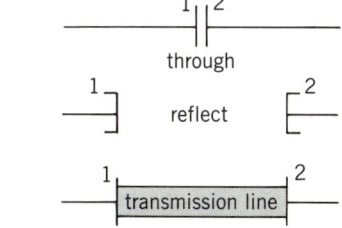

Fig. 3. Measurement of all of the scattering parameters of a two-port device with two six-port reflectometers. (a) Basic circuit. A_1 and A_2 are variable attenuators, ϕ is a variable phase shifter, a_1 and a_2 are the amplitudes of the voltage waves incident on the two-port device, and b_1 and b_2 are the amplitudes of the voltage waves emerging from the two-port device. (b) Steps required to calibrate the two six-ports.

can be obtained from four or more equations like Eq. (6). These extra measurements give a second level of redundancy (the first is in the measurement of each Γ) to further indicate the quality of the measurement.

The measurement system shown in Fig. 3a differs from most other network analyzers in that the six-port reflectometers can be made small enough and portable enough to connect directly to the device under test, eliminating errors caused by flexible cables or arms between the measurement system and the device being measured. Flexible cables "behind" each six-port do not enter into the system calibration except in the measurement of the phase shift through nonreciprocal two-ports. The system is simpler than most other network analyzers in that no local oscillator is needed and no phase detectors are required.

Typical performance. A six-port reflectometer which uses thermistor-type power detectors can have a resolution of about 2×10^{-5} in measuring Γ_2. If each reflectometer in the dual six-port automatic network analyzer of Fig. 3a has this resolution, the resolution in measuring S_{12} will be about 10^{-5}. This fixed resolution in S_{12} translates into a resolution of 0.001 dB for a 20-dB attenuator, 0.01 dB at 40 dB, and 0.1 dB at 60 dB.

The systematic errors in measuring S_{12} appear to be less than the resolution in measuring S_{12}. The systematic errors in measuring S_{11}, S_{22}, or Γ are determined primarily by the systematic errors in the standards used to calibrate the six-ports.

Calibration. A six-port reflectometer must be calibrated to determine the $|A_i|$ and q_i in Eq. (1), or the f_i, g_i and h_i in Eqs. (2) and (3). It turns out that it is actually easier to calibrate two six-ports than it is to calibrate a single six-port, because more equations can be generated with fewer standards. In fact, a dual six-port automatic network analyzer having mating connectors, such as 14-mm (0.6-in.) or 7-mm (0.3-in.) connectors, requires only a single standard to calibrate the complete system. The standard can be a length of precision transmission line, which is the most accurate impedance standard available at microwave frequencies. A single six-port or a dual six-port with mating but not identical connectors, such as type N or SMA, requires more than one impedance standard to complete the calibration.

The steps in calibrating the dual six-port automatic network analyzer with sexless connectors are shown in Fig. 3b. The measurement planes are connected together, then a highly reflecting termination such as a short or open is connected to one six-port and then to the other six-port. Finally a length of precision transmission line is connected between the two six-ports. This "through-reflect-line" calibration technique assumes that the scattering parameters S_{11} and S_{22} of the line are zero. The scattering parameter S_{12} of the line and the reflection coefficient Γ of the highly reflecting termination do not need to be known. Their parameters are determined along with the parameters describing the two six-ports.

If the connectors at the measurement ports are sex-type, the same calibration procedure is used except that the highly reflecting termination cannot be connected to both six-ports. In this case, one procedure is to connect an open-circuit termination to each measurement port and assume that the magnitude of their reflection coefficients are equal. The calibration is completed by connecting to one measurement port a short whose reflection coefficient phase angle is known or assigned. The reflection coefficient magnitude of the short need not be known.

It is possible to calibrate a single six-port as well as a dual six-port by taking advantage of the fact that a six-port can be mathematically reduced to an equivalent four-port reflectometer. This is done by measuring the sidearm powers when a collection of five or more different terminations of unknown reflection coefficient Γ are connected to the measurement port. Conventional calibration techniques that apply to a four-port can then be used. For example, a short, open, and matched termination, all with known values of Γ, can be used to calibrate a four-port, and also to calibrate a six-port after the six-port has been reduced to a four-port. Sliding terminations have been used successfully to provide the measurements needed to reduce a six-port to a four-port, and then to calibrate the resulting four-port.

For background information see HIGH-FREQUENCY IMPEDANCE MEASUREMENT; LEAST-SQUARES METHOD; MICROWAVE; MICROWAVE TRANSMISSION LINES; REFLECTION AND TRANSMISSION COEFFICIENTS in the McGraw-Hill Encyclopedia of Science and Technology.

[CLETUS A. HOER]

Bibliography: G. F. Engen, Calibrating the six-port reflectometer by means of sliding terminations, *IEEE Trans. Microw. Theory Tech.*, MTT-26(12):951–957, December 1978; G. F. Engen, The six-port reflectometer: An alternative network analyzer, *IEEE Trans. Microw. Theory Tech.*, MTT-25(12):1075–1080, December 1977; G. F. Engen and C. A. Hoer, Thru-reflect-line: An improved technique for calibrating the dual six-port automatic network analyzer, *IEEE Trans. Microw. Theory Tech.*, MTT-27(12):987–993, December 1979; C. A. Hoer, A network analyzer incorporating two six-port reflectometers, *IEEE Trans. Microw. Theory Tech.*, MTT-25(12):1070–1074, December 1977.

Mycobacteria

In addition to the well-known *Mycobacterium tuberculosis*, which causes tuberculosis in humans, some rather obscure and opportunistic mycobacteria species have recently obtained medical notoriety because of their ability to colonize and subsequently infect the compromised human host.

Species of *Mycobacterium* share certain characteristics—notably, they are rod-shaped, are acid-fast, and have a relatively slow growth rate. Isolation on culture media takes up to 5 days for rapid growers and up to 8 weeks for the majority of slow growers. Growth of most species requires complex media compositions containing lipids, proteins, carbohydrates, and trace elements.

According to their pathogenicity in humans and animals, mycobacteria species can be divided into three groups: (1) obligate pathogens that in very small numbers will infect healthy individuals upon exposure and may cause fulminant tuberculosis; (2) potential pathogens that usually colonize only the compromised host, but subsequently may cause an infection called mycobacteriosis; (3) ubiquitous saprophytes that are not known to cause disease.

Obligate pathogens. In addition to *M. tuberculosis*, *M. bovis* and *M. africanum* belong to the group of obligate human pathogens. The three species are genetically indistinguishable by current lab techniques, and each can cause tuberculosis in all its manifestations.

Mycobacterium bovis is the agent of tuberculosis in cattle and other wild and domestic ruminants. It is transmitted to humans mostly by ingestion, and it used to be the prime cause of cervical lymphadenitis (scrofula) in children in North America before the eradication of bovine tuberculosis by slaughtering of tuberculous cows earlier in this century.

A bovine mycobacterium strain attenuated in its virulence for cattle and humans by French scientists in 1908 became the basis for the BCG (Bacillus Calmette-Guerin) vaccine that can be used for active antituberculous immunization. In North America and Europe this vaccine has been used with varying success to retard the metastatic growth of certain malignant carcinomas by unspecific stimulation of the reticuloendothelial system.

Mycobacterium africanum appears to be a central and west African variety of the tubercle bacillus that is biochemically similar to *M. bovis*, because the niacin and nitrate reductase tests are not consistently positive in both species and both are resistant to the antituberculous drug pyrazinamide.

Infections by both species or a BCG vaccination will result in a tuberculin skin test conversion in an individual who had no prior exposure to the tubercle bacillus or these related species.

Potential pathogens. Potential pathogens are mycobacteria causing skin ulcerations, often with a wide variety of associated complications in the susceptible individual. In addition to *M. leprae*, the causative agent of leprosy, *M. marinum* is a potential pathogen of humans and of fresh- and salt-water fishes. *Mycobacterium marinum* can infect even minor lacerations of the extremities to cause swimming pool granuloma in humans. Originally diagnosed in swimmers, the organism is isolated more often now from ornamental fish, from fish tanks, or from persons who have reached into such infected tanks with lacerated and unprotected hands. The organism prefers growth temperatures of 77–90°F (25–32°C) and is photochromogenic, that is, develops bright yellow pigment only after light exposure. It does not reduce nitrate to nitrite, but hydrolyzes polysorbate. A combination of antituberculous chemotherapy and surgery is usually successful in curing this rather protracted, chronic ulceration of the skin.

Mycobacterium ulcerans and *M. haemophilum* are fastidious mycobacteria that cause skin ulcerations prevalent in Africa (Buruli ulcer), Australia, and the Near East.

Mycobacterium fortuitum and *M. chelonei* are ubiquitous, rapidly growing mycobacteria that can infect soft tissue as well as bone after traumatic injuries sustained outdoors. Invasive procedures or hemodialysis can precipitate a disseminating infection if aseptic techniques are inadequate. Both species are biochemically very active; they are highly resistant to antituberculous drugs and most antibiotics, except perhaps certain aminoglycosides. Debridement of necrotic tissue and draining of purulent material are essential. Chemotherapy may help, but usually not in cases of pulmonary mycobacteriosis in the elderly. Fortunately, such cases are seen infrequently.

Scrofula (or mycobacterial infection of the cervical lymph nodes) of children of less than 12 years of age is caused by *M. avium-intracellulare* or *M. scrofulaceum* (rarely by *M. tuberculosis*). *Mycobacterium avium-intracellulare* is nonchromogenic; *M. scrofulaceum* is scotochromogenic because it has an orange-yellow pigmentation that is independent of light exposure. *Mycobacterium scrofulaceum* possesses urease activity. However, both species share serologic and chemical similarities, and both have a high degree of resistance to antituberculous drugs. Surgical excision of infected and draining lymph nodes is the treatment of choice.

In many parts of the world, including the United States, *M. avium-intracellulare* is also an important opportunistic pulmonary pathogen. It is notorious for colonizing chronic lung lesions and old scars of compromised or elderly patients who may have been successfully treated for tuberculosis. Because of the organism's drug resistance and the host's impaired defense at an anatomic site, ensuing extensive infiltrates or cavities of the lungs offer only a poor prognosis, and a high rate of morbidity and mortality.

Granulomata, in organs such as liver and bone marrow, resulting from disseminated *M. avium-intracellulare* disease spreading via the bloodstream, is extremely rare, but has been seen recently in individuals with acquired immunodeficiency syndrome (AIDS). *Mycobacterium avium-intracellulare* is the most important bacterial infection of those with AIDS; it is extremely difficult to treat because of the bacteria's drug resistance and the toxic side effects of multiple drug regimens.

Only virulence experiments in chickens can reliably differentiate between *M. avium* and *M. intracellulare*. Serologic tests (Schaeffer's serotyping) are helpful in many instances. However, the test often indicates a considerable number of so-called intermediate strains; known biochemical tests are useless for making a distinction between the two species. Except for strains of *M. avium* that are pathogenic for birds and rabbits, the majority of *M. avium-intracellulare* strains are ubiquitous in soil and water. Drinking water has been implicated as a source of *M. avium-intracellulare* colonization.

Other nonchromogenic, slow-growing mycobacteria that are responsible for pulmonary mycobacteriosis in humans are *M. xenopi*, and the two newly described species *M. malmoense* and *M. shimoidei*. *Mycobacterium xenopi* can be waterborne, grows in water temperatures of up to 115°F (46°C), and may cause nosocomial infections.

Photochromogenic mycobacteria (listed in decreasing order of preponderance for pulmonary mycobacteriosis) are *M. kansasii*, *M. szulgai*, *M. simiae*, and *M. asiaticum*. Important tests to identify these species are urease activity, nitrate reduction, and polysorbate hydrolysis. They are slow growers and prefer growth temperatures of 95–99°F (35–37°C). *Mycobacterium kansasii* and *M. xenopi* are not highly resistant to the major antituberculous drugs, and so mycobacterioses by both species can be managed therapeutically with similar drug regimens as are used to treat tuberculosis.

Mycobacteriosis. Pulmonary mycobacteriosis is radiologically and clinically indistinguishable from pulmonary tuberculosis. Even positive microscopic findings in pulmonary secretions will not reveal the identity of the acid-fast pathogen. Only the isolation of the organism on culture medium with subsequent identification by biochemical tests and drug susceptibility studies to antituberculous drugs will provide the definite diagnosis as well as invaluable information regarding effective chemotherapy.

The difference between mycobacteriosis and tuberculosis is of epidemiologic and etiologic nature. In contrast to tuberculosis, transmission of mycobacteriosis via human contact has not been reported. The etiology of mycobacteriosis is still somewhat of a mystery. It is clear, however, that many mycobacteria live in the environment and may colonize the susceptible human host by various routes. Exposure to mycobacteria and subsequent colonization may take place repeatedly or continuously by large numbers of organisms. Eventual mycobacterial superinfection of existing lesions or foci infected earlier by other microorganisms may be the result of such massive colonization, particularly in the immunocompromised patient. On the other hand, mycobacterial colonization without outright infection may merely aggravate other pathologic processes, such as chronic lung disease, neoplastic disease, or emphysema, that in turn may be accompanied by opportunistic fungal and bacterial infections, making it extremely difficult to establish a definite and precise diagnosis.

For background information *see* LEPROSY; MYCOBACTERIAL DISEASES; TUBERCULOSIS in the McGraw-Hill Encyclopedia of Science and Technology.

[KURT D. STOTTMEIER]

Bibliography: A. I. Braude (ed.), *Medical Microbiology and Infectious Disease*, 1981; A. M. Macher et al., *Ann. Intern. Med.*, 99:782, 1983; A. Timpe and E. H. Runyon, *J. Lab. Clin. Med.*, 44:202, 1954; E. Wolinsky, *Amer. Rev. Resp. Dis.*, 119:107, 1979.

Mycoplasma

Mycoplasma hominis is one of 10 distinct species within the microbial order Mycoplasmatales which are indigenous to the human host (Table 1). This organism has been isolated most commonly from the genitalia of both men and women, and more rarely from the oral cavity; it also has been found occasionally in primates and as a contaminant of cell cultures. The first isolates of *M. hominis* came from patients with genital lesions and were thought to represent etiologic agents of these lesions. In several studies of men with nongonococcal urethritis, the mycoplasma was recovered from about 25% of cases. However, as 75% of cases still required a cause, there was some doubt about the pathogenic association of the mycoplasma. Later, more extensive surveys showed that *M. hominis* could be recovered from normal people, suggesting that it was a member of the autochthonous microflora of the genital mucosa. It has been found about twice as frequently in the presence of inflammation, which suggests that this condition facilitates growth. Thus, despite the fact that this mycoplasma species was among the first ever isolated from humans, its true role as a pathogenic organism has remained controversial. In recent years, evidence for pathogenicity of *M. hominis* has been strengthened by improved methods of isolation and serologic study, coupled with the increased ability of scientists to recognize and exclude other pathogens from disease processes. This article will consider some of the properties of *M. hominis* and its potential as a cause of human disease.

Culture characteristics and biologic features. *Mycoplasma hominis* grows readily on many mycoplasma media formulations, and is less fastidious in nutritive requirements and incubation conditions than most of the other species affecting humans. A source of cholesterol, usually provided by serum, and factors supplied by yeast extracts are required. The organisms also will grow on standard sheep or horse blood agars used for diagnostic bacteriology, although small colony size (0.1–0.2 mm) makes them hard to recognize against the opaque media. On good-quality transparent media, growth usually can be recognized microscopically within 48–72 h of incubation. Growth occurs either aerobically or anaerobically over a broad pH range (6.0–8.0). Typical colonies have the "fried-egg" or nipplelike appearance characteristic of many mycoplasma species, wherein the central part of the colony is deeply embedded in the agar matrix while the periphery is more superficial. The colony surface may appear roughened due to the presence of small vesicles or "large bodies" in the peripheral zone, representing degenerative forms. In broth media, individual organisms appear as tiny (0.2-micrometer) coccobacilli which grow into thin filaments that suddenly

Table 1. Mycoplasmas indigenous to humans

Species	Pathogenic status	Biochemical pattern			G + C, mol %
		Dextrose	Arginine	Urea	
Mycoplasma buccale	Normal flora	−	+	−	25.0–26.4
M. faucium	Normal flora	−	+	−	NT*
M. fermentans	Normal flora	+	+	−	27.5–28.7
M. genitalium	Genital pathogen	+	−	−	31.8
M. hominis	Genital pathogen	−	+	−	27.3–33.7
M. lipophilum	Normal flora	−	+	−	NT
M. orale	Normal flora	−	+	−	24.0–28.2
M. pneumoniae	Respiratory pathogen	+	−	−	38.6–40.8
M. salivarium	Normal flora	−	+	−	27.3–31.4
Ureaplasma urealyticum	Genital pathogen	−	−	+	26.9–29.8

*Not tested.

septate, producing many progeny. Ultrastructurally, these are prokaryotic cells bound only by trilaminar membranes, like other species of mycoplasmas.

Biochemically, *M. hominis* falls into the mycoplasma group which hydrolyzes arginine with production of ammonia but does not cleave urea or ferment carbohydrates. The guanine-plus-cytosine content of its deoxyribonucleic acid (DNA) is estimated at 27.3–33.7 mol %, which is at the low end of the spectrum for various mycoplasma species (Table 1). The genome size of mycoplasmas generally is 5×10^8 daltons, roughly one-sixth that of other bacteria such as *Escherichia coli*.

Different strains of *M. hominis* show considerable genotypic and phenotypic diversity. There is evidence for at least seven serotypes, which complicates the serologic identification of clinical isolates as well as determination of antibody responses. Among the various serologic procedures that have been described, the enzyme-linked immunosorbent assays (ELISA) appear most promising. It is not known if strain variations play a part in the degree of virulence demonstrated by *M. hominis*. There is recent evidence that strains isolated from pathologic processes have the ability to attach to cells, which could represent a virulence factor or marker. Mycoplasma viruses (phages) have been seen in *M. hominis* cells by electron microscopy, which suggests another determinant sometimes confering virulence.

Epidemiology. Various surveys have revealed *M. hominis* colonization in about 10% of normal male urethras and 25–30% of normal vaginas. These figures double in sexually promiscuous individuals, indicating sexual transmissibility of the organism. Transmission from mother to newborn infant also has been documented. Surveys for oropharyngeal colonization with *M. hominis* show the mycoplasma to be present in 1–3% of adults, but in 15% of those engaging in orogenital sexual activity. The prevalence of *M. hominis* in normal people complicates interpretation of its role in disease states, and indicates the need for careful control studies in clinical research using subjects matched for age, sex, parity, and sexual experience.

Clinical manifestations. *Mycoplasma hominis* has been implicated in a wide variety of diseases (Table 2). Many studies are difficult to interpret because they lack suitable control groups, or because comprehensive microbiologic studies have not been done (for anaerobic bacteria, viruses, *Chlamydia trachomatis*, *Ureaplasma urealyticum*). It is clear, however, that there are instances when *M. hominis* has been isolated from normally sterile body fluids or internal tissues in pure culture; whether these situations reflect primary or opportunistic invasion of the mycoplasma is less certain.

In the newborn infant, there are a number of reports of *M. hominis* meningitis or meningoencephalitis, with fatalities, serious neurologic sequelae, or successful outcome of treatment. Skin abscesses and lymphadenitis also have been reported in which *M. hominis* was the only microorganism isolated. Some of these cases followed trauma to the skin, such as forceps abrasions or placement of scalp electrodes for fetal monitoring. Asymptomatic colonization of the neonatal pharynx, skin, and genital mucosa also has been described, at times in association with lower-than-average birth weight. The mycoplasma has been recovered from discolored (but not clear) amniotic fluid samples taken for antenatal diagnostic purposes, indicating that it can penetrate intact

Table 2. Pathogenicity of *Mycoplasma hominis*

Host	Disease associations
Neonate	Meningitis, conjunctivitis, subcutaneous abscesses, pericardial effusion
Children	None described
Adults	Postpartum sepsis, pyelonephritis, pelvic inflammatory disease, cervicitis, prostatitis, nonspecific vaginitis, interstitial cystitis, parametritis
	Wound infection, brain abscess
Immunodeficient	Perirenal abscess, septic arthritis
Volunteers	Pharyngitis
Experimental animals	Arthritis, subcutaneous abscesses, salpingitis, pyelonephritis

membranes and potentially produce congenital infections.

A wide variety of urogenital pathologic processes associated with *M. hominis* has been described in adult patients (Table 2). The most significant of the associations from the medical management viewpoint are postpartum sepsis, pelvic inflammatory disease, and pyelonephritis. Mycoplasma infection of bloodstream during childbirth appears to be a benign and self-limited type of disease; however, this morbid condition is economically significant in terms of extended hospitalization, laboratory diagnostic studies, and empirical therapy of mother and infant. The occurrence of *M. hominis* salpingitis has been well documented with cultures collected through the laparoscope and thus not subject to incidental contamination as are cultures obtained transvaginally. It is not known if mycoplasma salpingitis has the same long-term consequences as do gonococcal infections. The association of *M. hominis* as the only potential pathogen in pyelonephritis also has important implications for management of this condition. In one series, 10% of cases studied yielded the mycoplasma alone. The pathogenicity of *M. hominis* in some of the other conditions listed in Table 2 is less certain because of the problem of colonization levels in matched control subjects and the need to exclude many other microorganisms. Thus, the role of *M. hominis* in nongonococcal urethritis, prostatitis, nonspecific vaginitis, and cervicitis remains enigmatic; currently, *C. trachomatis* and *U. urealyticum* are more highly implicated in these conditions.

That some strains of *M. hominis* have pathogenic potential has been demonstrated through experimental inoculations of humans and animals. Volunteers given the organism by the intranasal and oropharyngeal routes developed acute pharyngitis; however, surveys conducted in patients with naturally occurring pharyngitis have not shown a significant increase in isolation of the mycoplasma above the normal colonization rate. Subcutaneous abscess formation in mice, and salpingitis, parametritis, pyelonephritis, and arthritis in primates have been produced.

Laboratory diagnosis and treatment. *Mycoplasma hominis* can grow in several standard bacteriologic media, and may be isolated in clinical laboratories. Generally, however, the organisms are not detected, since agar cultures are not kept sufficiently long. Exceptions are late subcultures from blood broth bottles or agar anaerobic plates. Species identification of isolates requires serologic methods, since there are other mycoplasmas in humans which share similar biologic properties (notably *M. salivarium*).

Generally, *M. hominis* is susceptible to the tetracycline group of antibiotics, clindamycin, and lincomycin, though development of resistance to each of these drugs has been reported. There is intermediate sensitivity to the aminoglycosides and chloramphenicol, and all strains are resistant to erythromycin, the penicillins, and polymyxin B. Administration of appropriate therapy without laboratory assistance is difficult, since the sensitivity pattern differs from the drugs of choice for other microbial pathogens that would be in the differential diagnosis of most of the conditions under consideration.

For background information *see* ANTIBIOTIC; CLINICAL MICROBIOLOGY; INFECTIOUS DISEASE TRANSMISSION; MYCOPLASMATALES; SEROLOGY in the McGraw-Hill Encyclopedia ot Science and Technology.

[WALLACE CLYDE, JR.]

Bibliography: P.-A. Mardh, B. R. Møller, and W. M. McCormack (eds.), International Symposium on *Mycoplasma hominis*—A Human Pathogen, *Sex. Transm. Dis.*, 10(Supp. 4):225–385, 1983; C. S. Nicol and D. G. ff. Edward, Role of organisms of the pleuropneumonia group in human genital infections, *Brit. J. Vener. Dis.*, 29:141–150, 1953; J. G. Tully and R. F. Whitcomb (eds.), *The Mycoplasmas*, vol. 2, 1979.

Natural language processing

Natural language processing involves computer analysis and generation of natural language text. The goal is to enable natural languages, such as English, French, or Japanese, to serve either as the medium through which users interact with computer systems such as data-base management systems and expert systems (natural language interaction), or as the object that a system processes into a more useful form such as in automatic text translation or text summarization (natural language text processing).

In the computer analysis of natural language, the initial task is to translate from a natural language utterance, usually in context, into a formal specification that the system can process further. Further processing depends on the particular application. In natural language interaction, it may involve reasoning, factual data retrieval, and generation of an appropriate tabular, graphic, or natural language response. In text processing, analysis may be followed by generation of an appropriate translation or a summary of the original text, or the formal specification may be stored to serve as the basis for more accurate document retrieval later. Given its wide scope, natural language processing requires techniques for dealing with many aspects of language, in particular, syntax, semantics, discourse context, and pragmatics. (Analysis and generation of spoken natural language, not discussed in this article, also involve techniques for dealing with acoustic phonetics, phonology, stress, and intonation.) *See* SPEECH RECOGNITION.

Parsing. The aspect of natural language processing that has perhaps received the most attention is syntactic processing, or parsing. Most current techniques for parsing an input string of words involve (1) a description of the allowable sentences of the language (the grammar); (2) an inventory of the words of the language with their inflectional, syntactic, and possibly semantic properties (the lexicon); and (3) a processor which operates on the grammar, the lexicon, and the input string (the parser). This processor (1) simply accepts the input string, if grammatically well formed, or rejects it (a recognizer); (2) associates the string, if well formed, with its structure (or structures, if ambiguous) according to the grammar (an analyzer); or (3) associates the string with some other representation, for example, a semantic characterization (a transducer). Syntactic processing is important because certain aspects of meaning can be determined only from the underlying structure and not simply from the linear string of words.

One of the oldest parsing techniques, called augmented transition networks (ATNs), grew out of a system for parsing context-free (CF) languages, called recursive transition networks (RTNs). A recursive transition network parser consists of a set of named graphs or networks, each consisting of a set of nodes or states connected by a possibly ordered set of directed labeled arcs. The labels correspond to (1) words or classes of words that can be recognized or "consumed" on the arc; (2) the empty symbol, indicating an arc that can be taken without consuming any input; or (3) the name of a network, which indicates that the next segment of the input string must be recognizable by that network. Each network has a start state and one or more end states. If the parser can move through a network from start to end by consuming a segment of the input string, that segment is said to be recognizable by that network.

An augmented transition network adds to the basic recursive transition network framework the ability to set and test variables or registers, thereby giving it the power to recognize a wider class of languages than a recursive transition network. An augmented transition network is appealing because its grammar is relatively easy to specify. Its weaknesses lie in the simple, uniform control structure provided by the basic augmented transition network (unguided backtracking) and in its power, felt to be more than is needed for recognizing a natural language.

Current trends are to construct parsers and grammars which appear to follow more closely human parsing strategies and which have less power. In particular, researchers have begun to give almost context-free descriptions of natural languages, thereby allowing them to use slightly extended versions of efficient context-free parsing techniques. Such descriptions include generalized phrase structure grammar (GPSG), immediate dominance/linear precedence (ID/LP) grammar, and tree adjoining grammar (TAG).

Semantic analysis. A second phase of natural language processing, semantic analysis, involves extracting context-independent aspects of a sentence's meaning. These include the semantic roles played by the various entities mentioned in the sentence. For example, in the sentence "John unlocked the toolbox," "John" serves as the agent of the unlocking and "the toolbox" serves as the object; in "This key will unlock the toolbox," "this key" serves as the instrument. Context-independent aspects of sentence meaning also include quantificational information such as cardinality, iteration, and dependency. For example, in the sentence "In every car, the mechanic checked to see that the engine was working," the checking is iterated over each car; the identity of the engine depends on the identity of the car, while that of the mechanic does not; and the cardinality of engines per car is one. Thus there are as many engines as cars, but possibly only one mechanic. The representational formalism used by the system for semantic analysis (for example, first-order predicate calculus, case grammar, conceptual frames, procedures, and so forth) is usually chosen for its ability to convey those aspects of semantics that the system requires for later processing. For example, if temporal position (past/present/future) is not significant, it will not be captured in the formalism.

Most semantic analysis is done by applying pattern-action rules either during parsing or afterward. The pattern part of a rule consists of clauses, each of which specifies the presence of a particular lexical item, usually the head of some syntactic substructure (for example, the main verb of a clause or sentence or the head noun of a noun phrase); or a particular syntactic substructure (for example, a relative clause, to be interpreted as a restriction on the class described by the rest of the noun phrase). A pattern clause may also specify a test on another part of the current substructure. The action part of a rule usually calls for building a piece of semantic representation, often requiring the semantic analysis of some other part of the syntactic substructure. For example, there may be a pattern-action rule associated with "unlock" as the main verb of a clause. A test in the pattern may require that the subject of the clause be interpretable as an animate agent. The rule's action may call for the inclusion of a conceptual frame for the concept "unlock" as part of the semantic representation of the sentence. The rule's action may further specify that the agent role of the frame be filled by the semantic interpretation of the subject of the clause, and that the object role be filled by the semantic interpretation of the direct object. In some systems, a rule can have optional pattern clauses and actions: thus, the rule pattern for "unlock" might optionally specify a "with" prepositional phrase whose noun phrase object can be interpreted as a tool. If so, the rule action might additionally call for instantiating the instrument role in the frame with the semantic interpretation of the prepositional phrase object.

Contextual analysis. Given that most natural languages allow people to take advantage of discourse context, their mutual beliefs about the world, and their shared spatio-temporal context to leave things unsaid or say them with minimal effort, the purpose of a third phase of natural language processing, contextual analysis, is to elaborate the semantic representation of what has been made explicit in the utterance with what is implicit from context. Two major linguistic devices that contextual analysis must deal with are ellipsis and anaphora.

Ellipsis. Ellipsis involves leaving something unsaid. To handle ellipses computationally, techniques are required for recognizing that something is indeed missing and for recovering the ellipsed material. When the utterance is a sentence fragment and not a complete sentence, it is fairly easy to recognize that something is missing. An example appears in the following sequence:

User: What is the length of the JFK?
System: <some number of feet>
User: The draft?

On the other hand, since parsers are usually designed for well-formed input, either the system's grammar must be revised to accept sentence fragments or a special error-recovery routine must take over after the parser fails. When an utterance is syntactically well formed, it may still be elliptic in that some needed conceptual material is missing, as in the following example:

User: What maintenances were performed on plane 3 in May 1971?
System: <list of maintenances>
User: What maintenances were performed on plane 48?

In the user's second question, the time period of interest is missing, and the question should not be answered until it is recovered. (It is clearly May 1971.)

The primary technique for recovering ellipsed material is a simple one, based on semantic features. For sentence fragment ellipses, the previous discourse is searched for the most recent utterance containing a constituent with the same features as the fragment. The utterance minus that constituent is taken as the ellipsed material. For conceptual ellipses in a syntactically well-formed sentence, the previous discourse is searched for the most recent utterance with a constituent having the required semantic features. That constituent is taken to be the ellipsed material. In each case, a new well-formed sentence is then constructed and processed as if the ellipsis had never occurred. For instance, in the first example, both "length" and "draft" are properties of ships. Thus, given the fragmentary utterance "The draft?," What is ——— of the JFK?" is found as the ellipsed material. The question "What is the draft of the JFK?" is then interpreted and answered normally. This technique works often, but does not constitute a general solution. A more powerful solution has been developed based on recognizing a user's goals in producing an utterance, but has been found computationally efficient only in very narrow task-oriented domains. This will be discussed below.

Anaphora. Anaphoric expressions are very simple words or phrases which cospecify something previously evoked by the discourse or are strongly associated with something so evoked. Instances of anaphora include definite pronouns such as "he," "she," "it," "they," and definite noun phrases such as "the mechanic" and "the cars." The problem is that anaphora can be interpreted only in context, and the semantic intepretation of a sentence is not complete until all anaphoric expressions are resolved and the cospecified entities identified.

Early computational approaches to anaphora resembled those for dealing with ellipses: entities described in previous sentences were searched for the most recently mentioned one with appropriate semantic features. Now recency has been replaced by the notion of focus as a basis for anaphora resolution. Immediate focus reflects the particular thing the speaker is talking about; global focus involves things associated with it or in which it participates, and gives a sense of what may be talked about next. Techniques have been developed for tracking immediate focus, projecting ahead from the current utterance what may be focused on in the next one. This is useful for resolving anaphora, in that it predicts what entities are likely to be respecified anaphorically.

Pragmatics. A fourth phase of natural language processing, pragmatics, takes into account the speaker's goal in uttering a particular thought in a particular way—what the utterance is being used to do. In an interaction, this will influence what constitutes an appropriate response. For example, an utterance which has the form of a yes/no question or an assertion may have the goal of eliciting information (for example: "Do you know how to delete a control-Z?," "I can't get the set file protection command to work."). Because it is inappropriate (and possibly at times dangerous) to take a user's utterances literally or to assume that the user will take those of the system literally, computational techniques must be devised for relating the syntactic shape and semantic content of an utterance to its pragmatic function.

Plan recognition. One important approach to this problem has been to view language understanding as plan recognition. The actions (either communicative or physical) that constitute the plan may be motivated in one of two ways: goals in the world that the person wants to accomplish, for which he or she needs to elicit or offer aid or information; or aspects of an already ongoing interaction that need attention—for example, confusion over the speaker's foregoing utterance may lead the listener to seek clarification.

A user seeking particular information from the system illustrates clearly a plan-recognition approach to language understanding. The user's utterance—a well-formed sentence or an ellipsed fragment—is taken as a request for information that the user believes he or she needs in order to accomplish some goal, a goal which is not as yet presumed to be known to the system. Just as a medical diagnosis system uses rules which link findings back to those diseases which commonly manifest them, the plan-recognition system uses rules which link utterances back to those domain goals which need the intended information in order to be achieved. For example, consider the utterance "The train to Windsor?" made to a system serving as train information clerk.

The system interprets this as a request: the user wants to know some property of that train in order to fulfill his or her goal. The system then tries to figure out what that goal is, in order to figure out what information the user might be requesting. There are only two possible goals considered: meeting a train and boarding one. The description "train to X" does not match that of incoming train, so the user's goal is taken to be boarding the Windsor train. To board a train, one needs to know its departure time and track. Since the system does not have evidence of the user's knowing either of these properties, it responds with both: "It leaves at 3:15, from track 7." Currently, it is only by limiting the domain, and hence the range of possible goals the system needs to consider, that such a plan-based approach to pragmatics and natural language processing becomes feasible.

Cooperative principle. Because pragmatics acknowledges language use, it also acknowledges expectations that speakers and listeners have about the form and content of utterances, based on normal conventions of use. This has been well described by the philosopher Paul Grice, who noted that speakers acknowledge a "cooperative principle" of conversation (by either upholding it or purposely flouting it), which he further specified in terms of conversational maxims of quantity, quality, manner, and relation. For example, the quantity maxim states: "Make your contribution as informative as is required (for the current purposes of the exchange). Do not make it more informative than is required."

The cooperative principle and its maxims are important to natural language processing because, if a system does not behave in accord with normal conventions of use, the user is likely to be confused or misled by the system's behavior. Conversely, if the system does not interpret the user's behavior in terms of normal conventions of use, the system is unlikely to understand the user correctly, if at all. In particular, the cooperative principle and its maxims reveal a method of implicit communication which Grice termed implicature. An implicature is basically an aspect of an utterance's interpretation which makes no contribution to its truth value (that is, semantics) but constrains its appropriateness in discourse. For example, consider the following discourse:

Q: Is there a gas station on the next block?
R: Yes.

The simple "yes" answer implicates to Q that, as far as R knows, the gas station is able to provide its normal services and hence fulfill Q's probable goal. Q reasons that if R knew that the gas station was closed and hence could not fill Q's needs, R would have said so; that is, R would have said "Yes, but it's closed." Thus a system must be as aware of implicatures (both the user's and its own) as it is aware of what is communicated explicitly.

Overall organization. As for fitting the pieces together, there is no single way that natural language analysis is done. Some systems have a single processor for syntactic, semantic, contextual, and pragmatic analysis, with no distinction made as to the source of that knowledge. Some systems keep the knowledge sources separate but apply them simultaneously, extracting whatever can be derived at the moment and using whatever information is available. Other systems are very modular, separating the knowledge sources and specifying when they should be applied. Efficiency, extensibility, and transportability are some of the important issues to be considered when evaluating a system for natural language analysis.

Natural language generation. Until recently the bulk of research in natural language processing has been directed at natural language analysis. Now researchers have begun to take up seriously the task of natural language generation. Generation is not just the reverse of analysis because the status of user and system are fundamentally different. Systems can be developed which tolerate users' mechanical errors (for example, spelling, typing, and grammatical mistakes), treating them as insignificant variations. Users, on the other hand, may not be able to figure out which aspects of the system's natural language behavior reflect simple nonfluencies (for example, those due to limited lexical or grammatical options) and which embody significant aspects of communication. Moreover, the system's sense of language must be more highly developed for generation, lest it confuse or mislead the user by what it communicates or how. Work on explanation is also a significant aspect of natural language generation.

For background information *see* ARTIFICIAL INTELLIGENCE in the McGraw-Hill Encyclopedia of Science and Technology.   [BONNIE WEBBER]

Bibliography: *Proceedings of the Association for Computational Linguistics*; *Proceedings of the International Conference on Computational Linguistics*; H. Tennant, *Natural Language Processing*, 1981; T. Winograd, *Language as a Cognitive Process*, 1982.

Neoplasia

The correct number of human chromosomes was established in 1956. Just 4 years later investigators succeeded in relating a chromosomal abnormality with a type of neoplasia. The development of banding techniques for metaphase chromosomes in 1970 enabled scientists to undertake detailed studies of chromosome structure and prompted an international meeting to establish a standardized nomenclature. Banding techniques have led to the precise identification of a large number of associations between chromosomal defects and different types of malignancies (see table).

Before 1980, it was generally thought that most chromosomal defects were random and represented secondary aberrations in the neoplastic process. This was in part due to the fact that early banding techniques for tumor cells were not ideal for defining structural defects. Only about 150–250 bands per haploid set were identified, and the chromosomes often had a "fuzzy" appearance. The application of a methotrexate cell-synchronization technique and brief exposure of cells to Colcemid were introduced in 1976, and this method was applied in 1981 for the large-scale study of leukemia, lymphoma, and carcinoma cells at the 320–1200 band stages. With this high-resolution technique, it has become possible to detect chromosomal defects in approximately 95% of patients with acute leukemia and non-Hodgkin's lymphoma. Also, the recent use of mild tissue disaggregating enzymes, selective growth factors and feeder fibroblast layers for cultured tumor cells has facilitated the discovery of different solid tumors with recurrent defects.

Consistency of chromosomal lesions. At present, there are over 40 types of neoplasias known to have specific chromosomal lesions (see table). These lesions fall into two general categories: reciprocal translocations, commonly seen in leukemias and lymphomas; and deletion of a specific band, commonly seen in solid tumors. In a reciprocal translocation, segments from two chromosomes break and exchange places. In a deletion, a piece of a chromosome breaks off and is lost. Less common are a trisomy or extra copy of a chromosome and an inversion in which a segment flips over in the same chromosome. Banding nomenclature allows the breakpoints involved in chromosomal rearrangements to be uniformly designated by chromosome number, arm ("p" meaning short arm and "q" long arm), region, and band and subband numbers. For example, 14q32.3 indicates chromosome number 14, long arm, region 3, band 2, subband 3 (see illustration).

Prior to the development of high-resolution banding of methotrexate-synchronized cells, only 50% of patients with acute leukemia were found to have a clonal chromosomal abnormality in their malignant cells. Also, only a few patients with acute leukemia or non-Hodgkin's lymphoma showed a specific chromosomal defect. With the new technology, it is now possible to find specific defects in at least two-thirds of patients with these diseases. In acute myelogenous leukemias, for example, the most common specific defects include a translocation 8;21 [t(8;21)], a trisomy 8, and an inversion 16, each conferring a different prognosis. In non-Hodgkin's lymphomas, they include a t(8;14) and a t(14;18), also with different prognoses.

A given chromosomal defect can be shared by different related disorders. For example, a translocation t(9;22) is found in most patients with chronic myelogenous leukemia and in some patients with either acute myelogenous leukemia or acute lymphocytic leukemia. Likewise, a t(8;14) by itself can indicate that the patient has either Burkitt's lymphoma, small-cell non-Burkitt's lymphoma, acute lymphocytic leukemia, or immunoblastic lymphoma (see table). Indeed, it is not uncommon that two to five related malignancies share a primary chromosomal defect, suggesting the need of additional steps in the development of a given neoplasia. Also, in the evolution of neoplasia, secondary chromosomal defects may appear, some of them heralding a more rapid progression of the disease process.

Before 1980, only half of solid tumors could be analyzed and, even though most reported cases showed chromosomal abnormalities, no specific lesions were detected. With the new techniques, chromosomal defects are now found in nearly all carcinoma patients (for example, cancer of the lung, ovary and retina). In retinoblastoma, a well-studied malignant childhood tumor of the eye, a small piece of the long arm of a chromosome 13 may be deleted involving loss of band q14. This loss is thought to predispose an individual to neoplasia but may not by itself cause it. For malignancy to develop, the gene locus in the other chromosome 13 must be affected by a second deletion or mutation. Interestingly, in one variety of the disease individuals are born with all the cells of the body missing this segment, while in another only the tumor cells are affected. The absence of the small chromosome segment suggests an association between retinoblastoma and loss of a specific DNA sequence in chromosome 13. This sequence has been identified through restriction enzyme polymorphism and cloned. It is possible that in malignancies with a primary specific chromosomal loss, the missing DNA segment may contain a regulatory "repressor" sequence that normally prevents abnormal cell proliferation.

Oncogenes and fragile sites. The knowledge accumulated on chromosomal abnormalities and genes important in cellular proliferation or development has just begun to merge. In neoplasias with a recurrent reciprocal chromosomal translocation, a ge-

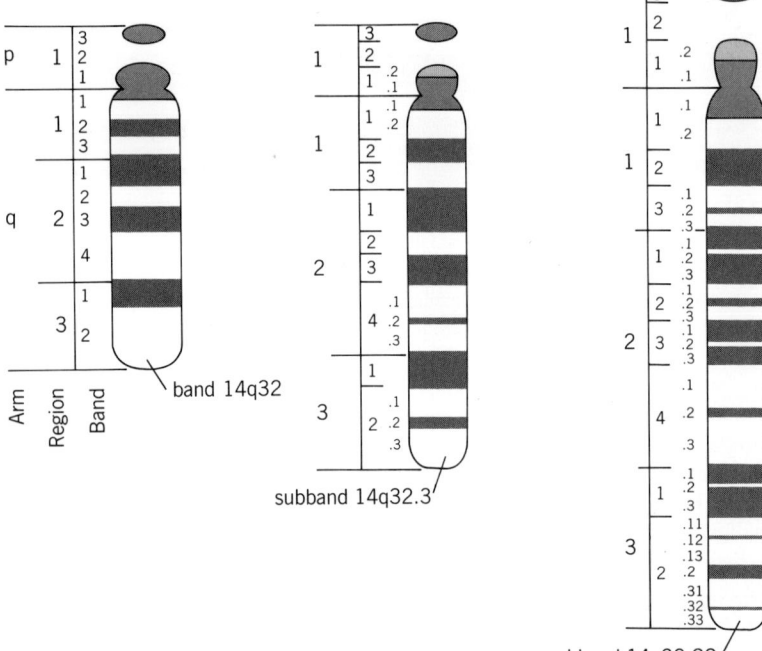

Schematic representation of chromosome 14 at the 400 (left), 500 (center), and 850 band stages (right), illustrating human chromosome nomenclature to designate chromosome arms, regions, bands, and subbands. (*After J. J. Yunis, Chromosomes and cancer: New nomenclature and future directions, Human Pathol., 12(6):494–503, 1981*)

Neoplasms with a known specific recurrent chromosomal defect*

Disease	Chromosome defect†	Breakpoints‡
Leukemias		
Chronic myelogenous leukemia	t(9;22)	9q34.1 and 22q11.21
Acute nonlymphocytic leukemia	inv 3	3q21 and 3q26.2?
	del 5q	5q13 and 5q31
	t(6;9)	6p22 and 9q34
	del 7q	7q22 and 7q32
	+8	
	t(8;21)	8q22.1 and 21q22.3
	t(9;11)	9p22 and 11q23
	t(9;22)	9q34.1 and 22q11.21
	t(15;17)	15q22 and 17q11.2
	inv 16	16p13.2 and 16q22.1
Chronic lymphocytic leukemia	t(11;14)	11q13 and 14q32
	+12	
Acute lymphocytic leukemia	t(1;19)	1q23? and 19q13?
	t(4;11)	4q21 and 11q23
	t(8;14)	8q24.13 and 14q32.33
	t(9;22)	9q34.1 and 22q11.21
	t(11;14)	11p13? and 14q11?
Lymphomas		
Burkitt's, small noncleaved cell (non-Burkitt), large-cell immunoblastic	t(8;14)	8q24.13 and 14q32.33
Follicular small cleaved, follicular mixed, and follicular large-cell	t(14;18)	14q32.3 and 18q21.3
Small-cell lymphocytic, B-cell	t(11;14)	11q13 and 14q32
	+12	
Small-cell lymphocytic, T-cell	inv(14)	14q11.2 and 14q32.3
Diffuse large cell lymphoma	del(16)	6q21 and 6q25
Carcinomas		
Neuroblastoma, disseminated	del 1p	1p31p36
Small-cell lung carcinoma	del 3p	3p14p23
Papillary cystadenocarcinoma of ovary	t(6;14)	6q21 and 14q24
Aniridia-Wilms' tumor	del 11p	11p13
Wilms' tumor	del 11p	11p13
Ewing sarcoma, neuroepithelioma	t(11;22)	q24 and q12
Adenocarcinoma of the colon	del 12q	q22q24
Constitutional retinoblastoma	del 13q	13q14.13
Retinoblastoma	del 13q	13q14
Benign solid tumors		
Mixed parotid gland tumor	t(3;8)	3p25 and 8q21
Meningioma	−22	

*Modified from J. J. Yunis in V. F. Fairbanks (ed.), *Current Hematology*, vol. 3, John Wiley & Sons, 1984.
†A plus or minus sign before a number indicates gain or loss of chromosome, respectively; del denotes deletion, inv inversion, and t reciprocal translocation. Parentheses in the translocation include the two chromosomes involved.
‡The breakpoint nomenclature is as follows: chromosome number, arm p or q, region, band number, and subband when preceded by decimal.

nomic rearrangement is visualized as the main event that sets a stem cell toward a malignant path. This process is thought to involve a "cancer gene" or "oncogene" and a cell-differentiation gene. More than 20 oncogenes have been identified and 17 localized to specific chromosome bands.

Early evidence of the relationship between chromosomal rearrangements and genes in human malignancy was established by the discovery in 1982 of the activated cellular myc oncogene (c-myc) in Burkitt's lymphoma. In this disease of B lymphoid cells, a t(8;14) with breakpoints at bands 8q24.1 and 14q32.3 is often found. In such instance, the oncogene c-myc moves from its normal location at subband 8q24.1 and becomes activated when rearranged with the constant genes of the immunoglobulin heavy chain at subband 14q32.3. As a result of the translocation, c-myc may have increased transcription, though the exact mechanism by which the chromosomal rearrangement triggers malignant growth remains unknown. The very active cell-differentiation genes of a given cell type are somehow involved, helping to activate a rearranged oncogene and possibly also helping to determine the stage of cell differentiation that a given malignancy manifests.

Besides oncogenes and cell-differentiation genes, 16 heritable chromosomal fragile sites have been identified, all inherited in a simple mendelian fashion. Fragile sites do not undergo normal chromosomal condensation and often appear as a gap, displacement, or break of a chromosomal segment. Most but not all of them are expressed when cells are deprived of folate and thymidine in culture. Eight heritable fragile sites are localized at or near the breakpoints of one or two inversions and 6 of 13 specific reciprocal translocations found in leukemias and non-Hodgkin's lymphomas.

A larger class of 51 constitutive fragile sites has

been found in homologous chromosomes of the human and primate genomes. Like most heritable fragile sites, they are expressed in conditions of cellular folic acid-thymidine deprivation. Their expression is dramatically enhanced by caffeine, a DNA repair inhibitor. Twenty of the 51 constitutive fragile sites map at or close to one of the breakpoints found in 24 of 31 specific structural chromosome defects known so far in human malignancy. In addition, eight fragile sites have been found at the same chromosome band where eight oncogenes have been mapped. These findings suggest that fragile sites may represent regulatory sequences of oncogenes and may play an important role in the origin of cancer.

Clinical significance of chromosomal defects. Specific chromosomal defects are clonal in nature and are present in the malignant tissue throughout the lifetime of the disease. They are central to the neoplastic process and have a direct bearing on patient prognosis and survival. For example, patients with a t(9;22) and chronic myelogenous leukemia have a better prognosis than those without this defect. Patients with acute myelogenous leukemia showing an inversion 16 tend to have a uniform and sustained complete remission and may be cured with standard chemotherapy. In contrast, patients with acute myelogenous leukemia and complex chromosomal abnormalities carry a very poor prognosis. In acute lymphocytic leukemia, chromosomal defects and immunologic cell markers are already widely used to establish an accurate diagnosis.

With the availability of different types of treatment for patients with acute leukemia, such as those employing new cancer-fighting drugs or bone marrow transplantation, chromosome analysis is becoming a valuable asset in selecting specific types of therapy for groups of patients with differing prognoses. Chromosome analysis is also emerging as an important prognostic tool in non-Hodgkin's lymphomas. At present little is known about specific chromosomal defects in the majority of carcinomas.

For background information see CHROMOSOME ABERRATION; HUMAN GENETICS; ONCOLOGY in the McGraw-Hill Encyclopedia of Science and Technology. [JORGE J. YUNIS]

Bibliography: A. A. Sandberg (ed.), *The Chromosomes in Human Cancer and Leukemia*, 1980; J. J. Yunis, The chromosomal basis of human neoplasia, *Science*, 221:227–236, 1983; J. J. Yunis, Clinical significance of high resolution chromosomes in the study of acute leukemias and non-Hodgkin's lymphomas, V. F. Fairbanks (ed.), *Current Hematology*, vol. 3, 1984; J. J. Yunis et al., High-resolution chromosomes as an independent prognostic indicator in adult acute nonlymphocytic leukemia, *New Engl. J. Med*, 311:812–818, 1984; J. J. Yunis and A. Lee Soreng, Constitutive fragile sites and cancer, *Science*, 226:1199–1204, 1984.

Neuron

The relationship between the neuronal cytoskeleton and axonal transport has been the subject of neurotoxicologic studies, studies in developing animals, and biochemical studies describing the interactions between cytoskeletal elements. In the optic axons of a newborn animal, the slow transport of the cytoskeletal elements, the neurofilaments and microtubules, is considerably more rapid than in the adult animal. This slowing can be correlated with the appearance of a cross-linking protein, which may provide additional rigidity to the mature axon. The cross-linking of these cytoskeletal elements has also been studied biochemically. The relationship between fast axonal transport and the cytoskeleton is less clear. Studies involving a variety of neurotoxins have indicated the absence of a correlation between neurofilaments and fast transport, and only a possible link between microtubules and fast transport. With a number of these neurotoxins, as well as in some neurological diseases, an accumulation of neurofilaments occurs, indicating an impairment of slow axonal transport.

Neuronal morphology. All neurons consist of a cell body or soma and two distinct types of extensions or processes called axons and dendrites. The cell body contains the nucleus and all the apparatus necessary for protein synthesis. Each neuron has a single axon and a number of dendrites. Axons can attain lengths of up to several meters, and maintain a relatively uniform diameter along this length. Dendrites are much shorter than axons, are highly branched, and become progressively narrower in diameter. The processes of other nerve cells terminate on both the dendrites and the cell soma at specialized regions of contact called synapses. Axons may branch, and the axon and its branches are the main transmitting channels through which nervous impulses are conducted, and by which the neuron communicates with other neurons, as well as muscles and other tissues.

Neuronal cytoskeleton. All cells are thought to be held together by a scaffolding known as the cytoskeleton. The cytoskeleton is composed of different structural elements, which can be seen by use of electron microscopy as long, thin filaments. These filaments pervade the cytoplasm of the cell, as well as the axonal and dendritic processes of a neuron. On the basis of their thickness, three types of filaments can be distinguished: microtubules, intermediate filaments, and microfilaments.

Microtubules are 24 nanometers in diameter and are composed of tubulin, an evolutionarily conserved protein present in all cell types. The microtubules have side arms extending from them, which are believed to be composed of microtubule-associated proteins (MAPs). Intermediate filaments, so called because they have a 10-nm diameter, which is between those of the other two types of filaments, are more varied than the microtubules, since their subunit composition changes with cell type. The proteins that make up the intermediate filaments in these different cell types are interrelated both structurally and genetically. A large number of homologies can be detected in both the protein and deoxyribonucleic acid (DNA) sequences. In the nerve cell, the intermediate filaments are called neurofilaments. The neurofilament is made up of a core filament protein (NF70) and two higher-molecular-weight proteins (NF150 and NF200), which are peripherally located, apparently winding around the core filament, and are responsible for cross-bridges between filaments. The microfilaments are composed of a protein called actin, which is also present in the contractile system of muscle. These thinnest filaments provide the most dynamic aspect of the cytoskeleton, since they are found directly connected to the cell membrane and to the nerve terminal.

The three filament systems are not independent of each other, but provide an interlinking network, which is presumably connected through various cytoskeletal associated proteins. In the nerve cell, the neuronal processes are filled with these filamentous structures, although not always in the same proportion. Dendritic processes are filled with microtubules, and for the most part are devoid of neurofilaments. There appears to be a relationship between the axon diameter and the ratio of neurofilaments to microtubules: the smaller-diameter axon consists mostly of microtubules, whereas the larger-diameter axonal processes are composed mostly of neurofilaments.

Axonal transport. Neurons can extend their axons over great distances. Since the axons do not have the capacity to synthesize proteins, all proteins are synthesized in the cell body and subsequently transported into the axon. The rate of axonal transport of different elements is dependent on the type of protein transported: rapidly transported proteins tend to be membrane proteins, secretory proteins, and smaller peptides; matrix proteins and cytoskeletal proteins are transported much more slowly. Fast transport is bidirectional: in addition to the orthogradely (toward the axon) transported material, there are membranous materials transported retrogradely (toward the cell nucleus). Retrogradely transported material also includes extracellular molecules that are obtained by internalization at the nerve terminals. The slowest transported proteins are the microtubule and neurofilament proteins, which are thought to be transported as a network of interconnected polymerized filaments. The microfilaments move faster than this network, and may have additional functions in terms of axonal growth, since only microfilaments are found all the way into the growth cones of the axons.

Axonal neuropathies. In an effort to understand the relationship between fast axonal transport and the cytoskeleton, studies have been conducted with drugs which affect either the microtubules or the neurofilaments. Fast transport is affected by colchicine, an antimitotic drug, which causes the disassembly of microtubules. However, this effect can be observed at concentrations below those necessary to disassemble the microtubules. Furthermore, calcium, which also disrupts microtubules, does not affect fast transport. β,β'-Iminodiproprionitrile (IDPN), a toxin which is known to produce accumulations of neurofilaments in the proximal regions of large axons throughout the nervous system, does not alter the rate of fast transport. These studies imply that neurofilaments are not essential for fast transport, but that microtubules may be. In addition to IDPN, an accumulation of neurofilaments occurs in a number of different toxicological and pathological states. For example, aluminum and acrylamide poisoning cause an accumulation of neurofilaments in the cell bodies of large neurons. In the disease amyotropic lateral sclerosis (Lou Gehrig's disease), large swellings appear in the axons, which are filled with neurofilaments. Thus in these cases the transport of neurofilaments is blocked, causing a buildup of neurofilamentous material. These swellings do not cause a complete physical blockage, since fast transport still occurs, although retrogradely transported organelles are retained within the swellings.

Developing animals. In the developing animal, both the composition of the cytoskeleton and the rate of axonal transport change. In the optic nerve of a newborn rat, neurofilaments are sparse, but microtubules are already abundant. Later in development, neurofilaments are much more numerous than microtubules. The biochemical composition of the cytoskeleton also changes. The neurofilaments in the newborn animals are composed solely of the NF70 and NF150 polypeptides. The NF200 polypeptide does not appear until the animal is nearly 3 weeks of age. In the optic nerve of a newborn rabbit, the neurofilament and microtubule proteins are transported at nearly 10 times the rate of the adult animal. The change in transport rate can be reasonably correlated with the appearance of the NF200 protein, which appears to be an interfilamentous cross-linker. Antibody studies have in fact shown that this particular protein can be found in the cross-bridges between filaments. The rate of axonal transport may be further affected by changes in the composition of other cytoskeletal components, such as the microtubule-associated proteins, or various intermediate filament–associated proteins, which have as yet not been identified.

Interactions between cytoskeletal elements. As mentioned earlier, the neuronal cytoskeleton is a complex of various interacting elements that varies in composition during development and as a function of axon diameter. In addition to the axonal transport studies, which have indicated the existence of a complex of microtubules and neurofilaments transported together in the slow component, biochemical data have indicated the presence of interactions between various cytoskeletal elements. The primary agents of these interactions are the cytoskeletal-associated proteins. By antibody localization studies, the NF200 protein has been shown to be an interfilament cross-linker. Microtubule-associated proteins have been implicated in the interconnection between neurofilaments and microtubules, as well as between microfilaments and microtubules. These studies have been conducted by direct binding studies and viscosity measurements. In the binding studies, MAPs were found to bind directly to the core NF70 protein. A MAP-dependent increase in viscosity has also been observed when neurofilaments and microtubules are mixed. In

IDPN intoxication studies, where the neurofilaments and microtubules segregate, the MAPs were found to colocalize with the neurofilaments. Phosphorylation may also play a role in the interactions of the various cytoskeletal elements. The MAPs and the two higher-molecular-weight neurofilament proteins are heavily phosphorylated. Phosphorylation appears to occur before assembly and transport, and may regulate the interactions between these various molecules. The presence of phosphokinases and phosphatases in the axons, and the regulation of these enzymes, may be important in understanding the dynamics of the interactions between the various cytoskeletal elements.

Summary. The neuronal cytoskeleton is composed of three types of filamentous structures. Two of these, the microtubules and the neurofilaments, interact with each other and move as a unit in the slowest component of axonal transport. The third type of structure, the microfilaments, are present in the growing tips of axons and are also involved in interactions with membrane components. The microtubules may be connected to fast transport, but the neurofilaments are not. Various toxic and pathological states cause the neurofilaments to accumulate in the proximal part of the axons, indicating a failure of their transport.

For background information see CYTOSKELETON; NEUROBIOLOGY; NEURON in the McGraw-Hill Encyclopedia of Science and Technology.

[RONALD K. H. LIEM]

Bibliography: R. J. Lasek, Translocation of the neuronal cytoskeleton and axonal locomotion, *Phil. Trans. Roy. Soc. Lond.*, B299:313–327, 1982; J. S. Pachter, R. K. H. Liem, and M. L. Shelanski, The neuronal cytoskeleton, in S. Fedoroff (ed.), *Advances in Cellular Neurobiology*, vol. 5, pp. 113–143, 1984.

Neutrino

An intense beam of neutrinos, produced by a particle accelerator, may be a useful tool with which to probe the remote interior of the Earth. The most powerful neutrino beams that are now available (those generated at the Fermilab National Accelerator Center near Chicago, or at the European Center for Nuclear Research in Geneva) are not sufficiently energetic for this purpose. However, the next generation of high-energy accelerators can produce beams capable of surveying thousands of square miles up to depths of tens of miles. They will serve as prototypes for future machines which may truly provide whole-earth tomography, and will allow the exploration of the last great frontier, Earth's interior.

Characteristics. Neutrinos constitute one of the four classes of particles currently regarded as elementary. These are the force carriers (like the photon), quarks (which make up the atomic nucleus), charged leptons (like the electron), and three varieties of neutral leptons which are known collectively as neutrinos. While neutrinos are not physically a component of atoms and molecules, they play a key role in the working of the universe. They are an essential ingredient of the beta process, by means of which primordial hydrogen was transmuted into the various chemical elements found on Earth. Furthermore, neutrinos are essential to the nuclear fusion reactions which power the Sun and other stars. Without neutrinos, the Earth, should it exist at all, would consist of a solid ball of frozen hydrogen.

Neutrinos were first detected experimentally in 1953. Today, they are as commonly studied in experiments as any other variety of so-called elementary particle. They are studied as by-products of cosmic-ray collisions, as a form of waste energy emerging from nuclear reactors, as decay products of naturally radioactive elements, as an invisible component of the Sun's radiation, and as intense beams produced by high-energy particle accelerators. In order to demonstrate the extreme elusiveness of the neutrino, it must be considered that 10^{11} solar neutrinos pass through each square centimeter of Earth in each second, both day and night. On the average, a solar neutrino could pass through 10^9 earths without being absorbed or deflected. Even though solar neutrinos are so numerous, their experimental detection was accomplished only in the 1970s. Solar neutrinos, impinging upon thousands of gallons of cleaning fluid in an underground tank, convert an occasional chlorine atom into radioactive argon. A few such events take place each month and are counted by sophisticated radiochemical techniques.

Earth exploration. If neutrinos, or for that matter any other particles or waves, are to be used as a probe of the deep earth, then two conditions must be satisfied: the probe must penetrate to great depths, but it must also interact with the underground material so as to produce a recognizable and interpretable signal. This is why only very high-energy neutrinos can be used for purposes of geological exploration. A neutrino with an energy of 1 TeV (1 TeV is 10^{12} eV, about 1 erg or 10^{-7} joule) stands about a 10% chance of making a nuclear collision as it passes through the Earth. Such neutrinos are produced in large numbers only by proton accelerators with a beam energy of 10 TeV or more, much larger than is now available.

In any scheme to probe the Earth's interior, protons have to be accelerated to great energies by a large proton synchrotron. The extracted beam must be carefully aimed toward the distant geological site of interest. The beam of protons is then allowed to collide with a target of heavy metal, and thus to generate a secondary beam of highly collimated energetic mesons. The mesons are directed through an evacuated or helium-filled decay tube which must be several miles in length. In this environment, the mesons avoid nuclear interactions and are enabled to decay in flight. Among their decay products are the neutrinos which constitute the underground probe.

Three scenarios have been suggested by which the neutrino beam could be used to provide information about the makeup of the interior of the Earth.

Project GENIUS. In Project GENIUS (Geological Exploration by Neutrino-Induced Underground Sound), the neutrino beam is used in order to introduce a controlled source of sound waves at great depth in the Earth. The sound waves propagate to the Earth's surface, where they are detected by an array of geophones (Fig. 1). In conventional reflective sonic seismology, sound is produced at the Earth's surface by explosions or by truck-mounted vibrators. The sound propagates to the interior of the Earth and is reflected to the surface by interfaces of different materials. Project GENIUS has three evident advantages over this procedure: simplicity of analysis, sensitivity to oil or gas deposits, and depth capability.

In reflective seismology, the sonic signal received at the surface is a complex mixture of reflected and multiply reflected waves coming from many sources. Great computational power is needed to reconstruct the geometry of subsurface strata. In Project GENIUS, the sound is injected into the Earth along a predetermined and precisely known trajectory, and the detected signal is direct and not reflected.

In addition to studying the subsurface geometry, Project GENIUS can reveal the physical constitution of the underground medium. Rock which is permeated with oil or gas produces a louder noise when traversed by the neutrino beam than does dry rock. This enhancement mechanism was discovered (in a different context) by Alexander Graham Bell a century ago. Project GENIUS offers a specific means of detecting the presence of trapped underground fluids: they literally scream in response to the neutrinos.

The greater depth sensitivity of Project GENIUS results from the direct, rather than reflected, nature of the sonic signal. If the procedure can be implemented at all, it will be sensitive to oil or gas deposits at any depth accessible to present or future drilling capabilities.

However, there are serious technical and financial obstacles to the implementation of Project GENIUS. Large accelerators are expensive, and a facility dedicated to geological exploration could cost more than a billion dollars. Feasability studies at smaller machines are a necessary prerequisite toward determination of the cost effectiveness of the procedure.

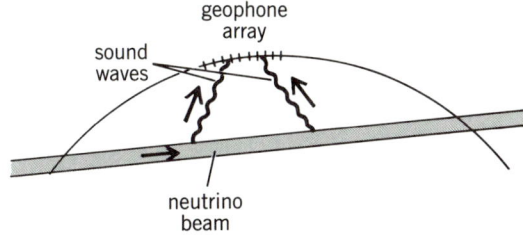

Fig. 1. Neutrino beam traversing the Earth in Project GENIUS. (*After A. De Rújula et al., Neutrino exploration of the Earth, Phys. Rep., 99:341–396, 1983*)

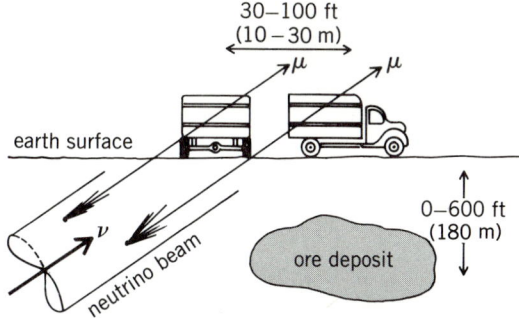

Fig. 2. Arrangement for detecting muons with truck-mounted counters in Project GEMINI. (*After A. De Rújula et al., Neutrino exploration of the Earth, Phys. Rep., 99:341–396, 1983*)

The neutrino beam must be capable of being redirected toward different geological sites. Nothing like this has ever been attempted before. While Project GENIUS may not become practicable until the next century, much preliminary research remains to be done today.

Project GEMINI. A neutrino beam traversing the Earth, as well as producing sound, generates a co-moving beam of parasitic muons which result from nuclear collisions of the neutrinos. These secondary particles provide the basis to Project GEMINI (Geological Exploration by Muons Induced by Neutrino Interactions). An array of truck-mounted muon counters (Fig. 2) is deployed at the location at which the neutrino beam emerges from the Earth. (This may be located thousands of miles from the accelerator itself.) The secondary muons provide a means of "x-raying" the last few miles of Earth through which the neutrinos pass. The GEMINI procedure is sensitive to the presence of ores of heavy metals like copper, lead, or uranium. It would be particularly useful as a means of mapping known deposits in order to expedite ore recovery, but it might also be used to discover new deposits.

Project GEOSCAN. A third implementation of a powerful neutrino beam is Project GEOSCAN. In this case, the neutrino beam must be directed in a vertical or nearly vertical direction. (One possible method of accomplishing this involves deploying the accelerator and beam at sea; Fig. 3.) Large ship-mounted muon detectors are placed at the antipodes of the Earth. In this fashion, the vertical profile of the density of the Earth can be measured with great precision. These experiments can complement and confirm indirect studies based upon the observations of seismic waves, and remaining ambiguities may be resolved.

An even more ambitious possibility, Project LUNASCAN, involves directing the neutrino beam toward the moon, and detection of its sonic signal by remote lunar seismic sensors.

Prospects. All of this, of course, sounds like science fiction. However, developments in accelerator technology are taking place rapidly. High-energy physicists in the United States have designated as their highest priority the construction of a new ma-

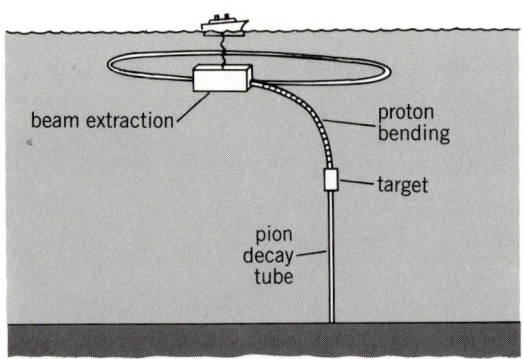

Fig. 3. One possible method of directing the neutrino beam in a vertical direction in Project GEOSCAN, involving accelerator and beam deployment at sea. (After A. De Rújula et al., Neutrino exploration of the Earth, Phys. Rep., 99:341–396, 1983)

chine called the Superconducting Super Collider (SSC). With relatively minor additions, this machine may serve as an effective test of the utility of neutrino prospecting. Its neutrino beam can be directed at known oil or gas deposits at distances up to several hundred miles from the accelerator. It would offer a unique facility for collaboration between academic and commercial science. Neutrinos may provide the key to the eventual discovery and recovery of Earth's remaining fossil energy reserves and vital minerals, and may lead to a deeper understanding of the planet and its satellite.

For background information see NEUTRINO; PARTICLE ACCELERATOR; SEISMOLOGY; SOLAR NEUTRINOS in the McGraw-Hill Encyclopedia of Science and Technology.

[SHELDON L. GLASHOW]

Bibliography: A. De Rújula et al., Neutrino exploration of the Earth, Phys. Rep., 99:341–396, 1983.

Nobel prizes

For 1984 nine recipients of the Nobel prizes were announced by the Swedish Royal Academy.

Medicine or physiology. This prize was awarded to three researchers for their work in immunology. Niels K. Jerne of the Basel Institute of Immunology, in Switzerland, was recognized for his formulation of three important theories. The first explained how the body produces specific antibodies to fit specific bacteria, viruses, or other foreign substances. The second gave an overall view of the process whereby the immune system develops and matures; and the third, the network theory, described how the many interrelated aspects of the immune response are coordinated by the body.

Cesar Milstein of the British Medical Research Council laboratory in Cambridge and Georges J. F. Köhler of the Basel Institute of Immunology shared the prize for their discovery of a laboratory technique for producing antibodies of remarkable uniformity and selectivity in responding to target substances. These monoclonal antibodies have enabled the creation of new tools for diagnosis and treatment of disease.

Physics. Carlo Rubbia and Simon van der Meer received this award for work in the detection of three subatomic particles—a negative and a positive W particle and a neutral Z particle—that theorists had postulated as the transmitters of the weak force. This discovery is an important step in confirming the modern theory of the unified quality of all the forces in nature. The Italian-born Rubbia, a Harvard professor as well as senior physicist at the European Center for Nuclear Research (CERN), was the principal architect of the experiment that effected the release of the particles through high-energy collisions of protons and antiprotons in CERN's circular synchrotron, and as the designer of the particle detector. Van der Meer, a Dutch research physicist at CERN, devised a method to ensure frequent and efficient collisions of the accelerated protons and antiprotons by preventing them from scattering.

Chemistry. R. Bruce Merrifield, a professor of biochemistry at the Rockefeller University in New York City, was recognized for his innovations in the field of protein synthesis, especially for his development of solid-phase peptide synthesis for making proteins by assembling amino acids sequentially into peptide chains. Originated by him in the 1950s, the process has greatly facilitated study of proteins and peptides, as well as practical applications in the development of medicines.

Economics. Richard Stone, a retired professor at Cambridge University, was chosen for his work in creating an accounting system for nations, wherein the respective incomes and spending of households, businesses, and government are weighed against each other to obtain a comprehensive profile of a nation's economy. The Nobel committee also cited his analyses of the details of national growth, and his studies aimed at assessing the costs of social problems in economic terms.

Literature. This prize was awarded to the Czechoslovak poet Jaroslav Seifert. He was praised for works which display an uninhibited and highly individual response to the everyday life in contemporary Prague.

Peace. Bishop Desmond Tutu, general secretary of the South African Council of Churches, was honored for his nonviolent campaign to end apartheid in South Africa. The committee also cited the "courage and heroism shown by all black South Africans in their use of peaceful methods in the struggle against apartheid."

Nuclear fusion

The purpose of nuclear fusion research is to find out whether or not the vast potential resource of energy in the nuclei of the light elements can be released in a controlled fashion, for the generation of electricity in central power stations and possibly other applications. The essential fuels are the hydrogen isotopes deuterium (D), occurring naturally, and tritium (T), manufactured from lithium, also widely

Table 1. Fusion reactions envisaged for controlled energy release

Solar reactions
  $H + H \to D + e^+ + \nu + 0.9$ MeV
  $D + H \to {}^3He + \gamma + 5.5$ MeV
Possible reactions for practical terrestrial thermonuclear fusion
  $D + D \to {}^3He + n + 3.3$ MeV 0.08*
  $D + D \to {}^3H + p + 4.0$ MeV 0.09
  $D + T \to {}^4He + n + 17.6$ MeV 5.0
  $T + T \to {}^4He + 2n + 11.3$ MeV 0.1
  $D + {}^3He \to {}^4He + p + 18.3$ MeV 0.8
  $D + {}^6Li \to {}^4He + {}^4He + 22.4$ MeV 0.026
  $p + {}^{11}B \to 3{}^4He + 8.7$ MeV 0.8
  $p + {}^6Li \to {}^4He + {}^3He + 4.0$ MeV 0.25
Tritium breeding reactions
  $n + {}^6Li \to {}^4He + T + 4.8$ MeV
  $n + {}^7Li \to {}^4He + T + n - 2.5$ MeV
Neutron multiplicator reactions
  $n + {}^9Be \to 2He + 2n + e - 1.85$ MeV

*Values are maximum cross section in barns (10^{-24} cm^2).

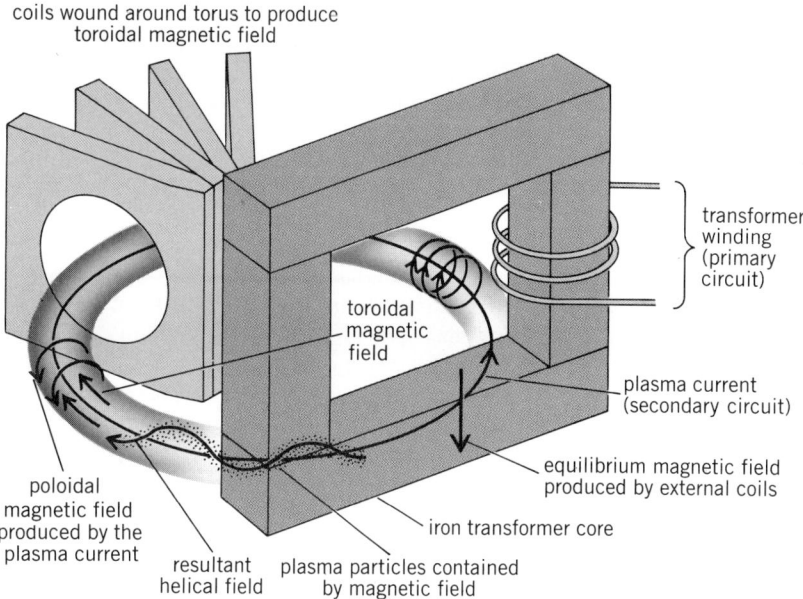

Fig. 2. Tokamak confinement system.

available. The principal nuclear fusion reactions envisaged are listed in Table 1. For net energy to be released, it is necessary to heat a mixture of hydrogen isotopes forming a plasma to a temperature of at least 10^8 K, and keep it thermally isolated according to the confinement requirement indicated in Fig. 1. The thermal insulation parameter is the product of the fusion fuel particle density n and the confinement time τ. In order for thermonuclear reactions to produce a net energy yield, it is necessary to reach the reactor plasma region indicated in the upper right corner of Fig. 1.

In Europe, the main line of research is to confine the high-temperature plasma with magnetic fields and heat it by the current flowing through it, supplemented by additional inputs such as the injection of beams of energetic neutral particles and the absorption of radio-frequency (rf) radiation. The achievements of some present European experiments are shown in Fig. 1.

Western Europe program. Research on controlled nuclear fusion in Europe began in 1946, when G. P. Thompson and M. Blackman patented a toroidal discharge as a means of reaching the required high temperature and magnetic thermal insulation. The research is now carried out in western Europe by a consortium of nations consisting of the European Economic Community together with Sweden and Switzerland. This Euratom research program is coordinated by the Fusion Directorate of the Commission of the European Communities. The commission provides over 40% of the funds, the remainder being provided by the national authorities carrying out the research.

Joint European Torus. The largest single project in the Euratom program is the Joint European Torus (JET), at Culham, United Kingdom, the world's most advanced fusion experiment. JET utilizes the tokamak configuration, a toroidal discharge stabilized by a very strong additional toroidal magnetic field (Fig. 2). This configuration, pioneered by workers in the Kurchatov Institute, Moscow, has been adopted as the most promising way of reaching fusion conditions.

In JET, currents of over 3×10^6 A lasting up to 10 s are induced through low-pressure hydrogen gas in a large toroidal vessel encircled by a magnetic circuit and electrical windings. The gas is ionized into a plasma which provides the secondary turn in a transformer mode when the current in the primary winding is pulsed on or off. The discharge is stabilized by a 3.4-tesla toroidal magnetic field provided by a ring of D-shaped water-cooled coils. The peak electrical power drawn by a JET pulse from the grid

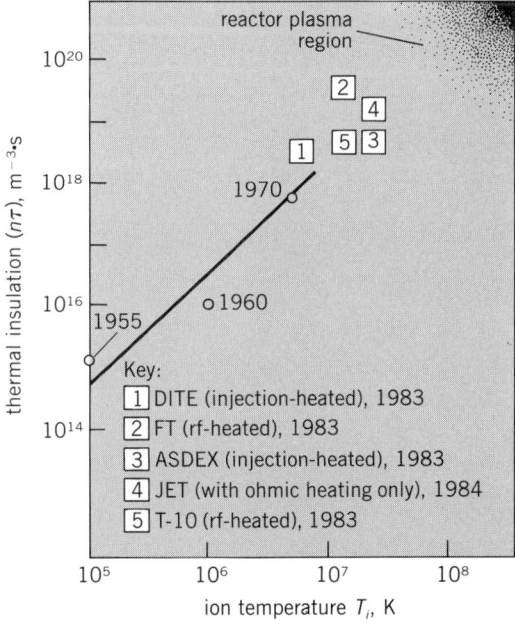

Fig. 1. Plasma parameters in current European tokamaks.

will exceed 500 MW. The sizes of the plasma current and of the machine were chosen so that, with the aid of additional heating, the 10^8 K temperature conditions needed for the envisaged energy-producing reactors could be approached. The flow of current provides about 3 MW of basic ohmic heating, but this power will decrease with the effective resistance as the temperature increases, and as the hydrogen gas becomes less contaminated with impurities from the walls of the chamber.

Additional heating apparatus (10 MW of neutral injection and 15 MW of ion cyclotron resonance heating) is being completed and will be installed in the period 1985–1987. If this additional heating is effective in eventually raising the temperature of a 50–50 deuterium-tritium mixture to 10^8 K, then thermonuclear reactions might contribute to the heating, maintaining the temperature without further input, until the end of the current pulse. Such operation would produce short bursts of fusion energy of tens of megawatts for a few seconds. There would then be some neutron irradiation of the surrounding apparatus, and so the apparatus incorporates some remote handling facilities. However, the theory of magnetic thermal insulation is very imperfect, and JET is solely a physics experiment which aims to find out whether or not the thermal insulation will in fact be sufficient for such a result to be achieved. JET cannot generate electricity.

Alternate configurations. Magnetic field confinement systems as alternates to the tokamak configuration are studied in western Europe, notably the stellarator configuration in West Germany, and the reverse-field pinch (RFP) configuration at Culham Laboratory, United Kingdom, and at Padua in Italy. Temperatures and confinement times in stellarators are now comparable to those achieved in tokamaks for apparatus of the same size. The reverse-field pinch produces high temperatures at relatively high gas pressures, but does not yet have such long confinement times as the tokamaks, so that the thermal insulation is not good enough to achieve a net energy yield. Heating methods being developed in Europe include: neutral beam injection (in which beams of accelerated ions are neutralized to pass into the magnetic fields); ion cyclotron resonance heating; electron cyclotron resonance heating; lower hybrid resonance heating; and Alfvén waves. (The resonance heating methods all involve input of rf energy.) The highest temperatures achieved in Europe by these methods are $3–5 \times 10^7$ K, compared to the 10^8 K ultimately needed.

Fusion reactor studies. The program development has centered on examination of the use of tokamaks

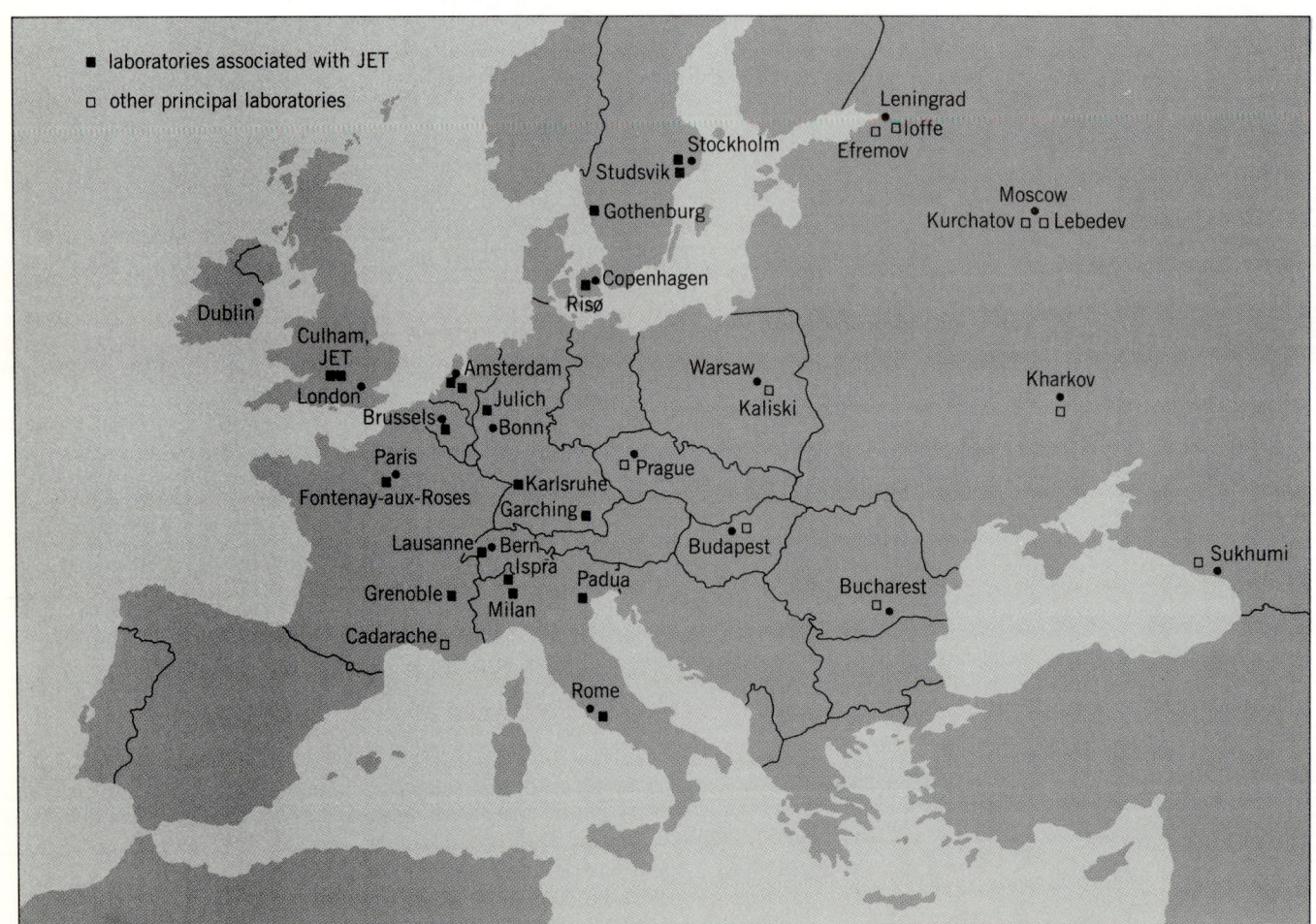

Fig. 3. Location of nuclear fusion research centers in Europe.

Table 2. Topics and physical parameters of some European tokamaks (indicative values)

Country and location	Device	Date of first operation	Radius, m (ft) Minor	Radius, m (ft) Major	Toroidal field, T	Plasma current, MA	Pulse length, s	Confinement time (τ), s	Electron density (n_e), m^{-3}	Peak temperatures* (T_i/T_i), keV	Average β_p/β_T†	Remarks
Commission of the European Communities U.K. Culham,	JET	1983	1.2/2.1 (4/7)	2.96 (9.7)	3.5	3.2	4	0.6	3×10^{19}	2.9/2	0.15/0.002	D-shaped cross-section, ohmic heating
France: Cadarache	TORE-SUPRA	1989	0.7 (2.3)	2.25 (7.4)	4.5	4.8	20	0.5–2	$0.3–1 \times 10^{20}$	5–10	1.0/0.03	rf + neutral beam heating, possible ignition in D+T Superconducting coils, rf current drive
France: Fontenay-aux-Roses	TFR	1986 1973	0.22 (0.7)	0.98 (3.2)	6	1.7 0.6	30 0.5	0.2–0.5 0.035	10^{20} 2×10^{20}	1–4 1–1.9/0.8–2	0.5/	rf, electron cyclotron, and neutral beam hearing; rf antenna development
West Germany: Garching	ASDEX	1980	0.4 (1.3)	1.65 (5.4)	2.8	0.5	3	0.04	7×10^{19}	1.2/2.2	1.8/0.016	2-null poloidal divertor, rf + neutral beam heating
West Germany: Garching	ASDEX upgrade	1987	0.5 (1.6)	1.65 (5.4)	4	2.0	10		1.2×10^{20}	≤ 7		Reactorlike 1-null poloidal rf + neutral beam heating
West Germany: Julich	TEXTOR	1981	0.5 (1.6)	1.75 (5.7)	2	0.5	3	0.05	5×10^{19}	1.3/0.8	0.2–0.4/	rf + neutral beam heating, plasma-wall interaction
United Kingdom: Culham	DITE	1975	0.26 (0.85)	1.17 (3.8)	2.8	0.34	1.0	0.04	1.2×10^{20}	1.6/1.0	2/0.01	Bundle divertor, electron cyclotron + neutral beam heating
United Kingdom: Culham	COMPASS	1987	0.22 × 0.37 (0.7 × 1.2)	0.56 (1.8)	2.1	0.4	5	0.02	10^{20}	7/1–2	2/0.1	Circular and D-shaped cross section, electron cyclotron heating
Italy: Frascati	FT	1977	0.23 (0.75)	0.83 (2.7)	10	0.8	1	0.06	7.5×10^{20}	1.3/1.2	0.5/0.003	rf heating and current drive, high $n\tau$
Italy: Frascati	FTU	1986	0.31 (1.0)	0.92 (3.0)	8	1.6	1.5	0.2	6.6×10^{20}	~ 6	1.5/	rf heating and current drive, high $n\tau$
Soviet Union: Ioffe (Leningrad)	Tuman-3	1978	0.23 (0.75)	0.55 (1.8)	1	0.17	0.04	$2–5 \times 10^{-3}$	$2–4 \times 10^{19}$	0.3–0.5/0.3	0.3/0.002	Adiabatic compression
Soviet Union: Kurchatov (Moscow)	T-7	1980	0.35 (1.15)	1.22 (4.0)	3	0.24–0.4	1	0.02	2×10^{19}	/0.35		Superconducting coils, rf current drive
Soviet Union: Kurchatov (Moscow)	T-10	1975	0.39 (1.3)	1.50 (4.9)	5	0.65	1	0.06	10^{20}	4.0/1.1	0.3/	Electron cyclotron heating and current drive
Soviet Union: Kurchatov (Moscow)	T-15	(1987)	0.75 (2.5)	2.4 (8)	3.5–4.5	1.4	5	0.3	10^{20}	5–7/		Superconducting coils

*When two numbers are given; number preceding solidus is electron temperature T_e, and number following solidus is ion temperature T_i. 1keV = 11.6×10^6°K.

†β_p is the ratio of the average plasma pressure to the magnetic field pressure associated with the plasma current. β_T is the ratio of the average plasma pressure to the magnetic field pressure associated with the toroidal magnetic field.

for reactors. However, the technological development which will be needed if power-generating reactors are to become a reality is still at a comparatively early stage. The Euratom countries have participated in the worldwide international INTOR workshop to establish what might be the next experiment after the successful operation of JET and of the equivalent experiments in the Soviet Union, the United States, and Japan; and have set up a European team specifically to develop the concepts for NET (Next European Torus). Decisions to proceed or not with NET will await the substantial completion of the experimental work on JET, that is, into the early 1990s.

Eastern Europe program. No figures are available for the size of the research effort in eastern Europe. The major part of the work is carried out in the laboratories of the Soviet Union. It would appear that the resources devoted there are roughly comparable to those in western Europe. Soviet research has been of major importance in a number of areas such as the development of magnetic mirror machines, the pioneering development of the tokamak system, theoretical developments, and the use of microwaves for plasma heating.

The research results are extensively interchanged, not only through the International Fusion Research Council of the International Atomic Energy Agency, but also, for example, through United States–Soviet and United Kingdom–Soviet agreements. No formal agreement between the fusion research program of the European Economic Community and of the eastern European bloc exists, but individual western nations have information and personnel exchange agreements with the Soviet Union.

Tokamaks. The Soviet Union's current research features a series of tokamak experiments at the Kurchatov Institute, including the world's first tokamak with superconducting coils, T-7; an extensive program of electron cyclotron resonance heating on their largest operating tokamak, T-10 (over 1 MW at 89 GHz has been achieved); and the construction of a major superconducting tokamak, T-15. At Leningrad, there is a compression-heated tokamak, Tuman 3.

Alternate confinement systems. Alternate magnetic confinement systems under study include mirror machines (at Kurchatov, Novosibirsk, and Kharkov), and stellarators at the Lebedev Institute (Moscow) and at institutes at Kharkov and Sukhumi. Laser-based inertial confinement schemes are developed at the Lebedev and Kurchatov Institutes (Moscow); and electron beams and ion beams are respectively developed at the Kurchatov Institute and in Kharkov. The first experiments to establish extensive agreement with the theory of the stability of magnetic confinement were carried out on the mirror machines PR5 and PR6 at the Kurchatov Institute and were of great general importance. All mirror machines were converted to the so-called magnetic well version now used. Current research at Novosibirsk Institute concentrates mainly on the so-called tandem mirror machines, which feature special arrangements to restrict by electrostatic potentials the loss of particles and energy along the lines of force; but the experiments have not yet reached a conclusive stage.

Among the smaller countries of eastern Europe, much work is done on the straight pinch discharge (the so-called Z pinch), where currents up to about 1 MA are passed through gas between two electrodes. The so-called plasma focus version of this is a form of magnetic inertial confinement producing temperatures of $2-3 \times 10^7$ °C and very high densities (10^{21} cm^{-3}) but for very short times (10^{-8} to 10^{-9} s). Poland, Romania, and Turkey all have examples of this development which, however, has been much reduced in western Europe. These pinches provide very interesting physical conditions, and a high power-density of fusion reactions; but the configuration is subject to rapidly growing instabilities which limit the duration and hence the apparent potential of the system.

Location of research centers. The main centers of fusion research in Europe are shown in Fig 3. At most of these laboratories, smaller and more specialized tokamaks are in operation or under construction (Table 2).

For background information *see* LAWSON CRITERION; NUCLEAR FUSION; PLASMA PHYSICS in the McGraw-Hill Encyclopedia of Science and Technology.

[R. S. PEASE]

Nuclear physics

Recent developments in nuclear physics include the incorporation of relativistic effects in the description of nuclei, and the study of dipole collectivity in nuclei.

RELATIVISTIC EFFECTS

The special theory of relativity is generally accepted as correct; therefore, dynamical systems such as nuclei should satisfy its requirements. Specifically, nuclei should be described by a relativistic many-body theory which combines the ideas of Lorentz covariance with the rules of quantum mechanics. However, almost all descriptions of nuclei use the Schrödinger equation and incorporate relativity in a minimal fashion, through kinematic corrections. A recent confluence of new theoretical work and accurate experimental data appears to challenge time-honored treatments of nuclear structure and reactions which use nonrelativistic approaches. There is now compelling evidence that the Dirac equation, long known to be the correct equation for describing relativistic spin-½ particles such as electrons, should also be used as the dynamical equation for systems involving protons and neutrons.

While complete relativistic descriptions of nuclear matter and finite nuclei using a relativistic quantum field theory are absent, considerable progress has been made by using approximate descriptions. Calculations are rapidly reaching the level of sophistication of the nonrelativistic treatments developed over the past 40 years.

Relativistic Hartree approximation. In the simplest approach, doubly magic nuclei such as ^{16}O, ^{40}Ca, and ^{208}Pb are described in the Hartree approximation, a treatment which neglects the identical-particles exchange effects due to the Pauli principle. The resulting single-particle wave functions are solutions of a Dirac equation containing two large covariant potential terms. One, a Lorentz scalar, transforms like the particle mass; the other, a Lorentz vector, behaves like the static Coulomb potential. In spite of their simplicity, relativistic Hartree calculations (in contrast to the corresponding nonrelativistic Hartree results) successfully reproduce level orderings, spacing, and major shell closures of the nuclear shell model. In addition, the predicted distribution of charge in these nuclei is in good agreement with the empirical distributions obtained from precise electron scattering measurements. Figure 1 shows that relativistic Hartree calculations agree as well with experiment as do the complicated nonrelativistic treatments which include the effects of the Pauli principle and of the nuclear medium on the effective nucleon-nucleon interaction, and other treatments which consider, in addition, ground-state correlations. The relativistic Hartree approach provides realistic nuclear ground-state proton and neutron densities as well as new proton and neutron scalar densities (densities formed by the difference between the squares of the large and small components of the relativistic wave function). The success of this approximation is attributable to the use of the Dirac equation as the relevant single-particle wave equation.

Comparison of wave equations. Some essential differences between relativistic and nonrelativistic treatments are made apparent by an examination of the wave equations used. For spin-½ particles, the one-body Schrödinger equation contains central and spin-orbit potentials and the single-particle wave function is a two-component spinor in Pauli space. These potentials, which are usually taken to be local and spherically symmetric, are independent of each other and, in the description of nucleon-nucleus scattering, they are complex. These two complex potentials form the heart of the nonrelativistic treatment of nuclear processes.

In the relativistic case, the one-body Dirac equation with local and energy-independent potentials may have as many as five different types of Lorentz terms. These potential types, distinguished by their properties under transformations of the Lorentz group, are scalar, vector, pseudoscalar, axial-vector, and tensor. It appears that two of these, the Lorentz scalar and the Lorentz vector, play a crucial role in the description of nuclear phenomena. In the nonrelativistic case, the question of Lorentz covariance is, of course, never raised. The Dirac single-particle wave function is a four-component, rather than two-component, spinor; and it is these other components which are the important feature of relativistic descriptions.

Reduction of the Dirac equation. A revealing comparison with the nonrelativistic treatment may

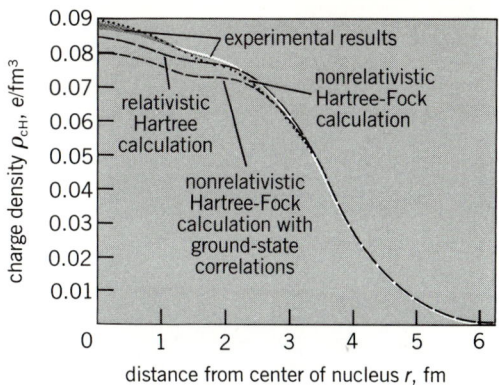

Fig. 1. Calculations and measurements of charge distributions for the nucleus of ^{40}Ca. Charge density is measured in units of electronic charge e per cubic femtometer. (*After C. Horowitz and B. D. Serot, Self-consistent Hartree description of finite nuclei in a relativistic quantum field theory, Nucl. Phys., A368:503–528, 1981*)

be made by performing a reduction of the Dirac equation containing scalar and vector potentials. This is done by solving for the two lower (negative-energy) components of the wave function in terms of the upper two (positive-energy) components. When this procedure is carried out, the resulting second-order differential equation resembles the Schrödinger equation with central and spin-orbit terms. However, these two potentials exhibit a number of important differences from the usual nonrelativistic case. First, the central potential, which depends largely on the sum of the scalar and vector terms, contains an explicit dependence on the single-particle energy. It also contains nonlinear (square) terms in each potential. In addition, a nonlocal (momentum-dependent) term called the Darwin term appears. The nonlinearity and nonlocality introduce complex nuclear density dependence of the central potential as well as cross terms between the nuclear and Coulomb interactions. Second, and as a consequence of the relativistic formulation, a spin-orbit term arises which depends on the difference between the scalar and vector potentials. Such a term is responsible for the spin-orbit splittings in atoms where the major interaction reflects the Coulomb force.

As early as 1956, it was pointed out that the explicit energy dependence and the addition of attractive scalar and repulsive vector interactions could be responsible for nuclear saturation. In fact, this energy dependence produces a central potential which changes from attraction for low-energy scattering processes to repulsion at higher energies, even when energy-independent scalar and vector potentials such as those from relativistic Hartree calculations are used. Such behavior is impossible to achieve by using energy-independent potentials in a Schrödinger description. As pointed out in 1936, the scalar and vector potentials contribute in different ways to the central and spin-orbit terms. In 1972 it was noted that the experimentally observed

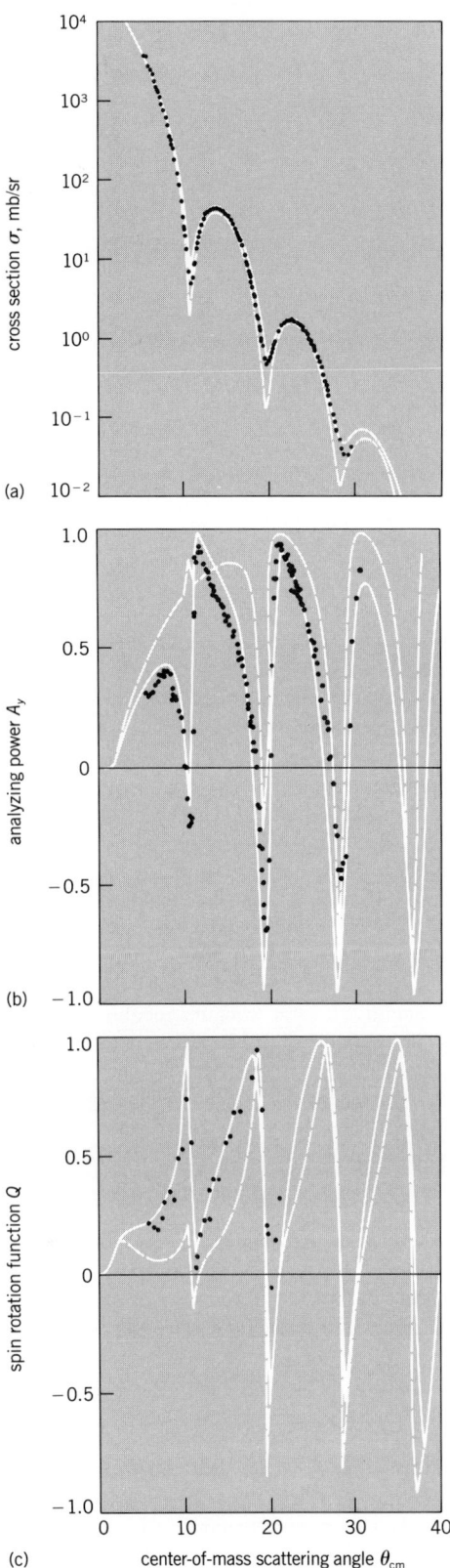

Fig. 2. Comparison of calculations using relativistic impulse approximation (solid curve) and nonrelativistic impulse approximation (broken curve) with experimental data for elastic scattering of 500-MeV protons from ^{40}Ca. (a) Cross section. (b) Analyzing power (left-right asymmetry). (c) Spin rotation function. (*After B. C. Clark, et al., Dirac-equation impulse approximation for intermediate-energy nucleon-nucleus scattering, Phys. Rev. Lett., 50:1644–1647, 1983*)

strong nuclear spin-orbit interaction results from this fact.

Proton-nucleus scattering. Clear phenomenological evidence favoring relativistic treatments of nuclear scattering has come from the analyses of intermediate-energy proton-nucleus scattering. These analyses, using the Dirac equation containing scalar and vector optical potentials, have been under way since the early 1970s. However, sophisticated new experimental facilities at the Los Alamos National Laboratory have made possible accurate measurement of spin observables heretofore not investigated. The results of the analyses using Dirac phenomenology were reported in 1982. They showed that it is possible to obtain precise fits to the new experimental data even when the standard Schrödinger equation–based phenomenology fails. They also showed that the potentials determined from Dirac phenomenology exhibit a smooth energy variation in both the central and the spin-orbit terms, a feature required in optical model analyses. This behavior is absent in the usual nonrelativistic approaches. The scalar and vector potentials determined from fitting the data are large (several hundred MeV) and opposite in sign, with the vector repulsive and the scalar attractive just as in the relativistic Hartree case. The imaginary potentials are also large and opposite in sign, with the vector corresponding to absorption and the scalar to production. The overall effect of the imaginary potentials is to produce absorption cross sections in agreement with experiment.

The most striking feature of this work was the ability, based on fits to two of the measured observables, cross section and left-right asymmetry (analyzing power), to predict the newly measured spin observable, the spin rotation function. These critical measurements, made possible by a new focal plane polarimeter at Los Alamos, also signaled an apparent inadequacy in the usual nonrelativistic approaches for microscopic derivation of optical model potentials. The measurements were made on a number of target nuclei using a beam of polarized 500-MeV protons. At this energy the elementary nucleon-nucleon amplitudes are experimentally well known, and the nucleon-nucleus elastic scattering was expected to be adequately represented by the first-order impulse approximation. In this approximation the multiple scatterings of the incident proton by the target nucleons are neglected so that, when constructing the overall scattering optical potential, each scattering may be considered to behave as if it were a free scattering event. The effects of the surrounding nuclear medium are ignored. This multiple scattering approach, elucidated in the 1950s, is a mainstay of nuclear reaction theory. In its first-order approximation, expected to be viable at these energies, the nuclear optical potential is constructed from the empirical free nucleon-nucleon amplitudes folded with the matter densities of neutrons and protons in the target nucleus. When used in the Schrödinger equation, this optical potential allows the prediction of the elastic scattering observables, cross sections, analyzing powers, and spin rotation functions. Surprisingly, the predictions of this theory were in severe disagreement with the measured spin observables, causing the use of the impulse approximation to be questioned. It appears, however, that the fault is not with the reaction model, but with the Schrödinger treatment employed.

Relativistic impulse approximation. Striking evidence that this is the case comes from the recently developed relativistic impulse approximation (RIA). Although a complete theory is currently not available, substantial progress has been made by a number of groups. The first series of results were reported in 1983. In these treatments the experimental free nucleon-nucleon scattering amplitudes, written in an explicitly invariant Dirac representation, are folded with the nuclear densities. The resulting nuclear optical potentials retain their Lorentz transformation properties. The two most important terms for scattering from spin-zero targets are a large attractive scalar potential and a large repulsive vector potential, just as from the phenomenological work. In addition, the imaginary parts of the scalar and vector optical potentials were large and opposite in sign, again in agreement with Dirac fits to the data. A small tensor component also occurs in the relativistic impulse approximation treatment; however, it produces little effect on the scattering observables. The calculations shown by solid lines in Fig. 2 use scalar and vector relativistic Hartree densities in calculating the relativistic-impulse-approximation optical potentials; thus relativistic effects in both the target and projectile are included. The broken lines give the nonrelativistic impulse approximation results. Comparison with the Los Alamos experimental data clearly shows the relativistic impulse approximation to be superior, especially in predicting the spin observables.

Effect of negative energy states. The explicit inclusion of negative energy states, omitted in the nonrelativistic calculations, plays the critical role in the success of the relativistic impulse approximation. In fact, the results are extremely sensitive to the presence of these negative energy components. This can be appreciated by looking at the differences between the scattered wave functions for the 500-MeV ^{40}Ca case shown in Fig. 3. Although the negative energy part of the wave function scattered by the target nucleus is very small, roughly 2% of the total, its inclusion in the scattering process produces a radial and angular pattern different from that in the nonrelativistic calculation. The wave functions shown in Fig. 3 were calculated using a Cray supercomputer. The differences in the relativistic and nonrelativistic scattered wave functions is mainly in the greater asymmetry shown by the relativistic case. The fact that the patterns, which look quite similar, produce such different spin observables is an indication of the sensitivity to relativistic effects.

Improvements in theoretical treatments incorpo-

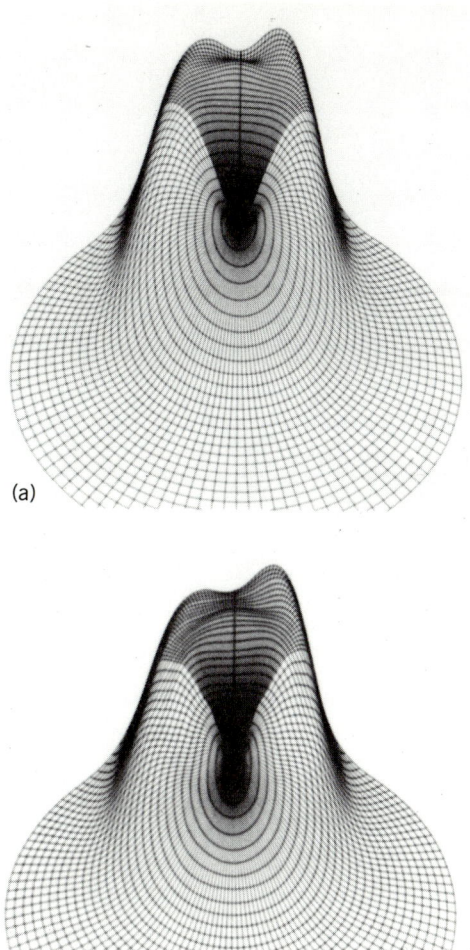

Fig. 3. Calculations of scattered wave functions for 500-MeV protons from ^{40}Ca. (a) Calculation from nonrelativistic impulse approximation. (b) Calculation from relativistic impulse approximation, which includes effects of negative energy states. (After M. V. Hynes, A. Picklesimer, P. C. Tandy, and R. M. Thaler)

rating relativity in the descriptions of both nuclear structure and nuclear reactions are under way. The next few years should tell nuclear physicists the extent to which it is necessary to use relativistic formulations in treating nuclear phenomena.

[BUNNY C. CLARK]

DIPOLE COLLECTIVITY IN NUCLEI

Low-lying collective nuclear states exhibit surface oscillations of several different shapes such as an (American) football shape ($\lambda = 2$, quadrupole mode), a (western) pear shape ($\lambda = 3$, octupole mode), and the more intricate hexadecapole ($\lambda = 4$) mode of excitation. In the early 1950s, it was pointed out that these collective states are characterized by enhanced electromagnetic decay rates, with so-called reduced transition rates $B(E\lambda)$, that exhaust a significant fraction of their respective sum

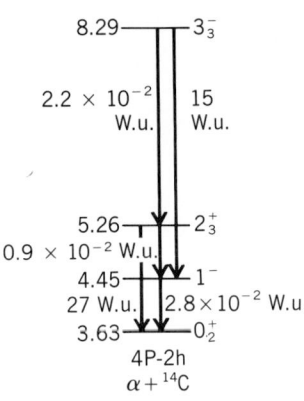
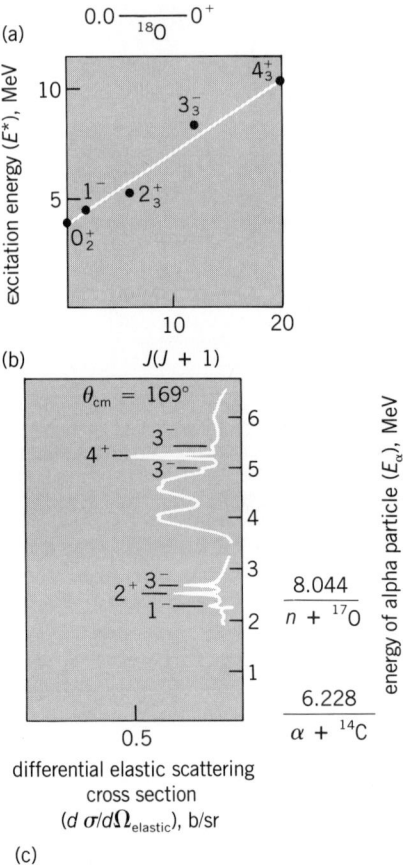

Fig. 4. The proposed $\alpha + {}^{14}C$ dipole molecular band in ${}^{18}O$. (a) Energy levels. The numbers on the left of each level are excitation energies in MeV, and the symbols on the right are spins J and parities π. $J^\pi = 0_2^+$ stands for the second 0^+ state, and so forth. W.u. is a measure of so-called single-particle reduced transition rate. 4P-2h indicates a state of 4-particle, 2-hole nature. (b) Excitation energy as a function of $J(J + 1)$. (c) Characteristic data on $\alpha + {}^{14}C$ backward-angle elastic scattering that exhibits characteristic molecular behavior. Data are shown for center-of-mass scattering angle θ_{cm} of 169°. 1 b = 10^{-28} m². The thresholds for dissociation of ${}^{18}O$ to $\alpha + {}^{14}C$ and $n + {}^{17}O$ are indicated.

rules, a measure of the maximum transition rate allowed by quantum mechanics. Such collective states can be related to the shell model of nuclei through the use of more recent theories developed in the 1970s, the theory of generalized seniority and the interacting boson model. The later theories utilize second-quantization formalism and group theory to give a description of collective shape variables that arise in collective models based on first quantization. In all theories of low-lying nuclear collective states, the $\lambda = 1$ electric dipole mode is omitted and is correctly assumed to represent motion of the center of mass. A similar conclusion was drawn in the nineteenth century by Lord Rayleigh, who studied the shape oscillations of a charged liquid drop, which experience similar surface oscillations. It is also well known that the $\lambda = 1$ electric dipole giant resonance, characteristically occurring at high excitations (approximately 15 MeV) exhausts the energy-weighted-isovector electric dipole sum rule, thereby hindering low-lying dipole collectivity in nuclei. In the giant dipole state, the neutron and proton fluids oscillate against one another, creating a dynamical displacement of the center of charge from the center of mass, and yielding enhanced electric dipole deexcitation of the giant dipole state. In most other nuclear states, the center of charge exactly or almost coincides with the center of mass, and this behavior hinders electric dipole transitions and results in reduced transition rates $B(E1)$ of the order of 10^{-5} to 10^{-7} single-particle units. Therefore it had been widely believed since the early 1950s that nuclei do not exhibit low-lying collective dipole excitations.

Dipole degree of freedom. Interest in the existence of nuclear cluster molecular states was generated by the observation in the early 1960s of such states in the ${}^{12}C + {}^{12}C$ system. Molecular states differ from ordinary nuclear states in that they involve at least two separate clusters, and are characterized by the vector separating the centers of the two clusters. Recently, such states have been associated with a dipole degree of freedom, in a model that utilizes the S^+ and P^+ boson creation operators defined in second-quantization formalism, and the dynamical symmetries of the group U(4). This model, referred to as the Vibron model, is analogous to the interacting boson model that uses S^+ and D^+ boson operators and the symmetries of the group U(6) to describe the collective quadrupole degree of freedom. Clearly if the charge-to-mass ratios of the two participant clusters are different, the dinuclear molecular state will exhibit polarization (that is, the center of charge will be displaced from the center of mass), and the model predicts enhanced E1 deexcitations for such cluster states. Furthermore, since cluster formation is enhanced at the nuclear surface region, in most cases it gives rise to a displacement of the center of charge from the center of mass, and the cluster oscillations may thus be considered to correspond to surface dipole oscillations. Since the oscillations involve only the surface region, they are low-lying, as opposed to the giant dipole resonance

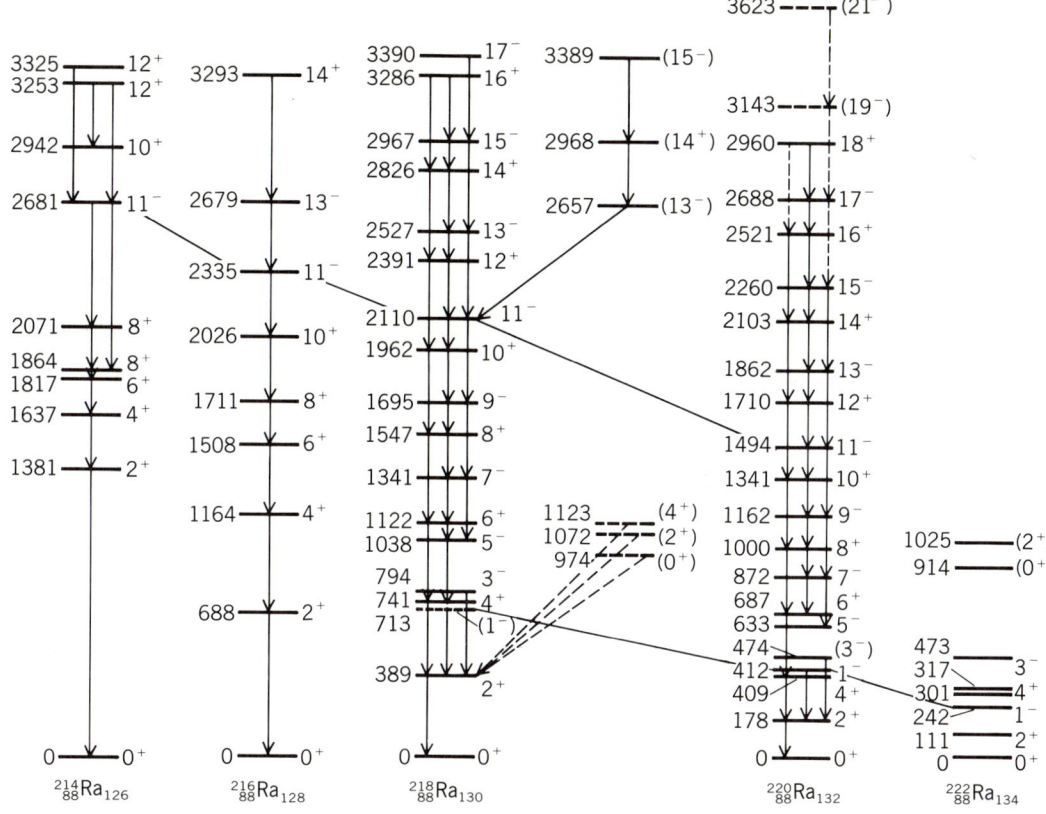

Fig. 5. Energy systematics of light even-even radium isotopes. The numbers on the left of each level are excitation energies in keV and the symbols on the right are spins and parities. The relation between the low-lying 1^- and high-lying 11^- states of neighboring nuclei is indicated. The ^{218}Ra and ^{220}Ra data appear at high spin with alternating parity sequences and enhanced E1 deexcitations are expected for a collective dipole band.

state which involves the entire nuclear volume and is thus high-lying. The Vibron model therefore predicts low-lying states with spin and parity $J^\pi = 1^-$, with enhanced $B(E1)$ deexcitations. In addition, the model predicts collective bands having spin parities $J^\pi = 0^+$, 1^-, 2^+, 3^-, 4^+, and so forth, as generally observed for diatomic molecules.

The E1 enhancement of deexcitations of nuclear molecular states clearly cannot be as large as the single-particle unit. Indeed a molecular dipole sum rule was recently derived for such deexcitations. An enhanced transition rate, $B(E1)$ of the order of 10^{-2} single-particle units, was predicted, which is still a factor of the order of 100 more than usually observed for E1 deexcitations in nuclei. It was also demonstrated that such cluster states, in addition to polarization (dipole moment), acquire deformation (quadrupole moment) and pear shape (octupole moment). Thus, simultaneous enhancement of $B(E1)$, $B(E2)$, $B(E3)$, and so forth is expected. Clearly the E1 enhancement is the most striking prediction. The nonvanishing octupole deformation of cluster molecular states suggests that the complete description of these states may require the inclusion of the octupole degree of freedom, together with the dipole degree of freedom.

Results from the ^{18}O nucleus. The $^4\text{He} + {}^{14}\text{C}$ system is composed of two tightly bound units, the doubly magic alpha particle and the semimagic ^{14}C nucleus. Thus molecular formation is expected to influence the spectrum of the ^{18}O nucleus. The alpha particle and ^{14}C nucleus have different charge densities (isospins) and thus, as discussed above, the predicted E1 enhancement should, in principle, be observed for $\alpha + {}^{14}\text{C}$ molecular states. Extensive study of the ^{18}O nucleus reveals a band of states of alternating parities having the $J^\pi = 0_2^+$, 1^-, 2_3^+, 3_3^-, and possibly the 4_3^+ states (where $J^\pi = 0_2^+$ stands for the second 0^+ state, and so forth) in ^{18}O (Fig. 4). The observed values of $B(E1)$ are very large, equal to or greater than 10^{-2} W.u., where W.u. is a measure of the so-called reduced transition rate, and represent some of the most enhanced E1 deexcitations identified in even-even nuclei. These values of $B(E1)$ correspond to a large fraction of the E1 molecular sum rule. The states involved are well recognized as collective cluster states, as indicated by the large observed so-called reduced alpha-decay widths of these states. In addition, the band appears to have a rotational character, as indicated by the almost linear dependence of the excitation energy on $J(J+1)$ (Fig. 4b). The observed moment of inertia appears to be close to that of two touching spheres of an alpha particle and a ^{14}C nucleus. The band-head is the $J^\pi = 0_2^+$ state at 3.63

MeV, which is a well-recognized collective state; in fact, it is the very first state suggested to exhibit coexistence between collective and single-particle degrees of freedom in nuclei. Thus, in ^{18}O the shell model single-particle degrees of freedom coexist with the well-known quadrupole collective degree of freedom and the newly suggested $\alpha + {}^{14}$C dipole cluster degree of freedom. The band observed in ^{18}O appears to be the first and, so far, the best example of dipole collectivity in nuclei. Microscopic calculations also appear to confirm the collective dipole character of this band in the ^{18}O nucleus.

Results in radium isotopes. Ever since pioneering work in the early 1950s, it has been known that radium isotopes display low-lying 1^- states. These are in fact the lowest (and also the first) collective negative parity states observed so far in even-even nuclei. They were suggested as a signature for nuclei having a pear shape (octupole deformation). Such a model could indeed account for many properties of these heavy nuclei. However, anomalies concerning large alpha-decay reduced widths, and unexpectedly small alpha-particle hindrance factors for decay into the 1^- and 0_2^+ excited states, were left unexplained. These anomalous alpha-decay rates led to the suggestion of clustering in heavy nuclei. As noted above, such cluster states also have a pear shape component (stable octupole moment), and thus the two seemingly different models of heavy nuclei are necessarily related.

Experimental research using heavy-ion beams to study high spin states in radium and thorium isotopes, carried out simultaneously in several laboratories, has revealed that the high spin states in the light radium and thorium isotopes appear with the alternating parity sequence $J^\pi = 4^+, 5^-, 6^+, 7^-, 8^+$, and so forth, and exhibit unusually enhanced E1 deexcitations. Figure 5 displays new data on ^{218}Ra and ^{220}Ra, together with the well-established level schemes of the neighboring even-even isotopes. In ^{214}Ra and ^{216}Ra, the lowest negative parity state now known is a noncollective two-particle $J^\pi = 11^-$ state. In contrast, in ^{218}Ra, ^{220}Ra, and ^{222}Ra a rich structure of low-lying negative parity states including a low-lying $J^\pi = 1^-$ state is observed, together with a structure that appears to be collective.

Furthermore, detailed experimental study of E1 deexcitations at high spins in ^{218}Ra indicates that the noncollective, two-particle $J^\pi = 11^-$ state (that is clearly identified in ^{216}Ra) interferes with the collective high-spin structure of ^{218}Ra, and leads to a reduction in the E1 deexcitation rates. This fact indicates that the origin of the E1 enhancement (at lower spins) is of a collective nature which is different than the nature of two-particle state. Thus, the level structure of the radium isotopes, with low-lying $J^\pi = 1^-$ states and bands of states of alternating parities and enhanced E1 deexcitations, together with observation of the reduction of E1 rates associated with the noncollective $J^\pi = 11^-$ state, clearly suggests that low-lying dipole collectivity is observed in the radium isotopes.

This dipole collectivity can be interpreted as arising from alpha-particle clustering, since in this model the dipole degree of freedom is naturally included. The model can also account for the displacement of the 1^- state upward above the 2^+ states. It arises from the fact that these nuclei appear to be nonspherical. Such a cluster model, however, also includes other degrees of freedom, such as the octupole one. The octupole model, on the other hand, associates pear shapes with these collective states, and such a shape has been shown to give rise to dipole polarization. Both models thus give rise to low-lying dipole collective states, but the latter does not readily encompass the observed enhancement of the E1 transitions without additional assumptions. More involved and detailed data are now being collected to study the overlap and differences of these two models.

The data on ^{18}O and the radium isotopes thus indicate the importance of the (low-lying) dipole degree of freedom in nuclei. The dipole degree of freedom appears related to a molecularlike structure. Low-lying cluster dipole states seem to appear in nuclei close to the doubly magic ^{16}O and ^{208}Pb nuclei, and they may be expected to occur in ^{212}Po and ^{52}Ti where states of $\alpha + {}^{208}$Pb and $\alpha + {}^{48}$Ca configurations are expected to be low-lying and play a role in their structure. The low-lying dipole degree of freedom represents an entirely new form of collectivity in the nuclear many-body system.

For background information see GIANT NUCLEAR RESONANCES; MULTIPOLE RADIATION; NONRELATIVISTIC QUANTUM THEORY; NUCLEAR MOLECULE; NUCLEAR REACTION; NUCLEAR STRUCTURE; RELATIVISTIC QUANTUM THEORY; SCATTERING EXPERIMENTS (NUCLEI) in the McGraw-Hill Encyclopedia of Science and Technology.

[MOSHE GAI]

Bibliography: Y. Alhassid, M. Gai, and G. F. Bertsch, Radiative width of molecular cluster states, *Phys. Rev. Lett.*, 49:1482–1485, 1982; M. R. Anastasio et al., Relativistic nuclear structure physics, *Phys. Rep.*, 100:327–401, 1983; B. C. Clark, S. Hama, and R. L. Mercer, Dirac Phenomenology, in H. D. Meyer (ed.), *The Interactions Between Medium Energy Nucleons in Nuclei-1982*, AIP Conf. Proc. 97, p. 260, 1983. P. D. Cottle et al., Level structure and deexcitations in ^{220}Ra and their systematic behavior as a function of neutron number, *Phys. Rev.*, C30:1768–1771, 1984; M. Gai et al., Coexistence of single-particle, collective-quadrupole, and $\alpha + {}^{14}$C molecular-dipole degrees of freedom in ^{18}O, *Phys. Rev. Lett.*, 50:239–242, 1983; M. Gai et al., Molecular alpha-particle clustering in ^{218}Ra: Dipole collectivity in the vicinity of nuclear shell closures, *Phys. Rev. Lett.*, 51:646–649, 1983; F. Iachello, Algebraic approach to nuclear quasi molecular spectra, *Phys. Rev.*, C23:2778–2780, 1981; F. Iachello and A. D. Jackson, A phenomenological approach to α-clustering in heavy nuclei, *Phys. Lett.*, 108B:151–154, 1982; B. M.

Schwarzschild, Relativistic treatment of low-energy nuclear phenomena, *Phys. Today*, 37(3):20–22, March 1984; B. D. Serot and J. D. Walecka, The relativistic many body problem, *Advances in Nuclear Physics*, vol. 16, 1985.

Nuclear reaction

The study of nuclear structure and nuclear spectroscopy is now a well-developed field, having been pursued with a variety of techniques for well over 30 years. The equilibrium shapes of nuclei at relatively low spins and excitation energies are now known to range from spherical to mildly deformed ellipsoids with axis ratios of order 4:3. These results are quite well understood in terms of models which have been developed in parallel with the experimental studies such as the nuclear shell model and the collective model.

To date, the experimental probes used in the study of nuclear structure have been restricted to those which do not appreciably disturb nuclei from their equilibrium shapes. This situation has changed, however, with the development of a new generation of accelerators capable of producing beams of heavy nuclei. Studies of the scattering and reactions of such heavy-ion beams with similar target nuclei have shown that it is possible to form nuclei in highly excited states with shapes corresponding to that of an ellipsoid with an axis ratio approaching 2:1. What is especially surprising is the observation that such superdeformed or quasimolecular states are unusually long-lived. This longevity must betray the existence of some special symmetry which stabilizes nuclei of this shape.

Observation of heavy-ion resonances. Historically, the first observation of quasimolecular states was made in a series of experiments in which the scattering and reactions of a ^{12}C beam with a ^{12}C target were measured as a function of the beam energy. The novel result of these experiments was that, instead of varying smoothly with energy as was expected, the cross sections for these processes varied rapidly with energy (Fig. 1). This rapid variation

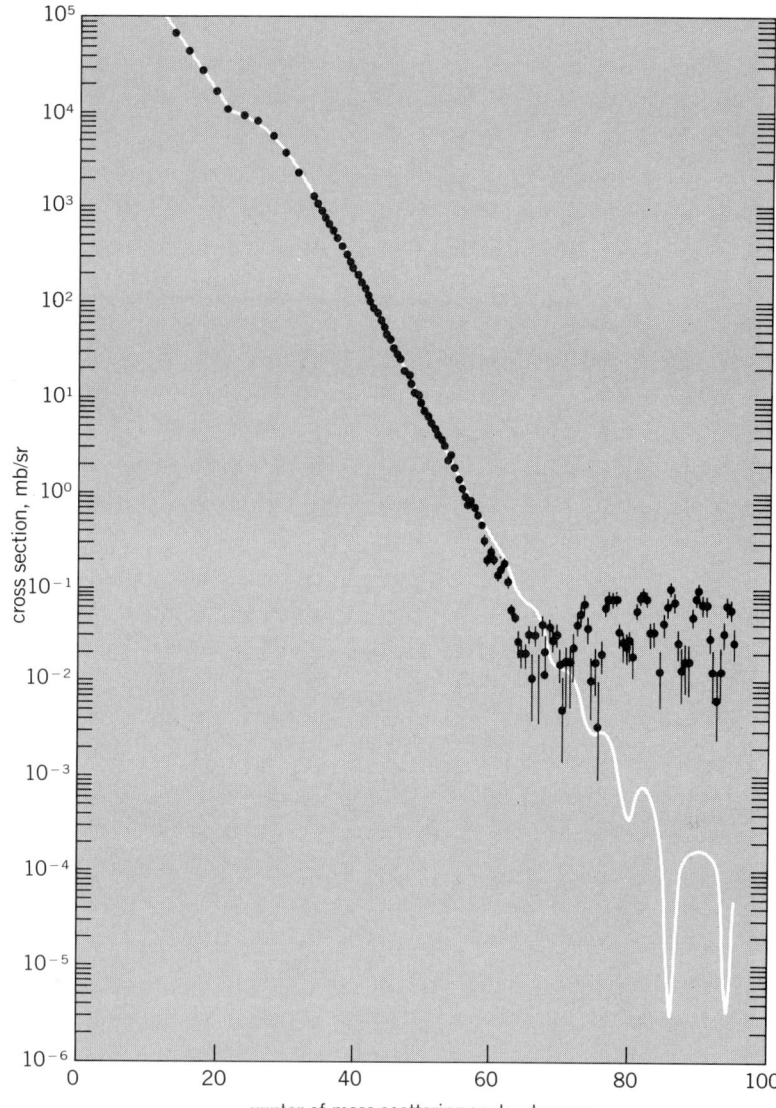

Fig. 2. Elastic scattering cross section for ^{28}Si + ^{28}Si measured at a center-of-mass bombarding energy of 60 MeV plotted as a function of scattering angle. The curve shows the result of a calculation of this quantity, excluding the possibility of resonance formation. 1 mb = 10^{-31} m^2.

with energy indicates the formation of a number of resonances corresponding to highly excited states of the compound nucleus ^{24}Mg. The unusual features of these results are, first, the relatively small number of these resonances when compared to the total number of available states in ^{24}Mg in this range of excitation energy and, second, the narrow width and therefore long lifetime of the resonant states.

These results remained as an isolated example for many years, and it was felt that the ^{12}C + ^{12}C system was therefore rather special in showing resonance behavior. Similarly, no clear understanding emerged of the nuclear structure underlying this resonance behavior, although an early suggestion was that the resonances perhaps corresponded to something akin to a nuclear molecule.

In recent years, a number of experiments have

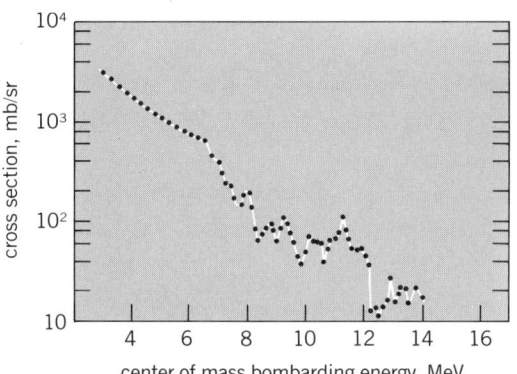

Fig. 1. Elastic scattering cross section for ^{12}C + ^{12}C measured at a center-of-mass scattering angle of 90° as a function of bombarding energy. 1 mb = 10^{-31} m^2.

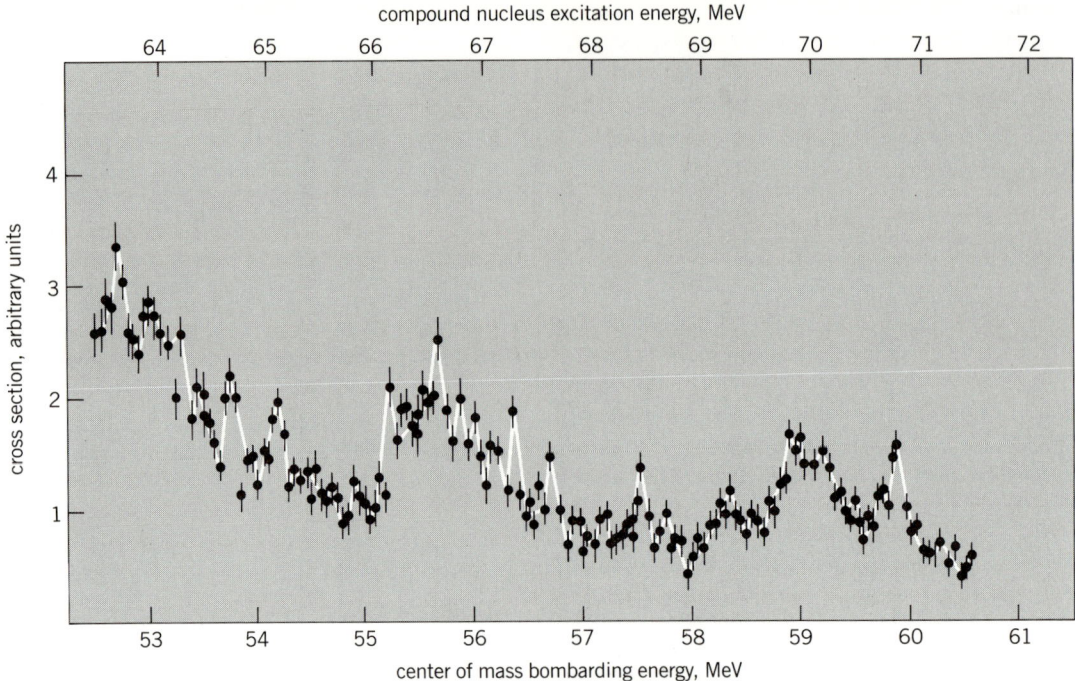

Fig. 3. Average elastic scattering cross section for ^{28}Si + ^{28}Si over the center-of-mass angular range of 60 to 90° plotted as a function of bombarding energy.

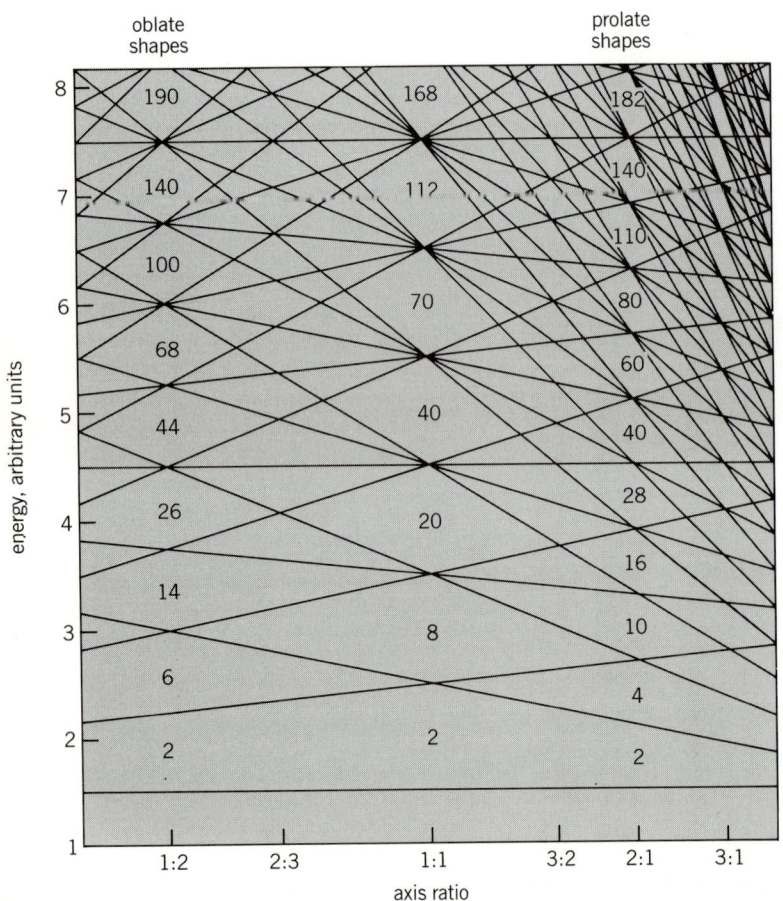

Fig. 4. Energy levels of a single nucleon moving in a deformed harmonic oscillator potential plotted versus the ratio of the length of the long axis to the other two. Numbers within gaps indicate number of energy levels below the gap.

shown that the resonances in the ^{12}C + ^{12}C system are in fact far from unique. Studies of much heavier systems such as ^{24}Mg + ^{24}Mg and ^{28}Si + ^{28}Si have shown that resonances in heavy-ion reactions are a much more general phenomenon which may well occur over the entire range of nuclear species that it is possible to produce.

An example of the kind of experimental observation for the heavier systems is shown in Fig. 2. This figure shows the cross section for elastic scattering of ^{28}Si + ^{28}Si measured at a center-of-mass bombarding energy of 60 MeV plotted as a function of scattering angle. The smooth behavior observed for scattering angles less than 60° is in agreement with the expectations based on conventional theories of heavy-ion scattering shown by the curve. The highly oscillatory behavior observed for scattering angles greater than 60° is completely different from the predictions of the conventional theory and corresponds to the formation of a resonance state in the compound nucleus (^{56}Ni) with angular momentum 40. Another set of data which displays the formation of resonances is shown in Fig. 3, where the elastic scattering cross section averaged over the angular range from 60 to 90° is shown plotted as a function of bombarding energy. The narrow peaks observed in these data again indicate the existence of resonant states in ^{56}Ni at excitation energies in the vicinity of 65 to 70 MeV. The special nature of these resonances is emphasized by the observation that in this excitation energy range, the nucleus ^{56}Ni is expected to have between 10^4 and 10^5 states per MeV whereas the number of resonance states observed is only the order of 1–10 per MeV.

Superdeformed nuclear states. Another important piece of information in the understanding of the nuclear structure underlying heavy-ion resonances is that their probability for decay into two large fragments, such as back into the entrance channel, is many times larger than that expected for a normal compound nucleus state. This, together with the narrow width of the resonances, has led to the speculation, which is in all likelihood correct, that they correspond to states of the nucleus which are extremely deformed, much more so than the normal states of the compound nucleus.

There is theoretical justification for assuming this hypothesis to be correct. Calculations of the equilibrium shapes of a number of nuclei have shown that, in addition to the normal spherical or slightly deformed shapes, there exist semistable states of extreme deformation. The origin of the special stability at extreme deformations arises from the existence of new shell gaps in the single-particle spectrum at particular values of the deformation. This feature is most graphically displayed by the results of a simplified calculation in which the energies of a single nucleon moving in a deformed harmonic oscillator potential are plotted versus axis ratio (Fig. 4). Gaps appear for integer ratios of the axis lengths. The gaps which appear in the spectrum at zero deformation (spherical shapes) are those which, in more detailed calculations, are responsible for the exceptional stability of nuclei with magic numbers of neutrons and protons corresponding to the number required to fill the single particle levels up to the shell gaps. The new gaps appearing at deformations corresponding to prolate ellipsoids with a 2:1 axis ratio will lead to stability of nuclei with this shape and with neutron and proton number corresponding to the filling of the new deformed shells. This feature is illustrated schematically in Fig. 5, where the total energy of nuclei with $N = Z = 20$ (^{40}Ca) and with $N = Z = 28$ (^{56}Ni) is shown plotted versus axis ratio. For the nucleus ^{40}Ca, the shell at axis ratio 1:1 stabilizes the spherical shape, whereas for ^{56}Ni, although the equilibrium shape is still spherical, there now exists a quasistable minimum at an axis ratio of 2:1 corresponding to the appearance of the $N = Z = 28$ shell. More detailed calculations employing a more realistic potential and allowing for more complex variations of the nuclear shape also lead to the same qualitative conclusions and do indeed predict the occurrence of such superdeformed minima in nuclei in which heavy-ion resonances are observed. Additionally, it is also found that for a number of cases in which no resonances are observed the calculations predict the absence of a superdeformed minimum.

Interaction process. The picture of heavy-ion resonances that emerges is therefore the following. The two colliding nuclei approach each other, interact, and form the configuration of the compound nucleus in the superdeformed minimum. This ex-

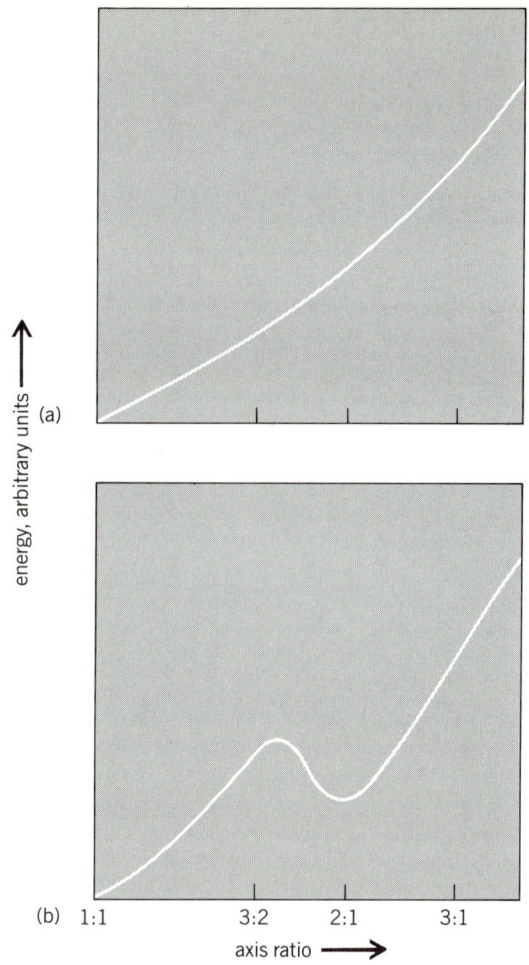

Fig. 5. Schematic representation of the total energy of (a) ^{40}Ca and (b) ^{56}Ni as a function of the axis ratio of each nucleus.

tended object then rotates several times before decaying into two large fragments resembling the incoming nuclei. The evidence for the validity of this picture is at present circumstantial, but there is a growing body of experimental data which point to its essential correctness. In that case, future experiments in which searches are made for resonances in even heavier systems are likely to prove fruitful and thus demonstrate the general nature of this phenomenon.

For background information see NUCLEAR MOLECULE; NUCLEAR REACTION; NUCLEAR STRUCTURE; SCATTERING EXPERIMENTS (NUCLEI) in the McGraw-Hill Encyclopedia of Science and Technology.

[R. RUSSELL BETTS]

Bibliography: R. R. Betts, B. B. Back, and B. G. Glagola, Intermediate structure resonances in ^{56}Ni, *Phys. Rev. Lett.*, 45:23–26, 1981; Aa. Bohr and B. Mottelson, *Nuclear Structure*, 1975; P. Braun-Munzinger (ed.), *Nuclear Physics with Heavy Ions*, 1984; D. A. Bromley, J. A. Kuehner and E. Almquist, Resonant elastic scattering of ^{12}C by carbon, *Phys. Rev. Lett.*, 4:365–367, 1960.

Nutrition

Research concerning essential elements can be divided into several broad areas, including (1) research to identify new essential elements; (2) studies to ascertain the functions and processes that known essential elements perform in life and their interactions with other nutrients; and (3) investigations designed to delineate the amounts of essential elements required to support optimum growth and performance of an organism. This article describes a few examples of research currently of interest to plant and animal scientists and human nutritionists for several elements in each of these categories.

Thirty-one elements are thought to be essential or potentially essential for higher plants (that is, angiosperms) or higher animal life (that is mammals; see table). The accepted criteria for establishing an element as essential differs between the botanical and zoological sciences. In plant science the criteria are much more rigid and conservative than those used by many animal scientists and human nutritionists. For plants, an essential element is defined as one with a specific role in the completion of the life cycle, and no other element can substitute for its function in plant metabolism. For animals, a dietary deficiency of the element is defined as consistently resulting in a suboptimal biological function that is preventable or reversible by physiological amounts of the element. Therefore, some elements that are considered essential for higher animals may not be considered essential for higher plants.

Problems arise in classifying elements as generally required for higher plants when the element is required by only one family (for example, sodium is essential only for plants having the C_4 dicarboxylic acid photosynthetic pathway), or when it is required only under a specific set of conditions (for example, molybdenum may be required by higher plants when they are dependent upon nitrate-nitrogen for growth but not when adequate sources of ammonium-nitrogen are available).

The philosophy and limitations of criteria of essentiality for plant growth are undergoing further examination, and the concept of "beneficial" or "functional" elements is emerging. Under this classification scheme, many of the elements listed in the table as being not required by higher plants could be classified as beneficial (that is, iodine, selenium, chromium, vanadium, fluorine, arsenic, lithium, and bromine) for the growth of certain plant species because under specific environmental conditions these elements have been reported to stimulate growth.

New essential elements. Not since 1954, when chlorine was recognized as being generally required for plant growth, has another element been recognized as essential for higher plants. New evidence suggests that nickel may be added to the list of essential elements. Soybean (*Glycine max*) and cowpea (*Vigna unguiculata*) plants develop necrotic lesions on their leaflet tips when grown without nickel in highly purified environments as a result of the accumulation of toxic levels of urea (the necrotic tissue contains 3–4% dry weight urea). This abnormality is not observed during vegetative growth but develops soon after flowering and pod set.

The enzyme urease, which hydrolyzes urea to ammonium and water, contains nickel as an essential component of its structure. The hydrolysis reaction was thought to be important in plant metabolism only when plants were supplied nitrogen solely as urea, but the new evidence suggests that a substantial amount of nitrogen is being cycled through urea as plants switch from vegetative to reproductive growth. Plants deprived of nickel accumulate urea to toxic levels because they do not contain a nickel-activated urease which is necessary to degrade urea. These findings must be substantiated in nonleguminous species before nickel is recognized as being generally required by higher plants.

Evidence is accumulating that silicon may also be essential for higher plant growth. Silicon treatments can increase the tolerance of plants to high manganese and iron levels, curtail fungus infection in several plant species, and effectively prevent lodging (that is, crops are beaten down by wind or rain) of some plant species, especially in crops supplied with high rates of nitrogen. Recently abnormalities, such as leaf curling in new leaves of cucumber (*Cucumis sativas*) plants, developed after flowering, and were attributed to silicon deficiency. Likewise, tomato (*Lycopersicon esculentum*) plants grown in nutrient culture without silicon developed malformed new leaves after the bud-flowering stage of growth and, in many cases, pollination and fruit formation were disrupted. Silicon additions to the culture media prevented or reversed the development of the abnormalities. Unfortunately, some doubt exists as to the adequacy of the chlorine level supplied to tomato and cucumber plants used in these studies. Further research is required to firmly establish silicon as an essential element for higher plant growth.

Since the early 1960s, animal scientists have proposed that 11 elements, including arsenic, boron, bromine, cadmium, fluorine, lead, lithium, nickel, silicon, tin, and vanadium, be added to the list of essential nutrients for mammals. They have been called ultratrace elements because they are usually found in very low concentrations in normal animal diets (that is, less than 1 microgram per gram dry weight of diet). Of these elements, only arsenic, nickel, and silicon meet the criteria of essentiality as defined for mammals. Preliminary findings provide some evidence that boron, lithium, and vanadium may be essential. Only very limited scientific evidence suggests that bromine, cadmium, fluorine, lead, and tin are essential. Fluorine has several beneficial pharmacological properties which can affect animal health, including a role in preventing dental caries.

The essential functions of arsenic, nickel, and silicon in animal metabolism are not known with any certainty. Arsenic may be necessary for the normal

Nutrition

Comparisons between essential elements for angiosperms and mammals

Element	Angiosperms	Mammals	Comments
Carbon	Required	Required	
Hydrogen	Required	Required	
Oxygen	Required	Required	
Nitrogen	Required	Required	
Sulfur	Required	Required	
Phosphorus	Required	Required	
Potassium	Required	Required	
Calcium	Required	Required	
Magnesium	Required	Required	
Sodium	Required by some	Required	Sodium is considered a micronutrient for plant species having the C_4 dicarboxylic acid photosynthetic pathway
Chlorine	Required	Required	
Iron	Required	Required	
Manganese	Required	Required	
Zinc	Required	Required	
Copper	Required	Required	
Molybdenum	Required	Required	
Boron	Required	Required	Further research needed to establish essentiality with certainty for mammals
Cobalt	Required by some species	Required	Essential for legumes dependent on nitrogen fixation for their nitrogen supply
Iodine	Beneficial effects	Required	Required by some lower plant forms (such as some algae)
Silicon	Required by some species	Required	Silicon is required for the growth of *Equisetum arvense*, a silicon accumulator species; beneficial effects of silicon on plant growth are quite common
Selenium	Beneficial effects	Required	
Chromium	Beneficial effects	Required	
Nickel	Required by some species	Required	Nickel is required by plants supplied urea as their sole source of nitrogen and by legume plants which accumulate ureides
Arsenic	Beneficial effects	Required	
Vanadium	Beneficial effects	Probably required	
Lithium	No evidence for essentiality	Probably required	
Fluorine	Beneficial effects	Possibly required	
Tin	No evidence for essentiality	Possibly required	
Cadmium	No evidence for essentiality	Possibly required	
Lead	No evidence for essentiality	Possibly required	
Bromine	Required by some species	Possibly required	Required by some lower plant forms (such as some algae)
Tungsten	Beneficial effects	No evidence for essentiality	
Aluminum	Beneficial effects	No evidence for essentiality	

functioning of certain amino acids or protein degradation products that are part of the urea cycle in animal metabolism. Nickel may be a cofactor or structural component in a metalloenzyme, or it may function as a cofactor in a metal-binding compound required for the efficient absorption of ferric iron by cells lining the animal's digestive tract. Silicon appears to play a role in biological cross-linkages in connective tissue in animals. More research is required before the physiological functions and required dietary levels of these ultratrace elements are defined.

New functions of essential elements. There is evidence that supports a direct involvement of boron in the functioning of plant cellular membranes. The absorption rates of both rubidium and chloride ions by root tips of corn (*Zea mays*) are depressed in the absence of boron in the absorption media but return to normal within an hour after small amounts of boric acid are added to the uptake solution. Also, evidence suggests that zinc may play a role in the structure and integrity of root cell membranes. Zinc-deficient wheat roots have depressed rates of phosphate and chloride absorption and retention. The effect is evident within 6 days after germination on zinc-free media, even though seedling tissues contain adequate concentrations of zinc (that is, from 30 to 40 micrograms per gram dry weight of tissue). If substantiated, these findings suggest that zinc must be supplied continuously to plant roots to assure optimum root cell membrane integrity and plant performance.

In animal studies, current research has stressed the importance of zinc in cellular immune functions and in brain development. Zinc-deficient rats have less developed hippocampal areas in their brains and demonstrate learning and memory impairments when compared to rats fed zinc-adequate diets. Interest has also focused on the role of zinc in stabilizing animal cell membranes. One hypothesis suggests that zinc may play a role in cell membranes by forming mercaptide bonds with membrane-protein sulfhydryl groups, thus preventing oxidation of the sulfhydryl by free organic radicals or by transition metals such as copper. Additionally, zinc may be involved in the regulation of a messenger ribonucleic acid (RNA) responsible for the synthesis of metallothionein (a low-molecular-weight metal-binding protein) in intestinal cells. Metallothionein may play a role in the regulation of zinc, copper, and cadmium absorption or excretion in the intestine.

Human health and essential elements. Human nutritionists have become increasingly concerned that calcium, magnesium, selenium, copper, iron, and zinc intakes by certain groups within the United States may not be sufficient for optimal health. Several government surveys show that significant segments of the American population have calcium, zinc, copper, and iron intakes below the Recommended Dietary Allowances (RDA) established for these elements.

Recently, interest in calcium nutrition has been stimulated by the controversial findings that a calcium intake which meets the RDA of 800 mg may help maintain normal blood pressure and while intakes in excess of 800 mg may lower both systolic and diastolic blood pressure in hypertensives. Also, it has been reported that zinc supplementation at a level of 5 mg per day in children between 2 and 6 years of age, with heights below the tenth percentile, has resulted in significant increases in their height over unsupplemented control children. This has intensified interest in zinc nutrition in the United States. Research on selenium has increased since the report of the prevention of Keshan disease (an endemic cardiomyopathy which previously resulted in the deaths of thousands of Chinese children each year) by dietary selenium supplementation in people living in the Keshan region of China. Interest in copper nutrition has been stimulated by research showing that copper plays a role in the metabolism of the sugar fructose. In the United States, fructose, particularly in the form of high-fructose corn syrup, is used a great deal as a substitute for sucrose to sweeten soft drinks and processed foods. Thus American diets are increasingly higher in fructose and marginally low in copper. This trend may increase the risk of cardiovascular problems because of difficulty in regulating blood sugar in people. This is so because copper is necessary for insulin binding, sugar transport, and breakdown of fats by adipose tissue.

The question of the available amounts of essential mineral elements in human foods has received much attention because of the possibilities of the borderline undernutrition in people for several of the essential elements. Research in identifying food components (so-called antinutritive substances) which can either promote or inhibit the absorption or utilization (bioavailability) of mineral nutrients by humans has increased greatly because of these concerns. In order to establish realistic RDAs, the bioavailability of mineral nutrients in varied human diets must be determined. However, little is known about the identity of antinutritive factors in foods and their mode of action in relation to the bioavailability of mineral nutrients to higher animals and humans.

For background information *see* NUTRITION; PLANT GROWTH; PLANT MINERAL NUTRITION in the McGraw-Hill Encyclopedia of Science and Technology.

[ROSS M. WELCH]

Bibliography: D. Bould, E. J. Hewitt, and P. Needham, *Diagnosis of Mineral Disorders in Plants*, vol. 1: *Principles*, 1984; A. Läuchli and R. L. Bieleski (eds.), *Inorganic Plant Nutrition*, 1983; F. H. Nielsen, Ultratrace elements in nutrition, *Annu. Rev. Nutr.*, 4:21–41, 1984; R. M. Welch and W. H. Gabelman, *Crops as Sources of Nutrients for Humans*, Amer. Soc. Agron. Spec. Publ. 48, 1984.

Oncogenes

Oncogenes are genes that cause cancer. Oncogenes were first noted in certain viruses, but the revelation that the cells of most animals have similar genes, termed proto-oncogenes, was a turning point in the understanding of carcinogenesis. Investigations have centered on identifying the proteins or enzymes coded for by these genes.

In the past few years several proto-oncogenes have been identified. It is emerging that these genes are involved in upsetting the delicate regulatory mechanisms of cell growth and division; both positive and negative regulators are involved. That is, similar responses can be evoked by an excess of a positive regulator (one that encourages growth or division) or insufficient amounts of a negative regulator. An ordinary animal cell may express several thousand genes, and the products of any of these can be made in wrong amounts or in defective forms. A large number of these genes are critical to the cell's normal activities, and therefore damage to them may kill the cell or interfere with its function, but very few (a few dozen) are actually involved in the regulation of cell growth and division. However, interference with these genes can result in uncontrolled cell growth. It is now thought that these relatively few genes correspond to oncogenes.

Oncogene concept. It has long been known that certain specialized viruses called retroviruses can cause cancer in specific host animals, including birds, cats, rodents, and monkeys. The best known of these is the Rous sarcoma virus discovered by P. Rous in 1910, which when injected into chickens invariably results in the development of tumors called sarcomas. As in the case of other retroviruses, the Rous sarcoma virus has a small genome

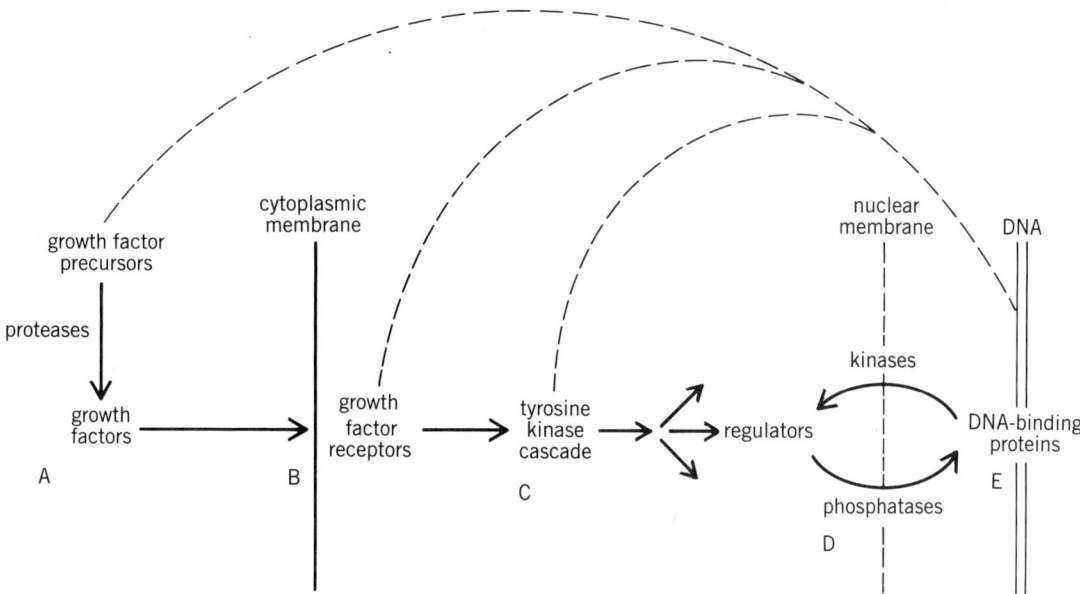

Schematic depiction of some events that control growth and cell division in a typical animal cell. A growth factor (A), ordinarily made in some cell in a precursor form and activated extracellularly by a proteolytic enzyme, combines with a receptor (B) lodged in the cellular membrane. The activated receptor, by various means, activates certain cellular kinases (C) that set in motion a series of reactions leading to cell growth and division. Each of these three agents (A, B, and C) is now known to correspond to a particular oncogene. Other oncogenes may be the counterparts of other cell regulators, including hypothetical phosphatases (D) and some DNA-binding proteins (E).

composed of ribonucleic acid (RNA) only a few thousand nucleotides in length. The virus has only four genes. One of these codes for the unique enzyme reverse transcriptase, which functions to copy RNA into deoxyribonucleic acid (DNA) that can then become integrated into the DNA of the host cell. The virus can thus propagate passively when the host cell divides, or actively as the result of DNA transcription back into (multiple) copies of viral RNA. While transiently resident in the genome of the host cell, the virus may inadvertently pick up some of the host cell's genetic material. It is likely that most of the time these genetic exchanges are of minor consequence. Occasionally, however, the acquired gene may be capable of inflicting great changes on the development of subsequently infected cells. These genes are called transforming factors because they transform cells into a cancerous state.

Modern recombinant DNA techniques allowed researchers to determine the sequences of bases that make up the genomes of these acutely transforming retroviruses and the oncogenes, or transforming factors, that they carry. These techniques also allowed the oncogene itself to be excised from the virus. When isolated oncogenes were radioactively labeled and incubated with fragmented DNA from normal cells, it was found that normal cellular DNA has regions that can hybridize with the oncogenes. This implied that the virus was not transforming the cell by some foreign agent, but rather by introducing an ordinary gene either in an altered form or in a manner that would allow its excessive production.

Types. Altogether about two dozens oncogenes have been discovered in retroviruses. The amino acid sequences of the hypothetical proteins for which the genes code can be inferred from the nucleic acid sequences of the genes. About half of these proteins are quite similar to each other and clearly belong to the same family. They are now known to belong to a group of enzymes called kinases, and most of them are able to phosphorylate certain tyrosine side chains in some critical proteins, including themselves. The consensus is that phosphorylation of certain pivotal proteins allows the cell to divide and grow without restraint. Alternatively, an altered form of an enzyme might be interfering with the action of the normal form. In either case the exact cellular enzymes to which the oncogenes correspond have not yet been identified.

The situation was significantly clarified, however, when it was discovered that the product of one of the oncogenes corresponded to a growth factor protein normally found in minute amounts in blood platelets. This protein, called platelet-derived growth factor (PDGF), is released into the immediate surroundings only when blood clots in response to an injury; it causes neighboring fibroblasts to divide and grow over the injured area. In this case the retrovirus, originally isolated from a woolly monkey and known as simian sarcoma virus, had been resident in a monkey chromosome where it picked up the monkey's PDGF gene. Ordinarily PDGF is under very tight regulatory control and is expressed only in blood platelets, and there only in very tiny amounts. However, upon transfer to the retrovirus, ordinary regulation of the gene is lost. Also the virus may provide a vehicle for release of the gene from cells to an extracellular milieu where it stimulates cell growth and division (see illustration).

An additional piece of information was provided by the discovery that the receptor for another growth factor (epidermal growth factor; EGF) is partially encoded by a retrovirus oncogene known as erb-B. It is known that receptors for several growth factors, including those of PDGF and EGF, are tyrosine kinases. Thus a connection was established between the two types of oncogene (see illustration).

Currently about six oncogenes remain that do not belong to any of the groups described above. Moreover, these oncogenes include several that have been associated most closely with known human cancers. In many cases human tumors have increased amounts of RNA transcripts that correspond to a family of oncogenes called ras (originally for "rat sarcoma"). There is reason to believe that this gene may give rise to a coupling factor that is instrumental in conveying extracellular stimuli to the internal machinery. Finally there is a set of widely differing oncogenes called myc and myb which occur in the nuclei of cells infected with certain retroviruses. In the view of some researchers, the problem in these cases is an altered protein that can no longer bind appropriately to DNA. As a consequence, some other agent may be deregulated, and this can lead to uncontrolled cell growth and division. The possibility also exists that some of the still-unidentified oncogenes might affect specific phosphatases, which act in the opposite way to kinases.

It is important to understand that transformation can result from many different interruptions in the normal cycle of cell division. It must also be recognized that while retroviruses have been extremely useful in providing a clear understanding of oncogenes and have led to the identification of proto-oncogenes, it is unlikely that retroviruses are the cause of most human cancers, although some carcinogenic disorders may arise in this fashion. It is much more probable that mutations in those genes that are negative regulators (governors or dampeners) are the major causes of cancer. In some cases augmented expression of a proto-oncogene may result from the viral introduction of positive regulators (enhancer elements) in adjacent regions of host DNA. It is also known that enhanced expression of a proto-oncogene can be effected by chromosomal rearrangements that may let the gene escape its normally stringent controls.

For background information see GENE ACTION; ONCOLOGY in the McGraw-Hill Encyclopedia of Science and Technology.

[RUSSELL F. DOOLITTLE]

Bibliography: J. M. Bishop, Cellular oncogenes and retroviruses, *Annu. Rev. Biochem.*, 52:301–354, 1983; R. F. Doolittle et al., Simian sarcoma virus onc gene, v-sis, is derived from the gene (or genes) encoding a platelet-derived growth factor, *Science*, 221:275–277, 1983; J. Downward et al., Close similarity of epidermal growth factor receptor and V-erb-B oncogene protein sequences, *Nature*, 307:521–27 (1984); T. Hunter, The proteins of oncogenes, *Sci. Amer.*, 251:70–79, 1984; M. D. Waterfield et al., Platelet-derived growth factor is structurally related to the putative transforming protein P28sis of simian sarcoma virus, *Nature*, 304:35–39 (1983).

Optical fiber sensors

In recently developed instruments for the optical measurement of position, rotation, pressure, temperature, and numerous other physical quantities, optical fibers are used (1) to provide particularly safe, reliable, or flexible transmission of the measuring signal between the sensor head and a remote evaluation unit, or (2) to serve as the sensing elements, promising higher sensitivity or otherwise better performance than conventional sensors in specific industrial, medical, and military applications. Important examples are fiber-optic displacement sensors, fiber-optic thermometers, and fiber-optic gyroscopes.

Advantages, applications, principles. Optical fibers have a number of properties which make them ideally suited for the collection and transmission of sensor signals, such as from a critical process in an industrial plant to a central control room, or from the interior of a patient's vein to an indicator observed by the operating surgeon. Electrical insulation of fibers permits their use in sensors in high-voltage applications (such as in electric power installations) and in medical applications (where they pose no danger of electric shock, and create no interference with electronic sensors such as those for electrocardiograms and electroencephalograms, or with diathermal heating). Reliability of transmission exists in fibers even in the presence of strong electromagnetic fields, extreme temperatures, corrosive atmospheres, or ionizing radiation (such as in chemical and nuclear reactors or oil well exploration). Safety of transmission through explosive environments (as in the mining and petrochemical industry) can be guaranteed by the use of low optical-power levels (microwatts). High flexibility of very thin fibers is important in sensors for endoscopic medical applications. Furthermore, fibers are key elements of distributed sensors and sensor systems, collecting information through one fiber simultaneously from many different locations.

Most fiber-optic sensors operate on some form of modulation principle, with fiber-guided light acting as the information carrier (Fig. 1). From a near-infrared light-emitting diode or laser diode (0.8–1.3-micrometer wavelength) in the source unit, the light is guided to the sensor head and back to a photodiode detector in the evaluation unit. In the head, the quantity x to be measured (which may be a displacement, velocity, acceleration, force, pressure, temperature, electric or magnetic field, or other physical or chemical parameter) modulates the light; that is, the momentary value of x is encoded into the intensity, the spectrum, or the phase of the light flux, so that x can be determined in the evaluation unit from the detector signal by suitable decoding electronics. Three groups of optical fiber sensors are discussed below, corresponding to those three encoding possibilities.

Intensity-encoding sensors. In the simplest class of fiber-optic sensors, the intensity or power of the light flux is modulated by a shutter or attenuator, controlled by x, and the evaluation unit is essentially an optical-power meter. As an example, Fig. 1 illustrates intensity modulation by moving a flexible fiber end. Incremental position and angular sensors use masks with clear and opaque regions moving across the path of light, and in fiber-optic proximity sensors a reflector moves in front of two parallel fiber ends. In other displacement sensors, modulation results from the loss of light at sharp fiber bends, controlled by x. Loss variations by changes of the refractive index outside a bent fiber are the basis of fiber-optic refractometers, thermometers, and liquid-level sensors.

A large class of fiber-optic sensors operates indirectly, employing suitable transducer elements in connection with a direct sensor. A simple lever, for example, may convert a linear displacement sensor to an angular one, or vice versa. Similar transducer elements are thermal expansion bodies (to measure temperature), springs (to measure force or torque), membranes (to measure pressure), a probe mass on a spring (to measure acceleration), and piezoelectric or magnetostrictive bodies (to measure electric and magnetic field strengths).

Many of the shutter-type sensors operate highly reliably with digital evaluation. An electronic circuit following the detector produces a logic "high" or "low" output signal, depending on whether the measured quantity x and the corresponding detector current exceed a set threshold value or not.

With analog evaluation, the detected optical power is indicated, representing directly the momentary value of x. Extremely high absolute sensitivities have been demonstrated or predicted in such sensors at audio frequencies (above about 1 kHz), paralleling those of the best conventional hydrophones or magnetometers. In the measurement of slowly varying quantities, however, the accuracy of analog shutter-type sensors is severely limited, in the order of 10% of full scale, because fiber losses may fluctuate that much by bending, stressing, and temperature variations. Better accuracy (3%) results in analog intensity-modulating sensors which combine two modulators to encode x as the power ratio of two light fluxes. This encoding scheme is employed with elasto-, electro-, and magnetooptic modulation for pressure-, voltage-, and current-sensors, respectively.

Sensors with spectral encoding. By using more complex modulators and filters in the sensor head, the information on x can be encoded into the optical spectrum of the transmitted light in order to reduce the influence of variable losses in the transmission fibers. Some sensors use the power ratio of two or more spectral lines to characterize x. Others use the spectral position of a sharp absorption edge or of a single spectral line, or the density of lines in a comb spectrum. Still another form of spectral encoding is the use of temporal modulation frequencies for the characterization of x.

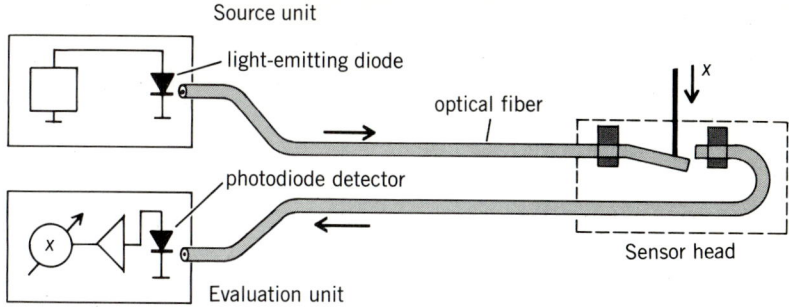

Fig. 1. Configuration of a typical fiber-optic sensor. The quantity x, which is to be measured, is here the displacement of a flexible fiber end.

In particular, the sensors based on pairs of fiber-coupled spectrometers and fiber-coupled interferometers show excellent independence of their operation from the properties of the transmission fibers. With such sensors, fibers of different quality and length may be used, fiber connectors may be inserted, and sensor heads may be interchanged without affecting the reading of the instrument. These sensors, however, are still under development.

In one widely used fiber-optic thermometer (Fig. 2), the key element is a tiny crystal of fluorescent material at the tip of the sensor fiber. Light from a mercury lamp is guided to the crystal and excites fluorescence at a number of distinct spectral lines (red, yellow, green). The fluorescent light is guided back to the evaluation unit, where the lines are separated by spectral filters and their powers are individually measured. As these three powers vary differently, but reproducibly, with the crystal temperature, it is possible to evaluate that temperature from the ratios of these powers. A recently proposed alternative is to evaluate the crystal temperature from a measurement of the decay time of the fluorescence after pulsed excitation.

Interferometric sensors. The most sophisticated sensors encode x into the optical phase of a light wave guided in a fiber, for example, by elastic ex-

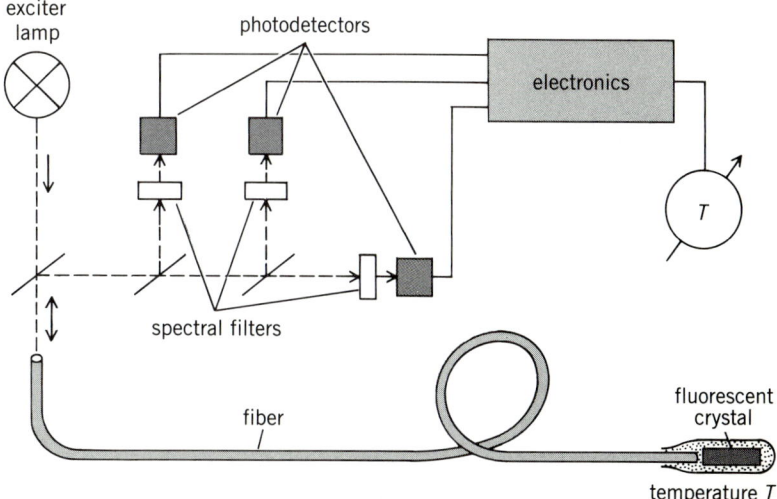

Fig. 2. Fiber-optic thermometer that operates by evaluating power ratios of fluorescent lines.

tension of the fiber itself. For evaluation of the resulting phase changes, a second light wave serves as a reference. It is guided along a second fiber which is not affected by x, or along the same fiber as the first wave but in an orthogonal polarization or in the opposite direction. These two light paths, together with beam-splitting and beam-combining elements at their ends, form a fiber-optic interferometer. For good interference contrast, the fibers must be single-mode fibers, preferably of the polarization-maintaining kind.

Interferometric sensors are most promising because they offer extremely high absolute sensitivities (dilatations below 10^{-10} m are measurable), high dynamic ranges (by digital counting of interference fringes and analog interpolation of fractional fringes), and great flexibility in deploying the sensing fiber (for example, in specially shaped fiber coils to tailor the directivity of a fiber-optic hydrophone).

In interferometric hydrophones for the detection of underwater sound, the pressure changes act directly or indirectly (through elastic transducer bodies) to deform a fiber and thus produce a phase change. Similarly, laboratory versions of fiber-optic magnetometers use a magnetostrictive body which deforms under the influence of a magnetic field. The fiber is wrapped around it and converts the deformation into an optical phase change. In fiber-optic current sensors, the magnetic field associated with the current changes the phase or polarization of a light wave by the Faraday effect, permitting current measurement at high-voltage lines.

By far the most advanced interferometric sensor is the fiber-optic gyroscope (Fig. 3). Light from a superluminescent diode is split at a fiber-optic directional coupler (DC 2) into two light waves which travel in opposite directions around a long fiber, coiled with, say, 2-in. (5-cm) diameter. Returning to DC 2, they are combined and guided through another directional coupler (DC 1) to the photodiode. When the coil is rotated with angular velocity Ω, a phase change $\Delta\phi$ results due to the Sagnac effect, given by the equation below, where A is the area of

$$\Delta\phi = 8\pi m A\Omega/\lambda c$$

the fiber coil, m is the number of turns in the coil, λ is the wavelength of the light, and c is the speed of light. With the help of a phase modulator and detection electronics, $\Delta\phi$ is measured and Ω evaluated. Industrial development of this sensor is progressing rapidly, as it promises to become a small, rugged, all-solid-state rotation sensor without moving parts, which may substitute for classical spinning-mass gyros in modern inertial navigation systems.

For background information *see* GYROSCOPE; INTERFEROMETRY; OPTICAL FIBERS in the McGraw-Hill Encyclopedia of Science and Technology.

[REINHARD ULRICH]

Bibliography: R. A. Bergh et al., An overview of fiber-optic gyroscopes, *IEEE J. Light-wave Technol.*, LT-2:91–107, 1984; T. G. Giallorenzi et al., Optical fiber sensor technology, *IEEE J. Quant. Electr.*, QE-18:626–665, 1982; *Optical Fibre Sensors*, IEE Conf. Publ. 221, London, 1983.

Parity (quantum mechanics)

The search for parity violation in atoms aims primarily at testing the new unified theories of electroweak (that is, electromagnetic and weak) interactions in a regime not accessible to high-energy experiments. Atomic physics experiments started in 1973, triggered by the theoretical discovery that the effects to be expected, although tiny, were much larger than previously anticipated in heavy atoms. In spite of considerable experimental difficulty, there is now evidence of parity violation effects in several heavy atoms, and the results complement those that are obtained at high energy. Parity violation in atoms is the first atomic physics effect to be observed that is not described by quantum electrodynamics.

Gauge theories predict electroweak interactions to be mediated by (at least) four gauge bosons: the photon (γ) for electromagnetic interactions; the charged W^+ and W^- bosons for the charged-current weak interactions such as those involved in β decay; and the neutral Z^0 boson for the neutral-current weak interactions, a new type of weak interaction first predicted by these theories and first observed in 1973 in neutrino experiments. Charge conservation commands a particle which emits or absorbs a W^+ or W^- boson to change its charge. Such a process, which takes place, for example, in β decay, is excluded by definition in a stable atom. However, just like photons, neutral Z^0 bosons may be exchanged between electrons and nucleons inside the atom without affecting its stability.

An important specific property of weak interactions is that they do not respect mirror symmetry. This means that two experiments, prepared in corresponding left-handed and right-handed configurations, do not in general produce corresponding left-handed and right-handed results. In 1956, T. D. Lee and C. N. Yang speculated that this peculiar property, known as parity violation in weak interactions, existed, and it was observed, beginning in

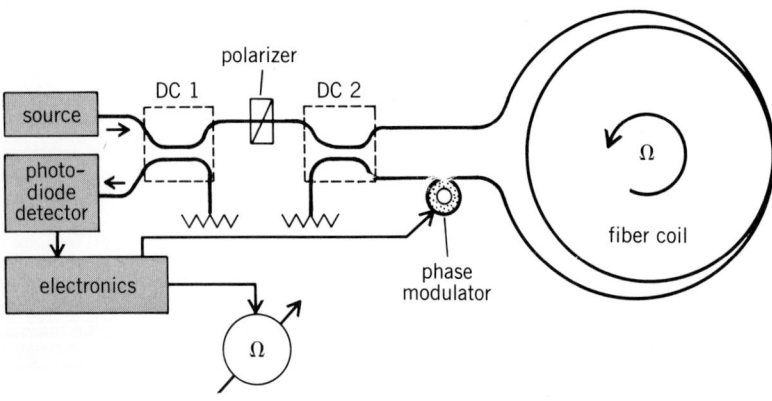

Fig. 3. Fiber-optic gyroscope.

1957, in all weak processes then known, corresponding to exchanges of $W^{\pm}$ bosons. Parity violation was predicted also in exchanges of Z^0 bosons by certain electroweak models, in particular the current standard Weinberg-Salam model. It was observed in 1978 in a high-energy experiment. Finally, in the standard model, parity violation (that is, violation of mirror symmetry) is expected in stable atoms. The search for this parity violation provides the basis for a test of the new electroweak theories. *See* ELEMENTARY PARTICLE.

Observed atomic properties. The Z^0 exchange, which violates mirror symmetry but preserves time reversal invariance, induces in atoms new transition amplitudes A_{pv}, otherwise forbidden by the parity selection rule. In particular, an electric dipole moment amplitude becomes possible between two states of the "same" parity—that is, the same parity in the absence of weak interactions. (The reason for considering a transition electric dipole moment rather than a static one is that a static electric dipole moment is not compatible with time reversal invariance.) The new transition amplitude A_{pv}, associated with Z^0 exchange, adds to the usual amplitude A_{em} associated with photon exchanges only, so that the transition probability becomes $|A_{em} + A_{pv}|^2$.

The fact that A_{em} respects mirror symmetry while A_{pv} violates it implies that associated right-handed and left-handed configurations lead to unequal transition probabilities $|A_{em} + A_{pv}|^2$ and $|A_{em} - A_{pv}|^2$. The electroweak interference term $\pm 2A_{em}A_{pv}$ gives rise to a contribution whose sign changes under reversal of the handedness of the experiment. The observed right-left asymmetry $\mathcal{A}_{RL}$, defined as the difference of these two probabilities divided by their sum, is practically $2A_{pv}/A_{em}$ (since in practice A_{pv} is much smaller than A_{em}).

Orders of magnitude. Because the mass M of the Z^0 boson is very large, the range ρ of the weak interaction, typically $\hbar/Mc$, where $\hbar$ is Planck's constant divided by 2π, is considerably smaller than the distance scale characteristic of atomic processes, the Bohr radius a_0 which is approximately 5×10^{-11} m or $10^7 \rho$. As a result, this interaction is expected to lead only to very small asymmetries (10^{-7} to 10^{-5}), even when very special experimental conditions are combined to maximize the effects.

Several methods are used to enhance $\mathcal{A}_{RL}$ up to this level:

1. Experiments are conducted with heavy atoms in order to benefit from the large increase of the asymmetry which grows slightly faster than the cube of the atomic number Z, as pointed out by M. A. Bouchiat and C. Bouchiat in 1973. Experimental results presently exist only in heavy elements: cesium (Cs), thallium (Tl), lead (Pb), and bismuth (Bi), with atomic numbers 55, 81, 82, and 83 respectively.

2. Forbidden transitions are observed, in order to reduce the electromagnetic amplitude A_{em}. This was the case for the experiments on cesium and thallium.

3. Experiments are conducted on the metastable $2S$ state of atomic hydrogen. The quasidegeneracy of the $2S$ and $2P$ states, of opposite parity, enhances the parity-violating amplitude.

Choice of elements. Whichever element is chosen, an atomic physics calculation is necessary in order to connect the measured quantity with the electroweak coupling constants. The need for a reliable atomic model obviously favors atoms with a single valence electron (that is, alkali atoms), especially that of largest atomic number Z (cesium, with $Z = 55$) in view of the Z^3 increase law. Yet there is considerable incentive for measuring effects in hydrogenic systems, since the atomic physics calculation introduces here virtually no theoretical uncertainty.

Interaction properties to be determined. Parity violation in atoms reflects the parity-violating weak neutral current interaction between electron and nucleon. A description of this interaction involves four fundamental constants: C_p^1 and C_p^2 describe the electron-proton interaction, respectively independent of and dependent on the proton spin; C_n^1 and C_n^2 are similar constants for the electron-neutron interaction. Each gauge theory makes definite predictions for these four constants, but ultimately they have to be determined from experiment.

Experiments in heavy atoms yield the so-called weak vector charge Q_w of the atomic nucleus. Q_w is the sum of the individual weak vector charges $-2C_p^1$ and $-2C_n^1$ of each of the Z protons and N neutrons that constitute the nucleus. Equivalently, it is the sum of the individual weak vector charges (denoted by $-2C_u^1$ and $-2C_d^1$) of the u and d quarks that constitute the protons and neutrons, as given by the equation below. This additivity origi-

$$Q_w = -2[ZC_p^1 + NC_n^1]$$
$$= 2[(2Z + N)C_u^1 + (Z + 2N)C_d^1]$$

nates in the conservation of weak vector charge. In high-energy experiments the nucleons are broken, so that the quarks act incoherently. As a result, although the only high-energy experiment sensitive to the same electron-nucleon interaction as the heavy-atom experiments also measures the value of a linear combination of $-2C_u$ and $-2C_d$, this linear combination turns out to be nearly orthogonal to the combination measured in heavy atoms. That is, when the ranges of values of C_u and C_d allowed by the two experiments are graphed (Fig. 1), the two domains are strips which are nearly at right angles to each other.

Each point of segment AB in Fig. 1 represents the prediction of the standard model for one value of its parameter $\sin^2\theta$. The intersection of the two strips is consistent with the standard model prediction for a value of $\sin^2\theta$ of approximately 0.2, in agreement with determinations from high-energy experiments of totally different types.

If they reach sufficient accuracy, experiments in hydrogen (and its isotope deuterium) can in principle determine the four constants separately. Since

328 Parity (quantum mechanics)

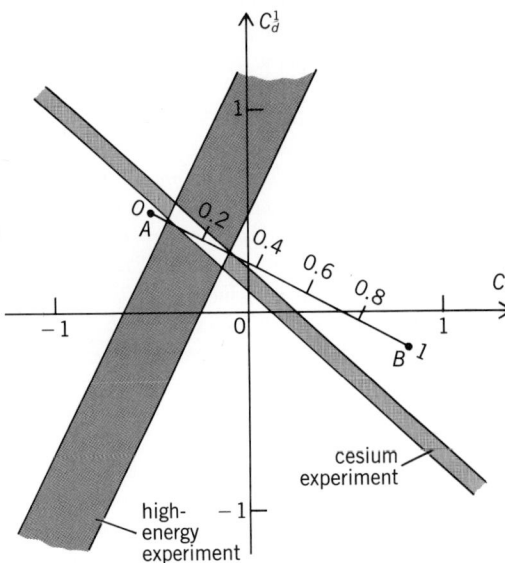

Fig. 1. Domains of values of the weak charges $-2C_u^1$ and $-2C_d^1$ of the up and down quarks allowed by the results of the cesium experiment and a high-energy experiment sensitive to the same weak electron-nucleon interaction (including experimental uncertainty at 90% confidence level and theoretical uncertainty).

three of these are expected to be very sensitive to higher-order corrections, this could provide refined tests of the theory. In addition, atomic physics tests the weak interaction in a distance range not accessible to high-energy experiments, which explore it at distances much closer to its range ρ.

Systematic effects. The observed asymmetry is the sum of the true asymmetry and of a systematic (that is, spurious) asymmetry. The latter originates in the experimental imperfections of the handedness reversal. Because of these imperfections, the two experiments which are compared are not exactly mirror images of each other. Consequently, parity-conserving signals may spuriously appear parity-violating. Since they are so overwhelming, even a small fraction of them may be comparable in size with the expected parity-violating signal. The main advantage of forbidden transitions is precisely to attenuate this difficulty by making the electromagnetic signals somewhat less overwhelming. Yet even in this case, reducing and controlling systematic effects is the major difficulty of these experiments.

Types of experiments. Three types of experiments have been, or are being, conducted involving atomic optical rotation in heavy atoms, forbidden transitions of heavy atoms, and transitions between sublevels of the 2S state in hydrogen.

Atomic optical rotation in heavy atoms. Since atoms have no geometrical handedness, optical rotation in atomic vapor violates mirror symmetry. The effect originates in a (parity-violating) difference between the refractive indexes for right and left circular polarization.

Atomic optical rotation has been observed in two transitions of bismuth and one of lead. In practical conditions the observed rotation is typically 10^{-7} radian. The first observation of parity-violating optical rotation was reported in bismuth in 1978. A factor-of-2 discrepancy between the original result and subsequent results from other groups remains. Quoted experimental uncertainties are typically 15–25%, essentially determined by potential systematic effects. Due to the complex structure of lead and bismuth, different atomic models yield theoretical predictions dispersed over a range of about ± 30%. Nevertheless, parity violation is clearly present with the expected sign and order of magnitude.

Forbidden transitions of heavy atoms. In one type of experiment (Fig. 2), the atomic transition is excited in a dc electric field $\vec{E}$ by a resonant, circularly polarized laser beam. The measured quantity is the electronic spin polarization P of the excited atoms. The handedness of the experimental configuration is determined by the right or left circular polarization of the beam. (The polarization modulator must be specially designed for high purity.) Reversing this circular polarization ($\xi = \pm 1$) amounts to reflecting the configuration in a mirror containing the beam and the field. The image of the spin component normal to this mirror is the component itself (it represents some sort of rotation within the mirror plane, and this rotation coincides with its own image). In fact, under reversal of the beam polarization a slight variation of this component is observed, so that mirror symmetry is violated. In Fig. 2b, the small component P^{pv} of the electronic polarization reverses; $P^{(1)}$ and $P^{(2)}$ are parity-preserving components of the electronic polarization. This experiment has been performed in thallium and cesium. In the cesium cell, mirrors M_1 and M_2 reflect the beam

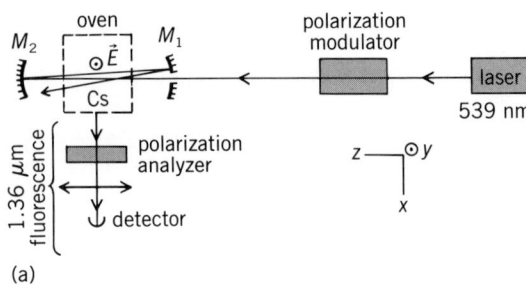

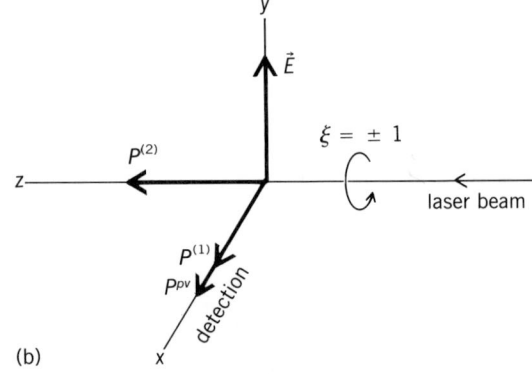

Fig. 2. Experiment to detect parity violation in cesium with a circularly polarized laser beam. (a) Layout of experimental equipment. (b) Relation of experimental parameters.

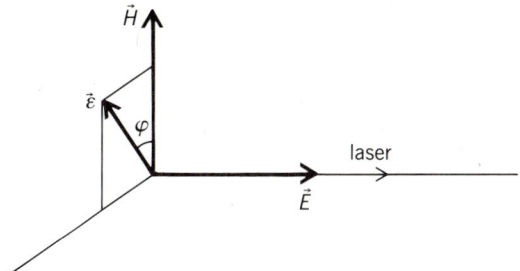

Fig. 3. Handed geometrical configuration used to detect parity violation in thallium, using a linearly polarized laser beam in crossed electric and magnetic fields.

back and forth many times through the vapor (beam multipass). The electronic polarization of the atoms is deduced from the measured polarization of the fluorescence light emitted at a different wavelength when the atoms excited by the laser deexcite. The experimental uncertainties are about 30 and 15%, and the uncertainties added by theoretical interpretation are about 30 and 10% for thallium and cesium respectively. In the cesium experiment, each potential source of systematic effect has been checked to be less than about 1% of the observed effect. The results agree with the standard electroweak theory.

In a second type of experiment, the transition is excited in crossed dc electric and magnetic fields, $\vec{E}$ and $\vec{H}$, and the laser is polarized linearly (Fig. 3) and directed along $\vec{E}$. The beam polarization $\vec{\epsilon}$ and the fields define a handedness. The measured quantity is the absorption of the laser light by the vapor. When the handedness is reversed (by reversing, for example, the electric field), a small difference in absorption is observed (in practical conditions typically 10^{-5}). Mathematically, the existence of a term proportional to $(\vec{\epsilon} \cdot \vec{H}) \cdot (\vec{\epsilon} \times \vec{E} \cdot \vec{H})$ in the absorption cross section manifests parity violation. This effect has been measured in thallium with an experimental uncertainty of about 20%.

Hydrogen atom. Because of their extreme difficulty, the three experiments in progress (microwave transitions between two sublevels of the metastable 2S state) have produced no results yet.

For background information *see* ATOMIC STRUCTURE AND SPECTRA; FUNDAMENTAL INTERACTIONS; PARITY (QUANTUM MECHANICS); WEAK NUCLEAR INTERACTIONS in the McGraw-Hill Encyclopedia of Science and Technology.

[MARIE-ANNE BOUCHIAT; LIONEL POTTIER]

Bibliography: L. M. Barkov, M. S. Zolotorev, and I. B. Khriplovich, Observation of parity nonconservation in atoms, *Sov. Phys. Usp.*, 23:713–731, 1980; M. A. Bouchiat and L. Pottier, An atomic preference between left and right, *Sci. Amer.*, 250(6):100–111, June 1984; M. Davier, Electroweak neutral currents, *J. Phys.* (Paris), 43 C3:471–511, 1982; E. N. Fortson (ed.), *Atomic Physics 9*, pp. 246–271, 1984; E. N. Fortson and L. L. Lewis, Parity violation in atoms, *Phys. Rep.*, 113:289–344, 1984.

Particle accelerator

The design and construction of higher-energy particle accelerators continues as physicists probe the fundamental structure of matter. This article discusses four machines at the forefront of this accelerator technology: The Tevatron is now in operation. The SLAC linear collider (SLC) is under construction and is scheduled for completion in 1986. A proposal for a continuous electron-beam accelerator facility (CEBAF) has been prepared. Lastly, a proposal for a superconducting supercollider (SSC) has also been prepared, which would have the highest energy of any accelerator.

TEVATRON

The Tevatron is the largest and highest-energy proton synchrotron in the world. It is the first high-energy particle accelerator to use superconducting magnets as the main guide field dipoles (bending magnets), quadrupoles (focusing magnets), and correction elements.

Features. This machine has many noteworthy features. The first is its energy. It is capable of accelerating protons to 1 teraelectronvolt (TeV)—hence the name Tevatron. (1 TeV = 10^{12} electronvolts = 1.6×10^{-7} joule.)

The second is its size. The accelerator is located in a tunnel almost 4 mi (6 km) in circumference. This tunnel is 30 ft (9 m) below the surface of the earth at the Fermi National Acceleratory Laboratory near Batavia, Illinois. An earthen mound above the tunnel is a prominent feature of the landscape (Fig. 1).

The third unique feature is that all of the 2000 electromagnets in this machine are cooled to liquid-helium temperature, $-451°F$ ($-268°C$). At this temperature the magnet coils, which contain niobium-titanium filaments, are superconducting; in other words they exhibit no electrical resistance. This means that the magnets can be excited to very high fields without the expenditure of a large amount of electic power. Thus, as the protons are accelerated to higher energy, the magnets can be excited to higher magnetic fields to confine the protons to the interior of a small beam tube as they pass around and around the machine.

Accelerating cycle. The Tevatron is the final device in an array of machines. The protons pass sequentially from a 750-keV Cockcroft-Walton machine to a 200-MeV linear accelerator, then are transferred to a GeV booster-synchrotron. From the booster they go to a 150-GeV proton synchrotron (the old 400-GeV Fermilab main ring tuned down to 150 GeV) and finally into the Tevatron (Fig. 2).

As the protons enter the Tevatron at 150 GeV, the superconducting magnets are set to a magnetic field of only 0.7 tesla (7000 gauss). As the protons circulate around the machine, they encounter a set of four radio-frequency accelerating cavities placed at one location on the 4-mi (6-km) circle. These cavities, working together, use a high electric field oscillating in phase with the beam to increase the pro-

Fig. 1. Aerial view of Fermilab. (*Fermilab*)

ton energy by 1.2×10^6 eV on each turn. As the proton energy increases turn by turn, the magnetic field must be raised to keep the beam on the same circular path. Finally after about 750,000 trips around the 4 mi circle, the protons have an energy of 10^{12} eV. The magnetic field in the dipoles has been raised to 4.5 T (45,000 G) by raising the magnet excitation current. This same current excites the focusing quadrupoles proportionally, keeping the proton beam comfortably inside the beam tube.

Precision. Every element of the Tevatron has to be constructed and controlled with extreme precision in order to make the accelerator work at all. The beam of protons must not stray more than 0.5 in. (1.25 cm) from the center of the beam tube or the protons will be lost into the magnet walls. This means that each turn of superconducting cable in each magnet has to be placed to an accuracy of about 0.001 in. (25 micrometers) and held there even under the tremendous forces (up to 3000 pounds per linear inch or 5×10^5 newtons per meter) generated by the pressure of the magnetic field. The machine contains almost 1000 mi (1600 km) of superconducting cable.

The control of the high-current power supplies and the correction magnets has to be correspondingly precise. For example, the twelve 4500-A 1000-V power supplies used to excite the magnets must be regulated to better than 1 part in 10^4.

Complexity. The complexity of the accelerator is also surprising. For example, the magnets are cooled with a helium refrigeration system which contains twenty-four 1000-W satellite refrigerators.

These refrigerators utilize liquid helium from a central helium liquefier capable of producing more than 1000 gallons (4 m^3) of liquid helium per hour. The compressors for this refrigeration use a total of 16,000 hp (12 MW). These devices, in fact, are the major consumers of electric power in this superconducting machine, because the energy used to excite the magnets is returned to the power grid at the end of each acceleration cycle.

Another example of complexity is the control and monitoring system. This system contains more than 500 microcomputers which oversee the power supplies, refrigerators, beam monitors, radio-frequency accelerating cavities, beam injection, extraction and abort systems, and so forth. All of these computers communicate with a central computer and through it with the operators working at consoles in the main control room.

Quench protection. Generally, the system operates smoothly but occasionally something goes wrong. For example, electromagnetic noise pulses can confuse the microcomputers; or there can be a loss of beam into the magnet walls. The lost protons will heat the magnets beyond the superconducting temperature. This causes a phenomenon known as a quench.

A quench involves the superconducting wire from which the magnet coils are made. The Fermilab magnet wire is made of 2100 niobium-titanium filaments, each only 3×10^{-4} in. (7.5 μm) in diameter. These filaments are embedded in a copper matrix making up the rest of the 0.027-in.-diameter (0.8-mm) wire. Twenty-three of these wires are

Fig. 2. The two large proton synchrotrons at Fermilab. The Tevatron is on the bottom. The upper machine is the old Fermilab main ring, now used as an injector for the Tevatron. (*Fermilab*)

made into a cable (Fig. 3). The cable is wound through 112 turns to form the coils of the 21-ft-long (6.4-m) magnets.

When the magnet temperature rises above −440°F (−262°C), the niobium-titanium filaments no longer have zero resistance. Electric current will rapidly heat the cable to room temperature, turning the liquid helium surrounding the coils into high-pressure vapor.

This quench would quickly melt the coils. However, in the Tevatron they are protected by a microprocessor-controlled quench protection system. This system switches current around the quenched magnet, dumps the stored energy through external resistors, and shuts off the power supplies. The microcomputer must diagnose the trouble and respond quickly, in less than a few thousandths of a second.

Experiment. There are many other examples of systems which must react promptly, reliably, and with precision in the Tevatron. However, all this engineering and technology are not ends in themselves. When the protons have finally been accelerated and extracted, they are sent to many experiments. It is these experiments, probing deeply into the structure of matter, that this accelerator was designed to serve.

Proton-antiproton collider. In the near future the facility will be used in a manner allowing experimenters to obtain even higher energy probes. Antiprotons will be injected into the ring, accelerated to 1 TeV, and made to collide with protons circulating in the opposite direction. This colliding mode will allow physicists to observe interactions at 2 TeV in the center of mass. This is over 40 times the useful energy available to them in the currently operating fixed-target mode.

[J. RICHIE ORR]

LINEAR COLLIDERS

The SLAC linear collider (SLC) is the first of a new type of electron-positron colliding-beam device. It is designed to be both a research tool for elementary particle physics experiments and a facility in which this new technique can be further developed and tested.

A true linear collider would consist of two linear

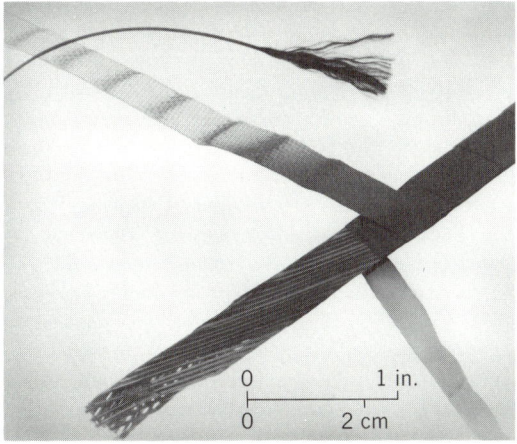

Fig. 3. Fermilab superconducting cable and one of the 23 cable strands with the copper matrix etched away to show the 2100 niobium-titanium filaments. (*Fermilab*)

accelerators (linacs) firing intense bunches of electrons and positrons at each other. During the collision, the beams in such a machine would be disrupted by the very strong electromagnetic fields in the collision region, and would thereafter be disposed of and not used again. Since the beams would not have to be deflected into a circular path (as they are in the storage-ring technique that has been used up to now for electron-positron colliding-beam experiments), no synchrotron radiation would be produced in the true linear collider. This makes the scaling laws for the size and cost of linear colliders as a function of energy much more favorable for extension to considerably higher energies than are the scaling laws for storage rings. Since the beams in linear colliders interact only once, the disruptive effect of the collision of the two beams can be much stronger than could be allowed in a storage ring, where the beams must be kept circulating in the magnetic aperture of the ring for periods of hours (billions of collisions).

SLC design. The SLC (Fig. 4) is not a true linear collider, since it uses one linac rather than two, and the beams coming out of the linac are magnetically deflected through curved arcs to reach the collision point. The linac is the existing SLAC (Stanford Linear Accelerator Center) 2-mi-long (3.2-km) machine, and in this variation of the linear collider technique intense bunches of electrons and of positrons are both accelerated in a single pulse of the linac, are separated by a magnet at the end of the machine, and are then injected into two beam-transport systems that guide the beams to the collision point. At the design energy of the SLC, the synchrotron radiation emitted in this transport system is comparatively unimportant, although at much higher energies it would become so severe as to make this one-linac variant impractical.

The main design goals fo the SLC are as follows. The center-of-mass energy at the collision point is to be 100 GeV, and the luminosity (reaction rate per unit cross section) at this energy is to be 6×10^{30} cm^{-2} s^{-1} at a repetition rate of 180 pulses per second. To reach these goals the linac must be upgraded in energy; a positron-production facility must be built to produce the 5×10^{10} positrons required for each pulse; a beam-conditioning system (damping rings) must be built to produce the low-emittance beams required for operation; tunnels must be bored to house the beam transport magnets required to bring the particles from the end of the linac to the final focus system; a highly corrected magneto-optical system must be built to focus the beam down to the 1.4-μm radius required at the collision point; and an experimental hall must be built to house the experimental apparatus. Construction of this facility began in October 1983, and the expected startup date is October 1986.

Operation cycle. To discuss the operation cycle, it may be assumed that the machine is running in equilibrium. The two damping rings near the front of the linac (one for electrons and one for positrons) are each filled with two bunches of particles. At the start of the cycle, one positron and two electron bunches are extracted from the damping rings and injected into the linac. At injection each bunch has an energy of 1.2 GeV, is 0.08 in. (2 mm) long, and has an emittance (radius times angular divergence) of 5×10^{-7} radian-inch (1.3×10^{-8} radian-meter). The bunches travel down the linac about 55 ft (17 m) apart, gaining 5.3 MeV for each foot (17.5 MeV for each meter) traveled.

At the two-thirds point of the linac, the positron and the first electron bunch pass the positron source. The second electron bunch is diverted from the linac to the positron source. The first two bunches continue down the linac, reaching 51 GeV at the end of the machine with an energy spread of ±0.4%. The bunches have a cross-sectional radius of approximately 90 μm at this point. A magnet separates the electron and positron bunches and directs them into the two transport systems. In these transport systems (the collider arcs), the beams lose about 1 GeV of energy by synchrotron radiation. The focusing must be very strong to prevent quantum fluctuations in synchrotron radiation from in-

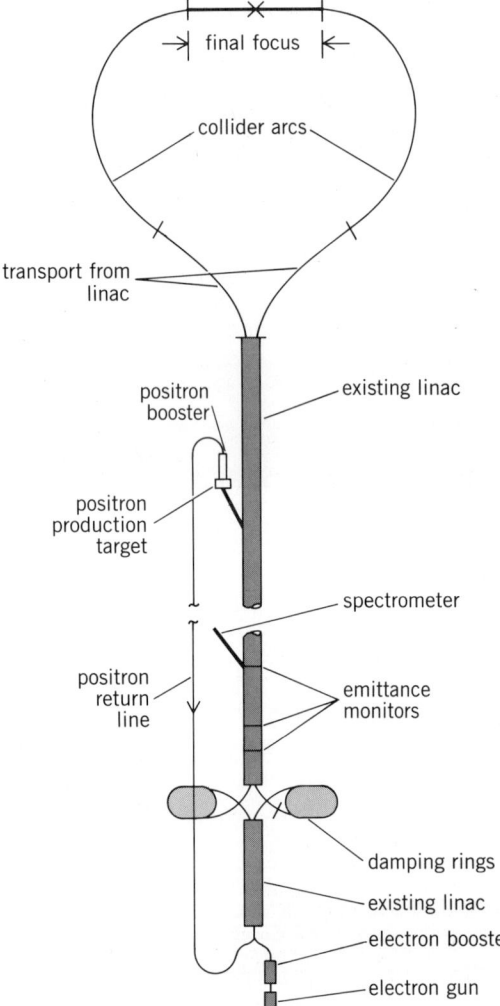

Fig. 4. Layout of SLAC linear collider (SLC).

creasing the emittance, which is 1.2×10^{-8} radian-inch (3×10^{-10} radian-meter) at this point. Small strong-focusing magnets (0.3-in. or 8-mm aperture) are required. The bunch radii are about 30 μm in the arcs.

Final focus. The final focus is a complex system that must reduce the beam size at the collision point to a radius of 1.4 μm and must also correct for both chromatic and geometric aberrations. During the actual collision of the two bunches, the extremely strong fields within the bunches, equivalent to about 100 teslas (10^6 gauss), reduce the mean beam cross-sectional area during the collision by about a factor of three (a pinch effect). After the collision the nonlinear part of these intense fields has so badly distorted the phase space of the beams that it is impossible to refocus them to a small size again; thus they are disposed of and not used further.

Positron acceleration. Meanwhile, the bunch that was extracted at the positron target has struck the target, producing about 10^{11} positrons within the acceptance of the positron production system. The average energy of these positrons at the production target is only 2 MeV. This intense low-energy bunch is boosted to 200 MeV by a small linac, then turned around and brought back to the front of the linac by a transport system built in the existing linac housing. It is then turned around again and injected into the first section of the existing linac, boosted to 1.2 GeV, and injected into the positron damping ring. Two electron bunches are produced and injected into the electron damping ring. In the damping rings the transverse beam sizes decrease through synchrotron radiation. The positron bunch, which is initially much larger than the electron bunches, must remain in the damping ring for a longer time than the electron bunch. Then in the next cycle, which starts about 5 milliseconds later, the "old" positron bunch and the two electron bunches are extracted to begin the cycle all over again.

Development, testing and application. The energy upgrade of the SLAC linac requires the development of a new 50-MW (peak power) S-band klystron to replace the existing 36-MW tube. These new tubes must operate at a higher voltage and they have twice the pulse length of the old tubes. A pulse compression scheme is used at the output of the kystron to raise the effective peak power of the new tubes at the linear accelerator by a factor of 2.5 over that provided by the old tubes. Successful prototypes of these new tubes have been tested, and full-scale production has begun.

Testing of the front end of the machine with the required two electron bunches has successfully demonstrated the full cycle of electron production, acceleration, damping, reinjection, and acceleration to the one-third point of the linac. The electron bunches meet the emittance specification at this point. The full SLC should begin tests in the last quarter of 1986, and the experimental physics program should begin early in 1987.

The experimental program is aimed at the study of the production and decay of the Z^0 boson, the transmitter of the weak force. The Z^0 production rate of 800 per hour expected at full luminosity should give much new information on the unification of the electromagnetic and the weak interactions and on the structure of the basic building blocks of matter.

Future colliders. In parallel with the construction of the SLC, work is proceeding in the United States, the Soviet Union, Europe, and Japan on the technology and optimization procedures required for the construction of a practical and economical large linear collider in the TeV energy range. These studies have already demonstrated accelerating gradients in conventional structures of more than 30 MeV per foot (100 MeV per meter). New types of high-efficiency, high-power radio-frequency sources are under development. Novel accelerating schemes capable of still higher accelerating gradients are also under investigation. A key element in progress toward these large machines will be the demonstration by the SLC that the dynamics of the acceleration and beam-beam interaction process are understood as well as is now believed.

[BURTON RICHTER]

CONTINUOUS ELECTRON BEAM ACCELERATOR FACILITY (CEBAF)

As a probe of nuclear matter the electron is unsurpassed. Its lack of internal structure and its well-understood interactions through the relatively weak electromagnetic interaction make the extraction of precise information about the probed system possible. Consequently, electrons have been used to study solids, atoms and molecules, atomic nuclei, and most recently, elementary particles.

Pioneering studies of nuclear sizes and configurations conducted in the 1960s opened up a field of subatomic studies which has continued to grow and diversify. Studies of nuclear properties using electrons with energies of a few hundreds of MeV have yielded, and continue to yield, some of the most precise information available about nuclear systems. Experiments conducted in the 1970s using electrons with energies of tens of GeV made essential contributions to the discovery of quarks and to the determination of their properties and distribution within nucleons. Throughout its growth, electronuclear physics has maintained the attributes of precision and clarity which characterized the early experiments.

It has long been understood that the traditional picture of nuclei as collections of neutrons and protons interacting through the exchange of mesons must break down. The nucleons and mesons are themselves composed of quarks, so at some level of detail their composite nature must manifest itself directly. Recent experiments at CERN underscored this point when they revealed that the distributions of quarks were different in nucleons in nuclei than in free nucleons. To further examine these effects requires a continuous electron beam to permit coincidence experiments, in which two or more outgoing

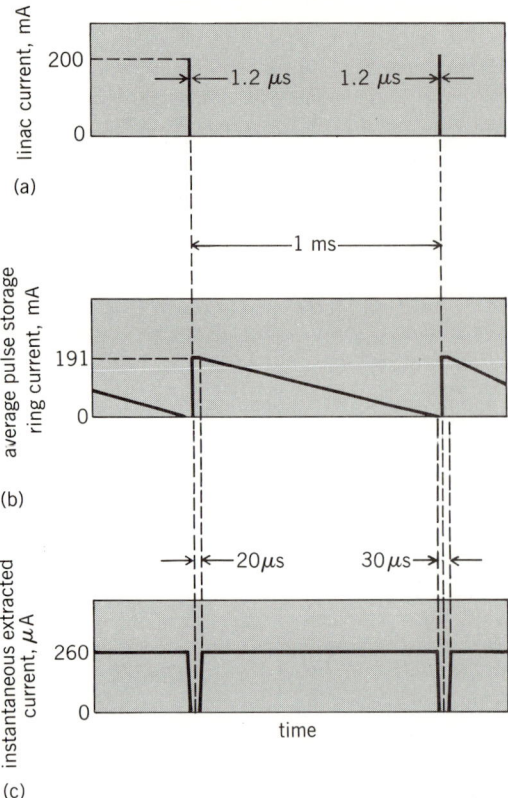

Fig. 5. Linac-PSR (pulse-stretcher-ring) time structure. (a) 200-mA, 1.2-μs linac beam pulses are injected into the pulse stretcher ring every 1.0 ms. (b) The average stored current in the ring is constant for approximately 20 μs after injection. (c) Current extracted from the ring rises to steady-state about 20 μs after injection, falling to zero about 10 μs before the next injection.

particles from one reaction are detected, with energies in the multi-GeV range. The need for an accelerator capable of producing such a beam was realized in the early 1970s; recent developments have brought this need into sharper focus.

In 1982 the Nuclear Science Advisory Committee solicited proposals for high-duty-factor electron accelerators in the GeV energy range. Five submissions were received, including one from the Southeastern Universities Research Association (SURA), a consortium of over 30 institutions in the southeastern United States. The SURA proposal was chosen and recommended for funding by the U.S. Department of Energy.

Description. The proposed Continuous Electron Beam Accelerator Facility will be located in Newport News, Virginia, on the former site of the NASA Space Radiation Effects Laboratory. The accelerator will be a combination of a linear accelerator followed by a pulse stretcher ring (PSR). As such, it represents a union of well-established linac and synchrotron technologies. A linac is used to efficiently accelerate high-current pulses of electrons which are then injected into a pulse stretcher ring. These electrons are subsequently extracted from the pulse stretcher ring by using the technique of slow extraction common to synchrotrons (Fig. 5). The present design was based originally on a suggestion by G. A. Loew.

Figure 6 shows a layout of the proposed facility. The linac will produce a beam of 1.2-microsecond pulses at a variable repetition rate of up to 1 kHz. The beam energy will be variable between 0.5 and 4.8 GeV on a pulse-to-pulse basis. Pulsed elements in the beam switch yard (BSY) will permit successive pulses to be directed either to the pulse stretcher ring or to one of the end stations (A or C). Pulses directed to the pulse stretcher ring will be injected vertically during a single turn and then extracted during the period before the next pulse is injected. The extracted beam may be split if desired, and all or part of it directed to each end station.

Linear accelerator. The accelerator will be composed of 40 traveling-wave, 2856-MHz, disk-loaded, accelerating waveguides structurally similar to those in use at SLAC. Each 10-ft (3-m) section will be powered by a 40-MW klystron/modulator-transmitter. For final energies below 2 GeV the electrons will pass once through the linac; for energies above 2 GeV they will be recirculated through 35 sections. Thus, 4-GeV electrons will be produced by acceleration through effectively a 75-section machine.

The length of the recirculation path was selected so that the head of a pulse being recirculated will reenter the linac immediately following the tail of the pulse. This "head-to-tail" recirculation will minimize the effects of transient beam loading and the potential for beam breakup. An energy compression system will be located immediately following the linac to reduce the energy spread.

Pulse stretcher ring. The pulse stretcher will consist of two 180°-bend regions joined by straight, achromatic insertions. Both injection and extraction will take place in the same insertion (Fig. 6).

The beam from the linac will be injected vertically onto the pulse-stretcher-ring closed orbit during a single turn. The circumference (1190 ft or 363 m) was chosen to be just greater than the length of the linac pulse so the pulse stretcher ring is essentially filled by the injected pulse. Beam will be extracted from the pulse stretcher ring by using achromatic, half-integral extraction in the horizontal plane. The extraction will be controlled by ramped quadrupole-octupole pairs in each insertion. A feedback system between the external-beam monitoring devices and the ramped multipoles will be used to

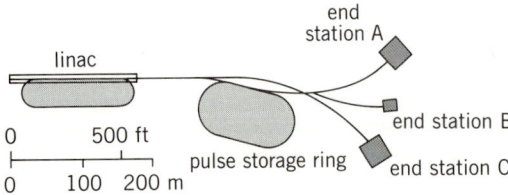

Fig. 6. Layout of Continuous Electron Beam Accelerator Facility (CEBAF).

ensure the quality and stability of the extracted beam. There will be two radio-frequency accelerating systems in the pulse stretcher ring, a continuous one and a pulsed one on only during the first few microseconds after injection. Combined, they will effect a reduction in the energy spread of the beam of a factor of four.

Beam switch yard. Beams can be directed from the linac to the pulse stretcher ring or to the end stations and from the pulse stretcher ring to the end stations. Pulsed elements will be provided in the beam switch yard to permit variation on a pulse-to-pulse basis. The beam extracted from the pulse stretcher ring can be split into three parts. High-duty-factor beams of differing intensity (high intensity to a primary beam experiment, low intensity to the tagged-photon facility) can be delivered simultaneously. Provision has been made for the addition of elements to permit the delivery of arbitrarily polarized beams to the end stations.

Experimental facilities. Primary experimental facilities will be housed in three end stations. Two of these (labeled A and C in Fig. 6) will be heavily shielded and capable of receiving full beam intensities. The third will be lightly shielded and, therefore, able to receive only low intensity (less than 1 microampere) beams. It will house a locally shielded system for producing tagged photons.

Upgrade and expansion. Flexibility for future upgrade and expansion has been built into the design from the beginning. In particular, the possibility of an increase in the beam energy has not been precluded.

With a linac in place an obvious avenue of expansion is by the addition of other rings. A second pulse stretcher ring would permit the delivery of simultaneous high-duty-factor beams of different energies, and a ring dedicated to experiments with internal targets could be added.

[JAMES McCARTHY; BLAINE NORUM]

SUPERCONDUCTING SUPERCOLLIDER

High-energy physicists are proposing the construction of a very large colliding proton-beam accelerator using superconducting magnets, called the superconducting supercollider. A facility of the proposed capacity has been made feasible by recent major advances in superconducting magnet technology, culminating in 1983 with the successful operation of the first full-scale high-energy superconducting accelerator, the Tevatron, discussed above. The proposed facility would address outstanding fundamental issues in understanding the elementary forces of nature.

Supercollider design. The proposed collider will use superconducting magnets to accelerate circular beams of protons (Fig. 7) to an energy of 20 TeV in each beam. The use of superconducting magnets permits increased magnetic field strength with substantially less energy required for operation. For example, the installation of superconducting magnets in the Fermilab accelerator ring reduced power consumption while doubling the accessible energy over that attainable by using conventional magnets in the same ring.

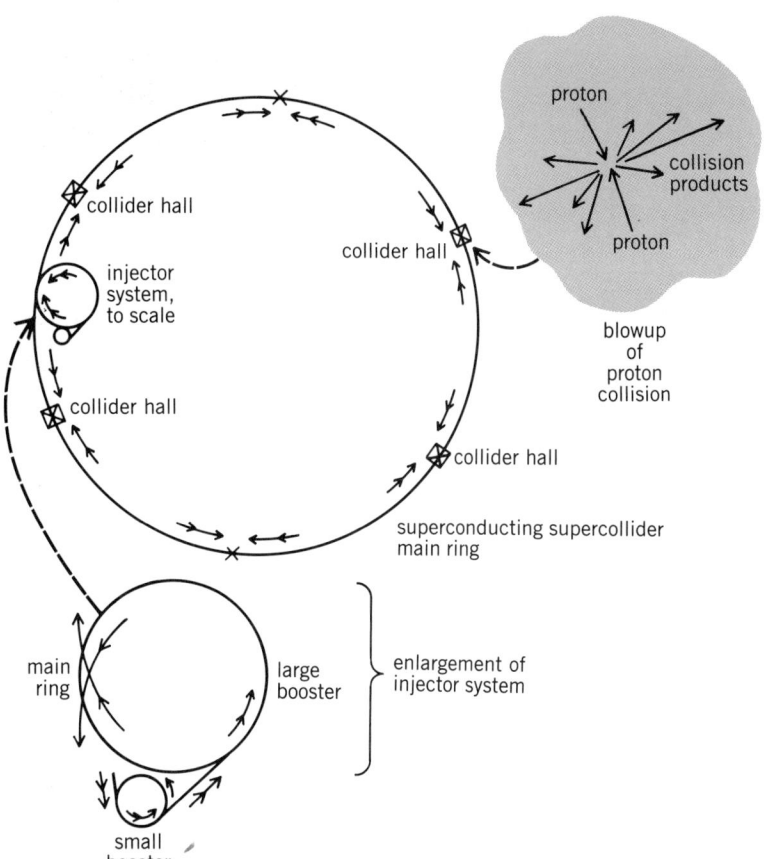

Fig. 7. Prototype design of the superconducting supercollider (SSC). Arrows indicate direction of protons. Crosses indicate points where protons are made to collide.

The Fermilab accelerator has a circumference of 4 mi (6 km) and can accelerate protons to an energy of 1 TeV. The proposed supercollider will have a radius of 56 to 103 mi (90 to 165 km), depending on the choice of magnetic field strength. In a prototype design (Fig. 7) from the reference design study prepared for the U.S. Department of Energy under the leadership of Maury Tigner, protons are accelerated to an energy of 10^{-3} TeV before injection into the small booster ring that accelerates them to 0.07 TeV. They are then injected into the large booster (the size of the Fermilab ring), accelerated in counterrotating bunches to 1 TeV, and injected into the main ring where the two counterrotating beams are accelerated to 20 TeV per beam and made to collide at six points around the ring. Collider halls can house detectors to record the passage of particles created in collisions. With an energy of 20 TeV in each beam, each collision will provide a total energy of 40 TeV that could in principle be converted into new forms of matter.

However, protons, as now understood, are not elementary particles, but rather "bags" containing more elementary constituents (quarks and gluons), each of which carries, on average, a sixth to a tenth of the proton energy. In order to meet the scientific

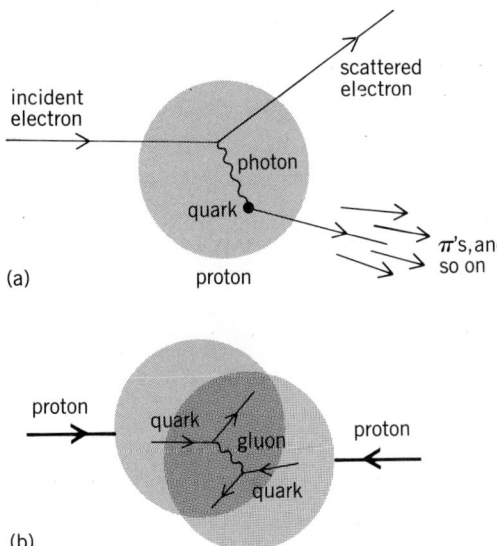

Fig. 8. High-energy collisions that resolve elementary interactions among quarks and gluons. (a) Penetration of a proton by an energetic electron that scatters from a single quark. (b) Collision of energetic protons with collision between constituent quarks.

goals of the superconducting supercollider, which require the study of individual collisions among quarks and gluons that can liberate energies up to several TeV, proton beams with tens of TeV energy are needed.

The energy stored in the colliding particles can be converted into new matter only when they come very close together; the creation of a particle of mass m requires an interaction distance less than $h/2\pi mc$, where h is Planck's constant and c is the velocity of light. Conversely, to study a new interaction that has a very short range r requires a collision energy greater than $ch/2\pi r$. For example, the constituent nature of protons and neutrons was revealed, and elementary interactions among quarks and gluons resolved, only when (Fig. 8) sufficient energy, namely billions of electronvolts, was available to probe distances much smaller than the nucleon radius, about 10^{-15} m. Thus, with sufficient energy an electron can penetrate the proton (Fig. 8a) and resolve its structure by scattering from a single quark; the struck quark converts its energy into a jet of π mesons and other particles. The composite structure of the nucleon was first revealed by a high-energy electrons scattering from nucleons at the Stanford Linear Accelerator Center. Energetic colliding protons (Fig. 8b) can also interact at sufficiently short distances that elementary collisions between their constituent quarks may occur. The most frequent collisions arise from gluon exchange. The weak interaction, which has a range of about 10^{-18} m, in contrast, requires tens to hundreds of billions of electronvolts for efficient study. Thus probing more deeply into the elementary forces of nature involves resolving phenomena over increasingly small distances.

The rate at which a particular type of interaction occurs is governed by the rate at which protons collide, and also by the area over which the interaction is effective. To attain a sufficiently high interaction rate for the study of new phenomena in the TeV energy region, or, equivalently, to resolve distances less than 10^{-19} m, the small effective interaction area must be compensated for by a high proton collision rate. The standard figure of merit for a particle collider is the number of crossing protons per square centimeter of area per second, called luminosity. The proposed supercollider could attain a luminosity of 10^{33} cm^{-2} s^{-1}, providing a significant number of events for the study of new phenomena.

Status of elementary particle physics. There has been a profound advance since the mid-1960s in understanding the basic forces of nature. The nuclear force is now attributed to elementary interactions among quarks and gluons, with colored quarks bound together inside the nucleon by a force transmitted by gluons, the quanta of the color electromagnetic field; this force is very similar to the force transmitted by photons, the quanta of the ordinary electromagnetic field, that binds electrons with nuclei to form atoms. An important difference is that the gluons are carriers of the color charge (analogous to ordinary electric charge) as well as the transmitters of force between color-charged particles. A consequence of this dual role of the gluons is that the force binding the quarks and gluons together becomes increasingly strong as the separation between them increases. This is thought to underlie the observed phenomenon known as confinement: it appears impossible to separate the constituents of the nucleon no matter how much energy is available.

The weak interaction, responsible for radioactive decays of unstable elements, is now understood as arising from a force that is also very similar to the ordinary electromagnetic force and the strong color force described above. The difference is that while the photon and gluons have no rest mass, the quanta that mediate the weak force, called W^+, W^-, and Z^0, are 80 to 95 times more massive than the proton. Nevertheless, the weak interaction is understood within an intricate framework with weak and

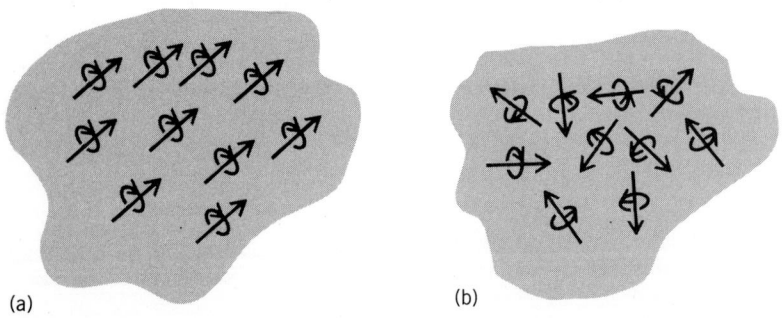

Fig. 9. Orientation of particle spins in a ferromagnetic substance. (a) Alignment of spins along a common direction in a cold substance, breaking the rotational symmetry of the system. (b) Random orientation of spins at high temperature, restoring rotational symmetry.

electromagnetic phenomena viewed as different manifestations of the same underlying forces. The formulation of this theory by Sheldon Glashow, Abdus Salam, and Steven Weinberg led to the prediction and subsequent observation of neutral current phenomena, and it was most strikingly confirmed by the discovery in 1983 at CERN in Geneva of the W and Z particles, with masses and other properties as predicted by the theory. See ELEMENTARY PARTICLE.

Spontaneous symmetry breaking. The large masses of the W and Z are attributed to a property of the vacuum, or state of lowest energy (ground state), known as spontaneous symmetry breaking. This phenomenon is most easily visualized by analogy with a ferromagnet. The laws of nature are invariant under rotations in space, which means that the energy of an isolated spinning particle is independent of its axis of rotation (spin). In a ferromagnet the total energy is lowest when all particle spins are aligned (Fig. 9a); this defines the ground state. Alignment of spins implies an axis of alignment, or direction along which all spins are pointing; the specification of this direction destroys the rotational invariance of the system. Because any direction would have been equally energetically favorable, the breaking of rotational symmetry is said to be spontaneous.

According to present understanding, an analogous phenomenon is responsible for the large masses of the W and Z particles. The analog of spin alignment is the energetically favored alignment of weak charges (similar to electric or color charges) in the vacuum state.

Little is known about the mechanism responsible for the vacuum charge alignment that triggers electroweak symmetry breaking, but very general principles imply that some manifestation of this mechanism must appear in collisions releasing trillions of electronvolts of energy. Uncovering its origin is a primary objective of the superconducting supercollider.

The simplest form this new phenomenon could take is that of a single spinless particle, called the Higgs particle, whose mass might even be considerably less than 1 TeV/c^2. However, attempts to formulate a fully plausible theory indicate that a more complex structure is involved; specific models generally predict a rich spectrum of states in the 1 TeV/c^2 mass region, possibly entailing new strong interactions of W and Z particles (Fig. 10).

Other physics issues. There are a number of other questions that can be addressed by experimentation at the superconducting supercollider. New phenomena observed may provide clues to understanding the observed quark and lepton spectrum. The similarity of the strong, electromagnetic, and weak forces, as presently understood, suggests a unifying principle. This idea is supported by the relative strengths of the observed interactions, but it predicts the instability of the proton with a decay rate that may exceed current experimental limits. The same unified theory offers an understanding of

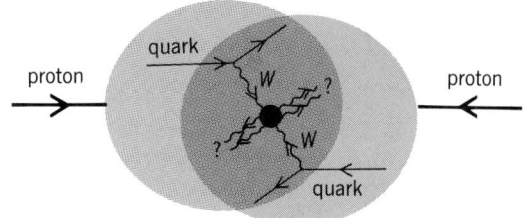

Fig. 10. Collision in which extremely energetic quarks radiate W and Z particles that can then collide to create a heavy Higgs particle or other new forms of matter related to the origin of electroweak symmetry breaking.

the predominance of matter over antimatter in the universe. All these observations could be reconciled by the existence of additional particles observable only at energies to be made accessible by the superconducting supercollider.

Answers to questions regarding the mechanism responsible for galaxy formation and the possible existence of "dark matter" in galactic halos may also involve new particles. Among the conjectured new states are objects called photinos and gravitinos, fermionic partners of photons and gravitons. These occur in the supersymmetry and supergravity theories that may provide a link between gravity and other forces, and could also play a role in electroweak symmetry breaking.

Vacuum charge alignment phenomena may have been important in cosmological history. Just as heating a ferromagnetic substance restores rotational invariance by random spin orientation (Fig. 9b), the electroweak symmetry of the vacuum should have been restored in the very hot early universe. The details of subsequent cooling and phase transitions to the asymmetric vacuum could have affected the expansion rate of the universe at various stages of its history, and also the relative densities of matter and radiation now observed. Understanding the origin of vacuum charge alignment phenomena like the one responsible for electroweak symmetry breaking can have far-reaching consequences for both particle physics and cosmology.

For background information see ELEMENTARY PARTICLE; FUNDAMENTAL INTERACTIONS; GLUONS; PARTICLE ACCELERATOR; QUARKS; SUPERCONDUCTING DEVICES; SYMMETRY LAWS (PHYSICS); WEAK NUCLEAR INTERACTIONS in the McGraw-Hill Encyclopedia of Science and Technology.

[MARY K. GAILLARD]

Bibliography: T. Appelquist, M. K. Gaillard, and J. D. Jackson, Physics at the superconducting supercollider, *Amer. Sci.*, 72:151–155, 1984; E. Eichten et al., *Supercollider Physics*, *Rev. Mod. Phys.*, 56:579–707, 1984; M. K., Gaillard, Toward a unified picture of elementary particle interactions, *Amer. Sci.*, 70:506–514, 1983; *A Long Range Plan for Nuclear Science*, Report of the DOE/NSF Nuclear Science Advisory Committee, December 1983; J. S. McCarthy, B. E. Norum, and R. C. York, *Proceedings of the 12th International Conference on*

High Energy Accelerators, Batavia, Illinois, August 1983; *The Physics of High Energy Particle Accelerators*, American Institute of Physics Conference Series, to be published; *The Role of Electromagnetic Interactions in Nuclear Science*, Report of the DOE/NSF Nuclear Science Advisory Committee, April 1982; *Superconducting Accelerator*, Fermilab Design Report, 1979; 12th International Conference on High Energy Accelerators, Fermilab, August 11–16, 1983.

Perthite

Members of the alkali feldspar group of minerals are a major constituent of the continental crust of the Earth, making up about 30% by volume. They are familiar as the pink, white, or cream-colored tabular crystals conspicuous in most granites. Although they are chemically simple (they are largely solid solutions between $NaAlSi_3O_8$ and $KAlSi_3O_8$), their internal atomic structural arrangements are extraordinarily complex, and in recent years mineralogists have made great strides in understanding how these structures develop and in using them to outline the geological history of rocks from which they come.

In almost all cases, what appears externally to be a single crystal of alkali feldspar proves, on microscopic examination, to be an intergrowth of two feldspars, one nearly pure sodium feldspar (albite), the other potassium-rich feldspar (sanidine, orthoclase, or microcline, depending on the exact structure). These intergrowths are called perthites, more specifically microperthites, when an optical microscope is required to observe them, or cryptoperthites, when x-ray diffraction or electron microscopy is required to characterize them (Figs. 1 and 2). Cryptoperthites may be iridescent, as a result of diffraction of light, and are familiar in the polished slabs surrounding many shop fronts.

Many of the recent advances in understanding perthites have come from the application of modern electron microscope techniques to samples whose geological histories are relatively well understood, and it is now possible to make estimates of cooling rates from these intergrowth textures. Textural differences also arise when alkali feldspars interact with the hot, hydrous fluids which circulate in cooling igneous rocks (deuteric fluids), and these interactions can often be detected by optical microscopic examination. They are geologically important because they may well represent the stage at which radioisotopic "clocks" are set in rocks.

Chemical and structural basis. Relationships in alkali feldspars which lead to perthite formation are best understood by means of a phase diagram (Fig. 3). The diagram shows phase relations at a relatively low water-vapor pressure (P_{H_2O}), appropriate for rocks emplaced at high levels in the crust. An important simplifying omission is of the calcium feldspar component, $CaAl_2Si_2O_8$, present in all alkali feldspars at low but not insignificant levels. This has some detailed effects but does not change general relationships. At high temperatures, feldspars can crystallize as solid solutions of continuously variable composition ranging from pure sodium feldspar to pure potassium feldspar. At lower temperatures, beneath the dome-shaped solvus curve, these solid solutions become unstable and break down into two coexisting feldspar phases whose compositions with respect to temperature lie on the Na- and K-rich limbs of the solvus.

Two solvus curves are shown in Fig. 3: the strain-free solvus defines the composition of separate discrete feldspar crystals; the coherent solvus shows compositions of feldspars in intimate intergrowths sharing a common, continuous, Al-Si-O framework. As temperature falls, the compositions of the intergrown feldspars approach pure $NaAlSi_3O_8$ and pure $KAlSi_3O_8$. The breakdown of the high-temperature solid solution occurs in a way which produces intimate perthitic intergrowths of the two feldspars on scales which range from as little as about 7 nanometers up to more than 1 mm (Figs. 1 and 2). The proportions of the two phases depend on the bulk composition of the parent crystal. The examples given consist of intergrowths of nearly pure K- and Na-feldspar in the approximate proportions 4:6 (by weight or volume), and are characteristic of syenites and certain high-level granites. More common, higher-pressure granites contain feldspars with much more K-rich bulk compositions, with K:Na > 7:3. The segregation process is called exsolution and is analogous to the separation of two immiscible liquids, but in this case occurs entirely in the solid state, by diffusion of Na and K ions through the framework of silicon, aluminum, and oxygen tetrahedra which form the basis of the feldspar structure.

Whereas immiscible liquids form globular mixtures, in solids the shape of the intergrowth reflects the crystal structure of host and subordinate phase, and the bulk composition. There are two types of structural variation in the framework. The first is the

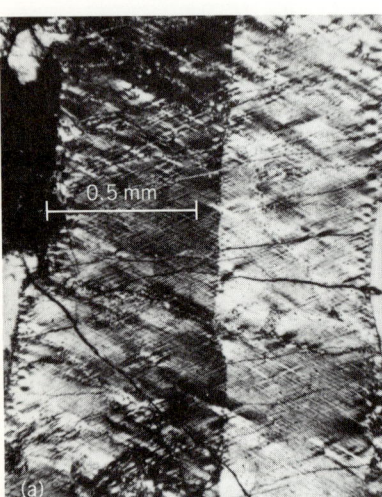

Fig. 1. Optical micrographs of microperthites. (a) Strain-controlled braid microperthite. The crystal is divided into two halves by a Carlsbad twin. The perthite consists of intersecting lamellae of Na-rich and K-rich feldspar. (b) Deuteric patch perthite. The dark areas are K-feldspar (microcline), the paler, boldly striped material is Na-feldspar (albite). The stripes are Albite twins. Note the coarseness and irregular shape of the exsolution texture. (*From I. Parsons, Feldspars and fluids in cooling plutons, Miner. Mag., 42:1–17, 1978*)

degree of regularity of distribution of aluminum and silicon atoms on similar sites in the framework. At high temperatures the arrangement is disordered, but becomes ordered with falling temperature. This diffusive transformation (Fig. 3) gives rise to the range of K-feldspar polymorphs from high sanidine to microcline, and the Na-feldspars high and low albite. In K-feldspar the ordering is accompanied by a symmetry change from monoclinic to triclinic. In addition, a second type of structural change occurs in Na-rich feldspars which involves only twisting of the framework and leads to a displacive monoclinic-triclinic symmetry change (Fig. 3). Both these symmetry changes are accompanied by twinning which produces a structure based on alternating left- and right-handed triclinic slabs (Figs. 1b and 2). Thus the interplay of three solid-state processes—exsolution, framework ordering, and displacive twisting of the framework—forms the basis of the structural variety of perthites.

Although the textures which result from exsolution may be extremely complex (Fig. 2a), they may nevertheless form in a structurally continuous Al-Si-O framework. Textures of this type are said to be coherent intergrowths, and recent high-resolution electron microscopy has shown that perthites may be structurally continuous both across boundaries between Na- and K-rich feldspar and across twins (Fig. 2b). Other perthites may be semicoherent with regularly distributed dislocations in the structure, while coarser perthites (Fig. 1b) are commonly thought to be noncoherent although this is by no means certain. Coherent intergrowths tend to have relatively well-defined orientations, either as thin, more or less straight lamellae parallel to the crystallographic plane $(\bar{6}01)$ or as intersecting lamellae parallel to $(\bar{6}\bar{6}1)$ and $(\bar{6}61)$ [Figs. 1a and 2a]. Research has shown that the planes along which the two compositionally different feldspars are united are orientated so that the amount of distortion (strain) needed to make the two structures fit together (Fig. 2b) is minimized. Structures of this type are elastically strained so that the crystal cell dimensions in the two structures are not the same as they would be in the same materials in isolation.

Coarser perthitic intergrowths are usually much more irregular (Fig. 1a), and the shapes are not controlled by the minimization of coherent elastic energy. These intergrowths have been called deuteric perthites to distinguish them from the strain-controlled intergrowths, because there is petrological evidence that their development depends on interaction with a water-rich fluid at relatively low temperatures, and they are very commonly associated with the development of turbidity in the feldspar.

Mechanisms and petrological implications of exsolution. Recently there have been considerable advances in the understanding of strain-controlled intergrowths, and some promising quantitative applications of the intergrowths have been devised. Crystals with bulk compositions in the central region of the phase diagram probably unmix by a process called spinodal decomposition, first described in

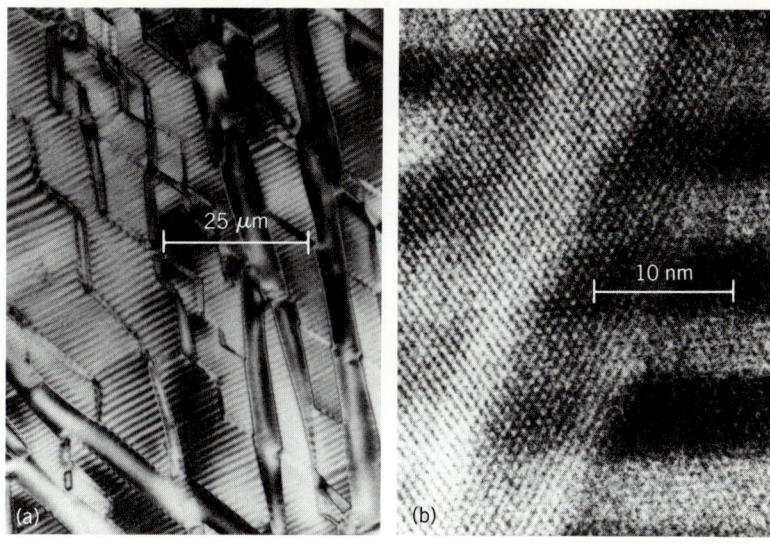

Fig. 2. Electron micrographs of cryptoperthites. (a) Bright-field micrograph shows relatively featureless, sometimes zigzag, continuous bands of microcline, the broader of which outline striped lozenges of Albite-twinned albite, which correspond with the braid texture of Fig. 1a (from W. L. Brown, S. M. Becker, and I. Parsons, Cryptoperthites and cooling rate in a layered syenite pluton: A chemical and TEM study, Contrib. Mineral. Petrol., 82:13–25, 1983). (b) High-resolution micrograph illustrating the continuity of structure between microcline (pale) and albite (darker with stripes which are Albite twins). The lattice fringes outline C-centered unit cells (from W. L. Brown and I. Parsons, Exsolution and coarsening mechanisms and kinetics in an ordered cryptoperthite series, Contrib. Mineral. Petrol., 86:3–18, 1984).

metals. If exsolution does occur by this process, a feldspar of median composition (Fig. 3), crystallized initially as a homogeneous crystal at say 850°C (1560°F), will cool through the strain-free solvus without unmixing. This is because the system has insufficient excess Gibbs free energy to overcome the strain which develops as soon as compositional fluctuations appear. However, when the crystal has reached a temperature just inside the coherent solvus, a regular wavelike compositional variation develops, first as long-wavelength, low-amplitude variation in Na:K, rapidly giving way to short-period, higher-amplitude fluctuations with compositions on the coherent solvus. The development of continuous compositional fluctuations constitutes spinodal decomposition. With time, the initially fine-scale (about 7-nm) exsolution lamellae can coarsen, without losing coherency, up to at least 1-micrometer scale (Figs. 1a and 2a). Spinodal decomposition is, by geological standards, a rapid process (it is observable after annealing of under 1 h at 550°C or 1020°F) so that only extremely quickly chilled sanidines (in volcanic ejecta) can remain homogeneous on the Earth's surface. Coarsening, however, is much slower, and the periodicity of the resulting exsolution textures depends on cooling rate.

Coarsening rates have been calibrated in laboratory experiments, and the rates have been successfully related to periodicities in relatively rapidly cooled igneous bodies (lavas and dykes). In such rocks, perthites appear to be promising "cooling-rate meters," provided cooling rates are sufficiently rapid to inhibit significant Al-Si framework ordering, in which case lamellae retain a simple planar

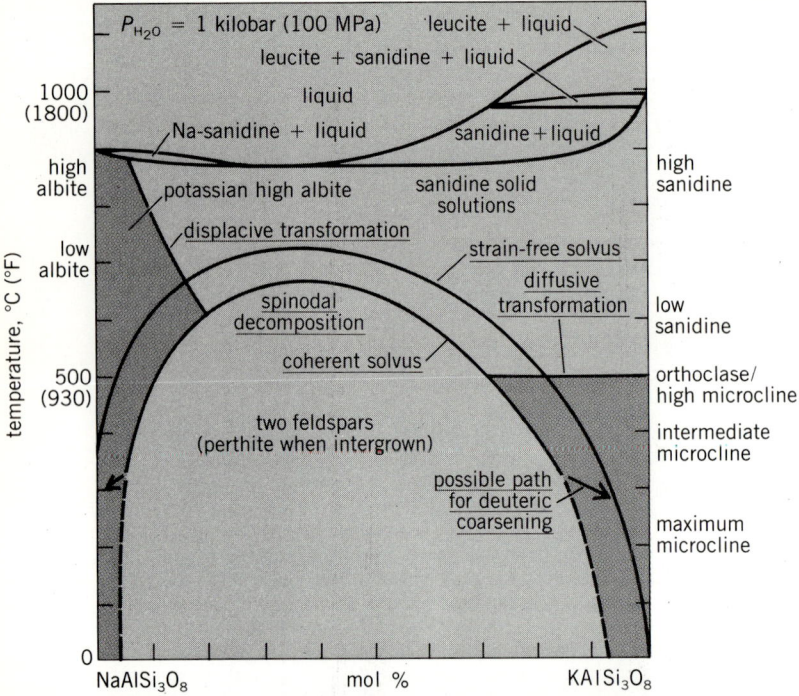

Fig. 3. Phase diagram for the alkali feldspar system at slow cooling rates. Feldspars in the shaded areas have triclinic symmetry, others are monoclinic. Mineral names beside the vertical axes show the approximate stability ranges of the polymorphs of the end members. Orthoclase is an ordered potassium feldspar structurally equivalent to microcline but with a fine-scale twinlike texture which gives overall monoclinic symmetry. It frequently persists in the stability range of microcline. Spinodal decomposition occurs at temperatures just below the coherent solvus, but at temperatures which depend on cooling rate.

morphology. In more slowly cooled rocks, ordering occurs and the coarsening process becomes very much more complex, leading to textures like those shown in Figs. 1a and 2a. Variation in cryptoperthites has been studied in a small syenite stock, and it has been demonstrated that perthite coarseness is systematically related to distance from the contact (roof) of the intrusion and hence most probably to cooling rate. Periodicities are much less, however, than would be predicted from the work on laboratory coarsening rates extrapolated to calculated cooling rates for the intrusion, and it seems that ordering (and probably the accompanying twinning; Fig. 2a) greatly slows the coarsening process. The method has promise, however, for locating relative cooling-rate surfaces in plutons even if a laboratory calibration proves impossible.

Deuteric exsolution, although very commonly observed in granitic and allied rocks, is less well understood as a mechanism, although it probably occurs by small-scale solution-redeposition steps, leading to loss of coherency and relaxation of elastic strain. The intergrowths are therefore of much more irregular shape (Fig. 1b). Qualitatively the presence of such textures is a good guide to the interaction of felsic rocks with a hydrous fluid, and there is evidence that these interactions often occur at low temperatures (less than 400°C or 750°F; Fig. 3), because within individual crystals the irregular deuteric perthites commonly cross-cut strain-controlled perthites which have also ordered. Island relics of cryptoperthite are common in rocks which have otherwise been extensively affected by deuteric exsolution and alteration, and may provide a valuable window into the earlier postmagmatic history of felsic plutonic rocks.

For background information see ALBITE; CRYSTALLOGRAPHY; FELDSPAR; LEUCITE; MICROCLINE; ORTHOCLASE; PETROLOGY; PHASE EQUILIBRIUM; SOLID-STATE CHEMISTRY in the McGraw-Hill Encyclopedia of Science and Technology.

[IAN PARSONS]

Bibliography: W. L. Brown, S. M. Becker, and I. Parsons, Cryptoperthites and cooling rate in a layered syenite pluton: A chemical and TEM study, Contrib. Mineral. Petrol., 82:13–25, 1983; I. Parsons and W. L. Brown, Feldspars and the thermal history of igneous rocks, in W. L. Brown (ed.), Feldspars and Feldspathoids: Structures, Properties and Occurrences, 1984; C. Willaime and W. L. Brown, A coherent elastic model for the determination of the orientation of exsolution boundaries: Application to the feldspars, Acta Crystallogr., A30:316–331, 1974; R. A. Yund, Alkali feldspar exsolution: Kinetics and dependence on alkali interdiffusion, in W. L. Brown (ed.), Feldspars and Feldspathoids: Structures, Properties and Occurrences, 1984.

Phonon

The conduction of heat through nonmetallic solids is caused by the high-frequency vibration of atoms. In a crystalline solid, these vibrations are conveniently described as elastic waves. The quantum-mechanical theory of solids dictates that a vibrational wave of given frequency and propagation direction (wave vector) can have only discrete values of energy or amplitude. Specifically, the energy in an elastic wave is given by $h\nu(n+\frac{1}{2})$, where h is Planck's constant (6.6×10^{-34} joule·second), ν is the frequency of the wave, and n is an integer. It is useful to think of the elastic wave as consisting of n discrete packets of elastic energy, called phonons. The elastic wave can undergo a change in energy only by emitting or absorbing a phonon of energy $h\nu$. This quantum description of elastic waves is not usually apparent in the thermal properties of crystals at room temperature because the number of phonons in a vibrational mode of given wave vector is generally much larger than 1. However, when the temperature of the crystal is reduced toward absolute zero, the effects of energy quantization on the thermal properties of the solid (for example, heat capacity and thermal conductivity) become apparent. A. Einstein showed in 1907 that the precipitous drop in the heat capacity of solids as the temperature is lowered could be explained by this quantization of elastic energy.

Phonon propagation. The transport of thermal energy in a crystalline solid can be described by the propagation and scattering of phonons. A phonon is

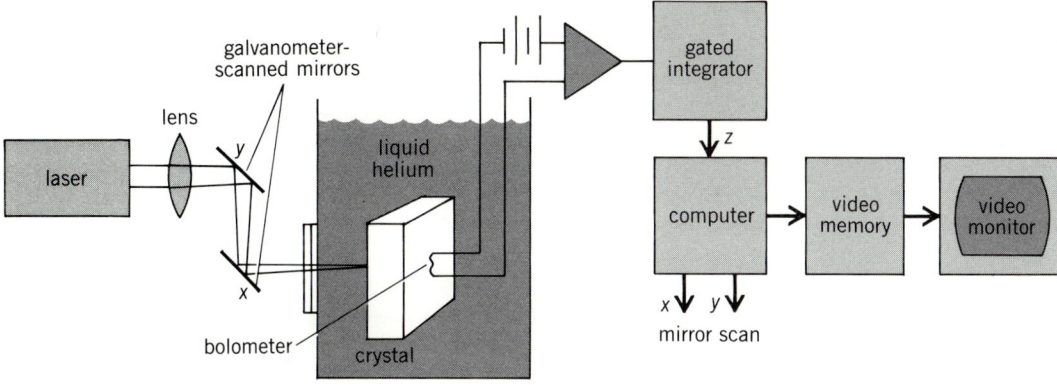

Fig. 1. Diagram of phonon-imaging apparatus.

said to be scattered when one unit ($h\nu$) of elastic energy is transferred from one vibrational mode to another. Scattering of phonons is caused by defects or impurities in the crystal lattice or interactions between the phonons themselves. The existence of different atomic isotopes in the crystal also causes the scattering of high-frequency phonons. At room temperature, the mean free path of a thermal phonon through a pure crystal is only about a micrometer, due to phonon-phonon interactions. But as the temperature is lowered, the number of phonons in the crystal diminishes and the mean free path between scattering events becomes much larger. Indeed, at temperatures below about 10 K (18°F above absolute zero), high-frequency thermal phonons can propagate completely across a centimeter-sized crystal without scattering. This nondiffusive type of phonon propagation is termed ballistic.

Ballistic phonon propagation occurs at the speed of sound and is found to display dramatic effects which are not seen at higher temperatures due to the many scattering events between phonons. At room temperature, phonon propagation is diffusive, and the corresponding thermal energy emitted from a point source of heat typically expands in a nearly spherical (or isotropic) pattern. In contrast, at low temperatures the propagation of ballistic heat flux is highly anisotropic, whereby thermal energy is strongly channeled along certain crystalline directions. In 1979 a new method was introduced to study this highly anisotropic ballistic propagation of phonons at low temperatures. The method is called phonon imaging, and in the ensuing period it has been applied to the problems of phonon propagation and scattering in crystals at low temperatures.

Technique of phonon imaging. Phonon imaging is based on a general heat-pulse method introduced in 1964. A thermal pulse of energy that locally raised the temperature of the crystal a few degrees was injected at one surface of a quartz crystal, and the arrival of this heat pulse was detected on the opposite face of the crystal. The crystal was immersed in superfluid helium, and the detector consisted of a superconducting metal film deposited on the crystal. The bath temperature was precisely adjusted to the superconducting transition of the metal film, providing a fast and extremely sensitive detector of heat (a bolometer). Following thermal excitation, several pulses were detected at the bolometer. The pulses corresponded to compressive (longitudinal) and shear (transverse) waves propagating at their respective sound velocities through the quartz.

In 1968 it was discovered that the intensity of these ballistic heat pulses depended strongly on the direction of propagation in the crystal, an effect called phonon focusing. Ten years later, heat-pulse experiments indicated that the thermal flux from a point source of heat exhibited mathematical singularities along certain crystalline directions. To study this anisotropy in detail, a phonon-imaging scheme was devised whereby the propagation direction of the phonons could be varied continuously in two dimensions. A fixed bolometer detector was employed in conjunction with a movable heat source in the form of a focused laser beam (Fig. 1). The laser beam was electronically deflected by two mirrors across one surface of the crystal. A gated integrator determined the heat-pulse intensity at a given time of flight after the laser pulse. A computer controlled the scanning of the laser beam and collected the measured heat-pulse intensities as a function of phonon propagation direction in the crystal. The intensity of ballistic phonons arriving at the detector was measured to be extremely anisotropic and showed a complex pattern characteristic of the particular crystal (Fig. 2).

This phonon-focusing effect arises simply from the basic elastic anisotropy in the crystal. That is, the velocity of a phonon depends upon the direction of its wave vector with respect to the crystal axes. The degree of this anisotropy depends upon the microscopic structure of the atomic lattice and the form of the interatomic forces. Basically, the restoring force upon compression of the crystal is greater along certain crystalline directions than along other directions.

An interesting consequence of elasticity theory in such an anisotropic medium is that the energy propagation direction is not along the wave vector direction (that is, the direction perpendicular to the wavefronts), and this effect gives rise to the extreme variations of heat-pulse intensity with propagation angle, as seen in the phonon images. In other words, an isotropic distribution of wave vectors emitted from an incoherent heat source can produce a highly anisotropic energy flux. In fact, the bright lines in the phonon-focusing pattern correspond to mathematically infinite phonon flux, although truly singular behavior cannot be observed in practice because of the finite size of heat sources and detectors. The term focusing is used here to describe the channeling of heat energy along certain directions in the crystal; it does not imply that the phonon trajectories are curved as in the geometrical optics use of the term.

Research applications. Since this phonon-imaging method was introduced in 1979, it has been applied to several major problems in the area of propagation and scattering of phonons. Phonon-focusing patterns have been observed for a number of crystals—germanium, silicon, gallium arsenide (GaAs), lithium fluoride (LiF), sapphire, quartz, diamond, and so forth (Fig. 2)—and for the most part, the resulting patterns can be predicted by the low-frequency elastic constants of the crystals. Typical frequencies of the phonons contained in a heat pulse are in the range 0.2–1 terahertz, which correspond to wavelengths in the range 30–5 nanometers. Thus the highest-frequency phonons have wavelengths approaching the lattice spacing between the atoms, typically about 0.5 nm. Since it is physically impossible to have a phonon wavelength shorter than the atomic spacing, interesting changes occur in the elastic waves as they approach this wavelength. For example, the propagation velocity becomes much smaller, and thus these high-frequency waves are said to be dispersive. Elaborate theories have been devised to understand wave propagation in this dispersive regime, and this comprises a major field in the study of condensed matter known as lattice dynamics. The general aim is to extract information about interatomic forces (that is, those which bind the crystal together) from the propagation of high-frequency elastic waves. The most productive means of studying dispersive phonons has been by neutron scattering. It appears now that phonon imaging will also provide additional new information about dispersive phonon propagation. By using special superconducting detectors, experimenters have observed shifts in the phonon-focusing patterns of germanium and gallium arsenide at frequencies of about 0.7 THz, as expected for dispersive phonons. This method may eventually give a global view of acoustic-phonon dispersion in a crystal.

The second major area of study is phonon scattering from defects and surfaces. Dislocations of atoms can be introduced into crystals by plastically deforming them, and such defects are known to greatly affect the mechanical strength and thermal conductivity of the solid. Phonon images of a lithium fluoride crystal before and after plastic deformation have shown that certain subsets of phonons are scattered more strongly from the dislocations than other subsets. Thus a quantitative measure of the scattering of phonons with particular wave vector and polarization has been obtained. In this way, the scattering mechanism was identified to be associated with a "fluttering-string" oscillation of a dislocation line. By selectively introducing dislocation lines of a particular orientation, it was found that the deformed crystal acted as an effective phonon polarizer.

The transmission and reflection of high-frequency phonons at surfaces has also been examined by phonon imaging. In 1941 P. L. Kapitza observed a large temperature discontinuity between a copper heater and the superfluid helium bath in which it was immersed. Numerous studies on solid-liquid and solid-solid interfaces have since characterized this thermal boundary resistance in terms of the acoustic mismatch between the contacting materials. It is found, however, that there exists an anomalously high transmission of phonons from a solid crystal into a liquid helium bath, and a microscopic description of this anomaly still remains as a major problem in phonon physics. It has been recently postulated that diffusely scattered phonons are the principal contributors to the anomalous Kapitza conductance. Phonon imaging has provided a means for positively identifying phonons which are reflected specularly or diffusely from a crystal boundary. It was discovered, for example, that about 80% of the heat-pulse energy incident on an optical-grade sapphire surface was specularly reflected, implying that the surface irregularities were smaller than the wavelength of the phonons, approximately 20 nm. In

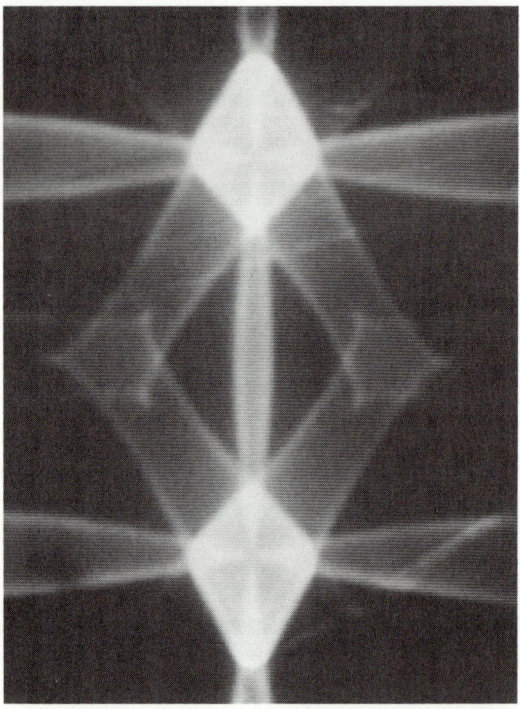

Fig. 2. Phonon image of gallium arsenide (GaAs) obtained by scanning the laser beam across a (110) face of the crystal. Sharp phonon-focusing structures are observed. (*G. A. Northrop and J. P. Wolfe*)

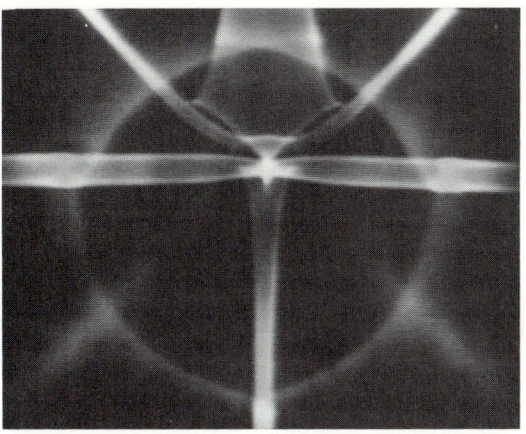

Fig. 3. Phonon image of sapphire (Al_2O_3). The heat pulses are produced by laser-heating a 100-nm copper film evaporated onto the crystal surface. The circular structure represents a selective channeling of phonon wave vectors across the metal-crystal interface. The remaining sharp structures are due to phonon focusing in the bulk of the crystal. (*From A. G. Every, G. L. Koos, and J. P. Wolfe, Ballistic phonon imaging in sapphire: Bulk focusing and critical cone channeling effects, Phys. Rev., B29:2190–2209, 1984*)

a related experiment, phonons generated in a metal film evaporated onto a sapphire crystal were found to be transmitted into the crystal primarily along certain crystalline directions (Fig. 3). This selective transmission of phonons with certain wave vectors provides information about the quality of the bonding between evaporated metal and the crystal substrate. Much work remains to be done to characterize the boundary scattering of phonons from surfaces of various preparation conditions.

In summary, the phonon-imaging method, which is generally applicable to a wide variety of nonmetallic crystals, provides an important new means for characterizing the propagation and scattering of thermal phonons at low temperatures in crystals.

For background information *see* CRYSTAL DEFECTS; ELASTICITY; HYPERSONICS; KAPITZA RESISTANCE; LATTICE VIBRATIONS; LOW-TEMPERATURE ACOUSTICS; PHONON in the McGraw-Hill Encyclopedia of Science and Technology.

[JAMES P. WOLFE]

Bibliography: A. G. Every, G. L. Koos, and J. P. Wolfe, Ballistic phonon imaging in sapphire: Bulk focusing and critical cone channeling effects, *Phys. Rev.*, B29:2190–2209, 1984; G. A. Northrop et al., Anisotropic phonon-dislocation scattering in deformed LiF, *Phys. Rev.*, B27:6395–6408, 1983; G. A. Northrop and J. P. Wolfe, Phonon reflection imaging: A determination of specular versus diffuse boundary scattering, *Phys. Rev. Lett.*, 52:2156, 1984; J. P. Wolfe, Ballistic heat pulses in crystals, *Phys. Today*, 33(12):44–50, December 1980.

Phylogenetic systematics

Systematics, the science of discovering and organizing the diversity of organisms, is an important biological discipline because its results provide a framework for comparing populations and species and higher taxa. Similarities and differences of interest to systematists may include characters drawn from morphology, biochemistry, ecology, behavior, or biogeography. Because the results of systematic analysis provide the basis for comparisons in many fields of inquiry, it is important to understand how systematists arrive at their conclusions.

Three approaches. Biologists are faced with three different approaches to systematic analysis. Each is characterized by a different philosophical outlook on the goals of systematics and taxonomy, which leads to differing opinions on how species should be classified. Each claims that its methods and results better serve the general needs of the biological community. Since many conclusions about the process of evolution may be affected by classification of organisms, these differences of approach are of some significance even to nonsystematists who use classifications for comparative purposes. This article is about one of the three approaches, phylogenetic systematics, but the goals of the other two approaches must be understood in order to put the phylogenetic approach into perspective.

"Phylogenetic systematics," often called cladistics, was developed by the German entomologist Willi Hennig, who first published a detailed account of his system in 1950. His system provides a method for reconstructing phylogenetic trees and a logical framework for constructing biological classifications that group species and higher taxa by evolutionary descent. The logic is based on using only characters that indicate unique common ancestry (derived characters). Phylogenetic classifications are directly derived from the phylogenetic tree reconstructed from this character information. Thus, there is a one-to-one correspondence between the classification and the phylogenetic tree, permitting the reconstruction of the branching structure of the tree directly from the classification.

Evolutionary taxonomy, the second approach, uses the same methods for reconstructing trees that are used by phylogenetics, but evolutionary taxonomists frequently classify in a different manner. Evolutionary taxonomists consider that the overall distinctiveness of a group is a sufficient criterion to give it a higher rank than its closest relative; thus the classification does not necessarily show a one-to-one correspondence with the evolutionary tree. This central difference between phylogenetic systematics and evolutionary taxonomy revolves around different concepts of monophyly, which are discussed below.

Phenetics, the third approach to systematics, asserts that while some aspects of phylogeny may be recovered from phylogenetics analysis, the result of the exercise is not certain enough to be used as the basis for classifications that serve the entire biological community. At its extreme, phenetics rejects the reconstruction of phylogenetic trees altogether. Pheneticists suggest that species should be recognized and classified on the basis of a measure of overall similarity using all characters (primitive, derived, and convergent) rather than on the basis of the special similarities (derived characters) used by phylogeneticists. Classifications, when proposed by pheneticists, are based on percent similarity between groups.

Central concepts. Given this characterization of the three schools, phylogenetic systematics may now be considered in detail. Phylogenetic systematics is based on several central principles. First, the evolutionary relationships that exist between species are genealogical relationships ("blood" relationships). Species are related to each other through ancestral species. While two species may appear similar in terms of overall resemblance, one may actually share a more recent common ancestor with another species which, because of evolutionary modifications, is less similar. For example, the Australian lungfish, *Neoceratodus forsteri*, may resemble the trout, *Salmo trutta*, more than it resembles the frog, *Rana pipiens*. However, the lungfish shares a more recent common ancestor with the frog than with the trout. Thus, the lungfish and the frog are more closely related. Second, characters used to show common ancestry relationships are restricted to those that are evolved in the most recent common ancestor of the species while characters evolved in earlier ancestors are disregarded. Lungfishes and frogs have a two-chambered auricle in the heart, which is presumed to have arisen in their immediate common ancestor (which is also the common ancestor of all lungfishes and all tetrapods). Such a homologue, termed a derived character or synapomorphy, is evidence for a genealogical connection between lungfishes and frogs (and cows for that matter). In contrast, lungfishes and trout have a caudal fin whereas frogs and cows do not. The presence of a caudal fin does not provide evidence for a unique common ancestry relationship between lungfishes and trout. Sharks and lampreys also have caudal fins. Thus the caudal fin originated before the origin of the common ancestor of lungfishes and trout and is a primitive, or symplesiomorphic, homologue.

Of course, most systematists know that lungfishes are more closely related to frogs and cows than to trout and other bony fishes. Where phylogenetic systematists differ from other systematists is that phylogeneticists classify lungfishes in the same higher taxon (Sarcopterygii) with frogs and cows while placing trout in another taxon (Actinopterygii). Phylogeneticists do so because of a third basic concept—only monophyletic groups are recognized as natural taxa. The phylogenetic concept of monophyly is specific and restricted: a monophyletic taxon is one that includes all of the descendants of a common ancestor (as well as the ancestral species if it is known). This monophyletic group is also known as a clade. A taxon which includes some, but not all, of the descendants of an ancestral species is termed a paraphyletic taxon, while a taxon that includes descendants of two different ancestral species is termed polyphyletic. Paraphyletic taxa are considered unnatural because the concept of grouping by descent is violated by excluding one or more descendant groups; polyphyletic taxa are also considered unnatural, since the same concept is violated by including two descendant groups. Examples of paraphyletic groups include Reptilia (mammals and birds excluded), Pisces (tetrapods excluded), Pongidae (*Homo sapiens* and fossil relatives excluded), and Invertebrata (vertebrates excluded). As a consequence of grouping by descent, phylogenetic classifications do not contain many familiar groups, and they contain certain familiar groups that have rather different ranks than have been traditionally assigned. Aves, rather than a class, becomes a taxon with the same rank as its nearest genealogical relative, Crocodilia. This points to the fourth principle: sister groups (nearest genealogical relatives) are classified together rather than apart. It is this principle that sets phylogenetic classifications apart from other forms of classification.

Reconstructing phylogenies. First, it is important to note that in reconstructing phylogenies, phylogeneticists do not treat fossils, no matter what their age, in any way differently from extant species. Fossils are certainly used in the analysis of a group, but assigned no special status—the idea is that the characters of fossil species should speak for themselves in determining how that species is placed in the phylogeny of the taxon.

Much recent effort has gone into examining the criteria by which characters are determined to be primitive, as in the single auricle in the heart of sharks, or derived, as in the double auricle of frogs. One major and one auxiliary criterion are now favored. The major criterion is the outgroup criterion, which states that of two homologous characters occurring within a group, the homologue that also oc-

Classification of tetrapod vertebrates

Current	Phylogenetic
Class Amphibia	Class Tetrapoda
Class Reptilia	Subclass Lissamphibia (Recent amphibians)
Order Anapsida (turtles)	Subclass Amniota
Order Lepidosauria (lizards, and so on)	Infraclass Mammalia
Order Crocodilia	Infraclass Reptilomorpha
Class Aves (birds)	Division Anapsida
Class Mammalia	Division Diapsida
	Subdivision Lepidosauria
	Subdivision Archosauria
	Cohort Crocodilia
	Cohort Aves

curs outside the group is the primitive (plesiomorphic) character while that restricted to the group is the derived (apomorphic) character. This criterion is relatively unambiguous when the group analyzed is monophyletic and the closest relatives are known with some precision. Single auricles are found in sharks and lampreys, two outgroups of the clade composed of lungfishes, frogs, cows, and trout. Thus, the double auricle is derived.

The auxiliary criterion is the ontogenetic criterion, which states that given an ontogenetic series of character changes (that is, a sequence of changes over the course of an organism's development), those characters that occur earlier in the ontogenetic series are primitive relative to those that follow. However, the outgroup criterion remains the primary criterion because changes in developmental timing and in embryonic development may obscure the original, primitive developmental patterns. But even the outgroup criterion may be ambiguous. Independent parallel development of characters, or character reversals (characters that revert to primitive expression) may cloud the phylogenetic picture. Rather than a single, fully corroborated phylogenetic tree, one may find several trees that have some support. In such cases the investigator may seek more characters, but must eventually pick the best-supported hypothesis on the basis of parsimony, that is, the hypothesis that assumes the fewest number of evolutionary changes for the characters examined.

Another major development in reconstructing phylogenetic relationships has been the application of computer algorithms to data sets. Although several classes of computer algorithms have been suggested and debate over which is best is not over, those algorithms employing a parsimony criterion seem best for duplicating the results obtained by more traditional Hennigian methods.

Classification. Phylogenetic systematists argue that classifications that exactly reflect the genealogical relationships among taxa are superior to nonphylogenetic classifications, and are superior to classifications that attempt to adjust the ranks given taxa by the magnitudes of difference between groups. Taxonomic classifications, which are hierarchical, can only portray group membership. Diagnoses (the naming of characters unique to a group) are necessary to convey character information concerning group members. A phylogenetic classification associated with a diagnosis for each monophyletic group produces a system that is more informative than other types of classifications. For example, phylogeneticists would abandon the current classification of higher tetrapod vertebrates (see table) and substitute a phylogenetic classification which groups birds with crocodiles (see illustration). Such a classification would exclude the paraphyletic Reptilia, but would include a monophyletic Archosauria. Characters such as the complex pneumatic spaces that occur in the skulls of both birds and crocodiles, and the same mode of tooth replacement in both crocodiles and Mesozoic birds, have no place in

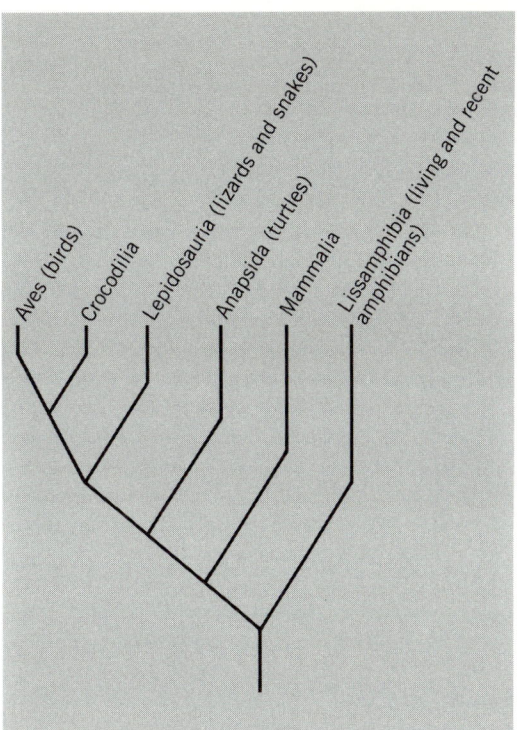

Phylogenetic relationships among living tetrapod vertebrates. Drawing corresponds to phylogenetic classification outlined in the table.

conventional classifications, or their inclusion implies that these characters are convergent (which they are not).

Phylogenetic classifications are logically consistent with the phylogenetic trees they attempt to summarize, while classifications that include paraphyletic groups such as Reptilia are not logically consistent with the phylogenetic trees they supposedly summarize. And conventions can be adopted which preserve the consistency and informativeness of phylogenetic classifications while keeping them relatively simple. Thus, phylogenetic systematists assert that abandonment of familiar classifications that include such paraphyletic groups as Reptilia and Pongidae (apes) is necessary if biological classifications are to serve the needs of evolutionary and comparative biology.

Biogeography and other fields. Hennig proposed various criteria for reconstructing the biogeographic history of groups, including the general principle that the oldest species were nearest the center of origin of the group. Lars Brundin, an important early supporter of phylogenetic classification, applied many of Hennig's ideas in his seminal monograph on southern chironomid midges. Beginning in the early 1970s, Gareth Nelson began applying some of the principles of Leon Croizat's panbiogeographic methods along with Hennig's systematic methods to problems in biogeography. The result has been the emergence of vicariance biogeography, which consists of reconstructing phylogenetic histories for several groups inhabiting the same geographic region or regions and then comparing their distributions in time and space to search for common historical patterns. Common patterns are assumed to denote common history. The method has proven useful for establishing historical connections between areas of the Earth and in studying speciation.

Phylogenetic systematics and vicariance biogeography provide a powerful combination for studying evolution because the results of such studies provide a firm base for making evolutionary comparisons. Phylogeneticists are moving toward applying their ideas to evolutionary theories, both in testing current ideas and in developing new ways of viewing the evolutionary process. Some (like D. R. Brooks) are also applying the methods of phylogenetic analysis to studying phenomena such as coevolution in an effort to sort out the historical and proximal factors involved in the phenomenon.

[E. O. WILEY]

Bibliography: N. Eldredge and J. Cracraft, *Phylogenetic Patterns and the Evolutionary Process*, 1980; W. Hennig, *Phylogenetic Systematics*, 1966, reprint 1979; G. J. Nelson and N. I. Platnick, *Systematics and Biogeography: Cladistics and Vicariance*, 1981; E. O. Wiley, *Phylogenetics: The Theory and Practice of Phylogenetic Systematics*, 1981.

Physiological ecology (plant)

Carbon dioxide (CO_2) concentration of the Earth's atmosphere has been carefully measured continuously since 1958. The concentration has risen from about 0.0315% or 315 parts per million (ppm) in 1958 to more than 345 ppm currently. Most of this recent rise in CO_2 concentration is attributed to burning of fossil fuels. Some scenarios of use of fossil fuels lead to predictions of a doubling of atmospheric CO_2 concentration in 100 years or less. Whereas most of the recent rise in global atmospheric CO_2 is attributed to fossil fuel combustion, other studies of tree ring data and critical examination of earlier air sample measurements suggest that CO_2 concentration could have been about 260–280 ppm before the rapid pioneering expansion of the eighteenth and nineteenth centuries. It is possible that large quantities of biotic and soil organic matter sources of carbon may have been oxidized and CO_2 returned to the atmosphere during those centuries of global population expansion. Furthermore, analyses of gas bubbles trapped in Antarctic and Greenland ice core samples indicate that atmospheric CO_2 concentration may have decreased to as low as 160–200 ppm 15,000–20,000 years ago in the coldest part of the Wisconsin Ice Age. Atmospheric CO_2 concentration appears to have varied in the past due to both natural and human causes.

Carbon makes up about 40% of the dry weight of plants. Carbon dioxide from the atmosphere provides both the structural material and the metabolic energy source for green plants through the process of photosynthesis. Life on Earth actually depends on photosynthetic fixation and assimilation of CO_2. In higher terrestrial green plants, CO_2 from the atmosphere diffuses into leaves through small pores called stomata. The CO_2 concentration in the atmosphere is below the optimum for maximum photosynthesis rates, but turbulence and molecular diffusion are nevertheless quite effective in supplying CO_2 from the atmosphere to the surface of leaves. Indeed growing, high-producing crops and other vegetation can take up CO_2 in a growing season equivalent to the entire content of the atmosphere in a vertical column above the field (about 0.72 lb CO_2/ft^2 or 3.5 kg CO_2/m^2). However, only a small 6-ppm peak-to-peak annual cycle of CO_2 concentration has been observed at Mauna Loa, Hawaii, an isolated island location. Obviously, recycling of sources of CO_2 due to decay, and low-to-nil CO_2 uptake rates over much of the Earth's surface, prevents the average global CO_2 concentration from decreasing seriously during each growing season.

Recently, there has been a growing scientific awareness of and public interest in the practical implications of increased atmospheric CO_2 concentration on agricultural productivity and ecosystem responses. The implications of direct effects of rising atmospheric CO_2 on vegetation are perhaps more important than the predicted effects of CO_2 on global climate (greenhouse effect). Direct photosynthetic increases caused by rising CO_2 levels is actually the real greenhouse gas effect.

Plant responses to carbon dioxide. Several plant responses to increased CO_2 have been observed under experimental conditions, including (1) increased leaf and vegetative canopy photosynthesis rates (under low light as well as high light conditions); (2) increased stomatal resistance to gaseous exchange of CO_2 and water vapor; (3) decreased transpiration rates and water use; (4) increased photosynthetic water-use efficiency (that is, larger ratios of photosynthesis to transpiration); (5) increased starch accumulation; (6) increased plant turgor; (7) increased plant size, biomass accumulation, and crop yield; (8) enhanced biological nitrogen fixation; (9) changes in weed-crop competition; (10) changes in pest-crop interactions; (11) increases in ecosystem net primary production; and (12) changes in ecosystem species competition which may lead eventually to changes in ecosystem species composition.

Most of these observed and expected responses of plants to rising atmospheric CO_2 concentration depend primarily on increased photosynthesis rates. However, decreased leaf transpiration rates are mediated primarily by the fact that stomata partially close as CO_2 concentration increases, so that stomatal resistance to transpiration increases as CO_2 concentration increases.

C_3, C_4, and CAM plants. Plants are classified by their characteristic carbon metabolism. Plants with different categories of the CO_2 photoassimilation biochemical pathway behave somewhat differently from each other. C_3 plants (such as soybean, wheat, potato, and many trees) assimilate CO_2 by direct reaction with ribulose-1,5-bisphosphate (RuBP) in the presence of the enzyme RuBP carboxylase to form phosphoglyceric acid, a 3-carbon compound. C_3

plants have a greater increase in photosynthesis rates and a smaller increase in stomatal resistance than C_4 plants as the CO_2 concentration is elevated. C_4 plants (such as maize, sugarcane, and crabgrass) trap CO_2 in their leaf mesophyll cells by a reaction with phosphoenolpyruvate (PEP) in a reaction catalyzed by the enzyme PEP carboxylase, transport 4-carbon malic or aspartic acid to leaf bundle sheath cells, and concentrate CO_2 in these cells by decarboxylation reactions for sucrose or starch synthesis to proceed by the C_3 cycle. C_4 plants have a smaller increase in photosynthesis rates and a larger increase in stomatal resistance than C_3 plants as the CO_2 concentration is elevated.

Photosynthetic CO_2 uptake in C_3 plants is hampered because the enzyme RuBP carboxylase can also function as an oxygenase. Under present-day levels of atmospheric CO_2 and O_2, oxygen competes with CO_2 for binding sites on the enzyme, and drains away potential photosynthetic energy. As atmospheric CO_2 rises, more of the photosynthetic potential of the C_3 plants will be realized as CO_2 competes more effectively with O_2 for binding sites on the enzyme, and C_3 species may compete more effectively with C_4 species.

Plants such as pineapple, a crassulacean acid metabolism (CAM) plant, may respond to increased CO_2. CAM plants typically open their stomata at night to take up and store CO_2 as malic acid (and thus avoid high transpiration rates during the day). However, their CO_2 uptake is usually limited more by malic acid storage capacity than by CO_2 concentration.

Photosynthesis and carbon dioxide. The most important direct response to rising atmospheric CO_2 concentration is increased photosynthesis rate. An example of the photosynthetic response to CO_2 in bright sunlight of soybean, a C_3 legume crop that is important worldwide, is shown in the illustration. These data were collected in closed, controlled-environment plant growth chambers located outdoors (phytorhizotrons) so that below-ambient as well as above-ambient CO_2 concentration effects could be measured.

The illustration shows that soybean-plant relative photosynthetic rates increase as CO_2 increases above a reference level of 330 ppm, at least up to 1000 ppm and perhaps well beyond that level. Also, photosynthetic rates decrease sharply as CO_2 decreases below the reference CO_2 concentration. Therefore, response to a small incremental increase in CO_2 is much more pronounced at low levels of CO_2 than at high levels. The best mathematical formulation to fit the curve is probably a form of the rectangular hyperbola equation shown below.

$$P = \frac{P_{max} \times (C - \Gamma)}{(C - \Gamma) + (K_m + \Gamma)}$$

In this formulation, which is a modified Michaelis-Menten equation, P is the net photosynthesis rate, C is the CO_2 concentration, P_{max} is the asymptotic upper limit of P at large values of C, Γ is the crop canopy CO_2 compensation point, or the con-

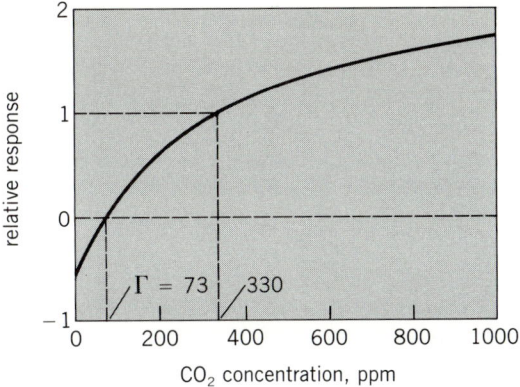

Relative response of soybean crop canopy photosynthesis rates to carbon dioxide concentration as measured in bright midday sunlight. Responses are shown relative to photosynthesis rates measured at a CO_2 concentration of 330 ppm. The CO_2 compensation point (Γ = 73 ppm) is also indicated. Below this CO_2 concentration, no net photosynthesis can occur.

centration of C at which P would be reduced to zero, K_m is the Michaelis constant of this rectangular hyperbola with respect to C, and $(K_m - \Gamma)$ is an apparent Michaelis constant with respect to $(C - \Gamma)$. This response curve does not pass through the origin of the axes (see illustration), but crosses the zero value of relative response at a CO_2 concentration of about 73 ppm, the value of Γ for this specific crop canopy. This value of Γ can be regarded as the canopy CO_2 starvation point because photosynthesis cannot proceed at CO_2 levels below this value. The value of P_{max} was about 2.45 greater than the reference P value at 330 ppm CO_2, which indicates the maximum photosynthesis rates that would be expected under very high CO_2 levels relative to photosynthesis rates at 330 ppm CO_2. The value of K_m for this response curve was 305 ppm.

The response curve to CO_2 can be applied to estimating increases in photosynthesis rates from prepioneer times to present conditions, and from present conditions to an expected doubling of atmospheric CO_2. The curve shows that a change in CO_2 concentration from a prepioneer value of about 260 ppm to a 1958 measured value of 315 ppm should give an 18% increase in photosynthesis rates. From 1958 to 1982, when global CO_2 concentrations rose from 315 to 340 ppm, photosynthesis rates rose 6% based on the response curve shown in the illustration. Furthermore, photosynthesis rates rose about 25% from the prepioneer concentration base of 260 ppm to 340 ppm in 1982. Extrapolating to a doubling of CO_2 concentration to 680 ppm expected in the future within the next century, the rectangular hyperbola formulation predicts a 48% increase in photosynthesis. Similar but slightly smaller responses are expected for crop biomass accumulation and seed yields.

Stomata, transpiration, and CO_2. Stomata of most higher terrestrial green plants partially close in response to elevated levels of CO_2. All other conditions being equal, a reduction of transpiration and water use will result. However, two factors partially mitigate against the full potential for reduction in transpiration. First, under CO_2 enrichment, plants grow slightly larger so they have slightly more leaf area exposed for transpiration. Second, as partial stomatal closure tends to reduce evaporation, leaf temperature will tend to rise. As leaf temperature rises slightly, the vapor pressure of water inside the leaf will increase, which will in turn tend to increase transpiration. The balance between increased stomatal resistance (which tends to decrease transpiration) and increased vapor pressure of water inside the leaves (which tends to increase transpiration) will determine the actual net effect of increased CO_2 on transpiration of individual leaves. This balance, plus more leaf area, will determine the direct effect of increased CO_2 on whole-plant and crop or ecosystem transpiration.

Efficiency of water use by plants (defined as the ratio of photosynthetic CO_2 uptake to transpiration) has been demonstrated to increase with increased CO_2. However, most of the increase is due to increased photosynthesis and growth rates, with only a minor contribution from decreased transpiration.

Source-sink effects. In a few cases, increasing the CO_2 concentration of plants that have been grown at atmospheric levels has been observed to temporarily increase photosynthesis rates, followed by an inhibition of photosynthesis. This response may be due to accumulation of starch and other photoassimilates, because the plants lacked sufficient growing organs, such as seed, to utilize the extra new photoassimilate. In other cases, although starch has been observed to accumulate in leaves, it apparently did not create a severe source-sink imbalance in the plants and hence did not cause photosynthetic inhibition.

Complex ecosystems. Since ecosystems are more diverse and complex than single agricultural crops, it is more difficult to conduct research to predict direct effects of rising CO_2 and competitive interactions of species. However, most investigators expect that C_3 plants will become more competitive with respect to C_4 plants as CO_2 increases, especially when water is not severely limiting. Under conditions where water is limiting, C_4 plants may fare better because they would have higher stomatal resistances which would restrict transpirational water losses.

Summary. The global CO_2 concentration is clearly rising and may double within the next 100 years. Based on experimental information, the photosynthesis, growth, biomass accumulation, and yield of the Earth's vegetation should increase as CO_2 rises. Direct responses of C_3 plants should be greater than those of C_4 plants. Crop responses all appear beneficial, but long-term ecosystem composition responses are harder to predict and impacts are more difficult to evaluate than for managed crops.

For background information see GREENHOUSE EFFECT, TERRESTRIAL; PHOTOSYNTHESIS; PHYTOTRONICS; PLANT METABOLISM; PLANT RESPIRATION in the McGraw-Hill Encyclopedia of Science and Technology.   [LEON H. ALLEN, JR.]

Bibliography: W. C. Clark (ed.), *Carbon Dioxide Review: 1982*, 1982; B. A. Kimball, Carbon dioxide and agricultural yield: An assemblage and analysis of 430 prior observations, *Agron. J.*, 75:779–788, 1983; E. R. Lemon (ed.), *CO_2 and Plants: The Response of Plants to Rising Levels of Atmospheric Carbon Dioxide*, AAAS Selected Symp. 84, 1983; S. H. Wittwer, Rising atmospheric CO_2 and crop productivity, *HortScience*, 18:667–673, 1983.

Plant metabolism

Secondary plant products are natural chemical compounds which are made by plants but are not normally involved in primary metabolic processes such as photosynthesis and cell respiration. These compounds, also called secondary metabolites, are often produced and sequestered in specialized tissues (epidermal glands, hairs), individual cells (idioblasts), or specific locations within a cell (vacuoles) where they do not interfere with primary metabolism. Indeed, less than 25 years ago, most of these secondary plant compounds were routinely assumed to be waste products which the plant excreted or sequestered as metabolic excesses. However, since then intensive study by biologists and chemists has shown that many of these compounds have a biological function which adds to the long-term survival of the plant. In addition to their immediate biological relationships, many of these compounds have already proven useful in commercial and economic development in modern medicine, in agriculture, and even socially in human cultural development and geographic expansion.

Types of secondary plant products. The classes of secondary plant products are diverse in both occurrence and structure (Figs. 1 and 2). They include flavonoid pigments such as cyanidin, a water-soluble polyphenolic pigment responsible for most of the red coloration seen in fruits and flowers. Other compounds such as the alkaloids codeine (used as a cough suppressant) and quinine (used as an antimalarial analgesic) have proven useful in medicine, while alkaloids such as strychnine are deadly toxins or hallucinogenic (lysergic acid). Other secondary metabolites called monoterpenes impart the pleasant aromas of the essential oils of the spice plants used in cooking and flavorings. The taste and smell of peppermint is due to the natural monoterpene menthone, and spearmint flavor is due to the compound carvone.

Most of these secondary plant products are of low molecular weight (about 1000 or less); others with individual low molecular weight also occur as complex polymers of these simpler units, often possessing a molecular weight of several thousand or more. These polymers include large carbohydrates such as natural gums (used in foods as thickeners), polyterpenes such as natural rubber and chicle (used in chewing gum), and some small proteins such as serendip, a natural sweetener from the seredipity berry (*Dioscoreophyllum cumminsii*).

What is most striking about secondary compounds is that they have been an integral part of everyday

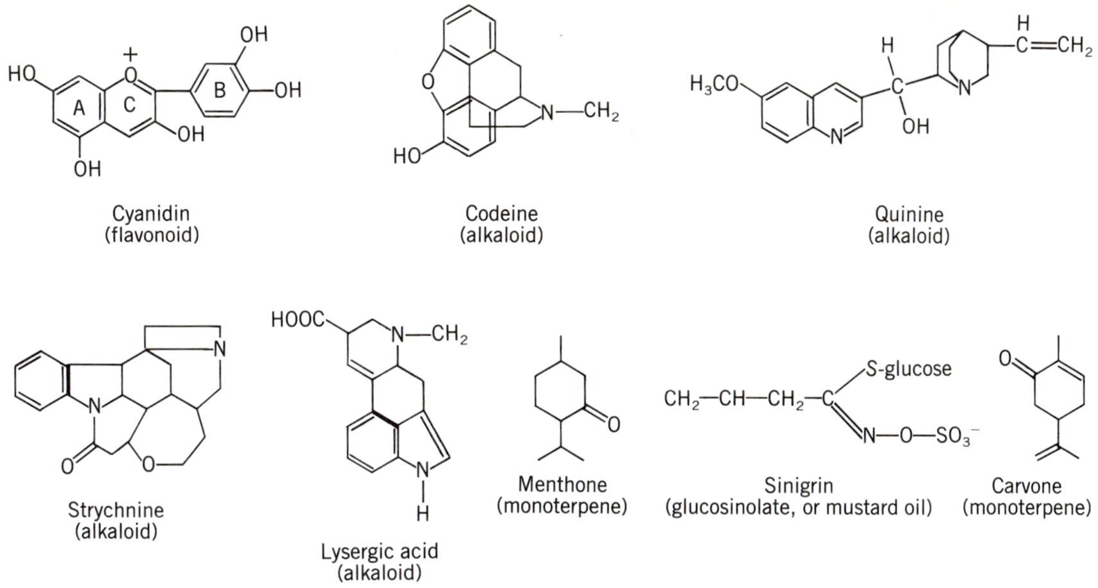

Fig. 1. Structures of typical secondary plant products.

life in foods, medicine, and cultural development for centuries (and often millennia) and yet until recently little knowledge has been available as to why the plant produces them. A survey of the biosynthetic pathways of secondary compounds, as elucidated by chemists, explains much of their diversity.

Biosynthesis. Generally, secondary plant products fall into two major biosynthetic groups (Fig. 3). The first group is derived from the biosynthesis of aromatic amino acids (phenylalanine, tyrosine, and tryptophan) via the shikimic acid pathway. This pathway gives rise to secondary plant products with a phenolic or aromatic ring in their structure and includes cinnamic acids, coumarins, lignins, hydrolyzable tannins, and other phenolics. The second major group of secondary plant products is derived directly or indirectly from acetate, an end product of glycolysis, the initial step of sugar metabolism in intracellular respiration. Products from this pathway include fatty acids and polyacetylenes, numerous types of terpenoids (via mevalonic acid), and other secondary compounds derived from nonaromatic amino acids resulting from the citric acid cycle. The last group includes many nitrogen-containing secondary products such as alkaloids, cyanogenic glycosides, glucosinolates (mustard oils), and nonprotein amino acids (Figs. 1 and 2).

In some cases, secondary plant products may involve precursor constituents from both major biosynthetic lines in their final synthesis. The flavonoid pigments are good examples. They all have the basic three-ring structure like that of cyanidin (Fig. 1), though they may differ in color, in number of external hydroxyl groups, and in the presence or ab-

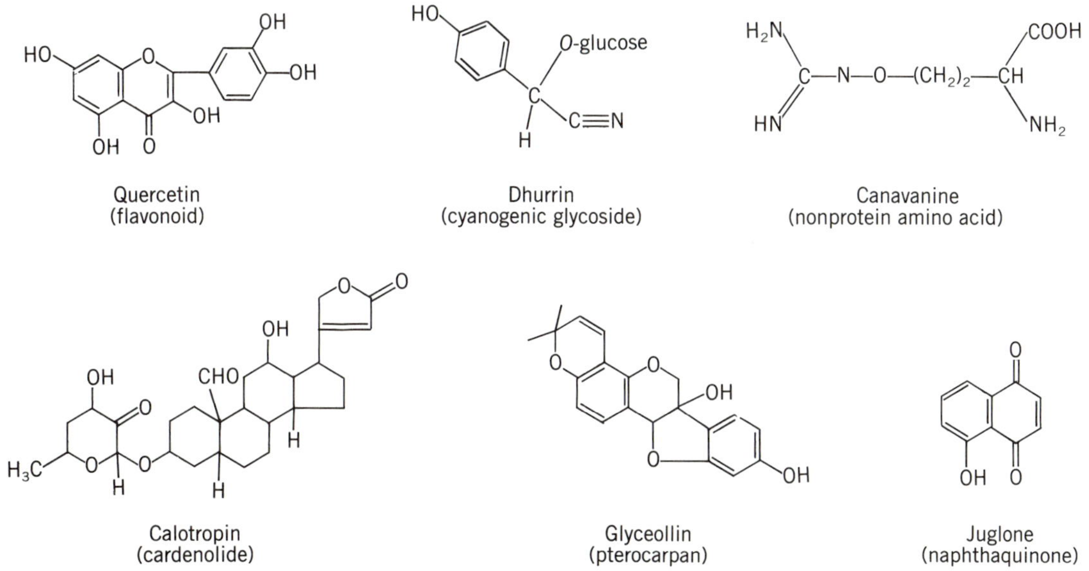

Fig. 2. Structures of secondary plant products with known biological functions.

Plant metabolism

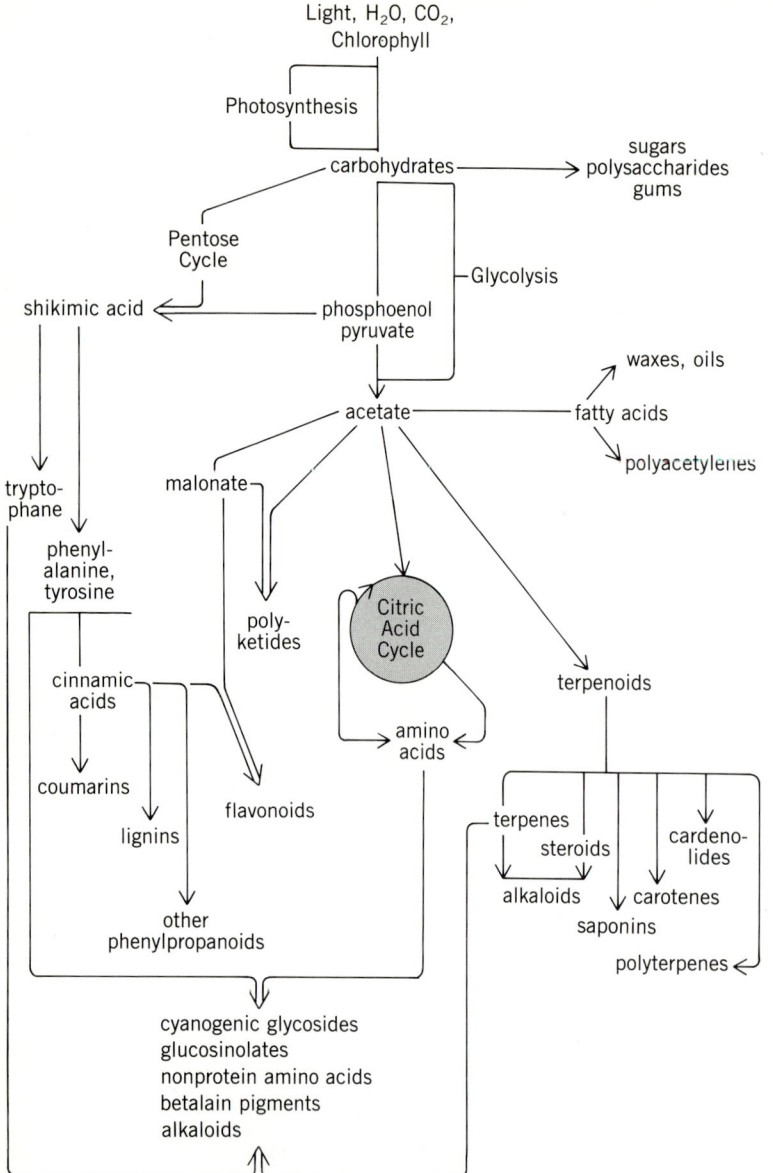

Fig. 3. Outline of secondary plant product metabolism.

derived from the condensation of acetate and the amino acid ornithine (also from the acetate line) and a molecule of phenylalanine (shikimic acid pathway). In some instances, however, alkaloids arise from single sources. Benzylisoquinoline alkaloids (codeine, morphine) all originate from tyrosine, while the lupine alkaloids from the legume plant *Lupinus* all originate from multiple units of lysine. In the case of alkaloids then, these secondary plant products are lumped together on the basis of several general features (they are charged ions, possess a heterocyclic nitrogen-containing ring, and exhibit some physiological activity in animals) rather than a unified biosynthetic origin.

Functions. Although secondary plant products were at first thought to be excess metabolic waste products, it seems unreasonable that a plant would consume much valuable energy to produce substances of little value. Certainly the complexity of some of these secondary products would suggest an inordinate effort on the part of a plant to excrete waste metabolites when simpler systems are often available. Some biologists believe that many of these secondary products probably have (or had or will have) an important biological function adding to the overall survival of a plant.

For example, the flavonoid pigment cyanidin (Fig. 1) is responsible for the red colors in flowers. It has been shown that red flowers are visually attractive to hummingbirds whereas bees are not attracted to red flowers because their vision is more sensitive in the ultraviolet range. Flowers that attract bees often produce another kind of flavonoid called an anthoxanthin, such as quercetin (Fig. 2). These compounds are colorless or nearly so to the human eye, but absorb ultraviolet light and thus are colored or at least dark in the ultraviolet range where the bee sees. Under ultraviolet the patterns of light and dark, and possibly color, act as visual attractants and nectar guides in these flowers.

In other situations a secondary plant product may act as a feeding deterrent to an animal. Cyanogenic glycosides (Fig. 2), for example, are toxic substances which are stored in specialized cells called idioblasts. When the plant tissue is disturbed, as by a grazing animal, the cyanogenic glycoside is released, a natural enzyme cleaves off the sugar, and the cyanogen breaks down giving off toxic hydrogen cyanide which discourages the animal, or perhaps kills it if it eats such plants.

Nonprotein amino acids are not involved in protein synthesis as are the 20 common "essential" amino acids. However, canavanine (Fig. 2) and several other nonprotein amino acids produced in legumes, and especially in their seeds, are often lethal to mammals and insects which feed on them, or at least interfere with normal development in these herbivores, especially insects.

Some animals can use secondary plant products to their advantage. For example, the larvae of the monarch butterfly feed on several species of milkweed (*Asclepias*) which produce the bitter-tasting

sence of various sugars or methyl groups attached to the hydroxyl groups. The left-hand A ring, although a phenolic ring, is actually derived from the condensation of several acetate units via polyketide synthesis. Only the right-hand B ring and 3-carbon chain making up the central C ring are actually derived (as a unit) from the aromatic shikimic acid pathway via tyrosine. Thus the flavonoid pigments are of mixed synthesis with portions from both biosynthetic lines.

Nowhere is this mixed-origin synthesis of secondary products more obvious than in the case of the alkaloids (Fig. 3). Alkaloids such as strychnine, quinine, lysergic acid, and ergotamine are all indole alkaloids derived from the amino acid tryptophan (shikimic acid pathway) and its condensation with one or more terpene units (acetate pathway). Similarly, cocaine is an alkaloid, one portion of which is cardiac glycosides called cardenolides (Fig. 2). These steroidal glycosides are sequestered in various body parts of the larva and are concentrated in the adult butterfly. When its common enemy, the blue jay, attempts to eat it, the cardenolides on and in the butterfly cause a powerful vomiting reaction so that the bird regurgitates the insect. So successful is this chemical deterrent that several other butterflies which do not sequester the cardenolides (such as the viceroy butterfly) mimic the color pattern of the monarch. This, of course, is effective only if the bird has "learned" not to attack the monarch.

Secondary plant products may also function as natural "antibiotics," or phytoalexins, to protect the host plant from fungal and bacterial pathogens. This phenomenon has been described for soybean, *Glycine max*, in which a common fungal pathogen, *Phytophthora megasperma* var. *sojae*, causes root and stem rot. When the fungus infects the host soybean, it releases a substance called an elicitor as it destroys the host cells. This elicitor is actually a series of polysaccharides with a molecular weight from about 5000 to 200,000. The elicitor evokes a phytoalexin response in the host, in this case the production of the phytoalexin called glyceollin (Fig. 2), a pterocarpan (a flavonoid compound). The phytoalexin then retards or stops the growth of the fungus. In some plants such phytoalexins are successful antibiotics, while in others the fungus has overcome the phytoalexin. This biochemical struggle between host and pathogen is a dynamic one which goes on continuously. Such biochemical contests may determine which organism (host or pathogen) will survive to evolve further.

Finally, a plant may also biochemically affect other plants with which it competes for the same ecological niche and available resources. This phenomenon is called allelopathy. An individual plant or a species releases into the air, or more often the soil, one or more substances called allelopaths which are toxic to seedlings and adult plants of competing species. A well-known example is juglone (Fig. 3), a toxic naphthaquinone released into the soil by the black walnut (*Juglans nigra*). This substance is highly toxic to competing plants and even grass. In some cases such allelopaths will inhibit the nitrogen-fixing bacteria in the soil so that nutrients (such as NO_3^-) are not available to competing species, although the producer plant is still able to obtain nitrogen (as NH_3^+) for its own growth.

Potential. Not only are secondary plant compounds commercially useful to humans, but the study of their actions in biological relationships among plants and between plants and animals may eventually provide clues toward similar developments in agriculture and in natural resources. The development of natural biodegradable insecticides and herbicides in the increasingly polluted environment is desirable. The ability to insert genes for useful secondary plant products into crops or other useful plants is an exciting possibility, made all the more plausible by recent developments in genetic engineering.

For background information *see* ALKALOID; ALLELOPATHY; CHEMICAL ECOLOGY; CHEMOTAXONOMY; FLAVONOID; GENETIC ENGINEERING; PLANT METABOLISM; PLANT RESPIRATION; PLANT TAXONOMY; TERPENE in the McGraw-Hill Encyclopedia of Science and Technology.

[DAVID E. GIANNASI]

Bibliography: P. Albersheim and B. S. Valent, Host-pathogen reactions in plants, *J. Cell Biol.*, 78:627–643, 1978; T. W. Goodwin and E. I. Mercer, *Introduction to Plant Biochemistry*, 2d ed., 1983; J. B. Harborne, *Introduction to Ecological Biochemistry*, 2d ed., 1982; M. L. Vickery and B. Vickery, *Secondary Plant Metabolism*, 1981.

Plant morphogenesis

Polyamines are among the oldest biochemicals known. Antony van Leeuwenhoek, studying human sperm on a microscope slide in 1678, noticed the gradual appearance of beautiful stellate crystals during prolonged storage. He named these crystals spermine. He was unaware of the nature of the material, which was much later shown to be the insoluble phosphate salt of a nitrogenous base (I); inor-

$$HN-(CH_2)_4-NH$$
$$H_2N-(CH_2)_3 \qquad (CH_2)_3-NH_2$$
$$(I)$$

ganic phosphate made available by the gradual hydrolytic breakdown of creatine phosphate, a cellular storehouse and carrier of phosphate, complexed with spermine to form the insoluble precipitate. Later on, a related material, spermidine (II), was discovered, along with putrescine (III), abundant in decaying materials, and cadaverine (IV), found in corpses. It

$$H_2N-(CH_2)_4-NH$$
$$(CH_2)_3-NH_2$$
$$(II)$$

$$H_2N-(CH_2)_4-NH_2 \qquad H_2N-(CH_2)_5-NH_2$$
$$(III) \qquad (IV)$$

is now known that polyamines occur in all living cells and that, as positively charged polycations, they can complex with the important cellular polyanions such as deoxyribonucleic acid (DNA), ribonucleic acid (RNA), and the phospholipids of cell membranes. Through such complex formation, they appear to control the configuration of membranes and nucleic acids, and are thus in a position to regulate key processes in cellular maintenance and growth. Spermine is restricted to eukaryotic organisms, where it is largely localized in the nucleus.

It has been known for several decades that polyamine titers are generally highest where cell proliferation is most active. Thus, in microbial systems grown synchronously, one can detect a rise in spermidine and spermine titer in concert with premitotic synthesis of DNA. The same can be seen in animal

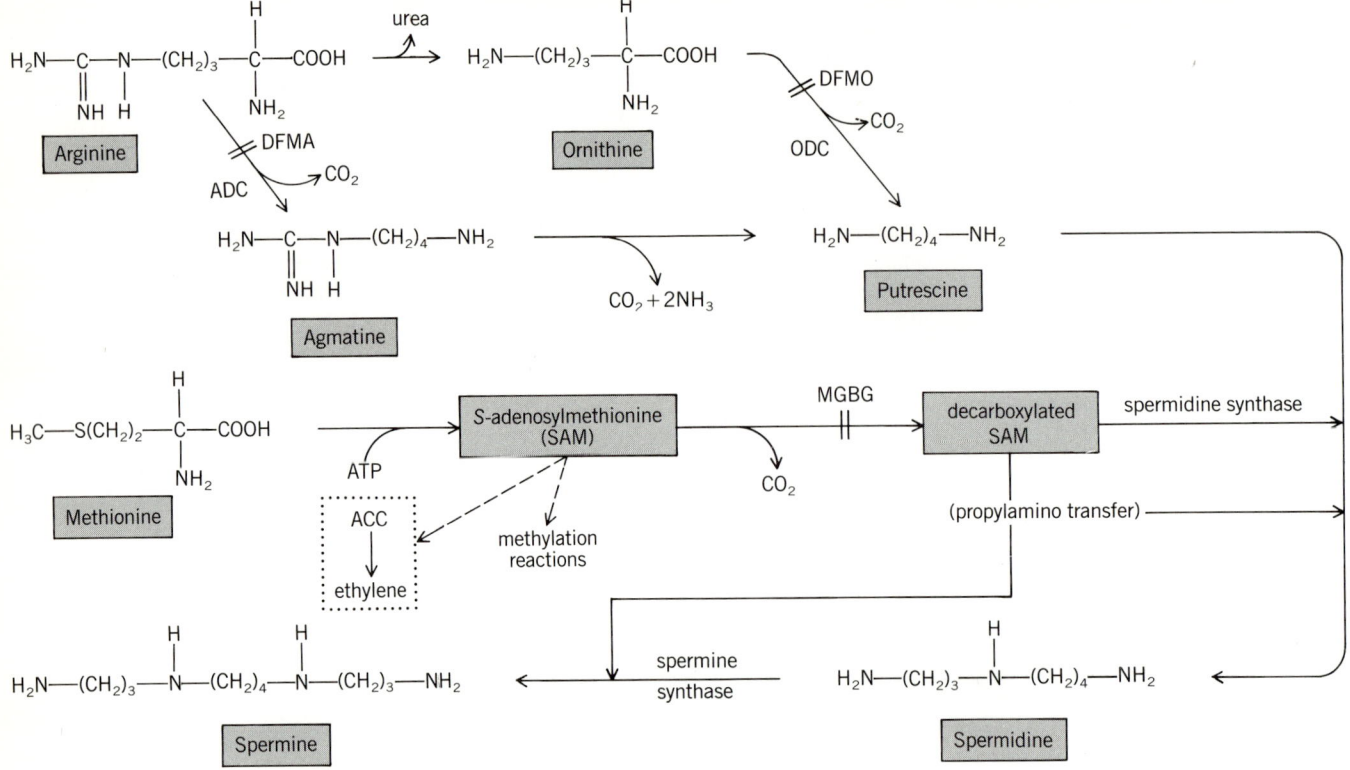

Pathways of biosynthesis of the major plant polyamines. Note that SAM is a precursor not only of the senescence inhibitors spermidine and spermine, but also of the senescence inducer, ethylene (dotted box). Chemical abbreviations are explained in text. (*Information from Terance Smith*)

tissues caused to enter vigorous mitosis by injury, hormonal stimulation, or exposure to carcinogens. Because chemical or genetic blockage of polyamine synthesis simultaneously retards DNA synthesis and cell division, antagonists of the major polyamine biosynthetic enzymes have been used as anticancer drugs. Two such prominent chemicals are α-difluoromethylornithine (DFMO), an analog of ornithine, an intermediate in the formation of putrescine from arginine, and methylglyoxal-bis-(guanylhydrazone) [MGBG], an inhibitor of the decarboxylation of S-adenosylmethione, involved in the further transformation of putrescine to spermidine and spermine (see illustration).

In plants, as in animals and microorganisms, polyamine titer tends to be highest where cell division is most active. This holds most for spermine and spermidine; putrescine titer, by contrast, tends to increase as the cell elongates and ages. Since the higher polyamines are derived from putrescine, it appears that during aging, the transformation of putrescine to spermidine and spermine is diminished.

Polyamines and plant stress. The putrescine content of plant cells rises sharply during exposure to stress. For example, young oat leaves, with a putrescine content of about 10 nM per gram fresh weight of putrescine, raise this level approximately 50-fold in 4–6 h when exposed to water stress, such as an osmoticum of 0.4–0.6 M sorbitol, to acid stress (pH below 5.0), or to nutritional stress, exemplified by nutrient solutions with potassium levels below 0.6 mM. This rapid rise in putrescine in stressed plant tissues is correlated with a rapid induction of the enzyme arginine decarboxylase (ADC), as shown by the inhibitory effect of α-difluoromethylarginine (DFMA) on the stress-induced changes. Like DFMO, which attacks ornithine decarboxylase, DFMA is an enzyme-activated, irreversible, "suicide inhibitor"; when it couples to the active site of ADC, it forms a firm bond that cannot be easily broken, and thus effectively removes the enzyme from cellular activity.

Despite the rapid and massive stress-induced accumulation of putrescine due to ADC induction, there are no similar changes in the higher polyamines, spermidine and spermine. Thus, it appears that stress induces two separate effects: first an enhancement of the rate of putrescine synthesis through ADC induction, and second a partial blockage of the further transformation of putrescine to spermidine and spermine.

The physiological significance of the stress-induced rise in putrescine levels is not yet known. It may be part of the plant's self-protective response to a danger imposed from without, and thus beneficial; alternatively, it could be part of the biochemical mechanism through which injury is produced; finally, it could be irrelevant for stress. A. Galston and colleagues have investigated this question with experiments in which DFMA is applied to or withheld from stressed and unstressed plants, yielding individuals in both populations with high or low putrescine titers. These individuals are then tested to see whether their ability to synthesize protein and RNA from radioactive precursors is impaired or enhanced. Such experiments suggest that the stress-induced rise in putrescine enhances a plant's ability to synthesize protein and RNA, and thus may be beneficial to the plant.

Other nitrogenous components of plant cells may also rise in concentration in response to stress. Such compounds include the amino acid proline, and a class of methylated low-molecular-weight compounds called betaines. The synthesis of these compounds can be related to metabolic pathways yielding putrescine; thus, the stress-induced rise in putrescine titer may represent only one facet of a metabolic disturbance induced by stress conditions.

Regulation of plant growth and development. Experiments suggest that polyamines are essential components of cells. In work with the bacterium *Escherichia coli*, it has been shown that blockage of polyamine synthesis prevents or depresses growth, which can then be restored if putrescine or spermidine is added. In work with the yeast *Saccharomyces cerevisiae*, it was found that blockage of the ornithine decarboxylase (ODC) locus strongly inhibits growth, which can be restored if putrescine is added to the medium. If the blockage is between putrescine and spermidine, vegetative growth is permitted, but no sporulation occurs. If spermidine is added, the life cycle can be completed normally. Other similar evidence, implicating individual polyamines in the control of specific aspects of the life cycle of higher plants, is also becoming available.

Individual plant cells can divide and grow on chemically defined media, producing an undifferentiated tissue mass called a callus. Calli, under some conditions, can differentiate many microscopic embryoids, which have the capacity for regenerating whole plants. Researchers have shown that shortly before embryogenesis in tissue cultures of carrot, polyamine titer and ADC activity rise remarkably. Working with the same system, others have shown that DFMA applications, which block ADC activity, substantially block embryogenesis without affecting growth. Such inhibition could be overcome by the application of putrescine or spermidine.

Similar experiments have been conducted with protoplast-derived tissue cultures of *Vigna aconitifolia* and *Petunia*. Embryogenesis is effectively blocked by DFMA; this can be reversed by putrescine or spermidine. On the other hand, DFMO paradoxically increases embryogenesis. This indicates that the two putrescine-synthesizing pathways, via ODC and ADC, have different functions in the cell. Attempts to localize these enzymes in the cell by radioautography, by coupling them with tritiated DFMA and DFMO, are now in progress.

Tobacco mutants were developed with altered polyamine biosynthetic patterns. One strain, a spermidine overproducer, has large, swollen ovaries with stamenlike organs instead of ovules. This indicates a possible role for polyamines in sex differentiation processes in higher plants, as in yeasts.

Mediation of hormonal effects. Researchers in the 1950's showed that explants of dormant tubers of *Helianthus tuberosus* could be made to grow in tissue culture by the addition of either indole-3-acetic acid (auxin), a plant hormone, or the simple diamine putrescine. Since then, many investigators have been able to show that administration of plant hormones to plant tissues leads to a greatly increased polyamine titer, and an enhanced activity of biosynthetic enzymes. Sometimes the main enzyme affected is ADC, sometimes ODC. In the tomato, pollination or auxin application to the pistil results in ovary enlargement to form a fruit. Fruit enlargement and development, which is accompanied by a massive rise in ODC activity, can be blocked by DFMO, and this inhibition can be reversed by applied putrescine or spermidine. Experiments showing similar polyamine rise following hormonal application have been described for cytokinins and gibberellins, the other two stimulatory plant hormones.

Special interest surrounds the interaction of ethylene, another plant hormone, and polyamines. Both ethylene and polyamines arise entirely or in part from decarboxylated *S*-adenosylmethionine (SAM; see illustration). The inhibition of one pathway promotes the flow of carbon from decarboxylated SAM to the other. Furthermore, the application of polyamines inhibits the evolution of ethylene, and vice versa. Since ethylene is a senescence inducer, polyamines act to retard senescence. This can be seen in fruits, leaves, and even in single isolated protoplasts. Polyamine titers and ADC activity decline in senescing leaves, which suggests that normal aging phenomena in plants may be controlled by their content of polyamines. The stimulation of polyamine synthesis and titer by applied plant hormones has led to the proposal that polyamines may serve as intracellular second messengers for plant hormone action. The reported control by polyamines of the activity of kinases (enzymes that phosphorylate nonhistone nuclear proteins) in the slime mold *Physarum polycephalum* supports this view.

Polyamine titer is also controlled by the status of phytochrome, the photomorphogenic receptor pigment of plants, in a manner suggesting control of growth patterns and possibly of photomorphogenesis itself. Polyamines have also been shown to oscillate in synchrony with circadian rhythms of organic acids in *Bryophyllum*, a fleshy crassulacean plant. Such circadian rhythms have been shown to be a critical part of the photoperiodic timing mechanisms.

Recently, a commercial U.S. patent was awarded for the use of polyamines in retarding senescence in plants. It appears that their practical use may grow.

For background information *see* PLANT GROWTH; PLANT HORMONES; PLANT MORPHOGENESIS in the McGraw-Hill Encyclopedia of Science and Technology.

[ARTHUR W. GALSTON]

Bibliography: U. Bachrach, *Function of Naturally Occurring Polyamines*, 1973; F. M. Dumortier et al., Gradients of polyamines and their biosynthetic enzymes in coleoptiles and roots of corn, *Plant Physiol.*, 72:915–918, 1983; H. Flores and A. W. Galston, Polyamines and plant stress: Activation of putrescine synthesis by osmotic shock, *Science*, 217:1259–1261, 1982; A. W. Galston, Polyamines as modulators of plant development, *BioScience*, 33:382–388, 1983; N. D. Young and A. W. Galston, Putrescine and acid stress: Induction of arginine decarboxylase activity and putrescine accumulation by low pH, *Plant Physiol.*, 71:767–771, 1983.

Plant pathology

Biological control of plant diseases never received major attention until 1972, when H. Wells, D. Bell, and C. Jaworski first demonstrated control of *Sclerotium rolfsii*, the soil-borne fungal pathogen of many crop plants, with the fungus *Trichoderma harzianum* (Fig. 1). Since that time there has been an enormous increase in the amount of research on biological control of plant diseases. In the following discussion, biological control agent is restricted to mean an identifiable biological entity, such as a fungus or bacterium, that can be removed from its natural setting and reintroduced as a constraint on disease. Parameters discussed cover modes of action, biological control agents, and variability in biological control.

Modes of action. Major modes of action of biological controls of plant diseases are competition, parasitism, antibiosis, and hypovirulence. Two or three modes are often associated with one biological control agent but, for clarity, are treated individually in this article.

Competition. Competition results from the biological control agent removing food required by the plant pathogen, or from the agent taking up space required for survival and infection by the plant pathogen. The biological control of apple scab caused by the fungus *Venturia inequalis* showed that biological agents controlled this disease by preemptive use of nutrients in fallen leaves on the orchard floor where ascocarps of the plant pathogen must mature and produce ascospores for initiating the disease in the spring. Preemptive utilization of natural food and space is accomplished by treating wounds of trees and shrubs after pruning, stump treating after forest thinning, and seed treating with biological control agents. Site occupation is a highly versatile component of biological control. The preempting may be general or specific, in which a species, subspecies, or strain of the biological control agent (for example, *Fusarium*) is closely related to the plant pathogen.

Parasitism. Parasitism as used here is the action of the antagonist in obtaining nutrients partially or entirely from living cells or thalli (Fig. 2). Parasites may be obligate or facultative, though most parasitic biological control agents being studied for plant disease control are facultative. A number of fungi, including *Trichoderma* and *Gliocladium*, have demonstrated biological control through parasitism of plant pathogens.

Antibiosis. Antibiosis includes all antagonistic chemical products produced by the living biological control agent and released into the environment. Antibiosis involves enzyme systems and antibiotic compounds that damage the plant pathogens. Antibiosis does not depend on direct contact of the control agent and the pathogen, but enzymatic antibiosis usually requires the antagonist to be physically close to the pathogen. Antibiotic antibiosis is exhibited in genera of a number of organisms used for biological control, such as the fungi *Trichoderma* and *Gliocladium*, and the bacteria *Pseudomonas* and *Agrobacterium*. Continuing growth of the agent is essential for long-lasting antibiotic biological control.

Hypovirulence. Hypovirulence, caused by viruslike genetic material, has potential for biological control of plant diseases. Hypovirulent indicates that the plant pathogen biotype will infect and cause slight but not appreciable damage. The major example of biological control using the hypovirulence concept is the control of chestnut blight caused by the fungus *Endothia parasitica*. A double-stranded ribonucleic acid (RNA) occurs in the cytoplasm of the hypovirulent isolates. This RNA can be transferred into virulent biotypes by simple hyphal fusion; the entire thallus of the virulent strain is then converted into a hypovirulent type containing the double-stranded RNA. Thus tree cankers can be cured by bringing the hypovirulent and virulent types in contact. Major limiting factors are that the hypovirulent and subsequent converted-to-hypovirulent types have reduced vigor and are poor sporulators. The converted types may also sector or produce spores that leave the double-stranded RNA behind and revert to virulent types. A crucial factor in converting the virulent types to hypovirulent types is lack of compatability between fungal isolates. The first hypovirulent type was discovered in Europe, but additional hypovirulent types have been found in Europe and the United States. This biological control system may return the American chestnut as a major forest tree in the United States. Genetic engineering techniques should facilitate the movement

Fig. 1. Biological control with *Trichoderma harzianum* of *Sclerotium rolfsii* in tomato transplant seedlings in naturally infested soils. Left: efficacy of one application; right: disease development in the nontreated control. (*From H. D. Wells et al., Efficacy of Trichoderma harzianum as a biocontrol for Sclerotium rolfsii, Phytopathology, 62:442–447, 1972*)

of the hypovirulent double-stranded RNA to many plant pathogens.

Biological control agents. Agents used in the biological control of plant diseases include four classes of fungi, bacteria, and double-stranded RNA.

Fungi. The basidiomycete *Corticium*, which offers excellent biological control of a number of plant pathogens, is parasitic, is a competitor, and has antibiosis.

The largest number of biological control agents are ascomycetes, but are generally designated by their deuteromycete Latin binominals in the context of biological control. In this group, *Trichoderma*, *Gliocladium*, *Penicillium*, *Fusarium*, *Chaetomia*, and *Sporidesmum* are widely used as biological control agents. These have representatives that are competitors, are mycoparasites, and that provide antibiosis through enzymatic and antibiotic activities. They control plant pathogens belonging to the basidiomycetes, ascomycetes, deuteromycetes, and phycomycetes.

Many phycomycetes are mycoparasites, but most have not been exploited for biological control. *Pythium* and *Phytophtora*, however, are phycomycetes that have recently shown promise as biological control agents.

Bacteria. Genera of bacteria most promising as biological control agents include *Pseudomonas* (primarily the fluorescent types), *Bacillus*, and *Agrobacterium*. Bacteria appear effective as a result of specific antibiotic activities which, in the fluorescent pseudomonads, are related to their fluorescent properties. Bacteria, used as seed protectant, multiply and follow root development, and thus afford extended control of the pathogen. In take-all, a fungal disease of wheat, the bacteria are effective against both the saprophytic and parasitic phases of the pathogen.

Double-stranded RNA. The genetic components of hypervirulence are double-stranded RNAs. Some controversy, however, exists about their nomenclature. The conservative view is that the term double-stranded RNA should only be used in reference to hypovirulence because some of the parameters usually associated with viruses, such as a protein coat, are absent or unidentified. Therefore, most do not call it a virus, though the double-stranded RNA appears to be its own messenger for duplication and to that extent meets the criteria of being a biological entity. There are at least three double-stranded RNA components in some hypovirulent fungal isolates.

Genetic information carried in double-stranded RNA differs widely, and has many loci in which changes affect phenotypic expression of the plant pathogen. Double-stranded RNA has been identified in association with hypovirulence in *Rhizoctonia*, and this system may be operating with many other pathogens. Most plant pathologists discard weak and atypical isolates because they may be studying host resistance or chemical actions and are primarily in-

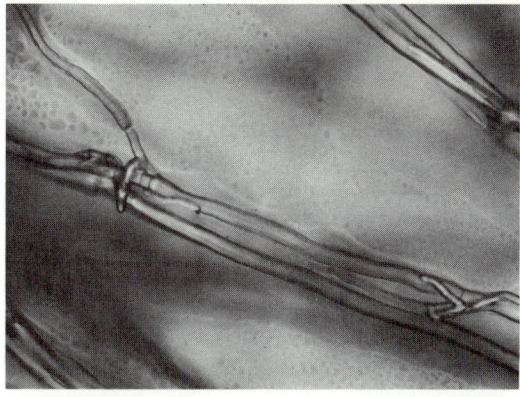

Fig. 2. Mycoparasitism of *Sclerotium rolfsii* by *Trichoderma harzianum*. The smaller strands of *T. harzianum* mycelia are shown clamping onto the larger strands of *S. rolfsii* mycelia.

terested in evaluating normal-appearing plant pathogen isolates. Weak isolates, however, may contain valuable material for biological control of pathogens.

Variability in biological control. The factors that most frequently cause variability in the efficacy of biological control agents are the formulation in which they are applied, the environment in which the biological control is used, and the genetic variability within both the biological control agent and the plant pathogen.

Formulation. The environment is a highly buffered system where many types of organisms compete. The efficacy of the biological control formulation is often enhanced by adding a suitable food base. This apparently gives the control agent sufficient energy to make it temporarily dominant. In some instances, however, the addition of a supplemental food has been counterproductive because some plant pathogens may be enhanced by leachates and become more aggressive in causing disease, or because other antagonistic organisms in the environment also may utilize the food base and reduce the biological control inoculum potential by antibiosis and competition. It is essential that the proper formulation be investigated for each biological control system.

Environment. General environmental effects are often important, since an excellent biological control system developed under one set of conditions may be ineffective in others. Soils that have high levels of antibiotic-producing fluorescent pseudomonads may reduce the control efficacy of some biological control agents that have good efficacy in other soils. Soil moisture and some minor elements are critical factors affecting efficacy of various biological control agents. In fluorescent bacteria, the production of pathogen-suppressing antibodies is negatively correlated with iron availability in the soil. Chemical pesticides used to control specific diseases often reduce efficacy of biological control agents. Research indicates that biological control agents can be selected with tolerance to some fungicides.

Genetic variability. Probably the largest factor affecting efficacy of biological control of plant disease is genetic variability within both the control agent and the pathogen. These systems evolved over millions of years, and diversity within each has occurred. A biological control agent may have good efficacy against biotypes of a given plant pathogen from widely scattered parts of the world. For example, the ATC 24274 isolate of *Trichoderma harzianum* has shown this ability on *Sclerotium rolfsii*. Other *Trichoderma* biotypes have not shown this consistency against numerous pathogens. A highly significant interaction often exists in which the best biological control for one pathogen biotype may be noxious to a different biotype of the same pathogen, which in turn is highly susceptible to a second biotype of the same species. Work with chestnut blight indicates that it may be essential to use many double-stranded RNA sources in an attempt to overcome interactions between *Endothia parasitica* and the double-stranded RNA components of hypovirulence. It has been shown that induced mutations are possible with biological control agents.

For background information *see* ECOLOGICAL INTERACTIONS; PLANT PATHOLOGY in the McGraw-Hill Encyclopedia of Science and Technology.

[HOMER D. WELLS]

Bibliography: R. J. Cook and K. F. Baker, *The Nature and Practice of Biological Control of Plant Pathogens*, American Phytopathological Society, 1983; C. G. Griffin et al., Survival of American chestnut trees: Evaluation of blight resistance and virulence in *Endothia parasitica*, *Phytopathology*, 73:1084–1092, 1983; D. M. Weller and R. J. Cook, Suppression of take-all of wheat by seed treatments with fluorescent pseudomonads, *Phytopathology*, 73:463–469, 1983; H. D. Wells, D. K. Bell, and C. A. Jaworski, Efficacy of *Trichoderma harzianum* as a biocontrol for *Sclerotium rolfsii*, *Phytopathology*, 62:442–447, 1972.

Pneumocystis

Since the discovery of Legionnaire's disease, microbiologists and clinicians alike have recognized that the human pulmonary flora may be changing and may not be as yet fully defined. *Pneumocystis carinii*, an organism of unsettled taxonomy, represents a case in point. It has been recognized as a far more common infectious agent than formerly thought because of the development of modern technology that has allowed researchers to conduct immunologic surveillance for both *P. carinii* antigen and antibody, and because of the association of *P. carinii* with the acquired immunodeficiency syndrome (AIDS).

AIDS and P. carinii pneumonia. *Pneumocystis carinii* has been recognized as the most common opportunistic infectious disease agent found in association with AIDS. AIDS is characterized by a loss of normal resistance to opportunistic microorganisms, that is, organisms that are harmless to healthy persons but attack individuals whose immunity to infectious disease is seriously impaired by cancer or other debilitating disorders.

Evidence is mounting that AIDS may be caused by one of the human T-cell leukemia viruses (HTLV) which severely damage the body's cell-mediated immune system and render the individual vulnerable to infectious diseases. Healthy individuals have resistance to these microorganisms. AIDS appears to be spread by intimate contact and specifically by blood or blood products. Support for this means of transmission is offered by the fact that a small proportion of AIDs victims are hemophiliacs who require treatment with factor VIII, which is a blood component that assists in clotting.

Nearly 60% of all AIDS patients initially need medical attention as a result of symptoms associated with *P. carinii* pneumonia (PCP). Symptoms include shortness of breath, unplanned weight loss, fever, and night sweats. Multiplication of the organisms in the alveolar spaces of the lung inhibits respiration, and death occurs by asphyxiation in 90–100% of acute untreated cases.

Two drugs have been demonstrated to be effective in the treatment and prevention of PCP: trimethoprim-sulfamethoxazole (TMP-SMX), and pentamidine isethionate or its methane sulfonate derivative. In patients with AIDS, however, 40% do not respond to prompt and appropriate therapy and thus the survival rate for PCP in AIDS patients is only 60%.

Until recently, the only means for demonstrating *P. carinii* infection was to resort to some invasive surgical procedure such as open-lung biopsy or needle aspiration of the lung. In 1978, however, a technological advance made it possible to detect *P. carinii* by means of an immunologic test on serum separated from blood. The technique is termed counterimmunoelectrophoresis (CIE) and is based upon the binding of *P. carinii* antigen(s) from a patient's serum to antibody prepared against *P. carinii* in rabbits (see table). A second-generation test is under development that should make noninvasive diagnosis possible in any routine hospital laboratory. It is hoped that very early detection of antigen in the blood may reduce the morbidity and mortality rate due to PCP in newborns, transplant recipients, and cancer and AIDS patients.

Biology of P. carinii. *Pneumocystis carinii* is a parasitic protozoan, probably a member of Sporozoa, and exhibits two fundamental stages, the trophozoite (2–3 micrometers across) and the cyst (about 7 micrometers). *Pneumocystis carinii* has been reported in the lungs of nearly all species of mammals investigated. Although it is apparently ubiquitous and most children (80%) have antibody to it by age 2–4, a source of contagion in nature has never been identified. It may in fact be a commensal or saprophyte of mammalian lung that poses a threat only when the immunologic defenses of the host are impaired. All that is required to induce fulminant infection in young healthy laboratory rats is a low-protein diet and twice-weekly injections with cortisone acetate. This drug suppresses resistance to infectious diseases, and within 8–12 weeks nearly 100% of the animals will develop acute PCP.

Incidence of *Pneumocystis carinii* antigenemia in several categories of patients

Group	Number tested	Number of positive/total	% positive
Normal children	184	0/184	0
Children with malignancy			
Tissue-documented PCP	20	19/20	95%
PCP undocumented, randomly selected	100	15/100	15%
PCP prophylaxis study*			
TMP-SMX	8	0/8	0%
Placebo	11	3/11	27%
Infant pneumonitis			
Healthy infants	64	0/64	0%
Mixed-agent pneumonitis†	104	19/104	18%
Normal adults	208	6/208	3%
Adults with malignancy			
Hematologic malignancy (afebrile)	172	5/172	2%
Solid tumors (afebrile)	75	0/75	0%
Lung cancer (afebrile)	36	1/36	3%
Malignancy (febrile)‡	19	6/19	32%
Bone marrow allograft study			
Recipients			
Documented PCP	28	22/28	79%
Viral pneumonia	26	18/26	69%
Idiopathic pneumonia	26	17/26	65%
No pneumonia	25	11/25	44%
Others			
Normal marrow donors	50	1/50	2%
Undiagnosed pulmonary infiltrates (clinical suspicion of Legionnaire's disease)	28	22/28	79%
Other adults			
Pulmonary infection	28	3/28	11%
Nonpulmonary infection	22	2/22	9%
Nonmalignant pulmonary disease	33	3/33	9%
Healthy homosexual males	20	0/20	0%
Renal allograft recipients			
Tissue-documented PCP	9	7/9	78%
No evidence of PCP, not biopsied	16	1/16	6%
AIDS patients with tissue-documented PCP	12	9/12	75%

*These patients were at high risk for developing PCP, hence their entry into the prophylaxis study.
†*Pneumocystis carinii* was present in the single open-lung biopsy performed on antigen (+) infant.
‡Hospitalized or terminally ill.

In much the same manner, organ transplant recipients, individuals who have some other immune defect, and others who undergo tumor chemotherapy and other treatment are at risk for developing PCP. The newer diagnostic test for *P. carinii* will likely make it possible to monitor these individuals and to therapeutically intervene when antigen first appears in the blood. It is theorized that antigen is liberated into the blood as a result of the enzymatic action of alveolar macrophages and other phagocytic cells in the lung.

Cyst-phase *P. carinii* organisms are about the size of a red blood cell; the trophozoites resemble blood platelets. Cysts are round, oval, or irregular in shape and are highly resistant to physical and chemical disruption. The stains of choice for demonstrating *P. carinii* cysts are Gomori's methenamine silver (GMS) method and toluidine blue "0". These stains are not, however, absorbed by the trophozoites, which require a Giemsa stain and a detergent. The eight sporozoites within cysts that ultimately give rise to eight trophozoites also require a Giemsa stain for visualization. Pathology laboratories employing only Gomori's method or toluidine blue "0" stain may miss the diagnosis if trophozoites predominate.

Culture of P. carinii. The ability to grow *P. carinii* in embryonic chick epithelial lung-cell cultures made possible developments which led to the noninvasive serologic test for PCP. Cell culture propagation permitted the production of large numbers of pure organisms that were used as antigen for the preparation of antibody in the CIE test.

The life cycle of *P. carinii* in cell culture requires about 6–8 h for completion and involves the attachment of trophozoites to the host cell surfaces. Experiments employing radioactively labeled monolayers of host cells have shown that the "trophs" derive labeled metabolic precursors (nucleic acid components and amino acids) from the host cells. When the trophs mature, eight comma-shaped sporozoites develop within the parent cyst, and these emerge at maturation through the cyst wall.

Antibody to P. carinii. Antibody titers to *P. carinii* as determined by indirect immunofluorescence (IFA) and enzyme-linked immunosorbent assay (ELISA) have provided valuable epidemiologic data, but have not been reliable from a diagnostic standpoint. Since most individuals have a substantial antibody titer to *P. carinii*, the mere presence of antibody is diagnostically meaningless. To further complicate the issue, those individuals who develop PCP do so because their immune response is inadequate. Therefore, they are seldom capable of mounting the fourfold rise in antibody titer that is usually accepted in the case of most infectious agents as a valid indication of infection. Recent data suggest, however, that antibody titers may be predictive of susceptibility to PCP. Consequently, antibody test results may prove useful in the surveillance of cancer, AIDS, and chronic diseases for evidence of impending problems with this infectious disease.

For background information *see* CELLULAR IMMUNOLOGY; IMMUNOELECTROPHORESIS; IMMUNOFLUORESCENCE; OPPORTUNISTIC INFECTIONS; SPOROZOA in the McGraw-Hill Encyclopedia of Science and Technology.

[LINDA PIFER]

Bibliography: L. L. Pifer, *Pneumocystis carinii*: A diagnostic dilemma, *Ped. Inf. Dis.*, 2:177–183, 1983; L. L. Pifer, *Pneumocystis carinii*: A misunderstood opportunist, *Eur. J. Microbiol.*, 3:169–173, 1984; L. L. Pifer et al., Propagation of *Pneumocystis carinii in vitro*, *Ped. Res.*, 11:305–316, 1977; Update on acquired immune deficiency syndrome (AIDS): United States, *Morbid. Mortal. Week. Rep.*, 31:507–514, 1982.

Pogonophora

Pogonophores are marine tube worms that lack mouth and gut, yet are not parasites. Many species live partly buried in the mud of the sea floor. They have been known for more than a century and recognized for more than 60 years, but the mystery of their way of feeding is only now nearing a solution.

During the 1960s and 1970s research was directed mainly to the measurement of the animals' ability to absorb and utilize dissolved organic compounds from their surroundings. It was concluded that some small species may be able to sustain themselves by absorption through the epidermis, but that larger species are unlikely to get sufficient nourishment in this way. An attractive hypothesis, recently developed, is that they all make use of symbiotic chemoautotrophic bacteria, which live inside the animals and contribute to their nutrition.

Symbiotic bacteria. The symbiosis hypothesis was developed after the discovery, in 1977, of giant pogonophoran tube worms surrounding hot springs, known as hydrothermal vents, in the Pacific Ocean floor. The water from these vents is hotter than the surrounding deep-ocean water and contains a high concentration of sulfides. Where it mixes with oxygenated ocean water, it supports an abundant population of free-living bacteria that use energy derived from oxidation of sulfide to fix carbon dioxide and build up organic compounds necessary for their life. Such chemoautotrophic bacteria also occur in boundary systems between oxic and anoxic conditions in sediments or stagnant water. The interesting feature of the hydrothermal vent tube worms (*Riftia pachyptila*) is that they lack mouth and gut but contain vast numbers of internal bacteria that are able to fix carbon dioxide, apparently using reduced sulfur compounds as their energy source.

Examination of the more widely distributed small pogonophores has shown that they also contain endosymbiotic bacteria capable of fixing carbon dioxide. Their energy source is not necessarily the same as that of the vent pogonophores, since they live in environments with much less free sulfide, but sulfur-oxidizing enzymes can be detected in some species. The symbiotic nature of the association between animal and bacteria is deduced from the anatomy of the tissues which support the bacteria, termed the trophosome; and from stable-carbon-isotope studies which show that the organic carbon in the animal tissue is probably derived from autotrophic bacteria, rather than from photosynthetically produced organic matter formed near the sea surface.

Classification. Both giant and small pogonophores have bacteria-containing tissue in the middle (trunk) region of the body, but this tissue is more bulky and complicated in the giant pogonophores; a study of the small pogonophores has been helpful in understanding the anatomy of both groups. The goups show many differences in both external and internal anatomy, and have been separated into the subphyla Obturata and Perviata. The obturates have an anterior pluglike structure, termed the obturaculum, with which they can close the top end of their tube. The giant pogonophores (*Riftia* spp.) and some medium-sized ones (*Lamellibrachia* spp.) belong to this group. The Perviata have no plug, but some can close the end of their tube in other ways. They are fairly small pogonophores, all very narrow in proportion to their length. *Siboglinum fiordicum*, for example, is about 0.2 mm (0.008 in.) wide and 100–200 mm (4–8 in.) long.

Perviata trophosome. In perviate pogonophores, the cylindrical trophosome lies between the two longitudinal blood vessels in the postannular part of the trunk (Fig. 1). In females and immature animals it extends from in front of the girdles back to the septum between trunk and opisthosoma. In males the gonads and sperm ducts occupy the front part of this region and the trophosome is restricted to the hind part. Transverse sections show that the trophosome is made up of two concentric layers of cells (epithelia) separated by a space filled with blood. There is also a central fluid-filled hollow in some species (Fig. 2). The outer epithelium is part of the lining of the coelom and has a storage function, while the inner epithelium is composed of cells termed bacteriocytes, each of which contains numerous bacteria enclosed individually in vacuoles. The blood space is bounded by the basal lamellae of the two epithelia, but connects at intervals with the longitudinal blood vessels. Blood does not circulate but moves in a "tidal" fashion and, according to the relative pressures in the two vessels, may flow in either direction.

In most of the perviate pogonophores examined, the symbiotic bacteria are rod-shaped, 0.1–0.3 μm wide and 2–5 μm long. Probably there is a different species in each species of pogonophore. *Siboglinum poseidoni* has thicker bacteria with many parallel intracytoplasmic membranes, like those of methane bacteria (Fig. 3). *Sclerolinum brattstromi*, which lives in rotting wood, has large oval bacteria with a few intracytoplasmic membranes.

Obturata trophosome. In the Obturata the trophosome is a bulky, much branched tissue, mingled with the gonads and filling most of the large trunk (Fig. 1). Its many lobes are penetrated by fine branches of the blood vessels, and its cells are of two types, with and without bacteria. More detailed studies are needed, but it seems likely that the trophosome of the obturates could have evolved by growth in volume together with branching, from the simple tubular structure in the perviates. The bacterial symbionts in *Riftia* and *Lamellibrachia* are large and rounded and contain many intracytoplasmic membranes (Fig. 3).

Function. The housing of the symbionts in cell vacuoles allows the host cell to control the symbionts' immediate environment. Apart from growth requirements for carbon dioxide, oxygen, and hydrogen sulfide or other reduced inorganic energy substrate, there are probably accessory factors required from the host. What these may be cannot be determined until the bacteria can be isolated in culture.

In the Perviata, oxygen is taken up by the blood at the tentacles from the oxygenated water above the sediment; carbon dioxide can diffuse in from the water above and in the sediment, and respired carbon dioxide can be reused; reduced sulfur compounds, present in the anoxic layers of the sediment, can diffuse through tube and epidermis all along the postannular region, which can make up about two-thirds of the total length of the animal. The blood is rich in hemoglobin with a high affinity for oxygen, and the circulatory system takes the hemoglobin into close contact with the bacteriocytes in the trophosome. The transfer of nutrients from bacteria to host cells can probably take place as exudates from the bacteria and after intracellular digestion of bacteria in lysosomes. The blood circulating over the bacteriocytes can then transport nutrients to the rest of the body, depositing some compounds for storage in the outer trophosome epithelium. The system seems well adapted for the nurture of the bacteria and the transport of their products. The hydrothermal vent Obturata can acquire sulfide via their tentacles from the vent water; a sulfide-transporting blood protein has been reported from *Riftia*.

Evolution. The development of the trophosome has been paralleled by the loss of the gut. Since the trophosome spatially takes the place of the midgut, it seems likely that it is a modified vestige of a gut that existed in the ancestor of the Pogonophora. The developmental origin of the trophosomal lining epithelium from embryonic endoderm is suspected but not yet confirmed. The original establishment of chemoautotrophic symbionts in the place of normal gut symbionts, such as are found in many invertebrates, would depend on the colonization of habitats where appropriate energy sources were available.

Zoologists have been unable to agree about the relationship between pogonophores and other invertebrate phyla, since both embryonic development and adult anatomy can be interpreted in more than one way. It now seems to be accepted that pogonophores are protostomians, rather than deuterostomians. The presence of annelid-type chaetae and septa (particularly in the opisthosoma) suggests annelidan ancestry. One can postulate an annelidan ancestor, living in nearly anoxic conditions, whose gut was colonized by sulfur-oxidizing bacteria from its surroundings so that the organism was enabled by their activity to penetrate further into sulfide-rich environments; the organism then gradually became dependent on the symbionts. The loss of foregut and

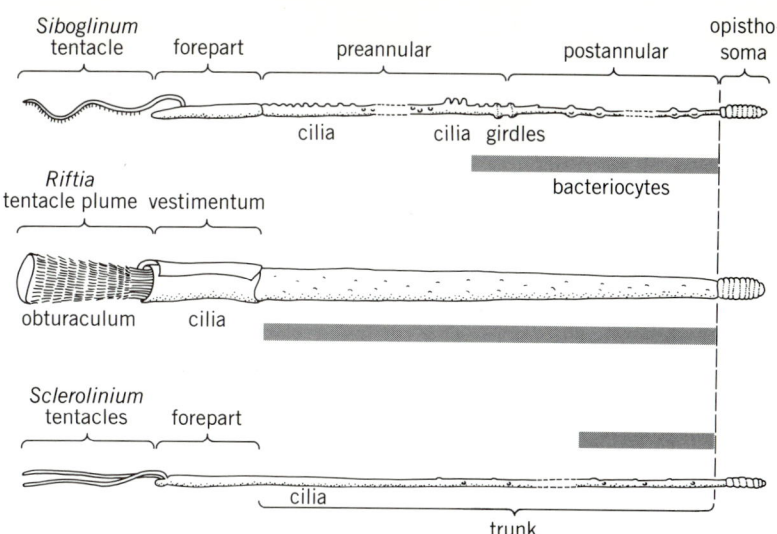

Fig. 1. Body divisions in Pogonophora (much shortened): *Siboglinum*, pattern found in most Perviata; *Riftia*, pattern found in Obturata; *Sclerolinum*, pattern found in one family of Perviata (Sclerolinidae). The bars show the extent of the trophosome in each. (*After E. C. Southward, Bacterial symbionta in Pogonophora, J. Mar. Biol. Ass., 62:889–906, 1982*)

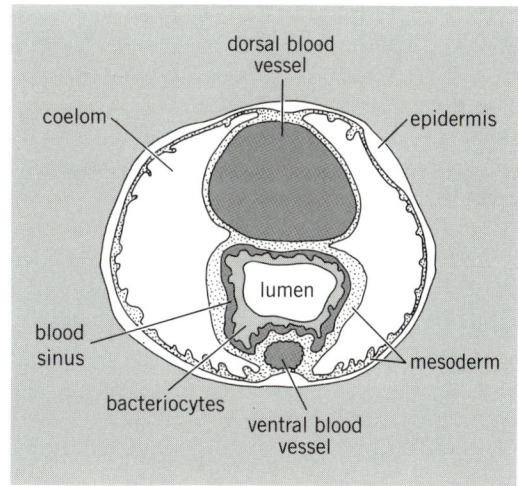

Fig. 2. Transverse section of postannular part of trunk of *Siboglinum fiordicum*.

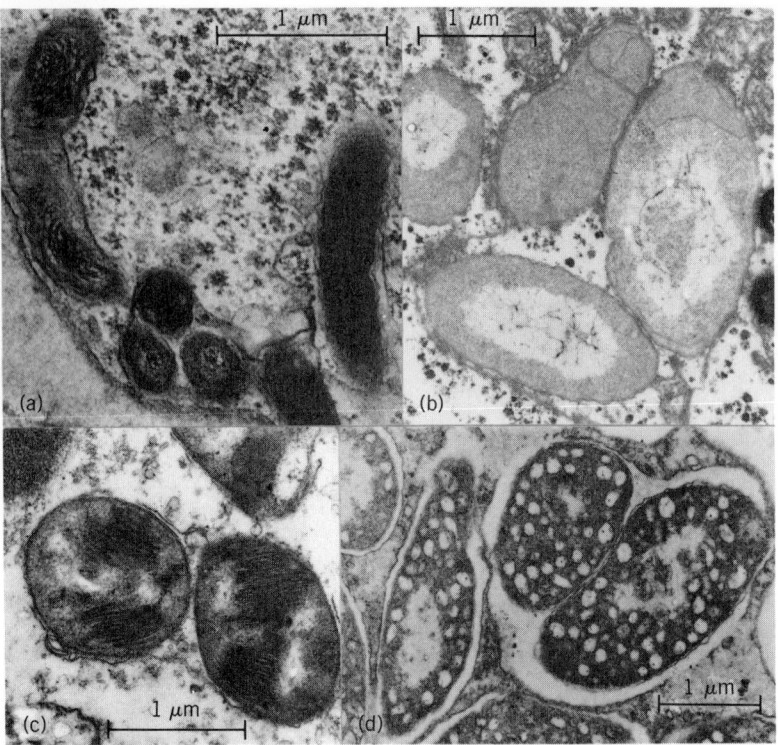

Fig. 3. Sections of symbiotic bacteria in pogonophoran trophosomes: (a) *Siboglinum fiordicum*, note blood space at lower left (*E. C. Southward*); (b) *Sclerolinum brattstromi* (from E. C. Southward, J. Mar. Biol. Ass., 62:889–906, 1982) (c) *Siboglinum poseidoni* (*H. J. Flügel*); and (d) *Riftia pachyptial* (from C. M. Cavanaugh, Nature Lond., 302:58–61, 1983).

hindgut could be viewed as an economy of effort. There are certain annelids (Oligochaeta) which have lost the entire gut and acquired epidermal sulfur-oxidizing symbionts. These live in sulfide-rich lagoon sediments and show a different approach to nutrition by chemotrophic procaryotes.

For background information *see* ANIMAL EVOLUTION; ANNELIDA; POGONOPHORA in the McGraw-Hill Encyclopedia of Science and Technology.

[EVE C. SOUTHWARD]

Bibliography: C. M. Cavanaugh, Symbiotic chemoautotrophic bacteria in marine invertebrates from sulphide-rich habitats, *Nature Lond.*, 302:58–61, 1983; H. J. Flügel and I. Langhof, A new hermaphroditic pogonophore from the Skagerrak, *Sarsia*, 68:131–138, 1983; A. J. Southward et al., Bacterial symbionts and low $^{13}C/^{12}C$ ratios in tissues of Pogonophora indicate unusual nutrition and metabolism, *Nature Lond.*, 293:616–620, 1981; E. C. Southward, Bacterial symbionts in Pogonophora, *J. Mar. Biol. Ass.*, 62:889–906, 1982.

Polarity (embryology)

Most cells and tissues in living organisms exhibit a highly organized structure that is critical for their function. For example, cells in secretory organs generally secrete only at one end where secretory vesicles are localized, and their nucleus is often slightly displaced toward the opposite end. Such cells are said to be polarized because they have a distinct apical-basal axis and exhibit an asymmetrical distribution of cytoplasmic organelles. Another example would be a cell in the intestinal wall which must collect nutrients on the luminal side and transport them through the cell to the opposite end where they can be delivered to the blood supply for transport to the rest of the body. Clearly, this cell must exhibit a polarized distribution of membrane proteins so that those necessary for sugar uptake are concentrated in the membrane facing the lumen and those needed to move the sugar out of the cell are located at the opposite side. Therefore, this cell polarity is extremely important for proper cell and tissue function.

Most organisms begin life as a rather symmetrical, spherical, single cell called an egg. During early development this egg divides many times and forms an embryo which exhibits much more intricate patterns (such as the polarized intestine) than were initially expressed by the egg. Determining how the egg controls the development of such patterns has been an area of active research over the past decade; there is evidence that the plasma (outer) membrane can influence cell polarity by driving ion currents through the cell.

Plant eggs. Much knowledge about this early development of cell polarity results from studies of plant eggs which exhibit no preformed axis. Among the most useful have been the eggs of brown algae *Fucus* and *Pelvetia*, common rockweeds found in the intertidal zone on both the Atlantic and Pacific coasts. The eggs develop by first secreting wall-softening and wall-building enzymes at one end. They then pump in potassium and chloride ions, and water follows these ions into the eggs to generate a large turgor pressure of about 5 atm (500 kilopascals). The eggs bulge (germinate) in the region where the wall has been softened by the secreted enzymes, thereby producing a structure which will develop into the rhizoid. This is the first morphologically obvious axis of polarity. The establishment of the axis of secretion, and thus the morphological polarity, can be influenced by a variety of environmental vectors, including light, temperature, and pH. The response to unilateral light is to germinate on the dark side; this is useful since the rhizoid outgrowth will form the holdfast for the plant and should form opposite the sun so that the plant can be anchored to the ground.

Transcellular ion currents. One can use this light response to orient the germination site while studying the transcellular ion current pattern associated with it by using an extracellular electrode technique called the vibrating probe. This instrument measures the small voltage gradients in the fluid just outside the egg which are generated by the movement of ions into or out of the egg's plasma membrane. Currents are detected around this egg as early as 30 min after fertilization and tend to enter at the dark hemisphere. The early spatial current pattern is unstable and shifts position, often with more than one inward current region. However, cur-

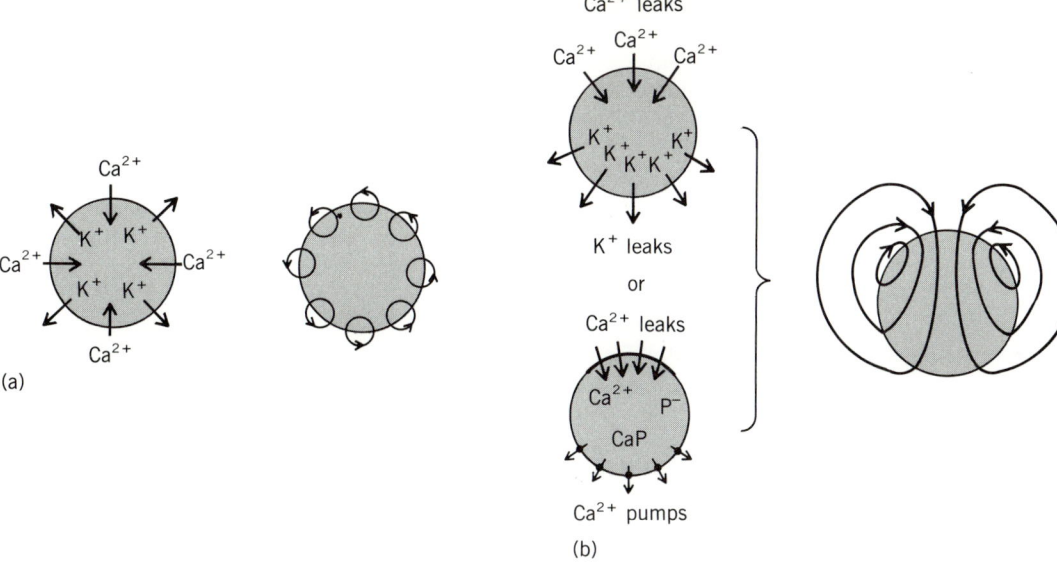

Fig. 1. Possible distributions of ion channels and pumps in a cell's plasma membrane with the resulting current pattern. (a) Uniform distribution of calcium (Ca^{2+}) and potassium (K^+) channels will result in many highly localized current loops. (b) Separation of the Ca^{2+} and K^+ channels (above) or Ca^{2+} channels and pumps (below) will result in the transcellular ion current pattern shown on the right.

rent enters mainly on the side where germination will occur and is usually largest at the prospective cortical clearing region where the rhizoid forms.

The current pattern observed during the 2-h period prior to germination is more stable and looks like the pattern illustrated in Fig. 1 on the right side. The site of inward current always predicts the germination site, even when the axis is reversed by light-direction reversal. The most likely hypothesis is that light receptors in the cell's plasma membrane control the distribution of open ion channels so that current enters the dark end.

In order to understand how such a transcellular current might be affecting the egg, one must first

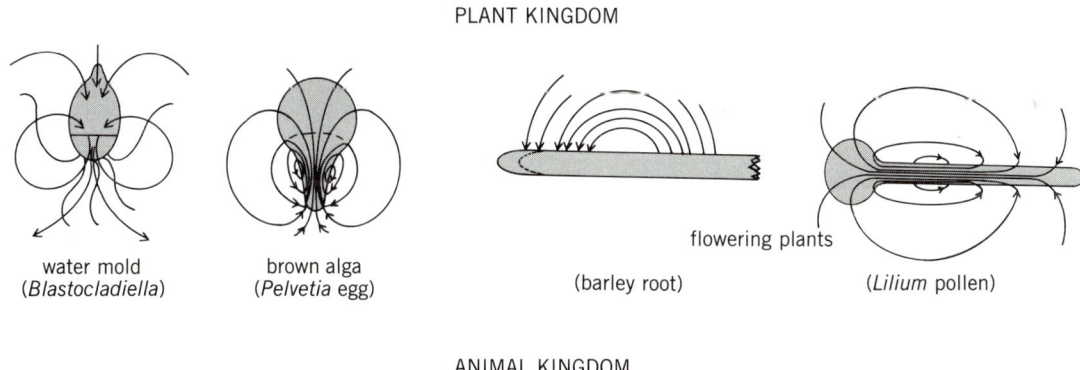

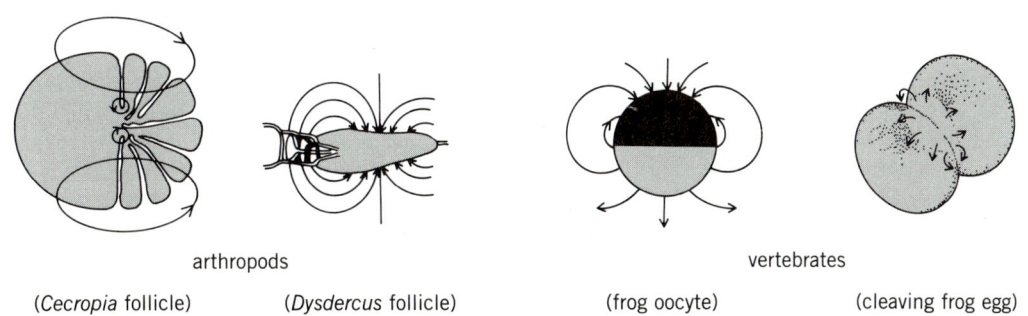

Fig. 2. Transcellular ion current patterns measured in plant and animal cells.

know what ions are carrying the current. Studies using artificial sea water with certain ions removed indicate that calcium influx and chloride efflux carry most of the inward current, with potassium efflux as the probable outward-current carrier. This has also been confirmed by directly measuring the movement of radioactive calcium tracer through the egg. This calcium influx will increase the local intracellular concentration of calcium, and high calcium has been demonstrated to stimulate vesicle secretion. Hence, the plasma membrane can directly influence the germination direction by determining the site of maximum calcium influx and thereby controlling the cell's polarity.

These transcellular ion currents are not a unique property of brown algae, but have been detected in every plant and animal cell and tissue investigated with the vibrating probe technique (Fig. 2). The transcellular current pattern is usually closely correlated with the axis of polarity. It would therefore appear that most cells do not have a uniform distribution of ion channels and pumps (Fig. 1a), but segregate channel types to varying degrees (Fig. 1b).

Moth egg. One of the most intriguing examples from the animal kingdom is the developing egg of the moth *Hyalophora cecropia*. This egg is connected to seven other cells called nurse cells which produce important molecules that are transported into the egg. This transport is polarized since most proteins will move from the nurse cells into the egg, but will not move from the egg back to the nurse cells. This polarized transport requires a voltage difference between the two cell types that is generated by a transcellular current which moves through the thin cytoplasmic connection between the nurse cells and egg. It has recently been shown that the movement of proteins between these two cells depends only on the net electrical charge of the proteins, and that the same protein can be made to reverse its transport direction by simply reversing its charge. This suggests that the egg uses the transcellular ion current to generate the voltage gradient which causes proteins to move according to their charge (in a manner similar to electrophoresis) from the nurse cell into the egg. Thus, the current controls the polarity of the protein movement, and this determines which cell will receive the products of the nurse cells.

For background information see DEVELOPMENTAL BIOLOGY; EMBRYOLOGY in the McGraw-Hill Encyclopedia of Science and Technology.

[RICHARD NUCCITELLI]

Bibliography: L. F. Jaffe and R. Nuccitelli, An ultrasensitive vibrating probe for measuring steady extracellular currents, *J. Cell Biol.*, 63:614–628, 1974; R. Nuccitelli, Ooplasmic segregation and secretion in the *Pelvetia* egg is accompanied by a membrane-generated electrical current, *Dev. Biol.*, 62:13–33, 1978; R. Nuccitelli, Transcellular ion currents: Signals and effectors of cell polarity. in J. R. McIntosh (ed.), *Modern Cell Biology*, vol. 2, pp. 451–481, 1983; R. I. Woodruff and W. H. Telfer, Electrophoresis of proteins in intercellular bridges, *Nature*, 286:84–86, 1980.

Primates

Although remarkably varied in their anatomy, locomotor patterns, diet, and social behavior, the living lemurs of Madagascar are only a remnant of a substantially more diverse fauna that disappeared some 2000 years ago, around the time the island was first colonized by humans. Some extinct lemurs differed little from their closest extant relatives, but others have no modern counterparts in either size or morphology. *Indri*, the largest living lemur, weighs about 13 lb (6 kg), or only as much as a medium-sized dog. By contrast, *Megaladapis edwardsi*, the largest extinct lemur for which a reliable weight estimate exists, may have been between 110 to 220 lb (50 to 100 kg)—the size of a large chimpanzee or small gorilla. Surviving lemurs are primarily arboreal, but the extinct forms were more diversely adapted. *Megaladapis*, for example, may have been an arboreal leaper, like the living koala bear of New Guinea, if its limb morphology is a good guide. *Archaeolemur* and *Hadropithecus* were convergently similar to ground-dwelling Old World monkeys in several respects, and may have been the ecological equivalent of baboons.

The oddest of the extinct lemurs, however, is *Palaeopropithecus ingens*. Previously one of the least known of the extinct primates of Madagascar, *Palaeopropithecus* is now represented by the most complete remains of an individual giant lemur ever discovered—an entire skeleton, discovered in 1983 at the cave site of Anjohibe by a paleontological expedition jointly organized by Duke University and the University of Madagascar.

Size. In the illustration, the Anjohibe specimen is shown mounted in a position that may have been habitually assumed by this species during some locomotor activities—hanging by three of its four extremities under a branch. Judging from this individual, *P. ingens* measured slightly more than 3 ft (about 1 m) from snout to rump and weighed from 88 to 110 lb (40 to 50 kg), which is the size of a rather large chimpanzee. Dental traits indicate that this lemur subsisted largely on fruit, leaves, twigs, and other plant material.

Arboreal activity. Limb disproportion is the single most striking characteristic of *Palaeopropithecus*. Among living primates in which the forelimb is markedly longer than the hindlimb only the gibbons and the orangutan display as great a degree of disproportion. In living primates very long forelimbs are highly correlated with arboreal activity, and in particular with the frequent use of brachiation, a method of progression in which the body is suspended from and propelled by the arms alone. Close study of the hip and ankle joints of *Palaeopropithecus* reveals that they were specialized for extreme mobility, and that the hindlimb could be used in a great variety of postures. Postural flexibility is important for arboreal animals, since tree trunks and branches are positioned at all angles to the pull of gravity. In sum, there is little doubt that this giant lemur was fully committed to life in the trees.

Features of the hands and feet set *Palaeopropithecus* apart from other arboreal primates, and suggest that it may have practised an unusual mode of locomotion in the trees. The hands and feet were permanently bent: the bones in the palms, soles, and digits are curved to such an extent that full straightening of the fingers and toes would not have been possible. Some other primates (apes, for example) have moderately curved digits, but none express the degree of curvature seen in *Palaeopropithecus*. In addition, the digits themselves appear to have had only limited independent movement, for the knuckle joints are so modified that it would have been difficult for the animal to spread its fingers or toes. Lastly, *Palaeopropithecus* may not have had functional thumbs, although firm evidence for this is lacking. Despite careful searching, no bones assignable to the first digital ray of the hand were recovered from Anjohibe. The bone that supports the big toe (the first metatarsal) was present in *Palaeopropithecus*, but the digit itself must have been short and very delicate since no large, spatulate phalangeal bones were recovered. Vestigial thumbs are found in several primates that frequently brachiate, but reduction of the big toe in both length and robusticity is unprecedented.

Locomotion. Whatever the precise form and degree of function of the first digits of *Palaeopropithecus*, the other characteristics discussed above imply that the hands and feet were chiefly adapted for use in positions where the natural partial flexion and relative immobility of the digits could be used to advantage. The most likely positions involve suspension of the body, with the extremities acting as grappling hooks. In principle, this adaptation could be useful for engaging in energetic brachiation, although the estimated body weight of *Palaeopropithecus* makes it improbable that this lemur would have moved at great speeds.

An alternative locomotion is slower, deliberate movement, mostly along the underside of tree branches, advancing one limb at a time. Locomotor patterns involving this style of progression are often seen in the Afro-Asian lorises and, among nonprimates, the tree sloths of South America. Some years ago Alan Walker showed that lorises, sloths, and *Palaeopropithecus* share several important anatomical traits that can be linked with locomotor activity. To these may now be added some special vertebral features, brought to light for the first time during investigation of the Anjohibe specimen. In comparison to other members of their taxonomic groups, lorises and sloths have an unusually high number of thoracolumbar vertebrae (22–23 in most lorises, compared to a primate average of 19–20; 20–24 in tree sloths, compared to 14 for terrestrial edentates). In addition, the bony dorsal projections for muscle attachment (spinous processes) are small in this region, instead of being long as in most of their relatives. These and related adaptations favor spinal flexibility over stability and strength, especially in the area above the pelvis. Lower-back flexibility seems to be important for slow-moving climbers and

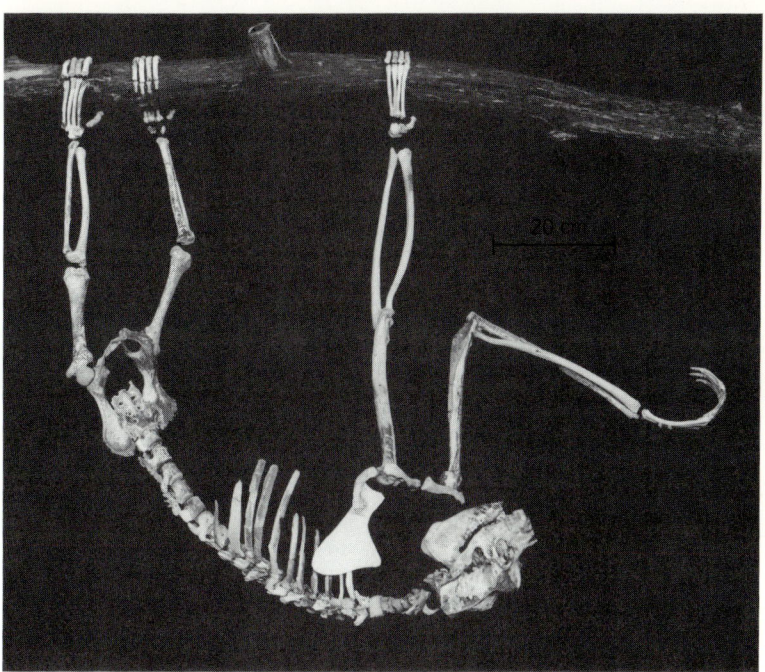

Anjohibe specimen of *Palaeopropithecus ingens*, mounted in probable habitual posture. Note limb disproportion and curvature of hands and feet. (*American Geological Institute*)

hangers—no doubt because it facilitates throwing the hindlimb into the odd angles required for this form of progression. By contrast, primates that frequently brachiate have comparatively few elements (16–19), an adaptation for bringing the limbs closer together and increasing spinal rigidity. This has the effect of both raising the center of gravity of the body and stabilizing the lower spine and hips, thereby reducing drag and twisting of the body during brachiation.

Palaeopropithecus is like the lorises rather than the apes in having a thoracolumbar column consisting of at least 22 elements, a record for lemurs. (Only 17 elements are shown in the illustration because this photograph was taken prior to the discovery of 5 more thoracolumbar vertebrae during additional excavations at Anjohibe in 1984.) The uniformly oriented spinous processes are comparatively large on cervical vertebrae, small on thoracic elements, and diminutive on lumbars—precisely the pattern found in lorises and sloths, and very different from that in either arm-swingers or quadrupedalists.

Although the evidence presented here shows that *Palaeopropithecus* exhibited a number of important similarities to slow-moving climbers and hangers, it should always be borne in mind that the locomotor repertoire of every living primate species is large, and varies as occasion and substrate demand. Even a species as specialized as *P. ingens* must have climbed, sat, scrambled, swung, and even walked on its hindlegs from time to time.

Extinction. What *Palaeopropithecus* and the other great lemurs could not do was survive the destruction of much of Madagascar's forest cover in the centuries following initial human occupation of the island. This ecological disaster affected all of the island's land vertebrates, but was most severely felt by the largest species, all of which disappeared. The present land vertebrate community is dominated by small, nocturnal species that reproduce well and live in isolated patches of forest and bush. But whether they will be able to persist in the face of uncontrolled land clearing and other destructive practices is a matter for grave concern.

For background information see LORIS AND ALLIES; PRIMATES; SLOTH in the McGraw-Hill Encyclopedia of Science and Technology.

[R. D. E. MacPHEE]

Bibliography: Alan Walker, Locomotor adaptations of past and present prosimian primates, in F. A. Jenkins (ed.), *Primate Locomotion*, pp. 349–381, 1974; W. L. Jungers, Hindlimb and pelvic adaptations to vertical clinging and leaping in *Megaladapis*, a giant subfossil prosimian from Madagascar, *Yearbook of Physical Anthropology*, 20:508–524, 1976; R. D. E. MacPhee et al., Team finds giant lemur skeleton in Madagascar, *Geotimes*, 29:10–11, 1984; E. L. Simons, *Primate Evolution*, 1972.

Printing

Recent advances in graphic arts include the extension of flexography to newspaper printing and the development of plate scanners for premakeready.

Flexography for newspaper printing. The simplicity of the flexographic process as well as its economic and technical advantages offers much to newspapers, and it is expected that this process will radically change the method of newspaper production over the next 20 years.

Background. When American newspapers expanded their operations into the suburbs in the middle 1970s, it became desirable to establish suburban printing plants for printing newspapers in the areas in which they were to be distributed. However, although the offset process was capable of providing the desired quality, the cost of new presses was too high. To deal with the problem, the American Newspaper Publishers Association (ANPA) Research Institute was asked to develop an inexpensive press design that could provide good printing quality at reasonable cost.

The ANPA Research Institute concluded that the flexographic process could be adapted to produce an economical design for newspaper use. Protypes were built, tests were made, and by 1977 a design concept was evolved. From this beginning, manufacturers have developed a variety of flexographic variations that are presently being used by newspapers and may provide, in the coming years, a radical change in newspaper production methods.

Advantages. The flexographic printing process seemed to have a number of potential advantages for newspaper printing. They included reduction in capital investment, operating costs, newsprint waste, operating noise levels, ink consumption, and maintenance costs; reduction in elimination of ink mist; and improvement in printing quality. With such promising advantages, the newspaper industry continued investigating the process in the hope that most of the benefits could be achieved in a reasonable period of time, and that flexography could provide the newspaper industry with a more efficient means of production at a lower cost.

A variety of flexographic variations have been developed and installed in newspaper plants. In general, there are two design types: a retrofit inking unit applied to an existing older press, and a complete press unit.

Technical aspects. For many years, two basic printing processes, letterpress and offset, have been used for printing newspapers. Although offset provided a better-looking image, the majority of metropolitan newspapers still printed with the older letterpress method. The major reason was the extremely high cost of replacing press equipment.

Letterpress uses a printing plate with the image in a raised relief. An inking system with a multitude of rollers transfers the ink fountain to the printing plate. The inked image then directly contacts the newsprint to transfer the image.

Offset uses a flat plate of minimal thickness that is chemically treated to accept either ink or a watery fountain solution. Two separate sets of rollers carry either the ink or the fountain solution to the plate. The chemistry of the system allows the plate image areas to accept the ink while repelling the fountain solution, and the nonimage areas to accept the fountain solution and repel the ink.

Flexographic printing for newspapers uses a resilient relief letterpress plate and a very fluid solvent ink. Therefore, rather than a long train of rollers to carry ink to the plate, the system can consist of a single intermediate roller, known as an anilox roller, to transfer the ink from the ink fountain to the plate. The simplicity of the system requires fewer moving parts, permits lighter frame construction, and yields many other engineering developments that make it simpler to operate and reduce the cost of the press and the energy to run it.

Inks. Flexographic printing has generally used thin inks for applications such as package printing. With this background, it seemed inevitable that experiments would proceed with water-based inks when flexographic printing was used in a newspaper mode. Early experiments showed that water-based inks improved the image quality considerably. Often, it was impossible to tell from the printed result if the newspaper had been printed by flexography or offset when water-based inks were used. In addition to the quality improvement, the quick-drying water-based inks also eliminated or considerably reduced ink setoff or ruboff. This is the cause of a major complaint of newspaper readers: the ink of the paper rubbing off on their hands.

Plates. The printing plates used in flexographic printing are modified versions of the photosensitive polymer plates that have long been used for newspaper letterpress printing. Changes have been made in the hardness of the polymer as well as in its thickness. However, these factors have not yet been fully standardized. The polymer thickness has been reduced from about 0.020 in. (0.51 mm) to about 0.014–0.016 in. (0.36–0.41 mm). It is believed that this thickness will probably be further reduced to 0.010–0.012 in. (0.25–0.30 mm) as more experimental printing continues. The advantages of a thinner polymer are twofold. It produces a better printing image due to the formation of a better dot structure and, as the cost of the plate is proportional to the polymer thickness, it reduces the final cost of the plate to the newspaper.

Future development. While flexographic printing has been adopted by some newspapers, its success has not been as rapid as had been predicted. However, many of the initially claimed advantages have already proven to be valid. The presses are considerably less expensive than other newspaper presses; there is a minimization of newsprint waste; maintenance is much cheaper; image quality can approach offset printing; and ink ruboff is almost completely eliminated.

Problems still exist, however. The majority of them have to deal with a lack of standardization and consistency. Excellent results, while possible, are not regularly produced, even in a single plant on a single flexographic press unit. Thus many newspapers have chosen to postpone using flexographic presses. However, a number of flexographic press units are being installed, the majority for major newspapers with skilled personnel and a commitment to reap the benefits of flexography. If most of the remaining problems are solved and if the testing newspapers find they are satisfied, it is anticipated that flexographic printing will become a major publication method for newspapers.

[ERWIN JAFFE]

Plate scanner premakeready. A plate scanner is a device for measuring the image area of a printing plate relative to the total area available. This information is used to preset the ink feeds on a printing press in order to achieve an acceptable printed result with a minimum amount of wasted material and time.

One of the important factors of makeready in color printing is the adjustment of the ink feed to satisfy the requirements of local areas on the printed sheet. Substantial adjustment in the amount of ink across the width of the press is possible by virtue of control of the gap spacing between the ink fountain blade and roller. The ink fed to the press locally is adjustable across the width of the press; however, there is no control around the cylinder. Therefore color separations must be produced so that the desired visual effect is achieved with a standard amount of ink. This does not preclude the necessity of adjusting the amount of ink to values beyond the standard to compensate for incorrect separations, last-minute editorial changes, and unfavorable distribution of the printed patterns on the printing plate.

In conventional makeready, an experienced press operator first views the printing plate and then adjusts the ink fountain zone controls to correspond with the area of coverage that is estimated to be needed in each key zone. For example, if the operator estimates that about 50% of the plate is covered with image in a given zone, the blade-roller gap is opened to about twice that for a zone that appears to have 25% coverage.

The press operator runs the press sufficiently to achieve dynamic stability, then examines the result with densitometer readings of a color bar being printed along with the job, or evaluates the appearance compared with a proof sheet or previously run copy. Individual ink feed adjustments are made as needed until the job is considered acceptable.

The estimate of area coverage when done visually is quite inaccurate, especially since solids, halftones, and type may be involved, and the corresponding key adjustments made are, in turn, highly approximate. This contributes substantially to the time and cost of makeready.

The plate scanners make use of the differences in reflectance between the image and the nonimage areas of the press plate. For example, if the background or nonimage has a reflectivity of 100 units and the image is 30 units, a 50% coverage at any given reading would be 50% between 30 and 100, or read as 65 units. Generally, many so-called slices or readings may be taken to cover the active area of the plate and the results averaged to give the effective image coverage of the plate for each key zone.

A plate scanner reads the image coverage of the plate relative to the key zones with a good level of accuracy. In addition, either in the plate scanner itself or in the interface to the press controls, the coverage data are manipulated so that various printing parameters are considered.

The factors affecting the relationship between coverage on the plate and the ink key setting are: paper absorption; ink strength; densitometric standard of the ink color; relationship between numerical setting of the key and the actual amount of ink fed, which may not be linear due to the hydrodynamics of the inking system; sweep adjustment or the overall control of the ink being fed from the fountain roller down into the ink roller train; and the amount of sideward vibration or oscillation of the drums which is used to prevent streaking, or a localized, undesired unevenness of the press inking system. Many plate scanners establish their operating parameters by the initial running of test plates and the statistical evaluation of data to relate the scanned information to the setting finally used under various conditions.

Unfortunately, the printing quality of the plate may have no relationship to the contrast between the image and nonimage of the plate or with the uniformity of the reflectance of the background or image areas of the processed plate. As a result, there is generally a substantial reduction in the accuracy of the reading when the image coverage is below about 10% in any given key zone. Up to the present, most offset plate manufacturers have not considered the optical appearance and uniformity important factors in their quality control.

Another problem that may exist at low coverage is the possible lack of reproducibility in the setting of the fountain key for very low settings. If a postage stamp–sized image is being printed next to a large solid-area coverage, it may not be possible to adjust the keys to achieve the proper ink film needed for each zone. A visual compromise must then be made.

It is not possible to predict the exact amount of savings resulting from the use of a plate scanner. Predicting the point in makeready at which a printed sheet is considered acceptable has a two-tiered level of indeterminacy. First, not all color combinations are equivalent in the ease with which they can be obtained and maintained. Second, criteria for acceptability vary with the function of the printed piece. For example, a color swatch of a beige tile in a manufacturer's catalog would impose a different order of complexity of color control compared to a landscape illustrating a travel brochure. Under the most favorable conditions, however, a plate scanner may reduce makeready time to that required for register, positioning, and in-line bindery and adjustments.

For background information *see* INK; PRINTING; PRINTING PRESS in the McGraw-Hill Encyclopedia of Science and Technology.

[PHILIP E. TOBIAS]

Bibliograpy: J. Kaivosoja, K. Simomaa, and T. Lehtonen, Intelligent plate scanner for offset presses, *TAGA (Technical Association of the Graphic Arts) Proceedings*, pp. 474–503, 1984; J. Kaivosoja, K. Simomaa, and T. Lehtonen, Plate scanner for presetting of an offset press, pt. 2: Measuring logics and experience, *Graph. Arts Finland*, 12(3):18–24, 1983; P. Tobias, R. Karch, and H. Mangen, The PLS plate scanner for the Baker Perkins press, *TAGA (Technical Association of the Graphic Arts) Proceedings*, pp. 104–114, 1983; D. Toth, Sheetfed press plate scanners, *Graph. Arts Mon.*, pp. 34–35, July 1984.

Protein

Green plant leaves are largely untapped resources of protein suitable for animal and human diets. In 1942, in response to the need for self-sufficiency in time of war, the use of fractionated leaves was proposed as a source of protein and other nutrients. Although basic fractionation methods were published in 1773, the rationale for studying green leaves as a protein source has only recently been accepted by research scientists and the agribusiness community. Leaf protein concentrates are now the subject of international symposia and vigorous research activity.

Forage crops are the major plants for which leaf protein concentrate production methods have been developed, but other plants are also potential sources. Depending on whether animals or humans are the intended consumer, many different varieties of leaf protein fractions can be prepared which range from mixed leaf proteins to homogeneous powders of certain proteins. Leaf protein concentrate production involves subjecting harvested leaf material to mechanical forces which disrupt the cell walls and membranes so that the cellular contents (the protein-containing cytoplasm) can be obtained in the form of a crude leaf juice. This juice is then processed further to yield products with desired purity, pigment composition, and nutritional value. Leaf protein concentrate and by-products of its production can be used for animal feed, in human diets as a protein source, or in specialized applications in clinical situations.

Sources of leaf protein Although the forage crop alfalfa has been used for a significant proportion of research and development on leaf protein concentrate production, recent studies have examined an array of plant species, including other legumes such as clover, lupines, and beans. In Australia the potential of mixed herbage, such as found in meadows, permanent pastures, and native grasslands, has been examined. Herbage from these sources has a protein content of 20–30% of total plant dry weight. However, less than 15% of the protein available from these sources is converted to milk or meat by conventional agricultural practices, the rest being undigested or ultimately excreted as urea by ruminants. Other crop plants whose leaves should be suitable for leaf protein concentrate isolation are those from which the fruits or roots are harvested. Estimates suggest that in Great Britain alone, as much leaf protein is wasted as is present in the harvested parts of pea, beet, potato, and other vegetables.

Numerous tropical species have been studied as potential sources of leaf protein concentrate, and several plants are known which yield leaf protein concentrate of equivalent quality to that of alfalfa. In principle, fields of common weeds and plants whose leaves may be highly toxic to animals are viable sources of leaf protein concentrate, provided that appropriate processing steps which remove the toxic substances are included during leaf protein concentrate production.

Products. Several products are obtained in the process of isolating leaf protein concentrate, and each of these products is of some economic value. Juice containing cytoplasmic components is separated from the fibrous leaf residue, which may be further processed to yield animal feed. The press cake obtained may be used directly for feed or as silage, field-dried for hay, or dehydrated to yield stable feed pellets. Depending on the details of the extraction procedures, leaf protein concentrate is removed from the leaf juice as crude green leaf protein concentrate or more refined white protein fractions which lack pigmentation. The deproteinized leaf juice, generally brown in color, has potential for use as fertilizer or as a medium for microbial growth to produce protein or other by-products.

Extraction. All types of leaf protein concentrate are isolated after first grinding leaves and expressing the leaf juice under pressure. The best current methods involve a two-stage process using a machine to macerate the leaves into a pulp which is then fed to a screw press to separate the juice from solid leaf residues. A variety of processes exist for subsequent treatment of the juice to extract the protein (2–5% of the juice). These processes involve steps such as heating the juice to different temperatures, acid treatment, or ultrafiltration. In the Pro-Xan process for alfalfa, heating the leaf juice to 85°C (185°F) for a short time coagulates about 50% of the dry matter present in the juice as a green curd, which is separated from the remaining brown juice by centrifugation. This product (Pro-Xan) and that obtained from heating the juice to only 60°C or 140°F (Pro-Xan II) are rich in protein (62% and 45%, respectively), chlorophylls, and xanthophyll. After removal of the Pro-Xan II product, the juice can be heated to 85°C (185°F) or acid-treated to yield a second protein fraction—a white leaf protein concentrate consisting primarily of fraction I protein, 90% of which is ribulose 1,5-bisphosphate carboxylase, the enzyme responsible for all photosynthetic CO_2 fixation, and the single most abundant protein in nature. The white leaf protein concentrate is of high quality and is comparable to soy protein isolate in protein content (about 90%).

Composition. Leaf protein concentrate products are a complex mixture of proteins and other components of the leaves from which they are extracted. Green leaf protein concentrate from a one-step high-temperature process contains most of the soluble leaf proteins as well as those insoluble proteins associated with the green photosynthetic membrane systems. The soluble proteins are partially fractionated by two-step heating processes, the higher-temperature extract enriched in fraction I protein while the lower-temperature fraction consists primarily of other soluble proteins and green chloroplast membranes.

The main classes of nonprotein compounds present in leaf protein concentrate are pigments, carbohydrates, lipids, and phenolics. Chlorophylls and their breakdown products are the most abundant pigments in green leaf protein concentrate. In addition, heat-stable carotenoids (β-carotene) and heat-labile xanthophylls (lutein, violoxanthin, and neoxanthin) are found in leaf protein concentrate; these pigments are easily degraded by oxidative and enzymatic reactions, with alkaline pH during processing favoring their preservation. Phenolic compounds, such as tannins and quinonoids, are also found. Leaf protein concentrate contains 1–5% carbohydrates, both free and covalently linked in glycoproteins. The most abundant reducing sugars found are arabinose, xylose, galactose, and traces of glucose. Fraction I leaf protein concentrate preparations generally contain some carbohydrates, but the crystalline protein from tobacco is carbohydrate-free. Ionic polar, nonionic polar, and nonpolar lipids are present in green leaf protein concentrate at 10–30% by weight. Simple lipids and glycolipids are more abundant than phospholipids.

Toxicity. Plant leaves of many species contain natural products which are toxic to animals and humans. Successful extraction and ultimate utilization for food or feed of leaf protein concentrate from a given plant depends on removal or inactivation of such compounds. In the case of alfalfa leaf protein concentrate products, saponins are present in amounts up to a few percent by weight. Among the reported effects of saponins in leaf protein concentrate are: depressed rates of animal growth, reduced palatability and thereby food intake, reduction of serum cholesterol levels, production of eggs with reduced cholesterol content, erythrocyte hemolysis, inhibition of smooth muscle activity, inhibition of cellular enzymes, and reduced cation absorption. Major classes of toxic compounds which could affect utilization of leaf protein concentrate include cyanogenic glycosides, sulfur-containing glucosinolates, nitrates, nitrites, oxalates, estrogenic isoflavonoids such as coumestrol, rare amino acids like canavanine and mimosine, solanine, tomatine, solasonine, and a variety of toxic alkaloids.

Nutritional value. The overall nutritional value of a given leaf protein concentrate depends on interactions between the protein and the leaf protein concentrate's other constituents. The protein quality (that is, amino acid balance and relative amounts of essential amino acids) is similar for leaf protein concentrate from most plants examined; however, differences in palatability and digestibility determine its food value. The protein quality of alfalfa leaf protein concentrate products are typical, being well balanced. Slight deficiencies in sulfur amino acids can be supplemented with free methionine and cysteine or with other protein sources. Green leaf protein concentrate could become an important source of vitamin A in some parts of the world as a result of its content of β-carotene, a precursor to vitamin A.

High temperature during leaf juice processing is probably the most deleterious aspect of current methods for leaf protein concentrate production, since it can cause color changes, breakdown of certain amino acids, and decreased solubility of leaf protein concentrate. Acid-precipitated leaf protein concentrate is a more soluble product, while reducing agents added during processing provide the greatest improvement in nutritional quality of leaf protein concentrate. Animal feeding experiments show that the best white leaf protein concentrate and fraction I protein preparations are of at least equal quality to casein as a dietary source of protein.

Utilization. Crude green leaf protein concentrate has been used as a feed supplement for poultry, swine, goats, and fishes. The press cake obtained during leaf protein concentrate production is also excellent for silage and ruminant feed. Green, decolorized, or white leaf protein concentrates have been used to supplement human diets in nations where protein deficiencies are common. Leaf protein concentrate can be used as a milk substitute in beverages or mixed in food dishes (such as soups and stews), assuring potential acceptance in different cultures. Research in India, Pakistan, and Nigeria shows that leaf protein concentrate is effective as a protein source in children's diets and can alleviate symptoms of kwashiorkor (protein-calorie malnutrition). Palatability of white leaf protein concentrate products permits its use in foods where avoidance of flavor alteration is desirable.

Fraction I protein is of very high quality and has many potential uses since it is a pure protein with little or no contamination by other leaf components. Tobacco fraction I protein is equivalent to casein as a protein source in rat diets. Its high solubility permits use as a solute in beverages in which it could be an effective milk substitute for infants, a component of weight-control formulations for adults, or a well-defined protein source for parenteric therapeutic use. Fraction I hydrolysates may also be suitable for use in nutrient mixtures for intravenous feeding.

For background information see FOOD ENGINEERING; FOOD MANUFACTURING; PROTEIN in the McGraw-Hill Encyclopedia of Science and Technology.

[DON P. BOURQUE]

Bibliography: N. W. Pirie, Green leaves as a source of proteins and other nutrients, *Nature*, 149:251, 1942; N. W. Pirie, *Leaf Protein and Other Aspects of Fodder Fractionation*, 1978; L. Telek and H. D. Graham (eds.), *Leaf Protein Concentrates*, 1983.

Quantum mechanics

Quantum mechanics describes the behavior of microscopic particles (electrons, protons, photons, and so forth) at the atomic and subatomic scales. A typical example of a well-known quantum-mechanical process is the tunneling of a microscopic particle through a potential barrier, when the particle has insufficient energy to jump over the barrier. Such a process has no analog in classical mechanics, where it is forbidden by conservation of energy. It is an intriguing question as to whether in certain circumstances such phenomena can be extrapolated to objects of macroscopic dimensions, when all the complex interactions of the constituent particles with the environment are taken into account. Usually the calculated tunneling probabilities are very small so that observation of such a process is impossible. There are systems, however, in which macroscopic quantum tunneling might be observed.

Interrupted superconducting ring. Recently it became clear that experiments relevant to macroscopic quantum tunneling could be performed at very low temperatures in superconducting rings interrupted by a low-capacitance Josephson junction. If an external magnetic field of appropriate strength is applied perpendicularly to the ring and given suitable junction parameters, this system has two metastable states corresponding to supercurrent states with a clockwise and counterclockwise direction of the current. These states, usually called magnetic flux states, are separated by a potential barrier. The height of the barrier is determined by the maximum supercurrent of the junction, the self-inductance of the ring, and the external magnetic field. The total magnetic flux passing through the ring can be measured with a very sensitive flux meter, a SQUID (superconductive quantum interference device), as a function of time. Intrinsic magnetic flux transitions are observed from one metastable flux state to the other, and vice versa; in other words, a weak persistent supercurrent switches stochastically from one direction to the opposite direction. At relatively high temperatures, the rate at which these flux transitions occur is dominated by thermal activation over the barrier. Thermal activation, however, decreases very rapidly as the temperature is lowered. If the junction interrupting the superconducting ring has a very small capacitance, such a stochastic flux transition behavior is observed even at low temperatures, where the product of the absolute temperature and Boltzmann constant is very much smaller than the barrier height. At sufficiently low temperatures, the transition rates become temperature-independent. This flux-transition phenomenon at very low temperatures can be interpreted as due to macroscopic quantum tunneling. Each of these flux states is macroscopic in the sense that it involves the cooperative motion of all the microscopic conduction electrons around the ring.

An electrical analysis of the superconducting ring system can be made for the case that the system approaches the stationary state. If the system is not in equilibrium, it is necessary to take into account, besides the contribution of the supercurrent, also the small contributions of the displacement current and normal current. This leads to behavior which resembles the motion of a classical particle with damping in a double potential well. The dynamic variable in this analysis is the total magnetic flux passing through the ring. The capacitance of the junction is equivalent to the mass of the pseudoparticle, and the damping is due to the motion of flux which gives rise to normal currents (Foucault damping). The potential consists of two terms: the coupling energy of the Josephson junction, which is determined by the supercurrent through the junction; and the magnetic energy of the circulating current due to the self-inductance of the ring. Macroscopic quantum tunneling will be favored by a very small capacitance of the junction (very small effective mass), when the de Broglie wavelength is large compared with the width of the barrier. There is, however, an important difference between microscopic and macroscopic quantum tunneling, since the macroscopic systems are intrinsically dissipative even at absolute zero temperature. The generally accepted conclusion is that dissipation reduces the tunneling rate for macroscopic systems, although not necessarily to the extent of making it unobservably small.

Current-biased Josephson junction. Macroscopic quantum tunneling has also been observed in single low-capacitance Josephson junctions biased by an external current. In this case the electrical analysis leads to behavior which resembles the motion of a macroscopic particle in a tilted periodic sinusoidal potential, the so-called washboard potential. If the system is underdamped, a transition out of the potential well will lead to continuous motion of the system down the washboard potential that will result in the switching of the junction to a finite voltage state. If such a junction is driven repeatedly by an external current source, the transition probabilities for switching out of the metastable superconductive state to a finite voltage state can be measured. Again at high temperatures the transition rates are dominated by thermal activation over the barrier, and at low temperatures the transition rates become temperature-independent. By studying junctions with different effective shunting resistances, it is even possible to show that dissipation reduces the tunneling rate for macroscopic systems.

Macroscopic quantum superposition. Experiments with a superconducting ring interrupted by a weak junction can also be used to answer the question whether a macroscopic system can be described by a wave function in the form of a linear superposition of macroscopically distinguishable flux states. A well-known analog in the microscopic physical world is the nitrogen inversion in the ammonia molecule (NH_3). The two metastable states of the ammonia molecule, with the nitrogen atom above and below the plane of the hydrogen atoms, are comparable to the two metastable flux states in the ring device. In both cases there is a double potential well separated by a potential barrier which is impenetrable in classical physics. For the ammonia molecule, this leads to a splitting of the ground-state energy into a doublet: two stationary energy states with different wave functions and a small energy difference. If the nitrogen atom is initially prepared, for example, in the position state above the plane determined by the hydrogen atoms, the nitrogen atom moves regularly backward and forward between the two potential wells with a frequency equal to this energy splitting divided by Planck's constant. The system is in a superposition of both energy states.

If the same laws are valid on a macroscopic level for the superconducting ring system with a weak junction, the magnetic flux in the ring can be expected to hop regularly between the clockwise and anticlockwise direction of the supercurrent with a frequency corresponding to the energy splitting in the ring device. However, experiments on this effect of macroscopic quantum superposition, also referred to as macroscopic phase coherence, have to be performed at very low temperatures in the millikelvin range or even lower, where the thermal energy is smaller than the splitting of the ground-state energy levels. The externally applied flux to the ring has to be precisely half a flux quantum in order to have a symmetrical potential double well. Furthermore, the system must be underdamped so that the interaction with the environment is sufficiently reduced. The experiment must be performed in such a way that the flux is not monitored continuously, as this will destroy phase coherence, but instead is measured during short time intervals which are interrupted by longer time intervals during which no measurements are performed. It is generally accepted that these conditions are very stringent.

So far, all attempts at observing macroscopic quantum superposition have failed. Macroscopic quantum tunneling is a necessary but by no means sufficient condition for macroscopic phase coherence. Experiments have demonstrated that macroscopic systems can show stochastic quantum tunneling with a total loss of phase coherence due to the interaction with the environment.

For background information *see* JOSEPHSON EFFECT; NONRELATIVISTIC QUANTUM THEORY; QUANTUM MECHANICS; TUNNELING IN SOLIDS in the McGraw-Hill Encyclopedia of Science and Technology.

[R. DE BRUYN OUBOTER]

Bibliography: *Proceedings of the International Symposium on the Foundations of Quantum Mechanics in the Light of New Technology*, Physical Society of Japan, 1984.

Radar meteorology

In the past decade there have been rapid advances in Doppler weather radar. A number of government laboratories have been actively testing this new technology to remotely measure air velocities both within storms and within clear air. As a result of this research, a wide variety of Doppler radar applications have emerged that can improve weather warnings and very short-period (0–6 h) weather forecasts. Because of the expected improvement in severe storm warnings, the United States government is funding the development of a national network of over 100 Doppler radars to be installed during the late 1980s. This program is called NEXRAD (next-generation weather Doppler radar).

Doppler effect. Radars presently used by the National Weather Service are able to measure only the intensity of the backscattered signal from precipitation particles. This returned power is called the radar reflectivity factor and is a measure of the precipitation rate. Doppler radar has the additional capability of measuring the differences between the frequency of the transmitted and received microwave signal. A frequency shift results when the scattering particles move relative to the radar. This is called the Doppler effect. This frequency shift can easily be converted to a speed value. The scatters are precipitation particles, insects, seeds, and changes in the atmosphere's index of refraction. These scatters essentially move with the wind, tracing the actual wind flow. Thus, a Doppler radar can also measure winds.

Doppler wavelength. Meteorological Doppler radars operate in the microwave frequency range, typically in the 1–10-cm wavelength range. Lower frequencies or longer wavelengths have the advantage of a larger velocity interval (Nyquist limits) over which velocities can be measured unambiguously. This velocity interval is given by Eq. (1), where PRF is

$$V_{max} = \pm PRF\, \lambda/4 \qquad (1)$$

the radar pulse repetition frequency (such as number of radar pulses emitted in 1 s) and λ is the radar wavelength. Since longer wavelengths make it possible to operate with relatively lower PRFs to obtain appropriately large Nyquist limits, a second advantage of longer wavelengths results in a relatively longer unambiguous radar range. The maximum distance that targets can be unambiguously observed is given by Eq (2), where C is the speed of light.

$$R_{max} = C/2\, PRF \qquad (2)$$

When targets are located beyond R_{max}, they return echoes after the next transmitted pulse, and echoes from close targets can be received simultaneously with those from distant targets. This overlaying of echoes makes interpretation difficult. A third advantage of longer wavelengths is less attenuation of the radar signal by precipitation. The primary disadvantage of longer wavelengths is that the diameter of the antenna required to maintain equivalent-size beam width increases proportional to the wavelength. The NEXRAD's will have wavelengths of 10 cm.

Wind measurement. Doppler radar can measure only wind velocity components that are parallel to the pointing direction of the radar beam, referred to as the Doppler radial velocity. In principle, two or more Doppler radars are required to measure the actual wind at any point in space and time. Determination of the actual wind is simply a matter of combining the radial velocities from the individual radars. Utilizing such techniques, it is possible to generate, in great detail, the evolving wind-flow pattern within a storm. Such studies during the past 10 years have done much to advance meteorologists' understanding of the structure and development of many different types of storms.

Multiple-Doppler techniques are presently impractical for routine weather forecasting because of the large number of Doppler radars that would be required for a national network. However, research field programs are showing that significant, short-period forecasting information can be obtained from a single Doppler radar.

A scanning Doppler radar will display a field of radial velocities representing velocity components of the wind in the direction that the antenna is pointing. Figure 1 shows a Doppler velocity display for a nearly uniform southeast wind. The radial velocities are at maximum while approaching and at maximum while receding when the antenna is pointed southeast and northwest, respectively, and are zero when pointing at right angles to the wind; that is, to the northeast and southwest.

When the radial velocities are plotted as a function of antenna azimuth angle, they trace a sine curve. It is easy to determine from the Doppler radar the wind direction and speed of a uniform wind field. When the wind flow is disturbed, as is the case in most storms, the Doppler display will be correspondingly complicated. Much of the success of Doppler radar for weather forecasting lies in its ability to measure strong velocity gradients. Divergence, convergence, and rotation of air exhibit distinctive signatures on a Doppler velocity display. Thus, even though the full-velocity vectors are not directly observable, the analyst is able to infer from the Doppler scalar field many important wind patterns. Typically, the Doppler velocity displays are in color, which greatly facilitates this interpretation.

Severe storm warnings. A most important application of Doppler radar, the detection and warning of severe thunderstorms, has been primarily developed and tested over the past 15 years by the National Severe Storms Laboratory in Norman, Oklahoma. This work indicated that many severe storms, that is, those with large hail, strong surface winds, or even tornadoes, showed large rotations at mid-heights within the storms. These circulations, called mesocyclones, had diameters of roughly 10 km (6 mi). It was found that the mesocyclone produced a distinctive signature on a Doppler radar display that was often visible many minutes before severe weather actually occurred at the ground. It was also discovered that tornadoes at relatively close radar range produced a much smaller but very intense rotation signature on the Doppler display, which is a direct indication of a tornado circulation. This signature was named the tornado vortex signature. Figure 2 is an example of both a mesocyclone and a

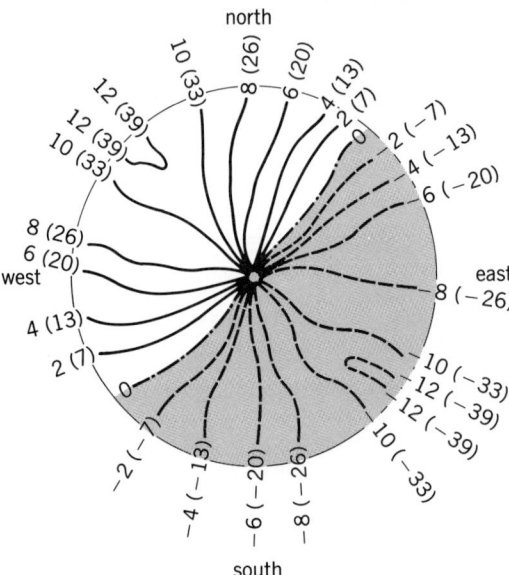

Fig. 1. Doppler velocity display of a relatively uniform wind field associated with a southeast wind of about 12 m/s (39 ft/s). The contours are radial velocities in m/s (ft/s). The broken-line contours and shading indicate approaching velocities, and the solid contours receding velocities.

tornado vortex signature observed with a tornado-producing storm in Oklahoma. The tornado vortex signature is evident by the very closely spaced couplet of 30 m/s (67 mi/h) approaching flow and the 50 m/s (112 mi/h) receding flow. The distance between these centers is approximately 1 km (0.6 mi). The mesocyclone has the same magnitude of approaching and receding velocities as the tornado vortex signature. However, the receding velocity center is located 8 km (5 mi) east of the approaching center. Utilizing Doppler radar, an operational test was performed in Oklahoma to forecast severe weather. The results were very encouraging. In fact, it was found that tornado warnings based on a Doppler radar had an average lead time of 20 min before actual touchdown. These findings, coupled with the need to replace the aging national network of conventional radar, resulted in NEXRAD, a cooperative effort by the Air Force, National Weather Service, and Federal Aviation Agency to develop and deploy a national network of Doppler radars.

Wind shear warnings. A recently discovered important application of Doppler radar is to warn airports of dangerous wind shear situations. Since 1964 there have been at least 27 civil airline accidents and incidents involving strong wind shears, resulting in 491 persons killed and 206 persons injured. It is believed that the wind shears were associated with strong vertical downdrafts of air from thunderstorms and even smaller cumulus clouds not producing lightning or heavy rain. The wind shear is produced when the downdraft hits the Earth's surface and spreads horizontally in many directions, just like water from a garden hose when it is pointed downward at the pavement. This meteorological event is called a microburst and is particularly dangerous to aircraft just at the time of takeoff or landing. An airplane will first experience an increase in airspeed followed by a drastic loss in airspeed and critical loss of altitude.

Experiments in 1982 and again in 1984 conducted by the National Center for Atmospheric Research in Boulder, Colorado, have shown that Doppler radar is an excellent tool for identifying and quantifying microbursts. A microburst produces a very distinctive diverging wind signature that is easy to identify on a Doppler radar display. Figure 3 is an example of the type of wind information that a Doppler radar could provide if it was located at an airport; it also shows the strength of the head wind and tail wind that an airplane would experience during a landing. In this example, the rapid change from a head wind of 7 m/s (16 mi/h) to a tail wind of 19 m/s (43 mi/h) is caused by a microburst and would produce a very serious situation for an aircraft on landing.

In addition to improving aircraft safety, the Denver experiments have shown that Doppler radar can be used to reduce airport delays and increase overall efficiency by providing forecasts of wind changes that will allow air-traffic controllers to anticipate wind-caused runway changes.

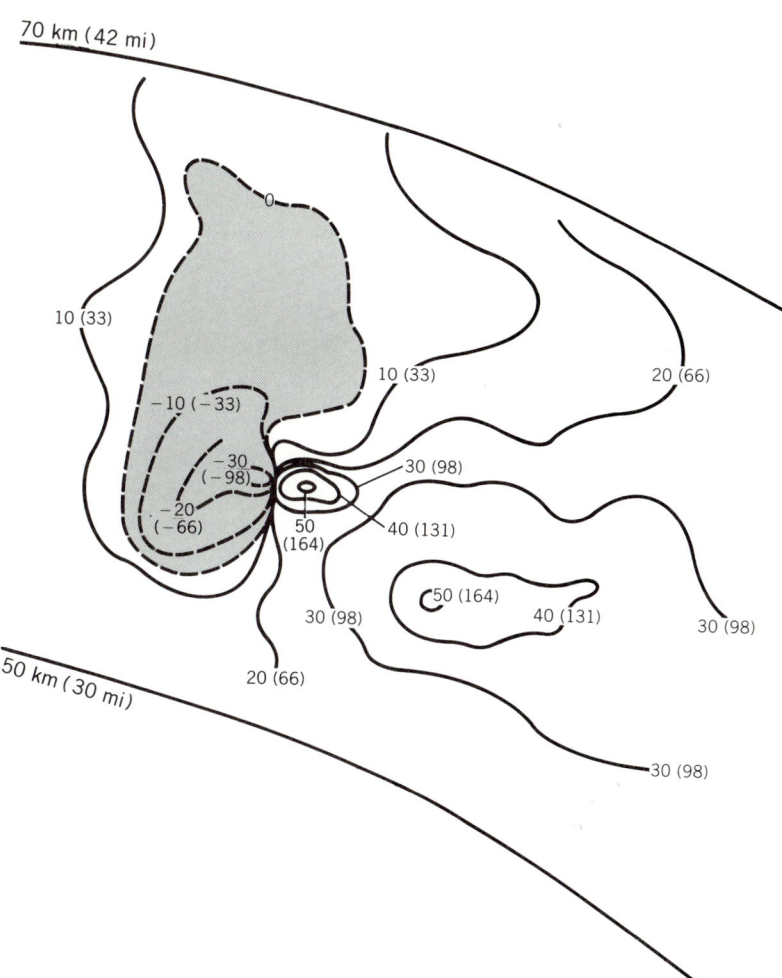

Fig 2. Doppler velocity display of a tornado-producing storm. The shading and broken-line contours indicate approaching velocities, and the solid contours receding. A tornado vortex signature is indicated by the closely spaced couplet (~1 km or 0.6 mi) of −30 m/s (−98 ft/s) (approaching) and 50 m/s (160 ft/s) (receding). The mesocyclone is represented by the larger-scale closed contours of approaching and receding radial velocities. Distances from the radar are indicated by the 50- and 70-km (30- and 42-mi) range marks. Contours are given in m/s (ft/s).

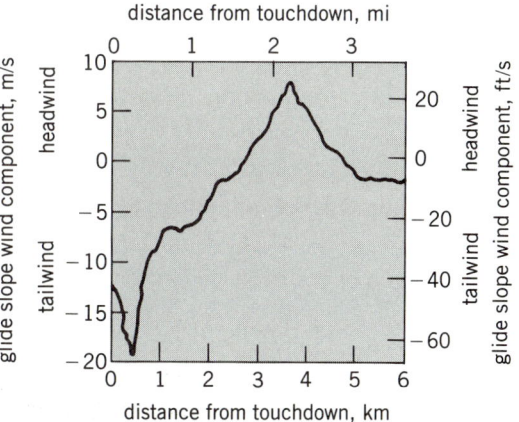

Fig. 3. Example of glide-slope wind component versus the distance from aircraft touchdown location that can be provided by a Doppler radar located at the airport. The wind is the component parallel to the plane's path, that is, the head wind or tail wind. The indicated 24-m/s (79-ft/s) wind shear in about 3 km (2 mi) was produced by a microburst.

It is expected that by the end of the decade Doppler radars will be deployed near major airports to improve aircraft safety and airport efficiency.

Thunderstorm forecasting. Previously, it was impossible to forecast with any precision the location and time that new showers and thunderstorms would develop. New findings show that Doppler radar is able to observe in the optically clear air wind-shift lines that trigger where and when showers and thunderstorms will form. These wind-shift lines represent regions where air is converging and being forced upward. When temperature and moisture conditions are suitable and the air convergence is sufficiently strong, storms will develop along these wind-shift lines. Storms are particularly likely to occur when two of these boundaries collide. These wind-shift lines are created by a large variety of mechanisms. The important point is that the Doppler radar can be used to monitor their movement and strength. The combined use of Doppler radar and cloud imagery data from satellites is particularly useful in forecasting thunderstorm formation. After storms have developed, the Doppler radar is very useful in predicting their movement, strength, and dissipation.

Winter storm forecasts. The short-period forecasting of winds, precipitation rate, and precipitation type during large-scale, winter-type storms can be improved with Doppler radar. A single Doppler radar can monitor the vertical profile of the wind above the radar during such episodes. Since these winds are critical to changes in storm strength, the forecaster can use this information to predict changes in precipitation rate. The Doppler radar can also be used to locate and monitor the movement of frontal systems, thus making it possible to make detailed forecasts of temperature, wind, and precipitation changes.

For background information see DOPPLER RADAR; FRONT; METEOROLOGY; RADAR METEOROLOGY; THUNDERSTORM; WEATHER FORECASTING AND PREDICTION; WIND in the McGraw-Hill Encyclopedia of Science and Technology.

[JAMES WILSON]

Bibliography: L. J. Battan, *Radar Observation of the Atmosphere*, 1973; K. Browning (ed.), *Nowcasting*, 1982; R. J. Doviak and D. S. Zrnić, *Doppler Radar and Weather Observations*, 1984.

Radioactive waste management

The United States and several European and Asian nations are currently conducting research to determine the feasibility of isolating high-level nuclear waste in subseabed geological formations. The concept is burial of solidified high-level waste or spent fuel, contained in high-integrity canisters, tens of meters below the deep-ocean floor within stable geologic formations. The canister, the clay sediments, and the ocean water act as barriers and delay migration of radionuclides while they decay to innocuous levels. If the concept proves feasible, a subseabed repository may be operated in the early twenty-first century, although no nation has as yet proposed to construct one.

High-level nuclear wastes are generated as part of the nuclear fuel cycle for commercial power reactors. By definition, these wastes are either the discarded fuel assemblies from power reactors or the wastes generated by the reprocessing of these fuel assemblies to recover ^{235}U, ^{238}U, and ^{239}Pu. Contained within high-level waste is short-lived ^{137}Cs and ^{90}Sr and longer-lived transuranic elements with half-lives exceeding 10,000 years, including ^{241}Am, ^{239}Pu, and ^{240}Pu. In the United States, fuel assemblies are in temporary storage awaiting establishment of permanent repositories for high-level wastes. Because the toxic lifetime of the radionuclides within high-level wastes exceeds the recorded span of human civilization, repositories using geological barriers and requiring no long-term monitoring have been given the greatest attention.

Within the United States, mined repositories on land will be the first disposal facilities considered for high-level nuclear waste. This is mandated by the Nuclear Waste Policy Act (NWPA), Public Law 97–425. NWPA also provides for further investigations of alternatives, one of which is subseabed disposal. Thus, the Department of Energy Office of Civilian Radioactive Waste Management manages a program to determine the feasibility of the subseabed disposal concept. Because of significant interest in the concept by various other nations, the Nuclear Energy Agency of the Organization of Economic Cooperation and Development has coordinated data exchange and cooperative research through the Seabed Working Group (SWG). The present membership of the SWG includes Canada, the commission of European Communities, France, West Germany, Japan, the Netherlands, Switzerland, the United Kingdon, and the United States.

Characteristics of deep seabed. The decision to consider seabed areas, which comprise 60% of the Earth's surface, as potential repositories was made because of several favorable characteristics of the ocean and the geological formations beneath the ocean floor. The abyssal clay sediment deposits under the deep ocean are some of the most stable and predictable geologic formations on Earth. Vast areas of sea floor are distant from the active edges of tectonic plates and, thus, not threatened by volcanic or earthquake activity. Lying deep beneath gyral, oceanic surface currents, these areas are covered with sediments formed by the slow, continuous accumulation of fine particles. The geologic record shows that neither the ice ages nor earlier climatic changes have altered the stability and uniformity of the deep-sea areas under consideration. In portions of the North Pacific, this slow accumulation has been shown to be continuous and uniform for 7×10^7 years. These areas have been termed midplate, midgyre (MPG) regions.

The characteristics of low strength, low permeability, and high adsorbancy combined to make deep-sea clay sediments excellent geologic barriers to the release of radionuclides. If geologic disturbances do occur within a subseabed repository, the lesser strength of the sediments will prevent the formation of cracks or voids. The very small, uniformly sized grains which compose the clay sediments found within midplate, midgyre regions tend to strongly absorb many radionuclides and are relatively resistant to the natural flow of enclosed pore waters.

The areas under investigation are remote from human activities and contain few, if any, useful resources. The potential repositories lie at water depths of 3–3.6 mi (5–6 km) and are distant from exploitable fisheries. In the areas of interest, the only minerals present are contained in manganese nodules which have been found to contain small amounts commercially important elements of copper, nickel, and cobalt. For these reasons, accidental disturbance to a repository by future generations is unlikely.

Dispersion and dilution of noxious materials occurs naturally within the oceans. While oceanic circulation is much slower than atmospheric circulation, it does act to disperse tracers into the enormous volume of the world ocean on time scales of 10^2–10^3 years.

Present status. While the characteristics of the deep seabed suggest that suitable repositories may be found beneath the ocean bottom, many related scientific and engineering questions have to be answered before the concept can be judged feasible. First among these questions is whether or not sites with no significant geological or geophysical flaws can be located in order that more site-specific studies can be carried out. Three locations in the North Pacific and two in the North Atlantic (Fig. 1) have been identified through a systematic evaluation of geological, geophysical, oceanographic, and resource data. Based on generic site-selection criteria, each of these locations appears to be geologically stable and predictable, as well as structurally simple and remote from human activities.

Studies have shown that oxidized deep-sea clay sediments within these locations should constitute a major barrier to radionuclide migration. Calculations based on radionuclide movement by diffusion, taking into account radionuclide adsorption onto sediments, suggest that these sediments form an extremely effective barrier to the migration of cationic radionuclides. The sediments would also serve as a slow release valve to oceanic waters for some very long-lived radionuclides which occur in low concentrations in high-level wastes. Further calculations have shown that the rate of pore waters convection within certain deep-sea sediments would not impair a subseabed repository.

Although deep-sea clay deposits in their natural state appear to be excellent barriers to the release of many radionuclides, the emplacement of waste canisters which generate heat through radioactive decay may alter conditions within the sediments of the repository sufficiently to increase the predicted release of radionuclides. For example, the heat generated by radioactive decay may speed up the rate of corrosion of the waste canister. It is estimated that each canister will generate about 1.8 kW of heat 5 years after emplacement, raising its surface

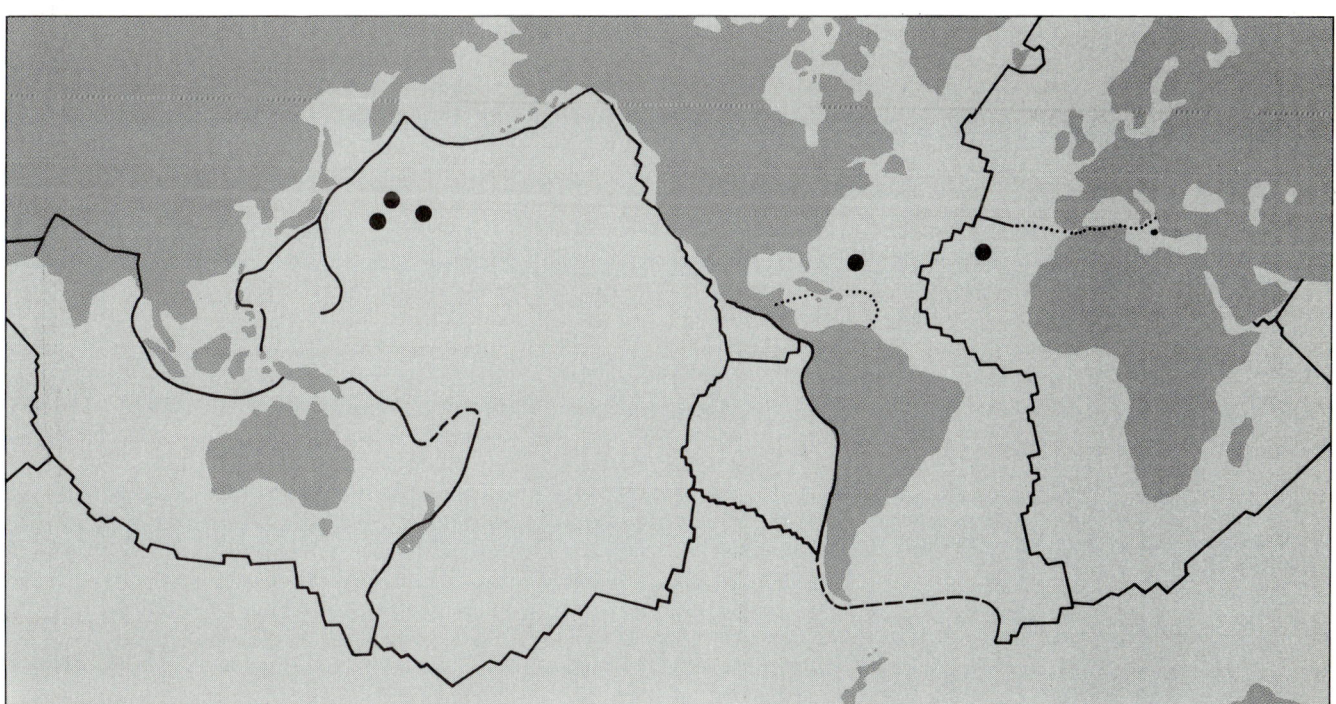

Fig. 1. Locations (solid circles) within the North Atlantic and North Pacific oceans selected for further consideration. Each location is distant from tectonically active plate boundaries within geologically stable and predictable clay sediments.

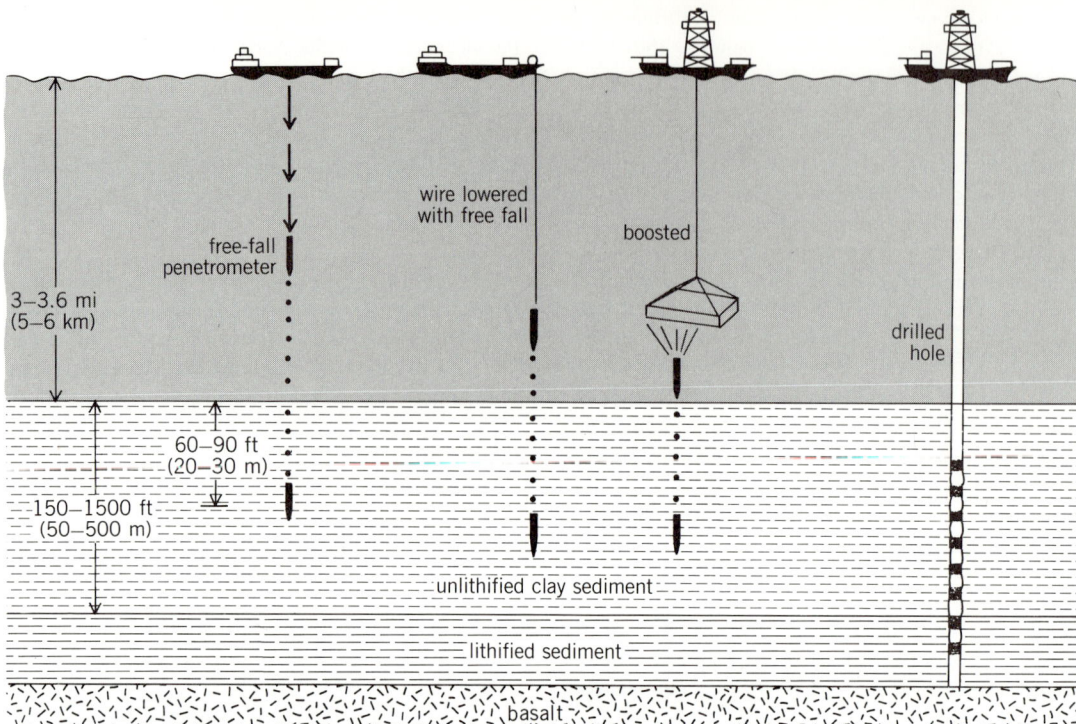

Fig. 2. Techniques that might be used for the emplacement of radionuclide-bearing waste canisters within subseabed geological formations. The clay sediments which compose these sediments will be a primary barrier to the release of radionuclides to the environment.

temperature to about 480°F (250°C). Also, the hole caused by emplacement may not close completely, thereby forming an artificial pathway for release of radionuclides to oceanic waters.

The decision of whether the concept of subseabed disposal is feasible will be made by the late 1980s. This decision will be based upon the properties and characteristics of sea-floor sediments, ocean waters, and the marine ecosystem. These properties and characteristics will be embodied within complex models describing the behavior of the barriers within a subseabed repository and the pathways to humans which might be followed by radionuclides. The final decision of feasibility will be based upon comparisons of the predictions from these models with safety and health guidelines established by national and international regulatory agencies. Verification of the soundness of these models is a fundamental part of both the United States and SWG research efforts.

To understand the effects that a subseabed repository operation, including accidents or faulty emplacement, may cause on both the marine ecosystem and human health, detailed modeling of the marine food web and the deep-sea carbon flux is necessary. To support this modeling, field investigations are being carried out to determine the structure and rates of metabolism of the bottom-dwelling communities in the oceans, the biology of deep-sea mobile scavengers, the faunal composition of the midwater nekton, the rates of microbial processes, and the radiation sensitivity of deep-sea fauna. Important to this biological modeling is accurate modeling of how released radionuclides would be stirred and mixed by the oceans. Linked to these biological models are models of the physical dispersion of radionuclides by oceanic circulation and mixing. These models are local, regional, and global in scale. Site-specific parameters are just now being gathered to support this modeling of oceanic circulation.

If a subseabed repository is constructed, it will likely occupy an area of 40–4000 mi^2 (10^2–10^4 km^2) on the sea floor. Waste canisters will be evenly spaced about 300 ft (100 m) apart to avoid interaction of their thermal fields. Canisters will be emplaced in a controlled manner (Fig. 2). Working within a network of acoustic navigation beacons moored to the sea floor, a waste disposal ship will lower each canister to a height several hundred meters above the sea floor. After release, the canister will accelerate to terminal velocity in sea water and penetrate the sediments. Alternatively, a hydrodynamically stable waste canister may be released at the sea surface and allowed to fall the entire 3–3.6 mi (5–6 km) before entering the sea floor. Modeling studies as well as in-place experiments indicate that, in using either method, penetration to 90-ft (30-m) depth beneath the sea floor is feasible for the sediments being studied. Deeper burial may be achieved by providing an energy source to boost each canister into the sediments or by drilling deep holes into the abyssal clays, emplacing several canisters within each hole, and backfilling each hole to restore the sediment barrier.

A titanium alloy (TiCode 12) has been under study for the waste canister because of its strength and resistance to corrosion. Calculations indicate that such a canister will remain intact for 100–300 years when emplaced in deep-sea sediments. The waste form within the canister will probably be a borosilicate glass, similar to that chosen for land-based repositories; however, the proportions of heat-generating radionuclides within the borosilicate glass may differ depending upon the sensitivity of deep-sea sediments to heat.

Major problems. Several major problems have been identified which, at a later time, could prove the concept of subseabed disposal unfeasible. The safe and reliable emplacement of waste canisters and their recovery in the event of accidents may not prove possible. Greater-than-expected velocities of natural pore water flow may make it impossible to find suitable sites. Once the canister is emplaced in the sea floor, chemical reactions between the waste form and the heated sediments surrounding the canister may alter the chemical nature of the radionuclides. This may render invalid the predictions of radionuclide migration based on the chemical species within the original waste form. Verification of the validity of all the parts of the complex numerical models upon which the ultimate decision of feasibility will be based may prove impossible. Use of the subseabed as a repository by one or more nuclear energy–generating countries may prove unacceptable to nonnuclear states regardless of the technical feasibility. Under these circumstances, it may be impossible to establish an international governing agency to approve and regulate the use of the subseabed for these purposes. Given these major problems, research continues on the feasibility of the concept of subseabed disposal, and no nation has yet proposed the construction and use of such a repository.

For background information see RADIOACTIVE WASTE MANAGEMENT; TITANIUM METALLURGY in the McGraw-Hill Encyclopedia of Science and Technology.

[EDWARD P. LAINE]

Bibliography: W. MacLeigh (ed.), High-Level Nuclear Wastes in the Seabed issue, *Oceanus*, vol. 20, no. 1, Winter 1977; P. K. Park et al., (eds.), *Wastes in the Ocean*, vol. 3: *Radioactive Wastes in the Ocean*, 1983; A. J. Silva and R. C. Chaney (eds.), Subseabed Disposal Program issue, *Mar. Geotechnol.*, vol. 5, no. 3/4, 1984; *The Subseabed Disposal Program, 1983: Status Report*, Sandia National Laboratories, SAND 83–1387, October 1983.

Radionuclides, cosmic-ray–produced

Cosmic rays, mostly high-energy protons that come from stars, constantly bombard the Earth and the rest of the solar system and induce nuclear reactions in a minute fraction of the atoms they strike. The radioactive products of the nuclear reactions, for example, ^{14}C, are called cosmogenic radionuclides.

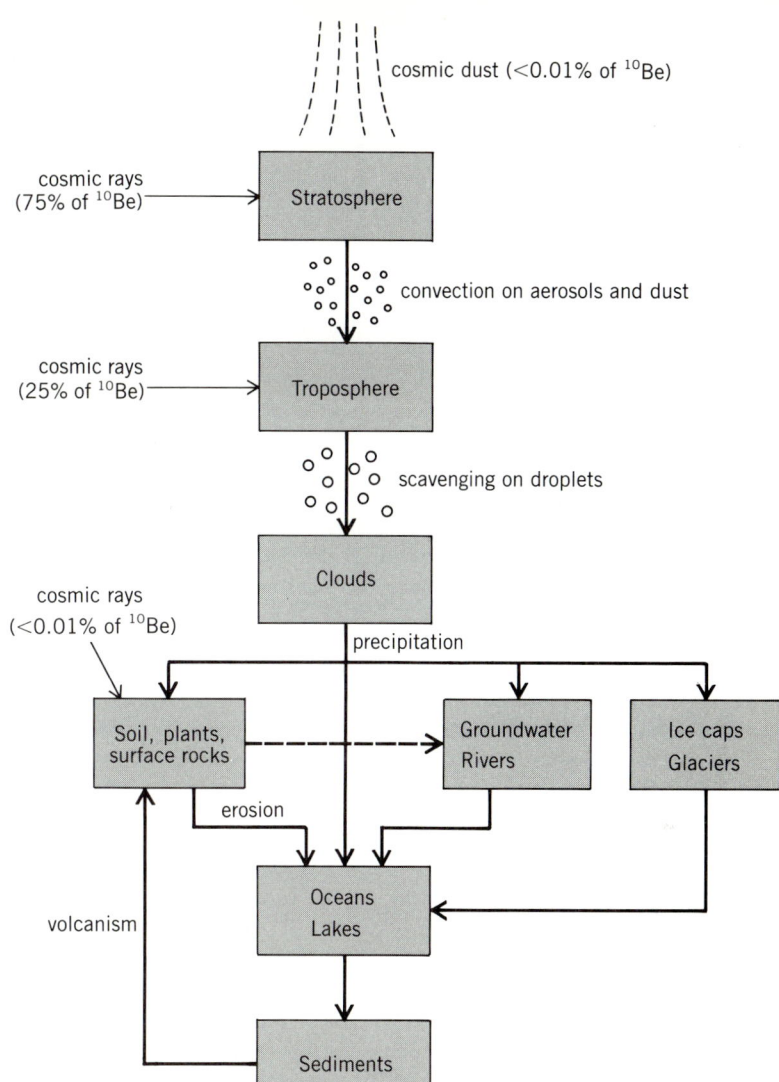

Production and transport of ^{10}Be through terrestrial reservoirs.

The illustration shows schematically the terrestrial reservoirs in which one of the less familiar cosmogenic radionuclides, ^{10}Be, can be found.

The cosmogenic radionuclides listed in the table have two main applications, as clocks and as tracers. Use of the longer-lived isotopes has been limited because of the experimental difficulties associated with measuring their extremely low decay rates. The recent development of an alternative measure-

Long-lived cosmogenic radionuclides

Isotope	Half-life, years	Representative samples
^{14}C	5.7×10^3	Trees, bones, shells
^{36}Cl	3.0×10^5	Groundwater, ice
^{26}Al	7.2×10^5	Impact glasses, sediments, meteorites
^{10}Be	1.6×10^6	Sediments, igneous rocks, manganese nodules
^{53}Mn	3.5×10^6	Extraterrestrial materials
^{129}I	1.6×10^7	Extraterrestrial materials

ment technique known as accelerator mass spectrometry has dramatically improved the sensitivity with which the longer-lived cosmogenic radionuclides can be detected. The accuracy of the new method now approaches or surpasses that obtainable by conventional decay counting.

Clocks. Samples may accumulate the cosmogenic radionuclides directly as a result of cosmic-ray bombardment or indirectly as a result of transport from a source region, usually the atmosphere. In the former case the amount of a cosmogenic radionuclide, R, will grow slowly toward a maximum or saturation value, R_0, according to Eq. (1), where λ is the decay constant for R and t is the exposure age. The

$$R = R_0 (1 - e^{-\lambda t}) \qquad (1)$$

derivation assumes the constancy of the cosmic-ray flux. If R_0 is known from analyses of samples in which R has reached saturation or from theoretical considerations, an exposure age can be calculated.

Meteorites contain relatively large quantities of the cosmogenic radionuclides as a consequence of exposure to cosmic rays for periods lasting up to several hundred million years. Exposure ages may help to link a meteorite to its parent body. Measurements of the ^{26}Al and ^{10}Be contents of a lunar meteorite recovered from the Antarctic indicate an unusually low exposure age compared to those of other meteorites thought to come from the asteroid belt. The short exposure age is consistent with the nearby, lunar origin. The shergottites, nakhlites, and Chassigny meteorites (SNC) compose a rare group of stony meteorites which some meteorite specialists believe came from Mars. The ^{10}Be and ^{21}Ne exposure ages of the shergottite, nakhlite, and Chassigny meteorites form three groups, a result that buttresses other evidence for their common origin and implies that three independent collisions brought them to Earth. Iodine-129 has a half-life of 1.7×10^7 years. By determining ^{129}I in meteorites, a feat possible at present only with accelerator mass spectrometry, it will be possible to look further back into their exposure histories.

The cosmic rays that completely traverse the Earth's atmosphere can produce appreciable quantities of the cosmogenic radionuclides in surface matter. Libyan Desert glass, which may have formed in an impact-related event, contains 3×10^6 and 9×10^6 atoms per gram of ^{10}Be and ^{26}Al, respectively. These results imply a total surface exposure of 0.5×10^6 years. In principle, tumbling ages for rocks overturned by landslides or earthquakes can be obtained with similar measurements.

The dating of soils millions of years old poses a problem toward whose solution the measurement of ^{10}Be and ^{26}Al may prove useful. Although production occurs mainly in the atmosphere, precipitation transports the two isotopes more or less quantitatively and continuously to the soil column, thereby satisfying, in effect, the assumptions built into Eq. (1). Certain soils from Merced, California, have ^{10}Be ages estimated between 0.04 and 3×10^6 years. Forthcoming studies of cosmogenic radionuclide geochemistry will help to identify other soils amenable to dating.

A second kind of age can be obtained for entities that collect the cosmogenic radionuclides produced elsewhere until death or burial stops the uptake. Equation (2) assumes that the sequestration of the

$$R = R_0 T e^{-\lambda t} \qquad (2)$$

system ends all losses and gains of R except those due to radioactive decay, where t represents the time elapsed after the sample became isolated, R_0 represents the production rate of R in the source region and is assumed constant, and T is the efficiency with which R was transported to the sample. To circumvent the need to specify T, Eq. (2) is usually divided by a second equation in which T or a quantity almost equal to it also appears. The second equation may apply to a stable isotope of R, to another cosmogenic radionuclide that resembles R geochemically, or to R itself but in another sample. In the case of ^{10}Be in manganese nodules, for instance, the second equation refers to the most recent deposit of ^{10}Be, that is, that of the outermost layer, as shown in expression (3a), which yields Eq. (3b).

$$[^{10}\text{Be}_{\text{sample}} = {}^{10}\text{Be}_{\text{atmosphere}} T e^{-\lambda t}] \div$$
$$[^{10}\text{Be}_{\text{outermost}} = {}^{10}\text{Be}_{\text{atmosphere}} T \qquad (3a)$$

$$\frac{^{10}\text{Be}_{\text{sample}}}{^{10}\text{Be}_{\text{outermost}}} = e^{-\lambda t} \qquad (3b)$$

The general area of ^{14}C dating has profited from the greater sensitivity afforded by accelerator mass spectrometry. Scarce fragments of cultivated plant remains have been dated to help evaluate the controversial evidence for early horticulture. The Yuha *Homo sapiens sapiens* skeleton has recently been shown to have a ^{14}C age of 4000 years before present (BP). This new result may supersede several age estimates of 20,000 years B.P. inferred from other data. Flecks of charcoal from the Wasatch Fault in Utah presumably formed in forest fires. By dating the charcoal associated with different faulting events, a history of earthquake recurrence intervals may be reconstructed.

The isotopes of ^{10}Be and ^{36}Cl have been used in tandem to obtain relative ages for polar ice. Results for Antarctic samples indicate that ice from the Yamato Mountains is probably 4×10^5 years younger than that from the Allan Hills.

The ^{36}Cl present in groundwater comes from three main sources: spallation of atmospheric nuclei by high-energy cosmic rays; neutron activation of ^{36}Ar, again in the atmosphere; and neutron activation of ^{35}Cl, mainly in rocks. The neutrons are secondary products of the cosmic-ray interactions. In selected cases, one source predominates. The ^{36}Cl contents of water samples from the Great Artesian Basin, for example, decrease regularly with distance from the recharge region, a result which suggests major importance for the atmospheric component. Flow rates inferred from the ^{36}Cl contents agree with estimates based on hydrodynamic models.

A number of marine sediments have been analyzed for ^{10}Be, with a view toward dating. The results for manganese nodules and crusts show a steady, exponential decrease with depth, consistent with constant deposition rates for periods lasting several million years. Hiatuses in the rates of ^{10}Be deposition evidently correlate with major changes in oceanic circulation patterns.

Tracers. Historically the shorter-lived cosmogenic radionuclides have served to monitor a variety of atmospheric, terrestrial, and oceanic transport processes and, to a lesser extent, recent short-lived fluctuations in the cosmic-ray flux. The uses of longer-lived cosmogenic radionuclides reflect a slightly different emphasis. Their applications as tracers can be divided into three broad classes: those in which the cosmogenic radionuclide contents of samples relate to the long-term behavior of the cosmic-ray flux; those in which the patterns of cosmogenic radionuclide abundances help to reveal the samples' places of origin; and those in which the cosmogenic radionuclide contents give mixing rates or residence times.

Cosmic-ray flux. The flux of cosmic rays incident on the solar system may change in time. Such changes would violate the assumption of constancy underlying Eq. (1), used to calculate the exposure ages of lunar rocks and meteorites; inconsistencies in the exposure ages based on different cosmogenic radionuclides would follow. In fact, all the cosmogenic radionuclides but ^{26}Al give a self-consistent set of exposure ages, a result suggesting that the average galactic cosmic-ray flux varied by less than 50% over the last $0.5-1 \times 10^7$ years. In this application the cosmogenic radionuclide contents of extraterrestrial materials suffer in that they integrate production rates over relatively long times and do not record sensitively any shorter-term fluctuations. As many terrestrial cosmogenic radionuclides fall from the sky to sediments within 100 years of so of formation, sediments may faithfully represent more rapid fluctuations of the galactic cosmic rays. The constant ^{10}Be deposition rates observed in certain manganese crusts from the deep sea indicate that 1000-year variations in the flux of galactic cosmic rays did not exceed 10–20% during the last $0.5-1 \times 10^7$ years.

The Sun's magnetic field modulates the flux of galactic cosmic rays on time scales less than 1000 years. Lower fields allow more cosmic radiation to penetrate further into the solar system, thus enhancing cosmogenic radionuclide production rates. Reliable, direct records of solar magnetic activity such as sunspot numbers go back no more than 1000 years. The cosmogenic radionuclides provide indirect measures that in principle extend this information considerably further back in time. The observation of increased ^{14}C contents in tree rings that grew about 300 years ago gave independent evidence for the Maunder minimum, a period of unusually low sunspot activity. Measurements of ^{10}Be in polar ice and perhaps lake sediments confirm the ^{14}C results. These studies point the way toward an examination of much more ancient solar history, and perhaps toward establishing a relationship between cosmogenic radionuclide deposition rates and climate.

In this context a connection has already been observed between the ^{10}Be and oxygen isotopic composition of polar ice. When the climate warmed at the end of the last ice age, both quantities underwent marked, correlated changes. The abrupt decrease of ^{10}Be contents may reflect a change in cosmic-ray intensity, or one in meteorological conditions. If the former proves true, a new tool linking the Earth's climate and the Sun's behavior is in hand.

The Earth's magnetic field influences the flux of cosmic rays that interacts with the Earth's atmosphere. Again, lower fields lead to higher cosmogenic radionuclide production rates. Vigorous searches for the effects of secular variation of the geomagnetic field, of geomagnetic reversals, and of geomagnetic excursions on cosmogenic radionuclide deposition rates are under way.

Abundance patterns. Cosmic rays produce different patterns of the cosmogenic radionuclides in different target materials. The bombardment of rocks which consist mostly of elements heavier than oxygen generates relatively more of the heavier cosmogenic radionuclides such as ^{26}Al and ^{36}Cl than does the bombardment of the atmosphere. Within rocks, the detailed patterns of cosmogenic radionuclide abundances depend on the sample's composition and on the geometrical conditions under which the exposure to cosmic rays took place. Two examples illustrate how these differences have been exploited.

Deep-sea sediments contain a small fraction of submillimeter spherules first recognized and suspected to be extraterrestrial 100 years ago. Confirmation of extraterrestriality has come from analyses of the cosmogenic radionuclides. Further, the cosmogenic radionuclide ratios observed seem to be closer to those expected from the irradiation of a subcentimeter-sized body than from the irradiation of a much larger body, that is, a small asteroid or comet. The cosmic spherules may have orbited the Sun as small bodies.

The naturally occurring impact glasses called tektites occur over large portions of the Earth's surface. Those from Australasia and the Ivory Coast retain appreciable amounts of ^{10}Be but apparently never had much ^{26}Al or ^{53}Mn. The cosmogenic radionuclide signature looks like that of a terrestrial soil or sediment. It does not resemble the patterns characteristic of extraterrestrial matter and is, in particular, inconsistent with a lunar origin. The uniformity of the ^{10}Be contents of certain Australasian tektites has been interpreted as reason to place the site of the originating impact in a region of rapid sedimentation.

Mixing rates and residence times. Volcanism is a recurrent feature at convergent plate margins, that is, where one plate burrows beneath another. An important, unresolved question is how much each of crust- and mantle-derived fluids go into the igneous rocks associated with the volcanism. Where pelagic sediments cover the subducted plate, they may accompany it downward. Thus, the cosmogenic radionuclides that accumulate in pelagic sediments could act as tracers for the crustal component. Appreciable amounts of ^{10}Be have been found in lavas collected near convergent plate margins. In comparison, rocks produced by other forms of volcanism contain much less ^{10}Be. It appears that the cosmogenic radionuclides provide direct evidence for the recycling of crustal materials.

Finally, there have been suggestions of some possible, new applications of ^{14}C, in particular to oceanographic problems. One such application is dating the remains of individual foraminifera, marine invertebrates that leave behind shells or tests rich in calcium carbonate ($CaCO_3$). By comparing the ages of surface- and bottom-dwelling species deposited in the same sediment sample, the rates at which various portions of the ocean exchanged water with each other in the past may be estimated. A second application concerns the ocean water heated at certain ridge crests on the sea floor. Complete carbon isotopic analyses—^{14}C, ^{13}C, and ^{12}C—may establish a reference point suitable for tracing the circulation of the heated water throughout the oceans.

For background information *see* METEORITE; RADIOCARBON DATING; RADIOISOTOPE (GEOCHEMISTRY); RADIONUCLIDES, COSMIC-RAY–PRODUCED; TEKTITE in the McGraw-Hill Encyclopedia of Science and Technology. [GREGORY F. HERZOG]

Bibliography: L. Brown, Applications of accelerator mass spectrometry, *Annu. Rev. Earth Planet. Sci.*, 12:39–59, 1984; G. F. Herzog and T. H. Kruse, Cosmogenic radionuclides, *Eos: Trans. Amer. Geophys. Union*, 64:594–596, 1983; D. Lal and B. Peters, Cosmic-ray produced radioactivity on the earth, *Handb. Phys.*, 46:551–612, 1967; R. C. Reedy, J. R. Arnold, and D. Lal, Cosmic-ray record in solar system matter, *Annu. Rev. Nucl. Part. Sci.*, 33:505–537, 1983.

Remote sensing

The soil layer covering the surface of the Earth is a vitally important reservoir for water, upon which all higher life depends. Historically people have had to accept what nature provided in the form of rain and its stored form of soil moisture. Recent advances in remote sensing technology have demonstrated that soil moisture can be measured from airborne or space platforms. Further development of this technology may provide agriculturists with an important management tool that could be as valuable as the daily weather reports and maps.

Importance of soil moisture. Soil moisture is important for various reasons. It is a factor in partitioning rainfall into streamflow or infiltration and groundwater recharge. Adequate soil moisture is essential for seed germination, growth, and maturation of all agricultural crops. Soil moisture affects the rate and amount of evapotranspiration and evaporation which, in turn, affect the global water balance and weather circulation patterns. Plant diseases and population levels of pests often are regulated by the soil moisture content. Erosion of soil by both wind and water depends on the soil moisture content. Thus soil moisture and its changes in both space and time are part of a very complex environment upon which plant life and the ultimate food source for humans depend.

The historical development of humans from nomadic to agrarian societies has in part been dependent upon adequate soil moisture. It has been hypothesized that many of the lost civilizations disappeared because climatic changes and droughts resulted in insufficient soil moisture to support the required crop production. Management of soil moisture in the form of irrigation has, to an extent, freed humans from total dependency on the whims of nature. However, there exists great competition for water resources, and agriculture faces continuing pressure to improve its efficiency of water use. Better management of soil moisture is one approach to increase the efficiency of water use while maintaining high crop productivity.

Measuring soil moisture. In spite of the obvious importance of soil moisture, it remains very difficult and costly to measure. Many methods are available for measuring soil moisture at a point. These include weighing and drying samples, techniques based on electrical properties, nuclear instruments, pressure gages and transducers, and many others. The main disadvantage of these techniques is that they determine the moisture only at a specific point. Because of heterogeneity of soil properties, crop cover, and rainfall distribution, this measurement may not be representative of soil even a few meters away. The remote sensing approach has an advantage over other techniques because it provides a spatial measurement.

Remote sensing techniques measure emitted or reflected energy from specific regions of the electromagnetic spectrum. Four regions of the spectrum have been the focus of soil moisture research: the gamma radiation region, the visible and near-infrared region, the thermal region, and the microwave region.

Gamma radiation. Since 1978 the National Weather Service (U.S. Department of Commerce) has been using an airborne gamma radiation snow survey program to gather snow water equivalent data for their river forecast centers. This technique relies on the attenuation of natural gamma radiation from the soil as a measure of the snow water content. The gamma radiation attenuation technique has also been used to make airborne soil moisture measurements for the upper 6–8 in. (15–20 cm). The National Weather Service has demonstrated that remotely sensed soil moisture is useful for hydrologic forecasting. Because of atmospheric attenuation of the natural gamma energy, this approach is limited to relatively low-level flights (about 1000 ft or 300 m). This limits the use of gamma radiation techniques to relatively flat terrain. In addition, the area sampled is a rather narrow strip of the surface directly beneath the aircraft.

Visible and near-infrared. People have long recognized that wet soil has a different reflectivity than dry soil. This difference can be described by the albedo, the ratio of reflected solar radiation to incident solar radiation. Researchers have developed good relationships between soil moisture and albedo for a specific soil. However, in addition to its dependency on soil moisture, the albedo depends upon particle size, color, stones, organic matter, surface roughness, angle of incidence, and measurement angle. These last two factors can be compensated for, but the other factors are site-specific, decreasing the usefulness of this technique. In addition, crop canopy shading and shielding limit this approach to bare soils.

Thermal. Thermal remote sensing is based on measuring emitted long-wave radiation from a surface. Long-wave radiation is proportional to the fourth power of the absolute temperature of the surface and its emissivity. A soil's emissivity is a measure of its efficiency as an emitter, which in turn depends on its moisture content. Direct measurement of temperature can give a qualitative indication of soil moisture. However, a technique based on thermal inertia appears most promising for quantitative measurement of soil moisture. Thermal in-

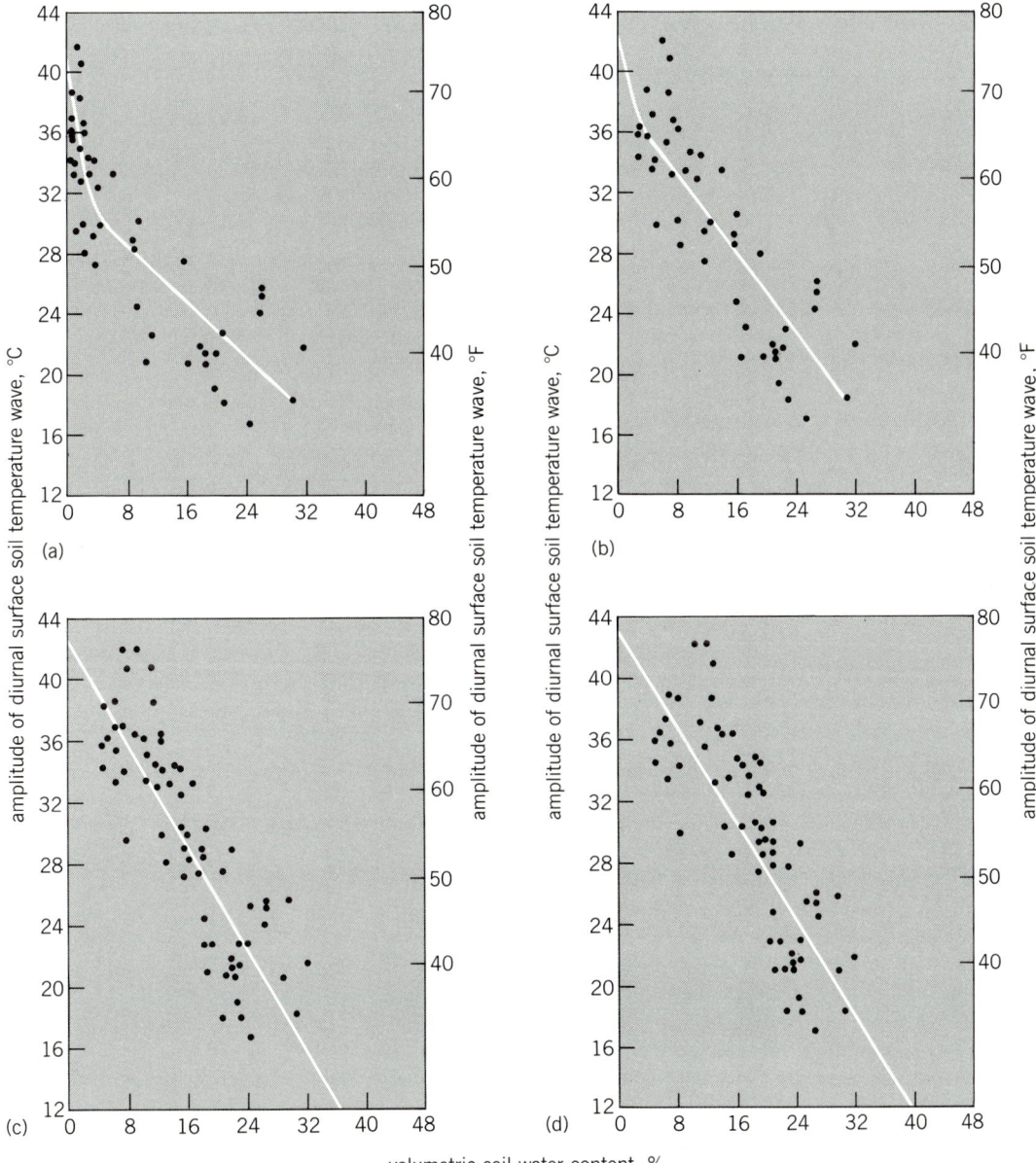

Fig. 1. Relationship between the amplitude of diurnal temperature and volumetric soil moisture at (a) soil surface and (b) upper 0.4 in. (1 cm), (c) upper 0.8 in. (2 cm), and (d) upper 1.6 in. (4 cm) of the soil. (*After S. B. Idso et al., The utility of surface temperature measurements for the remote sensing of soil water status, J. Geophys. Res. 80:3044–3049, 1975*)

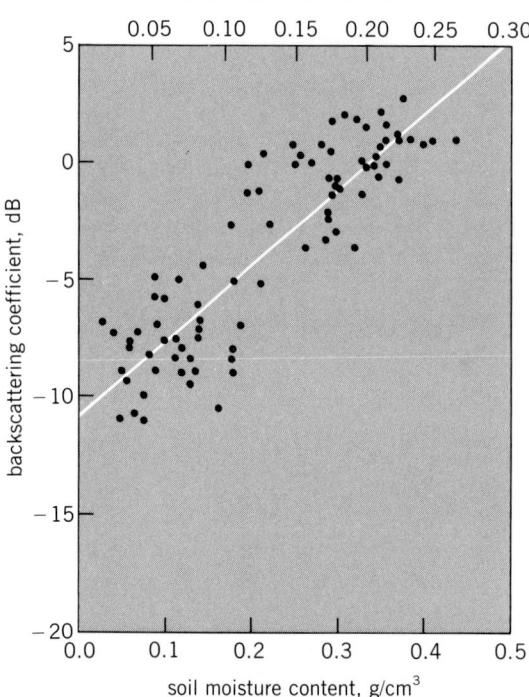

Fig. 2. Relationship between backscattering coefficient and volumetric soil moisture. (After F. T. Ulaby, P. P. Batlivala, and M. C. Dobson, Microwave backscatter dependence on surface roughness, soil moisture and soil texture, part 1, Bare soil, Trans. Geosci. Electr., IEEE, GS16(4):286–295, 1978)

ertia depends on the thermal conductivity and the volumetric heat capacity of the material which, in the case of soil, are both dependent on the moisture content. The difference between day and night temperature is a measure of the thermal inertia. Evaporation tends to reduce the amplitude of the day-night diurnal temperature cycle. Thus the difference between day and night temperatures is an indication of both soil moisture and evaporation. Field studies have shown good relationships between soil moisture in the top 0.8–1.6 in. (2–4 cm) and the amplitude of the diurnal surface temperature wave (Fig. 1).

The thermal infrared approach is limited to bare soils and is dependent on soil type. In the case of vegetated fields, the canopy temperature would be measured, not the soil surface temperature. However, the difference between canopy temperature and the air temperature has been shown to be an indication of plant stress. Thus, thermal measurements may reflect soil moisture throughout the plant's rooting depth. In considering crop production, this type of index may provide valuable threshold information for managing the soil moisture levels.

Microwave. Water has unique dielectric properties at microwave frequencies which make soil moisture measurement feasible. The dielectric constant for water is an order of magnitude larger than the dielectric constant for dry soils. Because of this, the soil surface emissivity and reflectivity in the microwave regions are strongly dependent upon the soil moisture content. These quantities can be remotely sensed with active microwave (radar) and passive microwave (radiometry) systems. Both of these approaches have been demonstrated with truck-mounted instruments (16–32 ft or 5–10 m above the soil) and with aircraft measurements. High correlations between soil moisture and microwave brightness temperature (radiometry) and backscatter coefficient (radar) are common. These studies have shown that it is possible to measure soil moisture in the upper 2–4 in. (5–10 cm) of soil (Fig. 2).

Additional research has shown that vegetation canopies typical of agricultural crops do not decrease the sensitivity of microwave techniques significantly for specific frequencies. Surface roughness such as furrows and tillage effects may result in misleading measurements unless compensated for. This is generally more serious for radar systems than for radiometers. All microwave systems have an additional advantage in that they are all-weather instruments and can make measurements through cloud cover.

Future applications. The application of space technology for remote sensing of soil moisture on a daily basis is technically feasible. Currently there are no satellites in orbit to provide this information; Research is progressing to specify the best of several approaches to be used. Data from a soil moisture satellite could be made available to farmers via television or through personal computers and telephone modems. Farmers will be able to use this information in their day-to-day management of water, fertilizers, and pesticides to maximize crop production and minimize energy use and possible environmental hazards.

For background information see GAMMA RAYS; MICROWAVE; RADAR; REMOTE SENSING; TERRESTRIAL WATER in the McGraw-Hill Encyclopedia of Science and Technology.

[EDWIN T. ENGMAN]

Bibliography: S. B. Idso et al., The utility of surface temperature measurements for remote sensing of soil water status, *J. Geophys. Res.*, 80:3044–3049, 1975; R. D. Jackson, R. J. Reginato, and S. B. Idso, Wheat canopy temperature; a practical tool for evaluating water requirements, *Water Resources Res.* 13(3):651–656, 1977, T. J. Jackson, T. J. Schmugge, and J. R. Wang, Passive microwave sensing of soil moisture under vegetation canopies, *Water Resources Res.* 18(4):1137–1142, 1982; T. J. Schmugge, T. J. Jackson, and H. L. McKim, Survey of methods for soil moisture determination, *Water Resources Res.* 16(6):961–979, 1980.

Ribozymes

Living organisms depend on a complex set of biochemical reactions that must be accelerated or catalyzed in order to occur within the time scale necessary for life. A biological catalyst is called an enzyme.

In 1926 J. B. Sumner isolated the first enzyme in crystalline form and showed it to be a protein. Since then, each new enzyme that was discovered proved to be a protein, and the word enzyme came to signify a protein. A few protein enzymes were shown to require a ribonucleic acid (RNA) component for activity, but in all cases the catalytic site of the ribonucleoprotein enzyme was thought to reside on the protein moiety. Recently, however, the definition of an enzyme has been challenged by the discovery of RNA molecules that, like protein enzymes, can catalyze, or lower the activation energy for, specific biochemical reactions. These catalytic RNAs are called ribozymes.

Self-splicing rRNA. The first example of a catalytic RNA molecule came during the study of RNA processing in the protozoan *Tetrahymena thermophila*. RNA processing, which refers to the way that a newly synthesized RNA molecule is made into its mature form, may involve removal of sequences at the ends of the nascent RNA molecule and, in eukaryotes, the removal of internal sequences that actually interrupt the mature RNA domains. The interruptions are called intervening sequences and are removed by a process called splicing. Splicing involves bond breakage to excise the intervening sequence and bond formation to re-ligate the mature sequences.

The 26S ribosomal RNA (rRNA) gene of *T. thermophila* contains a 413-base-pair intervening sequence. When the deoxyribonucleic acid (DNA) is transcribed into RNA, the resulting RNA precursor (pre-rRNA) includes a copy of the intervening sequence which must be deleted by RNA splicing. If the pre-rRNA is incubated in a test tube at physiological temperature and pH with only a monovalent cation, a divalent cation, and a guanosine molecule, the intervening sequence is excised and the mature sequences ligated. No protein or protein enzyme is required; the catalytic activity is intrinsic to the RNA molecule itself. The guanosine molecule is a substrate in the reaction and becomes covalently linked to the 5′ end of the intervening sequence during excision.

It appeared from early studies of the *Tetrahymena* self-splicing reaction that the catalytic activity was dependent on the three-dimensional structure of the RNA. For example, the denaturant urea completely destroyed the splicing activity. Structure is also very important for catalysis by a protein enzyme, and the similarity led researchers to suggest that the ribozyme was catalytic in the same way that a protein is catalytic. By assuming a precise three-dimensional structure, a protein creates an active site which binds its substrate in such a way that the activation energy is lowered and the reaction facilitated. The formation of a binding site is central to the lowering of the activation energy by a protein enzyme. A binding site brings the reacting molecules together, increasing their effective concentration and, in essence, approximating an intramolecular reaction. In addition, the binding may cause certain bonds to be strained and more reactive, that is, closer in energy to the transition state. Ribonucleic acids also have the capacity to fold into three-dimensional shapes as demonstrated crystallographically for transfer RNA (tRNA).

At least two experimental results suggest that the *Tetrahymena* rRNA folds to create a binding site for its guanosine substrate. First, there is a hyperbolic relationship between the concentration of guanosine and the reaction velocity. At low concentrations of guanosine the reaction rate depends on guanosine concentration in a first-order manner, while at higher concentrations the reaction rate becomes independent of guanosine concentration and approaches a limiting velocity. Such kinetics are typical of enzyme proteins and can be described by the Michaelis-Menten equation. In analogy to proteins, the approach to limiting velocity signals the approach to saturation of a binding site.

The second piece of evidence supporting a binding site is that the ribozyme is quite specific for its guanosine substrate. For example, guanosine, inosine, and 2-aminopurine ribonucleoside are all substrates for the splicing reaction, but adenosine, cytidine, uridine, and deoxyguanosine splice poorly, if at all. The number of 5′ phosphates does not affect the reaction; guanosine and guanosine triphosphate splice equally well. Many guanosine analogs were tested, and the results indicate that the substituents on the six-membered ring are very important for binding to the ribozyme. Specifically, the best substrate has, on its six-membered ring, a hydrogen bond acceptor at carbon 6 and a hydrogen bond donor at carbon 2. A model for the putative binding site is illustrated.

A model for the nucleoside binding site on the ribozyme. The mature RNA sequences (exons) are shown with a straight line and the intervening sequence with a curved line. The guanosine is shown binding to the RNA with four hydrogen bonds. The splicing reaction is initiated by a nucleophilic attack of the guanosine 3′ hydroxyl at the phosphodiester bond between the exon sequences and the intervening sequence. (*After B. Bass and T. R. Cech, Specific interaction between the self-splicing RNA of Tetrahymena and its guanosine substrate: Implications for biological catalysis by RNA, Nature, 308:820–826, 1984*)

RNase P. RNase P is also an RNA-processing enzyme; it is responsible for maturation of the 5′ end of tRNA molecules. It has been known for some time that RNase P required an RNA as well as a protein for its activity in a living cell. The standard conditions for assay of this enzyme include magnesium at a concentration of 2–5 mM. Under these conditions both protein and RNA components are required for activity. It has been discovered that, in test-tube preparations, in the presence of high magnesium (60 mM), the RNA component alone is able to catalyze the maturation of the tRNA precursor. By itself the protein moiety has no activity under the same conditions.

RNase P has been studied extensively in the two bacteria *Escherichia coli* and *Bacillus subtilis*. In *E. coli* the RNA subunit is referred to as M1 RNA and the protein as C5 protein. The analogous nomenclature for *B. subtilis* is P-RNA and P-protein. Cross-hybridization experiments with the cloned M1 RNA gene and *B. subtilis* genomic DNA indicate that there is no extensive primary sequence homology between the two RNA molecules. However, an active RNase P complex can be reconstituted with the protein of one bacterium and the RNA of the second. One interpretation of this result is that, like the *Tetrahymena* ribozyme, RNA structure is important for catalysis by RNase P. That is, although the M1 RNA and P-RNA differ in primary sequence, they may assume similar structures in solution.

Kinetic assays of *E. coli* RNase P also implicate structural requirements. A short lag is seen in the kinetics of M1 RNA catalysis. This lag disappears when the C5 protein is added or when the M1 RNA is dialyzed out of 7M urea. The lag may represent the time the denatured M1 RNA requires to assume its active structure. The addition of the protein may shorten the time required for this structural change.

Ribozymes versus enzymes. The discovery of biologically active RNA molecules has changed the way that biochemists think about RNA. Before the discovery of ribozymes, RNA was considered to play a passive role, if any, in catalysis. For instance, the rRNA of a ribosome was thought to function as a scaffold on which the proteins involved in translation assembled. Future studies on the ribosome may demonstrate a more active role by rRNA. Indeed, it is now known that mutations which affect the catalytic activity of ribosomes occur in the rRNA as well as in proteins.

It is interesting to compare the catalytic efficiency of protein enzymes and ribozymes. The two parameters of the activity of a biological catalyst are the K_m and the k_{cat}. The Michaelis-Menten constant, K_m, measures the affinity that the catalyst has for its substrate, while the k_{cat}, or catalytic rate constant, is a measure of how efficiently the catalyst accelerates the reaction. The *Tetrahymena* ribozyme has a K_m of 21 μM with its guanosine substrate; the *E. coli* M1 RNA has a K_m of 0.5 μM with a tRNA[tyr] precursor. Both values are within the range observed for the K_m's of protein enzymes.

It is not surprising that ribozymes and protein enzymes are able to bind their substrates with similar affinities. Both rely on the same three types of interactions for folding and substrate binding: hydrogen bonding, hydrophobic interactions, and charge-pairing interactions. These interactions allow formation of the precise three-dimensional shape, and furthermore can be positioned in the active site for specific binding of the substrate. Proteins are synthesized from a pool of 20 different amino acids, whereas RNA molecules are polymers derived from a pool of only 4 nucleic acids. In this respect proteins have more versatility in the reactions they catalyze and the substrates they bind. However, RNA catalysts have one capability that proteins do not have: they can form specific hydrogen bonds with a complementary sequence of nucleotides. Thus, RNA may be particularly well suited for catalysis of reactions that involve nucleotide or polynucleotide substrates.

Although the K_m values observed for ribozyme catalysis are similar to those of protein enzymes, the k_{cat} values of ribozyme reactions in test-tube preparations are an order of magnitude lower than observed in intact cells. Both protein and RNA are required for RNase P catalysis in cells, and in test tubes the protein moiety enhances the rate of M1 RNA catalysis. It may be that there is also a protein that enhances the rate of the *Tetrahymena* splicing reaction in intact cells.

It seems likely that the function of a protein cofactor in RNA catalysis would be to aid in the formation of the active RNA structure. As discussed previously, there is evidence that the C5 protein aids in the formation of the active M1 RNA structure. This view is also supported by the observation that in the fungus *Neurospora crassa*, if ribosomal proteins are not present, the rRNA precursor forms a structure that is believed to be inactive in splicing.

Primordial catalysts. Although most biochemists agree that the primordial genetic material was a nucleic acid, they disagree as to whether it was DNA or RNA. The ability of RNA molecules to catalyze biological reactions supports those who argue that RNA was the primordial molecule. It has been suggested that nucleotide coenzymes might be vestiges of these early ribozymes. Perhaps ribozymes were the primordial catalysts and served well to catalyze the crude reactions of a primordial biosphere. As life evolved, proteins began to fine-tune the catalysis until eventually the complexity of life demanded that a protein enzyme replace the primordial ribozyme. The only remaining ribozymes may be those few that were particularly well suited for carrying out reactions that involve interactions with other RNA molecules.

For background information *see* ENZYME; GENE ACTION; RIBONUCLEIC ACID (RNA) in the McGraw-Hill Encyclopedia of Science and Technology.

[BRENDA BASS]

Bibliography: B. L. Bass and T. R. Cech, Specific interaction between the self-splicing RNA of

Tetrahymena and its guanosine substrate: Implications for biological catalysis by RNA, *Nature*, 308:820–826, 1984; G. Garriga and A. M. Lambowitz, RNA splicing in Neurospora mitochondria, *J. Biol. Chem.*, 258:14745–14748, 1983; C. Guerrier-Takada et al., The RNA moiety of ribonuclease P is the catalytic subunit of the enzyme, *Cell*, 35:849–857, 1983; K. Kruger et al., Self-splicing RNA: Autoexcision and autocyclization of the ribosomal RNA intervening sequence of *Tetrahymena*, *Cell*, 31:147–157, 1983.

Rice

Rice vies with wheat in importance as a human food. It supplies about 80% of the calories for the 2 billion people of Asia, and about 33% of the caloric needs of the 1 billion people in Africa and Latin America. It also supplies a major portion of the protein in Asia. Among the two cultivated species, *Oryza sativa* is the cosmopolitan and major crop; *O. glaberrima* is limited to Africa.

Rice is a semiaquatic plant, and is the only cereal that can tolerate continuous flooding. Its caloric production can support a larger number of people per unit of land than any other cereal in the humid tropics. Because the microorganisms associated with the rhizosphere collaborate with the rice plant in the biological fixation of nitrogen, subsistence rice farmers are able to reap some harvest without fertilization. Rice is often the sole crop for subsistence farmers of monsoonal regions and supports numerous landless workers in such areas.

Sources and diversity of germplasm. The 20 wild species of *Oryza* are pantropical across Africa, South America, South and Southeast Asia, Madagascar, Australia, and the major islands of Oceania. The original home of *Oryza* may be traced to the Gondwana supercontinent, before its component plates drifted apart. Early progenitors of the 22 species of *Oryza* had differentiated before the Early Cretaceous Period, when the supercontinent began to fracture. Thus, the antiquity of *Oryza* exceeds 130 million years.

Progenitors of Asian rice underwent rapid differentiation in new environments after the South Asian plate collided with the original Asian mainland, and some of the species became widely distributed on the northern slopes of the Himalaya and associated mountain ranges—inside present-day China. During the Neothermal Period, about 15,000–20,000 years ago, the annual prototype of *O. sativa* quickly differentiated into early-maturing forms in the area along the foothills bordering India, Burma, and China. Since prehistoric times, the continuous and varied movements of people in Asia were instrumental in extending the geographic distribution of the primary ecogeographic race of rice called indica, and in the formation of two other races, javanica in Indonesia and sinica in China. Archeological findings in Chekiang Province of China clearly show that rice cultivation existed at least 7000 years ago. The oldest rice remains found in India date back to about 4530 B.C. Rice cultivation in Japan dates back to 300 B.C or earlier.

With the rapid growth in human population, accelerated migrations of people, and expanded trade, the tropically based cultivated species (cultigen) was brought into diverse climatic, hydrologic, edaphic, and seasonal environments. Extreme regimes led to the specialization into such types as upland and deepwater rices. Cultivars capable of tolerating salinity, alkalinity, or cool night temperatures also evolved under the combined forces of natural and human selection. There were probably more than 100,000 cultivars of Asian rice before genetic erosion set in shortly after World War II.

Although the African cultigen has a shorter history of cultivation in West Africa and is less genetically diverse than its Asian counterpart, the two cultigens have parallel evolutionary pathways. The wild relatives of each cultigen in adjoining primitive cultivation sites usually consist of the perennial wild, annual wild, and weed races. Geographic isolation has barred free gene flow between the two cultigens.

The genetic diversity in *Oryza* is indeed rich and remarkable. Both diploid ($2n = 24$) and tetraploid forms are found in some of the wild species. The full spectrum of germplasm in *Oryza* has developed: (1) from the centers of diversity—related genera, wild relatives, natural hybrids between cultigen and wild relatives, and primitive cultivars of the cultigen (land races); (2) from the areas of cultivation—commercial varieties, minor varieties, special-purpose types, and obsolete varieties; and (3) from breeding programs and genetic research—pure-line selections from farmers' varieties, elite hybrid varieties, F_1 hybrids, breeding lines, breeding stocks, mutants, genetic markers, polyploids, aneuploids, cytoplasmic sources, and intergeneric and interspecific hybrids.

History of conservation efforts. Most Asian countries exerted efforts to conserve their cultivars, both improved and unimproved, shortly after World War II. A number of foreign introductions of some fame were included in each national collection, but coverage on the land races of the remote areas was generally poor. Few wild species were conserved.

Although rice is not indigenous to the United States, rice workers in the Department of Agriculture (USDA) have assembled a world collection of rice. The USDA has also funded field-collection projects in India and Pakistan. The conservation efforts were materially aided by a medium-term cold-storage facility constructed at Beltsville, Maryland, in 1957 and the National Seed Storage Laboratory established at Fort Collins, Colorado, in 1958–1959. The United States rice collection continued to grow from about 6000 entries in 1960 to about 13,000 in 1981, of which more than 1000 are breeding lines.

During the early 1950s, the International Rice Commission under the Food and Agriculture Organization (FAO) of the United Nations attempted to set up three regional collections for the ecogeographic races: indicas in India, javanicas in Indonesia, and japonicas (Japanese varieties of the sinica race) in Japan. Another set of deepwater rices was maintained in Bangladesh. The four collections totaled 1344 varieties, but the three centers in the tropics could not adequately maintain seed viability without refrigerated storage facilities. Among the Asian countries, only Japan has had medium- and long-term storage facilities since 1965.

When the International Rice Research Institute (IRRI) began its research operations in 1961–1962, it took on the role of a global depository and exchange center. Because of its international character and dedication to service, IRRI was able to acquire a duplicate set of most national collections. When the high-yielding semidwarf rices spread quickly in the Asian tropics and began to replace the traditional varieties, 14 Asian countries collaborated with IRRI in launching systematic field-collection operations in both the threatened areas and other unexplored remote areas. Beginning in 1971, such collaborative efforts were systematically planned during international workshops. The pooled funds and worker power, together with the volunteering inputs of missionaries, service groups, and anthropologists, have added 38,000 samples to national collections and the base collection maintained at IRRI. The dwindling wild rices are also a target of conservation. Similar efforts by regional centers and national programs in West Africa have added 7700 African samples to the world's rice gene pools. A substantial number of the samples came from ecological niches where specific climatic, edaphic, or biotic stresses are endemic; rices from these habitats may have high levels of tolerances to such environmental factors. Since 1978 the International Board for Plant Genetic Resources (IBPGR) has joined the campaigns by providing some of the funds. During 1984, the IRRI collection of cultivars and wild species totaled 74,000 samples. Thus, the collective efforts of all concerned workers not only have saved the rice crop from a genetic wipeout but also have greatly enriched the gene pools available for further crop improvement.

Exchange and evaluation. Though short-lived, the FAO projects on conservation and interracial hybridization in the early 1950s aroused the interest of rice breeders in foreign rice germplasm. Shortly after IRRI was established, its seed distribution service and the promising sources of desired characters identified by its staff greatly stimulated the national rice research programs to expand their systematic evaluation projects, which also included increasing segments of foreign rices. Since 1962, the International Rice Germplasm Center at IRRI has supplied nearly 100,000 seed packets to rice researchers around the world in response to more than 3000 requests. The number of requests also indicates the magnitude of research experiments being conducted by rice scientists in different countries.

IRRI's systematic evaluation operations for a large number of desired traits were expanded and streamlined in 1974 under the Genetic Evaluation and Utilization (GEU) Program, which markedly augmented the combined inputs of the rice breeders and of the problem-area scientists such as plant protectionists, plant physiologists, and soil scientists. The multidisciplinary teams were organized to seek rice improvement in eight research areas: agronomic characters (especially grain yield), disease resistance, insect resistance, grain and nutritive quality, drought resistance, adverse soil tolerance, deepwater and flood tolerance, and extreme-temperature tolerance. The magnitude of IRRI's GEU Program is indicated by the 30,000–50,000 seed samples drawn from the Germplasm Center each year. Its major thrust is to raise yield levels and to stabilize crop production in vast rain-fed areas where the numerous subsistence farmers have not benefited from the semidwarf rices that dominate the irrigated areas. In recent years, several Asian countries have also organized national GEU programs.

On an international scale, exchange and evaluation were greatly expanded by the establishment of the collaborative International Rice Testing Program in 1976. Rice scientists of 76 countries are entering their promising parents and hybrid progenies in the international nurseries for broad-based testing. Rice breeders are also encouraged to release locally promising selections identified from foreign sources.

Use. Rice scientists have used most profitably the diversity present in the two cultigens and their wild relatives. The first breakthrough was the development of the high-yielding semidwarf varieties (HYVs) that raised grain yields to new heights in the tropics and subtropics during the 1960s. The quick spread and adoption of the semidwarf rices and wheats gave rise to the Green Revolution. The semidwarfs which constitute about 25% of the world's rice acreage, are now grown in temperate zone areas such as east-central China, South Korea, and California. The semidwarfing gene (sd_1) in the Asian HYVs probably came from spontaneous mutations in Chinese varieties; the same gene in the California varieties was artificially induced.

Further improvement of the semidwarfs' disease and insect resistance was based on many resistance sources present in the agronomically poor land races and a few wild relatives of *O. sativa*. The wild rices are the sole sources of resistance to two destructive virus diseases. A strain of *O. nivara* from north-central India has prevented the loss of yields worth untold millions of dollars to ravages of the grassy stunt virus. Other sources of resistance or tolerance present in the low-yielding land races are being incorporated with desirable agronomic backgrounds to produce new varieties able to cope with the environmental stresses in less favored production areas.

A sterile wild rice plant found on Hainan Island of China has provided breeders with the most usable source of cytoplasmic male sterility and has made hybrid rices possible. Since the program began in 1974, the area planted to hybrid rice has increased to over 19.7×10^6 acres (8×10^6 hectares).

Agricultural economists of IRRI have estimated

that during 1980–1981 the HYVs occupied 81.4×10^6 acres (33×10^6 ha), or about 40%, of the planted rice areas in 11 Asian countries (excluding China, Japan, and the two Koreas). Between 1965 and 1980, rice production in eight Asian countries (Bangladesh, Burma, China, India, Indonesia, Philippines, Sri Lanka, and Thailand) increased by 65% to 129.1×10^6 tons (117.1×10^6 metric tons) of grain worth $19.4 billion. The varietal contribution to the increase amounted to 30.1×10^6 tons (27.3×10^6 metric tons) or $4.5 billion.

Genetic vulnerability. Seed viability of rice during open-shelf storage in the humid tropics seldom lasts more than 2 years. Consequently, the staffs of many national gene banks have been burdened with the task of having to grow the entire collection every year or once in 2 years for seed rejuvenation when refrigerated storage facilities are lacking. Such stopgap measures have led to loss of stocks, errors in labeling, and overburdening of small staffs. Only in recent years did a few major national centers acquire medium-term storage facilities, but maintenance of the refrigerating equipment poses difficulties. As of 1984, only the United States, Japan, South Korea, Thailand, Soviet Union, and IRRI have both medium-term and long-term storage rooms. Most national conservation centers are also handicapped by insufficient funding, small or untrained staff, and lack of proper equipment for the hermetic sealing of stored seed.

The rapid spread and large-scale adoption of the HYVs and hybrid rices have inevitably led to an accelerated increase in genetic uniformity among the commercial varieties of South, Southeast, and East Asia. All the semidwarfs carry the sd_1 gene and are photoperiod-insensitive. Moreover, the early group of tropical semidwarfs shared the cytoplasm from Cina variety (China), which entered into the parentage of Peta, the female parent of IR8 and many IR varieties and lines. The hybrid rices of the indica type in China all have the Wild Abortive cytoplasm.

Along with the dwindling genetic base, the continuous multicropping of only one or a few HYVs under staggered and overlapping planting dates in the humid tropics has increased the genetic vulnerability of the principal HYVs in major production areas. Examples of uniformity-related epidemics are (1) the quick shifts in the biotype composition of the brown planthopper in the Philippines, Indonesia, and Vietnam during 1975–1976 and again in 1982; (2) the surge of virus diseases in the early 1970s; and (3) the blast epidemic and cold injury in South Korea in 1980. On the other hand, rice breeders are becoming increasingly aware of the need to reinstate genetic diversity in the commercial varieties, and are taking steps to remedy the situation.

Conservation necessity and cost. While most of the rice cultivars in readily accessible areas have been salvaged, thousands of minor varieties and many wild rices remain to be explored and collected in less accessible areas because of either physical remoteness or politicomilitary strife. Such germplasm-rich areas may be found in northeast India, high elevations on the southern slopes of the Himalayas, the northern borders of Indochina, and southwest China. Many areas in West Africa also remain to be canvassed. Acquisition of the remaining germplasm would provide rice researchers with gene pools that may cope with unexpected epidemics and shifts in pest populations or major climatic changes, and extend rice production into new areas. Moreover, rice yields have reached a plateau since the late 1960s. A continuous supply of diverse germplasm is needed to lend impetus to improved genetic potentials.

The total cost of conserving rice germplasm is about $1.25 million per year for the whole world. The crop amounts to about 440×10^6 tons (400×10^6 metric tons) a year and feeds more than 2.3 billion people, a number that is increasing by nearly 2.5% per year in the tropics. In the developing countries, where the population increases faster than in the developed nations, the need for rice will eventually top that for other cereals. The investments in rice germplasm will prove to be one of the most crucial and profitable research inputs. Developed and developing nations should collaborate on this area of common concern.

For background information see AGRICULTURAL SCIENCE (PLANT); BREEDING (PLANT); RICE in the McGraw-Hill Encyclopedia of Science and Technology.

[TE-TZU CHANG]

Bibliography: T. T. Chang, Conservation of rice genetic resources: Luxury or necessity?, *Science*, 224:251–256, 1984; T. T. Chang, The origin, evolution, cultivation, dissemination, and diversification of Asian and African rices, *Euphytica*, 25:425–441, 1976; R. W. Herdt and C. Capule, *Adoption, Spread, and Production Impact of Modern Rice Varieties in Asia*, IRRI, 1983; International Rice Research Institute, *Rice Improvement in China and Other Asian Countries*, Los Baños, Philippines, 1980.

River

Rivers have been an important area of study in the past few years. Recent work on the hydraulics, sediment transport, and chemistry of the Amazon and Orinoco rivers of South America has shown that older concepts are wrong. There has been some fundamental research on the origin and hydraulics of meandering river systems on Earth and on Mars.

THE AMAZON AND THE ORINOCO

The Amazon and the Orinoco rivers together drain approximately 2.7×10^6 mi^2 (7×10^6 km^2), and share a common basin divide along 600 mi (1000 km) of their watersheds (Fig. 1). They are the first and third largest rivers of the world in terms of average flow to the oceans. The estimated average flow of the Amazon at its mouth is 7×10^6 ft^{3}s^{-1} (200,000 m$^3 \cdot$s^{-1}); the average flow of the Orinoco is in excess of 1.3×10^6 ft^{3}s^{-1} (36,000 m$^3 \cdot$s^{-1}). The flows are roughly proportional to drainage areas,

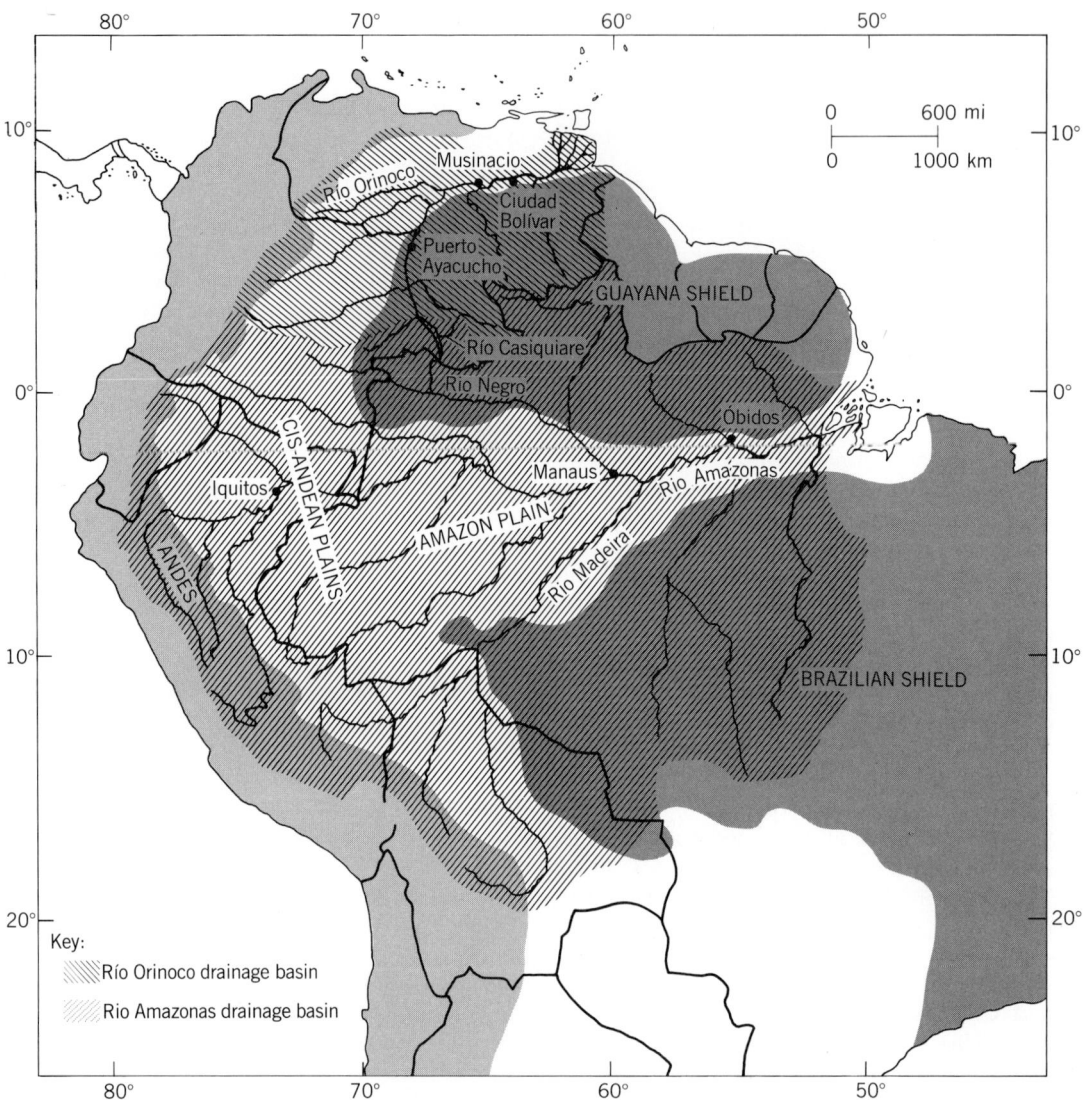

Fig. 1. Location map of the Amazon and Orinoco drainage basins. The downstream gaging stations are at Obidos and Musinacio.

the area of the Amazon basin being six times as large as the area of the Orinoco basin. The annual runoff from the basins is greater than 40 in. (1000 mm) per year, more than twice the average runoff from the global land surface to the oceans.

The two river basins are connected by the channel Rio Casiquiare (Fig. 1) through which almost 25% of the Orinoco headwater flows, perhaps as much as 25,000 ft^{3}s^{-1} (700 m$^3 \cdot$s^{-1}), and are diverted to the Rio Negro in a classical case of stream piracy. This channel provides a basis for transcontinental fluvial transport using small shallow-draft boats, although there are nonnavigable rapids on the Orinoco just upstream of Puerto Ayacucho that necessitate a 36-mi (60-km) portage by road.

The two rivers share many similarities. Both possess major tributaries that head along the east slope of the Andes and cross the dry llanos and deliver substantial flows and large sediment loads to the main rivers. Both drain large shield areas (Fig. 1) with streams that are relatively clear of suspended sediments and dilute in their chemical constituents, contributing large flows and small suspended and dissolved loads, but substantial quantities of bed load composed of medium sands and, in the case of Río Orinoco, coarse sands and fine gravels. In both basins, substantial areas are covered with tropical moist forests that are rich biologically in diversity of species but not very well understood, and that are being converted rapidly by deforestation to alternative patterns of land use. Rivers draining only these jungle areas are the classic black-water rivers like Río Negro and Río Atabapo, the characteristic dark color of the water reflecting the high levels of humic acids and corresponding small pH, generally around 4. The beds of these rivers usually are composed of pure white quartz sands, as all other minerals of the sediments in the bed load are leached by the acid waters to dissolved components.

Hydrology. Not much is known about the hydrology of these two rivers. A few long, continuous records of daily stage exist. For example, water-surface

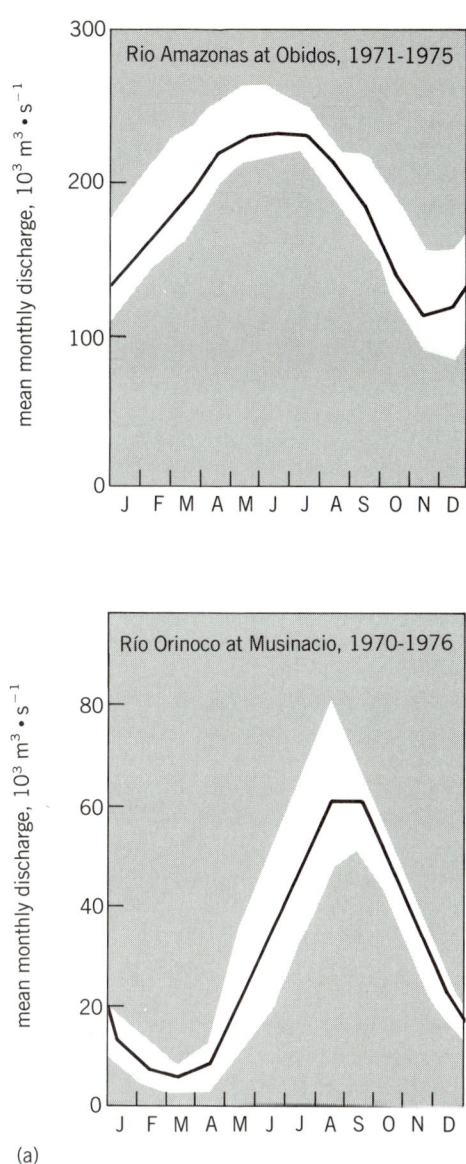
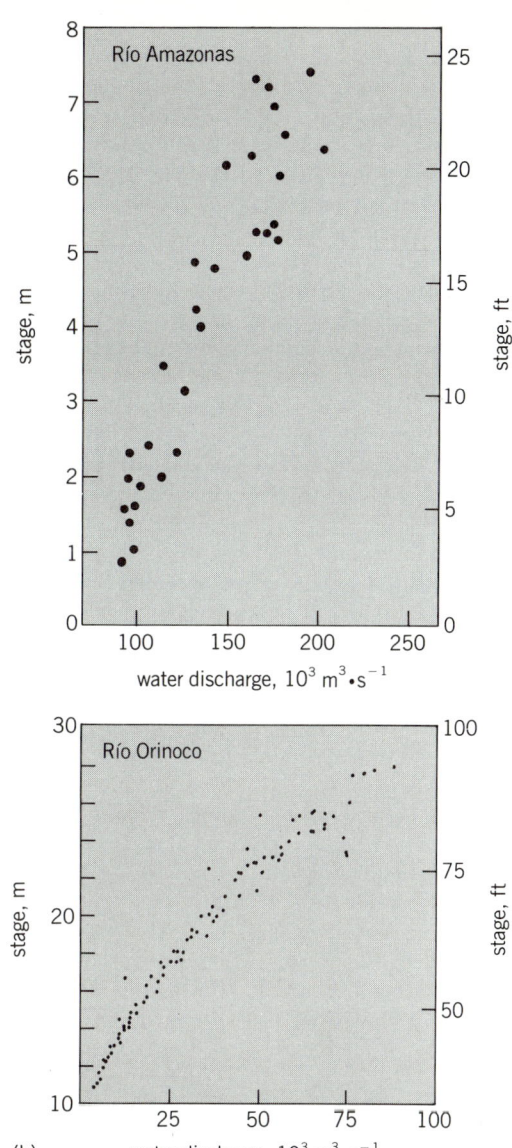

Fig. 2. Comparison of the hydrologic characteristics of the Amazon and Orinoco rivers. (a) Hydrographs of mean monthly discharges. The band about the mean curve shows the range of the monthly values. (b) Relations between stage and discharge. 1 m³ = 35.3 ft³.

elevations have been recorded continuously at the port of Manaus on the Rio Negro since 1902 and at Ciudad Bolívar on the Río Orinoco since 1924. However, systematic streamflow measurements were not undertaken until the mid 1960s, the first comprehensive sampling of suspended and bed sediment was conducted on the Amazon during 1976 and 1977, and the first comprehensive sampling of bed sediments and water chemistry of the Orinoco and its tributaries was undertaken in 1982.

Despite many similarities of the rivers, the hydrologic characteristics are markedly different. The Rio Amazonas and its tributaries straddle the Equator, so the peak flows from major left-bank tributaries of the Northern Hemisphere are offset from the peak flows of tributaries entering the right-bank from the Southern Hemisphere. The end result is a smoothing and damping of the annual flow cycle so that the relative difference between low and high flow is never very great, a factor of 3 or 4 at most in the Amazon mainstem (Fig. 2). The only other large river with this characteristic is the Zaire, in Africa, which also straddles the Equator and shares major tributaries entering from both Northern and Southern hemispheres.

The basin of the Río Orinoco, on the other hand, is completely north of the Equator and possesses distinct dry and wet seasons that lead to great differences between low and high flows. The ratios of highest-to-lowest flow each year for records from 1970 to 1981 range from 8 to 60 and average 26. Average monthly flows show a strong annual cycle (Fig. 2).

In addition to major climatic differences between the basins, the flow attenuation due to floodplain storage is considerably different for the two rivers.

The Amazon possesses a vast floodplain inundated during high flows that stores large volumes of the annual flow, reducing the flood peak and smoothing the flows through the year. The Río Orinoco is confined by rock at a number of locations where there are no floodplains. Between controls are large alluvial reaches with extensive floodplains that are inundated during the high-flow season, but the volume of floodplain storage relative to the annual flow volume is much smaller than for the Amazon, so floods are not attenuated much by storage.

The main gaging station of the Amazon, and the one closest to its mouth, is the station at Obidos (Fig. 1). This is a narrow, deep section of the river, about 8000 ft (2400 m) wide and 200 ft (60 m) deep at flood stage, and it shows the peculiar characteristic that most of the increase in discharge is accommodated by an increase in velocity rather than an increase in the cross-sectional area. The bed of the river here is mostly sand, so the relation between stage and discharge tends to shift from time to time and the individual measurements scatter considerably about a mean curve (Fig. 2). Deviations of plus or minus 10% about a mean relation are typical of rating curves for many sand-bed streams; in the case of the Amazon, this amounts to plus or minus one Mississippi.

The only rated gaging station of the lower Río Orinoco is at Musinacio, located 400 mi (650 km) upstream of the river mouth. The section is about 8500 ft (2600 m) wide and 59 ft (18 m) deep at bank-full stage and contains a sand bed. The rating curve here is looped, with higher flows for a given stage on the rising hydrograph. The relative scatter of measurements around a mean relationship is about the same as for the Amazon, approximately 10% (Fig. 2). The range of stage between low and high flow for the Orinoco is 52 ft (16 m), about twice that of the Amazon at Obidos.

The slope along the Orinoco is well established, and there are many gages along the river tied to a common datum. The average slope is 3.0 in./mi (4.6 cm/km). Slope along the Amazon is not known because the gages are not referenced to vertical control. Probably, it ranges from 1–2 in./mi (2 or 3 cm/km) near Iquitos to less than 0.7 in./mi (1 cm/km) below Obidos.

Bed sediments and sediment discharge. In most alluvial streams, the mean size and the range of sizes of particles in the bed material decrease in the downstream direction due to hydraulic sorting and abrasion. This tendency is not observed in either the Amazon or the Orinoco (Fig. 3). Along the lower 2000 mi (3200 km) reach of the Amazon, the bed is composed mostly of fine sand with a median diameter of about 0.24 mm. Along the Orinoco the median diameter is 0.4 mm, and although there is large variation from one sample site to the next, there is no pronounced trend for decreasing size downstream. As a general rule, there is more cross-channel variation in particle sizes at any given section than there is in the average size distributions along the river. The general uniformity of sizes along the Orinoco is attributed to the addition by tributaries draining the Precambrian shield areas of coarser bed sediments that just offset the reduction of size due to sorting and attrition of the finer Andean sediments. A similar mechanism may also be responsible for the lack of any downriver change in the mean size of the bed sediments of the Amazon.

The main bed configuration in both rivers is large dunes, with heights averaging 13–16 ft (4–5 m) and extremes up to 33 ft (10 m) in the Amazon and averaging about 7–10 ft (2–3 m) in the Orinoco. The ratios of average length to average height vary from 30 to 40.

The relation between sediment discharge and water discharge for the Río Orinoco at Musinacio is well established from measurements undertaken during 1969–1975 and 1982–1983 (Fig. 3). The estimated average sediment discharge is 2.48×10^8 short tons (2.25×10^8 metric tons) per year, the seventh or eighth largest sediment load delivered to the oceans from rivers of the world. Fewer measurements have been made for Rio Amazonas at Obidos (Fig. 3), its annual sediment load is estimated to be slightly more than 10^9 tons (10^9 metric tons) per year, ranking it second or third largest in sediment load. For both rivers, the relations between water discharge and sediment discharge are looped, with greater sediment discharges on rising stages than at equivalent water discharges on falling stages. Complex patterns of storage and reentrainment of sediments along the channels have been identified. In a 420-mi (700-km) reach of the channel of Rio Amazonas below Manaus, as much as 10^6 tons (10^6 metric tons) per day of suspended sediment are deposited into storage during some seasons and resuspended during other seasons. This pattern of seasonal storage and resuspension in Rio Amazonas, which is one of storage during rising river stages and resuspension during falling river stages, is related more directly to changes in water-surface slopes than to changes in water discharge. Seasonal storage and remobilization of suspended sediment also has been observed in the 120-mi (200-km) reach of Río Orinoco between the mouths of the tributaries Río Meta and Río Apure. The relation of the storage pattern to either slope or discharge in this reach of the Orinoco, however, is not as clearly defined as it is in the lower Amazon.

Geochemistry. Geochemical investigations of the Amazon River began in the 1950s, and today a great deal is known about the system. Especially significant among the many studies of the river were the comprehensive investigations undertaken from the research vessel *Alpha Helix* during its 1976–1977 Amazon expedition. Comparable studies of the Orinoco basin were undertaken during 1982–1985, and although final results are not yet available, some generalities about the geochemistry of the two basins can be drawn.

The waters of the Amazon and Orinoco are dilute in their dissolved constituents, but because of their large flow volumes they contribute large dissolved

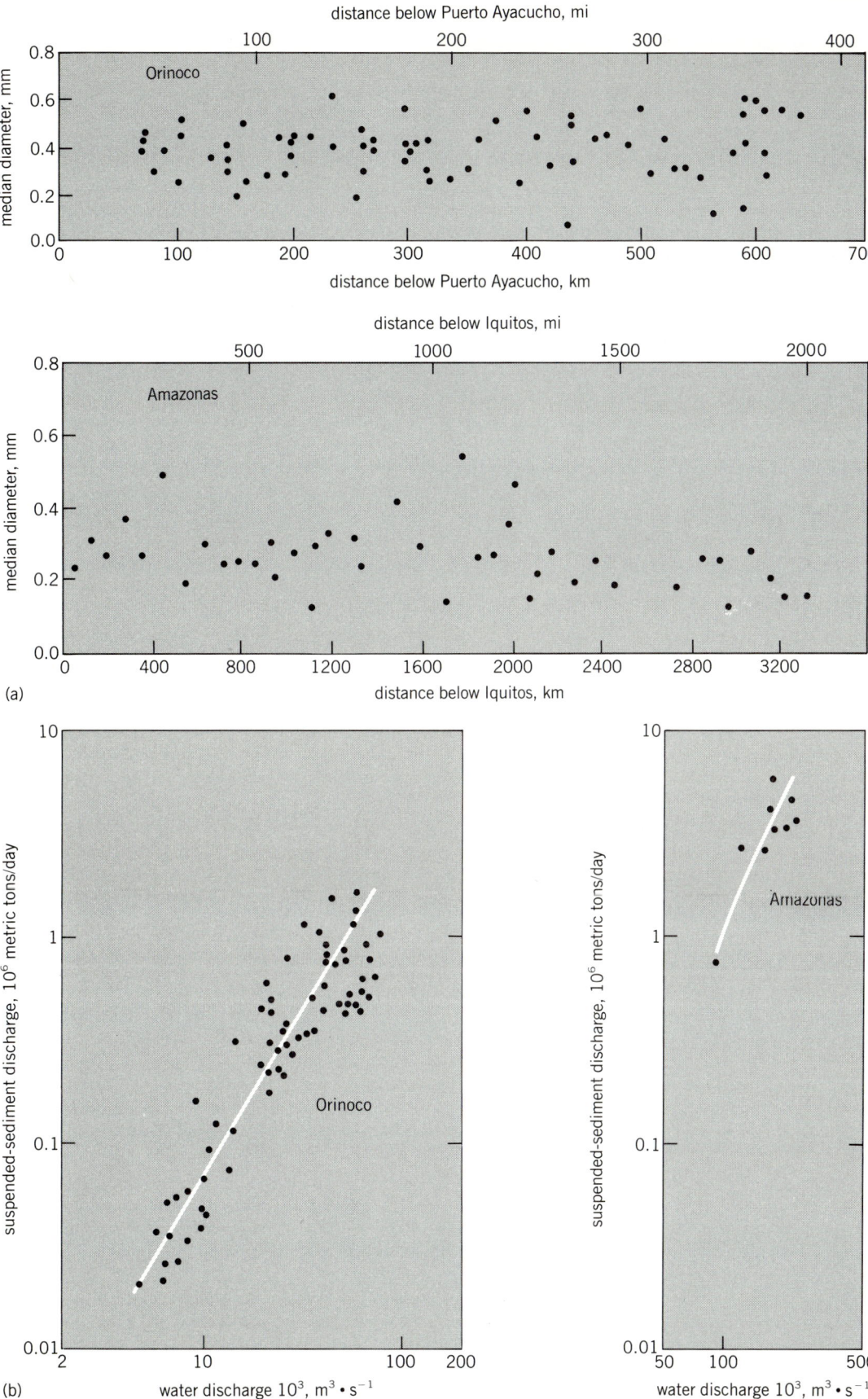

Fig. 3. Comparison of bed sediments and sediment discharge of the Amazon and Orinoco rivers. (a) Relations between distance and median particle size of bed material samples. Distances are measured downstream from Iquitos for the Amazon and downstream from Puerto Ayacucho for the Orinoco. (b) Relations between water discharge and sediment discharge. 1 m^3 = 35.3 ft.^{3} 1 metric ton = 1.1 short tons. 1 mm = 0.04 in.

loads to the ocean. Weathering and erosion processes over the basins are similar, so the dissolved loads, like the suspended loads, are roughly proportional to drainage areas. The dissolved load of the Amazon is estimated to average 3.19×10^8 tons (2.90×10^8 metric tons) per year, that of the Orinoco is 5.5×10^7 tons (5.0×10^7 metric tons) per year.

Rivers draining the shield areas are mostly dilute and acid, with concentrations of dissolved solids ranging about 1–10 mg/liter for lowland shield areas, such as the areas drained by the Rio Negro, and about 5–50 mg/liter in the highland shields, the higher concentrations associated with the drier basins. These waters are enriched generally in potassium and magnesium relative to other rivers of the basin. Andean streams drain marine sediments, shales, and occasionally massive evaporites. These streams possess high levels of calcium, magnesium, alkalinity, and SO_4 generally, and those draining evaporites also are rich in sodium and chloride. Concentrations range from about 100 up to several thousand milligrams per liter, the largest concentrations associated with the evaporites. A number of Andean streams drain mostly siliceous rocks; their concentrations range up to 100 mg/liter, and their waters are rich in silica relative to other species.

In a general way, the chemistry of the rivers in both basins can be related to the geology of the catchment areas when geomorphic factors are taken into account. Erosion processes occupy a continuum between weathering-limited processes in areas with steep slopes and thin soils, and transport-limited processes in areas of thick soils and small slopes. In the former, denudation depends mostly on lithology, and the cation composition of waters is similar to that of the weathering rocks. In the latter, denudation depends on regional lowering of the landscape and lithology is a minor factor, the composition of dissolved and solid river loads depending on soil chemistry. Both the Amazon and Orinoco discharge to the ocean waters that are dilute in their dissolved concentration relative to world averages, but the chemical compositions of their dissolved, suspended, and bed load are not greatly different from world average compositions.

[C. F. NORDIN, JR.; R. H. MEADE]

CHANNEL MEANDERING

Stream channels display a variety of patterns in map view. There are relatively straight stretches, braided streams where the flow divides into two or more intertwined channelways, and river courses which follow a regular meandering path. The regularity of river meandering, in particular, raises questions as to its origin and how it relates to the discharge of the river, the volume of water carried by the river per unit time. For more than a century scientists and engineers have been attempting to answer such questions, first by simple empirical correlations between the measured meander characteristics and the river flow, but more recently by analyses of the water flow processes and how they might initiate meander development.

This work has received renewed impetus by the discovery of sinuous channels on Mars, now dry but probably formed by water flows. There are also meandering channels in the deep sea formed by turbidity currents, the best example being the channels located on the sediment fan deposited in the sea near the mouth of the Amazon River. Finding sinuous channels in these very different environments offers additional opportunities to gain a better understanding of the basic cause of channel meandering. There is also the possibility of employing the empirical relationships for meandering of terrestrial rivers to evaluate what the discharges of water must have been in these now dry Martian channels and to estimate flows of deep-sea turbidity currents.

Controlling factors. A number of studies have been undertaken to determine the controlling factors which govern the type of channel pattern and, in the case of meandering channels, what determines the geometry of these meanders. Such studies have involved both direct measurements on rivers and streams as well as controlled experiments in laboratory flumes where there can be better examination of how a particular channel pattern develops. The most obvious attribute of a meandering channel is its wave length, L_m (Fig. 4a), and the goal of many past investigations was to determine what flow and sediment conditions control this parameter. All such studies find a close correlation with the flow discharge. A compilation of such measurements is shown in Fig. 4b, where the meander wave lengths are measured in meters and the flow discharge Q is given as cubic meters of water flow per second. The available data are seen to be consistent with Eqs. (1) and (2), two empirical relationships between L_m

$$L_m = 58.4 Q^{0.454} \qquad (1)$$

$$L_m = 19 Q^{0.62} \qquad (2)$$

and Q, depending on which data sets are considered best and, in particular, whether the laboratory flume measurements which form the chief basis for Eq. (1) are included. The flow discharge is known to be the principal controlling factor in determining the meander wave length, but it is seen that even with this dominant factor there is a problem in establishing a definite empirical relationship. Part of this problem arises from the great irregularity of river patterns and their discharges.

The discharge of a river varies from one day to the next, so that the question becomes which flow stage and value of Q should be employed in a correlation with the meander wave length. It has been concluded that the bank-full flood discharge should be used. This larger flow rate does the most work in establishing the channel geometry, and this choice has been verified in laboratory flume studies where the flow was changed through time, just as in natural rivers. There is the additional problem of the variability of the channel meandering in natural rivers, so that it is often difficult to select a dominant

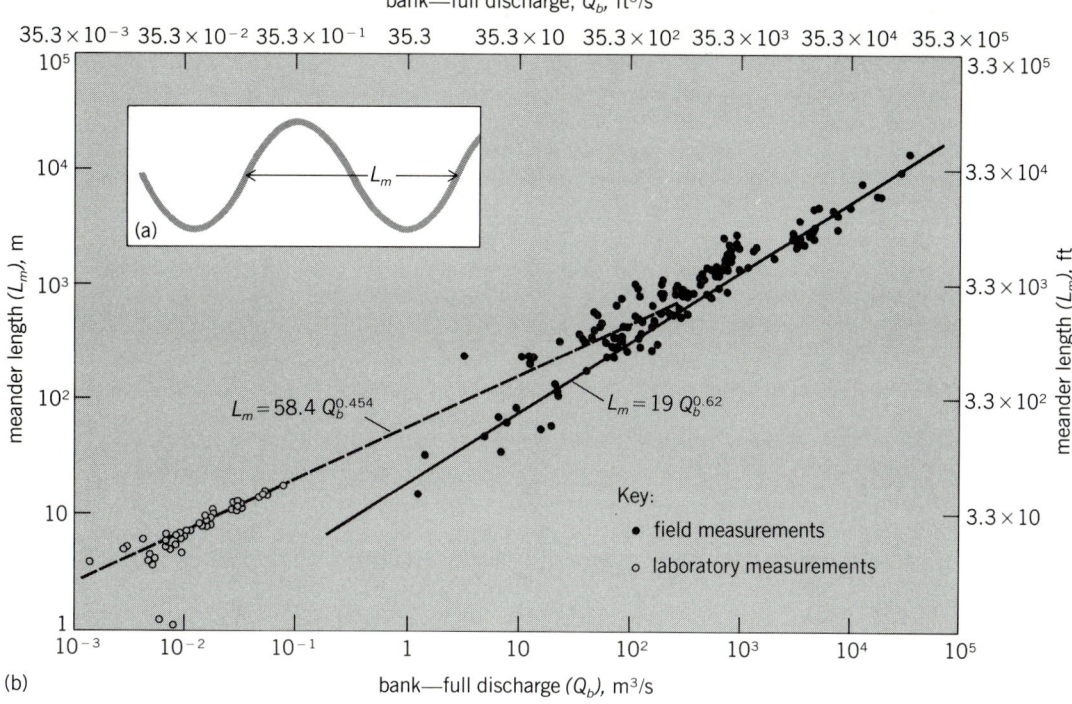

Fig. 4. Meandering channel measurements. (a) Diagram showing how the wave length L_m is determined. (b) Graph showing correlation of meander length and flow discharge.

wave length, that is, to determine one value for L_m. Recent studies have employed spectral analysis techniques to examine the meander patterns. This technique determines a number of sinusoidal waves of differing wave lengths which yield the observed channel meandering when summed. In such studies a number of L_m values for any given stretch of river is obtained, and an equal number of correlations with the river discharge.

Analysis of other channels. Relationships between the meander wave length and flow discharge, such as those of Eq. (1) and (2), are potentially useful in the analyses of the Martian and deep-sea channels. In both applications the meander wave length can be measured, and such relationships can be used in reverse to determine approximately the discharges of the water flows which eons ago carved the Martian channels, and the discharges of turbidity currents in the deep sea. But now the question arises as to whether such relationships based on terrestrial rivers are in fact applicable to such vastly different environments as Mars and the deep sea. Although the Martian channels must have been somewhat akin to rivers, gravity on Mars is roughly one-third that of Earth's (372 versus 981 cm/s², or 146 versus 386 in./s²). Turbidity currents in the deep sea show a comparable change in gravity; because these currents are immersed in the sea, the buoyancy of the surrounding water on the low-density turbidity current in effect reduces the gravity field that the flow experiences to a level even lower than that on Mars. Before the relationships between river meander wave length and flow discharge can be applied to these unusual environments, it must become possible to account for gravity differences. This in turn requires a better understanding of the actual physical causes and origin of channel meandering, considerations that did not enter into the simple empirical correlations of Eqs. (1) and (2).

Unfortunately, in spite of many hypotheses and analyses, there is still no firm understanding as to the basic cause of channel meandering. One approach considers the potential energy of the water as it exists in the headwater of the river, hypothesizing that the meandering comes about by the river adjusting in such a way as to facilitate the expenditure of this potential energy; a difference of opinion exists, however, as to whether this adjustment maximizes or minimizes this rate of potential energy loss. Similar to such approaches are attempts to consider the entropy of the river system, drawing on concepts from thermodynamics. Such hypotheses have tended to be more philosophical than mechanical and, most critically, do not yield predictive equations by which the hypotheses can be tested. More promising are the instability analyses which examine the flow processes themselves. These analyses determine which fluctuations of the flow are stable or unstable and thus which flow conditions lead to braided versus meandering channels, and what will be the dimensions of the resulting meanders. Instability analyses provide a more quantitative understanding and predictive capability of channel meandering, a growing understanding which may eventually be applicable to analyses of the Martian and deep-sea channels as well as to terrestrial rivers.

For background information *see* FLUVIAL EROSION LANDFORMS; FLUVIAL SEDIMENTS; MARINE SEDIMENTS; RIVER; STREAM TRANSPORT AND DEPOSITION; TURBIDITY CURRENT in the McGraw-Hill Encyclopedia of Science and Technology.

[PAUL D. KOMAR]

Bibliography: V. R. Baker, *The Channels of Mars*, 1982; J. E. Damuth et al., Distributary channel meandering and bifurcation patterns on the Amazon Deep-Sea Fan as revealed by long-range side-scan sonar (GLORIA), *Geology*, 11:94–98, 1983; R. H. Meade et al., Sediment and water discharge in Río Orinoco, Venezuela and Colombia, *Proceedings of the 2d International Symposium on River Sedimentation*, Beijing, 1983; R. H. Meade et al., Sediment loads of the Amazon River, *Nature*, 278(5700):161–163, 1979; G. Parker, On the cause and characteristic scales of meandering and braiding in rivers, *J. Fluid Mech.*, 76:457–480, 1976; K. Richards, *Rivers: Form and Process in Alluvial Channels*, 1982; S. A. Schumm, *The Fluvial System*, 1977; R. F. Stallard, River chemistry, geology, geomorphology, and soils in the Amazon and Orinoco basins, in J. I. Drever (ed.), *The Chemistry of Weathering: Proceedings of the NATO Advanced Research Workshop*, Reidel, France, 1985; R. F. Stallard and J. M. Edmond, Geochemistry of the Amazon, 2. The influence of geology and weathering environment on the dissolved load, *J. Geophys. Res.*, 88(C14):9671–9688, 1983.

Sclerochronology

Dendrochronology is the study of the annual rings in trees. The counterpart of dendrochronology in the marine realm is termed sclerochronology, the study of periodic growth structures in the skeletonized portions of marine organisms. Investigation of these periodic structures can provide information about the growth history and longevity of the organisms and about the environment in which they lived. This information is preserved as physical and chemical changes within the skeleton, and the interpretation of sclerochronological records has recently become an active area of interdisciplinary research among both marine biologists and paleontologists. Whereas the former are interested in modern organisms, the latter have been able to successfully interpret the skeletal records of fossil organisms and reconstruct their growth patterns, as well as the paleoenvironmental variability of ancient times.

Though applicable to any taxon possessing hard parts, the term sclerochronology has most frequently been associated with the study of light and dark density banding seen in cross sections through reef-building corals. The alternating light and dark growth bands are revealed when a thin cross-sectional slab is x-rayed. Radiochemical and field investigations have shown that these coral bands form with an annual periodicity in response to seasonal changes in the environment, analogous to the formation of yearly rings in trees. Thus, it is theoretically possible to develop sclerochronologies for the marine environment in the same manner that dendrochronologies are constructed for the terrestrial realm.

Dendrochronology has been responsible for extending knowledge of climatic change over the past 100 centuries. The size of yearly rings is largely a function of climate, so that variations in the size and chemical properties of rings reflect the history of climatic change. California bristlecone pines, the oldest known living organisms, provide chronologies for the past 5000 years. These can then be cross-dated with dead trees and wood so that a continuous record can be extended back to well over 9000 years. It does not appear that coral sclerochronologies will ever approach dendrochronologies in length. Sclerochronologies based upon multiple corals have been constructed for spans of up to 100 years, while individual corals with 200-year growth records are known. It has been predicted that coral sclerochronologies may eventually be extended back more than 500 years.

The calcium carbonate ($CaCO_3$) skeletons of corals provide the best sclerochronologies in the tropical and subtropical oceans. However, reef-building corals are environmentally limited. Temperature restrictions confine them to low-latitude zones, where the mean annual temperature remains above 64°F (18°C). The presence of symbiotic photosynthetic algae further restricts them to the shallow waters of the photic zone, where light levels are high and where water turbidity must also be low. For these reasons, other organisms which are not subject to the same environmental restrictions and whose skeletonized tissue also could provide sclerochronologies were sought. The bivalve mollusks have received the most attention in this regard. *See* ECOLOGICAL INTERACTIONS.

Bivalve shell growth increments. Marine bivalve mollusks (clams, mussels, and so on) have been shown to form shell growth increments which reflect a variety of environmental periodicities. These increments are best observed in radial shell cross sections, where they appear as alternating light and dark bands. The largest of these increments can easily be viewed with the unaided eye (see illustration). Such increments typically form with a yearly periodicity, and are the best documented and most pervasive of periodic shell fabrics. They have been noted to form in response to such environmental stimuli as low winter temperatures, high summer temperatures, and annual salinity cycles, as well as to internal stimuli such as the annual spawning cycle. The dark and light increments represent changes in the ratio of calcium carbonate to organic matrix (conchiolins) used to construct the shell. Often, the size and arrangement of the calcium carbonate crystals also vary between the two repeating increments.

Environmental and physiological periodicities of smaller magnitudes have also been documented in certain mollusk shells. For instance, many species are known to form daily increments in response to the 24-hour, day-night cycle. Certain intertidal bivalves inhabiting coasts that experience bidaily tides have been shown to record this phenomenon in their shells with subdaily increments, two sets of dark-light increments per daily increment. Lunar monthly tidal cycles have also been noted in bivalve shells through the fortnightly clustering of wide and narrow daily increments. The annual shell growth increments, however, remain the best documented and most widely used in sclerochronology.

Theoretically, modern bivalves should form 365 daily increments per annual increment. The same could be said for modern corals. Paleontologists have found these same types of increments in fossil organisms. By counting the number of daily increments per annual increment in fossil shells from several periods in Earth history, it has been estimated that there were once many more days in a solar year and that throughout geologic time the Earth's rate of axial rotation has been slowing. Sclerochronological records have thus played an important role in helping geophysicists calculate the change in orbital periodicities over geological time.

Growth rates and longevities. Counting the number of annual growth increments in a mollusk shell yields the age of the organism at the time of its death or capture. By measuring the size of each increment, it is possible to construct a growth curve indicating the increase in size associated with each year's growth. In this way, the growth records of both modern and fossil shells can be studied.

The use of annual growth increments exposed in shell cross sections to calculate age and growth rate has resulted in a reassessment of life history parameters for most species investigated. For example, the over-30-year life-spans calculated for the Atlantic surf clam, *Spisula solidissima*, and the soft-shell clam, *Mya arenaria*, are approximately twice previous estimates. Even more startling are the shells of certain species such as *Tindaria callistiformis* and *Arctica islandica*, which often possess over 100 growth increments that appear to be annual. The former is a deep-sea species and apparently grows very slowly, attaining a shell length of only 0.3 in. (8.4 mm) in about 100 years. The latter is a common, commercially harvested clam on the continental shelf of the northeastern United States and northwestern Europe. Specimens have been recovered which contain over 200 annual increments, ranking this species among the longest-lived invertebrates.

The use of incremental shell growth records in reconstructing life history patterns has resulted in very accurate estimates of age and growth rate. Most species investigated have been shown to live longer than previously imagined; over half exhibit longevities of 20 years, while a small percentage commonly attain ages of over 100 years. Investigations of the life history patterns of fossil shells have necessarily lagged behind studies of their modern counterparts. However, those species studied to date do not seem to differ significantly from their modern counterparts.

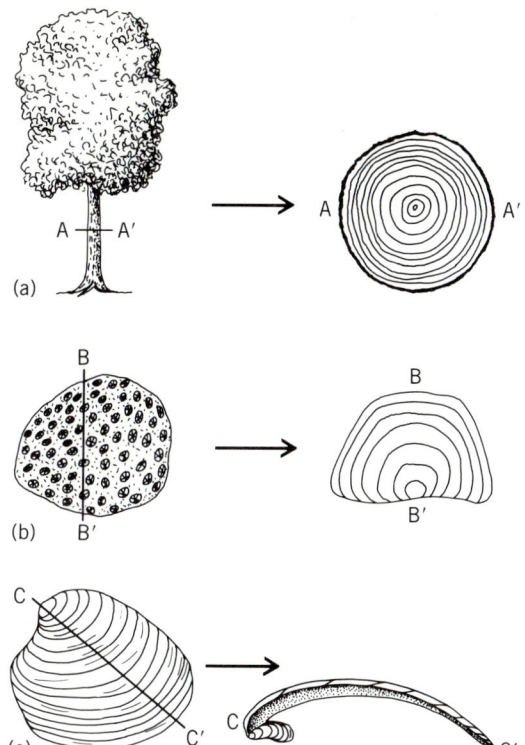

Yearly growth bands visible in cross sections of (a) tree, (b) reef-building coral, and (c) bivalve mollusk shell.

Climate records in shells. Marine climatic information is stored in annual increments as both physical and chemical changes across the shell. Perhaps the physical changes are easier to understand. Just as with tree rings, the normalized size of individual yearly increments can be related to environmental variables known to have a strong influence on growth. Paramount among these variables is temperature. In a recent study of the Atlantic surf clam from the east coast of the United States, the shell records of over 300 specimens were considered. Because the clams were collected alive, annual shell increments could be related to specific calendar years. By comparing the sclerochronological data with historically measured oceanographic variables, it was determined that mean annual temperatures were reflected in the widths of yearly shell increments. Cooler-than-average years yielded wide increments, whereas warm years resulted in narrow ones. A 25-year sclerochronological record of temperature change was produced.

Mollusk shells are composed of $CaCO_3$ in the form of calcite, aragonite, or a combination of the two. In the last two decades, many studies have shown that the stable isotope ratios of oxygen ($^{18}O/^{16}O$) and carbon ($^{13}C/^{12}C$) in shell carbonate are largely a function of water temperature and salinity. Therefore, by sampling across a shell at very fine intervals and determining the isotopic ratios of the samples on a mass spectrometer, it is possible to read the climatic history recorded by chemical variations in the shell. Furthermore, sampling techniques now permit many samples to be recovered from a single annual increment so that within-year as well as between-year variations can be studied.

To date, the isotopic studies have corroborated the results obtained from increment size analyses. Isotopic changes within annual increments track the seasonal cycle of temperature and salinity variation of the water in which the mollusks form their shells. Mollusk shells can thus be viewed as high-resolution recorders of seasonality and climate change.

Even though some mollusks have very long life-spans, they are probably not sufficiently long-lived to permit cross-dating from shell to shell back in time as dendrochronologists do with trees. Sclerochronologies, therefore, are not likely to approach the length of dendrochronologies. However, they will still remain useful for calculating age and growth rate of corals and mollusks. Sclerochronologies can also serve as after-the-fact records of the effects of environmental disturbances, whether natural such as storms or anthropogenic such as thermal pollution, on marine organisms. The shell climate records may also be used to interpret modern ocean conditions at sites where there are no historical data, or to trace the climatic variations of ancient oceans back through geologic time.

For background information *see* BIOSPHERE; CLIMATIC CHANGE; DENDROCHRONOLOGY; MOLLUSCA in the McGraw-Hill Encyclopedia of Science and Technology.

[DOUGLAS S. JONES]

Bibliography: D. Jones, Sclerochronology: Reading the record of the molluscan shell, *Amer. Sci.*, 71:384–391, 1983; D. Rhoads and R. Lutz (eds.), *Skeletal Growth of Aquatic Organisms*, 1980; K. Turekian, On reading seashells, *Discovery*, 13:2–11, 1978.

Sea water

Recently scientists reported a freshening, or reduction in salinity, of the deep North Atlantic Ocean waters that appeared to have occurred over the last 20 years; indeed, the change observed was found to have occurred principally within the last decade. The effect is of interest principally through the role of the ocean as a moderator of the Earth's climate. The heat capacity of the ocean is so large that the surface mixed layer of the ocean alone has a greater heat content than the atmosphere above it. The ocean is the largest reservoir of water on Earth, and changes in its water content can have consequences for lesser reservoirs that are important to humankind. The long-term geochemical stability of the oceans is also of fundamental interest to those concerned with the Earth's geochemical history.

An era of climate change is predicted through the greenhouse effect, generated by the buildup of long-wavelength-radiation–absorbing gases (CO_2, CH_4, N_2O, and so on) in the lower atmosphere. Thus a change in such a property as ocean salinity should be investigated for the revelations it may bring of natural or anthropogenic climatic variations.

Ocean salinity. The salinity of deep ocean water throughout the world averages close to 34.9 parts per thousand (‰, or grams of salt per kilogram of sea water). The oceans contain some 3.6×10^{20} gal (1.4×10^{21} liters) of sea water. The salinity may be determined by conductometric techniques to an accuracy and precision of about ±0.003‰, and thus small changes in the salinity of deep ocean water may be readily observed.

Unfortunately, a long and accurate temporal record of ocean salinity does not exist. The classical determination by silver nitrate titration held sway until the advent of the conductometric salinometer during the International Geophysical Year (IGY) expeditions in 1957–1958. Thus data of the required accuracy have been available only for a short time compared to that normally considered necessary for the observation of climatic variability.

Observations of change. The observation of freshening comes principally from the results of the 1981 TTO (Transient Tracers in the Ocean) cruise on the research vessel *Knorr*, which carried out an extensive survey of the North Atlantic Ocean to observe the penetration of fossil fuel carbon dioxide into the ocean abyss by using correlation of the CO_2 with tracers produced from weapons testing. The cruise track for this expedition is shown in Fig. 1. Among the apparently confusing lines left by the vessel shown on this diagram, two tracks are of principal interest: the east-west line at 59°N (Labrador-Greenland-Scotland), which is a repeat of a classic section from the research vessel *Erika Dan* in 1962, and the north-south section running from Iceland to the east of the Antilles, which may be pieced together, in the western Atlantic basin. This latter section repeats the track of the 1972 GEOSECS expedition on the *Knorr*.

A comparison of the 1981, the 1972, and earlier results suggests a freshening of the deep water of about 0.02‰ or greater north of about 50°N. The observation is strongly supported; it occurs over a wide geographical area and is based upon hundreds of measurements. The mean change is some 10 times the standard deviation of a single measurement. The change may be seen in the data shown in Fig. 2, which compares the 1962 and 1981 salinities for the 59°N east-west section.

There can be no doubt that the change is real. The calibration of ocean salinity determinations rests upon the standard–sea water service, provided in 1962 by F. Hermann and in 1981 by F. Culkin. The lineage of the standardization is impeccable.

Recently this change has been further examined, and it has been shown that the signal clearly originates in the Norwegian and Greenland seas. The source of the deep Atlantic waters lies in these basins, and the salinity change is greatest there. The scheme of deep-water formation and flow is such that warm saline surface water is transported northward past the British Isles. Winter cooling in the Norwegian and Greenland seas greatly increases the density of this surface water, whereupon it becomes vertically unstable and overturning of the water col-

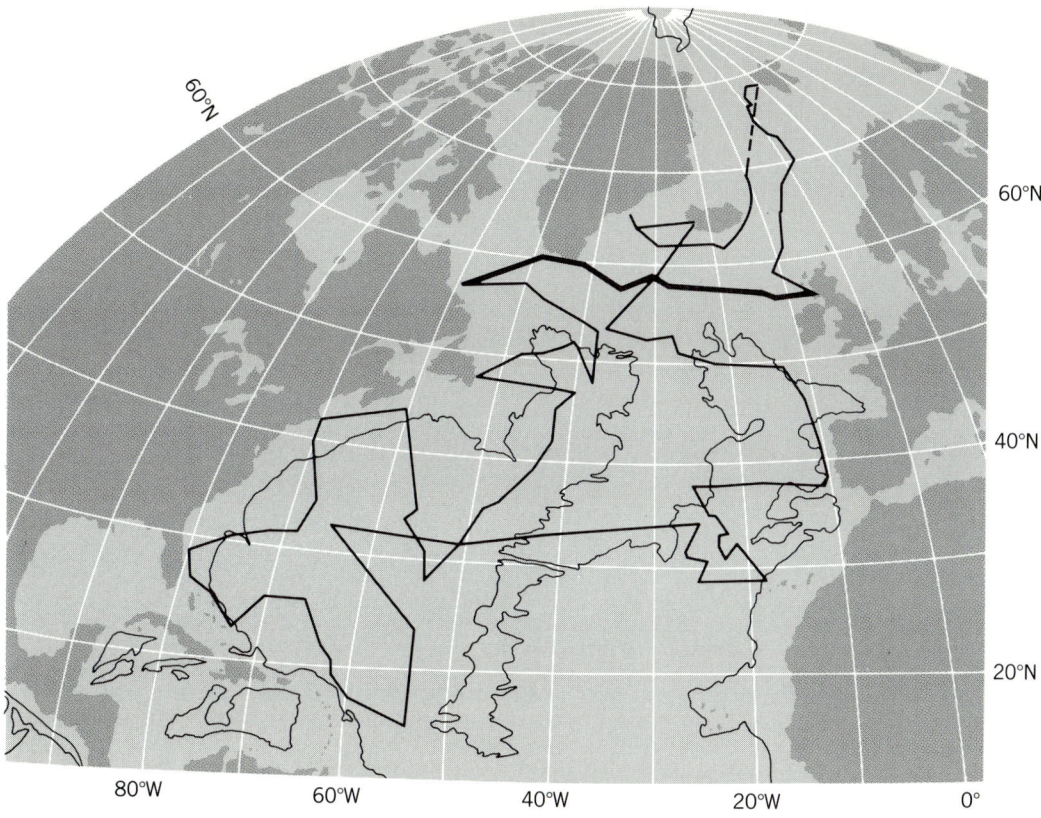

Fig. 1. The cruise track followed by the research vessel *Knorr* in 1981. The bold east-west line at 59°N corresponds to a repeat of a set of observations first made in 1962. (*After P.G. Brewer et al; A climatic freshening of the deep Atlantic north of 50°N over the past 20 years, Science, 222:1237–1239, December 16, 1983*)

umn occurs. The cold, dense water filling these basins spills out, entraining intermediate-depth waters from above, through the Denmark Straits as a deep current. Forced to the west by the Earth's rotation, this deep jet ventilates the waters of the abyssal western Atlantic basin.

The results shown in Fig. 2 reflect the emphasis on the density of ocean water masses. The ordinate in Fig. 2a is temperature (corrected for adiabatic effects), which in combination with salinity accurately prescribes density. The ordinate in Fig. 2b is density, expressed as sigma 2, that is, the density that sea water would possess if moved adiabatically to a reference level of 200 bars (20 megapascals), with the notation being such that an absolute density of 1.00370 g/cm^3 is expressed as 37.0. Because flow in the ocean takes place dominantly along surfaces of constant density, a representation of this kind eliminates noise associated with topography and the varying depth of density surfaces.

Causes of change. It is not known whether such a change is a random, aperiodic event driven by climatic instabilities of decade scale that are inherent in such a geophysical system, or whether this is a periodic event such as might be associated with cyclic changes due to a phenomenon such as insolation. It may be a manifestation also of a secular or long-term trend toward a fresher ocean. It is this last hypothesis that has evoked the most controversy.

Proponents of secular change observed that sea level is rising. A rise of 0.1 in. (3 mm) per year worldwide was reported for the period 1935 to 1975. In 1982 a more carefully edited data set was examined, and it was concluded that sea level had risen by about 5 in. (12 cm) worldwide in the last century. The situation is complicated by problems such as continental uplift in Scandinavia and the lack of representative Southern Hemisphere record. However, the rise in sea level is now widely accepted.

One group of researchers correlated the rise with climate change, the trace-gas greenhouse warming of the troposphere producing a thermal expansion of sea water, and the possible small net melting of the ice sheets. Others pressed this connection further, pointing out that thermal expansion alone could not, in their calculations, account for the observed rise; they suggested that significant discharges of polar ice must also be occurring. Moreover, they reported that the redistribution of mass on the Earth's surface from polar regions to lower latitudes as a result of melting should have increased the Earth's moment of inertia and have reduced the speed of the Earth's rotation. In fact, the Earth's angular velocity has been decreasing. They suggested that 75% of the observed decrease, or 1.5 parts in 10^8, was due to this effect.

Oceanic perturbations do indeed possess the capacity to affect the Earth's rotation. In 1984 it was demonstrated that the El Niño event of 1982–1983 was intense enough to have changed atmospheric

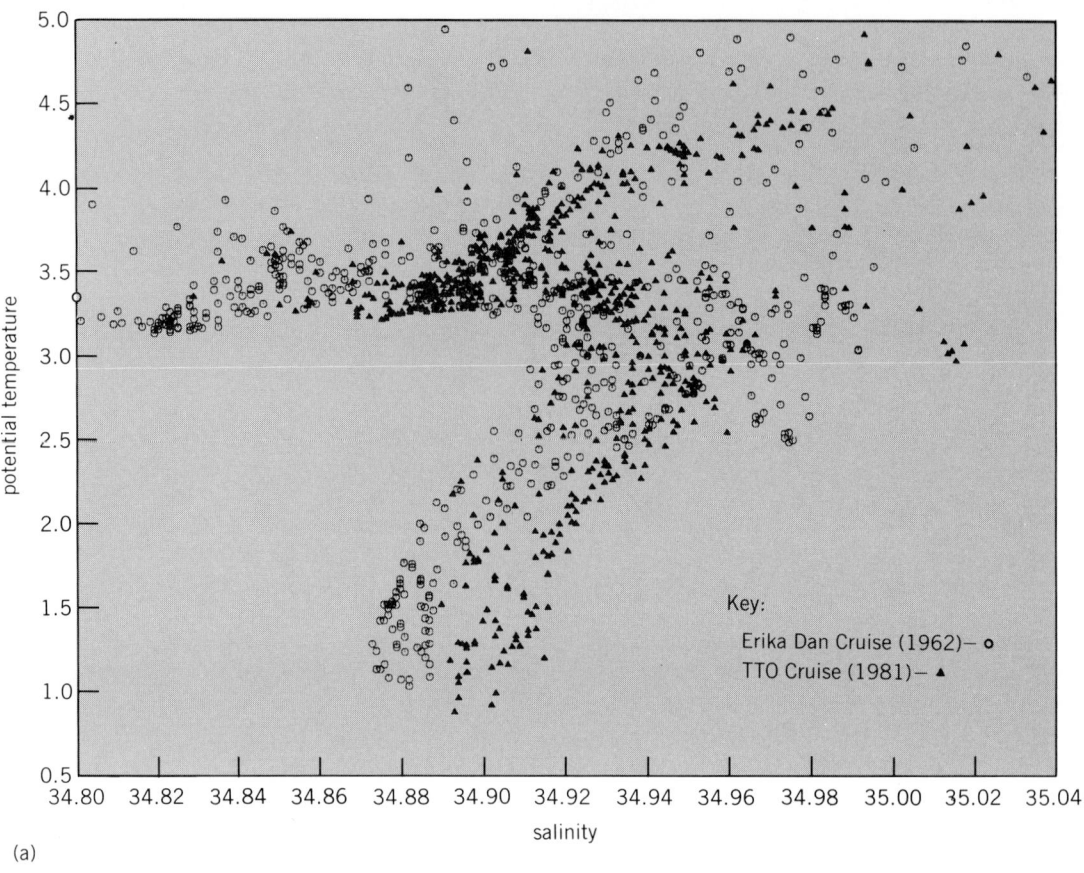
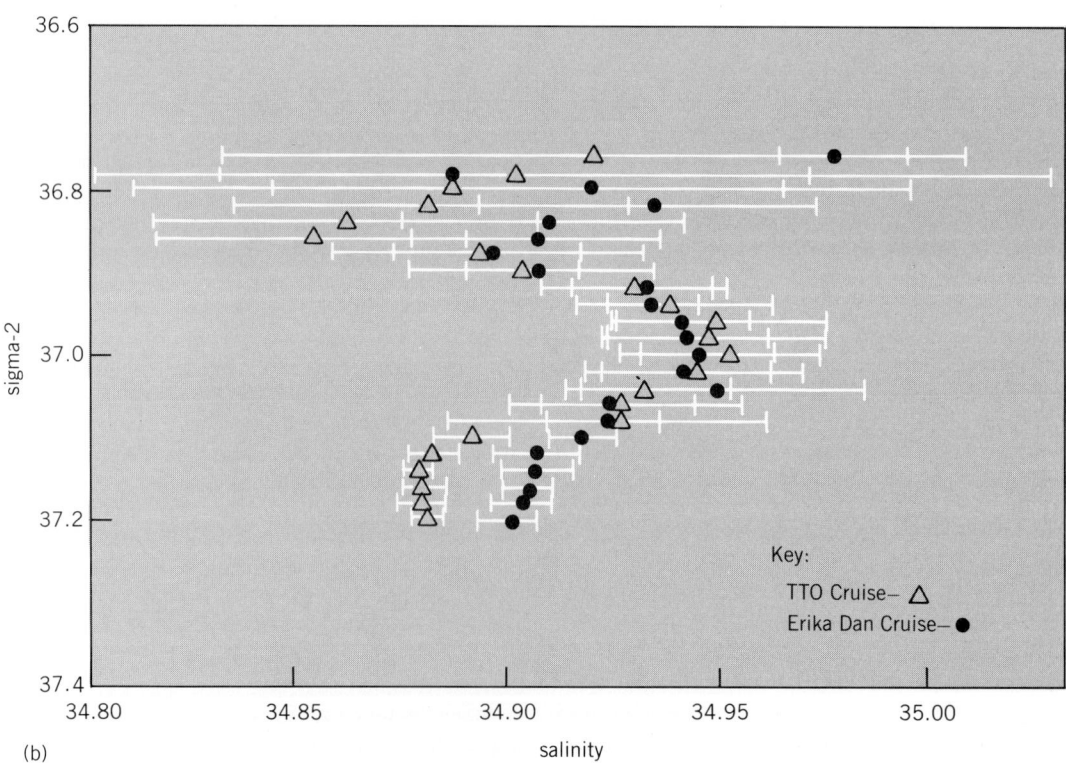

Fig. 2. Diagrams expressing the change in salinity of North Atlantic Ocean waters from 1962 to 1981 against (a) temperature and (b) the mean and standard deviation of observations on surfaces of constant density. (*After P.G. Brewer et al., A climatic freshening of the deep Atlantic north of 50°N over the past 20 years, Science, 222:1237–1239, December 16, 1983*)

angular momentum and the length of day. However, the provocative hypothesis of the occurrence of significant discharges of polar ice is not unchallenged. It has been pointed out that the ambiguous nature of ocean temperature and sea-level trends, together with large unknowns in the trace-gas–climate connection, could readily explain all observations without recourse to the drama of melting ice sheets. However, some researchers have countered that although uncertainty in each property observed indubitably existed, the physical linkage of the systems could not be ignored. They cited the freshening of the North Atlantic Ocean as further evidence of this causal chain. According to their calculations, the observed freshening between 1972 and 1981 would be equivalent to a melting of the Greenland ice cap of about 4 in. (10 cm) per year.

Conclusions. Although the curiosity surrounding such an observation is intense, there seem too many uncertainties to give a categorical explanation. The freshening described here is not an isolated oceanic event, but may well have partners in other areas of the ocean not under observation in this time period. It has recently been reported that the freshening of the North Atlantic was accompanied by a warming of the deep (2300–9800 ft or 700–3000 m) Atlantic waters south of 30°N by about 0.18°F (0.1°C) over the same time period.

The freshening may simply be a result of a shift in the locus of the deep-water formation areas in the Norwegian-Greenland seas toward a lower salinity area. It may be a glimpse of global change. Without systematic observational programs to document these events, it is hard to tell the difference. The creation and implementation of these strategies will be a key scientific concern in the near future.

For background information see CLIMATIC CHANGE; CLIMATOLOGY; GREENHOUSE EFFECT, TERRESTRIAL; OCEAN-ATMOSPHERE RELATIONS; SEA WATER in the McGraw-Hill Encyclopedia of Science and Technology.

[PETER G. BREWER]

Bibliography: P. G. Brewer et al., A climatic freshening of the deep Atlantic north of 50° N over the past 20 years, *Science*, 222:1237–1239, 1983; K.O. Emery, Relative sea levels from tide gauge records, *Proc. Nat. Acad. Sci. U.S.A.*, 77:6968–6972, 1980; R. Etkins and E. Epstein, The rise of global mean sea level as an indication of climate change, *Science*, 215:287–289, 1982; R. Etkins and E. S. Epstein, Response to criticism, *Science*, 219:997–998, 1983; V. Gornitz, S. Lebedeff, and J. Hansen, Global sea level trend in the past century, *Science*, 215:1611–1614, 1982; A. Robock, Global mean sea level: Indicator of climate change?, *Science*, 219:996–997, 1983.

Shipbuilding

The United States shipbuilding industry has been experiencing significant reorganization and redirection since the mid-1970s. These changes have been primarily the result of the shifting of the production function from a technology based on human labor, in the physical sense, to one with emphasis on advanced construction systems. This transition has been facilitated by the actions of two governmental agencies: the U. S. Maritime Administration and the U.S. Navy.

The Maritime Administration's main thrust has been in supporting a cooperative shipbuilding research program, which has existed since the early 1970s. The National Shipbuilding Research Program is a collaborative, cost-sharing effort with the shipbuilding industry, including universities and design agents. It has built a nationwide industry organization with several hundred participants involved from all of the major United States shipyards. The early stages of the program focused on short-term projects and concentrated on hardware development. In recent years the focus has broadened to include organizational and management technologies, education, and issues related to improved material flows and assembly techniques.

The Navy's activities have been in two areas: basic research and development, and shipyard expansion and modernization. The basic research programs have been concentrated in the physical and engineering sciences, and have been performed both

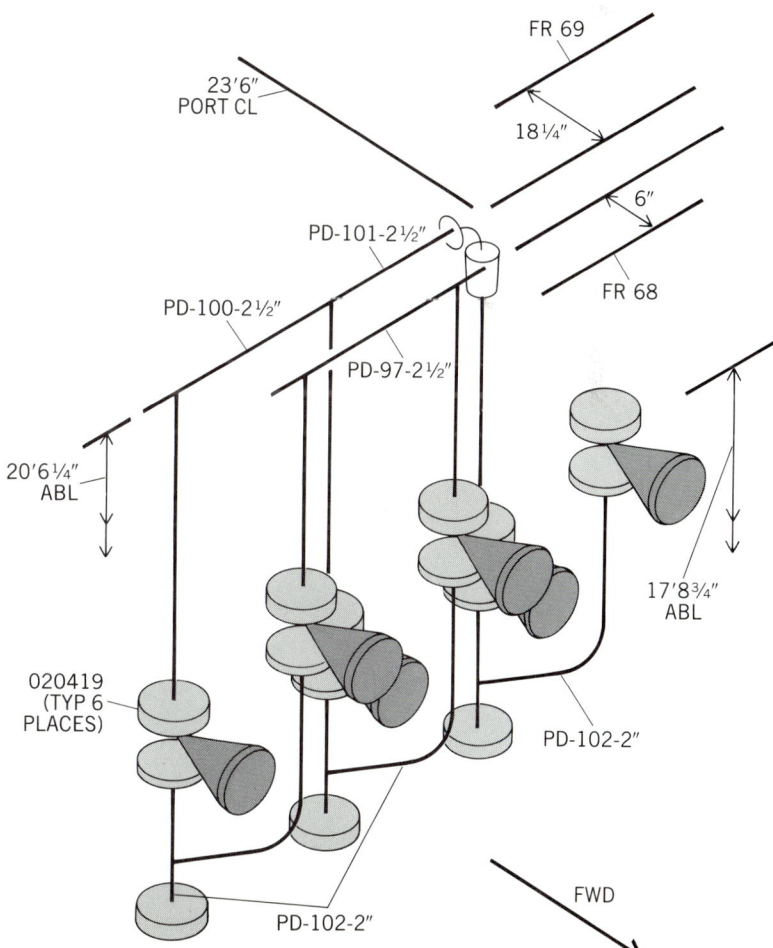

Fig. 1. Computer-developed isometric drawing of piping details for a lubricating-oil fill, transfer, and purification system.

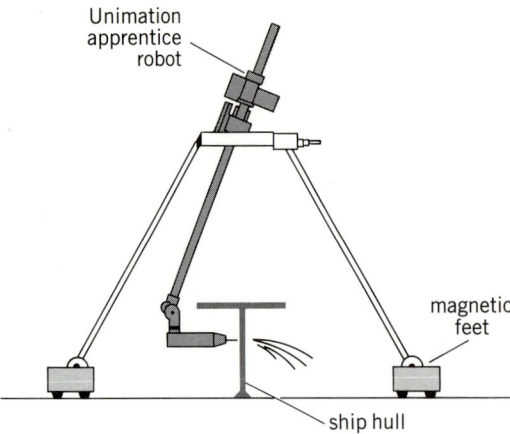

Fig. 2. Portable-robot cutting system.

sembly technology, new materials, and surface preparation and coatings.

Computer technology and robotics. The key to successful application of computer technology in shipbuilding is the development of a three-dimensional computer-data-base model of all significant items on the ship. The model is then used to form the basis for engineering analysis programs and design drafting, and to supply the data required for manufacture. The model is used to expedite the processes of performing arrangement studies, verifying compliance with the applicable design criteria, and confirming that producibility considerations are incorporated into the design. These details are accomplished by computer programs which refer to the data in the model and analyze the data for structural interferences and alignment, thermal stress, and weight distribution. Also, the geometry is compared with the limitations of the bending and fabricating machines. Finally, the three-dimensional data base can be used to generate the information needed for manufacturing purposes (Fig. 1).

A frontier of computer technology in the advanced-concept shipyard is in the field of robotics. While robot systems have seen limited application in United States shipyards, it is anticipated that their utilization will be expanded in future years. The main areas of application are in surface preparation and painting, and in fabrication and assembly. Robots are attractive in the surface preparation and painting areas because of their greater speed (5 to 10 times faster than manual blasting), better quality control, and increased safety.

Shipyard management finds robots attractive in fabrication and assembly because of their ability to enhance accuracy, to increase speed (25–50% faster than manual or semiautomatic welding methods), to reduce scrap, and to work in hazardous areas. The most elementary robots found in the industry are portable units, most often used in cutting or simple welding operations (Fig. 2). More exotic applications of robots may be found in welding operations where the robot has feedback control systems utilizing advanced infrared or radar technology. An optical system in which the feedback control mechanism is based on light reflection (Fig. 3) is utilized in Japan and is being critically examined by United States shipyards. This relatively simple system is appropriate where the fillet welds are applied to perpendicular plates. The light sources are arranged so that brightness of the vertical plate is different than that of the horizontal plate; the robot detects this light difference and follows a path at the intersection of the two reflections.

Welding and fabrication. Welding research has long been one of the major research areas of shipyards, with use of robots being but one area of interest. Major investments have been made in research in one-sided welding, whereby butt joints can be joined with only one pass. Shipyard management is also vitally interested in improving distortion control of welded joints. This interest has led to

at Navy laboratories and at private research centers (including universities). With the commitment to an expanded naval force, the Navy is supporting, with incentive-based, joint-investment programs, the shipbuilding industry's significant expenditures in plants and facilities. A large number of yards are improving (or expanding) their ship berthing, drydocking, and mechanical handling facilities, and industrial plants. Additional investments are being made by the yards, with Navy assistance, in computer graphics, numerically controlled cutting and welding machines, and advanced types of robots.

Productivity thrusts in the shipyards have been in the major areas of advanced technology, group technology, standardization, and institutional barriers.

Advanced technology. Advanced technology concepts have been applied in four areas: computer-aided design and manufacturing, fabrication and as-

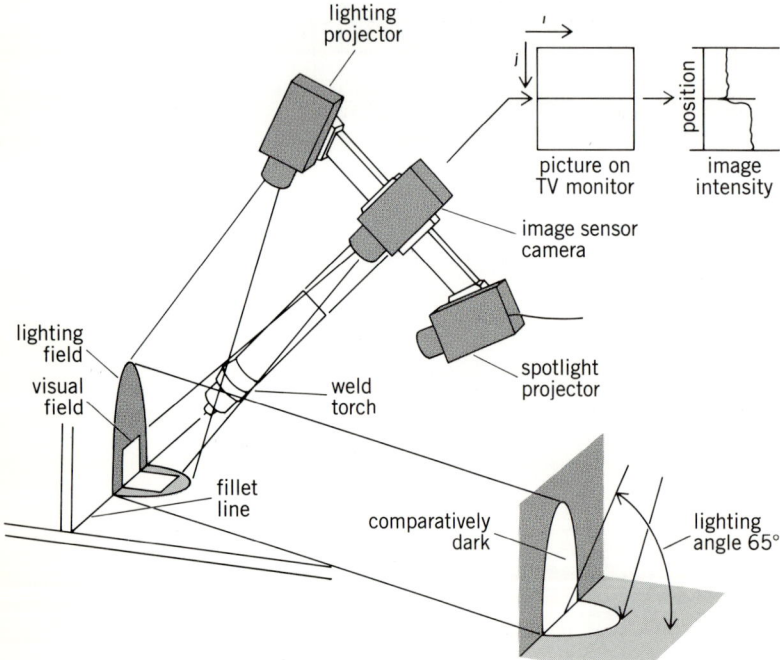

Fig. 3. Mitsui imaging and illumination system.

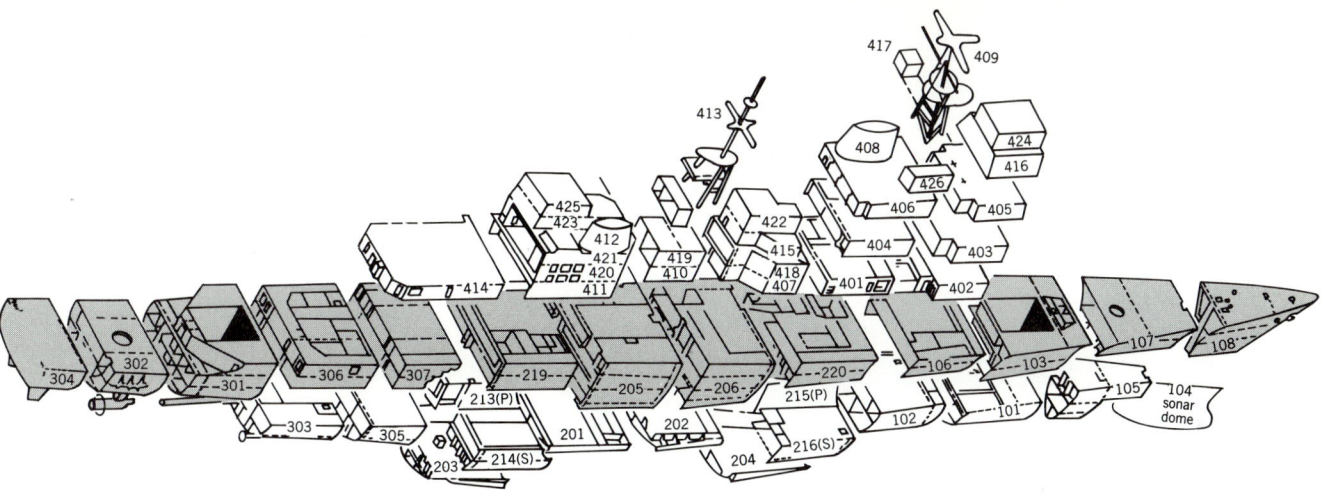

Fig. 4. Assembly block view of a U.S. Navy frigate.

efforts to understand the concept of line heating, the process of applying heat in controlled conditions to assist the shaping and forming of plate and structural shapes. As the physics of this process becomes better understood, it will become possible to automate the production system for fabrication of complicated shapes (by applying robots, for example). This automation will lead to improved processing of curved structures found in a ship, and will reduce a major cost impediment, especially in construction of high-performance ships.

Materials. The major thrust in materials research focuses on improved fabrication and lighter weight. The most significant program, sponsored by the Navy, has been the development of high-strength low-alloy (HSLA) steels, which will ultimately replace the Navy's class of high-strength (HY-80) steels that were developed in the 1940s. The HSLA steels are both lower in acquisition cost and better in welding characteristics than the HY-80, and they have superior strength characteristics.

In addition to metals, the Navy has investigative programs under way that are examining various applications of fiberglass-reinforced plastic. These programs are seeking to improve and expand on applications that presently exist for the material: piping, graphite-reinforced hydrofoil struts, foils and control surfaces, propeller-shaft coverings, and radomes.

Surface preparation and coatings. In the area of surface preparation and coatings, there have been major changes in the manner of preparation and application as well as the materials that are utilized. Steels are subjected to blasting materials that have higher specific gravity (which translates into greater kinetic energy), and are more easily reclaimable for repeated use than previously.

The major changes in coatings have been in the development of more fire-retardant, water-base coatings, and in the application of more friction-free material to the ship's underwater, outer-shell plating. The use of the former coatings results in meaningful cost savings because of the reduction of interference with welding and burning. The benefit of the latter coatings is that the ship moves with less resistance in the water; the result is an increase in fuel economy.

Group technology. Perhaps the most economically significant change in United States shipyards has been the application of group technology to the production process. Group technology is the system of manufacturing by grouping together parts (or pieces) that require common production processes. In shipbuilding, this concept has been applied to the interim products, the blocks that are put together to build the ship (Fig. 4). The blocks are placed in a specific manufacturing lane on the basis of material, work content, and processes required to fabricate and assemble the complete unit. Similar blocks are processed through the same construction lanes.

The new technology is based on complete assembly of the blocks, including the interior outfitting components, in assembly sheds away from the erection slipway. To the greatest extent possible, all blocks are moved at approximately the same time, and the assigned work is performed so as to optimize production efficiency. Figure 5 shows one of the completed blocks ready for movement to the erection site.

Standardization. In 1980 the National Shipbuilding Research Program initiated a major project to support the development and implementation of standardization in the United States shipbuilding industry. The program categorizes standards into national, industry-wide, and in-house, following the structure that exists in the Japanese shipbuilding industry. Development priority is given to the industry-wide standards, and includes basic materials, fittings, and equipment that compose the ship's various systems.

There are, in fact, many standards that already

Fig. 5. Completed block, ready for erection.

exist and are utilized by the United States shipbuilding industry. Unfortunately, many of these were issued by different organizations for the same area, resulting in overlap, duplication, and even contradiction. Therefore the project is addressing, as a first-priority goal, the consolidation of these standards. The most significant effort is the consolidation of existing Navy standards (12,000) with their commercial equivalents.

Institutional barriers. United States shipbuilding has long been burdened with institutional barriers to the improvement of its competitive position. These barriers include labor and management conflict, poor vendor relationships, separation of the design and production functions, and government policies, including contracting practices of the Navy and of other government agencies. Progress is being made in reducing the impacts of these factors, primarily by greater cooperation by all elements of the industry: government, management, and labor.

For background information see COMPOSITE MATERIAL; COMPUTER-AIDED DESIGN AND MANUFACTURING; HEAT TREATMENT (METALLURGY); INDUSTRIAL ROBOTS; SHIP DESIGN; SHIPBUILDING; WELDING AND CUTTING OF METALS in the McGraw-Hill Encyclopedia of Science and Technology.

[HOWARD McRAVEN BUNCH]

Bibliography: Bath Iron Works Corp., *Recommended U.S. Shipbuilding Standards Program, Long Range Plan, Final Report*, February 1982; L. D. Chirillo, *Scientific Shipbuilding: The Challenge*, SNAME Spring Meeting, Los Angeles, April 1984; R. L. Harrington, *The Application of Computer Technology to Shipbuilding*, Newport News Shipbuilding, speech delivered in May 1984; R. Ramsey, A time for shipbuilding renaissance, *Nav. Eng. J.*, 95(5):33–63, September 1983; U.S. Department of Defense, *Annual Report on the State of the Shipbuilding and Ship Repair Industry of the United States*, 1982.

Sickle cell anemia

Sickle cell anemia is a severe, genetically determined disease of the red blood cells resulting from an abnormality of the hemoglobin molecule. Hemoglobin (Hb) is a protein consisting of two polypeptide chain pairs (α_2 and β_2) with a total of 574 amino acid residues. In the abnormal hemoglobin of sickle cell anemia (Hb S), the hydrophilic glutamic acid residue in the sixth position of the normal β chain is replaced by the hydrophobic valine. This single change on the surface of the molecule results in diminished solubility of hemoglobin after the release of oxygen (deoxyhemoglobin). Oxyhemoglobin S is completely soluble. This difference in the behavior of Hb S is due to the fact that in the deoxystate hemoglobin has a different allosteric conformation from that of the oxy state which allows intermolecular reactions between deoxygenated hemoglobin molecules leading to polymerization of deoxyhemoglobin and the formation of fibers. These fibers can be considered liquid crystals of deoxyhemoglobin S which distort the erythrocytes containing Hb S into the sickle shape. The explanation of the mechanism of sickling, beginning with the discovery of the abnormal charge of Hb S, as revealed by electrophoresis, to the elucidation of the mechanism of the formation of the paracrystalline gels is one of the triumphs of molecular pathology. However, the hope that this understanding of the molecular basis of sickling would lead to a rational therapeutic approach has not yet been fullfilled.

Polymerization. The kinetics of polymerization of Hb S molecules has been studied by birefringence, light scattering, viscosity, calorimetry, nuclear magnetic resonance, and other methods. At first the Hb S molecules associate with each other to form aggregates of two, three, and successively larger groupings (nucleation). This is the rate-limiting step of gelling, the duration of which mainly depends on the concentration of deoxyhemoglobin S and the temperature. After this phase (delay time of sickling) longitudinal growth and alignment rapidly follow.

The process of gelling of deoxygenated concentrated Hb S solutions in laboratory preparations is an important model for studies of the kinetics of sickling and also of the conditions for the inhibition of sickling. These cell-free conditions also permit the reproduction of the environment of hemoglobin within the red cells where deoxyhemoglobin and oxyhemoglobin are in an equilibrium state expressed by the well-known sigmoid shape of the oxygen dissociation curve in which the degree of oxygen saturation is plotted against oxygen pressure. Under these equilibrium conditions the oxygen affinity of the red cells depends on the pH and the concentration of the affinity modulator 2,3-diphosphoglycerate. A decrease in pH and an increase of 2,3-diphosphoglycerate increases the relative concentration of deoxyhemoglobin, which is the most important variable determining the rate of gelling or sickling on deoxygenation. In individuals with sickle cell anemia, the concentration of Hb S is usually 30–34 grams per deciliter of red blood cells which corresponds to 28–30 picograms per cell.

Approximately 8% of the black population in the United States is heterozygous for the sickle globin gene. In the red cells of these usually asymptomatic individuals, Hb S constitutes 40–45% of the total hemoglobin. When mixtures of Hb A and S are deoxygenated, some of the Hb A copolymerizes with Hb S. This is also true for some other hemoglobin variants, in which heterozygosity for two hemoglobins is associated with sickling and vasoocclusive phenomena. On the other hand, fetal hemoglobin inhibits polymerization of Hb S.

Sickled cell characteristics. The survival of the red cells in the circulation of individuals with sickle cell anemia (those homozygous for the Hb S gene) is markedly shortened from the normal 120 days to 10 days or less, because the repeated sickling-unsickling cycles that the erythrocytes undergo in the circulation damage the red cell membrane, which loses its deformability so that the cells become irreversibly sickled. These cells either lyse within the circulation or are removed by phagocytic cells. Before this final stage occurs, there is an increased influx of sodium (Na^+) and calcium (Ca^{2+}) ions, potassium ions (K^+) leave the cell, the regeneration of adenosinetriphosphate (ATP) is impaired, the cell becomes dehydrated, the intracellular concentration of Hb S increases, and the cell becomes denser. When the red cells are separated by a density gradient centrifugation, the sickled cells are in the densest layer, which also includes still-round cells containing Hb S polymers. It has been suggested that the magnitude of this layer correlates with the occurrence of vasoocclusive pain crises, but just prior to and during the pain crisis this layer becomes smaller, probably because the dense cells are sequestered in capillary beds. The interaction of the Hb S fibers with the membrane appears also to affect the external surface of the red cell, which becomes stickier and shows increased adherence to the vascular endothelium.

Modification of gel formation. The investigations aiming at molecular therapy have been mainly directed at the inhibition of gelling by agents which affect the intermolecular contacts for polymer formation, or increase oxygen affinity, and thus maintain the molecule in the liganded conformation which does not lead to polymerization. The early attempts to increase the percentage of hemoglobin in the liganded conformation with carbon monoxide or nitrites proved the point, but had no practical utility. Inhibition of gelation by disrupting hydrophobic and hydrogen bonds with urea, which in the early 1970s created a flurry of enthusiasm, was ill conceived. When administered in dosages to accomplish this disruption, urea is highly toxic; in tolerable concentrations it is ineffective. Cyanate (—CNO), which is present in urea solution, was thought to have an antisickling effect by carbamylation of the α-amino groups of the amino-terminal valine residues of the two polypeptide chains interfering with polymer formation of deoxyhemoglobin S, but this reaction appears to cause mainly an increase in oxygen affinity. Although the initial therapeutic results seemed beneficial, later careful crossover studies revealed no essential benefits. Moreover, the occurrence of reversible peripheral neuropathies and cataracts led to discontinuation of all therapeutic trials with orally given sodium cyanate. Extracorporeal carbamylation and reinfusion of the carbamylated blood is not likely to find wide application.

Despite its lack of success, this early experience suggested that agents forming covalent or other bonds with crucial amino groups of the hemoglobin molecule might maintain the liganded conformation of Hb S to such degree as to lower the relative deoxyhemoglobin concentration by shifting of the O_2 equilibrium curve to the left. Examples of such agents are the above mentioned cyanate salts, carbamyl phosphate, the diaspirins, or pyridoxal phosphate which forms Schiff-base adducts with deoxyhemoglobin. Other potential agents with an antisickling effect are those that interefere directly with polymerization, such as aromatic amino acids or small peptides, which might exert their effect by stereospecific inhibition of binding sites. Covalent reagents such as nitrogen mustard or dimethyl adipimate, which are thought to inhibit contact sites essential for gelling, have also been used. The latter compound, an imido ester which is capable of crossing the membrane, is thought to have a bifunctional effect, increasing the oxygen affinity as well as inhibiting polymer formation by crosslinking with crucial binding sites. The inhibiting effect of these agents on gelling of concentrated Hb S solutions has been demonstrated, but their clinical use has not yet been found to be practical for various reasons, including the need for extensive toxicity testing, and the lack of proof of clinical effectiveness because of the absence of an animal model.

Alteration of cell membrane. Despite the fact that changes in the red cell membrane due to sickling are secondary to the polymerization of hemoglobin, the membrane has been the target of therapeutic attempts. Cetiedil, which is being used in Europe as a vasodilator, has been found to reduce the number of irreversibly sickled cells, and also to reduce the rigidity and increase the filterability of Hb S cells. The drug decreases calcium influx and increases the water and sodium content of the cell, thus decreasing the mean corpuscular hemoglobin concentration of Hb S. Some other agents that have been found to improve filterability of deoxygenated cells are L-phenylalanine, benzyl ester, monensin (an antibiotic for veterinary use), pirozecam, and potassium tellurite. Their end effect is dilution of deoxyhemoglobin S. This dilution has also been achieved in the recent attempt to induce swelling of the sickle cell by hyponatremia of the plasma produced by intranasally administered antidiuretic hormone, but this method is too risky to become practical.

Laboratory experiments with these agents, and the

limited experience in living organisms, has elucidated a great deal about the molecular physiology of sickling. These studies helped to identify the gelation delay phase of sickling during nucleation prior to the formation of larger polymers as the phase which probably is the most suitable target of antisickling agents. The length of this phase is indirectly proportional to the concentration of deoxyhemoglobin in an exponential relationship. Thus, small decreases in concentration may prolong the delay phase and enhance the chance for the inhibition of sickling.

Fetal hemoglobin. Clinical observations of the variable severity of sickle cell anemia have shown that fetal hemoglobin (Hb F, $\alpha_2 \gamma_2$), if present in the red cells in critical proportions of approximately 20% of total hemoglobin, inhibits sickling. This phenomenon is demonstrable in laboratory preparations by the increase in the minimal concentration of Hb S required for gelling of a mixture of Hb S and Hb F. This inhibition of sickling is likely to be due to the weak interaction of those amino acid residues of the γ chain ($\gamma 80$ and $\gamma 87$) which would face the valine of β^6 of Hb S if the polymer were formed. Apart from this effect, the increase in fetal hemoglobin is accompanied, for reasons which are not yet completely clarified, by a reciprocal decrease in the corpuscular concentration of Hb S so that the total intracellular hemoglobin concentration does not increase.

The main blood pigment during fetal life is Hb F. Toward the end of normal gestation, a poorly understood switch occurs, in which synthesis of γ chains decreases and the production of β chains increases, until at the end of approximately 9 months of life adult hemoglobin becomes the predominant blood pigment and fetal hemoglobin constitutes only 0.5–2% of the total. In most individuals with sickle cell anemia, Hb F is only slightly elevated to 4–6%. If it were possible to increase the transcription of the γ gene by pharmacologic means, so that Hb F would constitute approximately 20% of total hemoglobin, sickling would be inhibited. This has recently been accomplished by the use of 5-azacytidine, a cytotoxic agent which also causes hypomethylation of deoxyribonucleic acid (DNA) by inhibiting the enzymatic formation of methylcytosine. In the baboon, Hb F levels have been found to increase dramatically after the administration of this drug. Although the results of its use in patients with sickle cell disease were quite encouraging, the possibility that 5-azacytidine might be carcinogenic has restricted its use. Other cytotoxic drugs, such as hydroxyurea, also have been found to induce Hb F synthesis. Here the main mechanism is not hypomethylation of DNA, but probably a shift toward greater immaturity of erythropoietic progenitor cells.

Although no method of molecular therapy has become practical, the hope seems justified that these studies have provided important clues for future developments.

For background information see HEMOGLOBIN; HUMAN GENETICS in the McGraw-Hill Encyclopedia of Science and Technology.

[PAUL HELLER; JOSEPH DeSIMONE]

Bibliography: J. Dean and A. N. Schechter, Sickle cell anemia: Molecular and cellular bases of therapeutic approaches, *N. Engl. J. Med.*, 299:752–763, 804–811, 863–870, 1978; J. DeSimone et al., 5-Azacytidine stimulates fetal hemoglobin synthesis in the baboon, *Proc. Nat. Acad. Sci. U.S.A.*, 79:4428–4431, 1982; I. M. Klotz, D. N. Haney, and L. C. King, Rational approaches to chemotherapy: Antisickling agents, *Science*, 213:724–731, 1981; P. Heller and J. DeSimone, 5-Azacytidine and fetal hemoglobin, *Amer. J. Hematol.*, 16:439–447, 1984.

Simian AIDS

Simian acquired immunodeficiency syndrome (SAIDS) is a spontaneous disease of rhesus monkeys apparently caused by a type D retrovirus. This virus causes immunosuppression with clinical signs that are similar to acquired immunodeficiency syndrome (AIDS) in humans. SAIDS is an important animal model of the human disease AIDS, since they are so similar clinically and both are apparently caused by members of the retrovirus family. The information gained by studying SAIDS in a carefully controlled environment has direct benefit for both humans and monkeys. By understanding the retroviruses' mechanism of action, advances in therapy and prevention of both SAIDS and AIDS will be made.

Clinical syndrome. The hallmark of the clinical syndrome is generalized enlarged lymph nodes. In addition, enlarged spleen, diarrhea, fever, and weight loss occur. Chronic or opportunistic infections include organisms such as disseminated cytomegalovirus, *Mycobacterium avium intracellulare*, intestinal cryptosporidiosis, esophageal candidiasis, *Staphylococcus aureus*, or *Klebsiella pneumoniae*. *Pneumocystis carinii*, common to AIDS patients, is not seen in SAIDS. Neoplasia in animals with SAIDS includes cutaneous fibrosarcoma, cutaneous sarcomas (closely resembling Kaposi's sarcoma in AIDS patients), lymphosarcoma, and retroperitoneal fibromatosis. Severe lymphopenia, neutropenia, and anemia often occur early in the disease and predispose the animal to infections. The most susceptible animals are between 6 months and 3 years of age. Onset of the spontaneous disease is 3 to 6 months after exposure to infected animals. The prognosis is poor for animals with the disease, with mortality rates greater than 50%. Animals often die because of opportunistic infections, severe diarrhea, and dehydration, all unresponsive to intensive therapy.

Virology. It was important to determine if SAIDS was an infectious disease and, if so, what was the infectious agent. To answer these two questions, several sequential experiments were done. Experimental transmission of SAIDS to healthy rhesus monkeys was accomplished by inoculating tissue or whole blood taken from animals with SAIDS, thus demonstrating that SAIDS was an infectious disease.

Next, plasma was found to transmit the disease; plasma passed through a 0.45-micrometer filter was also infectious. The filter would allow passage of only virus-sized particles; that plasma passed through the filter was infectious provided strong evidence that a virus was responsible for the disease. Ether-treated plasma failed to transmit the disease, which demonstrated the virus was coated by a lipid envelope. The envelope was dissolved by the ether, thus removing the infectivity of the virus.

Simultaneous to animal inoculations, rhesus monkey kidney cells in tissue culture were infected with blood or plasma and observed for changes indicative of viral infection. When these cells were observed by electron microscopy, a type D retrovirus was seen (see illustration) and later confirmed by radioimmunoassay and reverse transcriptase assay.

The Retroviradae family consists of enveloped viruses with a nucleic acid core which is bar-shaped in type D retroviruses. The core contains linear plus stranded ribonucleic acid (RNA) and the virus has an enzyme, reverse transcriptase, which allows transcription of the RNA into deoxyribonucleic acid (DNA). The viral DNA then inserts into the infected host cell's DNA and is replicated at each subsequent cellular division. Some other retroviruses known to cause disease in humans and animals are human T-cell leukemia virus types I, II, and III, feline leukemia virus, bovine leukosis virus, and mouse mammary tumor virus.

The type D retrovirus was confirmed as the causative agent of SAIDS by inoculating rhesus monkeys with this virus grown in tissue culture cells in which no other virus could be detected. These animals developed SAIDS with clinical signs identical to the spontaneous disease in 4 to 8 weeks after inoculation. Investigators at the California, New England, Oregon, and Washington Regional Primate Research Centers have all isolated similar type D retrovirus from animals with SAIDS or related diseases. This increases evidence of the causative role of this virus in SAIDS.

Immunology. Defects in immune function are seen after injection with the SAIDS retrovirus. These immune defects allow opportunistic infections to occur, and eventually result in the death of the animal. One major feature of this immune defect is a decrease in the number of lymphocytes found in the peripheral blood. In SAIDS, it appears that the retrovirus infects both T and B lymphocytes as well as macrophages. This is in contrast to AIDS, in which the T helper lymphocyte appears to be the only cell type infected.

Response to nonspecific lymphocyte stimulators is decreased, but is partially to completely restored by adding interleukin 2, a product secreted by normal stimulated lymphocytes. Interleukin 2 and gamma interferon production by SAIDS lymphocytes is decreased. Immunoglobulin or antibody secretion by B lymphocytes is also markedly decreased in SAIDS. Neutrophils, another white blood cell responsible for immune defense, are decreased in number in peripheral blood, and neutrophil function is also abnormal in the terminal stages of the disease.

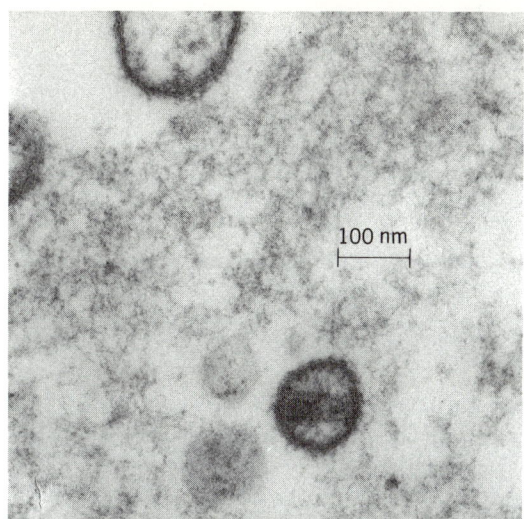

Transmission electron microscopy of rhesus monkey kidney cells infected with the SAIDS type D retrovirus. A typical virion with the cylindrical core characteristic of type D retroviruses is shown. (*From P. A. Marx et al., Simian AIDS: Isolation of a type D retrovirus and transmission of the disease, Science, 223:1083–1086, 1984*)

Pathology. The most characteristic histopathologic changes in SAIDS are seen in the lymph nodes. Very early in the disease, lymph nodes show hyperplastic T- and B-lymphocyte regions. As the disease progresses, the T- and B-lymphocyte regions become depleted of lymphocytes and are replaced by histiocytes, a type of macrophage. This severe lymphoid depletion stage corresponds to the lack of lymphocytes in the peripheral blood, abnormal cellular immune functions, decreased serum immunoglobulins, and severe opportunistic infections. These lymph node changes are nearly identical to AIDS, except the early hyperactive lymph node of SAIDS progresses to cellular depletion more rapidly than in AIDS.

Epidemiology. SAIDS has been described in at least six species at four primate research centers in the United States. This disease is undoubtedly more widespread and probably exists in most large colonies of Old World monkeys in this country. In addition, it is probable that zoo and wild populations of nonhuman primates have the disease which may go unrecognized. The origin of the disease is unknown, but further work will identify the source as wild or domestic.

The natural route of spread has been defined in a captive population at the California Primate Research Center. Several animals with no clinical illness but with large quantities of the SAIDS virus in blood and saliva have been identified. These healthy carrier animals likely spread the virus through saliva or perhaps blood to susceptible animals. Rhesus monkeys in a group cage are very social and close contact by licking, grooming, fighting, and biting is the most likely means of spread.

Outlook. This animal model of a human disease (AIDS) should provide benefits for both human and veterinary medicine. Knowledge gained in understanding the mechanism of action of the SAIDS retrovirus on the immune system may provide valuable information for developing drugs or other methods of treatment of AIDS, SAIDS, and other retrovirus-induced disease such as human T-cell leukemia or feline leukemia. Work is currently under way to develop a vaccine for the prevention of SAIDS in unexposed animals in research facilities or zoos. This vaccine could be very important for use in treatment of endangered species of nonhuman primates. Hopefully, these techniques will aid in developing a vaccine against AIDS and other human retrovirus diseases.

For background information see CELLULAR IMMUNOLOGY; ONCOLOGY; OPPORTUNISTIC INFECTIONS in the McGraw-Hill Encyclopedia of Science and Technology.

[DONALD H. MAUL]

Bibliography: M. B. Gardner et al., Simian acquired immunodeficiency syndrome: An overview, in *Acquired Immunodeficiency Syndrome*, pp. 9–27, 1984; R. V. Henrickson et al., Clinical features of simian acquired immunodeficiency syndrome (SAIDS) in rhesus monkeys, *Lab Anim. Sci.*, 34:140–145, 1984; P. A. Marx et al., Simian AIDS: Isolation of a type D retrovirus and transmission of the disease, *Science*, 223:1083–1086, 1984.

Soil chemistry

The uppermost layer of the Earth's crust, the soil, is an environment in which there is constant chemical activity. This article discusses the mechanisms by which soil mineral colloids bind humic substances, the processes of soil mineral degradation by organic acids, and the measurement and control of cation activities around roots.

Mechanisms of binding. Humic substances are the principal organic components of soils and waters, in which they interact with colloidal inorganic soil constituents, primarily hydrous oxides and clay minerals, to form associations of widely differing chemical and biological properties and stabilities. The products of these interactions affect the moisture and aeration regime, exchange capacity, nutrient availability, chemical and biological degradation, as well as many other reactions which occur in these systems.

Interactions with hydrous oxides. In soils and sediments, hydrous oxides of iron and aluminum provide large surfaces for the absorption of humic substances. At low pH values, these compounds are positively charged and can bind negatively charged humic substances by electrostatic bonding. More stable complexes, that is, inner-sphere complexes between the humic substances and the iron (Fe) and aluminum (Al) of the hydrous oxides, are formed when oxygen-containing functional groups of the humic materials penetrate the coordination shells of the metal ions and displace water of coordination.

Freshly precipitated iron hydroxides and aluminum hydroxides absorb humic acid and fulvic acid from aqueous solutions, with aluminum hydroxides having higher adsorption capacities than iron hydroxides.

In addition to the modes of bonding mentioned above, hydrogen bonding and van der Waals interactions between adsorbed humic acid and fulvic acid molecules or between adsorbed molecules and mineral surfaces are also operative. The van der Waals interactions are likely to be significant at high ionic strength or when the system is allowed to dry, which establishes intimate contact between solute and surface. The adsorption of humic substances on hydrous oxides changes their surface charge and tends to make them negative.

Humic substances dissolve soil minerals. This has significant effects on the weathering cycle and on soil genesis. For example, a 0.2% weight per volume solution of fulvic acid dissolves about 26% of the initial weight of thuringite after shaking at room temperature for 360 h, compared to distilled water which dissolves only 6% under the same experimental conditions. If a mineral is relatively rich in iron, it is more readily decomposed because of the strong complexing power of fulvic acid for iron. Through their ability to solubilize and complex metals from soils and rocks, humic substances play an important role in the migration of metals in the hydrosphere. Recent reports suggest that even gold is transported as a humic complex. The considerable quantities of metals leached by humic substances from the Earth's surface over wide areas and long periods of time and deposited in particular environments lead to the formation of ore bodies. Thus, nature has provided a very efficient leaching and transporting agent for metals. See GOLD, GEOCHEMISTRY OF.

Increasing concentrations of humic acids and fulvic acids inhibit the crystallization of iron oxides (goethite, lepidocrocite, and hematite) and of aluminum oxides (gibbsite, boehmite, bayerite, and nordstrandite), favoring instead the formation of amorphous ferrihydrite and pseudoboehmite, respectively. Interferences of humic acids and fulvic acids with the crystallization of these oxides are due to strong complexing of iron and aluminum by humic materials. As far as the crystallization of alu-

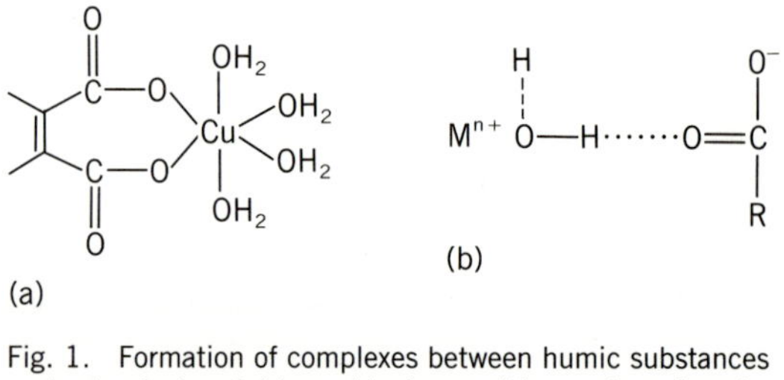

Fig. 1. Formation of complexes between humic substances and polyvalent metal ions of hydrous oxides or clay minerals. (a) An inner-sphere complex. (b) Water-bridge between the water of coordination of metal ion M^{n+} with a carboxylate group of humic substance R.

minum hydroxides is concerned, the occupation of coordination sites of aluminum by humic ligands disrupts the hydroxyl-bridging mechanism which is indispensable for the polymerization of aluminum ions. The stronger the humic ligand complexes aluminum, the greater is the disruptive effect on the crystallization of the aluminum. It is significant that substantial concentrations of crystalline iron oxides and aluminum oxides are absent in soils formed under cool climates, relatively rich in humic substances, although the total concentrations of iron and aluminum in these soils are high.

Humic substances can influence the types of iron oxides which are formed in soils, particularly the ratio of goethite to hematite. In the presence of humic materials, goethite or lepidocrocite concretions are converted by forest fire to maghemite. Increasing concentrations of humic substances favor the formation of the following iron compounds in soils: hematite → goethite → ferrihydrite → iron-humic complex.

Catalytic effects of iron oxides with high surface areas are known to exist, and may be operative in the synthesis and polymerization of humic materials.

Interaction with clay minerals. Appreciable adsorption of humic substance on clay minerals is observed when the clay is saturated with polyvalent cations which act as bridges between negatively charged oxygen-containing functional groups (mainly COO^-) of humic substances and mineral surfaces. Inner-sphere complexes (Fig. 1a) are often formed between humic substances and iron and aluminum on clays. In hydrated clay systems, anionic groups of humic substances are normally associated with exchangeable cations by hydrogen-bonding to water molecules in primary hydration shells of the metals. This mechanism (Fig. 1b) is referred to as the formation of water bridges. The resulting complexes resemble outer-sphere complexes, which are less stable than inner-sphere complexes. The formation of microaggregates in soils is visualized to involve the binding of organic matter (OM), consisting mainly of humic substances, to the polyvalent cation (P) on the clay (C) via water bridge (H_2O) to produce the complex C—P—H_2O—OM.

The adsorption of humic substances by clays is proportional to the crystal area and geometry of the surface of the clay. At a comparable pH, the extent of adsorption decreases in the following order: montmorillonite > vermiculite > kaolinite > chlorite > biotite > muscovite (see Fig. 2).

pH effect on adsorption of humic substances. The adsorption of humic acid and fulvic acid on external surfaces of sodium (Na) montmorillonite decreases with increases in pH. Also, the interlayer penetration of fulvic acid into Na-montmorillonite (Fig. 3) decreases between pH 2 and 5, with the steepest drop occurring between pH 4 and 5, which is close to the overall pK value of the fulvic acid. Thus, the magnitude of the d_{001} spacing is governed by the degree of ionization of the functional groups, particularly —COOH groups, of the fulvic acid. At pH <

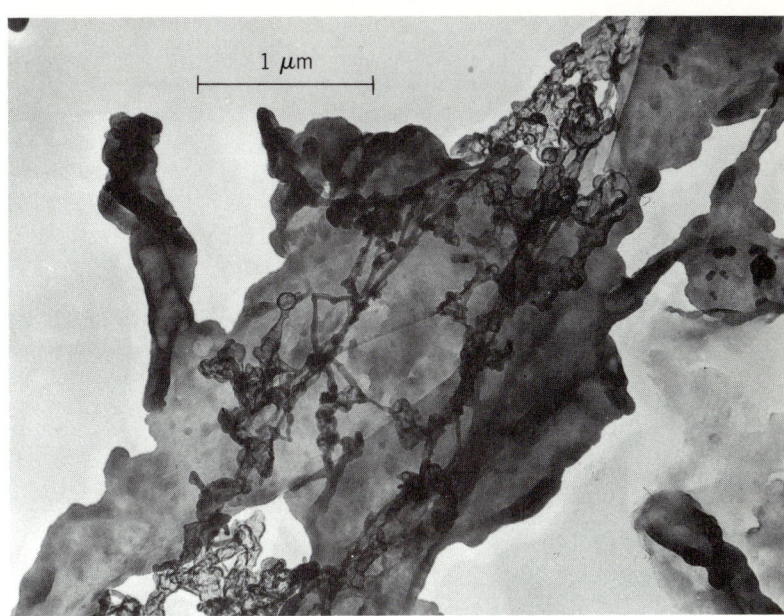

Fig. 2. Adsorption of fulvic acid (fibrous material) on sodium montmorillonite.

4, relatively few of these groups are ionized, so that the fulvic acid behaves like an uncharged molecule which competes with water for adsorption sites around exchangeable Na^+ ions. The bonding mechanism appears to be an ion-dipole interaction. The formation of some inner-sphere complexes could conceivably also occur because some Al^{3+} ions, released from the silicate interior at low pH, could migrate to exchange sites and interact with the fulvic acid. As the pH rises, more and more functional groups ionize, which results in a negative charge on the fulvic acid. The latter is then repulsed by the negatively charged clay. At pH 2, two layer thicknesses of fulvic acid are adsorbed in the interlayers of montmorillonite.

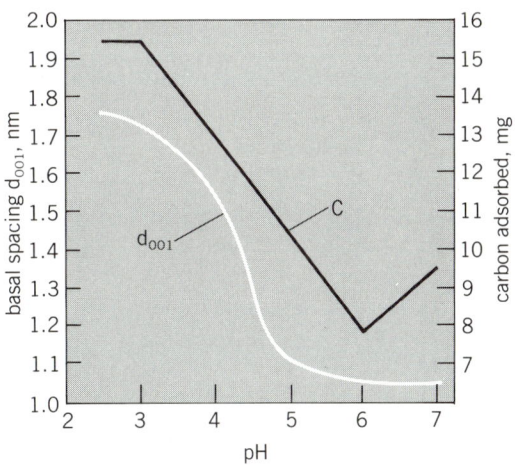

Fig. 3. Effect of pH on the d_{001} spacing of sodium montmorillonite and the adsorption of fulvic acid, which contains 50% carbon by weight and thus is double the weight of carbon adsorbed. (*After M. Schnitzer and H. Kodama, Montmorillonite: Effect of pH on its adsorption of a soil humic compound, Science, 153:70–71, 1966*)

Stability of complexes formed. Humic substances are resistant to biodegradation compared to such biopolymers as proteins and polysaccharides. This stability is related to the random manner in which constituent units of humic acid and fulvic acid are joined together. Therefore, a large variety of enzymes would be required at any given time to extensively hydrolyze such a "heteropolycondensate." The presence of polyvalent cations in the system will contribute to a still greater stabilization of the structures by forming strong complexes through the functional groups of the humic materials. Adsorption on clay minerals will lead to an increase in attractive interactions between the solute and the surface. These factors, along with the physical and steric inaccessibility of the adsorbed material to microorganisms or extracellular enzymes, will enhance even more the biostability of humic substances in soils and sediments.

[M. SCHNITZER]

Decomposition by organic acids. Soil minerals are inorganic compounds which are derived from rocks, and which are broken down constantly by physical, chemical, and biochemical forces to serve as the parent materials for soils. Under the influence of temperature differences and freezing water, minerals expand and contract, causing rock to fracture. Plants, both alive and dead, and organic waste produce various organic and inorganic acids, creating a chemical environment that contributes to the breakdown, or decomposition, of the soil minerals. Part of the soil minerals is dissolved, and part remains

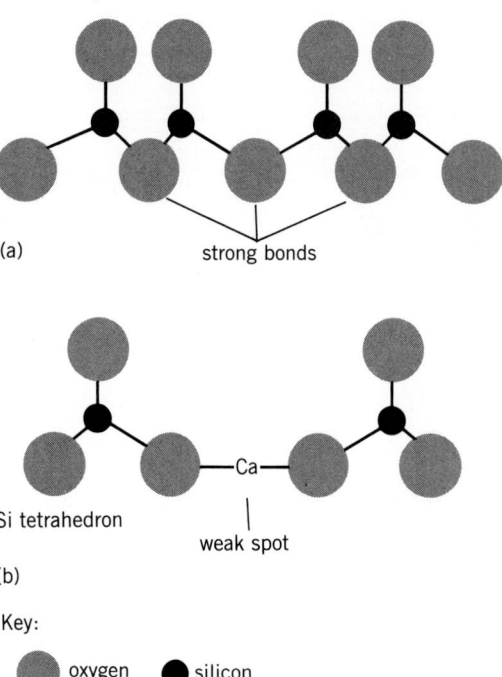

Fig. 4. Linkages of silica tetrahedra. (*a*) Linkage of several tetrahedra by mutually sharing oxygen atoms. The Si—O—Si bond, called the siloxane bond, is the strongest bond in nature. (*b*) Two silica tetrahedra joined together by a calcium (Ca) atom. This cationic bond is a weak spot in the mineral structure.

undissolved. Still another part of the soil minerals is transformed into clay minerals. For this reason, clay is frequently classified as a secondary mineral, in contrast to feldspars, micas, and the like which are called primary minerals. The undissolved part makes up the sand and silt fractions, which together with the clay form the inorganic soil matrix. All of them are subject to further decomposition and alteration processes. The dissolved part contains the inorganic elements that nourish plant growth.

Soil minerals. The inorganic fraction of soils is subject to decomposition processes and is composed of rock fragments (>0.08 in. or 2.0 mm), sand (0.08–0.002 in. or 2.0–0.050 mm), silt (0.002–0.00008 in. or 0.050–0.002 mm), and clay (<0.00008 in. or 0.002 mm). Despite the variability in composition, these fractions are composed mostly of silicate and oxide minerals. The silicate minerals are classified in terms of increasing complexity of the SiO_4 tetrahedra linkages into neso-, soro-, cyclo-, ino-, phyllo-, and tectosilicates. The silicate clays belong mainly to the phyllosilicates, although some of the minerals in the sand and silt fractions can also be in the phyllosilicate group, such as micas. Pyroxenes and amphiboles are inosilicates, whereas quartz and feldspars are minerals of the tectosilicate group.

Crystal chemistry and stability of soil minerals. Mineral stability depends to a large degree on the strength of the atoms binding neighboring atoms in the crystal structure. The four major types of bonding between atoms in crystals are ionic, homopolar, metallic, and van der Waals; ionic and homopolar bonds yield, in general, hard crystals with high melting points. Most of the bonds in soil minerals are ionic in nature. Single or several units of SiO_4 tetrahedra can be linked together by mutually sharing oxygen atoms (Fig. 4). The Si—O—Si linkage, called the siloxane bond, is the strongest bond in nature and will be the most resistant to weathering attack. Quartz is an example of a mineral with such a bond. The SiO_4 tetrahedra can also be linked through cation bonds, as in inosilicates, where double chains of SiO_4 tetrahedra are joined together by Ca or Mg (Fig. 4*b*). The cations acting as the connecting linkage are considered nonframework atoms, and form the weakest spots in the crystal. An example is amphibole. On the basis of a progressive increase in siloxane bonds, the silicate minerals have been ranked for stability as follows: olivine < pyroxene < hornblende < biotite < quartz.

Soil organic acids. Soil organic acids comprise two broad groups, nonhumified and humified. The nonhumified acids range from simple aliphatic acids [for example, acetic acid (vinegar), formic acid (an acid secreted by ants), and oxalic acid] to complex aromatic and heterocyclic acids, (for example, vanillic acid, tannic acid, and benzoic acid). Plant and animal tissue contain a large number of these acids as intermediate products of metabolism. They are released in the soil either as root exudates or as products of organic matter decay. The concentration of these acids is generally very small (0.5 to 0.9 millimole/100 g soil). The humified acids are essential soil constituents beneficial for the physical, chemical, and biological conditions of soils. Because of their differences in solubility in alkali, acid, and alcohol, they are separated into fulvic acid, humic acid, hymatomelanic acid, and humin. The concentration of fulvic and humic acids varies considerably from soil to soil, but its magnitude, ranging from 500 to 2000 mg/100 g soil, indicates concentrations far exceeding those of the nonhumified acids.

Effectiveness of organic acids in mineral decomposition. The nonhumified organic acids may be as effective in mineral decomposition as humic acids; however, since they are present in very small concentrations, their effect is masked by that of fulvic and humic acids. On the basis of chemical reactivity, the organic acids can be divided into two groups: (1) organic acids in which the acidic characteristics are attributed only to the presence of —COOH groups; and (2) organic acids in which the acidic characteristics are attributed to —COOH and phenolic OH groups. Acetic acid, formic acid, and oxalic acid belong to the first group of acids. They exhibit a complexing capacity, but seldom a strong chelation reaction, and their effect on mineral decomposition is more through the acidic (H^+ ions) effect. Chelation produces stronger bonds than complexation. The second group of acids includes fulvic and humic acids. By virtue of the presence of —COOH and phenolic OH groups in their molecules, they have the advantage over the simple or-

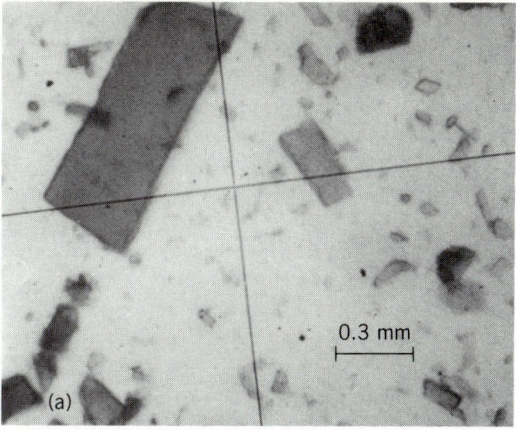

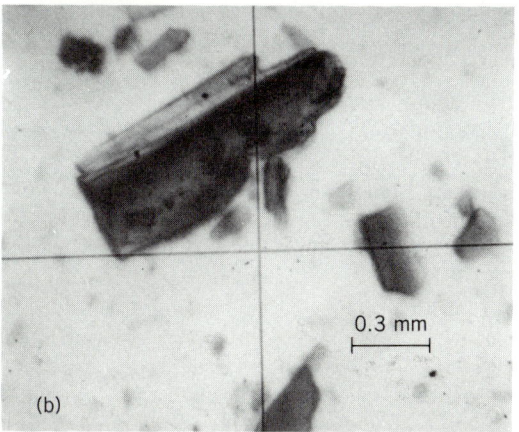

Fig. 6. Petrographic micrographs of the mineral hornblende, $Ca(Fe,Mg)_2 Si_4O_{12}$ (a) before treatment with humic acid, and (b) after treatment with 400 ppm humic acid. In nature, dissolution does not take place as rapidly as in the laboratory. Because of the prevailing dilute condition, it takes approximately 5 years or more under natural conditions to completely dissolve the mineral particle shown.

Fig. 5. Three types of interactions between humic acid and a metal cation. (a) Electrostatic attraction. (b) Chelation. (c) Coadsorption (H_2O bridge). R = the remainder of the humic acid structure, and M = cation with valence $n+$. Note that in chelation the M^{n+} is grabbed by both the carboxylic (—COOH) and the phenolic-OH groups.

ganic acids by being able to exert an acidic effect and an interaction effect (Fig. 5). By forming organometal chelates, these organic acids bring about the dissolution of soil minerals. The greater the affinity of the metal atoms in the mineral for interaction with humic acids, and the more stable the metal-humic chelates formed, the easier the dissolution of the cation from the minerals.

Dissolution process. The rate of mineral dissolution occurs rapidly during the first 24 h, and slows down thereafter to reach near-constant values after 5 days or longer. Aluminum and iron are frequently listed as exceeding other elements in amounts extracted from the minerals by the acids. Analysis with the petrographic microscope yields evidence for dissolution to proceed from the periphery toward the inner layers of the mineral (Fig. 6). Scanning electron microscopy shows the development of a diffusion layer on the surface of the dissolving mineral, which acts as a diffusion barrier; this accounts for the decrease in dissolution rate with time.

Dissolution products. The dissolution products have an important bearing on soil formation and fertility. Not only will they form the different horizons in soils, but they will also promote or inhibit development of new soil minerals. The formation of spodic horizons in spodosols is an example and is caused by the interaction of aluminum or iron with humic acid and the subsequent migration of the chelates to the subsoil. The formation of mollic and andic epipedons in mollisols and andosols, respectively, offers additional examples.

Many of the cations chelated are essentially micronutrients for plant growth, such as iron, manganese, copper, and zinc. As cations they tend to be converted rapidly into insoluble hydroxides, but as metal-humic chelates they may remain soluble, increasing the diffusion and mass flow of micronutrients to roots. Depending on soil pH and the stability of the chelates, they can then be made available to plants by an exchange mechanism.

[KIM H. TAN]

Cation activities around roots. Plant-available (labile) cations may exist in soils as the dissolved free ions, as dissolved complexes with organic or inorganic ligands, as exchangeable or specifically adsorbed ions, or in some cases as precipitates. Current evidence indicates that plant roots absorb cations as the free ion but that rate of transport to the root surface by either diffusion or mass flow is a function of the total dissolved species. In addition, diffusion-controlled transport depends upon the buffer power, which is the change in the concentration of total labile species (usually dissolved plus adsorbed) per unit change in concentration of total dissolved species. Therefore, estimation of nutrient absorption rates using a mechanistic approach involving simulations of transport by diffusion and mass flow requires methods for determining the concentrations of the various forms of the nutrient present in the soil (speciation) and an understanding of the reactions that determine relationships between species.

Cation-exchange and adsorption sites. Bonding of cations to cation-exchange sites is essentially ionic, and the relatively low degree of specificity for different cations is a function of ion charge, hydration energy, and polarizability. Cation-exchange sites arise from negatively charged sites on surfaces of mineral or organic particles. They may be subdivided into permanent sites (sites that do not form stable coordinate covalent bonds with hydrogen ions over the normal range of soil pH) and pH-dependent sites (sites that form stable coordinate covalent bonds with hydrogen ions at low pH but exist as negatively charged sites when the hydrogen dissociates at higher pH). Specific adsorption sites are either electron-pair donor sites which favor cations that form strong multiple coordinate covalent bonds (Fe^{3+}, Cu^{2+}, Zn^{2+}, and so on), or they may be sites with dominantly ionic bonding where the specificity depends on ion hydration energy or steric factors (K^+ "fixation" in micaceous minerals or vermiculite). Bonding of cations in dissolved complexes is similar to that for specific adsorption.

In the mass-action reactions below, describing reactions involving soluble complexes [reaction (1)], cation exchange [reaction (2)], and specific adsorption [reaction (3)], $M^{(m)}$ represents a cation of charge

$$M^{(m)} + NL^{(n+l)} \rightleftarrows N^{(n)} + ML^{(m+l)} \quad (1)$$

$$nM^{(m)} + mNX_n \rightleftarrows nMX_m + mN^{(n)} \quad (2)$$

$$M^{(m)} + NY^{(y+n)} \rightleftarrows N^{(n)} + MY^{(y+m)} \quad (3)$$

m, $N^{(n)}$ a competing cation of charge n, L a ligand (or group of similar ligands) of overall charge l, X a cation-exchange site of unit negative charge, and Y a specific adsorption site whose charge y depends on site characteristics and possibly pH. In reactions (1) and (3), an ion-for-ion exchange is assumed, with any charge difference between m and n being expressed as a change in charge on the dissolved ligand or the specific adsorption site. In reaction (2), equivalent amounts of charge are exchanged.

Equilibrium constants can be calculated for reactions (1–3), but they will not be true constants, partly because in soils there are many types of sites in each of these categories with each type having different selectivities for the different cations, and partly because treating an adsorbed cation and one or more surface adsorption sites as a molecular species is not valid. Because of this variability, the term selectivity coefficient is used instead to describe the relationship between the ratio of adsorbed or complexed cations and the ratio of cations in solution (each activity term raised to the appropriate power), as determined under specified experimental conditions.

Exchangeable cations. In most soils, concentrations of dissolved constituents are low compared with the total labile cations (exchangeable + specifically adsorbed + dissolved), and the number of cation-exchange sites far exceeds that of the specific adsorption sites. The most abundant cations in most soil solutions and on most cation-exchange sites (Ca^{2+}, Mg^{2+}, Na^+, K^+) have little affinity for the specific adsorption sites, except that K^+ (and NH_4^+) is preferentially adsorbed by micaceous minerals and vermiculite. Many metal ions such as Cu^{2+}, Zn^{2+}, Cd^{2+}, and Fe^{3+}, however, are strongly bonded to electron-pair donor sites on organic and mineral surfaces and can compete successfully against Ca^{2+} for those sites, even though the Ca^{2+} may be present at more than 10^5 times their concentrations. High selectivity for these metals is also the case with many dissolved organic ligands. Hydrogen ions, on the other hand, compete strongly for electron pair donors so that metal ions are progressively displaced from these sites as the pH is decreased.

Trace metals normally constitute only a very small proportion of the exchangeable cations because they show little, if any, specificity over the major cations on these sites. In acid soils, Al^{3+} and Mn^{2+} become soluble enough to compete for exchange sites, and Fe^{2+} and Mn^{2+} become important under anaerobic conditions.

Measurement of cation activity. Activities of major cations in the soil solution can be measured or estimated as follows: (1) activity measurements in displaced soil solution with ion-selective electrodes; (2) concentration measurements in displaced soil solution by various methods with activity corrections based on estimates of ionic strength and complex formation; and (3) concentration measurements in water extracts at a specific solution/soil ratio and back calculation of activities at an assumed soil water content. This calculation requires that the selectivity coefficients for the cation-exchange reactions be known or assumed, that the proportions of cations in the adsorbed phase are unchanged or changed to a known extent, and that no solid-phase dissolution occurs on dilution.

Determination of trace cation activities at realistic soil solution concentrations presents more of a challenge. In most cases, activities are below detection limits of specific-ion electrodes. However, a cation-exchange membrane has been used to separate the free cation from complexed forms, which permitted determination of concentrations of free metal ions down to the detection limits of the graphite-furnace atomic absorption spectrometer.

In other research metal-ion activities have been estimated from measurements of total concentration. This normally requires that the concentrations and selectivity coefficients for all complexing ligands be known. Although selectivity coefficients for most inorganic complexes are available, very little is known concerning the composition and selectivity coefficients of many of the organic ligands. A computer simulation program, GEOCHEM, has been developed which uses total dissolved carbon to estimate total organic ligand and a mixture of organic compounds with known selectivity coefficients to simulate complexing properties. Recently a method for directly determining gross selectivity coefficients of dissolved ligands was proposed. It involves equilibrating the solution containing the ligands with a metal-bearing chelating resin which supports known activities of the metal ions in solution. The gross selectivity coefficient for cadmium versus zinc was calculated from the known metal-ion activity ratio and the ratio of the complexed metals (total metal minus free ion).

In another recent approach to determining trace-metal activity ratios in soil suspensions, a small amount of ligand with known selectivity coefficients was added and the increase in dissolved metal was determined. Activity ratios of the free ions were estimated from the ratios of the concentration increases (assumed to equal metal complexed by the added ligand) and the known selectivity coefficients. The activity of Cd^{2+} was then determined from the increase in Cd concentration on adding Cl^- ($CdCl_n^{(2-n)}$ complex formation), and activities of other metals were then calculated from the previously determined Cd^{2+}/N^{2+} ratios. For activity estimates to be accurate, activities of the metals should not be affected by addition of ligands; that is, addition of the complexing agent should not dissolve so much of the adsorbed metal that a significantly lower activity of free metal ion in solution results.

Buffer power. The methods described here are mainly concerned with determining speciation and activities of nutrients in the solution phase. Methods are also required for accurately determining change in buffer power as nutrient concentrations are depleted to very low levels near the soil-root interface. To date, most buffer power estimates have been based on adsorption curves, whereas plant uptake is a desorption process. For most elements, adsorption and desorption curves are very different, and more emphasis must be placed on developing methods for determining the change in buffer power in a desorbing system at realistic soil solution concentrations.

For background information *see* ACTIVITY (THERMODYNAMICS); ADSORPTION; BUFFERS (CHEMISTRY); CHELATION; CHEMICAL EQUILIBRIUM; CLAY MINERALS; COMPLEX COMPOUNDS; COORDINATION CHEMISTRY; HUMUS; SOIL; SOIL CHEMISTRY in the McGraw-Hill Encyclopedia of Science and Technology.

[RICHARD B. COREY]

Bibliography: P. W. Birkeland, *Pedology, Weathering and Geomorphological Research*, 1974; F. De Coninck, Major mechanisms in formation of spodic horizons, *Geoderma*, 24:101–128, 1980; J. B. Dixon and S. B. Weed (eds.), *Minerals in Soil Environments*, 1977; R. C. Evans, *An Introduction to Crystal Chemistry*, 1939; R. Fujii, L. L. Hendrickson, and R. B. Corey, Ionic activities of trace metals in sludge-amended soils, *Sci. Total Environ.*, 28:179–190, 1983; D. J. Greeland and M. H. B. Hayes (eds.), *The Chemistry of Soil Processes*, 1981; L. L. Hendrickson and R. B. Corey, A chelating-resin method for characterizing soluble metal complexes, *Soil Sci. Soc. Amer. J.*, 47:467–474, 1983; H. Kodama, M. Schnitzer, and M. Jaakkimainen, Chlorite and biotite weathering by fulvic acid solutions in closed and open systems, *Can. J. Soil Sci.*, 63:619–629, 1983; J. K. Lampert, *Measurement of trace cation activities by Donnan membrane equilibrium and atomic absorption analysis*, Ph.D. thesis, University of Wisconsin-Madison, 1982; S. V. Mattigod and G. Sposito, Chemical modeling of trace metal equilibria in contaminated soil solutions using the computer program GEOCHEM, in E. A. Jenne (ed.), *Chemical Modeling: Speciation, Sorption, Solubility and Kinetics in Aqueous Systems*, Amer. Chem. Soc. Symp. Ser. 93, 1979; B. K. G. Theng, *Formation and Properties of Clay-Polymer Complexes*, 1977.

Soil fertility

Due to concern for soil erosion, fuel costs, machinery inventories, and timeliness, the amount and degree of tillage for seedbed preparation is being reduced by crop producers. Conservation tillage is the broad term used to categorize tillage systems that cause less soil disturbance and leave plant residues from the previous crop on the soil surface.

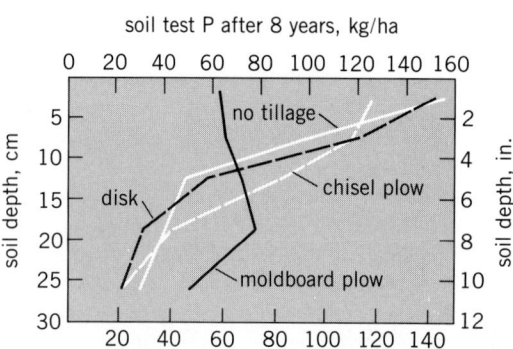

Fig. 1. Effect of tillage system on the distribution of soil test phosphorus (P) in the upper 12 in. (30 cm) of the soil profile after 8 years.

Other properties of the soil surface (6 in. or 15 cm) affected by various conservation tillage systems include: increased soil moisture content due to less evaporation; reduced soil temperature; increased organic matter level; increased microbial activity; changed soil pH; and greater density. These factors along with reduced soil disturbance affect soil fertility as well as the options available for most efficient fertilizer application.

Immobile nutrients. Because soils are not inverted with most conservation tillage systems, the top 2 in. (5 cm) of soil often becomes more acidic. This is a result of surface application of nitrogen (N) over a period of many years. Additions of limestone (calcium or magnesium carbonate) will neutralize this surface acidity. Since limestone is immobile, incorporation of limestone to raise soil pH below the top 4 in. (10 cm) of soil is difficult with most conservation tillage systems and is not possible with the no-tillage systems. Therefore, neutralization of soil acidity below the surface requires a tillage system that either inverts the soil or mixes the soil sufficiently so that the liming material is incorporated. Conservation tillage systems can then be used in the following years because of the residual, long-lasting effect of the limestone.

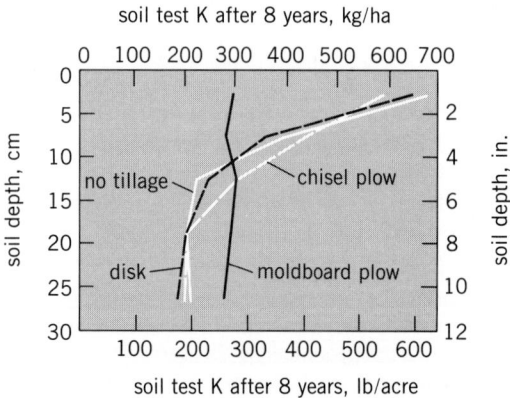

Fig. 2. Effect of tillage system on the distribution of soil test potassium (K) in the upper 12 in. (30 cm) of the soil profile after 8 years.

Immobile elements also accumulate in the top 4 in. (10 cm) of soil when conservation tillage practices are used. Sources of these nutrients are surface-applied fertilizers and recycled nutrients which have been accumulated from the subsoil by plant roots and translocated to the plant top. The degree of accumulation or downward movement depends on the exchange capacity of the soil, rainfall, and the type of tillage used. A crop production system with soils of high cation-exchange capacity (fine-textured soils), low rainfall conditions, and no-tillage will accumulate greater amounts of immobile nutrients close to the soil surface than will systems with low cation-exchange capacity, high rainfall, and some degree of tillage.

Phosphorus (P) and potassium (K) are two essential elements commonly applied as fertilizers which tend to accumulate near the surface with conservation tillage. Substantial accumulation of phosphorus (Fig. 1) and potassium (Fig. 2) occurred in the top 4 in. (10 cm) of soil with continuous no-tillage or an annual disking after an 8-year period on a Webster clay loam in Minnesota. Chisel plowing, which tends to invert some soil, resulted in incorporation to 6 in. (15 cm). Moldboard plowing, which almost completely inverts the soil and is not a conservation tillage method, incorporates and distributes the plant residues and fertilizers found on the soil surface to the depth at which it is operated.

Accumulation of phosphorus and potassium near the soil surface can result in reduced plant uptake and possible nutrient deficiency. This is especially true under dry conditions when root activity is greatest in the moist soil zones found deeper in the soil. Reduced potassium availability has been related to decreased aeration in the more densely compacted surface soils with no-tillage and some other methods of conservation tillage. Soils with a very low cation-exchange capacity generally do not accumulate large amounts of potassium at the surface because the potassium tends to move downward with rainfall. Micronutrients, for example, iron, manganese, zinc, and copper, are attached to both the clay colloids and organic matter and consequently accumulate close to the soil surface. However, cases of micronutrient deficiency associated with their surface accumulation and conservation tillage have rarely been documented.

Mobile nutrients. Microbial populations are increased in surface soil (0–4 in. or 0–10 cm) horizons with conservation tillage but are often lower at depths below 4 in. (10 cm). Increased nitrifier populations in the surface are largely due to increased soil water content associated with the surface residues. Denitrifier populations in the soil surface generally increase at a faster rate than the aerobic microorganisms with conservation tillage due to both the availability of water and carbon and energy sources.

Nitrogen is the primary mobile nutrient that is affected by conservation tillage. Accumulation of plant residues on the soil surface increases the potential for greater immobilization of broadcast fertilizer nitrogen, creates conditions conducive for ammonia (NH_3) volatilization of surface-applied ammoniacal nitrogen sources, and also provides a condition for greater potential denitrification losses. Surface application of nitrogen to a residue-covered soil surface, which is necessary with no tillage, can lead to serious losses of nitrogen by either volatilization or denitrification. Also, tie-up of the nitrogen by immobilization into the soil organic biomass results in less fertilizer nitrogen available for the plant. These potential nitrogen losses necessitate use of specific placement methods, substitute nitrogen sources, nitrification inhibitors, or improved application timing techniques to reduce the potential of these losses and thereby increase nitrogen efficiency.

Injecting fertilizer nitrogen below this zone of high microbial activity resulted in an average corn yield increase of 970 lb/acre (1090 kg/ha) in seven experiments conducted in Indiana. In those studies urea–ammonium nitrate solution was injected below the residue-covered soil surface and compared to a surface broadcast application. Nitrification inhibitors which delay the conversion of ammonium (NH_4^+) to nitrate (NO_3^-) reduce the potential for denitrification losses in soils with greater amounts of water or less air porosity. Application of nitrate sources of nitrogen (ammonium, calcium, or potassium nitrate) to a residue-covered soil surface reduces the volatilization potential compared to ammoniacal sources (urea or ammonium sulfate). Urea is hydrolyzed to ammonium carbonate. The rate of this hydrolysis is increased in the presence of urease, an enzyme commonly found in soil surface residues. Free NH_3 can then be lost (volatilized) from the ammonium salts, especially under high pH conditions. Delaying nitrogen application until the period of greatest plant uptake may also increase nitrogen efficiency on soils where loss potential is greater.

Application methods. Injection or knifing-in of nitrogen is the most efficient method of application in most conservation tillage systems. At present, however, application methods have not been developed that allow farmers to inject nitrogen into a no-tillage system without disturbing the soil or residues.

Application methods for the immobile elements, for example phosphorus and potassium, are not as critical as for the mobile nutrients which need to be applied annually. Phosphorus and potassium have been built to high levels throughout the plow layer on many soils by previous fertilizer applications before conservation tillage became widely practiced. Plants have been able to feed on this residual fertility in addition to the fertilizer applied directly for the crop grown. Starter fertilizers, which contain nitrogen, phosphorus, and potassium placed close to the row at planting, are often recommended to assist in the early growth of the plant and to maintain adequate fertility for sustained growth and yield in conservation tillage systems.

Elevated phosphate concentrations in runoff waters from no-tillage systems have been observed where phosphorus fertilizers have been broadcast-applied. Consequently, subsurface applications of the immobile nutrients may also be necessary to reduce environmental concerns where they occur and to increase fertilizer efficiency.

For background information see AGRICULTURAL SOIL AND CROP PRACTICES; FERTILIZER; FERTILIZING; PLANT MINERAL NUTRITION in the McGraw-Hill Encyclopedia of Science and Technology.

[GYLES RANDALL]

Bibliography: J. W. Doran, Soil microbial and biochemical changes associated with reduced tillage, *Soil Sci. Soc. Amer. J.*, 44:765–771, 1980; D. B. Mengel, D. W. Nelson, and D. M. Huber, Placement of nitrogen fertilizers for no-till and conventional till corn, *Agron. J.*, 74:515–518, 1982; G. W. Randall, Fertilization practices for conservation tillage, in *Proceedings of the 32d Annual Fertilizer and Agricultural Chemical Dealers Conference*, Iowa State University, 1980.

Soil microbiology

Aggregation of soil and the strength and stability of the aggregates formed are fundamental soil properties. Soil aggregation plays a large role in water infiltration, the resistance of soil to water and wind erosion, and provision of a good physical environment for plant root growth. It is controlled by many soil components and properties, but the role of microorganisms is especially important.

Soil aggregation process. It has long been established that microorganisms can increase soil aggregation and that the process can be influenced by the types of microorganisms and substrates present in the soil. Microorganisms influence soil aggregation by several mechanisms: they physically bind soil particles together, especially filamentous microorganisms such as actinomycetes and fungi; they hold soil particles together by the ionic charges present on their cell walls and by the cells' causing the particles to stick together; and they produce extracellular products such as gums and polysaccharides. The types and numbers of microorganisms, their extracellular products, the cellular components released upon death, and the different substrates available for microbial utilization are all factors that affect the aggregation process.

Inoculation of soil with microorganisms to promote aggregation does not appear to be a particularly feasible approach at this time; thus if aggregation can be improved, other approaches may be more feasible. The major influence of microbial cells and extracellular products on soil aggregation in cropland likely occurs during decomposition of crop residues. Whether or not the microbial cells or extracellular products are the primary contributors to soil aggregation during residue decomposition probably depends on the nitrogen content of the residue. However, it has been difficult to separate these processes in order to develop residue management practices that would encourage maximum production of microbial biomass or polysaccharides and gums. Separation of these processes and an understanding of their importance to aggregation may allow the design of residue management systems for specific situations and conditions to gain maximum beneficial effect on the soil.

Recent studies have indicated that the initial decomposition of wheat straw can be carbon-limited as well as nitrogen-limited. They also have predicted that a nitrogen limitation during this phase of decomposition would result in polysaccharide synthesis. The readily available carbon during the initial residue decomposition is used to produce the microbial biomass, gums, and polysaccharides, depending on nitrogen and carbon availability. Experiments give credence to the prediction that decomposition of straw under nitrogen-limited conditions would lead to increased polysaccharide production and resultant soil aggregation. Winter wheat straw with the lowest nitrogen content caused the greatest aggregation of soil and volcanic ash, with evidence that this treatment produced the largest amounts of polysaccharide and gum during the decomposition period. Although it may be presumed that the high-nitrogen–containing straw would generate greater microbial biomass, the straw caused less aggregation, indicating polysaccharide production had the most significant role under these conditions. The data from these studies indicate that two mechanisms play a primary role in microbially mediated soil aggregation as affected by residues. Aggregation from high-nitrogen–containing residues such as alfalfa and from other leguminous residues seems to be affected by the type and quantity of organisms present in the soil, while that from low-nitrogen residues results primarily from microbial extracellular products produced when nitrogen is limited. Thus, surface management of low-nitrogen–containing crop residues would offer the greatest potential for improving soil aggregation from these sources. Incorporation of low-nitrogen crop residues into the soil would greatly decrease the opportunity for polysaccharide formation, because nitrogen, available from the soil, would no longer be limiting.

Plant roots and their excretory products also influence soil aggregation by producing gums and by binding the soil with roots and root hairs. Roots are also colonized by microorganisms which use root excretions as a source of carbon and energy. As the microbial cells grow on root surfaces, the organisms and their extracellular products influence soil aggregation, as well as microbial cells and products resulting from senescence and decomposition of the roots.

Green manures. Incorporation of high-nitrogen–containing residues such as alfalfa for green manure, especially for shallow incorporation, may improve soil aggregation because the large microbial biomass generated would have more opportunity to interact and bind the soil particles. There is evidence that soils which utilize green manures and pasture in the rotation are less erodible than those that do not. Soil from a farm using green manures and native soil fertility for crop production was compared with the soil from an adjacent farm receiving recommended amounts of commercial fertilizers. Generally, the soil from the first farm had a higher potential soil biological activity, indicating a greater soil microbial biomass. The resultant increases in soil microflora should increase soil aggregation.

Fungal hyphae. When wet soil aggregates are taken from a field and dissected, hyphae are found encrusted with fine clay particles. The presence of both living and dead hyphae is considered to be responsible in part for the stabilization of the larger macroaggregates. While individual hyphae are not very strong, many hyphae, along with fine plant roots, develop a strong network that will hold an aggregate together and protect it against dissolution when subjected to rapid wetting. The network formed by the fungal hyphae and roots physically binds the particles. Also, the fungi as well as the roots produce gums that seem to function as cementing or binding agents. Laboratory studies indicate that aggregate stabilization by hyphae depends on access to readily available substrates. When the fungi utilize the substrates, there is a spurt of growth and an increase in the number of hyphae present in the soil.

Mycorrhizae, a group of fungi that form a symbiotic association with plant roots, tend to be most abundant in soils where the nutrients are low or unbalanced. Such soils support smaller microbial populations than fertile soils. Despite lower numbers, microbially mediated aggregation occurs as the mycorrhizal fungi produce an extensive network of hyphae. In these soils, aggregation is correlated with the total length of hyphae and roots present.

For background information *see* FUNGI; MYCORRHIZAE; POLYSACCHARIDE; RHIZOSPHERE; SOIL MICROBIOLOGY; SOIL MICROORGANISMS in the McGraw-Hill Encyclopedia of Science and Technology.

[L. F. ELLIOTT; D. E. STOTT]

Bibliography: H. Bolton, Jr., et al., The effect of fertilization and cropping practices on soil microbial biomass and selected soil enzyme activities, *Soil Biol. Biochem.*, vol. 17, 1985; L. F. Elliott and J. M. Lynch, The effect of available carbon and nitrogen in straw on soil and ash aggregation and acetic acid production, *Plant and Soil*, 78:335–343, 1984; C. M. Gilmour, O. N. Allen, and E. Truog, Soil aggregation as influenced by the growth of mold species, kind of soil and organic matter, *Proc. Soil Sci. Soc. Amer.*, 13:292–296, 1948; J. P. Martin, Decomposition and binding action of polysaccharides in soil, *Soil Biol. Biochem.*, 3:33–41, 1971; J. M. Tisdall and J. M. Oades, Organic matter and water-soluble aggregates in soils, *J. Soil Sci.*, 33:141–163, 1982; S. A. Waksman and J. P. Martin, The role of microorganisms in the conservation of soil, *Science*, 90:304–305, 1939.

Soil science

Soil acidity is a problem throughout the world. Acid mineral soils of less than pH 5 contain aluminum as the principal exchangeable cation. The aluminum is toxic to plant growth and inhibits uptake of a number

Table 1. Amounts of aluminum and calcium in exchangeable form and in soil solution and the percent saturation of the cation exchange sites of several acid soils

Soil no.	Soil pH	Exchangeable, cmol (p$^+$) kg^{-1}*		Saturation, %		Soil solution, mmol (p$^+$) L^{-1}*	
		Ca	Al	Ca	Al	Ca	Al
1	4.9	0.06	1.43	5	95	0.75	0.55
2	4.1	0.44	5.35	6	77	3.24	0.31
3	4.4	1.00	3.67	23	77	5.00	0.53
4	4.5	0.45	1.15	26	70	1.14	0.16

*The symbol p$^+$ represents the charge on the proton.

of essential plant nutrients. Inactivation of the exchangeable aluminum by liming eliminates toxicities due to soil acidity and enhances the uptake of plant-nutrients and the activity of beneficial bacteria.

Chemical properties of acid soils. In humid regions, rainwater moving through the soil removes the basic cations calcium and magnesium which are present on cation exchange sites. In the process the hydrogen ion concentration of the soil solution increases. The hydrogen ions react with the soil clay minerals, which are aluminosilicates, and release aluminum ions. The aluminum ions then occupy the cation-exchange sites which formerly held calcium and magnesium ions. At pH 4.8 to 5.0, about half of the cation-exchange sites are occupied by aluminum ions and the other half by calcium and magnesium, while at pH 4 aluminum is on most of the cation-exchange sites. Thus, in very acid soils a high proportion of the exchangeable cations are aluminum and only a very small proportion consist of calcium and magnesium (Table 1).

Soils with exchangeable aluminum are acid because of the chemical reaction of water with aluminum ions. Water reacts with aluminum ions to form hydroxyaluminum ions and hydrogen ions: $Al^{3+} + HOH \rightleftharpoons AlOH^{2+} + H^+$.

Acid soils can also have fairly high concentrations of manganese ion in the soil solution. If large amounts of manganese-containing minerals are present in soils, the high concentration of hydrogen will dissolve some of the minerals and the manganese concentration of the soil solution will increase, particularly at pH 5 or less.

Soils which have developed under an acid weathering regime contain hydrated oxides of iron and aluminum. The amounts of these oxides in soils tend to increase as the clay content increases. The hydrated oxides of iron and aluminum are very reactive with phosphorus and form sparingly soluble phosphorus compounds. Thus soils with a high clay content have a high capacity to absorb phosphate, and on such soils large amounts of phosphorus fertilizers have to be applied before plants are able to take up adequate amounts of phosphorus. Generally the red clayey soils are high fixers of phosphorus. *See* SOIL CHEMISTRY.

Detrimental effects of soil acidity. Plant growth is affected adversely by soil acidity. Detrimental effects include aluminum toxicity, hydrogen toxicity, manganese toxicity, and calcium deficiency.

Aluminum toxicity. Acid soils contain concentrations of aluminum which are detrimental to root growth. High concentrations of aluminum decrease cell division and elongation. In severe toxicity, roots are thickened and stubby in appearance. Decreased root growth results in less exploration of the soil for nutrients and water, which can adversely affect top growth of plants. The aluminum held on root surfaces or on cell walls is very reactive with phosphate ions. This can result in phosphorus not being translocated in adequate amounts to the aboveground portion of plants and results in reduced plant growth. Aluminum also inhibits uptake of calcium and can result in calcium deficiency of plants. Growth of rhizobia, particularly those associated with temperate legumes, are severely reduced by high concentrations of solution aluminum. This results in reduced nitrogen fixation and decreased growth of legumes.

Hydrogen toxicity. Hydrogen ions can affect plant growth in several ways. Below pH 4.2, membranes of plant roots become leaky and nutrients leak out of the root back into the soil. Growth is severely impaired in such situations. At pH 4.5 to 5, hydrogen ions have an inhibitory effect on the plant uptake of calcium and magnesium, nutrients essential for growth.

Manganese toxicity. Acid soils containing large amounts of manganese minerals have a high concentration of manganese in the soil solution, and plants acccumulate large concentrations of manganese in their leaves. When this occurs, physiological processes in the leaves are impaired and plant growth is reduced.

Calcium deficiency. Certain acid soils may have very low amounts of exchangeable calcium. In such instances, plants may exhibit calcium deficiency symptoms because of the lack of adequate available soil calcium. Calcium deficiency symptoms may also be induced by high concentrations of soil solution aluminum, which inhibits the uptake of calcium by plants and translocation to the leaves.

Beneficial effects of liming. Increased availability of nutrients and thus improved plant growth can be achieved by the application of lime to the soil. Beneficial effects include elimination of aluminum, hydrogen, and manganese toxicities; improvement of the availability of calcium, magnesium, and phosphorus; increase in potassium retention on cation-exchange sites; and increases in the number and activity of bacteria.

Elimination of aluminum, hydrogen, and manganese toxicities. Application of limestone ($CaCO_3$) results in the presence of hydroxyl ions (OH) and calcium ions in the soil solution. The hydroxyl ions combine with the hydrogen ions to form water and also with the aluminum ions to form aluminum hydroxide, a nontoxic form of aluminum. The calcium ions, which are supplied by the limestone, then occupy the cation-exchange sites vacated by the neutralized aluminum. Neutralization of the hydrogen ions causes the pH of the soil to rise. Below pH 5.6, the concentration of manganese in the soil solution is reduced to nontoxic levels because of the formation of manganese compounds with a low solubility. In mineral soils, essentially all of the exchangeable aluminum is neutralized at pH 5.8 to 6 and calcium is the predominant exchangeable cation (Fig. 1). Thus liming of acid soils eliminates toxicities caused by high concentrations of aluminum, hydrogen, and manganese, and provides a soil chemical environment favorable for plant growth.

Because the solubility of limestone in water is very low, limestone must be mixed throughout the volume of soil which is to be amended in order for the lime to react with the soil acidity. In order for the limestone to be in contact with as much soil surface area as possible, it should be finely ground. Limestone which passes a 100-mesh sieve is considered to be 100% reactive.

Calcium and magnesium availability. Application of calcitic lime supplies calcium, while dolomitic lime supplies both calcium and magnesium. When lime is applied to acid soils, the exchangeable aluminum is replaced from cation-exchange sites by calcium or calcium and magnesium, nutrients es-

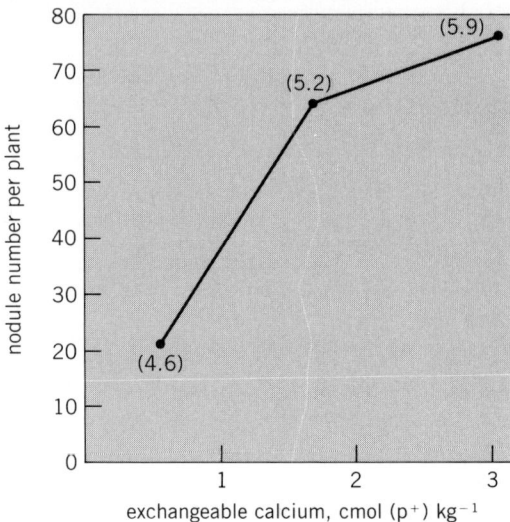

Fig. 2. Relationship between exchangeable calcium level of soil and number of nodules on soybean plants. Numbers in parentheses are soil pH values. The symbol p+ represents the charge on the proton. (*After J. B. Sartain and E. J. Kamprath, Effect of liming a highly aluminum saturated soil on the top and root growth and soybean nodulation, Agron. J., 67:507–510, 1975*)

sential for plant growth. Acid soils are relatively low in these two essential nutrients, and the presence of soil solution aluminum reduces the uptake of calcium and magnesium by plants. An adequate supply of calcium is particularly important for the growth of nitrogen-fixing plants such as soybeans, peanuts, clovers, and alfalfa. Calcium is very important in initiation of nodule development on the roots of nitrogen-fixing plants. When the exchangeable soil calcium level and pH of acid soil were increased by liming, the number of nodules on soybean roots was markedly increased (Fig. 2). Nitrogen fixation takes place in the nodules, and the amount of nitrogen fixed by legumes is increased by liming acid soils.

Availability of phosphorus. Liming has a beneficial effect on availability of phosphorus in soils which have a relatively high amount of exchangeable aluminum in comparison with exchangeable calcium. Addition of lime to an acid soil increased soybean yields substantially (Table 2). Neutralization of the exchangeable aluminum resulted in a twofold increase in the phosphorus concentration of the soybean plant. This was probably brought about by more root growth and greater root exploration of the soil volume and accessibility to soil phosphorus when the soil was limed.

Molybdenum availability. Molybdenum is especially important in the growth of nitrogen-fixing plants because it is required by rhizobia, the nitrogen-fixing bacteria. Liming of acid soils increases the availability of soil molybdenum compounds. When acid soils are limed to pH 6, the solubility of the soil molybdenum compounds is increased sufficiently so that plants can take up the required amounts of molybdenum for optimum plant growth.

Potassium retention on cation-exchange sites. Lim-

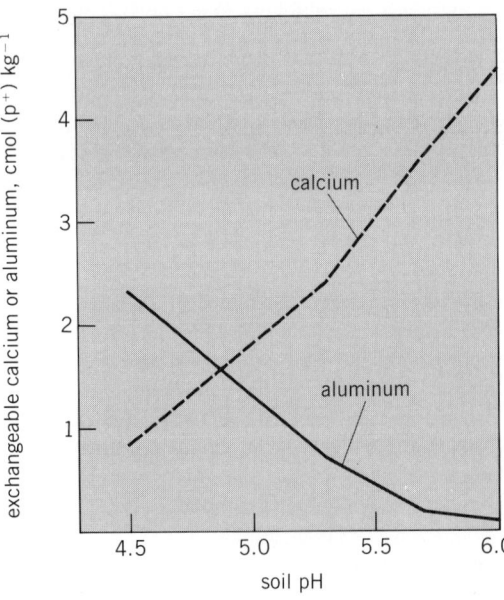

Fig. 1. Exchangeable aluminum and calcium contents of a mineral soil at various soil pH's. The symbol p+ represents the charge on the proton; cmol = centimole. (*After E. J. Kamprath, Exchangeable aluminum as a criterion for liming leached mineral soils, Soil Sci. Soc. Amer. Proc., 34:252–254, 1970*)

Table 2. Effect of aluminum on phosphorus (P) availability and yield of soybeans

Soil pH	Exchangeable, cmol (p^+) kg^{-1}*		Available P, μg/g	Soybean grain, kg/ha	P concentration in whole plant, g/kg
	Al	Ca			
4.9	2.00	0.79	33	1008	1.10
6.0	0.10	2.76	31	1882	2.10

*The term p^+ represents the charge on the proton.

ing of acid soils increases the amount of fertilizer potassium held on cation-exchange sites and thereby reduces leaching losses of potassium. The monovalent potassium ions can more easily replace the divalent calcium ions from the exchange sites of limed soils than it can the trivalent aluminum ions of acid soils. Thus additions of potassium are used more efficiently in crop production when acid soils are limed.

Activity of bacteria. Liming of acid soils to remove the detrimental effects of aluminum and hydrogen ions increases the number and activity of bacteria. Bacteria play an important role in the conversion of ammonium to nitrate and nitrogen fixation by legumes. The more favorable environment for soil bacteria created when soils are limed increases the rate at which these reactions occur.

For background information *see* AGRICULTURAL SOIL AND CROP PRACTICES; PLANT MINERAL NUTRITION; PLANT PATHOLOGY; SOIL CHEMISTRY in the McGraw-Hill Encyclopedia of Science and Technology.

[EUGENE J. KAMPRATH]

Bibliography: F. Adams (ed.), *Soil Acidity and Liming*, Agron. Monogr. 12, 2d ed., American Society of Agronomy, Crop Science Society of America, and Soil Science Society of America, 1984; C. S. Andrew and E. J. Kamprath (eds.), *Mineral Nutrition of Legumes in Tropical and Subtropical Soils*, CSIRO, 1978; R. A. Olson, Fertilizer Technology and Use, 2d ed., Soil Science Society of America, Madison, Wisconsin, 1971. J. T. Sims and B. G. Ellis, Adsorption and availability of phosphorus following the application of limestone to an acid, aluminous soil, *Soil Sci. Soc. Amer. J.*, 47:888–993, 1983.

Soliton

Solitons are stable, localized, pulselike motions that emerge from a collision with similar solitons with unchanged shapes and speeds. In various branches of scientific research, solitons have been described as bumps, vortices, kinks, domain walls, inhomogeneous states, discommensurations, order parameter defects, monopoles, instantons, and so forth.

The concept of solitary wave was introduced by J. Scott Russell in 1844, based upon his observations of the motion of water in a canal after a sluice gate was suddenly opened. The theory of those solitons was developed by D. J. Korteweg and G. de Vries, and thus they are generally referred to as KdV solitons. Until recently the KdV and envelope solitons were the only known solitary waves in fluids. The envelope soliton can be pictured as a rapidly oscillating wave cut off by a smoothly modulating envelope (Fig. 1). Both these solitons propagate at a nonzero group velocity determined by their amplitude and shape. They are also limited to motion in one dimension.

The search for new types of solitons is an important area of research in nonlinear physics. The recent experimental discovery of a nonpropagating hydrodynamic soliton has stimulated soliton research.

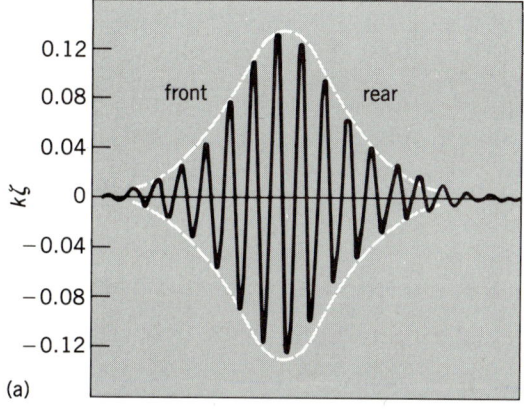

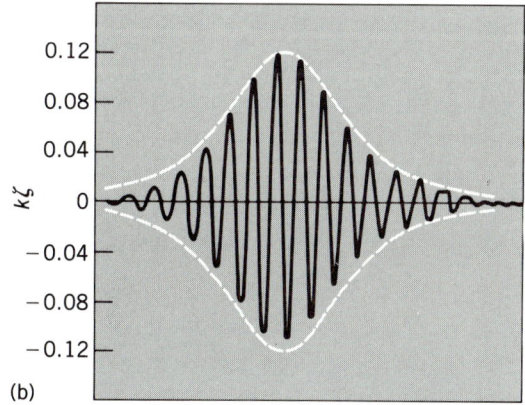

Fig. 1. Measured displacement of water surface showing evolution of envelope soliton at (a) 20 ft (6 m) downstream of wave maker and (b) 98 ft (30 m) downstream of wave maker. The quantity k is the wave number (equal to $2\pi/\lambda$ where λ is the wavelength), and ζ is the displacement, so that $k\zeta$ measures the displacement in units of wavelength. (After M. J. Ablowitz and H. Segur, Solitons and the Inverse Scattering Transform, Society for Industrial and Applied Mathematics, 1981)

Nonpropagating hydrodynamic solitons. Recently it was observed that a fluid can be driven sinusoidally and vertically from below so that all the fluid accelerates up and down in phase. A localized surface pulse could be observed. The motion of the pulse appeared to be so stable and robust that it is probably reasonable to regard it as being a soliton. Furthermore, since this soliton can appear anywhere in the fluid and since it can easily be at rest, it is fundamentally different from other types of solitons. It is called a nonpropagating hydrodynamic soliton. Although the motion is stable, it is rather complex and involves sloshing of fluid back and forth across the width of the channel in which the soliton is produced.

The typical experimental apparatus is a Plexiglas channel, 15 in. (38 cm) long and 1 in. (2.54 cm) wide, filled with water to a depth of 0.8 in. (2 cm). Waves are generated by placing the horizontal trough on a shake table which is driven vertically at an amplitude (peak to peak) of about 0.025 in. (0.06 cm) and at a frequency of about 10 Hz, which is the frequency at which the wavelength of the surface water wave is equal to twice the width of the trough. A stable bump, which is stationary and highly localized in length direction of the channel and oscillating across the width of the trough at half-frequency of the drive, will form on the surface of water. The profile of the bump can be well described by Eq. (1) [Fig. 2], where z is the height of

$$z = \text{sech}\frac{x}{1.12}$$
$$\cdot [2.8 \exp(-1.1y) - 0.70] \quad (\text{cm}) \quad (1)$$

the water surface, x is the coordinate in the direction along the length of the channel, and y is the coordinate in the direction across the channel.

The interaction of two such solitons can also be observed. Two solitons of the same polarity attract and pass through each other, turn around, and then attract each other again. This oscillation can be stable for at least 24 hours (Fig. 3). Two solitons of opposite polarity in close proximity to each other repel each other and slowly move until they are approximately 5 in. (12 cm) apart, and then maintain

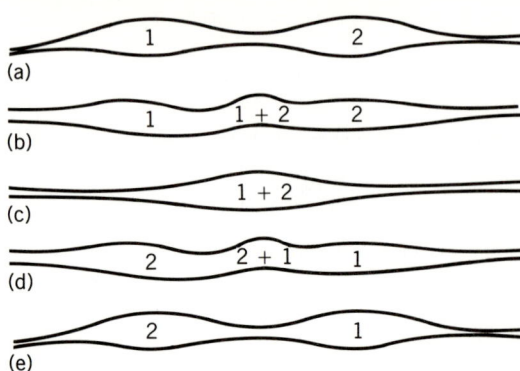

Fig. 3. Stages in the resulting oscillation of two like-phase solitons. (a) Significant overlap between soliton 1 and soliton 2. (b) Sufficient overlap to create an amplitude at the center comparable to the amplitude of solitons 1 and 2. (c) The solitons completely overlap. (d) Solitons 1 and 2 have passed through each other and exchanged places. (e) The original configuration; solitons 1 and 2 have exchanged places. (After J. Wu, R. Keolian, and I. Rudnick, Observation of a nonpropagating hydrodynamic soliton, Phys. Rev. Lett., 52:1421–1424, 1984)

this separation indefinitely. The polarity is defined as the phase of the motion relative to the drive.

A theory which explains this new soliton has recently been developed for the case of an incompressible, inviscid fluid. The essence of the theory is the concept of a cutoff frequency. This is the frequency below which waves of a particular mode do not propagate down a waveguide. Instead they are exponentially cut off. This concept, however, applies to small-amplitude waves (the linear theory). For the high-amplitude case (the nonlinear theory) the cutoff frequency can be a function of the amplitude of motion. It can happen that in some localized region, high-amplitude wave motion reduces the local cutoff frequency below the linear cutoff frequency. If the disturbance in some region has a frequency higher than the nonlinear cutoff frequency but lower than the linear cutoff, this mode will exist in that region, but toward the edges where its amplitude is small it will be self-trapped by an exponential decay. This is why the soliton has a profile in the direction along the length of the channel of the form of the hyperbolic secant function [Eq. (1)], which has tails of exponential decay at both sides. Theoretical interpretations suggest that in order for the soliton to avoid losing energy to higher harmonics (that is, to prevent the formation of shock waves and other irreversible effects), the system must have dispersion (the phenomenon whereby the phase velocity of a traveling plane wave varies with the frequency of the wave) that exceeds a certain minimum value.

Acoustic solitons. A unified theory of the three above-mentioned solitons (KdV solitons, envelope solitons, and nonpropagating solitons) has recently been proposed. It is based upon a nonlinear, dispersive wave equation for a continuum. A system is dispersive when group velocity and phase velocity are not equal, as would be the case for the dispersion law of Eq. (2), where ω is the frequency, K is

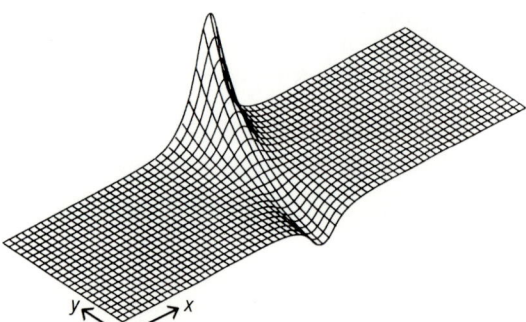

Fig. 2. A plot of Eq. (1), which is a profile of the nonpropagating hydrodynamic soliton based on curve fits to measurements. (After J. Wu, R. Keolian, and I. Rudnick, Observation of a nonpropagating hydrodynamic soliton, Phys. Rev. Lett., 52:1421–1424, 1984)

$$\omega^2 = C_0^2 K^2 + \gamma K^4 \qquad (2)$$

the wave number, C_0 is the speed of long-wavelength small-amplitude sound propagating in the medium, and γ is the dispersion coefficient of the medium. Thus, a system is nonlinear and dispersive when the speed of sound depends upon both amplitude and frequency.

Two asymptotic solutions exist in the one-dimensional case. One is KdV soliton solution which has the form of Eq. (3), where $\delta\rho$ is the density fluctua-

$$\delta\rho = \frac{3}{2}\left(\frac{u^2 - C_0^2}{\alpha}\right) \operatorname{sech}^2\left[\frac{1}{2}\sqrt{\frac{C_0^2 - u^2}{\gamma}} \cdot (x - ut)\right] \qquad (3)$$

tion caused by the soliton, u is the speed of the soliton, and α is the nonlinear coefficient of the medium. The other possible solution is the so-called envelope soliton, which is a traveling wave with slow modulation of its amplitude. This kind of soliton can be expressed as Eq. (4), where b_0 and b_1

$$\delta\rho = b_0 \operatorname{sech} b_1 (x - ut) \exp(ikx - i\omega t) \qquad (4)$$

are two constants depending upon the nonlinearity and dispersion of the medium as well as the frequency of the wave.

If one extra spatial dimension is involved in the motion, it is possible to look for another type of soliton solution with its wave vector $\vec{K}$ almost perpendicular to the phase velocity $\vec{C_0}$. Cutoff frequencies again play an essential role just as they do in the case of the nonpropagating surface-water-wave soliton. Since the new solution has the same profile as Eq. (1), it is called a nonpropagating acoustic soliton. It is sometimes also called a hydrodynamic polaron, since the self-trapping mechanism of the soliton is very much like that of polarons—electrons self-trapped by a lattice in crystal. It has been found that this polaronlike soliton is generally coupled to its radiation field. This kind of coupling has been described by V. E. Zakharov's equations in plasma physics.

It is surprising that this unusual solitary wave, which can exist in a simple system such as a pure one-component fluid, was only recently discovered. Future investigations will show whether these structures might play a role in membrane physics or turbulence, and whether they might exist in higher-dimensional geometries.

For background information *see* POLARON; SOLITON in the McGraw-Hill Encyclopedia of Science and Technology.

[JUNRU WU]

Bibliography: A. Larraza and S. Putterman, Theory of nonpropagating hydrodynamic solitons, *Phys. Lett.*, 103A:15–18, 1984; A. Larraza and S. Putterman, Theory of nonpropagating surface wave solitons, *J. Fluid Mech.*, 148:443–449, 1984; J. W. Miles, Parametrically excited solitary waves, *J. Fluid Mech.*, 148:451–460, 1984; J. Wu, R. Keolian, and I. Rudnick, Observation of a nonpropagating hydrodynamic soliton, *Phys. Rev. Lett.*, 52:1421–1424, 1984.

Somatic cell genetics

Cells from various parts of a plant can be stimulated to dedifferentiate and divide in chemically defined media. With hormones of the proper concentrations and ratios, the dividing cells can reorganize among themselves to form embryos or meristems that will subsequently differentiate into flowering plants. In the late 1960s, technologies were developed to strip plant cells of their cell walls to produce protoplasts which could fuse with the protoplasts of a different plant and absorb foreign molecules such as deoxyribonucleic acid (DNA) or chromosomes. These features of the plant tissue culture make it possible to temporarily treat complex organisms, such as flowering plants, like microorganisms which are more accessible for genetic manipulation and biochemical analysis. Thus began the field of plant somatic genetics.

Plant somatic genetics has been useful in studying fundamental problems such as primary metabolic processes, secondary metabolite accumulation, and hormone-regulated gene expression and differentiation. For agricultural economics, there has been much interest in deriving new variability from cultured cells either through spontaneous somatic variation or through deliberate genetic manipulation of cultured cells.

Sexual reproduction via seed produces progeny which are similar but not identical to the parents because the genetic complements segregate during meiosis. Reproduction via meristem or tissue culture, however, preserves the genetic makeup and produces identical clones of the parent. Once a new variety is selected or bred, this asexual propagation can be used for mass production of the same plant. This technique has been routinely practiced in horticultural nurseries.

Mutants and hybridization. Breeding for new varieties has traditionally depended on the introduction of new genes from wild germ plasms, a process that involves many generations of crosses. Cell culture, however, offers an opportunity to induce and select mutants effectively and thus obtain new breeding varieties more quickly. Furthermore, plants which are incompatible by sexual crosses may be combined to generate hybrids via protoplast fusion. In the past 15 years, mutants which are difficult to obtain in whole plants, such as those that lose the ability to synthesize certain amino acids, have been selected in culture. These biochemical mutants provided genetic markers for the selection of hybrids after protoplast fusion. In addition, agriculturally useful strains have been selected in culture, such as herbicide-resistant and salt-tolerant plants. Hybrid plants between *Arabidopsis* and *Brassica*, potato and tomato, and other combinations have been created via fusion of somatic cells.

Haploid culture. In addition to cells derived from vegetative organs, germ cells such as immature pollens and megasporocytes can be stimulated to divide in culture. Since the germ cells have undergone meiosis and contain half of the chromosomes of the somatic cells, the cultures derived from them are made of haploid cells. The dividing immature pollens sometimes develop into embryos, and subsequently into haploid plants. The major benefit of the haploid plants in agriculture is the reduction in time required for producing homozygous, inbred lines necessary for subsequent breeding hybrids to exhibit heterosis. Also, new rice and barley varieties have been produced from plants derived from haploid culture. It is believed that a true haploid exposes recessive traits, some of which can produce new and beneficial phenotypes.

Somaclonal variation. Plant cells are totipotent—they possess the potential to develop into a full plant from a single cell. The different cell types within a plant result from the selective expression of different sets of genes; no permanent changes in the genome are involved in cellular differentiation. Recently, however, it has been found that certain tissue-specific phenotypes are heritable.

Plants regenerated from tissue culture exhibit an unusually high rate of variation which in some cases has been shown to be stably inherited. The sources of variation originate from the explant or in culture. Thus, contrary to previous belief, permanent alterations can occur in somatic cells of the plant as well as in cells grown in culture. The alterations are masked until the full genetic potential of the cell is elaborated in plants regenerated from the cloned single cell. This somaclonal variation is characterized by a high frequency of occurrence and by somatic origin. Not only simple genetic traits such as disease resistance but also quantitatitive traits such as high kernel yields and large tuber sizes have been found. While little is known regarding the mechanism of this type of variation, it has generated tremendous enthusiasm because it is by far the most efficient way to create new crop varieties.

Plant regeneration from culture. While plant cultures contain relatively homogeneous populations of cells, variations do occur. The previous sections discussed the rare mutational events that occur at a frequency less than 10^{-5} per cell. In some cases, stable and heritable variations occur at frequencies as high as 1 in 100 cells. In addition to these permanent changes, epigenetic changes occur as a result of diverse tissue origins, changing microenvironment, or cellular interactions. Under the suitable physiological and chemical conditions, the proliferating tissues can reorganize and form meristems (organogenesis) or embryos (embryogenesis) which subsequently develop into plants.

In the absence of the whole plant, cell-culture organogenesis and somatic embryogenesis provide simple and manipulatable systems for studying plant development. Development consists of two steps: the induction of a developmental state and the execution of the developmental process. Little is known about the genetic and biochemical basis of either step, particularly the induced state which does not involve morphological changes. Yet, it is possible to distinguish a tissue that is competent or determined for organogenesis from one that is not by grafting or cell-culture experimentation. *Convolvulus* tissue culture can become competent to respond to conditions which will induce shoot formation. While the tissue is competent, it is not determined until after several days of incubation in the shoot-inducing medium. Once the determined state is achieved, the tissue no longer needs to be cultured on the shoot-inducing medium to carry out morphogenetic differentiation—in this case, shoot formation. The induction process requires approximately 14 days, within which, in addition to the competence and determination, there are a series of points which appear in sequence with respect to each other. These phenocritical points can be defined operationally by the use of mutations or by the developmental periods that are transiently sensitive to stage-specific inhibitors. The cell-culture system has enabled the dissection of the induction process into detailed and discrete steps which will facilitate the characterization of its biochemical and cellular basis.

Embryogenesis. One of the fundamental differences between organogenesis and embryogenesis lies in ontogeny. A group of cells is required for the induction of meristem formation, whereas embryos usually derive from a single cell. The embryogenic potential of a culture is usually acquired as the explant starts to dedifferentiate. The cells in culture thus are determined to an embryonic fate, but cannot carry out embryogenesis until transferred into embryogenic medium. Embryo development is characterized by distinct morphogenetic stages; the transition from one stage to the next can be interrupted by mutation or by specific inhibitors.

Because mass quantity of embryos of uniform stages can be obtained in culture, somatic embryogenesis is amenable for genetic, biochemical, and molecular analysis, which involves the identification of developmentally regulated biochemical phenotypes that provide the targets for studying the mechanism of gene regulation in development. A comparison between the mutants impaired in development and the wild type (normal) may reveal the genetic and biochemical basis of morphogenesis. The ultimate objective aims to identify genes and molecules that play a causal role in development. In this respect, the interesting candidates are biochemical phenotypes exhibiting a transient pattern of expression. For instance, the cellular concentrations of polyamines are transiently elevated during early embryogenesis. The prevention of this elevated synthesis by specific inhibitors blocks embryogenesis. Polyamine involvement in embryo development has been confirmed by the fact that mutants blocked in development no longer exhibit the transient increase of polyamine levels.

For background information *see* BREEDING (PLANT); PLANT MORPHOGENESIS; REPRODUCTION (PLANT); SOMATIC CELL GENETICS; TISSUE CULTURE in the McGraw-Hill Encyclopedia of Science and Technology.

[RENEE SUNG]

Bibliography: M. L. Christianson and D. A. Warnick, Phenocritical tissues in the process of in vitro shoot organogenesis, *Dev. Biol.*, 101:382–390, 1984; P. Maliga, Isolation and characterization of mutants in plant cell culture, *Annu. Rev. Plant. Physiol.*, 35:519–542, 1984; F. Meins, Heritable variation in plant cell culture, *Annu. Rev. Plant Physiol.*, 34:327–346, 1983.

Space flight

Astronauts and cosmonauts living and working above the Earth were central to United States and Soviet Union space activity in 1984. During the year, 26 Americans and one Canadian flew aboard shuttle orbiters (one American astronaut flew twice); 8 Soviet and 1 Indian cosmonauts traveled into space. Both countries demonstrated on-orbit repair and servicing of space hardware, broadening the role of humans in low Earth orbit, with a trio of Soviet cosmonauts setting a new long-duration space flight record aboard their *Salyut 7* space laboratory.

President Ronald Reagan, during his 1984 State of the Union address to Congress, directed the National Aeronautics and Space Administration (NASA) to develop a permanently manned space station within a decade. The station, to be operating by 1992 and to involve participation by other countries, is to provide a base for carrying out protracted operations and moving humans and machines between low and high Earth orbits.

Now 3.3×10^8 mi (5.3×10^8 km) from its next target, *Voyager 2* continued to chalk up space mileage heading toward the distant planet Uranus. Launched in 1977, the interplanetary probe began imaging Uranus to take readings on the globe's brightness from different angles of illumination. On January 24, 1986, *Voyager 2* will pass within 64,000 mi (102,000 km) of Uranus, relaying photographs to Earth of the planet.

Another celestial object, Comet Halley, was spotted by Earth-based astronomers in late September, 50 to 100% brighter than first estimated. As it approaches for its periodic (every 76 years) swing-by of the Sun, preparations are under way in the Soviet Union, Japan, and Europe to launch spacecraft to the vicinity of the comet. First was the December 15 and December 21 liftoff of the Soviet Union's *Vega 1* and *2* probes, both equipped to drop off a lander and upper atmospheric research balloon in the atmosphere of Venus, and then to travel onward to a rendezvous with Comet Halley in 1986.

Space as an arena of economic competition was heightened as the uncrewed Ariane launcher scored a series of successful flights, placing several communications satellites into precise orbits. In separate missions, Ariane boosted two American-built satellites into space; the customer had rejected use of the United States shuttle launch vehicle.

United States experts predicted that, given the right kind of investment decisions and program direction, the creation of a space-based economy yielding a half trillion dollars is possible in the future.

Significant space launches in 1984 are listed in the table.

UNITED STATES SPACE ACTIVITY

In 1984, 21 United States launches carrying a total of 35 payloads were placed in orbit. Launches were grouped generally as follows: science, 5; communications, 15; weather, 1; crewed, 5; reconnaissance, 3; electronic intelligence, 2; navigation, 3; and earth resources, 1.

Space Transportation System (STS) missions. The United States space shuttle effort accelerated throughout 1984 with the maiden voyage of orbiter *Discovery*, the latest addition to the shuttle fleet. *Discovery* flew twice in 1984, with veteran *Challenger* used in three missions.

Significant space launches in 1984

Payload	Date	Payload country or organization	Purpose and comments
Challenger, mission 41-B	2/3/84	U.S.	*Palapa* and *Westar* satellites reach improper orbits; first untethered spacewalk
Soyuz T-10	2/8/84	Soviet Union	Three-person crew takes off for *Salyut 7*, then board station for record 237 days
Landsat 5	3/1/84	U.S.	Remote-sensing satellite orbited to replace failed *Landsat 4*
Challenger, mission 41-C	4/6/84	U.S.	In-space repair of *Solar Max* satellite; deployment of *Long Duration Exposure Facility* (*LDEF*)
China 15	4/8/84	People's Rep. of China	Experimental communications satellite
SPACENET 1	5/23/84	Arianespace Inc.	First commercial flight of European Ariane booster orbits United States satellite
AMPTE	8/16/84	International	United States, West Germany, and United Kingdom satellites launched to measure solar wind and interaction with magnetosphere
Discovery, mission 41-D	8/30/84	U.S.	Test of large solar array; deployment of *LEASAT 2*, *Telstar 3*, and *SBS 4*
Challenger, mission 41-G	10/5/84	U.S.	First American spacewalk by woman; release of *ERBS* scientific satellite; demonstration of space refueling
Discovery, mission 51-A	11/8/84	U.S.	Space salvage of two errant satellites, *Palapa* and *Westar*; *Anik-D* and *LEASAT 4* spacecraft deployed

Despite several shuttle launch postponements, including the first space shuttle abort, five missions were successfully completed, yielding increased confidence in America's space prowess. Use of a shuttle orbiter as a launch pad for satellites became routine, as did the recovery, in-orbit maintenance, and return to Earth of uncrewed spacecraft. For the first time, an untethered astronaut floated free of the space vehicle, cruising the orbital environment via a propulsive backpack.

A new identification system was enacted by NASA. For example, 41-B is interpreted as: 4 = fiscal year 1984; 1 = launch site for Kennedy Space Center (a 2 would denote Vandenberg Air Force Base, California); B = numerical order of launch that fiscal year.

As the third space-capable orbiter off the production line, *Discovery* is lighter in weight than its two sisterships, *Challenger* and *Columbia*, and is designed to fend off greater heat loads during its fiery plunges back to Earth after orbital flight. Low temperature (white-colored) tiles throughout most of the upper wings and fuselage of the previous shuttle orbiters have been replaced with Advanced Flexible Reusable Surface Insulation (AFRSI). The advanced insulation is also installed on the payload bay doors on *Discovery*'s vertical stabilizer. *Discovery*'s Orbital Maneuvering System (OMS) pods—providing thrust for the space plane's orbit insertion, orbit change, orbit transfer, rendezvous operations, and deorbit—have also been covered with the thicker insulation. Graphite epoxy has replaced some internal aluminum spars and beams in the wings and in the payload bay doors.

Use of the new quiltlike "skin," as well as manufacturing changes to *Discovery*, enabled shuttle engineers to trim the dry weight of the vehicle to about 150,000 lb (67,500 kg), which is about 1000 lb (450 kg) less than the *Challenger* vehicle. This weight reduction permits heavier payloads or additional passengers. *Discovery*'s onboard systems have been updated and are of more advanced construction than those of either of its predecessors.

Shuttle mission 9. The close of 1983 saw the inaugural flight of Spacelab, a $1 billion module which is secured in the cargo bay of an orbiter and serves as a versatile, general-purpose laboratory for crewed and automated activities. The Spacelab, built in Europe under the guidance of the European Space Agency (ESA) with its own funds, remains attached to the orbiter during all phases of a mission. The *STS 9/Spacelab 1* mission was originally designed for an October 28, 1983, liftoff, but slipped due to analysis of a faulty booster nozzle on the preceding mission, *STS 8*.

The *STS 9* delay resulted in far from optimum use of several scientific experiments manifested aboard the *Spacelab 1*. NASA and ESA management agreed to offer reflight opportunities for one joint United States–Japan, two United States, and four ESA experiments on subsequent shuttle missions.

Orbiter *Columbia* lifted off from the Kennedy Space Center, Florida, at 11:00 A.M. EST on November 28, 1983. Tucked within its cargo bay, the 13-ft-wide (4-m) by 23-ft-long (7-m) 33,252-lb (14,964-kg) *Spacelab 1* and its attached instrument pallet held 38 instruments, representing some 70 experiments from 14 countries in physiology, astronomy, biology, materials science, Earth observations, and atmospheric studies. Racks filled with the experiments and auxiliary equipment line the inner walls of Spacelab.

As a joint NASA/ESA venture, the flight was the first time that six persons were orbited by a single vehicle: John Young, a veteran of five previous space flights, was in command, with Brewster Shaw, Jr., as his pilot. Mission specialists Owen Garriott (a former United States *Skylab* astronaut) and Robert Parker were joined by scientists Byron Lichtenberg and Ulf Merbold, a West German physicist and the first non-American to fly on a United States spacecraft. Lichtenberg and Merbold were Spacelab payload specialists, a new class of crew member who are not career astronauts and whose training is tailored to one particular mission.

Once in orbit, *Columbia* circled Earth in a 150-mi-high (250-km) orbit, and at a 57° inclination to the Equator (a higher inclination than any other United States crewed spacecraft). The six-person crew divided into two teams for 24-h continuous work.

Tests in biology and physiology occupied most of the crew's time during their first days in orbit. Studies focused on two factors: the coordination between vision and the vestibular system (the balance organs of the inner ear), and the shifting of body fluids that occurs in the weightless state. Investigations demonstrated that caloric nystagmus can be provoked even in weightlessness and that a change in gravity environment causes considerable fluid shifts inside the human body. An experiment on human lymphocytes, a parameter controlling the immune response, proved that lymphocyte proliferation is weaker in microgravity than in an Earth-type, 1-g environment. It is anticipated that Spacelab data may lead to understanding and possibly controlling space adaptation syndrome (space motion sickness); this malady affects nearly 50% of both Soviet cosmonauts and American astronauts.

In the field of materials sciences, the astronauts successfully produced large protein crystals; a crystal of lysocyme, close to 1000 times larger than producible on Earth, was grown. The crystals created in space were of sufficient size to allow structural analysis by x-ray diffraction techniques. In the discipline of Earth observations, a Metric Camera photographed millions of miles of Earth surface, with a ground resolution of 66 ft (20 m). Three spectrometers located on the Spacelab pallet, directly exposed to space, performed analyses of the gases of the Earth's atmosphere in visible, infrared, and ultraviolet wavelengths. In one of the atmospheric physics experiments, a Grille Spectrometer revealed the existence of methane up to 43 mi (70 km) altitude. In addition, deuterium—a heavy form of hydrogen—was discovered in the thermosphere with a peak in density around 68 mi (110 km) altitude, and the first observation of a narrow region of hot hydrogen emission in the upper atmosphere was made.

Astronomy experiments led to the detection of puzzling sources of x-rays while Spacelab's Very Wide Field Camera produced pictures of the sky at ultraviolet wavelengths and revealed a connecting structure between the small and the large Magellanic Clouds. One space physics instrument designed to generate a powerful beam of electrons to create an artificial aurora in the charged gas surrounding *Columbia* did not operate as planned, leading to cancellation of the experiment. The hardware did operate however, in a low-power mode and, with additional instruments, probed how the Earth's atmosphere and magnetic fields react to controlled artificial stimulation.

Throughout 166 orbits of Earth, Spacelab astronauts conducted their multidisciplinary experiments while in touch with scientists on the ground at the Payload Operations Control Center (POCC) at Johnson Space Center, Houston, Texas. This interchange between Earth-situated scientists who designed experiments for Spacelab and the astronauts who actually carried out the investigations in space, although unique, did prove troublesome on occasion. In more than one instance, tempers flaired between seemingly overworked astronauts and impatient scientists on the ground. Largely, however, the communications link proved extremely valuable in troubleshooting equipment problems, making additions to experiments, and going over procedures.

An extension of an extra day for the mission was granted, made possible by an unexpectedly low use of the spacecraft's fuel, electric power, and oxygen. After 10 days of highly productive science in orbit, and husbanding erratic guidance computers and a failed Inertial Measurement Unit, one of three such devices for vehicle navigation, *Columbia* touched down on December 8, 1983, at the Edwards Air Force Base in California. Later inspection of the space plane revealed one additional anomaly: several small fires, caused by fuel leaks, had started shortly after landing around two of *Columbia*'s three hydraulic power units located in the aft fuselage. Although these problems demanded further investigation by shuttle engineers, the international Spacelab mission was pronounced an unqualified success and the opening of a new era in scientific space research.

Shuttle mission 41-B. *Challenger*'s fourth flight, and the shuttle program's tenth mission, began at 8:00 A.M. EST on February 3, 1984. Onboard the orbiter was commander Vance Brand, on his second shuttle flight, aided by pilot Robert Gibson. Bruce McCandless, Robert Stewart, and Ronald McNair served as mission specialists.

The orbital inclination of mission 41-B was 28.5°, in orbits varying from 174 to 202 mi (278 to 323 km). Circuiting the Earth 127 times, *Challenger*'s crew deployed from the shuttle bay a *Palapa B-2*, an Indonesian national telecommunications satellite, and a *Westar 6* advanced Western Union communications spacecraft.

A central aspect of the mission was evaluation by astronauts McCandless and Stewart of two Manned Maneuvering Units (MMU), self-contained backpacks with nitrogen gas propulsion that permit crew members to move outside the payload bay to other parts of the orbiter or to nearby spacecraft. Much of the flight focused on demonstrating techniques important for the successful retrieval and repair of a disabled *Solar Maximum Mission* (*Solar Max*) spacecraft on a later (41-C) shuttle flight.

A variety of payloads were carried aboard *Challenger*. These included two Cinema 360 cameras to yield 35-mm motion picture photography in a unique format especially suited for planetarium viewing (films completed on the flight and subsequent shuttle missions are being used in the production of a motion picture about the space shuttle program); a Monodisperse Latex Reactor to manufacture tiny identical-sized latex beads with major medical and industrial research applications; the Acoustic Containerless Experiment System for materials processing; and an IsoElectric Focusing Experiment to evaluate the effect of electroosmosis on an array of eight columns of electrolyte solutions as direct-current power is applied and pH levels between anodes and cathodes increase. A shuttle Student Involvement Project experiment, the eighth to be flown, studying the effects of weightlessness on the development of arthritis using six rats as test subjects, was also carried into orbit.

Numerous experiments were carried in five Getaway Special (GAS) canisters, such as an assay of the effects of microgravity on capillary waves in liquids by a Utah State University student; the configuration of an arc lamp in gravity-free surroundings by GTE Laboratories, Inc.; and an Air Force investigation to evaluate the potential for cosmic rays to upset or change the logic state of a memory cell.

Mission 41-B comprised historical firsts as well as disappointing failures. In one incident, an ejected-into-space target balloon was to have been used to evaluate the navigational ability of *Challenger*'s onboard radar systems, as well as the interaction between the spacecraft, flight crew, and ground control. The aluminum-coated Mylar balloon ruptured during inflation, although target debris was tracked out to a distance of about 52,000 ft (16,000 m), verifying the orbiter's radar.

Both the ejected *Palapa B-2* and *Westar 6* satellites suffered similar failures of their commercially made Morton Thiokol Star-48 solid-rocket motors. Instead of the craft reaching a transfer orbit, the misfiring motors placed the two spacecraft into low orbits of about 190 × 750 mi (304 × 1200 km). In orbits negating their usefulness, the satellites were declared losses.

Two spectacular spacewalks marked the first time astronauts traveled away from a spacecraft without a connecting tether. Attached to the 300-lb (136-kg) MMU, McCandless and Stewart took turns flying free of *Challenger* out to 300 ft (91 m) distance. Radar-clocked at a speed of 0.7 mi/h (0.3 m/s), Stewart was cautioned to slow down by shuttle commander Brand. Using hand controllers to pulse combinations of the 24 dry nitrogen gas thrusters mounted on the MMU, the astronauts spent 12 h jetting outside the orbiter, stretched over two periods of extravehicular activity (EVA). The thorough checkout of the MMU, and the completion of additional EVA tasks, cleared the way for bolder objectives in the future.

A plan to lift out of the cargo bay a German *Shuttle Pallet Satellite* (*SPAS*), as a target for MMU-equipped astronauts to perform docking tests, had to be called off when the Canadian robot arm developed problems.

Completing a 127-orbit trek, *Challenger* cut through a slight early morning ground fog on February 11, touching down at 7:16 A.M. EST on the 15,000-ft long (4500-m) runway at Kennedy Space Center, Florida. The landing occurred within sight of the gantry from which *Challenger* had departed Earth nearly 8 days earlier. *Challenger* thus became the first spacecraft of any country to embark from and return to the same launch area after circling Earth. In doing so, it made possible a quick refurbishment of the orbiter, allowing it to return to space a little over 2 months later.

Shuttle mission 41-C. On April 6, 1984, *Challenger* was off again. The five-person mission was commanded by Robert Crippen on his third shuttle flight. The mission 41-C crew were pilot Francis (Dick) Scobee and mission specialists Terry Hart, James van Hoften, and George Nelson. Rocketing away from Kennedy Space Center at 8:58 A.M. EST, *Challenger* was lifted by its three main engines directly to an orbital high point of 285 mi (460 km). Smaller Orbital Maneuvering System (OMS) engines normally used to reach apogee were fired only once to circularize the orbit. This fuel-saving, direct-insertion technique, a first for the shuttle program, put the space plane in its highest orbit so far, at an altitude of 250 mi (400 km). The trajectory was intended to place the orbiter on a course for retrieval, repair, and relaunch of a disabled *Solar Maximum Mission* satellite. The inclination to the Equator was 28.5°.

On the morning of the mission's second day, a 21,400-lb (9707-kg) *Long Duration Exposure Facility* (*LDEF*) was eased out of the orbiter's bay by the robot arm. Left floating 250 mi (400 km) above the globe in a gravity-gradient stabilized attitude, inclined approximately 28.5° to the planet, *LDEF* holds 57 separate experiments on its 30-ft-long (9.14-m), 14-ft-wide (42-m), 12-sided cylindrical frame. Dozens of *LDEF* trays hold equipment and test samples developed by 194 researchers from nine countries. During 1985, a space shuttle is to retrieve *LDEF* from orbit, returning the structure to Earth for distribution of its experiments, as well as reuse. *LDEF* trays carry samples of foils, tapes, composite materials, and other hardware being evaluated for use in future spacecraft. Several experiments are studying interstellar gas, dust, and micrometeorites, and one investigation is analyzing how 1.2×10^7 tomato seeds are affected by 10½ months in space. Once the seeds are returned to the ground, they are to be distributed to elementary through university classrooms around the country as a cooperative educational project with the George Park Seed Company.

Besides the five-person astronaut crew, mission 41-C also carried thousands of other living "passengers": a Shuttle Student Involvement Project from Tennessee Technological Institute comprised a study of honeycomb structures fashioned by 3300 bees confined in a specially designed "bee-box" mounted in *Challenger*'s middeck. Documenting the work of mission 41-C, 70-mm IMAX and 35-mm Cinema 360 cameras were operated by shuttle astronauts, with the resulting film footage to be publicly shown.

By far the most complex aspect of mission 41-C was the first on-orbit repair of the crippled *Solar Max*. Sent into orbit in February 1980 to observe solar flares over a wide range of wavelengths, from visible light to gamma rays, the uncrewed spacecraft blew three fuses in its attitude control box 10 months after launching. Since the craft was unable to point itself accurately, four of its seven onboard instruments were rendered useless. Another instrument, the Coronagraph/Polarimeter, subsequently developed electrical problems and also failed. *Solar Max* drifted in a slow, stable spin of 1° per second with its winglike solar panels constantly pointed at the Sun to maintain battery charge for more than 3 years. Because the satellite was the first designed for on-orbit replacement of parts, planning for a rescue mission began almost immediately after the satellite failed.

On the morning of *Challenger*'s third day in space, the orbiter parked about 200 ft (61 m) away from the slowly turning satellite. Untethered astronaut George Nelson, using his MMU backpack—first tested on mission 41-B—propelled himself away from the orbiter and in a 10-min trip through space attempted to cancel out *Solar Max*'s turning by docking to a pin protruding from the satellite (Fig. 1a). A special Trunnion Pin Acquisition Device (TPAD), fixed to the arms of his MMU, failed to clamp to the pin; the unsuccessful docking introduced unwanted oscillations into *Solar Max*. After three tries, and with the spacecraft now wobbling, commander Crippen, observing from the orbiter's cockpit, suggested that Nelson try to steady the spacecraft by holding onto one of its solar panels. The situation worsened. Low on nitrogen fuel to power his MMU backpack, Crippen and Nelson called off the rescue.

With *Solar Max* tumbling unpredictably around all three axes, the crew attempted four times to snare the spacecraft with the mechanical arm, but without success. Pulling away from the satellite, the frustrated crew considered various options along with ground control. At NASA's Goddard Space Flight Center, Greenbelt, Maryland, the command station for *Solar Max*, engineers struggled to slow the satellite's rotation through day- and night-long sessions. Battling weakening batteries aboard the satellite, Goddard technicians utilized *Solar Max*'s magnetic torquer bars in a never-before-used mode, regaining control of the disabled craft. *Solar Max* also righted itself in orbit to position critical solar panels to begin recharging drained batteries.

Early on flight day five, *Challenger* was once again positioned nearby the rejuvenated and stable *Solar Max*. The satellite's spin reduced to half its original rate, Hart extended the robot arm upward and locked onto *Solar Max* on a first try, and set the spacecraft gently down into a cradle.

Working within the cargo bay, space-suited astronauts van Hoften and Nelson serviced *Solar Max* for less than an hour (Fig. 1b). The repairmen replaced the satellite's Coronograph/Polarimeter's electronic box, changed fuses in the attitude control box, and placed a baffle cover over the satellite's x-ray Polychromator to vent its exhaust gas away from other spacecraft instruments. When repairs and checkout were complete, *Solar Max* was picked up by the mechanical arm, and held up and away from *Challenger* while engineers verified that the new attitude control system and electronic box worked perfectly. Released from the arm's grip, *Solar Max* began its second life in orbit.

The long-distance repair call brought back to service a $77 million satellite, saving an estimated $150 million to replace the spacecraft. The cost to repair *Solar Max* in orbit was estimated at $40 million. To build a new spacecraft and launch it would cost $235 million in 1984 dollars. Technological benefits also resulted from closeup looks at components of the Sun-watching satellite. The nearly 500 lb (225 kg) of returned-from-space parts showed surprisingly little wear due to impinging of micrometeorites and atomic oxygen degradation. Space scientists now see the orbital altitude at which *Solar Max* travels as a benign environment. One item, the used *Solar Max* attitude control system, is to be repaired for reflight on another satellite later in the decade.

With the mission's tasks completed in just under 7 days, *Challenger* returned to Earth on April 13, touching down at the Edwards Air Force Base. Mission 41-C landed 1 day later than originally planned; the shuttle was scheduled to land at the Kennedy Space Center but bad weather at the prime landing site forced the change.

Shuttle mission 41-D. The maiden voyage of *Discovery* orbited a six-person crew on August 30, 1984. Liftoff of the twelfth shuttle mission was delayed several times. On June 25, 1984, an apparent malfunction of a backup flight computer led to a 24-h postponement. On the following day, when a fuel-line valve failed to open correctly on one of *Discovery*'s engines, computers signaled an abort of the mission, seconds prior to liftoff and with one of

Fig. 1. In-space repair of *Solar Max* satellite. (a) Astronaut George Nelson making initial excursion to the damaged satellite in an unsuccessful docking attempt. (b) *Challenger* cargo bay with astronaut James van Hoften and satellite almost ready for release into space after servicing. Robot arm is at upper right.

three engines ignited. It was the second time a United States spacecraft had been aborted after engines began to operate; the first time occurred on December 12, 1965, when *Gemini 6* engines were automatically shut down.

Due to the delay, called to understand more fully the abort incident, NASA combined cargo elements from mission 41-D and the next shuttle flight in line: 41-F. (41-E, a Defense Department flight, had been previously canceled.) A third launch try was foiled on August 29, 1984, by an electronic timing problem. Finally, the following day, *Discovery* lifted off at 9:38 A.M. EDT. Hauling the heaviest payload to date, 47,000 lb (21,150 kg), the orbiter ascended into a 184-mi-high (294-km) orbit, inclined 28.5° to the Equator.

Veteran shuttle astronaut Henry Hartsfield, pilot on the STS 4 flight, was commander of the mission, joined by pilot Michael Coats and three mission specialists, Judith Resnik, Steven Hawley, and Richard Mullane. The mission 41-D crew also included the first passenger ever to be launched into space for private industry; McDonnell Douglas engineer Charles Walker served as a payload specialist operating the firm's Continuous Flow Electrophoresis System (CFES)—hardware which churns out proprietary samples of a pharmaceutical product planned to be commercially sold in the late 1980s. Operating in microgravity, the apparatus permits 500 to 700 times as much drug to be processed per hour, in purer quantities, as is possible on Earth.

A record three satellites were deployed from *Discovery*: *LEASAT 2* (*Syncom IV-2*) for Hughes Communications Services, Inc., and leased to the Navy; *SBS 4* for Satellite Business Systems; and a *Telstar 3* for American Telephone and Telegraph (AT&T). Two of the spacecraft utilized solid propellant boost motors, similar to those which had failed during shuttle mission 41-C. The satellites achieved their desired orbits, to the relief of commercial users of the shuttle system, as well as the space insurance community. The *LEASAT* propelled itself into a 22,300-mi (35,680-km) geostationary position, but made use of two built-in liquid-fuel engines that burn hypergolic propellants. The *LEASAT* series are also the first satellites designed exclusively for launch from a shuttle. Measuring 14 ft (4.2 m) across, the spacecraft are too large to fit in the protective nose cone that sits on top of an expendable booster, such as a Delta or Atlas Centaur. Laying on its side within the shuttle cargo bay, *LEASAT* was dispatched from *Discovery* in a "frisbee" motion.

Also tucked inside *Discovery*'s payload section was a triangular-shaped assembly on which was mounted NASA's Office of Aeronautics and Space Technology (OAST 1) experiment. The primary aspect of this payload was a solar array wing. Other OAST 1 equipment evaluated solar cell calibration techniques and rated various types of solar cell.

On flight day three, Judith Resnik unfurled to varying lengths the experimental solar panel. Stretched to its maximum, the 102-ft-tall (31.5-m) by 13-ft-wide (4-m) array held a thin blanket of plastic material called Kapton, hoisted up and down on an epoxy-fiberglass mast. The blanket consisted of 84 panels that folded accordion style when the structure was retracted. The test not only proved that the array would deploy as designed but demonstrated the stability of a large structure similar to those planned for use in the projected United States space station effort. During maneuvering of the orbiter, with the array fully stretched, the solar panel displayed no more than 1 or 2 in. (2 or 5 cm) of oscillation.

McDonnell Douglas employee Walker processed about 85% of the pharmaceutical materials he had on board through the CFES equipment. Due to hardware breakdown within the CFES, the machine operated less on-orbit hours than anticipated. In addition, samples processed in space, when tested on the ground, proved contaminated by bacteria. The contamination voided 0.4 gallon (1.5 liters) of the secret drug made for clinical testing by the Food and Drug Administration. NASA and McDonnell Douglas have agreed to fly Walker on a second shuttle mission in 1985 to obtain additional CFES-generated samples.

Also running into trouble, a student experiment, crafted to increase the size of a 10-in.-long (25-cm) rod of gallium impregnated with thallium, metals used in the production of electronic components, failed due to an internal electrical short circuit.

One problem with *Discovery* required astronaut attention: an unwanted buildup of ice blocked waste outlets for the orbiter's toilet. Mission managers fretted about potential damage to the shuttle's delicate tiles if the ice were to break off during reentry. After trying to melt the estimated 25-lb (11-kg) chunk of ice by turning the orbiter's hull sunward, the crew finally managed to dislodge the material with a controlled swipe of the shuttle's robot arm.

Completing its 6-day maiden flight, *Discovery* ended its nearly flawless mission 41-D at 6:37 A.M. PDT, September 5, at Edwards Air Force Base.

Shuttle mission 41-G. For its third return to space in 1984, shuttle orbiter *Challenger* left its Florida-based launch pad on October 5, 1984, at 7:03 A.M. EDT. A string of notable achievements marked mission 41-G, including a record seven individuals orbited in one liftoff, the first flight of a Canadian payload specialist, the first spacewalk by an American woman, the first demonstration of a satellite refueling technique in space, and the first flight with a reentry profile crossing the eastern United States.

Leading the record-size crew was veteran astronaut Robert Crippen, on his fourth shuttle mission. Jon McBride was pilot, and mission specialists were Sally Ride (America's first woman into orbit on the shuttle program's seventh mission), Kathryn Sullivan, and David Leestma. The two payload specialists were U.S. Navy oceanographer Paul Scully-Power and Canadian Marc Garneau. Payload specialist Garneau carried out a series of space science, technology, and life science experiments for the National Research Council (NRC) of Canada.

The orbital inclination was 57° in a circular orbit at 218 mi (350 km). One of the first activities after reaching orbit was the deployment of an *Earth Radiation Budget Satellite* (*ERBS*). The 5087-lb (2289-kg) *ERBS* was lifted out of the orbiter's payload bay by using *Challenger*'s robot arm, and then set free to measure the amount of solar energy absorbed in different regions of Earth and the quantities of thermal energy emitted to space. Understanding this interaction will lead to enhanced climate predictions. Release of *ERBS* was accomplished, however, only after astronaut Ride shook the spacecraft with the mechanical arm to help deploy stuck solar panels. A set of hydrazine thrusters mounted on *ERBS* gradually raised the satellite to 329 nautical miles (609 km) to begin its mission to probe the ebb and flow of solar energy on Earth.

Additional payloads included eight Getaway Special canisters, one of which held a Japanese experiment to form crystals of three metal alloys and two glass composites in five small electrical furnaces. Another produced the first works of art using vacuum deposition techniques to coat eight glass spheres with gold, platinum, and other metals to create lustrous space sculptures.

As part of an Office of Space and Terrestrial Applications payload (*OSTA 3*), a modified version of the shuttle imaging radar (SIR-B) created two-dimensional images of the Earth's surface. The astronauts had difficulty in opening and folding panels which constituted the 35 × 7 ft (11 × 2 m) SIR-B antenna. Plans for lengthy radar data takes were shortened, resulting in achievement of just 40% of the mission objectives. Despite the problems, images of Bangladesh jungle were taken to measure radar penetration capabilities, as were scans of the

Rift Valley of Africa to search for remnants of early humans by detecting sedimentary basins. SIR-B studies encompassed investigations by over 200 scientists from 13 countries in geology, cartography, oceanography, vegetation studies, and archeology.

A 900-lb (405-kg) Large Format Camera (LFC) also captured images of terrain on five types of Kodak film, to assist in worldwide exploration for oil and mineral resources, mapping, and monitoring Earth's environment. The unique space eye took photographs with resolutions below 70 ft (21 m). Due to *Challenger*'s orbital inclination, images were also taken over Soviet territory, including the site of a serious Soviet nuclear accident that occurred over 25 years ago. A censorship of both radar and optical images was enforced by United States Defense/intelligence agencies to restrict certain security-sensitive data from public viewing.

Astronaut/oceanographer Scully-Power photographed thousands of miles of interconnected spiral eddies covering the world's oceans. Originally believed to be an isolated phenomenon, the linked eddies are a significant new ocean dynamics finding.

Astronaut Sullivan became the first American woman to spacewalk. In a 3-h stint of extravehicular work, Sullivan and Leestma installed a valve assembly into simulated satellite propulsion plumbing. Working in the orbiter's cargo bay, they checked out key aspects of the Orbital Refueling System (ORS) which is to permit replenishment of propellant and other liquid consumables on Earth-orbiting satellites in the future. After completing their ORS tasks, Sullivan and Leestma returned to the safety of *Challenger*'s cabin for the actual transfer of hydrazine, a highly toxic and corrosive fuel, controlled from the shuttle's aft flight deck.

Dropping from orbit, the 101-ton (92-metric-ton) *Challenger* glided across 13,000 mi (21,000 km) of Alaskan, Canadian, and United States midwest and southeast territory. Shuttle commander Crippen performed eleven different flight test maneuvers and four hypersonic and supersonic S-turns before landing at Kennedy Space Center on October 13, 1984.

Shuttle mission 51-A. The second flight of *Discovery* began at 7:15 A.M. EST on November 8, one day late due to high winds in the launch area. On board the orbiter were commander Frederick Hauck on his second shuttle mission, pilot David Walker, and mission specialists Anna Fisher, Dale Gardner, and Joseph Allen, both men on their second shuttle voyage. Mission 51-A was the product of lessons learned from previous shuttle experiences, putting to use orbital rendezvous techniques, satellite deployment and recovery methods, and extensive astronaut spacewalking. Altitudes between 220 and 97 mi (352 and 155 km) were achieved by the orbiter during its mission 51-A.

An *Anik D2* (*Telesat*) communications satellite, owned and operated by Telesat Canada, Ottawa, was deployed from *Discovery* on November 9. An 85-s burn of *Anik*'s Payload Assist Module (PAM) upper stage propelled the 9-ft-tall (2.7-m) 2727-lb (1227-kg) satellite into a geosynchronous transfer orbit. A later boost from a smaller engine within the satellite pushed *Anik* into geosynchronous altitude of roughly 22,300 mi (35,680 km). The following day, a 20-ft-long (6-m) *LEASAT* (*Syncom IV-1*) satellite, leased by the Department of Defense, was ejected from the orbiter's cargo bay in a "frisbee" fashion. This deployment technique gave the satellite a separation velocity from *Discovery* and its own gyroscopic stability. The *LEASAT*, using an internal engine, later transferred itself into a geosynchronous position. By unleashing the two satellites from the orbiter's cargo bay, room was made to attempt the boldest shuttle task so far.

As it traveled around the Earth in a 28.45° orbital inclination, a series of maneuvers allowed *Discovery* to rendezvous on November 12 with the *Palapa B-2*, a satellite deployed during shuttle mission 41-B in February and placed into a useless orbit, as discussed above. As the satellite drifted in a 207 × 220 mi (331 × 352 km) orbit, commander Hauck and pilot Walker eased *Discovery* 35 ft (11 m) away from the slowly spinning satellite. Propelled by his jet-powered backpack, astronaut Joe Allen approached the *Palapa* and inserted a 5-ft-long (1.5-m) "stinger" device into the nozzle of the satellite's spent rocket motor. Using backpack thrusters, Allen canceled out the spin of *Palapa*, while Fisher, manipulating the orbiter's mechanical arm from inside the *Discovery*'s cabin, locked onto the satellite and nudged it into the cargo bay.

Awaiting the satellite arrival, co-spacewalker Dale Gardner ran into an unexpected snag when a specially designed bracket would not fit onto the satellite, preventing additional mechanical arm handling to lower *Palapa* into the cargo bay. While Allen held the 1200-lb (540-kg) spacecraft for some 90 min, Gardner attached a berthing device to *Palapa*'s bottom and, together, Allen and Gardner gently hand-lowered the satellite and locked it to a cradle in the cargo bay. During 6 h of spacewalking, the two had completed the first satellite salvage in history.

The recovery of a second satellite, which suffered a similar failure, was accomplished 2 days later. After catch-up maneuvers, *Discovery* pulled-up alongside Western Union's *Westar 6*, which had been drifting 700 mi (1120 km) distant from the already rescued *Palapa*. Gardner rode the propulsion backpack to accomplish the snag of the errant second satellite. Improvised procedures, based on experience gained in wrestling the first spacecraft into the cargo bay, were used in securing the 1000-lb (450-kg) *Westar* into the cargo bay. Positioned on the end of the robot arm, Allen held onto the plucked-from-orbit spacecraft while Gardner completed berthing and bolting-down measures (Fig. 2). With the work going more smoothly than the first satellite retrieval, the second rescue was completed in less than 6 h.

Having captured the stray *Palapa* and *Westar*,

Fig. 2. Westar 6 retrieval. Astronaut Joseph Allen (right) holds onto spacecraft while on a mobile foot restraint attached to Discovery's mechanical arm. Astronaut Dale Gardner works to remove stinger device from the now stabilized satellite.

satellite insurers, who spent $5.5 million to help offset the costs of the salvage plan, hope to refurbish the spacecraft for sale and relaunch. When the two lost-in-space satellites were first declared failures, $180 million worth of insurance claims were paid to Indonesia and Western Union.

As one of the most trouble-free shuttle flights to date, Discovery returned to Earth on November 16, and at 6:59 A.M. EST touched down at the Kennedy Space Center's 3-mi-long (5-km) shuttle landing strip.

Space science flights. The 1-ton Infrared Astronomical Satellite (IRAS) ended its scientific life on November 21, 1983, when its supply of supercold helium, needed for operating telescope sensors, slowly evaporated. The 2370-lb (1075-kg) IRAS concluded 10 months of surveying the sky in the infrared region of the electromagnetic spectrum. Telescope sensors onboard IRAS had been maintained at 2.5 K ($-455°F$) to detect the faint thermal emissions of stellar objects, and other phenomena. As a joint project of the United States, the Netherlands, and the United Kingdom, the astronomical spacecraft had mapped 95% of the sky, recording over 250,000 infrared sources, encompassing the discovery of never-before-observed comets and asteroids.

As scientists began sifting through 2×10^{11} bits of relayed data from IRAS, it was announced in late December 1983 that the orbiting infrared eye had found rings of cool material encircling Fomalhaut, the brightest star in the constellation Piscis Austrinus. Estimated to be the size of grains of sand, the cool matter is similar to the matter discovered by IRAS early in 1983 around the star Vega. It is a reasoned possibility that such rings could eventually condense into planetary systems, akin to the one surrounding the Sun. Results of a systematic search led space scientists to report in 1984 that similar rings of grit and dirt may circle more than 40 other nearby stars. If so, the observed phenomenon provides the first evidence that planets around other stars might be a common occurrence. This possibility was seemingly corroborated by data released in October 1984. By employing special ground-based optical and computer techniques, a vast swarm of solid particles was photographed around Beta Pictoris, a star just 50 light-years (3×10^{14} mi or 5×10^{14} km) from Earth.

Additional new IRAS results were presented in June 1984: An extensive infrared tail of ice and dust as long as 2.2×10^7 mi (3.5×10^7 km) was found on Comet Bowell; a very young star roughly the mass and brightness of the Sun has formed in the last 100,000 years in a relatively nearby cloud of molecular hydrogen gas called Lynds 255, about 450 light-years (2.5×10^{15} mi or 4×10^{15} km) away; an unexpectedly high percentage of galaxies seen by IRAS are merging, colliding, or otherwise interacting with one another.

Pioneer Venus Orbiter. Other space science events included the February 1984 announcement that the Pioneer Venus Orbiter, orbiting cloud-covered Venus since 1978, had relayed data suggesting the planet is the site of massive volcanic eruption. The American probe also detected repeated clusters of lightning bolts, very similar to lightning discharges generated in terrestrial volcanic plumes.

In a different task, the Pioneer Venus Orbiter was commanded by NASA mission controllers on April 13, 1984, to tilt away from Venus for a period of time, scan across the solar system, and view Comet Encke. Pioneer's Ultraviolet Spectrometer found, to the surprise of scientists, that Comet Encke was losing water at a rate approximately three times higher than expected for its distance from the Sun. It was speculated the phenomenon could be due to the particular arrangement of ice and dust that the comet is made of, or to crumbling of "mesas and hills" that may cover the 1.2-mi-diameter (2-km) surface of the comet nucleus.

Comet probe. From a space locale 900,000 mi (1,440,000 km) from Earth, the International Sun-Earth Explorer (ISEE 3) was nudged out of its orbit and redirected through interplanetary space. Months of precision navigation permitted the 1050-lb (473-kg) ISEE 3 to make a December 22, 1983, swing-by of Earth's Moon, passing only 70 mi (112 km) above the lunar surface, and onward to a scheduled September 11, 1985, interception with Comet Giacobini-Zinner. Discovered in 1900, Giacobini-Zinner has been extensively observed from Earth on each of its 13-year returns to the inner solar system. Now renamed International Cometary Explorer (ICE) for its new mission, the spacecraft is the first targeted toward a comet. ICE instruments are to study the solar wind by measuring magnetic fields, particles, and gas ions that form the comet's wispy tail.

The comet probe should yield information regarding the composition of comets and their surrounding environment. In doing so, *ICE* would relay invaluable data useful for European, Japanese, and Soviet missions to Comet Halley in 1986.

AMPTE project. An "artificial comet" is one of the objectives of the Active Magnetospheric Particle Tracer Explorers (AMPTE), a research project involving satellites from the United States, West Germany, and the United Kingdom. A single Delta rocket launched the three-satellite probe on August 16, 1984, with each spacecraft placed into differing orbits. The international scientific venture, to be continued through 1985, is designed to gain new knowledge about the physical interactions of the solar wind and Earth's magnetic field.

To accomplish this task, the 1554-lb (705-kg) German satellite, the *Ion Release Module* (*IRM*), is to inject rapidly ionizing clouds of lithium and barium from both outside and inside the boundary of the Earth's magnetosphere. Located in close proximity to the *IRM*, the 172-lb (78-kg) *United Kingdom's Subsatellite* (*UKS*) will study the chemical releases as they develop, while the United States satellite, the 529-lb (240-kg) *Charge Composition Explorer* (*CCE*), is to analyze the activity from within the magnetosphere. A series of planned chemical releases embraces the creation of an artificial comet to be formed inside the bow-shock region but outside the magnetosphere, directly in the orbital path of Earth. As sunlight and charged particles hit the barium cloud, a visible tail is expected to form so that AMPTE investigators can study the behavior of an artificial comet under synthetic conditions. Yet another series of releases will occur in 1985 behind Earth, in the trailing magnetotail.

Scientists expect the triad of AMPTE spacecraft to shed light on how and where the solar wind particles pierce the magnetopause which surrounds the magnetosphere, how energies of the moving solar particles are increased by factors of up to a million, and whether they are the source of the Van Allen radiation belts.

Remote sensing. *Landsat 5* was rocketed into a 434-mi (700-km) Sun-synchronous polar orbit from Vandenberg Air Force Base on March 1, 1984, to replace an ailing *Landsat 4* which had fallen victim to poor workership and quality control inadequacies during its design and construction.

Images of the Earth taken by *Landsat* series spacecraft are in use by both national and international concerns for oil and mineral exploration; agriculture, forestry, and water management; map making; industrial plant site identification and location; and general land use planning.

Modified to assure the same faults would not befall it, *Landsat 5* carries similar instruments to those of its predecessor. Orbiting the Earth every 100 min at a 98.3° inclination to the Equator, the *Landsat* completes 14.5 orbits a day and repeats the cycle every 16 days, allowing the spacecraft's Thematic Mapper (TM) and Multispectral Scanner (MSS) instruments to sense the Earth below. The Thematic Mapper collects radiometric data in seven spectral bands, the Multispectral Scanner in four bands. Images with a resolution of 99 × 99 ft (30 × 30 m) in six bands of reflected sunlight and 396 × 396 ft (120 × 120 m) in the 10.4–12.5-micrometer thermal band are achievable with the Thematic Mapper. The Multispectral Scanner generates images with resolutions of 264 × 264 ft (80 × 80 m) in four reflective bands.

SOVIET SPACE ACTIVITY

In 1984, 97 Soviet launches carrying a total of 115 payloads were placed in orbit. Launches were grouped, in general, as follows: science, 2; communications, 30; weather, 1; crewed or crew-related, 9; reconnaissance, 36; electronic intelligence, 7; navigation and geodetic, 16; natural resources/oceanographic, 1; early warning, 7; radar calibration, 4; unknown, 2.

The year saw the Soviet Union establish a world record in human long-duration space flight, the pushing of humans and machine to new limits, expanded use of spacewalking cosmonauts to maintain the viability of the *Salyut 7* space station, scores of successful uncrewed launches, one of which was an international effort involving the United States, and another test of a mini-space shuttle–like device.

Space cooperation. On December 14, 1983, *Cosmos 1514* reached space carrying the first Soviet monkeys along with a complement of pregnant rats, fishes, and plants. The spacecraft remained in orbit for 5 days and then returned its biological cargo safely to Earth. United States–supplied biomedical instrumentation implanted into the monkeys collected data on blood flow under a weightless environment. The rats were used to study space effects on embryos and later gave birth to apparently normal young. Other countries participating in the successful biosatellite project included Czechoslovakia, Bulgaria, Hungary, East Germany, France, Poland, and Romania.

Shuttle developments. A third flight of the Soviet Union's mini-shuttle occurred on December 27, 1983, but differed from earlier tests. Launched under the guise of *Cosmos 1517*, the 2000-lb (900-kg) vehicle orbited Earth about one revolution before making a "controlled descent" to a recovery zone within the Black Sea. Previous mini-shuttle shots had landed in the Indian Ocean, but the Soviets apparently wanted to prevent uninvited onlookers. Royal Australian Air Force planes had monitored retrieval operations of the small device on the earlier tests.

Data released by the Pentagon in 1983 stated that Soviet space engineers are engaged in building two reusable vehicles, one similar in size to, but able to lift heavier cargo than, the American space shuttle; the other, a smaller spaceplane. It is believed the mini-shuttle may be serving as a test prototype for the two reusable craft. It was reported in early January 1984 that photographs taken by United States shuttle astronauts in November 1983, high above the Soviet Tyuratam launch complex for crewed flights, show a new 15,000-ft-long (4550-m) runway and support buildings, possibly associated with the Soviet shuttle program.

Space station activity. Dropping to Earth under billowing parachute, *Salyut 7* space station cosmonauts Vladimir Lyakhov and Aleksandr Aleksandrov returned on November 23, 1983, after 149 days in space. During their mission which had begun on April 20, 1983, the two carried out more than 250 technical, technological, and astrophysical experiments aboard the *Salyut*, a 42,000-lb (19,000-kg) habitat about the size of a compact house trailer. Prior to departing *Salyut 7*, the cosmonauts conducted two spacewalking sessions, installing new solar panels onto the station's larger center solar array. These panels, brought to the facility in the cargo hold of an uncrewed *Progress 18* resupply spacecraft, boosted the electrical output of the original array by 50%. By the end of 1983, it was clear the Soviet Union intended to perpetuate use of the *Salyut 7*.

A study reviewing the Soviet Union's space station effort was released in late December 1983 by the U.S. Congressional Office of Technology Assessment (OTA). The analysis reported that the *Salyut* program's "is the cornerstone of an official policy which looks not only toward a permanent Soviet human presence in low-Earth orbit but also toward permanent Soviet settlement of their people on the Moon and Mars. The Soviets take quite seriously the possibility that large numbers of their citizens will one day live in space."

Underscoring the OTA findings was the long duration sojourn of cosmonauts Leonid Kizim, Vladimir Solovyev, and Oleg Atkov, a heart specialist from the Cardiological Research Center in Moscow. Launched in their *Soyuz T-10* craft from the Baikonur space center at Leninsk in Kazakhstan on February 8, 1984, the space trio remained aboard *Salyut 7* for a world record of 237 days, surpassing a previous endurance record of 211 days established by two cosmonauts in 1982.

Circling Earth in a 149-mi-high (240-km) orbit, at an inclination of 51.6°, the three *Salyut 7* cosmonauts were visited by two other Soviet space crews and five uncrewed *Progress* supply vessels. *Progress 19* was rocketed spaceward on February 21, 1984, providing the long-duration team with food, oxygen, water, spare parts, and new equipment.

On April 3, 1984, a three-person *Soyuz T-11* crew lifted off from Earth, docking with the occupied *Salyut 7* one day later. Onboard the craft were India's first man in space, Rakesh Sharma, and Soviet cosmonauts Yuri Malyshev and Gennadi Strekalov. During an 8-day stay aboard *Salyut*, Sharma conducted yoga exercises as an adaptive measure to the effects of weightlessness. The Soviet-Indian team carried out 12 experiments, including mapping of the Indian subcontinent and creating samples of silver germanium alloy in a *Salyut* furnace. Using the *Soyuz T-10*, the spacecraft which transported the three long-duration cosmonauts to *Salyut*, Sharma, Malyshev, and Streakalov returned to Earth on April 11, 1984. The *Soyuz T-11* (piloted by the three remaining cosmonauts) was undocked from one *Salyut* airlock and reattached to another.

Continuing on their trek, Atkov, Solovyev, and Kizim received three *Progress* spacecraft to replenish food supplies, unload new hardware, and recharge *Salyut* fuel and oxygen tanks: *Progress 20* on April 17, 1984; *Progress 21* on May 10; and *Progress 22* on May 30. During that time span, a series of spacewalks by Kizim and Solovyev led to repair of a section of the station's propulsion system which developed a leak in September 1983. Cardiologist Atkov monitored his spacewalking colleagues from inside the Earth-circuiting laboratory.

Soviet cosmonauts Vladimir Dzhanibekov, Svetlana Savitskaya (on her second space mission), and Igor Volk reached *Salyut 7* on July 18, 1984, on *Soyuz T-12*. Docking to one of the station's airlocks, the *Soyuz T-12* crew brought mail with them, as well as additional supplies, and other material for the three men who had been aboard *Salyut* for some 160 days. Becoming the first woman to walk in space, Savitskaya left the space station with Dzhanibekov on July 25, 1984, working outside *Salyut 7* for nearly 4 h. Her tasks included six metal-cutting experiments, six soldering tasks using lead and tin, and two other experiments where a silver coating was applied to an aluminum surface. A hand-operated electron beam tool was evaluated during the cutting and welding tasks. Returning after a 13-day mission, Dzhanibekov, Savitskaya, and Volk parachuted to Earth in *Soyuz T-12* on July 29, 1984.

A *Progress 23* docked to *Salyut 7* on August 16, 1984. The engines of the resupply vehicle were later used to raise the station's orbit to a 234 × 218 mi (377 × 351 km) altitude. The three Soviet cosmonauts ended their mission on October 2, 1984. Completing 237 days in space, Kizim, Solovyev, and Atkov rode the *Soyuz T-10* descent capsule to a safe landing on the steppe in Soviet Kazakhstan, about 90 mi (144 km) from the launch center from which they departed nearly 8 months prior.

While the cosmonauts were orbiting Earth, six spacewalks totaling 22 h 50 min were accomplished, extending the spacewalk mark set by America's *Skylab* space station crew in 1973–1974, which saw four walks equaling 22 h 13 min. During their long-term space stint, months of astrophysical, biological, medical, and technological experiments were performed onboard *Salyut 7*. Weeks of adaptation to Earth gravity were required by the cosmonauts, but their overall medical status was deemed satisfactory at the conclusion of their flight.

At the close of 1984, Western space authorities were awaiting the first test flight of an uncrewed Soviet booster capable of orbiting payloads of 100 tons (90 metric tons) or more, roughly equivalent in capability to the now discarded United States Saturn V launcher. Such a heavy-lift rocket, it is theorized, has been constructed to orbit large modules for space station purposes, perhaps permitting dozens of cosmonauts to live and work in space at one time.

Uncrewed satellite utilization. For communications purposes, the Soviet Union continued to rocket into space *Molniya* satellites, normally maintaining 12 of this class satellite in highly elliptical orbits. *Molniya* satellites continue as the mainstay of the Soviet space-based communications network, providing domestic and international radio, telephone, and television services. Geostationary-positioned *Ekran*, *Raduga*, and *Gorizont* communications satellites were also orbited. Together with *Molniya* spacecraft, the satellites beam central television programming into the homes of 90% of the populace of the Soviet Union.

Meteorological satellites of the *Meteor* series were orbited in 1984 for keeping a television eye on the distribution of clouds, ice, and snow cover. Data on temperature fields, cloud-top heights, as well as on water surface temperatures, are also relayed from *Meteor* satellites.

ASIAN SPACE ACTIVITY

Three satellites were launched by Japan in 1984. As Japan's first operational direct broadcast satellite, the *BS-2A* was rocketed from the National Space Development Agency's (NASDA) launch facilities on Tanegashima Island on January 24 aboard a Japanese N-2 booster. The 770-lb (350-kg) satellite, built to provide two color television channels to 420,000 households, reached a stationary orbit 22,300 mi (35,680 km) above the Equator. However, the *BS-2A* suffered failures in several key electronic components, making unusable its direct-to-home satellite television capacity.

Japan's Institute of Space and Astronomical Science launched its ninth satellite, *Ohzora*, formally known as *EXOS-C*, on February 14. *Ohzora* is one of the Japanese contributions to a Middle Atmospheric Program (MAP), for infrared spectroscopy of the Earth's middle atmosphere.

Himawari 3 was orbited by Japan's NASDA on August 3, and later positioned into a geosynchronous slot. As Japan's third geosynchronous meteorological spacecraft, *Himawari 3* is being used for typhoon storm monitoring over Asia and the western Pacific.

China orbited its fourteenth satellite on January 29, 1984, making use of a new booster. The spacecraft was eventually maneuvered into a 4017 × 223 mi (6479 × 359 km) orbit.

Announced by China as its first experimental communications satellite, that country's fifteenth spacecraft was placed in space on April 8, reaching a geosynchronous orbit 8 days later. The nearly 1-ton satellite was boosted by a new launcher, a CZ-3 three-stage liquid propellant rocket, exhibiting greatly increased capabilities over earlier Chinese carrier rockets.

China launched its sixteenth satellite, a military reconnaissance spacecraft, on September 12 into an 108 × 247 mi (175 × 399 km) orbit. The satellite ejected a capsule to Earth 5 days later.

Events in 1984 indicate China's burgeoning role in space, including the making of payload reservations aboard the United States space shuttle as well as Europe's Ariane booster. The launch slots were requested by China for its domestic direct broadcast satellite network to be orbited within a 1987–1988 time period.

EUROPEAN SPACE ACTIVITY

The business of orbiting commercial satellites heated up in 1984, fueled by a string of successful launches of Europe's Ariane booster. Lifting off from a jungle site in Kourou, French Guiana, four Ariane launches placed six commercial satellites into their proper orbits over the year. An *INTELSAT V (F-8)* communications satellite was orbited by an Ariane on March 5, 1984, marking the booster's eighth "developmental" flight, and the last under the auspices of the European Space Agency.

Ariane is now property of Arianespace, formed in 1980 by companies and banks in 11 European countries which manufacture, market, and launch the European booster. The first commercial Ariane flight under the responsibility of Arianespace occurred on May 23, 1984, when a three-stage Ariane 1 placed into space an American-built *GTE SPACENET 1* communications satellite. That liftoff marked the first time a United States–owned satellite was rocketed spaceward by a non–United States booster. Orbiting an American satellite by Ariane also signaled to space customers that they need not depend solely on the United States shuttle for a relatively economical boost into space.

On August 4, 1984, both a *European Communications Satellite (ECS 2)* and the French satellite *TELECOM 1* were boosted by an Ariane 3 class vehicle. On its maiden flight, the Ariane 3 featured two first-stage strap-on solid propellant boosters added to the basic Ariane 1–type launcher, thereby augmenting the ability of the basic rocket to place heavier payloads into orbit.

Again using an Ariane 3, the European Space Agency Maritime Communications Satellite, *MARECS-B2*, and a *GTE SPACENET 2* satellite were orbited on November 10, 1984.

The spate of successful Ariane launches helped to promulgate, in essence, a trade war in outer space. A United States private firm, Transpace Carriers Inc., filed a trade practices complaint against Arianespace, contending that the French government is subsidizing Ariane launch services to American companies at unfairly low prices. Transpace, hoping to sell launches to customers with its Delta series of boosters, asked the Office of the U.S. Trade Representative to prohibit Arianespace from marketing services in America. An investigation was undertaken concerning the allegations.

For background information *see* APPLICATIONS SATELLITES; COMET; COMMUNICATIONS SATELLITE; MAGNETOSPHERE; REMOTE SENSING; SPACE BIOLOGY;

SPACE FLIGHT; SPACE PROBE; SPACE PROCESSING; SPACE SHUTTLE; SPACE STATION; VENUS in the McGraw-Hill Encyclopedia of Science and Technology.   [LEONARD DAVID]

Bibliography: *Aviat. Week Space Technol.*, issues from November 21, 1983, through November 19, 1984; *NASA Activities*, December 1983 through November 1984; *Science News*, November 19, 1983, through November 24, 1984; *Space World*, December 1983 through November 1984.

Speech recognition

To discuss progress in speech recognition, it is first necessary to clearly separate the area into specific tasks because of the large differences in technical difficulty that the respective tasks represent. Word recognition systems accept spoken input as sequences of isolated or connected words without significant reliance on the syntax or grammar that might restrict the acceptable word sequences. They embody only lexical information, generally a list of words that may be used in generating such sequences. In contrast, speech recognition systems deal with some subset of a spoken natural language and employ, in addition to the lexical information, syntactic or grammatical information to predict the likelihoods of specific words following each other.

Word recognition systems may be speaker-trained or speaker-independent, indicating whether each speaker is expected to train the system by pronouncing each word one or more times before it is ready to recognize his or her spoken words, or whether such training takes place beforehand by recording many speakers' productions of those words. Speaker-trained recognizers are generally simpler, less expensive, and prone to fewer recognition errors than speaker-independent systems.

Recent progress, both in terms of the algorithms or signal-processing steps that lead to reliable recognition, and of the computing hardware that implements those algorithms, has allowed the incorporation of speaker-trained isolated-word recognizers into personal computers to permit their control by means of selected spoken commands. So far, the most useful applications have been those of controlling some other apparatus when the speaker's hands and eyes are otherwise occupied. Results reflecting gradually improved recognition capabilities for vocabularies in excess of 1000 words are being achieved in research laboratories, but such techniques have not yet been incorporated into practical, cost-effective recognition systems.

Word recognition procedures. Practically all currently marketed word recognition systems represent each word in terms of a set of features describing aspects of the speech signal as it evolves over time corresponding to the production of the word. Underlying the recognition procedure is a high-speed comparison process that computes the similarity between a newly spoken unknown word and the stored representations of the previously spoken vocabulary words. Subsequently, a decision rule identifies the unknown word with the vocabulary word to which it is most similar or rejects it as not being sufficiently similar to any of the stored words. Recently introduced techniques recognize the need to overcome variations in how even a single speaker may pronounce a specific word by collecting sufficient training samples until a robust representation is judged to have been attained. Further enhancements have allowed the achievement of relatively reliable recognition even in the presence of significant amounts of background acoustic noise such as may be found in offices and some factories.

Hardware requirements. The most significant advances for cost-effective recognition result from the rapid growth of computing power available on special-purpose or programmable signal-processing integrated circuits. Such chips can execute both the signal analysis and the comparison functions. When augmented with additional circuitry to convert the analog speech signal to a frequently sampled digital form and adequate memory chips to store the reference information, inexpensive recognition systems can be achieved. The availability of chips with multiply times as low as 300 nanoseconds has made one-chip real-time word recognition possible. More recent designs allow extension of this capability to vocabularies up to 1000 words. Such units can be used in both speaker-trained and speaker-independent modes, depending on whether the reference data represent multiple forms of the same word from the same speaker or whether data from a wide variety of speakers have been analyzed so as to arrive at a few selected forms for each word.

Factors affecting recognition performance. The approach to achieve speaker independence by separately analyzing many speakers' productions of each individual word works best when the potential users belong to the same dialect or accent groups as the training speakers. True speaker independence is unlikely to be achieved without building into the recognizer a capability to learn continuously from newly input words. No commercially available recognition system manifests such learning capabilities.

In natural connected speech, pauses are not apparent between successive words. This introduces significant uncertainty as to where one word ends and another starts, leading to complex search strategies to explore all legal sequences of words. By requiring the speaker to pause between words, the task of the recognizer is eased considerably. Not only are the end points of the words located more easily, but the requirement to pause between words slows the speaking rate, clarifies the pronunciation, and reduces the variability between different appearances of the same word. Isolated-word recognition systems can be looked upon as short-term solutions that require the speaker to compensate for the limited capabilities of current recognizers.

Systems based on phonetic models. A vocabulary of 5000 words is considered adequate for rough-copy text input on correspondence tasks relating to specialized topics. One such system has been operated in real-time on a large laboratory-based computer system. The approach taken is to incorporate phonetic models for each word, that is, a representation in terms of sequences of sounds for all the alternative ways the word may be pronounced. The sounds in turn are represented by acoustic models, quantitative descriptions reflecting the evolution in time of the analyzed features. The speakers' representation of the sounds is inferred from spoken words consisting of known sound sequences by matching segments of the signal with the respective sounds. This avoids the need for training the recognizer on every word, an impractical task for large vocabularies. However, there is the risk that the same speech sounds may differ significantly from the form in which they were produced in the training word to the form in which they are produced in an unknown word to be recognized. Additionally, large bodies of text comparable to the type of text which the recognizer is expected to encounter must be analyzed to develop an adequate model for the likelihoods with which words follow each other. Statistical models which incorporate information on the likelihood of occurrence of specific three-word sequences (trigrams) are preferred at this time over grammatical models, due to the lack of adequately developed models for finite word subsets of natural languages.

Prospects. For many years, practical exploitation of word recognition was limited by the inability to provide the required processing capability in a cost-effective fashion. It appears that processing capability has now caught up with algorithmic knowledge, at least for small-vocabulary, isolated-word recognizers. To achieve further progress, additional basic research is needed on how to represent the individual speech sounds so as to best differentiate them from each other despite the significant speaker and context-dependent variations. Some researchers believe that this is best done through automatic learning techniques; others are focusing on human speech perception in an attempt to model computer-based recognition after known psychoacoustic processes. Effective speech recognition is carried out by large research teams working with sizable data bases and significant computing power to test the improved techniques that are continually proposed. Over the next 10 years a renewed effort can be expected to enhance the performance of algorithms to allow the tackling of more complex tasks. Although practical connected-word speech recognition systems are unlikely to be achieved before the end of this century, gradual enhancements in performance can be expected as the technology advances step by step.

For background information see SPEECH RECOGNITION in the McGraw-Hill Encyclopedia of Science and Technology.

[PAUL MERMELSTEIN]

Bibliography: L. R. Bahl et al., Some experiments with large-vocabulary isolated-word sentence recognition, *Proceedings of the International Conference on Acoustics, Speech and Signal Processing*, 26.5.1–26.5.2, 1984; F. Bucy et al., Ease-of-use features in the Texas Instruments Professional Computer, *Proc. IEEE*, 72:269–282, 1984; R. Kavaler et al., A dynamic time warp IC for a one thousand word recognition system, *Proceedings of the International Conference on Acoustics, Speech and Signal Processing* 25B6.1–25B6.4, 1984; D. S. Miller, A compact speech recognition chip set, *Electro '83*, 1983.

Staphylococcus

Staphylococci are one of the major groups of bacteria inhabiting skin. Staphylococci together with other resident microflora may form a defensive barrier to invasion by more serious pathogens and metabolize waste products secreted by the skin and its glands. When natural barriers and defense systems are compromised, staphylococci can pose a serious threat to health. Their versatility in producing antibiotic-resistant populations has stimulated bacteriologists, geneticists, and biochemists to explore strategies to outmaneuver their attack.

Historically, staphylococci have been divided into two major groups on the basis of their coagulase activity. Staphylocoagulase is an extracellular enzyme, and its occurrence is generally accepted as a distinguishing property of potentially pathogenic staphylococci, though the role of staphylocoagulase in infection is not yet clear.

The coagulase enzyme exerts a clotting effect on the plasmas derived from several animal species. More specifically, it aids in the activation of prothrombin to staphylothrombin, which in turn promotes the conversion of a soluble plasma protein, fibrinogen, into insoluble fibrin (clot formation).

Coagulase-positive species (especially *Staphylococcus aureus*) have been known for their ability to produce acute and pyogenic infections in humans and animals, whereas, up until recently, coagulase-negative species were regarded as part of the normal skin flora and thought to be nonpathogenic. Interest in the coagulase-negative species has increased dramatically in the last decade with the discovery of new species (now totaling 17) and the finding that under certain circumstances several of them may produce serious and even fatal infections.

Structure. Members of the genus *Staphylococcus* are gram-positive cocci (0.5–1.5 μm in diameter), occurring singly, in pairs, in tetrads, and in irregular grapelike clusters. They are nonmotile, nonsporeforming, and catalase-positive. Most species are facultatively anaerobic. Their cell wall contains a peptidoglycan and teichoic acid. The diamino acid present in the peptidoglycan is L-lysine, a feature which distinguishes them from many bacterial genera. Unsaturated menaquinones and cytochromes *a* and *b* (and *c* in certain species) form the electron transport system. Most staphylococci will grow well in the presence of up to 10% sodium chloride and between 64 and 104°F (18 and 40°C). *Staphylococcus* can be distinguished from the genus *Streptococ-*

cus on the basis of microscopic morphology, cell-wall composition, and the presence of catalase and cytochromes. It can be distinguished from the genus *Micrococcus*, often sharing the same habitat, on the basis of cell-wall composition, cytochrome type, intrinsic antibiotic susceptibilities, susceptibility to lysis by the endopeptidase lysostaphin, and deoxyribonucleic acid (DNA) base composition.

Hosts. Currently, 20 different species of *Staphylococcus* are recognized (see table), of which 17 are coagulase-negative (*S. aureus* and *S. intermedius* are coagulase-positive, while *S. hyicus* is coagulase-variable). Species living on humans include *S. aureus*, *S. epidermidis*, *S. capitis*, *S. hominis*, *S. haemolyticus*, *S. warneri*, *S. saccharolyticus*, *S. auricularis*, *S. saprophyticus*, *S. cohnii*, *S. xylosus* (uncommon), and *S. simulans* (uncommon). Other primates are colonized by the species *S. aureus*, *S. saprophyticus*, *S. xylosus*, *S. sciuri* (uncommon), and *S. simulans* (uncommon), and a different subspecies of *S. haemolyticus*, *S. warneri*, *S. cohnii*, and *S. auricularis* than that found on humans. The species *S. intermedius* mainly colonizes carnivora, and has also been isolated from certain other mammals and birds. *Staphylococcus hyicus* ssp. *hyicus* is found frequently living on pigs; whereas *S. hyicus* ssp. *chromogenes* is found predominantly on cattle. *Staphylococcus sciuri* is found living on rodents and certain other mammals. *Staphylococcus lentus* colonizes sheep and goats, *S. caprae* colonizes goats, and *S. gallinarum* has been isolated from poultry (chickens and pheasants). The species *S. caseolyticus* has been isolated from milk and dairy products, and *S. carnosus* is being used as a starter culture in the processing of fermented meats such as sausages and salamis. Both of these latter species may be found living on domestic artiodactyls.

Habitats. Most *Staphylococcus* species are common inhabitants of the skin, skin glands, and mucous membranes of warm-blooded animals, and many exhibit habitat preferences (see illustration). For example, *S. capitis* produces large populations

Differentiation of *Staphylococcus* species*

Character	*S. aureus*	*S. epidermidis*	*S. capitis*	*S. caprae*	*S. warneri*	*S. haemolyticus*	*S. hominis*	*S. saccharolyticus*	*S. auricularis*	*S. saprophyticus*	*S. cohnii*†	*S. xylosus*	*S. simulans*	*S. carnosus*	*S. intermedius*	*S. hyicus*‡	*S. sciuri*	*S. lentus*	*S. caseolyticus*	*S. gallinarum*
Colony size (large)	+	−	−	+	d	+	−	−	−	+	d+	+	+	+	+	++	+	−	−	+
Colony pigment	+	−	−	−	d	d	d	−	−	d	−d	d	−	−	−	−	−+	d	d	d
Anaerobic growth	+	+	(+)	(+)	+	(+)	−	+	(±)	(+)	d(+)	d	+	+	(+)	++	(+)	(±)	(±)	(+)
Aerobic growth	+	+	+	+	+	+	+	−	+	+	++	+	+	+	+	++	+	+	+	+
Coagulase	+	−	−	−	−	−	−	−	−	−	−−	−	−	−	+	d−	−	−	−	−
Hemolysis	+	−	−	(d)	(d)	(+)	−	−	−	−	(d)(d)	−	(d)	−	d	−−	−	−	−	(d)
Nitrate reduction	+	+	d	+	−	d	d	+	(d)	−	−−	d	+	+	+	++	+	+	+	+
Acetoin	+	+	d	+	+	d	d	ND	d	+	dd	d	−	+	−	−−	−	−	−	−
Cytochrome *c*	−	−	−	−	−	−	−	−	−	−	−−	−	−	−	−	−−	+	+	+	−
Phosphatase	+	+	−	(+)	−	−	−	d	−	−	−+	d	(d)	+	+	++	+	+	+	(+)
Urease	+	+	−	+	+	−	+	ND	−	+	−+	++	−	+	+	dd	−	−	ND	+
Arginine utilization	+	+	d	+	d	+	d	+	d	−	−−	−	+	+	d	++	−	−	ND	−
β-Glucosidase	+	(d)	−	−	+	d	−	ND	−	d	−−	+	−	−	d	dd	+	+	ND	+
β-Glucuronidase	−	−	−	−	d	d	−	ND	−	−	−+	d	d	−	+	d−	−	−	ND	−
β-Galactosidase	−	−	−	−	−	−	−	ND	(d)	d	−+	+	+	+	+	−−	−	−	ND	−
Novobiocin resistance	−	−	−	−	−	−	−	−	−	+	++	+	+	−	−	−−	+	+	−	+
Acid (aerobically) from:																				+
Maltose	+	+	−	d	(+)	+	+	−	(+)	+	(d)(+)	+	−	−	(±)	−d	(d)	d	+	
D-Trehalose	+	−	+	+	+	d	−	(+)	+	++	+	d	+	d	+	++	+	+	d	+
D-Mannitol	+	−	+	−	d	d	−	−	−	d	dd	d	+	+	(d)	−d	+	+	−	+
D-Xylose	−	−	−	−	−	−	−	−	−	−	−−	+	−	−	−	−−	−	(d)	−	+
Xylitol	−	−	−	−	−	−	−	−	−	d	(d)(d)	(d)	−	−	−	−−	−	−	−	d
D-Cellobiose	−	−	−	−	−	−	−	−	−	−	−−	−	−	−	−	−−	+	+	ND	+
Sucrose	+	+	(+)	−	+	+	(+)	−	d	+	−−	+	+	−	+	++	+	+	d	+
D-Turanose	+	d	−	−	d	d	d	ND	(d)	+	−−	d	−	d	d	−d	−	−	−	+
D-Mannose	+	(+)	+	+	−	−	(+)	−	−	−	(d)+	d	+	d	+	++	(d)	(+)	−	+
D-Ribose	+	d	−	−	d	d	−	ND	−	−	−−	d	d	ND	+	++	+	+	+	+
Raffinose	−	−	−	−	−	−	−	−	−	−	−−	−	−	−	−	−−	+	+	ND	+
α-Lactose	+	d	−	+	d	d	d	−	−	d	−+	d	+	d	d	++	(d)	d	+	d
β-D-Fructose	+	+	+	−	+	d	+	(+)	+	+	++	++	+	+	+	++	(+)	+	+	+

*+ = 90% or more strains positive; d = 11–80% strains positive; ± = 90% or more strains weak positive; − = 90% or more strains negative; ( ) = delayed reaction; ND = not determined.

†Characteristics of the human *S. cohnii* subspecies are noted on the left and those of the primate *S. cohnii* subspecies are noted on the right.

‡Characteristics of *S. hyicus* ssp. *hyicus* are noted on the left and those of *S. hyicus* ssp. *chromogenes* are noted on the right.

on the adult human head, especially the scalp and forehead where sebaceous glands are numerous and very active. In preadolescent children, where sebaceous glands are much less active, *S. capitis* is a relatively minor species. *Staphylococcus auricularis* has a strong preference for the human external auditory meatus, where it is in contact with ceruminous gland secretions (earwax). The nonhuman primate subspecies of *S. auricularis* prefers to colonize the external auditory meatus and also regions of skin in contact with the oily secretions of specialized glands or scent glands of monkeys and prosimians. The species *S. hominis* and *S. haemolyticus* produce large populations in areas of the skin where apocrine glands are numerous (for example, the axillae and pubic areas). These species also colonize the drier regions of skin more successfully than others. *Staphylococcus aureus* prefers the mucous membranes of the anterior nares in the human adult. *Staphylococcus saprophyticus* is usually found in small populations on human skin, but may also colonize the urinary tract. It has a higher capacity to adhere to uroepithelial cells than to buccal or skin cells, and does so better than other species. *Staphylococcus epidermidis* is usually the predominant species on human skin. It produces very large populations on the face, axillae, inguinal and perineal areas, feet, and the mucous membranes of the anterior nares. It is probably the most versatile of the resident species. Individual strains of *S. epidermidis* may demonstrate habitat preferences, though the species itself is widespread over the human body.

Certain species such as *S. xylosus* and *S. sciuri* have been isolated frequently from environmental sources such as beach sand and estuaries. Based on their ability to grow on inorganic sources of nitrogen, these species are perhaps more primitive and free-living than other staphylococci.

Clinical significance. The coagulase-positive species *S. aureus* is a major opportunistic pathogen in humans and a variety of other mammals and birds. Infections produced by this species are often pyogenic (pus-producing) and acute and, if untreated in some individuals, may spread contiguously to surrounding tissue or via bacteremia to metastatic sites of infection (that is, to other organs). The most common infections caused by *S. aureus* involve the skin and surrounding tissues and include pustules, boils, carbuncles, impetigo, cellulitis, and postoperative wound infections. A common community-acquired disorder is food poisoning caused by thermostable enterotoxins produced by some strains of *S. aureus*. *Staphylococcus aureus* may also have a relationship to toxic shock syndrome.

The coagulase-positive species *S. intermedius* is a major opportunistic pathogen of carnivora. It has been implicated in otitis externa, abscesses, pyoderma, reproductive tract infections, and mastitis in dogs and cats and is therefore of concern to veterinary medicine. *Staphylococcus hyicus* (a coagulase-

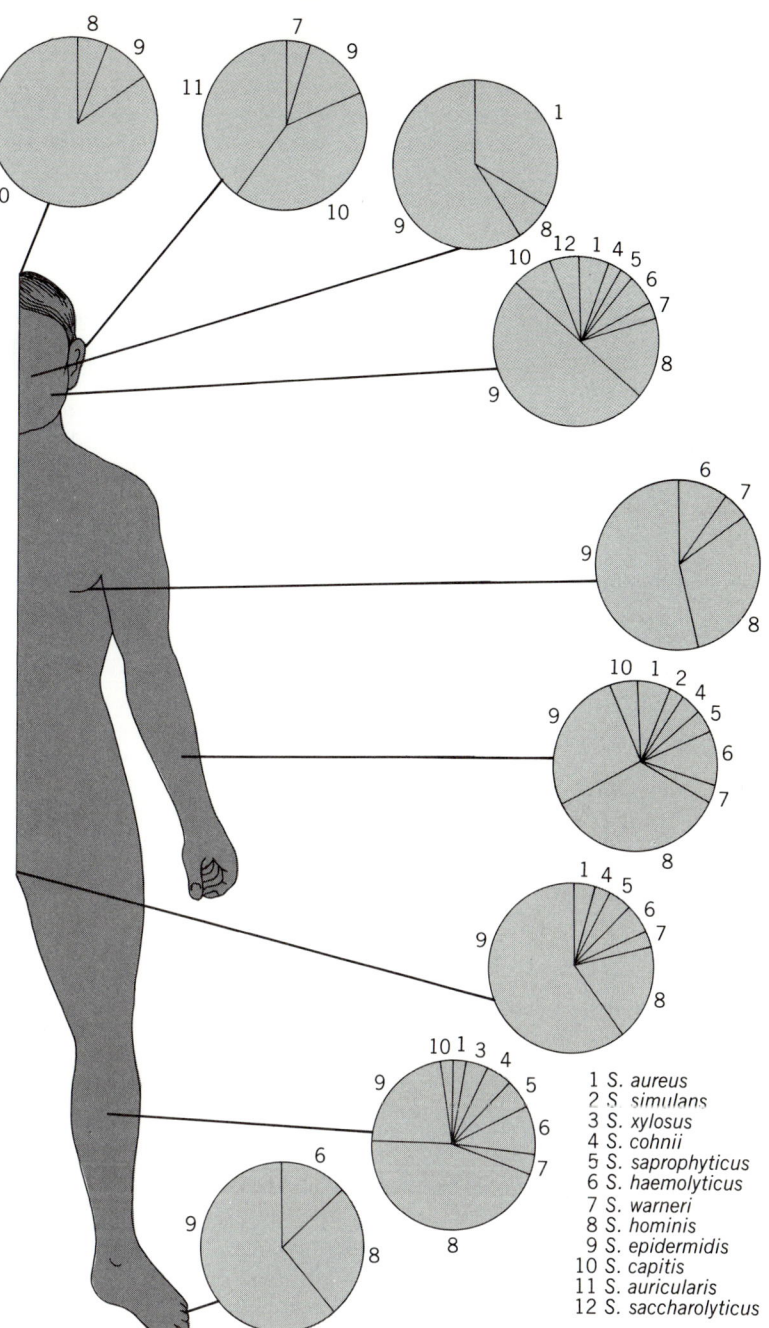

Distribution of *Staphylococcus* species on different regions of adult human skin. The percent species of total staphylococci is directly related to the percent area of sectors of the circle. (*After W. E. Kloos and J. H. Jorgensen, Staphylococci, in E. H. Lennette et al., (eds.), Manual of Clinical Microbiology, 4th ed., American Society for Microbiology*)

variable subspecies) has been implicated in infectious exudative epidermitis and septic polyarthritis of pigs. This subspecies and *S. hyicus* ssp. *chromogenes* (a coagulase-negative subspecies) have been isolated from the milk of cows with mastitis, though *S. aureus* remains the major species causing mastitis.

For many years the coagulase-negative *Staphylococcus* species indigenous to humans were generally regarded as contaminants when isolated from clinical specimens, but this view is rapidly changing. It is becoming clear that modern medical practices leading to the compromise of patients, such as the insertion of prosthetic devices or catheters and immunosuppressive therapy, have greatly enhanced the risk of infection by resident coagulase-negative species. *Staphylococcus epidermidis* and *S. saprophyticus* are the best-documented opportunistic pathogens. *Staphylococcus epidermidis* has been frequently implicated in prosthetic valve endocarditis, with mortality reaching as high as 70%, and in infections of intravascular catheters, cerebrospinal fluid shunts, and orthopedic appliances. Slime-mediated adherence may be an important factor in the pathogenesis of *S. epidermidis* infections of these medical devices. *Staphylococcus hominis*, *S. haemolyticus*, and *S. warneri* are occasionally implicated in similar infections. *Staphylococcus saprophyticus* appears to be the predominant species causing acute urinary tract infections in young adult women. This species has been implicated in cystitis, urethritis, and pyelonephritis, and is usually accompanied by a significant bacteriuria (bacteria present in the urine).

Antibiotic resistance. Resistance to the penicillins (for example, penicillin G, penicillin V. ampicillin) is quite frequent in most of the *Staphylococcus* species. Resistance to tetracycline and erythromycin is also common. Multiple resistance to penicillin, tetracycline, or erythromycin is especially prevalent in the species *S. epidermidis*, *S. hominis*, *S. warneri*, and *S. haemolyticus*. Methicillin- and gentamicin-resistant *S. aureus* and *S. epidermidis* strains are increasing in numbers in many localities of the world and are of growing concern, especially in hospitals using these antibiotics. Resistance to antibiotics is usually determined by genes located on plasmids (extrachromosomal DNA). The management of antibiotic therapy for coagulase-negative staphylococcal infections is often very difficult and complicated by the appearance of high levels of multiple antibiotic resistance. Coagulase-negative staphylococci may serve as a large reservoir of antibiotic-resistant plasmids capable of being transferred to susceptible strains. As more knowledge is gained of the strategies used by staphylococci for protecting themselves and their communities, scientists will be in a better position to compromise their defenses and improve antibiotic treatment management.

For background information *see* ANTIBIOTIC; DRUG RESISTANCE; OPPORTUNISTIC INFECTIONS; STAPHYLOCOCCUS in the McGraw-Hill Encyclopedia of Science and Technology.

[WESLEY E. KLOOS]

Bibliography: L. A. Devriese and V. Hajek, A review: Identification of pathogenic staphylococci isolated from animals and foods derived from animals, *J. Appl. Bacteriol.*, 49:1–11, 1980; C. S. F. Easmon and C. Adlam (eds.), *Staphylococci and Staphylococcal Infections*, 1983; W. E. Kloos, Natural populations of the genus *Staphylococcus*, *Annu. Rev. Microbiol.*, 34:559–592, 1980; W. E. Kloos and J. H. Jorgensen, *Manual of Clinical Microbiology*, 1985.

Staurolite

Staurolite is a common hydrous silicate of medium-grade pelitic schists. While considerable progress has been made in understanding the chemical reactions and pressure-temperature conditions under which it forms and is destroyed, the details of its crystal chemistry are not fully known. A complete understanding of the role of staurolite in metamorphic rocks will not be possible until its crystal chemistry has been worked out. The crystal chemistry, structure, and formula of staurolite are being solved through a combination of techniques, including x-ray structure determination, neutron diffraction, Mössbauer spectroscopy, transmission electron microscopy, and careful, complete chemical analysis. The model presented here represents a synthesis of the available data.

A simplified formula for staurolite consistent with the model discussed in this article is shown below.

$$H_4(Fe,Mg,Li,Zn,Mn)_4Al_{18}Si_{7.5}O_{48}$$

(Elements grouped in parentheses may exhibit solid solution with each other.) The first two subscripts are variable, while the third and fourth are nearly constant. Iron (Fe) is much more abundant than the other cations in almost all natural staurolites, but pure Fe, magnesium (Mg), and zinc (Zn) staurolites have been synthesized.

X-ray diffraction. The crystal structure of staurolite was determined by using x-ray diffraction methods. The positions of the hydrogen (H) atoms were located by using a combination of neutron diffraction and nuclear magnetic resonance. The structure consists of kyanite (Al_2SiO_5) and iron hydroxide ($Al_{0.7}Fe_2O_2(OH)_2$) units interlayered along the b crystallographic axis (Fig. 1). The kyanite layers contain silicon (Si) in tetrahedral coordination with oxygen (O) and aluminum (Al) in octahedral coordination with O. Details of the iron hydroxide layer are shown in Fig. 2, where zigzag chains are composed of Al octahedra and Fe tetrahedra. It has been determined that Si is partly replaced by Al in the kyanite layer; the Al(3A) and Al(3B) sites are occupied 50% or less by a combination of Al and Fe; tetrahedral sites in the iron hydroxide layer are about 90% occupied, and Fe is the principal ion; and the U(1) and U(2) sites average about 6% occupancy by ions such as Fe and manganese (Mn). Because of the similar scattering factors of Mg and Al, the positions for Mg, the most abundant of the nonessential ions, were not located with certainty. The H atoms are located in three or more of the eight sites per formula unit disposed as shown in Fig. 3, about midway between the tetrahedral and octahedral positions in the iron hydroxide layer. Where H is absent from an H site, the nearby O remains, leading to an $OH^- \leftrightharpoons O^{2-}$ substitution on the O(1A) and O(1B) sites (Fig. 2), which must be

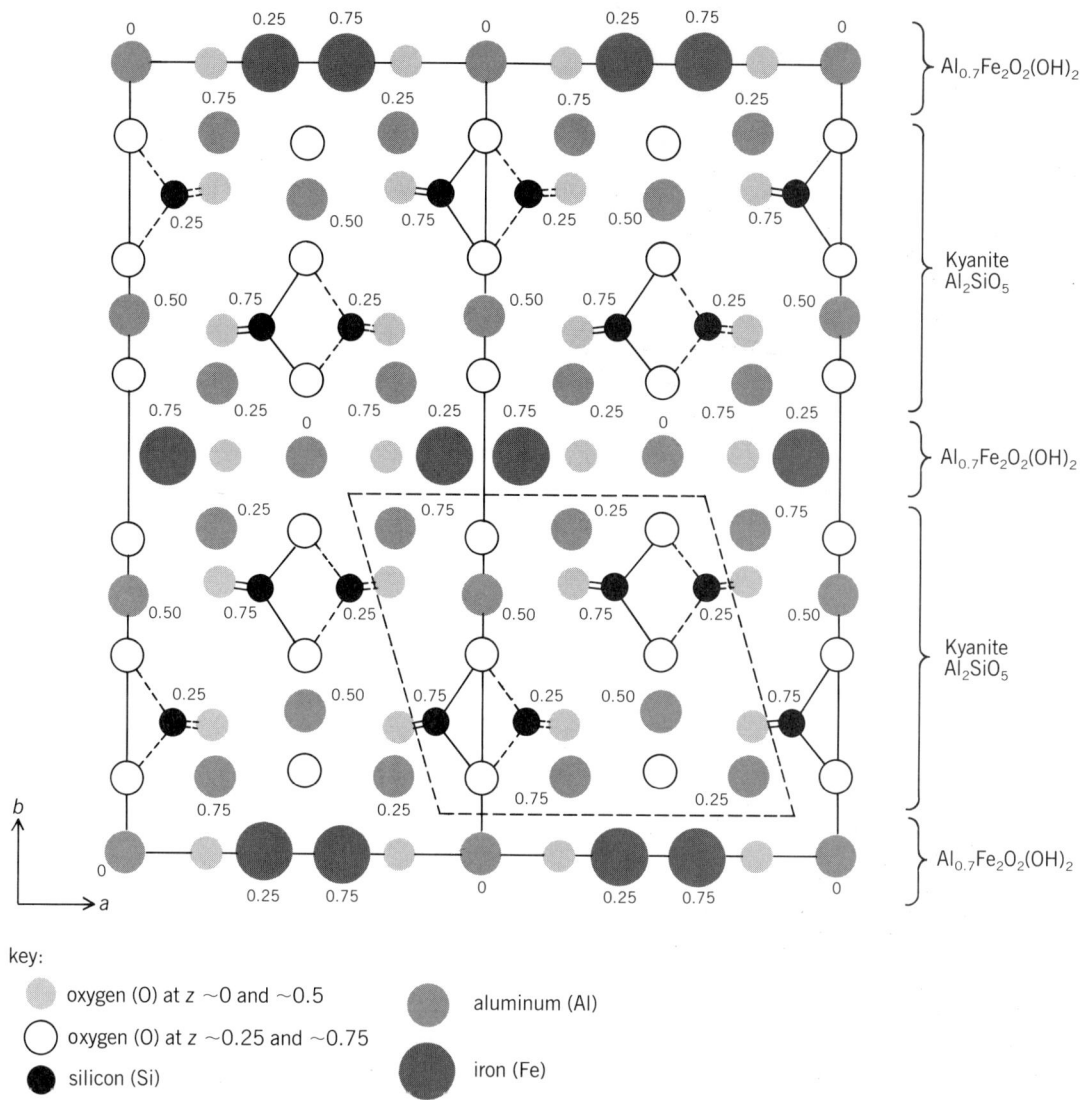

Fig. 1. Structure of staurolite on (001). Two unit cells are outlined by solid lines, and the kyanite unit cell is outlined by broken lines. The kyanite and iron hydroxide layers are indicated to the right of the diagram. Numbers on and below the diagram refer to the proportion of the unit cell, in the vertical sense (Z axis), that the ion is above the base. Thus each oxygen position relates to two oxygens in the unit cell. (After P. H. Ribbe, Staurolite, in P. H. Ribbe, ed., Reviews in Mineralogy, vol. 6: Orthosilicates, Mineralogical Society of America, pp. 171–187, 1980)

charge-compensated elsewhere in the structure.

X-ray diffraction structure determination has shown that the occupancy of Al(3A) is 1.5 times that of Al(3B), while the occupancy of U(1) is 2 times that of U(2). It has also been suggested that in fact Al(3B) and U(2) may be empty while Al(3A) is nearly full, and the apparent occupancy of Al(3B) and U(2) may result from submicroscopic twinning on (001). More recent transmission electron microscopy has demonstrated antiphase domain boundaries in some staurolites which could produce a similar result. The monoclinic (pseudo-orthorhombic) symmetry of staurolite results from the unequal occupancy of Al(3A) and Al(3B) and of U(1) and U(2). It is possible that staurolite in its most ordered form is monoclinic, having Al(3B) and U(2) empty, while disordered staurolite is orthorhombic, having equal occupancy of Al(3A) and Al(3B) as well as U(1) and U(2). An orthorhombic zincian staurolite has been described with equal occupancy of Al(3A) and Al(3B). The designations used in this context are names given to the sites by the original authors and are in general use; the letter U refers to a site which is largely unoccupied.

Mössbauer spectroscopy. Mössbauer studies indicate that most of the iron is ferrous and it exists in two or more crystallographically distinct sites. Some Mössbauer studies suggest minor ferric iron. Wet chemical analyses average about 2% Fe_2O_3, corresponding to about 0.45 Fe^{3+} on a 48-oxygen basis.

Chemical analysis It has recently shown by chemical analysis that the H content of staurolite is indeed variable. Because of this variability in H, it has become clear that an adequate understanding of the stoichiometric relations cannot be achieved without high-quality analyses of water content.

In one recent study, 31 staurolites were analyzed for elements heavier than fluorine (F) by electron microprobe, and for light elements by ion microprobe. H was analyzed by both ion microprobe and the stable isotope H extraction line. [Most recent studies have involved only electron microprobe analysis of elements heavier than F.] Two interesting results of this work are that staurolites have very little F replacing OH and that lithium (Li) is a ubiquitous constituent of all staurolites with a distribution not unlike that of Zn. Most staurolites have 0.1 to 0.4 Li atom, but a few reach values as high as 1.3 Li atoms. The average and standard deviation for each element (48-oxygen basis), and the analytical precision are given in the table. The elements Si, Al, and to a certain extent titanium (Ti) are more or less fixed in amount, while all the other elements are variable, their standard deviations being at least 2.5 times the analytical precision.

Certain elements may be grouped together to cause a reduction of their combined standard deviation while the analytical precision must necessarily become larger. The combination of Si and Al brings the standard deviation down to analytical precision, suggesting that these sums are indeed constant or nearly so. The grouping of Fe + Mg + Li + Zn

Average, standard deviation, and precision of 31 staurolite analyses*

Element	Average	Standard deviation	Precision
Si	7.62	0.07	0.05
Ti	0.11	0.02	0.01
Al	17.93	0.17	0.11
Si + Al	25.55	0.13	0.12
Fe	3.02	0.35	0.04
Mg	0.74	0.29	0.02
Li	0.22	0.26	0.02
Zn	0.14	0.30	0.01
Mn	0.05	0.03	0.01
Fe + Mg + Zn + Mn	3.94	0.36	0.04
Fe + Mg + Li + Zn + Mn	4.16	0.24	0.05
H	3.15	0.38	0.17

*Number of atoms based on 48 oxygen atoms.

+ Mn has a significantly lower standard deviation than any element in the group except Mn. However, the total number of cations in the group still has a standard deviation five times that of the analytical precision indicating that the total occupancy of the sites involved is variable. One possible interpretation of these two groups of cations (Si, Al, and Fe, Mg, Li, Zn, Mn) is that each of them represents groups of elements which substitute for each other on one or more sites. Other combinations of elements do not reduce the standard deviation.

Idealized crystal model. The following idealized model of staurolite crystal chemistry is consistent with all the available evidence. In the kyanite layer, the eight Si tetrahedra are fully occupied with 95% Si and 5% Al disordered over all the positions. The sixteen octahedral kyanite layer positions are occupied fully by Al. In the iron hydroxide layer the two Al(3A) positions are filled (or nearly filled) with 75% Al and 25% Fe, perhaps as ferric iron, disordered over the two positions. The two Al(3B) positions are empty, but on their edges near the O(1B) positions, the hydrogen sites [P(1B), where P is a proton] are approximately filled. Maximum charge balance is maintained by the P(1A) sites remaining largely empty against occupied Al(3A) sites. The four tetrahedral Fe sites and the two U(1) sites are occupied by an average of 3.7 ions of Fe, Mg, Li, Zn, and Mn. The two U(1) sites are each about 15% occupied, leaving about 3.4 ions for the four tetrahedral Fe positions. The tetrahedral vacancies probably result from the fact that wherever a U(1) site is filled, the adjacent tetrahedral sites are empty. Charge balance for the variable H content comes from some combination of $Si^{4+} \leftrightharpoons Al^{3+}$ substitution; deficiencies in the tetrahedral Fe sites which are only half compensated by U(1) site occupancy; substitution of Li^+ for Fe^{2+}; and possible $Al^{3+} \leftrightharpoons Fe^{2+}$ substitution in the Al(3A) sites.

Given the model described above and the stoichiometry from microprobe analyses, a more precise chemical formula for staurolite is as shown below.

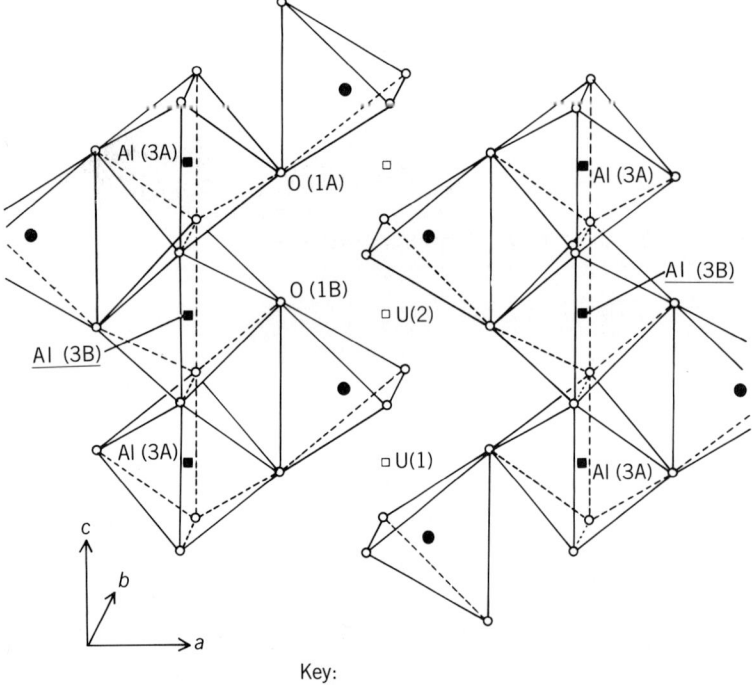

Key:
- ○ oxygen
- ● tetrahedral Fe positions
- ■ octahedral Al sites
- □ largely unoccupied octahedral sites

Fig. 2. Iron hydroxide layer of staurolite. Oxygen atoms occupy the corners of octahedra and tetrahedra, and the iron atoms occupy tetrahedral positions. O(1A) and O(1B) positions have associated hydrogen ions as shown in Fig. 3. Al(3B) octahedra (underlined) are presumed to be empty in the present model. (*After C. M. Ward, Magnesium staurolite and green chromian staurolite from Fiordland, New Zealand, Amer. Mineralog., 69:536, 1984*)

$$H_{2.7-4.2}(Fe,Mg,Li,Zn,Mn)_{3.2-4.2}(Al,Fe)_2Al_{16}(Si,Al)_8O_{48}$$

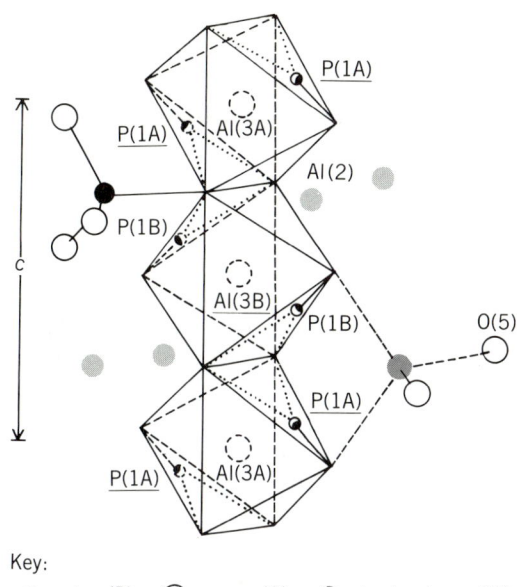

Fig. 3. Location of the partially filled H sites P(1A) and P(1B) in staurolite. The hydrogen ion (proton) lies nearly in the face of the Al(3B) octahedron near the O(1B) position (Fig. 2), and about midway between the tetrahedral iron and octahedral aluminum sites in the iron hydroxide layer. Underlined sites are presumed to be unoccupied in the present model. (*After Y. Takéuchi, N. Aikawa, and T. Yamamoto, The hydrogen locations and chemical composition of staurolite, Z. Kristallog., 136:1–22, 1972*)

Based on this model, ionic substitution occurs on five different types of crystalline sites, but site assigment for individual staurolites is still straightforward, given an accurate and complete chemical analysis and the assumption that partial occupancy occurs only on the tetrahedral Fe sites, the U(1) sites, and the P(1B) sites.

For background information *see* COORDINATION CHEMISTRY; CRYSTAL STRUCTURE; CRYSTALLOGRAPHY; MINERALOGY; SPECTROSCOPY; STAUROLITE in the McGraw-Hill Encyclopedia of Science and Technology.

[M. J. HOLDAWAY]

Bibliography: P. H. Ribbe, Staurolite, in P. H. Ribbe (ed.), *Reviews in Mineralogy*, vol. 6: *Orthosilicates*, pp. 171–187, 1980.

Stellar evolution

The formation of stars is one of the most intensively investigated topics in modern astronomy. This interest is due in large part to the rapid rise of both infrared and molecular radio astronomy in the late 1960s and early 1970s. More recently, observational infrared astronomers have advanced on two fronts: NASA's very successful *Infrared Astronomical Satellite* (*IRAS*), launched in 1983, has obtained a uniquely sensitive picture of essentially the entire far-infrared sky. Many objects have been revealed which are, almost certainly, sunlike stars that are now forming from clouds of gas and small dust particles. At the same time, new techniques have enabled ground-based infrared astronomers to partially overcome the effects of the Earth's atmosphere and examine some young, sunlike stars (called T Tauri stars) in unprecedented detail. These studies have revealed an unusual, hitherto-unknown, infrared companion to T Tauri itself (which is the prototype of this class of young stars) as well as the presence of substantial quantities of small dust particles orbiting other T Tauri–type stars. These particles may represent signposts for the formation of planets.

As exciting as these discoveries by infrared astronomers have been, the biggest surprise during the past decade was uncovered by radio and, to a lesser extent, infrared and optical astronomers. This is the realization that the process of star formation is accompanied by massive outflowing winds. These winds, which may be quite sporadic, are millions of times stronger than the solar wind and are generated by most, perhaps all, young stars either shortly before or, more probably, shortly after they initiate the burning of hydrogen in their interiors. Since it is natural to think of star formation as a process of collapse, that is, infall, the ubiquitous occurrence of an accompanying outflow comes as a bit of a shock.

T Tauri. The life histories and properties of stars in the Galaxy are determined almost entirely by their initial masses. Stars that have masses roughly equal to the mass of the Sun and that exhibit active and variable optical spectra similar to that of the prototypical star T Tauri (Fig. 1) are classified as T Tauri stars. They are found in the vicinity of clouds

Fig. 1. The star T Tauri (center) and, to its right, Hind's reflection nebulosity. Burnham's emission nebula is located just to the south of T Tauri and blended with its image on this photograph. The infrared companion is not visible. (*Courtesy of George Herbig*)

of interstellar molecules and dust grains. It is believed that these stars have very recently formed by the gravitational collapse of a small part of a molecular cloud. "Very recently" in this context means within the past several hundred thousand years, which is but an instant when compared to the 10^{10} years that these stars will live much as the Sun, peacefully burning hydrogen into helium in their interiors.

T Tauri itself has turned out to be a remarkable object and, therefore, perhaps not a good prototype for the class. In the optical part of the electromagnetic spectrum, both emission and reflection nebulosities can be seen near T Tauri (Fig. 1). In its radio and infrared spectrum, evidence from carbon monoxide and molecular hydrogen emission lines implies rapid mass loss at rather high velocity from T Tauri. Hind's reflection nebulosity (Fig. 1) may also be due to material ejected from T Tauri in the not too distant past. T Tauri has also been measured as an x-ray emitter by NASA's orbiting *Einstein Observatory*.

But the most unusual aspect of T Tauri is its close companion which emits most of its energy in the infrared and which was discovered only in 1981. Because the infrared companion is only 0.6 arc-second away (south) from T Tauri, it required a special observing technique called speckle interferometry (described below) to resolve the two stars. At the likely distance to T Tauri, about 500 light-years (3×10^{15} mi or 5×10^{15} km), 0.6 arc-second corresponds to about 100 astronomical units (1 AU is the distance between the Earth and the Sun, equal to 9.3×10^7 mi or 1.50×10^8 km, and the planet Pluto is located 40 AU from the Sun).

In spite of its cool infrared temperature, roughly 900°F (800 K), and apparently low luminosity (approximately that of the Sun), the companion to T Tauri is a rather intense radio source at centimeter wavelengths. This radio emission implies the existence of substantial quantities of ionized hydrogen gas. The source of energy that keeps the gas ionized is unclear—models of both outflow and infall have been proposed. In the infall models the infrared companion has been suggested to be either a sunlike protostar or, possibly, even a giant protoplanet orbiting T Tauri. Although most astronomers would be inclined to reject the latter interpretation, if it should be correct, then the companion is the first planet ever discovered outside the solar system.

Outflowing winds. Outflowing material, such as that observed toward T Tauri, has recently been discovered in the direction of many T Tauri stars. Even more powerful winds are generated by numerous young stars that are more massive and, therefore, more luminous than T Tauri. Indeed, because there is a rough proportionality between the strength of the outflowing wind and the luminosity of the young star that drives the wind, it was initially suggested that the underlying physical mechanism that powers the wind is radiation pressure on dust grains located in the outflowing gas. That is, photons in the stellar radiation field push on dust grains which, in turn, collide with gas particles (primarily hydrogen molecules and helium atoms) which are then driven away from the star along with the dust.

Unfortunately, this simple and, therefore, appealing mechanism probably does not work because the winds are often so powerful that the momentum they carry is hundreds of times greater than the momentum carried by the stellar radiation field. Therefore, other, more exotic driving forces have been suggested. A popular one, for T Tauri star winds, involves an outward-directed flux of magnetic (that is, Alfvén) waves which are generated near the stellar surface. Dissipation of these waves is envisioned to drive the outflowing gas. The underlying energy source for the waves might be convection and rapid differential rotation, both of which occur in the Sun, albeit to a lesser degree.

Whatever the underlying driving force might be, it seems clear observationally that most, and possibly all, stars undergo at least one episode of rapid mass loss during their formative years. This mass loss may be the critical reason why only relatively few stars grow to masses much greater than the mass of the Sun. In the absence of rapid mass loss, young

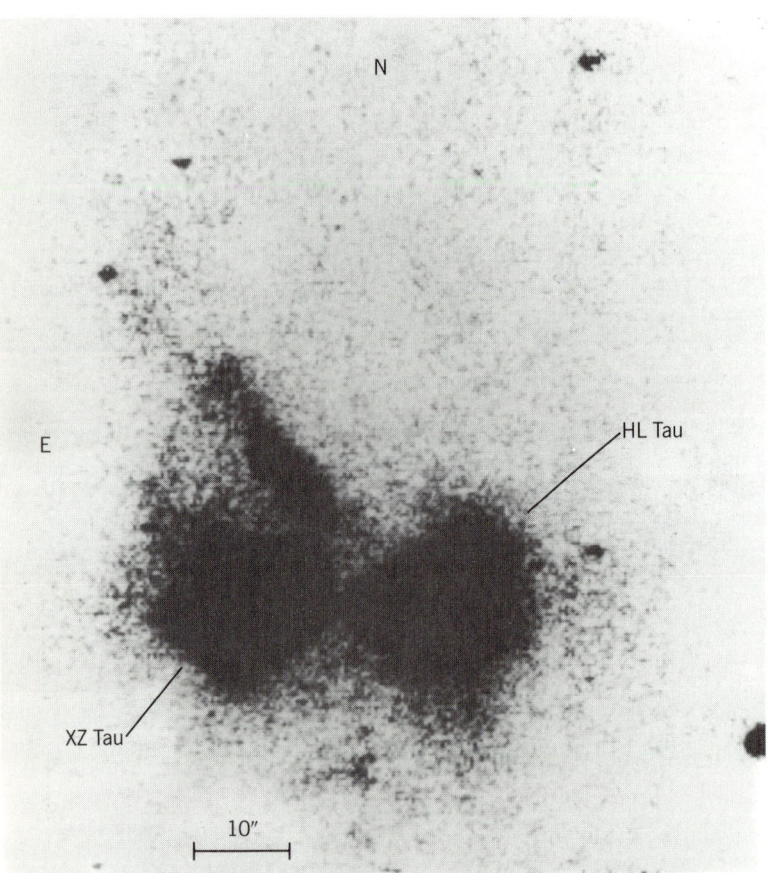

Fig. 2. Region near the star HL Tauri, located within the bright blob of scattered visible light to the right. The small cloud of dust particles that scatters infrared radiation is not visible. The jet of gas located just above the star XZ Tauri may have been ejected from HL Tauri. (*From R. Mundt and J. W. Fried, Jets from young stars, Astrophys. J., 274:L83–L86, 1983; © 1983 American Astronomical Society*)

stars located in molecular clouds might be expected to continue to gravitationally accrete surrounding material for an extremely long time. However, once the massive outflowing wind turns on, it overpowers the gravitational accretion and clears out a local cavity in the cloud. The key to the trigger that turns on the wind is still uncertain. One interesting possibility is the onset of the nuclear burning of deuterium to helium which happens at relatively low temperatures and might be regarded as the demarcation line between protostars shining by energy generated by gravitational infall, and true stars generating energy by nuclear reactions. The onset of deuterium burning should generate strong convective motions inside T Tauri stars which, combined with the differential stellar rotation, may then power the winds as outlined above.

An additional observational clue to the nature of the winds is that the flows appear bipolar. That is, in most, perhaps all, such outflows, the material is ejected in two opposite directions at velocities between a few 100 and 600 mi/s (a few 100 and 1000 km/s). Sometimes these high-velocity jets appear so narrowly collimated that the initial collimation must be taking place fairly near to or even at the stellar surface. One plausible source of the collimation is a dusty gaseous disk of material in orbit around the young star. The jets are envisaged as emerging from the poles of the disk. People have speculated that such a disk was the birthplace of the planets and other debris in the solar system. If this model is correct, then disks and possibly solar systems must be quite common. Infrared astronomers have recently found evidence for substantial material close to some young stars.

Formation of new planetary systems. The diffraction-limited angular resolving power of a large optical or infrared telescope is much better than the actual resolution obtained in ordinary astronomical photographs. The reason is the Earth's atmosphere, which distorts the incoming (plane) wave from the star of interest and smears out its image. (This is essentially the same effect that causes stars to twinkle when observed with the naked eye.) To circumvent this problem, astronomers are prepared to spend substantial sums of money on instruments such as the Space Telescope to be launched by NASA.

But there is a way to partially overcome the deleterious effects of the atmosphere while still using ground-based telescopes. This technique, called speckle interferometry, relies on very short exposure images that "freeze" the atmosphere and eliminate much of its smearing effect. It is thereby possible to resolve much finer details than with conventional astronomical imaging. Recently, speckle interferometry was employed to reveal the existence of halos of infrared light that had been scattered toward the Earth by tiny dust particles surrounding two young stars, HL Tauri and R Mon (Figs. 2 and 3).

The radius of the dust halo around HL Tauri, which is a T Tauri–type star, is roughly 100 AU. The total

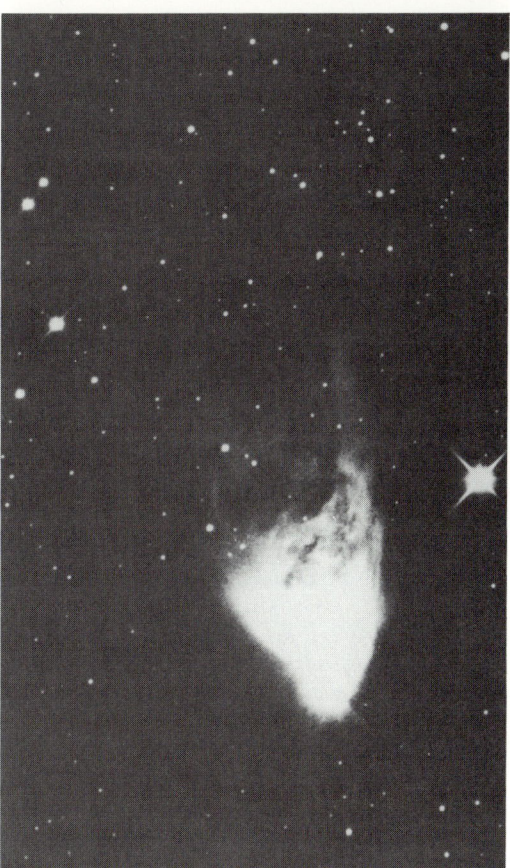

Fig. 3. Region near the star R Mon, located near the tip of the bright fan-shaped cloud of gas (bottom center). The cloud is probably material that was ejected from the vicinity of the star a few thousand years ago. The cloud of dust grains that scatters infrared radiation is not visible. (*Lick Observatory photograph; from B. F. Jones and G. H. Herbig, Proper motion of Herbig-Haro objects, II. The relationship of HH39 to R Mon and NGC 2261, Astron. J., 87:1223–1232, 1982*)

mass of dust detected in the halo is roughly the same as the mass of the Earth. When this dusty material (composed primarily of silicon and oxygen) is augmented with appropriate quantities of hydrogen and helium gas that is not seen directly but must almost certainly accompany the dust, then the minimum implied total mass of material orbiting around HL Tauri is nearly as great as the total amount of mass in the giant planets, Jupiter and Saturn. However, there is as yet no direct evidence for the presence of large (that is, Moon or planet size) objects orbiting this star or R Mon. Unfortunately, techniques that rely on either scattered or emitted light cannot easily detect such large objects. The reason has to do with surface-to-volume considerations. Large objects such as the Earth contain a lot of volume (and therefore mass) but relatively much less total surface area than a collection of tiny particles that have the same total mass as the Earth. It is surface area that produces scattered light.

Both HL Tauri and R Mon possess collimated outflows. Their existence raises the possibility that the dusty halos are actually disks that produce the collimation. In HL Tauri, the distribution of scattered light suggests that if the halo is a disk its plane lies in a roughly east-west direction. Unfortunately, the jet which may be associated with HL Tauri (Fig. 2) is oriented at a 45° angle to east-west and cannot be explained in any obvious way as being due to collimation by the dusty halo observed in the infrared. Likewise, in R Mon there is optical evidence for two oppositely directed jets oriented approximately in a north-south direction. However, the infrared halo around this star shows no north-south/east-west asymmetry, so once again the optical and infrared morphologies cannot be connected in any obvious way.

These halos were discovered late in 1983. Earlier that year, NASA startled the astronomical community by announcing that its *IRAS* satellite had detected evidence for particulate clouds around two older, more evolved stars, Vega and Fomalhaut. More recently, astronomers working with the *IRAS* data have announced the discovery of many low-luminosity infrared sources in molecular clouds. Some of these objects may be sunlike protostars, but additional infrared and microwave observations will be required to establish this.

For background information *see* INFRARED ASTRONOMY; INTERSTELLAR MATTER; SPECKLE; STELLAR EVOLUTION in the McGraw-Hill Encyclopedia of Science and Technology. [BEN ZUCKERMAN]

Bibliography: C. A. Beichman et al., The formation of solar type stars: IRAS observations of the dark cloud Barnard 5, *Astrophys. J.*, 278:L45–L48, 1984; J. P. Emerson et al., IRAS observations near young objects with bipolar outflows, *Astrophys. J.*, 278:L49–L52, 1984; Infrared evidence for protoplanetary rings around seven stars, *Phys. Today*, 37:17–20, 1984.

Superconductivity

The growing use of new materials in modern electronic technology has been encouraged by the superconductivity exhibited by some metals and metallic compounds when they are cooled below a certain critical temperature. A material in the superconducting phase is able to carry a current without any loss of energy from Joule heating, but it also exhibits many other properties which can be used for the development of future technologies. The expulsion of a magnetic field from the inside of a superconducting material (the Meissner effect) produces magnetic levitation which can be used for high-speed transportation. The characteristic response of two superconductors in close contact on either side of an insulating barrier (the Josephson effect) is used in extremely sensitive detectors of very low voltages, small magnetic fields, and so forth, and in components of high-speed logic circuits. Some of these components are also being evaluated for the construction of superfast computers, but these have not yet reached the development level, primarily because of the very low temperature conditions required to maintain the material in the superconducting state.

In recent years, two new types of superconductors have been discovered. One type is found among organic compounds which have one-dimensional molecular structures. The other type is found among heavy-fermion systems, metals in which the electrons have an unusually large effective mass.

Organic superconductors. In accordance with the BCS theory, proposed by J. Bardeen, L. N. Cooper, and R. Schrieffer in 1957, it has become well established that superconductivity in metals is based on the existence of an attractive interaction between conducting electrons which is mediated by the ionic crystalline lattice and leads to pair formation of electrons with opposite momenta and spins at low temperature. However, the theoretical understanding of superconductivity leaves little hope that the superconducting transition temperature T_c can be raised very much above 23 K ($-418°$F), which is attained in niobium-germanium compounds.

In hopes of achieving a significant enhancement of T_c, theoreticians have tried to image conductors in which the attractive pairing could be stronger than in ordinary metals. In 1964, W. A. Little proposed investigating organic compounds to discover a class of new materials in which superconductivity might be observed at room temperature. This proposal resulted in an intense search for high electronic conductivity in organic matter, which led to the discovery of many interesting properties of low dimensional conductors.

Molecular structure. The first organic superconductors, discovered in 1980, are admittedly far from Little's initial model, but the investigation of their physical properties has aroused great interest because their behavior deviates from that of other superconducting materials. The unique properties of organic superconductors are linked to a special feature: they are built from isolated molecules. These molecules, when brought together in a solid (a crystal), interact relatively weakly, thus allowing intermolecular charge delocalization. The interactions, however, are strongly directional, so that the resulting conductor is strongly anisotropic.

The materials consist of uniform stacks of rather flat molecules, tetramethyl tetraselenafulvalene (TMTSF). A substance with this structure is expected to have strongly interacting molecular orbitals primarily along the stacking direction (Fig. 1). The valence bands formed by the highest occupied molecular orbitals are opened by partial charge transfer from the electron-releasing stack to an electron-attracting inorganic anion, leading to formation of the salt $(TMTSF)_2X$ (where X is a monovalent anion). Single crystals are obtained by electrochemical oxidation of neutral TMTSF in a solvent containing an excess of X^- (Fig. 2). Due to the insolubility of the salt, the crystals separate on the electrode according to reaction (1). Inspection of the crystal

$$2\text{TMTSF} + X^- \xrightarrow{(-e^-)} (\text{TMTSF})_2X \text{ (insoluble)} \quad (1)$$

structure (Fig. 1) indicates that the material will exhibit anisotropic conducting properties.

Superconducting properties. Superconductivity in these materials was first observed in 1980 in the material $(TMTSF)_2PF_6$, which at ambient pressure undergoes a transition from a metallic state above 12 K ($-438°F$) to a magnetic insulating one below 12 K. When subjected to hydrostatic pressure above about 8.5 kilobars (850 megapascals), the material remains metallic down to about 1 K ($-458°F$), below which it becomes superconducting (Fig. 3). The investigation of the $(TMTSF)_2X$ series showed that the proper choice of the anion X, in particular ClO_4 (which is smaller than PF_6, ReO_4, and so forth), makes it possible to stabilize superconductivity below 1.2 K ($-457.5°F$) at ambient pressure. This discovery made the experimental investigation of organic superconductors readily accessible by usual means in physics laboratories. In fact, at 1.2 K $(TMTSF)_2ClO_4$ presents all the characteristics of superconductivity observed in metallic superconductors. The electrical resistance of the sample drops to zero value. A magnetic field is expelled from the inside of the sample when it is cooled below 1.2 K under a moderate magnetic field (Meissner effect). Finally, an anomaly in the heat capacity at T_c is the typical signature of the superconducting transition.

However, a question arises as to whether the quasi-one-dimensional nature of these new materials affects so deeply the phenomenon of superconductivity that the picture of superconductivity in metals cannot be straightforwardly extended to organic conductors. This question has stirred a great deal of interest, but it now seems established that one-dimensional physics must be taken as the starting point for the interpretation of these new conductors.

One-dimensional properties. Of particular importance is the existence of a wide temperature domain above T_c where cooperative phenomena give rise to simultaneous enhancement of magnetic and conducting properties. Both features are observed up to about 40 K ($-388°F$) in nuclear magnetic resonance experiments and also in far-infrared data that show a highly conducting narrow collective mode at zero frequency. The coexistence of magnetic and superconducting fluctuations at temperatures far above the onset of long-range order is the signature of the one-dimensional properties of the electron gas. One-dimensionality is also illustrated by the interplay between various long-range orders (magnetism and superconductivity) that occur at low temperature, depending on the value of external parameters. For instance, the magnetic ground state of $(TMTSF)_2PF_6$ is converted into a superconducting state under pressure (Fig. 3). Moreover, the converse is observed (superconductivity→magnetism) when only a few percent of nonmagnetic impurities is added to $(TMTSF)_2ClO_4$ or when the pure compound is subjected to a large magnetic field.

Higher transition temperature. In the framework of a predominantly one-dimensional interpretation of organic superconductors, the onset of bulk one-dimensional superconductivity is greatly suppressed by the weak interchain coupling. Therefore, the possibility of superconductivity up to 20 or 30 K (-424 or $-406°F$) has been suggested for future organic superconductors with increased interchain coupling. This possibility opens the prospect of superconductivity at temperatures higher than those obtained in metals at present. In this respect, the discovery of another superconducting family based on the molecule bis(ethylenedithiolo)tetrathiafulvalene (BEDT-TTF) is very encouraging. Even more exciting is the recent observation of superconductivity at 2.5 K ($-455.2°F$) and possibly up to 7 K

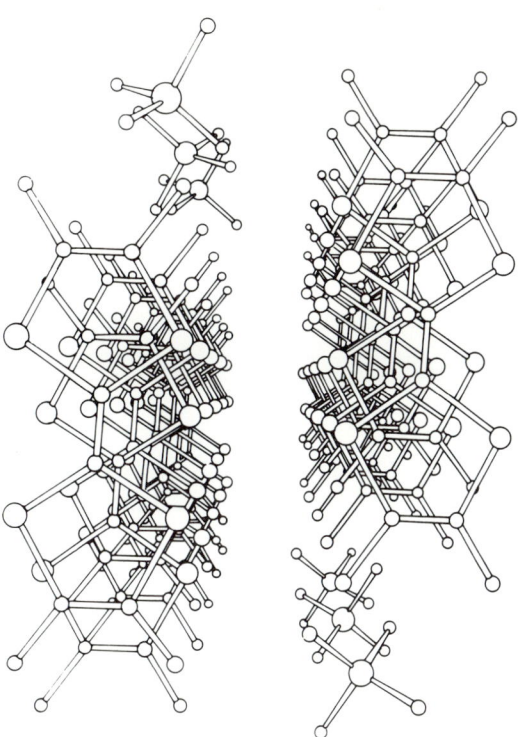

Fig. 1. Structure of $(TMTSF)_2X$ conductors showing the stacking of TMTSF molecules along the direction of high conductivity.

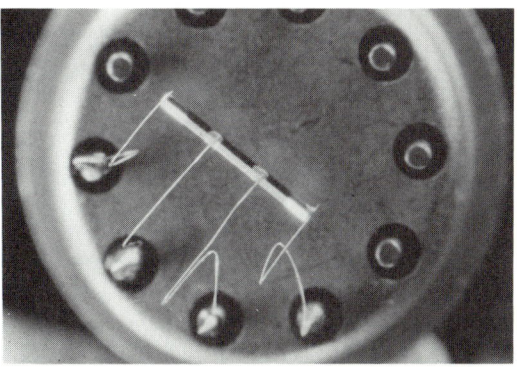

Fig. 2. $(TMTSF)_2ClO_4$ single crystal mounted with four electrodes for the purpose of conductivity measurements. The length of the crystal is about 0.4 in. (10 mm). The shiny parts are gold-evaporated contacts.

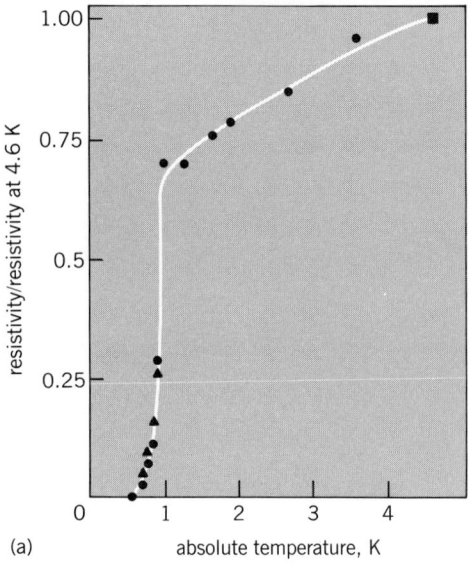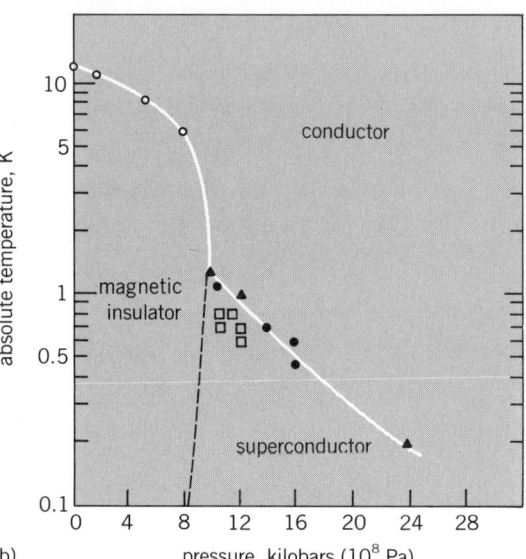

Fig. 3. Low-temperature phases of $(TMTSF)_2PF_6$. (a) The superconducting transition observed by resistivity measurements at 1 K under a pressure of 12 kilobars (1.2 gigapascals). (b) Temperature-pressure phase diagram showing the interplay between the magnetic-insulating state and superconductivity.

($-447°$F) at ambient pressure by a Soviet team in the BEDT-TTF-iodine system.

Applications. Several properties (electrochemistry, sensitivity to irradiation or magnetic field, and so forth) are very specific to organic superconductivity. Because of these properties, organic superconductors are of special interest for a number of potential applications in which they would be of significantly greater value than inorganic conductors.

In the realm of basic research, these materials have provided a broad range of new phenomena for study—for example, the quantization of the Hall effect and the sequence of magnetic phase transitions which are observed for $(TMTSF)_2X$ under high magnetic field at low temperatures. Furthermore, the flexibility of organic chemistry makes feasible a link between various domains: organic conductors, doped polymers, and thin-film preparation. The development of this field promises to be fruitful for potential applications in molecular electronics. *See* HALL EFFECT.

The primary unanswered question is whether the electron pairing in organic superconductivity is of the BCS type or whether it is mediated by a purely electronic process resembling the mechanism proposed by Little, in spite of the modest values of critical temperatures obtained thus far in organic matter.

[D. JEROME]

Heavy-fermion superconductors. The discovery in 1979 of superconductivity in the compound $CeCu_2Si_2$, where the conduction electron wave functions are derived from atomic f orbitals, resulted in a rapid expansion of the search for and study of other so-called heavy-fermion systems. The term heavy fermion refers to the fact that these systems are characterized by an unusually large effective electronic mass m^*, more than 100 times the mass of the free electron. At present, eight heavy-fermion systems are known of which three, UBe_{13}, UPt_3, and $CeCu_2Si_2$, are superconducting.

Free-electron model. Electrons in a metal constitute a many-body interacting system. In the face of strong Coulomb interactions between the electrons as well as interactions between the electrons and lattice ions, it seems a hopelessly difficult task to understand the system. However, most solids have a periodic lattice which extends over considerable distances. As researchers showed in the early days of quantum mechanics, conduction electrons in a metal can be modeled as a gas of noninteracting particles obeying Fermi statistics.

In this simple picture of the free-electron Fermi gas, the specific heat associated with the electrons is proportional to the temperature. The proportionality constant γ, determined from low-temperature heat-capacity experiments, is usually not too far from the value of γ calculated for the free-electron gas. Since the free-electron mass m is proportional to γ, the effective mass m^* is defined by Eq. (2),

$$m^* = m \times [\gamma(\text{experiment})/\gamma(\text{free electron})] \quad (2)$$

where m is the free-electron mass. The given metal can be viewed as a gas of free electrons of mass m^*.

The departure from the free-electron mass is due to the interactions that exist in a real solid. In the simple free-electron theory, an interacting system is approximated by transforming it into an equivalent system of noninteracting particles. This is done by renormalizing (that is, changing) the electron mass, and the new particles are called quasiparticles. Fermi-liquid theory is a semiphenomenological theory which retains the quasiparticle concept to describe an interacting system in its low-energy excited states. An example of how interactions between the electrons can renormalize the mass is provided by an electron moving through a gas of electrons. As it moves, it will drag along with it a region of void since the similarly charged electrons repel each other. This picture can be modeled by an electron accompanied by a compensating cloud of oppositely charged particles which will provide additional inertia to the original electron.

In most metals and alloys, including superconducting ones, m^* is of the order of the bare electron mass m. Hence, heavy-fermion systems with m^* hundreds of times greater are very unusual Fermi liquids. The discovery of superconductivity in them adds more complexity and interest to an already interesting theoretical problem.

Nature of superconductivity. In an ordinary superconductor, the ground-state wave function is a coherent superposition of bound electron pairs (called Cooper pairs); that is, the ground state which consists of many particles can be correctly described by a single wave function. The Cooper pairs in BCS theory are bound in states of zero total spin and angular momentum, and are formed through a phonon-mediated electron-electron attraction. Even though electrons normally repel each other, the paired state becomes possible in the lattice because the electrons traverse one lattice spacing in much less time than the vibration period of a lattice ion. A positively charged ion (whose energy comes in quantized packets called phonons) can transmit the vestige of its attractive interaction with the first electron of a pair to the second electron that comes along before the ion relaxes back into its original position. The normal repulsion of electrons does not matter because the first electron is already far away. This retarded nature of the Coulomb interaction is very important in the formation of Cooper pairs.

Since heavy-fermion systems are so different from previously known superconductors, the nature of superconductivity in heavy fermions is of primary concern. More specifically, these superconductors are being examined for the possibility of being paired in states which have orbital momenta with unit angular momentum, generally referred to as p-state superconductivity. Such states might be favored when the electrons are not much faster than the phonons so that electron-electron repulsion is less retarded. Electron-electron repulsion would then be comparable to or larger than the phonon-induced attraction, making the conventional BCS state which has zero angular momentum (that is, s-state superconductivity) no longer energetically favorable. On the other hand, p-state superconductivity has pair wave functions with p parity or, more generally, odd parity. In odd-parity states, electron density goes to zero at the origin of the relative coordinates, thus guaranteeing less Coulomb repulsion than in s states.

There are several experimental ways to distinguish between s-state and p-state superconductors. One is to observe how sound decays as a function of temperature below the superconducting transition temperature T_c. Another involves measuring inelastic neutron scattering. Also, the sensitivity of T_c to nonmagnetic and magnetic impurities is expected to be quite different in the two types of superconductors. Experiments with tunnel junctions that look for electron-pair tunneling through a barrier between an electrode made with a conventional superconductor and one made with a heavy-fermion superconductor can provide conclusive results under some circumstances. If Josephson tunneling is observed, the heavy-fermion superconductor is not likely to be p-state since the change in symmetry prevents Josephson tunneling between the two types of superconductors. On the other hand, if no Josephson tunneling is observed, the results will not guarantee that the heavy-fermion system under investigation is a p-state superconductor since nonoptimal sample preparation also suppresses Josephson tunneling. Many of the above experiments are being undertaken, but conclusive results are difficult to establish and it may be some time before heavy-fermion superconductors are well understood.

Properties. All three known heavy-fermion superconductors contain electrons in unfilled atomic f orbitals and have temperature-dependent magnetic susceptibilities (of the order of 10^{-2} emu/mol) that are much larger than that of nonmagnetic metals (of the order of 10^{-4} emu/mol) [emu = electromagnetic units). Their susceptibilities obey Curie-Weiss laws from 100 to 300 K (-280 to $+80°$F) with large effective magnetic moments. Magnetism and s-state superconductivity have been thought to be mutually incompatible because magnetism prefers electrons to have their spins line up in parallel while conventional BCS-state superconductivity prefers them antiparallel. Since the p state is formed by two parallel electrons, the strong spin correlations responsible for magnetism might favor p-state superconductivity.

In $CeCu_2Si_2$, a transition into a superconducting state is observed at $T_c = 0.5$ K ($-458.8°$F) both in resistance and specific-heat measurements. The jump in heat capacity at the superconducting transition temperature is very close to the value calculated in BCS theory for the measured γ, implying that the Cooper pairs in $CeCu_2Si_2$ are formed by all the heavy-fermion quasiparticles which cause the large γ. Properties, including γ and T_c, of $CeCu_2Si_2$ are very sample-dependent. They are sensitive to small changes in the concentration of copper, for example.

Therefore, the discovery in 1983 of superconductivity in UBe_{13} was of great value, because it demonstrated that heavy-fermion superconductivity was not just accidental. Since UBe_{13} does not exhibit the sample variation found in $CeCu_2Si_2$, more reproducible samples upon which to experiment are provided. Its transition temperature is found to be approximately 0.9 K ($-458.0°$F). $CeCu_2Si_2$ and UBe_{13} have comparable electronic specific heat; the γ values are identical (approximately 1.1 Jmol^{-1} K^{-2}). Recently, it has been pointed out that the similarity of the temperature dependence of the specific heat for UBe_{13} between 1 and 8 K (-458 and $-445°F$) with that of the specific heat of liquid helium-3 between 0.1 and 0.6 K (-459.5 and $-458.6°F$) suggests the existence of p-state superconductivity in UBe_{13}.

Liquid helium-3 is a fermion system because the nuclei have spin $\frac{1}{2}$. It condenses into a p-state paired superfluid with anisotropic magnetic properties at a very low temperature (approximately 2 millikelvins). Superfluidity is the analog of superconductivity for neutral systems. In the dependence of its low-temperature specific heat upon absolute temperature T, there is a $T^3 \ln T$ term, which is indicative of the tendency toward ferromagnetism. This tendency is caused by long-range spin fluctuations that are sometimes called paramagons. See LIQUID HELIUM.

UPt_3 is the most recently found heavy-fermion superconductor. Its superconducting behavior and its other properties are different from the other two heavy-fermion superconductors, suggesting that it might be a different kind of p-state superconductor. The value of γ (0.45 J mol^{-1} K^{-2}) is somewhat lower than that of the other two systems. In its specific heat, a $T^3 \ln T$ dependence is observed which strongly indicates the presence of spin fluctuations. In analogy with helium-3, superconductivity in UPt_3 might be due to coupling of the electrons through spin fluctuations.

In applied magnetic fields, UPt_3 shows anisotropic properties which can be modeled as one of the possible p-type ground states. However, more experiments and theoretical work need to be done before any one of the three heavy-fermion superconductors is understood well enough so that it can be decided whether a new type of superconducting paired state has been observed and whether it is the result of a new pairing mechanism. Either a new pairing mechanism or a new type of pairing would present a new frontier for superconductivity.

For background information see FREE-ELECTRON THEORY OF METALS; LIQUID HELIUM; SUPERCONDUCTIVITY in the McGraw-Hill Encyclopedia of Science and Technology.

[THEODORE H. GEBALLE; SUNG IL PARK]

Bibliography: D. Jerome and H. J. Schulz, Organic superconductors, *Adv. Phys.*, 31:299, 1982; H. R. Ott et al., UBe_{13}: An unconventional actinide superconductor, *Phys. Rev. Lett.*, 50:1595–1598, 1983; Proceedings of the Colloque International du CNRS sur la Physique et la Chimie des Métaux Synthétiques et Organiques, les Arcs, December 1982, *J. Phys. (Paris)*, vol. 44, Colloque C3, 1983; Proceedings of the International Conference on Synthetic Metals, Abano-Terme, June 1984, *Mol. Crys. Liq. Crys.*, 1985; F. Steiglich et al., Superconductivity in the presence of strong Pauli paramagnetism: $CeCu_2Si_2$, *Phys. Rev. Lett.*, 43:1892–1896, 1979; G. R. Stewart, Heavy fermion systems in perspective, *Rev. Mod. Phys.*, 56:755–787, 1984; G. R. Stewart et al., Possibility of coexistence of bulk superconductivity and spin fluctuations in UPt_3, *Phys. Rev. Lett.*, 52:679–682, 1984.

Surgery

One of the greatest difficulties in reconstructive surgery of the upper extremities is the return of good function to injured nerves. When a nerve is cut, the portion of the nerve cell that does not contain the cell nucleus dies. The peripheral nervous system can regrow these severed parts, called axons, to a remarkable degree, but the severed axons need to be reconnected properly for nerve function to be restored. This is no easy task, and over the years many methods have been tried, with mixed results. A new method of repairing severed nerves involves a combination of microsurgery and argon laser technology. This method has been used, with excellent results, to repair sciatic nerves in rats and median nerves in primates.

Nerves consist of a multitude of single cells, each stretching from the spinal column to the end organs in the skin or muscle. The axons serve as electrical conduits, and there are tens of thousands of them in each major nerve. It is technically impossible to realign each axon when a nerve has been cut; even with the most sophisticated repair the cells will always be somewhat disorganized.

In recent years, microsurgical techniques employing tiny sutures 25 micrometers in diameter have been used in attempts to obtain better nerve repair. A variety of cuffs or "sleeves" have been used in an attempt to direct the regenerating axons toward their distal counterparts, prevent escape of the sprouting axons from the juncture site, and prevent ingrowing of scar formation between the severed ends of the nerve. These have not proved practical in the clinical setting.

Argon laser. The advent of the argon laser has made it possible to form autogenous blood, that is, the subject's own blood, into an encompassing, nonreactive, adherent tubule around the severed nerve, resulting in a nerve repair that would seem to approximate a technical ideal.

The argon laser produces a blue-green light of 488- to 521-millimurons wavelengths. These wavelengths are absorbed almost completely by red-colored substances, such as blood, and are reflected almost completely by white surfaces, such as nerves. Thus, when an argon laser beam is applied to red blood cells on a white nerve, the blood cells absorb the light and become an adherent coagulant, while the white nerve repels the light and is not affected by the laser beam. This process creates an adherent minitubule around the two ends of severed nerve and binds them together, theoretically without causing further damage to the nerve.

Laser repair of injured nerves was conceived in 1976. In its earliest form, the system consisted of an argon laser attached to a microscope and focused through prisms and mirrors. It was possible to repair major nerves with this system, but the process was too awkward to be practical. This problem was overcome by development of a fiber-optic system capable of the finest "pinpoint" accuracy and tissue specificity.

The system developed by E. E. Almquist and co-workers consists of an argon ion laser fitted with a mirror shutter, a timer, and a foot switch. The shutter opens when the foot switch is depressed, and remains open for the period preset on the timer. The laser energy is focused through uncoated 10 × 0.25 microscopic objectives to the end point of the fiber. The optical fiber, consisting of fired quartz, is 4.5 m (14.85 ft) long; its core diameter is 400 μm and the outside diameter is 850 μm.

The method for using this technology in both rats and primates is to dissect the severed nerve into its fascicle components, visually matching corresponding axons. The nerve ends are held together with forceps, and a drop of the animal's blood is placed around the nerve ends. The blood is coagulated by several brief applications of the argon laser and thus becomes a cohesive mass around the adjoining fascicles, which then cannot be teased apart easily (see illustration). The wound is closed conventionally, and the limb or digit is immobilized to prevent damage to the repair site.

Evaluation of nerve repair. These nerve repairs have been evaluated extensively by scanning electron microscopy, transmission electron microscopy, and traditional light microscopy. The results have shown what appears to be very satisfactory nerve regeneration without any further damage to the perineurium due to laser application. The small cuff or tubule of blood seems to shunt the growing axons toward their appropriate destinations and to prevent the ingrowth of scar tissue. Use of the subject's own blood eliminates the foreign-body reaction. The blood eventually dissolves and leaves what appears to be a very normal sheath around the nerve repair.

Scanning electron microscopy reveals that when the nerve repair is a few days old, the coagulum appears as a globular, homogeneous structure at the surface. The globular mass is soon intermeshed with collagen fibrils, and the structure matures rapidly into a fibrous sheath that has many characteristics of an epineurial structure, the normal sheath that covers a nerve. The laser method thus appears to be safe and particularly useful for grafting or splicing nerves.

The most important issue for nerve repair by any method is the return of function. This outcome is difficult to assess in experimental animals because the results of electromyography, nerve conduction velocity studies, and nerve axonal counts do not always correlate with the return of function. Therefore, laser nerve repairs have been evaluated only histologically to date. Scanning electron microscopy shows that the typical laser repair allows early effusive axonal budding proceeding distally from the repair site. Transmission electron microscopy also shows relatively healthy axons of the appropriate density. Thus, the argon laser yields a technically satisfactory repair compared with those obtained by traditional methods. Laser-repaired nerves seem to show less escapement of axons and smaller neuroma than surgically repaired nerves, although this observation has not been quantitated. Rats with laser-re-

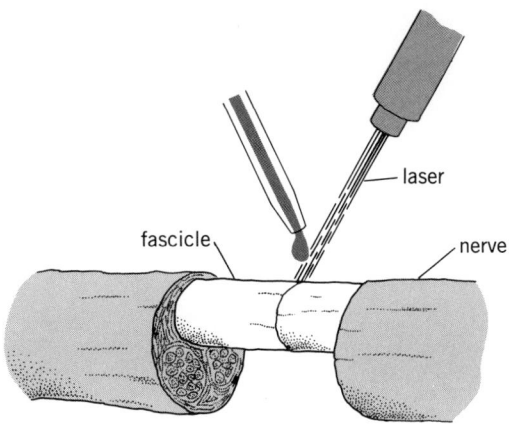
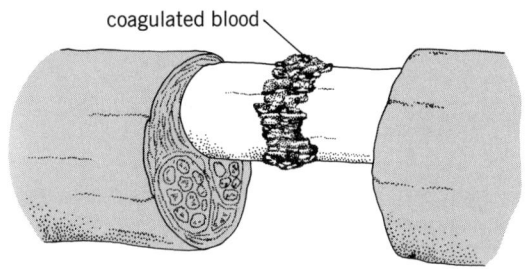

Laser energy is applied to a nerve fascicle with a cuff of blood. The laser-coagulated blood forms an adherent minitubule cuffing the fascicle.

paired nerves show a return of muscle function at three months; this observation must be interpreted cautiously, however, as studies have shown that almost any type of nerve repair in rats results in return of function. Whether the argon laser technique can actually produce an improved return of nerve function awaits long-term evaluation following application to human nerve repairs.

For background information see LASER; MICROSCOPE; NERVE; SURGERY in the McGraw-Hill Encyclopedia of Science and Technology.

[EDWARD E. ALMQUIST]

Bibliography: E. E. Almquist et al., Argon laser repair of rat and primate nerves, *J. Hand Surg.*, 1984; E. E. Almquist et al., Evaluation of the use of the argon laser in repairing rat and primate nerves, *J. Hand Surg.*, 1984; E. E. Almquist et al., Nerve conduction velocity, microscopic, and electron microscopy studies comparing repaired adult and baby monkey median nerves, *J. Hand Surg.*, 8:406–410, 1983; L. Goldman, *Biomedical Aspects of the Laser: The Introduction of Laser Applications into Biology and Medicine*, 1967.

Synchrotron radiation

Synchrotron radiation is light emitted by charged particles (usually electrons) as they circulate in high-energy accelerators or storage rings. The light is very bright and covers the entire electromagnetic spectrum from the x-ray region up to energies of several tens of kilovolts, through the soft x-ray, ultraviolet, visible, and infrared region.

Fig. 1. National Synchrotron Light Source electron storage ring, 164 ft (50 m) in circumference, at Brookhaven National Laboratory, New York, before installation of shielding and beam lines. One of the 16 beam ports on the ring is indicated.

Sources. When charged particles are accelerated, they emit electromagnetic radiation. If the charged particles are of low mass, such as electrons, and if they are traveling relativistically, the emitted radiation is very intense and highly collimated, with opening angles of less than 1 milliradian. Circular accelerators used in high-energy physics such as synchrotrons or storage rings, in which bunches of up to 10^{12} electrons continuously circulate in vacuum, guided by magnetic fields, are sources of synchrotron radiation. Recently, dedicated light sources have been constructed in the United States, United Kingdom, West Germany, and Japan (Figs. 1 and 2). These dedicated sources run at higher circulating beam currents, but their most important difference from other accelerators is that they have a magnetic focusing system which concentrates the electrons into bunches of very small cross section. Thus, when the electrons radiate as they are accelerated in their circular paths, the radiation originates from a very small source. The combination of the high intensity with small opening angle and small source dimensions (0.008 × 0.024 in. or 0.2 × 0.6 mm) results in synchrotron radiation sources being several orders of magnitude brighter than any other source (Fig. 3), even than the most powerful x-ray tubes with water-cooled rotating anodes. Even in the infrared region where the output is falling off with wavelength, synchrotron radiation is two to three orders of magnitude brighter than that from a 2000 K (3100°F) blackbody.

Most of the work done so far with synchrotron radiation has used that radiation which is emitted by the electrons during their passage through the dipole bending magnets. A typical modern synchrotron storage ring will have several dipoles to bend the paths of the electrons and generate a closed orbit. Between the dipoles are straight sections where focusing magnets are placed. In the future, however, these straight sections will increasingly be used in addition for devices which modulate the electron beam with rapidly alternating magnetic dipole fields, causing continuous radiation emission over a longer path length to increase intensity. The devices can either have higher fields to increase the hard x-ray flux for a given electron beam energy, or have many periods whose wavelength is chosen to given coherent emission. These insertion devices are called wigglers and undulators respectively (Fig. 3). The first wiggler beam line was commissioned at Stanford University, and the first undulator beam line is being commissioned at Brookhaven National Laboratory. In undulators with 30 periods, gains in brightness of three orders of magnitude are theoretically possible, on account of the coherence of these sources. The spectrum of undulators is not smooth but peaks in harmonics. Some tunability may be achieved by varying the gap and hence the field in the device.

Applications. Because of its strong interaction with matter, synchrotron light has tremendous potential for studies of physical structures using x-ray scattering techniques, or electronic structures using absorption techniques. These general techniques are applicable to studies from a broad range of disciplines, including physics, chemistry, biology, metallurgy, and materials science.

Since the emitted radiation covers a wide spectral range, it is usually necessary to be able to select a wavelength of interest and to focus it onto a sample. Beam lines which do this involve new frontiers in precision ultrahigh-vacuum engineering, in optical

Synchrotron radiation 447

Fig. 2. National Synchrotron Light Source after installation of shielding and some of the beam lines. The latter fan out radially from the ports and are used for a variety of physics, chemistry, biology, and materials science experiments.

design and fabrication, and in coping with thermal loads on mirrors as high as several hundred watts per square centimeter. Wavelengths are selected by using crystal diffraction below 0.5 nanometer and gratings above 2 nanometers. A large research effort involving new materials for crystals and new grating spectrometer designs is being implemented to obtain wavelength selection for the difficult region between 0.5 and 2 nm.

Apart from being very bright, synchrotron radiation is pulsed with on-times of 0.2 to 1 nanosecond and off-times of 10 to 1000 nanoseconds, depending on the ring circumference. It is also polarized parallel to the orbit plane. Some applications, though relatively few to date, take advantage of these unique characteristics.

Scattering experiments. Synchrotron radiation in the 0.15–0.20-nm wavelength range, which is preferred for scattering, is five orders of magnitude brighter than previously existing x-ray sources. Experimentally, this has been translated into several opportunities for new studies. For example, the beam can be focused down so that very small crystals such as those of proteins can be studied for the first time. The collimation of the beam allows high-resolution scattering to be performed so that samples with structure on the scale of 1 micrometer can be accurately studied.

With conventional samples, the brightness of the synchrotron radiation can be used to make measurements as a function of time. One demonstration of time-resolved scattering involved the study of the laser annealing of a silicon crystal. From the x-ray scattering data taken during the 50–100-ns annealing process, the temperature profile of the surface as a function of time was derived. This helped to solve the controversy about whether classical melting really occurred or not. It was found that melting-point temperatures were achieved during laser annealing.

Synchrotron radiation has also been successfully used to study dilute samples or their equivalent such as surfaces, where the number of scatterers can be a few orders of magnitude lower than in bulk samples. Studies of this type has been made on germanium reconstructed surfaces and on liquid-crystal thin films.

An important technique involving the preparation of two-dimensional protein single crystals by adsorption to a lipid monolayer is opening up a new field of structure determination for biological molecules. Preliminary work has shown that it is possible to obtain well-defined diffraction peaks from adsorbed lipid monolayers on a cystalline substrate. The extension to crystals of macromolecules, such as proteins, has not yet begun. In the near future, it should be possible to repeatedly prepare 10×10 μm monolayers of immunoglobin crystals. The use of x-ray diffraction from such samples could allow the determination of structure at high resolution for molecules that will not crystallize as three-dimensional structures.

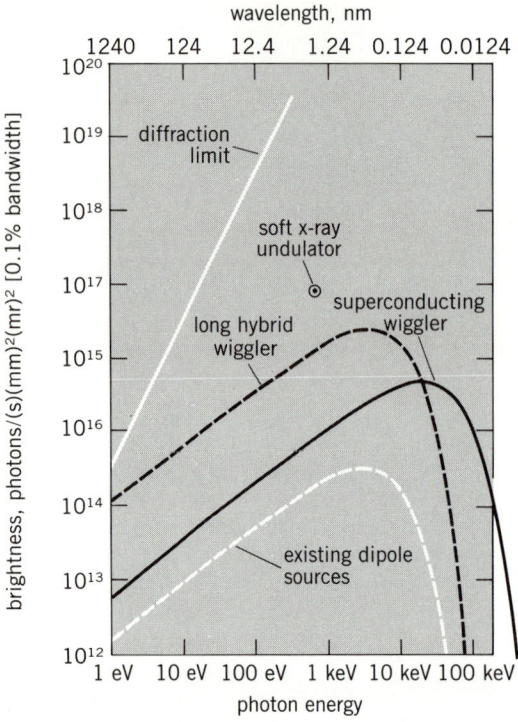

Fig. 3. Average brightness of typical synchrotron radiation sources. The spectrum of the soft x-ray undulator is not smooth but peaks in harmonics, the brightest of which is indicated.

Absorption experiments and surface studies. So far the majority of these experiments have used photoelectric emission to study electronic structure. The incident photons are typically in the wavelength range 2–200 nm, and the emitted photoelectrons are energy analyzed with a resolving power of 200 meV. There have been several recent developments. New toroidal-grating, grazing-incidence monochromators with a resolving power of 1000 or more for photons around the carbon and oxygen K edges have made it possible for the first time to measure the vibrational energies of the carbon-oxygen stretch in carbon monoxide. A very high-resolution electron spectrometer has been developed which has also been coupled to a high-resolution, toroidal-grating monochromator, resulting in observations of new electronic bulk and surface states in metals such as copper and semiconductors such as silicon.

A series of carbon and oxygen K-edge studies have been carried out on hydrocarbons, including benzene, methane, and carbon monoxide, on surfaces of the metals nickel and platinum which are important catalytically. From the peaks in the photoemission spectra which result from unoccupied orbitals being filled in the photoabsorption process, and in particular from their polarization and energy dependence, molecular orientations and distances have been deduced. *See the feature article* SURFACE SCIENCE.

Photon-stimulated desorption has been used to study surfaces in a number of research endeavors. The technique really requires the intensity provided by undulators due to the low yields, but it affords unique opportunities for studying surface species, including hydrogen, and surface chemistry, including bonding.

X-ray microscopy and holography. Recently several in-line holograms have been made by using light of 3.2-nm wavelength. They were reconstructed by using visible light. The new high-brightness synchrotron radiation sources are sufficiently coherent to permit these early experiments, but in the long term undulators will have a significant advantage which is necessary if this field is to progress.

X-ray microscopy has made significant strides with the availability of the new sources, and the field is under rapid development. In principle it is possible to study living cells by using elemental contrast techniques and at submicrometer spatial resolution. In scanning x-ray microscopy, the beam has to be monochromatized, focused to a submicrometer spot, and brought through a window in front of which the sample is scanned. A detector behind the sample allows a picture to be created similar to that in a television tube. Elemental contrast is achieved by recording a picture just above and then just below a principal absorption edge and then taking the difference electronically. The available edges of interest biologically include carbon, oxygen, and calcium. Cardiac angiograms offer a very important frontier in visualizing the cardiac circulation without the risk of invasive catheterizations.

EXAFS studies. The extended x-ray absorption fine-structure (EXAFS) technique makes it possible to measure interatomic distances with an accuracy of 1 picometer. It is based on the fact that the absorption cross section above a core electron absorption edge exhibits oscillations in strength because the density of available electron final states is determined by scattering from neighboring atoms. It is possible to measure these oscillations and deduce distances to the neighboring atoms. This is a general technique which has been applied to a number of systems. The technique is element-selective and therefore affords the opportunity to study dilute impurities such as iron in hemoglobin.

For background information *see* SYNCHROTRON RADIATION in the McGraw-Hill Encyclopedia of Science and Technology.

[GWYN P. WILLIAMS]

Bibliography: E. E. Koch (ed.), *Handbook on Synchrotron Radiation*, 1983; H. Winick and S. Doniach (eds.), *Synchrotron Radiation Research*, 1980.

Tectonic geomorphology

Tectonic geomorphology involves studying the shapes of hills and streams in order to learn more about the origins of mountains. Early researchers identified tectonic landforms such as piedmont fault scarps and triangular facets that are indicative of rising mountain fronts. Recent studies focused on

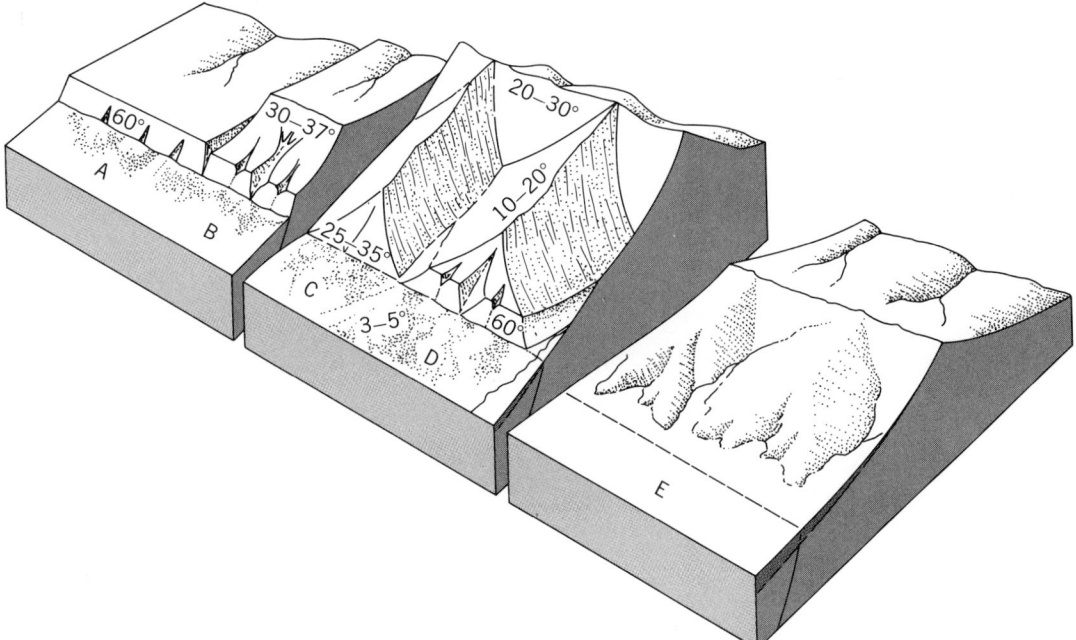

Fig. 1. Block diagrams depicting the sequence of mountain landscapes resulting from uplift along a single range-bounding fault; initial uplift stages (A, B), maximum relief, and development of drainage basins (C, D), and dominance of erosional processes during subsequent stage of tectonic quiescence (E). (*After R. E. Wallace, Geometry and rates of change of fault-generated range fronts, north-central Nevada, USGS J. Res., 6:637–650, 1978*)

rates of uplift and horizontal displacement that are associated with different landscape assemblages. By emphasizing the time factor, tectonic geomorphology provides a new tool for evaluation of seismic risk in areas of faulted landforms and helps determine the times of last surface rupture by specific faults.

Fault-block mountain ranges. The landscapes of fault-bounded mountains are controlled mainly by the type and rate of earth deformation, climate-controlled weathering and erosion, and resistance of different rock types to erosion. Hundreds of mountain ranges within the Basic and Range Province of the western United States, and elsewhere, have been raised relative to their adjacent valleys by bounding faults whose component of movement has been dominantly vertical. The mountains are uplifted gradually during the course of geologic time by 3–10-ft (1–3-m) seismic events that are repeated roughly every 10,000 years. In the structures shown in the block diagram in Fig. 1, the initial faulting (stage A) creates a linear uplifted scarp crest, but erosion causes the crest to migrate from the fault zone, and eventually it becomes the main watershed divide separating the two sides of the mountains. Erosional power is concentrated along stream courses, so valleys that are notched into the rising fault block (stage B) are the sites of most rapid downcutting. Areas of less rapid downcutting become gently sloping spur ridges that separate the valleys (stages C and D). Triangular facets at the ends of the spur ridges reflect the various episodes of mountain-range uplift. The junction between the mountains and the adjacent valley tends to be straight because most faults are straight. Thus, tectonically active mountain ranges typically show straight mountain-piedmont junctions and narrow floored V-shaped canyons. Upon decline of tectonic activity, erosion rates will exceed uplift rates. The relief of the mountains will become less (stage E), the mountain-piedmont junction more sinuous. and the cross-section topography of the valley floors more U-shaped as erosion lowers spur-ridge crests adjacent to valley floors that are progressively widening. Flowing water is responsible for shaping most of the surface of the earth. Uplift profoundly affects the slope and thereby the energy of streams, so fault-bounded mountain ranges are an excellent example of the effects of uplift on fluvial landscapes.

Landforms of strike-slip faults. Horizontal displacements are typical for tectonic movement along strike-slip faults. Whereas vertical fault displacements tend to accentuate fluvial downcutting, horizontal displacements tend to disarrange fluvial landscapes. Linear fluvial landforms such as stream channels and spur ridges that cross active strike-slip faults are offset horizontally. The offset is termed right-lateral if streams and ridge crests on the opposite side of the fault appear to have been displaced to the right, and left-lateral if displaced to the left. The diagnostic landforms that result from such horizontal movements are shown in Fig. 2a, and include offset ridges (which are termed shutter ridges if they have been moved in front of a valley so as to deflect streamflow), offset stream channels, and offset piedmont landforms such as alluvial fans.

Ruptures along strike-slip fault zones commonly

occur along interrelated, nearly parallel faults. A small vertical component of displacement due to local compression or tension is common even where the fault zone is vertical. The result of dominant lateral movement with minor vertical movement is the formation of landforms such as parallel ridges and elongate depressions (some of which may be undrained to form sag ponds). The ridges have undergone slight uplift relative to the adjacent depressions. Other features such as scarps and faceted ends of spur ridges result more from disarrangement of the fluvial landscapes than by vertical uplift; the vertical sense of movement for such landforms is only apparent. Features such as notches and linear valleys reflect the presence of weaker and more easily eroded rocks that have been crushed by move-

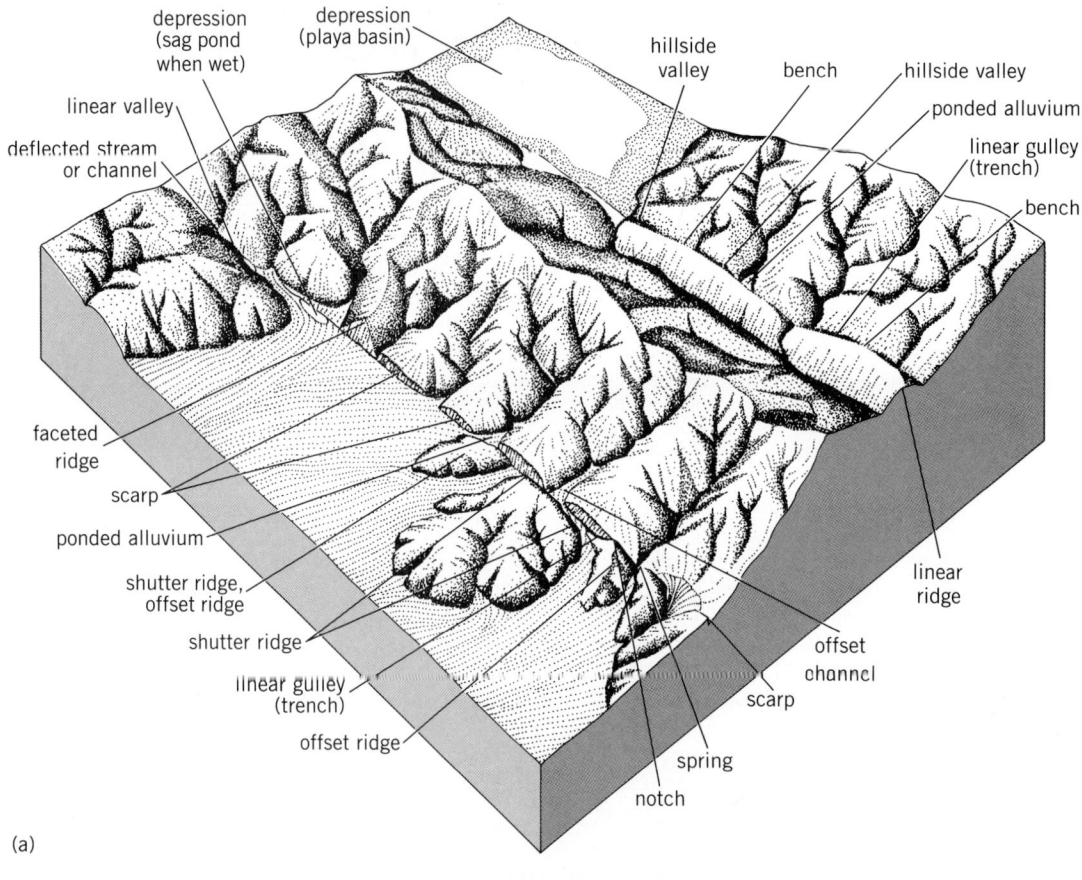

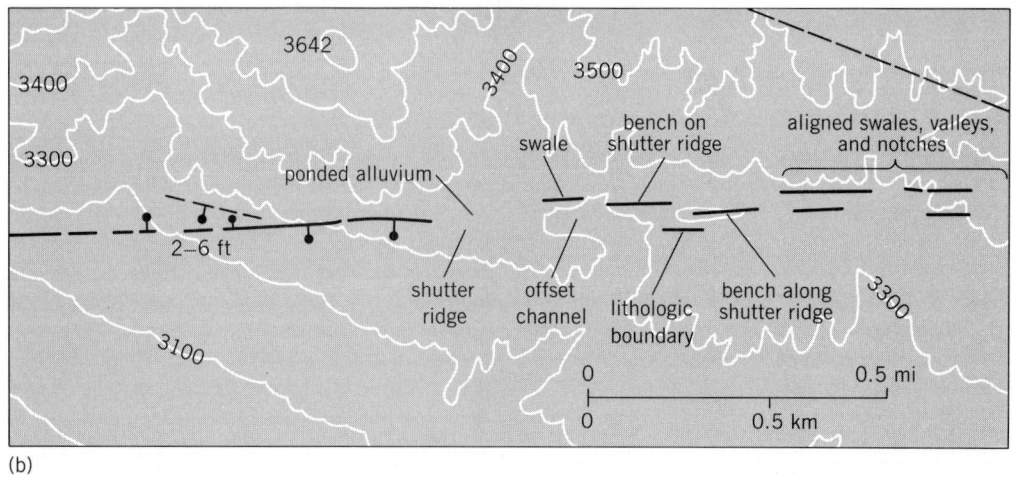

Fig. 2. Landforms associated with horizontal fault movements. (a) Block diagram showing the variety of diagnostic landforms produced by fluvial processes along a hypothetical zone of active strike-slip faulting. (b) Topographic map with notes showing the locations of landforms indicative of active left-lateral strike-slip displacements along a section of the Garlock Fault, California. The black dots indicate the relatively downthrown side of scarps along segments of faults where movement has been predominantly horizontal. (After M. M. Clark, Map showing recently active breaks along the Garlock and associated faults, California, USGS Miscellaneous Geol. Investigat. Map I-741, 1973)

ments along the fault zone.

The recent surface structures along the Garlock Fault of southeastern California have been mapped (Fig. 2b). Several active faults are present within the fault zone, as is indicated by parallel sets of notches, valleys, and benches. The stream channels in the center and right side of Fig. 2b have been offset in a left-lateral sense, and are preserved because of permanent entrenchment of small valleys both upstream and downstream from the fault zone. In contrast, the stream channel at the left side is not offset; this stream is not entrenched and can easily return to its former course when offset by the fault. The ridge in the center has been offset to form a shutter that partially blocks streamflow, which results in accumulation of ponded alluvium.

Landforms as time lines. Although hills and mountains are formed during time spans of millions of years, certain landforms show abrupt terminations of formative processes and therefore may be regarded as time lines within the overall landscape. The retreat of a glacier leaves glacial moraines. The renewed incision of a valley that has been partially backfilled with stream gravel results in the formation of terrace risers and treads. Where aggradation and degradation alternate, successive risers and treads may ascend from the active stream channel like a flight of stairs. Each terrace within the flight records changes in fluvial processes. Shore platforms are coastal landforms that are formed typically during high stands of sea level when wave erosion and other processes produce gently sloping bedrock platforms that may be more than 1600 ft (500 m) wide. Subsequent sea-level decline terminates the constructional phase of the shore-platform time line, and incorporates these coastal landforms into preexisting fluvial watersheds. Each type of time line may be faulted or folded subsequent to its formation. Vertical deformation results in fault scarps, and where episodic uplift has affected a flight of terraces, the older fluvial terraces will have higher fault scarps. The magnitudes and locations of both faulting and folding may be determined by tracing a shore platform for a given age of marine terrace along a coastline.

Careful dating of landforms that serve as time lines is essential in order to ascertain rates of horizontal and vertical tectonic movements and to estimate the time of last movement. Most researchers seek volcanic and organic materials which can be dated by using potassium-argon and radiocarbon isotopes. Another approach is to first make detailed studies of the characteristics of soil profiles and weathering rinds on cobbles, which tend to be ubiquitous features that are affected by the passage of time. The next step is to use radiogenic dating at selected localities to date such pedogenic features, and thereby establish a relative-absolute dating relation for flights of stream terraces that have different time-dependent cobble weathering rinds and soil-profile characteristics such as the amounts and distribution of calcium carbonate, iron oxyhydroxides, and phosphate. Shore platforms associated

Fig. 3. The Hope Fault separates the alluvial-fan piedmont from the rugged Seaward Kaikoura Range, New Zealand. A medium and a large triangular facet is present in the center of the view, and the flat skylined ridge at an altitude of 5100 ft (1600 m) has widely scattered beach pebbles on it.

with marine terraces may be dated by radiogenic dating of coral, wood, volcanic ash, and other materials that occur on the platform or in the capping deposits. They may also be dated indirectly by comparing the spacing of flights of undated terraces with flights of dated late Pleistocene global marine terraces at New Guinea and elsewhere. The high stands of sea level associated with each marine terrace vary considerably, as does the time interval between episodes of shore-platform formation. Such diversity results in unique altitudinal spacings of terraces for each uniform uplift rate—spacings that provide a basis for terrace correlations. Local identification of dated global terraces allows estimates to be made of inferred uplift rate.

Many aspects of tectonic geomorphology are summarized in the structures shown in Fig. 3. The mountains rise abruptly from the piedmont because of the location of the strongly active Hope Fault at the mountain-piedmont junction. Alluvial fans and stream terraces have been displaced both vertically and laterally, and the soil profiles on these surfaces indicate that several fault movements have occurred during the last 12,000 years. The sag ponds and shutter ridges that occur elsewhere along this mountain front indicate substantial horizontal movements during the late Pleistocene, and the triangular facets and V-shaped canyons are indicative of substantial rapid uplift. Notches in the spur ridges, and the flat summit-ridge crest, are remnants of former shore platforms that still possess widely scattered beach pebbles on them. The spacing of these marine terrace remnants infers that the rate of uplift of this structural block during the last 300,000 years has been uniform and is about 14 ft (4.5 m) per 1000 years. The rate of horizontal movement probably has been at least twice the vertical component of tectonic deformation.

For background information *see* FAULT AND FAULT STRUCTURES; FLUVIAL EROSION LANDFORMS; GEOMORPHOLOGY; OROGENY; STRUCTURAL GEOLOGY in the McGraw-Hill Encyclopedia of Science and Technology.

[WILLIAM B. BULL]

Bibliography: W. B. Bull, Tectonic geomorphology, in M. E. Kauffman and D. E. Ritter (eds.), Recent advances in geomorphology: Implications for instruction, *J. Geol. Educ.*, 32:310–324, 1984; M. M. Clark, Map showing recently active breaks along the Garlock and associated faults, California, USGS Misc. Geol. Investig. Map I-741, 1973; M. Morisawa and J. Hack (eds.), *Tectonic Geomorphology: Proceedings of the 15th Annual Geomorphology Symposium*, State University of New York at Binghamton, 1984; R. E. Wallace, Geometry and rates of change of fault-generated range fronts, north-central Nevada, *USGS J. Res.*, 6:637–650, 1978.

Telephone service

Teleconferencing service systems connect two or more locations and allow the multipoint transmission of audio, graphic, or video information. The rising cost of travel and the escalating demand for higher productivity have increased the need for groups of people to have electronically connected meetings or teleconferences. Recent applications of digital technology have brought forth systems which are better able to meet these needs. Multipoint digital systems provide superior audio quality, connect high-resolution protocol graphic terminals (or other, nonprotocol graphic devices) and provide full-motion video transmission at a variety of bandwidths.

Audio teleconferencing systems. Teleconferencing systems have a spectrum of capabilities, including audio, graphics, freeze-frame television, and full-motion video. Audio teleconferencing is the oldest technology and constitutes the backbone of all teleconferencing services; many teleconferences use audio transmission only. Teleconferencing users consider improved audio quality to be the most important property desired of new teleconferencing systems.

A variety of teleconferencing systems and services are offered today. The basis for one widely used operator-assisted teleconferencing service is the Mitchell bridge, which uses analog technology over 50 years old. Other systems use more sophisticated technology, but most are still analog. A new bridge uses digital technology to provide audio quality on a teleconference which is subjectively equivalent to a two-point direct-distance-dialing (DDD) connection.

This digital bridging capability is provided by a newly developed system called the Network Services Complex (NSC), which is a set of program-controlled equipment units loosely coupled to a 4ESS (No. 4 electronic switching system) switch that is located at each of the bridging locations. The Network Services Complex contains the audio bridges, digital tone receivers, announcements, and data storage capability to allow conferences to be set up and to keep track of calls in progress and billing details (Fig. 1). When conference legs are to be added to a conference, the Network Services Complex directs the host 4ESS switch to establish the needed connections through the national and international networks. The Network Services Complex's position in the heart of the national telephone network further enhances the audio quality.

The Network Services Complex bridge uses state-of-the-art technology to achieve superior audio quality. This technology includes echo cancellation, speech detection, signal processing, and automatic gain control. These capabilities, taken together, are designed to equalize sound levels, reduce noise, minimize clipping, and allow multiple simultaneous speakers.

Network considerations. The location of a conference bridge with respect to conferees greatly influences the quality of transmission. Bridges are either centrally located in the network or located on a customer's premises (Fig. 2).

On-premises bridges may have up to twice the loss of a network bridge, but even network bridges incur additional loss over a two-point connection because of the variety of lengths of different conference legs in any conference. To compensate for this additional loss, bridges commonly add gain or amplification to incoming signals.

In addition to the added loss encountered in bridged connections, the variability of loss among the connections in a teleconference can cause large differences in received-signal levels from different conferees. This loss contrast can be a major performance problem in a conference with legs of many different lengths.

Multipoint teleconferences have the potential to sum not only the signals generated by the conferees but the echoes of these signals as well. In two-point connections, echoes can reflect only along a single path, whereas bridged connections permit many re-

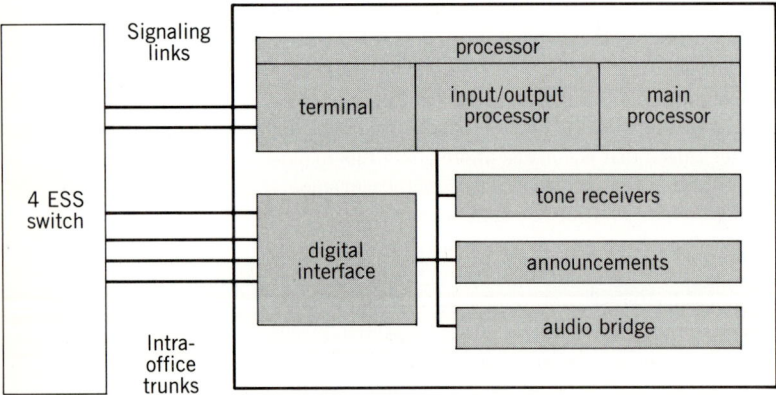

Fig. 1. Basic organization of Network Services Complex.

flection paths. It is this echo summation that makes it necessary for conference bridges to contain their own echo control systems.

Since most bridges are designed to sum the signals from multiple locations, they can also sum the noise. When no one is talking, any echo suppression will be inactive and all the noise from all legs connected can add together to become quite loud. Noise gating and speech detection techniques are used to reduce noise when no one is speaking on the conference bridge.

Audio terminals. Audio terminals are a critical component in the overall teleconferencing system performance. In some teleconferences, there may be just a two-point connection between two groups. In this case, there is no bridge but rather audio terminals in each room which allow everyone to hear and anyone to speak.

The speakerphone is the most commonly used audio terminal. About half of the persons participating in operator-assisted teleconferencing use a speakerphone frequently. Most speakerphones use voice switching, which provides audio stability by allowing only speaking or only listening at any one time. However, the voice switching often clips speech and does not easily allow interruptions. A full-duplex speakerphone has been manufactured, but such speakerphones are quite difficult to develop because room echoes can have delays of 200 milliseconds and greater.

Speakerphones generally are useful only in office-sized rooms. For larger conference rooms, directional microphones provide serviceable results. There must be one directional microphone per person, or, if not, the single microphone must be handed to each speaker. Recent developments in voice-activated directional microphones mitigate this problem somewhat. A single linear array of microphones provides another alternative for conference rooms. The solution which provides the best audio quality with little or no impairment is the use of telephone handsets, or individual headsets with earphones and a mouth mike for each participant. However, such arrangements provide an encumbrance which tends to reduce the naturalness of the meeting. On a large conference with many locations, different audio terminals may be used at each location.

Nonprotocol graphics teleconferencing. There are many graphics devices now available that can be used with audio teleconferencing systems. These devices can operate within the telephone bandwidth and do not use a communication protocol. Nonprotocol graphics devices include electronic blackboards, telewriter devices, and slow-scan or freeze-frame television. These devices must be used in the broadcast mode; that is, only one device can transmit at a time. Some bridges have a broadcast mode, but similar results should be possible if there is an agreement as to who will transmit and when the transmission will begin and end. Graphics usually supplement a voice conference. Two separate con-

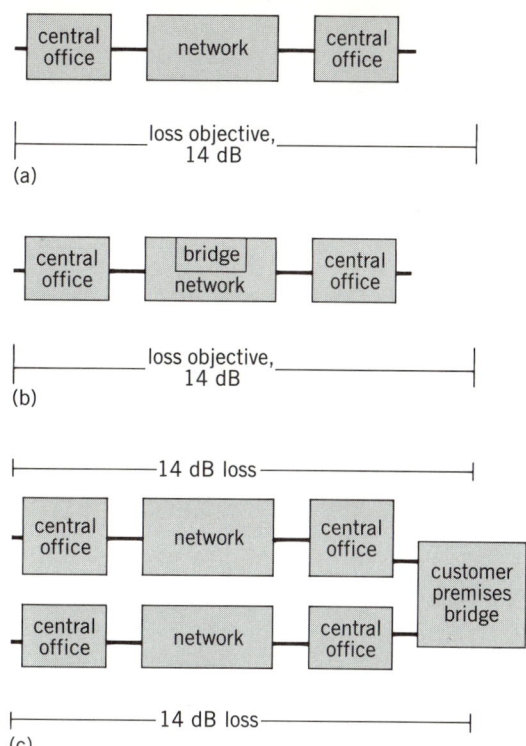

Fig. 2. Network teleconferencing configurations. (a) Two-point (direct distance dialing). (b) Multipoint with network bridge. (c) Multipoint with customer-premises bridge.

ferences, one for voice and one for graphics, must be created for an audiographics conference. Coordinating the transmission of graphics can be done through the voice conference.

Protocol graphics systems. These systems provide a significant step forward in that the transmission negotiations are intrinsic to the system. There are two architectural configurations for protocol graphics systems. One configuration is the star pattern, where the bridge is at the center and does the multipoint transmission to each leg or ray of the star. The ring is another architecture. In this case, there is no bridge. Each graphics terminal has a communications package that receives data from the terminal on one side and retransmits the data to the terminal on the other side.

The digital 3000 bridge uses protocol graphics in a star configuration. It is based on the group-four facsimile protocol, now being approved by the International Telegraph and Telephone Conservative Committee. Bit rate is also a consideration for protocol graphics systems. The 3000 bridge is planned to operate at 4.8 or 9.6 kb/s over normal telephone lines and at 56 kb/s using dedicated lines. Another system, which uses a ring configuration, operates at 9.6 kilobits per second.

Because the group-four facsimile protocol is so new, there are only prototype terminals available. Interand manufactures terminals for their conference arrangement which have a number of state-of-the-art terminal design features. Graphics terminal features

Fig. 3. Room arrangement for teleconferencing.

include high resolution (text transmission), zoom, color, interactive symbols, light pens, and color palette.

Teleconferencing setup. The advent of sophisticated graphics terminals should eventually make it easier to set up a conference. The greater the number of locations in a conference, the more difficult the teleconference setup. On the average, setting up a teleconference takes about 1 min per leg. There are a variety of teleconferencing setup options. Operators are most commonly used by teleconferencing services, and some use multiple simultaneous operator setup to reduce the setup time for large conferences. The new digital Alliance bridges offer self-dial setup as well. Meet-me teleconferencing is also used, and it reduces setup time by having each conferee dial the bridge at the start of or during the conference. Autodialers can be used to speed dialing for self-dial and meet-me conferences. Another setup capability is called blast, in which the conferees are automatically called at once.

Once the conference is set up, there is a certain probability of a leg being dropped. Most bridging services provide tones to indicate to the conferees when a party is added or dropped from the conference. The ability for the dropped conferee to be added back is a necessary feature for any bridging service.

Video teleconferencing systems. Video teleconferencing tends to be from one location to another or from one location to many locations (broadcast). Systems are being developed which provide multipoint video teleconferencing. These systems conserve bandwidth by using coding, multiplexing, and switching techniques. Advances in coders are providing better pictures at bandwidths as low as 56 kb/s. Still, the primary video conferencing services operate at 1.5 megabits per second. Control and display of the multiple locations to every other location is the next key issue beyond bandwidth and transmission costs. Voice-actuated systems with multiscreen or split-screen displays are currently being developed.

Conference rooms. Conference room design is receiving careful attention by teleconferencing system providers and developers (Fig. 3). Microphone, speaker, camera, screen, and terminal placement are all critical issues involving complex acoustic and optical tradeoffs. Room and table size, shape, and placement are also key issues. Tradeoffs often improve audio or video transmission at the expense of ease of local communication. Acoustic and optical room treatments are essential for improving audio and video quality.

Future directions. Teleconferencing technology has reached a point of sophistication, penetration, price, quality, and ease of use which will allow it to be relied upon more regularly by the business community. As people use teleconferencing more, they will become more sophisticated and demand higher quality and more features. End-to-end digital connectivity and end-to-end bandwidth increases due to the use of fiber optics should be, eventually, the basis for improvements in transmission quality and increased use of graphics and video. Stereo audio teleconferencing is probably not far off. Satellites have advanced the use of teleconferencing (especially international) but at the expense of increased transmission delays.

Computer conferencing has been only experimental. The proliferation of personal computers could spur their use in graphics conferencing.

For background information *see* COMPUTER GRAPHICS; SWITCHING SYSTEMS (COMMUNICATIONS); TELEPHONE; TELEPHONE SERVICE; VIDEO TELEPHONE in the McGraw-Hill Encyclopedia of Science and Technology.

[D. J. EIGEN]

Bibliography: D. J. Eigen, R. J. Jaeger, Jr., and E. G. Sable, Customer-dialed audiographics teleconferencing service description and evaluation, *International Teleconferencing Symposium*, Philadelphia, April 1984; D. R. Fischell, Performance issues in multipoint audiographics teleconferencing, *International Teleconferencing Symposium*, Philadelphia, April 1984.

Teleterminals

The communications and computing disciplines are merging. This confluence is most apparent in the corporate and regulatory events in the industries that have grown so rapidly around each discipline. The electronic revolution is driving the confluence because the systems in each discipline are made from the same components. Since the respective systems are virtually indistinguishable in their appearance, technology, and operational features, their convolution is beginning to carry into the services they provide to their respective users. Research in integrated services and instruments began only about 1979, and commercial products, both hardware and software, are just becoming available. The convolution of services is more than a simple union of two sets because of attempts to present the services to the user in a unified fashion. But it is also more than integration of services because they combine synergistically and new services are possible.

Consider two services: (1) Computers provide a means for a user at a conventional terminal to retrieve a record of information from a file by typing in one field of the desired record. If the file is the English dictionary and the input is an English word, the output would be the meaning of the word. (2) Some modern telephones or telephone adjuncts allow their users to store telephone numbers and assign them to extra buttons on the instrument. Some telephone systems allow the subscriber to simulate this service from a regular telephone. The user perceives that one of these special calls may be placed without having to dial the entire telephone number.

In the synergy of these services, the user accesses a telephone directory as a computer file and types in the name of the desired party. Instead of a telephone number appearing on a terminal screen, however, the call is placed automatically. Such a service is logical to the naive user who does not know that it is half computer discipline and half communications discipline, and it is now feasible in the convolved superdiscipline. The missing ingredient is the integrated instrument that delivers the services. A teleterminal, a combination of a telephone and a terminal, is such an instrument.

Fig. 1. Prototype teleterminal.

User interface. Figure 1 shows a research prototype of a teleterminal. It has the handset, ringer, modular connection, and size of a telephone. It has a keyboard and video screen like a conventional computer terminal, but in miniature. Since the terminal features conform to the size of a telephone, the use perceives the instrument as a telephone that doubles as a terminal. The services provided through the instrument are consistent with this perception. The user would perceive an instrument that is the size and shape of a conventional terminal, but with a handset on the side, as a terminal that doubles as a telephone.

In the software interface to the user, services are chosen by menu selection. The user need not know all the available services and need not have previously memorized the detailed name of a desired service. On the teleterminal of Fig. 1, function buttons mounted adjacent to the each side of the video screen provide the user with a mechanism for pointing that further simplifies the selection of a service from the menu. The program dynamically labels the buttons.

Since the list of services is larger than the number of buttons, a means of partitioning the list is required. Scrolling and paging are useful, but organ-

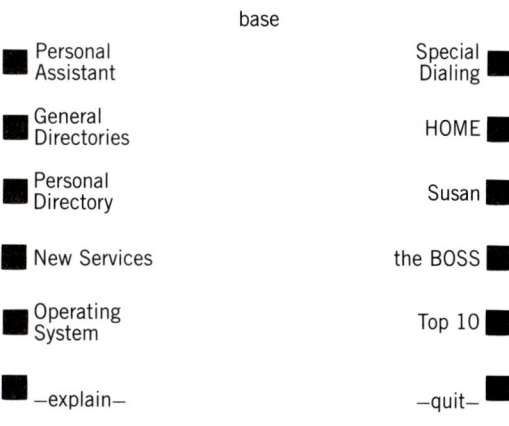

Fig. 2. Base of an access tree.

izing the list into a tree structure has proven to be friendlier, that is, easier for the user to handle. All services are assigned as leaves in a tree. Starting from the base, the user climbs the tree by selecting appropriate branches at successive nodes. The video screen displays only the labels of those branches and leaves attached to the current node of the tree. Because there are only 12 buttons, the teleterminal never confronts the user with more than 12 choices. After selecting a branch, the user perceives the traversal by a new display of button labels.

When the user reaches the node containing the desired leaf, the corresponding service is selected. The tree is customized for each user and, even, by the user.

Figure 2 shows a typical base of such a tree structure. "Personal Assistant" is a branch leading to the node in Figure 3a that contains access to electronic mail and calendar services. "General Directories" is a branch leading to a node that contains access to white-page, yellow-page, and organization-chart directories. "Personal Directory" is a branch leading

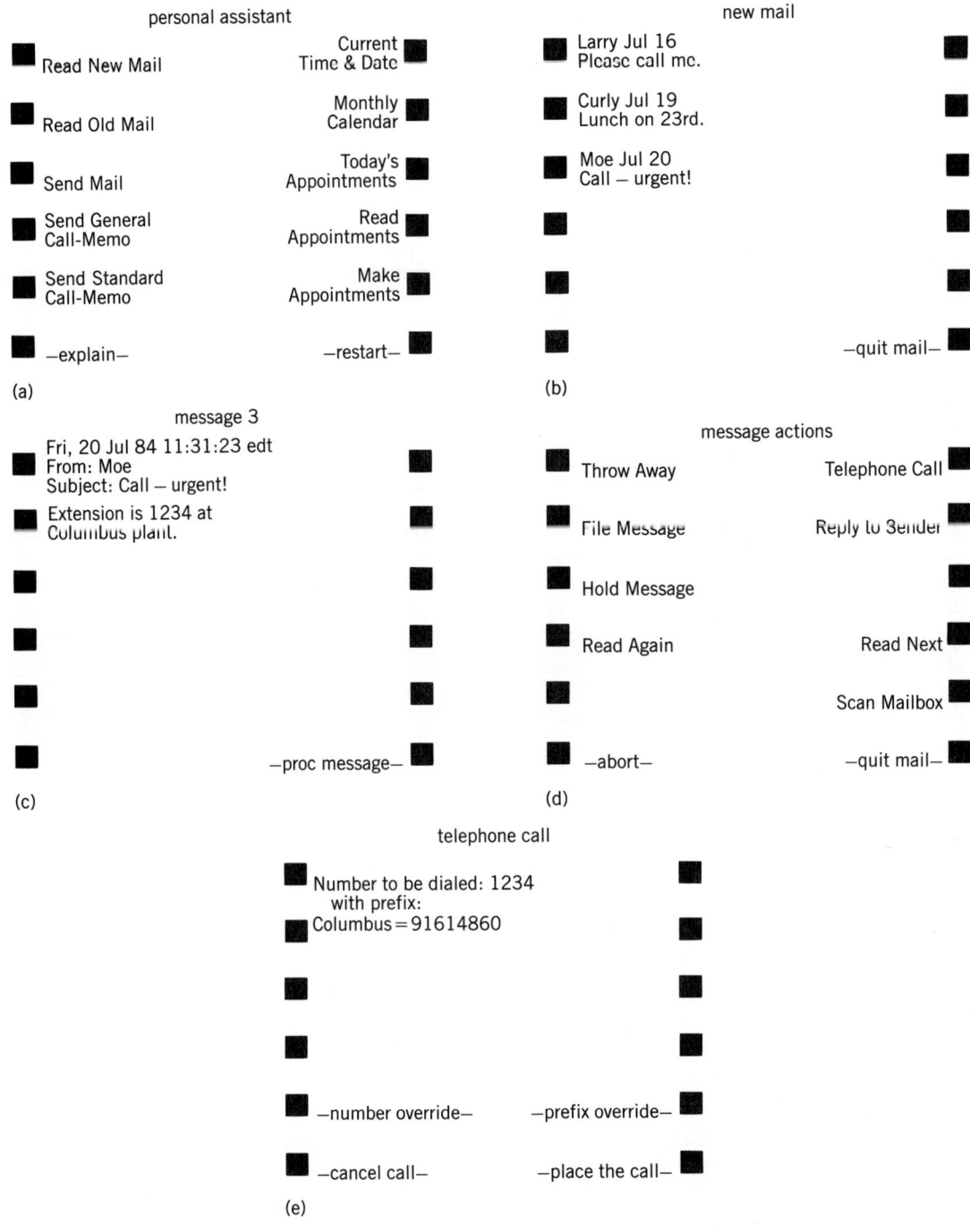

Fig. 3. Successive nodes in an access tree. (a) "Personal Assistant" node. (b) "New Mail" node. (c) "Message 3" node. (d) "Message Actions" node. (e) "Telephone Call" node.

to a node that allows access to a special personal directory service. From this node the user makes new entries in the personal directory or retrieves existing entries by traditional alphabetic sort or by matches on key words. "New Services" is a branch leading to a node that is the repository for newly created or improved services. The user tries one at his leisure and then discards it or moves it to a more appropriate place in the tree. "Operating System" is a leaf providing direct access into the host computer's command interpreter. When the user pushes this button, the video screen is cleared and the user perceives the teleterminal as a regular computer terminal. When the user pushes the "–explain–" button, its text is capitalized to "–EXPLAIN–"and the next button push is explained on the video screen instead of executed. "Special Dialing" is a branch leading to a node that contains translation services from mnemonics into area codes, tie-line codes, and dialing sequences for services like transfer and call forwarding. The next three buttons are leaves that cause telephone calls to be placed to the respective parties. These are three parties that the owner of this tree apparently calls frequently because the owner placed these leaves directly on the base of the tree to reduce the access time. The user installed 10 other parties that are called almost as frequently in the node reached by the button labeled "Top 10" (more precisely, the last 10 of the top 13). The button marked "–quit–" is the means for the user to shut off the software.

Service example. Suppose Moe tried to call someone while the callee was away from the telephone. Assume that either someone else answered the phone and left a message by electronic mail or Moe sent the mail directly because no one answered the phone. The callee returns to find the teleterminal screen as in Fig. 2 except that the bottom line of the video screen has the message: "You have new mail." By pressing the "Personal Assistant" button, the "Personal Assistant" node (Fig. 3a) can be reached. In similar fashion, the buttons "Read New Mail," "Moe," "–proc message–," and "Telephone Call" can be pressed in succession to reach, in succession, the nodes "New Mail" (Fig. 3b), "Message 3" (Fig. 3c), "Message Actions" (Fig. 3d), and "Telephone Call" (Fig. 3e). After the callee presses "–place the call–" and lifts the receiver, Moe's telephone number is called and the screen returns to Moe's message (Fig. 3c). If Moe answers, while speaking with him the callee would probably press "–proc message–" to get message actions, and then "Throw Away" and "–quit mail–" to return to the "Personal Assistant" node.

Implementation. The particular teleterminal described here was initially designed for a research environment. The research objective is to investigate synergistic services and the user interface. The teleterminals are deliberately provided with little local intelligence and must be permanently connected to a host computer during operation. They have two independent communications ports, one for connection to the telephone network and one for connection like a terminal to the computer. Each user executes the same program in the time-shared host. The user perceives customization because the program reads an access tree file that each user owns and maintains.

For a real market a different environment might be logical. More local computing power in the teleterminal might be practical to reduce the data communications requirements. Access through the tree file and some of the services, like calendars and directories, are provided well by the teleterminal directly. Other services, like centralized directories and mail, are best implemented in a centralized computer. These price-performance issues will be resolved in the marketplace as teleterminals become commercially available.

As voice communications become increasingly digital, voice and data will use the same communications media. The need for two separate ports will be removed, and the distinction between remote and local computing will become more blurred. In the more distant future, as photonics provides extremely high bandwidth, this exercise will be repeated because of a convolution with the video discipline. The instrument will then be a synergy of the telephone, the computer terminal, and the television set.

For background information *see* MULTIACCESS COMPUTER; TELEPHONE EQUIPMENT, CUSTOMER in the McGraw-Hill Encyclopedia of Science and Technology.

[RICHARD A. THOMPSON]

Bibliography: D. W. Hagelbarger and R. A. Thompson, Experiments in teleterminal design, *IEEE Spectrum*, 20(10):40–45, October 1983.

Toxicology

Dioxins, or polychlorinated dibenzo-*para*-dioxins, are highly toxic, lipophilic, unwanted contaminants found in a number of chemical products; they are also found in the food chain in fish, meat, eggs, poultry, and milk. Dioxins are chemically and toxicologically similar to the chlorinated dibenzofurans and polychlorinated biphenyls (PCBs; Fig. 1). The term dioxin often refers specifically to 2,3,7,8-TCDD, or simply TCDD, the most toxic of the dioxins in many species, including monkeys, rats, mice, rabbits, and guinea pigs.

TCDD is found as an impurity or contaminant in chlorinated phenols, which are frequently used as wood or paper preservatives, in the herbicide 2,4,5-trichlorophenoxy-acetic acid (2,4,5-T), and in hexachlorophene, an antibacterial agent. These compounds have also been reported in fly ash as well as in stack effluent from municipal incinerators. PCBs and polychlorinated benzene–containing electrical transformers can release furans and dioxins when involved in fires by pyrolytic conversion of PCBs to furans and of the chlorinated benzenes to dioxin isomers.

The finding of trace amounts, in the parts per mil-

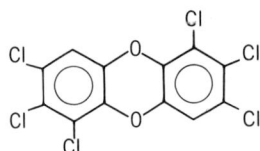

Fig. 1. Chemical structure of several dioxins, furans, and polychlorinated biphenyls.

lion range, of TCDD in Agent Orange, a mixture of 2,4-D and 2,4,5-T, used extensively as a defoliant in Vietnam during the 1960s as well as in the United States, has caused great concern about health effects of dioxin.

Dioxin exposure. In reviewing current evidence regarding the health effects of dioxins, several different methodological approaches must be considered. Because some of the 75 possible dioxin isomers are extremely toxic to sensitive species of animals such as monkeys and guinea pigs, no deliberate dosing of humans is performed, although some deliberate human dosing studies have been performed in the past. However, there have been accidental occupational or environmental exposures of humans upon which to base some conclusions, though usually such incidents are not followed with the methodologic rigor typical of prospective medical research projects: sample size is often too small to be certain of detecting increasing incidents of rare events, such as cancer of a given type or birth anomalies; frequently several chemicals are involved; and estimation of dose of dioxin exposure is usually difficult.

One of the earliest accidents was reported after an explosion of a chemical plant in Nitro, West Virginia, in 1949; other explosions or incidents with dioxin exposure have since been reported in England, Germany, Italy, the Netherlands, Czechoslovakia, and elsewhere. Dioxin contamination of soil has been reported in parts of Missouri, including the town of Times Beach, where the United States government purchased the homes of the inhabitants after a finding of from one to several hundred parts per billion of TCDD in soil samples. In Japan and Taiwan, rice oil contaminated with PCBs and furans led to serious illness in persons consuming food containing it. Because transplacental transfer occurs, babies born to women who ingested the rice oil were also affected. Recent PCB transformer incidents in San Francisco and Binghamton, New York, have also provided human health data, especially fat biopsy measurements of furans and dioxins, as well as liver ultrastructural alterations after PCB, furan, and dioxin exposure.

As a result of exposure to dioxins, a transient skin rash may be seen; on some occasions, a lesion known as chloracne, a type of acne, is found on the face, especially the nose and cheeks, and in more severe cases on the neck, back of ears, chest and back, and, sometimes, genitalia or legs. Chloracne, which is also induced by other chlorinated chemicals, may resolve in the course of time or persist for years. Comedones and cysts may be noted, but keratosis rather than excess oil secretion is characteristic.

Signs and symptoms of nerve damage such as weakness and pain in the lower extremities with difficulty coordinating movement have been reported after exposure to dioxins, as have alterations in nerve conduction velocities. Arthritis, hyperirritability, sleep disorders, and decreased libido have also been reported and psychiatric pathology is frequently observed. Respiratory pathology has sometimes been well documented after exposure.

Liver damage is noted (Fig. 2): Mitochondrial alterations, the presence of lipid droplets and myelin figures, as well as changes in both smooth and rough endoplasmic reticulum have been described after dioxin, PCB, or chlordecone (Kepone) exposure. Necrosis sometimes occurs, and porphyria cutanea tarda, an acquired defect of porphyrin metabolism in liver, is sometimes seen.

A variety of less common effects of exposure have also been found. Elevated serum lipids are sometimes seen, suggesting alteration of lipid metabolism. In some DDT and PCB incidents, increased blood pressure has been noted. An excess of cardiovascular deaths was reported in one European incident. Hemorrhagic cystitis was seen in a child exposed to dioxin in Missouri. Alterations of sperm morphology, motility, and quantity were reported in a court proceeding.

A Swedish investigator reported a marked increase in the incidence of soft tissue sarcomas as well as non-Hodgkins lymphoma in phenoxy herbicide–exposed workers. But a similar New Zealand study did not note increased pathology in phenoxy herbicide–exposed workers.

In the Yusho contaminated–rice oil incident in

Japan and the similar Yu Cheng incident in Taiwan, discoloration of gingival membranes in the mouth persisted for years as did some chloracne. Some babies whose mothers had been exposed to the rice oil were born with cola-colored or brown skin which gradually returned to normal. Biopsies of fat and other tissues were found to contain elevated levels of some of the PCB and furan, especially penta- and hexachlorinated furan, isomers and congeners to which patients had been exposed. Persistence of polybrominated biphenyls (PBBs) for decades in human adipose tissue has been reported, and dioxins and furans have been found in control human adipose tissue in the United States and Canada, suggesting possible contamination of the food chain.

After the Yu Cheng incident, reports of attempts to reduce the body burden of these chemicals through dietary changes have recently appeared. It is unclear whether these attempts have in fact decreased the body level of chemicals below the level which would be consistent with the natural, but slow, unassisted rate of elimination. Oral ingestion of sucrose polyesters as well as certain other compounds have been proposed as a means of removing these chemicals from the gastrointestinal tract, where they are found as part of the circulation between intestine and liver.

Laboratory studies. There are two types of nonhuman data that have been used to study the health effects of dioxins: First is evidence from studies of cells in culture. There is some evidence of mutagenicity of TCDD in at least one mammalian cell culture study. This is compatible with the possibility that the compound might be carcinogenic in humans as well as in animals. In addition, in liver or skin cultures, some dioxins, as well as some of the furan and PCB isomers, induce the activity of an enzyme that serves to metabolize dioxins (aryl hydrocarbon hydroxylase), and also induce the activity of cytochrome P450. The activity of these chemicals parallels the LD_{50} (lethal dose for 50% of the animals) of a given dioxin isomer and hence is useful in screening the many dioxin and dioxinlike chemical isomers for toxicity.

The extreme toxicity of the dioxins and dioxinlike chemicals has been demonstrated and characterized from animal dosing or toxicologic studies. When there is chlorination of the 2,3,7,8 positions on a dioxin or furan molecule, that molecule will be much more toxic than if those lateral positions were not chlorinated. Guinea pigs, monkeys, rats, mice, and rabbits are more sensitive than hamsters or bullfrogs, and different genetic strains of mice or rats vary in their sensitivity. It is not clear from these studies how sensitive humans are to these compounds.

The effects found in animal dosing studies are dramatic. Dioxins and related compounds cause a delayed death after a wasting (weight loss) syndrome. They also cause thymic involution, splenic atrophy, bone marrow atrophy, liver necrosis, also gastrointestinal tract hyperplasia, especially of

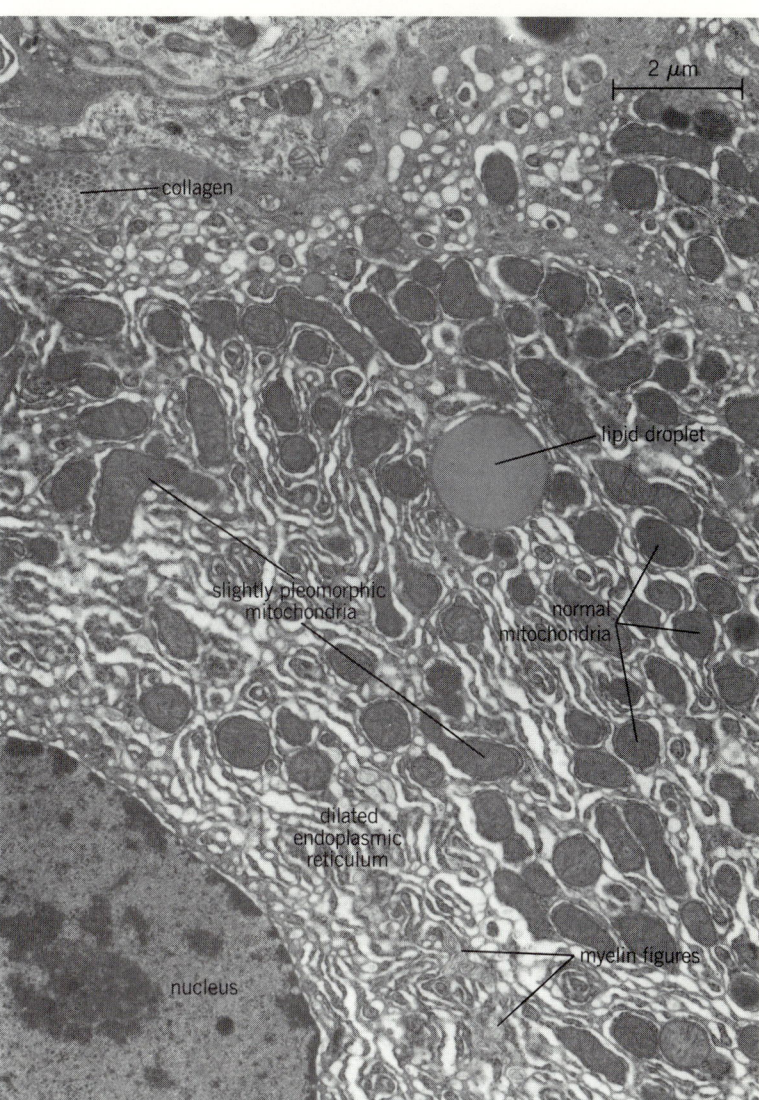

Fig. 2. Electron micrograph of a rat liver after 14 days of first dosing with 2,3,7,8-TCDD. Marked dilation of endoplasmic reticulum can be seen along with some membrane hypertrophy as well as the beginning of myelin body formation. A small lipid droplet is seen. Mitochondria demonstrate mild pleomorphism. (*Toxicologic preparation in the laboratory of Thomas Gasiewicz; electron microscopy by Arnold Schecter*)

stomach and colon, and intrahepatic bile duct hyperplasia, gallbladder hyperplasia, urinary tract hyperplasia, hyperplasia of skin and sebaceous glands, atrophy of the testes, and neoplasms of various organs including liver, oral cavity, nasal turbinates, and lung. In addition, there is an effect on the reproductive system such that inability to conceive, early abortions, or resorbtion of embryos may result. Teratogenicity has also been found. Acne (chloracne) caused by a hyperkeratosis and alterations of the Meibomian glands of the eyelids are noted in some species. Hyperkeratosis is noted in certain species, as are alopecia and edema.

Summary. Some of the polychlorinated dibenzodioxins and closely related chemicals including certain of the polychlorinated dibenzofurans and some polychlorinated biphenyls, as well as some polybrominated biphenyls, are very toxic for numerous animal species, affecting multiple organ systems. The mechanism of this toxicity is unknown. They may be mutagenic and are certainly carcinogenic and teratogenic to many species, including monkeys, rats, and guinea pigs. For a number of reasons, including difficulty in measuring or estimating dose as well as finding a nonexposed control group, the extent of their toxicity in humans is not yet clear. Toxicity of the dioxins is not in doubt; the extent of this toxicity still remains to be defined.

For background information see LETHAL DOSE 50; POISON; TOXICOLOGY in the McGraw-Hill Encyclopedia of Science and Technology.

[ARNOLD SCHECTER]

Bibliography: R. Kimbrough (ed.), *Topics in Environmental Health*, 4: *Halogenated Biphenyls, Terphenyls, Naphthalenes, Dibenzodioxins, and Related Products*, 1980; R. Kimbrough and A. Poland (eds.), *Biological Mechanism of Dioxin Action*, Banbury Rep. 18, Cold Spring Harbor, 1984; W. W. Lawrence (ed.), *Public Health Risks of the Dioxins*, William Kaufmann, Inc., Los Altos, California, 1984; W. J. Nicholson and J. A. Moore (eds.), *Health Effects of Halogenated Aromatic Hydrocarbons*, New York Academy of Sciences, vol. 320, 1979; Polychlorinated biphenyls, *Environmental Health Perspectives*, vols. 59 and 60, 1984.

Toxin

Within the last two decades there has been a great increase in the reorganized number of microbial agents associated with gastrointestinal infections and food-borne intoxications. Pioneer studies in India in the 1950s showed that *Vibrio cholera* produces an extracellular protein toxin that causes fluid secretion when injected into intestinal loops of young rabbits were soon confirmed. These discoveries also led to a new definition of enterotoxin as a protein or peptide that causes net fluid hypersecretion from the intestinal mucosa by mechanisms that may or may not involve cytotoxic damage to the epithelium. Enterotoxins that are not cytotoxic are now designated cytotonic because of the passive efflux of watery stools and the active secretion of chloride into the intestinal lumen. In general, cytotonic enterotoxins seem to act by affecting the cyclic nucleotide system at the level of the cell adenylate and guanylate cyclase system.

Cholera. *Vibrio cholerae* enterotoxin [also known as cholera toxin (CT) or choleragen] consists of one A subunit (mol wt 28,000 daltons) and of five binding (B) subunits (mol wt 11,500 daltons each) responsible for the binding of the toxin to specific glycoconjugate cell receptors (the GM_1 ganglioside; see illustration). Together they are referred to as AB_5. Any such toxin molecules that have lost the enzymatically active A subunit are natural nontoxic products called choleragenoids. Recent immunological studies in animals and humans have shown that nontoxic B subunits are promising cholera vaccine candidates, while conventional vaccines based on heat-killed cholera vibrios give an incomplete short-lasting immunity upon vaccination. It is now known that the active AB_5 toxin can only be neutralized by antibodies (antitoxin) for up to 60 s after binding. In the first 10 to 15 min after binding, the A subunit is transported through the cell membrane, either enclosed within a vesicle and released into the cytosol, or directly through the plasma membrane through a channel formed by the B subunits with the A_2 fragment of the A subunit as the leading sequence (see illustration). The A_1 fragment then catalyzes transfer of adenosinediphosphate (ADP) ribose from nicotinamide adenine dinucleotide (NAD) to membrane-bound proteins through a guanosinetriphosphate (GTP) binding component with an inherent feedback regulatory mechanism by which guanosinetriphosphate (GTP) is hydrolyzed and cyclic adenosinemonophosphate (cAMP) accumulates intracellularly. Interestingly, most other eukaryotic cells in culture, except certain GM_1-deficient cell lines, also bind cholera toxin and trigger cAMP-regulated events such as production of steroid hormones. Cholera toxin has become an important probe for studies of cAMP-mediated events in cell biology.

The accumulation of intracellular cAMP in intestinal epithelial cells stimulates the active secretion of chloride and bicarbonate by villus crypt cells and inhibits the normal absorption of chloride-coupled sodium ions by villus cells, the net result being diarrhea with ricewater-like stools of up to several liters per day for more than a week in severe cholera. Recent studies have shown that the secretion mechanism is more complex, probably involving intramural neurons and the release of local neural transmittors such as serotonin. Furthermore, prostaglandin inhibitors such as acetylsalicylic acid (aspirin) and indomethacin, as well as other drugs such as chlorpromazin, inhibit the net secretion caused by cholera toxin measured as a decrease in stool volume. Other promising antisecretory drugs have

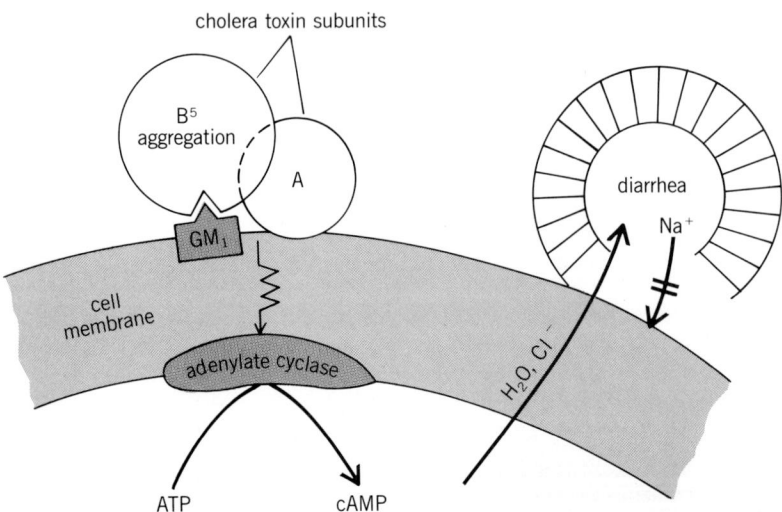

Binding of cholera toxin–binding subunits to a specific cell receptor in the epithelium of the small intestine (ganglioside GM_1) triggers transport through the cell membrane of the A (active) subunit. After activation of the A subunit, and after a series of intracellular events including increase in cellular cyclic adenosinemonophosphate (cAMP), chloride, water, and ions are excreted and reabsorption of sodium is inhibited; the net result is severe diarrhea.

also been reported recently. Finally, the discovery that glucose and certain amino acids, such as glycine, stimulate net absorption in cholera and other enterotoxic enteropathies has made it possible to develop an oral rehydration therapy, initiated at the World Health Organization, which has, in the last few years, drastically decreased the mortality of cholera as well as pediatric diarrhea in some developing countries.

Escherichia coli. Through pioneer studies on diarrhea in young pigs and calves, in the late 1960s a heat-labile protein toxin (LT) and a heat-stable (ST) peptide were both shown to cause net fluid secretion in intestinal loop experiments in young animals. However, fully virulent strains also produce specific cell-surface proteins which enable the organism to colonize the small bowel and to release enterotoxins close to specific cell receptors on the epithelial surface. Despite the fact that heat-labile toxin was found to be very similar in structure as well as in action to cholera toxin, heat-labile toxin and heat-stable toxin were found to be regulated by extrachromosomal genes on plasmids while cholera toxin is regulated by chromosomal genes.

Epidemiological studies showed quite early that such enterotoxigenic *Escherichia coli* (ETEC) strains producing heat-labile toxin or heat-stable toxin, or both toxins in combination, are the most important pathogens to cause diarrhea in calves and piglets in Europe and the United States, in human infants in tropical and subtropical countries, and among adult travelers of the Western Hemisphere. These epidemiological findings have encouraged a rapid development of vaccines based on detoxified heat-labile toxin and toxin subunits (inoculation of pregnant sows). Promising results with such vaccines as well as vaccines based on bacterial colonization factors are a breakthrough in preventive veterinary medicine, now followed by attempts to develop new gene technology-based products for both human and veterinary medicine. This recent development also encouraged studies to define new enterotoxins when it was found that animal ETEC strains can produce at least two different heat-stable peptides (ST_a and ST_b). In brief, ST_a is very active in the intestine of infant mice and rabbits, while ST_b is active only in the intestine of weaned piglets.

Recently, ST_a was found to be a peptide with an identical amino acid sequence in human, calf, and pig strains. This toxin binds to a unidentified cell receptor and causes activation of the guanylate cyclase leading to an increased cell cyclic GMP level. More recently the toxin was also found to activate cell metabolism of prostaglandins and release of free radicals and to cause activation of a protein kinase in the cell, leading to impaired chloride absorption and isotonic fluid secretion. The mode of action of ST_b is much less well understood, since ST_b has not yet been purified.

Enterotoxins in other bacterial diarrheas. The clinical disease in diarrhea caused by cytotonic enterotoxins is of a noninflammatory type with typical watery diarrhea. However, diseases caused by or-

Cytotonic and cytotoxic bacterial enterotoxins

Cytotonic enterotoxins	Cytotoxic enterotoxins
Intestinal infections	
*Vibrio choleae**	
Escherichia coli	
Yersinia enterocolitica	
Salmonella species	*Salmonella* species
Aeromonas hydrophila	*Campylobacter* species
	Shigella species
Food poisoning or intestinal infections	
Bacillus cereus	*Clostridium perfringens*
	Clostridium difficile

*Other related vibrios, which are also aquatic organisms isolated from water, fishes, and other aquatic organisms as well as human stools, produce choleralike enterotoxin as well as cytotoxic enterotoxins (for example, *Vibrio parahaemolyticus*).

ganisms elaborating cytotoxic enterotoxins and probably other cell-destructive factors such as endotoxins cause tissue inflammation with mucoid bloody stools. *Shigella dysenteriae* organisms produce an AB-subunit protein toxin and release of an A_1 fragment into the enterocyte cytoplasm which binds to 60S ribosomal subunits and blocks protein biosynthesis. Interestingly, certain strains of *E. coli* produce a very similar toxin now generally called the Shiga toxin. Injection of this purified toxin in experimental animals causes hemorrhages in the intestinal mucosa. More recently, strains of certain *Salmonella* species were found to produce both a cholera toxin and heat-labile toxin like cytotonic enterotoxin as well as cytotoxic proteins which may act in a very similar way to Shiga toxin (see table). It seems likely that salmonellosis, a disease with a wide spectrum of symptoms in animals and humans, may be explained by the production of different combinations of toxins. Moreover, other newly discovered enteropathogens such as certain *Campylobacter* and *Aeromonas* species also cause intestinal infections with symptoms similar to classical shigellosis or salmonellosis.

Enterotoxins and food poisoning. *Staphylococcus aureus* is the classical food poisoning agent able to produce several different protein toxins upon growth in certain food vehicles. More than two decades before the discovery of cholera toxin these toxins were already called enterotoxins. Purified staphylococcal enterotoxin A and B, when given to animals and human volunteers, cause rapid onset of nausea and vomiting sometimes followed by diarrhea. Such symptoms develop usually within 12 h after consumption of the contaminated food which contains preformed toxin. These toxins, however, do not cause fluid accumulation in intestinal loop assays and probably act after uptake into the local nervous system in the gut wall. They should thus probably better be called neurotoxins. Interestingly, purified staphylococcal enterotoxin, when injected into the bloodstream of monkeys, causes a disease similar to the toxic shock syndrome in young female monkeys. In both syndromes, bacterial endotoxin seems to act in synergism with toxic shock syndrome toxins to cause shock. Other food poisoning agents, such as *Clostridium perfringens* and *Bacillus cereus*, have also been found to produce cytotoxic as well as cytotonic enterotoxins.

For background information *see* CHOLERA VIBRIO; DIARRHEA; FOOD POISONING; TOXIN in the McGraw-Hill Encyclopedia of Science and Technology.

[TORKEL WADSTRÖM]

Bibliography: H. J. Binder, *Mechanisms of Intestinal Secretion*, 1979; H. L. Du Pont and L. K. Pickering, *Infections of the Gastrointestinal Tract: Microbiology, Pathophysiology and Clinical Features*, 1980; C. G. Gemmell, Comparative study of the nature and biological activities of bacterial enterotoxins, *J. Med. Microbiol.*, 17:217–223, 1984.

Transportation engineering

A number of alternative modes of transportation are under development or have come into increasing use in recent years. These include automated driverless mass transit systems and new types of bicycles, tricycles, and quadricycles.

Advanced technology. The city of Vancouver, British Columbia, is constructing a modern light rail rapid transit system employing advanced transportation technology. A fully automated rapid transit system designed to serve 15 stations along an approximately 13-mi (21-km) corridor will serve 10,000 passengers per hour when full revenue service begins in 1986.

The Vancouver system (Advanced Light Rapid Transit, or ALRT) will incorporate four unique features that distinguish it from the other modern transit systems of the world: (1) a totally automated, driverless train control system; (2) linear induction

Fig. 2. Linear induction motor (undercar view), a truck-mounted vehicle propulsion system. (*Courtesy of K. Knight, Metro Canada, Ltd.*)

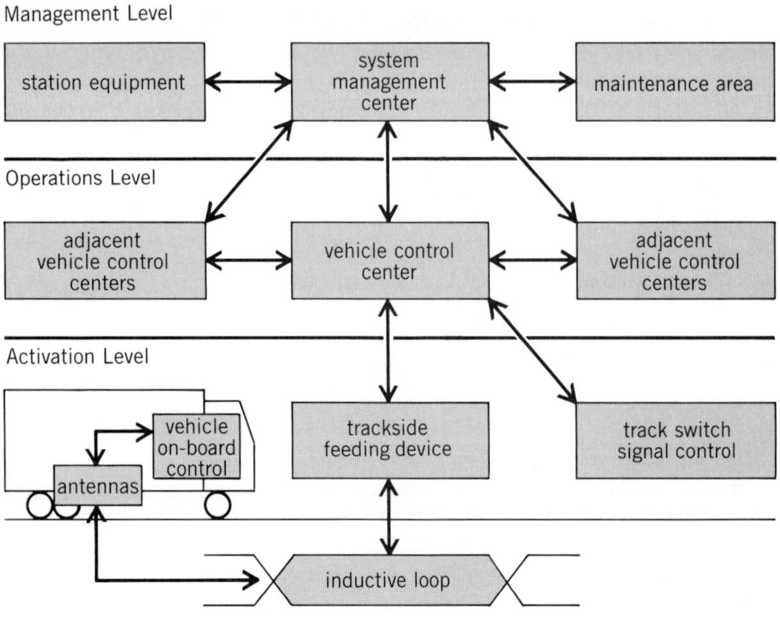

motors as vehicle propulsion; (3) steerable-axle trucks to reduce wheel-rail wear and produce quieter trains; and (4) track elements supported directly on guideways requiring special quality-control measures.

Automated train control system. A totally automated, driverless train control system becomes a key choice to operate at very short headways while minimizing operating costs. The Vancouver system uses a computerized automatic train control system to direct train movement by checking the speed, direction, and location of all cars at all times. The control system also provides for high levels of efficiency, safety, and schedule adherence. The system selected for Vancouver utilizes the moving block principle to attain short headways and is structured in a rigid three-level hierarchy (Fig. 1).

The three vital elements of this control system are the vehicle control center, the inductive loop, and the vehicle on-board controller. The vehicle control center is a threefold redundant computer complex which exchanges data with an on-board vehicle control unit (one per train). The inductive loop, a cable laid between the tracks, transmits a continuous flow of digital data. The vehicle on-board controller is a microprocessor to transmit accurate position, velocity, and system status data to the vehicle control center. It acts on validated commands to control the train safely and precisely within the limits imposed by the control center.

Linear induction motor. This is essentially a conventional motor with its stator stretched out flat and attached beneath each of the two trucks of each vehicle (Fig. 2). In a conventional induction motor there is also a movable solid iron rotor, spun by the force of the electromagnetic field in the outer stator. In the linear induction motor, the rotor is unrolled along the track, forming a stationary, continuous, laminated iron and aluminum reaction rail, mounted on the guideway between the rails, to complete the magnetic circuit in conjunction with the vehicle's stator. The power rails energize the stator. The linear induction motor is of single-sided axial flux design and consists of a laminated core and a three-

Fig. 1. Diagram of an automated train control system, a three-level hierarchy that ensures overall system safety.

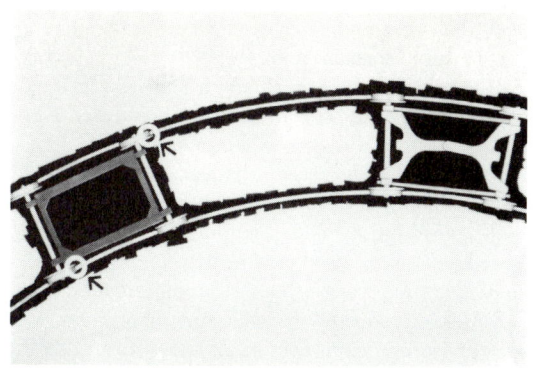

Fig. 3. A steerable-axle truck that ultilizes swivel-thru mechanical linkage as compared with a fixed-axle truck. The arrows indicate the rubbing of flanges to the steel rail. (*Courtesy of J. Eagen, Metro Canada, Ltd.*)

Steerable axle truck. Characteristic sounds associated with many trains are primarily caused by the contact between the wheel and the running rail; curve squeal results when wheel flanges rub the side of the rail. On a conventional truck running on a straight track, each axle is held parallel to the other, so the wheels are lined up straight. Rounding a curve, however, causes the flanges to rub the rail and sometimes develop a squeal. The Vancouver vehicle utilizes axles with swivel-thru mechanical linkage. This steerable truck makes the wheels follow the rails through curves, resulting in a reduced flange-rail contact and a much quieter system (Fig. 3).

Guideway surface and track elements. Assuring optimal mechanical interfacing between vehicle and guideway is vital to the linear induction motor system. The vehicle subsystems interface with the track elements at the reaction rail, the running rails, and the power rails (Fig 4). Because of these critical contacts between subsystem elements and guideway surface–mounted elements, strict quality control measures must be implemented during construction. The trackwork installation—which follows the guideway construction—must accommodate deviations in the guideway surface to meet the tight tolerance requirements for proper vehicle performance.

The power pick-up mechanism needs careful alignment. The Vancouver system uses brushes for a horizontal power pick-up from the power rail, in

phase six-pole winding system with necessary electrical subsystem features.

In the linear induction motor, the field is continually induced in front of the pole, and the pole pursues, in contrast to the conventional rotary induction motor, in which the rotor is continually attracted to a moving field. In the conventional motor, mechanical linkage is required to produce a linear force to propel the vehicle. In the linear induction motor there are no linkages, no moving parts, and therefore no routine maintenance. There is no drive train and, consequently, no gears or related bearings to require lubrication and periodic replacement.

Operating in the acceleration mode, the linear induction motor provides thrust in the direction of travel in response to the input voltage, frequency, and phase sequence supplied by the power conversion unit. In the deceleration mode, the linear induction motor operates as a generator and feeds power back to supply lines under the control of the power conversion unit. This regenerated power is used directly by other vehicles in the system, or if the system is insufficiently receptive, transferred to wayside resistors at the traction power substations to be dissipated as heat.

Another feature of the linear induction motor mode of propulsion system is that it does not rely much on wheel-on-rail friction. Because of this, a steel-wheeled vehicle on steel rails, such as used in Vancouver can negotiate 6.5% grades. This capability is substantially in excess of the capability of a system using friction between the wheels and rails. Similarly, dynamic braking down a grade is virtually unlimited by any friction factor. These advantages become even more important when friction is reduced by ice and snow.

Optimum efficiency of the linear induction motor is dependent upon a close air gap between the linear induction motor rail and the reaction rail, which should be maintained between 0.4 and 0.6 in. (9 and 15 mm). This requirement of very close tolerance for the vehicle-track-guideway interface is an extremely critical factor in the successful operation of Vancouver's linear induction motor.

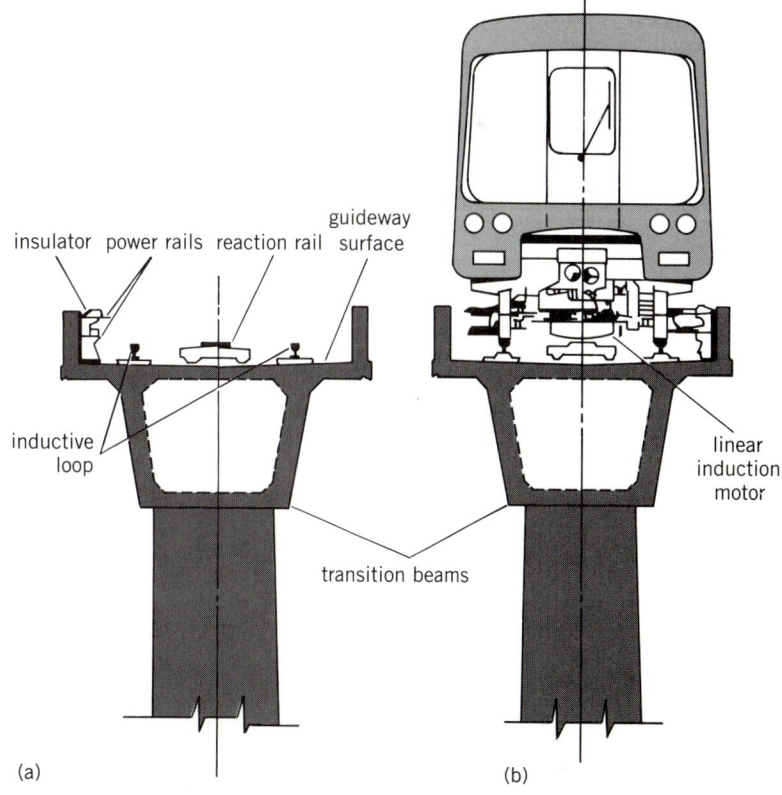

Fig. 4. Vehicle-guideway interface: (*a*) guideway surface–mounted track elements and (*b*) their interface with vehicle-mounted linear induction motor.

contrast with the usual pick-up contact with a vertical power rail.

Also of significant concern is the relative location of the top of running rail with respect to reaction rail, which greatly affects the air gap between the linear induction motor and the reaction rail. This gap is very critical to linear induction motor operational efficiency and may necessitate periodic adjustments to maintain desired air gap, depending upon the running rail wear and the power rail wear.

[A. A. DORMAN; D. B. DESAI]

Human-powered land vehicles. Human-powered land vehicles can be divided into two categories: bicycles, representing the predominant mode for the last 100 years, and three- and four-wheeled vehicles, posing a significant challenge for high-speed vehicles and for practical commuting vehicles. Bicycles have developed in three new directions in the last few years: BMX (bicycle motor cross) bicycles for youngsters; and mountain or city bikes and recumbents for adults. A large variety of tricycles and quadricycles are being developed for commuting and for attempts on speed records.

BMX bicycles. The BMX bicycle has developed into a small and extremely strong machine, having generally 20-in. (500-mm) wheels, large-cleat (knobbly) tires, upright but not high-rise handlebars, and a seat positioned more toward the rear wheel than on a conventional bicycle. The wheels are often a complete break with bicycle tradition in being made from injection-molded glass-reinforced nylon in bright colors. Either a single-speed transmission is fitted or a five- or six-speed derailleur gear. Another break with previous youth bicycles is the use of caliper brakes operated from the handlebars.

These vehicles are used for stunt riding and tricks. These bicycles can be ridden off a ramp at high speeds, covering long distances while airborne and usually coming down on the rear wheel. They can also be ridden on the back wheel alone under full control and can jump curbs or other obstacles. The bicycles are equipped with pads on the crossbars and handlebars, as a result of government bicycle-safety regulations.

Mountain bikes. The BMX bikes can also be used to go off-road, over building sites in cities and along trails in the country, and they undoubtedly had some influence in the development in California in the late 1970s of the so-called mountain bike for adults. This development was perhaps also a reaction to the trailing off of the boom in lightweight racing or ten-speed bicycles. New riders appreciated the liveliness and easier pedaling of the lightweights, but found them fragile in potholed city streets and almost useless in soft ground or gravel. The earlier heavy American bicycle was a single-speed machine with large wheels and large-diameter tires that could be ridden over gravel, rocks, and street curbs, but could not be manipulated by body dynamics over rocks, and could not be ridden up more than gentle inclines because of the weight and the inflexible transmission.

The mountain bike is simply a marriage of lightweight multispeed bicycle technology with large-diameter tires on full-size wheels. Unlike the lightweight bicycles, however, a straight handlebar giving a comfortable semiupright riding position, with well-padded grips, and sometimes a padded contoured saddle are used, rather than the racing-style handlebars requiring a back-straining crouch and a narrow seat producing discomfort in many riders and sometimes injury. Fifteen- or eighteen-speed derailleur-gear transmissions are usually fitted, with the lowest gears giving an approximately 1-to-1 ratio with the rear wheel, and a consequent ability to climb very steep grades. The large-cleat or knobbly tires, similar to large-size versions of the BMX type, give traction and braking on mountain paths. Skilled riders can use body dynamics, as do the youngsters on the BMX machines, to perform striking maneuvers such as crossing logs almost 3 ft (1 m) in diameter.

The mountain bike is inevitably heavier than the racing-style lightweight, but the penalty is not large—29 versus 22 lb (13 versus 10 kg) for high-quality versions of the two types respectively. There is some concern about the effect of large numbers of bikers on the stability of mountain trails, but little evidence as yet that low-pressure tires are more damaging than hiking boots. Meanwhile, both the users and the makers of mountain bikes have discovered their suitability for use in poor-surface city streets, where their robustness and greater comfort are considered by many to be ample compensation for slightly lower speeds.

Recumbent bicycles. In contrast to the widespread use of BMX bicycles, and the growing use of the mountain and city bikes, the recumbents are relatively insignificant.

The name "recumbent" covers a far wider range of possible configurations than the two types of bicycles discussed above, but the common feature is that the rider is fully supported in a seat and the pedals are out in front of the rider, rather than below as in all other machines (Fig. 5). Generally the seat back is at an angle similar to that in an automobile, in which case the bicycle is perhaps more correctly termed a semirecumbent, to distinguish it from the fully recumbent position adopted by some vehicle constructors when aiming for speed records.

Some recumbents follow the parallel with the automobile seat by having the handlebars, sometimes in the shape of a steering wheel, in front of the rider. In others, the handlebars are beneath the seat, a position which, it is felt by some, reduces the danger of injury in certain accident situations. Recumbents can also be classified into two types based on the position of the front wheel in relation to the pedal cranks: long-wheelbase models have the front wheel ahead of the cranks; short-wheelbase recumbents have a small wheel (16-in. or 400-mm diameter or less) behind the cranks and below the rider's knees. Most manufacturers seem to be choosing the long-wheelbase style with handlebars under the seat, but there are some major exceptions.

The modern recumbent is partly a revival, because bicycles that could fall into this category have been made at intervals since before the turn of the century. The revival has had two driving forces: safety and speed. The quite dramatic improvement in safety over conventional bicycles comes largely from the greatly reduced danger of spine and skull fractures that are frequently a result of over-the-handlebar falls. Spills from recumbent bicycles are usually made feet first. Other safety features are greatly improved braking, the near impossibility of catching pedals on the ground, and improved view of traffic situations because the rider's head is comfortably balanced in a natural position instead of cantilevered forward (in racing-style bicycles) in a position so uncomfortable for many that they relax frequently by looking down on the ground. The increased speed results principally from the lower frontal area. There is some disagreement about whether or not the recumbent pedaling position is ergonomically better or worse than the upright or crouched positions of conventional bicycles. There is no doubt that the position is far more comfortable, and speed records for bicycles have been held by recumbents since the late 1970s.

The most likely cause for the slow acceptance of recumbents seems to be the perception by potential manufacturers that the lower riding position would lead to a lack of visibility of the bike by other road users. Existing manufacturers of recumbent bicycles deny that there is a visibility problem, so long as the recumbent carries a bike flag; in fact it is maintained that with this flag the recumbent is far more easily seen than a conventional bicycle.

Tricycles and quadricycles. As with recumbents, the impetus behind the development of these alternative human-powered vehicles is sometimes the need for an improved practical or commuting vehicle, and sometimes the desire to reach record speeds. Both movements have been stimulated by speed trials and practical-vehicle contests conducted by the International Human-Powered-Vehicle Association.

The fastest speeds yet attained by human-powered vehicles were both attained with tricycles called Vector: the single-rider Vector achieved 58.89 mi/h (26.32 m/s) and the two-rider Vector 63 mi/h (28 m/s). The latter also traveled 42 mi (67 km) on a freeway at 50.5 mi/h (22.6 m/s). The success of the Vectors has resulted in the development of a large number of vehicles having similar specifications: an almost-recumbent rider facing forward, very close to the ground, and driving, through a chain routed below the cockpit, a standard bicycle rear wheel, with front wheels within the fairing having a limited degree of steering ability. The fastest single vehicle in the 1983 speed trials, Dragonfly, was, however, quite different: a prone, head-first rider operating linear slides with hands and feet. The speed was

Fig. 5. Avatar 2000 recumbent bicycle. (*NASA*)

below that of the Vector, but the conditions were different. (A streetworthy, short-wheelbase recumbent bicycle, Lightning X2, having flaps through which the feet could be thrust for starting and stopping, reached almost the same speed at the same trials.) At the Hull, Ontario, festival of 1984, the winner of the practical-vehicle competition was a tricycle called Windcheetah that has been approved in Britain for street use.

The battle between recumbent bicycles and tricyles for what their proponents hope will be the title as the successor to the conventional commuting bicycle has been joined. In the future, new human-powered vehicles can be expected that are easier to use, are more comfortable, have increased protection from the weather, have much improved safety, and are faster for the same effort.

For background information *see* INDUCTION MOTOR; TRANSPORTATION ENGINEERING in the McGraw-Hill Encyclopedia of Science and Technology.

[DAVID GORDON WILSON]

Bibliography: *1st* and *2d Scientific Symposium on Human-Powered Vehicles*, Proceedings of the International Human-Powered-Vehicle Association, Seal Beach, CA, 1981 and 1983; *Greater Vancouver ALRT System: Design Criteria Manual*, Metro Canada Limited, Urban Transportation Development Corporation, Ltd., Toronto, May 1982; C. R. Kyle, *Human Powered Vehicles*, 1983; F. R. Whitt and D. G. Wilson, *Bicycling Science*, 1982; D. F. Williamson, *A Transit Application for the Linear Induction Motor*, APTA Conference, Pittsburgh, June 1983.

Trichothecenes

Fungal metabolites such as alcohols, organic acids, and aldehydes are important natural sources for industry, being used in the production of chemical products in food processing, and in medicine. Few fungal metabolites exhibit toxic effects to animals

Structure of major trichothecenes.

and humans. One exception is aflatoxins, which are dihydrobisfuranoids of *Aspergillus flavus* and are presumed to be one cause of liver cancer. Toxic fungal metabolites are generally called mycotoxins.

Trichothecenes are a group of chemically related sesquiterpenoid compounds produced by fungi such as *Fusarium*, *Trichoderma*, *Myrothecium*, and *Stachybotrys*, and by higher plants such as *Baccharis*. During the past 20 years, more than 70 kinds of trichothecenes have been isolated from their metabolites. In the early stage of the trichothecene research, the first compound, trichothecin, an antifungal metabolite, was isolated from *Trichothecium roseum*. Crotocin, verrucarins, and roridins were also detected as antifungal agents. Diacetoxyscirpenol (DAS) was isolated from *F. scirpi* as a potent phytotoxic compound, and the scirpen ring system was adopted for these related compounds. The name trichothecane was proposed for the group of spiroepoxy-containing sesquiterpenoids.

Chemical characteristics. Trichothecenes possess the tetracyclic 12,13-epoxy-trichothecene skeleton belonging to sesquiterpenoids. Biogenetically, the ring system arises from cyclization of farnesyl pyrophosphate followed by two 1,2-methyl group shifts. More than 70 kinds of naturally occurring trichothecenes are tentatively divided into four categories according to the similarity of functional groups and producing fungi (see illustration). The first category is characterized by a functional group other than a ketone at C-8 (type A). This is represented by T-2 toxin, DAS, and others. The second category has a carbonyl function at C-8; nivalenol (NIV), deoxynivalenol (DON), and their esters are representatives of this group (type B). The third category is characterized by a second epoxide at C-7,8 (crotocin) or C-9,10 (type C). The last category includes those compounds containing a macrocyclic ring between C-4 and C-15 with two ester-linkages (type D). Verrucarins, roridins, satratoxins, and baccharinols are included in this category.

Food-borne diseases. The role of trichothecenes as an etiological agent in food-borne mycotoxicosis was first elucidated when T-2 toxin, a toxin metabolite of *F. tricinctum* (= *F. sporotrichioides*), was found in moldy corn associated with illness and death of cows in the United States in 1972. Alimentary toxic aleukia, a disease reported in Russia in the nineteenth century is also caused by T-2 toxin and related trichothecene mycotoxins. Another food-borne disease caused by the trichothecenes is scabby wheat intoxication found in the United States, Canada, Japan, Korea, China, and other countries. When *F. graminearum* colonizes wheat, trichothecenes such as NIV and DON are produced on the grains; these grains induce vomiting, refusal of feed, diarrhea, and hemorrhage in the intestine. A recent epidemiolgeical surveys, have revealed that NIV and DON are the major toxic pollutants in wheat, barley, corn, and other cereal grains.

Yellow rain. Since 1975, the United States government has received consistent reports detailing chemical attacks in Southeast Asia (Laos and Kampuchea). Some reports described the use of lethal agents which induced symptoms in humans that included vomiting, diarrhea, hemorrhage, and skin irritation, and even led to death. Since these symptoms best fit those induced by the trichothecenes, several chemists searched for the possible presence of trichothecenes in the samples of "Yellow Rain," which were presumed to have been used as chemical warfare agent in these areas. As a result of these studies, three trichothecenes (T-2 toxin, DAS, 4-acetyl-DON) and another *Fusarium* mycotoxin, zearalenone, were detected along with a synthetic material, polyethylene glycol. Furthermore, T-2 toxin and HT-2 toxin (a metabolite of T-2 toxin) were detected in human blood, urine, and body tissues.

While this evidence may seem strong, an alternative proposal is that "Yellow Rain" can be explained by natural phenomena. This purported agent of chemical warfare may be the feces of wild Asian honeybees, a good substrate for bioproduction of trichothecenes by *Fusarium* spp. Neither the chemical warfare hypothesis nor the bee-feces hypothesis can be conclusively shown to be correct.

Sources. The trichothecenes are secondary metabolities of a small number of genera of fungi, including (but not restricted to) *Fusarium*, *Trichothecium*, *Trichoderma*, *Myrothecium*, *Verticimonosporium*, *Stachybotrys*, and *Cylindocarpon*. Most of these fungi come from soil and are parasitic to plants. *Fusarium* is the most important genus: *F. sporotrichioides* produces T-2 toxin, HT-2 toxin, neosolaniol, acetyl-T-2 toxin, and other toxins; *F. graminearum* produces DON, 3-acetyl-DON, 15-acetyl-DON, NIV, 4-acetyl-NIV, and 4,15-diacetyl NIV.

Myrothecium verrucaria produces verrucarin A and B; *M. roridum* produces roridin A and D, and E. *Stachybotrys atra* (*S. alternance*) produces satratoxin F, G, and H.

The most important food-contaminating fungus of those listed above is *F. graminearum*. This fungus is widely distributed in wheat, barley, corn, millet, and other grains, and produces both the DON and NIV trichothecene families. In laboratory conditions with synthetic liquid culture media, 4-acetyl-NIV and 3-acetyl-DON are the major products; their alcohols (DON and NIV) are of minor significance. In natural conditions on cereal grains, however, this fungus produces DON and NIV. This difference in products can be explained by the initial production of the acetyl derivatives of DON and NIV; under natural conditions, some esterases originated from the fungus and cereal grains deacylate the esters into the parent alcohols, DON and NIV, which accumulate in the grains. Single-spore isolation techniques have revealed that *F. graminearum* can be subdivided chemotaxonomically into two groups: the producer of DON and 3-acetyl-DON, and the producer of NIV and 4-acetyl-NIV. There are some regional differences in the distribution of these two chemotypes.

While most of the trichothecene compounds are produced by fungi, several macrocyclic trichothecenes, the baccharins, are isolated from a higher plant, the Brazilian shrub *Baccharis megapotamica*. While researchers were screening extracts of higher plants for anticancer agents, an extract of *B. megapotamica* was found to contain a series of sesquiterpenes, the baccharinoids, which exhibited very high activity against a specific mouse leukemia. The active principle has been identified as baccharinol.

Extract of *B. megapotamica* also contains roridins D and E, which are the products of fungi *Myrothecium* spp. Therefore, there is a possibility that roridins produced by the soil fungus *Myrothecium* spp. are being taken up the plant, translocated to the leaves, and metabolized to baccharinoids. Thus these compounds may be products of an interaction between fungi and higher plants.

Toxicology. The major trichothecenes have been implicated in food-borne diseases throughout the world. Experimental studies of toxicology in laboratory animals have revealed the following symptoms as a common feature of the trichothecenes: vomiting, emesis, burning sensation in the mouth and esophagus, diarrhea, a marked decrease in leukocytes, petechial hemorrhages in chest, arms, face, intestine, and brain, and bleeding from the nose, throat, and gums. Histopathologically, cellular destruction and karyorrhexis are evident in the thymus, spleen, bone marrow, and the epithelia of the small intestine. Trichothecenes also cause immunological disorders: depletion of lymphoid cells in thymus, spleen, and bone marrow is often seen. Antibody formation, cell-mediated immune responses, graft rejection, delayed hypersensitivity, and mitogen-dependent proliferation of T cells are also greatly affected by the trichothecenes. Thus, the trichothecenes may be considered immunosuppressants.

Trichothecenes administered to animals are converted into several metabolites, and several enzyme systems participate in these transformations. In the liver, microsomal carboxyesterases attack the acyl residues of trichothecenes. For example, T-2 toxin is transformed to T-2 tetraol via HT-2 toxin, and DAS is metabolized into monacetoxyscirpenol (MAS). The hydroxylation of T-2 toxin and HT-2 toxin results in the formation of 3′-hydroxy T-2 toxin and 3′-hydroxy HT-2 toxin, respectively. In the case of DON, about 50% of DON in swine serum was conjugated as glucuronide.

Trichothecenes such as T-2 toxin and DON are teratogenic to mice, but short-term tests on mammalian cell cultures reveal no mutagenicity. Long-term feeding experiments with T-2 toxin and 4-acetyl-NIV (fusarenon-X) have resulted in some tumors in heart and several other tissues, but the incidence was too low to define the carcinogenicity of these trichothecenes in mice and rats.

The trichothecene mycotoxins are cytotoxic to eukaryotic cells, including fungi, protozoa, and cultured mammalian cells. Biochemical analysis has revealed that the trichothecenes inhibit protein synthesis by binding to the large subunits of eukaryotic ribosomes, and destroy the polyribosomal structure. This is one reason why the trichothecenes cause cellular damage in the actively dividing cells of thymus, small intestine, bone marrow, and testis.

For background information *see* AFLATOXIN; MYCOTOXIN; TOXICOLOGY in the McGraw-Hill Encyclopedia of Science and Technology.

[YOSHIO UENO]

Bibliography: L. R. Ember, Yellow rain controversy remains unresolved, *Chem. & Eng.* pp. 25–28, June 25, 1984; H. Kurata and Y. Ueno (eds.), *Developments in Food Science*, vol. 7; *Toxigenic Fungi: Their Toxins and Health Hazard*, 1984; R. T. Rosen and J. D. Rosen, Presence of four *Fusarium* mycotoxins and synthetic material in "Yellow Rain": Evidence for the use of chemical weapons in Laos, *Biomed. Mass Spectrosc.*, 9:443–450, 1982; Y. Ueno (ed.), *Developments in Food Science*, vol. 4: *Trichothecenes: Chemical, Biological and Toxicological Aspects*, 1983.

Turbine

Straight-flow turbines, which have the advantage of great simplicity of design, have reached an advanced stage of development. They are based on conventional machine components, so that any experienced bulb-turbine manufacturer should be able to produce them safely. Recent operating experience with these machines has been encouraging.

Concept. The straight-flow concept, proposed by L. F. Harza in 1919, makes possible horizontal-axis low-head hydro installations. Constructional problems have prevented straight-flow turbines from coming into widespread use in spite of their simplic-

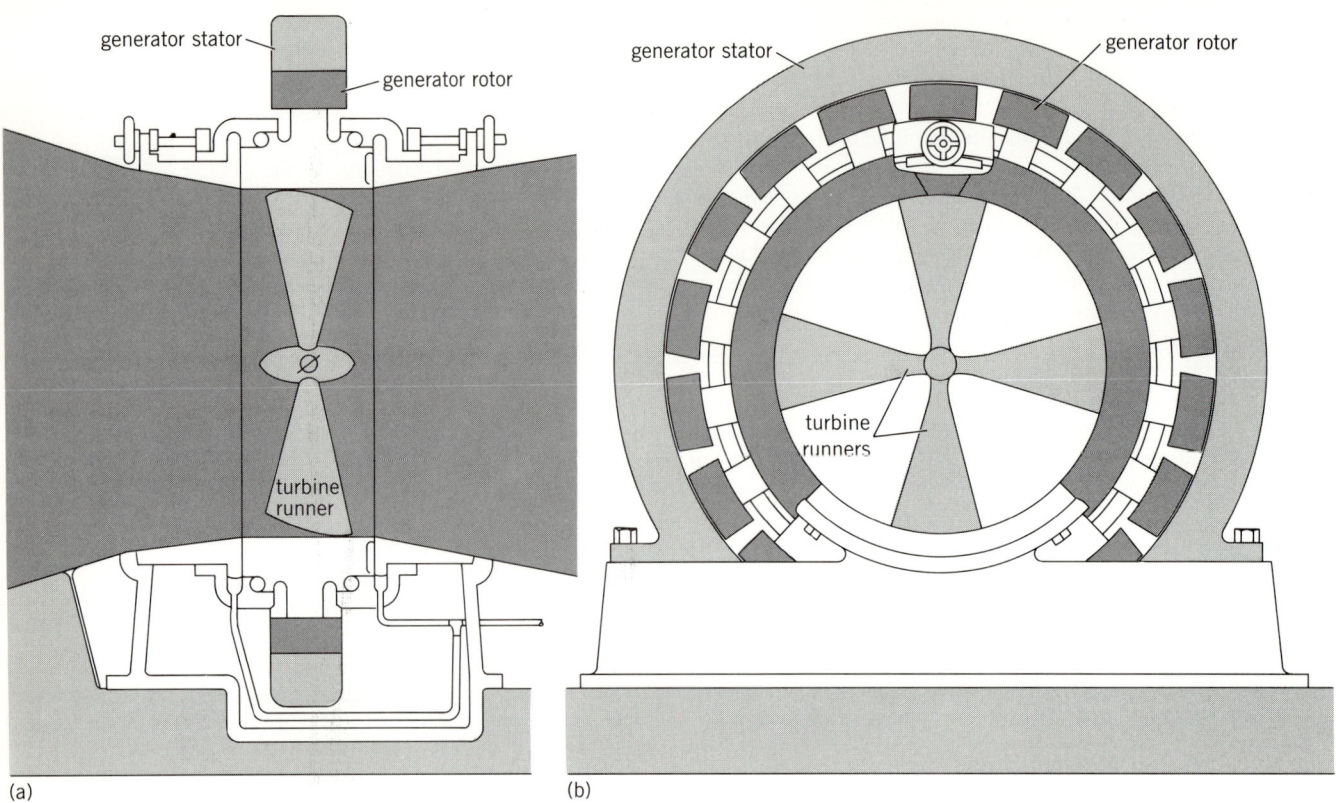

Fig. 1. Diagrams from patent application of L. F. Harza for a turbine in which the generator is placed directly on the periphery of a runner rotating in the tubular water passage. (a) Side view. (b) Cross section.

ity, although the concept has given rise to many alternative solutions.

As the runner blades of a turbine become more axial, its discharge capacity increases, and therefore the dimensions of the hydraulic passage become smaller. A design proposed by V. Kaplan utilized this idea, but his main contribution was to propose double-regulation between the guide vanes and runner blades, so that the flow, both upstream and downstream of the runner, is optimally guided over the whole range of operation. In Harza's straight-flow concept, not only the runner but the whole turbine unit is axial. That is, the upstream and downstream reservoirs are connected by a straight tube into which the turbine runners is integrated. This results in a compact low-head machine with a simple hydraulic passage, free from complicated configurations, such as spiral casings and elbow-bend draft tubes.

In Harza's concept, the generator is placed directly on the periphery of a runner rotating in the tubular water passage. A ring installed on the tips of the runner blades carries the generator poles. The rotor is supported on this ring radially with sleeve bearings and axially with water-fed hydrostatic bearings (Fig. 1). In order to protect the generator from humidity, seals are placed on both sides of the ring. The ring can also be used as a belt or gear drive.

Harza made further improvements by proposing regulated runner blades and transferring the rotor bearings inside the water passage with an overhung configuration.

Realization attempts. The turbine industry did not take up Harza's ideas immediately but "rediscovered" them many years later.

Iller, Lech, and Saalach turbines. Between 1937 and 1951, 73 turbines based on Harza's proposals were installed in 14 power stations on the Iller, Lech, and Saalach rivers (in Germany and Austria). The turbines have a runner diameter of 77–85 in. (1950–2100 mm) and are designed for a discharge of 700–900 ft^3/s (20–25 m^3/s). The rated heads vary between 26 and 30 ft (8 and 9.2 m). Rated outputs vary between 1000 and 1900 kW. The turbines have bearings on both sides of the runner.

All machines have one-piece steel-cast runners with fixed runner blades connecting the hub to the outer rim. A one-piece generator rotor is shrink-fitted to the outer rim.

Sealing between the rotating outer rim, in the water, and the fixed turbine casing is realized with lip seals. These are pressed by the water pressure against the outer rim whose surface is protected by an overlay of wear-resisting material (Remanit). Radial grooves on the seals through which river water passes ensure lubrication and cooling of the surface in contact with the seal. The duration of these seals is more than 20,000 h. They can be replaced from inside the water passage of the dewatered turbine.

The compactness of the turbine-generator unit allows the spillway to be built above the powerhouse. These submerged power stations are well integrated in the lanscape and were subsequently declared under environmental protection.

An attempt to install a runner with movable blades on one of the machines was not successful. The weight of the rotating outer rim and the internal forces created through thermal dilatation developed such forces on the connection between the blades and the outer rim that blade movement was impossible.

Since these power stations were to produce peak power, and water storage between barrages was possible, regulation of the runner blades was not necessary. In spite of this, the problems discussed above led to the abandonment of the outer-rim concept and to the development of tubular turbines with generators inside the water passage (bulb turbines).

Soviet turbines. In 1953, at Kura, Soviet Union, three straight-flow turbines with outer-rim generators were installed in the Ortatschlaskoj power station. Each had a runner diameter of 130 in. (3300 mm) and an output of 6300 kW under a head of 34.5 ft (10.5 m). The turbines were designed to have double-regulation. The concept resembled that of the Iller, Lech, and Saalach units. The runner had upstream and downstream supports, and the generator rotor was supported by the runner. Further, the outer-rim seals were of the same type as the earlier seals that could not cope with the axial movements of the runner. These machines were a total failure and had to be replaced by conventional ones.

Engineering study. From 1962 to 1970, a wide-ranging study was undertaken on the hydraulics, seals and bearings of straight-flow turbines. This work resulted in recommendations that the problems encountered with these turbines could be alleviated by using lip seals with wear-resisting plates on the rotor, bearings inside the hub, and a nonrotating shaft anchored on a concrete pier located in the turbine inlet. The proposed solution also included movable runner blades.

Straflo turbines. Beginning in 1973, the increasing importance of unused low-head potential and the physical limits of bulb turbines led to further development of straight-flow turbines with outer-rim generators. The first prototypes of this development are already in operation under the trade mark Straflo.

The experience with the units on the Iller, Lech, and Saalach rivers and at Ortatschlaskoj had led to the belief that there were very narrow limits on the diameter, head, and output of outer-rim-generator turbines. This limitation was overcome by equipping the generator rotor with a separate support and sealing system using hydrostatic bearings and seals.

A prototype featuring these elements was installed at Höngg, Zurich, and has been in operation since 1982. The runner diameter is 118 in. (3000 mm), the head 11.5 ft (3.5 m), and the output 1.5 MW. The unit is equipped with conventional bearings for the turbine rotor inside a bulb in the water passage and both axial and radial hydrostatic bearings on the outer rim of the generator. In order to prevent large forces that would result from even a minimal eccentricity between the turbine and generator rotors, the connection between the runner blades and the outer rim is made through tangential servomotors that secure an even distribution of load on the blades. In order to minimize radial deformations of the generator rotor surface that result from centrifugal forces and thermal influences, the position of the surface is controlled, and the necessary corrections are automatically made, through the hydraulically supported (hydrostatic) bearing elements.

The 15 hydrostatic support elements are supplied with a constant volume of filtered river water. Redundancy as well as a continuous energy supply are secured. The same is true of the hydrostatic seal of the rotor. An identical seal has been in operation since 1981 on one of the 16 Iller units.

Parallel to the so-called high-output Straflo with hydrostatic generator bearings, a concept with conventional inner bearings was further developed with special attention to the rotor dynamics and simple rotor seals. Corresponding prototypes have been in operation since 1980 in Andenne and Lixhe (Belgium), since 1983 in Weinzödl (Austria), and since 1984 in Annapolis Royal (Canada) [see table].

The Andenne, Lixhe, Annapolis Royal, and Aboisso units are equipped with turbines with adjustable wicket gates and fixed runner blades, whereas the turbines at Weinzödl have movable runner blades. With the exception of Aboisso, the bearings are on both sides of the runner with the upstream bearing acting as a combined thrust and guide bearing. In the case of Aboisso, the bearings are inside the runner hub, similar to the design proposed by the 1962–1970 study. The downstream support thus disappears with corresponding cost savings, improvement in efficiency, and shortening of the erection time.

With a diameter of 25 ft (7.6 m), the Annapolis Royal turbine is twice as large as any other straight-flow turbine with outer-rim generator; furthermore, this plant is the first tidal power plant in North America (Fig. 2). Special corrosion protection is

Properties of Straflo turbines with conventional bearings

Plant	Andenne	Lixhe	Weinzödl	Annapolis Royal	Aboisso
Runner diameter, in. (mm)	140 (3550)	140 (3550)	146 (3700)	299 (7600)	126 (3200)
Output, kW	3200	5530	8000	20,000	6980
Head, ft (m)	16.1 (4.9)	24.0 (7.3)	36.1 (11.0)	23.0 (7.0)	20.0 (6.1)
Speed, revolutions per minute	107.1	120.0	150.0	50.0	125.0
Number of units	3	4	2	1	2

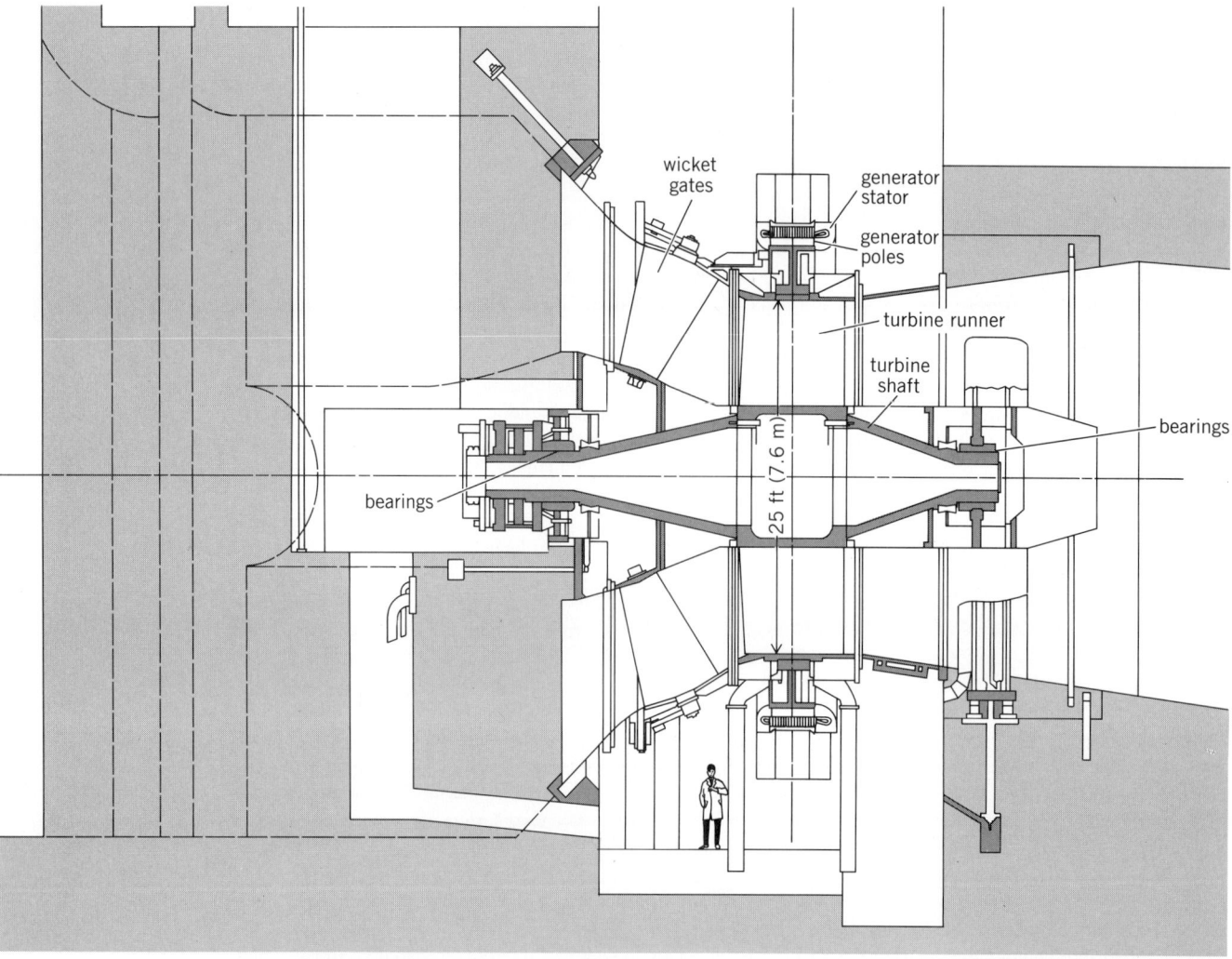

Fig. 2. Section through the turbine at Annapolis Royal. (*After Sulzer-Escher Wyss*)

necessary for operation with sea water. Treatment of cooling water and sea water also poses a considerable challenge.

The extremely low rotational speed of 50 revolutions per minute entails a generator with 144 poles. This large number of poles is also a first. Another characteristic of the Annapolis Royal turbine is the fact that the upstream turbine support, including the bulb, is made from concrete.

The Weinzödl turbines have movable runner blades. Unlike other turbines with axial movable runner blades, their runner blades are not held in the hub with a thrust bearing, but in the outer rim. This rim varies in diameter due to rotational speed and thermal effects, but the blades follow these radial movements.

In order to transmit the forces to the bearings inside the turbine, a connection between the outer rim, runner blades, hub, shaft, bearing, and support is necessary. This is achieved by means of a conical piece inside the hub which, through prestressed springs, presses the blades concentrically radially outward. The blades are less loaded than in an axial machine with movable runner blades. The blade-hub transition is unloaded through centrifugal forces and moments, and the blades are constantly under compression. This arrangement reduces the danger of fatigue cracks, a problem that is present with runner-regulated machines.

Experience with prototypes. Much has been learned from the operation of Straflo turbines, particularly with regard to the performance of the bearings and rotor seals.

Bearings. The Höngg installation has shown that turbines with the combination of conventional inner bearings, hydrostatic outer bearings, and position control are capable of operating. But the hydrostatic bearings do not operate satisfactorily. The deformation through age of the bearing supports made out of Robadur is considerably larger than what was expected. Because of the sophisticated protection system, however, no damage has occurred. The Höngg bearing system is expensive and requires considerable maintenance because of the necessary auxiliaries and control and protection systems.

The rotor dynamic calculations made in connection with the Andenne, Lixhe, Weinzödl, and Annapolis Royal turbines, as well as the measurements executed at site, have shown that with conventional inner bearings the rotor speed is well under the critical value, with a large margin on safety. Since the generator is not present in the turbine intake, the concrete intake pier can be extended up to the distributor, which results in a direct and very stiff bearing support. The bearing inside the runner hub renders a second bearing support in the draft tube redundant (Fig. 3).

In conclusion, direct support of the outer rim, which is possible only with hydrostatic bearings, has proven unnecessary. Simple solutions with inner bearings and an overhung design without downstream support are possible and economical within the foreseen range of application of Straflo turbines.

Runner regulation. The concept utilized at Höngg and at Weinzödl is fully operable. Blade regulation has caused no problems.

Rotor seals. The hydrostatic seals at Iller VII, Höngg, and Annapolis Royal are fully operable. Their disadvantage lies in the use of filtered water.

Experience with the considerably cheaper lip seals has shown that the correct form of the lip profile, efficient cooling of the friction surface, and the choice of material for the protection rings are important factors in determining success. While friction rings made from a normal stainless steel material (with 5% chromium and 13.4% nickel) have not yet proven themselves, friction rings made of Remanit stainless steel with 20% chromium and 14% nickel are still in operation in the Lech and Iller turbines after nearly 50 years.

An overlay with chromium oxide in the Lixhe turbines has proved itself, whereas in the Weinzödl turbines the overlay has partially disappeared.

Operating experience with aluminum oxide (Al_2O_3) and silicon carbide ($SeC+Si$) is satisfactory in bulb turbines, and corresponding results are available from the Weinzödl turbines, where the water of the Mur River is especially corrosive and erosive. Tests have shown that the resistance to wear of these so-called ceramic protection rings is 50–100 times higher than with a normal stainless steel material.

In conclusion, through correct material combinations it is possible to reach an acceptable lifetime, even with large machines and heads. By using silicon carbide, aluminum oxide, or chromium oxide segments, the wear is reduced to acceptable levels,

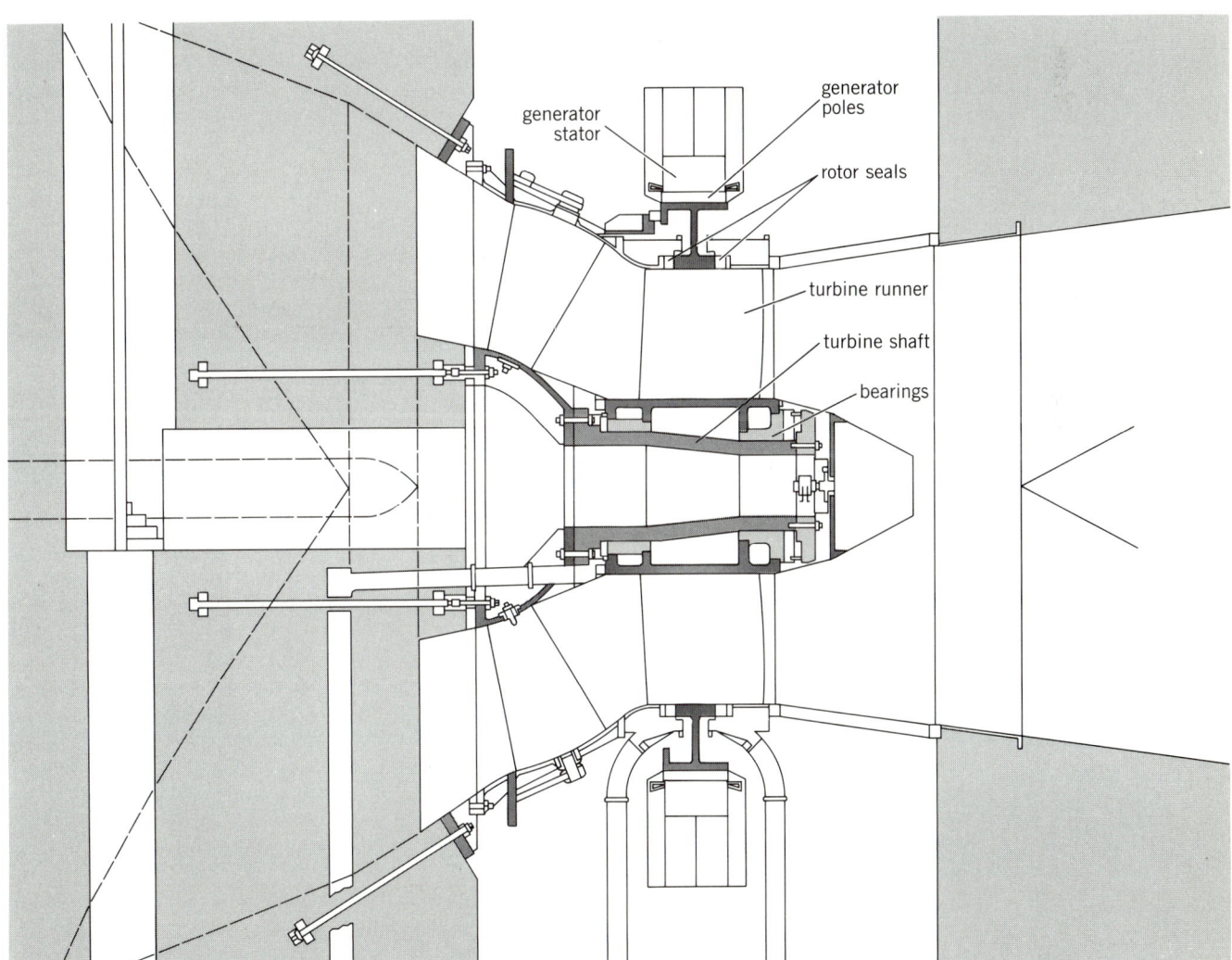

Fig. 3. Section through a Straflo turbine with bearing inside the runner hub. (*After Sulzer-Escher Wyss*)

even in water with unusually high sand content.

The lip itself must be accepted as a wearing element with a lifetime of 2–5 years. Care must be taken that the lips are easily accessible and can be quickly replaced.

Generator. With the exception of the Annapolis Royal unit, no problems have been encountered in generator construction. The problems with the Annapolis Royal generator were connected with the choice of insulation material for the poles.

For background information *see* HYDRAULIC TURBINE; TIDAL POWER in the McGraw-Hill Encyclopedia of Science and Technology.

[HELMUT MILLER]

Bibliography: H. Cardinal von Widdern, The tubular turbine, *Escher Wyss News*, 25/26:22–30, 1952/1953; M. Braikevitch, *The Straight Flow Turbine with Rim Generator: 8th World Energy Conference*, Bucharest, 1971; A. Douma, G. D. Stewart, and W. Meier, *Straflo-Turbine at Annapolis Royal, First Tidal Power Plant in the Bay of Fundy*, Escher Wyss Print 21.34.30, Zurich, 1982.

Turbocharger

Turbocharging has long been used to boost diesel engine power, and recently has proved a popular addition to passenger cars. But it has two drawbacks: turbo lag, the delay between depressing the accelerator and feeling the power that indicates turbocharger boost; and an inability to match the wide speed range of a car or truck engine. Engineers are now developing variable-geometry turbochargers as one solution.

Disadvantages. Conventional turbochargers consist of a tiny rotor inside a sturdy cast housing with a turbine wheel at one end of the rotor and a compressor wheel at the other. Exhaust gases leaving the engine are routed through a spiral passage in the turbo housing and onto the turbine blades, making the rotor spin. The compressor wheel then sucks in air-fuel mix (or simply air in the case of a diesel or fuel-injected gasoline engine) and propels it under pressure through another volute passage and into the engine. Squeezing more air and fuel into the engine increases power and torque, often by more than half.

However, since a turbocharger is essentially a steady-state device, working best at constant speed, it has difficulty meeting the varied demands of a typical engine operating cycle. If it is matched to low engine speeds so that low-velocity exhaust flow gives high turbo rotor speed and strong compressor boost, it will provide excessive boost at high engine speeds because pressure from the compressor increases with the square of the rotor's speed. Overboost can blow up an engine by imposing dangerous structural loads on reciprocating parts, or destroy a turbocharger by overspeeding. Additionally, at high engine speeds overboost can prevent the exhaust from escaping as freely as it would from a nonturbocharged engine, creating back pressure that makes the turbo absorb power instead of providing it. The conventional but inefficient answer to over-

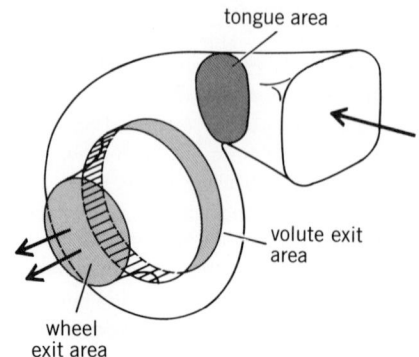

Fig. 1. Potential control areas in the turbine flow path of a turbocharger. (*After D. Flaxington and D. T. Szczupak, Variable area radial-inflow turbine, I.Mech.E. Tech. Pap. C 36/82, 1982*)

boost is to spill excess exhaust through a wastegate once a predetermined boost limit is reached. Conversely, a turbocharger engineered for high exhaust-gas speed and pressure provides negligible boost at low engine speeds.

Allied to the mismatch problem is poor transient response, or turbo lag. It arises because bearing friction and the rotor's own inertia prevent it from rapid acceleration or deceleration when needed; spinning often above 100,000 revolutions per minute (rpm) it is not easy to slow down or speed up. In trucks, where vehicle acceleration is slow, and in marine diesels and most construction equipment, this is of little consequence. But in cars, lag is quite definite, and in power generation sets it can be severe enough to make electrical output frequency jump from a steady 60 Hz during sudden loads. Optimized for the engine's midspeed operation, the turbo will be spinning too slowly at low engine revolutions to be able to pick up speed when required.

Variable geometry. One solution undergoing serious development is a ceramic rotor which is 60% lighter, and therefore has less rotating inertia, than the metal unit. Studies on air bearings show promise too. However, the approach that seems to offer most potential is variable-geometry turbocharging, where control elements sit in the entry passage to the radial-inflow turbine wheel.

The idea of variable-geometry turbocharing is to govern the speed of exhaust flow through the turbo so the rotor spins at a constant high speed irrespective of engine rpm. This is achieved by closing down the control elements at low engine speeds so the meager exhaust flow is accelerated through them, and opening up at high engine rpm so the flow impinges more slowly on the turbocharger turbine blades. A crude but simple analogy is pinching the end of a garden hose to force the water out faster.

Three areas of the turbo can be controlled: the tongue of the volute (the scrolled spiral passage feeding exhaust into the turbine wheel); the wheel exit, from which exhaust gas exits axially; and the volute exit, where incoming exhaust starts to hit the turbine blades (Fig. 1).

Wheel exit area control is equivalent to throttling and produces unacceptable losses; and tongue area elements, like flaps and sliding gates that cut off flow, result in lower reliability and controllability. Additionally, flaps can set up destructive vibrations in the turbine blades because they create asymmetric entry conditions for the exhaust gas.

Volute control, using a ring of 11 (or up to 17) pivoting aerofoil-shaped vanes around the periphery of the turbine wheel, is the most practical (Fig. 2). At low engine speeds, a control rod pulls on the unison ring connected to each vane; the vanes bunch up like venetian blinds to present a series of tiny throats, or nozzles, that accelerate the exhaust flow. At high engine speeds, a push on the control rod swivels the vanes open to prevent overboost. The control rod can be linked to engine speed or boost pressure. The optimum would be microprocessor control with inputs from engine speed, fueling rate, ignition timing, and manifold pressure.

The variable-area turbine nozzle is not a new design; it is borrowed from water turbine control in hydroelectric power stations, and from high-pressure-ratio centrifugal refrigeration compressors. It can be extended to the compressor wheel also, but added cost and complexity make this prohibitive for all but the most demanding jobs, such as in military tank engines, where power/size ratio is critical.

A prototype variable-geometry turbocharger has been installed in a diesel automobile and is reported to have cut lag—the time required to reach 1.6-bar (160-kilopascal) boost pressure—from 3.2 to 1.1 s. In addition, a 23% gain in power, from 175 to 215 hp (130 to 100 kW), has been reported. The U.S. Army's research and development wing, TARADCOM, has sponsored a program to investigate variable-geometry turbocharger performance gains on diesel tank engines. Fitted with variable nozzles on both turbine and compressor sides, a 750-hp (560-kW) 12-cylinder air-cooled diesel almost doubled peak torque ratings: from about 1770 ft lb (2400 newton-meters) at 2000 rpm to 3319 ft lb (4500 N-m) at 1600 rpm. Gross power output climbed to over 1208 hp (900 kW); specific fuel consumption fell by about 9% (Fig. 3). In Europe, a variable-area turbine nozzle unit is under development. Studies point to torque peaking at over 738 ft lb (1000 N-m) at 1500 rpm against the 664 ft lb (900 N-m) obtainable at 1400 rpm from the standard fixed-geometry turbo engine. The specific fuel consumption graph bottoms out lower too: 0.34 lb/(hp/h) or 205 g/(kW/h) for the variable-area turbine nozzle turbocharger against 0.35 lb/(hp/h) or 210 g/(kW/h) for the standard unit. Noise is also reduced. Variable-area turbine nozzle turbochargers have also been shown to cut down diesel smoke.

Future development. Among the disadvantages of variable-geometry turbochargers are that the moving vanes bring a drop in efficiency because they augment exhaust-gas leakage past their tips and that pivot points are prone to seize, particulary during heat transients. These and other defects must be eliminated before problems of cost can be resolved.

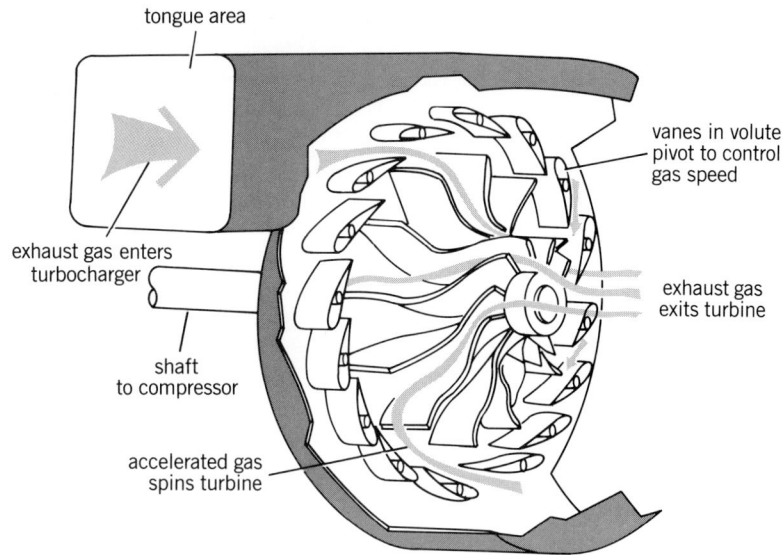

Fig. 2. Cutaway of turbine side on variable-geometry turbocharger, showing exhaust-gas flow through the movable vanes. (*From H. John, Advanced turbos poised to hit the streets, High Technol., 4(5):21–22, 1984*)

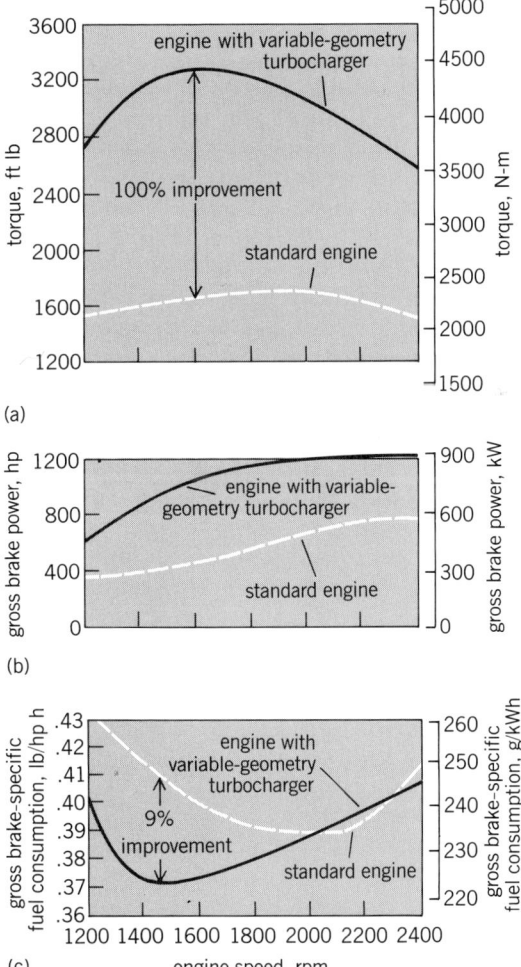

Fig. 3. Performance improvements in (*a*) torque, (*b*) brake power, and (*c*) gross on a military tank diesel engine fitted with variable-geometry turbocharger. (*After S. G. Berenyi and C. J. Raffa, Variable Area Turbocharger for High Output Diesel Engines, SAE Tech. Pap. Ser. 790064, 1979*)

It is anticipated that the auto industry will soon want more advanced turbochargers. In a few years, the car market will be able to absorb three to four times the number of turbochargers supplied to the relatively static commercial diesel sector; in 1983 the split was 50/50. There is some discussion over which type of advanced turbo will be adopted. If no breakthroughs in mechanical supercharging take place, variable-geometry turbochargers would seem to have the advantage over other designs. Considerable laboratory testing has been carried out, but little on-highway, where 2 years is a minimum.

Prospects may be best in the truck industry, where torque back-up is more important than lack of turbo lag. This is particularly true in Europe, where duty cycles call for considerable gear shifting. While a variable-geometry turbocharger may be accepted in a European truck by 1987, it appears that no one advanced turbocharger will suit all applications.

For background information *see* DIESEL ENGINE; GAS TURBINE; INTERNAL COMBUSTION ENGINE; TURBINE in the McGraw-Hill Encyclopedia of Science and Technology.

[JOHN KERR]

Bibliography: J. R. Arvin and N. L. Osborn, *Design Features and Operating Experiences of the Aerodyne Dallas VATN Turbocharger*, SAE Tech. Pap. Ser. 830013, February-March 1983; S. G. Berenyi and C. J. Raffa, *Variable Area Turbocharger for High Output Diesel Engines*, SAE Tech. Pap. Ser. 790064, 1979; H. John, Advanced turbos poised to hit the streets, *High Technol.*, 4(5):21–22, May 1984; F. J. Wallace, R. J. B. Way, and A. Baghery, *Variable Geometry Turbocharging: The Realistic Way Forward*, SAE Tech. Pap. Ser. 810336, February 1981.

Turbulence

Liquid helium (^{4}He) has two phases. The higher-temperature phase, which is called helium I, exists at the saturated vapor pressure above 2.172 K (−455.76°F). Helium II, which exists from 2.172 K to absolute zero (−459.67°F), exhibits superfluidity: the fluid can pass easily through microscopic cracks without resistance; it can conduct immense quantities of heat; and it supports second sound, a wave motion corresponding to temperature fluctuations. From a hydrodynamical point of view, helium II can be considered a mixture of a normal fluid of density ρ_n and a superfluid with density ρ_s, which can flow independently.

Observation of quantum turbulence. Consider the channel immersed in helium II shown in Fig. 1*a*. If the heater at the closed end is turned on and generates a small heat flux W, the superfluid will flow at velocity v_s toward the heater and the normal fluid will flow at velocity v_n toward the exit: no net mass of fluid flows, but a strong jet of normal fluid coming out the exit can turn a paddle wheel. Experiments show that the entire heat content of helium II is carried by the normal component, and a channel such

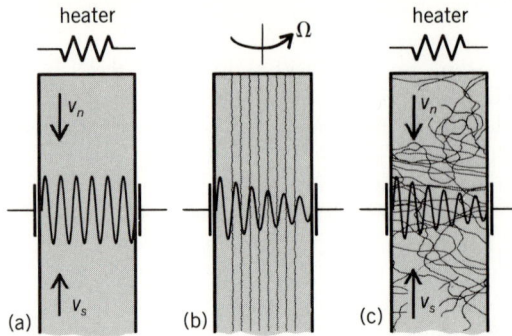

Fig. 1. Wide-channel counterflow experiments. (*a*) Heat flux produced by heater is subcritical, producing smooth counterflow. (*b*) Heater is turned off and channel is rotated at an angular velocity Ω to produce a uniform array of quantized vortices. (*c*) Heat flux is supercritical, producing tangle of quantized vortices.

as that shown has an enormous thermal conductivity because of this internal counterflow. If the heater is switched on and off rapidly, the periodic counterflow of the two fluids becomes a standing second-sound wave.

Another property of the superfluid is shown in Fig. 1*b*, where the heater is off and the channel is rotated about its long axis. The superfluid is known to rotate by creating a regular array of quantized vortex lines. Each line has a core size of atomic dimensions, and each atom of helium circulating around a vortex has angular momentum $h/2\pi$, where h is Planck's constant. For an angular velocity of 1 revolution per second, there are about 12,600 lines in a channel of cross section 0.4 × 0.4 in. (1 × 1 cm). A second-sound wave transmitted across the channel will be greatly attenuated by the presence of these quantized vortices.

The simple counterflow of Fig. 1*a* ceases when a certain critical heat input W_c is exceeded (about 14 mW/cm^2 at 1.65 K or −456.70°F). The situation is illustrated in Fig. 1*c*, where second-sound attenuation reveals that quantized vortices are present in the channel even without rotation. Experiments by W. F. Vinen in the 1950s established that the superfluid is filled with a tangled mass of quantized vortex lines, and the study of the properties of this unique flow is now known as quantum turbulence. The turbulence problem in classical fluids is a very important and largely unsolved problem. Quantum turbulence is a rather different phenomenon which is becoming a research field in its own right.

Channels such as are shown in Fig. 1 become the "wind tunnel" to produce quantum turbulence. The range of sizes of channels under study is enormous: from about 0.4 in. (1 cm) to 4 × 10^{-5} in. (1 micrometer), or 10,000 to 1. More general flows can also be established, including cases where only the normal fluid or only the superfluid is in motion relative to the channel walls.

Nature of quantum turbulence. The nature of the turbulent flow in wide channels is beginning to be understood. The density of vortex lines is measured in terms of length of line L per unit volume; that is, the dimensions of L are cm/cm^3 = cm^{-2}. In equilibrium, L is observed to be uniform down the channel and over at least 80% of the width of the channel. The normal fluid velocity v_n is also uniform across the channel, in contrast to the parabolic flow profile of ordinary viscous flow. The mean superfluid flow must also be, on average, uniform across the channel.

The variation of line density with applied heat flux is shown in Fig. 2. One way to represent the flow is in terms of the average relative velocity V_{ns} of the two fluids produced by a given heat flux W. This flux is given by the equation below, where S is

$$V_{ns} = (v_n - v_s) = \frac{W}{\rho_s ST}$$

the entropy of the fluid and T the absolute temperature measured in kelvin degrees (K). There is no line density when V_{ns} is less than a critical value V_c, equal in this case to 0.08 in./s (0.2 cm/s). At higher heat fluxes, L is proportional to V^2_{ns}. These features were first noted by Vinen.

Vinen assumed that the vortex tangle, being created from the superfluid, would move with it toward the heater. Various experiments have been designed to measure the drift of the tangle; the results are still somewhat in doubt, although measurements support a slow drift of the tangle toward the heater. Since there is no way to visualize the flow of liquid helium, the methods used are of necessity rather indirect.

It is possible to measure the average line density L both across the flow and along the channel axis. Measurements of this type indicate that the line density is considerably larger as seen up and down the tube, compared to across it. Thus the line-density distribution is flattened in a direction perpendicular to the axial flow.

Measurement of local properties. One of the most important measurements in classical turbulence is the local velocity field and its fluctuations. This is facilitated by local probes, such as hot-wire anemometers, which are usually small compared to the length scales being studied. In quantum turbulence, however, no comparable local probe has yet been devised. Second-sound detectors average the line density over a cross section of the channel, and cannot detect the direction of individual vortex lines. Thus, in some sense, second sound measures the absolute value of the vorticity averaged over a section of the channel, and hence is a measure of the enstrophy (the enstrophy is the square of the vorticity). With such a large-scale probe, it is clear that intrinsic fluctuations of the turbulence due to individual vortex lines will tend to average out and to depend upon the size of the volume studied. In spite of the difficulties, intrinsic fluctuations were observed in 1977 and confirmed later. The results are not entirely clear and not easy to interpret. An alternative technique giving different but decisive information in the induced-fluctuation method,

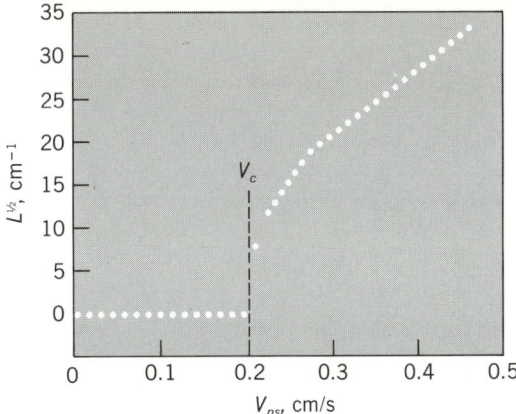

Fig. 2. Variation of square root of line density L as function of average relative velocity V_{ns} in a wide channel. Temperature of this experiment is $T = 1.65$ K ($-456.70°$F).

where second sound probes the response of the vortex tangle to white noise fluctuations imposed upon the heater power. In the 1950s, Vinen proposed simple phenomenological equations to describe the buildup, decay, and equilibrium of the vortex tangle. The induced-fluctuation technique verifies the predictions of Vinen's equations for the response of the vortex tangle to fluctuations of the counterflow velocity induced by noise on the heater.

Effects of rotation. Experiments in fluid mechanics are often strongly affected by rotation. Quantum turbulence displays a rich set of phenomena in rotation. The relative influence of heat and rotation can be characterized by the amount of vortex line generated by each alone: the line generated due to rotation L_R (Fig. 1b) and that due to heat L_H (Fig. 1c). In the limit of fast rotation and small heat flux, any rotation eliminates the critical velocity seen in Fig. 2. Instead, two new critical velocities appear. The first appears to be due to instabilities created on the vortex lines by the counterflow. The second appears to be due to a transition to turbulence. In the opposite limit of slow rotation and large heat, the effect of rotation is not simply to add line density to the tangle. It appears that the tangle polarizes itself in such a way as to rotate. In both limits, various experiments "scale" in terms of the dimensionless ratio L_R/L_H.

Prospects. There are many avenues to explore in quantum turbulence, some of which are just beginning to be investigated. For example, second-sound shock waves have been studied and observed to produce quantized vorticity after passage. These effects, too, are modified by rotation. A microscopic theory of the turbulent tangle has been developed. This computer simulation gives valuable insight into the dynamics of the vortex tangle.

For background information see LIQUID HELIUM; QUANTIZED VORTICES; SUPERFLUIDITY; TURBULENT FLOW in the McGraw-Hill Encyclopedia of Science and Technology. [RUSSELL J. DONNELLY]

Bibliography: J. G. M. Armitage (ed.), *75th Ju-*

bilee Conference on Liquid Helium, 1983; D. F. Brewer (ed.), *Progress in Low Temperature Physics*, vol. 8, 1982; K. W. Schwarz, Generation of superfluid turbulence deduced from simple dynamical rules, *Phys. Rev. Lett.*, 49:283–285, 1982; K. W. Schwarz, Turbulence in superfluid helium: Steady homogeneous counterflow, *Phys. Rev.*, B18:245–262, 1978; W. F. Vinen, Mutual friction in a heat current in liquid helium II, *Proc. Roy. Soc.*, A242:493–515, 1957.

Underwater sound

The speed of sound in the ocean is characterized by a weak but nevertheless important dependence on position. As a result of this inhomogeneous sound speed structure, acoustic waves propagate through the ocean along curved trajectories. A consequence of the curved trajectories is that in some regions acoustic-wave energy converges while in other regions it diverges. The focal regions (caustics) are of particular interest in studies of underwater acoustics since high intensities are predicted and observed in such regions. Catastrophe theory shows that caustics take on only certain forms and, when combined with certain mathematical expressions for the form of the acoustic wavefield, provide a quantitative description of the wavefield in the vicinity of the caustic.

The sound speed structure in most of the world's deep ocean is characterized by a minimum at a depth of roughly 1 km (about 0.6 mi). Above the minimum, sound speed increases because of the influence of increasing temperature. In the nearly isothermal water below the minimum, sound speed increases with depth due to the influence of pressure. A simple model that contains this important feature is one in which the sound speed increases linearly both above and below the sound channel axis and is independent of lateral position (Fig. 1). The real ocean is more complicated, of course, as its properties vary both vertically and laterally over a wide range of length and time scales. For the purpose of demonstration, however, a simple static model is helpful.

Formation of caustics. In many commonly encountered situations in underwater acoustics, variations in the sound speed over a wavelength of the acoustic wave are negligible. (Variations in the sound speed in Fig. 1 throughout the entire ocean depth are at most a few percent of the mean.) Under this condition the propagation of acoustic waves is governed by the rules of geometric acoustics. According to these rules, acoustic waves in an inhomogeneous ocean propagate along curved trajectories, called rays, in accordance with Snell's law (Fig. 1). Because the rays are curved, neighboring rays intersect in regions called caustics. More generally, a caustic is the envelope of a family of rays. According to the rules of geometric acoustics, the intensity of the acoustic wavefield is everywhere inversely proportional to the cross-sectional area between neighboring rays. At caustics this cross-sectional area vanishes, and infinite intensities are predicted. The latter does not happen in reality, of course, but high intensities are observed. The practical importance of caustics in applications stems from this fact.

Application of catastrophe theory. The introduction of catastrophe theory to underwater acoustics is useful for two reasons. First, it leads to an increased understanding of acoustic-wave propagation in focal regions on the conceptual level. It provides a framework within which such problems can be discussed and comparisons made to other fields of study. Second, due to the fact that classification of the catas-

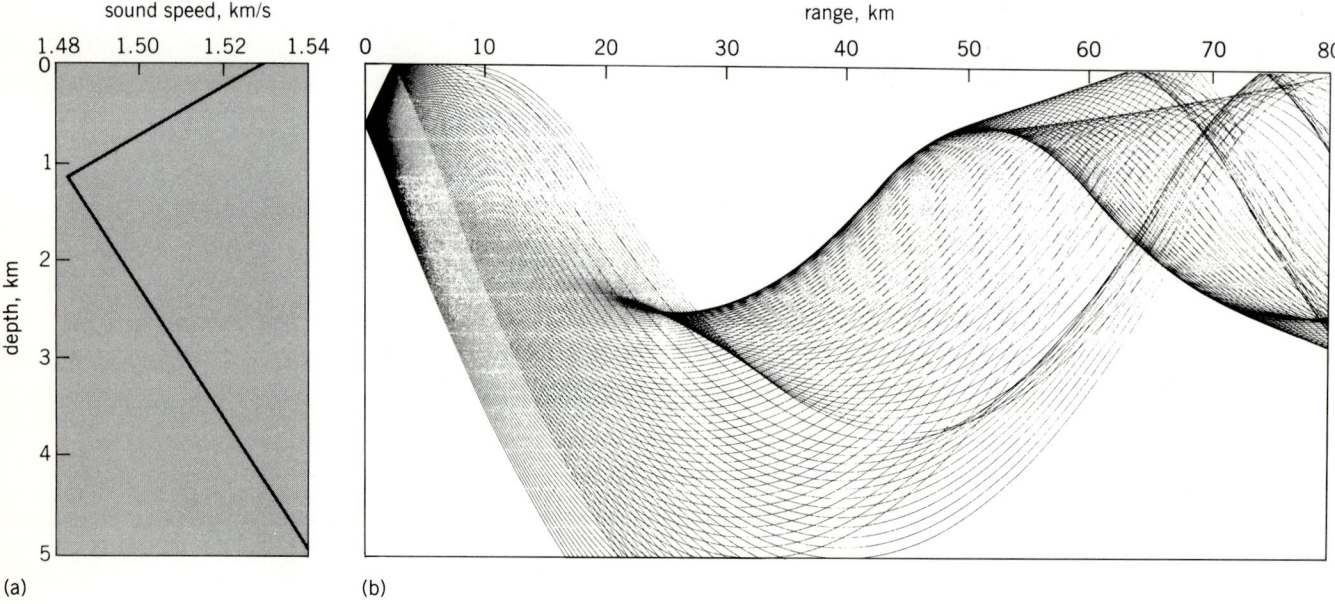

Fig. 1. Ocean sound propagation. (*a*) An idealized deep-ocean sound speed profile. (*b*) Corresponding acoustic-ray trace diagram for a shallow source. 1 km = 0.6 mi.

trophes is done quantitatively, the results of catastrophe theory may be combined with simple expressions for the acoustic wavefield to yield quantitative predictions about the structure of the wavefield in the vicinity of any caustic.

Classification of caustics. Catastrophe theory, a branch of mathematics concerned with structurally stable singularities of a certain type of mapping called a gradient map, provides a means to classify caustics. Fermat's principle, expressing travel time stationarity of the rays of geometric acoustics, defines such a map and thus provides the bridge between catastrophe theory and the caustics in acoustic-wave propagation. (These ideas hold for a much larger class of wave propagation problems.) Catastrophe theory shows that structurally stable caustics take on only certain forms—those of the catastrophes. Structural stability implies that the form of a caustic survives small perturbations to the ocean sound speed structure. Some of the most simple catastrophes are shown in Fig. 2; the similarity between the form of the caustics in Fig. 1 and the catastrophes in Fig. 2 is apparent. The cusp catastrophe is most easily identified.

Quantitative predictions. When combined with certain mathematical expressions for the form of the acoustic wavefield (these mathematical expressions, like geometric acoustics, are valid when the ocean's sound speed is approximately constant over a single wavelength), the results of catastrophe theory give testable, quantitative predictions about the wavefield structure in the vicinity of any caustic. Herein lies the utility and practical importance of catastrophe theory in underwater acoustics. An example of two different views of such a wavefield is shown in Fig. 3. These correspond to the fold caustic which separates a region in which there are two rays from another in which there are no rays. Many fold caustics can be seen in Fig. 1. By optical analogy, the two-ray and zero-ray regions are called the illuminated and shadow regions, respectively.

Figure 3a shows variations in the wavefield generated by an acoustic source of fixed frequency (such as a submarine propeller rotating at a constant speed) through a fold caustic. This figure shows only variations in the peak amplitude of the wavefield. The time dependence at all points is sinusoidal, with frequency identical to that of the source. In the illuminated region (negative x), waves corresponding to the two rays intefere with each other. Peaks and nulls correspond to places where constructive and destructive interference takes place, respectively. Peak amplitudes increase near the caustic ($x = 0$) as would be expected from the rules of geometric acoustics. The amplitude does not increase without bound, however, as predicted by geometric acoustics. Amplitudes decay smoothly in the shadow region (positive x).

A different view of the wavefield in the vicinity of the fold caustic is shown in Fig. 3b. This corresponds to the wavefield generated by an impulsive source (such as an underwater explosion). Here the

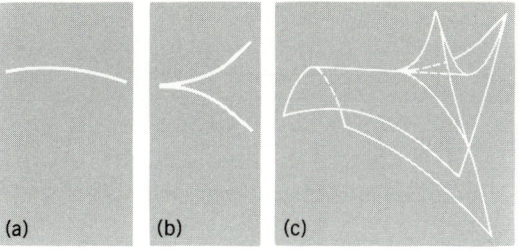

Fig. 2. Catastrophes. (*a*) Fold. (*b*) Cusp. (*c*) Swallowtail.

time dependence is shown explicitly for a series of values of x. On the illuminated side of the caustic the wavefield consists of two pulses with different shapes arriving at different times. Each is associated with a ray, one of which has touched the caustic. The pulse associated with this ray has been distorted (Hilbert-transformed) as a result of touching the caustic. The two pulses converge at the caustic ($x = 0$). Again, the wavefield is seen to decay smoothly in the shadow region. An interesting feature of this figure which has been appreciated only recently is that the times of arrival of the geometric rays on the illuminated side of the fold caustic form a cusp catastrophe in space and time, as may be seen by comparing Figs. 2b and 3b. The general result for waves in two space dimensions is that the times of arrival of the geometric rays in the vicinity of any caustic (catastrophe) form a catastrophe in space and time whose dimension is one higher than that of the caustic itself. The wavefields shown in

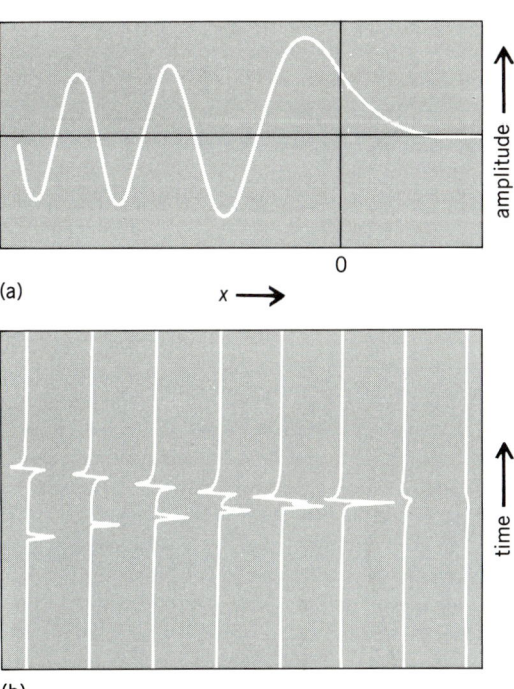

Fig. 3. Acoustic wavefield in the vicinity of a fold caustic for (*a*) a harmonic source and (*b*) an impulsive source. The *x* coordinate measures distance along a line perpendicular to the caustic, with the caustic itself located at $x = 0$.

Fig. 3a and b are closely related (by a Fourier transform) as both describe the acoustic wavefield in the vicinity of a fold caustic. The difference in appearance of the two is due to the different nature of the associated acoustic-wave–generating mechanisms.

Thus, the catastrophes appear naturally in underwater acoustics in two related contexts. The caustics of geometric acoustics take the form of the catastrophes as do the time histories of the acoustic wavefields in the vicinity of the caustics.

For background information see REFRACTION OF WAVES; UNDERWATER SOUND in the McGraw-Hill Encyclopedia of Science and Technology.

[MICHAEL G. BROWN]

Bibliography: M. V. Berry and C. Upstill, Catastrophe optics: Morphologies of caustics and their diffraction patterns, Prog. Opt., 18:257–346, 1980; Allan D. Pierce, Acoustics: An Introduction to Its Physical Principles and Applications, 1981.

Vaccine

The prevention of viral diseases by active immunization with inactivated or attenuated viruses has been critical to the control of such maladies as smallpox, rabies, yellow fever, and poliomyelitis. Unfortunately, many viral diseases remain beyond the reach of conventional vaccine technology. Further progress is predicated upon a thorough understanding of the molecular basis of antigenicity and virulence. However, the recent application of recombinant deoxyribonucleic acid (DNA) technology, computer simulation, and biopolymer sequencing and synthesis now permit the detailed analysis of viral structure, function, and host interactions. From these studies is emerging a new generation of vaccines, which are described below, using herpes simpex virus as an example.

Conventional vaccines. Conventional "killed" vaccines consist of virus particles whose infectivity has been destroyed with chemicals. These inactivated virus particles retain the ability to elicit immune responses and to protect the host from infection. The protective immunity that arises subsequent to vaccination results from host reactions to the virus surface proteins, which often represent only a fraction of all the proteins present in the virus particles. These host reactions include the synthesis of several classes of antibodies, proliferation and functional activation of various kinds of lymphoid cells and macrophages, and the production by these cells of interferons and other lymphokines.

It appears that there are distinct though overlapping groups of viral antigens responsible for stimulating antibodies and cell-mediated immunity. Neutralization of virus infectivity occurs when restricted molecular domains on the proteins in the external shell or surface of virus particles, called antigenic determinants or epitopes, are bound by antibodies. Neutralizing antibodies may interfere with absorption of a virus to the cell, or may induce conformational changes which interfere with virus penetration and uncoating. Antibodies induced by viral antigens may also bind to viral proteins on the surface of infected cells, thereby eliciting cytotoxicity, which may be antibody-dependent, cell-mediated or antibody-dependent, complement-mediated.

Structural analysis of antigenic determinants of viral proteins has shown that they are often hydrophilic, solvent-accessible regions where loops and turns in the polypeptide backbone occur. Furthermore, antigenic determinants correspond to those polypeptide segments with greater degrees of conformational flexibility. Defining at a molecular level those components of viral particles that elicit reactive antibodies may permit the design of vaccines free from residual surviving virus or other irrelevant components which may interfere with safety or efficacy.

Herpes simplex virus. Herpes simplex is a medium-sized virus, 150–200 nanometers in diameter, with one double-stranded DNA molecule of about 150 kilobases (kb) that is enclosed in a protective protein shell (capsid). The capsid together with the DNA, or nucleocapsid, is about 100 nm in diameter and is surrounded by a lipid-containing envelope. The viral envelope includes four major glycoproteins, designated gB, gC, gD, and gE, which induce neutralizing antibodies and cell-mediated immune responses.

Herpes simplex virus causes oral, ocular, genital, and neuronal infections. The incidence of genital herpes infections has been increasing in recent years. Studies of the sexual transmission of genital herpes virus reveal that many persons who transmit the disease are either asymptomatic or have unrecognized symptoms at the time of transmission. Since antiviral drugs cannot eradicate established latent herpes virus infections, herpes virus vaccines have a potential role in disease prevention. Concerns regarding the potential oncogenicity of herpes virus DNA, and the evidence that humoral and cell-mediated immune responses are directed largely against the virus glycoproteins, have prompted the development of glycoprotein subunit vaccines free of DNA as the first approach to prevention.

Subunit vaccine. One type of herpes virus subunit vaccine has been derived by stripping the viral envelope from the other constituents by detergent extraction. The combined virus envelope glycoproteins have been further purified by affinity chromatography and, when administered to human volunteers, elicited neutralizing antibodies and cell-mediated immunity. There appear, however, to be qualitative as well as quantitative differences in the immunity induced with the various glycoproteins, and a truly effective vaccine for as complex a virus as herpes simplex may require an experimentally tested balance of the individual envelope glycoproteins.

Considerable attention has been directed at characterizing the structure of one herpes virus glycoprotein, namely, glycoprotein D (gD), since both oral (type 1) and genital (type 2) strains of herpes virus have substantial antigenic cross-reactivity with this protein compared to the others. Purified gD's from either type 1 or 2 herpes virus stimulate high titers of type-common virus neutralizing antibodies that mediate cytotoxicity. Glycoprotein D (molecular weight 59,000 daltons) also plays a central role in absorption of the virus to the cell. The gene encoding gD has been mapped on the herpes virus DNA, and the entire DNA sequence has been determined.

Synthetic vaccine. Utilizing the universal genetic code, the amino acid sequences of the types 1 and 2 herpes virus gD polypeptides have been predicted from the DNA sequences. It appears that the gD polypeptides consist of 393 amino acids and that each polypeptide contains three glycosylation sites. By using a combination of proteolysis, computer predictions, synthetic peptides, and panels of type-common and type-specific monoclonal antibodies, major antibody binding sites (epitopes) have been determined. Amino acid residues 8–23 define one such major epitope. Computer model predictions of other antigenic determinants based on hydrophilicity and secondary structure as well as the use of synthetic peptides and antibody reactions has also identified residues 268–287 and 340–356 as conformational epitopes. Thus, peptides containing amino acid residues 8–23, 268–287, and 340–356 coupled to protein carriers are potential vaccines.

Synthetic polypeptides. The use of synthetic polypeptides as vaccines has been advocated because of their stability, purity, and low production costs. In addition, synthetic peptides may elicit neutralizing antibodies to antigenically conserved regions which are not, when part of the full viral protein, capable of stimulating antibodies. This could permit immunization against a wide variety of viral serotypes. Studies are currently under way to ascertain the level of protective immunity produced by synthetic herpes virus polypeptide vaccines. Analogous experiments with synthetic polypeptide vaccines for hepatitis B, foot-and-mouth disease, and influenza have demonstrated the potential usefulness and pitfalls of this approach.

Chimeric DNA. Cloning of the herpes virus gD gene in bacterial plasmids has facilitated the construction of chimeric DNA, which when placed under the control of high-efficiency gene expression systems results in the production of large amounts of gD, either as the authentic protein or as fusion of truncated forms. Recombinant DNA–derived gD produced in *Escherichia coli* has been found to be poorly immunogenic, probably due to incorrect protein folding occurring in the absence of a eukaryotic cell membrane. However, when a plasmid vector, pgD-DHFR, was used for transfer to the gD gene to Chinese hamster ovary cells, abundant quantities of an immunogenic form of gD were obtained. The vector pgD-DHFR consisted of *E. coli* plasmid pBR322 with inserted herpes virus gD and mouse dihydrofolate reductase genes under the control of simian virus 40 early promoters. In selective media containing aminopterin, the herpes virus gD and the mouse dihydrofolate reductase genes were stably cotransferred in an amplified form. The Chinese hamster ovary cells were sensitive to aminopterin toxicity under ordinary tissue culture conditions, but became aminopterin-resistant, provided that the mouse DHFR gene of the plasmid was expressed. Immunization with this animal cell–derived gD protected mice from lethal challenge with herpes virus.

Hybrid vaccines. Hybrid vaccines have also been constructed in which the gD gene of herpes virus is inserted into the genome of vaccinia, the smallpox vaccine virus. Glycoprotein D is produced in the host during the reproductive cycle of vaccinia by tricking the virus to recognize the herpes virus gene as its own. Vaccination has been shown to result in the production of neutralizing antibodies, and a protective immunity against subsequent exposure to herpes virus occurs. Also, because of the large size of vaccinia, up to a dozen other foreign genes might be incorporated into a single vaccine capable of providing protection to a wide variety of infectious diseases.

Modified live virus vaccine. Attenuated or modified-live virus vaccines are crippled or nonpathogenic viruses derived from virulent field strains which impart immunological cross-reactivity. Modified-live virus vaccines are generally regarded as superior to killed or subunit vaccines because they elicit more complete and balanced immunological responses. Conventional attenuated vaccines have been obtained by repeated passage of virus in cells of a species different than the natural host. During multiple passages in tissue culture cells, mutations accumulate as the virus adapts to its new environment. These mutations adversely affect viral reproduction in the natural host, resulting in lessened virulence. Unfortunately, this process may sometimes be reversed, so that vaccine virus may revert to a virulent state. Furthermore, attenuated virus vaccines may be unsafe in young, atopic, or immunocompromised individuals.

With regard to a specific modified-live virus vaccine for genital herpes, the additional concern exists that genital herpes infection increases the risk of cervical carcinoma. Besides numerous epidemiological studies correlating cervical carcinoma with herpes virus infection, tisue culture studies have demonstrated that herpes virus DNA and fragments of this DNA can transform rodent cells to neoplasia. Attempts to show that particular viral gene products are associated with transformed cells or tumors have failed to identify any tumor-specific viral proteins, as has been the case with adenoviruses and papovaviruses. These findings have led to speculation that the oncogenicity of herpes virus may operate through novel mechanisms. The "hit-and-run" hypothesis suggests that the virus alters the cell to a neoplastic or preneoplastic state that does not require the continued presence of the complete virus for its maintenance. An 0.8-kb fragment of herpes virus DNA (about 1% of the total DNA) can substitute for the transforming ability of intact viral DNA, and may function as an enhancer-insertion element which activates cellular oncogenes.

The suspected oncogenic potential of herpes virus does not necessarily preclude modified-live viruses from being used as herpes virus vaccines for three reasons. First, currently available genetic engineering techniques permit the selective excision of identifiable transforming DNA sequences from herpes virus, fulfilling a basic safety requirement. Second, a recent prospective study involving 10,000 women in Czechoslovakia has failed to confirm the direct association between genital herpes infection and cervical cancer. Third, the oncogenic risk due to vaccination has to be balanced against the oncogenic risk from natural infection with virulent herpes virus.

Introduced nonvirulence. A goal of contemporary research is to identify at the molecular level the herpes virus genes responsible for virulence, broadly defined as the ability to cause disease. As with DNA sequences associated with oncogenic potential, virulence genes can be prevented from causing disease by surgically introducing nonrevertible deletion mutations into the respective viral genes.

Virulence is a complex phenomenon that varies with each virus agent. For herpes virus, virulent manifestations of infection are ulcerative skin lesions, urinary retention, central nervous system disturbances such as hindleg paralysis in experimental animals, latent nerve cell infections, encephalitis, abortion, generalized infections in neonates and immunocompromised hosts, and cancer.

Virulence for herpes virus has been identified with several genetic loci, which fall into two broad categories: (1) genes determining products essential for herpes virus replication in tissue culture and in living animals; and (2) genes dispensable under ordinary conditions of virus growth in tissue culture but required for the life cycle of the virus in the living host. The first category of genes may be inappropriate for genetic engineering manipulations because their inactivation represents too drastic an attenuation of the virus. That the second class of genes provides a more promising target is evident from recent results utilizing animal models. One of these genes is the thymidine kinase gene. Replication of herpes virus in nondividing cell populations, such as nerve tissue, is dependent upon the virus bringing with it the metabolic machinery necessary for virus growth. The replication of viral DNA is dependent upon precursor pools of nucleotides formed by viral thymidine kinase. The gene for this enzyme has been characterized in detail. Attenuated herpes virus strains, constructed by genetic engineering techniques which delete portions of the thymidine kinase gene, fail to express enzyme activity and are markedly impaired in their ability to grow in nerve cell. These thymidine kinase–negative herpes virus strains less frequently establish ganglionic infections, either acute or latent, and are markedly less likely to ascend into the central nervous system to cause encephalitis. Vaccinated experimental animals develop mild skin lesions, but are solidly protected from lethal doses of virulent virus.

A second herpes virus gene dispensable for ordinary tissue culture growth but important for latency is $\alpha 22$ which encodes a protein that functions at the very early onset of viral replication. Although mutants lacking this gene replicate in tissue culture, they are defective in establishing latency. Vaccination by $\alpha 22$ mutants has likewise protected laboratory animals against lethal challenge by virulent virus.

Conclusion. It is important to emphasize that the herpes virus vaccines now being developed through genetic engineering techniques appear capable of decreasing the severity of initial infections and decreasing the likelihood of the establishment of recurrent infections. It is doubtful, however, that any of the various types of subunit vaccines can prevent infection of a vaccinated subject by virulent strains of herpes virus and, thus, the continued spread of virulent virus in the population. It is also unlikely that modified-live herpes virus vaccines can totally prevent reinfection of vaccinated subjects and the establishment of latency, but they have the greatest potential for doing so. Given the perplexing and paradoxical nature of herpes infections, the most rational approach appears to be continuation of efforts to develop modified-live virus vaccines that are acceptably safe.

For background information see ANIMAL VIRUS; CELLULAR IMMUNOLOGY; IMMUNITY; VACCINATION; VIRUS in the McGraw-Hill Encyclopedia of Science and Technology. [MALON KIT; SAUL KIT]

Bibliography: R. M. Chanock and R. A. Lerner (eds.), *Modern Approaches to Vaccines: Molecular and Chemical Basis of Virus Virulence and Immunogenicity*, 1984; L. A. Lasky et al., Protection of mice from lethal herpes simplex virus infection by vaccination with a secreted form of cloned glycoprotein D, *Biotechnology*, pp. 527–532, June 1984; G. J. Mertz et al., Herpes simplex virus type-2 glycoprotein-subunit vaccine: Tolerance and humoral and cellular responses in humans, *J. Infect. Dis.*, 150:242–249, 1984; E. Paoletti et al., Construction of live vaccines using genetically engineered poxviruses: Biological activity of vaccinia virus recombinants expressing the hepatitis B virus surface antigen and the herpes simplex virus glycoprotein D, *Proc. Nat. Acad. Sci. USA*, 81:193–197, 1984.

Vibration

In the analysis of vibrations, statistics may be used to study the temporal behavior of the response or to describe the system itself. The first topic falls in the realm of random vibration; the second is variously termed statistical energy analysis, power flow methods, or asymptotic analysis. Of course, there are situations in which descriptions of both the time behavior and the system parameters are statistical.

Many structures that vibrate are thought of as highly deterministic in their mechanical properties such as wall thicknesses, overall size, component weights, and rigidities. When these systems are excited by unpredictable forces such as earthquakes, ocean waves, or turbulent fluid flow, they vibrate in an unpredictable way. The object of random vibration theory is to predict various statistics of the response based on a statistical description of the excitation.

Some structures are so large and complicated that a precise description of them is impractical. Examples include buildings, ships, and aircraft. Acoustical spaces (concert halls, office complexes) are not structures, but have dynamical properties that are similar to those of structures. These complex systems can be described in statistical terms such as mean free path or distribution of their resonance frequencies. The response of these systems to even a deterministic excitation is describable in statistical terms, except that now the statistics describe the system parameters and the response of a population of systems is being described.

A major impetus to the statistical analysis of vibratory response has been some very significant changes in experimental procedures as a result of digital processing of data. Detailed information on the response of structures to excitation is now readily available from digital instruments and by computation that was either unavailable or available with very great effort in the past. This additional information has put great pressure on vibration analysts to understand details of the response that have been bypassed until now.

Random response of deterministic systems. The most commonly used descriptor of a random signal is the power spectrum, which is a measure of how the vibrational energy is distributed in frequency. As a statistic, it is a second moment of the Fourier transform of the signal. A major advantage of its use is that there is a simple relation between the power spectrum of the system excitation and its response; the relationship can be found by deterministic measurements or analyses of the system (Fig. 1).

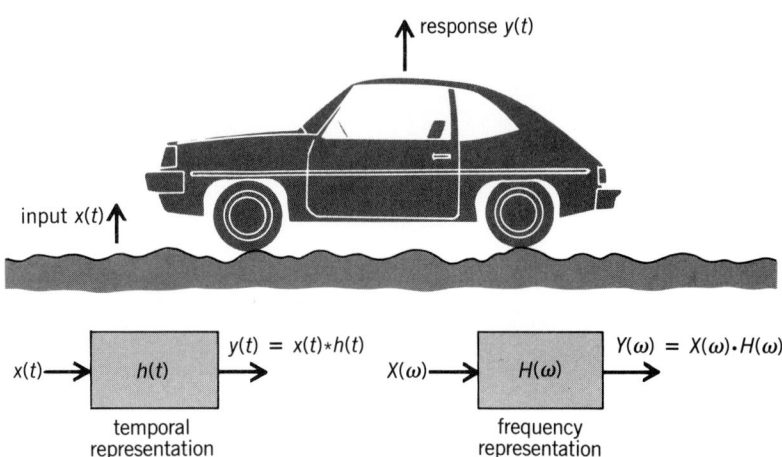

Fig. 1. Response of a deterministic system to a random input. In the temporal representation the input $x(t)$ and the response function of the system $h(t)$ must be combined in the convolution operation (symbol *) to obtain the response $y(t)$. In the frequency representation the power spectrum of the response $Y(\omega)$ is obtained from the power spectra of the input and the response function, $X(\omega)$ and $H(\omega)$, by simple multiplication.

There are occasions when the power spectrum of vibration does not provide enough information regarding the response or excitation. The likelihood that the suspension of an automobile will "bottom out" when one drives along a rough road is associated with the probability density of the response. Very often, this function is the bell-shaped probability curve, or gaussian distributional, of elementary statistics. In such an event, the second moments determine the probability curve, and the power spectrum continues to define the random process. But the response will not be gaussian in the case of strong system nonlinearities or a strongly nongaussian excitation. Although procedures exist for analyzing nongaussian cases, they tend to be very solution-specific and computationally complex.

An interesting phenomenon may occur when structures with geometric symmetries are excited by random noise sources at a point. Figure 2 shows a sand pattern on a plate for such a situation. The sand moves away from the regions of higher vibration and creates line patterns on the plate that connect at the excitation point and its images. The vibration amplitude along these lines is generally no more than 50% higher than at other points, but the sand can effectively show this difference.

Vibrating systems that have certain kinds of nonlinearity can show a random behavior which is termed chaotic. Even if the excitation is deterministic (a single frequency tone, for example) the vibrational response can be very erratic with a power spectrum that is nearly continuous, with the appearance of being totally random. The behavior of such systems is currently under active study. *See* CHAOTIC BEHAVIOR.

Response of probabilistic systems. A probabilistic system is a member of a group that has one or more of its parameters chosen randomly. For example, a set of flat steel plates, 0.04 in. thick and 9 ft^2 in area, forms such a population if the plate widths are defined by a probability law, for example, uniformly distributed between 1.5 and 2.25 ft. Any mass-produced product will form such a set because of the slight unintentional variations that occur, even among items made according to the same blueprints on the same production line. The computed response to excitation varies over the set of

Fig. 2. Chladni pattern for point excitation with a wideband random force.

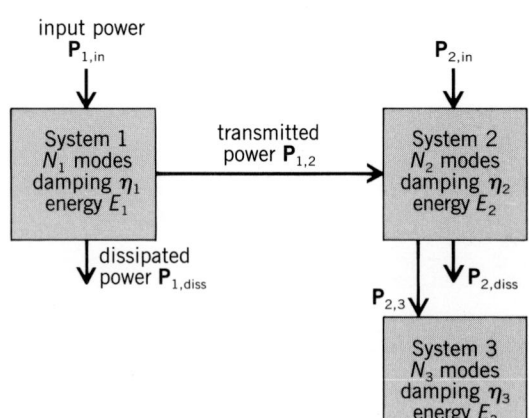

Fig. 3. Statistical energy analysis diagram showing energy interactions between systems.

systems, even though the response of each member may be deterministic.

There have been several approaches to dealing with this problem. The most common is to select an average or nominal system and treat it as a fully deterministic problem. Such an approach does not allow a calculation of how the response would vary as the system parameters vary over their range. The approach can also imply a precision in parameter knowledge that is unrealistic, and can produce unrealistic results in consequence.

Certain averaged response measures of a deterministic system tend to be independent of parameter choices. For example, the frequency-averaged response of a plate to a point force depends on the thickness and area of the plate, but not on its dimensions otherwise. There is the strong implication therefore that such an average would apply to a population of plates with the same thickness and area as that described above. Since frequency averaging is equivalent to calculating the average response to a band of noise, the insensitivity of random noise response to some structural parameters also is implied by these results.

A calculational approach called statistical energy analysis meets this issue directly by defining a system as a sample from a statistical population (Fig. 3). The average response, or indeed any statistic, applies to the population, and therefore only by inference to any particular member of the population. In basic viewpoint, statistical energy analysis is akin to statistical mechanics.

Many of the results of this analysis have certain similarities to those of statistical mechanics. The set of modal resonances of a complicated system on a specified frequency band is divided into subgroups of similar modes, shown as the boxes in Fig. 3. Within each box, equipartition (equal modal energy) is achieved. The mode groups or subsystems interact and energy flows between the boxes. This energy flow depends on the difference in modal energies (analogous to temperature) and the nature of the physical coupling between the modes. Energy is also lost by system damping. Balancing power sources against energy flows makes it possible to calculate an equilibrium of energy distribution among the boxes, much as in a heat flow calculation.

The question of how well the average over the population applies to any individual member is dealt with by computing the variance in the response. Calculating the variance is much more complicated than the average and has only been done for a few simple cases.

Measurement technology. Random vibration analysis has been of interest because some systems, like that in Fig. 1, are truly randomly excited, and also because random excitation and response can represent a simpler experimental test of a system. It is very time-consuming to painstakingly measure the excitation and response frequency by frequency and to plot out a system magnitude and phase response.

Digital processing of vibration data has changed this situation drastically and makes it possible to carry out these measurements quickly with great precision. Thus, it is now possible to consider the statistics of response or transfer functions, both magnitude and phase, and the use of these functions in source recovery, so that the excitation-force waveforms or spectra can be recovered from the response data for diagnostic purposes.

One effect of these developments has been to introduce the use of new statistical measures of the data such as the coherence and cepstrum functions. Coherence tells the degree to which the excitation and response are related by linear processes (nonlinearity and added noise reduce coherence). The cepstrum relates the excitation, the system, and the output response in an additive manner, and may allow the separation or extraction of various features of the excitation or of the system from the response data.

For background information see INTEGRAL TRANSFORM; STATISTICAL MECHANICS; STATISTICS; VIBRATION in the McGraw-Hill Encyclopedia of Science and Technology.   [RICHARD H. LYON]

Bibliography: J. S. Bendat and A. G. Piersol, *Random Data Analysis and Measurement Procedures*, 1971; S. Crandall and W. Mark, *Random Vibration*, 1963; R. H. Lyon, *Statistical Energy Analysis of Dynamical Systems*, 1975.

McGRAW-HILL YEARBOOK OF SCIENCE AND TECHNOLOGY

List of Contributors

List of Contributors

A

Allen, Dr. Leon H., Jr. Agronomy Department, Agronomy Physiology Laboratory, Institute of Food and Agricultural Sciences, University of Florida. PHYSIOLOGICAL ECOLOGY (PLANT).

Almquist, Dr. Edward E. Department of Orthopaedics, University of Washington, Seattle. SURGERY.

Anderson, Dr. J. Lawford. Associate Professor, Department of Geological Sciences, University of Southern California, Los Angeles. GRANITE—in part.

Appelquist, Prof. Thomas W. Physics Department, Yale University. ELEMENTARY PARTICLE—in part.

Ashton, Dr. Floyd M. Botanist, Agricultral Experiment Station, Professor of Botany, Department of Botany, University of California, Davis. HERBICIDE.

B

Bailey, Prof. Adrian G. Bill Bright Professor of Applied Electrostatics, Department of Electrical Engineering, University of Southampton, England. ELECTROSTATICS.

Barkworth, Dr. Mary. Associate Professor, Department of Biology, College of Science, Utah State University. GRASS.

Barry, Prof. Roger G. Professor of Geography and Director, WDC-A for Glaciology, Cooperative Institute for Research in Environmental Sciences, University of Colorado, Boulder. CRYOSPHERE.

Bass, Dr. Brenda L. Department of Chemistry, University of Colorado, Boulder. RIBOZYMES.

Bauer, John A. RCA Government Systems Division, Moorestown, New Jersey. ELECTRONICS—in part.

Bayfield, Prof. James E. Space Research Coordination Center, Department of Physics, University of Pittsburgh. HYDROGENIC IONS.

Berenbaum, Dr. May R. Assistant Professor, Department of Entomology, College of Liberal Arts and Sciences, University of Illinois at Urbana-Champaign. ECOLOGICAL INTERACTIONS.

Betts, Dr. R. Russell. Department of Nuclear Physics, Nuclear Physics Laboratory, University of Oxford, England. NUCLEAR REACTION.

Bocko, Dr. Mark F. Department of Physics and Astronomy, River Campus Station, University of Rochester. GRAVITATIONAL RADIATION—coauthored.

Bouchiat, Dr. Marie-Anne. Laboratoire de Spectroscopie Hertzienne de l'Ecole Normale Superiéure, Associé au C.N.R.S., Université de Paris VI, France. PARITY (QUANTUM MECHANICS)—coauthored.

Bourque, Dr. Don P. Associate Professor, Departments of Biochemistry, Nutrition and Food Science, and Molecular and Cellular Biology, University of Arizona. PROTEIN.

Brewer, Dr. Peter G. Department of Chemistry, Woods Hole Oceanographic Institution, Woods Hole, Massachusetts. SEA WATER.

Brookbank, Prof. John W. Professor of Microbiology and Cell Science, University of Florida. GENETICS.

Brown, Dr. Michael G. Division of Ocean Engineering, Dorothy H. and Lewis Rosenstiel School of Marine and Atmospheric Science, University of Miami. UNDERWATER SOUND.

Buchoff, Dr. Leonard S. Technical Director, PCK Elastomers, Inc., PCK Technology Division, Kollmorgen Corporation, Hatboro, Pennsylvania. ELECTRONICS—in part.

Bull, Prof. William B. Professor of Geosciences, Department of Geosciences, University of Arizona. TECTONIC GEOMORPHOLOGY.

Bunch, Prof. Howard McRaven. Navsea Professor of Ship Production, Department of Naval Architecture, University of Michigan. SHIPBUILDING.

Burgdorfer, Dr. Willy. Acting Chief, Epidemiology Branch, National Institute of Allergy and Infectious Diseases, Department of Health and Human Services, Public Health Service, National Institutes of Health, Rocky Mountain Laboratories, Hamilton, Montana. LYME DISEASE.

C

Cameron, Dr. Eion M. Energy, Mines and Resources Canada, Earth Sciences, Geological Survey of Canada, Ottawa, Ontario. GEOCHEMICAL PROSPECTING—in part.

Chang, R. Te-Tzu. Geneticist and Head, International Rice Germplasm Center, International Rice Research Institute, Manila, Philippines. RICE.

Clark, Prof. Bunny C. Department of Physics, Ohio State University. NUCLEAR PHYSICS—in part.

Clyde, Dr. Wallace A., Jr. Professor of Pediatrics and Microbiology, Departments of Pediatrics and Microbiology, Infectious Disease Division, Division of Health Affairs, School of Medicine, University of North Carolina at Chapel Hill. MYCOPLASMA.

Corey, Prof. Richard B. Department of Soil Science, College of Agricultural and Life Sciences, University of Wisconsin–Madison. SOIL CHEMISTRY—in part.

Cox, Peter A. Senior Partner, Rendel, Palmer & Tritton, Consulting and Designing Engineers, London, England. COASTAL ENGINEERING.

Craig, Prof. James R. Professor of Geology, Department of Geological Sciences, Virginia Polytechnic Institute and State University, Blacksburg, Virginia. GOLD, GEOCHEMISTRY OF—coauthored.

D

Dale, Prof. Alfred G. Department of Computer Sciences, College of Natural Sciences, University of Texas at Austin. DATA-BASE MANAGEMENT SYSTEMS.

David, Dr. Leonard. Editor/Programs Manager, National Space Institute. SPACE FLIGHT.

de Bruyn Ouboter, Prof. Dr. R. Kamerlingh Onnes Laboratorium, Der Rijksuniversiteit, Leiden, Nederland. QUANTUM MECHANICS.

Demianski, Dr. Marek. Professor of Physics and Astronomy, University of Warsaw, Poland, and Copernicus Astronomical Center. THE INFLATIONARY UNIVERSE—feature.

Desai, D. B. Associate Vice President, Daniel, Mann, Johnson & Mendenhall, Los Angeles, California. TRANSPORTATION ENGINEERING—in part, coauthored.

DeSimone, Dr. Joseph. Veterans Administration, Medical Center, West Side, Chicago, Illinois. SICKLE CELL ANEMIA—coauthored.

Di Fonzo, Daniel F. Communications Satellite Corporation, Washington, D.C. ANTENNA.

Donnelly, Prof. Russell J. Professor of Physics, Department of Physics, College of Arts and Sciences, University of Oregon. TURBULENCE.

Doolittle, Dr. Russell F. Department of Chemistry, University of California, San Diego, California. ONCOGENES.

Dorman, Albert A. Chairman of the Board and Chief Executive Officer, Daniel, Mann, Johnson & Mendenhall, Los Angeles, California. TRANSPORTATION ENGINEERING—in part, coauthored.

Douglass, Dr. David H. Department of Physics and Astronomy, River Campus Station, University of Rochester. GRAVITATIONAL RADIATION—coauthored.

E

Eigen, Dr. Daryl J. Siemen's Communications, Inc., Boca Raton, Florida. TELEPHONE SERVICE.

Elliott, Dr. Lloyd F. Microbiologist, United States Department of Agriculture, Agricultural Research Service, Western Region, Land Management and Water Conservation Research, Washington State University. SOIL MICROBIOLOGY—coauthored.

Engman, Dr. Edwin T. Agricultural Research Service, Hydrology Laboratory, United States Department of Agriculture, Beltsville, Maryland. REMOTE SENSING.

Ernst, Prof. W. G. Professor of Geology and Geophysics, Department of Earth and Space Sciences, University of California, Los Angeles. ACCRETION TECTONICS—feature.

F

Fairbridge, Dr. Rhodes W. Department of Geology, Columbia University, New York, New York. MOUNTAIN SYSTEMS.

Fedoroff, Dr. Nina V. Department of Embryology, Carnegie Institution of Washington, Baltimore, Maryland. GENE.

Flaim, Dr. Stephen F. Principal Scientist, Department of Biological Research, Section for Cardiovascular Research, McNeil Pharmaceutical, Spring House, Pennsylvania. HEART.

Ford, Prof. Joseph. Regents' Professor, School of Physics, College of Sciences and Liberal Studies, Georgia Institute of Technology. CHAOTIC BEHAVIOR.

Foster, Dr. John S. W. W. Hansen Laboratories of Physics, Edward L. Ginzton Laboratory, High Energy Physics Laboratory, Stanford University. ACOUSTIC MICROSCOPE.

Furth, Prof. Harold P. Director, Plasma Physics Laboratory, James Forrestal Campus, Princeton University. THE PATH TOWARD FUSION ENERGY—feature.

G

Gai, Prof. Moshe. Physics Department, Yale University. NUCLEAR PHYSICS—in part.

Gaillard, Prof. Mary K. Lawrence Berkeley Laboratory, University of California, Berkeley. PARTICLE ACCELERATOR—in part.

Gainer, Prof. John Lloyd. Chemical Engineering Department, University of Virginia. IMMOBILIZED CELL TECHNOLOGY—coauthored.

Galston, Prof. Arthur W. Professor of Biology, Department of Biology, Yale University. PLANT MORPHOGENESIS.

Geballe, Prof. Theodore H. Department of Applied Physics, Stanford University. SUPERCONDUCTIVITY—in part, coauthored.

Giacomo, Dr. P. Director, Bureau International des Poids et Mesures, Pavillon de Breteuil, Sèvres, France. METER (UNIT).

Giannasi, Dr. David E. Associate Professor, Botany Department, University of Georgia. PLANT METABOLISM.

Gill, Dr. J. M. Head of Research Unit for the Blind, Brunel Institute for Bioengineering, Brunel University, Uxbridge, Middlesex, England. BIOMEDICAL ENGINEERING.

Gilliam, Prof. J. W. Department of Soil Science, School of Agriculture and Life Sciences, Academic Affairs, Extension and Research, North Carolina State University. FERTILIZER.

Glashow, Prof. Sheldon L. Higgins Professor of Physics, Department of Physics, Lyman Laboratory of Physics, Harvard University. NEUTRINO.

Gollin, Dr. Susanne M. Director, Cytogenetics Laboratory, Arkansas Children's Hospital, Assistant Professor of Pathology and Pediatrics, University of Arkansas for Medical Sciences. GENETIC MAPPING.

H

Hanway, Dr. John J. Department of Agronomy, Iowa State University of Science and Technology. FERTILIZING.

Harbottle, Dr. Garman. Department of Chemistry, Brookhaven National Laboratory, Upton, New York. ARCHEOLOGY.

Harris, Dr. John W. K. Department of Anthropology, College of Letters and Science, University of Wisconsin–Milwaukee. ANTHROPOLOGY—in part.

Hastings, Prof. J. Woodland. Professor of Biology, Department of Cellular and Developmental Biology, Biological Laboratories, Harvard University. BIOLUMINESCENCE.

Hauptli, Dr. Holly H. Calgene, Inc., Davis, California. BREEDING (PLANT)—coauthored.

Hayes, William C. "Electrical World," McGraw-Hill Publications Company, New York, New York. ELECTRICAL UTILITY INDUSTRY.

Heller, Dr. Paul. Senior Medical Investigator, Veterans Administration, Medical Center, West Side, Chicago, Illinois, and Professor of Medicine, College of Medicine, University of Illinois. SICKLE CELL ANEMIA—coauthored.

Hemingway, Mark P. Department of Geological Sciences, New Mexico Institute of Mining and Technology. GEOCHEMICAL PROSPECTING—in part, coauthored.

Herzog, Dr. Gregory F. Department of Chemistry, Rutgers University. RADIONUCLIDES, COSMIC-RAY-PRODUCED.

Higgins, Dr. Robert P. Curator, Department of Invertebrate Zoology, National Museum of Natural History, Smithsonian Institution. LORICIFERA.

Hoer, Dr. Cletus A. Leader, Microwave Metrology Group, Electromagnetic Technology Division, United States Department of Commerce, National Bureau of Standards, Boulder, Colorado. MULTIPORT NETWORK ANALYZERS.

Hofstadter, Dr. Robert. Max H. Stein Professor of Physics, Varian Laboratory of Physics, Stanford University. MEDICAL IMAGING—in part.

Holdaway, Dr. M. J. Department of Geological Sciences, Southern Methodist University, Dallas, Texas. STAUROLITE.

Hopfinger, Dr. A. J. Director, Department of Medicinal Chemistry, Searle Research and Development, Division of G. D. Searle & Company, Skokie, Illinois. COMPUTER-ASSISTED DRUG DESIGN—feature.

Houck, Dr. Catherine M. Calgene, Inc., Davis, California. BREEDING (PLANT)—coauthored.

Hudson, Prof. John B. Materials Engineering Department, School of Engineering, Rensselaer Polytechnic Institute, Troy, New York. SURFACE SCIENCE—feature.

Hussong, Prof. Donald M. Professor of Geophysics, University of Hawaii at Manoa, Hawaii Institute of Geophysics. MUD VOLCANO.

J

Jaffe, Dr. Erwin. Newspaper and Graphic Arts Consultant, Easton, Pennsylvania. PRINTING—in part.

Jellison, Dr. J. L. Supervisor, Process Metallurgy, Sandia National Laboratories, Albuquerque, New Mexico. LASER WELDING.

Jérome, Prof. D. Associé au C.N.R.S., Laboratoire de Physique des Solides, Centre D'Orsay, Université de Paris-Sud, Orsay, France. SUPERCONDUCTIVITY—in part.

John, Prof. David T. Professor and Chairman, Department of Microbiology/Immunology, School of Medicine, Oral Roberts University. AMEBAS.

Johnson, Dr. David Linton. Schlumberger-Doll Research, Ridgefield, Connecticut. ACOUSTICAL MATERIALS.

Jones, Dr. David A. Director, Ewbank Preece Limited, Consulting Engineers, Brighton, United Kingdom. ELECTRICAL POWER ENGINEERING.

Jones, Dr. Douglas S. Associate Professor, Department of Geology, University of Florida. SCLEROCHRONOLOGY.

Junger, Dr. Miguel C. President, Cambridge Acoustical Associates, Inc., Cambridge, Massachusetts. ACOUSTIC TRANSIENTS.

K

Kamprath, Dr. Eugene J. Department of Soil Science, Academic Affairs, Extension and Research, School of Agriculture and Life Sciences, North Carolina State University. SOIL SCIENCE.

Kerr, John. "Electronic Business," Cahners Publishing Company, Irvine, California. TURBOCHARGER.

Kibble, Dr. Bryan P. Division of Electrical Science, National Physical Laboratory, Teddington, Middlesex, England. ELECTRICAL UNITS.

Kinzey, Dr. Bertram Y., Jr. Department of Architecture, University of Florida. ARCHITECTURAL ACOUSTICS.

Kit, Dr. Malon. Assistant Professor, Department of Virology and Epidemiology, Texas Medical Center, Baylor College of Medicine. VACCINE—coauthored.

Kit, Prof. Saul. Head, Division of Biochemical Virology, Department of Virology and Epidemiology, Texas Medical Center, Baylor College of Medicine. VACCINE—coauthored.

Kloos, Prof. Wesley E. Professor of Genetics and Microbiology, Department of Genetics, School of Agriculture and Life Sciences, North Carolina State University. STAPHYLOCOCCUS.

Komar, Prof. Paul D. Professor of Oceanography, College of Oceanography, Oregon State University. RIVER—in part.

Krider, Prof. E. Philip. Institute of Atmospheric Physics, University of Arizona. LIGHTNING.

Kucherlapati, Prof. Raju. Professor of Genetics, Center for Genetics, College of Medicine, University of Illinois at Chicago. GENE THERAPY—feature.

L

Laine, Dr. Edward P. Associate Professor of Oceanography, Graduate School of Oceanography, Narragansett Bay Campus, University of Rhode Island. RADIOACTIVE WASTE MANAGEMENT.

Lane, Dr. Kenneth D. Department of Physics, Ohio State University. ELEMENTARY PARTICLE—in part.

Liem, Dr. Ronald K. H. Associate Professor, Department of Pharmacology, School of Medicine, New York University Medical Center. NEURON.

Lyon, Prof. Richard H. Professor of Mechanical Engineering, Department of Mechanical Engineering, Massachusetts Institute of Technology. VIBRATION.

M

McCarthy, Prof. James. Department of Physics, University of Virginia. PARTICLE ACCELERATOR—in part, coauthored.

Mackowiak, Dr. Philip A. Assistant Chief, Medical Service, Veterans Administration Medical Center, and Associate Professor of Medicine, University of Texas Southwestern Medical School, Dallas. MICROBIAL INTERFERENCE.

MacPhee, Dr. Ross D. E. Assistant Professor of Anatomy, Department of Anatomy, Duke University Medical Center. PRIMATES.

Martin, Prof. James C. Professor of Chemistry, Roger Adams Laboratory, University of Illinois at Urbana–Champaign. HYPERVALENT SPECIES.

Maul, Dr. Donald H. California Primate Research Center, University of California, Davis. SIMIAN AIDS.

Meade, Dr. Robert H. United States Department of the Interior, Geological Survey, Denver Federal Center. RIVER—in part, coauthored.

Melvold, Dr. Roger W. Department of Microbiology-Immunology, Northwestern University Medical School. IMMUNOGENETICS.

Mermelstein, Paul. Manager, Speech Communications Systems, Recherche Bell-Northern Ltée, Verdun, Quebec, Canada. SPEECH RECOGNITION.

Meyer, Dr. Audrey Anne Wright. Ocean Drilling Program, Department of Oceanography, Texas A & M University. MARINE SEDIMENTS—in part, coauthored.

Miller, Dr. Helmut. Electrowatt Engineering Services, Ltd., Zurich, Switzerland. TURBINE.

Morris, Dr. Peter G. Medical Research Council Biomedical Nuclear Magnetic Resonance Centre, National Institute for Medical Research, London, England. MEDICAL IMAGING—in part.

N

Naney, Dr. Michael T. Geochemist, Chemistry Division, Oak Ridge National Laboratory, Oak Ridge, Tennessee. GRANITE—in part.

Nordheim, Dr. Alfred. ZMBH, Zentrum für Molekulare Biologie, Universität Heidelberg, Federal Republic of Germany. DEOXYRIBONUCLEIC ACID (DNA).

Nordin, Dr. C. F., Jr. (Retired) Hydrologist, United States Department of the Interior, Geological Survey, Denver Federal Center. RIVER—in part.

Norman, Dr. David I. Associate Professor of Geology, Department of Geosciences, New Mexico Institute of Mining and Technology. GEOCHEMICAL PROSPECTING—in part, coauthored.

Norrby, Prof. S. Ragnar. Department of Infectious Diseases, University of Umeå, Umeå Regional Hospital, Sweden. ANTIBACTERIAL AGENTS.

Norton, Dr. Roy A. Research Associate, Department of Environmental and Forest Biology, College of Environmental Science and Forestry, State University of New York, Syracuse. ACARINA.

Norum, Dr. Blaine. Associate Professor of Physics, Department of Physics, University of Virginia. PARTICLE ACCELERATOR—in part, coauthored.

Nuccitelli, Dr. Richard. Associate Professor, Department of Zoology, University of California, Davis. POLARITY (EMBRYOLOGY).

O

Okita, Dr. W. Blair. Chemical Engineering Department, University of Virginia. IMMOBILIZED CELL TECHNOLOGY—coauthored.

O'Neill, Dr. Gerard K. President and Chief Executive Officer, Geostar Corporation, Princeton, New Jersey. COMMUNICATIONS SATELLITE.

Orr, Dr. J. Richie. Fermi National Accelerator Laboratory, Batavia, Illinois. PARTICLE ACCELERATOR—in part.

P

Pal, Dr. R. Geneva, Switzerland. INSECT CONTROL, BIOLOGICAL.

Park, Sung Il. Department of Applied Physics, Stanford University. SUPERCONDUCTIVITY—in part, coauthored.

Parker, Prof. Eugene N. Laboratory for Astrophysics and Space Research, Enrico Fermi Institute, University of Chicago. MAGNETIC FIELD.

Parsons, Prof. Ian. Department of Geology and Mineralogy, Marischal College, University of Aberdeen, Scotland. PERTHITE.

Pasachoff, Prof. Jay M. Field Memorial Professor of Astronomy and Director, Hopkins Observatory, Williams College, and Visiting Colleague, Institute for Astronomy, University of Hawaii. ECLIPSE.

Pease, Dr. R. S. Authority Programme Director for Fusion, Culham Laboratory, United Kingdom Atomic Energy Authority, Abingdon, Oxfordshire. NUCLEAR FUSION.

Perkins, Dr. Lizabeth A. Department of Developmental Genetics and Anatomy, School of Medicine, Case Western Reserve University. CELL DIFFERENTIATION.

Pifer, Dr. Linda L. Associate Professor of Pediatrics, Department of Pediatrics, College of Medicine, Center for the Health Sciences, University of Tennessee. PNEUMOCYSTIS.

Pirrung, Dr. Michael C. Department of Chemistry, Stanford University. ETHYLENE.

Plaut, Dr. Andrew G. Professor of Medicine, Tufts University School of Medicine, and Physician, Gastroenterology Service, New England Medical Center, Inc. MICROORGANISM—in part.

Pottier, Dr. Lionel. Associé au C.N.R.S., Laboratoire de Spectroscopie Hertzienne de l'Ecole Normale Superieure, Université de Paris VI, France. PARITY (QUANTUM MECHANICS)—coauthored.

Purcell, Dr. Jennifer E. College of Oceanography, Oregon State University. CNIDARIA.

R

Rabinowitz, Dr. Mario. Senior Scientist, Electrical Systems Division, Electrical Power Research Institute, Palo Alto, California. NUCLEAR ELECTROMAGNETIC PULSE—feature.

Rainbow, Dr. Philip S. School of Biological Sciences, Queen Mary College, University of London, England. BARNACLE.

Randall, Prof. Gyles W. Southern Experiment Station, University of Minnesota. SOIL FERTILITY.

Rea, Dr. David K. Associate Professor of Oceanography, Department of Atmospheric and Oceanic Science, College of Engineering, University of Michigan. MARINE SEDIMENTS—in part.

Reed, Dr. Donald L. Ocean Drilling Program, Department of Oceanography, Texas A & M University. MARINE SEDIMENTS—in part, coauthored.

Richter, Prof. Burton. Director, Stanford Linear Accelerator Center, Stanford University. PARTICLE ACCELERATOR—in part.

Rimstidt, Dr. J. Donald. Department of Geological Sciences, Virginia Polytechnic Institute and State University. GOLD, GEOCHEMISTRY OF—coauthored.

Rosenfeld, Dr. Ron G. Assistant Professor of Pediatrics, Department of Pediatrics, Stanford University Medical Center. ADENOHYPOPHYSIS HORMONE.

S

Salomaa, Dr. Martti M. Low Temperature Theory Group Leader, Low Temperature Laboratory, Helsinki University of Technology, Espoo, Finland. LIQUID HELIUM—in part.

Schecter, Dr. Arnold. Professor of Preventive Medicine, Upstate Medical Center, Clinical Campus, State University of New York, Binghamton. TOXICOLOGY.

Schneidewind, Prof. Norman F. Computer Scientist, Computer Research, Pebble Beach, California. LOCAL-AREA NETWORKS.

Schnitzer, Dr. M. Principal Research Scientist, Chemistry and Biology Research Institute, Research Branch, Agriculture Canada, Ottawa, Ontario. SOIL CHEMISTRY—in part.

Smith, Prof. William H. Professor of Forest Pathology, Greeley Memorial Laboratory, School of Forestry and Environmental Studies, Yale University. ACID RAIN.

Southward, Dr. Eve C. The Laboratory, Marine Biological Association of the United Kingdom, Plymouth, England. POGONOPHORA.

Steen, Dr. Robert F. IBM Corporation, Research Triangle Park, North Carolina. COMPUTER NETWORKING

Stormer, Dr. H. L. AT & T Bell Laboratories, Murray Hill, New Jersey. HALL EFFECT.

Stott, Dr. Diane E. Microbiologist, United States Department of Agriculture, Agricultural Research Service, and Post-Doctoral Fellow, Washington State University. SOIL MICROBIOLOGY.

Stottmeier, Dr. Kurt D. Director, Medical Microbiology, Boston University School of Medicine, Maxwell Finland Laboratory for Infectious Diseases, Boston City Hospital. MYCOBACTERIA.

Sung, Dr. Renee. Department of Genetics, University of California, Berkeley. SOMATIC CELL GENETICS.

Suthers, Prof. Roderick A. Professor of Physiology, Physiology Section, Medical Sciences Program, School of Medicine, Indiana University. ECHOLOCATION.

Swartzel, Dr. Kenneth R. Assistant Professor, Food Engineering, Department of Food Science, School of Agriculture and Life Sciences, North Carolina State University. FOOD MANUFACTURING.

T

Tan, Prof. Kim H. Professor of Agronomy, Department of Agronomy, University of Georgia College of Agriculture. SOIL CHEMISTRY—in part.

Tannas, Lawrence E., Jr. Consultant, Orange, California. ELECTRONIC DISPLAYS.

Thompson, Dr. Richard A. Member of Technical Staff, Digital Systems, Research Department, AT & T Bell Laboratories, Murray Hill, New Jersey. TELETERMINALS.

Tobias, Philip E. President, Tobias Associates, Inc., Precision Equipment and Instrumentation, Ivyland, Pennsylvania. PRINTING—in part.

Trinkaus, Dr. Erik. Department of Anthropology, University of New Mexico. ANTHROPOLOGY—in part.

Turco, Dr. Richard P. Research Scientist and Program Manager, Atmospheric Chemistry and Physics, R & D Associates, Marina del Rey, California. NUCLEAR WINTER—feature.

Tuttle, Prof. Russell H. Professor in Anthropology, Evolutionary Biology, and The College, Department of Anthropology, University of Chicago. ANTHROPOLOGY—in part.

Tzipori, Dr. Saul. Veterinary Officer, "Attwood" Veterinary Research Laboratory, Department of Agriculture, Victoria, Australia. CRYPTOSPORIDIUM.

U

Ueno, Prof. Yoshio. Faculty of Pharmaceutical Sciences, Department of Toxicology and Microbial Chemistry, Science University of Tokyo, Japan. TRICHOTHECENES.

Ulrich, Prof. Dr. Reinhard. Technische Universität Hamburg-Harburg, Federal Republic of Germany. OPTICAL FIBER SENSORS.

V

Vogel, Prof. Steven. Professor of Zoology, Department of Zoology, Duke University. BIOMECHANICS.

W

Wadström, Prof. Torkel. Department of Veterinary Microbiology, Section of Bacteriology and Epizootology, College of Veterinary Medicine, Swedish University of Agricultural Sciences, Uppsala. TOXIN.

Webber, Dr. Bonnie Lynn. Department of Computer and Information Science, Moore School, University of Pennsylvania. NATURAL LANGUAGE PROCESSING.

Weinberg, Dr. Eugene D. Professor and Head, Microbiology Section, School of Medicine, Indiana University. MICROORGANISM—in part.

Welch, Dr. Ross M. Plant Physiologist, Plant, Soil, and Nutrition Laboratory, United States Department of Agriculture, Agricultural Research Service, Northeastern Region, North Atlantic Area, Ithaca, New York. NUTRITION.

Wells, Dr. Homer D. Research Plant Pathologist, United States Department of Agriculture, Agricultural Research Service, Southern Region, South Atlantic Area, Department of Plant Pathology, Coastal Plain Station, Tifton, Georgia. PLANT PATHOLOGY.

Westbrook, Dr. M. H. Manager, Body and Electrical Research, Technological Research, Research and Engineering Centre, Ford Motor Company Limited, Basildon, Essex, England. AUTOMOBILE.

Wienski, Robert M. District Manager, ISDN Architecture Planning, Bell Communications Research, Red Bank, New Jersey. INTEGRATED SERVICES DIGITAL NETWORK.

Wiley, Dr. E. O. Associate Curator, Museum of Natural History, Associate Professor, Department of Systematics and Ecology, University of Kansas. PHYLOGENETIC SYSTEMATICS.

Williams, Dr. Gwyn P. National Synchrotron Light Source, Brookhaven National Laboratory, Associated Universities, Inc., Upton, New York. SYNCHROTRON RADIATION.

Wilson, Dr. David Gordon. Department of Mechanical Engineering, Massachusetts Institute of Technology. TRANSPORTATION ENGINEERING—in part.

Wilson, Dr. James. Field Observing Facility, National Center for Atmospheric Research, Atmospheric Technology Division, Boulder, Colorado. RADAR METEOROLOGY.

Wilson, Dr. Timothy D. Assistant Professor of Psychology, Department of Psychology, University of Virginia. INFORMATION PROCESSING (PSYCHOLOGY).

Wolfe, Prof. James P. Professor of Physics, Physics Department and Materials Research Laboratory, Loomis Laboratory of Physics, University of Illinois at Urbana-Champaign. PHONON.

Wölfle, Prof. Dr. Peter. Physik-Department der Technischen Universität München, Theoretische Physik, München, West Germany. LIQUID HELIUM—in part.

Wright, Dr. Roger N. Acting Chairman, Materials Engineering Department, School of Engineering, Rensselaer Polytechnic Institute, Troy, New York. COMPUTER-AIDED DESIGN AND MANUFACTURING.

Wu, Dr. Junru. Department of Physics, University of California, Los Angeles. SOLITON.

Y

Yunis, Prof. Jorge J. Department of Laboratory Medicine and Pathology, Medical School, University of Minnesota. NEOPLASIA.

Z

Zierdt, Dr. Charles H. Microbiology Service, Clinical Pathology Department, Department of Health and Human Services, Public Health Service, National Institutes of Health, Bethesda, Maryland. BLASTOCYSTIS.

Zuckerman, Dr. Ben. Department of Astronomy, University of California, Los Angeles. STELLAR EVOLUTION.

McGRAW-HILL YEARBOOK OF SCIENCE AND TECHNOLOGY

Index

Index

Asterisks indicate page references to article titles.

A

Acanthamoeba 103
Acarina 89–91*
  oribatid cytogenetics 90
  Oribatida-Astigmata relationships 90–91
ACC *see* Amino cyclopropanecarboxylic acid
Accretion tectonics 81–88*
  evidence for far-traveled terranes 83–85
  inferred ancient allochthonous terranes 86–88
  modern allochthonous terranes 85–86
  true continental growth 88
Achipteria 90
Acid rain 91–93*
  air pollutants 92
  effect on forest soils 92–93
  effect on forest trees 93
Acid soil 412–415
  beneficial effects of liming 413–415
  chemical properties 413
  detrimental effects of acidity 413
Acoustic microscope 93–95*
  capability 95
  contrast 94–95
  cooling of liquid helium 94
  depth of focus 94
  imaging 94–95
  principles of operation 93–94
  sound attenuation problem 93–94
  ultimate limit of resolution 95
Acoustic solitons 416–417
Acoustic transients 95–97*
  hearing loss caused by sound pulses 96–97
  sound pulses as diagnostic tools 96
Acoustical materials 97–99*
  high-frequency behavior 98
  low-frequency behavior 97–98
  theoretical prediction of sound speeds 98–99
Acoustics: acoustic microscope 93–95*

Acoustics—*cont.*
  acoustic transients 95–97*
  acoustical materials 97–99*
  architectural acoustics 117–119*
  echolocation 158–160*
  phonon 340–343*
  underwater sound 476–478*
Acquired immune deficiency syndrome: association with pneumocystis 356
  simian AIDS 402–404*
Actinomyces viscosus 280
Actinopterygii 344
Active Magnetospheric Particle Tracer Explorers 427
Adams, J. W. 191
Adenohypophysis hormone 99–102*
  amino acid sequence 100
  biological activity of synthesized hormone 101
  purity of synthesized hormone 101
  synthesis by recombinant-DNA technology 100–101
  therapeutic use of synthesized hormone 101–102
Adsorption: soil chemistry 404–409*
Advanced Light Rapid Transit: automated train control system 462
  guideway surface and track elements 463–464
  linear induction motor 462–463
  steerable axle truck 463
Advanced Research Projects Agent's Network 148
Aeluropodeae 222
AEMP *see* Atmospheric heave electromagnetic pulse
Aeromonas 461
Aeromonas hydrophila 461
Agalma okeni 141
Agent Orange 458
Agricultural soil and crop practices: electrostatic pesticide spraying 182–183
  fertilizer 189–192*
  fertilizing 192–195*
  herbicide 231

Agricultural soil and crop practices—*cont.*
  insect control, biological 242–244*
  plant pathology 354–356*
  soil fertility 409–411*
Agrobacterium 354
Agrobacterium tumefaciens: plant cell transformation 133
Agropyron 222
AIDS *see* Acquired immune deficiency syndrome
Air pollution: acid rain 91–93*
  physiological ecology (plant) 346–348*
Air temperature: clouds and surface temperatures 2–3
Albrecht, A. 32
Aleksandrov, Aleksandr 428
Algebraic languages, relational 155
Alkaloid: secondary plant products 164, 348–351
Allelochemical 165, 166, 348–351
Allelopathy 351
Allen, Joseph 425
Allochthonous terranes: ancient 86–88
  Indonesian Archipelago 86
  modern 85–86
  oceanic plateaus of Pacific Basin 85
  salinian block of coastal California 85–86
  South Island, New Zealand 87
  Taiwan 87–88
  western North America 87
Almquist, E. E. 445
ALRT *see* Advanced Light Rapid Transit
Aluminum: toxicity in acid soils 413
Alvarez, L. 2
Amazon river 385–390, 386–388
  bed sediments and sediment discharge 388
  geochemistry 388–390
Amblyomma americanum 265
Amebas 102–104*
  nonpathogenic 102–103
  pathogenic 103–104

Amebas—*cont.*
  pathogenic free-living 103–104
Amebic meningoencephalitis 103
Amino cyclopropanecarboxylic acid: ethylene synthesis (plants) 188
Ampere (unit): direct determination 169–170
  indirect determination 169
AMPTE *see* Active Magnetospheric Particle Tracer Explorers
Andropogon 222
Andropogoneae 219, 222
Anemia, sickle cell 400–402
Angina pectoris: calcium channel blockers 229
Anik D2 (Telesat) 425
Animal: resistance to plant toxins 165–166
Animal communication: strategies using bioluminescence 125
Animal evolution: Pogonophora 359–360
  primates 362–364*
Animal growth: sclerochronology 392–394*
Animal virus: simian AIDS 402–404*
  vaccine 478–480*
Annular solar eclipse 162
Anopheles albimanus 243
Anopheles gambiae 242, 243
Anorogenic granite, Proterozoic 214–217
Antenna 104–107*
  designs for communications satellite 106–107
  optical resolution 106–107
  satellite frequency reuse 104–106
Anthropology 107–113*
  bipedal locomotion 107–109
  cultural evolution 113
  footprint trails 108
  modern anatomical changes 112–113
  modern subsistence pattern 112
  neurological evolution 113

Anthropology—cont.
  origin of modern humans 112–113
  stone artifacts 109–112
  timing of evolutionary transition 112
Antibacterial agents 113–116*
  carboxylic acid activity 114–115
  carboxylic acid chemistry 113–114
  clinical use of carboxylic acids 115
  mode of carboxylic acid action 114–115
  pharmocokinetics of carboxylic acids 115
  resistance to carboxylic acids 115
  safety of carboxylic acids 115
Antibiosis: plant disease control 354
Antibiotics: carboxylic acids 113–116
Antibody: class switching 239–240
  diversity of structure 237–238
  genetic specification of molecules 239–240
  IgA1 proteases 284–285
  immunogenetics 237–240*
Apicomplexa: *Cryptosporidium* 153–154*
Applications satellites: 1984 flights 426–427
Arabidopsis 417
Arachnida: Acarina 89–91*
Archaeolemur 362
Archeology 116–117*
  mass spectroscopy 117
  Mössbauer effect 117
  neutron activation analysis 116–117
  radiocarbon dating 116
  stone artifacts 109–112
Architectural acoustics 117–119*
Arctic sea ice 151–153
Arctica islandica 393
Argon laser: repair of injured nerves 444–445
Ariane satellite booster 429
Aristida 222
Aristideae 219, 222
Aristolochia 166
Aristolochiaceae 166
Aristolochic acid 166
ARPANET *see* Advanced Research Projects Agent's Network
Arrhenotokous parthenogenesis 90
Artificial intelligence: natural language processing 296–299*
  speech recognition 430–431*

Arunda 222
Arundinaria 222
Arundineae 222
Arundinoideae 221, 222
Asclepias 350
Aseptic food packaging 196
Aseptic food processing 195–196
  equipment operation 196
  equipment sterilization 196
  processing equipment 195–196
Asia: space activity (1984) 429
Astigmata: relationships with Oribatida 90–91
Astronomy: eclipse 160–164*
  gravitational radiation 222–225*
  inflationary universe 25–33*
  magnetic field 266–269*
  stellar evolution 437–440*
Athoryba lucida 141
Athoryba rosacea 141
Atkov, Oleg 428
Atmospheric electricity: lightning 249–252*
Atmospheric heave electromagnetic pulse 37
Atomic constants: electrical units 168–170*
Atomic, molecular and nuclear physics: elementary particle 184–188*
  nuclear physics 310–317*
  nuclear reaction 317–319*
  parity (quantum mechanics) 326–329*
  particle accelerator 329–338*
Atomic nucleus: dipole collectivity 313–316
  interaction process 319
  observation of heavy-ion resonances 317–319
  relativistic effects 310–313
  superdeformed nuclear states 319
Atomic structure and spectra: hydrogenic ion structure 232–233
Atomization: electrostatic spraying 182–184
Audio teleconferencing systems 452
Auger electron spectroscopy: surface chemical composition 17
Augmented transition network 297
Australian lungfish 344
Australian sheep blowfly 242
Australopithecus africanus 111
Australopithecus boisei 111
Austrochloris 219
Automobile 119–121*
  electronic engine controls 119–120
  electronic instrumentation 120

Automobile—cont.
  electronic traffic and road information 120–121
  electronic transmission control 120
  multiplex wiring 121–123*
  turbocharger 472–474*
Avdulov, N. 218, 219
Avena 222
Aveneae 222
Aves 344
Axon: morphology 302
  neuropathies 303
  transport 303

B

B-DNA 156
B lymphocyte: class switching in antibodies 239–240
Baccharis 466
Baccharis megapotamica 467
Bacillus cereus 461, 462
  geochemical prospecting 207
Bacillus subtilis: RNase P 382
Bacteria: activity in limed soils 415
  as control agents of plant disease 355
  luminous 125
  Staphylococcus 431–434*
  symbiosis with pogonophorans 358
  toxin 460–462*
  transposon structure 197–198
  use in geochemical prospecting 207
Bacteriocins 280
Bacteriophage: parasitism 281
Bacteroides: IgA1 protease 284
Balanus balanoides 123
Balanus perforatus 123
Bambusa 222
Bambuseae 222
Bambusoideae 219, 220, 221, 222
Band application of fertilizer 193
Bangladesh 167–168
Bardeen, J. 258, 440
Barley 222
Barnacle 121–123*
  cypris larvae 123
  feeding of nauplius larvae 122–123
  life cycle 121–122
  nauplius larvae 122–123
Barnyard grass 222
Bats: echolocation 158–160*
Battus philenor 166
Bell, Alexander Graham 305
Bell, D. 354
BEMP *see* Magnetic bubble electromagnetic pulse 36–37
Bermuda grass 222

Bicycles: BMX 464
  mountain bikes 464
  recumbent 464–465
Big bang theory 26–28
  decoupling of radiation and matter 27
  evolution from singular stage 26
  first 3 minutes 26
  formation of galaxies, stars, and planets 27
  future evolution 27
  properties of elementary particles 26
  radiation era 26–27
  spontaneous symmetry breaking 31
  standard model 26–28
  unanswered questions 27–28
Big brown bat: echolocation 158–159
Biocatalyst, immobilized 236
Biochemistry: ribozymes 380–383*
Biogeochemistry: mineral exploration 207–208
Biogeography: phylogenetic systematics 345–346
Bioluminescence 123–125*
  bacterial 125
  diversity and evolution 125
  functions of 124–125
  strategies for defense 124
  strategies for offense 124–125
  strategies in communication 125
Biomechanics 125–127*
Biomedical engineering 127–129*
  computer-based braille systems 128
  electronic mobility aids for blind people 128–129
  large-print reading material 128
  reading machines 128
  synthetic speech 127–128
  tape recorders 127
  television reading aids 127
Biophysics: biomechanics 125–127*
Biotite: granite crystallization 212
Bipedal locomotion (human): footprint trails 108
  gait analyses 108
  Hadar hominids 108–109
Birks, J. 1
Bivalvia: shell growth increments 392–393
Black smoke 5
Black walnut 351
Blackman, M. 307
Blastocystis 129–132*
Blastocystis hominis 129–132
  classification 130
  culture 131
  life cycle 131

Blastocystis hominis—cont.
   mitochondrial numbers 131
   morphology 129–130
   pathogenicity 131–132
   protozoan properties 130
Blindness: biomedical engineering 127–129*
   computer-based braille systems 128
   electronic mobility aids 128–129
   reading machines 128
   synthetic speech aids 127–128
   tape recorder aids 127
   television reading aids 127
Blood disorders: sickle cell anemia 400–402*
Bluegrass 222
Bluestem 222
BMX bicycles 464
Borrelia burgdorferi: Lyme disease 264–266*
Borrelia hermsii 266
Bouchiat, C. 327
Bouchiat, M. A. 327
Bouteloua 220, 222
Brand, Vance 421
Brassica 417
Breeding (plant) 132–134*
   cell transformation methods 133
   foreign gene expression in plants 133–134
   principles of genetic engineering 132–133
   prospects for plant improvement 134
   rice 383–385*
   somatic cell genetics 417–419*
Brevibacterium ammoniagenes: immobilized cell technology 237
Bristlegrass 222
Broadcast application of fertilizer 193
Bromegrass 222
Bromus 222
Brooks, D. R. 346
Brown algae: egg polarity 360–362
Brown smoke 5
Brownell, G. L. 276
Brundin, Lars 345
Bryophyllum 353
BS-2A (satellite) 429
Butterfly: pest control methods 243
   use of plant chemicals 166

C

Cabbage looper 243
CAD/CAM *see* Computer-aided design and manufacturing
Cadaverine 351

CADD *see* Computer-assisted drug design
Cadmium: accumulation in soils 192
Caenorhabditis elegans: cell differentiation 134–137*
   cell lineage 134–135
   mutation and cell fate 136–137
   selective cell ablation experiments 135–136
   transposable elements 197–198
Calcium: availability in limed soils 414
   deficiency in acid soils 413
   human nutrition 322
Calcium channel blockers: angina pectoris 229
   clinical agents 229
   experimental studies 230–231
   mechanisms of action 229–230
Calculus languages, relational 155
Calotropin 349
Calycophorae: nematocysts 140
Campbell, C. 221
Campylobacter 461
Canavanine 349
Cancer (medicine): neoplasia 299–302*
   oncogenes 322–324*
Candida albicans 280
Capacitor: derivation of SI value 169
Capnocytophaga: IgA1 protease 284
Capparaceae 164
Carbon dioxide: physiological ecology (plant) 346–348*
Carboxylic acid: antibacterial activity 114–115
   antibacterial agents 113–116*
   bacterial resistance 115
   chemical structures 114
   chemistry 113–114
   clinical use 115
   mode of antibacterial action 114
   pharmacokinetics 115
   safety 115
Carvone 349
Cassen, B. 277
Catalysis, heterogeneous 19–20
Catastrophe theory: underwater acoustics 476–478
CEBAF *see* Continuous electron beam accelerator facility
Cell (biology): immobilization technology 236
Cell differentiation 134–137*
   lineage of *Caenorhabditis elegans* 134–135

Cell differentiation—*cont.*
   mutation and cell fate 136
   selective cell ablation 135–136
Cell lineage: *Caenorhabditis elegans* 134–135, 136–137
Cenozoic: climate 273
Centosteca 222
Centosteceae 222
Centostecoideae 219, 220–221, 222
Ceratitis capitata 243
Challenger 420, 421–423, 424–425
Chaotic behavior 137–140*
   chaotic motion 139
   examples 138–139
   highly ordered motion 138
   noise 140
   ordered motion 138–139
   systems in the real world 139–140
   vibration 481
Charged particle beams: hydrogenic ion production 233
Chasmanthium 222
Chemical bonding: hypervalent species 233–235*
   surface electronic structure 18
Chemical ecology: ecological interactions 164–167*
   plant metabolism 348–351*
Chemistry: Nobel prize 306
China: space activity (1984) 429
Chip carrier (electronics) 178–180
Chlamydia trachomatis 295
Chlorideae 221
Chloridoideae 221, 222
Chloris 222
Cholera: enterotoxin 460–461
Chromatin: active versus inactive 203
   in the living cell 203
   structure 202–203
Chromosome: genetic mapping 199–202*
   histones 202–204
   neoplasia 299–302*
Chromosphere: eclipse studies 160
Chrysaora 140
Chthamalus 123
Chusquea 222
Circuit (electronics): elastomeric connectors 180–182
   microelectronic devices 22
   surface-mounted devices 177–180
Circumdehiscentiae 90, 91
Cladistics *see* Phylogenetic systematics

Classification (biology): phylogenetic systematics 343–346*
Clay minerals: interaction with humic substances 405
Climatic change: sea water 394–397*
Climatology: eolian sediments and processes 271–273
   nuclear winter 1–13*
Climatology *see* Meteorology and climatology
Clinopyroxene: granite crystallization 212
Cloning: mapping cloned genes 199–200
   use in plant breeding 132–134
Clostridium botulinum 281
Clostridium difficile 461
Clostridium perfringens 461, 462
Cloud: effect on surface temperature 2–3
   lightning 249–252*
Cnidaria 140–142*
Coastal engineering 142–144*
   characteristics of Thames Barrier 142–143
   flood-control studies of the Thames 142
   flood experience with Thames Barrier 143–144
Coats, Michael 423
Coccidia: *Cryptosporidium* 153–154*
Cochliomyia hominivorax 242
Codeine 349
Codling moth 243
Coelenterata *see* Cnidaria
Cole, R. D. 204
Coleman, S. 32
Color television: flat-panel displays 175–177
Columbia 420, 421
Comet: space probe 426–427
Communications: data *see* Data communications
   digital *see* Digital communications
   integrated services digital network 244–247*
   telephone service 452–455*
   teleterminals 455–457*
Communications satellite 144–146*
   antenna 104–107*
   antenna designs 106–107
   direct communication with mobile users 145
   frequency reuse 104–106
   1984 launches 429
   operation 104
   Radio Determination Satellite Services 145–146
   strengths and weaknesses 144–145

Computational chemistry: drug
    design 52
Computer: data-base
    management systems
    154–156*
  flat-panel displays 177
  local-area networks
    259–262*
  natural language processing
    296–299*
  speech recognition
    430–431*
  teleterminals 455–457*
Computer-aided design and
    manufacturing 146–147*
  use in shipbuilding 398
Computer-assisted drug design
    49–59*
  computer graphics 51–52
  definitions for three-
    dimensional design 51–52
  Hansch analysis 49–51
  molecular connectivity
    indices 51
  molecular modeling 51
  operational components for
    three-dimensional design
    52–53
  strategies for three-
    dimensional design
    53–59
  three-dimensional design
    51–59
Computer graphics: drug
    design 51–52
Computer networking
    147–150*
  development 147–149
  integrated services digital
    network 244–247*
  local-area networks
    259–262*
  principles 149
  prospects 149
Conalbumin 282
Conformational analysis: drug
    design 52, 57–59
Conservation tillage: effect on
    immobile nutrients 410
  effect on mobile nutrients
    410–411
  fertilizer application methods
    411
  soil fertility 409–411
Continent formation: accretion
    tectonics 88
Continental drift: accretion
    tectonics 81–88*
Continuous Electron Beam
    Accelerator Facility
    333–335
  beam switch yard 335
  description 334
  experimental facilities 335
  linear accelerator 334
  pulse stretcher ring 334–335
  upgrade and expansion 335
Cooper, L. N. 258, 440

Cooper pairs: superfluid helium
    258
  superfluid helium-3 253–255
Coordination chemistry:
    hypervalent species
    233–235*
Copper: human nutrition 322
Cordateria 222
Corn 321
Corn earworm 243
Corrosion: surface science
    techniques 20–21
Corynebacterium diphtheriae
    282
Cosmic rays: magnetic field
    266–267
  radionuclides, cosmic-ray
    produced 375–378*
Cosmogenic radionuclides *see*
    Radionuclides, cosmic-
    ray-produced
Cosmology: inflationary
    universe 25–33*
Cosmos 1514 427
Coulomb, C. A. 186
Cowpea 320
Crabgrass 222
Crested wheatgrass 222
Cretaceous: climate 273
Crick, J. F. 156
Crippen, Robert 422, 424
Crocodilia 344
Croizat, Leon 345
Crops: breeding (plant)
    132–134*
  rice 383–385*
Cruciferae 164, 166
Crustacea: barnacle 121–123*
Crutzen, P. 1
Cryosphere 150–153*
  Arctic sea ice 151–153
  borehole data 150–151
  glacier retreat 151
  numerical models of ice
    sheets and glaciers 151
  remote sensing 150
Cryptosporidiidae:
    Cryptosporidium
    153–154*
Cryptosporidiosis 153–154
Cryptosporidium 153–154*
  biology 153–154
  life cycle 153
  medical and veterinary
    significance 154
Crystal structure: surface
    structure 16–17
Ctenacaridae 89
Ctenophora 140
Cucumber 320
Cucumis sativas 320
Culkin, F. 394
Curcurbitaceae 165
Current balance: ampere
    determination 169
Current measurement: fiber-
    optic sensors 326
Cutgrass 222

Cyanidin 349
Cydia pomenella 243
Cylindocarpon 466
Cynodon 222
Cynodonteae 220, 222
Cyperales: grass 217–222*
Cypris larvae 123
Cystonectae: nematocyst 141
Cytogenetics: in oribatid mites
    90
Cytoskeleton: neuron
    302–304*
Czapek, F. 164

D

Dacus cucurbitae 243
Dacus dorsalis 243
Damaeus 90
Danthonia 222
Danthonieae 222
Data-base management systems
    154–156*
  development trends 156
  relational data models and
    languages 155
  relational hardware systems
    155–156
  relational software systems
    155
Data communications:
    computer networking
    147–150*
  integrated services digital
    network 244–247*
  local-area networks
    259–262*
Dating methods: cosmogenic
    radionuclide clocks
    376–377
de Laplace, P. S. 137
de Sitter, W.: space-time 31
de Vries, H. 415
Deep-penetration laser welding
    247
Deer mouse 265
Dendrocalamus 222
Dendrocalamus strictus 221
Dendrochronology 392
Deoxyribonucleic acid
    156–158*
  active versus inactive
    chromatin 203
  anonymous fragments in
    human genome 200
  chromatin in the living cell
    203
  function of Z-DNA 157–158
  gene 197–199*
  gene therapy 73–79*
  histones and chromatin
    structure 202–203
  human polymorphisms 200
  molecular structure of
    Z-DNA 156
  stabilization of Z-DNA
    156–157

Deoxyribonucleic acid—*cont.*
    see also Genetic engineering
Dermacentor variabilis 265
Deschampsia 219
Desmonomata 90
Determinism: chaotic behavior
    137–140*
Developmental biology: cell
    differentiation 134–137*
  polarity (embryology)
    360–362*
Devonian: Oribatida 89–90
Dhurrin 349
Diarrhea: *Blastocystis*
    129–132*
  Cryptosporidium 153–154*
  toxin 460–462*
Diesel engine: turbocharger
    472–474*
Digital communications:
    communications satellite
    144–146*
  integrated services digital
    network 244–247*
  telephone service 452–455*
Digiteria 222
Diltiazem 229
Dimopoulous, S. 187
Dioscoreophyllum cumminsii
    348
Dioxins 457–460
  chemical structure 458
  exposure 458–459
  laboratory studies of health
    effects 459
Discovery 420, 423–424,
    425–426
Distichlis 222
DNA *see* Deoxyribonucleic acid
Dopamine: drug design 54
Doppler effect: Doppler radar
    370–372
  echolocation 159–160
Doppler radar 370–372
  severe storm warnings
    370–371
  thunderstorm forecasting 372
  wind measurement 370
  wind shear warnings
    371–372
  winter storm forecasts 372
Drip irrigation 194
Dropseed 222
Drosophila melanogaster:
    transposable elements
    197–198
Drug design, computer-assisted
    49–59
Drug resistance: carboxylic
    acids 115
  Staphylococcus resistance to
    antibiotics 434
Dust (nuclear explosion):
    climatic impact 6–7
  light extinction by 5–6
  production in a nuclear war
    7
Dzhanibekov, Vladimir 428

E

Earth: magnetic field 267
Earth interior: neutrino 304–306*
Earth Radiation Budget Satellite (ERBS) 424
Echinochloa 222
Echolocation 158–160*
  bat ear structure 159
  target detection 158
  target direction 158–159
  target velocity 159
Eclipse 160–164*
  annular 162
  future eclipses 163
  paths of total eclipses 163
  recent expeditions 161–162
  safety in observing 162–163
  solar chromosphere and corona 160
Ecological interactions 164–167*
  animal resistance to plant toxins 165–166
  biological plant disease control 354–356
  cost of plant toxin production 165
  microbial interference 279
  plant-animal coevolution 164–165
  plant pathology 354–356*
  quantitative vs. qualitative plant toxins 165
  strategies using bioluminescence 124–125
  use of plant chemicals by insects 166
Ecology: ecological interactions 164–167*
  physiological ecology (plant) 346–348*
  plant metabolism 348–351*
  plant pathology 354–356*
Economics: Nobel prize 306
ECS *see* European Communications Satellite
Egg: polarity in moths 362
  polarity in plants 360–362
Ehrlich, P. 164
Eichter, E. 187
Eimeria 153
Eimeriina: *Cryptosporidium* 153–154*
Einstein, A. 29, 31, 340
Einstein Observatory: eclipse studies 160
Elastomeric electronic connectors: advantages 181
  composition 180
  future applications 181–182
  types 180
  uses 180–181

Electric power generation: engineering 167–168
  nuclear fusion 306–310*
Electrical distribution systems: electrical power engineering 167–168*
Electrical power engineering 167–168*
  availability of finance in third world countries 167
  environmental conditions in third world countries 167–168
  existing system in third world countries 167
  infrastructure in third world countries 168
  maintenance and spare parts in third world countries 168
  manufacture in third world countries 168
  nuclear fusion 306–310*
  prospects in third world countries 168
  training in third world countries 168
  urgency in third world countries 167
Electrical units 168–170*
  ampere 169–170
  derivation of kilogram 170
  direct determinations of ampere 169–170
  farad and derived units 169
  indirect determinations of ampere 169
  resistance unit and quantum Hall effect 169
Electrical utility industry 170–175*
  capacity additions 172–173
  capital expenditures 175
  combustion turbines 172–173
  distribution of electricity 174
  electricity usage 174
  fossil-fired capacity 172
  fuel consumption 174
  hydroelectric installations 173
  nuclear power 172
  ownership 171
  pumped storage 173
  rate of growth in demand 173–174
  renewable energy sources 173
  statistics (1984) 171
  transmission of electricity 174–175
Electricity: electrical power engineering 167–168*
  electrical units 168–170*
  electrical utility industry 170–175*
  nuclear fusion 306–310*

Electrodyn 182–183
Electroluminescence: flat-panel computer displays 177
Electromagnetic fields: lightning generation 249–250
Electromagnetic interactions 29
Electromagnetic pulse, nuclear 35–47
Electromagnetism: synchrotron radiation 445–448*
Electron-beam welding: compared with laser welding 248–249
Electron energy loss spectroscopy: surface atomic dynamics 18–19
Electronic displays 175–177*
  automobile instrumentation 120
  flat-panel computer displays 177
  flat-panel portable color TV displays 175–177
  liquid crystal displays 175–177
Electronic mail: computer networking 147–150*
Electronics 177–182*
  automobile 119–121*
  elastomeric connectors 180–182
  surface-mounted devices 177–180
Electrostatics 182–184*
  agricultural spraying 182–183
  liquid-insulator spraying 183–184
  liquid-metal spraying 183
  other applications 184
  paint spraying 182
Electroweak interactions: parity (quantum mechanics) 326–329*
Elementary particle 184–188*
  fundamental interactions 29
  masses of gauge bosons 186
  neutrino 304–306*
  possible extensions of standard model 185
  properties 26
  quantum chromodynamics 184–185
  quarks 184
  standard model 184–186
  successes of the standard model 185
  supercollider 187, 336–337
  symmetry breaking mechanisms 186–187
  technicolor forces 186–187
  technipions 187
Elements (chemistry): essential for plants and animals 320–322

Elements (chemistry)—*cont.*
  hypervalent species 233–235*
Elsasser, W. M. 267
Elymus 222
Embryology: cell differentiation 134–137*
  polarity 360–362*
EMP *see* Electromagnetic pulse
Enarthronata 90
Endocrine system (vertebrate): adenohypophysis hormone 99–102*
Endothia parasitica 354
Energy sources: path toward fusion energy 61–71*
  renewable 173
Engen, G. F. 289
Engineering: biomedical engineering 127–129
  coastal engineering 142–144*
  electrical power engineering 167–168*
  genetic *see* Genetic engineering
  transportation engineering 462–465*
Enterotoxin: bacterial diarrheas 461
  Escherichia coli 461
  food poisoning 461–462
  Vibrio cholera 460
Entomology, economic: insect control, biological 242–244*
Envelope soliton 415
Environmental toxicology: dioxins 457–460
Enzyme: IgA1 proteases 284–285
  ribozymes 380–383*
Eocene: climate 273
Eolian sediments: climate studies 271–273
Epidermal growth factor 324
Epidote: granite crystallization 213
Epilachna borealis 165
Eptesicus fuscus: echolocation 158–159
Eragrosteae 221
Eragrostis 222
Eragrostis pectinacea 221
Ericsson, A. 241
Erythrocyte: sickle cell membrane alteration 401–402
Escherichia coli: cloning of IgA1 protease gene 285
  enterotoxin 461
  immobilized cell technology 237
  polyamines and growth 353
  RNase P 382
Ethylene 188–189*
  substrates for ACC synthase 189

Ethylene—cont.
    synthesis in plants 188–189
Eucoccida: *Cryptosporidium*
    153–154*
Euptyctima 90
Europe: space activity 1984
    429
European Communications
    Satellite 429
Eutrophication: fertilizer
    189–192*
    nitrogen from fertilizers
        191–192
    phosphorus from fertilizers
        190–191
Euzetes 90
Evolution, animal *see* Animal
    evolution
Evolutionary taxonomy 343
Explosion: nuclear explosions
    4

F

Farad (unit) 169
    derived units 169
Fate maps (embryology):
    Caenorhabditis elegans
    136–137
Fault and fault structures:
    tectonic geomorphology
    448–452*
Fault-block mountains 449
Feeding mechanisms
    (invertebrate): barnacle
    121–123*
    Cnidaria 140–142*
Feeny, P. 164–165
Feldspar: perthite 338–340*
Fermi, Enrico 35
Fermion: heavy-fermion
    superconductors 442–444
Fertilizer 189–192*
    application methods 411
    cadmium accumulation 192
    combined with pesticides
        193
    fertilizing 192–195*
    immobile nutrients 410
    materials 193
    mobile nutrients 410–411
    nitrogen fertilizer and
        atmospheric effects 192
    nitrogen losses to
        environment 191
    phosphorus losses to
        environment 190
    soil application methods
        193–194
    soil fertility 409–411*
Fertilizing 192–195*
    fertilizer 189–192*
    fertilizer application methods
        193–194
    fertilizer materials 193
    foliar applications 194–195
Fescue 222

Festuca 222
Fetal hemoglobin: sickle cell
    anemia 402
Fiber optics *see* Optical fiber
    sensors
Field ion microscope: surface
    atomic structure 16–17
Fish-catching bat: echolocation
    159
Fisher, Anna 425
Flat-panel displays: color TV
    175–177
Fleming, A. 207, 280
Flexography: newspaper
    printing 364–365
Flood control: coastal
    engineering 142–144*
Flood irrigation 194
Flow cytometry: use in gene
    mapping 201–202
Fluid inclusion analysis
    (geochemistry) 204
    exploration methods
        206–207
    methods 205–206
    nature of inclusions 205
Foliar fertilization 194–195
Food-borne disease:
    trichothecenes 466
Food manufacturing 195–197*
    aseptic packaging 196
    aseptic processing 195–196
    plant leaf protein 366–368
Food packaging: aseptic 196
Food poisoning: enterotoxins
    461–462
Forest ecology: acid rain
    91–93*
    variability of ecosystems 92
Forest soil: acid rain effects
    92–93
Forest trees: effect of acid rain
    93
Fossil: Oribatida 89–90
Fossil fuel: electrical utility
    industry use 172
Fossil man: anthropology
    107–113*
    cultural evolution 113
    footprint trails 108
    gait analyses 108
    Hadar hominids 108–109
    modern anatomical changes
        112–113
    modern subsistence pattern
        112
    neurological evolution 113
    origin of modern humans
        112–113
    timing of evolutionary
        transition 112
Fourier series: acoustic
    transients 95–97*
Fractional quantum Hall effect
    227–228
    origin 227–228
    research areas 228
Fraenkel, G. 164

Free, S. U. 50
Freud, Sigmund 240
Friction: surface science
    techniques 21
Frog 344
Fructose: human nutrition 322
Fruit flies: eradication 243
Fucus 360
Fundamental interactions
    28–29
    grand unified theories 29
    parity (quantum mechanics)
        326–329*
    standard model (elementary
        particle) 184–186
    unification of 29
Fundamental particle *see*
    Elementary particle
Fungi: as control agents of
    plant disease, 355
    effect on soil aggregation
        412
    trichothecenes 465–467*
Fusarium 354
    trichothecenes 466
Fusarium graminearium 466,
    467
Fusarium sporotrichioides 466
Fusarium tricinctum 466
Fusion energy: alternative
    toroidal configurations 69
    fusion systems 63–66
    general magnetic
        confinement constraints
        64–65
    inertial confinement 65–66,
        69–71
    magnetic confinement 66–69
    mirror machines 69
    nuclear fusion development
        61–62
    open-ended magnetic bottles
        64
    plasmas and fusion reactions
        62–63
    practical utilization of fusion
        energy 71
    reactor wall materials 65
    tokamaks 66–69
    toroidal magnetic
        confinement 64
    see also Nuclear fusion

G

Galaxy, external: formation 27
    magnetic field 268
Galumna 90
Gamow, George 25
Gardner, Dale 425
Garneau, Marc 424
Garriott, Owen 420
Garwin, R. L. 36
Gauge bosons 184–187
    technicolor forces 186
Gauge theory: technicolor
    interaction 186

Gell-Mann, M. 184
Gene 197–199*
    breeding (plant) 132–134*
    deoxyribonucleic acid
        156–158*
    gene therapy 73–79*
    human gene mapping
        199–202
    oncogenes 322–324*
    selectable and nonselectable
        74–75
Gene action: expression of
    introduced genes 75–76
    histones 202–204
    oncogenes 322–324*
    regulation of introduced
        genes 76
Gene therapy 73–79*
    approaches 76–78
    DNA transfer into
        mammalian cells 73–74
    expression of introduced
        genes 75–76
    fate of foreign DNA 74
    future 79
    gene replacement through
        homologous recombination
        78–79
    gene transfer methods 73
    regulation of introduced
        genes 76
    selectable and nonselectable
        genes 74–75
Genetic engineering: breeding
    (plant) 132–134*
    gene therapy 73–79*
    herpes virus gD gene 479
    recombinant DNA mapping
        technology 199–200
    somatic cell genetics
        417–419*
    synthesis of adenohypo-
        physis hormone
        100–101
Genetic mapping 199–202*
    anonymous human DNA
        fragments 200
    flow sorting 201–202
    human DNA polymorphisms
        200
    Huntington's disease
        200–201
    in-situ hybridization 201
    somatic cell hybridization
        199
Genetics 202–204*
    active versus inactive
        chromatin 203
    chromatin in the living cell
        203
    deoxyribonucleic acid
        156–158*
    gene 197–199*
    gene therapy 73–79*
    genetic mapping 199–202*
    histones and chromatin
        structure 202–203
    human *see* Human genetics

Genetics—cont.
   immunogenetics 237–240*
   insect control, biological 242–244*
   somatic cell genetics 417–419*
Geobotanical indicators: mineral exploration 207
Geochemical prospecting 204–208*
   biological methods 207–208
   fluid inclusion analysis 204–207
Geochemistry: Amazon and Orinoco rivers 388–390
   geochemical prospecting 204
   gold, geochemistry of 208–210*
   sea water 394–397*
   soil chemistry 404–409*
   soil microbiology 411–412*
   soil science 412–415*
Geological Exploration by Muons Induced by Neutrino Interactions 305
Geological Exploration by Neutrino-Induced Underground Sound 305
Geology: accretion tectonics 81–88*
   cryosphere 150–153*
   granite 211–217*
   marine sediments 269–274*
   mountain systems 285–286*
   mud volcano 286–289*
   perthite 338–340*
   staurolite 434–437*
Geomorphology: accretion tectonics 81–88*
   tectonic geomorphology 448–452*
Geophysics: neutrino 304–306*
Geostationary communications satellite: antenna 104–107*
   antenna designs 106–107
   frequency reuse 104–106
   operation 104
Giant lemur: arboreal activity 362–363
   extinction 363–364
   locomotion 363
   size 362
Giant reed 222
Gibson, Robert 421
Glaciology: borehole data 150–151
   cryosphere 150–153*
   glacier retreat 151
   numerical models of ice sheets and glaciers 151
Glashow, Sheldon 29, 184, 337
Gliocladium 354
Global Positioning System 145
Glucosinolate: secondary plant products 164

Glyceollin 349
Glyceria 222
Glycine max 320, 351
Glycoside: secondary plant products 164
Gold, geochemistry of 208–210*
   economic gold-bearing deposits 209–210
   geochemical abundance and compositions 208–209
   gold-bearing deposits 210
   gold-bearing minerals 209
   gold transport and deposition 210
Goldstone, J. 258
GPS *see* Global Positioning System
Grain crops: rice 383–385*
Grama 222
Graminae *see* Poaceae
Grand unification theories: parity (quantum mechanics) 326–329*
   standard model (elementary particle) 185
   testing inflationary universe model 33
Granite 211–217*
   cooling experiments 213
   crystallization 211–214
   crystallization sequences 211
   nucleation and crystal growth 213–214
   phase equilibrium studies 211–213
   Proterozoic anorogenic 214–217
Granodiorite: granite crystallization 211–212
Graphics teleconferencing 453–454
Grass 217–222*
   anatomical taxonomic characters 219
   cytological taxonomic characters 219–220
   morphological taxonomic characters 219
   physiological taxonomic characters 220
   taxonomic history 218–219
   taxonomic overview 220–221
Gravitational radiation 222–225*
   binary pulsar source 223
   experimental detection efforts 223–225
   laser interferometer detectors 225
   nature 222–223
   resonant-bar detectors 224–225
Green manures: effect on soil aggregation 412
Greenhouse effect 3
   change in sea water salinity 395

Groundwater hydrology: soil moisture 378–380
Growth hormone: adenohypophysis hormone 99–102*
Gusella, James 200
Gypsy moth 243
Gyroscope: fiber-optic 326

H

Hadar hominids 108–109
Hadropithecus 362
Hall effect 225–228*
Hammer, M. 89
Hansch, C. 49
Hansch analysis (drug design) 49–51
Hardware systems, relational 155–156
Hart, Terry 422
Hartree approximation: doubly magic nuclei 311
Hartsfield, Henry 423
Harza, L. F. 467
Hauck, Frederick 425
Hawking, S. 32
Hawley, Steven 423
Hearing (human): loss caused by sound pulses 96–97
Heart 288–231
   calcium channel blockers 228–231
Heavy-fermion superconductors 442–444
   free-electron model 442–443
   nature of superconductivity 443
   properties 443–444
Helianthus tuberosus 353
Heliothis subflexa 242
Heliothis virescens 242
Heliothis zea 243
Helium: hydrogenic 233
   liquid *see* Liquid helium
   superfluid *see* Superfluid helium
Helle, W. 90
Hemoglobin: sickle cell anemia 400–402*
Hemophilus: IgA1 protease 284
HEMP *see* High-altitude electromagnetic pulse
Hennig, Willi 343
Henry (unit): derivation 169
Herbicide 231–232*
   mechanism of plant resistance 232
   potential benefits of resistance in plants 232
   prevention of plant resistance 232
   resistance in plants 231
Hermann, F. 394
Hermanniidae 90
Herpes simplex virus: vaccine 478–480*

Hertz, H. 223
Hessian fly 242
Heterogeneous catalysis: surface science techniques 19–20
Higgins larva (Loricifera) 263
Higgs, P. W. 186
Higgs bosons: symmetry breaking mechanisms 186–187
Higgs fields: spontaneous symmetry breaking 31
Higgs phenomenon: standard model (elementary particle) 185
High-altitude electromagnetic pulse 36
Himalayas: uplift 286
Himawari 3 429
Histone 202–204
   active versus inactive chromatin 203
   chromatin in the living cell 203
   chromatin structure 202–203
HL Tauri stars 439–440
Hoer, C. A. 289
Holography: synchrotron radiation application 448
Homo habilis 111
Homo sapiens 113, 344
Homologous recombination (genetics) 78–79
Hordeum 222
Hormone: adenohypophysis hormone 99–102*
   plant *see* Plant hormone
Hornblende: granite crystallization 213
Horseshoe bat: echolocation 159
Hubble, Edwin 25
Human evolution: anthropology 107–113*
   bipedal locomotion 107–109
   cultural changes 113
   footprint trails 108
   modern anatomical changes 112–113
   modern subsistence pattern 112
   neurological aspects 113
   origin of modern humans 112–113
   stone artifacts 109–112
   timing of transition 112
Human genetics: gene therapy 73–79*
   genetic mapping 199–202*
   sickle cell anemia 400–402*
Humerobates 90
Humus: binding mechanisms 404–406
Huntington's disease: gene mapping 200–201
Hybrid herpes virus vaccine 479

Hydroelectric power: electrical utility installations 173
Hydrogen: toxicity in acid soils 413
Hydrogenic ions 232–233*
   excited ion lifetimes 233
   hydrogenic helium 233
   partially ionized media 233
   production of highly charged ions 233
   relativistic effects 233
   structure 232–233
Hydrology: Amazon and Orinoco rivers 386–388
Hydrophone: fiber-optics 326
Hydrous oxide: interaction with humic substances in soil 404–405
Hypervalent species 233–235*
   first-row elements 235
   molecular orbital description of bonding 235
   transition states 234–235
Hypovirulence: plant disease control 354–355

I

Ice: borehole data 150–151
   cryosphere 150–153*
   remote sensing 150
ICE see International Cometary Explorer
IgA1 proteases: characteristics 284
   genetics and infectivity 285
   organization of mucosal immune system 284
   synthetic substrate analogs 285
   types 284–285
Iller turbine 468–469
Immobilized cell technology 236–237*
   industrial applications 236–237
   methods of immobilization 236
   rationale for whole-cell immobilization 236
Immunogenetics 237–240*
   class switching in antibodies 239–240
   diversity of antibody structure 237–238
   genetic specification of antibody molecules 238–240
Immunoglobulin: class switching 239–240
   diversity of structure 237–238
   genetic specification of molecules 239–240
   IgA1 proteases 284–285
Immunology: immunogenetics 237–240*
   vaccine 478–480*

Indri 362
Infection: alteration of iron-withholding host defense 283
   capture of host iron by invaders 283
   host emergency mechanisms of iron withholding 283
   host iron-withholding proteins 282–283
   IgA1 proteases 284–285
   normal flora as barrier 281–282
   role of iron 282
Inflationary universe 25–33*
   evolving universe 25
   fundamental interactions and particles 28–29
   inflationary scenario 31–32
   microwave background radiation 25–26
   resolution of big-bang problems 32–33
   spontaneous symmetry breaking 29–31
   standard big bang model 26–28
   testing of grand unified theories 33
   unification of interactions 29
Information processing (psychology) 240–242*
   inaccuracy of introspective reports 240
   lower-order versus higher-order mental processes 241
   mental processes versus mental states 241–242
   recent experimental work 241
   threatening versus nonthreatening thoughts 240–241
Infrared absorption spectroscopy: surface atomic dynamics 18–19
Infrared Astronomical Satellite 426, 437
Insect control, biological 242–244*
   available genetic techniques 242–243
   genetic control application and success 244
   genetically altered insects 242
   lepidopteran species 243
   Mediterranean fruit fly 243
   other fruit flies 243
   screwworm 243
   sterile-insect technique 242
Insecta: plant coevolution 164–165
   resistance to plant toxins 165–166
   use of plant chemicals 166
Insertion sequence (bacteria) 197–198

Integral quantum Hall effect 226–227
   applications 227
   origin 227
Integrated circuits: electronics 177–182*
Integrated services digital network 244–247*
   applications and prospects 246–247
   characteristics 244–245
   communications network 246
   digital technology 244
   interfaces 245–246
   service integration and standard interface 244
   signaling 244–245
   standards 245
INTELSAT: antenna 104–107
   antenna designs 106–107
   INTELSAT V (F-8) 429
Interferometry: fiber-optic sensors 325–326
   laser see Laser interferometer
Internal combustion engine: turbocharger 472–474*
International Cometary Explorer 426–427
International Rice Research Institute 384
International Sun-Earth Explorer 426
Introspective reports (psychology): inaccuracy due to lack of access 240
   inaccuracy due to lying 240
   threatening versus nonthreatening thoughts 240–241
Invertebrate zoology: Acarina 89–91*
   barnacle 121–123*
   biomechanics of soft, sessile marine organisms 125–127
   Cnidaria 140–142*
   Loricifera 262–264*
   Pogonophora 358–360*
   sclerochronology 392–394*
Ionization: hydrogenic ions 232–233*
IRAS see Infrared Astronomical Satellite
Iron: role in microbial infection 282–283
IRRI see International Rice Research Institute
Irrigation (agriculture): fertilizer application 194
Ischemia, myocardial: calcium channel blockers 228–231
ISDN see Integrated services digital network
ISEE see International Sun-Earth Explorer
Isospora 153
Ixodes dammini: Lyme disease 265

Ixodes ricinus: Lyme disease 264–266
Ixodes scapularis: Lyme disease 265

J

Japan: space activity (1984) 429
Jaworski, C. 354
Jellyfish: nematocysts 140
Jerne, Niels K. 306
JET see Joint European Torus
Joint European Torus 307–308
Josephson effect: current-biased Josephson junction 369
   interrupted superconducting ring 368–369
Jovin, T. M. 157
Juglans nigra 351
Juglone 349

K

Kapitza, P. L. 342
KdV soliton 415
Keshan disease 322
Kirchhoff, G. R. 97
Kizim, Leonid 428
Klebsiella pneumoniae 402
Knudsen, V. O. 118
Köhler, Georges J. F. 306
Kornberg, Roger 202
Korteweg, D. J. 415
Kristensen, R. M. 262
Kunst-Wilson, W. 241

L

Lactoferrin 282
Laetoli G hominids 108
Lamellibrachia 358, 359
Lamprey 344
LAN see Local-area networks
Landsat 5 427
Lane, K. 187
Langmuir, Irving 15
Languages see Programming languages
Laser: meter (unit) measurement 279
   repair of injured nerves 444–445
   speed of light determination 278
   surface science applications 22–23
   use in fusion reactor 69–71
Laser interferometer: gravitational radiation detector 224
   meter definition 279
Laser ranging: gravitational radiation 225

Laser welding 247–249*
   advantages 248
   applications 249
   comparison with electron-beam welding 248–249
   laser selection 247–248
   laser types 247
   weld types 247
Laughlin, R. B. 228
Lauterbur, Paul 274
Law, S. E. 182
Leaf protein concentrate 366–368
   composition 367
   extraction 367
   nutritional value 367
   products 366–367
   sources 366
   toxicity 367
   utilization 367–368
Leakey, Louis 109
Leakey, Mary 109
LEASAT (Syncom IV-1) 425
LEASAT 2 (Syncom IV-2) 424
Lech turbine 468–469
Lee, T. D. 326
Leersia 222
Leestma, David 424
Legionella pneumophila 281
LEM *see* Leukocytic endogenous mediator
Lemur: primates 362–364*
Lepas ruca 123
Lepidoptera: pest control methods 243
Leptasis 222
Leukocytic endogenous mediator 283
Lewis, G. N. 233
Lichtenberg, Byron 420
Life, origin of: role of ribozymes 382
Light: speed determination 278–279
Lightning 249–252*
   artificial triggering of 251–252
   cloud-to-ground 249
   compared to nuclear EMP disturbance 39–40
   electromagnetic fields 249–250
   future research 252
   hazards 249
   radio sources 250–251
Lime: effect on acid soil 413–415
Limestone: use in soil liming 414
Linde, A. 32
Linear colliders (particle accelerator) 331–333
Liquid crystals: electronic automobile instrument displays 120
   flat-panel computer displays 177
   flat-panel portable color TV displays 175–177

Liquid helium 252–259*
   acoustic microscope 93–95*
   collective modes in helium-3 257–259
   as heavy-fermion superconductor 444
   quantized vortices in helium-3 252–257
   turbulence 474–476*
Liquid metal: electrostatic spraying 183
Literature: Nobel prize 306
Little, W. A. 440
Local-area networks 259–262*
   hardware interconnection considerations 261–262
   operating system considerations 260
   software interconnection considerations 260–261
   status of interconnection technology 262
   strategy for interconnection 260
Logan, T. J. 191
Loricifera 262–264*
   external morphology 262–263
   Higgins larva 263
   internal morphology 263–264
   taxonomy and phylogenetic implications 264
Lovegrass 222
Low-energy electron diffraction: surface atomic structure 17
Low-temperature acoustics: phonon 340–343*
Low-temperature physics: acoustic microscope 93–95*
   superconductivity 440–444*
   turbulence 474–476*
Lubrication: microsprayer 184
Lucilia cuprina 242
Lucy (human fossil) 109
Luminescence: bioluminescence 123–125*
Lyakhov, Vladimir 428
Lycopersicon esculentum 320
Lyme disease 264–266*
   diagnosis 266
   epidemiology 265–266
   etiologic agent 264–265
   geographic distribution 265
   symptoms 264
   treatment 266
Lysergic acid 349

M

McBride, Jon 424
McCandless, Bruce 421
McClintock, Barbara 197
McDougall, J. 74
Mackay, D. 241
Maclurolyra 220

McNair, Ronald 421
Magma: granite crystallization 211–214
   Proterozoic anorogenic granite 214–217
Magnesium: availability in limed soils 414
Magnetic bubble EMP 36
Magnetic field 266–269*
   AMPTE project 427
   cosmic rays 266–267
   field of the Earth 267
   galactic fields 268
   quantitative study 268–269
   solar field 267
   solar wind 266
   stellar fields 267–268
Magnetic resonance imaging 274–276
   clinical applications 276
   image contrast 275–276
   principles 274–275
Magnetohydrodynamic electromagnetic pulse 36–37
Magnetometer: fiber-optic 326
Maitland, N. 74
Maize 222
Malyshev, Yuri 428
Manduca sexta 166, 243
Manganese: toxicity in acid soils 413
Mannagrass 222
Mansfield, Peter 274
Manufacturing systems: computer-aided design and manufacturing 146–147*
Manure: effect on soil aggregation 412
Mapping *see* Genetic mapping
Marine engineering: shipbuilding 397–400*
Marine geology: side-scan sonar sea-floor swathmapping 287–289
Marine organisms: barnacle 121–123*
   bioluminescence 123–125*
   biomechanics 125–127*
Marine sediments 269–274*
   drifts in active margins 269–271
   eolian sediments and processes 271–273
Mars: river channel meandering 390–391
Mass spectroscopy: use in archeology 117
Maxwell, J. C. 29, 138
Meadowgrass 222
Mechanical power engineering: turbine 467–472*
   turbocharger 472–474*
Medical bacteriology: Lyme disease 264–266*
   microbial interference 279–282*

Medical bacteriology—*cont.*
   microorganism 282–285*
   mycobacteria 292–294*
   Mycoplasma 294–296*
   Staphylococcus 431–434*
   toxin 460–462*
Medical imaging 274–278*
   magnetic resonance imaging 274–276
   nuclear medicine imaging 277
   positron emission tomography 276–277
Medical parasitology: amebas 102–104*
   Blastocystis 129–132*
   Cryptosporidium 153–154*
   pneumocystis 356–358*
Medicine and pathology: heart 228–231*
   human health and essential elements 322
   Lyme disease 264–266*
   Nobel prize 306
   pneumocystis 356–358*
   sickle cell anemia 400–402*
   surgery 444–445*
   toxicology 457–460*
Mediterranean fruit fly: eradication 243
Megaladapis edwardsi 362
Melic 222
Melica 222
Meliceae 222
Melon fly 243
Meningoencephalitis, amebic 103
Menthone 349
Merbold, Ulf 420
Merrifield, R. Bruce 306
Metabolism *see* Plant metabolism
Metallurgy: computer-aided design and manufacturing 146–147*
   laser welding 247–249*
Meteor (satellite) 429
Meteorite: cosmogenic radionuclides 376
   meteorite winter 2
Meteorology and climatology: climate records in mollusk shells 393–394
   lightning 249–252*
   radar meteorology 369–372*
   weather in nuclear winter 12
Meter (unit) 278–279*
   alternative definitions 278
   benefits of redefinition 279
   continuity of measurements 278–279
   need for redefinition 278
   new definition 279
Microbial interference 279–282*
   capitalizing on interference 281–282
   direct inhibition 280
   indirect inhibition 280–282

Microbial interference—cont.
  microbial assassins 281
  microbial synergism 282
  potential importance to humans 281
Microbiology: microbial interference 279–282*
  microorganism 282–285*
  soil microbiology 411–412*
Micrococcus 432
Microcomputer: biomedical engineering 127–129*
  data-base software 155
  local-area networks 259–262*
Microelectronics: surface science techniques 21–22
Microorganism 282–285*
  IgA1 proteases 284–285
  microbial interference 279–282*
  role of iron in infection 282–283
Microprocessor: biomedical engineering 127–129*
  electronic automobile controls 119–121
Microscope *see* Acoustic microscope
Microtubules: neuronal cytoskeleton 302–303
Microwave: background radiation 25–26
  multiport network analyzers 289–292*
Milankovitch, M. 273
Milkweed 350
Milstein, Cesar 306
Mineral: biological exploration methods 207–208
  decomposition in soil 406–408
  fluid inclusion analysis 204–207
  geochemical prospecting 204
  gold-bearing 209–210
  perthite 338–340*
  staurolite 434–437*
Miocene: climate 273
Mirror magnetic confinement 69
Mites: Acarina 89–91*
Mixonomata 90
Mobile communications: communications satellite 144–146*
  direct via satellite 145
  Radio Determination Satellite Services 145–146
Modified-live herpes virus vaccine 479–480
Molecular beam epitaxy: microelectronics 21–22
Molecular biology: deoxyribonucleic acid 156–158*
Molecular connectivity: drug design indices 51

Molecular modeling: drug design 51
Molecular orbital theory: hypervalent bonding 235
Molecular structure: drug design 52
Molinia 222
Mollusca: bivalve shell growth increments 392
  shell growth rates and longevities 393
Molniya 429
Molybdenum: availability in limed soils 414
Moon: eclipse 160–164*
Morphogenesis *see* Plant morphogenesis
Mössbauer effect: use in archeology 117
Moth: egg polarity 362
  pest control methods 243
Mountain bikes 464
Mountain systems 285–286*
  tectonic geomorphology 448–452*
MRI *see* Magnetic resonance imaging
Mucronothus nasalis 89
Mud volcano 286–289*
  overpressured sediments 287
  side-scan sonar sea-floor swathmapping 287–289
Muhlenbergia 222
Muhly 222
Mullane, Richard 423
Multiport network analyzers 289–292*
  calibration 292
  least-squares solution 291
  linear solution 291
  theory 289–291
  two-port measurements 291–292
  typical performance 292
Musical acoustics: architectural acoustics 117–119*
Mustard oils: secondary plant products 164
Mutation: effect on cell differentiation 136–137
Mya arenaria 393
Mycobacteria 292–294*
  mycobacteriosis 294
  obligate pathogens 293
  potential pathogens 293–294
Mycobacteriosis 294
Mycobacterium africanum 293
Mycobacterium asiaticum 294
Mycobacterium avium 293
Mycobacterium avium intracellulare 293, 402
Mycobacterium bovis 293
Mycobacterium chelonei 293
Mycobacterium fortuitum 293
Mycobacterium haemophilum 293
Mycobacterium intracellulare 293
Mycobacterium kansasii 294

Mycobacterium leprae 293
Mycobacterium malmoense 294
Mycobacterium marinum 293
Mycobacterium scofulaceum 293
Mycobacterium shimoidei 294
Mycobacterium simiae 294
Mycobacterium szulgai 294
Mycobacterium tuberculosis 281, 292
Mycobacterium ulcerans 293
Mycobacterium xenopi 294
Mycoplasma 294–296*
  indigenous to humans 295
Mycoplasma buccale 295
Mycoplasma faucium 295
Mycoplasma fermentans 295
Mycoplasma genitalium 295
Mycoplasma hominis 294–296
  clinical manifestations 295–296
  culture characteristics and biologic features 294–295
  epidemiology 295
  laboratory diagnosis and treatment 296
  pathogenicity 295
Mycoplasma lipophilum 295
Mycoplasma orale 295
Mycoplasma pneumoniae 295
Mycoplasma salivarium 295
Mycorrhizae: effect on soil aggregation 412
Mycotoxicosis: trichothecenes 466
Mycotoxin: trichothecenes 465–467*
Myocardial infarction: calcium channel blockers 228–231
Myrothecium 466, 467
Myrothecium verrucaria 467

N

Naegleria fowleri: primary amebic meningoencephalitis 103
Nannaloricus mysticus 262–264
Natural language processing 296–299*
  anaphora 298
  contextual analysis 297–298
  ellipsis 298
  natural language generation 299
  overall organization 299
  pragmatics 298–299
  semantic analysis 297
  syntactic processing (parsing) 296–297
Nauplius larvae: feeding mechanisms 122–123
Neisseria: IgA1 protease 284
Neisseria gonorrheae 281
Neisseria meningitidis 281
Nelson, Gareth 345
Nelson, George 422

Nemata: cell differentiation 134–137
Nematocysts: Cnidaria 140–142*
  effects on prey 140–141
  other functions 141–142
  structure 140
Neoceratodus forsteri 344
Neoplasia 299–302*
  chromosomal lesions 300
  clinical significance of chromosomal defects 302
  oncogenes and fragile sites 300–302*
  role of iron in infection 282–283
Nerve: surgery 444–445*
Nervous system (vertebrate): neuron 302–304*
Networking: computer 147
  teleconferencing 452–453
Neurofilaments 302–304
Neuron 302–304*
  axonal neuropathies 303
  axonal transport 303
  cytoskeleton 302–303
  developing animals 303
  interactions between cytoskeletal elements 303–304
  morphology 302
Neurospora crassa 382
Neutrino 304–306*
  characteristics 304
  Project GEMINI 305
  Project GENIUS 305
  Project GEOSCAN 305
  prospects for Earth exploration 305–306
  use in Earth exploration 304–305
Neutron activation: use in archeology 116–117
New Zealand: effects of accretionary tectonics 87
Newton, I. 137
Nickel: essential element for plants 320
Nicotiana tabacum 165
Nifedipine 229
Nisbett, R. 241
Nitrogen: fertilizer and the atmosphere 192
  fertilizer materials 193
  fertilizing 191–192
  as mobile soil nutrient 410–411
Nitrogen cycle: fertilizer effects 192
NMR *see* Nuclear magnetic resonance
Nobel prizes 306*
Noctilio leporinus: echolocation 159
Noise: relation to chaos 140
Nondestructive testing: sound pulses as diagnostic tools 96

Nonlinear physics: soliton 415–417*
Nonpropagating hydrodynamic soliton 416
Nonselectable genes 74–75
North America: effects of accretionary tectonics 87
Nothronata 90
Nozzle: electrostatic spraying 182–184
Nuclear electromagnetic pulse 35–47*
  atmospheric heave EMP 37
  definitions 36–39
  electron avalanche 42–43
  electron time of flight 41–42
  energy in the TEMP 40–41
  ensemble energy derivation in TEMP 44–45
  lightning compared to 39–40
  magnetic bubble EMP 36–37
  physical-insight derivation of TEMP 45–46
  power radiated per electron by TEMP 44
  tachy-electromagnetic pulse (TEMP) 37–39
  TEMP, nature of 43–46
  TEMP and the power line flashover 41–43
Nuclear explosion: nuclear electromagnetic pulse 35–47*
  nuclear winter 1–13*
  physical manifestations of 4
Nuclear fusion 306–310*
  alternative toroidal configurations 69
  alternatives to tokamak in Europe 308–310
  alternatives to tokamak in USSR 310
  development 61–62
  eastern Europe program 310
  fusion reactor studies 308–310
  general magnetic confinement constraints 64–65
  inertial confinement 65–66, 69–71
  Joint European Torus 307–308
  location of research centers in Europe 308
  magnetic confinement 66–69
  mirror machines 69
  open-ended magnetic bottles 64
  path toward fusion energy 61–71*
  plasmas and reactions 62–63
  practical utilization 71
  reactor systems 63–66
  reactor wall materials 65
  tokamaks 66–69
  tokamaks in USSR 310

Nuclear fusion—cont.
  toroidal magnetic confinement 64
  western Europe program 307–308
Nuclear magnetic resonance:
  medical imaging 274–276
  principles 274–275
Nuclear medicine imaging 277
Nuclear physics 310–317*
  comparison of wave equations 311
  dipole collectivity in nuclei 313–316
  dipole degree of freedom 314
  effect of negative energy states 313
  proton-nucleus scattering 312–313
  reduction of the Dirac equation 311–312
  relativistic effects 310–313
  relativistic Hartree approximation 311
  relativistic impulse approximation 313
  results from the ^{18}O nucleus 315–316
  results in radium isotopes 316
Nuclear power: electrical utility industry use 172
  path toward fusion energy 61–71*
Nuclear reaction 317–319*
  interaction process 319
  observation of heavy-ion sources 317–319
  radionuclides, cosmic-ray-produced 375–378*
  superdeformed nuclear states 319
Nuclear reactor: alternative toroidal configurations 69
  fusion reactor studies 308–310
  fusion systems 63–66
  inertial confinement 69–71
  magnetic confinement 66–69
  mirror machines 69
  radioactive waste management 372–375*
  tokamaks 66
Nuclear spectroscopy: nuclear reaction 317–319*
Nuclear structure: nuclear reaction 317–319*
Nuclear weapons 4–5
Nuclear winter 1–13*
  air circulation 9–12
  climate 2–3
  climatic impact of smoke and dust 6–7
  clouds and surface temperatures 2–3
  global circulation models of geographic distribution 9–12

Nuclear winter—cont.
  global surface temperatures 10
  greenhouse effect 3
  light attenuation 7–8
  light extinction by smoke and dust 5–6
  meteorite winter 2
  nuclear arsenals and scenarios 4–5
  nuclear explosions 4
  post–nuclear war environments 7–12
  smoke and dust 5–7
  temperature fluctuations 8–9
  types of smoke released by nuclear explosion 5
  uncertainty in 12
  weather effects 12
  winds 11
Nucleoprotein: histones 202
Nucleosome: chromatin structure 202–203
Nucleus see Atomic nucleus
Nutrition 320–322*
  human health and essential elements 322
  leaf protein concentrate 367
  new essential elements 320–321
  new functions of essential elements 321–322

O

Oatgrass 222
Oats 222
Obturata: characteristics 358
  trophosome 359
Oceanography: sea water 394–397*
O'Connor, B. 90–91
Odocoileus virginianus 265
Oersted, H. C. 29
Ohzora (satellite) 429
Oncogenes 322–324*
  central concept 322–323
  neoplasia 300–302
  types 323–324
Oncology: oncogenes 322–324*
Oppia 90
Optical fiber sensors 324–326*
  advantages, applications, principles 324
  intensity-encoding sensors 325
  interferometric sensors 325–326
  sensors with spectral encoding 325
Optics: biomedical engineering 127–129*
  optical fiber sensors 324–326*
Orbiting Solar Observatory: eclipse studies 160

Ore and mineral deposits:
  gold, geochemistry of 208–210*
Organ acoustics 117–119
  architectural design 119
  reflections 118–119
  reverberation time 117–118
Organic acid: decomposition of soil minerals 406–408
Organic superconductors 440–442
  applications 442
  higher transition temperature 441–442
  molecular structure 440–441
  one-dimensional properties 441
  superconducting properties 441
Oribatida: Acarina 89–91*
  chromosomes and parthenogenesis 90
  relationships with Astigmata 90–91
  slow early evolution 89–90
Oriental fly 243
Orinoco river 385–390
  bed sediments and sediment discharge 388
  geochemistry 388–390
Orogeny: fault-block mountain ranges 449
  landforms as time lines 451–452
  landforms of strike-slip faults 449–451
Orthogalumna 90
Orthopyroxene: granite crystallization 212
Oryza 222
Oryza glaberrima 383
Oryza sativa 383
Oryzeae 220, 222
OSO see *Orbiting Solar Observatory*

P

Packaging see Food packaging
Packet switching: computer networking 147–150*
Paint: electrostatic spraying 182
Palaeacaridae 90
Palaeopropithecus ingens 362–364
Palaeosomata 90
Palapa (satellite) 425
Palapa B-2 421
Paleocene: climate 273
Paleoclimatology 273
Paleomagnetism: accretion tectonics 84–85
Paleontology: Oribatida 89–90
Pampasgrass 222
Paniceae 219, 222
Panicgrass 222

Panicoideae 219, 221, 222
Panicum 219, 222
Panicum clandestinum 221
Parasitism: microorganisms 281
   plant disease control 354
Parasitology: *Cryptosporidium* 153–154*
   medical *see* Medical parasitology
Parhypochthonata 90
Parity (quantum mechanics) 326–329*
   choice of elements 327
   interaction properties to be determined 327–328
   observed atomic properties 327
   orders of magnitude 327
   systematic effects 328
   types of experiments 328–329
Parker, Robert 420
Parsing (natural language processing) 296–297
Parthenogenesis: evolution in mites 90–91
   in oribatid mites 90
Particle accelerator 329–338*
   CEBAF beam switch yard 335
   CEBAF description 334
   CEBAF experimental facility 335
   CEBAF linear accelerator 334
   CEBAF pulse stretcher ring 334–335
   CEBAF upgrade and expansion 335
   continuous electron beam accelerator facility (CEBAF) 333–335
   future colliders 333
   linear colliders 331–333
   SLC design 332
   SLC development, testing, and application 333
   SLC operation cycle 332–333
   spontaneous symmetry breaking in supercollider 337
   supercollider design 335–336
   supercollider in elementary particle physics 336
   superconducting supercollider 335–337
   synchrotron radiation 445–448*
   Tevatron 329–331
   Tevatron accelerating cycle 329–330
   Tevatron as proton-antiproton collider 331
   Tevatron complexity 330
   Tevatron experiment 331

Particle accelerator—*cont*.
   Tevatron features 329
   Tevatron precision 330
   Tevatron quench protection 330–331
Paspalum 222
Pathogen: microbial interference 281
Pathology *see* Medicine and pathology; Plant pathology
PCB *see* Polychlorinated biphenyl
Peace: Nobel prize 306
Pectinophora gossypiella 243
Pelvetia 360
Penicillium: geochemical prospecting 207
Penzias, Arno 25
Periodic table: hypervalent first-row elements 235
   molecular orbital description of hypervalent bonding 235
Peromyscus leucopus 265
Perthite 338–340*
   chemical and structural basis 338–339
   exsolution 339–340
Perviata: characteristics 358
   trophosome 358–359
Pesticide: combined with fertilizer 193
   electrostatic spraying 182–183
PET *see* Positron emission tomography
Petunia 353
pH: effect on humic substances 405
   precipitation (meteorology) 91
Phareae 222
Pharmacology: calcium channel blockers 228–231
   computer-assisted drug design 49–59*
Pharus 222
Phenetics 343–344
Phleum 222
Phonon 340–343*
   imaging technique 341–342
   propagation in crystalline solid 340–341
   research application of imaging method 342–343
Phonoreception: echolocation 158–160*
Phosphorus: availability in limed soils 414
   fertilizer materials 193
   fertilizing 190–191
   as immobile soil nutrient 410
Photoblepharon 124
Photosynthesis: physiological ecology (plant) 346–348*
Phragmites 222
Phragmites australis 221

Phylogenetic systematics 343–346*
   biogeography and other fields 345–346
   central concepts 344
   classification 345
   reconstructing phylogenies 344–345
Physalia 140, 141
Physarum polycephalum 353
Physical anthropology: anthropology 107–113*
Physical measurement: electrical units 168–170*
   meter (unit) 278–279*
   optical fiber sensors 324–326*
Physics: Nobel prize 306
Physiological acoustics: acoustic transients 95–97*
Physiological ecology (plant) 346–348*
   C_3, C_4, and CAM plants 346–347
   CO_2 source-sink effects 348
   complex ecosystems and rising CO_2 348
   photosynthesis and carbon dioxide concentration 347
   plant responses to carbon dioxide 346
   stomata, transpiration, and CO_2 347–348
Physiology: Nobel prize 306
   plant *see* Plant physiology
Physonectae: nematocysts 140
Phytophthora megasperma var. *sojae* 351
Pieris brassicae 166
Pink bollworm 243
Pioneer Venus Orbiter 426
Pipevine swallowtail 166
Pisces 344
Pituitary gland: adenohypophysis hormone 99–102*
Planet: formation 27
Plant: animal coevolution 164–165
   breeding 132–134*
   egg polarity 360–362
   grass taxonomy 217–222
   growth *see* Plant growth
   hormone *see* Plant hormone
   metabolism *see* Plant metabolism
   mineral nutrition 320–322
   morphogenesis *see* Plant morphogenesis
   pathology *see* Plant pathology
   physiology *see* Plant physiology
   quantitative vs. qualitative toxins 165
   secondary products 164

Plant—*cont*.
   somatic cell genetics 417–419
Plant growth: effect of soil acidity 413
   essential elements 320–322
   polyamine regulation 353
Plant hormone: ethylene 188–189*
   polyamine mediation of effects 353
Plant metabolism 348–351*
   biosynthesis of secondary products 349–350
   functions of secondary products 350–351
   potential for secondary products 351
   secondary products 348–351
Plant morphogenesis 351–354*
   polyamine mediation of hormonal effects 353
   polyamine regulation 353
   polyamines and plant stress 352–353
Plant pathology 354–356*
   biological control agents 355
   modes of action of biological control 354–355
   variability in biological control 355–356
Plant physiology: cost of toxin production 165
   ethylene 188–189*
   physiological ecology (plant) 346–348*
   plant metabolism 348–351*
   plant morphogenesis 351–354*
   quantitative vs. qualitative toxins 165
   secondary plant products 164
Plasma-display panels (computer) 177
Plasma physics: fusion systems 63–66
   plasmas and fusion reactions 62–63
Plasmid: gene replacement through homologous recombination 78–79
   plant cell transformation 133
Plasmodium berghei 51
Plasmodium falciparum 244
Plasmodium vivax 244
Plasmodium yoelii nigeriensis 243
Plate scanner (printing) 365–366
Plate tectonics 81
   marine sediment drifts 269–271
   tectonic geomorphology 448–452*
Platelet-derived growth factor 323
Pneumocystis 356–358*

Pneumocystis carinii: AIDS and pneumonia 356
   AIDS vs. simian AIDS 402
   antibody to 357–358
   biology 356–357
   culture 357
Pneumonia: *Pneumocystis carinii* 356
Poa 220, 222
Poa pratensis 221
Poaceae: grass 217–222*
Poeae 222
Pogonophora 358–360*
   classification 358
   evolution 359–360
   function of symbionts 359
   Obturata trophosome 359
   Perviata trophosome 358
   symbiotic bacteria 358
Pohl, F. M. 157
Poincaré, J. H. 138
Polarity (embryology) 360–362*
   moth egg 362
   plant eggs 360–362
Pollution *see* Water pollution
Polyamines: mediation of hormonal effects 353
   plant morphogenesis 351–354*
   plant stress 352–353
   regulation of plant growth and development 353
Polychlorinated biphenyls 457–460
Polychlorinated dibenzo-*para*-dioxins *see* Dioxins
Polymorphism (genetics): human 200
Pongidae 344
Pooideae 219, 220, 221, 222
Poroliodes 90
Porthetria dispar 243
Portuguese man-of-war 140
Positron emission tomography 276–277
Potassium: availability in limed soil 414–415
   fertilizer materials 193
   as immobile soil nutrient 410
Power-line flashover 41–43
Prat, H. 218
Precipitation (meteorology): acid rain 91–93*
Primary amebic meningoencephalitis 103
Primates 362–364*
   simian AIDS 402–404*
Printing 364–366*
   flexography for newspaper printing 364–365
   plate scanner premakeready 365–366
Probability: chaotic behavior 137–140*

Process models: computer-aided design and manufacturing 146–147*
Procyon lotor 265
Programming languages: relational data-base management 155
Progress 23 428
Project GEMINI 305
Project GENIUS 305
Project GEOSCAN 305
Prospecting *see* Geochemical prospecting
Prostigmata 90
Protein 366–368*
Proterozoic anorogenic granite 214–217
   conditions of crystallization 215–216
   essential features 214–215
   magma evolution 216–217
   tectonic considerations 217
   tectonic setting 214
Proto-oncogenes 322
Protozoa: amebas 102–104*
   Blastocystis 129–132*
   Cryptosporidium 153–154*
   pneumocystis 356–358*
Pseudomonas 354
Psychoacoustics: hearing loss caused by sound pulses 96–97
Psychology: information processing 240–242
Puccinellia 222
Pulmonary mycobacteriosis 294
Pulsar: gravitational radiation 222–225*
Pumped storage: electrical utility industry use 173
Putrescine 351

Q

QCD *see* Quantum chromodynamics
Quadricycles 465
Quantized vortices: superfluid helium-3 252–257
Quantum chromodynamics 184–185
Quantum Hall effect 225–228
   fractional 227–228
   Hall experiment 226
   integral 226–227
   resistance unit 169
   two-dimensional electron systems 226
Quantum mechanics 368–369*
   parity (quantum mechanics) 326–329*
Quantum turbulence: effects of rotation 475
   measurement of local properties 475
   nature of 474–475
   observation 474

Quantum turbulence—*cont*.
   prospects 475
Quarks 184
Quercetin 349
Quercus 165
Quinine 349

R

R Mon (star) 439–440
Raben, M. 99
Raccoon 265
Radar meteorology 369–372*
   Doppler effect 369
   severe storm warnings with Doppler radar 370–371
   wind measurement with Doppler radar 370
   wind shear warnings with Doppler radar 371–372
   winter storm forecasts with Doppler radar 372
Radiation: gravitational *see* Gravitational radiation
   sychrotron radiation 445–448*
Radio communications 144
Radio Determination Satellite Services 145–146
Radio ranging: gravitational radiation 225
Radioactive tracer: cosmogenic radionuclides 377–378
   medical imaging 277
Radioactive waste management 372–375*
   characteristics of deep seabed 372–373
   major problems of deep seabed 375
   present status of deep seabed 373–375
Radiocarbon dating: use in archeology 116–117
Radioisotopes: radionuclides, cosmic-ray-produced 375–378*
Radionuclides, cosmic-ray-produced 375–378*
   as clocks 376–377
   as tracers 377–378
Rana pipiens 344
Random process: chaotic behavior 137–140*
Rapid transit *see* Advanced Light Rapid Transit
Raven, P. 164
Ravennagrass 222
Rayleigh, Lord 314
RDSS *see* Radio Determination Satellite Services
Reactor *see* Nuclear reactor
Reactor physics: fusion systems 63–66
Recombinant DNA *see* Genetic engineering

Recombination (genetics): gene replacement through homologous recombination 78–79
Recumbent bicycles 464–465
Recursive transition network 297
Red blood cell: sickle cell membrane alteration 401–402
Reduced tillage systems: fertilizer application 193–194
Reedgrass 222
Relational algebraic languages 155
Relational calculus languages 155
Relational data-base theory 155
Relational hardware systems 155–156
Relational software systems 155
Relativity: gravitational radiation 223
Remote sensing 378–380*
   geochemical prospecting 208
   ice and snow surveys 150
   Landsat 5 427
   measuring soil moisture 378–380
Renewable energy sources: electrical utility industry use 173
Reptilia 344
Resedaceae 164
Resistance, electrical: derivation of unit 169
Resnik, Judith 423
Retrovirus: oncogenes 322–324
   simian AIDS 402–404*
   vector for gene therapy 77–78
Rhinolophus: echolocation 159
Ribonucleic acid: enzyme *see* Ribozymes
   plant disease control 355
Ribosome: self-splicing rRNA 381
Ribozymes 380–383*
   compared to enzymes 382
   as primordial catalysts 382
   RNase P 382
   self-splicing rRNA 381
Rice 383–385*
   conservation necessity and cost 385
   genetic evaluation 384
   genetic vulnerability 385
   germplasm sources and diversity 385
   history of conservation 383–384
   seed exchange 384
   taxonomic placement 222
   use of specific varieties 384–385

Rich, A. 156
Ride, Sally 424
Riftia 359
Riftia pachyptila 358
River 385–392*
    Amazon and Orinoco 385–390
    analysis of channel types 391
    channel meandering 390–392
    coastal engineering 142–144*
    factors controlling meandering 390–391
Robotics: use in shipbuilding 398
Rock: granite 211–217*
Rocket: artificial triggering of lightning 251–252
Root: cation activity in soil 408–409
Rous, P. 322
Rous sarcoma virus: oncogenes 322–324
Rubbia, Carlo 306
Russell, J. Scott 415
Rutaceae 164
Rutherford, Lord 62
Rye 222

S

Saalach turbine 468–469
Saccharomyces cerevisiae 353
Saccharum 222
Sakharov, A. 33
Salam, Abdus 29, 184, 337
Salmo trutta 344
Salmonella: enterotoxin 461
    inhibition by resident microorganisms 281
Saltgrass 222
Salyut 7 428
Samuel, A. L. 289
Sarcomastigophora: amebas 102–104*
Sarcopterygii 344
Satellite (spacecraft): antenna 104–107*
    communications satellite 144–146*
    space flight 419–430*
Savitskaya, Svetlana 428
Sawu-Timor Ridge 269
Scanning Auger microscopy: surface chemical composition 17–18
Scattering experiments: atoms and molecules 447
    nuclei 312–313
Schismus 222
Schrieffer, J. R. 258, 440
Scintigraphy 277
Sclerochronology 392–394*
    bivalve shell growth increments 392–393

Sclerochronology—*cont.*
    climate records in shells 393–394
    shell growth rates and longevities 393
Sclerolinum brattstromi 359
Sclerotium rolfsii 354
Scobee, Francis 422
Screwworm: eradication 243
Scully-Power, Paul 424
Sea ice: Arctic 151–153
    cryosphere 150–153*
    remote sensing 150
Sea nettle 140
Sea water 394–397*
    causes of salinity change 395–397
    observations of salinity change 394–395
    ocean salinity 394
SeaMARC II system 287–288
Secale 222
Secondary plant products: biosynthesis 349–350
    functions 350–351
    potential 351
    types 348
Seifert, Jaroslav 306
Selectable genes 74–75
Selenium: human nutrition 322
Semibalanus balanoides 123
Semiconductor devices: quantum Hall effect 226
Serendipity berry 348
Setaria 222
Sharks 344
Sharma, Rakesh 428
Shaw, Brewster 420
Shear, W. 89
Shigella 281
Shigella dysenteriae: enterotoxin 461
Shipbuilding 397–400*
    advanced technology 398–399
    computer technology and robotics 398
    group technology 399
    institutional barriers to improvement 400
    materials 399
    standardization 399–400
    surface preparation and coatings 399
    welding and fabrication 398–399
Siboglinum fiordicum 359
Siboglinum poseidoni 359
Sickle cell anemia 400–402*
    alteration of red cell membrane 401–402
    fetal hemoglobin 402
    modification of hemoglobin gel formation 401
    polymerization of hemoglobins 400–401
    sickled cell characteristics 401
    therapy research 401–402

Side-scan sea-floor mapping systems 287–289
Silicon: essential element for plants 320
Simian AIDS 402–404*
    clinical syndrome 402
    epidemiology 403
    immunology 403
    outlook 404
    pathology 403
    virology 402–403
Simon, E. 241
Sinigrin 349
Siphonophora: nematocysts 140–142
SLAC linear collider 331–333
    design 332
    development, testing, and application 333
    operation cycle 332–333
SLC *see* SLAC linear collider
Slime mold 353
Slipher, V. M. 25
Smoke (nuclear explosion): climatic impact 6–7
    light extinction by 5–6
    production in a nuclear war 7–8
Snow: cryosphere 150–153*
    remote sensing 150
Software systems, relational 155
Soil: aggregation 411–412
    chemistry *see* Soil chemistry
    fertility *see* Soil fertility
    forest *see* Forest soil
    microbiology *see* Soil microbiology
    moisture, importance of 378
    moisture measurements 378–380
    science 412–415
    use in geochemical prospecting 207
    see also Agricultural soil and crop practices
Soil chemistry 404–409*
    acid soils 412–415
    cation activities around roots 408–409
    decomposition of minerals by organic acids 406–408
    mechanisms of soil binding 404–406
Soil fertility 409–411*
    fertilizer application methods 411
    immobile nutrients 410
    mobile nutrients 410–411
    soil science 412–415
Soil microbiology 411–412*
    bacterial activity in limed soil 415
    fungal hyphae for soil stabilization 412
    green manures for aggregation 412
    soil aggregation process 411–412

Soil science 412–415*
Solar corona: eclipse studies 160
Solar eclipse 160–164
Solar energy: clouds and surface temperatures 2–3
    greenhouse effect 3
Solar Max 422–423
Solar wind: AMPTE project 427
    magnetic field 266
Soliton 415–417*
    acoustic solitons 416–417
    nonpropagating hydrodynamic 416
Solovyev, Vladimir 428
Somatic cell genetics 417–419*
    breeding (plant) 132–134*
    hybridization for mapping 199
    plant embryogenesis 418
    plant regulation from culture 418
    plant somatic and germ cell culture 417–418
    somaclonal variation in plants 418
Sonar: echolocation 158–160*
    side-scan sea-floor mapping systems 287–289
Sound *see* Acoustics
South Island: effects of accretionary tectonics 87
Soviet Union: space activity (1984) 427–429
Soybean: essential elements for growth 320
    response to CO_2 concentration 347, 351
Soyuz T-10 428
Soyuz T-11 428
Soyuz T-12 428
Space communications: antenna 104–107*
Space flight 419–430*
    Asian space activity 429
    European space activity 429
    remote sensing 427
    significant launches in 1984 419
    Soviet shuttle developments 427–428
    Soviet space activity 427–429
    Soviet space cooperation 427
    Soviet space station activity 428–429
    space science flights 426–427
    Space Transportation System missions 419–426
    uncrewed Soviet satellite utilization 429
    United States 1984 space activity 419–427
Space shuttle 419–426
    mission 9 420–421
    mission 41-B 421–422

Space shuttle—cont.
   mission 41-C 422–423
   mission 41-D 423–424
   mission 41-G 424–425
   mission 51-A 425
   Soviet developments (1984) 427
Space Transportation System: United States missions (1984) 419–426
Spacelab 1 420
SPACENET 429
Speakerphone: teleconferencing 453
Spectrometer: fiber-optic sensors 325
Speech recognition 430–431*
   factors affecting intelligibility 430
   hardware implementation 430
   prospects 431
   system based on phonetic models 430–431
   word recognition procedure 430
Speech synthesis: reading machines 127–128
Spermidine 351
Spermine 351
Spirochete: Lyme disease 264–266*
Spisula solidissima 393
Spontaneous symmetry breaking 29–31
Sporoboleae 221
Sporobolus 222
Sporozoa: *Blastocystis* 129–132*
   Cryptosporidium 153–154*
   Pneumocystis carinii 356–358
Sprinkler irrigation 194
Squash beetle 165
SSC *see* Superconducting Super Collider
Stachybotrys 466
Stachybotrys alternance 467
Stachybotrys atra 467
Stahl, E. 164
Stainless steel: corrosion 21
Standard model (elementary particle): quantum chromodynamics 184–185
   quarks 184
Staphylococcus 431–434*
   antibiotic resistance 434
   clinical significance 433–434
   habitats 432–433
   hosts 432
   species differentiation 432
   structure 431–432
Staphylococcus species: *aureus* 281, 402, 431, 432, 433, 434, 461–462
   auricularis 432, 433
   capitus 432, 433

Staphylococcus species—cont.
   caprae 432
   carnosus 432
   caseolyticus 432
   cohnii 432
   epidermidis 432, 433, 434
   gallinarum 432
   haemolyticus 432, 433, 434
   hominis 432, 433, 434
   hyicus 432, 433
   hyicus ssp. *chromogenes* 432, 433
   intermedius 432, 433
   lentus 432
   saccharolyticus 432
   saprophyticus 432, 433, 434
   sciuri 432, 433
   simulans 432
   warneri 432, 434
   xylosus 432, 433
Star: evolution *see* Stellar evolution
   formation 27
   magnetic fields 267–268
Starobinsky, A. A. 32
Staurolite 434–437*
   chemical analysis 435–436
   idealized crystal model 436–437
   Mössbauer spectroscopy 435
   standard deviation 436
   x-ray diffraction 434–435
Steinhardt, P. 32
Stellar evolution 437–440*
   formation of new planetary systems 439–440
   gravitational radiation 222–225*
   outflowing T Tauri winds 438–439
   T Tauri 437–438
Stellar magnetic field 267–268
Sterility: insect-control technique 242, 243
Sterilization: aseptic food processing 195–196
Stewart, Robert 421
Stinging cells *see* Nematocysts
Stinkgrass 222
Stipeae 219
Stochastic process: chaotic behavior 137–140*
Stone, Richard 306
Stone tool artifacts: age and geographical distribution 109–110
   behavioral implications 111–112
   form and function 110–111
Storm: Doppler radar warning and detection 370–371
   winter forecasts with Doppler radar 372
Straflo turbines 469–472
Straight-flow turbine: concept 467–469
   engineering study 469
   experience with prototypes 470–472

Straight-flow turbine—cont.
   realization attempts 468–470
   Soviet turbines 469
   Straflo turbines 469–470
Stream transport and deposition: Amazon and Orinoco rivers 388
Strekalov, Gennadi 428
Strength of materials: computer-aided design and manufacturing 146–147*
Streptococcus: distinguished from *Staphylococcus* 431
   IgA1 protease 284
Streptococcus faecalis 281
Streptococcus pneumoniae 280, 281
Streptococcus pyogenes 280
Stress (biology): polyamines and plant stress 352–353
Stress and strain: biomechanics 125–127*
Strike-slip fault: landforms of 449–451
Strong nuclear interactions 29
Structure-insensitive catalysis 20
Structure-sensitive catalysis 20
Strychnine 349
STS *see* Space Transportation System
Sugarcane 222
Sullivan, Kathryn 424
Sumba Ridge 269
Sumner, J. B. 380
Sun: eclipse 160–164*
   magnetic field 267
   solar wind 266
Superconducting supercollider 335–337
   design 335–336
   other physics issues 337
   spontaneous symmetry breaking 337
   status of elementary particle physics 336–337
   supercollider design 335–336
   technicolor forces 187
Superconductivity 440–444*
   heavy-fermion superconductors 442–444
   interrupted superconducting ring 368–369
   organic superconductors 440–442
Superfluid helium: A phase 259
   acoustic microscope 93–95*
   B phase 258–259
   collective modes 257–259
   continuous vortices in the A phase 254
   Cooper pairing 253
   macroscopic order 258
   magnetic vortices in B phase 254–256
   point vortices in momentum space 256–257

Superfluid helium—cont.
   quantized vortices in 252–257
   sound attenuation in 258
Superfluidity: turbulence 474–476*
Superlattice structures 22
Supersymmetry: standard model (elementary particle) 185–186
Surface-mounted electronic circuits 177–180
   chip carrier 178–180
   hybrids 178
   through-the-board attachment method 178
Surface science 15–23*
   applications 19–22
   atomic dynamics 18–19
   atomic structure 16–17
   chemical composition 17–18
   corrosion 20–21
   electronic structure 18
   friction and wear 21
   future 22–23
   heterogeneous catalysis 19–20
   microelectronics 21–22
   tools 15–19
   use of lasers 22–23
Surgery 444–445*
   argon laser nerve repair 444–445
   evaluation of nerve repair 445
Susskind, L. 187
Switching systems (communications): telephone service 452–455*
Symbiosis: Pogonophora 358–360*
Symmetry breaking: technicolor forces 186–187
Synchrotron radiation 445–448*
   absorption experiments and surface studies 448
   applications 446–448
   EXAFS studies 448
   scattering experiments 447
   sources 446
   x-ray microscopy and holography 448
Synergism, microbial 282
Syntactic processing (natural language processing) 296–297
Systematics: phylogenetic systematics 343–346*

T

T Tauri stars 437–438
   HL Tauri 439–440
Tachy-electromagnetic pulse 37–39

Tachy-electromagnetic pulse—cont.
  electromechanics and electrodynamics 43–46
  energy in 40–41
  ensemble energy derivation 44–45
  high-altitude burst 38–39
  low-altitude burst 37–38
  physical-insight derivation 45–46
  power radiated per electron 44
  surface burst 38
Taiwan: effects of accretionary tectonics 87–88
Tape recorder: use for blind people 127
Taxonomy: phylogenetic systematics 343–346*
TCDD see Dioxins
Technicolor forces 186–187
  masses of gauge bosons 186
  symmetry breaking mechanisms 186–187
  technipions 187
Technipions 187
Tectonic geomorphology 448–452*
  fault-block mountain ranges 449
  landforms as time lines 451–452
  landforms of strike-slip faults 449–451
Tectonics see Accretion tectonics; Plate tectonics
TELECOM 1 429
Telecommunications: integrated services digital network 244–247*
  telephone service 452–455*
  teleterminals 455–457*
Teleconferencing: audio teleconferencing systems 452
  audio terminals 453
  conference rooms 454
  future directions 454–455
  network considerations 452–453
  nonprotocol graphics teleconferencing 453
  protocol graphics systems 453–454
  setup 454
  video teleconferencing systems 454
Telephone service 452–455*
Teleterminals 455–457*
  implementation 457
  service example 457
  user interface 455–457
Television: color flat-panel displays 175–177
  reading aids for blind people 127
Telstar 3 424

TEMP see Tachy-electromagnetic pulse
Temperature inversion: post-nuclear war effect 8
Termites: use in geochemical prospecting 207–208
Tertiary: climate 273
Tetrahymena thermophila 381
Tetramethyl tetraselenafulvalene: organic superconductors 440–442
Tevatron: accelerating cycle 329–330
  complexity 330
  experiment 331
  features 329
  precision 330
  proton-antiproton collider 331
  quench protection 330–331
Thames Barrier (London):
  characteristics 142–143
  experience in flood control 143–144
  feasibility studies 142
  historical background 142
Thelytokous parthenogenesis 90
Themeda 222
Thomasson, J. 218
Thompson, G. P. 307
Three-awn grass 222
Three-dimensional drug design 51–59
  connection table 52
  definitions 51–52
  lead discovery 54–56
  lead optimization 56–59
  operational components 52
  strategies 53–59
Thunderstorm: Doppler radar warning and detection 370–371
Tick: Lyme disease 264–266*
Tillage: soil fertility 409–411
Timothy 222
Tindaria callistiformis 393
Tissue culture: breeding (plant) 132–134*
  plant regeneration from culture 418
  plant somatic and germ cell culture 417–418
  somaclonal variation in plants 418
Tobacco 353
Tobacco hornworm 166, 243
Tokamak 62, 66–69, 307, 309, 310
Tomato 320
Tomography see Positron emission tomography
Total solar eclipse 163
Toxicology 457–460*
Toxin 460–462*
  cholera 460
  enterotoxins and food poisoning 461–462

Toxin—cont.
  enterotoxins in bacterial diarrheas 461
  Escherichia coli 461
  nematocysts 140
  plant 165–166
  trichothecenes 465–467*
Toxoplasma 153
Transcellular ion current: patterns 361
Transferrin: role in infection 282–283
Transients see Acoustic transients
Transmission, automotive: electronic control 120
Transmission lines: flashover and TEMP 41–43
  utility industry capacity 174–175
Transportation engineering 462–465*
  advanced technology 462–464
  human-powered land vehicles 464–465
Transposable elements:
  function 198–199
  history 197
  structure 197–198
Transposons see Transposable elements
Tree: effect of acid rain on growth 93
  use in geochemical prospecting 207
Treponema phagodenis 266
Triazine: plant resistance 231–232
Trichoderma 354, 466
Trichoderma harzianum 354
Trichoplusia ni 243
Trichothecenes 465–467*
  chemical characteristics 466
  food-borne diseases 466
  sources 466–467
  toxicology 467
  yellow rain 466
Trichothecium roseum 466
Trickle irrigation 194
Tricycles 465
Triticeae 222
Triticum 222
Trout 344
Tsvelev, N. N. 220
Tube worms see Pogonophora
Tuberculosis: mycobacteria 293–294
Tunneling in solids: current-biased Josephson junction 369
  interrupted superconducting ring 368–369
  macroscopic quantum superposition 369
  quantum mechanics 361–369*

Turbine 467–472*
  electrical utility industry use 171–172
  straight-flow concept 467–468
  straight-flow prototypes 470–472
  straight-flow realization attempts 468–470
  turbocharger 472–474*
Turbocharger 472–474*
  disadvantages of conventional type 472
  future development 473–474
  variable geometry type 472–473
Turbulence 474–476*
  effects of rotation 475
  local properties in quantum turbulence 475
  nature of quantum turbulence 474–475
  observation of quantum turbulence 474
  prospects 475
  quantum turbulence in rotation 475
Turco, R. 2
Tutu, Desmond 306

U

Ultrahigh-vacuum technology: surface science 15–16
Ultraviolet photoelectron spectroscopy: surface electronic structure 18
Umbelliferae 165
Umbelliferone 164
Underwater sound 476–478*
  application of catastrophe theory 476–478
  formation of caustics 476
Unified electroweak theory 184
Universe, inflationary 25–33
Ureaplasma urealyticum 295
Utility industry see Electrical utility industry

V

Vaccine 478–480*
  conventional 478
  herpes simplex virus 478
  herpes virus glycoprotein 478–479
  herpes virus subunit 478
  introduced nonvirulence for herpes virus 480
  modified-live herpes virus 479–480
van der Meer, Simon 306
van Hoften, James 422

van Leeuwenhoek, Anton 351
Variable geometry turbocharger 472–473
Venturia inequalis 354
Venus: *Pioneer Venus Orbiter* 426
Verapamil 229
Verticimonosporium 466
Vibration 480–482*
   measurement technology 482
   random response of deterministic systems 481
   response of probabilistic systems 481–482
Vibrio cholerae 281
Video teleconferencing systems 454
Vigna aconitifolia 353
Vigna unguiculata 320
Vinen, W. F. 474
Virus: plant cell transformation 133
   vector for gene therapy 77–78
Visible absorption spectroscopy: surface atomic dynamics 18–19

Vision aids: biomedical engineering 127–129*
Volcano *see* Mud volcano
Volk, Igor 428
von Klitzing, K. 169

W

Walker, Charles 423
Walker, David 425
Wallwork, J. 89
Wang, A. H.-J. 156
Water pollution: acid rain 91–93*
   fertilizer 189–192*
Weak nuclear interactions: parity (quantum mechanics) 326–329*
Weapon, nuclear 4–5
Wear: surface science techniques 21
Weather *see* Meteorology and climatology
Weinberg, E. 32
Weinberg, Steven 29, 184, 337

Weinberg-Salam model 29
Welding and cutting of metals: laser welding 247–249*
   shipbuilding welding technology 398–399
Wells, H. 354
Westar 6 421, 425
Wheat 222
Wheatgrass 222
White-tailed deer 265
Whittaker, R. 164–165
Wigler, M. 74
Wildrice 222
Wildrye 222
Wilson, J. W. 50
Wilson, Robert 25
Wilson, T. 241
Wind: Doppler radar measurement 370
   shear warnings with Doppler radar 371–372
Windmillgrass 222

X

X-ray absorption, fine-structure: synchrotron radiation application 448

X-ray microscopy: synchrotron radiation application 448
Xenillus 90

Y

Yang, C. N. 326
Yellow rain 466
Yersinia enterocolitica 461
Young, John 420

Z

Z-DNA: function 157–158
   molecular structure 156
   stabilization 156–157
Zajonc, R. 241
Zakharov, V. E. 417
Zea 222
Zea mays 321
Zinc: human nutrition 322
Zizania 222
Zweig, G. 184